ORGANIC REACTION MECHANISMS · 1978

ORGANIC REACTION MECHANISMS · 1978

An annual survey covering the literature dated December 1977 through November 1978

Edited by

A. C. KNIPE and **W. E. WATTS**,
The New University of Ulster,
Northern Ireland

An Interscience® Publication

JOHN WILEY & SONS
Chichester · New York · Brisbane · Toronto

An Interscience® Publication

Library of Congress Catalog Card Number 66–23143

ISBN 0 471 27613 8

Printed in Great Britain by
John Wright & Sons Ltd.,
at the Stonebridge Press, Bristol

Contributors

M. S. BAIRD	Department of Organic Chemistry, The University, Newcastle-upon-Tyne
C. BROWN	Chemical Laboratory, University of Kent, Canterbury
A. R. BUTLER	Department of Chemistry, The Purdie Building, University of St. Andrews
B. CAPON	Department of Chemistry, Glasgow University
D. J. COWLEY	School of Physical Sciences, New University of Ulster
M. R. CRAMPTON	Department of Chemistry, Durham University
G. W. J. FLEET	Dyson Perrins Laboratory, South Parks Road, Oxford University
I. GOSNEY	Department of Chemistry, University of Edinburgh
M. C. GROSSEL	Dyson Perrins Laboratory, South Parks Road, Oxford University
A. F. HEGARTY	Chemistry Department, University College, Cork, Ireland
A. J. KIRBY	University Chemical Laboratory, Cambridge
A. W. MURRAY	Department of Chemistry, University of Dundee
D. C. NONHEBEL	Department of Pure and Applied Chemistry, University of Strathclyde
J. SHORTER	Department of Chemistry, University of Hull

Preface

'The Road goes ever on and on
  Down from the door where it began.
Now far ahead the Road has gone,
  And I must follow, if I can.
Pursuing it with weary feet,
  Until it joins some larger way,
Where many paths and errands meet.
  And whither then? I cannot say.'

The Fellowship of the Ring, J. R. R. TOLKIEN

[*Published by permission of George Allen & Unwin (Publishers) Limited*]

We have little doubt that the above passage will have particular significance for the dedicated team of contributors to *Organic Reaction Mechanisms*. As editors, we undertake the task of scanning the entire literature of organic chemistry in search of papers of mechanistic interest; we are consequently very much aware of the manifold development of the subject and of the difficulty in keeping up with its relentless progress. In keeping with previous volumes we have allocated references to fourteen chapter areas. It has subsequently been the task of each author to provide an expert and comprehensive review of his subject while taking care to highlight those papers of particular interest. Thus, we hope that successive volumes of *Organic Reaction Mechanisms* will effectively map the major developments in this diverse area while also describing the surrounding fields of activity.

The present volume, the fourteenth of the series, surveys research on organic reaction mechanisms described in the literature dated December 1977 to November 1978. In order to limit the size of the volume we must necessarily exclude or restrict overlap with other publications which review related specialist areas (e.g. photochemical reactions, biosynthesis, electrochemistry, surface chemistry, organometallic chemistry, heterogeneous catalysis). Furthermore, we try to avoid duplication between chapters. Thus, while a particular reference may occasionally be allocated to more than one chapter, we do assume that readers will be aware of the alternative chapters to which a borderline topic of interest may have been preferentially assigned.

We would once again like to express our particular thanks to our very experienced contributors, to the publications staff of John Wiley and Sons, and to Dr. R S. Cahn whose expertise as technical editor has ensured maintenance of the high standard of presentation expected of this series.

August 1979

A. C. K.
W. E. W.

Contents

CHAPTER 1

Reactions of Aldehydes and Ketones and their Derivatives

B. CAPON

Chemistry Department, Glasgow University

Formation and Reactions of Acetals and Ketals

There have been reviews entitled "Transition States of Hydrolysis of Acetals, Ketals, Glycosides, and Glycosylamines",[1] "Some Mechanistic Studies on the Hydrolysis of Acetals and Hemiacetals",[2] and "Stereoelectronic Control in Hydrolytic Reactions".[3]

Hemiacetals have been detected at quite high concentrations in the hydrolysis of some benzaldehyde diethyl acetals by making use of the fact that the hydrolysis of the hemiacetal is base-catalysed but that of the acetal is not.[4] Addition of sodium hydroxide to a hydrolysing solution of a benzaldehyde acetal, therefore, produces a sharp increase in the concentration of benzaldehyde (and hence UV absorbance) which is proportional to the concentration of hemiacetal present. The ρ-value for the hydrolysis of benzaldehyde acetals is -3.4 and of hemiacetals is *ca.* -2; hence

this effect is most noticeable in the reactions of *p*-methoxybenzaldehyde diethyl acetal. Also, as the hydrolysis of the hemiacetal is base-catalysed and that of the acetal is not, the effect is most noticeable at low pH. Under certain conditions, the hemiacetal is present in a hydrolysing solution of *p*-methoxybenzaldehyde diethyl acetal to the extent of 40% of the starting concentration.

The ethoxytropylium ion (**1**) has been detected as an intermediate in the hydrolysis of tropone diethyl ketal, and methoxy- and ethoxy-diphenylcyclopropenium ions have been detected in the hydrolysis of the acetals (**2**). However, no intermediate could be detected in the hydrolysis of 4-ethoxyacetophenone dimethyl acetal. Similar experiments were carried out with orthoesters and the intermediate ion could be detected in the hydrolysis of trimethyl orthomesitoate; by taking advantage of the rate-decreasing effect of electrolytes on the hydrolysis

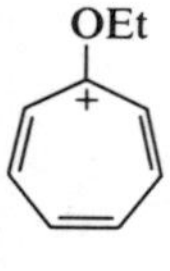

OEt (**1**)

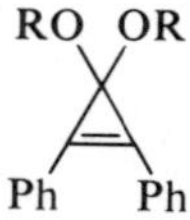

RO OR, Ph Ph (**2**) R = Me or Et

$Ar_2\overset{+}{C}—XMe$

Ar = *p*-MeO_6CH_4

(**3**) X = O

(**4**) X = S

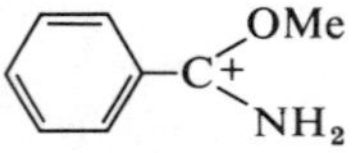

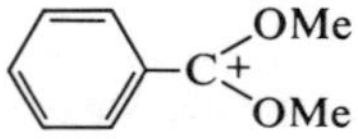

OMe, NH_2 (**5**)

OMe, OMe (**6**)

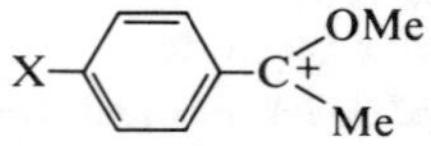

X, OMe, Me (**7**)

of carbocations, the intermediate ion could also be detected in the hydrolysis of trimethyl 4-methoxyorthobenzoate and trimethyl orthobenzoate.[5] The rate constants for the reactions of these and several other ions with water were also determined, the ions being generated in sulphuric acid solutions. Values of k_{H_2O} were obtained by extrapolation to zero acid concentration. It was shown *inter alia* that the methoxy-substituted cation (**3**) reacts over 1000 times faster than the methythio-analogue (**4**) and that the α-amino-substituted ion (**5**) reacts almost 10^9 times more slowly than ion (**6**). The ρ^+-values for the reactions of ions (**7**) with water is +2.1.[6] This was compared with the ρ^+-value of +1.6 estimated by Young and Jencks[7] using a different method as described below.

Young and Jencks studied trapping, by sulphite dianion, of the ions (**7**) which were generated by hydrolysis of the corresponding acetophenone dimethyl ketals. The fraction trapped increases with increasing sulphite concentration up to a maximum value ($f_{max.}$) which increases with increasing electron-releasing power of substituents in the ketal. The overall rate of reaction is unaffected by sulphite. Since the fraction of α-methoxysulphonic acid formed reaches a limiting value with increasing sulphite concentration, sulphite must also catalyse the reaction of water with cations (**7**) as well as attack them as a nucleophile. General base catalysis of attack of water by acetate was also detected as reflected by a decrease in the yield of α-methoxysulphonic acid with increasing acetate concentration. It was concluded from the low ratio of the rate constant for reaction with sulphite to that with water that the reaction with sulphite was diffusion-controlled. On this basis, on the assumption that the same values apply for the rate constants for reaction of all the ions with sulphite ($k_s = 5\times10^9\ M^{-1}\ s^{-1}$), a series of values for reactions of the ions with water was calculated which, at 25 °C, ranged from $7\times10^6\ s^{-1}$ for *p*-methoxy to *ca.* 4×10^8 s^{-1} for *m*-bromo, leading to the value of $\rho^+ = 1.6$ mentioned in the last paragraph. The value obtained by McClelland and Ahmad for the *p*-methoxy ion, by quite a long extrapolation from solutions in concentrated sulphuric acid to water (1.4×10^6 s^{-1}), was in quite good agreement. A linear plot of these values against the rate constants for reactions of sulphite with the corresponding carbonyl compounds was obtained. When this was extrapolated to isobutyraldehyde, *p*-nitrobenzaldehyde, and formaldehyde, very high values of rate constants for the reaction of the corresponding ions with water were estimated, *e.g.* 10^{15} s^{-1} for $CH_2{=}\overset{+}{O}Me$. This suggests that reactions which might involve these ions as intermediates, *e.g.* hydrolysis of the corresponding acetals, probably proceed by S_N2 processes.[7]

The rate constant for nucleophilic attack by water on the stabilized oxonium ion (**8**) is 8.5×10^{-2} s^{-1} at 25 °C.[8]

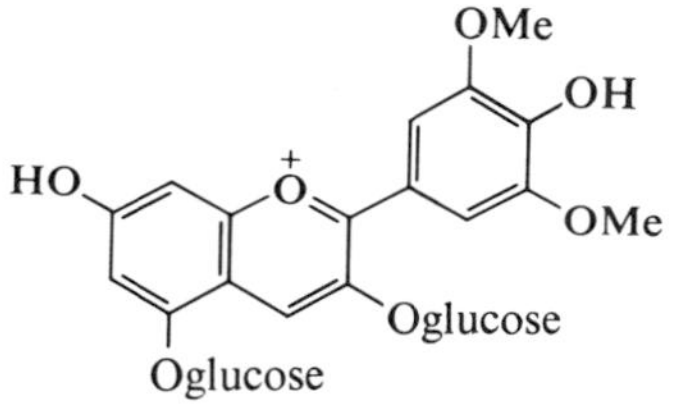

(**8**)

A discussion of the previously determined solvent deuterium isotope effects and Brønsted α-coefficients for the general acid catalysed hydrolysis of acetals and orthoesters has been given.[9] In agreement with earlier discussions,[10] it was considered that the rate-determining step was the concerted displacement of the alkoxycarbonium ion by the acid catalyst. In addition, however, it was thought necessary to invoke several additional intermediates to explain the difference in solvent isotope effects (and fractionation factors) between these reactions and the general acid catalysed hydrolysis of vinyl ethers.

The spontaneous hydrolysis of acetal (**10**) with an axial OAr group is 3·3 times slower than that of (**9**) with an equatorial OAr group.[11] This result contrasts with those previously reported which indicated that orthoester (**11**) should react with preferential cleavage of the axial carbon–methoxyl bond. It is possible that the acetal with an axial OAr group reacts *via* a half-chair transition state and cation, whereas that with an equatorial OAr group reacts *via* a boat or twist-boat transition state and cation. If these transition states were of similar energy, the rates of cleavage of axial and equatorial bonds would be similar, as found. With this structure, both the half-chair and boat (or twist-boat) conformations can accommodate the additional fused six-membered rings without additional strain. With the orthoester (**11**), however, the ring fusion is in a different position and a boat conformation would require the other six-membered ring to be fused *trans* across the base of the boat which is very unfavourable. In an investigation of the tricyclic acetals (**12**) and (**13**), evidence was obtained that, if the system is made sufficiently rigid, fusion of the axial C(2)—O bond of a 2-aryloxytetrahydropyran occurs faster than fission of an equatorial one.[12] It was found that the acid catalysed hydrolysis of (**13**) occurs over 3000 times faster than that of (**12**). The spontaneous hydrolysis of (**13**) is over 1000 times slower than that of (**10**); this was explained as resulting from the intermediate ion in the reaction of (**13**) undergoing recyclization faster than reaction with water. The rate-limiting step is therefore attack of water on the ion. Consistently, a solvent deuterium isotope effect, $k_{H_2O}/k_{D_2O} = 1.8$ at 100 °C, was found. On the other hand, (**12**) appears to undergo a spontaneous hydrolysis in which the rate-limiting step is cleavage of the C—O bond as

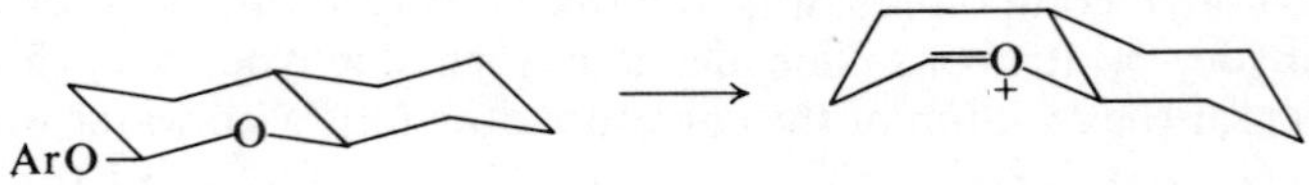

ArO (**9**)

ArO (**10**)

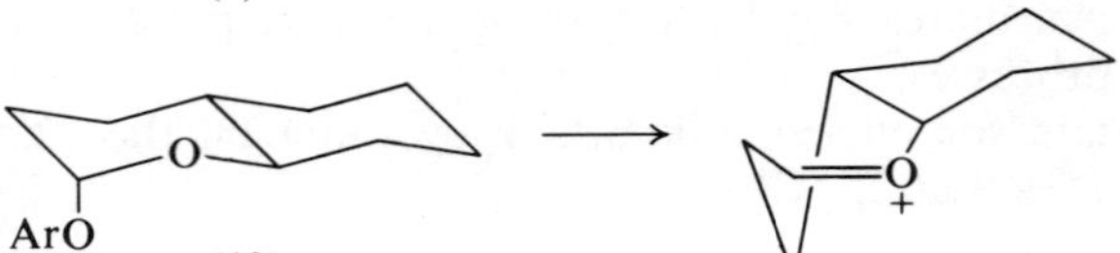

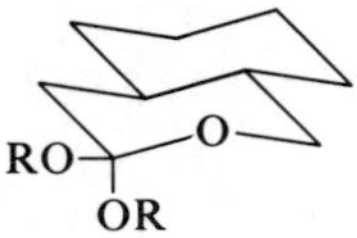

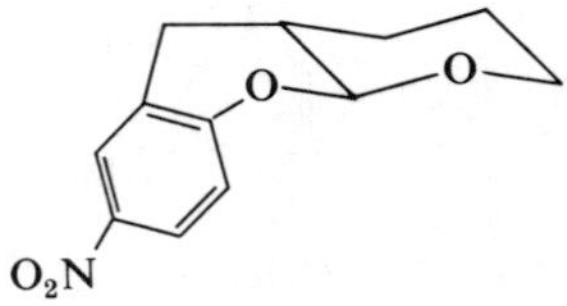

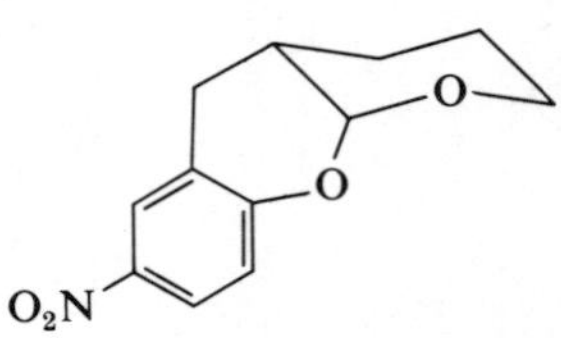

RO OR (**11**) O_2N (**12**) O_2N (**13**)

$k_{H_2O}/k_{D_2O} = 1.03$; this reaction occurs about 10^4 times more slowly than that of (**9**). The important difference between (**9**) and (**12**) is that (**9**) can achieve the boat or twist-boat conformation whereas (**12**) cannot.

Hydrolyses of the orthoesters (**14**) have strong negative entropies of activation and it was suggested that the reaction involved rate-limiting attack of water on a carbonium ion which is formed reversibly.[13]

The hydrolyses of 2,2-dialkoxy-tetrahydrofuran and -tetrahydropyran have been shown to be general acid catalysed.[14]

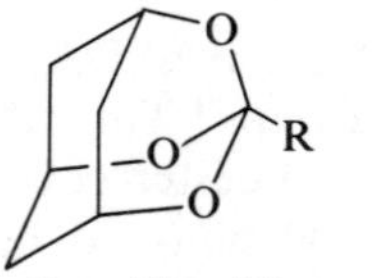

R = H or Me

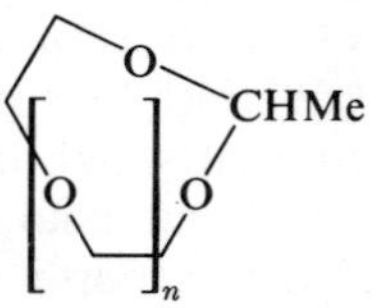

(**14**) CHMe n (**15**)

There have been further quantum-mechanical calculations on stereoelectronic effects in the breakdown of orthoesters and related species.[15]

Authentic intermolecular S_N2 reactions of acetals have been observed by Kirby and his co-workers.[16] The compound studied was methoxymethoxy-2,4-dinitrobenzene and the nucleophiles included AcO^-, F^-, N_3^-, I^-, CO_2^-, $S_2O_3^-$, HO_2^-, and HO^-. The α-deuterium isotope effect (k_H/k_D) falls in the range 1.05 to 1.16. It is clear that considerable caution must be used in interpreting α-deuterium isotope in enzymically catalysed reactions of glycosides.

The spontaneous hydrolysis of 2-aryloxytetrahydropyrans has a ρ^- value of 2.7.[17]

The effect of metal ions on the rates of hydrolysis of the "Crown Ether Acetals" (**15**) has been determined. The effects are slight when n = 0 to 3, but when n = 4 and 5 up to 11-fold rate decreases were observed, presumably as a result of cation binding.[18]

Halmann and his co-workers have reported further studies of the hydrolysis of phosphonated acetals.[19]

The acid-catalysed hydrolysis of *p*-nitrobenzaldehyde diethyl acetal in micelles of sodium lauryl benzoate has been investigated.[20]

The use of acetal protecting groups in nucleotide synthesis and their relative rates of hydrolysis have been discussed by Reese.[21]

Formation of acetals from *exo,cis*-2,3-dihydroxynorbornane and their subsequent equilibration has been studied.[22] The three-membered cyclic compounds (**16**) have been synthesized and tested as substrates for the enzyme epoxide hydrase.[23]

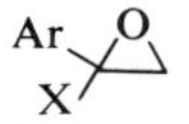

X = CH_3CO_2, CF, CH_2O, or F

(**16**)

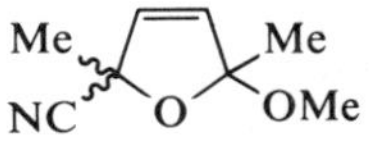

(**17**)

Dithioacetals and oxathiolones may be cleaved by isopentyl nitrite *via* the *S*-nitrosyl compound.[24] Electrochemical cleavage of 1,3-dithiones has also been investigated.[25]

Anodic oxidation of 2,5-dimethylfuran in methanol in the presence of cyanide ion yields a mixture of *cis*- and *trans*-isomers (**17**). The interconversion of these in methanol which contained trifluoroacetic acid was studied and thought to proceed *via* the cyclic carbonium-oxonium ion.[26]

Cycloheptatrienone reacts with ethane-1,2-dithiol or propane-1,3-dithiol to yield the products of 1,2-annelation rather than the thioketal.[27]

Other reactions that have been studied include photolysis of acetals,[28] reaction of cyclic acetals with ozone,[29, 30] homolytic isomerization of cyclic acetals,[31] isomerization of 1,3-dioxolanes,[32] reaction of 1,3-dioxolanes with Grignard reagents,[33] acid-catalysed hydrolysis of 2,4-disubstituted 1,3-dioxolanes,[34] hydrolysis of 2-(2,6-dichlorophenyl)-4,6-dimethyl-1,3-dioxane,[35] thiolysis of cyclic acetals,[36] neutral hydrolysis of 1,4-dioxaspiro[4,4]deca-6,9-diene-2,8-diones,[37] base-catalysed decomposition of 1-(2-hydroxyphenyl)-2-phenyl-3,3-dimethoxypropan-1-one,[38] reaction of 3-methylbut-3-en-1-ol with formaldehyde,[39] elimination reactions of acetals to yield vinyl ethers,[40] acetal formation,[41] methanolysis of acetals,[42] transketalization,[43] hydrolysis of 1,2-*O*-alkylidene-α-D-glucofuranoses,[44] acetal formation from 2-deoxyhexitols and aldehydes,[45] the reaction of mannose with acetone,[46] migration of the benzylidene group of methyl 3,4-*O*-(*R*)-benzylidene-β-D-ribopyranoside to yield methyl 2,3-*O*-(*R*)-benzylidene-β-D-ribopyranoside and the slower conversion of these compounds into their *S*-isomers,[47] and condensation of meso- and 2D,3L-dimethylbutane-1,2,3,4-tetraol with formaldehyde.[48]

The conformation of 1,3-dioxanes has been studied.[49]

Hydrolysis and Formation of Glycosides

Non-enzymic Reactions

The hydrolysis of sucrose in reversed micelles in benzene has been studied.[50]

It has been suggested that the thermal degradation of sucrose involves intramolecular nucleophilic attack by the hydroxyl group at C(1) of the fructosyl residue on C(2).[51]

Base-catalysed formation of 1,6-anhydro-β-D-glucopyranose from phenyl α-D-glucopyranoside has been studied.[52] Equilibrium constants for formation of 1,6-anhydropyranoses from aldoheptoses have been determined.[53]

The anomerization of 1,2-*trans*-glycofuranosides with Grignard reagents has been studied.[54]

Enzymic Reactions

(a) *Galactosidases.* The amino-acid sequence of β-galactosidase from *E. coli*, a tetramer of molecular weight 500 000, has been determined.[55] Methionine 500 is the site of reaction of the active-site directed irreversible inhibitor, β-D-galactopyranoxylmethyl *p*-nitrophenyltriazine.[56]

Inhibition of the β-galactosidase of *E. coli* has been investigated. Four classes of behaviour were found: (i) compounds that affected neither K_{m} not V_{max}; (ii) competitive inhibitors that modified K_{m} but not V_{m}; these bind to the free enzyme but not to the ES′ complex; (iii) non-competitive inhibitors that modify V_{m} but not K_{m}; these bind to free enzyme and ES′ complex; (iv) uncompetitive inhibitors that affect K_{m} and V_{m} but the ratio $K_{\mathrm{m}}/V_{\mathrm{m}}$ remains constant; these bind to the ES′ complex only. It was concluded that there are two binding subsites. It was also shown that there is conformational change when an inhibitor is bound to the enzyme.[57]

There have been investigations of the binding of phenyl 1-thio-β-D-galactopyranosides to the active site of β-D-galactosidase from *E. coli*[58] and of the pH-dependence of the kinetic parameters for reactions catalysed by this enzyme.[59]

There has been a comparison of wild-type β-galactosidase from *E. coli* and a defective β-galactosidase from deletion mutant strain M15.[60]

The reaction of glycerol with D-galactal-2-d catalysed by the β-galactosidase from *E. coli*, to yield 2,3-dihydroxypropyl 2-deoxy-β-D-galactoside, occurs by a *trans*-addition.[61]

2,6:3,4-Dianhydro-1-deoxy-D-talohept-1-enitol has been used as an irreversible inhibitor for the β-D-galactosidase from *E. coli*.[62]

The bind of Mn^{2+} and Mg^{2+} to the β-D-galactosidase from *E. coli* has been investigated.[63]

(b) *Glucosidases.* Good evidence that an isozyme of the glycosidase from almond emulsin, which catalyses the hydrolysis of β-D-galactosidase, β-D-glucosidase, and β-D-fucosidase, uses the same active site to do so has been reported by Walker and Axelrod.[64] Thus D-glucosylamine, D-galactosylamine, and D-fucosylamine each acts as a competitive inhibitor against each substrate, and each inhibitor has the same K_{i} value for whichever substrate is used.[64]

It has been reported that (**20**) is a by-product of the hydration of D-glucal catalysed by β-glucosidase. It was proposed that the reaction involved ionization with loss of the 3-hydroxyl group of the glucal (**18**) to yield an allylic cation (**19**) which was attacked by 2-deoxyglucose, the main reaction product.[65]

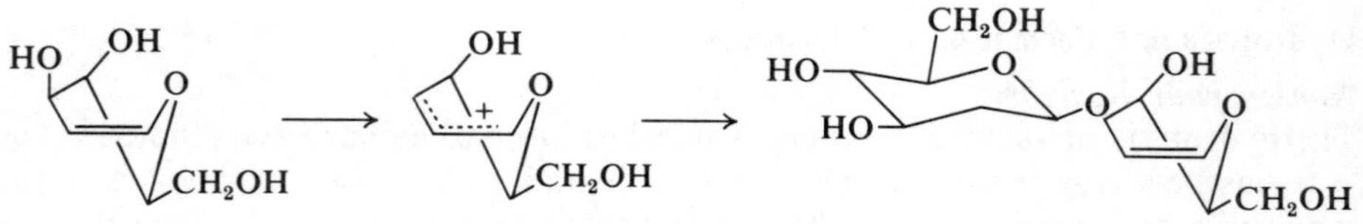

(**18**) → (**19**) → (**20**)

6-Bromo-3,4,5-trihydroxycyclohexene-2-t oxide (**21**) is an irreversible inhibitor for the β-glucosidase A from bitter almonds. Two peptides were obtained from the tryptic digest in which the inhibitor had become attached to different aspartic acid residues. It seems likely that the inhibitor binds in two different ways to the enzyme.[66]

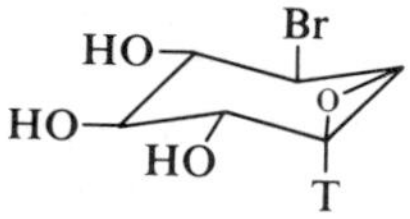

HO, HO, HO, Br, O, T

(**21**)

The β-D-glucosidase isoenzymes B from sweet and bitter almonds have been compared.[67]

There have been investigations of the β-D-glucosidase from *Pyricularia oryeza*.[68] the undiced β-D-glucosidase from *Stachybotrys atra*,[69] and of the β-D-glucosidases from baker's yeast,[70] *S. carlsbergensis*,[71] and sugar beet seed.[72]

(c) *Lysozymes*. Details have been published of an investigation by Capon and his co-workers of the hydrolysis of 3,4-dinitrophenyl glycosides of chitin oligosaccharides and of modified chitin oligosaccharides catalysed by HEW lysozyme. In order for these compounds to be substrates, it is essential that the CH_2OH group of the residue bound in subsite-D be unmodified; it was concluded that this residue must undergo an important interaction with the enzyme in the transition state and that there was no evidence of any distortion in the ground state of the enzyme–substrate complex.[73]

A PMR spectroscopic investigation of the bounding of the cell-wall trisaccharide NAM-NAG-NAM by HEW lysozyme indicated that the H(1)–H(2) coupling constant of the reduced NAM residue is unchanged when it is bound in subsite-D. It therefore seems probable that there is no formational change on bonding and hence little strain on bonding in subsite-D.[74]

There have been several other investigations of HEW lysozyme.[75]

An electron density map at 2.4 Å resolution has been reported for lysozyme from bacteriophage T4.[76]

There have been investigations on the lysozymes from *Pseudomonas aeruginosa*[77] and *Streptomyces erythraeus*.[78]

(d) *Amylases and Glucamylases*. Maltotrioses substituted at position 6 of the non-reducing terminal glucose residue have been synthesized and tested as substrates for Taka-amylase A. They were hydrolysed to glucose and a modified maltose. The relative values of k_0/K_m for maltotriose and the 6-deoxy-, 6-chloro-, 6-bromo, and 6-iodo-substituted compounds were 1, 0.07, 0.24, 0.19, and 0.22, respectively; this suggests that the 6-hydroxyl group of the terminal non-reducing residue undergoes an important interaction with enzyme in the transition state but that 6-halo-substituents are also able to do this.[79]

The binding of substrates to Taka-amylase A has been determined by UV-difference spectroscopy.[80]

D-Gluco-δ-lactone and maltobionic-δ-lactone have been compared as inhibitors for amylolytic enzymes.[81]

The hydrolysis of aryl β-maltotriosides by sweet potato β-amylase has been investigated.[82]

Other enzymes that have been studied include the saccharifying α-amylase from

Bacillus subtilis,[83] α-amylase from *Thermoactinomyces vulgaris*[84] and porcine pancreas,[85] Taka-amylase A,[86,87] and the glucoamylase from *A. niger*.[88]

(e) *Cellulases*. The steric course of the reactions catalysed by two *exo*-cellulolytic enzymes from *T. viride* has been determined. A β-glucosidase catalysed the hydrolysis of *p*-nitrophenyl β-D-cellobioside with retention of configuration and an *exo*-β-1,4-glucanase-hydrolysed cellopentitol with inversion of configuration.[89]

The cellulolytic enzymes from *T. koningii* have been investigated.[90]

(f) *Other Enzymes*. Other enzymes that have been investigated include β-D-glucuronidase,[91] dextranases,[92,93] α-D-mannosidases,[94,95] β-D-mannosidases,[96] α-L-arabinofuranosidases,[97,98] β-D-xylosidases,[99,100] *exo*-D-galacturonanase,[101] *endo*-galacturonanase,[102] trehalase,[103] and β-*N*-acetylhexosaminidase.[104]

Hydration of Aldehydes and Ketones and Related Reactions

Jencks and his co-workers[105] have made an important and detailed investigation of the breakdown of the hydrate and of some hemiacetals of formaldehyde ($HOCH_2OR$). Plots of $\log k_{gb}$ for the general base catalysed reaction against pK_{ROH} are curves that are convex downwards. The fact that electron-releasing substitutents are rate-enhancing is strong support for a class *n* mechanism (equation 1) rather than a class *e* mechanism (equation 2); if the reaction involved expulsion of RO^-, as in the latter case, no such enhancement would be expected.

$$H_2C(OH)(OR) + B \rightleftharpoons H_2C(O^-)(OR) + \overset{+}{B}H \rightleftharpoons H_2C(\cdots O^{\delta-})(\cdots O(R)\cdots H\cdots \overset{\delta+}{B}) \rightleftharpoons H_2C{=}O + B + ROH \quad (1)$$

$$H_2C(OH)(OR) \rightleftharpoons H_2C(\cdots O\cdots H\cdots \overset{\delta+}{B})(\cdots \overset{\delta-}{O}R) \rightleftharpoons H_2C{=}O + RO^- + HB^+ \rightleftharpoons H_2C{=}O + B + ROH \quad (2)$$

$$H_2C(OH)(OR) + HA \rightleftharpoons H_2C(OH)(\overset{+}{O}(H)R)\ A^- \rightleftharpoons H_2C(\cdots O\cdots H\cdots \overset{\delta-}{A})(\cdots \overset{\delta+}{O}(H)R) \rightleftharpoons H_2C{=}O + HA + ROH \quad (3)$$

$$H_2C(OH)(OR) + HA \rightleftharpoons H_2C(\cdots \overset{\delta+}{O}H)(\cdots O(R)\cdots H\cdots \overset{\delta-}{A}) \rightleftharpoons H_2C{=}O + A^- + ROH \rightleftharpoons H_2C{=}O + HA + ROH \quad (4)$$

Also the class *e* mechanism would require some of the individual rate constants to be larger than the rate constant for diffusion. The general acid catalysed reaction may also proceed by a class *e* (equation 3) or a class *n* (equation 4) mechanism. The value of α increases with decreasing pK_a of the leaving group, *i.e.* $\partial\alpha/-\partial pK_{lg} = 0.022$. This is consistent with both mechanisms but the fact that $k_{H_3O^+}$ for cleavage of formaldehyde methyl hemiacetal is 2600 times greater than for cleavage of formaldehyde dimethyl acetal strongly suggests that the removal of the proton has occurred or is occurring in the transition state for the former reaction. Therefore a class *e* mechanism is most likely. The water reaction appears to be a general base catalysed rather than a general acid-catalysed reaction.[105]

Young and Jencks have investigated the breakdown of substituted acetophenone bisulphite compounds (**22**) and of the corresponding α-methoxy-sulphonic acids (**23**). The latter compounds are hydrolysed with general acid catalysis and reaction is thought to involve partial proton donation to the leaving sulphite group. The breakdown of the corresponding bisulphite compounds (**22**) occurs only about

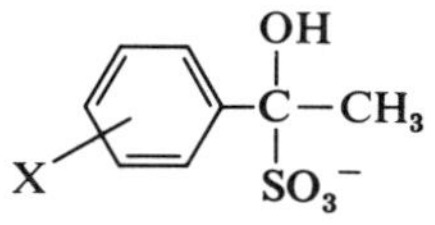

(**22**) 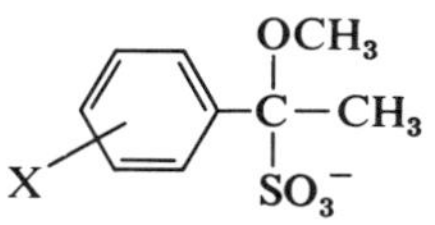 (**23**)

six-fold faster than that of (**23**) and hence probably proceeds by the same mechanism and not by one involving proton abstraction from the OH group. The variation of α with substituent (X = *p*-MeO, $\alpha = 0.6$; X = *p*-Cl, $\alpha =$ *ca.* 0.4; X = *p*-NO_2, $\alpha = 0.31$) is in the opposite sense to that found in the hydrolysis of benzaldehyde methyl phenyl acetals.[106] These results and those for other reactions have been discussed extensively in terms of three-dimensional free-energy contour diagrams.[107,108] An extensive investigation of the acid-catalysed addition of thiols to ketones in aqueous ethanol has been reported by Lamaty and his co-workers. The results were interpreted in terms of a rate-limiting attack of the thiol on the conjugate acid of the ketone with a relatively late transition state.[109]

The addition of methyl mercaptoacetate to acetaldehyde shows general acid catalysis but the plots of k_{obs} *versus* concentration of catalyst are curves that show a strong dependence of k_{obs} on [catalyst] at low concentrations and a weak dependence at high concentrations. This behaviour suggests that there is a change in the rate-determining step with change in catalyst concentration. It was proposed that, at low concentrations of catalyst, the rate-determining step was trapping of the anionic intermediate (**24**) which is formed reversibly; at high concentrations of catalyst, formation of this intermediate was rate-determining. In addition, in order

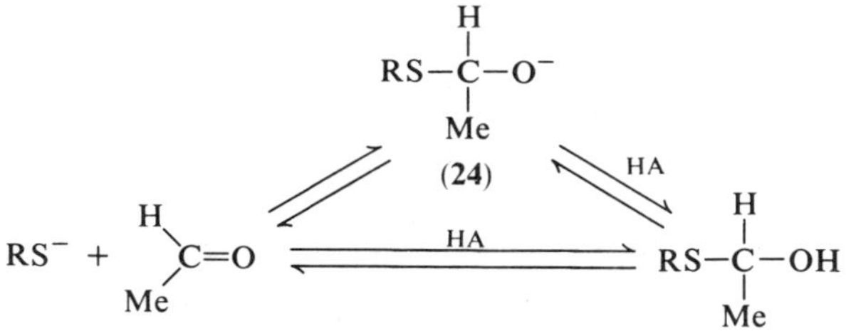

(**24**)

to explain the weak dependence of k_{obs}[catalyst] at high concentrations, it was proposed that there was a general acid catalysed reaction that did not proceed *via* the anionic intermediate but involved attack of thiolate on the carbonyl group which was hydrogen-bonded to the catalysing acid.[110,111]

The rate constants for hydration of 1,3-dichloroacetone in reverse micelles of Aerosol-OT (**25**) were 1808 to 111 times greater than those in aqueous dioxan with the same molar concentration of water. This may arise from a difference in structure between micelle-solubilized water and water in the dioxan–water mixture and/or from general base catalysis by the sulphite head-group of the surfactant.[112]

Benzils and ethyl phenyl glyoxylate are very effective inhibitors for chicken-liver esterases as a result of forming hemiacetals with the active-site serine. (*E*)-Benzil mono-oxime *O*-2,4-dinitrophenyl ether also forms a hemiacetal that undergoes a base-catalysed fragmentation by reaction with the enzyme.[113]

Several investigations of solvent effects on the mutarotation of glucose and tetramethylglucose have been reported.[114–116]

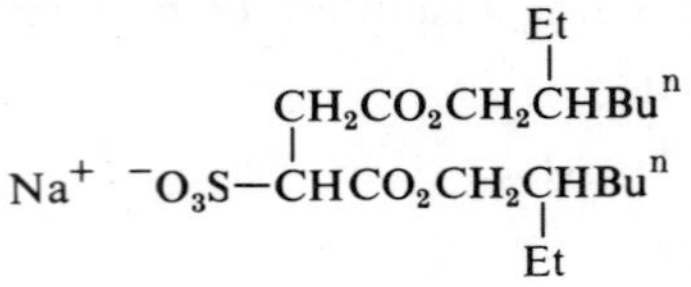

$$Na^+ \; {}^-O_3S-\underset{|}{\overset{CH_2CO_2CH_2CH(Et)Bu^n}{|}}CHCO_2CH_2CH(Et)Bu^n$$

(25)

Further work on the intramolecular catalysis of mutarotation of glucose 6-phosphate has been reported.[117]

Base-catalysed ring opening of 5-thio-D-glucopyranose and 5-thio-D-xylopyranose has been studied[118] by trapping the free thiol group with bis-(4-pyridyl) disulphide: it was found that base-catalysed mutarotation of these sugars is 500 times faster than ring opening as measured by this method; it was suggested that there is "a non-covalently bonded intermediate which can undergo ring closure to either α- or β-pyranose (thus causing mutarotation) without ring opening".[118]

The dimerization of tetramethylglucose and 2-pyridone in benzene has been studied by microcalorimetry.[119] The equilibrium between 2-hydroxypyridine and 2-pyridone has been investigated further.[120]

It has been shown by ^{13}C-NMR spectroscopy that D-idose contains 1.6% of a septanose isomer.[121] Complexing of carbohydrates with molybdenum has been studied.[122]

A review[123] on "Bifunctional Catalysis" and a monograph[124] on "Ring-chain Isomerism" have been published.

Dimerizations of lactaldehyde[125] and 3-hydroxy-2,2-dimethylpropanal[126] have been studied.

There have been several investigations of the hydration of aldehydes and ketones,[127–131] and MO calculations on the hydration of formaldehyde have been reported.[132] The chemistry of ninhydrin has been reviewed.[133]

The carbonyl hydrate has been detected as an intermediate in the cleavage of dimethylacetylphosphonate[134] (see Chapter 2). The cleavage of acylsilanes has been studied.[135]

A hemiacetal[136] and a hemiacylal[137] with the bicyclo[3.3.1]nonane skeleton have been prepared.

Aldehydes have been shown to inhibit papain through formation of hemithioacetals with the active-site thiol group.[138]

The reaction of formaldehyde with sodium sulphide has been studied.[139]

Reactions with Nitrogen Bases[140]

Schiff Bases

Fife[141] has extended his work on the hydrolysis of imidazolines to the 2-*tert*-butyl-*N,N'*-dimethyl and *N*-isopropyl-2-*p*-methoxyphenyl-*N'*-phenyl compounds **(26)** and **(27)**. At low pH, the reaction to form compound **(28)** shows kinetic general acid catalysis and the reaction was thought to involve general base catalysis of the hydrolysis of the intermediate protonated Schiff base which is present at a small concentration. At high pH, there is a pH-independent reaction which may involve either rate-determining water-catalysed or uncatalysed breakdown of an imidazoline or a hydronium ion-catalysed ring-opening of the imidazoline followed by a rate-determining hydroxide ion-catalysed hydrolysis of the Schiff base. The intermediate Schiff base could be detected in the hydrolysis of (**27**) was thought to be the *N*-alkyl compound (**29**) from its UV spectrum.[141]

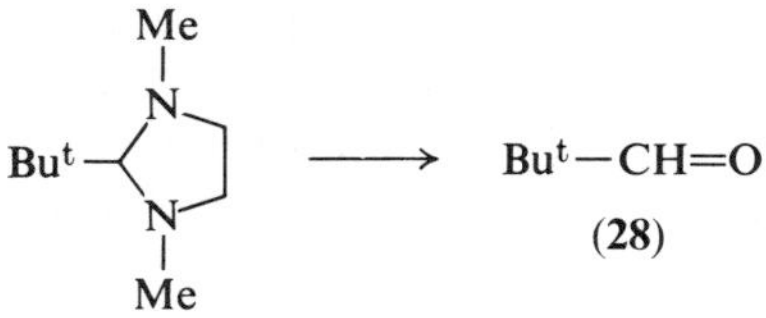

(**26**) $\longrightarrow$ Bu^t—CH=O (**28**)

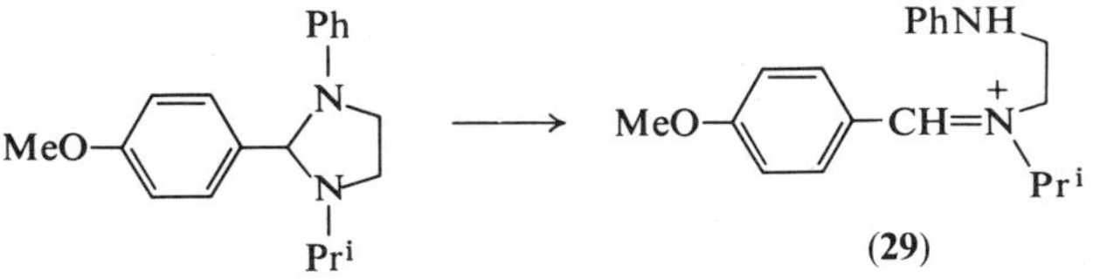

(**27**) $\longrightarrow$ (**29**)

The mode of breaking of unsymmetrical *gem*-diamines derived from formaldehyde has been studied. They were generated by reduction of the corresponding amidines with borohydride. Throughout the pH range studied (2–11), the preferred reaction was expulsion of the less basic amine.[142]

Pollack and his co-workers have reported further work on the hydrolysis of Schiff bases derived from cyclohexene-1-carboxaldehyde and amino-acids.[143]

Cyclization of the imidoyl azide (**30**) to the tetrazole (**31**) is slow in inert solvents

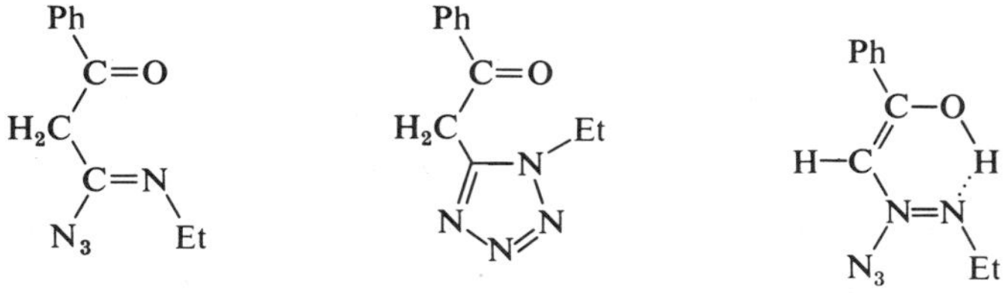

(**30**) (**31**) (**32**)

(**33**) RNH—C(Me)=CHCOMe (**34**)

and in aqueous solution at neutral pH. It shows acid and base catalysis and at $pH < 2$ acid inhibition; this was explained in terms of stabilization of the diazo-form (**32**) by intramolecular hydrogen bonding at neutral pH. The base catalysis was thought to arise from proton abstraction which destroys the hydrogen bond and allows rapid isomerization into the reactive form (**33**); the acid catalysis was attributed to protonation of the nitrogen which disrupted the hydrogen bonding and allowed rotation to occur. At high acidities, there was complete conversion into the conjugate acid which is unreactive.[144] The rate-decreasing effect of the intramolecular hydrogen bond is an example of deceleration due to stabilization of the initial state by interaction with a neighbouring group, and it could be referred to as "anchimeric deceleration".

The amine-catalysed eliminations of β-hydroxy- and β-acetoxy-ketones show large kinetic isotope effects ($k_H/k_D = ca.\ 5$) when the axial α-proton is replaced by deuterium. It was estimated (by determining the deuterium contained in the product) that the preference for axial proton abstraction rather than equatorial proton abstraction from the intermediate Schiff base is greater than 100.[145]

The hydrolysis of *N*-alkyl- and *N*-aryl-4-aminopent-3-en-2-ones (**34**) has been investigated: above pH 2, the rate-limiting step in the reaction of the *N*-alkyl compounds is base-catalysed attack of water on the substrate protonated at C(3); below pH 2, it is hydration of the carbinolamine. With the *N*-aryl compounds, base-catalysed attack of water on the protonated substrate is rate-limiting over the whole pH range.[146]

The effect of acid on the thermochromic system shown in equation 5 has been studied. Acid catalyses equilibration of the environments of the *gem*-methyl groups. This reaction involves ring opening to give the Schiff base (**35**) which

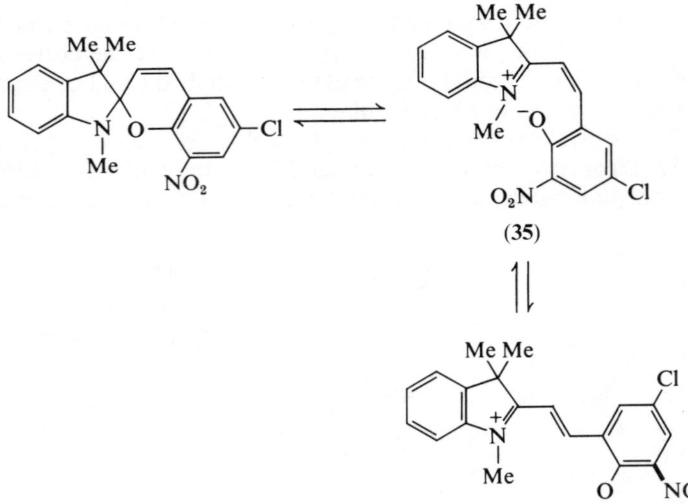

Me Me N Me O NO$_2$ Cl ⇌ Me Me N$^+$ Me $^-$O O$_2$N Cl (**35**) ⇌ Me Me N$^+$ Me Cl O NO$_2$ (5)

recyclizes. In addition, (**35**) is converted by a non-acid-catalysed pathway into its *trans*-isomer which is blue.[147]

Labelling experiments have demonstrated that the hydroperoxide formed on autoxidation of the Schiff base formed from 9-aminoanthracene and isobutyraldehyde undergoes base-catalysed decomposition in water–dimethyl sulphoxide *via* the dioxetan as shown in equation (6).[148]

It appears that α-proton abstraction from the Schiff bases formed from pyridoxal and amino-acids is partly controlled by stereoelectronic factors; Schiff bases in which a C(α)—H(α) bond is orthogonal to the π-system are especially reactive.[149]

Pyridoxal 5′-phosphate reacts with mixed micelles of hexadecyltrimethylammonium bromide and dodecylamine hydrochloride to form a Schiff base which has similar spectral characteristics to the Schiff base formed by pyridoxal 5′-phosphate with glycogen phosphorylase.[150]

The rate constant for Schiff base formation with pyridoxal-pyruvate transaminase and 5-deoxypyridoxal is $1.79 \times 10^7\ M^{-1}\ s^{-1}$. It was suggested that the rate-determining step in imine formation is production of the ES complex.[151]

There have also been investigations of the template effect in Schiff base formation,[152] Schiff base hydrolysis,[153] the effect of micelles on Schiff base hydrolysis,[154]

$$\text{ArN=CH-C(OOH)Me}_2 \longrightarrow \text{ArNH-CH(O-O)CMe}_2 \longrightarrow \text{ArNHCHO} + \text{O=CMe}_2 \quad (6)$$

condensation of methyl benzoylpyruvates with aniline,[155] alkaline hydrolysis of 3-(substituted phenylimino-oxindoles,[156] hydrolysis of "Etazolate hydrochloride",[157] reaction of α-amino-acids with phenalene-1,2,3-trione hydrate,[158] reaction of acetyl- and formyl-ferrocene with acetone cyanohydrin and amines to yield $FcCR(CN)NHC_6H_4R'$,[159] hydrolysis of methyl (α-acetoxybenzyl)nitrosamine,[160] ketimine–enamine tautomerism,[161] reaction of phthalaldehyde with ammonia and amines[162] and with thiols and primary amines,[163] transamination reaction between α-amino-acids and phthalaldehydic acid,[164] reduction of a pyridoxal 5′-phosphate Schiff base,[165] pyridoxyl-catalysed γ-elimination of α-amino-acids,[166] and *E–Z* isomerization of imines.[167,168]

Enamines

There have been investigations of the protonation of enamines,[169,170] and of the reaction of cyclohexanone enamines with diacyl di-imides.[171] MO calculations on vinylamine have been reported.[172]

Nucleosides

Cordes and his co-workers have reported α-deuterium isotope effects for the hydrolysis of several nucleosides and related compounds. The α-deuterium isotope effects for the hydrolysis of unprotonated, monoprotonated, and diprotonated inosine and of deprotonated adenosine (k_H/k_D = *ca.* 1.2 in each case) were interpreted in terms of rate-limiting unimolecular decompositions.[173]

The α-deuterium isotope effects for the hydrolysis of β-NAD^+ and β-NMN (1.103 and 1.135, respectively) also indicate unimolecular processes.[174] However, the enzymically catalysed hydrolysis of β-NAD^+ by the NAD-glycohydrolases from pig-brain and *Neurospora crassa* are 0.985 ± 0.021 and 1.003 ± 0.002, respectively, and it was concluded that, alternatively, (i) substrate binding to these enzymes is irreversible, (ii) C—N bond cleavage is not involved in the rate-determining step, or (iii) C—N bond cleavage occurs with participation of a nucleophile. Of these, (ii) was preferred and it was thought that a conformational change was the rate-determining step. In contrast, for the hydrolysis of β-NMN catalysed by the same enzymes, the α-deuterium isotope effects are 1.132 and 1.100, respectively. With this substrate, therefore, there appears to be a different rate-determining step which is probably one involving formation of a carbonium ion.[174]

The phosphorylation of adenosine and inosine by calf spleen purine nucleoside phosphorylase was also studied. Conditions were used where the measured isotope effect was for binding of the nucleoside to the enzyme. Values of k_H/k_D = 1.04–1.05

were obtained and interpreted as arising from distortion of the ribose ring in bonding to the enzyme.[175]

Hydrazones, Oximes, Semicarbazones and Related Compounds

The solvent isotope effect for the general acid catalysed reaction of *p*-methoxybenzaldehyde and methoxylamine varies from *ca.* 1 for catalysis by formic acid and cyanoacetic acid to *ca.* 3 for $NCCH_2CH_2NH_3^+$ and $O(CH_2CH_2)_2NH_2^+$, and is 2 for $MeOCH_2CH_2CH_2NH_3^+$ and $CH_3CH_2CH_2NH_3^+$. The results were interpreted in terms of a mechanism that involved rate-limiting protonation of the zwitteronic intermediate (**36**). In this process, it was proposed that the rate-limiting step changed from formation of the encounter complex with the strong

$$\mathrm{ArCHO} + \mathrm{CH_3ONH_2} \rightleftharpoons \mathrm{ArC(O^-)-\overset{+}{N}H_2OCH_3}\ (\mathbf{36}) \xrightarrow{\mathrm{HA}} \mathrm{ArC(OH)-\overset{+}{N}H_2OCH_3}$$

acid to proton transfer with $NCCH_2CH_2CH_2NH_3^+$ and $O(CH_2CH_2)_2NH_2^+$ and was starting to change to separation of the products with the weaker acids (cf. Ref. 177 and Chapter 2).[176]

The Brønsted coefficient for catalysis by phosphonate buffers of the dehydration of the intermediate carbinolamine in semicarbazone formation from substituted benzaldehydes is independent of the substituent ($\alpha = 0.80 \pm 0.02$; $\partial\alpha/\partial\sigma = 0$) and also the ρ-constant (-1.9) is independent of acid strength ($\partial\rho/-\partial pK_{HA} = 0$). On the other hand, the Bronsted α-coefficient increases as the leaving group becomes more basic; this apparent imbalance of the transition state has been discussed.[178]

The kinetics of the addition of nitrogen base to pyruvic acid have been studied by line-shape analysis of the PMR spectra of a flowing reaction mixture.[179]

The reaction of hydrazine with ethyl acetoacetate has been investigated by flow NMR methods. A time-averaged spectrum of the ethyl acetoacetate and carbinolamine was observed in addition to the spectra of the *trans*-hydrazone and 3-methylpyrazol-5-one which is the product of intramolecular displacement on the ester group by the NH_2 group of the hydrazone. The *cis*-hydrazone was not detected, so the *trans*-hydrazone must isomerize more rapidly than it cyclizes. In addition, cyclization probably takes place at the carbinolamine stage.[180]

The pH–rate profile for the formation of the phenylhydrazone of 2-hydroxyacetophenone shows a plateau in the pH-range 5–7. Under these conditions, formation of the carbinolamine is rate-limiting but no general acid catalysis by external acids was detected. It was suggested that the reaction involved intramolecular general acid catalysis by the hydroxyl group of the phenol.[181]

The kinetics of both steps in the formation of pyruvate oxime have been measured by continuous-flow microcalorimetry.[182]

The 2,4-dimethylsemicarbazones of some aromatic aldehydes exist as ring forms in deuteriotrifluoroacetic acid.[183]

The imine (**38**) has been detected (^{13}C NMR spectroscopy) as an intermediate in the Fischer indole synthesis of indomethacin (**39**) from the hydrazone (**37**).[184] Other investigations of the Fischer indole synthesis have also been described.[185, 186]

Other reactions which have been investigated include hydrolysis of α-nitrobenzaldehyde phenylhydrazone[187] and of erythromycin oxime,[188] formation of semicarbazones from heterocyclic ketones,[189] iodine-promoted cyclization of

γ,δ-unsaturated ketones,[190] reaction of ketoximes with phenyl isothiocyanate,[191,192] condensation of *o*-phthalaldehyde with urea and thiourea,[193] the Hoch–Campbell reaction,[194] formation of pyrazoles from 4,4-dimethoxybutan-2-one and methylhydrazine,[195] and the *E*–*Z*-isomerization of azines[196] and oximes.[197,198]

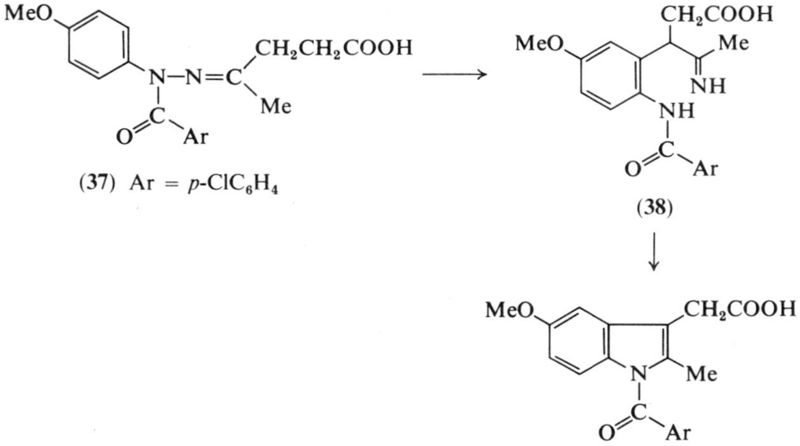

(37) Ar = *p*-ClC_6H_4

(38)

(39)

Hydrolysis of Enol Ethers

In a further attempt to find an enol ether whose hydrolysis involves a rapid and reversible protonation, Kresge and Chwang[199] have studied the hydrolysis of 1-cyclopropylvinyl methyl ether. However, hydrolysis of this compound showed all the characteristics of a reaction that involves a slow proton transfer (*e.g.* general acid catalysis and $k_{H_3O^+}/k_{D_3O^+} = 4.51$). It was also shown that *cis*- and *trans*-2-arylvinyl ethers are not interconverted during hydrolysis;[200] therefore, with these compounds too, the rate-determining step is protonation of the carbon–carbon double bond. It is interesting to note that it has been suggested that, under certain conditions, hydrolysis of ketene *S*-ethyl phenyl monothioacetal does involve the reversible proton transfer sought.[201]

An α-phenyl group has a smaller rate-enhancing effect in vinyl ether hydrolysis than an α-methyl group. This probably arises from the fact that the α-phenyl group is conjugated with the double bond and thus stabilizes the initial state as well as the transition state.[202]

The isotope effects of the hydronium ion catalysed hydrolysis of vinyl ethers have been discussed in terms of Marcus rate theory.[203]

Acid-catalysed hydrolysis of the enol ether (**40**) yields (*inter alia*) compound (**43**). This was thought to be formed by protonation and bromine participation to yield the bromonium ion (**41**) which then underwent attack from the *exo*-face to yield the hemiacetal (**42**) which subsequently reacted with neighbouring methoxyl participation as shown,[204]

Tracer studies have shown that the biosynthetic enol ether cleavage of thebaine to form codeinone involves methyl–oxygen bond cleavage and not the normal type of enol ether hydrolysis.[205]

The reactions of cyclic enol ethers (including glycals) with organopalladium compounds have been studied.[206]

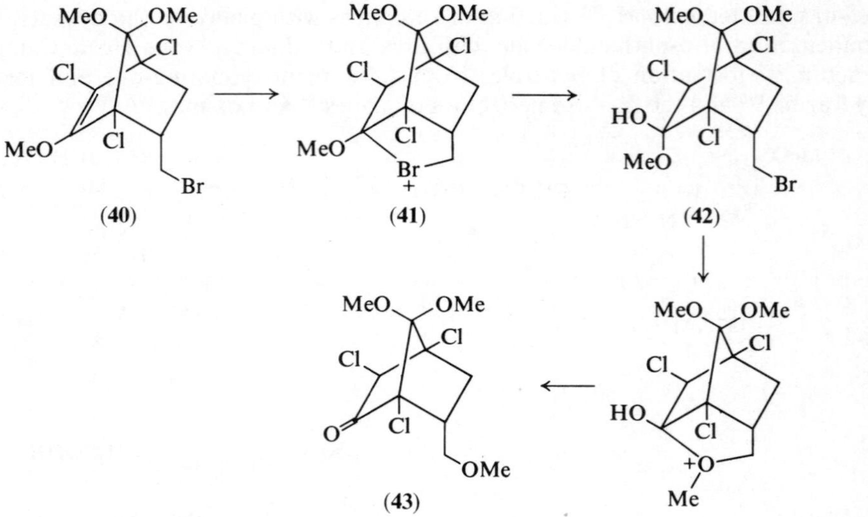

(40) → (41) → (42) → → (43)

There has been an extensive investigation of the ^{13}C-NMR spectra of vinyl ethers.[207]

Enolization and Related Reactions

There have been reviews on "Bifunctional Catalysis of α-Hydrogen Exchange of Aldehydes and Ketones"[208] and "Studies of Amine Catalysis *via* Iminium Ion Formation".[209]

Details have been published of Bruice and Bruice's demonstration that the enolization of oxaloacetic acid does not proceed by a concerted general acid general base catalysis.[210] It was found that tertiary amines with $pK_a > 8$ show a special type of kinetic behaviour: the intercept of the plot of $\log k$ *versus* $[Me_3N]$ does not lie in the pH–rate profile, determined by using other buffers. It was proposed that these tertiary amines catalyse the reaction through formation of a carbinolamine (equation 7).[210] Similar behaviour was reported for diethyl oxaloacetate.[211] The keto–enol tautomerism of oxaloacetic acid has also been investigated by PMR spectroscopy.[212]

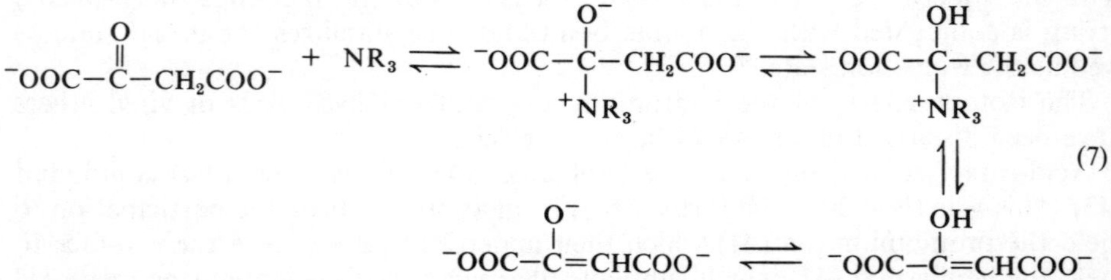

$$^{-}OOC{-}C(=O){-}CH_2COO^{-} + NR_3 \rightleftharpoons {}^{-}OOC{-}C(O^{-})(^{+}NR_3){-}CH_2COO^{-} \rightleftharpoons {}^{-}OOC{-}C(OH)(^{+}NR_3){-}CH_2COO^{-}$$

$$\rightleftharpoons {}^{-}OOC{-}C(OH){=}CHCOO^{-} \rightleftharpoons {}^{-}OOC{-}C(O^{-}){=}CHCOO^{-} \qquad (7)$$

Enol pyruvate has been generated by the treatment of phosphoenol pyruvate with phosphorylase. Tritiated phosphoenol pyruvate yielded the triply labelled pyruvate in D_2O. This was racemic when no other enzyme was present but was the (3*S*)-enantiomer when the reaction was carried out in the presence of pyruvate kinase. This was the same isomer as was formed in the presence of pyruvate kinase alone; hence these results are consistent with the reaction, catalysed by pyruvate kinase, proceeding *via* enol pyruvate since it was shown that pyruvate kinase catalyses the conversion of enol pyruvate into pyruvate.[213]

On the other hand, hydrolysis of *Z*- and *E*-phosphoenol-2-oxobutyrate by pyruvate kinase yields 2-oxobutyrate only in a stereoselective manner. It was suggested that protonation of the enol occurred partly through a non-enzymatic pathway.[214a]

It has been demonstrated that ionization of both enzyme and substrate contributes to the form of the pH–rate profiles for the interconversion of D-glyceraldehyde 3-phosphate and dihydroxyacetone phosphate catalysed by triose phosphate isomerase from chicken muscle.[214b]

Phosphomannose isomerase, which catalyses the interconversion of D-mannose 6-phosphate and D-fructose 6-phosphate, uses β-D-fructose 6-phosphate as its substrate.[214c]

Enolization is the rate-determining step for release of *p*-nitrophenol from 4-(*p*-nitrophenoxy)butan-2-one and base; the latter reaction is therefore easily studied with this substrate. Rate constants were measured for the reaction with 30 bases with pK_a values 4.70 to 15.75; the Brønsted plot, which was curved, was discussed in terms of a modified Marcus theory in which the curvature was ascribed to a variation of w_r.[215] Similar curvature was found in the reactions of pivalate esters.[216]

Carboxypeptidase A has been shown to catalyse deuterium exchange, presumably *via* an enol or enolate, of the substrate analogue 3-*p*-methoxybenzoyl-2-benzylpropanoic acid-3,3-d_2 (**44**). This reaction was inhibited by $\pm$-benzylsuccinic acid which is an inhibitor of hydrolyses, catalysed by carboxypeptidase A.[217]

As part of an extensive investigation of the volumes of activation of proton- and hydroxide-transfer reactions, Brower and Hughes[218] have determined the volume of activation for dedeuteriation of acetophenone-2,2,2-d_3 by hydroxide ion to be -1 ml mol^{-1}. This result was interpreted in terms of a late transition state. The $\Delta V^{\ddagger}$ for the Cannizzaro reaction was also determined. This was -27 ml mol^{-1} which was thought to be made up of *ca.* -12 ml mol^{-1} for the initial addition of HO^- and of *ca.* -15 ml mol^{-1} for the transfer of H^-.[218]

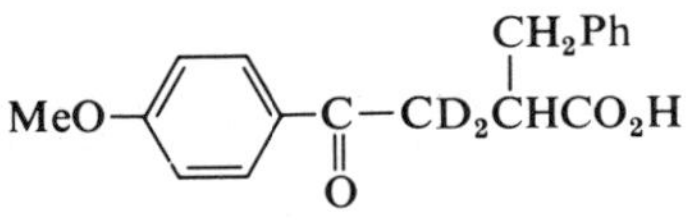

MeO–C$_6$H$_4$–C(=O)–CD$_2$CH(CH$_2$Ph)CO$_2$H

(**44**)

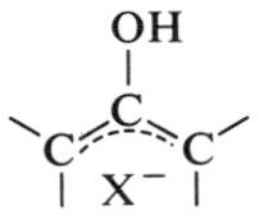

(**45**)

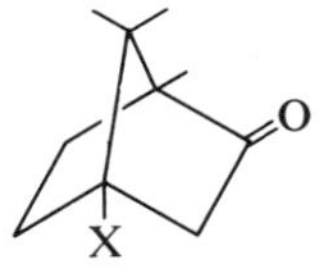

(**46**)

The α- to α'-rearrangement of α-halo-ketones may proceed by two mechanisms: (i) dehalogenation–rehalogenation, or (ii) *via* 2-hydroxyallylic cation (**45**). The former may be detected by carrying out the reaction in the presence of 2-naphthol which traps the halogen, and the latter by allowing a bromo-ketone to rearrange in the presence of HCl when halogen exchange can occur. By using these and other tests, it was found that, in acetic acid containing anhydrous hydrogen chloride, 2,2-dibromocholestan-3-one rearranges by mechanism (i); on the other hand, 1-bromobicyclo[5.3.1]undecan-11-one and 1-chloro-1-phenylbutan-2-one rearrange by mechanism (ii). 1-Bromo-1-phenylbutan-2-one rearranges by both mechanisms.[219]

The effect of 18-crown-6 and of cryptand[2.2.2] on the alkylation of potassium ethyl acetoacetate has been studied.[220a]

Hammett correlation of the acidities of *m*- and *p*-substituted acetophenones has been reported.[220b]

The rate of deuterium exchange of 4-substituted camphors (**46**) increases with increasing electron-withdrawing power of the substituent. The ratio of the rate constants for exchange at *exo*- and *endo*-positions decreases with increasing size of the substituent and a linear correlation was obtained between log $k_{2\text{-}exo}/k_{2\text{-}endo}$ and the A-value of the substituent; this was rationalized in terms of a steric interaction between the substituent and deuterium oxide in the transition state for *exo*-protonation of the enolate anion.[221a] Werstiuck and his co-workers[221b] have investigated deuterium exchange of norbornanones substituted in other positions; they also found that the rate of exchange at the 3-*exo*-position is more sensitive to the substituent than that at the 3-*endo*-position; the plot of $\log k_{exo}$ *versus* $\log k_{endo}$ has a slope of 1.67.[221b]

Proton exchange of norbornane-2,5-diones occurs over 100 times faster than that of the corresponding monoketones. This enhancement was attributed partly to inductive and strain effects and (by a factor of 13 to 32) to "homoenolic assistance".[222]

α-Thioenolization of thiocamphor occurs faster than enolization of camphor as determined by deuterium exchange. The *exo*- and *endo*-rates are, respectively, 23.2 and 12.3 times faster for the thioketone than for the ketone.[223]

Twistan-4-one, *i.e.* tricyclo[4.4.0.0^{3,8}]decan-4-one, undergoes exchange of one CH_2CO 270 times faster than that of the other. Examination of models shows that one C—H makes an angle of about 90° with the double bond of the carbonyl group while the other makes an angle of *ca.* 30°; presumably the former undergoes rapid exchange and the latter slow exchange.[224]

The conversion of phenylglyoxal into mandelic acid is catalysed by glutathione, which forms a hemithioacetal, and a base. Reaction proceeds with proton abstraction from the hemithioacetal *via* an enediol intermediate.[225]

Yeast enolase is inactivated by butanedione through reaction with its essential arginine residue.[226]

Other enzymes investigated include muscle pyruvate kinase,[227] yeast hexokinse,[228] and Δ^5-3-oxosteroid isomerase.[229]

Other reactions that have been investigated include photoenolization of crotonaldehyde,[230] amine-catalysed ionization of dihydroxyacetone phosphate,[231] isomerization of pentoses and 3-deoxyhexoses,[232] interconversions of glyceraldehyde and dihydroxyacetone,[233, 234] interconversion of D-glucose and D-fructose catalysed by sodium aluminate,[235] epimerization of 2-acetamido-2-deoxyhexoses,[236] catalysis by thiamine,[237, 238] iodination of acetophenones[239] and of pyruvic acid,[240] base-catalysed enolization of cyclopentanone,[241] keto-enol tautomerism of β-diketones,[242] β-thioxo-ketones,[243] phenol radical anion,[244] and 3-acetyl-5-isopropylpyrrolidine-2,4-dione,[245] alkylation of enolates,[246] and the gas-phase 1,3-hydrogen shift of the enol of acetic acid.[247]

MO calculations on keto–enol tautomerism have been reported.[248]

Homoenolization

The interconversion of the ketones (**47**) and (**48**) through homoenolization occurs 100 times faster than with the saturated analogues.[249] The compound (**48**) also undergoes migration of the double bond *via* an allylic anion, and both compounds undergo extensive deuterium exchange when the solvent is Bu^{t}OD.[250]

With potassium *tert*-butoxide in *tert*-butyl alcohol at 185°, bicyclo[2.2.2]-octenone undergoes a Claisen condensation followed by retro-Dields–Alder elimination of ethylene and Haller–Bauer cleavage.[251]

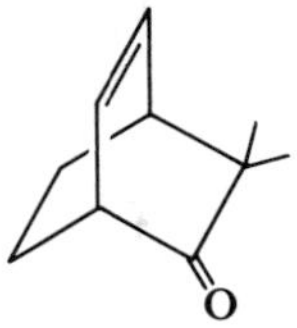

(**47**) 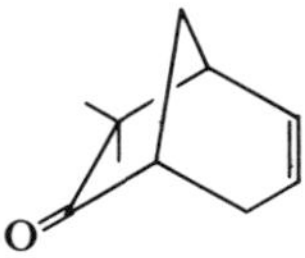 (**48**)

Aldol and Related Reactions

A linear free-energy relationship has been established between the equilibrium constants for aldol condensations and for other carbonyl additions.[252]

Ribulose 1,5-bisphosphate carboxylase from *Rhodospirillum rubrum* catalyses the conversion of D-ribulose 1,5-bisphosphate and carbon dioxide into 3-phosphoglycerate with retention of the oxygens at C(2) and C(3) of the substrate. This excludes any mechanism that involves exchange of these oxygens and is consistent with the mechanism of Calvin (equation 8).[253]

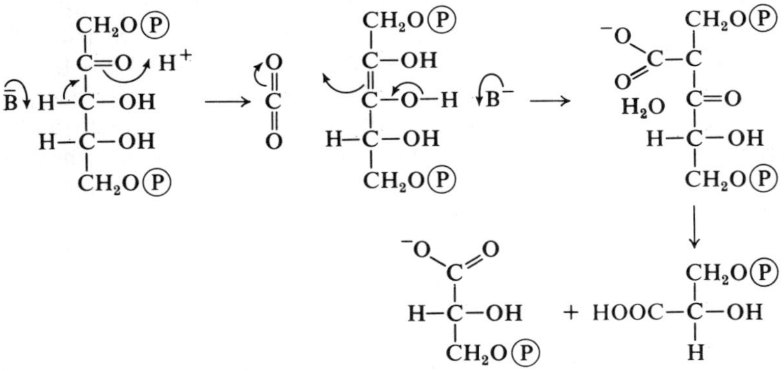

(8)

The cleavage of β-phenylserines by serine hydroxy methylase to yield benzaldehyde and glycine proceeds *via* a quinonoid intermediate (**49**).[254]

In the presence of pyridoxal and aluminium(III) or zinc(II) ions, β-phenylserine undergoes a retro-aldol reaction and a β-elimination. The metal ions promote the retro-aldol reaction more than the β-elimination.[255]

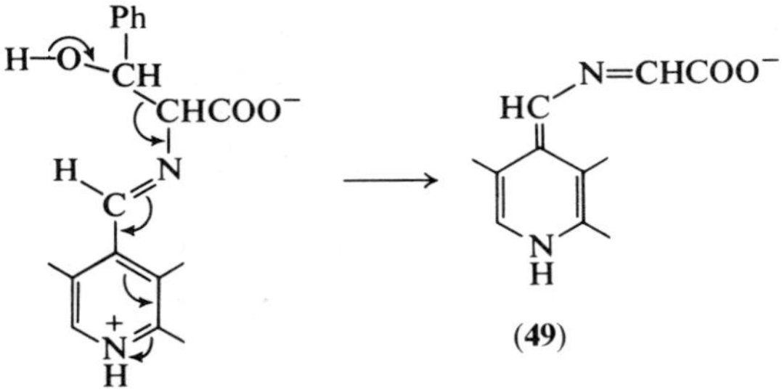

(**49**)

Studies have been reported of the cyclization of D-glucose 6-phosphate into *myo*-inositol 1-phosphate by L-*myo*-inositol 1-phosphate synthesase from rat testis, *Acer pseudoplatanus L.*, and *Oryza sativa*.[256]

The following condensations have also been studied: benzaldehyde with the anion of α-chloro-α-phenylacetate,[257] *o*-, *m*-, and *p*-hydroxybenzaldehyde with acetone,[258] sulgin with acetylacetone,[259] benzaldehyde with 3,5-dimethyl-4-nitro-isoxazole,[260] furfural with acetone,[261] chloroacetone with itself,[262] and propanal with itself catalysed by glycine.[263a]

Aluminium enolates, generated from α-bromo-ketones and diethylaluminium chloride and zinc, have been used in a regiospecific aldol reaction.[263–]

The stereochemistry of the addition of enolates to aldehydes and ketones has been investigated.[264]

Investigations of enzymically catalysed aldol reactions have been described.[265, 266]

Other Reactions

Ching and Kallen[267] have carried out a detailed investigation of the formation of cyanohydrins from aromatic aldehydes. The ρ^+-values for the addition of CN^- and HCN are 1.01 and 1.49, respectively. The reaction with HCN was thought to be stepwise with attack of CN^- followed by proton transfer. The ratio of ρ^+ for the rate constant (k_1') for attack of CN^- to the equilibrium constant (K_1') for addition of CN^- is 0.8, which indicates that bond formation is well advanced in the transition state.[267] This is in agreement with the previously determined α-deuterium isotope effect.[268] There have been two other investigations of cyanohydrins.[269–271]

The effects of ring size on the rate of reduction of cycloalkanones by sodium borohydride and horse liver alcohol dehydrogenase are very similar.[272]

L-Glyceraldehyde 3-phosphate is a poor substrate for D-glyceraldehyde 3-phosphate dehydrogenase with a $k_{cat.}/K_m$ ratio which is at least 3×10^4 times smaller than that for the D-isomer.[273] Another investigation with this enzyme has been reported.[274]

The reduction of several aromatic ketones by sodium borohydride in the presence of micelles of (+)-(*R*)-*N*-dodecyl-*N*,*N*-dimethyl-1-phenylethylammonium bromide shows a small enantio-selectivity.[275] The asymmetric reduction of ketones with sodium borohydride in the presence of a leather has been reported.[276]

There have been investigations of the Wolff–Kishner reaction[277] and of the reduction of ketones by metal hydrides.[278, 279] Other examples of reductions of carbonyl compounds are discussed in Chapter 4.

Micellar catalysis of the CN^--assisted isoalloxazine oxidation of aldehydes has been studied.[280]

There have been investigations of the oxidation of benzaldehyde[281] and of aldoses[282] by vanadium(V), of D-xylose by manganese(III) pyrophosphate,[283] of isobutyraldehyde by chromium(VI),[284] and of aliphatic ketones by chloramine T.[285] Other oxidations of carbonyl compounds are described in Chapter 4.

There have been several investigations of the reactions of aldehydes and ketones with Grignard reagents,[286] triphenylaluminium,[287] organolithium compounds,[288] and organosilanes.[289]

The stereochemistry of additions to carbonyl compounds has been investigated.[290]

There have been investigations of the following reactions: Schmidt reaction,[291] Darzens condensation,[292] reaction of aldehydes with acid chlorides to form α-chloroalkyl esters,[293] decarboxylation of aldehydes with $RhCl(PPh_3)_3$,[294] reaction between aldehydes, methyl diphenylphosphinite and carboxylic acids,[295] condensation of chloral with dialkyl phosphites,[296] isomerization of 2-bromocyclohexanones,[297] reaction of formaldehyde with chlorine,[298] and complex formation between formaldehyde and lithium.[299]

There have been several theoretical discussions of the protonation of carbonyl compounds.[300]

References

1 Cordes, E. H., and Bull, H. G., in "Transition States of Biochemical Processes", R. D. Gandour and R. L. Schowen, Eds., Plenum Press, New York, 1978, p. 429.

2 Capon, B., *Pure Appl. Chem.*, **49**, 1001 (1977).

3 Deslongchamps, P., *Heterocycles*, **7**, 1271 (1977).

4 Jensen, J. L., and Lenz, P. A., *J. Am. Chem. Soc.*, **100**, 1291 (1978).

5 McClelland, R. A., and Ahmad, M., *J. Am. Chem. Soc.*, **100**, 7027 (1978).

6 McClelland, R. A., and Ahmad, M., *J. Am. Chem. Soc.*, **100**, 7031 (1978).

7 Young, P. R., and Jencks, W. P., *J. Am. Chem. Soc.*, **99**, 8238 (1977).

8 Brouillard, R., and Delaporte, B., *J. Am. Chem. Soc.*, **99**, 8461 (1977).

9 Eliason, R., and Kreevoy, M. M., *J. Am. Chem. Soc.*, **100**, 7037 (1978).

10 Cf. *Org. React. Mech.*, **1968**, 350; **1969**, 399.

11 Chandrasekhar, S., and Kirby, A. J., *J. Chem. Soc., Chem. Commun.*, **1978**, 171.

12 Kirby, A. J., and Martin, R. J., *J. Chem. Soc., Chem. Commun.*, **1978**, 803.

13 Bouab, O., Lamaty, G., and Moreau, C., *J. Chem. Soc., Chem. Commun.*, **1978**, 678.

14 Bouab, O., Moreau, C., and Ako, M. Z., *Tetrahedron Lett.*, **1978**, 61.

15 Wipff, G., *Tetrahedron Lett.*, **1978**, 3269; Lehn, J. M., and Wipff, G., *Helv. Chim. Acta*, **61**, 1274 (1978).

16 Craze, G.-A., Kirby, A. J., and Osborne, R., *J. Chem. Soc., Perkin Trans.* 2, **1978**, 357.

17 Craze, G.-A., and Kirby, A. J., *J. Chem. Soc., Perkin Trans.* 2, **1978**, 354.

18 Gold, V., and Sghibartz, C. M., *J. Chem. Soc., Chem. Commun.*, **1978**, 507.

19 Yansai, S., Vofsi, D., and Halmann, D., *J. Chem. Soc., Perkin Trans.* 2, **1978**, 511, 517; cf. *Org. React. Mech.*, **1976**, 2.

20 Bunton, C. A., Romsted, L. S., and Smith, H. J., *J. Org. Chem.*, **43**, 4299 (1978).

21 Reese, C. B., *Tetrahedron*, **34**, 3143 (1976).

22 Bazbouz, A., Christol, H., Coste, J., and Plenat, F., *Bull. Soc. Chim. Fr. II*, **1978**, 305; cf. *Org. React. Mech.*, **1970**, 419.

23 Hanzlik, R. P., and Hilbert, J. M., *J. Org. Chem.*, **43**, 610 (1978); cf. *Org. React.Mech.*, **1972**, 389.

24 Fuji, K., Ichikawa, K., and Fujita, E., *Tetrahedron Lett.*, **1978**, 3561.

25 Porter, Q. N., and Utley, J. H. P., *J. Chem. Soc., Chem. Commun.*, **1978**, 255.

26 Yoshida, K., *J. Chem. Soc., Chem. Commun.*, **1978**, 20.

27 Cavazza, M., Morganti, G., and Pietra, F., *Tetrahedron Lett.*, **1978**, 2137.

28 Çetinkaya, E., Schuchmann, H.-P., and von Sonntag, C., *J. Chem. Soc., Perkin Trans.* 2, **1978**, 985.

29 Kovac, F., and Plesničar, B., *J. Chem. Soc., Chem. Commun.*, **1978**, 122.

30 Brudnik, B. M., Zlotshii, S. S., Imashev, U. B., and Rakhmankulov, D. L., *Dokl. Akad. Nauk SSSR*, **241**, 129 (1978); *Chem. Abs.*, **89**, 107496 (1978).

31 Kostyukevich, L. L., Zlotskii, S. S., Kravets, E. K., Martem'yanov, V. S., and Rakhmankulov, D. L., *Zh. Prikl. Chim. (Leningrad)*, **50**, 2124 (1977); *Chem. Abs.*, **88**, 21791 (1978).

32 Malardeau, C., and Mousset, G., *Bull. Soc. Chim. Fr. II*, **1977**, 988.

33 Westera, G., Blomberg, C., and Bickelhaupt, F., *J. Org. Chem.*, **43**, 285, 291 (1978).

34 Chernousova, N. N., *Deposited Doc.* VINITI 67 (1975); *Chem. Abs.*, **88**, 61761 (1978).

35 Schacht, E. H., St. Pierre, T., Desmarets, G. E., and Goethals, E. J., *Bull. Soc. Chim. Belg.*, **86**, 977 (1977).

36 Kostyukevich, L. L., Maksimova, N. E., Kantor, E. A., and Rakhmankulov, D. L., *Neftekhim. Sint. Tekh. Prog.*, **1976**, 34; *Chem. Abs.*, **88**, 189553 (1978).

37 Meyers, M. B., Gallagher, M. J., and Orr, F. F., *Proc. Roy. Irish Acad.*, **77B**, 507 (1977); *Chem. Abs.*, **89**, 146041 (1978).

38 Zsuga, M., Szabo, V., and Bulogh, L., *React. Kinet. Catal. Lett.*, **8**, 1 (1978); *Chem. Abs.*, **89**, 41861 (1978).

[39] Sharf, V. Z., Khardin, A. P., Val'dman, D. I., Panfilov, B. I., Val'dman, A. I., and Freidlin, L. K., *Kinet, Katal.*, **19**, 788 (1978); *Chem. Abs.*, **89**, 146053 (1978).
[40] Uzlyaner-Neglo, A. L., Kushina, V. V., Lesteva, T. M., and Katsnel'son, M. G., *Zh. Prikl. Khim.* (*Leningrad*), **51**, 1150 (1978); *Chem. Abs.*, **89**, 41915 (1978).
[41] Blaya, A. I., Pope, R. F., and Chirila, T., *Rev. Roum. Chim.*, **23**, 581 (1978).
[42] Giachin, G. G., *J. Chem. Educ.*, **55**, 201 (1971).
[43] Bauduin, G., Bondon, D., Pietrasanta, Y., and Pucci, B., *Tetrahedron*, **34**, 3269 (1978).
[44] van Heeswizk, W. A. R., Goedhardt, J. B., and Vliegenthurt, J. F. G., *Carbohydr. Res.*, **58**, 337 (1977); cf. *Org. React. Mech.*, **1965**, 241.
[45] Bonner, T. G., Gibson, D., and Lewis, D., *Carbohydr. Res.*, **66**, 265 (1978).
[46] Gelas, J., and Horton, D., *Carbohydr. Res.*, **67**, 371 (1978).
[47] Clode, D. M., *Can. J. Chem.*, **55**, 4071 (1977).
[48] Fredriksen, N., Jensen, R. B., Jørgensen, S. E., Nielsen, J. U. R., Norskov, L., and Schroll, G., *Acta Chem. Scand.*, **31B**, 694 (1977).
[49] Bailey, W. F., Connon, H., Eliel, E. L., and Wiberg, K. B., *J. Am. Chem. Soc.*, **100**, 2202 (1978).
[50] Arai, K., and Ogiwara, Y., *Bull. Chem. Soc. Jpn.*, **51**, 182 (1978).
[51] Richards, G. N., and Shafizadeh, F., *Aust. J. Chem.*, **31**, 1825 (1978).
[52] Lai, Y.-Z., and Ontto, D. E., *Carbohydr. Res.*, **67**, 500 (1978).
[53] Angyal, S. J., and Beveridge, R. J., *Carbohydr. Res.*, **65**, 229 (1978); cf. Paulsen, H., Landsky, G., and Koebernick, H., *Chem. Ber.*, **111**, 3699 (1978).
[54] Kawana, M., and Emoto, S., *Tetrahedron Lett.*, **1978**, 1561.
[55] Fowler, A. V., *J. Biol. Chem.*, **253**, 5484, 5499 (1978); Fowler, A. V., Brake, A. J., and Zabin, I., *ibid.*, pp. 5490, 5515; Fowler, A. V., and Zabin, I., *ibid.*, pp. 5505, 5510, 5521; cf. Hood, J. M., Fowler, A. V., and Zabin, I., *Proc. Natl. Acad. Sci. USA*, **75**, 113 (1978).
[56] Fowler, A. V., Zabin, I., Sinnott, M. L., and Smith, P. J., *J. Biol. Chem.*, **253**, 5283 (1978).
[57] Deschavanne, P. J., Viratelle, O. M., and Yon, J. M., *J. Biol. Chem.*, **253**, 833 (1978).
[58] Yde, M., and De Bruyne, C. K., *Carbohydr. Res.*, **60**, 155 (1978).
[59] Withers, S. G., Jullien, M., Sinnott, M. L., Viratelle, O. M., and Yon, J. M., *Eur. J. Biochem.*, **87**, 249 (1978).
[60] Celada, F., Fowler, A. V., and Zabin, I., *Biochemistry*, **17**, 5156 (1978); Jörnvall, H., Fowler, A. V., and Zabin, I., *ibid.*, p. 5160.
[61] Lehmann, J., and Zieger, B., *Carbohydr. Res.*, **58**, 73 (1977); cf. *Org. React. Mech.*, **1977**, 5.
[62] Lehmann, J., and Schwesinger, B., *Chem. Ber.*, **111**, 3961 (1978).
[63] Woulfe-Flanagan, H., and Huber, R. E., *Biochem. Biophys. Res. Commun.*, **82**, 1079 (1978).
[64] Walker, D. E., and Axelrod, B., *Arch. Biochem. Biophys.*, **187**, 102 (1978); cf. *Org. React. Mech.*, **1977**, 5.
[65] Lehmann, J., and Schröter, E., *Carbohydr. Res.*, **58**, 65 (1977).
[66] Legler, G., and Harder, A., *Biochim. Biophys. Acta*, **524**, 102 (1978).
[67] Legler, G., *Biochim. Biophys. Acta*, **524**, 94 (1978).
[68] Hirayama, T., Horie, S., Nagoyama, H., and Matsuda, K., *J. Biochem.* (*Tokyo*), **84**, 27 (1978).
[69] De Gussem, R. L., Aerts, G. M., Claeyssens, M., and De Bruyne, C. K., *Biochim. Biophys. Acta*, **525**, 142 (1978).
[70] Siro, M.-R. and Lövgren, T., *Acta Chem. Scand.*, **32B**, 447 (1978).
[71] Needleman, R. B., Federoff, H. J., Eccleshall, T. R. Buchferer, B., and Marmur, J., *Biochemistry*, **17**, 4657 (1978).
[72] Chiba, S., Inomata, S., Matsui, H., and Shimomura, T., *Agric. Biol. Chem.*, **42**, 241 (1978); Matsui, H., Chiba, S., and Shimomura, T., *ibid.*, p. 1855.
[73] Ballardie, F. W., Capon, B., Cuthbert, M. W., and Dearie, W. B., *Bioorg. Chem.*, **6**, 483 (1977); cf. *Org. React. Mech.*, **1972**, 396; **1974**, 5.
[74] Patt, S. L., Baldo, J. H., Boekelheide, K., Weisz, G., and Sykes, B. D., *Can. J. Chem.*, **56**, 624 (1978).
[75] Lenkinski, R. E., Agresti, D. G., Chen, D. M., and Glickson, J. D., *Biochemistry*, **17**, 1463 (1978); Atassi, M. Z., and Lee, C.-L., *Biochem. J.*, **171**, 419, 429 (1978); Johnson, E. R., Anderson, W. L., Wetlaufer, D. B., Lee, C.-L., and Atassi, M. Z., *J. Biol. Chem.*, **253**, 3408 (1978); Porubcan, R. S., Watters, K. L., and McFarland, J. T., *Arch. Biochem. Biophys.*, **186**, 255 (1978); Amano, K., and Ito, E., *Eur. J. Biochem.*, **85**, 97 (1978); Acharya, A. S., and Taniuchi, H., *Biochemistry*, **17**, 3064 (1978).

[76] Remington, S. J., Anderson, W. F., Owen, J., Ten-Eyck, L. F., Grainger, C. T., and Matthews, B. W., *J. Mol. Biol.*, **118**, 81 (1978); cf. Remington, S. J., and Matthews, B. W., *Proc. Natl. Acad Sci. USA*, **75**, 2180 (1978); Timchenko, A. A., Ptitsyn, O. B., Troitsky, A. V., and Denesyuk, A. I., *FEBS Lett.*, **88**, 109 (1978),
[77] Ochi, N., Azegami, M., and Ishii, S., *J. Biochem. (Tokyo)*, **83**, 727 (1978).
[78] Morita, T., Hora, S., and Matsushima, Y., *J. Biochem. (Tokyo)*, **83**, 893 (1978).
[79] Omichi, K., and Matsushima, Y., *J. Biochem. (Tokyo)*, **84**, 835 (1978); cf. Omichi, K., Fujii, K., Mizukami, J., and Matsushima, Y., *ibid.*, **83**, 1443 (1978).
[80] Kunikata, T., Nitta, Y., and Watanabe, T., *J. Biochem. (Tokyo)*, **83**, 1435 (1978).
[81] László, E., Holló, J., Hoschke, A., and Sávosi, G., *Carbohydr. Res.*, **61**, 387 (1978).
[82] Suegtsugu, N., Takeo, K., Sanai, Y., and Kuge, T., *J. Biochem. (Tokyo)*, **83**, 473 (1978).
[83] Fujimori, H., Ohnishi, M., and Hiromi, K., *J. Biochem. (Tokyo)*, **83**, 1503 (1978).
[84] Shimizu, M., Kanno, M., Tamura, M., and Suekane, M., *Agric. Biol. Chem.*, **42**, 1681 (1978).
[85] Kondo, H., Nakatani, H., Hiromi, K., and Matsuno, R., *J. Biochem. (Tokyo)*, **84**, 403 (1978).
[86] Suganuma, T., Matsuno, R., Ohnishi, M., and Hiromi, K., *J. Biochem. (Tokyo)*, **84**, 293 (1978).
[87] Matsuno, R., Suganuma, T., Fujimori, H., Nakanishi, K., Hiromi, K., and Kamikubo, T., *J. Biochem. (Tokyo)*, **83**, 385 (1978).
[88] Hinsch, B., and Kula, M.-R., *Biochim. Biophys. Acta*, **525**, 392 (1978).
[89] Nisizawa, K., Kunda, T., Shikata, S., and Wakabayashi, K., *J. Biochem. (Tokyo)*, **83**, 1625 (1978).
[90] Wood, T. M., and McCrae, S. I., *Biochem. J.*, **171**, 61 (1978).
[91] Brot, F. E., Bell, C. E., and Sly, W. S., *Biochemistry*, **17**, 385 (1978).
[92] Pulkownik, A., Thoma, J. A., and Walker, G. J., *Carbohydr. Res.*, **61**, 493 (1978).
[93] Richards, G. N., and Streamer, M., *Carbohydr. Res.*, **62**, 191 (1978).
[94] Paus, E., *Biochem. Biophys. Acta*, **526**, 507 (1978); **533**, 446 (1978).
[95] De Prijcker, J., De Bock, A., and De Bruyne, C. K., *Carbohydr. Res.*, **60**, 141 (1978).
[96] Bouquelet, S., Spik, G., and Montreuil, J., *Biochim. Biophys. Acta*, **522**, 521 (1978).
[97] Tanaka, M., and Uchida, T., *Biochim. Biophys. Acta*, **522**, 531 (1978).
[98] Schwabe, K., Grossmann, A., Fehrmann, B., Tschiersch, B., *Carbohydr. Res.*, **67**, 541 (1978).
[99] Deleyn, F., Claeyssens, M., Van Beeumen, J., and De Bruyne, C. K., *Can. J. Chem.*, **56**, 43 (1978).
[100] Saman, E., Claeyssens, M., and De Bruyne, C. K., *Eur. J. Biochem.*, **85**, 301 (1978); Kerslers-Hilderson, H., Von Doorslaer, E., and De Bruyne, C. K., *Carbohydr. Res.*, **65**, 219 (1978); Claeyssens, M., and De Bruyne, C. K., *Biochim. Biophys. Acta*, **533**, 98 (1978).
[101] Heinrichová, K., *Collect. Czech. Chem. Commun.*, **42**, 3214 (1977).
[102] Rexová-Benková, L., Mračková, M., Luknár, O., and Kohn, R., *Collect. Czech. Chem. Commun.*, **42**, 3204 (1978).
[103] Terra, W. R., Ferreira, C., and De Bianchi, A. G., *Biochim. Biophys. Acta*, **524**, 131 (1978).
[104] Mion, N., Herries, D. G., and Batte, E. A., *Biochim. Biophys. Acta*, **523**, 454 (1978).
[105] Funderburk, L. H., Aldwin, L., and Jencks, W. P., *J. Am. Chem. Soc.*, **100**, 5444 (1978).
[106] Young, P. R., and Jencks, W. P., *J. Am. Chem. Soc.*, **100**, 1228 (1978).
[107] Jencks, D. A., and Jencks, W. P., *J. Am. Chem. Soc.*, **99**, 7948 (1977).
[108] Cf. Cruickshank, F. R., Hyde, A. J., and Pugh, D., *J. Chem. Educ.*, **54**, 288 (1977); Kost, D., and Pross, A., *Educ. Chem.*, **14**, 87 (1977); Cruickshank, F. R., Hyde, A. J., and Pugh, D., *ibid.*, **15**, 133 (1978).
[109] Fournier, L., Lamaty, G., and Roque, J. P., *Tetrahedron*, **34**, 2901 (1978).
[110] Gilbert, H. F., and Jencks, W. P., *J. Am. Chem. Soc.*, **99**, 7931 (1977).
[111] Jencks, W. P., and Gilbert, H. F., *Pure Appl. Chem.*, **49**, 1021 (1977).
[112] El Seoud, O. A., da Silva, M. J., Barbur, L. P., and Martins, A., *J. Chem. Soc., Perkin Trans.* 2, **1978**, 331.
[113] Berndt, M. C., de Jersey, J., and Zerner, B., *J. Am. Chem. Soc.*, **99**, 8332, 8334 (1977).
[114] Kjaer, A. M., Sørensen, P. E., and Ulstrup, J., *J. Chem. Soc., Perkin Trans.* 2, **1978**, 51.
[115] Chin, C. S., and Huang, H. H., *J. Chem. Soc., Perkin Trans.* 2, **1978**, 474.
[116] Cox, B. G., and Natarajan, R., *J. Chem. Soc., Perkin Trans.* 2, **1977**, 2021.
[117] Schray, K. J., and Howell, E. E., *Arch. Biochem. Biophys.*, **189**, 102 (1978); cf. *Org. React. Mech.*, **1970**, 431; **1971**, 403; **1974**, 7.
[118] Grimshaw, C. E., *Fed. Proc., Fed. Am. Soc. Exp. Biol.*, **37**, 1806 (1978).
[119] Engdahl, K. Å., and Ahlberg, P., *J. Chem. Res.*, (*S*) **1977**, 340.

[120] Bensaude, O., Chevrier, M., and Dubois, J. E., *J. Am. Chem. Soc.*, **100**, 7055 (1978).
[121] Grindley, T. B., and Gulasekharam, V., *J. Chem. Soc., Chem. Commun.*, **1978**, 1073.
[122] Alföldi, J., Petruš, L., and Bílik, V., *Collect. Czech. Chem. Commun.*, **43**, 1476 (1978).
[123] Litvinenko, L. M., and Oleinik, N. M., *Usp. Khim.*, **47**, 777 (1978); *Russian Chem. Rev.*, **47**, 401 (1978).
[124] Valters, R., "Ring-chain Isomerism in Organic Chemistry", Riga, 1978.
[125] Nielsen, H., and Sørensen, P. E., *Acta Chem. Scand.*, **31A**, 739 (1977).
[126] Santoro, E., and Chiavarini, M., *J. Chem. Soc., Perkin Trans.* 2, **1978**, 189.
[127] Lewis, C. A., and Wolfenden, R., *Biochemistry*, **16**, 4886 (1977).
[128] Metzger, P., and Casadevall, E., *Bull. Soc. Chim. Fr. II*, **1978**, 109.
[129] Shibuya, H., Watarai, H., and Suzuki, N., *Bull. Chem. Soc. Jpn.*, **51**, 2932 (1978).
[130] Bogatkov, S. V., and Tur'yan, Y. I., *Zh. Org. Khim.*, **13**, 2486 (1977); *Chem. Abs.*, **88**, 73928 (1978).
[131] Rudnev, A. V., Kalyazin, E. P., Kalugin, K. S., and Kovalev, G. V., *Zh. Fiz. Khim.*, **51**, 2603 (1977); *Chem. Abs.*, **88**, 21649 (1978).
[132] Thang, N. D., Hobza, P., Panćiř, J., and Zahradnik, R., *Collect. Czech. Chem. Commun.*, **43**, 1366 (1978).
[133] Schönberg, A., and Singer, E., *Tetrahedron*, **34**, 1285 (1978).
[134] Kluger, R., Pike, D. C., and Chin, J., *Can. J. Chem.*, **56**, 1792 (1978).
[135] Brook, A. G., Vandersar, T. J. D., and Limburg, W., *Can. J. Chem.*, **56**, 1758 (1978).
[136] Bok, T. R., Kruk, C., and Speckamp, W. N., *Tetrahedron Lett.*, **1978**, 657.
[137] van Oosterhout, H., Kruk, C., and Speckamp, W. N., *Tetrahedron Lett.*, **1978**, 653.
[138] Lewis, C. A., and Wolfenden, R., *Biochemistry*, **16**, 4890 (1977).
[139] Pancheshnikova, R. B., Danilov, V. T., Komissarov, V. D., Maksimov, S. M., and Aleev, R. S., *Neftekhim. Smt. Tekh. Prog.*, **1976**, 48; *Chem. Abs.*, **89**, 5621 (1978).
[140] Barnett, R. W., "Mechanism of proton transfer in carbonyl and acyl group reactions" in "Bioorganic Chemistry", E. E. van Tamelen, Ed., Academic Press, New York, Vol. II, 1978, p. 39.
[141] Fife, T. H., Hutchins, J. E. C., and Pellino, A. M., *J. Am. Chem. Soc.*, **100**, 6455 (1978); cf. *Org. React. Mech.*, **1976**, 7.
[142] Moad, G., and Benkovic, S. J., *J. Am. Chem. Soc.*, **100**, 5495 (1978).
[143] Pollack, R. M., Kayser, R. H., and Damewood, J. R., *J. Am. Chem. Soc.*, **99**, 8232 (1977); cf. *Org. React. Mech.*, **1977**, 6.
[144] Hegarty, A. F., Digram, K. J., and Hickey, D. M., *Tetrahedron Lett.*, **1978**, 2121.
[145] Perran, H. E., Roberts, R. D., Jacob, J. N., and Spencer, T. A., *J. Chem. Soc., Chem. Commun.*, **1978**, 49.
[146] Kaválek, J., El Bahei, S., and Štěrba, V., *Collect. Czech. Commun.*, **43**, 1732 (1978).
[147] Menger, F. M., and Perinis, M., *Tetrahedron Lett.*, **1978**, 4653.
[148] McCapra, F., and Barford, A., *J. Chem. Soc., Chem. Commun.*, **1977**, 874.
[149] Tsai, M.-D., Byrn, S. R., Chang, C., Floss, H. G., and Weintraub, H. J. R., *Biochemistry*, **17**, 3177, 3183 (1978).
[150] Gani, V., Kupfer, A., and Shaltiel, S., *Biochemistry*, **17**, 1294 (1978).
[151] Gilmer, P. J., and Kirsch, J. F., *Biochemistry*, **16**, 5246 (1977).
[152] Drew, M. G. B., Rodgers, A., McCann, M., and Neilson, S. M., *J. Chem. Soc., Chem. Commun.*, **1978**, 415.
[153] Levina, S. G., Kulikova, T. M., Shishkina, V. I., and Kalmichenko, I. I., *Izv. Vyssh. Uchebn. Zaved., Khim., Khim. Tekhnol.*, **21**, 485 (1978); *Chem. Abs.*, **89**, 107263 (1978).
[154] Negoro, K., *Sen'i Kako*, **29**, 427 (1977); *Chem. Abs.*, **88**, 49924 (1978).
[155] Andreichikov, Y. S., Kozlov, A. P., Tokmakova, T. N., and Tendryakova, S. P., *Zh. Org. Khim.*, **14**, 163 (1978); *Chem. Abs.*, **88**, 151691 (1978).
[156] Talán, P., and Kosturiak, A., *Collect. Czech. Chem. Commun.*, **43**, 2600 (1978).
[157] Quar, M. S., Funke, P. T., and Cohen, A. I., *J. Pharm. Sci.*, **67**, 850 (1978).
[158] Zakhary, R. F., and Iskander, M. L., *Tetrahedron*, **34**, 339 (1978).
[159] Vittal, E. E., and Boer, V. I., *Zh. Org. Khim.*, **13**, 2059 (1977); *Chem. Abs.*, **88**, 36861 (1978).
[160] Baldwin, J. E., Scott, A., Branz, S. E., Tannenbaum, S. R., and Green, L., *J. Org. Chem.*, **43**, 2427 (1978).
[161] Macháček, V., Toman, J., and Klicnar, J., *Collect. Czech. Chem. Commun.*, **43**, 1634 (1978).
[162] DoMinh, T., Johnson, A. L., Jones, J. E., and Senise, P. P., *J. Org. Chem.*, **42**, 4217 (1978).
[163] Simons, S. S., and Johnson, D. F., *J. Org. Chem.*, **43**, 2886 (1978).

[164] Panetta, C. A., and Miller, A. L., *J. Org. Chem.*, **43**, 2113 (1978); see also Ando, H., and Emoto, S., *Bull. Chem. Soc. Jpn.*, **51**, 2366, 2433, 2435 (1978).
[165] Llor, J., and Cortijo, M., *J. Chem. Soc., Perkin Trans.* 1, **1978**, 409.
[166] Karube, Y., and Matsushima, Y., *J. Am. Chem. Soc.*, **99**, 7356 (1978).
[167] Boyd, D. R., Al-Showiman, S., and Jennings, W. B., *J. Org. Chem.*, **43**, 3335 (1978).
[168] Brown, C., Hudson, R. F., and Grayson, B. T., *J. Chem. Soc., Chem. Commun.*, **1978**, 156.
[169] Carlson, R., Nilsson, L., Rappe, C., Babadjamian, A., and Metzger, J., *Acta Chem. Scand.*, **32B**, 85 (1978); Carlson, R., and Nilsson, L., *ibid.*, p. 732.
[170] Hickmott, P. W., and Firrell, N. F., *J. Chem. Soc., Perkin Trans.* 1, **1978**, 340.
[171] Marchetti, L., *J. Chem. Soc., Perkin Trans.* 2, **1978**, 382.
[172] Müller, K., and Brown, L. D., *Helv. Chim. Acta*, **61**, 1407 (1978); Meyer, R., *ibid.*, p. 1418.
[173] Romero, R., Stein, R., Bull, H. G., and Cordes, E. H., *J. Am. Chem. Soc.*, **100**, 7620 (1978).
[174] Bull, H. G., Ferraz, J. P., Cordes, E. H., Ribbi, A., and Apitz-Castro, R., *J. Biol. Chem.*, **253**, 5186 (1978).
[175] Stein, R. L., Romero, R., Bull, H. G., and Cordes, E. H., *J. Am. Chem. Soc.*, **100**, 6249 (1978).
[176] Bergman, N.-Å., Chiang, Y., and Kresge, A. J., *J. Am. Chem. Soc.*, **100**, 5954 (1978).
[177] Cox, M. M., and Jencks, W. P., *J. Am. Chem. Soc.*, **100**, 5956 (1978).
[178] Funderburk, L. H., and Jencks, W. P., *J. Am. Chem. Soc.*, **100**, 6708 (1978).
[179] Malatesta, V., and Cocivera, M., *J. Org. Chem.*, **43**, 1737 (1978).
[180] Cocivera, M., Woo, K. W., and Livant, P., *Can. J. Chem.*, **56**, 473 (1978).
[181] Alves, K. B., Bastos, M. P., and do Amaral, L., *J. Org. Chem.*, **43**, 4032 (1978).
[182] Fischer, H. F., Stickel, D. C., Brown, A., and Cerretti, D., *J. Am. Chem. Soc.*, **99**, 8180 (1977).
[183] Uda, M., and Kubota, S., *J. Heterocycl. Chem.*, **15**, 807 (1978).
[184] Douglas, A. W., *J. Am. Chem. Soc.*, **100**, 6463 (1978).
[185] Miller, F. M., and Schinske, W. N., *J. Org. Chem.*, **43**, 3384 (1978).
[186] Glish, G. L., and Cooks, R. G., *J. Am. Chem. Soc.*, **100**, 6720 (1978).
[187] Kim, T.-R., and Choi, W.-S., *Taehan Kwakah Hoechi*, **22**, 30 (1978); *Chem. Abs.*, **89**, 107287 (1978).
[188] Lazarevski, T., Radoboja, G., and Djokić, S., *J. Pharm. Sci.*, **67**, 1031 (1978).
[189] Baliah, V., Chandrasekharan, J., and Natarajan, A., *Indian J. Chem.*, **15B**, 829 (1977); *Chem. Abs.*, **88**, 88737 (1978).
[190] Gevaza, Y. I., Kupchik, I. P., and Staninets, V. I., *Dopov. Akad. Nauk Ukr. RSR, Ser. B; Geol., Khim. Biol. Nauki*, **1977**, 703; *Chem. Abs.*, **88**, 5709 (1978).
[191] Hudson, R. F., and Dj-Forudian, H., *Phosphorus Sulfur*, **4**, 9 (1978); *Chem. Abs.*, **88**, 189533 (1978).
[192] Hudson, R. F., and Dj-Forudian, H., *J. Chem. Soc., Perkin Trans.* 1, **1978**, 12.
[193] Reynolds, D. D., Guanci, D. F., Neynaber, C. B., and Conboy, R. J., *J. Org. Chem.*, **43**, 3838 (1978).
[194] Alvernhe, G., and Laurent, A., *J. Chem. Res.*, (*S*) **1978**, 28.
[195] Lazaro, R., Mathieu, D., Phan Tan Luu, R., and Elguero, J., *Bull. Soc. Chim. Fr. II*, **1977**, 1163.
[196] Naulet, M., Leymarie-Beljean, M., and Martin, G. J., *Org. Magn, Reson.*, **11**, 16 (1978).
[197] Geneste, P., Durand, R., Kamenka, J.-M., Beierbeck, H., Martino, R., and Saunders, J. K., *Can. J. Chem.*, **56**, 1940 (1978).
[198] Buchanan, G. W., and Dauson, B. A., *Can. J. Chem.*, **56**, 2200 (1978).
[199] Kresge, A. J., and Chwang, W. K., *J. Am. Chem. Soc.*, **100**, 1249 (1978); cf. *Org. React. Mech.*, **1972**, 406.
[200] Chiang, Y., Kresge, A. J., and Young, C. I., *Can. J. Chem.*, **56**, 461 (1978).
[201] Heschfield, R., Yeager, M. J., and Schmir, G. L., *J. Org. Chem.*, **40**, 2940 (1975).
[202] Chiang, Y., Chwang, W. K., Kresge, A. J., Robinson, L. H., Sagatys, D. S., and Young, C. I., *Can. J. Chem.*, **56**, 456 (1978).
[203] Kresge, A. J., Sagatys, D. S., and Chen, H. C., *J. Am. Chem. Soc.*, **99**, 7228 (1977).
[204] van Bree, J., and Anteunis, M. J. O., *Tetrahedron*, **33**, 3331 (1977).
[205] Horn, J. S., Paul, A. G., and Rapoport, H., *J. Am. Chem. Soc.*, **100**, 1895 (1978).
[206] Arai, I., and Daves, G. D., *J. Am. Chem. Soc.*, **100**, 287 (1978).
[207] Taskinen, E., *Tetrahedron*, **34**, 353, 425, 429, 439 (1978).
[208] Hine, J., *Acc. Chem. Res.*, **11**, 1 (1978).

[209] Spencer, T. A., in "Bioorganic Chemistry", E. E. van Tamelen, Ed., Academic Press, New York, Vol. 1, 1977, p. 313.
[210] Bruice, P. Y., and Bruice, T. C., *J. Am. Chem. Soc.*, **100**, 4793 (1978).
[211] Bruice, P. Y., and Bruice, T. C., *J. Am. Chem. Soc.*, **100**, 4802 (1978).
[212] Cocivera, M., Kohesh, F. C., Malatesta, V., and Zinck, J. J., *J. Org. Chem.*, **42**, 4076 (1977).
[213] Kuo, D. J., and Rose, I. A., *J. Am. Chem. Soc.*, **100**, 6288 (1978).
[214a] Adlersberg, M., Dayan, J., Bondinell, W. E., and Sprinson, D. B., *Biochemistry*, **16**, 4382 (1977).
[214b] Belasio, J. G., Herlihy, J. M., and Knowles, J. R., *Biochemistry*, **17**, 2971 (1978).
[214c] Schray, K. J., Waud, J. M., and Howell, E. E., *Arch. Biochem. Biophys.*, **189**, 106 (1978).
[215] Hupe, D. J., and Wu, D., *J. Am. Chem. Soc.*, **99**, 7653 (1977).
[216] Hupe, D. J., Wu, D., and Shepperd, P., *J. Am. Chem. Soc.*, **99**, 7659 (1977).
[217] Sugimoto, T., and Kaiser, E. T., *J. Am. Chem. Soc.*, **100**, 7750 (1978).
[218] Brower, K. R., and Hughes, D., *J. Am. Chem. Soc.*, **100**, 7591 (1978).
[219] Warnhoff, E., Rampersad, M., Raman, P. S., and Yerhoff, F. W., *Tetrahedron Lett.*, **1978**, 1659.
[220a] Cambillau, C., Bram, G., Corset, J., Riche, C., and Pascard-Billy, C., *Tetrahedron*, **34**, 2675 (1978).
[220b] Bordwell, F. G., and Cornforth, F. J., *J. Org. Chem.*, **43**, 1763 (1978).
[221a] Brown, F. C., Casadevall, E., Metzger, P., and Morris, D. G., *J. Chem. Res.*, (*S*) **1977**, 335.
[221b] Werstiuk, N. H., Taillefer, R., and Banerjee, S., *Can. J. Chem.*, **56**, 1140 (1978).
[222] Werstiuk, N. H., Taillefer, R., and Banerjee, S., *Can. J. Chem.*, **56**, 1148 (1978).
[223] Werstiuk, N. H., and Andrews, P., *Can. J. Chem.*, **56**, 2605 (1978).
[224] Fraser, R. R., and Champagne, P. J., *J. Am. Chem. Soc.*, **100**, 657 (1978).
[225] Hall, S. S., Doweyko, A. M., and Jordan, F., *J. Am. Chem. Soc.*, **100**, 5934 (1978).
[226] Borders, C. L., Woodall, M. L., and George, A. L., *Biochem. Biophys. Res. Commun.*, **82**, 901 (1978).
[227] Dann, L. G., and Britton, H. G., *Biochem. J.*, **169**, 39 (1978).
[228] Viola, R. E., and Cleland, W. W., *Biochemistry*, **17**, 4111 (1978).
[229] Covey, D. F., and Talahay, P., *Fed. Proc., Fed. Am. Soc. Exp. Biol.*, **37**, 1296 (1978).
[230] Leforstier, C., *Nouveau J. Chim.*, **2**, 73 (1978).
[231] Gettys, G. A., and Gutsche, C. D., *Bioorg. Chem.*, **7**, 141 (1978).
[232] Bílik, V., and Petrus, L., *Collect. Czech. Chem. Commun.*, **43**, 1159 (1978); Bilik, V., Petruš, L., and Farkaš, V., *ibid.*, p. 1163.
[233] Lookhart, G. L. and Feather, M. S., *Carbohydr. Res.*, **60**, 259 (1978).
[234] Königstein, J., *Collect. Czech. Chem. Commun.*, **43**, 1152 (1978).
[235] Shaw, A. J., and Taso, G. T., *Carbohydr. Res.*, **60**, 327 (1978).
[236] Sallam, M. A. E., *Carbohydr. Res.*, **63**, 127 (1978).
[237] Gallo, A. A., Mieyal, J. J., and Sable, H. Z., in "Bioorganic Chemistry", E. E. van Tamelen, Ed., Academic Press, New York, Vol. 4, 1978, p. 147.
[238] John, F., and Mariam, Y. H., *J. Am. Chem. Soc.*, **100**, 2534 (1978).
[239] Satyanarayana, N., and Sunduram, E. V., *J. Indian Chem. Soc.*, **54**, 886 (1977); *Chem. Abs.*, **89**, 162711 (1978).
[240] Volkova, N. V. and Kanivets, N. P., *Ukr. Khim. Zh.* (*Russ. Ed.*), **43**, 1221 (1977); *Chem. Abs.*, **88**, 61745 (1978).
[241] Jagdale, M. H., and More, B. S., *Sci. J. Shivaji Univ.*, **15**, 131 (1975); *Chem. Abs.*, **88**, 5852 (1978).
[242] Neilands, O., and Kalnins, S., *Str. Tautomernye Prevrashch.* (*Batu*)-*Dikarbonil'nykh Sodein*, **1977**, 8, 381; *Chem. Abs.*, **88**, 49772 (1978); Adler, M., and Schank, K. *Chem. Ber.*, **111**, 2859 (1978).
[243] Carlsen, L., and Duus, F., *J. Am. Chem. Soc.*, **100**, 281 (1978).
[244] Russell, D. H., Gross, M. L., and Nibbering, N. M. M., *J. Am. Chem. Soc.*, **100**, 6133 (1978).
[245] Steyn, P. S., and Wessels, P. L., *Tetrahedron Lett.*, **1978**, 4707.
[246] Lefour, J. M., Sathou, P., Bram, G., Guibé, F., Loupy, A., and Seyden-Penne, J., *Tetrahedron Lett.*, **1978**, 3831; De Palma, V. M., and Arnett, E. M., *J. Am. Chem. Soc.*, **100**, 3524 (1978); Raban, M., and Haritos, D., *J. Chem. Soc., Chem. Commun.*, **1978**, 965; Groenewegen, P., Kallenberg, H., and van der Gen, A., *Tetrahedron Lett.*, **1978**, 491.
[247] Schwarz, H., Williams, D. H., and Wesdemiotis, C., *J. Am. Chem. Soc.*, **100**, 7052 (1978).

[248] Bouma, W. J., and Radom, L., *Aust. J. Chem.*, **31**, 1167, 1649 (1978); *J. Mol. Struct.*, **43**, 267 (1978).
[249] Cf. *Org. React. Mech.*, **1977**, 9.
[250] Cheng, A. K., and Stothers, J. B., *Can. J. Chem.*, **56**, 1342 (1978).
[251] Cheng, A. K., and Stothers, J. B., *Can. J. Chem.*, **55**, 4184 (1977).
[252] Guthrie, J. P., *Can. J. Chem.*, **56**, 961 (1978).
[253] Sue, J. M., and Knowles, J. R., *Biochemistry*, **17**, 4041 (1978).
[254] Ulevitch, R. J., and Kallen, R. G., *Biochemistry*, **16**, 5350, 5355 (1977).
[255] Tatsumoto, K., and Martell, A. E., *J. Am. Chem. Soc.*, **100**, 5549 (1978).
[256] Loewus, M. W., *J. Biol. Chem.*, **252**, 7221 (1977).
[257] Kyriakakou, G., Loupy, A., and Seyden-Penne, J., *J. Chem. Res.*, (*S*) **1978**, 8, 147.
[258] Kandlikar, S., Sethuram, B., and Rao, T. N., *Indian J. Chem.*, **15A**, 1076 (1977); *Chem. Abs.*, **88**, 189787 (1978).
[259] Markitanova, L. I., Nesterov, V. M., Drevina, V. M., and Teten'chuk, K. P., *Khim.-Farm Zh.*, **11**, 94 (1977); *Chem. Abs.*, **88**, 21612 (1978).
[260] Jagannadham, V., Kandlikar, S., Sethuram, B., and Rao, T. N., *Natl. Acad. Sci. Lett.*, **1**, 207 (1978); *Chem. Abs.*, **89**, 162906, (1978).
[261] Isǎcescu, D. A., and Avramescu, F., *Rev. Roum. Chim.*, **23**, 661, 865, 873 (1978).
[262] Awad, W. I., Abdel-Wahhab, S. M., Awad, B. M., Guirguis, N. R., and Wassef, W. N., *Indian J. Chem.*, **15B**, 737 (1977); *Chem. Abs.*, **88**, 61758 (1978).
[263a] Suyama, K., and Nakanishi, T., *Agric. Biol. Chem.*, **42**, 507 (1978); *Chem. Abs.*, **89**, 41826 (1978).
[263b] Maruoka, K., Hashimoto, S., Kitagawa, Y., Yamamoto, H., and Nozaki, H., *J. Am. Chem. Soc.*, **99**, 7705 (1977).
[264] Fellmann, P., and Dubois, J.-E., *Tetrahedron*, **34**, 1349 (1978); Mulzer, J., Segner, J., and Bruntrup, G., *Tetrahedron Lett.*, **1977**, 4651; Eichenauer, H., Friedrich, E., Lutz, W., and Enders, D., *Angew. Chem. Int. Edn.*, **17**, 206 (1978).
[265] Barnard, G. F., Itoh, R., Hohberger, L. M., and Shemin, D., *J. Biol. Chem.*, **252**, 8965 (1977).
[266] Wilde, J., Hunt, W., and Hupe, D. J., *J. Am. Chem. Soc.*, **99**, 8319 (1977).
[267] Ching, W.-M., and Kallen, R. G., *J. Am. Chem. Soc.*, **100**, 6119 (1978).
[268] Okaro, V., do Amaral, L., and Cordes, E. H., *J. Am. Chem. Soc.*, **98**, 4201 (1976).
[269] Gassman, P. G., and Talley, J. J., *Tetrahedron Lett.*, **1978**, 3773.
[270] McIntosh, J. M., *Can. J. Chem.*, **55**, 4200 (1977).
[271] Nadar, P. A., Granasekaran, C., and Chandrasekharan, J., *Bull. Chem. Soc. Jpn.*, **51**, 2187 (1978).
[272] Van Osselaer, T. A., Lemière, G. L., Lepoivre, J. A., and Alderweireldt, F. C., *Bull. Soc. Chim. Belg.*, **87**, 153 (1978).
[273] Byers, L. D., *Arch. Biochem. Biophys.*, **186**, 335 (1978).
[274] Meunier, J.-C., and Dalziel, K., *Eur. J. Biochem.*, **82**, 483 (1978).
[275] Goldberg, S. I., Baba, N., Green, R. L., Pandian, R., Stowers, J., and Dunlap, R. B., *J. Am. Chem. Soc.*, **100**, 6768 (1978).
[276] Doiuchi, T., and Minoura, Y., *Isr. J. Chem.*, **15**, 84 (1977); *Chem. Abs.*, **87**, 200492 (1977).
[277] Szmant, H. H., and Alciaturi, C. E., *J. Solution Chem.*, **7**, 269 (1978); *Chem. Abs.*, **89**, 146156 (1978).
[278] Alvarez-Ibarra, C., Goya, P., Lissavetzky, J., Perez-Ossorio, R., Perez-Rubalcaba, A., and Quiroga, M. L., *An. Quim.*, **73**, 760 (1977); *Chem. Abs.*, **87**, 200672 (1977).
[279] Arnaud, C., Accary, A., and Huet, J., *C. R. Hebd. Séances Acad. Sci., Sér. C*, **285**, 325 (1977).
[280] Shinkai, S., Ide, T., and Monabe, O., *Chem. Lett.*, **1978**, 583.
[281] Murty, G. S. S., Sethuram, B., and Rao, T. N., *Indian J. Chem.*, **15A**, 1073 (1977); *Chem. Abs.*, **88**, 189652 (1978).
[282] Haldorsen, K. M., *Carbohydr. Res.*, **63**, 61 (1978).
[283] Bhatnogar, R. P., and Fadnis, A. G., *Monatsh. Chem.*, **109**, 319 (1978).
[284] Bhalekar, A. A., Shanker, R., and Bakore, G. V., *Indian J. Chem.*, **15A**, 705 (1977); *Chem. Abs.*, **88**, 189640 (1978).
[285] Naidu, H. M. K., and Mahadevappa, D. S., *Monatsh. Chem.*, **109**, 269 (1978).
[286] Ashby, E. C., and Bowers, J. S., *J. Am. Chem. Soc.*, **99**, 8504 (1977); Ashby, E. C., and Wiesemann, T. L., *ibid.*, **100**, 189, 3101 (1978); Benkeser, R. A., Siklosi, M. P., and Mozdzen, E. C., *ibid.*, **100**, 2134 (1978); Garcia Marinte, A., Perez-Ossorio, R., and Plumet, J., *An.*

Quím., **72**, 873 (1976); Colantoni, A., Di Maio, G., Vecchi, E., Zeuli, E. and Quagliata, C., *Tetrahedron*, **34**, 357 (1978); Colantoni, A., Di Maio, G., Vecchi, E., and Zeuli, E., *ibid.*, **34**, 1235 (1978); Rubin, M. B., and Ben-Bassat, J. M., *J. Org. Chem.*, **43**, 1009 (1978); Makino, T., Shibata, K., Rohrer, D. C., and Osawa, Y., *ibid.*, **43**, 276 (1978); Anderson, N. H., Duffy, P. F., Denniston, A. D., and Grotjahn, D. B., *Tetrahedron Lett.*, **1978**, 4315.

[287] Palm, N., and Tuulmets, A., *Org. React.* (*Tartu*), **14**, 246 (1977).

[288] Lomas, J. S., Luong, P. K., and Dubois, J.-E., *J. Org. Chem.*, **42**, 3394 (1977).

[289] Evans, D. A., Hunt, K. M., and Takacs, J. M., *J. Am. Chem. Soc.*, **100**, 3467 (1978).

[290] Zioudrou, C., Moustakali-Mavridis, I., Chrysochou, P., and Karabatsos, G. J., *Tetrahedron*, **34**, 3181 (1978); Lefour, J.-M., and Loupy, A., *ibid.*, **34**, 2597 (1978); Aranda, G., Bernassau, J.-M., and Fetizon, M., *J. Org. Chem.*, **42**, 4256 (1977); Idriss, N., Perry, M., Maroni-Barnaud, Y., Roux-Schmitt, M.-C., and Seyden-Penne, J., *J. Chem. Res.*, (*S*) **1978**, 128; Dubois, J.-E., MacPhee, J. A., and Panaye, A., *Tetrahedron Lett.*, **1978**, 4099; Kobayashi, Y. M., Lambrecht, J., and Jochims, J. C., *Chem. Ber.*, **111**, 3442 (1978).

[291] Poplavskii, V. S., Ostrovskii, V. A., Koldobskii, G. I., Gidaspov, B. V., and Moskvin, A. V., *Zh. Org. Khim.*, **14**, 121 (1978); *Chem. Abs.*, **88**, 151690 (1978).

[292] Jończyk, A., Kwast, A., and Makosza, M., *J. Chem. Soc.*, *Chem. Commun.*, **1977**, 902.

[293] Bigler, P. and Neuenschwander, M., *Helv. Chim. Acta*, **61**, 2165, 2381 (1978); Bigler, P., Schönholzer, S., and Neuenschwander, M., *ibid.*, **61**, 2059 (1978); Neuenschwander, M., Bigler, P., Christen, K., Iseli, R., Kyburz, R., and Mühle, H., *ibid.*, **61**, 2047 (1978).

[294] Suggs, J. W., *J. Am. Chem. Soc.*, **100**, 640 (1978).

[295] Miller, J. A., and Stewart, D., *J. Chem. Soc.*, *Perkin Trans.* 1, **1977**, 2416.

[296] Vorontsova, N. A., Vlasov, O. N., Bogel'fer, L. Y., and Mel'nikov, N. N., *Zh. Obshch. Khim.*, **47**, 2698 (1977); *Chem. Abs.*, **88**, 120251 (1978).

[297] Bellucci, G., Ingrossu, G., and Mastrorilli, E., *Tetrahedron*, **34**, 387 (1978).

[298] Kudesia, V. P., and Mukherjee, S. K., *Acta Cienc. Indica*, **3**, 23 (1977); *Chem. Abs.*, **87**, 200511 (1977).

[299] Ha, T.-K., Wild, U. P., Kühne, R. O., Loesch, C., Schaffhauser, T., Stachel, J., and Wokaun, A., *Helv. Chim. Acta*, **61**, 1193 (1978).

[300] Del Bene, J. E., *J. Am. Chem. Soc.*, **100**, 1673 (1978); Del Bene, J. E., and Radovick, S., *ibid.*, **100**, 6936 (1978); Jost, R., Sommer, J., and Wipff, G., *Nouveau J. Chem.*, **2**, 63 (1978).

CHAPTER 2

Reactions of Acids and their Derivatives

A. J. Kirby

University Chemical Laboratory, Cambridge

CARBOXYLIC ACIDS

Tetrahedral Intermediates

Guthrie[1] has derived values of 4.7 ± 2.7 and 3.8 ± 1.6 kcal mol^{-1} for the free energy of hydration of *S*-ethyl thioformate and trifluorothioacetate, respectively. These figures, and estimated pK_a values of the tetrahedral addition intermediates (**1**; Z = SEt), allow the construction of detailed reaction co-ordinate diagrams for the reaction, and the calculation of full sets of rate constants. This detailed information is now available for a wide range of carboxylic acid derivatives (as well as aldehydes and ketones),[2] and it has been shown that rate and equilibrium constants referring to a given microscopic step are related by a common curve. The curves can be described by a simple one-parameter equation (1) derived from Marcus theory, and the correlations cover a very wide range of reactivity (up to 20 powers of 10 in equilibrium constant, and 10 powers of 10 in rate).

$$\log k = 10 - b(1 - \log K/4b)^2 \qquad (1)$$

For base-catalysed hydration, for example, b is about 6, though it varies a little for different classes of compound ($b = 5.35$, 6.68, and 7.41 for amides, esters, and thiolesters, respectively).

These correlations allow calculation of the equilibrium constant for the addition of hydroxide (and hence that for hydration) from the rate constant for the addition of hydroxide to any carbonyl compound. More important, the correlation provides a potentially powerful new criterion of mechanism, since the rate and equilibrium constants have to be calculated in terms of the correct rate-determining step of the reaction. When applied to the spontaneous hydration of carbonyl compounds, for example, application of this criterion leads to the clear conclusion that the rate-determining step is the simple addition of water to give the zwitterion (**2**). An

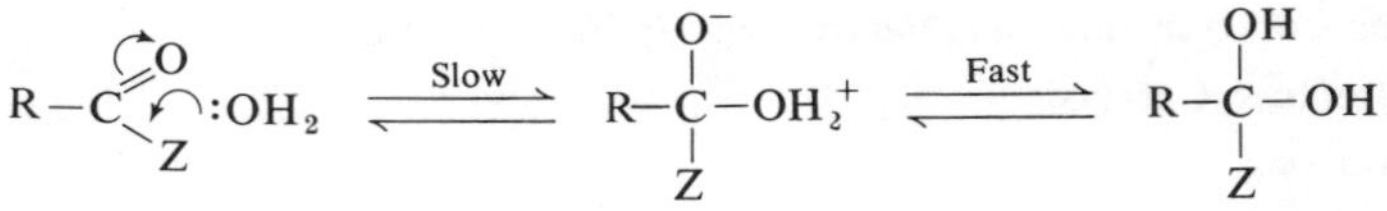

$$R-C(=O)(Z)\ {:}OH_2 \xrightarrow{\text{Slow}} R-C(O^-)(Z)-OH_2^+ \xrightarrow{\text{Fast}} R-C(OH)(Z)-OH$$

(**2**) (**1**)

explanation also emerges for the observation that the well-known acid-inhibition of the hydrolysis of thiolesters is found only for esters with electron-withdrawing acyl groups, particularly CF_3COSEt; the electron-withdrawing group favours loss of water, and slows loss of thiol from the tetrahedral intermediate (**1**; Z = SEt), so that the partitioning of (**1**) becomes important in the rate-determining step.

Theoretical calculations on tetrahedral intermediates are beginning to appear in increasing numbers, stimulated especially by Deslongchamps' theory of stereoelectronic control (see below). A self-consistent field study[3] (using a STO-3G basis set) of the stereochemistry of the hydroxide adduct of methyl formate (**3**) shows the usual three minima associated with rotation about the C—OMe bond, and two associated with rotation about O—H; the latter minima are complicated by H-bonding between the OH and the geminal O^- group (**3**). Calculations on tetrahedral intermediates from amides include a MINDO/3 study[4] of the hydroxide

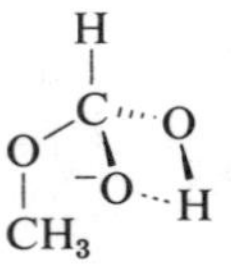

(**3**) 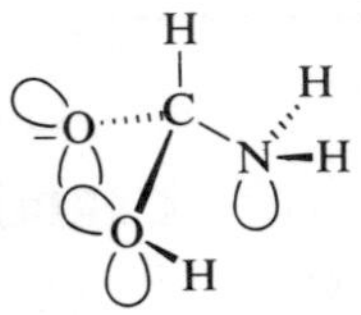 (**4**)

adduct of glycylglycine, as a model for the peptide-cleavage step in bacterial cell-wall synthesis inhibited by β-lactum antibiotics; and an *ab initio* study[5] on all 15 fully staggered conformations of the hydrate (**4**) of formamide. The results of Lehn and Wipff[5] are in agreement with a simple hyperconjugation picture in which lone pairs of electrons interact with the antibonding σ^* orbitals of *antiperiplanar*-oriented C—X bonds. Thus the weakest C—N bond is found for the conformation

(**4**) in which lone pairs on both oxygens are *antiperiplanar* to the C—N bond, and the nitrogen lone pair is *antiperiplanar* to the non-polar C—H bond. This form has the longest C—N and shortest C—O bond lengths, and hence is expected to show selective C—N cleavage: though the effects of substitution, solvation, prototropy, and general species catalysis will undoubtedly complicate the picture for a real reaction in solution.

Deslongchamps and his co-workers[6] have reported further results of their experimental study of stereoelectronic control of the base-catalysed hydrolysis of amides using the oxygen-exchange reaction of carbonyl-^{18}O-labelled tertiary amides.[7] Stereoelectronic theory predicts that tertiary amides will show no ^{18}O-exchange: the tetrahedral intermediate (**6**) formed by the addition of hydroxide ion to the amide (**5**) corresponds to a form (*e.g.* **4**, above) expected to show selective C—N cleavage, because lone pairs on both oxygens can be *antiperiplanar* to the C—N bond. What little experimental evidence is available in the literature is

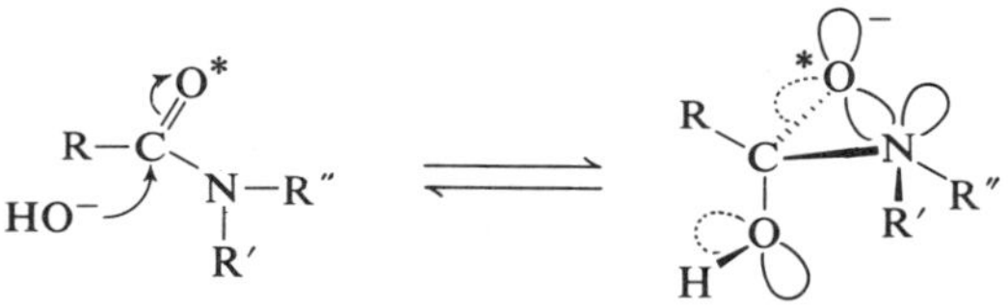

(**5**) (**6**)

consistent with this prediction, but the hydrolysis of *carbonyl*-^{18}*O*-*N*-benzyl-*N*-methylformamide is accompanied by substantial loss of label from the carbonyl group.

Paradoxically, this result, far from weakening the theory of stereoelectronic control, leads to an explanation that is convincing enough to strengthen it. The authors[6] suggest that where R = H the barrier for internal rotation (or inversion) of the amino-group is very low and becomes comparable with that for C—N cleavage of (**6**) (which is itself relatively high because of the minimal steric crowding in a formate derivative). When the corresponding acetamide is used, ^{18}O-exchange is sharply reduced, and it disappears altogether for the propionamide.[6] The stereoelectronic factor is thus seen, not as an all-or-nothing effect, but as competing with other factors. For example, the breakdown of (**6**) involves (at least) three competing processes, of which two (the cleavage reactions) are subject to primary stereoelectronic control; while the third, conformational, change will be subject to what might be termed secondary stereoelectronic control, since the energies of both conformers (**6** and **7**) depend on stereoelectronic interactions.[5]

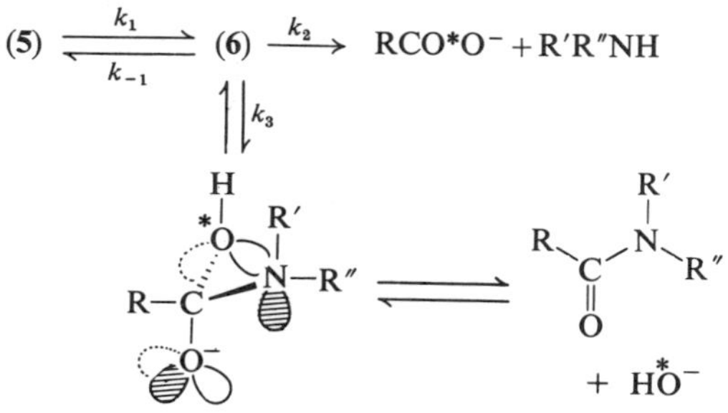

$$(5) \underset{k_{-1}}{\overset{k_1}{\rightleftharpoons}} (6) \xrightarrow{k_2} RCO^*O^- + R'R''NH$$

$$(6) \underset{}{\overset{k_3}{\rightleftharpoons}} (7) \rightleftharpoons RCONR'R'' + HO^{*-}$$

(**7**)

The alkaline hydrolysis of *N*,*N'*-dimethylformamidine (**8**) leads initially to the thermodynamically less favourable isomer (**9**) of formamide; here, too, stereo-electronic control is the preferred explanation.[8]

Ester aminolysis has come under very close scrutiny in recent years, particularly because the reaction offers such a wide range of variables, and because general

$$\underset{(\mathbf{8})}{\mathrm{H{-}C(H)({=}N^{+}HMe)(NHMe)}} \xrightarrow{\mathrm{HO^-}} \underset{(\mathbf{9})}{\mathrm{H{-}C({=}O){-}N(H)Me}} \xrightarrow{\text{Slow}} \mathrm{H{-}C({=}O){-}N(Me)H}$$

acid–base catalysis is readily observed. General acid catalysis is most important when weakly basic amines are involved, which in turn necessitates the use of esters with good leaving groups, at least for investigation of intermolecular reactions. Satterthwait and Jencks suggested some years ago[9] that the mechanism of the weak general acid catalysis of the addition of basic amines to aryl acetates involves the rate-determining transfer of a proton from the general acid to the initial tetrahedral addition intermediate $\mathbf{T}^{\pm}$. This is a case where general acid catalysis is

$$\mathrm{R'{-}C({=}O){-}OAr} + \mathrm{RNH_2} \underset{k_2}{\overset{k_1}{\rightleftharpoons}} \underset{\mathbf{T}^{\pm}}{\mathrm{R'{-}C(O^-)(OAr){-}\overset{+}{N}H_2R}} \overset{\mathrm{HA}}{\rightleftharpoons} \mathrm{R'{-}C(OH)(OAr){-}\overset{+}{N}H_2R} + \mathrm{A^-}$$

enforced[10] by the short lifetime of the intermediate: that is, the amine is expelled faster than $\mathbf{T}^{\pm}$ is trapped by protonation. The question then arises as to what happens when $\mathbf{T}^{\pm}$ reverts to starting materials faster than the catalyst (HA) can diffuse away from the intermediate; that is, when the lifetime of $\mathbf{T}^{\pm}$ is too short for the catalyst to diffuse to meet it in a separate step. Simple analysis suggests that under these circumstances catalysis must be accounted for by a preassociation mechanism (Scheme 1) by which $\mathbf{T}^{\pm}$ is formed in the presence of the catalyst, in a preformed ternary complex.

Cox and Jencks[11] have measured general acid catalysis of the methoxyaminolysis of phenyl acetate by a wide range of general acids, and have presented evidence that catalysis in this case does indeed involve the preassociation mechanism shown in the Scheme.

The Brönsted plot for general acid catalysis is curved and approaches unit slope for catalysis by weak acids ($\mathrm{p}K_a > 7$), for which the proton transfer is thermodynamically unfavourable and the diffusion away of A^- becomes rate-determining. For acids with $\mathrm{p}K_a$'s in the region 4–7, where the Brönsted plot is curved, the large deuterium isotope effects observed reach a maximum of almost 4 for a general acid of $\mathrm{p}K_a$ 6.8, close to the estimated $\mathrm{p}K_a$ of $\mathbf{T}^{+}$; this suggests that the proton transfer step k_p is partially rate-determining in this region.

The Brönsted plot does not level out to zero slope for catalysis by strong acids but approaches a line of slope $\alpha = 0.16$, consistent with catalysis of k_1' by hydrogen bonding in the association complex, but not with a rate-determining diffusion-controlled encounter step. Other features of reactions of strong acids are also inconsistent with a diffusion step: catalysis by H_3O^+ does not show the rate enhancement expected for the diffusion of a proton through water, and catalysis by chloroacetic acid is not inhibited by increasing solvent viscosity. Above all, the rate constants observed for general acid catalysis are indeed greater than those expected for a reaction involving the diffusion-controlled encounter of **T**± with a general acid, thus establishing fairly convincingly that the preassociation mechanism can provide the favourable pathway to products under the right conditions.[11]

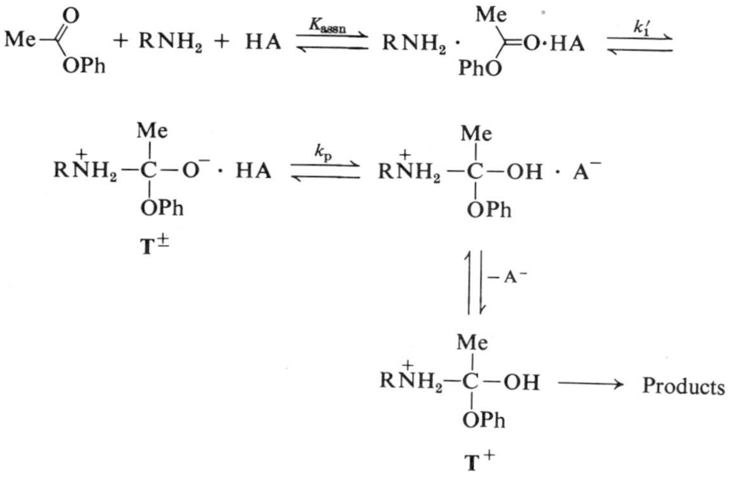

$$\mathrm{MeC(=O)OPh} + \mathrm{RNH_2} + \mathrm{HA} \underset{}{\overset{K_{assn}}{\rightleftharpoons}} \mathrm{RNH_2 \cdot MeC(OPh){=}O{\cdot}HA} \overset{k_1'}{\rightleftharpoons}$$

$$\mathrm{R\overset{+}{N}H_2{-}C(Me)(OPh){-}O^- \cdot HA}\ (\mathbf{T}^{\pm}) \overset{k_p}{\rightleftharpoons} \mathrm{R\overset{+}{N}H_2{-}C(Me)(OPh){-}OH \cdot A^-}$$

$$\mathrm{R\overset{+}{N}H_2{-}C(Me)(OPh){-}OH \cdot A^-} \underset{}{\overset{-A^-}{\rightleftharpoons}} \mathrm{R\overset{+}{N}H_2{-}C(Me)(OPh){-}OH}\ (\mathbf{T}^{+}) \longrightarrow \text{Products}$$

SCHEME 1

The very similar reaction of semicarbazide with *p*-methoxyphenyl formate probably involves the same mechanism.[12]

The partitioning of the tetrahedral intermediate generated by the addition of a basic tertiary amine to an aryl ester is a decidedly simpler problem, uncomplicated by general species catalysis. Breakdown of the zwitterionic tetrahedral intermediate is rate-determining when the amine is equally or less basic than the phenoxide leaving group, as suggested by the high sensitivity ($\beta_{nuc} = 0.9$) to the basicity of the amine. The sensitivity is much lower ($\beta_{nuc} = 0.2$) when the amine is 4–5 p*K* units more basic, and formation of the intermediate becomes rate-determining. A very similar picture has been found by Gresser and Jencks[13] for the reactions of substituted quinuclidines with diaryl carbonates, with a similar change in sensitivity to the leaving group $\beta_{LG} = (-1.3) \rightarrow (-0.2)$ at the point where β_{nuc} changes. The system is of interest because the presumed tetrahedral intermediate can be generated independently[14] (see Scheme 2). The relative rates of C—N and C—O cleavage, as measured by product analysis, do not depend on the way the tetrahedral intermediate is generated: they agree with the figures estimated from kinetic data, thus confirming the change of rate-determining step. These results allow a comparison of the leaving group capabilities of aryloxides with those of tertiary amines. A quinuclidine is as good a leaving group as an aryloxide anion that is 4–5 p*K* units less basic (*i.e.* an amine is a better leaving group than an oxyanion of the same basicity), though this difference depends on the absolute figures involved, increasing for more basic leaving groups (and for more electron-withdrawing "acyl" substituents in the ester).[14]

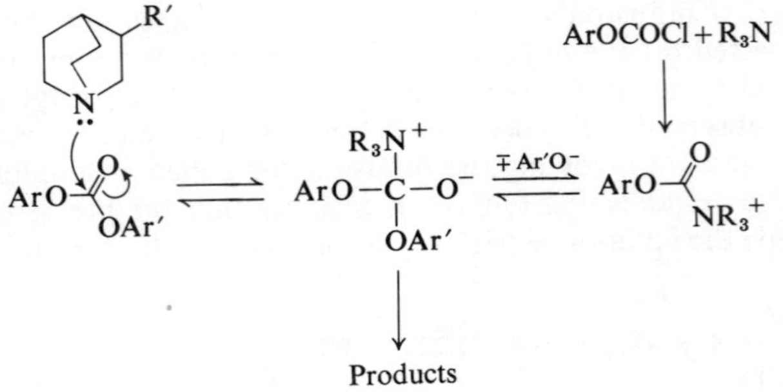

SCHEME 2

Very similar behaviour has been found in a study of the reactions of substituted pyridines with methyl 2,4-dinitrophenyl carbonate.[15]

Two reports concern gas-phase pathways for ester hydrolysis, as studied by ion-cyclotron resonance: the reactions of a range of esters with $^{18}OH^-$ show that the $B_{AC}2$ path is dominant, though $B_{AL}2$ is more important than in solution;[16] however, the ion–molecule reaction corresponding to acid-catalysed esterification may not involve tetrahedral intermediates.[17]

Hydrolyses of the trioxa-adamantanes (**10**; R = H or Me) show significant differences from the reactions of open-chain ortho-formates and -acetates;[18] In particular, the reactions are much slower, and replacement of H by Me causes a decrease of 50-fold in rate, though open-chain orthoacetates are more reactive than orthoformates. The explanation seems to be that the rate-determining step for the

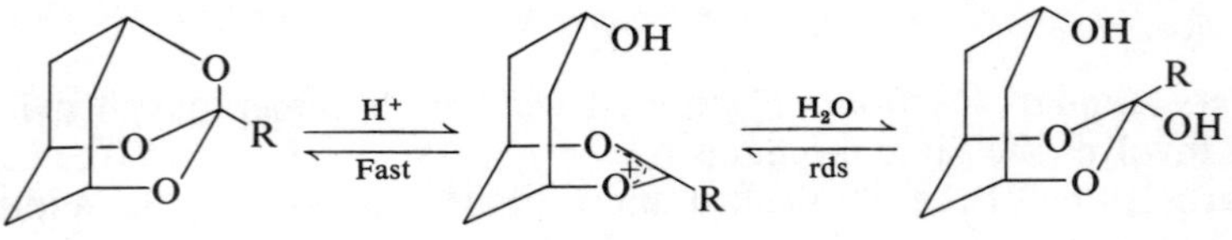

(**10**) (**11**)

hydrolysis of (**10**) is not C—O cleavage, as is normal for the hydrolysis of ortho-esters, but the hydration of the dioxocarbonium ion intermediate (**11**). This is eminently reasonable, requiring simply that intramolecular attack by the free OH group on the carbonium centre of (**11**) be faster than intermolecular attack by water, and it is supported by the substantial negative entropies of activation observed for the hydrolysis of (**10**; R = H or Me).[18]

McLelland and Ahmad have also found conditions under which hydration of the dioxocarbonium ion intermediate in orthoester hydrolysis becomes rate-determining;[19] thus, since the intermediate (**13**) formed upon reaction of (**12**) is particularly unreactive for steric reasons, its formation and decay can be monitored

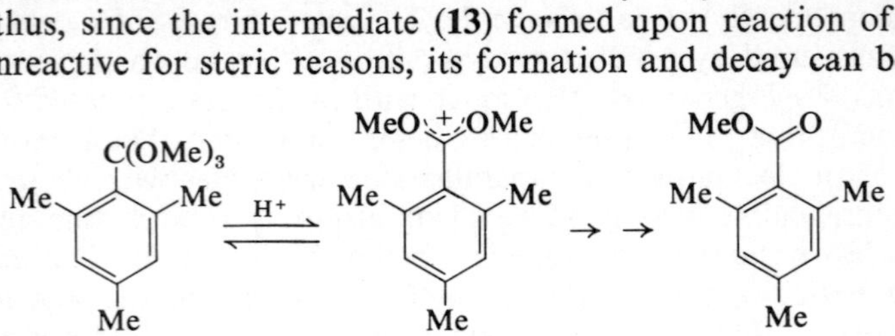

(**12**) (**13**)

spectroscopically. In fact (**13**) is less stable than it looks, because it is a prime candidate for classical steric inhibition of resonance. This can be detected even in the absence of the two *ortho*-substituents: Hammett's ρ for the H_3O^+-catalysed hydrolysis of a series of 2-aryl-2-methoxy-1,3-dioxolanes is -1.60, significantly larger than $\rho = -1.16$ found for the corresponding *p*-substituted trimethylorthobenzoates, suggesting that steric congestion in the transition state leading to the dioxocarbonium ion is reduced in the cyclic system (**14**), compared with (**15**).[20]

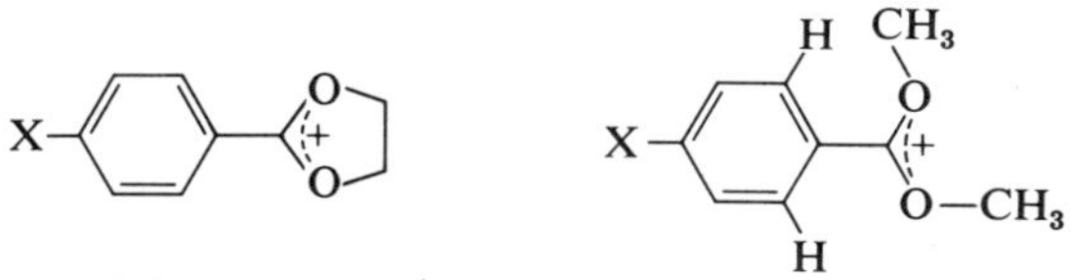

(**14**) (**15**)

The detailed mechanism of general acid catalysis of orthoester hydrolysis has been discussed[21] in the light of kinetic hydrogen isotope effects, and Brönsted α values have been measured for orthoesters derived from a few lactones.[22]

Two studies have dealt with the reactions of amide acetals.[23,24] Both C–O and C–N cleavage occurs when compounds (**16**) are hydrolysed in 0.1M-HCl;[23] the proportion of C–O cleavage increases with increasing electron withdrawal by Ar, with decreasing polarity of the solvent, with increasing concentrations of general acid, and, of course, with increasing pH. These are very reactive compounds at all

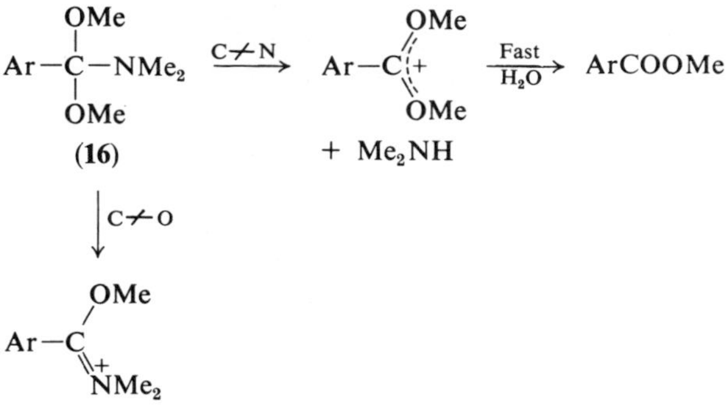

$$\mathrm{Ar{-}C(OMe)_2{-}NMe_2}\ (\mathbf{16}) \xrightarrow{\mathrm{C \not- N}} \mathrm{Ar{-}C^+(OMe)_2} + \mathrm{Me_2NH} \xrightarrow[\mathrm{H_2O}]{\text{Fast}} \mathrm{ArCOOMe}$$

$$\mathrm{Ar{-}C(OMe)_2{-}NMe_2}\ (\mathbf{16}) \xrightarrow{\mathrm{C \not- O}} \mathrm{Ar{-}C(OMe){=}N^+Me_2}\ (\mathbf{17})$$

pH's, because, the imidatonium ion (**17**) is stable enough to be produced in a spontaneous cleavage reaction of the neutral species (**16**). Acid-catalysed hydrolysis probably involves competing general acid-catalysed cleavage of the C—O bonds and the spontaneous breakdown of the *N*-protonated substrate.

In contrast to the similar rates of C–O and C–N cleavage in amide acetals such as (**16**), the hydrolyses of tetrahedral *intermediates*, with at least one OH group, favour C—N cleavage decisively. The hydrolysis of (**18**), for example, gives predominant C—N cleavage up to pH 7,[24] and here the proportion of C—N cleavage *increases* with increasing general acid concentration: even at high pH some C—N cleavage occurs. The *formation* of (**18**) is the rate-determining step in the hydrolysis of (**19**), but rate constants for buffer catalysis can be deduced from product studies; general acid catalysis of C—N cleavage involves a diffusion-controlled proton transfer, while the general base-catalysed breakdown of (**18**) may involve concerted protonation and C—N cleavage.[24]

$$H_3C-C(OEt)=\overset{+}{N}(Me)-Ph \ \underset{H_2O}{\overset{HO^-}{\rightleftharpoons}}\ H_3C-C(OEt)(OH)-NMePh \longrightarrow CH_3CONMePh + EtOH$$

(19) (18)

$$\mathbf{(18)} \longrightarrow CH_3COOEt + PhNHMe$$

Intermolecular Catalysis and Reactions

Reactions in Hydroxylic Solvents

A substantial review of the alkyl–oxygen cleavage (S_N2) reaction of esters has appeared in *Organic Reactions*.[25] Yates and Modro have summarized the transition-state activity-coefficient approach, with special reference to the acid-catalysed hydrolysis of esters and amides.[26] Reactions catalysed by folate cofactors, which include formyl transfer,[27] have been studied and aspects of bifunctional catalysis have been reviewed.[28]

Charton[29] has recently questioned the validity of Taft's polar substituent constants σ^*, concluding that they result almost entirely from the incomplete cancellation of steric factors. This view has now been vigorously challenged by Dubois and his co-workers. Charton analysed data for base-catalysed ester hydrolysis in terms of a multiple correlation equation and found no significant dependence on the inductive substituent constant σ_I for alkyl groups. MacPhee and Dubois[30] point out that σ_I is itself a dubious measure of polar effects, measuring as it does differences in pK_a of alkyl-substituted acetic acids in water. Since these are often entropy-controlled and enthalpies of ionization differ because of solvation differences (which are in variable part dependent on steric effects), the values, and even the signs, of the σ_I constants must vary with temperature.

A more satisfactory measure of inductive effects is available in the pK_a's of 4-substituted quinuclidinium ions, which Grob and his co-workers[31] have shown to be correlated by σ^* constants for 4-alkyl substituents. MacPhee, Panaye, and Dubois[32] suggest the use of the much more readily accessible infrared stretching frequency, ν_{CO}, of carboxylic acids, as a measure of the polar effects of alkyl groups; they have shown that ν_{CO} values for monomeric acids RCOOH are linearly related to σ^* in four non-polar solvents (though the correlation is far from perfect), and they suggest that steric effects make little if any contribution to this parameter.

The Paris group have also measured steric substituent constants for a group of highly branched alkyl groups for which E_s values are not available. They define[32] a modified constant, E_s', as the rate of esterification of RCOOH in methanol at 40 °C relative to that of acetic acid, and obtain a scale differing only in a few particulars from the E_s scale, where both parameters are available for comparison. An examination of the new E_s' values for highly branched alkyl groups shows an unexpected feature.[33] As is well known, the effect on steric compression of introducing one more carbon atom into a group close to a reaction centre increases sharply with the size of the group: the E_s values of Me, Et, Pr^i, and Bu^t, for example, are 0 (by definition), 0.07, 0.47, and 1.54, respectively. When the group is introduced further from the reaction centre, of course, the extra effect is reduced and the steric effect of the extra group levels out. But with very large, highly branched groups larger groups appear to have *smaller* steric effects; that is, the introduction of extra methyl groups actually *reduces* the rate of esterification of RCOOH in methanol. For example, the E_s' values for R = Bu^{t}Pr^{i}MeC and Bu^{t}Pr^{i}EtC are 7.56 and 6.62, respectively.[33] (One possible explanation is that the alkyl groups solvate each other better than does the solvent methanol, so that hydrophobic bonding has an effect similar to that of tying back the groups in a ring. In which case the E_s' values may turn out to be solvent-dependent.)

The Charton's, meanwhile, have measured the steric effects of different alkyl groups on the rates of alkaline hydrolysis of a series of thiolacetates, CH_3COSR. The values obtained are used in the correlation analysis of rate and equilibrium data for reactions involving alkyl groups attached to sulphur.[34] This complements similar work on *O*-alkyl[35] and *N*-alkyl[36] substituent effects.

The Kirkwood–Westheimer theory has been applied to the dissociation of the carboxyl group of the ω-amino-acids $H_3N^+(CH_2)_nCOOH$ (n = 1–10),[37] and of several 3- and 4-substituted cyclohexanecarboxylic acids.[38] The measured pK_a's of the series of acids (**20**; X = S) show that the transmission coefficients decrease in the series X absent > X = S > X = SO_2 (ρ = 0.41, 0.26 and 0.19, respectively).[39]

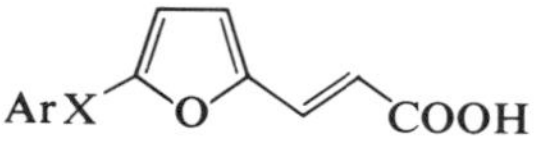

ArX COOH

(20)

Activation parameters have been measured for the hydrolysis of a series of simple esters in concentrated perchloric acid.[40] Activation enthalpy and entropy may be expected to change with decreasing water activity (and hence solvation), substrate protonation, or change in mechanism; all of these factors probably contribute to the results for hydrolysis of ethyl acetate.

The hydrolysis of chloromethyl pivalate involves S_N2 displacement of chloride at low acid concentrations, but in more concentrated $HClO_4$ the $A_{AC}2$ and subsequently the $A_{AC}1$ mechanisms take over,[41] partly because increasing electrolyte concentrations depress the rate of the S_N2 reaction.

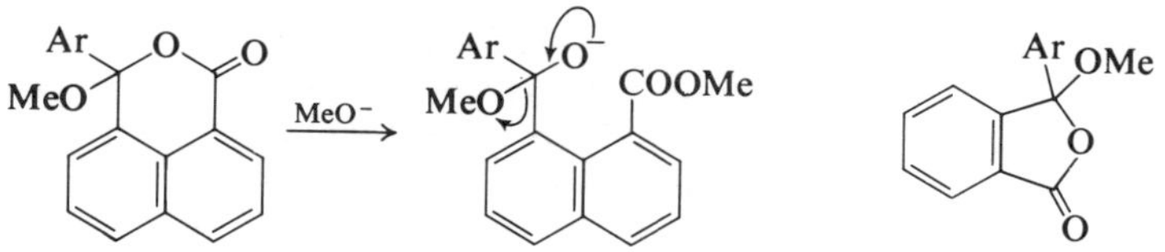

Ar MeO O O —MeO⁻→ Ar MeO O⁻ COOMe   Ar OMe O O

(21) **(23)** **(22)**

The effects of pressure on the acid(including sulphonic acid polymer)-catalysed hydrolysis of three *n*-alkyl acetates have been studied,[42] and ester hydrolysis catalysed by ion-exchange resins has been reviewed.[43]

The reactivities of a series of diastereoisomeric sugar-derived lactones, and their various OMe derivatives, towards acid hydrolysis have been compared in water,[44] and the acid- and base-catalysed rearrangements (in MeOH)) of pseudo-esters of general structure (**21**) and (**22**) to the open-chain esters have been studied;[45] ring-opening catalysed by methoxide is a simple reaction of $B_{AC}2$ type, but the very different Hammett ρ-values (2.1 for the naphthoates, 1.1 for the benzoates) suggest different rate-determining steps, with more negative charge in the bridging oxygen atom in the former case for which it has been suggested that breakdown of the tetrahedral intermediate (**23**) is rate-determining.

A double-labelling technique has been applied to measurement of the oxygen isotope effect in the alkaline hydrolysis and hydrazinolysis of methyl benzoate; k^{16}/k^{19} for the two reactions of the carbonyl-^{18}O-methyl-^{13}C-doubly-labelled ester were found to be 0.46‰ and 1.84‰, respectively.[46] Under the conditions employed oxygen-exchange is many times slower than hydrolysis—so the low value of k^{16}/k^{18} for alkaline hydrolysis implies that the transition state for this reaction is not close to the tetrahedral intermediate. The simplest explanation, consistent with a good deal of evidence of various kinds, is that the formation of a tetrahedral intermediate is rate-determining, and that the transition state is relatively early.

The opposite conclusion is drawn by Bogatkov, who has considered the effect of varying the leaving group on the rates of alkaline hydrolysis of a series of acetate, benzoate, *p*-nitrobenzoate, and methacrylate esters;[47] the Taft reaction constants ρ^* (for the variable leaving group) are closely similar for all four series, and the transition state, which is judged to be close to the tetrahedral intermediate, is correspondingly believed to remain at a fairly constant position on the reaction co-ordinate. The evidence that an OH group on a tetrahedral intermediate labilizes leaving groups very effectively compared with the corresponding OMe compound (discussed above for amide acetals, and well known for the hydrolysis of orthoesters and acetals, where the rate-determining step is clearly the hydrolysis of the first OR group), leaves little doubt that the formation of the tetrahedral intermediate *is* the rate-determining step in the alkaline hydrolysis of esters; but the position of the transition state *vis-à-vis* the tetrahedral intermediate is not well defined, although it presumably varies from earlier to later as the leaving group changes from very good (dinitrophenolate) to very poor (alkoxide).

Substituent effects on the alkaline hydrolysis of esters have also been measured for ethyl (7-substituted)fluorene-2-carboxylates,[48] aryl benzoates,[49] and aryl 2,4-dichlorophenoxyacetates.[50,51] The effects of the amino-group on the alkaline hydrolysis of the amino-alkyl esters, $RCOO(CH_2)_mNMe_2$ of acetic, benzoic, and methacrylic acids involve a combination of inductive and steric effects,[52] to which are added (small) electrostatic effects for the quaternary ammonium derivatives. When the rate constants for the hydrolysis of the conjugate acids $RCOO(CH_2)_{2,3}$–N^+HMe_2 are compared with those for the corresponding trimethylammonium compounds, a further rate increase of up to 40-fold is found, which has been ascribed to intramolecular catalysis by the NH^+ group. Clear-cut cases of intramolecular general acid catalysis of ester hydrolysis are hard to find; the example studied here was one of the earliest,[53] and although the analysis is now more rigorous, the mechanism involved is still not clearly established.

Activation parameters for the hydrolysis of several alkyl acetates in aqueous ethanol have been measured,[54–56] and the alkaline hydrolysis of monoethyl phthalate has been followed in mixtures of water and various alcohols[57,58] as well as in aqueous acetone. The effects of varying the solvent water : acetone ratio on the two steps involved in the alkaline hydrolysis of ethyleneglycol diacetate have been studied.[59] Data for the alkaline hydrolysis of a large number of substituted ethyl benzoates in various solvent mixtures at different temperatures have been fitted to a many-parameter equation;[60] the significant coefficients for cross-terms indicate non-additivity of the effects of the variables studied.

Alkaline hydrolyses of the ethyl esters of a large number of heteroaromatic carboxylic acids in aqueous dimethyl sulphoxide have been studied:[61] in most cases the usual acceleration by a dipolar aprotic solvent is observed, but for ethyl indole-2-carboxylate and methyl 4-pyridylacetate the observed rate decreases with increasing dimethyl sulphoxide concentration; since similar effects have been noted previously for the reactions of β-keto-esters, competition from the *E*1*cb* mechanism has been suggested in these cases.

Salt effects have been measured for the alkaline hydrolysis of monomethyl phthalate,[62] aryl benzoates,[63] and several alkyl benzoates,[64] as well as for the esterification of chloroacetic acid in ethanol.[65]

The self-catalysed esterifications of acetic and pivalic acids in ethanol have large negative volumes of activation.[66] The effect of pressure on the reaction of benzoic acid and PhCOOD with diphenyldiazomethane has also been studied;[67] the deuterium isotope effect does not change significantly in the range 0–2 kbar. The same reaction has been studied[68] in "a further 13 aprotic solvents" – making 43 solvents in all, and the data have been subjected to multiple correlation analysis with various solvent parameters. The substituent dependence of the reaction of diphenyldiazomethane with substituted pyridinecarboxylic acids is somewhat less[69] than for reaction with benzoic acids, or for alkaline hydrolysis of the corresponding esters.

The synthesis of macrocyclic lactones has been reviewed.[70]

Steric hindrance in aryl trimethylacetates reduces the rate of alkaline hydrolysis by nearly two orders of magnitude, compared with the corresponding acetates;[71] the factor is only a few times larger for nucleophilic attack by imidazole, which thus still acts as a nucleophilic catalyst, even for esters with *ortho*-substituents. It has been pointed out that trimethylacetylimidazole is only 2–3 times less reactive than acetylimidazole in these reactions, perhaps because of steric inhibition of resonance (**24**) in the ground state. The relative importance of bond-breaking in the transition state (**25**) evidently increases as steric congestion builds up: the Brönsted

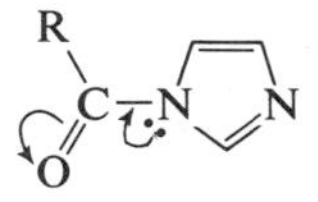

(**24**) 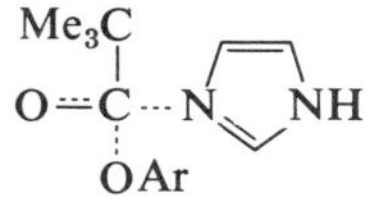 (**25**)

β_{LG} (for leaving group dependence) increases sharply from 0.4 for aryl acetates to 1.16 for aryl trimethylacetates.[71]

Steric hindrance is also observed when the imidazole carries an *ortho*-substituent: 2-*tert*-butylimidazole reacts over 10^4 times more slowly with *p*-nitrophenyl acetate than does imidazole itself, and rates for a series of 2-substituted compounds depend on E_s, with a slope of 1.33;[72] here there is some evidence for a change-over to general base catalysis for the most hindered compound (2-$HOCH_2$–CMe_2–imidazole), for which a significant solvent deuterium isotope effect, $k_H/k_D = 1.7$, and an increased entropy of activation are observed.

Imidazole-catalysis of the hydrolysis of *p*-acetoxybenzyl bromides (e.g. **26**; R = H) triggers the conjugate elimination of bromide and the formation of the unstable methylene quinone, which subsequently polymerizes. Steric hindrance in this case directs attack to the benzyl bromide centre, and for (**26**; R = Me) the product is the alkylated imidazole.[73]

OAc, R, R, CH_2Br **(26)** $\xrightarrow{Im}$ AcIm + (phenoxide, R, R, Br) $\longrightarrow$ (O, R, R, CH_2) + Br^-

Im = Imidazole.

Catalysis of acyl transfer reactions by the highly effective 4-(dialkylamino)-pyridines has been reviewed.[74]

The nucleophilic reactivity of amines (which bear β- or γ-methoxy or hydroxy-substituents) towards *p*-nitrophenyl acetate shows no enhancement that might be ascribed to the lone pairs on the substituent heteroatoms;[75] there is no β- or γ-effect corresponding to the α-effect. The α-effect is worth factors of 10^2–10^3 in rate for the reactions of peroxobenzoate ions, ArC(O)OO$^-$, with *p*-nitrophenyl acetate,[76] for which the value $\beta_{nuc} = 0.38$ approaches the limiting low sensitivity observed for normal oxyanion nucleophiles of high basicity and suggests that there is little bond formation to the nucleophile in the transition state.

2,4-Dinitrophenyl acetate and benzoate are cleaved by benzenethiolate anions in competing reactions involving both acyl–O and aryl–O cleavage (in 95% aqueous ethanol);[77] aromatic substitution is more sensitive to nucleophile basicity ($\beta_{nuc} = 0.44$ and 0.69, respectfully, for acyl-transfer and S_NAr reactions of the dinitrophenyl acetate), and this becomes the dominant mode of attack for both esters for the most basic thiolate anions. Steric hindrance, however, directs attack towards the carbonyl group: aromatic substitution is slowed by *ortho*-substituents on the nucleophile, as might be expected, but attack on the carbonyl group is actually accelerated;[78] the relative rates of attack on *p*-nitrophenyl acetate, for example, by thiophenoxide and its *ortho*-methyl and *tert*-butyl derivatives are 1 : 1.4 : 4.9. An explanation in terms of steric hindrance of solvation has been proposed.

The chemistry of the β-lactam antibiotics is the focus of a great deal of current synthetic work,[79] and is the subject of a *Special Publication*,[80] which includes a paper on the hydrolysis of some *N*-aryl-β-lactams.[81] The β-lactam ring of the penicillins is particularly reactive towards attack by nucleophiles because of ring-strain, and because the thiazolidine ring prevents full delocalization across the

Ac′NH, S, N, O, RNH_2 $\longrightarrow$ $PhCH_2CONH$, S, ^-O, N, $R\overset{+}{N}H_2$, COO^- **(27)** $\xrightarrow{HO^-}$ Ac′NH, S, RNH–C(=O), HN

amide bond. These factors should make tetrahedral addition intermediates more accessible than for open-chain amides, and kinetic evidence for an intermediate in the aminolysis of benzyl penicillin has been reported.[82] The reactions with glycine and 2-methoxyethylamine are general base catalysed by hydroxide at low pH but become independent of hydroxide concentration as this approaches 1M; the usual explanation of such behaviour is a change in rate-determining step, from the diffusion-controlled trapping of the zwitterionic tetrahedral intermediate **(27)** by hydroxide at low pH to its formation at high hydroxide concentration.[82] The aminolysis of mecillinam, which is a penicillin with an amidine side-chain on the β-lactam ring, is also general base catalysed.[83]

The NMR spectrum of *N*-(trifluoroacetyl)pyrrole, which is arguably an amide, shows in mixed aqueous solvents a new set of peaks which are attributed to the hydrate (**28**); dissociation constants in the region of 20M have been calculated for

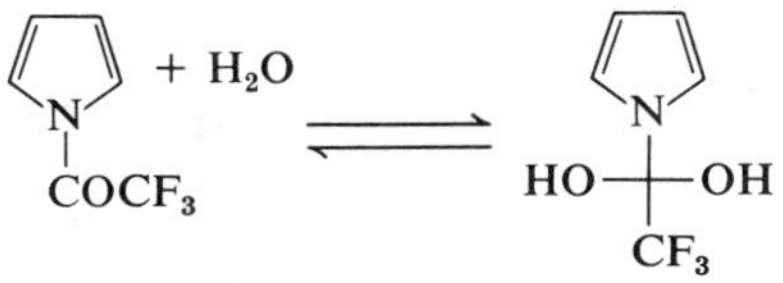

$$\text{C}_4\text{H}_4\text{N}\text{-COCF}_3 + H_2O \rightleftharpoons \text{C}_4\text{H}_4\text{N}\text{-C(OH)}_2\text{CF}_3 \quad (\mathbf{28})$$

(**28**) which is presumably one of the tetrahedral intermediates involved in the hydrolysis of the acylpyrrole.[84] The neutral hydrolysis of a series of 1-acyl-1,2,4-triazoles (**29**, also amides of a sort) appear to involve the classical general base catalysis mechanism shown,[85] and not an acylium ion route as proposed by earlier workers. Evidence includes proton inventory studies on three compounds and the effects of substituents on the rates of hydrolysis of a series of substituted benzoyl compounds (**29**; R = Ar, R′ = H) which are correlated by the Yukawa–Tsuno equation, with $\rho = 1.48$, $r = 0.42$.

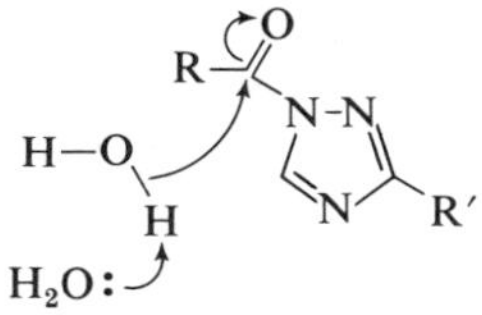

(**29**)

The acid-catalysed hydrolysis of acetamide has been studied at three temperatures in 1–6M-HCl;[86] so too has the hydrolysis of formamide in aqueous dioxan[87] and propan-2-ol[88] catalysed by both acid and base. In the alkaline hydrolysis of ethyl oxamate, EtOOC.CONH$_2$, measured at 65–85°,[89] the first step is hydrolysis of the ester group, but this is faster than the subsequent hydrolysis of the oxamate ion only at high base concentrations. It is odd, though the evidence seems clear on the point, that the amide group should be more reactive in NH$_2$CO.COO$^-$ than in NH$_2$CO.COOEt: the answer is presumably that the amide group is converted into its conjugate base at moderate hydroxide concentrations. With the number of variables available, a reasonable explanation of the complex dependence of rate on hydroxide concentration scarcely needs the authors' introduction of two kinetically significant tetrahedral intermediates.

The acid-catalysed hydrolysis of a series of *N*-acetyl-lactams[90] gives mainly *N*-acetyl cleavage for 6- and 7-ring compounds, but *N*-acetylpyrrolidone, which is significantly less reactive, shows a slight preponderance of ring-opening. The conjugate acid is probably significantly stabilized by intramolecular hydrogen-bonding (see **30**); this would explain the relatively low reactivity of the γ-lactam and is consistent with the very much slower hydrolysis of *N*-methylsuccinimide (**31**), for which such stabilization is not possible. The hydrolysis of a series of *N*-arylmaleimides has been measured at pH 8; it follows the Hammett equation with $\rho = 0.35$.[91]

No term of the second order in hydroxide is observed in the alkaline hydrolysis of several anilides of α-amino-acids[92] and oligopeptides[93] but catalysis by morpholine of the hydrolysis of the *p*-nitroanilide of L-proline shows saturation

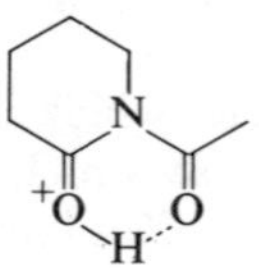

(30) 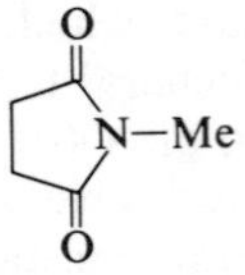 (31)

kinetics at lower pH[92], as if the rate-determining step changes from breakdown to formation of the tetrahedral intermediate with increasing catalyst concentration; there is some evidence that the partitioning of this intermediate may be subject to stereoelectronic control[93] when hydrogen bonding to polar substituents restricts free rotation of bonds to the central carbon atom.

A complete pH–rate profile for the hydrolysis of *p*-nitrofluoroacetanilide shows a pH-independent reaction between pH 1 and 4.5 ($k_{H_2O}/k_{D_2O} = 3.7$), which is catalysed by imidazole;[94] this is probably general base catalysis ($\rho = 0.84$ for a series of anilides[95]) and quite different from the imidazolium ion-catalysed alkaline hydrolysis, thought to involve general acid-catalysed breakdown of the tetrahedral intermediate (**32**),[95] for which ρ is large and negative (-1.56, similar to the value

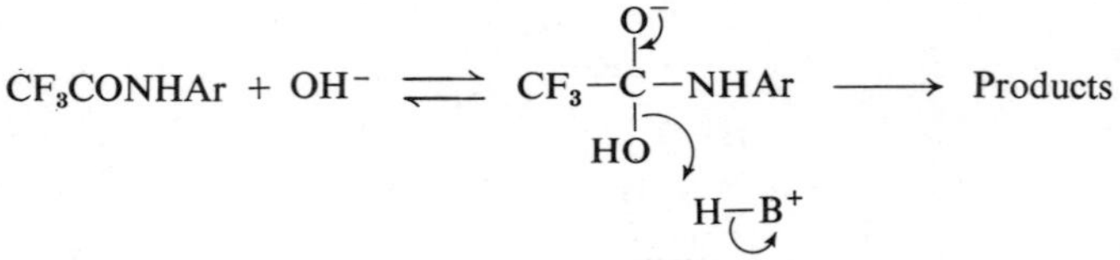

$$CF_3CONHAr + OH^- \rightleftharpoons CF_3-C(O^-)(HO)-NHAr \longrightarrow \text{Products}$$

H–B$^+$

(**32**)

of $\rho = -1.8$ found for the chymotrypsin-catalysed hydrolysis of *N*-acetyltyrosine-anilides). The imidazole-catalysed hydrolysis of *p*-nitrotrifluoroacetanilide has also been studied in acetonitrile in the presence of various water concentrations.[96]

With the completion of a study of the base-catalysed methanolysis of a series of (*N*-benzoyl-*N*-methylamino)pyridines, pyridine *N*-oxides, and *N*-methylpyridinium compounds,[97] data are now available for the reactions of *N*-aryl-*N*-methylbenzamides spanning a range of reactivity of 10^7; a single Hammett plot ($\rho = 3.2$) correlates the data for the complete range of amides, and rate-determining breakdown of the tetrahedral intermediate is believed to account for the reactions of all of the compounds studied, with the possible exception of the *p*-nitroanilide.[97]

The acid-catalysed hydrolysis of hydroxamic acids is similar to that of other amides, with a rate maximum in the region of 5M-acid; it almost certainly involves the $A_{AC}2$ mechanism,[98-100] with a suggestion that an *A*1 mechanism might be important in very strong acid.[101] The effects of *ortho*-substituents on the rates of hydrolysis of *N*-methylbenzohydroxamic acids in acid and in base have been examined.[102,103] The acid-catalysed hydrolysis of benzohydrazide has also been studied.[104]

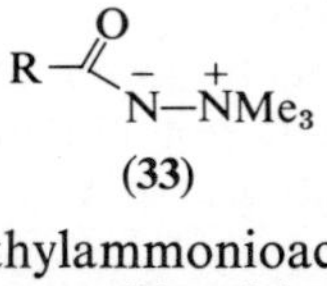

$$R-C(=O)-\overset{-}{N}-\overset{+}{N}Me_3$$

(**33**)

Less familiar are the *N*-trimethylammonioacylamidates (**33**). These are protonated on nitrogen to give conjugate acids with pK_a values that are comparable to those of the parent carboxylic acids,[105,106] but with a different response to substituents ($\rho = 1.50$ for the pK_a's of the benzamidates (**33**; R = Ph).

The acid-catalysed hydrolyses of thiol esters and thion esters involve mechanisms that are comparable with those for hydrolysis of oxygen esters, modified in detail because of the special properties of the third-row element and in particular by its reluctance to enter into pπ-conjugation. Thus a careful study by Edward and his co-workers, using the Yates–McClelland r and the transition-state activity-coefficient treatments, shows that *S*-ethyl thiobenzoate is hydrolysed by the $A_{AC}2$ mechanism at low H_2SO_4 concentrations (up to 60%), with bond-making to water well advanced; the $A_{AC}1$ mechanism then sets in, at considerably lower acidity than is required to elicit this mechanism from ethyl benzoate. *O*-Ethyl thiobenzoate, on the other hand, is hydrolysed by the $A_{AC}2$ mechanism over the complete concentration range of H_2SO_4;[107] evidently the thioacylium ion $PhC{\equiv}S^+$ is formed much less readily than the acylium ion. The complete rate–acidity profiles (30–99% H_2SO_4) show that the *O*-ester is very much more reactive at moderate acid concentrations, but that the rate passes through the $A_{AC}2$ maximum at about 70% acid and is soon overtaken by that of the *S*-ester in very strong acid. The effects of substituents on the reactions at high acidity have also been measured.[108]

The hydrolysis of thioacids resembles that of the *S*-esters under the same conditions: the conjugate acid (**34**) has both acylium and thioacylium ion routes available in competition with its reaction with water, and naturally is hydrolysed

$$R{-}C(SH)(OH)^+ \rightleftharpoons R{-}C(SH){=}O \rightleftharpoons R{-}C(\overset{+}{S}H_2){=}O \rightleftharpoons R{-}C{\equiv}\overset{+}{O} + H_2S$$

(**34**)

by the most favourable pathway available. Thus, the Hammett ρ value is approximately zero for the hydrolysis of substituted thiobenzoic acids in 36% H_2SO_4 (as found for oxygen esters) but it changes to about -2 in 65% acid.[109]

A similar study[110] of the hydrolysis of *p*-nitroacetanilide and the corresponding thioamide, in 22–96% H_2SO_4, is consistent with substantial bonding of water in the tetrahedral intermediate in the hydrolysis of the amide, but considerably less bond formation to the protonated thioamide, as in (**35**).

Results of a proton inventory study are consistent with the classical general base catalysis mechanism (see **29** above) for the neutral hydrolysis of *S*-ethyl trifluoroacetate.[111] The mechanism of aminolysis of *S*-acetyl-thioglycollic acid, CH_3-$COSCH_2COOH$, by a series of amino-acids and -alcohols has been studied as a function of pH.[112]

$$H_2O\cdots R{-}C(SH)(NH_2)^+$$

(**35**)

[Caprolactam ring: N(H)–C=O(S)]

(**36**)

Base-catalysed NH exchange (measured by ^{15}N-NMR)[113] is $>10^3$ times faster for the lactam (**36**) than for the corresponding thioamide, as it is in acyclic compounds.[114] On the other hand the alkaline hydrolysis of thioamides is slower than that of the corresponding amides, which are in fact the observed products of the reaction under most conditions. A complication is that primary thioamides, which are more reactive than tertiary amides, have pK_a's in the region of 12–13 and, even when ionization is allowed for, both classes of thioamide are found to be hydrolysed in reactions which are of intermediate (about 1.5) order in hydroxide;[115] the rate law is reasonably explained in terms of a reaction involving both mono- and di-anionic tetrahedral intermediates (**37**, **38**), which can eliminate HS^- and S^{2-} (a much better leaving group than RNH^-), respectively. But it is clear that C—N cleavage also occurs under some circumstances, and a study of product composition, combined with careful kinetic measurements, is needed before this reaction can be fully understood.

$$\text{Ar—C(=S)NR}_2 \underset{}{\overset{HO^-}{\rightleftharpoons}} \text{Ar—C(S}^-\text{)(OH)NR}_2\ \textbf{(37)} \rightleftharpoons \text{Ar—C(S}^-\text{)(O}^-\text{)NR}_2\ \textbf{(38)} \xrightarrow{-S^{2-}} \text{ArCONR}_2$$

$$\textbf{(37)} \rightleftharpoons \text{Ar—C(SH)(O}^-\text{)NR}_2 \xrightarrow{-SH^-} \text{ArCONR}_2$$

A kinetically much simpler reaction is the methoxide-catalysed methanolysis of *N*-methyl(thiobenzanilides) $Ar^1CSNMeAr^2$, which is of first order in base, and leads to a single thioacid product;[116] Hammett ρ-values ($\rho_1 = 1.46$, $\rho_2 = 2.16$) are consistent with rate-determining breakdown of the tetrahedral intermediate.

The alkaline hydrolysis of the nitrile of isonicotinic acid has been studied.[117] The acid-catalysed hydrolysis of 4-aryl-5-phenylisosydnones (**39**) involves (^{18}O)-attack by water at both ring carbon atoms of the conjugate acid.[118] The alkaline hydrolysis of 3-aryl-hydantoins (**40**) (R = Ar; R′ = H) is of the first order in hydroxide up to pH 12–13, and changes to zero-order in stronger base, as the

(**39**) (**40**) $\xrightarrow{HO^-}$ RNHCONR′CH$_2$COOH (**41**)

substrate is converted into the unreactive anion. (Saturation kinetics are not observed for the 1-*N*-alkyl derivatives, R′ = Me.[119–120]) The product hydantoic acids (**41**) arise from attack by hydroxide at the more reactive acylurea-carbonyl group. For the 3-*N*-methyl compound (**40**; R = Me, R′ = H) the reaction approaches *second* order in hydroxide at low base concentrations,[121] while for hydantoin itself (**40**; R = R′ = H), which is fully ionized in alkali (pK_a = 9.04), the picture is similar to that found for the 3-*N*-aryl compounds.[121] These results, and much supporting evidence, are consistent with a mechanism in which breakdown of a dianionic tetrahedral intermediate is rate-determining for compounds with poor leaving groups (R = H or Me). At high base concentrations the rate-determining step changes to formation of the tetrahedral intermediate (either first order in hydroxide and neutral substrate or, the kinetic equivalent, zero-order in hydroxide when the substrate is present as its conjugate base). For the 3-*N*-aryl compounds (**40**; R = Ar) the better leaving group causes rapid C—N cleavage, and consequent rate-determining formation of the tetrahedral intermediate, over a wider range of hydroxide concentrations; the observed value of $\rho = 0.8$[119,120] is consistent with this interpretation.

The hydrolysis of amidines is another multistage reaction where the rates of individual steps are finely balanced, and the rate-determining step readily changes with pH or substitution. The hydrolysis of 5,6-dihydro-1-methylcytosine (**42**), for example, is subject to general base catalysis in acid, suggesting that breakdown of the tetrahedral intermediate is rate-determining under these conditions; but no general species catalysis is observed at high pH, where the reaction becomes pH-independent above the pK_a of the substrate. The simplest explanation of this rate law is that the rate-determining step is the attack of hydroxide ion on the substrate conjugate acid, and most authors agree with this mechanism at high pH;[122] for

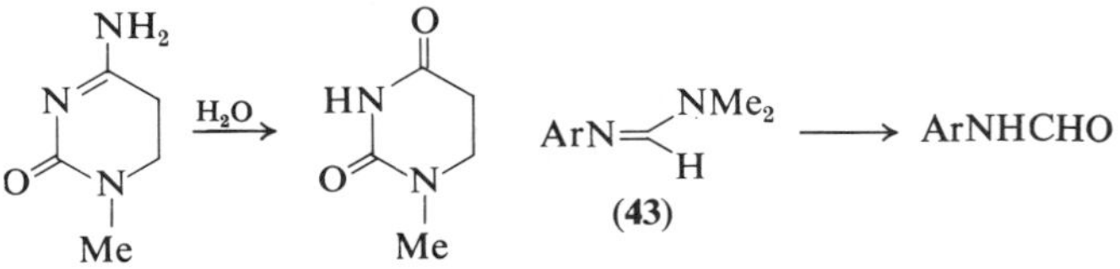

(**42**) $\xrightarrow{H_2O}$ ArN=C(NMe$_2$)H (**43**) ⟶ ArNHCHO

example, for the hydrolysis of 2-imidazolines.[123,124] A different interpretation is given by Večeřa and his co-workers,[125] who have studied the hydrolysis of a series of *N*′-aryl-*N*,*N*-dimethylformamidines (**43**). A complication here is that the two C—N bonds in the conjugate acid and tetrahedral intermediate are different. In this case the products are dimethylamine and the formanilide, so that the bond to the poorer leaving group is broken. Hammett ρ-values are -1.54 for protonation, -0.53 for hydrolysis in the pH-independent region at high pH, and almost zero below pH 8 (where the reaction is of the first order in hydroxide). Neither value appears incompatible with rate-determining attack of hydroxide on the conjugate acid at high pH, changing to rate-determining breakdown of the tetrahedral intermediate below pH 8.

The chemistry of cyanates and thiocyanates is the subject of a new treatise,[126] and has been reviewed in an article[127] that covers also carbodi-imide reactions.

Reactions in Aprotic Solvents

The dealkylation of esters by trimethylsilyl iodide,[128] which is of some preparative interest, and by superoxide anion,[129] which is of some biological interest, has been reported. The pyrolytic dealkylations of $PhCH_2CMe_2OAc$ and $PhCMe_2CH_2OAc$ give identical product ratios: evidently the acetates are interconverted under the reaction conditions.[130] The corresponding reaction of carbamates of 1,1-diphenylethanol has C—O bond breaking as the kinetically dominant process.[131]

The third-order term for the *n*-butylaminolysis of nitrophenyl acetates in dioxan is relatively smaller for *ortho*-nitro-compounds.[132] The benzylaminolysis of *S*-*p*-nitrophenyl thioacetate and the anilinolysis of 2,4-dinitrophenyl acetate have been followed in several organic solvents, and the rate constants (which decrease with solvent polarity) have been related to the Kirkwood equation.[133] The imidazole-plus-benzoate-catalysed cleavage of *p*-nitrophenyl acetate in acetonitrile 1M in H_2O, proposed as a model for the charge relay system of the serine proteases,[134] does not in fact lead to hydrolysis but occurs by direct nucleophilic attack by imidazole;[135] catalysis apparently involves imidazole–benzoate and imidazole–imidazole dimers, but studies of the deuterium isotope effect show that no kinetically significant proton-transfer step is involved.

The rate of cleavage of *p*-nitrophenyl propionate by imidazole–carboxylate mixtures in benzene is actually reduced by the addition of small amounts of water;[136,137] catalysis by carboxylate alone is much more effective than by imidazole alone, but very large rate increases occur when tetra-alkylammonium carboxylates are added to the solution of ester and imidazole.[136] Imidazole is most effective in the presence of an equimolar amount of carboxylate, and here too hydrogen-bonded imidazole–carboxylate pairs are thought to be involved (**44**).

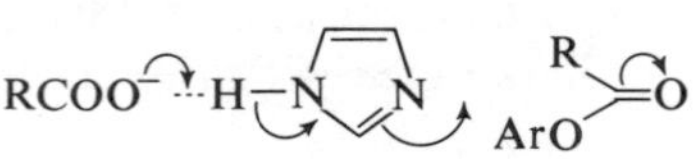

RCOO⁻ ··· H–N⌒N   R(ArO)C=O

(**44**)

The reaction of aniline with urethanes, PhNHCOOAr, catalysed by diazabicyclo-octane gives a Hammett ρ-value of 1.63;[138] the same author has also studied the benzylaminolysis of aryl phenyl carbonates.[139] The full paper has appeared describing the cleavage of an amide catalysed by Kunitake's *N*-methylmyristohydroxamate ion-pair in dry aprotic solvents.[140]

The exchange of ^{14}C-labelled acetic acid into *p*-nitrophenyl acetate catalysed by substituted pyridines has been studied.[140a] The benzoylation of 2- and 3-(dialkylamino)alcohols is a second-order reaction, suggesting that the tertiary amino-group acts as an intramolecular catalyst, since reaction with the linear $R_2NCH_2C{\equiv}C{-}CH_2OH$ is second-order in amino-alcohol.[141] The rate-determining step in these reactions is probably nucleophilic attack on the carbonyl group.[142] The reaction (with 3,5-dinitrobenzoyl chloride in toluene) is also catalysed by biphenylphosphoric acid (diphenyl hydrogen phosphate)[143] and naphthalene-1-sulphonic[144] acids; these are considered to be acting as bifunctional catalysts, on the grounds that HCl and 3,4-dinitrophenol are ineffective, but the evidence is not conclusive.

The mixed anhydrides $RCO{-}O{-}POCl_2$, presumed intermediates in acyl activation catalysed by $POCl_3$, have been synthesized by an anhydride exchange route.[145] The reaction of various picolinic or quinaldinic acids (ArCOOH) with PCl_5 in refluxing thionyl chloride (to give $ArCCl_3$) is second-order in the reactants and follows the Hammett equation with $\rho = -4.52$.[146] The reaction of aliphatic carboxylic acids in benzene with thionyl chloride itself is thought to go by way of an intial donor–acceptor complex between the reactants:[147] a Taft treatment shows separate lines for *n*-alkanecarboxylic acids and branched-chain acids, with $\rho^* -(2.5\text{–}3.0)$ and $+(3.5\text{–}3.8)$, respectively; other authors consider that formation of a chlorosulphite intermediate is the rate-determining step for the less hindered acids.[148]

The role of HF in the reaction of carboxylic acids with SF_4 has been discussed.[149]

The acetylation of aminopyridines in acetone involves direct attack by the NH_2 group for all but the 4-NH_2-compounds, where initial acylation of the ring nitrogen occurs if this is not sterically hindered by a 2-substituent.[150] Charge-transfer complexes are involved in the reactions of electron-rich anilines with pyromellitic anhydride.[151] Acylation of *p*-phenoxyaniline by phthalic and 1,2,3,6-tetrahydrophthalic anhydrides in several solvents is autocatalytic, because of the effects of the carboxylic acid produced.[152] The reaction of thiobenzoic anhydride with *p*-chloroaniline in dimethyl sulphoxide and dioxan is catalysed by heterocyclic bases.[153] Trithiocarboxylic anhydrides are produced by the lithium alkoxide- or dialkyl sulphide-catalysed disproportionation of RCS—S—COR′.[154]

Kinetic measurements have been reported for the reversible second-order reaction between a series of *N*-heterocycles and carboxylic acid chlorides;[155] the kinetics of reaction of various anilines with acid chlorides, catalysed by 1-methylimidazole[156] or acetic acid and an α-pyridone have been studied,[157] and the dependence on substituent[158–160] or solvent polarity has been ascertained.[160–162] Arylcarbamoyl chlorides, Ar′NHCOCl, as expected, are less reactive than benzoyl halides towards $ArNH_2$ in benzene and give Hammett ρ-values of about 1 (Ar′) and -3 ($ArNH_2$).[163] Selectivity and reactivity are not related in the reactions of substituted anilines with ArSCOCl[164] (ρ does not change for different substituted anilines). The intermediates ($Ph_2NCO^+NR_3$) involved in carbamoyl transfer catalysed by pyridine and quinoline have been made and their reactions with *p*-anisidine have been studied directly.[165]

The Hammett ρ-value for reaction of dimethylamine with substituted cyanates ArNCO in benzene is 3.26,[166] and the rate-determining step in the triethylamine-catalysed reaction of phenol with $Cl(CH_2)_6NCO$ in CCl_4 is thought to be attack on the cyanate group by the hydrogen-bonded complex of Et_3N and phenol.[167]

Intramolecular Catalysis and Neighbouring-group Participation

Neighbouring-group participation in intramolecular reactions has been reviewed.[168]

Solvation and microenvironmental effects on group reactivity may make important contributions to the high efficiency of enzymic catalysis, and intramolecular reactions provide perhaps the most suitable systems for testing the mutual reactivity of pairs of functional groups as a function of environment. If a particular group becomes enormously more reactive at an enzyme active site simply because it is held in a hydrophobic environment, for example, then one might expect to observe a similar effect on reactivity in an intramolecular reaction on changing to a less polar solvent.

Though this is an important question, very little work has been done along these lines; largely, no doubt, because an early search for such an effect proved negative.[169] Dafforn and Koshland have now examined the lactonization and thiolactonization of their celebrated set of hydroxy- and thio-acids,[170] catalysed by acid in water and by trifluoromethanesulphonic acid in sulpholane.[171] Solvent effects on the rates of the reactions are not large, and relative reactivities in the dipolar aprotic solvent differ little from those in water. Evidently the rates of these particular reactions would not be expected to be susceptible to large environmentally-induced accelerations. Other reactions, known to be more sensitive to solvation effects, make more plausible candidates; but the evidence so far suggests that solvation effects of this sort cannot be a factor of universal importance in enzyme catalysis.

The secondary deuterium isotope effect has been measured for intramolecular nucleophilic catalysis of the hydrolysis of *p*-bromophenyl succinate and its d_4-derivative (**45**) by carboxylate.[172] The effect ($k_{H_4}/k_{D_4} = 1.035$) is normal and clearly different from the value (0.91 per deuterium) expected for a fully tetrahedral transition state; a late transition state (**46**), close in structure to the anhydride product, is therefore likely and this is consistent with other evidence.

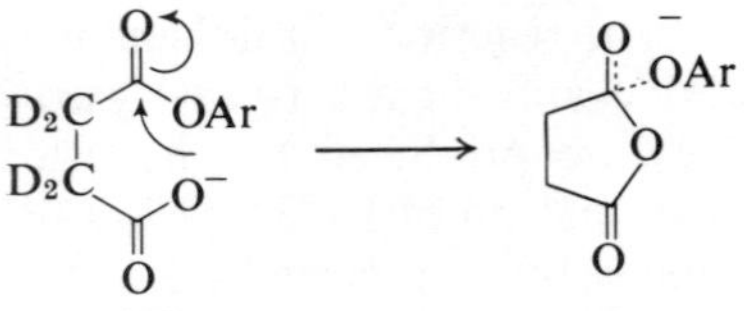

(45) (46)

Norbornene anilic acids (**47**), like other succinic and maleic acid derivatives,[173] are hydrolysed at low pH by intramolecular nucleophilic catalysis;[174] for the *m*- and *p*-nitro-compounds the reaction is general base catalysed, with Brönsted coefficients, $\beta = 0.66$ and 0.62, respectively. The H_3O^+- and H_2O-catalysed reactions depend on the basicity of the leaving group ($\beta_{LG} = 0.16$ and 0.38, respectively), and a mechanism (**48**) involving classical general acid catalysis, with proton transfer concerted with the cleavage of the C—N bond of the anionic tetrahedral intermediate (rather similar, in fact, to **46** above), has been proposed.[174]

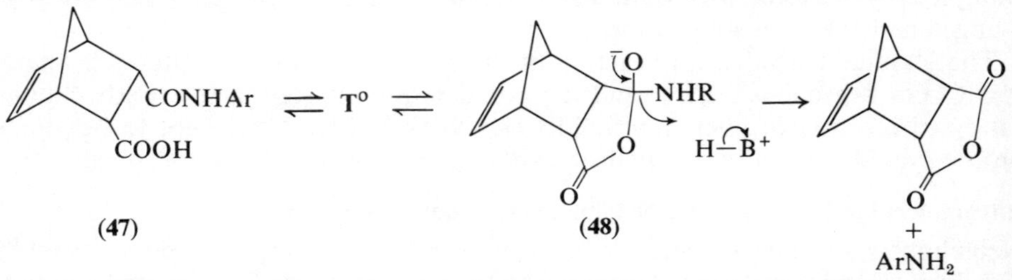

(47) (48)

The rates of hydrolysis of several substituted phthalanilic acids in dimethyl formamide,[175,176] and those of compound (**49**) and the corresponding monoglutarate ester (**50**) over a wide pH-range in water,[177] have been measured; compounds (**49**) and (**50**) have bell-shaped pH-rate profiles (with maxima near pH 4) which are attributable to nucleophilic catalysis (by the carboxyl group) of the

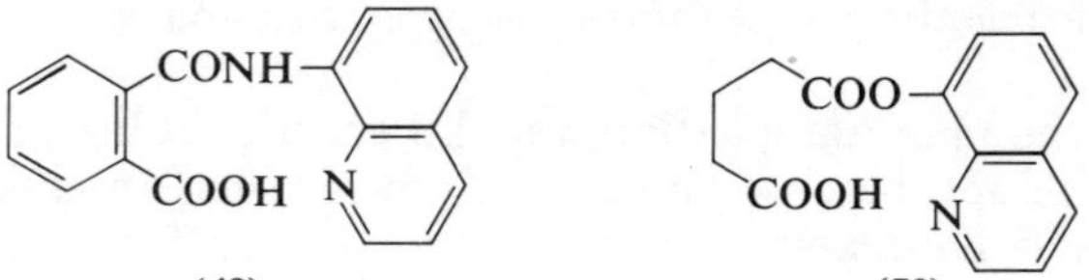

(49) (50)

hydrolysis of the neutral species/zwitterion, and slower reactions of the conjugate acid and base; divalent metal ions do not catalyse these reactions significantly.

The hydrolysis of (**51**) to the *N*-(acetylamino)benzene dicarboxylic acid (**52**) is apparently catalysed by the neighbouring carboxyl group, since in acid the reaction is 250 times faster than that of the 6-carboxy-compound;[178] a carboxyl group at position 5, by contrast slows the reaction. Given the geometry of the system the

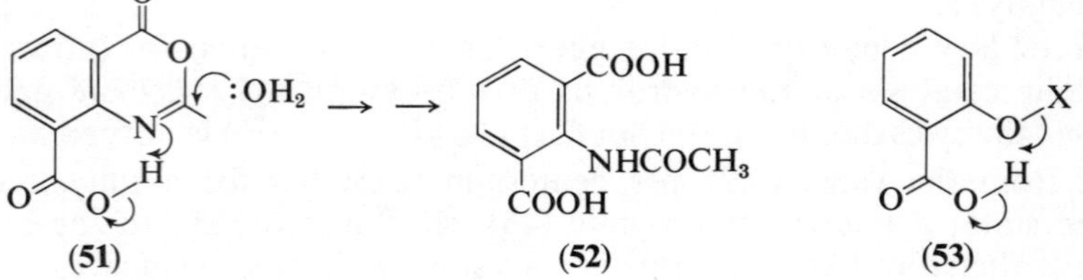

(51) (52) (53)

general acid catalysis mechanism depicted by (**51**) seems to offer the only reasonable explanation. Since intermolecular general acid catalysis appears not to be readily detectable in this system the effective concentration of the neighbouring COOH group must be substantial; this is not unexpected, because the analogy to salicylic acid derivatives is very close, and the cleavage of salicylyl phosphate and acetals [**53**; X = PO_3^{2-}, $CH(OR)_2$] is known to be subject to efficient general acid catalysis of this sort.

Capon *et al.* have reported the first mechanistic investigation of intramolecular catalysis of the hydrolysis of the cyano-group by COOH.[179] The hydrolysis of *o*-cyanobenzoic acid (**54**) is more than 10^5 times faster than that of the *para*-isomer in the pH-independent region near pH 1. The anion is unreactive, but the apparent pK_a (1.87) obtained from the sigmoid pH-rate profile is clearly different from the measured pK_a (3.30); a nucleophilic catalysis mechanism, in which the rate-determining step changes from the hydrolysis to the formation of the intermediate isophthalimide (**56**), as the pH falls, accounts satisfactorily for the observed results.

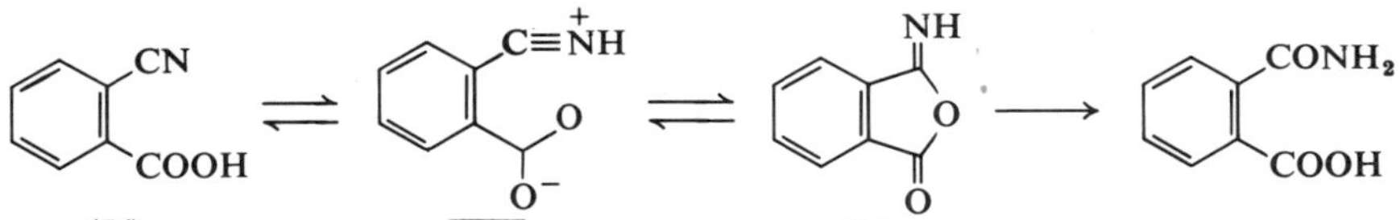

(**54**) (**55**) (**56**)

The subsequent hydrolysis of the phthalamic acid produced is subject to equally efficient intramolecular nucleophilic catalysis by the carboxyl group, but is almost an order of magnitude slower than the hydrolysis of (**54**) at low pH.

The hydrolysis of the cyano-group of α-amino-nitriles, of preparative importance in the Strecker synthesis of α-amino-acids is usually carried out in acid. Reaction in dilute base also leads to hydrolysis, in competition with decomposition to regenerate cyanide and the carbonyl compound: the hydrolysis reaction is autocatalytic,[180] the reaction being catalysed in the case of 2-amino-2-methylpropionitrile (**57**; R = Me) by the acetaldehyde produced by the decomposition reaction; added acetone has an even greater catalytic effect. The mechanism appears to involve the addition of the primary amino-group [compounds $RCH(CN)NMe_2$ are hydrolysed much more slowly under the same conditions]

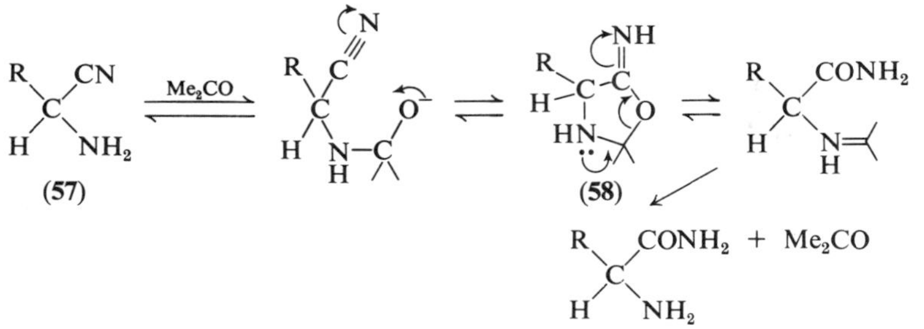

(**57**) (**58**)

to the carbonyl compound followed by intramolecular formation of a 5-imino-oxazolidine (**58**) which can break down with incorporation of the oxygen atom of the catalyst within the amide group of the product α-amino-amide.[181]

Neighbouring-group participation by the amide group is involved in the reactions of substituted *O*-benzoylglycol anilides (**59**); products from the hydrolysis of (**59**) include benzanilides, which have been shown to result from intramolecular attack of the amide anion nitrogen on the benzoyl group to give diacylamine intermediates (the *N*-methyl compounds do not undergo the reaction).[182]

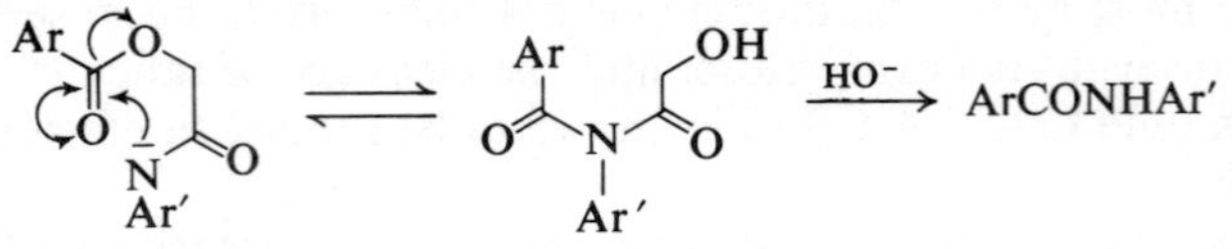

(59) $\rightleftharpoons$ $\xrightarrow{HO^-}$ ArCONHAr′

Intramolecular general base catalysis by thiolate ion is believed to assist intramolecular nucleophilic attack by amide-nitrogen, and thereby account for the enhanced rate of hydrolysis of several methyl esters (of which the most reactive is **60**): the SMe compounds are some two orders of magnitude less reactive.[183] This would be an interesting mechanism if confirmed, but in cases like this it is rarely a simple matter to rule out kinetically equivalent mechanisms, and no evidence specifically supporting mechanism (**60**) is available.

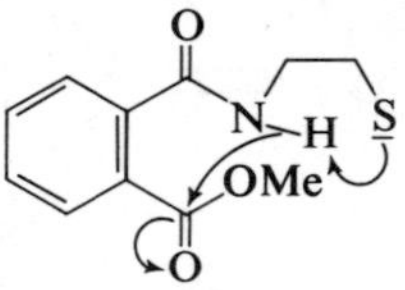

(60)

Intramolecular catalysis by an amide group has been suggested to account for the absence of a third-order term in the piperidinolysis of the ester (**61**) in acetonitrile and chlorobenzene.[184] The *n*-butylaminolysis of aspirin in acetonitrile also appears to be subject to intramolecular catalysis, in this case by the ionized carboxy-group, since the observed rate of reaction shows a sigmoid dependence on amine concentration in the region where ionization is expected.[185] Here too the reaction is first-order in amine at high amine concentration, and a mechanism in

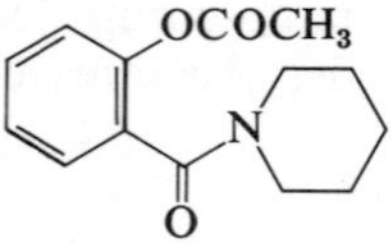

(61) 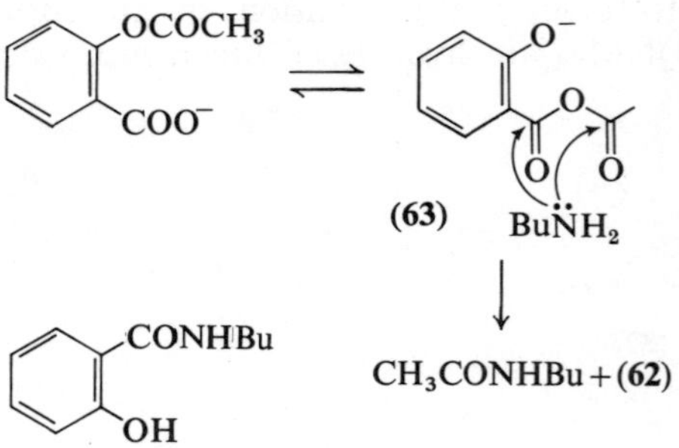 **(63)** $BuNH_2$ → $CH_3CONHBu$ + (**62**)

(62)

which the COO^- group acts as a general base has been suggested. Between 0.005M- and 0.1M-$BuNH_2$ the rate is independent of amine concentration, and 6% of *N*-*n*-butylsalicylamide (**62**) is produced; this can be reasonably explained only by the reaction of butylamine with the salicylic acetic anhydride (**63**), the formation of which is presumably rate-determining under these conditions.[185]

A free carboxyl group will also react with a neighbouring amino-group, usually under forcing conditions, to produce lactams. Hexanolactam, for example, is produced when 6-aminohexanoic acid is heated in hydroxylic solvents to temperatures in the region of 200 °C.[196] These reactions presumably involve the neutral species (in this case $H_2N(CH_2)_5COOH$) and are very slow because the difference in pK_a's of the amino and the carboxyl groups means that no more than about one part per million will be present as this form. A quite different reaction may have been observed in a study[187] of the cyclization of *o*-(1-aminoethyl)benzoate ion (**65**) to give (**64**); the amino-acid anion is the reactive species, and shows a pH-independent reaction near pH 11–12 which has been accounted for in terms of a hydroxide-catalysed reaction of the neutral compound, as discussed above. However, in

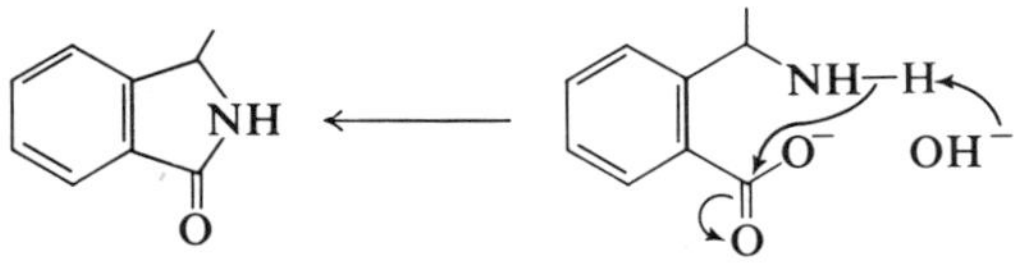

(**64**) (**65**)

0.1–1M-NaOH the cyclization occurs by a faster reaction which is of first-order in hydroxide and amino-acid anion. This can only reasonably involve attack by the amino-group on the ionized carboxyl; the kinetic parameters are consistent with the classical general base catalysis mechanism (**65**).[187]

Intramolecular aminolysis of *N*-(ω-aminoalkyl)amides (**66**) is irreversible when a tertiary amide is converted into a secondary amide.[188] An analysis in terms of stereoelectronic control suggests that the *N*-alkyl group (Me in **66**) will be equatorial

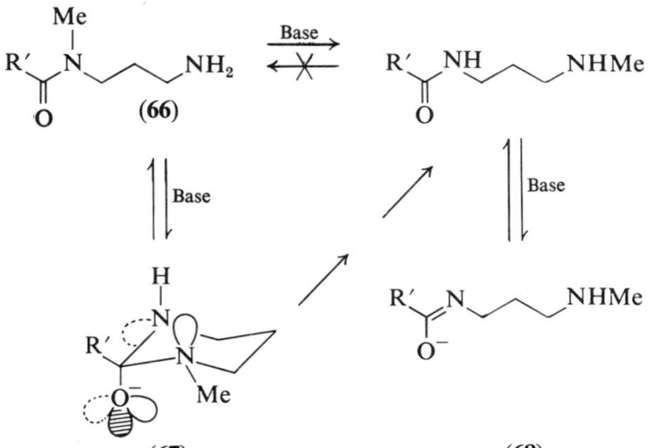

(**66**) (**67**) (**68**)

in the preferred conformation of the tetrahedral intermediate (**67**), in which case only the C—N(Me) bond has lone pairs on the two other heteroatoms antiperiplanar, so that this bond may be expected to be cleaved selectively. The observed reaction can in any case be fully accounted for, because the secondary amide can be converted into the conjugate base (**68**) which, under the conditions of the reaction, is therefore the thermodynamic product.[188]

Esters (**69**) of *o*-(2-imidazolyl)benzoic acid are hydrolysed with intramolecular catalysis by the imidazole group.[189] A general base catalysis mechanism fits the data for the reaction of the methyl ester, while hydrolysis of the phenyl ester clearly involves nucleophilic catalysis, since the acylimidazole intermediate (**70**) can be detected. The change over from nucleophilic to general base catalysis depends not only on the relative basicities of nucleophile and leaving group, but also on the system concerned: in the case of (**69**) the trifluoroethyl ester hydrolyses with nucleophilic participation also.[189]

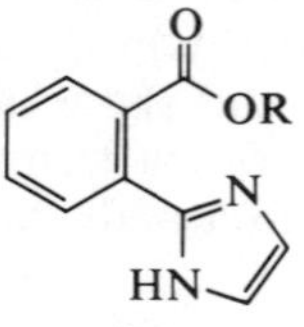

(**69**) 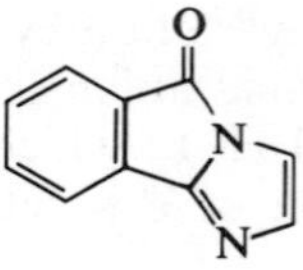 (**70**)

Intramolecular general base catalysis by imidazole also occurs in the hydrolysis of the ester group of α-*N*-(benzyloxycarbonyl)-L-histidyl-(*O*-acetyl)-L-serine amide,[190] and of several esters (*e.g.***71**) derived from benzimidazole.[191] Freezing out one internal rotation in this way produces a (corrected) four-fold increase in rate compared with (**72**). At the other extreme, intramolecular nucleophilic catalysis of the hydrolysis of the *p*-nitrophenyl esters of a series of polyoxyethylene derivatives

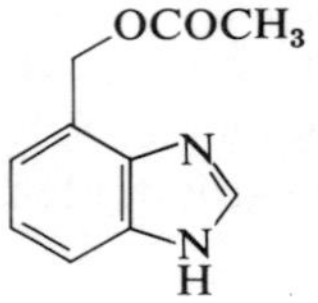

(**71**) 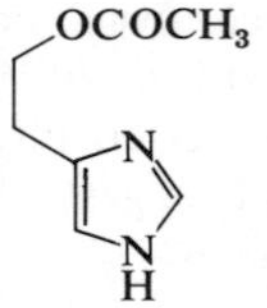 (**72**)

(**73**) still shows a detectable dependence on chain length, the oligomer with $n = 7$ being the most reactive.[192]

Pyrimidine-nitrogen acts as a nucleophile in the hydrolysis of the amide (**74**);[193,194] the reaction is general acid catalysed and provides another example of kinetic evidence for a tetrahedral (presumably) intermediate, based on a change of rate-determining step with pH.

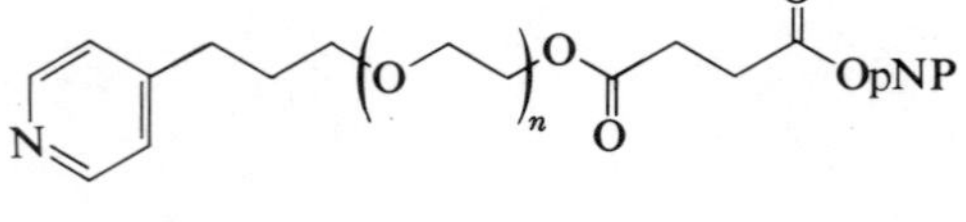

(**73**) 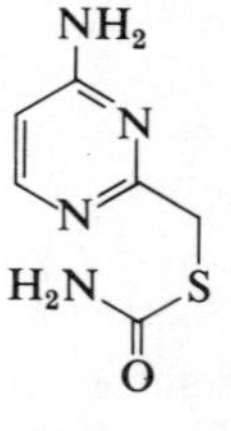 (**74**)

The hydrolysis of *O*-acetylsalicylaldehyde involves nucleophilic catalysis by the hydrated aldehyde group, which is so efficient that the hydration of the aldehyde is the rate-determining step.[195] This high efficiency is due largely to the independent mode of breakdown available to the hemiacylal (**75**), in which a rapid elimination of acetic acid regenerates the free aldehyde group. Salicylaldehyde can

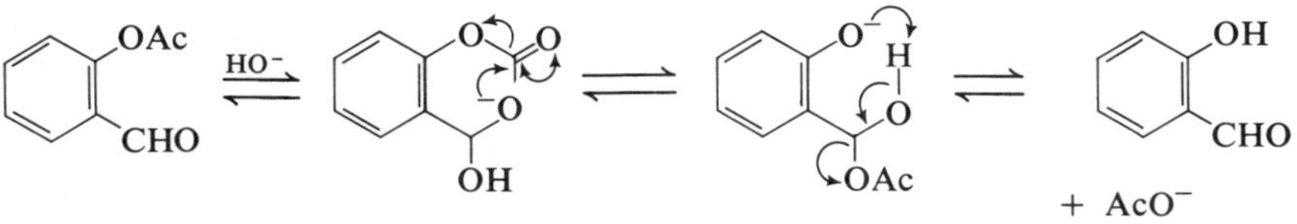

(75)

thus act as a true (nucleophilic) catalyst for the hydrolysis of acyl compounds, and the catalytic reaction has been demonstrated with *p*-nitrophenyl acetate.[196]

Nucleophilic catalysis, by the hydroxyl group, of the hydrolysis of amides (**76**) (to form the lactone) is known to be highly efficient:[197] the hydroxide-catalysed reaction of the amide derived from ethylenediamine ($R = CH_2CH_2NH_2$) shows a rate-enhancement of some 150-fold below pH 10, where the leaving group is present as the conjugate acid ($R = CH_2CH_2NH_3^+$); this has been attributed to intramolecular general acid catalysis of the breakdown of the tetrahedral intermediate involved and is characterized by the usual low effective molarity ($\simeq 1$M) for the intramolecular general species catalyst compared with that for the intramolecular nucleophile ($\simeq 10^8$M for the OH group of **76**).

(76) 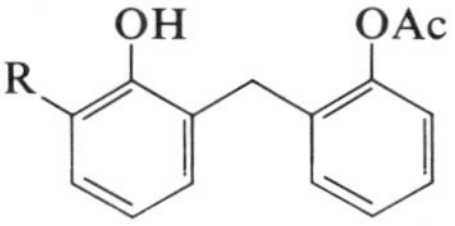 (77)

General species catalysis by phenolic OH groups is observed in the *n*-butylaminolysis of *S*-phenyl thiobenzoates[198] ($\rho = 1.45$) and 2-(*o*-hydroxyphenyl) benzoate[199] in acetonitrile, and of *o*-(2-hydroxybenzyl)phenyl acetates (**77**) in dioxan.[200] Suggestions[199,200] that the OH group may act as a general acid are supported by the observation of catalysis by added fluoro-alcohols with pK_a's comparable to that of phenol. Comparison with the observed effect of the phenolic OH group of 2-(*o*-hydroxyphenyl) benzoate suggests that the effective concentration concerned is unexceptional.[199,201]

The piperidinolysis of aryl acetates in acetonitrile and chlorobenzene is sterically hindered by *o*-halogeno-substituents, but not by *o*-nitro- or *o*-cyano-groups.[202] These derivatives react as fast as, or even slightly faster than, the corresponding *para*-substituted compounds, but the aminolysis still involves both second- and third-order terms; more evidence is required to substantiate the suggestion that these groups are acting catalytically.

Association-prefaced Catalysis

A strong upsurge of interest in this area is indicated by a sharp increase in the number of papers devoted to reactions that involve a more or less specific binding step before covalent bond-making or -breaking occurs. A substantial review of micellar reactions has appeared[203] together with brief reviews of cryptates[264] and of the kinetics of simple organic reactions in micelles,[205] and even a student's laboratory experiment on the determination of critical micelle concentration.[206] A series of important papers by Menger and his co-workers deal with fundamental properties of micellar systems.[207–209] A study of ^{13}C-spin–lattice relaxation times provides evidence for anisotropic motion of phenyl rings within micelles,[207] and another ^{13}C-NMR study of carbonyl "labels" within micelles suggests that water penetrates deeply into the structure (up to at least the first seven carbon atoms); however, the incorporation of the polar probe in the middle of $CH_3(CH_2)_7$-$CO(CH_2)_7NMe_3^+$ is a possible perturbing influence.[208] The preferred model has a hydrophobic core into which water does not penetrate, surrounded by regions where water penetrates deeply between the hydrophobic chains.

The water pools in inverse micelles of di-2-ethylhexyl sodium sulphosuccinate in hexane have intriguing acid–base properties.[209] In systems with inorganic buffers the pK_a of *p*-nitrophenol is raised by some 4.5 units, but imidazole–*p*-nitrophenol systems appear to involve not only this form of high pK_a but also a proportion of the phenol with a more normal pK_a. A possible explanation is that *p*-nitrophenol is normally held at the pool interface, from which it is progressively displaced by increasing concentrations of imidazole.[209] Work on micellar catalysis is largely inspired by the example of enzymic catalysis, and Piszkiewicz has extended his theoretical analysis[210] of micellar reactions with a treatment of the fall-off of rates in bimolecular micelle-catalysed reactions observed at high reagent concentrations, based on the analogy with substrate inhibition of enzyme reactions.[211]

Brief reviews of reactive polymers designed to immobilize enzymes[212] and to act themselves as association-prefaced catalysts[213,214] have been published as part of the proceedings of a symposium.

The reactivity of esters towards alkaline hydrolysis is well known to decrease with increasing chain-length. The reaction has been studied with a series of *p*-nitrophenyl alkanoates:[215] for esters that do not form micelles, and below the cmc for those which do, the decrease in reactivity of the monomeric ester with increasing chain-length has been explained in terms of self-coiling of the ester; the effect is reduced in solvents containing increasing amounts of an organic component. The reaction catalysed by micelles of CTAB (cetyltrimethylammonium bromide) has also been studied,[216] as has the alkaline hydrolysis of various tertiary anilides,[217,218] catalysed by CTAB and inhibited by sodium dodecyl sulphate. Micellar catalysis is more effective for amide hydrolysis than for esters.[218] The acid-catalysed hydrolysis of octanohydroxamic acid is catalysed by micelles of sodium dodecyl sulphate.[219] Micelles of CTAB also catalyse the attack of benzenethiolate ions on *p*-nitrophenyl acetate,[220] and the reactions of hydroxamic acids $H(CH_2)_{m-1}$-CONHOH with *p*-nitrophenyl alkanoates;[221] hydroxamic acids of longer chain-length ($m = 8$ or 10) themselves form micelles, which also enhance catalysis specifically of esters of similar chain-lengths, when these have a negatively charged leaving group; this may be because it inhibits self-coiling. The S to N acyl transfer reaction of *S*-2-(octanoylthio)ethylamine is also accelerated slightly in the presence of CTAB,[222] but is strongly inhibited (5×10^3 times) by the anionic detergent sodium dodecyl sulphate, largely because of an increase in the pK_a of the nucleophilic amino-group; even acetyl transfer is 100 times slower.

The reaction of imidazole with *p*-nitrophenyl dodecanoate is catalysed weakly by micelles of cholic acid.[223] When two cholic acid amide molecules are joined by a flexible oxyethylene chain $[2RCOOH \rightarrow (RCONHCH_2CH_2OCH_2CH_2)_2O]$ its solubility falls, so that solvents containing 40% of dioxan have to be used, and only a small catalytic effect at low concentrations is observed.

The hydrolysis of alkyl *p*-nitrophenyl carbonates in moist benzene or hexane is greatly accelerated by the addition of surfactant ammonium salts such as dodecylammonium propionate,[224] which organize the available water into reversed micelles and also catalyse the reaction by some form of general acid–base mechanism; a quaternary ammonium chloride has no catalytic effect. The exchange of ^{18}O into propionate–propionic acid in these reversed micelles is substantially slower than in bulk water.[225] On the other hand, Fendler has detected ultra-fast proton transfer processes involving excited-state pyrene-1-carboxylic acid in the same reversed micelle system. Proton transfers are 30–40 times faster than in bulk water, with a rate constant (2×10^{12} $\text{M}^{-1}\text{s}^{-1}$) for protonation of Py*COO$^-$ by propionic acid which is faster than any previously observed in solution.[226] These reactions almost certainly occur within the highly structured interface region, which constitutes the hydration shell of the surfactant headgroups; the substrate carboxylate group and the proton donor must be held in very close proximity.

Mixed micelles of CTAB and long-chain-acyl hydroxamic acids[227] and *N*-acylhistidines[228] show substrate selectivity. For example, the CTAB-decanohydroxamic acid co-micelles selectively hydrolyse *p*-nitrophenyl decanoate with very high efficiency;[227] reactions with both shorter- and longer-chain substrates are substantially slower. The *N*-acylhistidines catalyse the stereoselective deacylation of *p*-nitrophenyl *N*-acylphenylalanines. Both the rate and the selectivity increase with increasing chain-length of the catalyst: for example, *N*-acetyl-L-histidine mixed with CTAB hydrolyses *p*-nitrophenyl *N*-(phenylacetyl)-L-phenylaniline 1.4 times faster than the D-enantiomer; whereas the reaction with *N*-stearoylhistidine is 1000 times more effective in terms of rate, and hydrolyses the L-substrate 2.8 times faster than its D-isomer.[228]

Micelles of functionalized zwitterionic surfactants (**78**; $n = 1, 2,$ or 3) inhibit the alkaline hydrolysis of a series of esters, including *p*-nitrophenyl acetate and the

$$C_{10}H_{21}\overset{+}{N}Me_2(CH_2)_nCOO^- \qquad C_{10}H_{21}\overset{+}{N}Me_2(CH_2)_3COOpNP$$

(78) **(79)**

p-nitrophenyl ester (**79**), though small rate enhancements are observed below the cmc.[229] pH–rate profiles for hydrolysis in the presence of the micelles show a micelle-catalysed reaction at low pH, catalysed by the COO$^-$ form of the surfactant. The mechanism changes from intramicelle nucleophilic to general base catalysis in the usual way as the ester leaving group is changed from dinitro- to mononitrophenolate.[230] The corresponding functionalized micelles with OH in place of the COO$^-$ group of (**78**) also act as weak nucleophilic catalysts for the hydrolysis of neutral and positively charged esters.[231]

The effectiveness of nucleophilic catalysts of this sort is limited by the rate of deacylation of the catalyst; for example, (**80**) and (**81**) are two functionalized surfactants which are highly reactive in the first step (nucleophilic attack) but whose turnover rates are limited by their ease of subsequent deacylation.

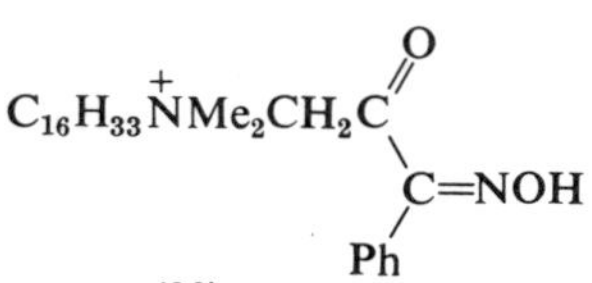

$$C_{16}H_{33}\overset{+}{N}Me_2CH_2C(=O)C(=NOH)Ph$$

(80)

$$C_{16}H_{33}\overset{+}{N}Me_2(CH_2)_2NHC(=O)CH(\overset{+}{N}H_3)CH_2SH$$

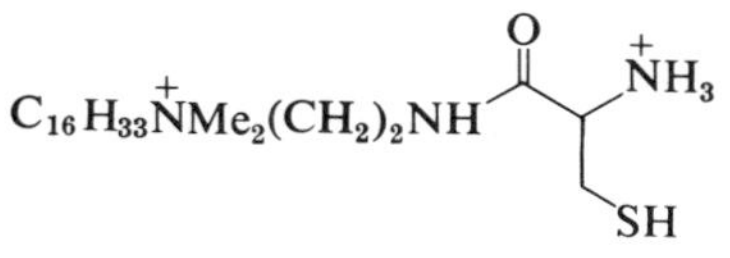

(81)

Mixed micelles of (**80**) and CTAB deacylate *p*-nitrophenyl hexanoate in a reaction which shows a very rapid "burst" of *p*-nitrophenolate release, followed by a much slower steady-state rate;[232] the value obtained for k_{cat} is over 10^4 $M^{-1}\,s^{-1}$ at pH 7.95 (25 °C), faster than that for any similar surfactant reaction or for the reaction of chymotrypsin with *p*-nitrophenyl esters; the turnover rate, on the other hand, is only $\sim 10^{-4}\,s^{-1}$.

The cysteine-derived surfactant (**81**) cleaves *p*-nitrophenyl acetate rapidly under micellar conditions $k_{cat} = 26\ \text{M}^{-1}\,s^{-1}$ at pH 8.0) by nucleophilic attack of the thiolate form, but the *S*-ester produced is deacylated by the neighbouring amino-group.[233] The resulting *N*-acetylcysteine reagent is actually more reactive than (**81**), and removes an acetyl group from a second molecule of ester with a k_{cat} of 36 $M^{-1}\,s^{-1}$, before reverting once again to a slow steady-state reaction limited by the rate of deacetylation.

The hydrolysis of acetyl phosphate is not much accelerated by several cationic and metal-chelating micelles, but dodecylammonium chloride enhances the cleavage by factors of up to 160 in a reaction which may be partly aminolysis.[234]

The polymerization of [^{14}C]-alanyl adenylate to form oligopeptides has been studied in the presence and absence of CTAB micelles, and in reversed co-micelles of CTAB and sodium di-2-ethylhexyl sulphosuccinate in benzene.[235] In the presence of micelles, and especially in the reversed micellar system, the yields of polyalanine are sharply increased, as is the degree of polymerization; high yields of polypeptide chains containing up to 40 amino-acid units are obtained.

The hydrolysis of *p*-nitrophenyl palmitate is catalysed by the cholesterol ester of imidazole-4-carboxylic acid in the presence of vesicles formed from bis(dodecyl)-ammonium bromide in water. Both catalyst and substrate are insoluble, and sonication is required to achieve incorporation into the surfactant aggregates.[236] Without sonication the palmitate reacts more slowly than *p*-nitrophenyl acetate, but incorporation on sonication leads to a rate increase of over 200-fold.

Micelles, and even the vesicles of Kunitake and Sakamoto,[236] are structurally relatively "soft", ill-defined structures, and increasing efforts are being made to design relatively small single-molecule hosts with catalytic capabilities. Cram's chiral binaphthalene-based hosts[237] begin to recognize enantiomers of amino-acids with NH_3^+-groups,[238,239] and this capability can be used for total separation of enantiomers when combined with chromatographic techniques.[240] Simple crown ethers also bind RNH_3^+ groups selectively (thus allowing, for example, selective acylation of secondary amines in the presence of primary amino-groups[241]), and functionalized crowns are proving to be remarkably effective and specific catalysts; for example, the dithiol (**82**; Y = H, X = $(CH_2)_3SH$) catalyses the release of *p*-nitrophenolate from α-amino-acid esters several thousands of times faster than does a mixture of 18-crown-6 and butanethiol[242] (in 20% ethanol–dichloromethane, buffered with acetic acid–pyridine); the related L-cysteine derivatives (**82**; X = Y = CO—CysOMe) bind and cleave *p*-nitrophenyl esters of dipeptide hydrobromides, also by nucleophilic attack by the S^- group in the side-chain.[243] The discrimination between secondary and primary ammonium is shown by the 15 000 times faster hydrolysis of the Gly–Gly ester than the Pro–Gly derivative, and chiral selectivity is also very high (*e.g.* a rate factor of 90 in favour of Gly–L-Phe, compared with the Gly–D-Phe ester).

Crown ether-type macrocycles modified by the introduction of 2,6-linked pyridine rings[244] are efficient catalysts for the *n*-butylaminolysis of *p*-nitrophenyl

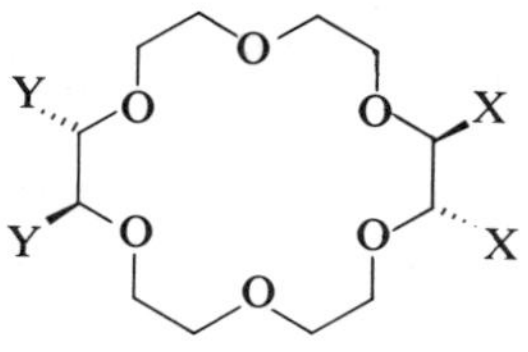

(82)

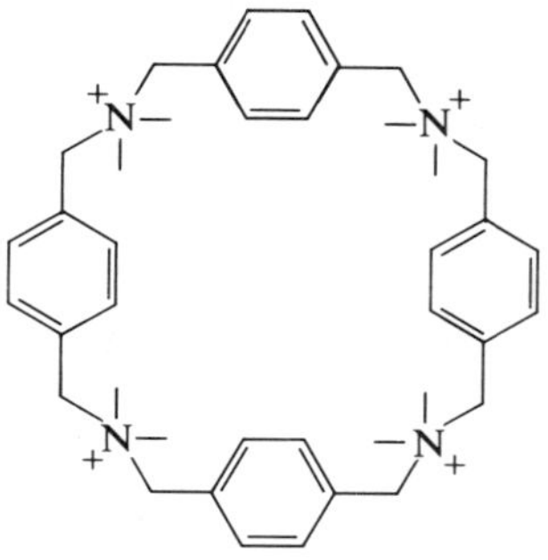

(83)

acetate in chlorobenzene.[245] The largest rings are the most efficient, but the pyridine-nitrogen is not essential for catalysis. The water-soluble heterophane **(83)**, on the other hand, catalyses the hydrolysis of aromatic esters of chloroacetic acid at pH 7–8 by way of inclusion complexes, in reactions that show kinetic specificity for the aromatic leaving group. Electrostatic stabilization of the transition state, a phenomenon difficult to demonstrate clearly in model systems, could be involved.[246]

The introduction of a histamine-imidazole group into Murakami's [20]-paracyclophanes[247] (*e.g.* **84**; X = O) results in nucleophilic catalysis of the cleavage of hydrophobic esters and in accumulation of the *N*-acylimidazole.[248] Very large

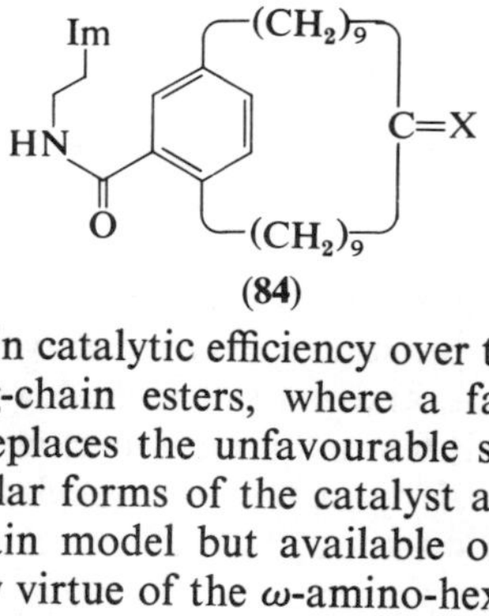

(84)

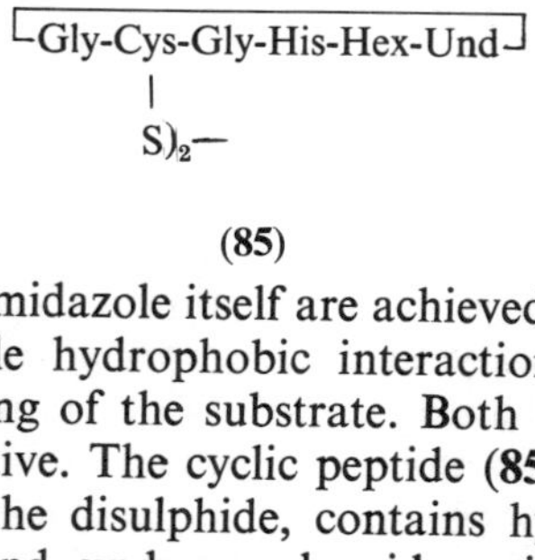

(85)

increases in catalytic efficiency over that of imidazole itself are achieved, especially with long-chain esters, where a favourable hydrophobic interaction with the catalyst replaces the unfavourable self-coiling of the substrate. Both monomeric and micellar forms of the catalyst are effective. The cyclic peptide **(85)**, designed as a papain model but available only as the disulphide, contains hydrophobic regions by virtue of the ω-amino-hexanoyl and -undecanoyl residues; it binds and hydrolyses long-chain *p*-nitrophenyl esters with Michaelis–Menten kinetics for which k_{cat}/K_M is comparable with that for chymotrypsin. The catalytic group is the imidazole anion, and effective molarities of over 10^4 M are achieved.[249]

Hydrolysis of *p*-nitrophenyl acetate catalysed by the anion of salicylaldehyde (see above) is accelerated in the presence of polyethylenimine quaternized with lauryl groups.[196] A graft copolymer of L-histidine on polyethylenimine is an efficient catalyst for the hydrolysis of nitrophenyl esters, and shows a bell-shaped pH–rate profile for the hydrolysis of 4-carboxy-2-nitrophenyl acetate.[250] The decarboxylation–fragmentation of 6-nitrobenzisoxazole-3-carboxylate, which is known to be very sensitive to solvent polarity,[251] is strongly catalysed by poly-(vinylbenzo-18-crown-6) in water:[252] decarboxylation of the bound carboxylate is 2300 times faster than in water, as expected for a benzene-like environment; binding and the rate of decarboxylation are increased substantially in the presence of K^+ and Cs^+ ions.

The binding of apolar substrates by α-cyclodextrin has been analysed,[253] and the binding of sodium benzoate and benzoic acid has been investigated by NMR spectroscopy.[254] From the X-ray structure of β-cyclodextrin-12 H_2O[255] it appears that no change of conformation is required to accommodate guest molecules; this is in contrast to the picture for α-cyclodextrin.[256]

Catalysis, by α-cyclodextrin, of the alkaline hydrolysis of aryl acetates is due to a more favourable enthalpy of activation;[257] the alkaline hydrolysis of trifluoroacetanilides is also catalysed, but not that of less activated amides.[258]

Of several functionalized cyclodextrins, two have been designed with phosphate ester substrates in mind. The 6-bisimidazole derivative (**86**) hydrolyses 4-*tert*-butylcatechol cyclic phosphate in a reaction that involves substrate binding and a bell-shaped pH–rate profile with a maximum rate near pH 7, indicating that the two imidazoles are involved as the free base and conjugate acid, respectively;

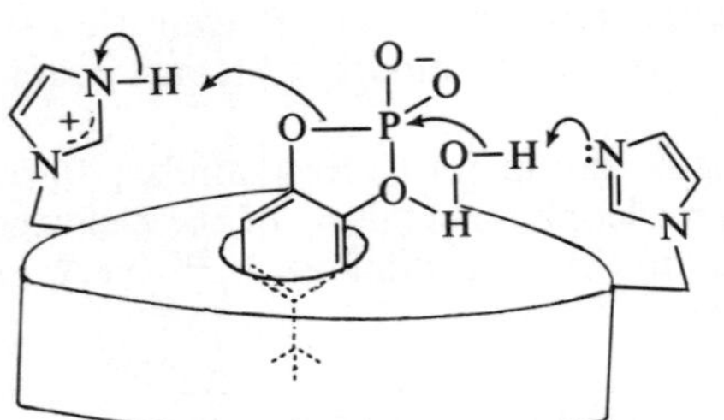

(**86**)

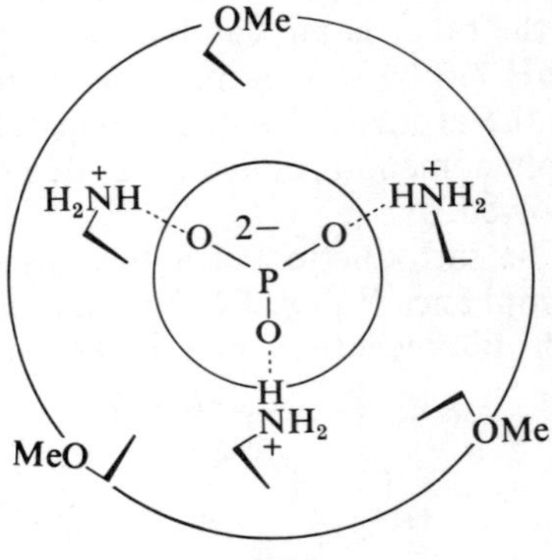

(**87**)

only one of the two possible isomeric monophosphates is produced by this regioselective reaction which qualitatively has many properties in common with similar reactions catalysed by the enzyme ribonuclease and may conceivably involve a similar mechanism (**86**).[259] The symmetrical tris-6-ammonium-α-cyclodextrin (**87**, in which the remaining fifteen 2-,3-, and 6-OH groups are methylated) binds *p*-nitrophenyl phosphate, as it was designed to do, though the effect on the rate of hydrolysis of the phosphate ester is small.[260]

The introduction of a single NMe_3^+ group at just one 6-position of β-cyclodextrin has only small effects on the rate of hydrolysis of the three acetoxybenzoates.[261] On the other hand, β-cyclodextrin functionalized with polyethylenepolyamine chains, which are powerful ligands for metal cations, binds hydrophobic anions especially effectively;[262] two recognition sites are operating, the hydrophobic cavity of the cyclodextrin, and the metal cation bound in the flexible "cap".

Metal-ion Catalysis

The rapid alkaline hydrolysis of α-amino-acid ester ligands in *cis*-$[CoX(en)_2]^{2+}$ complexes is not observed for co-ordinated ethyl 4-aminobutanoate, which is hydrolysed only 7 times faster than the free substrate,[263] but very efficient catalysis by metal ions is found when the ester contains the strongly bound 8-hydroxyquinoline group: for example, the alkaline hydrolysis of metal complexes of the methyl ester (**88**) is 10^3–10^6 times faster than that of the uncomplexed anion (with

Cu^{2+} most effective and $Zn^{2+} > Mn^{2+}$ least), and most probably involves simple electrophilic catalysis of the attack of hydroxide.[264] When the ligand is the leaving group, as in (**89**), catalysis is more efficient still;[265] the acetoxy-group of (**89**) is hydrolysed in the absence of metal by a mechanism involving intramolecular general base catalysis by the quinoline nitrogen atom. In the presence of metal ions

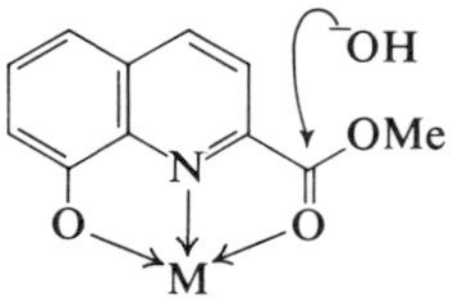

(**88**) 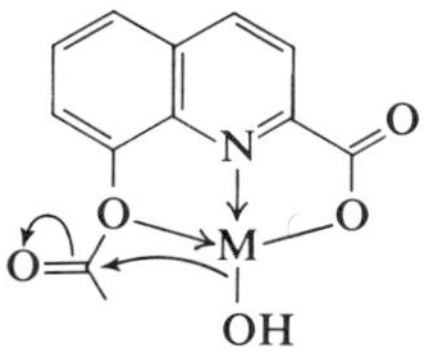 (**89**)

a substantial proportion of the substrate is complexed (with Cu^{2+} complete complexation is possible) and accelerations of over 10^8-fold in k_{OH} are observed for the Cu and Zn complexes. These large rate-enhancements have been explained in terms of intramolecular attack by bound hydroxide on the ligand complexed through the leaving-group oxygen (**89**).

Cu^{2+} ions also catalyse the hydrolysis and aminolysis of benzylpenicillin;[266] rate-enhancements as large as 10^7 are observed, and a mechanism involving attack of external nucleophile on the bidentate complex (**90**) has been proposed [the lactam-nitrogen is probably at least as basic as the ester-oxygen in (**89**), since full delocalization is not possible in the penicillin-amide group]; the reactions of

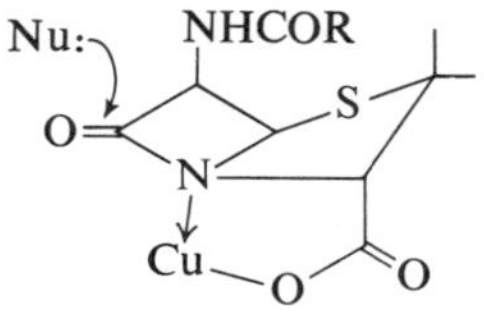

(**90**) 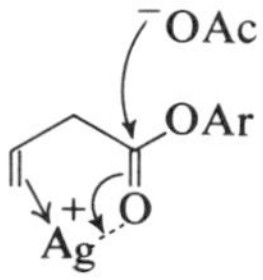 (**91**)

the methyl ester of benzyl penicillin are much less effectively catalysed. Intramolecular attack by metal-bound nucleophile cannot be ruled out, but the rate-determining step is probably breakdown of the tetrahedral intermediate, since the Brönsted β-coefficient for aminolysis is close to unity.[266]

Ag(I) specifically catalyses the hydrolysis of the $\beta\gamma$-unsaturated ester (**91**; Ar = dinitrophenyl) in acetate buffer, possibly by way of the complex shown.[267] The elimination of H_2S from thioacetamide to give acetonitrile is catalysed by Pb > Cd > Co and proceeds by rate-determining breakdown of the conjugate base of the 1 : 1 complex.[268]

Specific ester–cation interactions have been proposed to explain the observed effects of added salts on the competing acyl-oxygen and alkyl-oxygen cleavages of 1-phenylallyl chloroacetate in aqueous ethanol[269] and the methanolysis of phenyl acetate;[270] large effects are observed for the reaction of dimethyl-*n* propylmalonate, where there is IR and Raman spectroscopic evidence for the formation of ester–cation and ester–solvent complexes.[271]

In the presence of Cu^{2+} the paracyclophane (**84**; X = O) no longer acts as a nucleophile through the imidazole group, because this is complexed to the metal;[272] in contrast, the bifunctional oxime (**84**; X = NOH) has a higher residual activity in the presence of Cu^{2+}, possibly because ester substrates can be complexed by the metal and held within range of the nucleophilic oxime group.

Metal-ion catalysed esterifications of benzoate salts with ethylene glycol[273] and of chloroacetic acid[274] have been studied, as has the hydration of nitriles in the presence of Cu-containing catalysts.[275,276]

Enzymic Catalysis

Serine Proteinases

Recent reviews of enzyme mechanism have dealt with their determination by kinetic studies,[277] the use of kinetic isotope effects,[278] studies at sub-zero temperatures,[279] the structural basis of the activation and action of trypsin,[280] and the detection and properties of active-site histidine-imidazole groups.[281] A new NMR technique has been applied to the study of one of the protons in the charge-relay system of several serine proteinases.[282]

Two groups have examined the mechanism of the slow reactions of methyl chymotrypsin (methylated on histidine-57).[283,284] The modified enzyme is some 10^5 times less reactive than α-chymotrypsin, but this residual activity presents a considerable problem; the methylated nitrogen is the one which normally acts as a general base in the first step of the peptide cleavage reaction, but the observed differences [*i.e.* in leaving-group and acyl-group specificity, weaker binding of some protein inhibitors and transition-state analogues, and increased solvent deuterium isotope effect (k_H/k_D about 4, compared with 2.9 for deacylation)] are largely quantitative.[283] The reaction still involves an acyl-enzyme intermediate and appears to involve general acid–base catalysis. The enzyme is still a relatively efficient catalyst (2×10^7 times more effective than hydroxide at pH 7), but if the catalytic mechanism is essentially unchanged one is faced with two equally unlikely alternatives: either the methylated nitrogen atom acts as a general base (**92**), just as in the parent enzyme, or the imidazole residue rotates so that the other N becomes available to act. In the first case the methylated N is not a viable base (and even if it were it would have to act almost perpendicularly to the ring), but on the other hand the geometry of the system appears to prohibit the formation of the key H-bond to serine-195 upon rotation of the imidazole. The question remains unresolved.

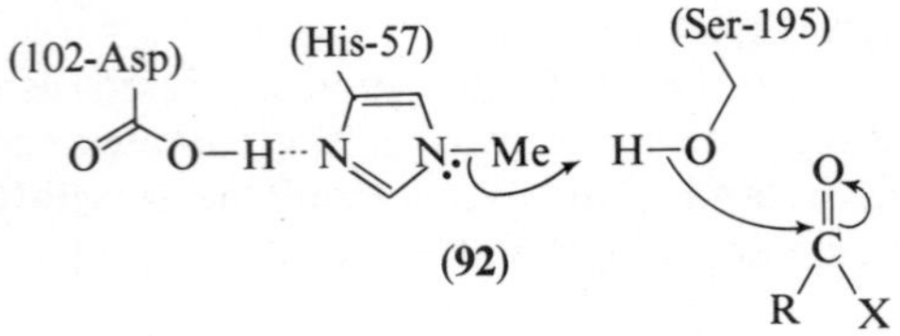

(**92**)

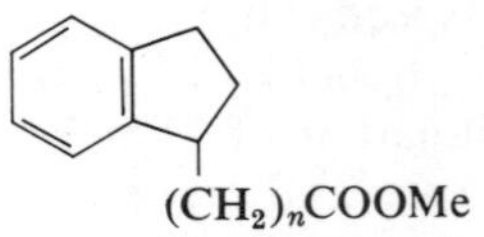

(**93**)

Taft's ρ^* value (2.18, for k_{cat}/K_m) for the chymotrypsin-catalysed hydrolysis of a series of acyl-substituted phenyl acetates is similar to that for alkaline hydrolysis (2.07) and is not consistent with differential hydrogen-bonding in Michaelis complex and transition state.[285] The enantiomeric specificity of chymotrypsin towards (**93**), and towards methyl cyclohex-3-enecarboxylate and its homologues, is reversed as n is increased.[286] Evidence[287] suggesting the intermediacy of a second acyl-enzyme has been refuted;[288] and the reaction of chymotrypsin with a modified trypsin–kallikrein inhibitor has been studied.[289]

Though trifluoroacetyl-dipeptides are more powerful reversible inhibitors of elastase than the *N*-acetyl compounds, the corresponding chloromethyl ketones (*e.g.* CF_3CO–Ala–Ala–CH_2Cl) are not better irreversible inhibitors; they bind better but react more slowly.[290]

The inefficient hydrolysis of several *p*-nitrophenyl esters by chymotrypsinogen and trypsinogen has been studied.[291] Results of a stopped-flow study of the trypsin-catalysed hydrolysis of two substituted anilide substrates are interpreted in terms of the rate-determining disappearance of a tetrahedral intermediate.[292]

A refinement of the X-ray structure of subtilisin shows that the hydrogen bond between Ser-221 and His-64, which is part of the charge-relay system, is either absent or severely distorted;[293] in contrast the buried Asp–His hydrogen bond is strong; a similar situation is found in other serine proteinases. It has always been clear that the imidazole group of the active-site histidine of the serine proteinases cannot be ideally placed to act both as a general base for the initial attack by serine (**92**, above) and as a general acid to assist the departure of the leaving group; it now appears that it is better placed to perform the latter function.

The hydrolysis of neutral esters by acetylcholinesterase is promoted by the binding of several quaternized heterocycles.[294]

Pantothenase (probably a serine protease) hydrolyses pantothenic acid to β-alanine and pantoic acid in a reaction that appears to be reversible at low concentrations of buffer and to involve an acyl-enzyme.[295]

Thiol Proteinases

Ab initio calculations on the interaction of papain with *N*-methylacetamide as a model substrate have been presented.[296] Resonance Raman spectra of three chromophoric acyl papains show a characteristic red shift and an unusually low-frequency double-bond absorption at 1570 cm^{-1}, tentatively associated with hydrogen bonding to the imidazole-NH^+ of histidine-159.[297]

Ingenious chemical conversions of the active-site cysteine-25 to give serine-25 and glycine-25 papains have been reported;[298] the key reaction is the photolysis of the phenacyl derivative (**94**; Ar = 2,4-dimethoxyphenyl) to give the thioaldehyde,

$$\text{Enz—CH(S)(HO)—Ar}\ (\mathbf{94}) \xrightarrow{h\nu} \text{Enz—CH=S} + \text{ArCOMe} \xrightarrow{H_2O} \text{EnzCHO} \begin{cases} \xrightarrow{BH_4^-} \text{Enz—CH}_2\text{OH} \\ \xrightarrow{OH^-} \text{EnzH} \end{cases}$$

which can be either hydrolysed to the aldehyde and subsequently reduced with borohydride, or hydrolysed in mild base (retro-aldol reaction); the modified enzymes show no activity towards good substrates of papain.

A study of the reaction of papain with α-*N*-(benzyloxycarbonyl)-L-lysine methyl ester at sub-zero temperatures in aqueous dimethyl sulphoxide at 276 nm reveals three reactions preceding the rate-determining step:[299] these have been identified as substrate binding, a conformational change involving the active site (His-159) imidazole, and the formation of the acyl enzyme. On the basis of these results a modified mechanism has been proposed which involves one conformation for the neutral imidazole form, in which it is H-bonded to Asn-175, and another for the protonated imidazole, interacting electrostatically with Asp-158–COO^- as well as Cys-25–S^-.

The deacylation of substituted-benzoyl papains depends on a single ionizing group of $pK_a = 4.2–4.3$, and follows the Hammett equation with $\rho = 2.74 \pm 0.32$; $k_H/k_D = 2.2$, and the reaction closely resembles the deacylation of benzoyl chymotrypsins; involvement of Asp-158–COO^- as a general base has been suggested.[300]

Careful analysis of the pH-dependence of k_{cat}/K_m for the papain-catalysed hydrolysis of several ester and amide substrates below pH 6.4 leads to the conclusion that two ionizations are involved in this region, with pK_a's of 3.78 ± 0.2 and 3.95 ± 0.1;[301] the groups likely to be involved are the imidazole of His-159 and the carboxyl of Asp-158, and it looks increasingly likely that this aspartic acid residue is involved in the catalytic process.

Rates of acylation and deacylation of papain by a range of specific ester substrates have been measured.[302]

Two $\alpha\beta$-unsaturated aromatic amino-acids (phenyldehydroalanine PhCH=C(NH_2)COOH and its vinylogue) have been used as spectroscopic probes of the changes that take place on acylation of papain.[303] Shipton and Brocklehurst have reported results with two thiol-specific reactivity probes, benzofuroxan (**95**)[304] and

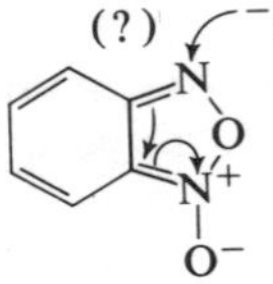

(?) $^-$SR

(**95**)

RS$^-$

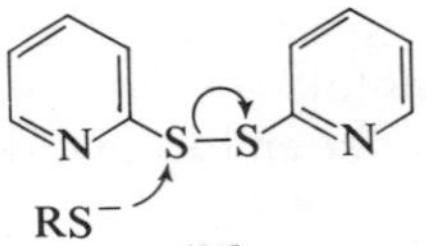

(**96**)

2,2′-dipyridyl disulphide (**96**).[305] The reactions of (**95**) with papain, ficin, and bromelain have been interpreted in terms of an ionization of pK_a 3–4 to produce a free thiolate at the active site of all three enzymes, and a second pK_a of about 9 associated with the imidazolium group of His-159.[304] Further work with (**96**), and with *n*-propyl 2-pyridyl disulphide,[305] shows a marked pH-optimum for the reaction with papain and provides further evidence for a difference between the reactions of papain, on the one hand, and ficin and bromelain, on the other; it has been suggested that this difference is explained by the absence, in the last-mentioned two enzymes, of an active-site carboxyl group equivalent to that of Asp-158 in papain.[305] Once again two ionizations with pK_a close to 4 are apparent and there is now fairly good agreement that active-site reactivity is dependent on three ionizations (two near $pK_a = 4$ and a third of pK_a close to 9), but no general agreement on which pK_a is associated with which group.

Alkyl aryl sulphates inhibit papain and ficin by non-specific methylation.[306] Alkylation of the active-site thiol is achieved by the specific inhibitors Z-Phe– and Z-Phe–Phe–$COCHN_2$,[307] but the reactions of these diazoketones are unusual in that the inhibitors do not react with simple thiols, clearly because they are not acidic enough to protonate the diazoketone. A mechanism involving initial addition of SH to the reagent-carbonyl group to give a monothioacetal (**97**) has been suggested,[308] and this would indeed account for the activation of the diazoalkyl group towards protonation; a mechanism (**98**), involving protonation by a general acid other than the SH group, would be a simpler explanation. The inhibition constant for the reaction of papain with the specific inhibitor $PhCONHCH_2CDO$ is 11% smaller than for the protio-aldehyde and is consistent with a mechanism of inhibition involving the formation of a monothioacetal.[309]

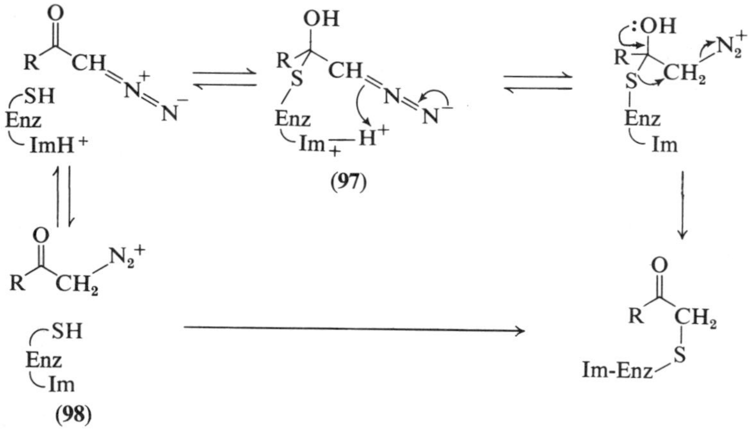

(97)

(98)

Acid Proteinases
The value of three-dimensional structural information in the discussion of enzyme mechanism is nowhere more apparent than in the cases of much-studied enzymes where it is not available. The acid proteinases have been a case in point, but now the crystal structure of penicillopepsin has become available,[310] and the active-site region is well enough defined to allow some mechanistic inferences to be drawn.[311] The carboxyl group of Asp-32 is surrounded by a network of H-bonds, which account for its low pK_a; these are formed to a main-chain amide-NH, a water molecule, and the carboxyl group of Asp-125, which is more exposed. Asp-32 in fact appears too buried to act as a nucleophile, while Asp-215 is probably protonated at the pH-optimum, and consequently less likely to act as a nucleophile. The first mechanistic proposals, based on the active-site structure, therefore involve a general base catalysis mechanism (**99**), with tyrosine-75 available as a general acid to assist the departure of the leaving group.[311] The similarities to recent proposals for the mechanism of action of carboxypeptidase are interesting.

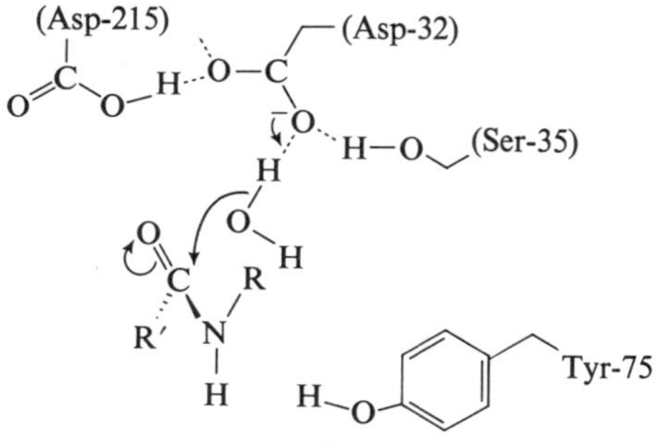

(99)

This mechanism is relevant to other acid proteases, including porcine pepsin, which is a closely similar enzyme. There is growing support for the belief that covalent acyl- and amino-enzyme intermediates, proposed to explain transpeptidation reactions, are not involved, at least in the hydrolysis of substrates at the pH optimum. Antonov *et al.*[312] have reported new ^{18}O-exchange evidence which is inconsistent with an acyl-enzyme intermediate. The transpeptidation reaction of L-leucine-L-tyrosine amide gives Leu–Leu containing one atom of oxygen-18 from labelled water in the peptide bond, and partial exchange of the peptide carbonyl-oxygen is also observed; under these conditions resynthesis of starting material from leucine and tyrosine amide is negligible, so transpeptidation involving the aminolysis of a first-formed leucyl-enzyme intermediate is ruled out. A general base catalysis mechanism is preferred by these authors also.

Metallo-proteinases

Co(III)-carboxypeptidase has been prepared and properly characterized. It is catalytically inactive, and binds amide but not ester substrates of the native enzyme, suggesting that inner-sphere co-ordination is necessary for ester binding.[313] Temperature-jump experiments, using the chromophore of arsanilazotyrosine-248 as a probe, allow the detection of discrete binding and conformational-change steps in the reactions of carboxypeptidase with products and poor substrates.[314] Substrate activation or inhibition is observed with hippuric acid ester substrates $PhCONHCH_2COOCHRCOOH$, depending on the geometry of the group R.[315]

Model experiments on the cleavage of *trans-p*-chlorocinnamic propionic anhydride by hydroxylamine show that very rapid attack coccurs at the propionic acid carbonyl group, so that a similar acyl enzyme would be trapped with release of the chromophore.[316]

Thermolysin catalysis of the cleavage of a series of peptide substrates terminating in Phe–Leu–Ala (the Phe–Leu bond is cleaved) shows little sensitivity to the *N*-terminal groups, unlike reactions of enzymes such as pepsin and papain, suggesting a more rigid active site region for the metalloenzyme.[317] Hydroxamic acids are potent irreversible inhibitors of thermolysin: *Z*–Gly–L-Leu–NHOH gives a K_i value of 13 M at pH 7.2, and an interaction of the hydroxamic acid oxygen with the active site Zn ion is consistent with the results.[318] The enzyme is also inactivated by related haloacetohydroxamic acid derivatives derived from leucine, which probably alkylate the carboxyl group of Glu-143.[319]

Other Enzymes

A review of the codon-specific activation of amino-acids has dealt with the high specificity of the aminoacyl-tRNA synthetases;[320] the three-dimensional structure of tRNA, and its functional implications, are the subject of a more extensive review.[321] The aminoacylation of arginyl-tRNA catalysed by arginyl-tRNA synthetase from yeast has been shown to be a stepwise process, like other similar reactions, although the aminoacyl adenylate complex cannot in this case be isolated in the absence of tRNA.[322] The rate-determining step is the release of Arg–tRNA from the enzyme, which is preceded by the formation of Arg–AMP and the transfer of arginine to tRNA.

Aspirin has been shown to inhibit prostaglandin synthesis by acetylating a serine-OH at the amino-terminus of prostaglandin synthetase.[323]

Clavulanic acid (**100**) is a potent inhibitor of β-lactamases from a variety of bacteria; the reaction with an *E. coli* enzyme involves both reversible and irreversible inhibition, the latter requiring a substantial number (about 100) turnovers and

(**100**) (**101**)

involving up to three different inactivated species;[324,325] various possible mechanisms have been suggested.

The pH-dependence of L-asparaginase from *E. coli* (bell-shaped pH–rate profile, pK_a's 6.58, 8.69) has been interpreted tentatively in terms of the ionizations of an active-site imidazole and of the α-amino-group of the substrate.[326]

Stopped-flow studies of the interconversion of CO_2 and bicarbonate catalysed by carbonic anhydrase show substantial solvent deuterium isotope effects.[327] The ratio $k_H/k_D = 3.3$ for the hydration of CO_2 in the high pH-independent region; for dehydration of HCO_3^- a ratio of 4.3 is found on the acid plateau. These figures (for k_{cat}) are consistent with the involvement of proton transfer in the turnover step. Since K_m also shows a substantial isotope effect (about 3 in both directions) much smaller isotope effects apply to k_{cat}/K_m.

Carbonic anhydrase (from bovine erythrocytes) also catalyses the hydration and hydrolysis of pyruvate esters (though the K_m's differ for the two reactions),[328] and the hydrolysis of dimethyl 2,4-dinitrophenyl phosphate.[328a]

Benzil inhibits chicken liver carboxylesterase by forming a hemiacetal adduct with an active-site serine-OH group;[329] the similar reaction with the 2,4-dinitrophenyl ether of benzil monoxime leads to fragmentation of the substrate (**101**) to give benzonitrile.[330]

The carbon-13 isotope effect on the decarboxylation of pyruvate catalysed by yeast pyruvate decarboxylase ranges from $k^{12}/k^{13} = 1.002$ at pH 7.5 to 1.011 at pH 5.0;[331] this compares with figures of 0.992–1·007 (depending on the solvent) for the reaction catalysed by thiamine,[332] and 1.051–1.058 for the decarboxylation of preformed thiamine-pyruvate addition intermediate. Evidently the decarboxylation step is not rate-limiting in the enzyme reaction. Close analysis of the pH–activity profile for the enzyme reveals a pH-independent region between pH 5.4 and pH 5.8, and two more (making four in all) apparent pK_a's.[333] For decarboxylation of arginine and homoarginine catalysed by the arginine decarboxylase from *E. coli* $k^{12}/k^{13} = 1.0144$ and 1.0535, respectively:[334] thus, decarboxylation appears to be rate-determining in the latter case, so that specificity is apparently manifested in the decarboxylation step.

Analysis of the chiral product of decarboxylation of L–glutamate in D_2O catalysed by glutamate decarboxylase from *E. coli* shows that the reaction proceeds with retention of configuration.[335]

Successful suicide inhibitors for mammalian ornithine decarboxylase have been reported.[336]

The carboxylation of propionyl-CoA catalysed by the eponymous carboxylase involves abstraction of a hydrogen atom from the α-carbon, and carboxylation by biotin–CO_2 to give methylmalonyl-CoA. When 2-fluoropropionyl-CoA is used, the carbanion eliminates F^- 6 times faster than biotin–CO_2 is formed; thus the formation of carbanion and carboxylation proceed at different rates and the reaction is not concerted.[337]

Decarboxylation

The decomposition of bicarbonate and several monoalkyl carbonate anions, $ROCOO^-$, at pH 6.7–10.8 have been studied:[338] the main mechanism is specific acid catalysis ($k_{H_3O^+}/k_{D_3O^+} = 0.6$) but weak general acid catalysis by phosphate and by water ($k_{H_2O}/k_{D_2O} = 1.8$) is observed; a late transition state is indicated.

MINDO/3 calculations for the transition state for the decarboxylation of but-3-enoic acid predict observed kinetic parameters rather well.[339]

Cuprous cyanoacetate undergoes reversible decarboxylation in the presence of tri-*n*-butylphosphine ligand, and is a potential CO_2-carrier soluble in organic solvents.[340]

The remarkable sensitivity of the decarboxylation-fragmentation of benzisoxazole-3-carboxylates[251] to solvent polarity disappears in the presence of a 4-hydroxyl group, as in (**102**; X = OH). The effect (on solvent sensitivity) of incorporating a pyridinium group into the 4-position (**102**; X = 2,4,6-trimethylpyridinium) is minimal (cf. **102**; X = H) and it has been concluded that a suitable hydrogen bond has far greater capacity for anionic stabilization than even a neighbouring cation.[341]

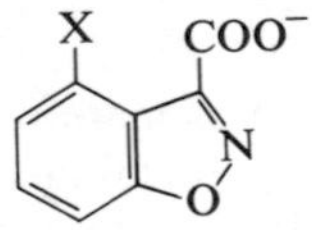

X COO$^-$ N O

(**102**)

$\ddot{N}H_2$ O O N + H

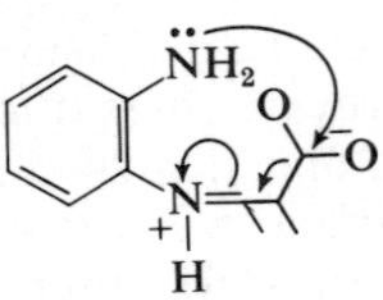

(**103**)

The decarboxylation of several mono-, di- and tri-phenyl acetate anions in THF is accelerated by factors of up to a few hundred in the presence of 18-crown-6; the rate of decarboxylation is correlated with the pK_a of the leaving carbanion, and cation effects (K > Na > Li) persist even in the presence of 1M crown.[342] Decarboxylation of mono- and di-phenylglycine is accelerated in dipolar aprotic solvents:[343] that of the monoamide monoanion of phenylmalonic acid proceeds at room temperature in acetone;[344] and those of trichloroacetate anion[345] and oxalic acid[346] in aqueous ethanol and acetophenone, respectively, have been studied.

Bromodecarboxylation of substituted cinnamic acids in water and methanol occurs cleanly when electron-donating substituents are present, but gives β-lactone and addition products as more electron-withdrawal is introduced into the system.[347]

The protodecarboxylation of solid 4-aminosalicylic acid hydrochloride follows initial loss of HCl;[348] a homolytic process is involved in the thermal decarboxylation of naphthylacetic acids.[349]

Phenylhydrazine and *o*-phenylenediamine are the most effective of a series of amine catalysts for the decarboxylation of methylacetoacetic acid. A mechanism (**103**) involving transfer of CO_2 to the *ortho*-amino-group has been suggested for the latter case.[350]

NON-CARBOXYLIC ACIDS

Phosphorus-containing Acids

Non-enzymic Reactions

Subjects covered by reviews include the chemical synthesis of polynucleotides by the triester route,[351] phosphorylation by means of enediol cyclic phosphates,[352] trivalent phosphorus thioesters,[353] and the hydrolysis kinetics of phosphate esters.[354] The Charton's have analysed steric and substituent effects on reactions at phosphorus centres;[355] calculations on high-energy phosphates and phosphoramidates underline the major importance of solvation;[356] and Guthrie[357] has constructed wide-ranging Brönsted plots for the hydrolysis and formation of methyl phosphates.

A major achievement this year is the synthesis of a phosphate monoester that is chiral by virture of the specific incorporation of the three isotopes of oxygen. Knowles and his co-workers[358] prepared [1(*R*)-^{16}O,^{17}O,^{18}O]phospho-(*S*)-propane-1,2-diol (**104**) by the relatively straightforward route shown, starting from (−)-ephedrine. The real problem is the stereochemical analysis of (**104**), and of the products of phosphate-transfer processes, for which an ingenious mass spectrometric analysis was developed.[358] Cullis and Lowe[359] report a related synthesis of chiral methyl phosphate that is also generally applicable; they find that their

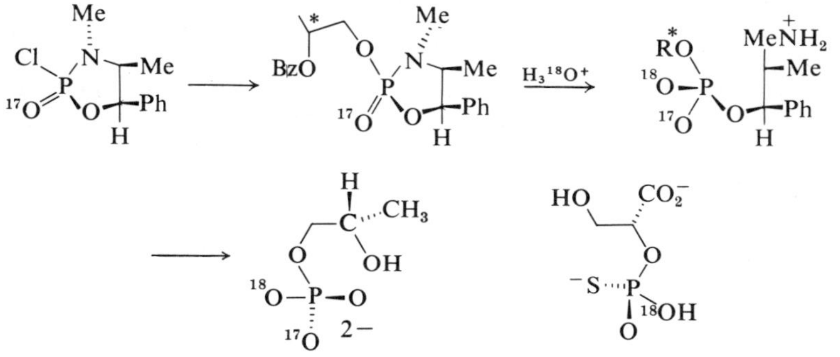

(**104**) (**105**)

product exhibits circular dichroism, thus making possible a relatively simple stereochemical analysis. These advances make it possible to answer fundamental questions about the stereochemistry of phosphate transfer.[359a]

Another important analytical advance is the discovery that ^{18}O-labelled phosphate undergoes a readily detectable isotope shift of about 0.02 ppm to higher field in the ^{31}P-NMR spectrum, for each atom of ^{18}O bound to phosphorus.[360]

ATP with a chiral γ-thiophosphate group has been prepared by the transfer of the phosphate group concerned, from chiral D-glycerate 2-(^{18}O)-monothiophosphate (**105**) to ADP, by using an enzyme system.[361]

Further results from Westheimer's laboratory testify to the difficulties involved in working with monomeric metaphosphates: methyl metaphosphate has been generated by both the pyrolysis of the cyclic phosphonate (**106**) and by the stereospecific fragmentation of (**107**) in base;[362] the highly electrophilic [$MeOPO_2$] so produced will carry out electrophilic aromatic substitution with dialkylanilines,

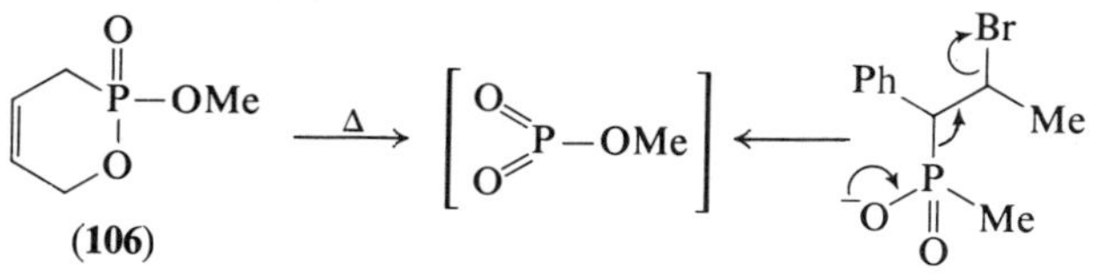

(**106**) (**107**)

albeit inefficiently. The fragmentation of (**107**) is slower than that of the phosphonate dianion, which is thought to cleave to give PO_3^-. Methyl metaphosphate may also be generated by fragmentation of a phosphate diester in the mass spectrometer.[363] The generation of monomeric mesityl metaphosphonate from the Diels–Alder reaction of (**108**) and dimethyl acetylenedicarboxylate does not occur under mild conditions, presumably because of the high activation energy for the initial Diels–Alder reaction.[364] Dimethyl phthalate is produced at 140 °C, as expected,

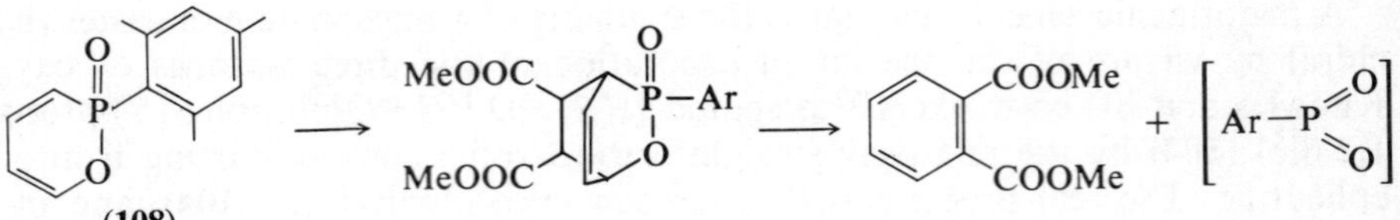

(108)

+ MeOOC≡CCOOMe → MeOOC / MeOOC → COOMe / COOMe + [Ar—PO_2]

but the metaphosphonate could not be trapped cleanly before it polymerized. The phosphorus centre of mesitylphosphonate esters is not subject to significant steric hindrance.

The striking effect of added cations on the stereochemistry of substitution at phosphorus in cyclic esters such as (**109**) in dipolar aprotic solvents[365] has been investigated further; the interpretation of the observed change to retention of configuration in the presence of Li^+ ions has been revised.[336] Displacement with inversion is retarded, and that with retention accelerated, by ion association. A mechanism has been proposed which involves rate-determining breakdown of a square-pyramidal intermediate (**110**). A Hammett plot for the displacement of

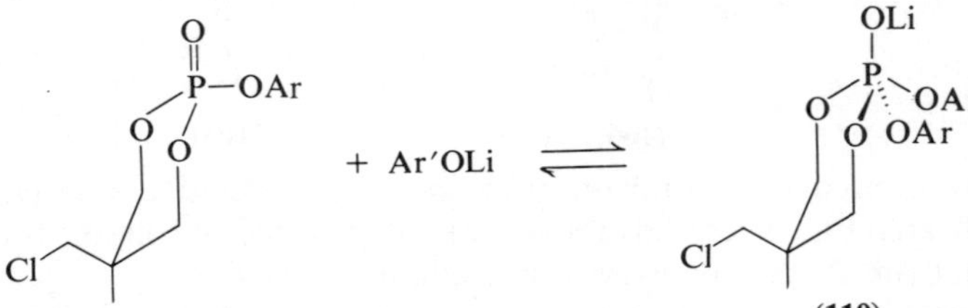

(109) + Ar′OLi ⇌ (110)

four *p*-substituted phenoxides by *p*-methylphenolate gives $\rho = 1.5$ when σ^- values are used (though a plot against σ appears to give a better correlation). In fact the breakdown of (**110**) to products cannot be cleanly rate-determining, since the leaving group is less basic than the nucleophile in most cases; there is no direct evidence for this sort of intermediate (though it has been suggested before to explain difficult stereochemical observations[367]). Nevertheless, we still have a great deal to learn about even apparently simple S_N2(P) reactions.

Phosphylated amide oximes (**111**) cleave in alkali to give the amide oxime when R = Ph, but phosphates (R = OEt) rearrange to the cyanamide R′NHCN, probably by a 1,2-shift of the anion (**112**); phosphonates (R_2 = Ph or OEt) rearrange in yet

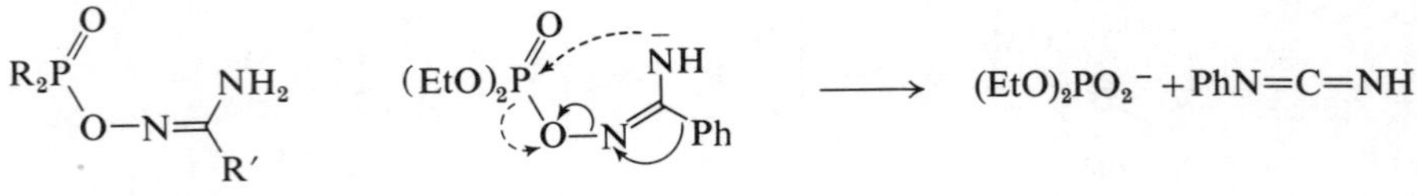

(111) (112) → $(EtO)_2PO_2^-$ + PhN=C=NH

a different way (broken arrows in **112**).[368] The selective displacement of *p*-nitrophenoxide from *p*-nitrophenyl diphenyl phosphate by the oximate anion of *p*-nitrobenzaldehyde suggests that the reagent could be useful for the deprotection of triesters; the oxime phosphate produced eliminates the phosphate diester, giving *p*-nitrobenzonitrile.[369] Disodium dodecyl phosphate is a better nucleophilic catalyst for the cleavage of *p*-nitrophenyl diphenyl phosphate in the presence of cationic micelles than is butyl phosphate, but micelles of the phosphate itself are poor catalysts.[370] Micelles of *N*-dodecylpyridinium-3-aldoximes are good catalysts for the hydrolysis of neutral esters.[371] Polycations catalyse the cleavage of 2,4-dinitrophenyl phosphate dianion by their electrostatic effects.[372]

The alkaline hydrolysis of aryl diphenylphosphinates ($\rho^0 = 1.4$)[373] and of *O,S*-alkyl, 2-dimethylaminoethyl methylthiophosphonates, and their trimethylammonium derivatives[374] have been studied. The rapid alkaline hydrolysis of dimethyl acetylphosphonate is of the first order in hydroxide and involves acyl–P cleavage of the anionic tetrahedral addition intermediate.[375] The hydrolysis of trichloroethyl phosphate shows the usual rate maximum near pH 4.[376]

The gas-phase reactions of hydroxide ion with cyclic phosphorus esters, studied by the pulsed ICR technique, show a pattern of reactivity similar to that observed in solution.[377] Extended Hückel calculations suggest that the surprisingly high heat of hydrolysis of 3′-5′-cyclic AMP is due to solvation, and specifically to extra stabilization of the product by a water molecule bridging the ribose-O(1′) and O(5′) by H-bonding.[378] The hydrolysis of α-D-glucose 1,2-cyclic phosphate does not lead cleanly to glucose 1-phosphate, but to a mixture of 1- and 2-phosphates, the former being hydrolysed at about the same rate as it is formed at pH 1. The formation of the 2-phosphate involves P—O cleavage.[379]

In dichloromethane in the presence of one equivalent of FSO_3H the phosphorane derivative (**113**) is in equilibrium with the open-chain enol (**114**), which is converted irreversibly into the keto-form only above 0 °C.[379a]

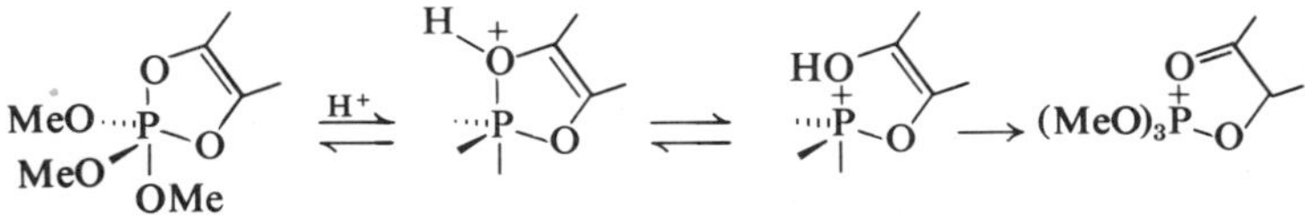

MeO···P(MeO)(OMe) $\underset{}{\overset{H^+}{\rightleftharpoons}}$ ⇌ → $(MeO)_3\overset{+}{P}$

(**113**) (**114**)

Benzilic acid reacts with $POCl_3$ to give the hydroxyphosphorane (**115**), isolated as the (pentacovalent) Et_3NH^+ salt.[380] There is some evidence that (**115**) forms a salt by protonation of dimethylformamide, in which case this hydroxyphosphorane would be an unexpectedly strong acid; in fact the best evidence yet concerning the acidity of hydroxyphosphoranes comes with the observation that (**116**) titrates as a weak acid in methanol, to give a phosphorane oxide.[381] (**117**; R = OH) has also

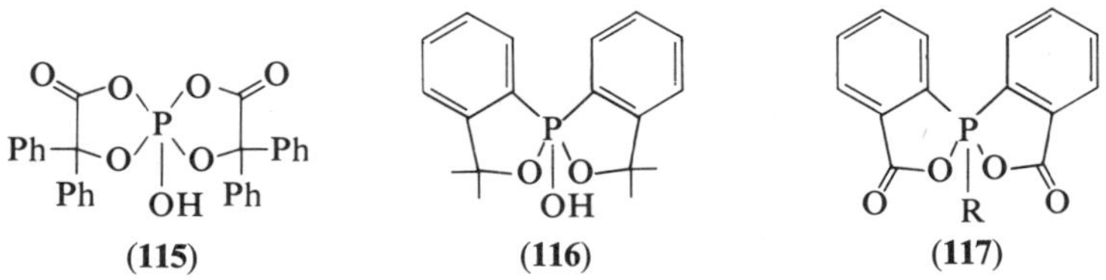

(**115**) (**116**) (**117**)

been prepared, and the crystal structure of (**117**; R = Ph) reveals an almost perfect trigonal bipyramid;[382] the crystal structure has been reported[383] for (**118**), which is present predominantly as the hydroxyphosphorane in solution[384] but not, alas, in the solid state.

Alkyl–oxygen cleavage reactions of phosphate derivatives reported involve acid-catalysed debenzylation,[385] neighbouring-group participation by phenyl[386] and dimethylamino-groups,[387] and the conversion of thiono-[388,389] and imino-phosphates[390] to the more stable P=O forms. Synthetically useful reactions are the demethylation of phosphonates by Me_3SiCl—NaI[391] and of pyro-phosphates by $MgBr_2$; the last reaction is significantly faster than demethylation by $R_4N^+Br^-$ and is thought to involve electrophilic catalysis by Mg^{2+} ions.[392]

The epimerization of methyl phenylphosphonochloridate catalysed by dimethyl formamide or acetamide is first-order in phosphonate and second-order in amide, suggesting a mechanism involving nucleophilic attack on a preformed phosphorane intermediate.[393,394] Polar substituent effects on the hydrolysis of phosphorus acid halides are not additive.[395] The reaction of *N*-methylanilines with dimethyl phosphorochloridate in nitrobenzene is catalysed by bases only.[396]

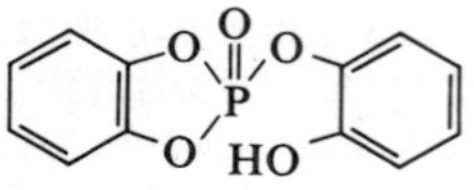

(118)

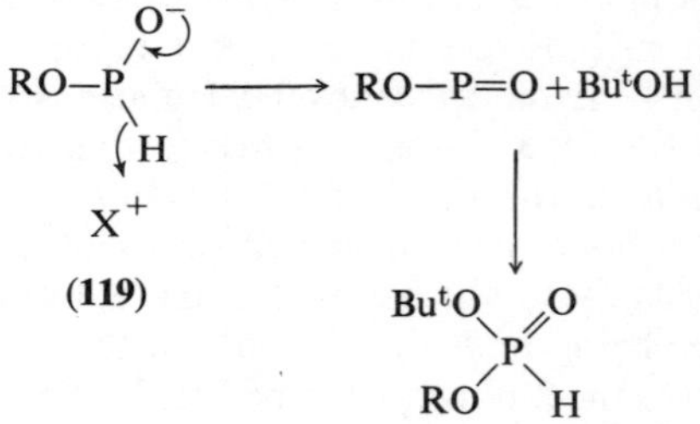

RO—P(O⁻)(H) + X^+ ⟶ RO—P=O + Bu^tOH ⟶ Bu^tO(RO)P(=O)H

(119)

Alkyl phosphinates $ROP(O)H_2$ undergo very ready transesterification with aliphatic alcohols,[397] no doubt because of the low steric demands of the substituents around P; oxidation to mixed phosphonate esters $Bu^tO(RO)POH$ occurs in Bu^tOH containing triethylamine and a mild oxidizing agent;[398] a mechanism (**119**) involving hydride transfer to give an alkyl phosphenite intermediate seems likely.

Oxidative addition of water to (**120**) to give the hydroxyphosphorane (**121**) appears to be the first step in its hydrolysis; because of the specially favourable bicyclic structure, (**121**) is stable enough to be observed by NMR spectroscopy.[399] An addition of this sort may be a common first step in the hydrolysis of P(III) compounds.[400]

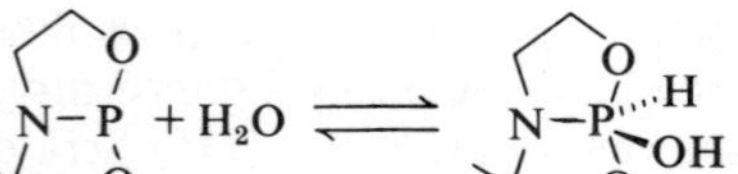

N–P + H_2O ⇌ N–P(H)(OH)

(120) **(121)**

RO^- + P_4 ⇌ RO–P_4^- + Cl–CCl_3

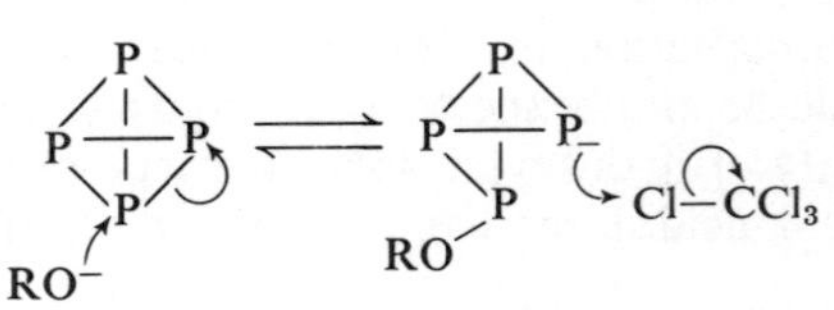

(122)

Good yields of trialkyl phosphites are obtained when alcoholic alkoxides react with elemental P_4 in the presence of CCl_4;[401] it has been suggested that the P—P cleavage is made irreversible by trapping of the phosphide anion (**122**) produced by the positive halogen compound.

Enzymic Reactions

The first determination of the stereochemistry of an intermolecular phosphate transfer has been reported by Knowles and his co-workers.[359a] The phosphate group from chiral phenyl [(*R*)-^{16}O-,^{17}O,^{18}O-]phosphate is transferred to propane-1,2-diol (see above), in a reaction catalysed by the alkaline phosphatase from *E. coli*, with retention of configuration. Since a phosphorylenzyme mechanism is well established for reactions catalysed by this enzyme, a double displacement mechanism is almost certainly involved, with both steps involving inversion at phosphorus;

however, no direct evidence about the stereochemistry of the individual steps is yet available.

Enzyme-catalysed group transfer (including phosphate transfer),[402] the involvement of enzymes (mostly kinases) in pyrophosphate metabolism in bacteria,[403] and isotope-exchange probes of enzyme mechanism (including the use of ^{18}O-exchange as a probe for phosphate transfer enzymes)[404] have been reviewed. Continuous monitoring of the exchange of ^{18}O with phosphate centres is now possible as a result of the discovery[360,405] of the isotopic shift of the ^{31}P-NMR resonance (see above). This technique has been applied to the exchange of the β-phosphate group of ADP with inorganic phosphate, which involves cleavage of the bond between the α-P and the pyrophosphate bridge oxygen;[405] to the phosphate–pyrophosphate exchange catalysed by yeast inorganic pyrophosphatase;[405] and to the phosphate–$H_2{}^{18}O$ exchange reactions catalysed by *E. coli* alkaline phosphatase,[406] myosin subfragment 1,[407] and prostatic acid phosphatase.[408] The myosin reaction has also been followed by the mass spectrometric method.[409] When ATP (**123**, labelled with ^{18}O in the β-phosphate group and the $\alpha\beta$-bridging oxygen) is incubated with pyruvate kinase the ^{31}P-resonance of the γ-phosphate-phosphorus develops an isotope shift, as label from the β-phosphate group finds its way into this position;[410] the simplest explanation is that the phosphate-transfer reaction catalysed by the enzyme involves an $S_N1(P)$ mechanism.

$$\text{AMP}-O^*-\overset{O^*}{\underset{O^*}{P}}-O-\overset{O}{\underset{O}{P}}-O^{3-} \rightleftharpoons \text{AMP}-O^*-\overset{O^*}{\underset{O^*}{P}}-O^{2-} + PO_3^-$$

(123)

$$\rightleftharpoons \qquad \xrightarrow{\text{Pyruvate}} \text{ADP}+\text{PEP}$$

$$\text{AMP}-O^*-\overset{O^*}{\underset{O}{P}}-O^*-\overset{O}{\underset{O}{P}}-O^{3-} \rightleftharpoons \text{AMP}-O^*-\overset{O^*}{\underset{O}{P}}-\overset{*}{O}{}^{2-} + PO_3^-$$

Circular dichroism has provided evidence that the formation of 5-phosphoribosyl α-1-pyrophosphate from ribose 5-phosphate and ATP, catalysed by PP-rib-P-synthetase from *Salmonella typhimurium* in the presence of Mn^{2+} or Mg^{2+}, involves an $S_N2(P)$ displacement on a metal-activated β-phosphorus, proceeding with inversion of configuration.[411]

Burgers and Eckstein have separated and identified the diastereoisomers of uridyl *O,O*-(3′, 5′)adenyl phosphorothioate and have determined the chirality of the α-thiophosphate centre of ATPαS, by enzymic methods.[412] This allows the assignment of the stereochemistry of the reaction catalysed by the DNA-dependent RNA-polymerase from *E. coli*, which occurs with inversion of configuration at the α-phosphorus centre.[412] So too does acetate activation by yeast acetyl-CoA synthetase,[413] which involves the formation of enzyme-bound acetyl adenylate. The reaction of UTPαS with α-D-glucose 1-phosphate, to give uridine-5′[1-thiodiphosphate]glucose, catalysed by UDPG-pyrophosphorylase, also goes with inversion at α-phosphorus.[414] The diastereoisomers of ATPβS have also been assigned.[415] A pH study of yeast hexokinase provides good evidence that an active-site carboxyl group, tentatively identified as an aspartic acid residue, is involved in the reaction:[416] the anionic group catalyses the transfer of the terminal phosphate of ATP to sugars, while the reverse reaction requires the COOH form; a mechanism for the phosphate transfer similar to one recently suggested for fructokinase[417] has been proposed.

Structural changes at the active site of pyruvate kinase have been monitored by following the relaxation time of the CH_3 protons of $CH_3NH_3^+$, which can act as the activating monovalent cation.[418,419]

The infrared spectra of the anions which specifically inhibit creatine kinase show characteristic changes when they are bound to the enzyme within the enzyme : M^{2+} : ADP : creatine complex;[420] it is thought that these anions, which include nitrate, thiocyanate, and azide, take the place of the terminal phosphate of ATP, perhaps mimicking transition-state metaphosphate.[421] Similar shifts occur in concentrated solutions of M^{2+} salts, suggesting that the anions are co-ordinated to the metal cation in the inhibited enzyme complex. This could mean that the phosphoryl group being transferred is also co-ordinated to M^{2+}, which would be surprising on chemical grounds and would require a revision of earlier pictures of the enzyme mechanism.[421] The mechanism and subunit behaviour of creatine kinase have been briefly reviewed by Bickerstaff and Price.[422] The transfer of the terminal phosphoryl group of ATP catalysed by cAMP-dependent histone kinase may involve a phosphoenzyme intermediate; apart from a nucleophilic histidine, the active site may also contain a lysine.[423]

A non-specific acid phosphatase isolated from a yeast appears to require three groups ionizing between pK_a 3 and 5 to hydrolyse *p*-nitrophenyl phosphate, all of which may be carboxyls.[424]

When stoichiometric amounts of 5′-nucleotidephosphodiesterase are incubated with tritium-labelled 3′,5′-cAMP covalently labelled enzyme is apparently produced;[425] this supports the earlier suggestion that a phosphoryl-enzyme intermediate is involved in the reactions catalysed by this enzyme.[426]

A complete pathway for the series of reactions catalysed by the glutamine-dependent carbamoyl phosphate synthetase from *E. coli* described by Powers and Meister assigns separate roles to the two bound ATP molecules[427] (their terminal phosphates are probably close to the bicarbonate binding site). The first reaction is the phosphorylation of bicarbonate to give ADP and carboxyphosphate, $^-O_2CPOP^{2-}$, which in the absence of glutamine can be trapped as the triester with diazomethane, or by reduction by borohydride to give formate.[428] When glutamine is present it is hydrolysed by the enzyme to provide ammonia, which reacts with the activated carbonate to give $H_2N{-}COO^-$ and inorganic phosphate. Finally, carbamate is phosphorylated by the second molecule of ATP, to give ADP and carbamoyl phosphate, NH_2COOPO^{2-}.[429]

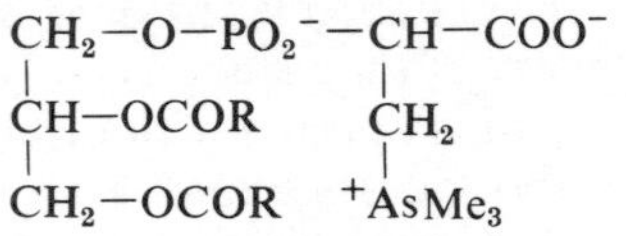

$$\begin{array}{ll} CH_2{-}O{-}PO_2^-{-} & CH{-}COO^- \\ | & | \\ CH{-}OCOR & CH_2 \\ | & | \\ CH_2{-}OCOR & ^+AsMe_3 \end{array}$$

(124)

Results of a pH-study of the transesterification of four dinucleotides catalysed by ribonuclease T_1 suggest that perhaps two carboxyl groups as well as two histidines are involved in binding and/or catalysis.[430] Conformational studies show that the C(2)-oxygen atom of pyrimidine nucleotides is not likely to be involved in the RNAse-catalysed hydrolysis of RNA;[431] the rates of H/D exchange of the six tyrosines of RNAse have been studied by a stopped-flow UV method.[432]

The arsenic-containing phospholipid (**124**), isolated from marine algae, may be a product of a detoxification mechanism used by organisms living in tropical waters where arsenate levels approach those of inorganic phosphate.[433]

Sulphur-containing Acids

It is still not entirely clear whether $S_N2(S)$ reactions at sulphonyl centres are concerted displacements or addition–eliminations involving pentacovalent intermediates. However, as the years go by, the continuing absence of evidence for the existence of addition intermediates makes it increasingly probable that these reactions are generally concerted processes; support for this presumption comes from a study[434] of the reversible reactions of substituted phenolates with substituted sultones (**125**).[435] The equilibrium is, not unexpectedly, favoured by

$$(\mathbf{125})\ \text{X-sultone}\ SO_2 + ArO^- \rightleftharpoons \text{X-C}_6\text{H}_3(O^-)CH_2SO_2OAr$$

electron-withdrawal in the sultone and by electron-donation in the phenolate; it is close to 1 : 1 for the reaction of *p*-nitrophenolate with the 5-nitro-sultone. The Brönsted coefficients, $\beta_{nuc} = -0.83$ (for reaction with the 5-nitro-sultone) and $\beta_{LG} = -0.85$ (for attack by PhO^-), when compared with equilibrium constant β-values, are consistent with an almost symmetrical transition state (with similar amounts of bonding to both nucleophile and leaving group) and not with the rate-determining formation or breakdown of an addition intermediate.

Thiolate and oximate show their usual enhanced nucleophilicity towards *p*-nitrophenyl toluene-*p*-sulphonate in aqueous ethanol, and reactions with weakly basic amines show general acid–base catalysis.[436] The alkaline hydrolysis of aryl toluene-*p*-sulphonates shows a negative salt effect with $NaClO_4$, similar to that observed with aryl benzoates.[437]

Trimethylammoniosulphate esters (**126**) are presumably intermediates in reactions of alkyl chlorosulphates catalysed by trimethylamine. The phenyl ester (**126**; R′ = Ph) has now been prepared,[438] and its properties underline the powerful leaving-group capabilities of sulphate derivatives. Not surprisingly, alkyl compounds (**126**; R′ = alkyl) are readily dealkylated, even by fluorosulphate anion, but even the phenyl ester is demethylated by trimethylamine, though secondary amines displace the amine.

$$Me_2NSO_2OPh + Me_4\overset{+}{N} \xleftarrow[R'=Ph]{Me_3N} Me_3\overset{+}{N}SO_2{-}OR' \xrightarrow[R'=Ph]{R_2NH} R_2NSO_2OPh$$

(**126**)

Extensive studies on the solvolysis of allyl arylsulphonates have been reported.[439–443] These reactions generally involve C—O cleavage, though weakly basic or sterically hindered amines may attack the sulphur centre of allyl 2,4-dinitrobenzenesulphonate.[444]

The reactions of nucleophiles with sulphonyl chlorides have been reviewed,[445] as have reactions of chloramine T and related reagents.[446]

Selectivity increases with reactivity in the reaction between substituted anilines and substituted benzenesulphonyl chlorides in a given solvent, but the slower reactions are more sensitive to solvent polarity.[447] An explanation has been presented in terms of variable transition states for a concerted displacement; electron-donating substituents on the sulphonyl halide give transition states with more S—Cl bond-breaking [more S_N1(S)-like] while more basic amines give transition states with more N—S bond-making (more addition–elimination-like).[447] The tosylation of anilines catalysed by pyridines may involve a different mechanism from the self-catalysed reaction.[448]

The solvolysis of sulphonyl halides in aqueous acetone is probably an S_N2(S) reaction.[449,450]

The activation of carboxylic acids by sulphonyl halides in the presence of tertiary amines involves acylammonium intermediates.[451]

The rate of hydrolysis of diethyl sulphite in $HClO_4$ has been measured at different temperatures.[40]

Ar–*S(=^{18}O)–OR + $(R'CO)_2O$ ⇌ $Ar-\overset{+}{S}(OAc)(OR)$ AcO^- ⇌ Ar–S(OAc)$_2$–OR

(127)

The rate of racemization of (−)methyl (−)*p*-toluene-sulphinate catalysed by trichloroacetic anhydride in benzene is twice that of ^{18}O-exchange, as expected for a reaction involving inversion at sulphur.[452] A mechanism involving the formation of N sulphurane (**127**) has been proposed, and the effects of substituents suggest that the initial acylation step is rate-determining.

Ph(Ar)S=NCl **(128)** $\xrightarrow{+HO^-}$ Ph(Ar)S$^-$(OH)(NCl) $\xrightarrow{-Cl^-}$ Ph(Ar)S(=O)=NH **(129)**

Alkaline hydrolysis of (−)-*N*-chloro-*S*-(*o*-methoxyphenyl)-*S*-phenylsulphimide (**128**) to the sulphoximide (**129**), on the other hand, proceeds with retention of configuration at sulphur; nonetheless, a mechanism involving attack by hydroxide at sulphur is preferred to several possible alternatives.[453] An *N*-acylsulphoximide is produced by the base-catalysed transannular migration of the acyl (or arene-sulphonyl) group of the phenothiazine derivative (**130**).[454]

N-COR phenothiazine S(=O)=NH **(130)** $\xrightarrow{HO^-}$ N-H phenothiazine S(=O)=NCOR

The hydrolysis of methyl toluene-*p*-sulphenate (**131**; Ar = *p*-$CH_3C_6H_4$) in solvents containing small amounts (> 1%) of water is an acid-catalysed reaction that produces diaryl thiolsulphonate (**132**) and disulphide, but at very low water concentrations the product is the thiolsulphinate ester (**133**).[455] The first step is presumably the hydrolysis of (**131**) to the sulphenic acid (**134**), which rapidly gives

$$4ArSOMe + 2H_2O \xrightarrow{H^+} ArSO_2SAr + ArSSAr + 4MeOH$$
(**131**) (**132**)

$$3ArSOMe + H_2O \xrightarrow{H^+} ArSO.OMe + ArSSAr + 2MeOH$$
(**133**)

diarylthiolsulphinate (**135**); in the absence of water this reacts with unchanged starting material to give sulphinate (**133**) and disulphide, thus accounting for the observed products at lowest water concentrations. If sufficient water is present the intermediate (**136**) is diverted as shown.

$$2\,ArS{-}OH\ (\mathbf{134}) \xrightarrow{-H_2O} ArS{-}SOAr\ (\mathbf{135}) \xrightarrow{(131),\ H^+} Ar\overset{+}{S}{-}SOAr\ (\text{with } ArS) + MeOH\ (\mathbf{136})$$

$$(\mathbf{136}) \xrightarrow{H_2O} ArSSAr + ArSO_2H \xrightarrow{(131)} (\mathbf{132})$$

$$(\mathbf{136}) \xrightarrow{MeOH} ArSO{-}OMe\ (\mathbf{133}) + ArSSAr$$

Sulphenic acids are generally very reactive species, and the isolation of the simplest, methanesulphenic acid, CH_3SOH (obtained by flash-vacuum pyrolysis of *tert*-butyl methyl sulphoxide at 400 °C), is a significant advance; the acid gives thiosulphinate (cf. **134**→**135**, above) in the condensed phase even at −196 °C but could be characterized by microwave spectroscopy.[456]

The reduction of biphenyl-2,2′-disulphonyl chloride by sulphite, which gives the cyclic thiolsulphonate (**137**) on acidic work-up, does in fact go through the expected disulphinic acid (**138**), but this cyclizes to the sulphinyl-sulphone (**139**), which is reduced by sulphite to the observed product.[457]

$$O_2S{-}S\ \text{(biphenyl-2,2'-diyl)}\ (\mathbf{137}) \qquad SO_2H,\ HSO_2\ \text{(biphenyl-2,2'-diyl)}\ (\mathbf{138}) \rightleftharpoons O_2S{-}SO\ \text{(biphenyl-2,2'-diyl)}\ (\mathbf{139})$$

Nucleophilic attack by cyanide and sulphite ions on (**137**) and similar compounds occurs at sulphide-sulphur and leads to ring-opening; the reactions are reversed in acid[458] but the similar reaction of the disulphone (**140**) is not reversed. The reactions of nucleophiles with the 5-membered cyclic disulphone (**141**) are not significantly faster than those with (**140**) or with diphenyl α-disulphone.[459] Apparently the ring-strain which accounts for the high reactivity of 5-membered cyclic esters with one ring sulphur atom is absent in (**141**).

Permutational isomerism in catechol-derived spirosulphuranes (*e.g.* **142**) has been studied by NMR spectroscopy.[460] Dialkoxysulphuranes also undergo ligand

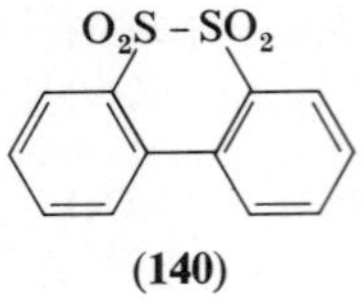

O_2S-SO_2 $O_2S—SO_2$

(140) 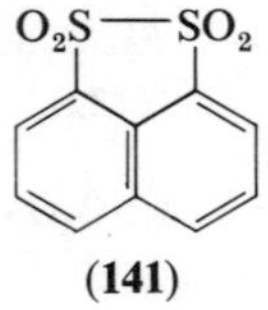 **(141)** 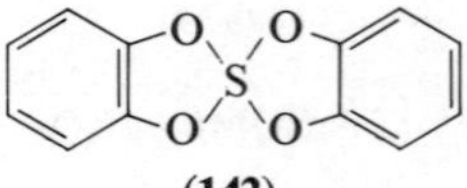 **(142)**

exchange, but in this case reaction is intermolecular.[461] The unsymmetrical sulphurane (**143**; $R^1 = R^2 = CF_3$) has two S—O bond-lengths differing by 0.24 Å, in the direction indicated; the estimated bond order for the bond to the fluoro-alcoholate is only 0.37.[462]

The barrier to inversion at the sulphur centre of an almost symmetrical sulphrane of this type has been estimated as >30 kcal mol^{-1} from the rates of interconversion of diastereomers (**143** R^1 = Me, R^2 = Et, and *vice-versa*).[463] Bicyclic sulphuranes (**144**) are attacked by nucleophiles at the carbonyl group, as shown.[464]

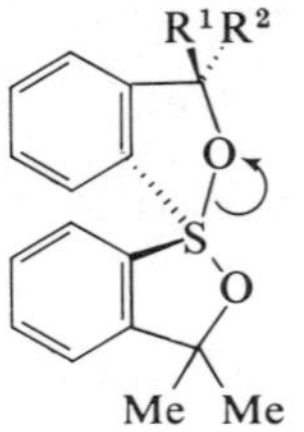

(143) 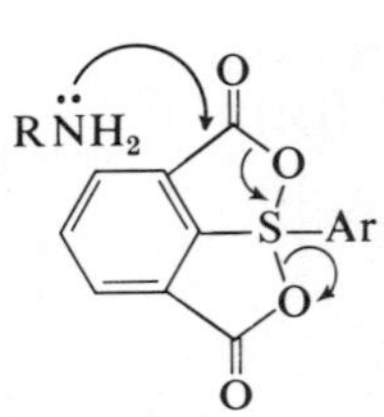 **(144)**

Other Acids

The rapid exchange of methoxyl groups between methyl borate and solvent methanol has been followed by NMR line-broadening measurements.[465] The pH-independent reaction shows a remarkably large solvent deuterium isotope effect, $k_{MeOH}/k_{MeOD} = 11.0$ at 25 °C, close to the theoretical maximum, and much higher than normally observed for solvolysis reactions. This, the large negative entropy of activation (about -40 eu), and an earlier observation of general base catalysis of a similar reaction are consistent with a general base catalysis mechanism (**145**); however, this does not explain the high isotope effect and may not account for the observed rate of the reaction.[465]

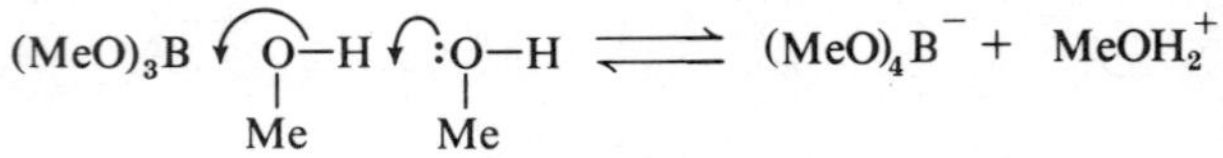

$$(MeO)_3B \leftarrow O(Me)-H \leftarrow :O(Me)-H \rightleftharpoons (MeO)_4B^- + MeOH_2^+$$

(145)

The hydrolysis of substituted 2-phenyl-1,3,2-benzodiazaboroles (**146**) (to the benzeneboronic acid and *o*-phenylenediamine) is catalysed by both general acids and bases.[466] In the acid region the mechanism appears to be *A*-2 for catalysis by $H_3O^+(k_D > k_H, \rho = -0.68)$, but may change to rate-determining breakdown of a tetrahedral addition intermediate for general acid catalysis (for catalysis by acetic acid, $k_H/k_D = 2.2$, $\rho = 0.81$). General base catalysis has both k_H/k_D and $\rho > 1$, and probably involves the kinetically equivalent general acid catalysed breakdown of the hydroxide adduct (**147**).[466]

The alkaline hydrolysis of nitrite esters has been compared with the same reaction of carboxylic esters:[467] the reaction is up to 10^5 times slower with nitrites (though they react the faster in acid), and no ^{18}O-exchange with solvent is observed;[468] the alkaline hydrolysis of nitrites is twice as sensitive to the leaving group, and steric effects are less important. It has been suggested that the addition–elimination route is less favourable for nitrite esters because the N is a softer centre (smaller difference in electronegativity across N=O than across C=O), and because of lone pair–lone pair repulsion between the nucleophile and the electrophilic nitrogen centre; a more or less concerted process is therefore preferred.

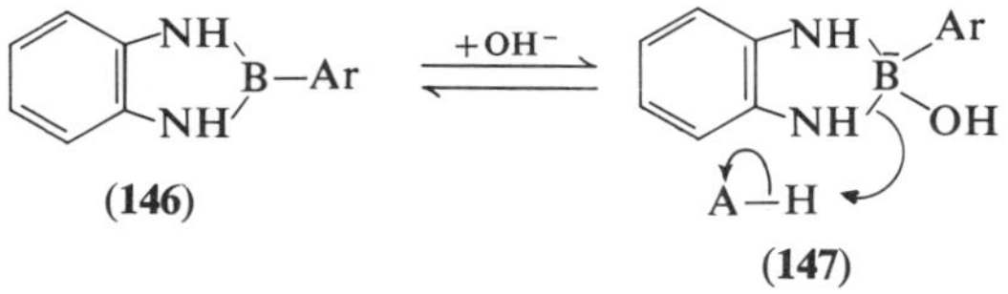

(146) ⇌ (+OH⁻) (147)

This is also the conclusion from a study of the aminolysis of 2-phenylethyl nitrite.[469] Lone pair–lone pair repulsions are ruled out for attack by nitrogen (**148**), and the aminolysis reactions are indeed relatively faster than attack by hydroxide, in fact faster than attack on esters. Another striking difference is that the aminolysis of nitrites is not general species catalysed (apart from a single case of general acid catalysis, by Me_3NH^+, of the reaction with trimethylamine), although substantial solvent deuterium isotope effects ($k_H/k_D \geqslant 2$) are found, suggesting that proton transfer is involved in the rate-determining step. The mechanism (**149**) which has been proposed to account for all this is stereoelectronically equivalent to an addition–elimination process, but with the adduct no longer stable enough to be an intermediate.[469]

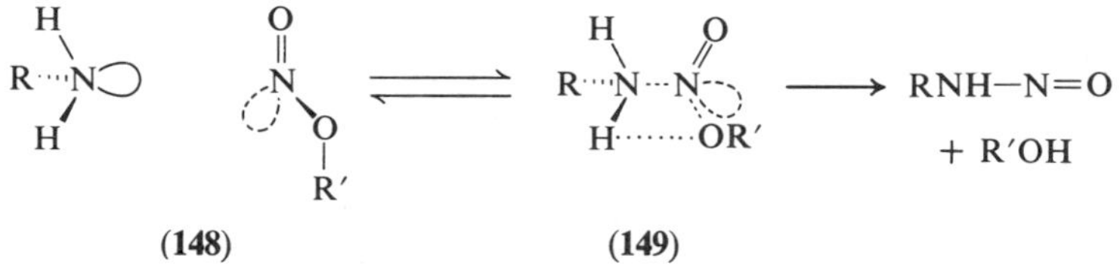

(148) ⇌ (149) ⟶ RNH—N=O + R′OH

The reaction of benzeneseleninic acid, $PhSeO_2H$, with thiols (RSH) in water gives PhSeSR and the disulphide, by way of an intermediate PhSe(O)SR and mechanisms involving nucleophilic attack by S on the Se centre of various protonated species.[470]

References

1 Guthrie, J. P., *J. Am. Chem. Soc.*, **100**, 5892 (1978).
2 See *Org. Reaction Mech.*, **1973**, 22; **1974**, 23; **1977**, 23.
3 Lien, M. H., Hopkinson, A. C., Peterson, M. R., Yates, K., and Csizmadia, I. G., *Prog. Theor. Org. Chem.*, **1977**, 162.
4 Boyd, D. B., *Proc. Natl. Acad. Sci. U.S.A.*, **47**, 5239 (1977).
5 Lehn, J.-M., and Wipff, G., *Helv. Chim. Acta*, **61**, 1274 (1978).
6 Deslongchamps, P., Gerval, P., Cheriyan, U. O., Guida, A., and Taillefer, R. J., *Nouveau J. de Chimie*, **2**, 631 (1978).
7 See *Org. Reaction Mech.*, **1977**, 23.
8 Halliday, J. D., and Symons, E. A., *Can. J. Chem.*, **56**, 1463 (1978).
9 See *Org. Reaction Mech.*, **1974**, 26.
10 See *Org. Reaction Mech.*, **1977**, 25.
11 Cox, M. M., and Jencks, W. P., *J. Am. Chem. Soc.*, **100**, 5956 (1978).
12 Ortiz, J. J., and Cordes, E. H., *J. Am. Chem. Soc.*, **100**, 7081 (1978).

[13] Gresser, M. J., and Jencks, W. P., *J. Am. Chem. Soc.*, **99**, 6963 (1977).
[14] Gresser, M. J., and Jencks, W. P., *J. Am. Chem. Soc.*, **99**, 6970 (1977).
[15] Castro, E. A., and Gil, F. J., *J. Am. Chem. Soc.*, **99**, 7611 (1977).
[16] Takashima, K., and Riveros, J. M., *J. Am. Chem. Soc.*, **100**, 6128 (1978).
[17] Pau, J. K., Kim, J. K., and Caserio, M. C., *J. Am. Chem. Soc.*, **100**, 3831 (1978).
[18] Bouab, O., Lamaty, G., and Moreau, C., *J. Chem. Soc., Chem. Commun.*, **1978**, 678.
[19] McClelland, R. A., and Ahmad, M., *J. Am. Chem. Soc.*, **100**, 7027 (1978).
[20] Chiang, Y., Kresge, A. J., and Young, C. I., *Finn. Chem. Lett.*, **1978**, 13.
[21] Eliason, R., and Kreevoy, M. M., *J. Am. Chem. Soc.*, **100**, 7037 (1978).
[22] Bouab, O., Moreau, C., and Zeh Ako, M., *Tetrahedron Lett.*, **1978**, 898.
[23] McClelland, R. A., *J. Am. Chem. Soc.*, **100**, 1844 (1978).
[24] Lee, Y.-N., and Schmir, G. L., *J. Am. Chem. Soc.*, **100**, 6700 (1978).
[25] McMurry, J. E., *Org. React.* (*N.Y.*), **24**, 187 (1976).
[26] Yates, K., and Modro, T. A., *Acc. Chem. Res.*, **11**, 190 (1978).
[27] Benkovic, S. J., *Acc. Chem. Res.*, **11**, 314 (1978).
[28] Litvinenko, L. M., and Oleinik, N. M., *Usp. Khim.*, **47**, 777 (1978); *Chem. Abs.*, **89**, 89771 (1978).
[29] See *Org. Reaction Mech.*, **1975**, 33; **1977**, 28.
[30] MacPhee, J. A., and Dubois, J.-E., *Tetrahedron Lett.*, **1978**, 2225.
[31] Grob, C. A., and Schlageter, M. G., *Helv. Chim. Acta*, **59**, 264 (1976).
[32] MacPhee, J. A., Panaye, A., and Dubois, J.-E., *Tetrahedron Lett.*, **1978**, 3293.
[33] Panaye, A., MacPhee, J. A., and Dubois, J.-E., *Tetrahedron Lett.*, **1978**, 3297.
[34] Charton, M., and Charton, B. I., *J. Org. Chem.*, **34**, 1161 (1978).
[35] Charton, M., *J. Org. Chem.*, **42**, 3531 (1977).
[36] Charton, M., *J. Org. Chem.*, **42**, 3535 (1977).
[37] Edward, J. T., Farrell, P. G., Job, J. L., and Poh, B.-L., *Can. J. Chem.*, **56**, 1122 (1978).
[38] Kirchnerova, J., Farrell, P. G., Edward, J. T., Halle, J.-C., and Schaal, R., *Can. J. Chem.*, **56**, 1130 (1978).
[39] Kada, R., Surá, J., Jurášek, A., Kováč, J., and Žvaková, A., *Coll. Czech. Chem. Commun.*, **43**, 621 (1978).
[40] Attiga, S. A., and Rochester, C. H., *J. Chem. Soc., Perkin Trans.* 2, **1978**, 466.
[41] Euranto, E. K., and Heinonen, U., *Finn. Chem. Lett.*, **1978**, 19.
[42] Taniguchi, Y., Sugiyama, N., and Suzuki, K., *J. Phys. Chem.*, **82**, 1231 (1978).
[43] Bhagade, S. S., and Nageshwar, G. D., *Chem. Petro-Chem. J.*, **8**, 15 (1977); *Chem. Abs.*, **89**, 89784 (1978).
[44] Helešic, L., Kefurt, K., and Jarý, J., *Collect. Czech. Chem. Commun.*, **43**, 1241 (1978).
[45] Bowden, K., and El-Kaissi, F. E., *J. Chem. Soc., Perkin Trans.* 2, **1977**, 1927.
[46] O'Leary, M. H., and Marliev, J. F., *J. Am. Chem. Soc.*, **100**, 2582 (1978).
[47] Bogatkov, S., *Org. React.* (*Tartu*), **14**, 153 (1977).
[48] Tsuno, Y., Tairaka, Y., Sawada, M., Fujii, T., and Yukawa, Y., *Bull. Chem. Soc. Jpn.*, **51**, 601 (1978).
[49] Mollin, J., and Baisa, A. A., *Collect. Czech. Chem. Commun.*, **43**, 304 (1978).
[50] Mitzner, R., Kramer, C.-R., Kempter, G., and Heilmann, D., *Z. Phys. Chem.* (*Leipzig*), **259**, 688 (1978).
[51] Mitzner, R., and Kramer, C.-R., *Z. Phys. Chem.* (*Leipzig*), **259**, 695 (1978).
[52] Bogatkov, S., and Cherkasova, E., *Org. React.* (*Tartu*), **14**, 165 (1977).
[53] Bruice, T. C., and Benkovic, S. J., "Bioorganic Mechanisms", W. A. Benjamin, Inc., New York, Vol. 1, 1966, pp. 134–136.
[54] Martinez, P., and Ramon, M. V., *An. Quim.*, **73**, 456 (1977); *Chem. Abs.*, **87**, 167086 (1977).
[55] Rodriguez, A. F., Gasper, I., and Martinez, P., *An. Quím.*, **74**, 193 (1978); *Chem Abs.*, **89**, 162699 (1978).
[56] Martinez, P., Celdran, M. S., and Ramon, M. V., *An. Quím.*, **74**, 212 (1978); *Chem Abs.*, **89**, 162700 (1978).
[57] Benko, J. ,and Holba, V., *Collect. Czech. Chem. Commun.*, **43**, 193 (1978).
[58] Holba, V., and Benko, J., *Z. Phys. Chem.* (*Leipzig*), **258**, 1088 (1977).
[59] Ševčík, S., Kubín, M., and Štamberg, J., *Collect. Czech. Chem. Commun.*, **34**, 1214 (1978).
[60] Istomin, B. I., Finkelstein, B. L., Sukhorukov, Yu. I., and Donskikh, V. I., *Org. React.* (*Tartu*), **14**, **476** (1977).
[61] Rao, G. V., Balakrishnan, M., Venkatasubramanian, N., Subramanian, P. V., and Subramanian, V., *J. Chem. Soc., Perkin Trans.* 2, **1978**, 8.

[62] Holba, V., Benko, J., and Okalova, K., *Collect. Czech. Chem. Commun.*, **43**, 1581 (1978).
[63] Nummert, V., and Piirsalu, M., *Org. React.* (*Tartu*), **14**, 263 (1977).
[64] Nummert, V., and Piirsalu, M., *Org. React.* (*Tartu*), **15**, 240 (1978).
[65] Khomutov, N. E., Eryshev, B. Ya., Yatsenko, B. F., and Smirnova, T. A., *Zh. Vses. Khim. O-va.*, **23**, 118 (1978); *Chem. Abs.*, **88**, 169285 (1978).
[66] Linton, M., *Chem. Abs.*, **88**, 104333 (1978).
[67] Isaacs, N. S., Javard, K., and Rannala, E., *Nature* (*London*), **268**, 372 (1977).
[68] Burden, A. G., Chapman, N. B., Duggira, H. F., and Shorter, J., *J. Chem. Soc.*, *Perkin Trans.* 2, **1978**, 296.
[69] Mišić-Vuković, M., Dimitrijević, D. M., Muskatirović, M. D., Radojković-Veličković, M., and Tadić, Z. D., *J. Chem. Soc.*, *Perkin Trans.* 2, **1978**, 34.
[70] Back, T. G., *Tetrahedron*, **33**, 3041 (1977).
[71] Douglas, K. T., Nakagawa, Y., and Kaiser, E. T., *J. Org. Chem.*, **42**, 3677 (1977).
[72] Akiyama, M., Hara, Y., and Tanabe, M., *J. Chem. Soc.*, *Perkin Trans.*, 2, **1978**, 288.
[73] Remuzon, P., and Wakselman, M., *Tetrahedron*, **33**, 3097 (1977).
[74] Höfle, G., Steglich, W., and Vorbrüggen, H., *Angew. Chem.*, *Int. Edn.*, **17**, 569 (1978).
[75] Klopman, G., and Evans, R. C., *Tetrahedron*, **34**, 269 (1978).
[76] Davies, D. M., and Jones, P., *J. Org. Chem.*, **43**, 769 (1978).
[77] Guanti, G., dell'Erba, C., Pero, F., and Cevasco, G., *J. Chem. Soc.*, *Perkin Trans.* 2, **1978**, 422.
[78] Guanti, G., dell'Erba, C., Cevasco, G., and Narisano, E., *J. Chem. Soc.*, *Chem. Commun.*, **1978**, 613.
[79] Mukerjee, A. K., and Singh, A. K., *Tetrahedron*, **34**, 1731 (1978).
[80] "Recent Advances in the Chemistry of β-Lactam Antibiotics", Chem. Soc. Special Publication No. 28, 1977.
[81] Butler, A. R., Freeman, K. A., and Wright, D. E., Ref. 80, p. 299.
[82] Gensmantel, N. P., and Page, M. I., *J. Chem. Soc.*, *Chem. Commun.*, **1978**, 374.
[83] Bundgaard, H., *Acta Pharm. Suec.*, **14**, 267 (1977); *Chem Abs.*, **88**, 21630 (1978).
[84] Cipiani, A., Linda, P., and Savelli, G., *J. Chem. Soc.*, *Chem. Commun.*, **1977**, 857.
[85] Karzijn, W., and Engberts, J. B. F. N., *Tetrahedron Lett.*, **1978**, 1787.
[86] Mittal, S., Gupta, K. S., and Gupta, Y. K., *Indian J. Chem.*, *Sect. A*, **15A**, 827 (1977). *Chem. Abs.*, **88**, 61878 (1978).
[87] Elsemongy, M. M., and Elnader, H. M. A., *J. Indian Chem. Soc.*, **54**, 1055 (1977); *Chem. Abs.*, **89**, 59428 (1978).
[88] Elsemongy, M. M., and Elnader, H. M. A., *J. Indian Chem. Soc.*, **55**, 60 (1978); *Chem. Abs.* **89**, 146247 (1978).
[89] Khan, M. M., and Khan, A. A., *J. Chem. Soc.*, *Perkin Trans.* 2, **1978**, 1176.,
[90] Laurent, E., Lee, S. K., and Pellissier, N., *J. Chem. Res.*, (*S*) **1978**, 440; (*M*) 5201.
[91] Machida, M., Machida, M. I., and Kanaoka, Y., *Chem. Pharm. Bull.*, **25**, 2739 (1977).
[92] Fischer, G., Küllertz, G., and Barth, A., *Z. Chem.*, **17**, 380 (1977).
[93] Fischer, G., Küllertz, G., and Barth, A., *Tetrahedron*, **34**, 2123 (1978).
[94] Komiyama, M., and Bender, M. L., *Bioorg. Chem.*, **7**, 133 (1978).
[95] Komiyama, M., and Bender, M. L., *J. Am. Chem. Soc.*, **100**, 5977 (1978).
[96] Henderson, J. W., and Haake, P., *J. Org. Chem.*, **42**, 3989 (1977).
[97] Broxton, T. J., Deady, L. W., and Pang, Y.-T., *J. Org. Chem.*, **43**, 2023 (1978).
[98] Mane, B. S., and Jagdale, M. H., *Acta Cienc. Indica*, **3**, 1 (1977); *Chem Abs.*, **87**, 200417 (1977).
[99] Mane, B. S., and Jagdale, M. H., *Indian J. Chem.*, *Sect. A*, **15A**, 1086 (1977); *Chem. Abs.*, **88**, 89532 (1978).
[100] Mane, B, S., and Jagdale, M. H., *J. Indian Chem. Soc.*, **54**, 615 (1977); *Chem Abs.*, **88**, 120272 (1978).
[101] Kucerova, T., and Mollin, J., *Collect. Czech. Chem. Commun.*, **43**, 1571 (1978).
[102] Ward, I. E., *Diss. Abstr. Int. B*, **38**, 2190 (1977); *Chem Abs.*, **88**, 49927 (1978).
[103] Berndt, D. C., and Ward, I. E., *J. Org. Chem.*, **43**, 13 (1978).
[104] Jagdale, M. H., and Nimbalkar, A. Y., *Sci. J. Shivaji Univ.*, **15**, 41 (1975); *Chem Abs.*, **88**, 5979 (1978).
[105] Beck, W. H., Liler, M., and Morris, D. G., *J. Chem. Soc.*, *Perkin Trans.* 2, **1977**, 1876.
[106] Beck, W. H., Liler, M., and Morris, D. G., *J. Chem. Soc.*, *Perkin Trans.* 2, **1978**, 1173.
[107] Edward, J. T., and Wong, S. C., *J. Am. Chem. Soc.*, **99**, 7224 (1977).
[108] Edward, J. T., Wong, S. C., and Welch, G., *Can. J. Chem.*, **56**, 931 (1978).

[109] Edward, J. T., Welch, G., and Wong, S. C., *Can. J. Chem.*, **56**, 935 (1978).
[110] Edward, J. T., Derdalt, G. D., and Wong, S. C., *J. Am. Chem. Soc.*, **100**, 7023 (1978).
[111] Venkatasubban, K. S., Davis, K. R., and Hogg, J. L., *J. Am. Chem. Soc.*, **100**, 6125 (1978).
[112] Boiko, T. S., and Yasnikov, A. A., *Ukr. Khim. Zh.* (*Russ. Ed.*), **43**, 1081 (1977); *Chem Abs.*, **88**, 21645 (1978).
[113] Yavari, I., and Roberts, J. D., *Tetrahedron Lett.*, **1978**, 2491.
[114] See *Org. Reaction Mech.*, **1977**, 37.
[115] Mollin, J., and Bouchalová, P., *Collect. Czech. Chem. Commun.*, **43**, 2283 (1978).
[116] Broxton, T. J., Deady, L. W., and Rowe, J. E., *Aust. J. Chem.*, **31**, 1731 (1978).
[117] Suvorov, B. V., Kagarlitskii, A. D., and Lebedeva, O. B., *Izv. Akad. Nauk SSR, Ser. Khim.*, **27**, 78 (1977); *Chem. Abs.*, **88**, 73826 (1978).
[118] Isukul, E. A., and Tillett, J. G., *J. Chem. Soc., Perkin Trans.* 2, **1978**, 908.
[119] Bergon, M., and Calmon, J.-P., *J. Chem. Soc., Perkin Trans.* 2, **1978**, 493.
[120] Blagoeva, I., and Pojarlieff, I., *Dokl, Bolg. Akad. Nauk.*, **30**, 1043 (1977); *Chem. Abs.*, **88**, 21623 (1978).
[121] Blagoeva, I. B., Pojarlieff, I. G., and Dimitrov, V. S., *J. Chem. Soc., Perkin Trans.* 2, **1978**, 887.
[122] Slae, S., and Shapiro, R. H., *J. Org. Chem.*, **43**, 1721 (1978).
[123] de Savignac, A., Kabbage, T., Dupin, P., and Calmon, M., *J. Heterocycl. Chem.*, **15**, 897 (1978).
[124] Kolomiets, B. S., German, V. K., and Suchkov, V. V., *Zh. Prikl. Khim.* (*Leningrad*), **51**, 1141 (1978); *Chem. Abs.*, **89**, 41914 (1978).
[125] Čegan, A., Šlosar, J., and Večeřa, M., *Collect. Czech. Chem. Commun.*, **43**, 134 (1978).
[126] "The Chemistry of Cyanates and their Thio Derivatives", S. Patai, Ed., Wiley, Chichester, 1977.
[127] Reichen, W., *Chem. Rev.*, **78**, 569 (1978).
[128] Ho, T.-L., and Olah, G. A., *Proc. Natl. Acad. Sci. U.S.A.*, **75**, 4 (1978).
[129] Niehaus, W. G., *Bioorg. Chem.*, **7**, 77 (1977).
[130] Taylor, R., *J. Chem. Res.*, (*S*) **1978**, 269.
[131] Thorne, M. P., *J. Chem. Res.*, (*S*) **1978**, 222. See also *idem*, *J. Chem. Soc., Perkin Trans.* 2, **1978**, 716; and Amin, H. B., and Taylor, R., *ibid.*, **1978**, 1090.
[132] Böhmer, V., Wörsdörfer, K., and Becher, U., *Tetrahedron*, **34**, 2737 (1978).
[133] Oleinik, N. M., Kurchenko, L. P., Sadovskii, Yu. S., and Litvinenko, L. M., *Zh. Org. Khim.*, **13**, 2118 (1977); *Chem Abs.*, **88**, 50090 (1978).
[134] See *Org. Reaction Mech.*, **1971**, 438.
[135] Hogg, J. L., Morris, R., and Durrant, N. A., *J. Am. Chem. Soc.*, **100**, 1590 (1978).
[136] Tonellato, U., *J. Chem. Soc., Perkin Trans.* 2, **1978**, 307.
[137] See *Org. Reaction Mech.*, **1973**, 24.
[138] Itoho, K., *Daiichi Yakka Daigaku Kenkyu Nempo*, **6**, 49 (1975); *Chem. Abs.*, **88**, 104334 (1978).
[139] Itoho, K., *Daiichi Yakka Daigaku Kenkyu Nempo*, **6**, 63 (1975); *Chem. Abs.*, **88**, 104335 (1978).
[140] Shinkai, S., Nakashima, N., and Kunitake, T., *J. Am. Chem. Soc.*, **100**, 5887 (1978). See also *Org. Reaction Mech.*, **1976**, 52.
[140a] Marton, A. F., Komives, T., and Dutka, F., *Radiochem. Radioanal. Lett.*, **23**, 1 (1978); *Chem. Abs.*, **88**, 151695 (1978).
[141] Bogatkov, S. V., Golovina, Z. P., Flid, V. R., and Cherkasova, E. M., *Zh. Org. Khim.*, **13**, 2271 (1977); *Chem. Abs.*, **88**, 61750 (1978).
[142] Bogatkov, S. V., Golovina, Z. P., and Cherkasova, E. M., *Zh. Org. Khim.*, **13**, 2262 (1977); *Chem. Abs.*, **88**, 88885 (1978).
[143] Litvinenko, L. M., Semenyuk, G. V., and Zhil'tsov, N. P., *Dokl. Akad. Nauk SSSR*, **239**, 129 (1978) [*Phys. Chem.*]; *Chem. Abs.*, **88**, 189555 (1978).
[144] Semenyuk, G. V., Litvinenko, L. M., and Nikitko, N. I., *Dopov. Akad. Nauk Ukr. RSR, Ser. B: Geol. Khim. Biol. Nauki*, **1978**, 342; *Chem. Abs.*, **89**, 23472 (1978).
[145] Effenberger, F., König, G., and Kleuk, H., *Angrew. Chem., Int. Ed.*, **17**, 695 (1978).
[146] Takahashi, K., Takeda, K., and Mitsuhashi, K., *J. Heterocycl. Chem.*, **15**, 893 (1978).
[147] Makitra, R., Marshalok, I., Pirig, Ya., and Sendega, R., *Org. React.* (*Tartu*). **15**, 25 (1978).
[148] Singh, H. N., Singh, S., and Mahalwar, D. S., *Acta Univ. Acad. Sci. Hung.*, **95**, 417 (1977); *Chem. Abs.*, **89**, 89867 (1978).

[149] Dmowski, L., and Kolinski, R. A., *Pol. J. Chem.*, **52**, 547 (1978); *Chem. Abs.*, **89**, 41906 (1978).
[150] Deady, L. W., and Stillman, D. C., *Aust. J. Chem.*, **31**, 1725 (1978).
[151] Kalnins, K., Svetlichnyi, V. M., Antonov, N. G., and Koton, M. M., *Dokl. Akad. Nauk SSSR*, **241**, 849 (1978) [*Chem.*]; *Chem. Abs.*, **89**, 162713 (1978).
[152] Solomin, V. A., Kardash, I. E., Snagovskii, Yu. S., Messerle, P. E., Zhubanov, B. A., and Pravednikov, A. N., *Dokl. Akad. Nauk SSSR*, **236**, 139 (1977); *Chem. Abs.*, **87**, 183680 (1977).
[153] Dadali, V. A., Tsupilo, I. A., and Litvinenko. L. M., *Zh. Org. Khim.*, **13**, 2111 (1977); *Chem. Abs.*, **88**, 36863 (1978).
[154] Kato, S., Sugino, K., Mizuta, M., and Katada, T., *Angew. Chem., Int. Edn.*, **17**, 675 (1978).
[155] Stupinkova, T. V., Zherebchenko, V. I., Serdyuk, A. I., and Sheinkman, A. K., *Dopov. Akad. Nauk Ukr. RSR. Ser. B: Geol. Khim. Biol. Nauki*, **1977**, 920; *Chem. Abs.*, **88**, 36853 (1978).
[156] Dadali, V. A., Zubareva, T. M., Litvinenko, L. M., and Simanenko, Yu. S., *Dokl. Akad. Nauk SSSR*, **239**, 868 (1978) [*Phys. Chem.*]; *Chem. Abs.*, **89**, 5632 (1978).
[157] Savelova, V. A., Skirpka, A. V., and Litvinenko, L. M., *Zh. Org. Khim.*, **14**, 1260 (1978); *Chem. Abs.*, **89**, 89864 (1978).
[158] Fisichella, S., and Alberghina, G., *J. Chem. Soc., Perkin Trans. 2*, **1978**, 567.
[159] Alberghina, G., and Fisichella, S., *J. Chem. Soc., Perkin Trans. 2*, **1978**, 81.
[160] Shpan'ko, I. V., Litvinenko, L. M., and Goncharov, A. N., *Zh. Org. Khim.*, **14**, 574 (1978); *Chem. Abs.*, **89**, 5626 (1978).
[161] Ivanova, N. S., Vorob'ev, N. K., and Zvereva, V. K., *Izv. Vyssh. Uchebn. Zaved., Khim. Khim. Tekhnol.*, **21**, 529 (1978); *Chem. Abs.*, **89**, 107264 (1978).
[162] Ivanova, N. S., Vorob'ev, N. K., and Frolova, I. N., *Izv. Vyssh. Uchebn. Zaved., Khim. Khim. Tekhnol.*, **21**. 371 (1978); *Chem. Abs.*, **89**, 107283 (1978).
[163] Litvinenko, L. M., Galusko, L. Ya., and Savchenko, A. S., *Zh. Org. Khim.*, **14**, 1036 (1978); *Chem. Abs.*, **89**, 107255 (1978).
[164] Litvinenko, L. M., Savchenko, A. S., and Galusko, L. Ya., *Zh. Org. Khim.*, **13**, 1657 (1977); *Chem Abs.*, **87**, 183799 (1977).
[165] Kostin, A. I., and Savchenko, A. S., *Deposited Doc.* **1976**, *VINITI* 490; *Chem. Abs.*, **88**, 73816 (1978).
[166] Chimishkyan, A. L., Dragolov, V. V., Kelekhsaeva, E. A., and Dvornikova, E. E., *Deposited Doc.* **1976**, *VINTI* 3300; *Chem. Abs.*, **89**, 128751 (1978).
[167] Ivanov, M. G., Subbotin, O. I., Golov, V. G., Vodop'yanov, V. G., and Gerega, V. F., *Kinet. Katal.*, **19**, 342 (1978); *Chem. Abs.*, **89**, 5642 (1978).
[168] Tanida, H., *Yuki Gosei Kagaku Kyokaishi*, **35**, 439 (1977); *Chem. Abs.*, **87**, 166928 (1977).
[169] See *Org. Reaction Mech.*, **1970**, 467.
[170] See *Org. Reaction Mech.*, **1970**, 468; **1971**, 434; **1972**, 434.
[171] Dafforn, G. A., and Koshland, D. E., *J. Am. Chem. Soc.*, **99**, 7246 (1977).
[172] Gandour, R. D., Stella, V. J., Coyne, M., Schowen, R. L., and Icaza, E. A., *J. Org. Chem.*, **43**, 1705 (1978).
[173] See *Org. Reaction Mech.*, **1976**, 42—43.
[174] Kluger, R., and Lam., C.-H., *J. Am. Chem. Soc.*, **100**, 2191 (1978).
[175] Nechaev, P. P., Mukhina, O. A., Kosobutskii, V. A., Belyakov, V. K., Vygodskii, Ya. S., Moiseev, Yu. V., and Zaikov, G. E., *Izv. Akad, Nauk SSSR, Ser. Khim.*, **1977**, 1750; *Chem. Abs.*, **87**, 200590 (1977).
[176] Lavrov, S. V., Kardash, I. E., and Pravednikov, A. N., *Dokl. Akad. Nauk SSSR*, **239**, 126 (1978) [*Phys. Chem.*]; *Chem. Abs.*, **88**, 189554 (1978).
[177] Fife, T. H., and Squillacote, V. L., *J. Am. Chem. Soc.*, **100**, 4787 (1978).
[178] Cremin, D. J., and Hegarty, A. F., *J. Chem. Soc., Perkin Trans. 2*, **1978**, 208.
[179] Capon, B., Ghosh, B. Ch., and Raftery, W. V., *Finn. Chem. Lett.*, **1978**, 9.
[180] Pascal, R., Taillades, J., and Commeyras, A., *Bull. Soc. Chim. Fr. II*, **1978**, 177.
[181] Pascal, R., Taillades, J., and Commeyras, A., *Tetrahedron*, **34**, 2275 (1978).
[182] Boudreau, J. A., and Williams, A., *J. Chem. Soc., Perkin Trans. 2*, **1977**, 2028.
[183] Petz, D., Schneider, F., and Loeffler, H. G., *Z. Naturforsch., C: Biosci.*, **33C**, 151 (1978).
[184] Komives, T., Marton, A. F., and Dutka, F., *React. Kinet. Catal. Lett.*, **4**, 43 (1976); *Chem. Abs.*, **88**, 21591 (1978).
[185] Guanti, G., Thea, S., Dell'Erba, C., and Pero, F., *J. Chem. Soc., Chem. Commun.*, **1978**, 886.

[186] Mares, F., and Sheehan, D., *Ind. Eng. Chem. Process Des. Dev.*, **17**, 9 (1978); *Chem. Abs.*, **88**, 49938 (1978).
[187] Camilleri, P., Cassola, A., Kirby. A. J., and Mujahid, T. G., *J. Chem. Soc., Chem. Commun.*, **1978**, 807.
[188] Guggisberg, A., Dabrowski, B., Heidelberger, C., Hesse, M., and Schmid, H., *Helv. Chim. Acta*, **61**, 1039 (1978).
[189] Fife, T. H., Bambery, R. J., and DeMark, B. R., *J. Am. Chem. Soc.*, **100**, 5500 (1978).
[190] Boudreau, J. A., Farrar, C. R., and Williams, A., *J. Chem. Soc., Perkin Trans. 2*, **1978**, 46.
[191] Utaka, M., Takatsu, M., and Takeda, A., *Bull. Chem. Soc. Jpn.*, **50**, 376 (1977).
[192] Sisido, M., Yoshikawa, E., Imnaishi, Y., and Higashimura, T., *Bull. Soc. Chim. Jpn.*, **51**, 1464 (1978).
[193] Kirby, A. J., and Mujahid., T. G., *Tetrahedron Lett.*, **1978**, 4081.
[194] See *Org. Reaction Mech.*, **1976**, 47.
[195] Walder, J. A., Johnson, R. S., and Klotz, I. M., *J. Am. Chem. Soc.*, **100**, 5156 (1978).
[196] Johnson, R. S., Walder, J. A., and Klotz., I. M., *J. Am. Chem. Soc.*, **100**, 5159 (1978).
[197] Morris, J. J., and Page, M. I., *J. Chem. Soc., Chem. Commun.*, **1978**, 591.
[198] Komives, T., Marton, A. F., Holly, S., and Dutka, F., *React. Kinet. Catal. Lett.*, **8**, 19 (1978); *Chem. Abs.*, **89**, 23571 (1978).
[199] Kirby, A. J., and Koh, L. L., *Tetrahedron Lett.*, **1978**, 4079.
[200] Boehmer, V., and Woersdoerfer, K., *Z. Naturforsch., B: Anorg. Chem., Org. Chem.*, **33B**, 439 (1978).
[201] See *Org. Reaction Mech.*, **1976**, 49.
[202] Komives, T., Marton, A., Dutka, F., Low, M., and Kisfaludy, L., *Z. Naturforsch., B: Anorg. Chem. Org. Chem.*, **32B**, 1359 (1977); *Chem. Abs.*, **88**, 49921 (1978).
[203] Bunton, C. A., *Tech. Chem. (New York)*, **10**, 731 (1976); *Chem. Abs.*, **88**, 36708 (1978).
[204] Lehn, J.-M., *Pure Appl. Chem.*, **50**, 871 (1978).
[205] Cordes, E. H., *Pure Appl. Chem.*, **50**, 617 (1978).
[206] Rujimethabhas, M., and Wilaivat, P., *J. Chem. Ed.*, **55**, 342 (1978).
[207] Menger, F. M., and Jerkunica, J. M., *J. Am. Chem. Soc.*, **100**, 688 (1978).
[208] Menger, F. M., Jerkunica, J. M., and Johnson, J. C., *J. Am. Chem. Soc.*, **100**, 4676 (1978).
[209] Menger, F. M., and Saito, G., *J. Am. Chem. Soc.*, **100**, 4376 (1978).
[210] See *Org. Reaction Mech.*, **1977**, 47.
[211] Piszkiewicz, D., *J. Am. Chem. Soc.*, **99**, 7695 (1977).
[212] Manecke, G., and Vogt, H.-G., *Pure Appl. Chem.*, **50**, 655 (1978).
[213] Overberger, C. G., Guterl, A. C., Kawakami, Y., Mathias, L. J., Meenakshi, A., and Tomono, T., *Pure Appl. Chem.*, **50**, 309 (1978).
[214] Manecke, G., and Storck, W., *Angew. Chem. Int. Edn.*, **17**, 657 (1978).
[215] Murakami, Y., Aoyama, Y., and Kilda, M., *J. Chem. Soc., Perkin Trans. 2*, **1977**, 1947.
[216] Funasaki, M., *J. Colloid Interface Sci.*, **64**, 473 (1978); *Chem. Abs.*, **89**, 59417 (1978).
[217] Gani, V., and Viout, P., *Tetrahedron*, **34**, 1337 (1978).
[218] Broxton, T. J., Deady, L. W., and Duddy, N. W., *Aust. J. Chem.*, **31**, 1525 (1978).
[219] Berndt, D. C., and Sendelbach, L. E., *J. Org. Chem.*, **42**, 3305 (1977).
[220] Cuccovia, I. M., Schröter, E. H., Monteiro, P. M., and Chaimovich, H., *J. Org. Chem.*, **43**, 2248 (1978).
[221] Ueoka, R., Shimamoto, K., Maezato, Y., and Ohkubo, K., *J. Org. Chem.*, **43**, 1815 (1978).
[222] Cuccovia, I. M., Schröter, E. H., de Baptista, R. C., and Chaimovich, H., *J. Org. Chem.*, **42**, 3400 (1977).
[223] Cairns-Smith, A. G., and Rasool, S., *J. Chem. Soc., Perkin Trans. 2*, **1978**, 1007.
[224] Kondo, H., Fujiki, K., and Sunamoto, J., *J. Org. Chem.*, **43**, 3584 (1978).
[225] Lomax, T. D., and O'Connor, C. J., *J. Am. Chem. Soc.*, **100**, 5910 (1978).
[226] Escabi-Perez, J. R., and Fendler, J. H., *J. Am. Chem. Soc.*, **100**, 2235 (1978).
[227] Ueko, R., and Ohkubo, K., *Tetrahedron Lett.*, **1978**, 4131.
[228] Ihara, Y., *J. Chem. Soc., Chem. Commun.*, **1978**, 984.
[229] Shiffman, R., Rav-Acha, Ch., Chevion, M., Katzhendler, J., and Sarel. S., *J. Org. Chem.*, **42**, 3279 (1977).
[230] Rav-Acha, C., Chevion, M., Katzhendler, J., and Sarel, S., *J. Org. Chem.*, **43**, 591 (1978).
[231] Wexler, D., Pillersdorf, A., Shiffman, R., Katzhendler, J., and Sarel, S., *J. Chem. Soc., Perkin Trans. 2*, **1978**, 479.
[232] Anoardi, L., de Buzzacarini, F., Fornasier, R., and Tonellato, U., *Tetrahedron Lett.*, **1978**, 3945.

[233] Moss, R. A., Bizzigotti, T. O., Lukas, T. J., and Sanders, W. J., *Tetrahedron Lett.*, **1978**, 3661; Moss, R. A., Lukas, T. J., and Nahas, R. C., *J. Am. Chem. Soc.*, **100**, 5920 (1978); *Tetrahedron Lett.*, **1978**, 507.

[234] Melhado, L. L., and Gutsche, C. D., *J. Am. Chem. Soc.*, **100**, 1850 (1978).

[235] Armstrong, D. W., Seguin, R., McNeal, C. J., MacFarlane, R. D., and Fendler, J. H., *J. Am. Chem. Soc.*, **100**, 4665 (1978).

[236] Kunitake, T., and Sakamoto, T., *J. Am. Chem. Soc.*, **100**, 4615 (1978).

[237] See *Org. Reaction Mech.*, **1976**, 51–52; **1977**, 48.

[238] Timko, J. M., Helgeson, R. C., and Cram., D. J., *J. Am. Chem. Soc.*, **100**, 2828 (1978).

[239] Kyba, E. P., Timko, J. M., Kaplan, L. J., de Jong, F., Gokel, G. W., and Cram, D. J., *J. Am. Chem. Soc.*, **100**, 4555 (1978).

[240] de Sousa, L. R., Sogah, G. D. Y., Hoffman, D. H., and Cram, D. J., *J. Am. Chem. Soc.*, **100**, 4569 (1978).

[241] Barrett, A. G. M., and Lana, J. C. A., *J. Chem. Soc., Chem. Commun.*, **1978**, 471.

[242] Matsui, T., and Koga, K., *Tetrahedron Lett.*, **1978**, 1115.

[243] Lehn, J.-M., and Sirlin, C., *J. Chem. Soc., Chem. Commun.*, **1978**, 949.

[244] See *Org. Reaction Mech.*, **1977**, 48.

[245] Gandour, R. D., Walker, D. A., Nayak, A., and Newkome, G. R., *J. Am. Chem. Soc.*, **100**, 3608 (1978).

[246] Tabushi, I., Kimura, Y., and Yamamura, K., *J. Am. Chem. Soc.*, **100**, 1304 (1978).

[247] See *Org. Reaction Mech.*, **1977**, 51.

[248] Murakami, Y., Aoyama, Y., Kida, M., and Nakano, A., *Bull. Chem. Soc. Jpn.*, **50**, 3365 (1977).

[249] Murakami, Y., Nakano, A., Matsumoto, K., and Iwamoto, K., *Bull. Chem. Soc. Jpn.*, **51**, 2690 (1978).

[250] Overberger, C. G., and Dixon, K. W., *J. Polymer Sci., Polym. Chem. Ed.*, **15**, 1863 (1977); *Chem. Abs.*, **87**, 167417 (1977).

[251] See *Org. Reaction Mech.*, **1976**, 66.

[252] Shah, S. C., and Smid, J., *J. Am. Chem. Soc.*, **100**, 1426 (1978).

[253] Tabushi, I., Kiyosuke, Y., Sugimoto, T., and Yamamura, K., *J. Am. Chem. Soc.*, **100**, 916 (1978).

[254] Bergeron, R. J., Channing, M. A., and McGovern, K. A., *J. Am. Chem. Soc.*, **100**, 2878 (1978).

[255] Lindner, K., and Saenger, W., *Angew. Chem., Int. Edn.*, **17**, 694 (1978).

[256] See *Org. Reaction Mech.*, **1976**, 54.

[257] Komiyama, M., and Bender, M. L., *J. Am. Chem. Soc.*, **100**, 4576 (1978).

[258] Komiyama, M., and Bender, M. L., *J. Am. Chem. Soc.*, **99**, 8021 (1977).

[259] Breslow, R., Doherty, J. B., Guillot, G., and Lipsey, C., *J. Am. Chem. Soc.*, **100**, 3224 (1978).

[260] Boger, J., and Knowles, J. R., in "Molecular Interactions and Activity in Proteins", CIBA Foundation Symposium No. 60, Excerpta Medica, Amsterdam, 1978, p. 225.

[261] Matsui, Y., and Okimoto, A., *Bull. Chem. Soc. Jpn.*, **51**, 3030 (1978).

[262] Tabushi, I., Shimizu, N., Sugimoto, T., Shiozuka, M., and Yamamura, K., *J. Am. Chem. Soc.*, **99**, 7100 (1977).

[263] Hay, R. W., Bennett, R., and Piplani, D., *J. Chem. Soc., Dalton Trans.*, **1978**, 1046.

[264] Hay, R. W., and Clark, C. R., *J. Chem. Soc., Dalton Trans.*, **1977**, 1866.

[265] Hay, R. W., and Clark, C. R., *J. Chem. Soc., Dalton Trans.*, **1977**, 1993.

[266] Gensmantel, N. P., Gowling, E. W., and Page, M. I., *J. Chem. Soc., Perkin Trans.* 2, **1978**, 335.

[267] Takeishi, M., Fujii, S., Niino, S., and Hayama, S., *Tetrahedron Lett.*, **1978**, 3361.

[268] Peeters, O. M., Blaton, N. B., and De Ranter, C. J., *J. Chem. Soc., Perkin Trans.*, **1978**, 23.

[269] Meyer, G., and Viout, P., *Tetrahedron*, **34**, 305 (1978).

[270] Loupy, A., Meyer, G., and Tchoubar, B., *Tetrahedron*, **34**, 1333 (1978).

[271] Corset, J., Froment, F., Deschamps, E., Loupy, A., Nee, G., Le Ny, G., and Tchoubar, B., *Tetrahedron*, **34**, 2645 (1978).

[272] Murakami, Y., Aoyama, Y., Kida, M., and Kikuchi, J., *J. Chem. Soc., Chem. Commun.*, **1978**, 494.

[273] Vejrosta, J., Zelena, E., and Malek, J., *Collect. Czech. Chem. Commun.*, **43**, 424 (1978).

[274] Eryshev, B. Ya., Yatsenko, B. P., Smirnova, T. A., and Khomutov, N. E., *Zh. Prikl. Khim. (Leningrad)*, **51**, 1406 (1978); *Chem. Abs.*, **89**, 107316 (1978).

[275] Zil'berman, E. N., Trachenko, V. I., Danov, S. M., and Shipunova, N. R., *Izv. Vyssh. Uchebn. Zaved., Khim. Khim. Tekhnol.*, **20**, 1141 (1977); *Chem. Abs.*, **87**, 183830 (1977).
[276] Elsemongy, M. M., and Onsager, D. T., *Acta Chem. Scand., Ser. B*, **32**, 167 (1978).
[277] Cleland, W. W., *Adv. Enzymol.*, **45**, 273 (1977).
[278] Klinman, J. P., *Adv. Enzymol.*, **46**, 416 (1977).
[279] Douzou, P., *Adv. Enzymol.*, **45**, 157 (1977).
[280] Huber, R., and Bode, W., *Acc. Chem. Res.*, **11**, 114 (1978).
[281] Schneider, F., *Angew. Chem., Int. Edn.*, **17**, 583 (1978).
[282] Markley, J. L., *Biochemistry*, **17**, 4648 (1978).
[283] Byers, J. D., and Koshland, D. E., *Bioorg. Chem.*, **7**, 15 (1977).
[284] Fastrez, J., and Houyet, N., *Eur. J. Biochem.*, **81**, 515 (1977).
[285] Campbell, P., and Nashed, N, T., *Bioorg. Chem.*, **7**, 69 (1977).
[286] Schwartz, H. M., Wu, W.-S., Marr, P. W., and Jones, J. B., *J. Am. Chem. Soc.*, **100**, 5199 (1978).
[287] See *Org. Reaction Mech.*, **1977**, 54.
[288] Riddles, P. W., de Jersey, J., and Zerner, B., *Proc. Natl. Acad. Sci. U.S.A.*, **75**, 172 (1978).
[289] Quast, U., Engel, J., Steffen, E., Tschesche, H., and Kupfer, S., *Eur. J. Biochem.*, **86**, 353 (1978).
[290] Renaud, A., Dimicoli, J.-L., Lestienne, P., and Bieth, J., *J. Am. Chem. Soc.*, **100**, 1005 (1978).
[291] Lonsdale-Eccles, J. D., Neurath, H., and Walsh, K. A., *Biochemistry*, **17**, 2805 (1978).
[292] Petkov, D. D., *Biochem. Biophys. Acta*, **523**, 538 (1978).
[293] Matthews, D. A., Alden, R. A., Birktoft, J. J., Freer, S. T., and Kraut, J., *J. Biol. Chem.*, **252**, 8875 (1977).
[294] Barnett, P., and Rosenberry, T. L., *J. Biol. Chem.*, **252**, 7200 (1977).
[295] Airas, R. K., *Biochemistry*, **17**, 4932 (1978).
[296] Bolis, G., Ragazzi, M., Salvaderi, D., Ferro, D. R., and Clementi, E., *Gazz. Chim. Ital.*, **108**, 425 (1978).
[297] Carey, P. R., Carriere, R. G., Phelps, D. J., and Schneider, H., *Biochemistry*, **17**, 1081 (1978).
[298] Clark, P. I., and Lowe, G., *J. Chem. Soc., Chem. Commun.*, **1977**, 923.
[299] Angelides, K. J., and Fink, A. L., *Biochemistry*, **17**, 2659 (1978).
[300] Zannis, V. I., and Kirsch, J. F., *Biochemistry*, **17**, 2669 (1978).
[301] Lewis, S. D., Johnson, F. A., Ohno, A. K., and Shafer, J. A., *J. Biol. Chem.*, **253**, 5080 (1978).
[302] Yuthavong, Y., and Suttimool, W., *Biochem. Biophys. Acta*, **523**, 198 (1978).
[303] Smolarsky, M., *Biochemistry*, **17**, 4606 (1978).
[304] Shipton, M., and Brocklehurst, K., *Biochem. J.*, **167**, 799 (1977).
[305] Shipton, M., and Brocklehurst, K., *Biochem. J.*, **171**, 385 (1978). See *Org. Reaction Mech.*, **1976**, 61.
[306] Buncel, E., Chuaqui, C., Lynn, K. R., and Raoult, A., *Bioorg. Chem.*, **7**, 1 (1977).
[307] Leary, R., Larsen, D., Watanabe, H., and Shaw, E., *Biochemistry*, **16**, 5857 (1977).
[308] Brocklehurst, K., and Malthouse, J. P. G., *Biochem. J.*, **175**, 761 (1978).
[309] Lewis, C. A., and Wolfenden, R., *Biochemistry*, **16**, 4890 (1977).
[310] Hsu, I. N., Delbaere, L. T. J., James, M. N. G., and Hofmann, T., *Nature* (*London*), **266**, 140 (1977).
[311] James, M. N. G., Hsu, I. N., and Delbaere, L. T. J., *Nature* (*London*), 808 (1977).
[312] Antonov, V. K., Ginodman, L. M., Kapitannikov, Yu. V., Barshevskaya, T. N., Gurova, A. G., and Rumsh, L. D., *FEBS Lett.*, **88**, 87 (1978).
[313] Van Wart, H. E., and Vallee, B. L., *Biochemistry*, **17**, 3385 (1978).
[314] Harrison, L. W., and Vallee, B. L., *Biochemistry*, **17**, 4359 (1978).
[315] Bunting, J. W., and Chu, S. S.-T., *Can. J. Chem.*, **56**, 2188 (1978).
[316] Sugimoto, T., and Kaiser, E. T., *J. Org. Chem.*, **43**, 3311 (1978).
[317] Morgan, G., and Fruton, J. S., *Biochemistry*, **17**, 3562 (1978).
[318] Nishino, N., and Powers, J. C., *Biochemistry*, **17**, 2846 (1978).
[319] Rasnick, D., and Powers, J. C., *Biochemistry*, **17**, 4363 (1978).
[320] Holler, E., *Angew. Chem., Int. Edn.*, **17**, 648 (1978).
[321] Kim, S.-H., *Adv. Enzymol.*, **46**, 279 (1977).
[322] Fersht, A. R., Gangloff, G., and Dirheimer, G., *Biochemistry*, **17**, 3740 (1978); see *Org. Reaction Mech.*, **1976**, 65.
[323] Roth, G. J., and Siok, C. J., *J. Biol. Chem.*, **253**, 3782 (1978).
[324] Fisher, J., Charnas, R. L., and Knowles, J. R., *Biochemistry*, **17**, 2180 (1978).

[325] Charnas, R. L., Fisher, J., and Knowles, J. R., *Biochemistry*, **17**, 2185 (1978).
[326] O'Leary, M. H., and Mattes, S. L., *Biochim. Biophys. Acta*, **522**, 238 (1977).
[327] Pocker, Y., and Bjorgquist, D. W., *Biochemistry*, **16**, 5698 (1977).
[328] Pocker, Y., Meany, J. E., Davis, B. C., Arrigoni, J., and Stein, J. E., *J. Am. Chem. Soc.*, **100**, 2883 (1978).
[328a] Pocker, Y., and Sarkanen, S., *Biochemistry*, **17**, 1110 (1978).
[329] Berndt, M. C., de Jersey, J., and Zerner, B., *J. Am. Chem. Soc.*, **99**, 8332 (1977).
[330] Berndt, M. C., de Jersey, J., and Zerner, B., *J. Am. Chem. Soc.*, **99**, 8334 (1977).
[331] Jordan, F., Kuo, D. J., and Monse, E. U., *J. Am. Chem. Soc.*, **100**, 2872 (1978).
[332] Jordan, F., Kuo, D. J., ans Monse, E. U., *J. Org. Chem.*, **43**, 2828 (1978).
[333] Jordan, F., Kuo, D. J., and Monse, E. U., *Anal. Biochem.*, **86**, 289 (1978).
[334] O'Leary, M. H., and Piazza, G. J., *J. Am. Chem. Soc.*, **100**, 632 (1978).
[335] Yamada, H., and O'Leary, M. H., *Biochemistry*, **17**, 669 (1978).
[336] Metcalf, B. W., Bey, P., Danzin, C., Jung, M. J., Casara, P., and Vevert, J. P., *J. Am. Chem. Soc.*, **100**, 2551 (1978).
[337] Stubbe, J., and Abeles, R. H., *J. Biol. Chem.*, **252**, 8338 (1977).
[338] Pocker, Y., Davison, B. L., and Deits, T. L., *J. Am. Chem. Soc.*, **100**, 3564 (1978).
[339] Dewar, M. J. S., and Ford, G. P., *J. Am. Chem. Soc.*, **99**, 8343 (1977).
[340] Tsuda, T., Chiyo, Y., and Saegusa, T. *J. Am. Chem. Soc.*, **100**, 630 (1978).
[341] Kemp, D. S., Reczek, J., and Vellaccio, F., *Tetrahedron Lett.*, **1978**, 741.
[342] Hunter, D. H., Hamity, M., Patel, V., and Perry, R. A., *Can. J. Chem.*, **56**, 104 (1978).
[343] Malherbe, M., and Chatelus, G., *Ann. Chim.* (*Paris*), **2**, 115 (1977).
[344] Hargreaves, M., and Khan, M., *Monatsh. Chem.*, **109**, 799 (1978).
[345] Diefallah, El-H. M., and El-Nadi, A. M., *Can. J. Chem.*, **56**, 2053 (1978).
[346] Adams, L. J., Franzus, B., and Huang, T. T.-S., *Int. J. Chem. Kinet.*, **10**, 669 (1978).
[347] Kingsbury, C. A., and Max, G., *J. Org. Chem.*, **43**, 3131 (1978).
[348] Lin, C. T., Siew, P.-K., and Byrn, S. R., *J. Chem. Soc., Perkin Trans.* 2, **1978**, 957.
[349] Bigley, D. B., and Weatherhead, R. H., *Can. J. Chem.*, **56**, 628 (1978).
[350] Boiko, T. S., Latenko, V. A., and Yasnikov, A. A., *Urk. Khim. Zh.* (*Russ. Ed.*), **44**, 74 (1978); *Chem. Abs.*, **88**, 210381 (1978).
[351] Reese, C. B., *Tetrahedron*, **34**, 3143 (1978).
[352] Ramirez, F., and Marecek, J. F., *Acc. Chem. Res.*, **11**, 239 (1978).
[353] Krasil'nikova, E. A., *Usp. Khim.*, **46**, 1638 (1977); *Chem. Abs.*, **87**, 200266 (1977).
[354] Bel'skii, V. E., *Usp. Khim.*, **46**, 1578 (1977); *Chem. Abs.*, **88**, 5640 (1978).
[355] Charton, M., and Charton, B., *J. Org. Chem.*, **43**, 2383 (1978).
[356] Hayes, D. M., Kenyon, G. L., and Kollman, P. A., *J. Am. Chem. Soc.*, **100**, 4331 (1978).
[357] Guthrie, J. P., *Can. J. Chem.*, **56**, 2342 (1978).
[358] Abbott, S. J., Jones, S. R., Weinman, S. A., and Knowles, J. R., *J. Am. Chem. Soc.*, **100**, 2558 (1978).
[359] Cullis, P. M., and Lowe, G., *J. Chem. Soc., Chem. Commun.*, **1978**, 512.
[359a] Jones, S. R., Kindman, L. A., and Knowles, J. R., *Nature*, (*London*), **275**, 564 (1978).
[360] Lowe, G., and Sproat, B. S., *J. Chem. Soc., Chem. Commun.*, **1978** 565.
[361] Orr, G. A., Simon, J., Jones, S. R., Chin, G. J., and Knowles, J. R., *Proc. Natl. Acad. Sci. U.S.A.*, **75**, 2230 (1978).
[362] Satterthwait, A. C., and Westheimer, F. H., *J. Am. Chem. Soc.*, **100**, 3197 (1978).
[363] Meyerson, S., Kuhn, E. S., Ramirez, F., Marecek, J. F., and Okazaki, H., *J. Am. Chem. Soc.*, **100**, 4062 (1978).
[364] Sigal, I., and Loew, L., *J. Am. Chem. Soc.*, **100**, 6394 (1978).
[365] See *Org. Reaction Mech.*, **1976**, 73.
[366] Bauman, M., and Wadsworth, W. S., *J. Am. Chem. Soc.*, **100**, 6388 (1978).
[367] See *Org. Reaction Mech.*, **1975**, 59.
[368] Hudson, R. F., and Woodcock, R. C., *Justus Liebeigs Ann. Chem.*, **1978**, 176.
[369] Reese, C. B., and Yau, L., *Tetrahedron Lett.*, **1978**, 4443.
[370] Bunton, C. A., Diaz, S., van Fleteren, G. M., and Paik, C., *J. Org. Chem.*, **43**, 258 (1978).
[371] Epstein, J., Kaminski, J. J., Bodor, N., Enever, R., Sowa, J., and Higuchi, T., *J. Org. Chem.*, **43**, 2816 (1978).
[372] Fernández-Prini, R., Baumgarten, E., and Turyn, D., *J. Chem. Soc., Faraday Trans.* 1, **1978**, 1196.
[373] Finkelstein, B. L., Sukchorukov, Yu. I., and Istomin, B. I., *Org. React.* (*Tartu*), **14**, 181 (1977).

[374] Minic, D., and Rakin, D., *Croat. Chem. Acta*, **49**, 819 (1977); *Chem. Abs.*, **89**, 107626 (1978).
[375] Kluger, R., Pike, D. C., and Chin, J., *Can. J. Chem.*, **56**, 1792 (1978).
[376] McRae, J. D., and Tadros, L. M., *J. Pharm. Sci.*, **67**, 631 (1978).
[377] Asubiojo, O. I., Brauman, J. I., and Levin, R. H., *J. Am. Chem. Soc.*, **99**, 7707 (1977).
[378] Scheffers-Sap, M. M. E., and Buck, H. M., *Recl. Trav. Chim. Pays-Bas*, **97**, 252 (1978).
[379] Kokesh, F. C., Cameron, D. A., Kukuda, Y., and Kwas, P. V., *Carbohydr. Res.*, **62**, 289 (1978).
[379a] Castlellijns, M. M. C. F., Schipper, P., and Buck, H. M., *J. Chem. Soc., Chem. Commun.*, **1978**, 382.
[380] Munoz, A., Garrigues, B., and Koenig, M., *J. Chem. Soc., Chem. Commun.*, **1978**, 219.
[381] Granoth, I., and Martin, J. C., *J. Am. Chem. Soc.*, **100**, 5229 (1978).
[382] Segall, Y., and Granoth, I., *J. Am. Chem. Soc.*, **100**, 5130 (1978).
[383] Sarma, R., Ramirez, F., McKeever, B., Nowakowski, M., and Marecek, J. F., *J. Am. Chem. Soc.* **100**, 5391 (1978).
[384] See *Org. Reaction Mech.*, **1977**, 62.
[385] Mhala, M. M., Jagdale, M. H., and Killedar, A. V., *Sci. J. Shivaji Univ.*, **15**, 195 (1975); *Chem. Abs.*, **88**, 5853 (1978).
[386] Mhala, M. M., and Sonawane, V. R., *Curr. Sci.*, **46**, 841 (1977); *Chem. Abs.*, **88**, 104319 (1978).
[387] Manninen, P. A., *Acta Chem. Scand., Ser. B*, **32B**, 269 (1978).
[388] Michalski, J., Mikołajczak, J., and Skowrońska, A., *J. Am. Chem. Soc.*, **100**, 5386 (1978).
[389] Markowska, A., and Nowicki, T., *Chem. Ber.*, **111**, 56 (1978).
[390] Challis, B. C., Challis, J. A., and Iley, J. N., *J. Chem. Soc., Perkin Trans. 2*, **1978**, 813.
[391] Morita, T., Okamoto, Y., and Sakurai, H., *Tetrahedron Lett.*, **1978**, 2523.
[392] Ramirez, F., Marecek, J. F., Chaw, Y., and McCaffrey, T., *Synthesis*, **1978**, 519.
[393] Corriu, R. J. P., Lanneau, G. F., and LeClerq, D., *J. Organomet. Chem.*, **153**, Cl (1978).
[394] Corriu, R. J. P., Lanneau, G. F., and LeClerq, D., *J. Chem. Soc., Chem. Commun.*, **1978**, 104.
[395] Gubaidullin, M. G., *Zh. Obshch. Khim.*, **47**, 2659 (1977); *Chem. Abs.*, **88**, 88923 (1978).
[396] Savelova, V. A., and Chentsova, N. M., *Ukr. Khim. Zh.*, **43**, 1194 (1977); *Chem. Abs.*, **88**, 61744 (1978).
[397] Gallagher, M. J., and Honegger, H., *J. Chem. Soc., Chem. Commun.*, **1978**, 54.
[398] Fookes, C. J. R., Gallagher, M. J., and Honegger, H., *J. Chem. Soc., Chem. Commun.*, **1978**, 324.
[399] Houalla, D., Sanchez, M., Wolf, R., and Osman, F. H., *Tetrahedron Lett.*, **1978**, 4675
[400] See *Org. Reaction. Mech.*, **1977**, 61.
[401] Brown, C., Hudson, R. F., Wartew, G. A., and Coates, H., *J. Chem. Soc., Chem. Commun.*, **1978**, 7.
[402] Wimmer, M. J., and Rose, I. A., *Ann. Rev. Biochem.*, **47**, 1031 (1978).
[403] Wood, H. G., O'Brien, W. E., and Michaels, G., *Adv. Enzymol.*, **45**, 85 (1977).
[404] Boyer, P. D., *Acc. Chem. Res.*, **11**, 218 (1978).
[405] Cohn, M., and Hu, H., *Proc. Natl. Acad. Sci. U.S.A.*, **75**, 200 (1978).
[406] Bock, J. L., and Cohn, M., *J. Biol. Chem.*, **253**, 4082 (1978).
[407] Webb, M. R., McDonald, G. G., and Trentham, D. R., *J. Biol. Chem.*, **253**, 2908 (1978).
[408] Van Etten, R. L., and Risley, J. M., *Proc. Natl. Acad. Sci. U.S.A.*, **75**, 4784 (1978).
[409] Sleep, J. A., Hackney, D. D., and Boyer, P. D., *J. Biol. Chem.*, **253**, 5235 (1978).
[410] Lowe, G., and Sproat, B. S., *J. Chem. Soc., Chem. Commun.*, **1978**, 783.
[411] Li, T. M., Mildvan, A. S., and Switzer, R. L., *J. Biol. Chem.*, **253**, 3918 (1978).
[412] Burgers, P. M. J., and Eckstein, F., *Proc. Natl. Acad. Sci. U.S.A.*, **75**, 4798 (1978).
[413] Jaffe, E. K., and Cohn, M., *J. Biol. Chem.*, **253**, 4823 (1978).
[414] Sheu, K.-F. R., and Frey, P. A., *J. Biol. Chem.*, **253**, 3378 (1978).
[415] Midelfort, C. F., and Sarton-Miller, I., *J. Biol. Chem.*, **253**, 7127 (1978).
[416] Viola, R. E., and Cleland, W. W., *Biochemistry*, **17**, 4111 (1978).
[417] See *Org. Reaction Mech.*, **1977**, 69.
[418] Nowak, T., *J. Biol. Chem.*, **253**, 1998 (1978).
[419] See *Org. Reaction Mech.*, **1976**, 82.
[420] Reed, G. H., Barlow, C. H., and Burns, R. A., *J. Biol. Chem.*, **253**, 4153 (1978).
[421] See *Org. Reaction Mech.*, **1976**, 84.
[422] Bickerstaff, G. F., and Price, N. C., *Int. J. Biochem.*, **9**, 1 (1978).
[423] Kochetkov, S. N., Bulargina, T. V., Sashchenko, L. P., and Severin, E. S., *Eur. J. Biochem.*, **81**, 111 (1977).

[424] Dibenedetto, G., and Mura, U., *Biochem. Biophys. Acta*, **522**, 122 (1977).
[425] Landt, M., and Butler, L. G., *Biochemistry*, **17**, 4130 (1978).
[426] See *Org. Reaction Mech.*, **1977**, 70.
[427] See *Org. Reaction Mech.*, **1977**, 69.
[428] Rowan, S. G., and Meister, A., *J. Biol. Chem.*, **253**, 1258 (1978).
[429] Powers, S. G., and Meister, A., *J. Biol. Chem.*, **253**, 800 (1978).
[430] Osterman, H. L., and Walz, F. G., *Biochemistry*, **17**, 4124 (1978).
[431] Lavallee, D. K., and Myers, R. B., *J. Am. Chem. Soc.*, **100**, 3907 (1978).
[432] Nakanishi, M., and Tsubio, M., *J. Am. Chem. Soc.*, **100**, 1273 (1978).
[433] Cooney, R. V., Mumma, R. O., and Benson, A. A., *Proc. Natl. Acad. Sci. U.S.A.*, **75**, 4262 (1978).
[434] Deacon, T., Farrar, C. R., Sikkel, B. J., and Williams, A., *J. Am. Chem. Soc.*, **100**, 2525 (1978).
[435] See *Org. Reaction Mech.*, **1977**, 71.
[436] Monjoint, P., Guillot, G., and Laloi-Diard, M., *Phosphorus, Sulphur*, **2**, 192 (1976); *Chem. Abs.*, **89**, 59391 (1978).
[437] Nummert, V., Piirsalu, M., and Alakivi, I., *Org. React.* (*Tartu*), **15**, 133 (1978).
[438] King, J. F., and Lee, T. M., *J. Chem. Soc., Chem. Commun.*, **1978**, 48.
[439] Sendega, R., Makitra, R., and Protsaylo, T., *Org. React.* (*Tartu*), **14**, 123 (1978).
[440] Sendega, R. V., Gorbachenko, N. G., and Pelekh, B. L., *Ukr. Khim. Zh.* (*Russ. Ed.*), **43**, 1291 (1977); *Chem. Abs.*, **88**, 73825 (1978).
[441] Protsailo, T. A., and Sendega, R. V., *Zh. Fiz. Khim.*, **51**, 2149 (1977); *Chem. Abs.*, **87**, 167386 (1977).
[442] Sendega, R. V., Makitra, R. G., and Protsailo, T. A., *Dopov. Akad. Nauk Ukr. RSR, Ser. B: Geol. Khim. Biol. Nauki*, **1977**, 711; *Chem. Abs.*, **87**, 200402 (1977).
[443] Sendega, R. V., *Ukr. Khim. Zh.* (*Russ. Ed.*), **44**, 89 (1978); *Chem. Abs.*, **88**, 104339 (1978).
[444] Mikhalevich, M. K., Ozdrovskaya, I. M., Sheiko, S. G., and Vizgert, R. V., *Izv. Vyssh. Uchebn. Zaved., Khim. Khim. Tekhnol.*, **21**, 323 (1978); *Chem. Abs.*, **89**, 107282 (1978).
[445] Vizgert, R. V., Skrypnik, Yu. V., Starodubtseva, M. P., Maksimenko, N. N., and Sheiko, S. G., *Deposited Doc.*, **1976**, *VINITI* 1237; *Chem. Abs.*, **88**, 61682 (1978).
[446] Campbell, M. M., and Johnson, G. J., *Chem. Rev.*, **78**, 65 (1978).
[447] Giese, B., and Heuck, K., *Chem. Ber.*, **111**, 1384 (1978).
[448] Maleeva, N. T., Savelova, V. A., Butko, O. I., and Solomoichenko, T. N., *Deposited Doc.*, **1975**, *VINITI* 3433; *Chem. Abs.*, **88**, 73814 (1978).
[449] Yang, K. Y., Chang, C. K., and Lee, K. A., *Hwahak Kyoyuk*, **4**, 68 (1977); *Chem. Abs.*, **87**, 200393 (1977).
[450] Uhm, T. S., Lee, I. C., and Lee, E. S., *Taehan Hwahak Hoechi*, **21**, 262 (1977); *Chem. Abs.*, **88**, 49907 (1978).
[451] Urbanska, H., and Urbanski, J., *Pol. J. Chem.*, **52**, 538 (1978); *Chem. Abs.*, **89**, 41905 (1978).
[452] Drabowicz, J., and Oae, S., *Tetrahedron*, **34**, 63 (1978).
[453] Akasaka, T., Yoshimura, T., Furukawa, N., and Oae, S., *Chem. Lett.*, **1978**, 417.
[454] Stoss, P., and Satzinger, G., *Chem. Ber.*, **111**, 1453 (1978).
[455] Ciuffarin, E., Gambarotta, S., Isola, M., and Senatore, L., *J. Chem. Soc., Perkin Trans. 2*, **1978**, 554.
[456] Penn, R. E., Block, E., and Revelle, R. K., *J. Am. Chem. Soc.*, **100**, 3622 (1978).
[457] Chau, M. M., and Kice, J. L., *J. Org. Chem.*, **42**, 3265 (1977).
[458] Chau, M. M., and Kice, J. L., *J. Org. Chem.*, **43**, 914 (1978).
[459] Chau, M. M., Kice, J. L., and Margolis, H. C., *J. Org. Chem.*, **43**, 910 (1978).
[460] Belkind, B. A., Denney, D. B., Denny, D. Z., Hsu, Y. F., and Wilson, G. E., *J. Am. Chem. Soc.*, **100**, 6327 (1978).
[461] Perozzi, E. F., and Martin, J. C., *J. Org. Chem.*, **42**, 3222 (1977).
[462] Adzima, L. J., Duesler, E. N., and Martin, J. C., *J. Org. Chem.*, **42**, 4001 (1977).
[463] Adzima, L. J., and Martin, J. C., *J. Org. Chem.*, **42**, 4006 (1977).
[464] Walter, W., Krische, B., and Voss, J., *J. Chem. Res.*, (*S*), **1978**, 332, (M) 4101.
[465] Hutton, W. C., and Crowell, T. I., *J. Am. Chem. Soc.*, **100**, 6904 (1978).
[466] Okuyama, T., Takimoto, K.. and Fueno, T., *J. Org. Chem.*, **42**, 3545 (1977).
[467] Oae, S., Asai, N., and Fujimori, K., *J. Chem. Soc., Perkin Trans. 2*, **1978**, 571.
[468] See *Org. Reaction Mech.*, **1972**, 461.
[469] Oae, S., Asai, N., and Fujimori, K., *J. Chem. Soc., Perkin Trans. 2*, **1978**, 1124.
[470] Kice, J. L., and Lee, T. W. S., *J. Am. Chem. Soc.*, **100**, 5094 (1978).

CHAPTER 3

Radical Reactions

D. J. Cowley

School of Physical Sciences, New University of Ulster

and D. C. Nonhebel

Department of Pure and Applied Chemistry, University of Strathclyde

---

Introduction

The third volume in the series "Free Radicals in Biology" has been published.[1] The role of singlet oxygen and related species was the subject of a recent symposium.[2] The proceedings of symposia including general free-radical chemistry,[3,4] radical processes in matrices, and reaction mechanisms (including radical processes),[5] models and computers[6] have been published. Other reviews have dealt with the chemistry of nitrogen-containing radicals[7] the structure and reactivity of amidyl radicals,[8] aminium radical cations[9], the chemistry[10] and thermochemistry[11] of sulphur-containing radicals, CIDNP,[12,13] and industrial applications of photochemical syntheses.[14] The influence of activation and reaction volumes in homolytic processes has been discussed.[15]

Structure, Stereochemistry, and Stability

Reviews on the application of Raman[16] and ESR spectroscopy[17] to the study of radical structure have appeared and long-lived radicals have also been reviewed.[18]

Carbon Radicals

A distinction can be made between σ- and π-radicals by plotting the difference in activation energies for radical reactions with CCl_4 and $BrCCl_3$ against their *F*-strain values[19,20] (π-radicals have greater *F*-strain than similarly substituted σ-radicals and hence are more sensitive to steric influences); on this basis Ph·, $CH_2{=}CH\cdot$, cyclopropyl, and 7-norbornyl radicals are σ-radicals while primary, secondary, and tertiary acyclic alkyl are π-radicals. The π-radical behaviour of *tert*-alkyl radicals contrasts with recent results for the *tert*-butyl radicals which tend to favour a pyramidal structure with a very low barrier to inversion.[21–24] IR studies indicate that primary alkyl radicals have an sp^2 radical centre which is easily distorted.[25]

Matrix IR studies indicate that the staggered conformation of the ethyl radical is the most stable form.[26]

MO studies indicate that the increasing removal from planarity along the series $\cdot CH_2F$, $\cdot CHF_2$, and $\cdot CF_3$ is due to a combination of inductive and conjugative effects involving a 2-orbital 3-electron bond.[27] The MO treatments of the conformations of $\cdot CH_2X$ radicals have been reviewed.[28]

The fluorine-containing radicals (**1–3**) exist in eclipsed, staggered and perpendicular conformations, respectively. Like their unfluorinated analogues they are very persistent.[29] The fluorinated α-ketoalkyl radical, $C_6F_5CO\dot{C}Bu^t_2$ has the

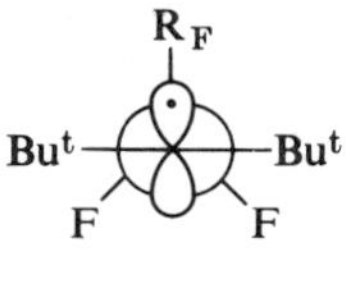

(**1**) 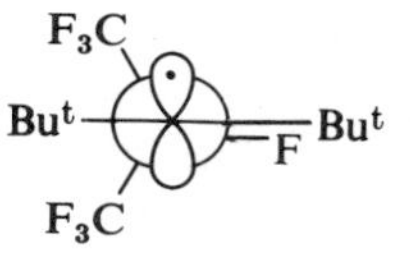 (**2**) 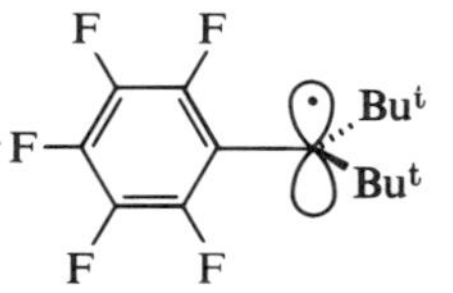 (**3**) 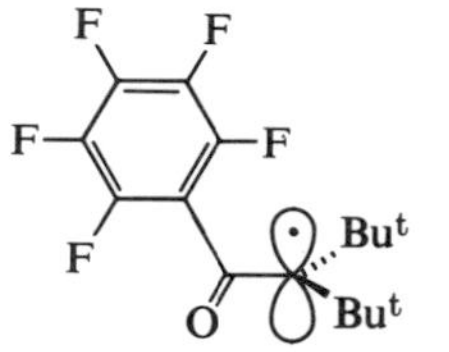 (**4**)

structure (**4**): other radicals of the type $R_FCO\dot{C}Bu^t_2$ have analogous structures. Equations, that consider conformational effects, have been derived for the calculation of the heats of formation of alkyl radicals.[30]

Radicals of the type $\cdot CMe(OR)_2$ are bent at the radical centre though $\cdot CPh(OR)_2$ radicals are planar.[31] Tris(organothiyl)methyl radicals, $R_nMS\dot{C}(SR')_2$ preferentially adopt the staggered conformation (**5**) in which conjugative delocalization between the unpaired electron and the *p*-type lone pair on the sulphur of the R_nMS group is maximized (cf. **6**).[32] In contrast, the related cyclic radicals (**7** and **8**)

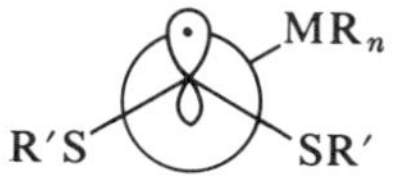

(**5**)  (**6**)

(**7**) 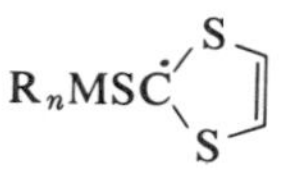 (**8**) 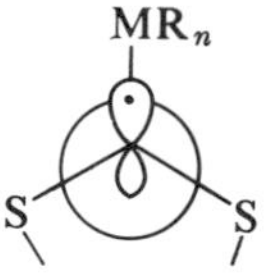 (**9**)

prefer the eclipsed conformation (**9**) in which steric repulsion between the MR_n group and α-substituents is minimized: in these cyclic radicals the sulphur atoms in the rings are well orientated for conjugative delocalization of the unpaired electron and hence the possibility of delocalization on to the R_nMS sulphur is much reduced.

The rate of inversion at the pyramidal carbon atom of cyclopropyl radicals with an α-hydrogen is very rapid ($k > 8 \times 10^7\ s^{-1}$), whereas the 1-fluorocyclopropyl radical is configurationally fixed on the ESR time scale.[33] A second study indicated that the rate of inversion of 2,3-dimethylcyclopropyl radicals is $\geqslant 5 \times 10^9\ s^{-1}$.[34] The first persistent cyclopropyl radicals (**10**; X = H, or F) have been reported.[35] They are also unusual in that they have planar and almost planar configurations, respectively.: this has been attributed to steric repulsion between the 1-substituent and the

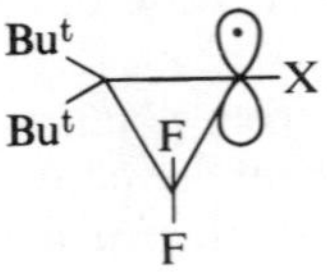

(10) 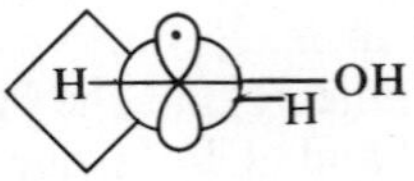 (11)

tert-butyl groups. The cyclobutylhydroxymethyl radical has the conformation **(11)** analogous to the corresponding cyclopropylhydroxymethyl radical.[36] The preferred configurations and barriers to inversion of 1-X cyclopentyl radicals have been determined: the radicals are planar when X = H or NCO but pyramidal with X = CH_3, CH_2F, or CH_2Cl. The barrier to inversion of the 1-methylcyclopentyl radical is 3.2 kcal mol^{-1}, *i.e.* greater than that of the analogous *tert*-butyl radical: this is probably due to the influence of the ring.[37] The tetrahydrofuran-2-yl radical has a C_4 envelope structure with rapid ring-flipping between the two alternative conformations. The tetrahydrothiophen-2-yl radical shows considerably greater conformational stability: INDO calculations indicate that there is significant delocalization of unpaired spin on to sulphur in this radical.[38] The *endo*-radical **(13)** is formed by reduction of (**12**; X = Br) with triethylsilyl radicals: the preferential formation of the *endo* radical has been attributed to a hyperconjugative interaction between the cyclopropane ring and the radical centre.[39] This accounts for the highly stereospecific transfer of a chlorine atom in the chlorination of (**12**; X = H).

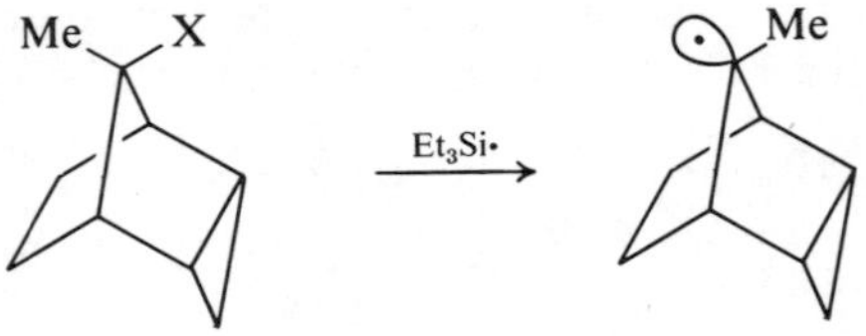

$Et_3Si\cdot$ →

(12) (13)

The relative rates of reduction of bridgehead halides by tri-*n*-butyltin radicals indicate that the stability of bridgehead radicals is in the order: 1-methylcyclohexyl > 1-adamantyl > 1-bicyclo[2.2.2]octyl > 1-norbornyl.[40] Hyperconjugative spin transfer in 1-norbornyl and bicyclo[2.1.1]hexan-1-yl radicals from the semi-filled orbital at the pyramidal radical centre decreases as this becomes more regularly pyramidal.[41] Reactions involving the 7,7-dimethyl-2-oxobicyclo[2.2.1]hept-1-yl radical proceed with partial retention of optical activity.[42]

Radicals with both electron-donating and electron-withdrawing substituents at the radical centre have enhanced stability.[43] A new example of a persistent radical of this type is (**14**);[44] other similar radicals include $Me_2N\dot{C}HCN$, $Me_2N\dot{C}(CN)_2$,[45] and $ArRN\dot{C}ArCN$.[46]

$$R^1R^2N\dot{C}(CN)_2 \longleftrightarrow R^1R^2\overset{+\cdot}{N}\overset{-}{C}(CN)_2 \longleftrightarrow R^1R^2NC(CN){=}C{=}N\cdot$$

(14)

Photo-electron spectroscopic studies indicate that the unpaired electron in the allyl radical is mainly non-bonding.[47] The (*Z,E*)-1,3-diphenylallyl radical undergoes isomerization even at −82 °C at a rate that competes with the coupling rate of the radicals.[48] The stabilization energy of the propargyl radical, $HC{\equiv}CCH_2\cdot$, has been estimated to be 8.6 and 8.8 kcal mol^{-1},[49, 50] which is slightly less than that of the allyl radical (11.4 kcal mol^{-1}).[51] The stabilization energy of HC≡CĊMe_2 is 11 kcal mol^{-1}.[52] The acetonyl radical has a very small stabilization energy of <2 kcal mol^{-1}.[53] In contrast to the acetonyl radical, in which the unpaired electron is centred mainly on carbon, the major canonical hybrid of the 1-thia-alkyl radical (**15**) is (**15b**) in which the unpaired spin is on sulphur.[32] This difference may be rationalized by consideration of the relative electronegativities of carbon, oxygen, and sulphur and/or the much greater bond strength of the C=O π-bond compared to the C=S π-bond.

(**15a**) (**15b**)

The barrier to ring flipping in the cyclohex-2-en-1-yl radical is 6.8 kcal mol^{-1}: most of the barrier can be attributed to eclipsing interactions present in the planar transition state.[54] ESR studies indicate that there is very little interaction between the HOMO of the allyl moiety and the LUMO of the double bond in (**16**).[55] A photoelectron-spectral study of the tropyl radical indicates that the unpaired electron is either localized on C(1) or delocalized over the allyl component C(1), C(2), and C(3) but is not delocalized over the whole ring.[56]

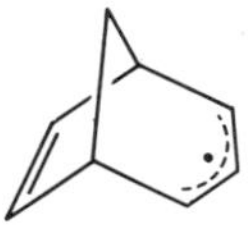

(**16**)

Ab initio calculations on the cyclopropenyl radical indicate that one C—C bond is shorter than the other two.[57] ESR studies of the cyclopentadienyl radical $C_5H_4\cdot$ and 1-(trialkylsilyl)cyclopentadienyl radicals indicate that Ph_2MeSi, $PhMe_2Si$, and Me_3Si groups are electron-accepting owing to p_π–d_π conjugation whereas Me_5Si_2 and Me_7Si_3 are electron-donating owing to σ–π mixing.[58]

The radical centre in the norbornadien-7-yl radical is more pyramidal than that in the cyclopropyl radical as a consequence of an electron-transfer interaction from the etheno-bridges to the radical centre as well as to the small bond angle about this carbon.[59] ESR studies indicate that there is little hyperconjugation in norborn-5-en-2-yl and benzonorbornen-2-yl radicals.[60]

In a study of 4-substituted 3-cyanobenzyl radicals, electron-withdrawing substituents, *e.g.* CH_3CO, PhN=N, CN, and NO_2, enhance the stability of the radical whilst fluoro- and methoxyl groups destabilize the radical; fluorine cannot expand its octet as the other halogens do, while the methoxyl group could only delocalize the unpaired electron if this were placed in a higher-energy orbital of oxygen.[61] A new fluorine-containing benzylic radical, F—C_6H_4—$C(CF_3)_2$—C_6H_4—Ċ$(CF_3)_2$, has been reported.[62] The spin density distribution in 4-substituted benzyl radicals has been obtained by INDO and PPP methods.[63] SCF-MO calculations on the properties of benzyl, phenoxy-, and anilino-radicals have been reported.[64] The extent of dissociation of *O*-silylated pinacols ArRC($OSiMe_3$)C($OSiMe_3$)RAr to the ketyl radicals ArRĊ$OSiMe_3$ is controlled mainly by the steric bulk of the groups Ar and R.[65,66] The first planar persistent triarylmethyl radical (**17**) has been reported; its persistence is due to the spatial requirements of the isopropylidene bridges.[67] The phenaleno[1,9-*cd*]dithiolyl radical is another example of a monomeric coplanar carbon radical.[68] Triarylmethyl radicals with bulky substituents at all the *meta* and/or *para* positions (**18**; R = H or Ph) dimerize to give the hexa-arylethanes rather than the more usual methylenecyclohexadienes.[69]

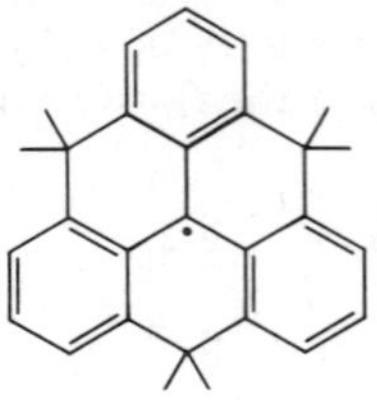

(**17**) 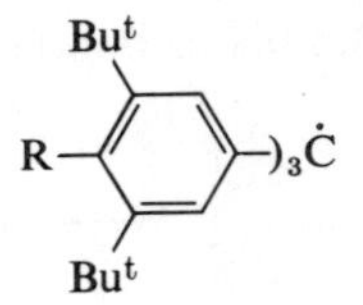 (**18**) 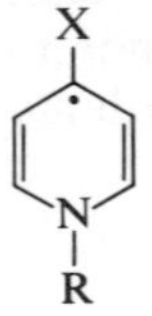 (**19**)

The 1-*tert*-butyl-4-methoxycarbonylpyridin-4-yl radical (**19**; R = Bu^t, X = COOMe) is a long-lived radical,[70] as are the adducts of Group IVB radicals with 4-substituted pyridines (**19**; R = SiR_3 or GeR_3); the corresponding adducts with $\cdot SnR_3$ radicals are only long-lived when X = CN, a reflection of the weak N–Sn bond.[71]

INDO calculations indicate that pyridyl radicals are more stable than phenyl radicals because of conjugation of the non-bonded electron pair with the unpaired electron in π-radicals.[72]

Nitrogen Radicals

The preferred conformation of aminyl radicals $Bu^t\dot{N}CH_2R$ depends on the nature of R; when R = alkyl the radicals are in the staggered conformation (**20**) whereas when R = Et_3Si the eclipsed conformation (**21**) is more favourable as the hyperconjugative interaction between the semi-filled *p*-orbital and the C_β—Si bond is maximized.[73] ESCA studies indicate that there is substantial unpaired spin density

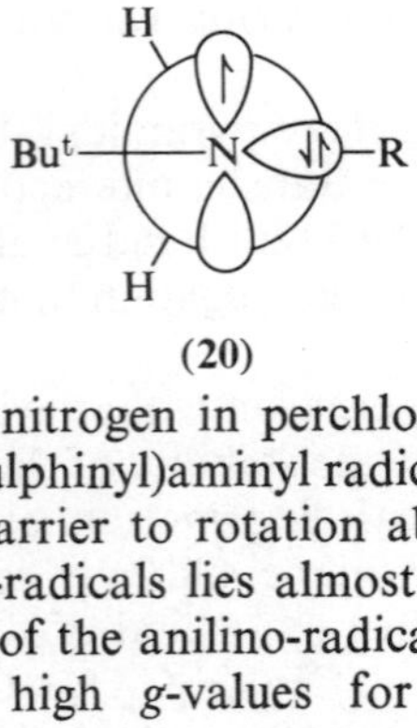

(**20**) 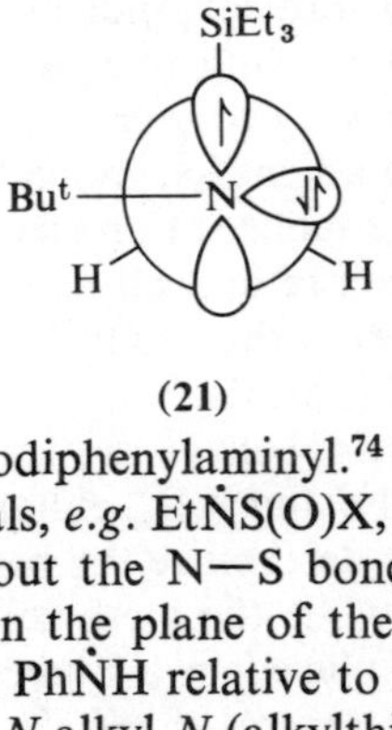 (**21**) 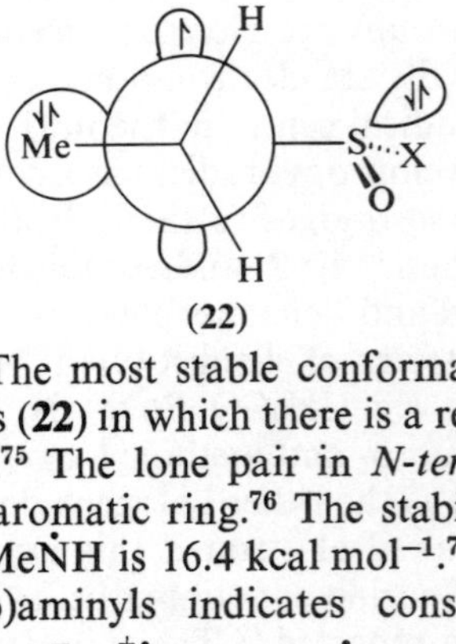 (**22**)

on the nitrogen in perchlorodiphenylaminyl.[74] The most stable conformation of alkyl(sulphinyl)aminyl radicals, *e.g.* $Et\dot{N}S(O)X$, is (**22**) in which there is a relatively large barrier to rotation about the N—S bond.[75] The lone pair in *N-tert*-butylanilino-radicals lies almost in the plane of the aromatic ring.[76] The stabilization energy of the anilino-radical $Ph\dot{N}H$ relative to $Me\dot{N}H$ is 16.4 kcal mol^{-1}.[77]

The high *g*-values for *N*-alkyl-*N*-(alkylthio)aminyls indicates considerable delocalization of unpaired spin onto sulphur, $R—\bar{N}—\dot{S}^{+}R' \leftrightarrow R—\dot{N}—SR'$.[78] The dipolar canonical form is much less important for $Bu^t\dot{N}SCCl_3$, as reflected by its lower *g*-value. A novel class of persistent radical is the 1,2-thiazet-2-yl type (**23**).[79]

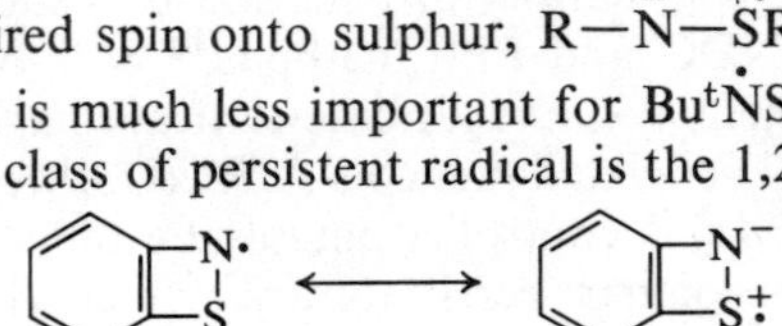

(**23**)

The unpaired spin in *N*-sulphinylaminyls, ArṄSOR,[80] and in *N*-sulphonylaminyls, $ArNSO_2R$, is located primarily on the nitrogen and in the *N*-aryl ring.[81]

1-*tert*-Butyl and 1-α-cumyl-urazole radicals (**24**; R′ = Bu^t or $PhCMe_2$) are the first examples of persistent hydrazyl radicals in which the hydrazyl-nitrogen atoms are not directly bonded to any aryl group.[82] ESR studies indicate these radicals are π-radicals in which the unpaired electron is delocalized over both hydrazyl-nitrogen atoms. It is also delocalized, though to a lesser extent, onto the imide-nitrogen, as reflected by the lower stability of the isoelectronic pyrazolidine-3,5-dione radicals (**25**). The persistence of (**24**; R = Bu^t or $PhCMe_2$) is due to steric shielding of the radical centre and to the fact that disproportionation cannot occur.

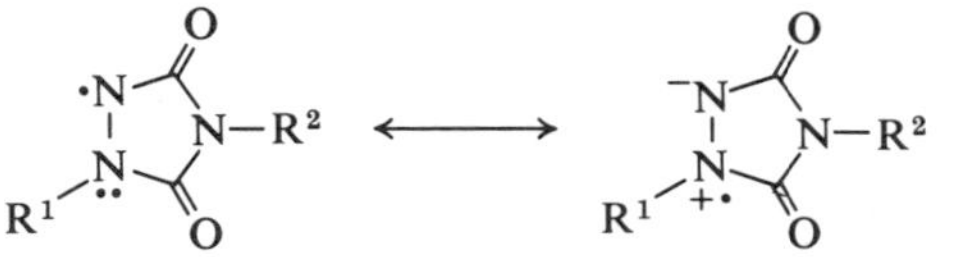

(**24**) 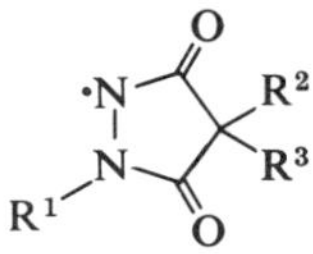 (**25**)

These radicals have a lesser tendency to dimerize in more polar solutions, reflecting the greater importance of dipolar canonical forms.

The values of the hyperfine splitting constants and INDO calculations favour a π-radical structure for the 1,3-dimethyltriaza-allyl radical, $CH_3\dot{N}{-}N{=}NCH_3$.[83] Both σ- and π-succinimidyl radicals have been recognized: they have very different patterns of reactivity (see p. 112).[84,85]

Oxygen Radicals

ESR studies show that the benzoyloxy-radical is a σ-radical in the ground state.[86] The values of the electron affinities of a series of alkoxy-radicals indicate that alkyl groups exert a stabilizing influence.[87] MO studies of phenoxy-radicals and radical anions from 1,2- and 1,4-dihydroxybenzenes demonstrate that the radicals have quinonoid-, and not quinol-like, geometries in which the C—O bond is short, the adjacent C—C is long and the next C—C bond short.[88] ENDOR spectral studies of *p*-phenylenebisgalvinoxyl have been reported.[89] An unusual phenoxy-radical with a hexa-co-ordinated phosphorus atom has been obtained.[90]

The iminoxy-radical derived from acetophenone oxime is a mixture of the *Z*- and *E*-isomers, and not exclusively the *Z*-isomer as had previously been reported.[91] ESR studies of phosphorus-containing iminoxyls (these also consist of both *Z*- and *E*-isomers) have been reported.[92,93]

Miscellaneous Radicals

The radical obtained from reaction of tri-*n*-butyltin radicals with oxygen has the unusual trigonal bipyramidal structure (**26**).[94] A novel persistent radical (**27**) with

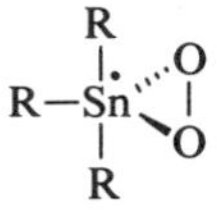

(**26**) 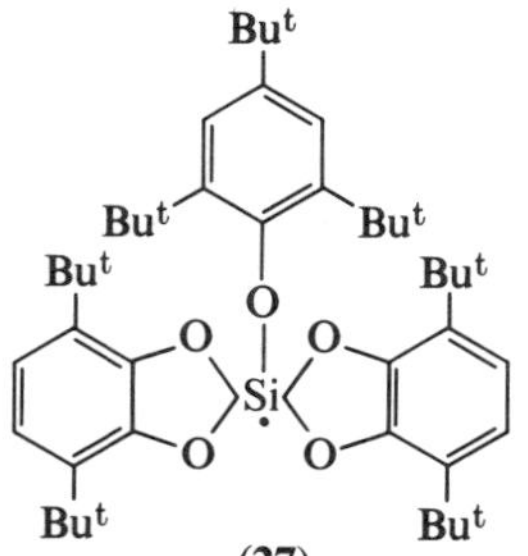 (**27**)

pentaco-ordinate silicon has been reported.[95] ESR studies indicate that the magnitude of the coupling constants for $Ph_2P\cdot$ are sensitive to the radical environment[96] and that there is increased pyramidality at the radical centre in the series $Ph_3C\cdot$, $Ph_3Si\cdot$, $Ph_3Ge\cdot$ and $Ph_3Sn\cdot$.[97]

Peroxides

MO-LCAO methods have been used to correlate the structure, stability, and reactivity of peroxides.[98] Both product and kinetic studies indicate that *tert*-butyl isopropyl peroxide undergoes induced decomposition, and this is reflected by the relatively low yields of propan-2-ol obtained from decompositions in alkylbenzenes:[99]

$$Bu^tO\cdot \text{ (or Me}\cdot) + Me_2CHOOBu^t \longrightarrow Bu^tOH \text{ (or } CH_4) + Me_2\dot{C}OOBu^t$$

$$Me_2\dot{C}OOBu^t \longrightarrow Me_2CO + Bu^tO\cdot$$

Di-isopropyl peroxides decompose predominantly by O—O homolysis, though an electrocyclic process leading to acetone and hydrogen becomes more important with increased solvent polarity.[100] The two peroxy-links in poly(peroxydisiloxanes), $R'OOSiR^2_2OOR^1$, undergo homolysis independently of each other.[101] The decomposition of the peroxide, $Bu^tOOSiMe_3$, proceeds homolytically whereas bis-trialkylsilyl peroxides decompose by a non-radical mechanism.[102] Simple O—O bond homolysis occurs in the thermolysis of *tert*-butylperoxyacetals.[103] Photochemical decomposition of the peroxide (**28**) leads to the rearranged radical (**29**);[104] this parallels the behaviour of the analogous cyclohexa-2,5-dienone peroxide.[105]

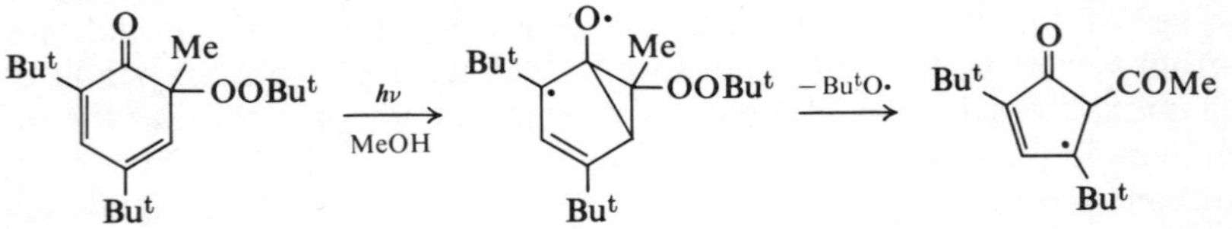

(**28**) $\xrightarrow[\text{MeOH}]{h\nu}$ → $\xrightarrow{-Bu^tO\cdot}$ (**29**)

Reactions of *tert*-butylperoxy-phosphates and -phosphonates with cyclohexene in presence of copper(I) give the corresponding cyclohex-3-enyl phosphate or phosphonate (*cf.* Scheme 1).[106, 107]

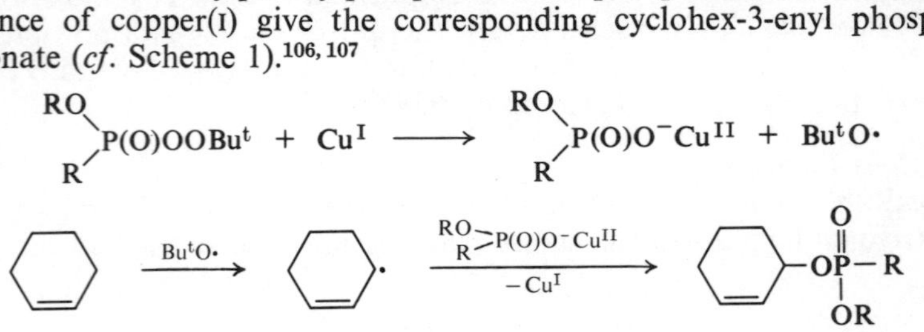

$$(RO)(R)P(O)OOBu^t + Cu^I \longrightarrow (RO)(R)P(O)O^-Cu^{II} + Bu^tO\cdot$$

cyclohexene $\xrightarrow{Bu^tO\cdot}$ cyclohexenyl radical $\xrightarrow[-Cu^I]{(RO)(R)P(O)O^-Cu^{II}}$ cyclohexenyl–OP(O)(R)(OR)

SCHEME 1

The polymerization of vinyl monomers can be initiated by using organic peroxides and chromium(II) acetate.[108, 109]

The chemistry of 1,2-dioxetanes has been reviewed.[110] The activation parameters for thermolysis of triphenyl-1,2-dioxetane suggest that decomposition occurs by O—O bond homolysis; substituents in the phenyl groups exert little influence on the rate of decomposition.[111] The same conclusions apply for the thermolyses of 3,3-bis-(*p*-methoxyphenyl)-1,2-dioxetane,[112] 3,4-diadamantyl-1,2-dioxetane,[113] and

3,3-dichloro-4,4-bis-(*p*-chlorophenyl)dioxetane.[114] In contrast, MINDO calculations indicate that a C—C biradicaloid mechanism is preferred for dioxetanes possessing alkoxy- or fluoro-substituents.[115] It has been suggested that dioxetanes (*e.g.* **30**) substituted by or reacting with compounds possessing strongly electron-donating groups may decompose by an electron-transfer mechanism.[116,117] An analogous mechanism is operative in the decomposition of 1,6-diaryl-2,5,7,8-tetraoxobicyclo[4.2.0]octanes.[118]

(**30**) $\xrightarrow{\Delta}$ [N-methylacridinium O· ; $R_2\bar{C}$—O·] ⟶ N-methylacridone* + $R_2C{=}O$

The chemistry of benzyl hydroperoxide has been studied: this decomposes in part by an induced process in styrene.[119] AIBN induces the decomposition of cumene hydroperoxide.[120] Decompositions of hydroperoxides catalysed by sulphides and disulphides,[121] chromium stearate,[122] cobalt naphthenoate,[123] and amines[124,125] have been studied, as has the cobalt octanoate-catalysed decomposition of *p*-xylene hydroperoxide.[126] Dimethylacetamide hydroperoxide decomposes by a radical process.[127] The radical polymerization of methyl methacrylate by *tert*-butyl hydroperoxide occurs at temperatures below 100 °C in presence of triethylamine and acetic acid.[128] Acetal hydrotrioxides (**31**) break down by two alternative radical pathways (and also in a concerted process).[129]

$R^1C(OOOH)(OR^2)_2$ (**31**) ⟶ $R^1C(OR^2)_2$—O· + HOO·

⟶ $R^1C(OR^2)_2$—OO· + HO·

⟶ $R^1COOR^2 + R^2OH + {}^1O_2$

The structure and properties of peroxy-radicals have been reviewed.[130] ESR spectroscopy has been used in the quantitative estimation of concentrations of $HO_2\cdot$ and $RO_2\cdot$.[131] MO calculations on the coupling of alkylperoxy-radicals support reaction *via* a six-membered transition state for primary and secondary alkylperoxy-radicals (and also for acylperoxy-radicals) while tertiary alkylperoxy-radicals form the tetroxide by coupling of the terminal oxygens and not by way of a four-centred transition state.[132,133] The rate constant for the coupling of isopropylperoxy-radicals is $0.81–0.97 \times 10^6$ l mol^{-1} s^{-1} *i.e.* between the reported values for primary and tertiary alkylperoxy-radicals.[134] α-Hydroxymethylperoxy-radicals, $HOCH_2OO\cdot$, undergo bimolecular decay *via* the tetroxide to give hydrogen peroxide and formic acid.[135] The radical derived from di-isopropyl ketone peroxide, $Me_2CHCOCMe_2OO\cdot$, undergoes intramolecular 1,5-hydrogen abstraction to give $Me_2\dot{C}COCMe_2OOH$.[136]

MO calculations have been carried out on the influence of substituents on the stability of dibenzoyl peroxides towards homolysis.[137] The photolytic decompositions of diacyl peroxides have been studied in argon matrices: the intermediate alkyl or aryl radicals have been detected by IR spectroscopy.[138] CIDNP studies point to an electron-transfer reaction between two acyloxy-radicals generated by thermolysis of certain diacyl peroxides.[139] Electron transfer occurs only when two radicals of very different electronegativities encounter each other, as in the decomposition of *tert*-butylacetyl *m*-chlorobenzoyl peroxide (**32**) (Scheme 2); in this

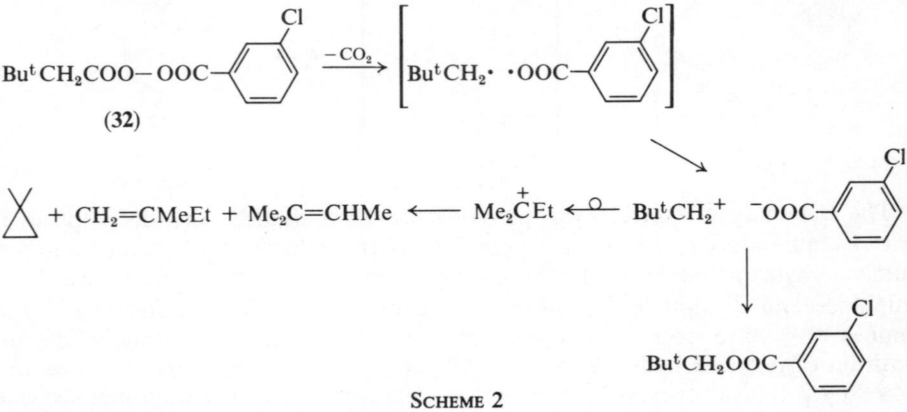

$$\mathrm{Bu^tCH_2COO{-}OOC{-}C_6H_4Cl\text{-}m}\ (\mathbf{32}) \xrightarrow{-CO_2} [\mathrm{Bu^tCH_2\cdot\ \cdot OOC{-}C_6H_4Cl\text{-}m}] \longrightarrow \mathrm{Bu^tCH_2^+\ {}^-OOC{-}C_6H_4Cl\text{-}m}$$

$$\mathrm{Bu^tCH_2^+\ {}^-OOC{-}C_6H_4Cl\text{-}m} \longrightarrow \mathrm{Bu^tCH_2OOC{-}C_6H_4Cl\text{-}m}$$

$$\mathrm{Bu^tCH_2^+} \longrightarrow \mathrm{Me_2\overset{+}{C}Et} \longrightarrow \text{1,1-dimethylcyclopropane} + \mathrm{CH_2{=}CMeEt} + \mathrm{Me_2C{=}CHMe}$$

SCHEME 2

example, products derived from the *tert*-pentyl cation are formed, thus implicating the neopentyl cation as an intermediate. CIDNP effects also support an electron-transfer process in the thermolysis of acetylbenzoyl peroxide in DMSO.[140] The formation of rearranged esters in thermolyses of cycloalkylacetyl peroxides is suggestive of an ion-pair rather than a radical mechanism, at least for the formation of these products.[141] The decomposition of dibenzoyl peroxide is greatly accelerated by *n*-butyl orthotitanate: the reaction proceeds *via* a complex that dissociates to give butyloxy-radicals.[142] The polymerization of methyl methacrylate initiated by dibenzoyl peroxide is retarded when carried out in the cathode, though not the anode, compartment of an electrolytic cell: this is attributed to the formation of oxygen in the cathode compartment.[143] Reaction of superoxide with diacyl peroxides leads to singlet oxygen.[144] Thermolysis of bis(diarylmethylenecarbonyl) peroxides (**33**) in toluene occurs very readily;[145] the peroxide is believed to break down to give diarylmethyleniminyl radicals which dimerize, abstract hydrogen from toluene or react with benzyl radicals to give (**34–36**), respectively; it is difficult to

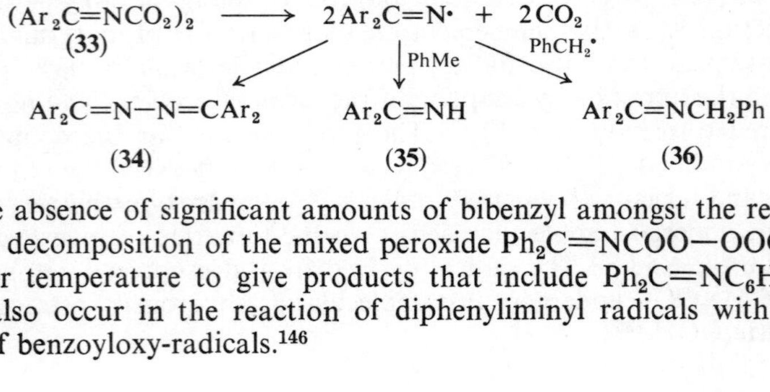

$$\underset{(\mathbf{33})}{\mathrm{(Ar_2C{=}NCO_2)_2}} \longrightarrow \mathrm{2\,Ar_2C{=}N\cdot} + \mathrm{2\,CO_2}$$

$$\mathrm{2\,Ar_2C{=}N\cdot} \longrightarrow \underset{(\mathbf{34})}{\mathrm{Ar_2C{=}N{-}N{=}CAr_2}}; \quad \xrightarrow{\mathrm{PhMe}} \underset{(\mathbf{35})}{\mathrm{Ar_2C{=}NH}}; \quad \xrightarrow{\mathrm{PhCH_2^{\bullet}}} \underset{(\mathbf{36})}{\mathrm{Ar_2C{=}NCH_2Ph}}$$

explain the absence of significant amounts of bibenzyl amongst the reaction products. The decomposition of the mixed peroxide $\mathrm{Ph_2C{=}NCOO{-}OOCPh}$ occurs at a higher temperature to give products that include $\mathrm{Ph_2C{=}NC_6H_4Me}$; such products also occur in the reaction of diphenyliminyl radicals with toluene in presence of benzoyloxy-radicals.[146]

Decomposition of *tert*-butyl cubanepercarboxylate is 4800-fold slower than that of *tert*-butyl trimethylacetate.[147] The competition constant for reactions of the cubyl radical with $BrCCl_3$ and CCl_4 is higher than for most bridgehead radicals. These observations are postulated to arise from greater *s*-character of the exocyclic C—CO bond which results in a stronger bond and also from the decreased possibility for the development of polar character in the transition state for such radicals. Decomposition of *tert*-butyl 1- and 2-peroxyadamantanecarboxylate in biacetyl leads to mixtures of adamantane and the corresponding 1- and 2-acetyladamantanes.[148] The decomposition of *tert*-butyl imidazolepercarboxylate has been studied.[149] *tert*-Butyl 3-(triethylgermyl)perpropanoate undergoes induced decomposition in carbon tetrachloride involving attack at the α-CH_2 and subsequent migration of the triethylgermyl group to give $CH_2{=}CHCOOGeEt_3$.[150]

The lack of solvent effects on the activation parameters for decarboxylation of β-peroxy-lactones (*e.g.* **37**), coupled with the absence of secondary deuterium isotope effects, supports a mechanism involving a 1,5-biradical intermediate

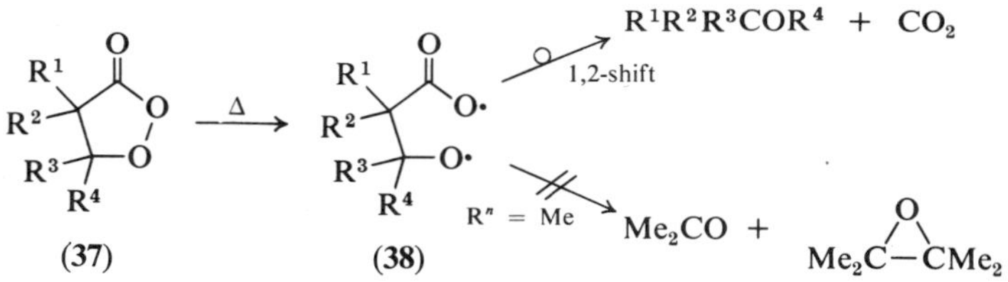

$R^1R^2R^3COR^4 + CO_2$ (1,2-shift); R″ = Me: Me_2CO + $Me_2C(O)CMe_2$

(**37**) (**38**)

(**38**).[151] The acetoxy-substituted γ-peroxy-γ-butanolactone (**39**) undergoes two-bond homolysis to give (**40**) and acetone but no carbon dioxide:[152] this contrasts with earlier results[153] for the lactone (**41**) which breaks down by a different two bond scission (O—O and C—CO) to give carbon dioxide, acetone, and ethylene.

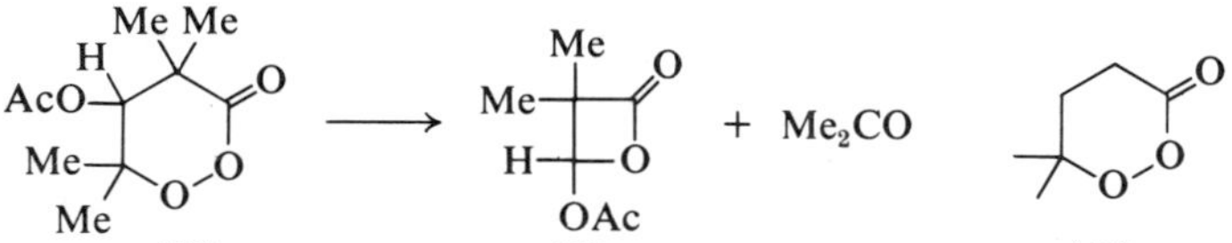

(**39**) ⟶ (**40**) + Me_2CO (**41**)

Peroxyacyl nitrates break down by both O—O and N—O homolysis; the first pathway occurs in the presence of NO_2, whilst the second is followed with added NO which scavenges the acylperoxy-radicals:[154]

$$RCOOONO_2 \longrightarrow [RCOO\cdot \quad \cdot NO_3]$$
$$RCOOONO_2 \longrightarrow [RCOOO\cdot \quad \cdot NO_2]$$
$$RCOOO\cdot + NO \longrightarrow [RCOO\cdot \quad \cdot NO_2] \longrightarrow RNO_2(\text{and } RONO) + CO_2$$

Peroxyalkyl nitrates are formed by reaction of alkylperoxy-radicals with nitrogen dioxide.[155]

The photoinduced decomposition of peracetic acid in both benzene[156] and toluene has been studied;[157] in toluene the main product is benzyl alcohol, resulting from attack by benzyl radicals on the peroxide bond.

$$MeCOOOH \longrightarrow Me\cdot + HO\cdot + CO_2$$
$$Me\cdot(\text{or } HO\cdot) + PhCH_3 \longrightarrow PhCH_2\cdot + CH_4 \ (\text{or } H_2O)$$
$$PhCH_2\cdot + MeCOOOH \longrightarrow PhCH_2OH + Me\cdot + CO_2$$

Ab initio calculations, together with analysis of thermochemical data for gas-phase ozonolyses of alk-1-enes, suggest that the intermediate CH_2OO can best be considered as a peroxy-biradical. In addition, solution-phase results on the stereospecificity of ozonolysis are consistent with a biradical mechanism;[158] evidence in favour of a biradical has also been obtained from the gas-phase ozonolyses of propene and isobutene.[159] These results are consistent with the postulate that the principal resonance hybrid of ozone is $\dot{O}{-}O{-}\dot{O}$.[160] *tert*-Butylperoxy- and hydroxyl radicals are formed in the molecule-assisted homolysis of *tert*-butyl hydroperoxide with ozone:[161]

$$Bu^tOOH + O_3 \longrightarrow Bu^tOO\cdot + HO\cdot + O_2$$

The reaction of ozone with peroxy-radicals has also been studied.[162]

Autoxidations

The relative reactivities of secondary and tertiary C—H bonds in 3-methylpentane towards *tert*-butylperoxy-radicals are 1 : 35 and product studies in this oxidation of 3-methylpentane indicate that termination involves both 3-methyl-2-pentylperoxy- and 3-methyl-3-pentylperoxy-radicals.[163] The reactivity of alkylbenzenes, PhR, towards PhCHMeOO· radicals increases in the order $R = Me_3CCH_2 < Me_2CHCH_2 < n\text{-}C_6H_{13} < BrCH_2CH_2 < Pr < Me_2CHCH_2CH_2 < Me_2CH < MeCHEt < c\text{-}C_6H_{11}$, thus reflecting the importance of steric factors.[164] Kinetic studies on the autoxidation of ethylbenzene indicate that at high hydroperoxide concentrations, O—O bond homolysis of the hydroperoxide gives PhCHMeO· radicals which initiate the chain process;[165] however, the alcohol PhCHMeOH formed in this step retards the rate of autoxidation.[166] The rates of initiation in the liquid-phase oxidation of ethylbenzene have been evaluated by using a computer modelling method.[167] In autoxidations of hydrocarbons catalysed by cobalt(II) tetra-(*p*-tolyl)porphyrin, the cobalt complex converts molecular oxygen to $O_2^{\cdot -}$ which then induces the autoxidation.[168] The rate-determining step in the metal-catalysed autoxidations of tetralin is the decomposition of the tetralin hydroperoxide.[169] The rates of oxygen absorption in the oxidation of nitroalkylbenzenes show deviation from a first-order dependence on substrate concentration due to interaction of the alkyl radicals with nitro-groups.[170] 1-Hydroxyfluoren-9-one is one of the major products from the photo-oxygenation of 9-isopropylidenefluorene; this is formed *via* homolysis of a peroxide intermediate.[171] The photo-oxygenation of alkenes proceeds as shown.[172]

$$^3R_2CO + {>}C{=}C{<} \longrightarrow R_2\dot{C}O{-}\overset{|}{\underset{|}{C}}{-}\dot{C}{<} \xrightarrow{O_2} R_2\dot{C}O{-}\overset{|}{\underset{|}{C}}{-}\overset{|}{\underset{|}{C}}{-}OO\cdot$$

$$2R_2\dot{C}O{-}\overset{|}{\underset{|}{C}}{-}\overset{|}{\underset{|}{C}}{-}OO\cdot \longrightarrow R_2\dot{C}O{-}\overset{|}{\underset{|}{C}}{-}\overset{|}{\underset{|}{C}}{-}OO{-}OO{-}\overset{|}{\underset{|}{C}}{-}\overset{|}{\underset{|}{C}}{-}O\dot{C}R_2$$

$$\swarrow$$

$$O_2 + 2R_2CO + 2\ {>}C{-}C{<}\ (\text{epoxide, }{-}O{-})$$

Liquid-phase oxidations of 2-alkoxytetrahydrofurans[173] and 1,1-dialkoxyalkanes[174] have been studied.

2,4,6-Trimethylphenol inhibits the autoxidation of aldehydes in acetic acid.[175] The benzoyl radical has been spin-trapped in the homogeneous oxidation of benzaldehyde catalysed by $(Ph_3P)_2PdO_2$.[176] Peroxidic intermediates are formed in oxidation of cyclohexanone catalysed by metal salts.[177,178] Dimethylketene reacts with oxygen by a radical mechanism whereas zwitterionic intermediates are formed in reactions of $Ph_2C{=}C{=}O$ and $PhCMe{=}C{=}O$.[179] The radical autoxidation of enolic 4-hydroxyphenylpyruvic acid[180] and of cobalt(III) acetylacetonate[181] have been studied. The rate-determining step in the autoxidation of *N*,*N*-diethylhydroxylamine involves formation of diethyl nitroxide:[182]

$$Et_2NOH + O_2 \longrightarrow Et_2NO\cdot + HO_2\cdot$$
$$HO_2\cdot + Et_2NOH \longrightarrow Et_2NO\cdot + H_2O_2$$

Aminoperoxy-radicals are intermediates in the oxidation of dimethylaminoethyl methacrylate:[183] this reaction is inhibited by cobalt and manganese acetates which react with the intermediate peroxy-radicals.[184] Hydroperoxides are formed in the photo-oxidation of amino-acid derivatives.[185]

Copper(II) complexes of dialkyldithiophosphoric and dialkyldithiocarbamic acid inhibit autoxidation of hydrocarbons. The results for the dithiocarbamates are best interpreted as involving attack by the peroxy-radical at the metal centre; an alkoxy-radical is formed subsequently.[186,187] Substituted 3-hydroxypyridines act as inhibitors of autoxidation.[188] Steric effects reduce the antioxidant ability of 4-alkyl-2,6-di-*tert*-butylphenols.[189] A cyclohexadienone is formed upon reaction of *tert*-butylperoxy-radicals with the antioxidant tetrakis[methylene 3-(3,5-di-*tert*-butyl-4-hydroxyphenyl)propanoate]methane.[190] Mixtures of aromatic amines and copper compounds are good low-temperature inhibitors of autoxidations: peroxy-radicals react with the amine which is activated by complexation with the copper(II) salt; the resultant amino-radical is within the co-ordination sphere of the metal, thus reducing the probability of its involvement in chain extension.[191]

Azo Compounds

A ^{15}N-CIDNP study has shown that the azo-compounds, $PhN{=}NCPh_3$, $PhN{=}NCMe_2CN$, $PhN{=}NCMe_2Ph$, and 1-norbornyl-$N{=}NCMe_2Ph$ all decompose by one-bond scission.[192] The diazenyl radicals from $PhN{=}NCMe_2Ph$ have been trapped with $Ph_3C\cdot$ radicals to give $PhN{=}NCPh_3$:

$$PhN{=}NCMe_2Ph \rightleftharpoons PhN{=}N\cdot + Ph\dot{C}Me_2$$
$$PhN{=}N\cdot + Ph_3C\cdot \longrightarrow PhN{=}NCPh_3$$

One-bond scission also occurs in the gas-phase thermolyses of azoalkanes, including symmetrical compounds;[193] hydrogen abstraction by the intermediate alkyl radicals also occurs:

$$CH_3CH_2N{=}NCH_2CH_3 \longrightarrow CH_3CH_2\cdot + CH_3CH_2N{=}N\cdot$$
$$CH_3CH_2\cdot + CH_3CH_2N{=}NCH_2CH_3 \longrightarrow C_2H_6 + CH_3\dot{C}HN{=}NCH_2CH_3$$
$$CH_3CH_2\cdot + CH_3\dot{C}HN{=}NCH_2CH_3 \longrightarrow CH_3CH{=}NN(CH_2CH_3)_2$$

A kinetic study of the decomposition of azomethane has been reported.[194] Unlike the case with saturated azoalkanes, triplet sensitization of $CH_2{=}CHCMe_2$-$N{=}NCMe_2CH{=}CH_2$ and $CH_3N{=}NCMe_2CH{=}CH_2$ occurs moderately efficiently without intervention of *cis*-azo-compounds.[195] Cage effects are more important for direct photolyses ($\sim$0.60) than for sensitized reactions ($\sim$0.10): the difference is due, not so much to spin-correlation effects, as to stepwise decomposition of the azo-compound. A CIDNP study of the decomposition of aryldiazenes, $ArN{=}NH$, has revealed the intermediacy of an aryl–arylhydrazyl radical pair, the logical precursor of which is an aryldiazenyl–arylhydrazyl radical pair:[196]

$$2\mathrm{ArN{=}NH} \longrightarrow [\mathrm{ArN{=}N}\cdot + \mathrm{Ar}\dot{\mathrm{N}}\mathrm{NH_2}] \longrightarrow [\mathrm{Ar}\cdot + \mathrm{Ar}\dot{\mathrm{N}}\mathrm{NH_2}] + \mathrm{N_2}$$

Ab initio calculations for the photolysis of di-imide indicate that successive rupture of the N—H bonds is the preferred pathway.[197]

Elimination of nitrogen from cyclic azo-compounds has been reviewed.[198] The spectroscopic properties of cyclic and bicyclic azoalkanes and their relation to modes of decomposition have been studied.[199] The influence of solvent polarity on the rates of decomposition of cyclic azo-compounds has been examined.[200] The rates of thermolysis of the five- and six-membered cyclic azo-compounds (**42** and **43**) are greater than that of *trans*-azo-*tert*-butane due to release of strain on decomposition: the sum of the ground-state strain energy and ΔH^* of thermolysis

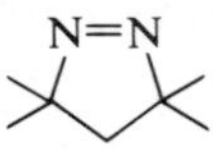

N=N (**42**) N=N (**43**) N=N (**44**)

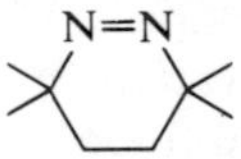

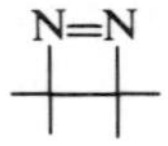

for the cyclic azo-compounds is approximately equal to ΔH^* for decomposition of the cyclic compound.[201] The four-membered compound (**44**) is, however, at least 11 kcal mol^{-1} more stable than would be expected and the following explanations have been suggested: (i) concerted $[2_s+2_s]$ decomposition is orbital-symmetry-forbidden and 11 kcal mol^{-1} represents the price of orbital-symmetry forbiddenness; (ii) decomposition occurs by a higher-energy $[2_s+2_a]$ pathway. The synchronous process $[2_a(\text{olefin})+2_s(\mathrm{N_2})]$ is excluded as thermolyses of *meso*- and (±)-3,4-diethyl-3,4-dimethyldiazetine occur with stereospecific loss of nitrogen;[202] a stepwise mechanism remains a possibility. Bicyclic 2,3-diazabicyclo-[2.2.2]octenes decompose by a concerted mechanism.[201] Thermolyses of 2-acetoxy-Δ^3-1,3,4-oxadiazolines[203] and photolysis of 3*H*-thienodiazepines[204] proceed by a stepwise loss of nitrogen *via* a diazenyl radical intermediate. Photolyses of (**45**) and

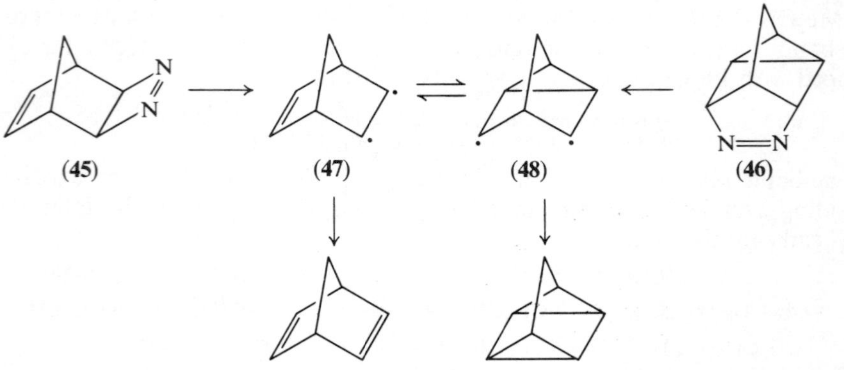

N N (**45**) (**47**) (**48**) N=N (**46**)

(**46**) lead to the same mixture of norbornadiene and quadricyclene via the equilibrating biradicals (**47**) and (**48**);[205] photolysis of (**49**) also proceeds via a biradical intermediate (**50**),[206] and such species are also formed in the thermal extrusion of nitrogen from 7,8-diazabicyclo[4.2.2]deca-2,4,7-triene,[207] and benzo[1,2-*c*: 4,3-*c*′]-dicinnoline.[208]

Several studies on the photolysis and thermolysis of 4-alkylidenepyrazolines have been designed to investigate the nature of the trimethylenemethane biradical

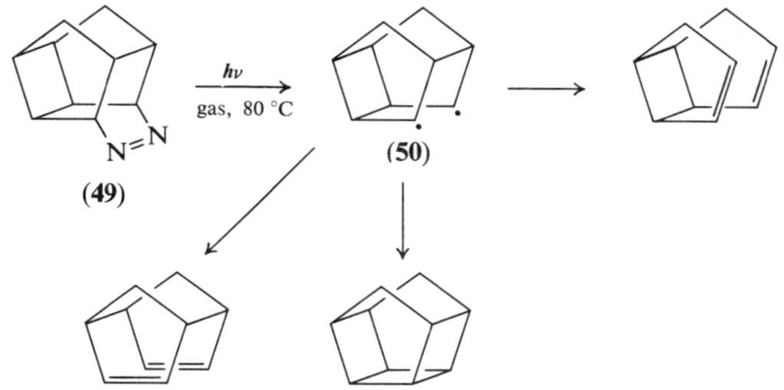

(49) $\xrightarrow[\text{gas, 80 °C}]{h\nu}$ (50)

intermediate.[209–212] Thermolysis of **(51)** leads to the rearranged product **(53)** *via* the diazenyl biradical intermediate **(52)**.[209]

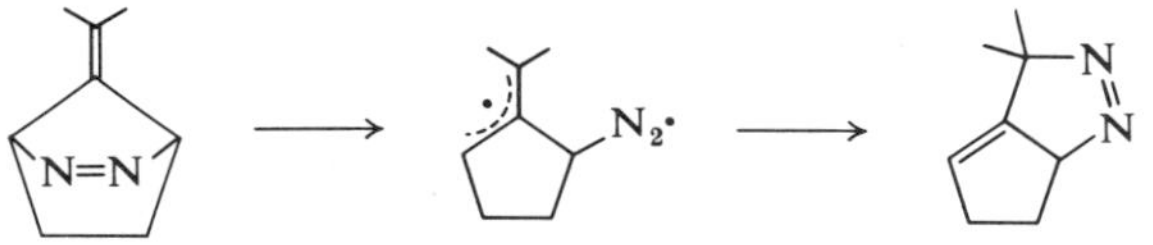

(51) (52) (53)

CIDNP studies indicate that aryltriazenes $ArN{=}NNR^1R^2$, on thermolysis, give $[ArN{=}N\cdot \ \ \cdot NR^1R^2]$ radical pairs.[213] Thermolysis of 3-aryl-1-heteroaryltriazenes Het—N=N—NHAr give products derived from the radical pairs generated from the less stable tautomeric form, HetNH—N=N—Ar.[214]

1,4-Diaryl-1,4-dimethyltetrazenes, which normally exist in the *trans*-planar configuration **(54)** in which the lone pair on nitrogen can overlap with the aromatic

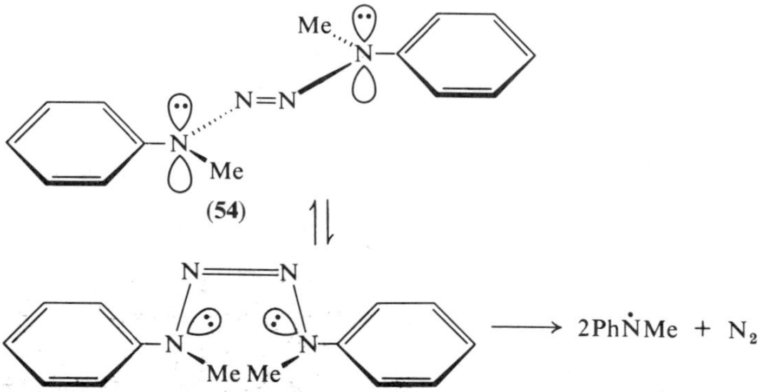

(54)

(55) $\longrightarrow$ $2PhN^{\bullet}Me + N_2$

π-cloud, decompose to *N*-methylanilino radicals *via* the *cis*-configuration **(55)**; this configuration is needed for maximum stabilization of the unpaired electron in the transition state.[215] The variation in rate of decomposition with viscosity of the solvent suggests that a stepwise decomposition of the tetrazene is operative; the rates of decomposition increase with increasing solvent polarity.[216] The rates of decomposition of the tetrazenes give a good Hammett plot; the negative ρ value indicates that the decomposition is facilitated by electron-donating groups.[217, 218]

1,4-Diaryl-1,4-dimethyltetrazenes act as chain-transfer agents in the vinyl polymerization of acrylonitrile.[219]

Under acidic conditions photolysis of 1,1,4,4-tetramethyltetrazene gives dimethylaminium radical cations.[220]

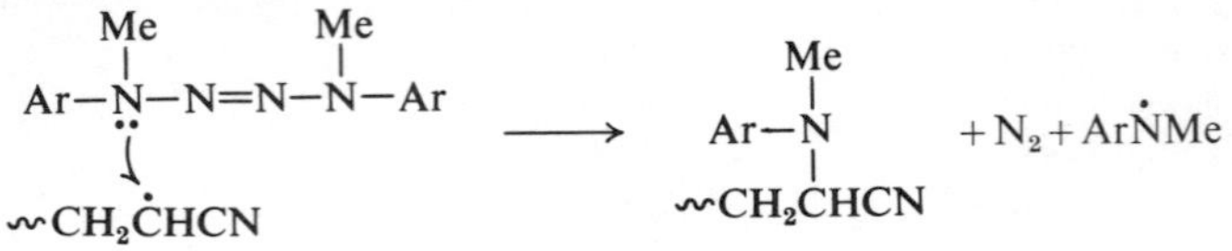

$$\mathrm{Ar{-}\overset{Me}{\underset{\cdot\cdot}{N}}{-}N{=}N{-}\overset{Me}{N}{-}Ar} \quad + \quad \sim\mathrm{CH_2\dot{C}HCN} \longrightarrow \mathrm{Ar{-}\overset{Me}{N}{-}CH_2CHCN}\sim \; + \mathrm{N_2} + \mathrm{Ar\dot{N}Me}$$

Diazonium Salts

The nature of the specific solvation of diazonium salts has considerable influence on whether these salts decompose by homolysis or heterolysis. A homolytic mechanism is favoured in solvents of high nucleophilicity, *e.g.* DMSO or HMPA, or by addition of added nucleophiles, *e.g.* diphenylhydroxylamine, whereas in solvents of low nucleophilicity a heterolytic mechanism predominates[221, 222] (cf. Scheme 3).

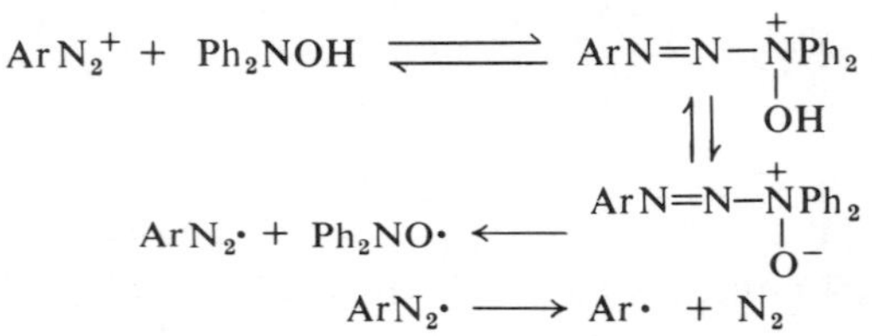

$$\mathrm{ArN_2^+ + Ph_2NOH \rightleftharpoons ArN{=}N{-}\overset{+}{N}Ph_2(OH)}$$

$$\rightleftharpoons \mathrm{ArN{=}N{-}\overset{+}{N}Ph_2(O^-)}$$

$$\mathrm{ArN{=}N{-}\overset{+}{N}Ph_2(O^-)} \longrightarrow \mathrm{ArN_2\cdot + Ph_2NO\cdot}$$

$$\mathrm{ArN_2\cdot \longrightarrow Ar\cdot + N_2}$$

SCHEME 3

Radical products are also formed more readily from diazonium salts possessing electron-withdrawing groups. Aromatic hydrocarbons have been obtained by the reductive amination of arylamines by alkyl nitrites in DMF, the final step probably involving hydrogen abstraction from the methyl groups of DMF:[223]

$$\mathrm{ArNH_2 + RONO \longrightarrow ArN{=}NOR}$$
$$\mathrm{ArN{=}NOR \longrightarrow ArN_2\cdot + RO\cdot}$$
$$\mathrm{ArN_2\cdot \longrightarrow Ar\cdot + N_2}$$
$$\mathrm{Ar\cdot + SH \longrightarrow ArH + S\cdot}$$

Aromatic hydrocarbons are also formed in the decomposition of diazonium salts in dioxane in presence of oxygen; the radical-chain pathway is catalysed by hydroquinone.[224] CIDNP effects confirm the formation of aryl radicals in the photoreactions of diazonium salts with pyrene[225] and benzoquinone.[226]

Combination and Disproportionation Reactions

A perturbational MO calculation on the model system, (Et· + Me·), has revealed that recombination and disproportionation pathways do not proceed through a common transition state. Recombination involves a maximization of the interaction between the singly-occupied MO's, whereas disproportionation involves significant charge-transfer from the doubly-occupied MO localized at the C—H bond into the singly-occupied MO of the other radical.[227] Termination rate constants of the $Bu^t\cdot$ radical in a range of solvents are well described by the Smoluchowski equation for diffusion control on the basis of a spin statistical factor, a temperature- and solvent-independent isotropic reaction diameter, and critically estimated diffusion coefficients D_R. Indeed, if reliable D_R values can be obtained empirically or semiempirically, the calculated termination rate constants may be better than currently available experimental values.[228] The solvent- and temperature-dependences of the disproportionation to combination ratio, P_d/P_c, for $Bu^t\cdot$ radicals have been explained very adequately on the basis of the anisotropic reorientational motions of the radicals during their encounter in the solvent cage. The rotational correlation times of the radical, treated as an oblate ellipsoid, were estimated by means of various models.[229]

P_d/P_c ratios for cyclohexyl and cycloheptyl radicals in the solid state, determined by a technique involving dissolution of solid irradiated samples in oxygenated liquid propane, differ appreciably from the value of 1.0 found for cycloalkyl radicals in the liquid phase at 298 K.[230]

A further examination[231] of the disproportionation of the substituted cyclohexyl radical (**56**) supports the hypothesis that homolysis of the C_β—H bond is subject to stereoelectronic control since (**57**), formed by loss of the *pseudo*-axial hydrogen atom, is the main product and (**58**) is not.

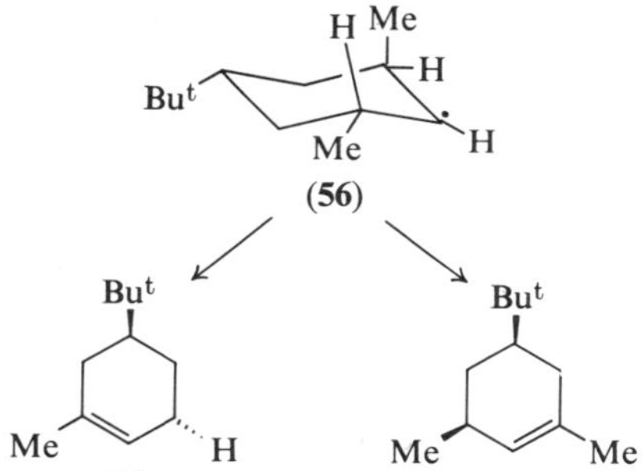

(**56**)

(**57**) Major product

(**58**) Minor product

Tetra-*tert*-butylethane, existing in the gauche conformation, dissociates to give two $Bu^t_2CH\cdot$ radicals and is the most thermolabile hydrocarbon known at present.[232] Unstable semibenzenes, formed *via* αo- and αp-coupling, constitute 19% of the products in the dimerization of benzyl radicals at 30 °C.[233] The combination of trityl radicals has been studied indirectly by means of equilibrations in the presence of thiophenol.[234] The reversible formation of stannyl radicals from the distannanes $R_3Sn—SnR_3$ (R = 2,4,6-trimethyl- or 2,4,6-triethyl-phenyl) has been demonstrated by ESR studies.[235] Dimerization of the radical 4-[4-$FC_6H_4C(CF_3)_2$]-$C_6H_4C(CF_3)_2\cdot$, generated from 4-$FC_6H_4C(CF_3)_2Br$ and $Fe(CO)_5$, has also been investigated.[236] The self-termination rate constants of various substituted alkyl radicals, generated by reductive dissociation of halo-compounds in propan-2-ol at 300 K, have been correlated with molecular mass.[237]

Discrepancies in the reports of the relative abundance of racemic and *meso*-diastereoisomers of 3,3′-diphenylbiphthalid-3-yl (**59**) formed on combination of the corresponding chemically and photochemically generated free radicals have recently been ascribed to thermal isomerization of the lower-melting diastereoisomer.[238]

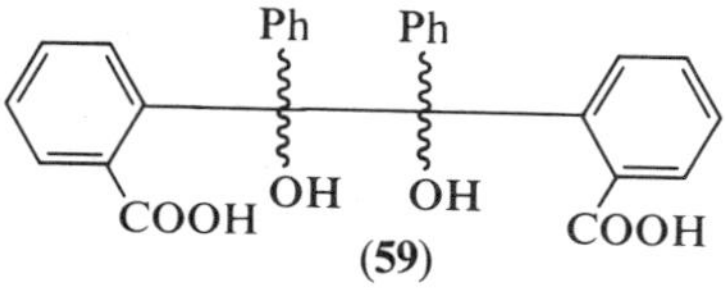

(**59**)

Rate constants for the combination of methyl and acetyl radicals, and for the self-combination of acetyl radicals, have been measured.[239]

The influences of binary solvent mixtures[240] and of copper ions[241] on the decay kinetics of substituted phenoxy-radicals have been investigated.

Reaction of Bu^{t}O· radicals with 2-(cyclohexyloxy)tetrahydropyran in solution involves a chain process with quadratic termination of linear radicals.[242]

The possible influence of magnetic field on recombination rates has been investigated in the liquid-phase oxidation of hydrocarbons[243] and in the (Br· + ·CCl_3) system;[244] in the latter case, the observed effect ($<2\%$) was much smaller than predictions based on current assumptions.

From perturbation MO calculations, the transition state in the termination of *tert*-alkylperoxy-radicals involves the terminal oxygen atom of each radical only; however, self-reaction of primary and secondary alkylperoxy-radicals requires the six-membered transition state (**60**) which is analogous to that for termination of acylperoxyl radicals also.[245]

The decay kinetics of isopropylperoxy-radicals in the gas phase are very similar to those in the liquid phase.[246] 3-Carbamoyl-1-methyl-pyridinyl radical (**61**) dimerizes in a pH-independent manner, whereas 4-carbamoylpyridinyl radical (**62**) is protonated before disproportionating to the corresponding dihydropyridine and pyridinium ion; the results of the study lend credance to the proposed 1e + H^+ pathway in NAD-catalysed enzymic reactions.[247]

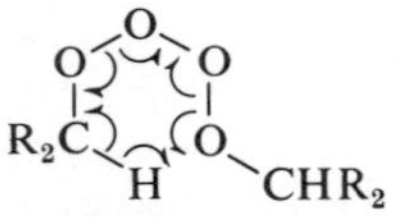

(**60**) 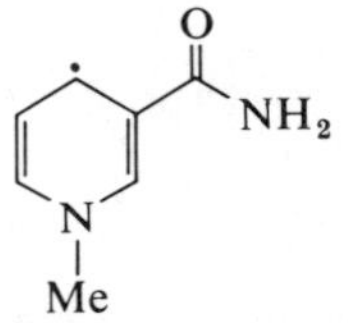 (**61**) 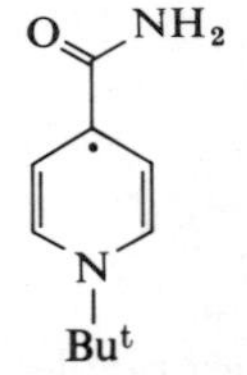 (**62**)

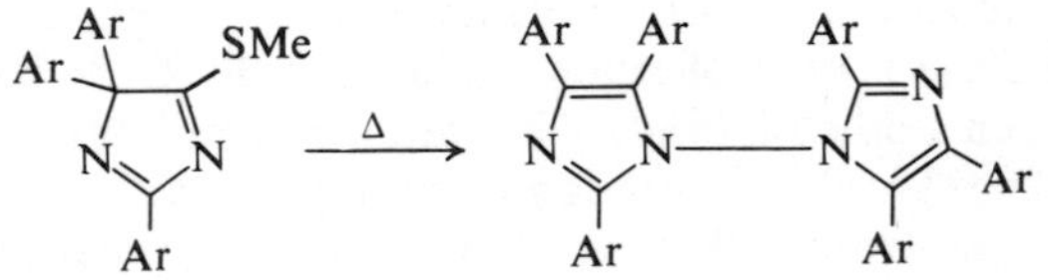

(**64**) (**63**)

The novel hexa-arylbi-imidazolyl (**63**) is formed by self-combination, following rearrangement, of the radical produced by thermal fission of the C(5)—S bond in (**64**).[248]

Good consistency is observed between gas-phase combination rate constants determined from comparative-rate single-pulse shock-tube experiments at 1000–1200 K and those found by the "radical buffer" method at 354–415 K. The latter method yields highly accurate values of k_r/k_s if the equilibrium constant K_{RS} of the reaction R· + SX = RX + S· can be found; 2R· → R_2 (rate constant k_r) and 2S· → S_2 (rate constant k_s). In the same study the heats of formation of Et·, Pr^i·, and Bu^t· radicals have been found to be higher by 10, 10, and 20 kJ mol^{-1}, respectively, than the currently accepted values obtained from metathesis reactions, and doubts have been expressed on the reliability of bromination/iodination as a method for the determination of, *inter alia*, radical enthalpies of formation.[249]

Recombination of methyl radicals behind shock waves[250] has been monitored by UV-absorption measurements.[251] The rate constant of the termination reaction of H· and ·CH_3 at 1100–1300 K has been found to be 80 times larger than the older value previously assumed.[252] The activation energy of the recombination of iodine atoms with ·CH_3/·CD_3 has been determined by the method of overlapping relaxations employing an iodine laser.[253] Rate constants have been measured for the gas-phase recombinations of amidogen radicals, ·NH_2, with Et·, Pr^i·, and Bu^t· radicals, and also for hydrogen abstraction from various alkanes by ·NH_2.[254] Radiofrequency plasmolysis of anisole yield cresols by cleavage and recombination.[255] Plasma pyrolysis of propane leads to higher acetylenes from combinations of HC≡C· radicals.[256] The unorthodox hypothesis of the disproportionation of $Me_3SiOSi(Me)_2$· radicals, formed by vacuum-UV-photolysis of hexamethyldisiloxanes, has been revived.[257]

Atom-abstraction Reactions

Isotope effects in hydrogen-atom transfer reactions have been reviewed.[258] Activation energies have been calculated for gas-phase hydrogen abstractions using bond indices and heats of formation derived for self-consistent electronic structures of all species lying on the reaction co-ordinate.[259]

Carbon-centred Radicals

The reaction of CH_3· radical with methane has been investigated by *ab initio* and semiempirical MO methods.[260] Rate constants of the reactions of CH_3· radical with dimethyl ether and formaldehyde, occurring in the pyrolysis of $(CH_3)_2O$, have been measured,[261,262] and so have rate constants for the hydrogen-atom detachment by CH_3· radicals from nitromethane[263] and 2-nitropropane.[264] The reaction of CH_3· radical with either O_2 or NO is not rapid at temperatures up to 1220 K.[265]

Undecyl radicals in PhCl at 100 °C attack the C—H bonds but not the O—H bond of monocarboxylic acids.[266] Partial rate constants, and activation energies, have been determined for the attack of undecyl radicals on toluene, ethylbenzene, and cumene.[267] Hydrogen-abstraction from phenols by polyvinyl acetate radical R· has been shown to proceed within the hydrogen bond to give the radical obtained by the addition of R· to the ester-carbonyl group of the solvent (*e.g.* EtOAc, $Me_2C{=}O$) which itself is hydrogen-bonded to the phenol:[268]

$$\mathrm{ArOH\cdots O{=}C(O\text{—})\text{—}} + \mathrm{R\cdot} \rightleftharpoons \mathrm{ArOH\cdots \dot{O}\text{—}C(O\text{—})(\text{—})\text{—}R} \longrightarrow \mathrm{ArO\cdot} + \mathrm{H\text{—}O\text{—}C(O\text{—})(\text{—})\text{—}R}$$

Chain transfer, and radical isomerization rate constants, together with activation-energy data, have been obtained for the radiation-induced telomerization of ethylene by bromobutane[269] and C_4 to C_8 *n*-alkanoic acids.[270] The radical cotelomerization of ethylene and vinyl chloride with CCl_4 results in the formation of homotelomers of ethylene with CCl_4 and of vinyl chloride with CCl_4, as well as cotelomers. The latter are sensitive to polar effects in the growth stage and in the chain-transfer step with CCl_4.[271] Telomerization of propene with methyl acetate is accompanied by rearrangement of the first and second growing radicals, with both 1,3- and 1,5-migration of hydrogen.[272] Investigation of the isobutene–methyl chloroacetate telomerization[273] has shown that only 1,5-migration of an H atom is realized in the radical $\cdot(CR^1R^2CH_2)_nCHClCOOCH_3$, where $R^1 = R^2 = CH_3$ and $n = 2$, as it is also in the case of ethylene, where $R^1 = R^2 = H$ and $n = 2$. Telomerization of propene with methyl dichloroacetate[274] includes chain transfer *via* C—H and C—Cl bonds, while with methyl trichloroacetate[275] the chain-transfer constants are an order of magnitude greater than in the reaction of ethylene and CCl_3COOMe.

1-Phenylethyl radicals are 100 times less reactive than peroxy-radicals towards phenols and amines.[276] The relative reactivities of scavengers such as I_2, CBr_4, and Me_2CHI towards phenyl radical in high-viscosity mineral oil tend to unity; the *apparent* unreactivity of Ph· towards O_2 in CCl_4 is a consequence of its high reactivity towards CCl_4.[277] The nucleophilic character of Ph· in hydrogen abstraction from *meta*-substituted toluenes is shown by the reaction ρ value (+0.18). Decomposition of phenylazotriphenylmethane (PAT) in the toluenes, for which a ρ value for −0.49 is found, suggests that Ph—N=N· radicals abstract hydrogen from the toluenes to contribute *ca.* 50% of the total benzene formed.[278] Vinyl radicals, generated during photolytic dechlorination of polycyclic chlorinated hydrocarbons, *e.g.* aldrin or dieldrin, show a clear preference for intermolecular reaction owing to the lack of correct intramolecular alignment of the vinyl radical-orbital and methylene-bridge hydrogen atoms.[279]

Hydrogen abstraction by $CF_3\cdot$ radicals from fluorine substituted alkanes shows a pattern similar to that from $CCl_3\cdot$ radicals with respect to sites of attack. An α-substituent has no effect on attack at the γ- and δ-positions. While the α-position in 1-fluorobutane is activated to attack (owing to a drop in activation energy) that in 1,1,1-trifluoropentane is strongly deactivated towards $CF_3\cdot$ radicals; in the latter compound, the reverse situation is found for attack by $CH_3\cdot$, making an explanation of the low reactivity of the β-position with respect to $CH_3\cdot$ difficult.[280] Activation energies have been determined for the reaction of $CF_2Cl\cdot$ radicals with Me_2O, Me_3N, and Me_4Si.[281] Relative rates of abstraction by $CCl_3\cdot$ radicals, produced on γ-irradiation of CCl_4, in cyclohexane–*n*-hexane liquid mixtures have been obtained with respect to self-recombination of $CCl_3\cdot$ radicals.[282]

Halogenation

Activation energies and Arrhenius factors have been calculated by the MO bond-index and group-contribution method for hydrogen- and chlorine-transfer reactions in the chlorination of paraffin hydrocarbons.[283] In the initial period vapour-phase chlorination of $CHCl_3$ involves chain-termination reactions Cl· + Cl· and Cl· + $CCl_3\cdot$ which do not appear to involve third bodies.[284] Regiospecificity in the chlorination of 1-chloropropane is influenced by the thermal complexing properties of *tert*-butylbenzene as solvent.[285] In contrast to the rate-retarding effect of a chlorine substituent on H abstraction by Br· from a secondary position in chloroalkanes, anchimeric assistance is observed for reaction at a vicinal tertiary position.[286]

Photo-induced chlorination and bromination of 1-fluoro-, 1-chloro-, or 1-bromo-adamantane proceed without loss or exchange of the 1-halo-atom; in the iodo-derivative only exchange occurs.[287] Deviations from the expected Goldfinger mechanism for chlorination of alkanes by *N*-chlorophthalimide have been attributed to equilibration of the initially formed alkyl radicals with HCl.[288] Bis(trimethylsilyl)bromamine is an effective brominating agent.[289] An unusual catalysis, by bromine, of the preparation of benzylidyne trichloride by photochlorination has been reported.[290] The participation of Cl· atoms in the chain chlorination of hydrocarbons by Bu^tOCl increases as the pressure is increased.[291] The products of the gas-phase photochlorination of tetramethylsilane and hexamethylsilane differ from those found in the solution reaction.[292]

Hydrogen Abstraction by Atoms

The reaction pathway and interaction terms in the abstraction of hydrogen from CH_4 by H· and Cl· atoms have been examined in an *ab initio* MO study,[293] and a re-evaluation of parameters in the empirical BEBO method, following criticisms, has restored the predictive ability of the method as applied, *inter alia*, to the calculation of activation energies and transition-state bond lengths in H-atom transfers.[294] Activation energies of atom-abstraction reactions have been correlated inversely with molecular polarizability.[295]

The phenomenological threshold energy[296] for the abstraction of deuterium by H· atoms from $CDCl_3$ is 34 ± 4 kJ mol^{-1}; the abstraction-to-substitution ratios per bond in the reactions of hot tritium atoms with Me_2O[297] and MeCl[298] are influenced markedly by the lone-pair electrons near the reaction site. The activation energies of the reactions, $H\cdot + RSH \rightarrow R\cdot + H_2S$ and $H\cdot + RSH \rightarrow RS\cdot + H_2$ differ by only 3.8 kcal mol^{-1} and this may account for some discrepancies in the literature on the photochemistry and radiation chemistry of alcohols when thiols are used as a source of H· atoms.[299] Addition of H· atom to thiolane in the gas phase leads to ring-opening; the resultant linear radical adds H·, to yield a chemically activated butane-1-thiol which can also be produced by initial H-abstraction from the thiol component of butane-1-thiol by H· followed by addition of H·.[300] Hydrogen abstraction is the sole reactive channel in the reaction of $O(^3P)$ atoms with ethylene oxide.[301] Fluorine-atom reactions with HCl, CH_4, $CHCl_3$, $CHCl_2F$, and $CHClF_2$, monitored by using atomic fluorine resonance absorption spectrometry in the far-vacuum-UV, have rate constants an order of magnitude greater than those for reactions of ·OH, and which correlate with the dipole moment of the substrate molecule rather than with C—H bond energy.[302]

The stratospherically important reaction of Cl· atoms with methane has received considerable attention. Rate constants determined by the resonance fluorescence technique[303, 304] are higher than those found from competitive chlorination studies[305] in the temperature range 220–300 K, and disagreement exists concerning possible curvature of the Arrhenius plots. Above 300 K, good agreement exists between the different methods. A new direct technique – a modification of the very low pressure pyrolysis (VLPP) method – has been applied to the $Cl\cdot + CH_4$ reaction to yield accurate measurements of rates and the equilibrium constant. Enthalpies of formation of free radicals, *e.g.* $CH_3\cdot$, accurate to ± 0.25 kcal mol^{-1}, are claimed.[306] Calculations on the reaction of Cl· with CH_2D_2 contradict the popular assumption that large kinetic isotope effects require symmetrical transition states.[307]

Abstraction of hydrogen, by Br atoms, from 2-substituted 5-methylbenzonitriles in benzene at 80 °C can be correlated with substituent σ^+ constants;[308] the abstracting reagent Br· involves a compromise between excessive and zero polar effects and enables a scale of Hammett free-radical substituent constants $\sigma\cdot$ to be constructed.[309] The logarithm of the relative rate constant of the bromination of 3-bromotoluene and 4-chlorotoluene by *N*-bromosuccinimide correlates with the Kirkwood solvent-polarity function $(\varepsilon - 1)/(2\varepsilon + 1)$ and confirms the expected separation of electric charges in the activated complex of the H-abstraction process.[310] Some kinetic evidence suggests that iodine atoms can abstract hydrogen from F_3SiH.[311]

Hydroxyl, Alkoxy-, and Peroxy-radicals

CNDO/2 calculations indicate that the electrophilicity of the hydroxyl radical dominates the H-abstraction from CH_4.[312] The partial reactivities of alcohols and carboxylate anions towards attack by ·OH have been related to the delocalizabilities of the hydrogen atoms, as calculated by the INDO method.[313]

Rate constants for H-abstraction from alkanes,[314] formaldehyde,[315] and other aldehydes[316, 317] by the ·OH radical in the gas phase have been determined by various methods in an effort to improve their accuracy. Abstractions by ·OH radicals from aliphatic alkanols in the gas phase,[318] from diethylene glycol dimethyl ether and its analogues in aqueous solution,[319] and from 1,3-dioxacycloalkanes in the liquid phase[320] have been studied.

Hydroxyl radical abstracts from the alkoxy-component of aliphatic esters in the vapour phase at 292 K.[321] In the reaction with aliphatic amines a ·OH radical behaves similar to $\cdot CH_3$ radicals.[322]

The kinetic isotope ratio k_H/k_D for the H-abstraction by Bu^{t}O· from Ph-CHDCH=CH_2 is constant at 2.90 ± 0.06 over the temperature range 25–115 °C, and is greater than the theoretical maximum of 1.2 for a linear transfer. The implied bent transition state (**65**) may be a consequence of overlap of the *p*-orbital in the *tert*-butoxyl radical with the olefin π-system.[323] The regioselectivity of

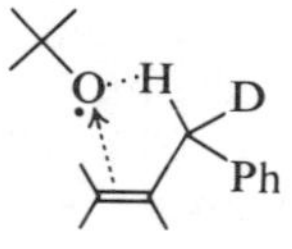

(**65**)

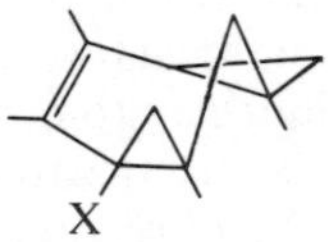

syn-(**66**)

H-abstraction by Bu^{t}O· from one of the equivalent allylic cyclopropyl positions in (**66**; X = H), and also from its *anti*-isomer, along with the retention of structural integrity, contrasts with the opening of the three-membered ring on solvolysis of the halide (**66**; X = Cl) *via* the carbocation.[324]

The relative reactivities of the α-hydrogen atoms in alkylbenzenes towards cumyloxy-radical are: primary 1.0; secondary, 4.5; tertiary, 7.5. The difference in the activation energies of hydrogen abstraction and fragmentation [$PhCHMe_2O\cdot \rightarrow Me\cdot + PhC(=O)Me$] is 26.1 kJ mol^{-1} in cyclohexane and 20.4 kJ mol^{-1} in toluene.[325] The chiral alkoxy-radical derived from photolysis of optically active 1-phenyl-1-methylpropyl hypochlorite abstracts hydrogen 1.42 times faster from (+)- than from (−)-2-phenylbutane, thus accomplishing a partial resolution.[326] Hammett correlations for the reaction of Bu^{t}O· with substituted anisoles monitored by kinetic ESR spectroscopy have been examined; the contribution of polar forms in the H-abstraction suggested by $\rho = -0.59$ in $C_2Cl_3F_3$ and $\rho = -0.74$ in 50/50 Bu^{t}OH–$C_2Cl_3F_3$ solvent at −11 °C is still in dispute since solvation of Bu^{t}O· may make the radical more electrophilic.[327]

After a nanosecond laser photolysis study of the reaction of Bu^{t}O· with Ph_2-CHOH and cumene, rate constants obtained by other techniques have been questioned.[328] The same workers have reaffirmed that alkoxyl-radicals are similar to carbonyl triplets in their reactions with alkylbenzenes, alcohols, ethers etc.; relative values of the rate constants obtained are in good agreement with those of other groups but the absolute values are much higher.[329] Reaction of MeO·, generated by azomethane photolysis in the presence of O_2, with CH_3CHO has a rate constant of 5.0 ± 10^6 M^{-1} s^{-1} at 25 °C in the gas phase.[330] Abstraction of hydrogen from the 1,3-diheterocycloalkanes (**67**), (**68**), and (**69**) by Bu^tO· in the

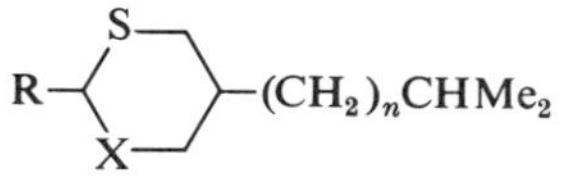

(**67**) (**a**) R = Pr^n; X = O; n = 2.
(**b**) R = Me; X = S; n = 1.

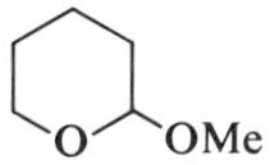

(**68**) 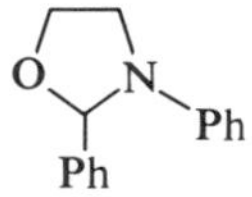 (**69**) 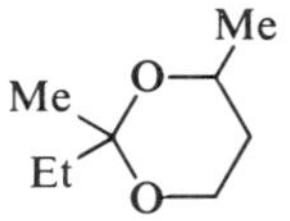 (**70**)

liquid phase occurs from the 2-position whereas abstraction from both 4- and 6-positions occurs with (**70**).[331] The reactivity of amines towards Bu^tO· in benzene parallels that in the quenching of carbonyl triplets but no charge-transfer quenching is possible when both reactants are in the ground state. However, the greater selectivity of Bu^tO· over Me· towards the amines indicates some charge-delocalization influence.[332]

The reactivity of *N,N*-diethylhydroxylamine towards Bu^tO· radical in benzene at 115 °C ($k = 7.7 \times 10^5$ M^{-1} s^{-1}), Et· radical in the gas phase at 20 °C ($k = 7.2 \times 10^5$ M^{-1} s^{-1}), and poly(peroxystyryl)peroxy-radical in styrene at 50 °C ($k = 2.9 \times 10^5$ M^{-1} s^{-1}) has been studied.[333] The H abstraction from Me_3SiH by Bu^tO· has been measured by a light-modulation ESR technique.[334] Abstractions of α- and β-H from Et_4Sn by Bu^tO· are fast; reversible elimination of alkene from the radical product after β-H-abstraction provides a useful source of trialkyltin radicals.[335]

The reaction of galvinoxyl radical with phenols, involving an initial slow H-abstraction step followed by a rapid cross-coupling of galvinoxyl and phenoxy-radicals, appears to be similar to that of DPPH with phenols in that galvinoxyl is electrophilic and polar resonance contributions in the transition state are important.[336] Rate constants and Arrhenius parameters have been determined for the reaction of cumylperoxy-radicals with benzoic acid esters[337] and for the reaction of Bu^tO_2· radicals with 3-methylpentane.[338]

Other Heteroradicals

Evidence has been adduced to support the counterclaim that radical chemistry can simulate diol dehydrase reactions;[339, 340] ethylene glycol is converted into acetaldehyde on anaerobic photolysis of organocobalamins in the presence of SH compounds such as dihydrolipoic acid amide, a thiyl radical being responsible for H-abstraction from the ethylene glycol.[340] Rate constants of H-abstraction by cyclohexanethiyl and benzenethiyl radicals from alkylbenzenes have been determined with a tritium-labelled thiol as solvent. Thiyl radicals are more selective than Br· or CCl_3· and show similar polar effects to Br· (attributed to similar polarizabilities).[341] Photolysis of C_6F_5SCl in toluene yields $C_6F_5SCH_2C_6H_5$, $(C_6F_5)_2S_2$, and $C_6H_5CH_2Cl$ arising from interactions of the initial C_6F_5S· and Cl· with benzyl radicals formed by H-abstraction.[342]

The chemistry of succinimidoyl radicals in the ground (**71**) and the excited (**72**) state has been reviewed.[343] It was suggested that a re-examination of the reactions of species such as Ph·, vinyl, acyl, acyloxyl, and amidyl radicals generated with different energetics would reveal possible differences between reactions of ground and excited doublet states.[343] In the presence of free halogen, imidyl radicals generated from *N*-haloimides are in the ground state (π_N) whereas in the presence of halogen scavengers such as olefins the excited state (σ_N) radicals are formed.

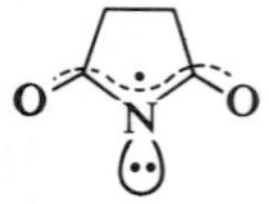

π_N ground state (**71**)

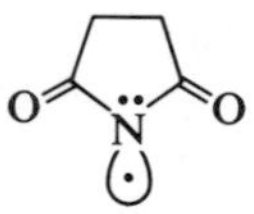

σ_N excited state (**72**)

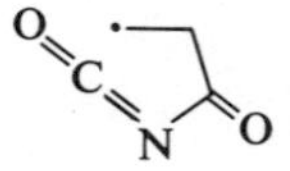

(**73**)

Ground-state imidoyl radicals are equally reactive towards neopentane and CH_2Cl_2 but excited state radicals (σ_N) are more reactive towards the latter.[344] A rapid and reversible cleavage of the σ_N radical (**72**) to give (**73**) has been postulated in order to rationalize many apparently contradictory observations.[345]

Bis(trimethylsilyl)aminyl radical, $(Me_3Si)_2N$·, is more reactive and more electrophilic than Me_2N·; H-abstraction from isobutane yields isobutyl radical as the main product rather than Bu^t·, due to steric effects.[346] Flash-photolysis of 1-chloro-2,2,6,6-tetramethylpiperidine gives the radical (**74**) (see p. 113) which on protonation abstracts H· from alcohols and adds to cyclohexene; the rates are diminished by steric crowding.[347] The cation radical of *N,N,N′,N′*-tetramethyl-*p*-phenylenediamine acts as a catalyst of the reduction of DPPH and phenoxy-radicals by MeCN.[348] A linear correlation between the rate constant of H-abstraction by DPPH and the ionization potential of the donor amine has been found.[349] Changes in the rate of reaction of DPPH with various hydroxylated derivatives of mesitylene are due to changes in the Arrhenius factors alone.[350]

The preferential reduction by $Bu^n{}_3SnH$ of certain α-chlorosilanes, *e.g.* Me_3SiCH_2Cl, over their all-carbon analogues, has been interpreted in terms of the greater stability of some α-silyl radicals due to $d\pi$–$p\pi$ overlap.[351] Below 950 K the thermolysis of Me_4Si and Me_3SiH proceeds *via* a short-chain mechanism,[352] thus: $Me_3Si\cdot + Me_4Si \rightarrow Me_3SiH + Me_3SiCH_2\cdot \rightarrow Me\cdot + Me_2Si{=}CH_2$, etc. The formation of disilacyclopropane from tetramethyldisilene is now thought to proceed *via* the intermediate (trimethylsilyl)methylsilylene rather than by double 1,2-migration of H from carbon to a silyl radical.[353]

Intramolecular Hydrogen Abstraction

At 690 K, in the gas phase, H-abstraction from the terminal methyl group of *n*-butoxy-radicals occurs with a rate constant of 2000 s^{-1} yielding $\cdot CH_2CH_2CH_2CH_2OH$ radicals.[354] The hydrogen-atom migration in ω-alkoxycarbonylalkyl radicals, $\cdot CH_2(CH_2)_nCH_2COOR$, to yield $CH_3(CH_2)_nCH(COOR)\cdot$ radicals has been studied with the aid of nitrosodurene as a spin trap. The ratio of rearranged to unrearranged radicals varied markedly with chain length, thus: $n = 4$, 0.25; $n = 3$, 1.0; $n = 2$, 0.10; $n = 1$, 0.03; it reveals the clear preference for a six-membered transition state.[355] The yield of *meso*- and racemic dibenzyl products $(PhR^1R^2C)_2$ formed by the action of $(Bu^tO)_2$ on the alkylbenzenes $PhCHR^1R^2$ (R^1 = H or Me;

R^2 = Et, Pr^n, Bu^n, pentyl, hexyl) decreases steadily with increase in chain length but rises again when the ($\omega-1$) alkyl radical (**75**) can readily undergo an intramolecular H-migration to yield a benzyl-type radical (**76**).[356]

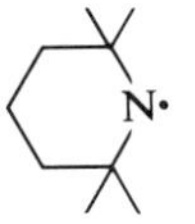

(**74**) 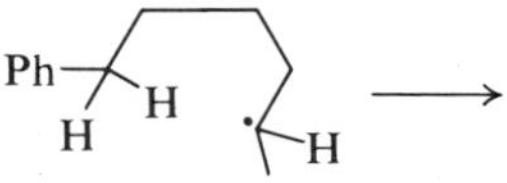 (**75**) 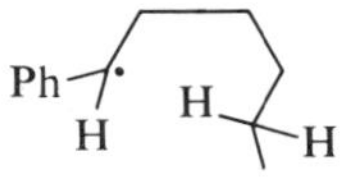 (**76**)

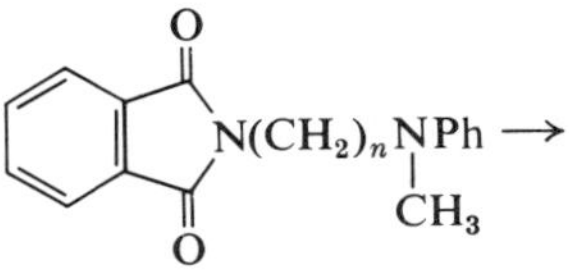

(**77**) 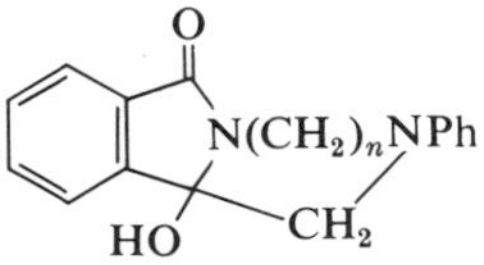 (**78**) 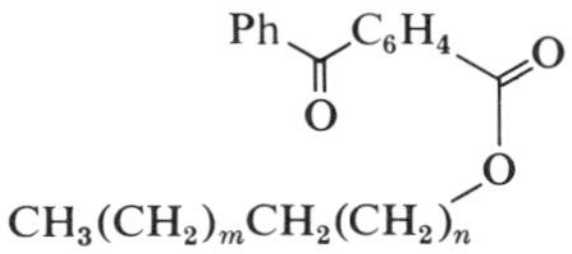 (**79**)

Intramolecular H-abstraction by an excited carbonyl group followed by C—C bond formation between the so-formed alkyl and ketyl radicals results in cyclization, *e.g.* (**77**)→(**78**).[357] The distribution of the sites of attack in the benzophenone derivatives (**79**) undergoing an analogous reaction provides a probe of the coiling of the alkyl chains, any theory of which must explicitly include the solvent.[358]

The kinetics of the intramolecular H-migration in the 3,5-di-*tert*-butyl-2-chloro-6-hydroxyphenoxy-radical have been examined.[359] Decomposition of *o*-di-*n*-propylaminosulphonylbenzenediazonium tetrafluoroborate in the presence of $CuBr_2$ and $Me_2S{=}O$ yields products arising from intramolecular 1,5-, 1,6-, and 1,7-H-transfer in the radical (**80**) as well as the normal "Sandmeyer-like" aryl bromide; the competition reflects the steric effects since (**81**) shows no intramolecular H-transfer.[360]

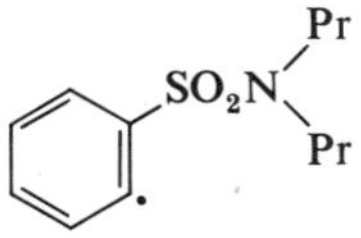

(**80**) 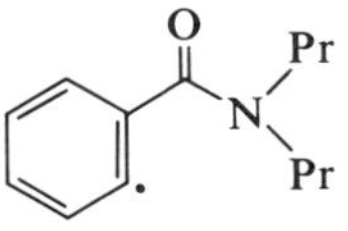 (**81**)

Halogen Atom Abstraction

A thorough study of the reaction of non-photolytically generated $CH_3\cdot$ radicals with Br_2, Cl_2, and BrCl has revealed that activation energies do not necessarily follow reaction exothermicities.[361] The ratio (r) of the rate constant for abstraction of Br· from CCl_3Br to that of Cl· abstraction from CCl_4 is 80 for the cubyl radical, surprisingly larger than that of 30 found for the 1-adamantyl radical, and indicative of less development of polar character in the abstraction by the cubyl radical.[362] Relative activation energies of radical–molecule metathesis reactions, including Cl-abstraction by cyclohexyl radicals, have been linearly correlated with bond dissociation energies and Taft polar (σ^*) and steric (E_s) substituent constants.[363] The α-methyl group of the penultimate methyl methacrylate (MMA) appears to obstruct chain transfer with CBr_4 in the copolymerization of styrene and MMA.[364]

The 2-tetrahydrofuryloxyl radical abstracts a halogen atom directly from *meso*-stilbene dibromide before scission to 3-formyloxypropyl radical occurs.[365] Removal of bromine from the dienone (**82**) by phenoxy-radicals is facilitated by electron-withdrawing substituents in the radical.[366] Inversion of stereochemistry in 4-*tert*-butyl-1-methyl-1-silacyclohexyl radicals competes with chlorine-abstraction from polychloroalkanes, configurational purity in the chlorosilane product being systematically lost as the halocarbon reactivity diminishes.[367] While $Me_3Ge\cdot$ radicals are less selective than $Me_3Sn\cdot$ towards monohalogen substituted alkanes, in Cl-abstraction from compounds of type $RCCl_3$ they are more selective: the role of polarity effects *vis-a-vis* bond strength effects is greater in the case of $Me_3Ge\cdot$ radicals.[368]

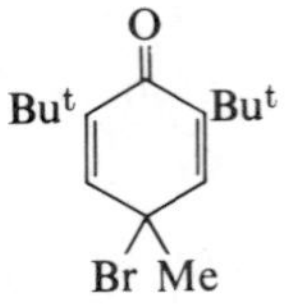

O, Bu^t, Bu^t, Br Me

(**82**)

R^2, R^3, R^1, R^4, Br_2

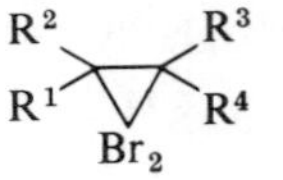

(**83**)

Attack of $Bu_3Sn\cdot$ radicals on *gem*-dibromocyclopropanes (**83**) occurs at the less hindered C—Br bond. The resulting pyramidal cyclopropyl radical can invert rapidly and the configurational stability depends on the β- as well as on the α-substituents.[369] Iodine-atom abstraction by $Br\cdot$ from C_6F_5I in the gas phase does not seem to take place but BrI is formed by addition to the benzene ring followed by decomposition of the adduct.[370] The reaction of I_2 with CH_3CF_2Br involves abstraction of Br by iodine atoms.[371] Rate constants for the reaction of ground-state lead atoms, $Pb(6^3P_0)$, with alkyl bromides have been obtained by atomic absorption spectroscopy.[372] Optimum conditions for the generation of alkyl radicals by the reaction of sodium with alkyl halides in cool flames have been explored.[373]

Addition Reactions

The kinetics of free-radical addition reactions in the gas phase, together with some theoretical calculations, have been reviewed.[374] Unrestricted MINDO/3 calculations of the addition of $Me\cdot$ to ethylene, acetylene, propene, and allene correctly predict activation energies and points of attack; in the cases of propene and allene the transition states are unusual in being very reactant-like, except for the attacking methyl which is pyramidal.[375]

Carbon-centred Radicals

The gradation of reactivity of the olefins (**84**) towards (substituted) alkyl radicals $X\cdot$ is no unequivocal proof of the greater stability of the secondary radical (**86b**) than of (**86a**) because the difference in the rate of addition of electron-deficient radicals $X\cdot$ can also be due to the polar effect of Me in the transition state (**85**).[376]

$$X\cdot + H_2C{=}CHY \longrightarrow [X\cdots CH_2 \overset{\dots}{=} CHY]^{\cdot\ddagger} \longrightarrow XCH_2{-}\dot{C}HY$$

(**84**) (**85**) (**86**)

(**a**), Y = H; (**b**), Y = Me.

A novel radical equilibrium method, $R^1\cdot \rightleftharpoons R^2\cdot + \text{olefin}$, involving determination of the recombination products of radicals $R^1\cdot$ and $R^2\cdot$, has been used to determine *inter alia* the enthalpy of formation of the *n*-propyl radical.[377]

Activation energies and Arrhenius factors for the addition of various alkyl radicals to ethylene in the liquid phase are largely independent of radical structure;[378] addition of Et• to alkyl-substituted olefins has an isokinetic temperature of 220 ± 15 K,[379,380] while that for $HOOCCMe_2$• radical is 290 ± 15 K.[380] The nucleophilic character of alkyl radicals shows itself in the greater reactivity towards acrylic than styryl monomers as determined by a competitive method involving electron-transfer oxidation by copper(II) acetate versus addition to double bonds, as well as by measurement of relative rates of alternating addition.[381] There is no evidence that 5-methylenecyclohexa-1,3-diene, used as a model for a postulated dimer in the radical polymerization of styrene, can initiate such a polymerization.[382]

Addition to the 6,7-alkene centre of 3,7-dimethylocta-1,6-diene by difunctional radical addends, *e.g.* ketone triplets, and addition to the 1,2-alkene centre by monofunctional addends, *e.g.* $Me_2C(OH)$•, can be rationalized by means of frontier-orbital control in some, but not in all, cases; reversible addition to the 6,7-bond may be a factor.[383] Products from the reaction of tetrahydrofuran with propadiene initiated by $(BzO)_2$ can be rationalized by means of addition of the tetrahydro-2-furyl radical to both central and terminal positions of the allene.[384] The rate of addition of a •CH_2COOCH_3 radical to dec-1-ene is reduced four-fold when an α-hydrogen atom is replaced by methyl but only two-fold by a chlorine substitution; in the latter case the steric effect is counterbalanced by a polar effect.[385] The radical adduct (**87**) is stable in hot mixtures when derived from the olefin (**88**) which bears both electron donor (X) and acceptor (Y) substituents.[386]

$$H_2C{=}C(X)(Y) + R\cdot \longrightarrow RCH_2{-}\dot{C}(X)(Y)$$

(**88**) (**87**)

(**a**) X = —N(morpholino)O , Y = —CN

(**b**) X = —SMe, Y = —CN.

R• = $Me_2\dot{C}CN$

Homolytic alkylation of benzoquinone in the presence of Cu^{2+} ions involves a polar transition state.[387] Addition of isomeric xylenes to methyl acrylate, initiated by $(Bu^tO)_2$, can be accomplished at 200–300 °C in an autoclave but not in an open reactor at 140–150 °C.[388] Ionization constants for alkynes have been correlated with the rate constants for the their reaction with chloromethyne radical.[389] The mechanisms of addition of CH_2Cl_2 to ethylene,[390] and of CH_2Cl_2 to olefins assisted by dichlorobis(triphenylphosphine)nickel(II)[391] have been investigated. Activation-parameter differences between intermolecular Cl-abstraction and transannular H-abstraction in the addition of CCl_4 to *cis*-cyclo-octene catalysed by dichlorotris(triphenylphosphine)ruthenium(II) have been determined.[392]

Addition of $ClCH_2$• radicals to vinyl fluoride occurs mainly at the β-position, due primarily to the higher enthalpy of activation for addition at the α (CHF) position: this is not attributable to resonance delocalization effects since addition to ethylene occurs more readily than α/β addition to $CH_2{=}CHF$; for the electrophilic radical, •CH_2Cl, addition is controlled by the olefin HOMO whereas addition of the nucleophilic •CH_3 is controlled by the olefin LUMO.[393] Rates of addition of methyl radical to fluorine-substituted ethylenes, in iso-octane at 65 °C, decreased in the order $C_2F_4 > CF_2{=}CHF > CH_2{=}CHF > CH_2{=}CH_2$.[394] Solvent effects on the kinetics and regioselectivity of the addition of cyclopropyl radical to fluoro-olefins are indicative of p–π repulsion-induced distortions in the π-clouds of the olefins.[395] Addition of fluorine-containing radicals to olefins has been reviewed.[396] After revision of the relative Arrhenius parameters for *n*-heptafluoropropyl radicals, a gradual increase in the selectivity of $CF_3\cdot$, $CF_3CF_2\cdot$, and $CF_3CF_2CF_2\cdot$ radicals in addition to fluoroethylenes is found; it is attributable to changes in the activation-energy term.[397] Attack of $CF_3\cdot$ radicals, generated by electrolysis of trifluoroacetate anion in aqueous MeCN, on the olefins $CH_2{=}CHX$ (where X = Pr^n, Bu^n, COOMe, CN, or CH_2COOMe) yields a mixture of the products $CF_3CH_2CH_2X$, $CF_3CH{=}CHX$, $CF_3CH_2CH(CF_3)X$, and $(CF_3CH_2\text{-}CHX)_2$.[398] Addition of $CF_3\cdot$ radicals to various vinyl monomers at 164 °C occurs exclusively at the unsubstituted end of the alkene, with little variation in rate.[399]

It is uncertain whether the reaction of phenyl radicals with thiourea to form $PhSC({=}NH)NH_2$ involves addition as the primary step.[400] Addition of substituted alkyl radicals X· to *N*-methylene-*tert*-butylamine provides a new route to aminyl radicals, $\cdot N(Bu^t)CH_2X$.[401] The reaction of methyl radicals with perfluorocarboxylic acid nitriles has been investigated.[402]

Halogen Atoms

Addition reactions of radioactive fluorine-18 atoms have been reviewed.[403] *cis–trans* Isomerization accompanies the reaction of 2,3-dichlorohexafluorobut-2-ene with recoil $^{38}Cl\cdot$ atoms.[404] The transition state in the addition to ethylene occurs later in the case of F· than of $CH_3\cdot$, with considerable charge polarization in the alkene arising mainly from the transfer of a β-electron to the fluorine.[405] Ionic and radical pathways in the halogenation of cyclopropylacetylene have been discussed.[406] The radical chain chlorination of 3,4-dimethyl-1,2,5-oxadiazole,[407] and the bromination of styrene in the dark,[408] have been studied. Dispiro[2.0.2.2]oct-7-ene (**89**) homolytically adds HBr to give an 80% yield of (**90**).[409]

Br

$\xrightarrow[\text{HBr}]{(BzO)_2}$

(**89**) (**90**)

The isomerization of *E,Z*- and *E,E*- 1,4-dialkylbutadienes in the presence of iodine atoms is reversible, but the *Z,Z* dienes do not isomerize at all; that iodine atoms add to the terminal atom in the dienes is confirmed by the lack of racemization of the chiral centres in *E,Z*-(3*S*,8*S*)-3,8-dimethyldeca-4,6-diene on isomerization.[410]

Heteroradicals

The reactions of H· atoms with carbon suboxide,[411] alkenes,[412] and acrylaldehyde[413] in the gas phase have been studied. Addition of oxygen (^3P) atoms to the terminal exocyclic carbon of methylenecycloalkanes (**91**) is more favoured as the stability of the resultant cycloalkyl radical site increases. Thus the observed product ratio (**92**)/[(**93**)+(**94**)] is 2.9, 4.6, and 21 for n = 3, 4, and 5, respectively.[414]

The kinetics of the addition of oxygen (^3P) atoms and hydroxyl radicals to 2-methylbut-2-ene have been monitored by the flash-photolysis—NO_2-chemiluminescence technique.[415] Of importance for tropospheric chemistry are the

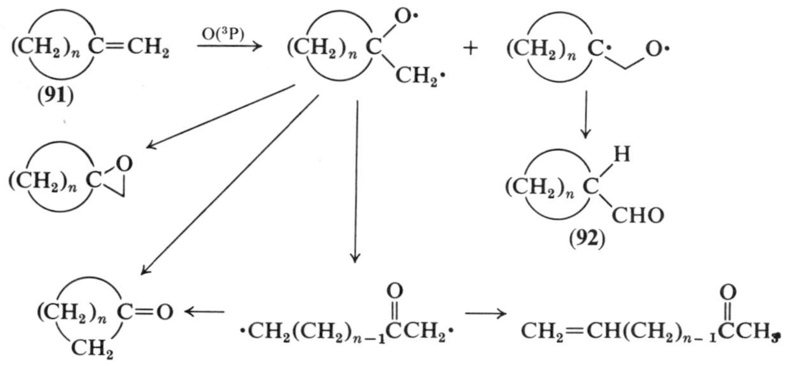

(91) (92) (93) (94)

reactions of •OH radical with CO and acetylene.[416] Hydroxyl radical reactions with allene, buta-1,3-diene, and 3-methylbut-1-ene,[417] with olefins in the presence of N_2O and H_2,[418] with olefin–nitric oxide mixtures,[419] and with nitric oxide[420] are all of interest in the study of polluted atmospheres and of photochemical smog formation. Photochemical oxidation of alkylbenzenes by nitrous acid involves addition of •OH to the aromatic ring and also H-abstraction from the side-chain.[421] A kinetic study, motivated by the need to devise appropriate control strategies for the degradation of aromatics in the troposhpere, has revealed that •OH radicals react with alkylbenzenes, C_6F_6, and pentafluoro-*n*-propylbenzene predominantly by addition; perfluorination of the aromatic ring lowers the ring electron density and reduces the rate of addition of the electrophilic •OH by six-fold.[422] The NIH shifts of deuterium during the oxidation of chlorobenzenes, anisoles, and naphthalene by iron(II) perchlorate and H_2O_2 in MeCN lead to the conclusion that reactions occur by addition of an oxidizing radical (*e.g.* "•OH" or an iron derivative FeO^{3+}/FeO^{2+}) followed by oxidation of the radical adduct to a cyclohexadienyl cation, and thence to phenols.[423]

Pulse radiolysis of aqueous solutions of acetylene leads to the enol radical (**95**) which rapidly forms the formylmethyl radical (**96**); addition of (**96**) to acetylene is at least 100 times slower than that of (**95**).[424] The very slow formation of •OH radical in the Haber–Weiss reaction ($O_2^{\cdot -} + H_2O_2 \rightarrow \cdot OH + OH^- + O_2$) can be detected by the addition of •OH to zone-refined sodium benzoate, to yield sodium salicylate which is readily determined fluorometrically.[425] Thiophene is more reactive than benzene towards addition of hydroxyl radical, and formation of the

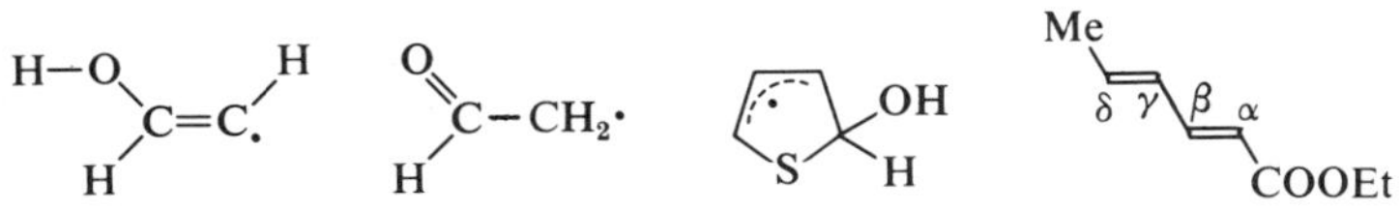

(95) (96) (97) (98)

2-hydroxythienyl radical (**97**) occurs at a near diffusion-controlled rate; (**97**), unlike the analogous radical in the benzene system (which gives rise to phenol, phenyl-phenol and biphenyl), undergoes radical–radical reactions with itself to yield 2-thiolactone and 2,2′-bithiophene, except in basic solution where C—S bond cleavage dominates.[426]

Ionic addition to ethyl sorbate (**98**) occurs essentially at the γ,δ bond whereas radical conditions with, *e.g.* MeOCl, MeOBr, or NBS, result in attack at the β- and δ-positions, the latter being preferred.[427] Similarly, free-radical addition of methyl and *tert*-butyl hypochlorite to cyclopentadiene yields mainly the *trans*-1,4 adduct, and the 1,2-adduct is the anti-Markonikoff product.[428] The probability of addition to olefins decreases in the order $R_3SiO\cdot > R_3GeO\cdot > R_3CO\cdot$ (where R = alkyl) while the probability of H-abstraction increases in the same series.[429]

Absolute rates of addition of sulphur atoms to alkynes have been measured by vacuum-UV kinetic spectroscopy.[430] In micellar solutions of sodium dodecane-1-sulphonate containing hydrogen sulphite ions γ-irradiation produces $\cdot SO_3^-$ radicals which add to dodec-1-ene at the terminal carbon; the resultant adduct radical can abstract H from HSO_3^- or combine with a further $\cdot SO_3^-$ radical.[431] Absorption bands attributable to transient radical adducts have been detected on flash-photolysis of $(Bu^tS)_2$ in the presence of styrene and 1,1-diphenylethylene, but the measured kinetics do not correspond well with rotating sector measurements.[432] Photochemical decomposition of *S*-phenyl and *S*-*n*-butyl thioacetate in the presence of acetophenone or benzophenone leads to the formation of thiyl radical adducts to the carbonyl group, as judged by spin-trapping experiments; other modes of generation of thiyl radicals yielded analogous results.[433]

Benzenethiyl radical regiospecifically approaches the sterically less hindered bridgehead position C(7) of the bicyclo[1.1.0]butane component in (**99**), forming

X　　H PhS X　　Cl (EtO)$_2$P(=O)—N

(**99**) 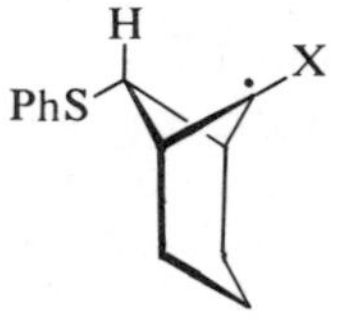　(**100**) 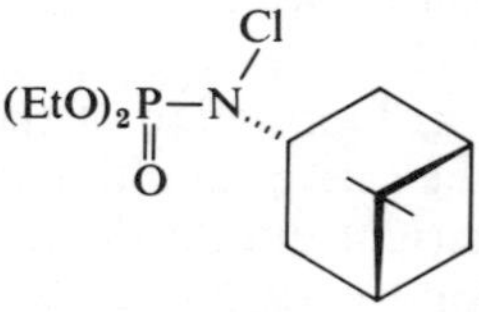　(**101**)

X = Me, Ph, SMe, Cl or CN

the more stable of the two possible cyclobutyl radicals (**100**) *via* opening of the strained C(6)—C(7) bond; steric factors in the bicyclobutane unit determine the outcome.[434] Thiocyanation of alkenes by thiocyanogen and UV light (previously believed to be a heterolytic reaction) has been shown to be a radical chain process; additions to acyclic alkenes are *trans*-stereoselective while additions to cyclohexenes are *trans*-stereospecific.[435] Addition of SF_5Br to both *cis*- and *trans*-1,2-difluoroethylene yields the same relative amounts of diastereomers; the radical adduct from $SF_5\cdot$ attack on the olefin abstracts bromine from SF_5Br.[436]

The initial adduct, $F_3CCF{=}NNF_2$, from photochemical reaction of N_2F_4 and CF_3CN, rearranges and decomposes.[437] Photolysis of tetramethyl-2-tetrazene, in acidic media only, leads to the formation of $Me_2NH^{+\cdot}$ and $Me_2N\cdot$ radicals; in the presence of oxygen and an olefin, *e.g.* cyclohexene or norbornene, 1-amino-2-hydroperoxides are formed by addition reactions.[438] Light- or $(BzO)_2$-initiated reaction of olefins and *N*-halosuccinimide in halogenated solvents involves excited state (σ_O or σ_N) imidyl radicals which behave similarly to Cl· and ·OH in addition.[439] Chromous chloride promotes a highly regiospecific addition of *N*-haloamides to enol ethers such as 1-methoxycyclohexene.[440] Addition of *N*-chlorophosphoramidate radical, $(EtO)_2P({=}O){-}N(Cl)\cdot$, to α-pinene occurs *trans* to the isopropylidene bridge to yield (**101**).[441]

$SiPh_3\cdot$ and $GePh_3\cdot$ radicals form adducts with 4-substituted pyridines by attachment at the nitrogen atom.[442] Addition of $X_3Ge\cdot$ radicals (X = Et, Ph, or mesityl) and the $Ph_2ClGe\cdot$ radical to oxaziranes occurs with cleavage of the N—O bond to give (α-aminoalkoxy)germanes.[443] ESR evidence has been adduced for the existence of a penta-co-ordinate stannylperoxy-radical formed from $Bu^n{}_3Sn\cdot$ and O_2.[444] $R_3Ge\cdot$ and $R_3Sn\cdot$ radicals add to aromatic ketones and 1,2 quinones at oxygen;[445] in the adducts from the latter substrate a tautomeric equilibrium exists, but this is not so in the adduct of $\cdot Mn(CO)_5$ with 1,2-diketones.[446] The hydrogenation of α-methylstyrene by pentacarbonylhydridomanganese(I) occurs by a free-radical addition mechanism, as shown by CIDNP studies.[447] A stereospecific insertion reaction of MeOOCC≡CCOOMe, (Acet), with a platinum hydride, *cis/trans* $PtHCl(PEt_3)_2$, involves the radical $\cdot Pt(I)Cl(PEt_3)_2$ (Acet) in a chain process.[449]

Intramolecular Addition

In 5,6-unsaturated hexyl radicals substituted at the 5-position, *e.g.* (**103**), with highly radical-stabilizing groups there is a competition between intermolecular H-abstraction from Bu–SnH and irreversible cyclization reactions; the ratio of six- (**102**) to five-membered (**104**) ring formation is an order of magnitude lower for X = Me than for X = Ph.[449]

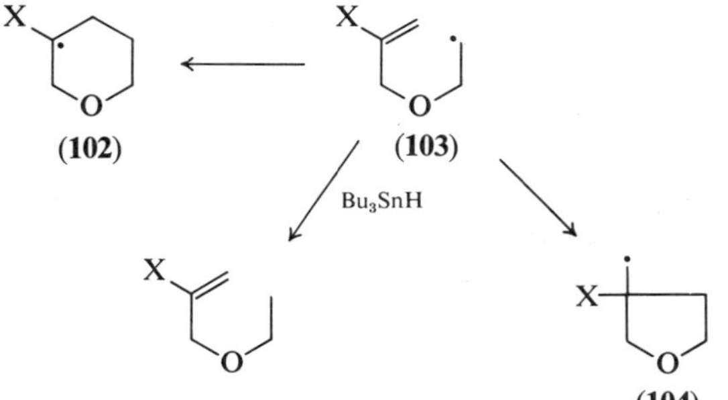

(**102**) (**103**) Bu_3SnH (**104**)

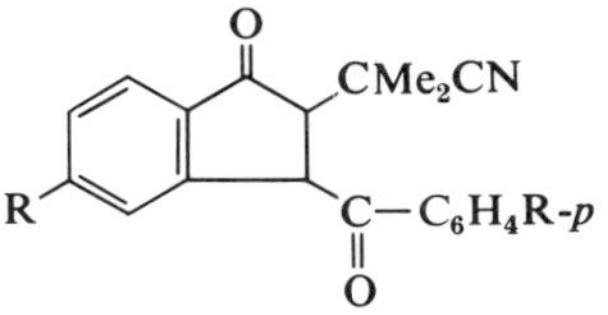

(**107**) R = H or Me

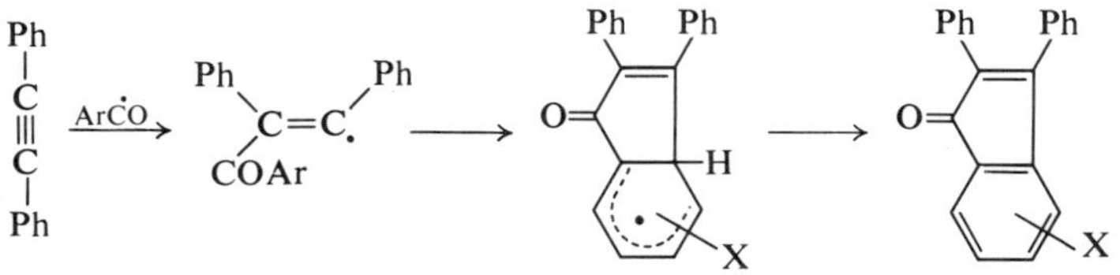

(**105**) (**106**)

Indenones (**106**) can be synthesized by radical addition of aromatic aldehydes to diphenylacetylene; this involves cyclization of the intermediate vinyl radical (**105**).[450] Reaction of $\cdot CMe_2CN$ radicals with *trans*-PhCOCH=CHCOPh yields the expected products of addition to the olefinic bond and also the cyclized product (**107**).[451] Intramolecular addition of *N*-methylcarboxamido-radicals, derived from olefinic *N*-chloroamides and probably in the excited σ_N state, is highly efficient whereas this reaction mode of the analogous acetamido-radicals occurs very inefficiently.[452] Chromous chloride promotes such cyclizations, *e.g.* that of *N*-chloro-*N*-methylcycloheptenecarboxamide,[453] acting as a radical chain inducer rather than in a redox fashion.[454]

Photolysis of (**108**) leads to an aminium radical (**109**), which cyclizes exclusively at C(6) rather than C(5), to give the thermodynamically less stable carbon radical (**110**), because of the good orbital overlap with the minimum of molecular motion.[455] A similar reaction of (**111**) forms carbon radicals (**112**) which can be intercepted cleanly to yield nitrate esters or halides having pyrrolidine rings.[456]

(**108**) $\xrightarrow[(H^+)]{h\nu}$ (**109**) ⟶ (**110**)

(**111**) $\xrightarrow[(H^+)]{h\nu}$ (**112**)

(**113**) ⟶ (**114**) $\xrightarrow{-N_2}$ Carbazolyl radical

(**115**)

Generation of radical (**113**) from the corresponding diazonium salt in the presence of iodide ion leads to the normal products, together with those derived from the carbazolyl radical formed by loss of N_2 from the cyclized radical (**114**). Evidence for cyclization was not found in the case of radical (**115**).[457]

Fragmentations

The relative rates of decarbonylation of cycloalkylcarbonyl and 1-methylcycloalkylcarbonyl radicals vary with ring size in the order: $6 \cong 5 > 4 > 3$, reflecting the stability of the radicals.[458] *tert*-Pentyloxy radicals, $Me_2CEtO\cdot$, fragment to give ethyl radicals and acetone, or methyl radicals and butanone in the ratio of 80 : 1.[459] Several studies on the fragmentation of radicals derived from acyclic acetals[460] and 2-alkoxytetrahydropyrans[461, 462] have been reported. β-(Phenylsulphinyl)-*sec*-butyl radicals (**116**) eliminate phenylsulphinyl radicals stereoselectively:[463] this stereoselectivity is the result of rapid loss of $Ph\dot{S}O$ radicals before the initial non-equilibrium conformations of the radical can undergo rotation about the C2—C3

$$\text{MeCH(Br)–CH(Me)SOPh} \xrightarrow{Bu_3Sn\cdot} \text{MeĊH–CH(Me)SOPh} \longrightarrow \text{MeCH=CHMe}$$

(116)

bond. C—O rather than C—N cleavage occurs in the fragmentation of *N,N*-dialkylaminoalkoxymethyl radicals:[464]

$$Et\cdot + EtN{=}CHOR \;\not\longleftarrow\; Et_2N\dot{C}HOR \longrightarrow HCONEt_2 + R\cdot$$

Methyl radicals are eliminated from 6-ethoxy-1,2-dihydro-2,2,4-trimethylquinolin-1-yl radicals.[465] α-Acyloxyalkyl radicals generated in the γ-radiolysis of esters fragment to give acyl radicals and ketones:[466] $R^1COO\dot{C}R^2R^3 \rightarrow R^1\dot{C}O + R^2COR^3$. Radicals anions (**117**) generated from esters undergo fragmentation by path (a) or path (b); when R^2 is bulky the former path is favoured owing to release of strain on fragmentation.[466–468]

$$R^1COOR^2 \xrightarrow[EtNH_2]{Li} R^1\text{–}\dot{C}(O^-)\text{–}O\text{–}R^2 \;(\mathbf{117}) \xrightarrow{(a)} R^2\cdot + R^1COO^- \quad ; \quad \xrightarrow{(b)} R^1\dot{C}O + R^2O^-$$

The radical $(CH_3O)_2\dot{C}CH_2Cl$ loses chloride to give the radical cation, $CH_3\overset{+}{O}{=}C(OCH_3)\dot{C}H_2$.[469]

Homolytic Aromatic Substitution

Homolytic aromatic substitution has been reviewed.[470] The rates of addition of phenyl radicals to benzene and chlorobenzene are 1.03×10^6 and 1.32×10^6 l mol^{-1} s^{-1};[471] this is very substantially *less* than the diffusion-controlled limit, the approach to which cannot therefore be invoked to account for the similar rates of reaction of benzenoid compounds with electron-donating or electron-withdrawing substituents. The rates of reaction are of the same order of magnitude (*ca.* 10^6 l mol^{-1} s^{-1}) for reactions of the *p*-carboxyphenyl radical with *p*-bromobenzoate anion and benzene[472] and for the reaction of $Cl_2^{\dot{-}}$ with benzene.[473] Rates of alkylation of benzene derivatives are much lower, $\sim 10^2$ l mol^{-1} s^{-1}.[474] Rates of alkylation of protonated heteroaromatic bases are much higher (10^5–10^8 l mol^{-1} s^{-1});[474] the latter reactions are much more selective[475] than the former and this failure of the reactivity–selectivity principle finds explanation both in the polar transition state for the alkylation of protonated heteroaromatics and in frontier-orbital theory. Thus, benzene derivatives have considerably higher HOMO and LUMO energies than protonated heteroaromatic compounds; the nucleophilic alkyl radicals have relatively high energy SOMO's and hence for protonated heteroaromatics the strong SOMO–LUMO interactions determine the high reactivity and high selectivity. The corresponding interaction for benzene derivatives is much weaker, giving lower reactivity and selectivity. There is some evidence for reversibility in homolytic alkylations of benzene derivatives.[474]

The acetoxyethylation of benzene derivatives has been accomplished by reaction of di-*tert*-butyl peroxide with 2-methyl-1,3-dioxolane in presence of the aromatic substrates: the positive value of the Hammett reaction constant for this process ($\rho = 0.24$) indicates that acetoxyethyl radicals, $AcOCH_2CH_2\cdot$, are slightly nucleophilic.[476] Aromatic nitromethylation has been effected by generation of nitromethyl radicals by oxidation of nitromethane with manganese(III) acetate; these radicals are electrophilic, as shown by the negative ρ value for the reaction ($\rho = -1.1$).[477] The homolytic alkylation of anthrone with 1-cyano-1-methylethyl radicals, $Me_2\dot{C}CN$ proceeds much more efficiently in basic media when alkylation of anthranol takes place.[478] Benzylation of quinoline in acid solution leads to 2- and 4-benzylquinolines, but when the reaction is carried out under reducing conditions, in presence of titanium(III), 2,4-dibenzyl-1,2,3,4-tetrahydroquinoline is formed;[479] the variation in the proportions of the reaction products with conditions suggests that the alkylation process is reversible.

The homolytic chlorination of 2-phenylnaphthalene has been found to occur by initial attack at the 1-position, to give the 1-chloro-1,2-dihydro-2-phenyl-2-naphthyl radical.[480] The initial step in the vapour-phase chlorination of benzene involves hydrogen abstraction by chlorine atoms, is endothermic by 6 kcal mol^{-1}, and proceeds *via* a transition state which closely resembles the free phenyl radical.[481]

$$PhH + Cl\cdot \longrightarrow Ph\cdot + HCl$$
$$Ph\cdot + Cl_2 \longrightarrow PhCl + Cl\cdot$$

Homolytic nitration of toluene with tetranitromethane at 300 °C gives an almost statistical mixture of *o*-, *m*- and *p*-nitrotoluene.[482]

The main products resulting from hydroxylation of aromatic compounds by $O(^3P)$ atoms are mixtures of the corresponding isomeric phenols but products resulting from migration of an alkyl group are also obtained (see Scheme 4).[483]

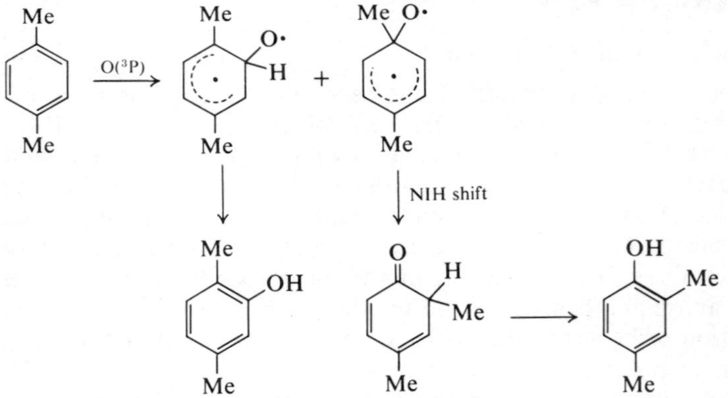

Me
O(³P)
O·
H
Me O·
NIH shift
OH
O
H
Me

SCHEME 4

NIH deuterium migrations occur in the analogous reaction with 2,4,6-trideuteriotoluene (see Scheme 5).[484] An NIH shift of deuterium also occurs in oxidations of aromatic compounds with iron(III) perchlorate and hydrogen peroxide in acetonitrile.[485] The active oxidant in these reactions is HO· or its equivalent FeO^{3+} or FeO^{2+}. The results have an important bearing on the mechanism of hydroxylation by cytochrome-P-450-dependent oxygenases. Hydroxylation of pyridine and pyridinecarboxylic acids takes place at the least electron-deficient positions of the pyridine ring, i.e. at the 3- or 5-position, consistently with the electrophilic character of HO·.[486]

The photochemical oxidation of benzene, toluene, or ethylbenzene with hydroxyl radicals or oxygen in presence of NO or NO_2 yields mixtures of phenols and nitro-compounds;[487–489] initially the aromatic substrate is attacked by $O(^3P)$ or HO·, and the resultant cyclohexadienyl radical is subsequently attacked by NO_2.

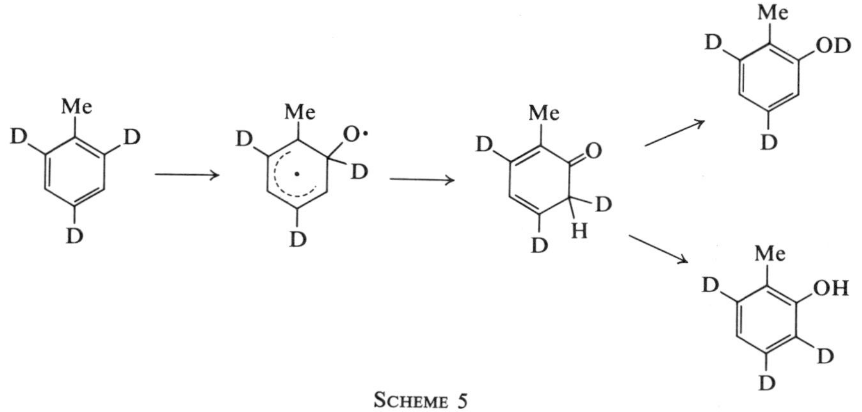

SCHEME 5

The cathodic reduction of 2-halobenzanilides affords aryl radicals which undergo intramolecular attack on the other aromatic ring to give the radicals **(118)** and **(119)** which subsequently afford **(120)** and **(121)**.[490, 491]

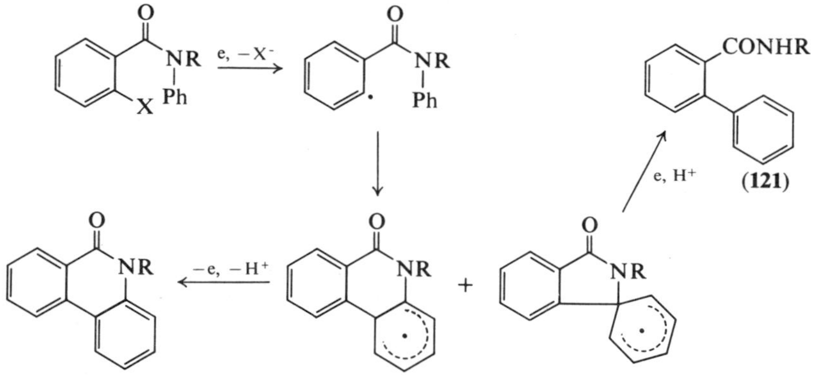

(120) **(118)** **(119)** **(121)**

Further examples of *ipso*-substitution have been reported;[492] an interesting one involves the alkyldenitration of benzene derivatives, containing electron-withdrawing groups in the *para*-position, by reaction with the nucleophilic 1-adamantyl radicals,[493] as shown in the annexed reaction sequence where Ad = adamantyl.

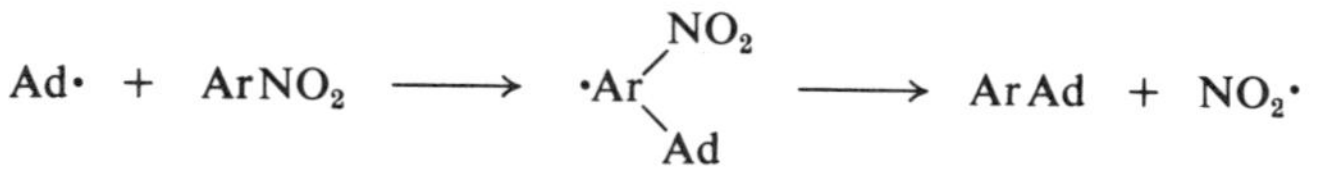

$$Ad\cdot + ArNO_2 \longrightarrow \cdot Ar(NO_2)(Ad) \longrightarrow ArAd + NO_2\cdot$$

Decomposition of dibenzoyl peroxide in polyfluorobenzenes gives biaryls resulting from displacement of both hydrogen and fluorine; the defluorinated products arise from *ipso*-substitution.[494] Replacement of fluorine also occurs in the photoreaction of hexafluorobenzene with cyclohexane in presence of benzophenone: the reaction involves radical displacement of fluorine by cyclohexyl radicals.[495] Other examples of *ipso*-substitution occur in the photo-initiated chlorination of *p*-chloronitrobenzene (which gives *p*-dichlorobenzene)[496] and in the formation of bromopentafluorobenzene by gas-phase bromination of pentafluoroiodobenzene.[497]

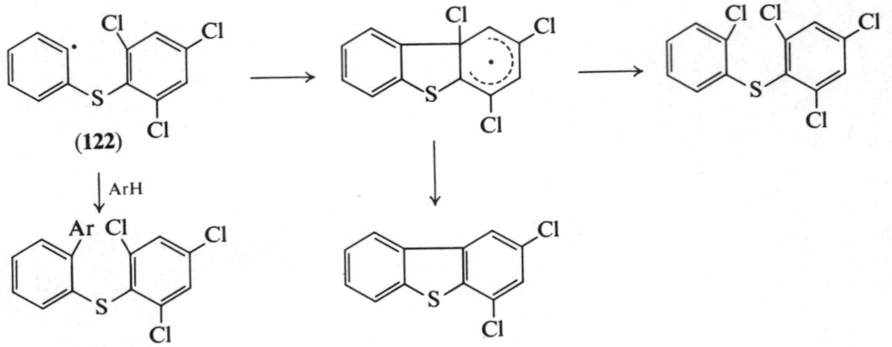

(122)

ArH

SCHEME 6

Intramolecular *ipso*-substitution is encountered in the replacement of chlorine in reactions of 2-(2,4,6-trichlorophenylthio)phenyl radicals (**122**); Scheme 6 depicts this and other reactions of these radicals.[498] A further report of *ipso*-substitution and intramolecular homolytic aromatic substitution of radicals derived from reactions of *N*-arenesulphonyl-1-iodomethylpiperidines with tributyltin radicals has been published.[499, 500] The *ipso*-substitution process is favoured with both electron-donating and electron-withdrawing substituents, emphasizing the importance of polar contributions to resonance stabilization of the transition state $[RCH_2 + YPhX] \leftrightarrow [R\overset{+}{C}H_2\dot{Y} : \overset{-}{Ph}X] \leftrightarrow [RCH_2^{-}\ \dot{Y}\overset{+}{Ph}X]$; the polar form contributing to the transition state depends on whether the attacking radical has nucleophilic or electrophilic character. The direct aromatic substitution process is facilitated by electron-releasing substituents ($\rho = -0.31$) for this reaction.

Deactivation of triplet benzophenone by aromatic compounds has long been believed to involve formation of the biradical (**123**) which decomposes to give ground-state benzophenone and the aromatic substrate.[501] Direct evidence for the

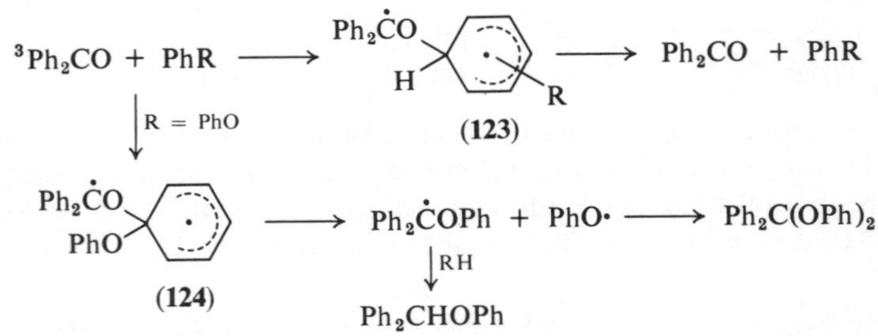

$$^{3}Ph_2CO + PhR \longrightarrow (\mathbf{123}) \longrightarrow Ph_2CO + PhR$$

R = PhO

$$(\mathbf{124}) \longrightarrow Ph_2\dot{C}OPh + PhO\cdot \longrightarrow Ph_2C(OPh)_2$$

$$Ph_2\dot{C}OPh \xrightarrow{RH} Ph_2CHOPh$$

formation of biradical species comes from the photolysis of benzophenone in diphenyl ether; *ipso*-substitution gives rise to (**124**) which then fragments to give $Ph_2\dot{C}OPh$ and PhO· radicals. *ipso*-Substitution is also an important process in the reaction of diphenyl ether with benzoyloxy-radicals.[502]

Rearrangements

Further studies of the rearrangement of alkoxy-radicals to 2-hydroxyalkyl radicals have been reported.[503, 504] Isomerization of the 2,4,6-tri-*tert*-butylphenyl radical to the 3,5-di-*tert*-butylneophyl radical occurs by quantum-mechanical tunnelling of a hydrogen atom from an *o*-*tert*-butyl group.[505]

A theoretical study has been carried out on the electrocyclic rearrangement of the bicyclo[3.2.0]heptadienyl radical.[506] ESR and ENDOR studies have been carried out on the electrocyclic ring-opening of 7-acenaphthyl radicals to phenalenyl radicals.[507] Intramolecular 1,2-migration of R_3Ge and R_3Sn but not R_3Si occurs in the radical adducts of these radicals to acenaphthoquinone;[508] the decreasing activation energy between Si and Sn is attributable to the decrease in the M—O bond strength; the ease of migration also depends on the internuclear O—O distance. Trialkylsilyl groups migrate in 3,6-di-*tert*-butyl-2-[(trimethylsilyl)-oxy]phenoxy-radicals,[509] and a 1,2-migration of the formyl group occurs in the biradical generated during photolysis of citral.[510]

The lack of the 1,2-aryl migration from silicon to carbon (reaction 1) has been attributed to stabilization of the α-silyl radicals by p_π–d_π delocalization, and to steric factors in the Ar_1-3 transition state.[511]

$$PhSiMe_2CH_2\cdot \not\longrightarrow Me_2\dot{S}iCH_2Ph \qquad (1)$$

The latter factor cannot be critical since the reverse rearrangement has been shown to occur at high temperatures[512] and thus it seems that the lack of rearrangement is due to the stability of the α-silyl radical. The first examples of aryl migration from silicon to carbon have been encountered in the 1,4- and 1,5-aryl migrations of the

$$\underset{\textbf{(125)}}{PhSiMe_2(CH_2)_{n-1}CH_2\cdot} \longrightarrow \underset{\textbf{(126)}}{[\text{spirocyclohexadienyl},\ Me_2Si\text{–}(CH_2)_n]} \longrightarrow \underset{\textbf{(127)}}{Ph(CH_2)_n\dot{S}iMe_2}$$

radicals (**125**; $n = 3$ or 4) to (**127**) *via* the transition states (**126**).[512] 1,2-Aryl migrations occur in 2,4,4-triaryl-4*H*-imidazol-5-yl radicals.[513] Intra-molecular 1,4-aryl migration from arsenic to carbon takes place during the pyrolysis of 9-aryl-10-benzyl-9,10-dihydro-9-arsa-anthracenes.[514] 1,4-Aryl migration from oxygen to nitrogen occurs in the rearrangement of aryl hydrazonates, $RC(OAr^1){=}NNHAr^2$, into *N*',*N*'-diarylhydrazides, $RCONHNAr^1Ar^2$.[515]

1,5-Migrations of tributyltin groups occur in reactions of the *o*-tributyltin derivatives (**128**) formed by reaction of glycidols with tri-*n*-butyltin radicals: the ring-opening process involves exclusive cleavage of the C—O rather than the C—C bond even in those cases where C—C cleavage would give a secondary radical.[516]

$$\underset{\textbf{(128)}}{H_2C\overset{O}{—}CH–CH_2OSnBu_3} \xrightarrow{Bu^tO\cdot} H_2C\overset{O}{—}CH–\dot{C}H(OSnBu_3) \longrightarrow [\text{cyclic } O\cdot \ \text{/}\ O\text{–}SnBu_3] \longrightarrow [\text{cyclic } \cdot\ \text{/}\ O\text{–}SnBu_3]$$

Ring-opening of the cyclobutylhydroxymethyl radical followed by 1,6-hydrogen transfer to the pent-1-en-1-oxy-radical can be detected in a matrix but not in solution (where the cyclobutylhydroxymethyl radicals combine before rearrangement);[517] cyclopropylhydroxymethyl radicals rearrange in solution as well as in a matrix, since ring-opening of the cyclopropyl ring is much faster than that of the cyclobutyl ring. Radical addition of halogens to cyclopropylacetylene gives *cis*- and *trans*-1,2-dihalocyclopropylethylenes, 1-halo-2-(halomethylene)cyclobutanes, and 1,5-dihalopenta-1,2-dienes; the initial vinylcyclopropylcarbinyl radical ring opens and the resultant 5-halopenta-3,4-dienyl radical cyclizes to the 2-(halomethylene)cyclobutyl radical.[518]

Rearrangement of the quadricyclyl radical (**129**) to the bicyclo[3.2.0]hepta-3,6-dien-2-yl radical (**130**) does not occur *via* the 7-norbornadienyl radical (**131**) (*cf.* the rearrangement of the analogous cation).[519]

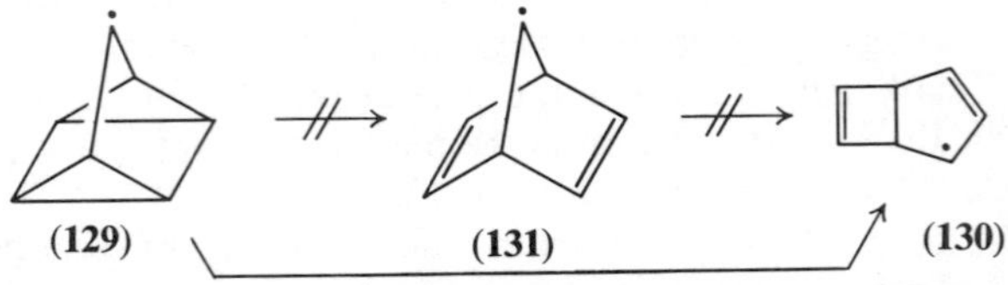

(**129**) (**131**) (**130**)

The bicyclo[3.2.1]octa-2,6-dienyl radical (**132**) is more stable than the isomeric tricyclo- and tetracyclo-radicals (**133**) and (**134**), both of which rearrange to (**132**).[520]

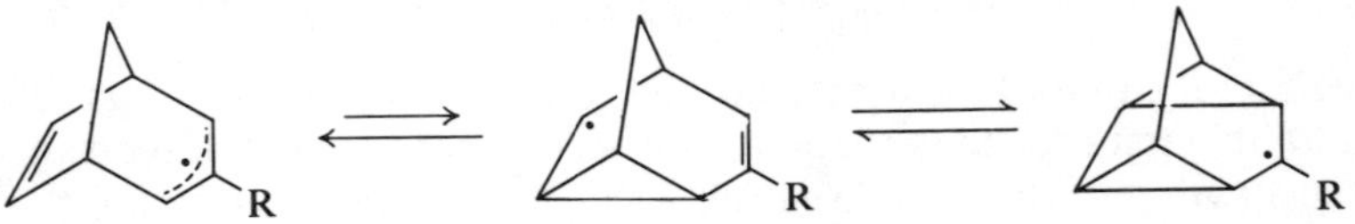

(**132**) (**133**) (**134**)

Substrates that give rise to either barbaralyl or bicyclo[3.2.2]nonatrienyl radicals invariably give products derived from the more stable bicyclononatrienyl radicals.[521]

The biradicals generated during addition of dibromocarbene to 1,2-dihaloethylenes undergo a 1,2-halogen shift to give 1,1-dibromo-3,3-dihaloethylenes.[522] In the rearrangements of polychloroalkyl radicals, *e.g.* $ClĊHCCl_3$ to $Cl_2CHĊCl_2$, the migrating chlorine is transferred to the $2p_z$ orbital of the carbon atom to which it migrates; rearrangements of this type have been followed by using spin-trapping techniques.[523]

Several studies on the rearrangements of radicals derived from 1,3-dioxanes[524, 525] and 1,3-dioxolanes,[526, 527] have been carried out; rearrangements of 1,3-dioxanes are more selective than those of 1,3-dioxolanes[526] owing to the greater stability of the six-membered ring. *trans*-2,5-Dialkyl-1,3-dioxanes isomerize to the *cis*-isomer as well as undergoing ring-opening; the isomerization process presumably involves ring-opening followed by recyclization to give the *cis*-isomer.[525] The alkoxy-radical (**136**) obtained from the base-catalysed decomposition of (**135**) rearranges to give

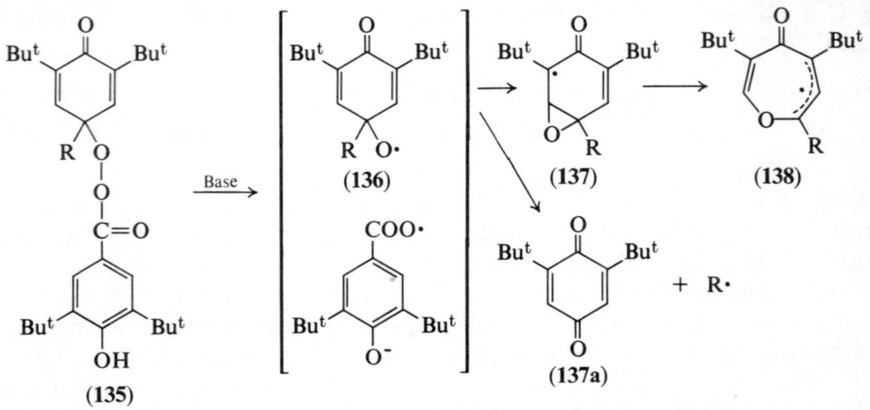

(**135**) (**136**) (**137**) (**138**) (**137a**)

136 → **137** when R = Me or Et; **136** → **137a** when R = Pr^i or Bu^t.

(**138**) when R = Me or Et; the reaction probably involves ring closure to (**137**) followed by C—C cleavage;[528] when R = Pr^i or Bu^t the radical (**136**) merely undergoes fragmentation to give 2,6-di-*tert*-butylbenzoquinone (**137a**).

Photolysis of the steroidal nitrite proceeds as shown (Scheme 7);[529] a related process involving ring opening of an alkoxy-radical generated by photolysis of a steroidal cyclopentyl nitrite leads to an α-hydroxy-nitrone.[530]

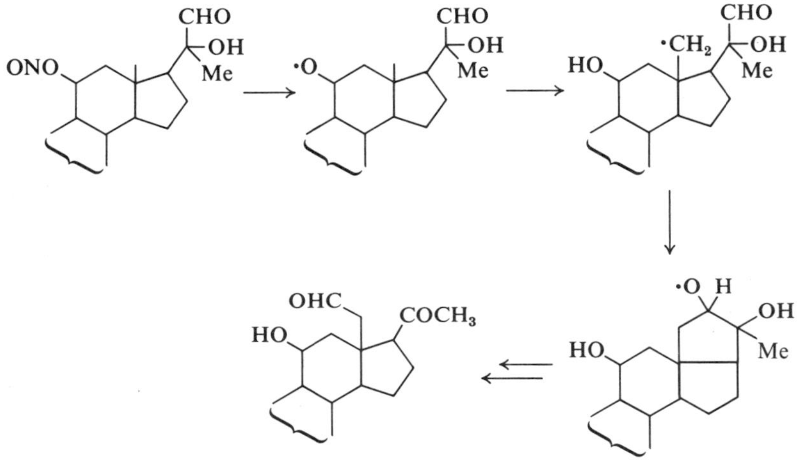

CHO OH Me ONO → CHO OH Me •O → CHO •CH_2 OH Me HO

↓

OHC $COCH_3$ HO ⇇ •O H OH Me HO

SCHEME 7

Spin-trapping studies have confirmed that the 1,2-migration of a phenylthio-group in acetophenone diphenyl ketal proceeds by a radical mechanism.[531] A 1,3-migration of a phenylthio-group occurs in the photolysis or thermolysis of trimethylsilyl enol ethers of α,α,-dialkyl-α-phenylthio-ketones by an addition–elimination pathway.[532] An addition–elimination mechanism has been invoked also for the photolytic rearrangement of 1,2,3-tris(trifluoromethyl)cyclopropenyl trifluoromethyl ketone to tetrakis(trifluoromethyl)furan.[533] The four methyl linoleate hydroperoxides (the *cis–trans*-9-hydroperoxides and the *trans–cis*- and *trans–trans*-13-hydroperoxides) are obtained upon thermal rearrangement of the *cis–trans*-9-hydroperoxide; labelling studies suggest that the reaction proceeds by reversible loss of molecular oxygen to give a pentadienyl radical which can isomerize.[534]

Radical processes for the oxy-Cope rearrangement have been shown to be thermodynamically feasible.[535] *N*-Benzhydryl-α,α-diaryl nitrones rearrange by a radical-pair mechanism to the substituted benzophenone *O*-benzhydryl oxime (Scheme 8); labelling studies indicate that free benzhydryl and iminoxy-radicals are generated to a major extent during the reaction.[536]

O-Trityl oximes, unlike other *O*-alkyl oximes, undergo *syn–anti* isomerization on being heated; an iminoxy-trityl radical pair is formed as intermediate; the developing trityl radical lowers the activation energy for homolysis.[537] CIDNP and trapping experiments support a radical-pair mechanism for the thiono–thiolo-rearrangement of *N*-aroyl-*O*-(*N*,*N*-dimethylthiocarbamoyl)-*N*-methylhydroxylamines:[538]

$$\mathrm{ArCONMeOCSNMe_2} \longrightarrow [\mathrm{ArCO\dot{N}Me}\quad \mathrm{Me_2NC(S\cdot){=}O}] \longrightarrow \mathrm{ArCONMeSCONMe_2}$$

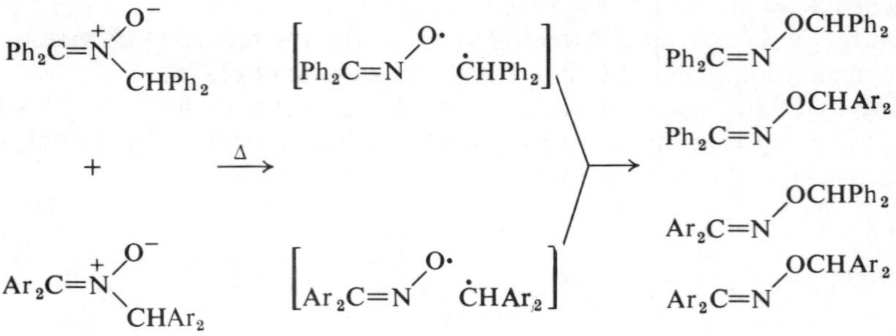

$Ph_2C{=}\overset{+}{N}(O^-)CHPh_2$ + $Ar_2C{=}\overset{+}{N}(O^-)CHAr_2$ $\xrightarrow{\Delta}$ $[Ph_2C{=}NO\cdot\ \dot{C}HPh_2]$ + $[Ar_2C{=}NO\cdot\ \dot{C}HAr_2]$ $\longrightarrow$ $Ph_2C{=}NOCHPh_2$ + $Ph_2C{=}NOCHAr_2$ + $Ar_2C{=}NOCHPh_2$ + $Ar_2C{=}NOCHAr_2$

[Ar = *p*-DC_6H_4]

SCHEME 8

The nitroalkane–alkyl nitrite rearrangement of $PhRC(CN)NO_2$ to PhRC(CN)ONO proceeds *via* the $[PhR\dot{C}CN\ \ \cdot NO_2]$ radical pair.[539] 1,2-Alkyl migration results from thermolysis of alkyl silylmethyl ethers (**139**; R = $PhCH_2$ or cyclo-$C_3H_5CH_2$):[540, 541] migration of the silyl group to oxygen initiates the reaction, with

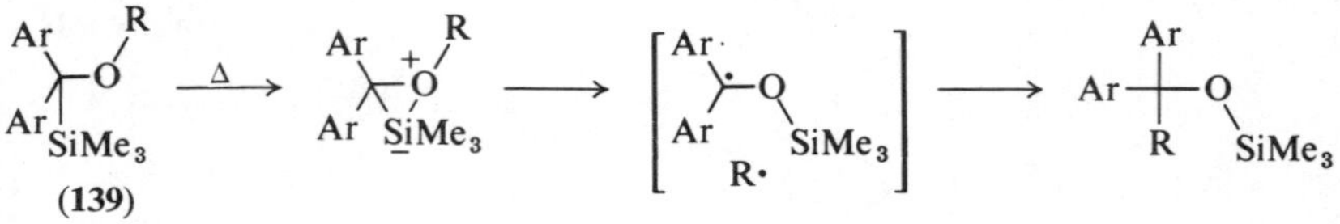

(**139**) $Ar_2C(SiMe_3)OR$ $\xrightarrow{\Delta}$ $Ar_2C(\underline{Si}Me_3)\overset{+}{O}R$ $\longrightarrow$ $[Ar_2\dot{C}OSiMe_3\ \ R\cdot]$ $\longrightarrow$ $Ar_2C(R)OSiMe_3$

the formation of a strong Si—O bond, but this proceeds only when the ketyl radical fragment is of the type $Ph_2\dot{C}OSiMe_3$: the homolytic nature of the reaction has been confirmed by cross-over experiments, CIDNP and ESR studies, and the ability to trap the migrating alkyl radicals, Photolysis of allyl aryl ethers proceeds by way of allyl-*O*-homolysis followed by combination of the geminate radical pair.[542] 4-Methyl-4-nitrocyclohexa-2,5-dienone rearranges by a radical pair process to 2-nitro-*p*-cresol.[543] The detection of bibenzyl in the 1,3-migration of a benzyl group in the rearrangement of 3,3′-dibenzyl-$\Delta^{2,2'}$-bibenzothiazoline to 2-(2-benzothiazolyl)-2,3-dibenzylbenzothiazoline suggests a radical mechanism.[544] The radical

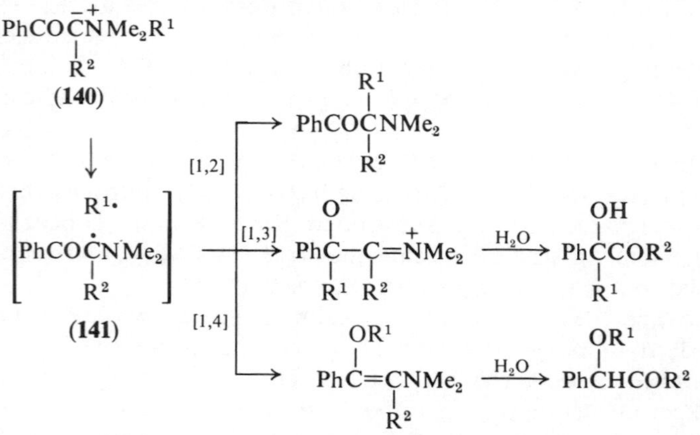

(**140**) $PhCO\overset{-}{C}(R^2)\overset{+}{N}Me_2R^1$ $\longrightarrow$ (**141**) $[R^1\cdot\ \ PhCO\dot{C}(R^2)NMe_2]$

[1,2] $\longrightarrow$ $PhCOC(R^1)(R^2)NMe_2$

[1,3] $\longrightarrow$ $PhC(O^-)(R^1){-}C(R^2){=}\overset{+}{N}Me_2$ $\xrightarrow{H_2O}$ $PhC(OH)(R^1)COR^2$

[1,4] $\longrightarrow$ $PhC(OR^1){=}C(R^2)NMe_2$ $\xrightarrow{H_2O}$ $PhCH(OR^1)COR^2$

pair (**141**) derived from ammonium ylides (**140**) gives products derived from 1,2-, 1,3-, and 1,4-rearrangements; deuterium-labelling studies indicate that all three processes are significantly intermolecular.[545] The intramolecular stereoselectivity for the 1,2-process is higher than that of the 1,3-process, as a consequence of the limited translational motion within the radical pair before coupling: the rather greater translational motion demanded for the 1,3-process permits more rotation and tumbling within the radical pair, leading to more of the racemic product.[546] Competing [1,2]-, [3,2]-, and [5,2]-rearrangements occur by a similar mechanism in reactions of pentadienyl ammonium ylides:[547] the ^{13}C-labelled bicyclic ylide **(142)** gives both **(144)** and **(145)**; the intermediate biradical **(143)** is apparently incompletely equilibrated as a result of steric and/or electronic hindrance to equilibration (or else the results are a consequence of competition by a concerted process).[548]

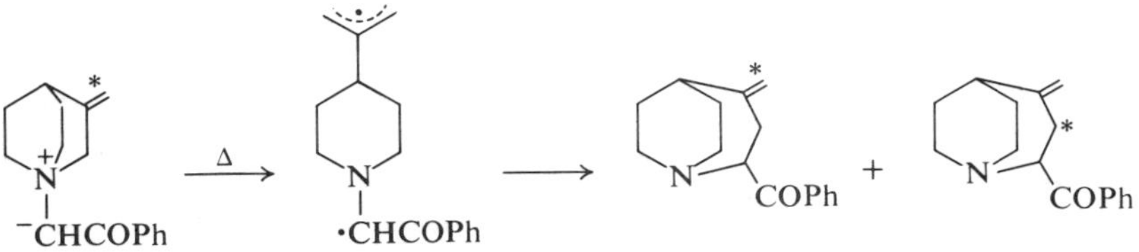

(142) (143) (144) (145)

1,4-Ylidic thianthrenes, *e.g.* 10-alkyl-9-mesityl(or duryl)-10-thia-anthracenes, undergo 1,4-alkyl migration to give the corresponding 3-alkyl-9-arylthioxanthenes; the intermediate 9-arylthioxanthyl radicals can be detected during the reaction.[549]

Biradicals

Theory

Electron spin resonance studies of triplet states and biradicals have been reviewed.[550] Geometry-optimization calculations on trimethylene diradical[551, 552] and 1,3-diradicals from ring-opened aziridine and ethylene oxide[553] indicate that a *trans*-bent conformation is most stable in all these cases. The low barriers to internal rotation in the singlet states of the isomeric bisallyl molecules **(146)** and **(147)** explain the racemizations observed in the thermal rearrangements of optically active methylenecyclopropanes and 1,2-dimethylenecyclobutanes.[554] Steric effects have been utilized as a tool to distinguish between pericyclic and biradical processes in the Cope rearrangement of bicyclo[3.2.1]octa-2,6-diene.[555] Topological considerations have been employed to show that a "bicycle or slither" mechanism for the movement of C(6) on the fulvene π-system, rather than a pivot mechanism, pertains in the transformation of **(148)** into **(149)**.[556]

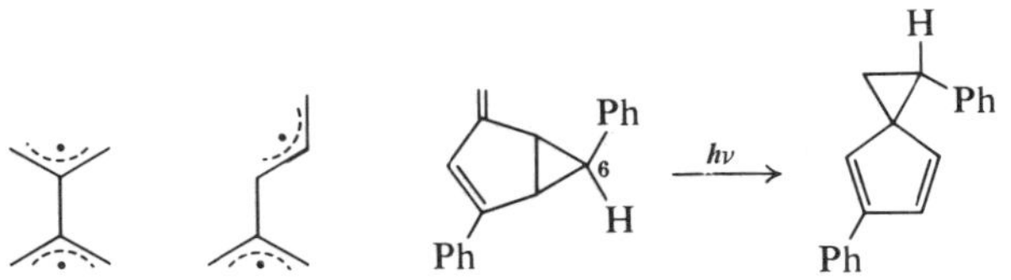

(146) (147) (148) (149)

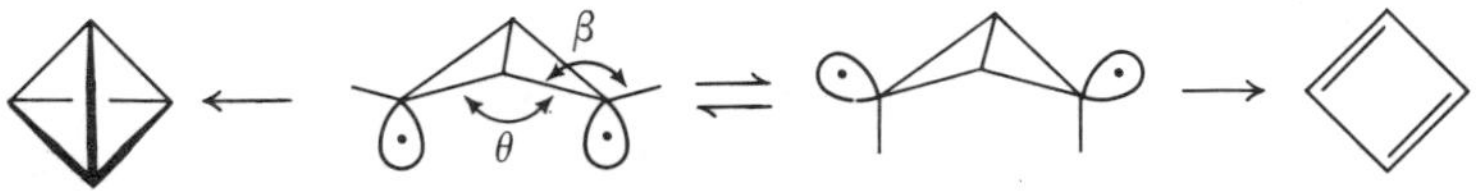

(150) (151) (152) (153)

A Different Orbitals for Different Spins-CI investigation of the 1,3-diradicals $\cdot CH_2NHO\cdot$, $\cdot CH_2OO\cdot$, and $\cdot CH_2CH_2O\cdot$ has been made and the results have been compared with restricted Hartree–Fock calculations.[557]

Extended Hückel calculations show that the formation of tetrahedrane (**150**) from (**153**) by direct $2\pi + 2\pi$ cycloaddition, or any concerted process, is thermally forbidden; however, the processes (**151**)→(**150**) and (**152**)→(**153**) are possible if the HOMO possesses A_1 symmetry. The important orbitals are strongly dependent on the angles θ and β by means of through-space interactions.[558]

Carbon–Carbon Centred Biradicals

Intersystem crossing and cyclization of thermalized and vibrationally hot triplet 2-methylbuta-1,3-diyl biradicals have been studied.[559, 560] The spectral characteristics of singlet and triplet 1,3-perinaphthadiyl are very different from those of alternant hydrocarbons. There is a large barrier to ring closure of the π,π-biradical to form a fused cyclopropyl ring.[561] A finite activation energy exists for cleavage of the weak 5,6-bond in (**154**), and thus (**154**) and the singlet trimethylenemethane derivative (**155**) are distinct.[562] The regioselectivity in the photochemical ring-opening of 3-phenyl- and 3-vinyl-substituted cyclopropenes, to afford indenes and 1,3-cyclopentadienes, favours a ring-opening mechanism involving a 1,3-diradical which transforms to a vinylcarbene, rather than one proceeding *via* π,π^* bridging to give a bicyclo[2.1.0]pentene diradical which then fragments.[563, 564] The photo-ene reaction of (**156**) is thought to involve the biradical (**157**).[565]

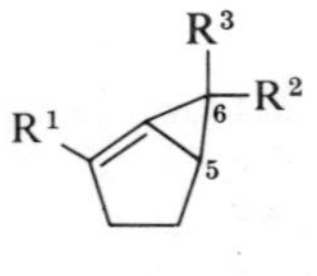

(**154**) 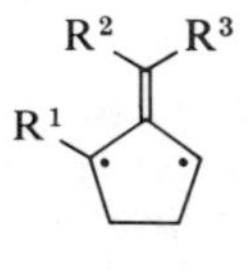 (**155**) 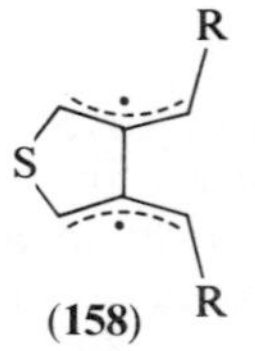 (**158**)

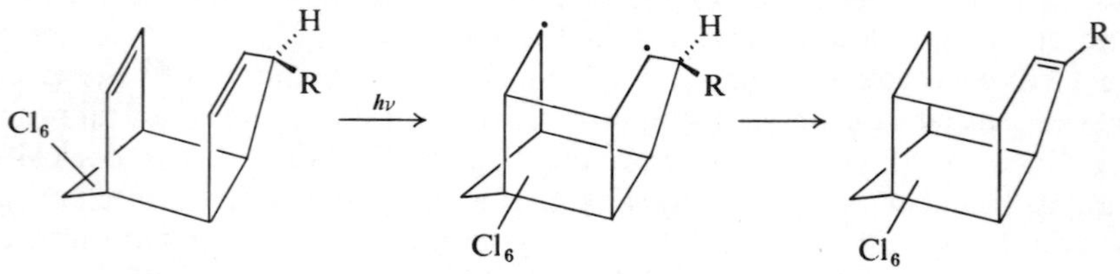

(**156**) (**157**)

The products arising from the treatment of bispropargyl sulphide with potassium *tert*-butoxide in THF can be rationalized by the intervention of the diradical(**158**).[566] The highly stabilized diradical (**159**) appears to be the major intermediate in the thermal rearrangement of 3-methylenebicyclo[3.2.2]nona-6,8-diene (**160**).[567] Rearrangement of (**161**) proceeds *via* a biradical intermediate but the radical anion of (**161**) rearranges 10^{15} times more rapidly.[568]

Rotational equilibration of diradicals (**162**) and (**163**) is incomplete before radical coupling occurs.[569] Elimination of CO on pyrolysis of fuchsone yields the biradical $Ph_2C{=}C(CH{=}CH\cdot)_2$ which forms 6,6-diphenylfulvene by intramolecular recombination; this, on dehydrogenation, ring-closure, and rearrangement, yields 1-phenylacenaphthylene.[570] The triplet sensitized photorearrangement of geranonitrile, $(CH_3)_2C{=}CHCH_2CH_2C(Me){=}CHCN$, involves a 1,3-shift of a nitrile

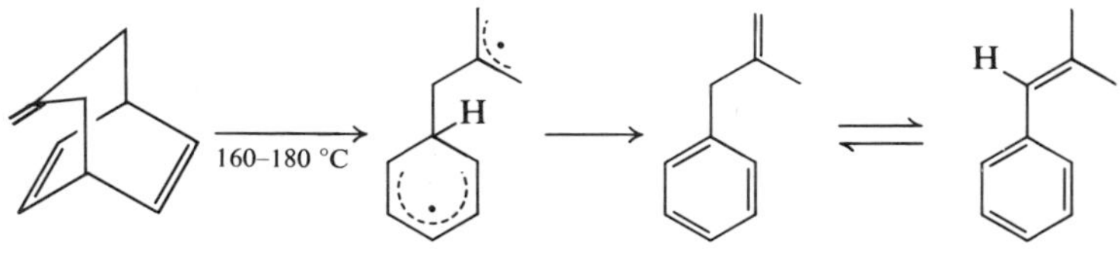

160–180 °C

H

H

(160) (159)

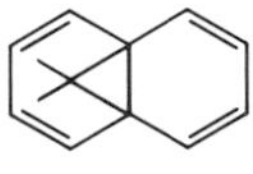

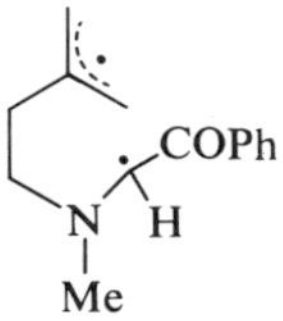

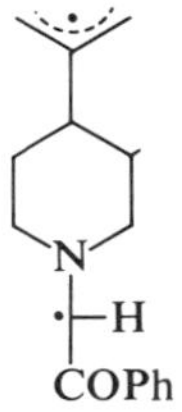

COPh

N H

Me

N

H

COPh

(161) (162) (163)

group in a biradical, analogous to the isomerization of citral.[571] Cyclopenta-1,3-diyls, formed on photolysis of cyclic azo-compounds, are trapped by O_2,[572] and by cyclopentadiene; frontier orbital control of regiospecificity is observed in the latter singlet cycloadditions.[573] Biradicals are involved in the dimerization of 2-fluoro- and 2-cyano-butadiene,[574] and in the Diels–Alder reaction of but-2-ene and cyclohexa-1,3-diene,[575] whilst theoretical support exists for a concerted biradical mechanism in 1,3-dipolar cycloadditions.[576] Interception of the biradical in the $Ph_2C{=}O$-sensitized reaction of phenanthrene with dimethyl fumarate by O_2 or $Bu^t_2NO\cdot$ enhances the probability of formation of the cyclobutane.[577] The formation and partitioning of biradicals in di-(9-anthryl)methane and other linked anthracene photo-isomerizations appear to control both excited singlet-state lifetimes and photodimerization yields; excimers appear to be subordinate.[578]

Hetero-centred Biradicals

The compound (**164**) isomerizes to the methylenepyrazoline derivative (**165**) *via* the diazenyl biradical (**166**).[579] Pyrolytic conversion of (**167**) into (**168**) involves the singlet diradical (**169**) rather than a synchronous transformation.[580]

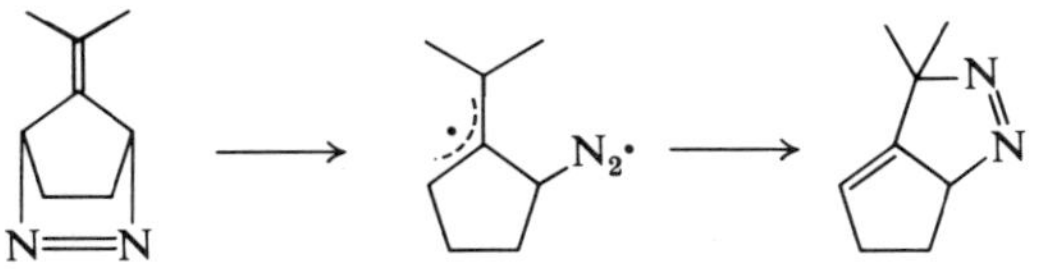

N=N $N_2\cdot$ N N

(164) (166) (165)

H H N COOEt O Ar O R

H H N COOEt O Ar O R

H COOEt O R Ar O

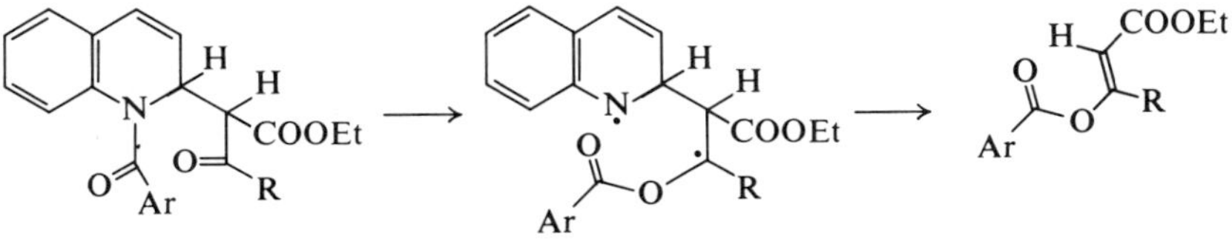

(167) (169) (168)

The photochemistry of the stable pyridinyl (py) diradicals, (**170**) and (**171**),[581] and the optical and ESR spectra of the diradicals, $\cdot py(CH_2)_n py\cdot$ (n = 3–10), complexed

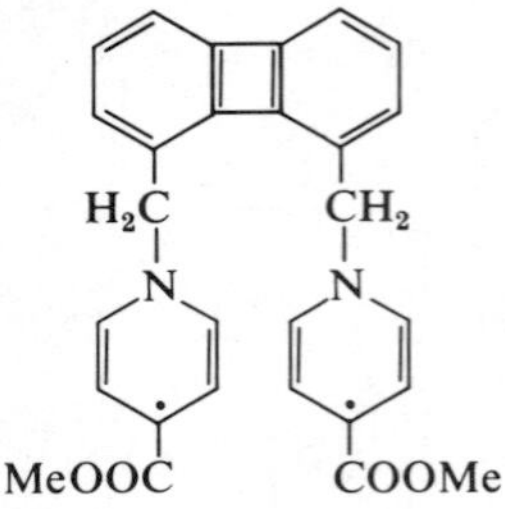

(170) 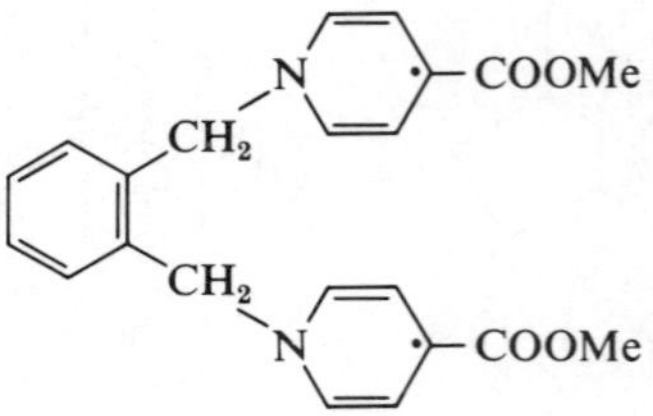 (171)

with metal halides,[582] have been documented. Experimental evidence on the stereoselectivity,[583] and a theoretical analysis[584] of the interactions (including anomeric effects) of the reaction of singlet oxygen with methoxy-substituted olefins are consistent with a non-concerted biradical mechanism but counter to one involving perepoxide intermediates. Mercury-photosensitized decomposition of thietane leads to thiatrimethylene and 4,5-dithiahexamethylene biradicals.[585]

Carbonyl-derived Biradicals

Competitive electron-transfer, to paraquat dication, from the 1,4-biradicals from the Norrish type II photoreaction of alkyl phenyl ketones enables the biradical lifetimes to be found (~100 ns); this has enabled a check to be made on the common assumption that biradical lifetimes approximate to those of free radicals with the same site configuration.[586] Only one biradical presursor, of lifetime ~300 ns, is obtained from the *syn*- or the *anti*-conformer of the *o*-methylacetophenone triplet, thus substantiating Wagner's proposal but *contra* Lindqvist.[587]

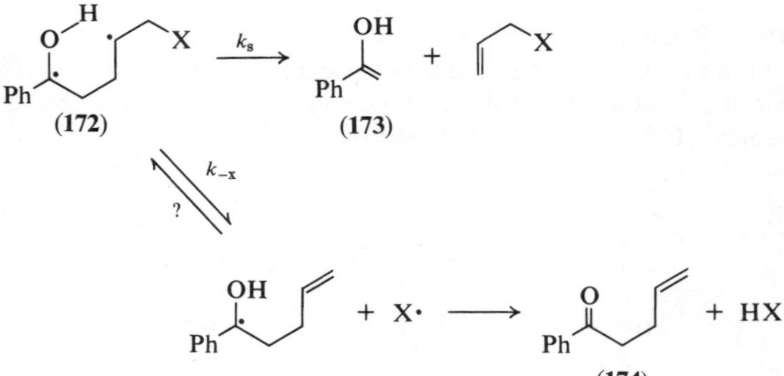

(172) (173) (174)

By using triplet quenchers to adjust the triplet-state lifetimes of alkyl phenyl ketones to a conveniently short value, the reaction of biradicals with oxygen can be differentiated; contrary to earlier reports an oxygen affect is observable.[588]

Relative rates of β-cleavage of the biradical (**172**) from δ-halovalerophenones, and also from the δ-benzoyl sulphides, sulphones, and sulphoxides, have been determined from the (**173**)/(**174**) ratios; the rate constant k_s is almost independent of the substituent X and k_{-X} follows the expected order I > Br > Cl and PhS ≫ Bu^{n}S except that, surprisingly, Cl· is eliminated more readily than thiyl radical.[589] Phosphine oxides and trisubstituted organic phosphates form complexes with the biradicals from the Norrish type II photolysis of alkyl phenyl ketones and enhance the yield of fragmentation product.[590] The activation volume for cyclobutanol formation by the 1,4-diradical derived from phenyl propyl ketone is less than 2 cc mol^{-1} smaller than that for elimination; this is consistent with a shallow energy minimum for the diradical and with little bond formation or breakage in the transition state.[591] The triplet state and Norrish II biradical generated from valeraldehyde in aqueous MeCN have lifetimes of 35 μs and 2 μs, respectively; the yield of biradical from the triplet state was estimated to be 0.32.[592] Biradicals from the type I photocleavage of cycloalkanones can be trapped by nitric oxide to yield long-lived nitroxide radicals whose ESR parameters contain information on the nature of the two ends of the biradical.[593] The competition between decarbonylation, disproportionation, and reclosure of the primary Norrish type I photoproduct, an acyl alkyl biradical, from 3-substituted alkyl 1-methyl-2-oxocyclohexane-1-carboxylate depends dramatically on the 3-substituent(s).[594]

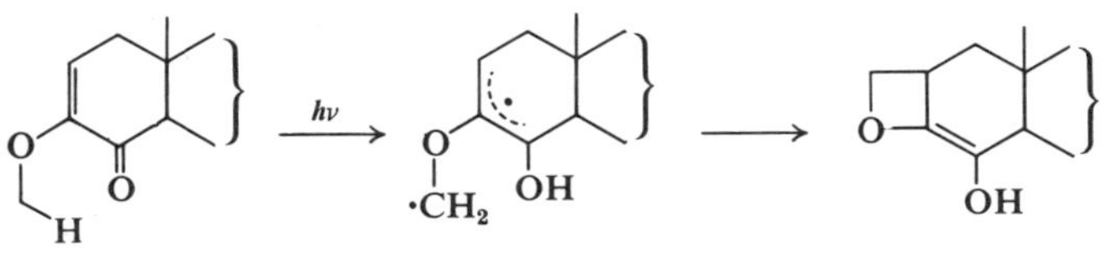

(176) (175) (177)

The biradical (**175**) derived by γ-H abstraction from the α-alkoxycyclohexenone component (**176**) cyclizes to yield the enol form (**177**) of the α-keto-oxetane product, as well as α-alkylideneoxetanols.[595] Following a Norrish type I cleavage of (**178**), a stereoselective transfer of the axial hydrogen from C(2) to C(4) occurs, with rotation about the C(1)—C(2) bond but without epimerization at C(4).[596]

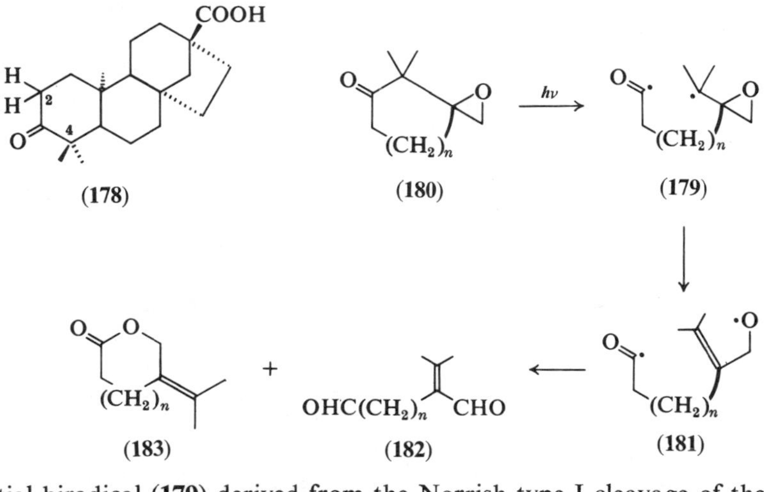

(178) (180) (179)

(183) (182) (181)

The initial biradical (**179**) derived from the Norrish type I cleavage of the 1-oxaspiro[2.*n*]alkan-5-ones (**180**) transforms to the oxirane ring-cleaved biradical (**181**) which gives the products (**182**) and (**183**).[597] In the photolysis of citral a 1,2-shift of a formyl group occurs *via* a 1,4-biradical.[598]

Stereoelectronic control in the fragmentation of the cyclopropylcarbinyl radical component of (**184**) leads to the *trans*-homoallylic species (**185**) and thus smoothly to the previously unknown *cis,trans*-cyclonona-1,4-diene as the kinetically controlled product, (**186**), from the photodecarbonylation of (**187**).[599]

(**187**) → (**184**) → (**185**) → (**186**)

Nitroxides and Spin Trapping

This section is deliberately brief on the synthetic and structural aspects of nitroxides and other stable radicals, concentrating rather on the kinetics of spin trapping and its applications.

Synthetic and Structural Aspects

The synthesis and chemistry of nitroxides spin labels have been reviewed.[600] Syntheses of acyl nitroxides,[601] azetidine nitroxides,[602] 2,2,4,6,6-pentamethyl-1,2,5,6-tetrahydropyrimidin-1-oxyl 3-oxide,[603] and a potential phenoxy-radical spin label from 3,5-di-*tert*-butyl-4-(hydroxyphenyl)glycine[604] have been reported. Synthesis and properties of a spin-labelled sodium dodecyl sulphate[605] have been announced. Nitroxides have been used to probe binding in micelles[606] and the segmental motions of poly(ethylene oxide) molecules in solution.[607] A new series of azethoxyl nitroxides involving the minimum steric perturbation on incorporation into lipids has been described.[608,609] The large dipolar ESR hyperfine splitting in a series of proximate dinitroxides based on the imidazolidine nucleus is very sensitive to molecular motion, thus making them valuable probes.[610] The synthesis of intercalating and carcinogenic nitroxide spin labels has enabled a study of the binding of carcinogens to DNA to be instigated.[611] Spin-labelled cyclic oxyphosphoranes[612] and phospholipids[613] have been prepared, and a transphosphorylation reaction has been studied in the case of phosphoryl imidazolides.[614] ESR evidence for phase transitions in bilayers of spin-labelled lipids[615] and for the surface characteristics of wood and cellulose[616] has been obtained.

The tetradecachloro-4-hydroxytriphenylmethyl radical can be coupled to alanine, valine, and proline, etc. by "mixed carboxylic anhydride methods".[617]

The magnitudes of the ^{14}N and ^{17}O coupling constants in a series of nitroxides have been used to establish the geometry of the nitroxide group.[618] Long-range ESR couplings[619] and an X-ray study[620] have shown the torsion angle of the nitroxide bond to be 30° in the *N*-oxy-2-azanoradamantane radical. An NMR study of [2,2]paracyclophan-4-yl *tert*-butyl nitroxide has shown that there is no substantial interaction of the unpaired electron with the transannular protons.[621] The ESR spectra of *N*-acyl-*N*-alkyl nitroxides suggest that the *trans*-conformation is preferred with hindered rotation about the *N*-alkyl bond.[622] Anomalies in the apparent spin-density distribution in PhC(=O)(O·)Bu^t radical have been attributed to non-planarity in the standard Bu^t_2NO· used as a model by which distributions are estimated;[623] thus, also in Bu^{t}N(O·)C(=O)R radicals the a_N values are higher than expected due to more efficient π–σ spin polarization.[624] In the radical $R^1R^2NC(=O)N(O\cdot)CH_2SO_2C_6H_4CH_3$-*p* there is slow rotation about the C_α—N(O·) bond with the S—C_α bond nearly eclipsed with respect to the unpaired electron *p* orbital in the most favoured conformation.[625] The very low barrier to motion (<2.7 kcal mol^{-1}) in (**188**) is ascribed to rotation of the five-membered rings.[626]

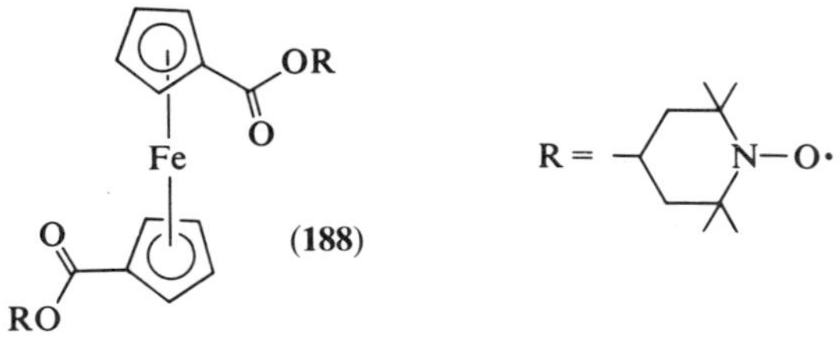

(**188**) R =

It has been claimed that up to five conformational isomers of some binitroxides can be detected.[627] Distances between radical functions in a nitroxide biradical have been estimated from the variation of ESR-signal line width with temperature.[628]

Reactions of Nitroxides

Reduction of (**189**) by H· or solvated electrons leads to (**190**) which undergoes disproportionation.[629, 630] HO_2· and $O_2^{\cdot-}$ radicals did not react with (**189**), but

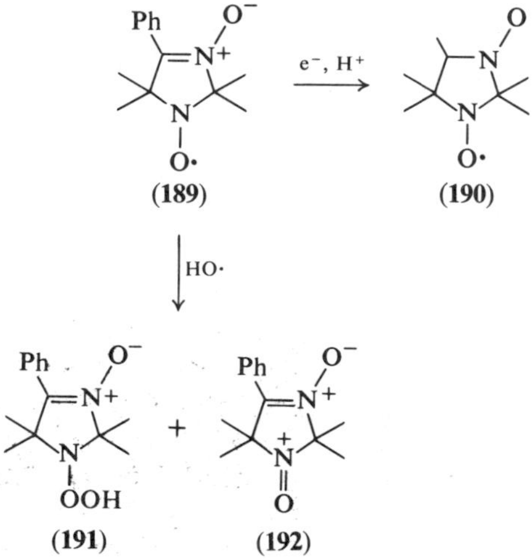

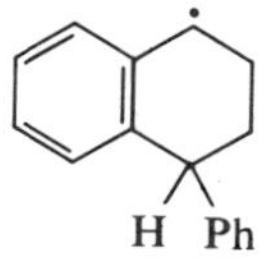

(**189**) → (**190**) [e$^-$, H$^+$]

(**189**) → (**191**) + (**192**) [HO·]

(**193**)

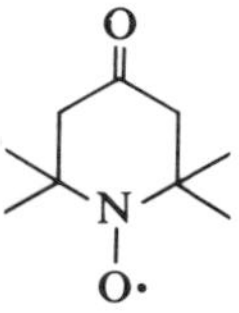

(**194**) 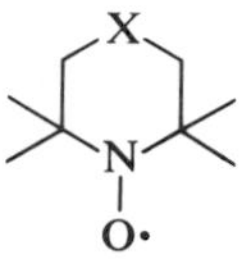 (**195**) 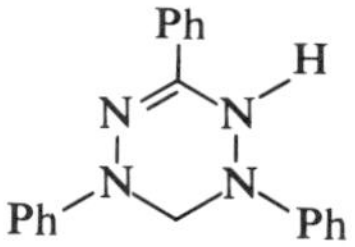 (**196**)

hydroxyl radicals reacted to give (**191**) and (**192**) as well as products hydroxylated in the Ph ring.[631] The photochemical reaction of (**189**) with Bu^{t}OH has been investigated.[632] Nitroxides have been used as scavengers in the photolysis of phenothiazines in solution[633] and of alkylpentacyanocobaltates[634] (thereby measuring rates of lateral diffusion in phospholipid bilayers). Fluorescence quenching, by nitroxides, of tetrasodium pyrenetetrasulphonate in micellar systems has been used as an environmental probe.[635]

The effects of pressure on the inhibition of styrene polymerization by 2,2,6,6-tetramethylpiperidin-1-oxyl radical[636, 637] are in accord with the involvement of the intermediate (**193**) in the thermal initiation (but see ref. 382). The nitroxide (**194**) has been used to model styrene peroxy-radical.[638] XeF_2 oxidizes (**195**; $X = CH_2$);[639] the nitroxide (**195**; $X = CHOH$) catalyses the oxidation of EtOH, PhOH, and H_2O_2 by $C(NO_2)_4$.[640] The hydrogen-transfer equilibrium reaction between leucoverdazyl (**196**) and (**194**) lies on the side of verdazyl radical and reduced (**194**) with $K = 10.0 \pm 1.0$.[641] The thionitroxyl radicals such as $PhCH_2N(CH_3)S\cdot$ decay by abstraction of benzylic hydrogen from the parent disulphide $[PhCH_2N(CH_3)S{-}]_2$.[642]

Spin-trapping

Reviews have appeared on the spin-trapping of organic radicals in the gas-phase[643] and on the kinetics of spin-trapping by nitrones in solution.[644] New, water-soluble, nitrones have been prepared.[645] Nitrosobenzenes continue to be attractive as spin traps. 2,6-Dichloronitrosobenzene is easily synthesized; in the presence of *o*-xylene at 50 °C the 2-methylbenzyl radical can be trapped and arises probably *via* $ArNO \rightarrow Ar\cdot$, $Ar\cdot + RH \rightarrow ArH + R\cdot$.[646] The dissociation constant of pentamethylnitrosobenzene dimer in benzene has been determined to be $10^{5.26} \exp(-50/\boldsymbol{RT})$; despite the small fraction of monomeric form in solution, the rate constant for the trapping of Bu$^t\cdot$ radicals by this form is $1.4 \times 10^8\ M^{-1}\ s^{-1}$ at 299 K.[647] By utilizing the competition between cyclization and spin-trapping (by common reagents) of the *n*-hexyl radical it has been shown that the rate constant for trapping increases as the solvent polarity increases, and therefore that alkyl radicals will be better trapped in an aqueous medium.[648] Extensions of this study, with 1-[^{13}C]-hex-5-enyl radical as a primary standard, to competitions between pairs of spin traps provide a very useful data base.[649] Relative spin-trapping rate constants towards Bu$^t\cdot$ radical (generated by photodissociation of Bu^{t}NO) for several common traps are: *C*-phenyl-*N*-*tert*-butylnitrone, <0.003; 2,4,6-tri-*tert*-butylnitrosobenzene, 0.07; 2-methyl-2-nitrosopropane, 1.0; pentamethylnitrosobenzene, 41; nitrosodurene, 63; nitrosobenzene, >50.[650]

New experimental evidence has confirmed conclusively the identity of the radical adduct from $\cdot OH$ radical and *C*-phenyl-*N*-*tert*-butylnitrone; the identity of that from $\cdot O_2H$ radical is less certain.[651] The hydroxyl radical adduct (**197**) derived from α-pyridyl 1-oxide–*N*-*tert*-butylnitrone in aqueous solution at pH 6–7 has been characterized by ESR spectroscopy.[652]

The absence of an observable stationary concentration of peroxyalkyl nitroxides in the photo-oxidation of *C*-nitrosoalkanes has been attributed to a rapid decay of the adduct (**198**) rather than to a failure of the trap to scavenge peroxyalkyl $R'O_2\cdot$ radicals.[653] 2,6-Di-*tert*-butyl-1,4-benzoquinone has been shown to give stable radical adducts with various alkyl radicals.[654] Acyl radicals are trapped ~50 times more efficiently by fumarate dianion and maleate anion than by maleate dianion.

$$^{-}O-\overset{+}{N}C_5H_4-\underset{OH}{\overset{H}{C}}-\overset{\dot{O}}{N}-CMe_3$$

(**197**)

$$R'O_2\cdot + RNO \rightleftharpoons R'O_2\overset{\dot{O}}{N}R \;(\textbf{198}) \rightarrow R\cdot + R'O_2NO$$

$$R'O_2\overset{\dot{O}}{N}R \rightarrow R'O\cdot + RNO_2 \xrightarrow{RNO} (R)(R'O)N-O\cdot$$

Observed radical

$$RCOCH(COO^-)\dot{C}HCOO^- \;(\textbf{199}) \longrightarrow RC(O^-)=CH\dot{C}HCOO^- \;(\textbf{200}) + CO_2$$

The acyl radical adducts (**199**) decarboxylate to yield 1-carboxy-3-oxyallyl radical dianions (**200**), the activation energy being ~9.5 kcal mol^{-1} for R = Me or Et.[655]

H· and HO· radicals generated during the electrolysis of water have been trapped with *C*-phenyl-*N*-*tert*-butylnitrone (PBN).[656] Superoxide ions, $O_2^{\dot{-}}$, have been trapped with 5,5-dimethyl-1-pyrroline 1-oxide.[657] Methyl and trityl radicals from the photo-Kolbe reaction of the appropriate acids have been trapped by using PBN.[658] Cumyloxy-radicals, produced in the reaction of cumene hydroperoxide and metmyoglobin, generate methyl radicals which can oxidize aminopyrine, as shown by trapping studies with Bu^tNO.[659] The $HOCMe_2CH_2\cdot$ radical is the sole product from the γ-radiolysis of Bu^tOH at −196 °C whereas $Bu^tO\cdot$, H·, and derived product radicals are detected at 30 °C.[660] The ESR spectra of the adducts of ^{17}O-labelled $Bu^tO_2\cdot$ and $Bu^tO\cdot$ with Bu^tNO and PBN have been recorded.[661] Spin trapping of Ph· by PBN in the radiolysis of benzene has been discussed.[662] The rearrangement of $Cl_3CCH_2\cdot$ radical (trapped with $MeCOCMe_2NO$) to $Cl_2C(CH_2Cl)\cdot$ (trapped by nitrosodurene) has been monitored.[663] Trichloromethyl radicals have been found in the reactions of CCl_4 and $CCl_3CH_2CH_2Cl$ with hex-1-ene photo-initiated by $Fe(CO)_5$ and nucleophilic co-catalysts.[664] The formation of the $\cdot CCl_3$–PBN adduct in the enzymic oxidation of NADPH by rat liver microsomes is itself an enzymic process.[665] The detailed mechanism of the homolytic rearrangement of acetophenone diphenyl ketal has been elucidated by ESR with Bu^tNO as a spin trap.[666] Aminoacyl[667] and aminium[668] radicals have been trapped, as have $\cdot NO_2$ and $\cdot CRR'NO_2$ radicals formed in the reaction of $C(NO_2)_4$ with α-nitrocarbanions in aprotic media.[669] The spin-trapping of short-lived chlorine- and sulphur-containing radicals has been reviewed.[670]

The nitroxide radical (**201**) can be detected by using nitrosodurene in the asymmetric hydrosilylation of acetophenone by 1-naphthylphenylsilane catalysed by a rhodium complex.[671] Germanes, $HGeR^1{}_2R^2$ ($R^1 = R^2$ = Et or Ph; $R^1{}_2R^2 = Ph_2Cl$ or $PhCl_2$), undergo a free-radical 1,3-addition to nitrones with the intermediacy of germyl radicals, $\cdot GeR^1{}_2R^2$, to yield *O*-germylhydroxylamines.[672] Phenyl radicals produced in the thermal decomposition of tetraphenylantimony mercaptides thus: $Ph_4SbSC_6H_4\text{—}X\text{-}p \rightarrow Ph_3Sb + Ph\cdot + p\text{-}X\text{—}C_6H_4S\cdot$ can be trapped by Bu^tNO but the thiyl radical does not give a stable adduct.[673]

The nitroxide radicals (**202**) are formed in the addition of aromatic *C*-nitroso-compounds to olefins.[674] Organophosphorus compounds such as $Bu^n{}_3P$ and $[(EtO)_2P]_2O$ yield nitroxide radicals with 2-methyl-2-nitrosopropane (Bu^tNO).[675]

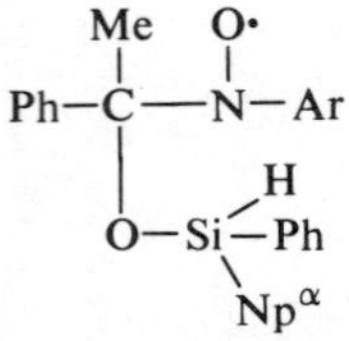

(201; Np$^{\alpha}$ = 1-naphthyl) 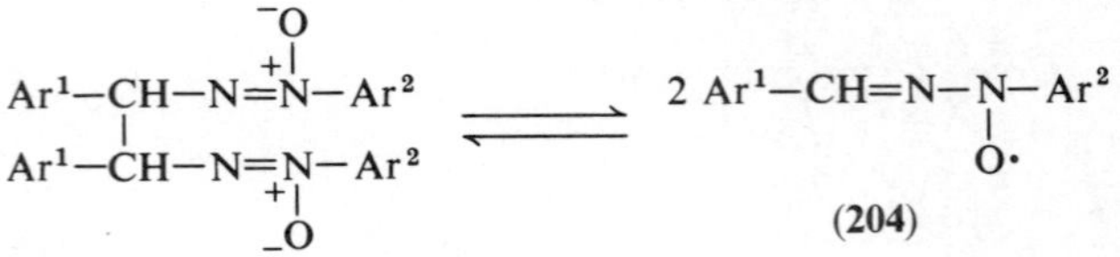 (203) (204)

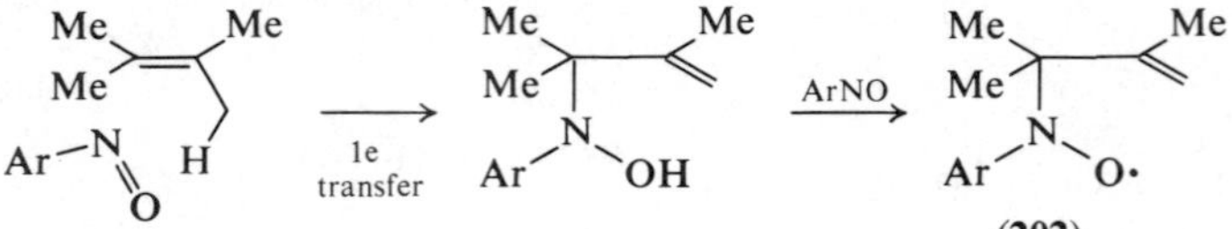

1e transfer ArNO (202)

Heating the "Bergman oxide" (203) at 90 °C in toluene leads to the nitroxide (204); the unpaired spin is delocalized over the whole molecule, except for Ar^1 = mesityl where a substantial twist occurs about the N—N bond; for Ar^1 = Ph, (204) reacts at its β-carbon atom with PhNO or nitrones.[676]

S_H2 Reactions

Rate constants have been measured for the reactions of formyl radical, $H\dot{C}O$, with nitric oxide and with oxygen in the gas phase.[677] Vinyl radical anion, $CH_2{=}\dot{C}^-$, has been shown by ion cyclotron resonance techniques to react with labelled nitrous oxide, $^{14}N^{15}NO$, at the terminal nitrogen atom to liberate $^{15}\cdot N{=}O$.[678] Small alkyl radicals (*e.g.* Et·) abstract $NF_2\cdot$ from nitrogen trifluoride but with the larger alkyl radicals (*e.g.* Pr^i, $Bu^t\cdot$ both F· and $\cdot NF_2$ abstractions occur; steric effects do not appear to be important.[679]

Intramolecular S_Hi attack of a carbon radical on the peroxide bond seems to require a co-linear alignment, *i.e.* back-side attack; thus, (205) readily undergoes S_Hi reaction but (206) does not react in this manner.[680]

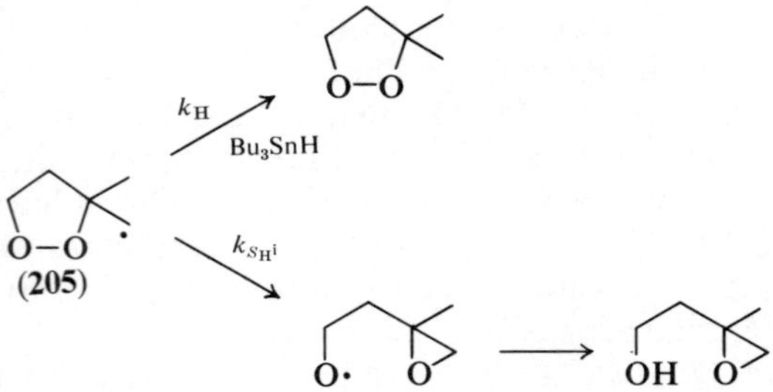

k_H Bu_3SnH k_{S_Hi} (205)

(206)

$R_3Sn\cdot$ (207) (208) (209)

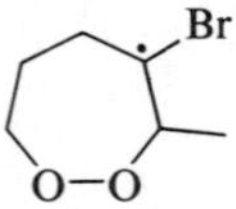

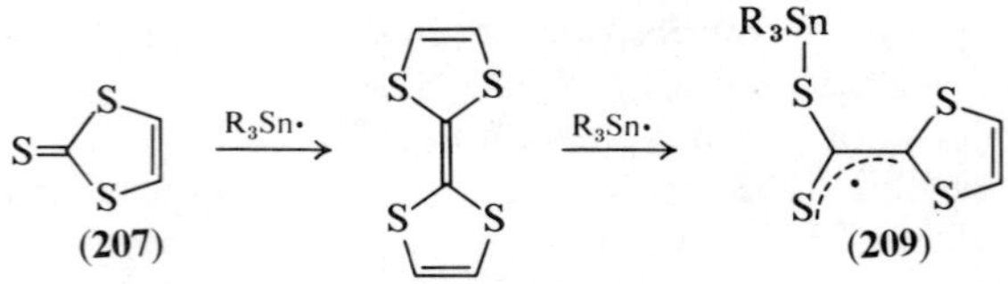

Aryl radicals, Ar·, can displace benzenethiyl, PhS·, from Ph_2S;[681] Bu^tO· radicals displace dialkylaminothiyl radicals, R_2NS·, from bis-(*N*,*N*-dialkylamino)-sulphides.[682] 1,3-Dithiole-2-thione (**207**) reacts with trialkyltin radicals, R_3Sn·, to yield (**208**) *via* the initial adduct; (**208**) undergoes an S_H2 reaction with R_3Sn· radical, giving acetylene and the persistent thiallyl radical (**209**).[683] A kinetic and ^{119}Sn-CIDNP study of the photoreaction of distannanes with alkyl halides, and their thermal reaction with diacyl peroxides, has shown, *inter alia*, that the effectiveness of the S_H2 reaction

$$R_6Sn_2 + R\cdot \longrightarrow R_4Sn + R_3Sn\cdot$$

varies with R, thus, $Bu^t < Pr^i \ll Pr^n < Et < Me \sim Ph$.[684]

Phosphoranyl and Sulphuranyl Radicals

The question of memory effects in the alkoxy-radical oxidations of tervalent phosphorus derivatives has been examined in some detail: on ESR and other evidence, the apical–equatorial ligand exchange in the intermediate tetra-alkoxyphosphoranyl radicals (**210**), generated by apical addition of alkoxy-radicals to

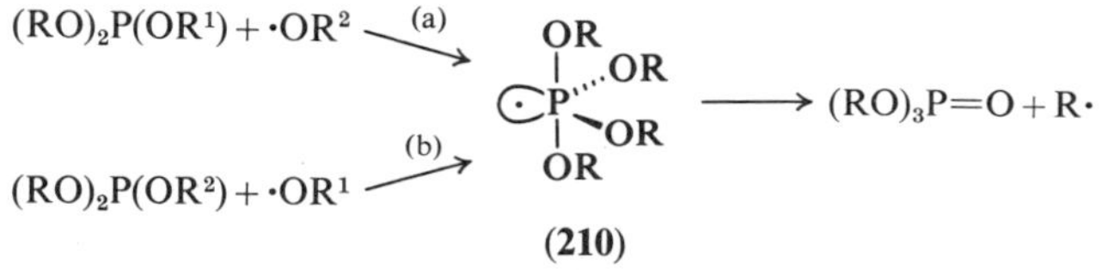

$$(RO)_2P(OR^1) + \cdot OR^2 \xrightarrow{(a)} \quad (RO)_2P(OR^2) + \cdot OR^1 \xrightarrow{(b)} \quad \dot{P}(OR)_4 \;(\mathbf{210}) \longrightarrow (RO)_3P{=}O + R\cdot$$

(**210**)

trialkyl phosphites (routes a and b), is faster than the β-scission; thus, memory effects are absent but some evidence was obtained for apical site selectivity in the β-scission process.[685] Bentrude and his co-workers have now confirmed the absence of memory effects in the product oxide ratio in systems where potentially common phosphoranyl radical intermediates are generated by two different pathways from five- and six-membered ring phosphites and five-membered ring phosphorodiamidites.[686] Attempts have been made to clarify the nature of the process(es) whereby alkoxy-ligands in $(RO)_4P$· radicals are made equivalent; relative rates of β-scission by $(Bu^sO)(EtO)_3P$· radicals, generated by two independent pathways, have been determined by ESR detection of the product butyl and ethyl radicals; a rapid, permutational isomerization in the phosphoranyl radical, exchanging apical and equatorial alkoxy-groups, seems most reasonable.[687] Addition of ethoxy(or *tert*-butoxy)-radicals to (**211**) occurs irreversibly at the

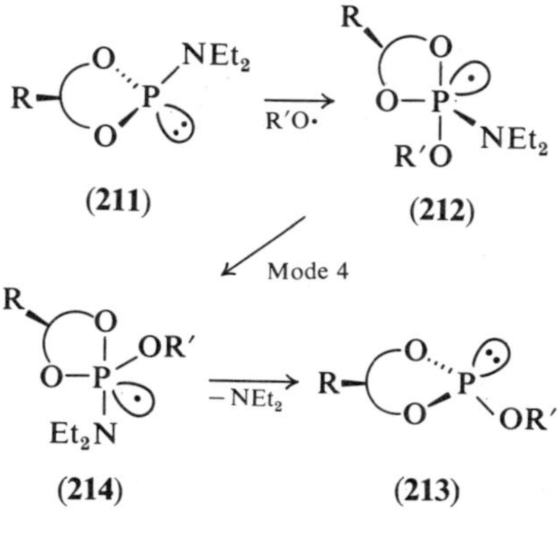

(**211**) $\xrightarrow{R'O\cdot}$ (**212**)

(**212**) $\xrightarrow{\text{Mode 4}}$ (**214**)

(**214**) $\xrightarrow{-NEt_2}$ (**213**)

apical position initially. Radical (**212**) derived from (*cis*-**211**) yields the *trans*-phosphite (**213**) on elimination of $Et_2N\cdot$. The observed inversion of configuration about phosphorus indicates that mode 4 rather than mode 1 is the most rapid permutation available to (**212**); however, it should be noted that mode 4, yielding (**214**), could consist of two successive mode 1 processes.[688]

Reaction of the four-co-ordinate $Ph_3(Bu^tO)P\cdot$ radical with molecular oxygen gives radical (**215**) in which the two oxygen atoms are magnetically inequivalent as determined from the ESR spectrum.[689] The phosphoranyl radicals, $R_n(Bu^tO)_{4-n}P\cdot$, derived from chlorophosphines and $Bu^tO\cdot$ radicals, have characteristic Arrhenius decay parameters: for R = Et or Bu, when $n = 3$, E_a is *ca.* 4 kcal mol^{-1} and A is *ca.* 10^5 s^{-1}; but, when $n = 2$, E_a is 12 kcal mol^{-1} and A is *ca.* 10^{13} s^{-1}.[690] The phosphonyl radical (**216**) has been shown to be pyramidal; this is supported by the observed retention of configuration upon addition of radical (**217**) to ethylene.[691]

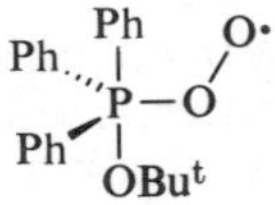

(**215**) 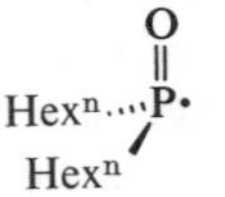 (**216**) 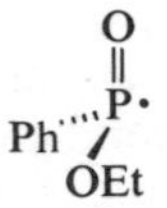 (**217**) 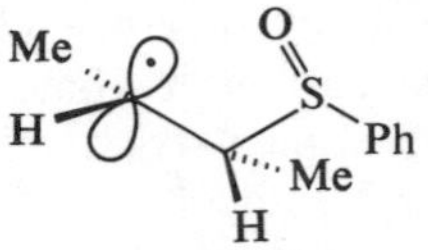 (**218**)

Addition of $CF_3S\cdot$ radical to dialkyl sulphides, R_2S, yields the sulphuranyl radicals $[CF_3S—SR_2]\cdot$ whose stability towards fragmentation is much enhanced compared with $[CH_3SSR_2]\cdot$ radicals.[692] Elimination from the phenylsulphinyl *s*-radicals, (**218**) to give but-2-enes has been shown to be faster than rotation about the C(2)—C(3) bond; the small differences in the rates of elimination from radicals, e.g. (**218**), generated from the four diastereomeric 2-bromo-3-(phenylsulphinyl)-butanes using Bu_3SnH, is consistent with a modest interaction between sulphur and the developing *p*-orbital of the radical but inconsistent with anchimeric assistance by the sulphoxide-oxygen.[693]

Homolytic Oxidation and Reduction Reactions

The mechanism of dehydrodimerization of aromatic substrates by aluminium chloride–copper(II) chloride involves initial formation of an arene radical cation; this generally attacks a second molecule of arene but on occasion dimerizes;[694, 695] dimerization occurs with fluorene. Oxidation of alkylbenzenes with metal acetates gives diarylmethanes as the main products; in presence of oxygen benzaldehydes are obtained owing to scavenging of benzyl radicals.[696] Mixtures of benzylic acetates and nitrates are obtained on oxidation of polymethylbenzenes with cerium(IV) ammonium nitrate (CAN): oxidation gives initially the radical cation which loses a proton to give a benzylic radical.[697] Oxidation of toluene with cobalt bromide involves attack by $Br\cdot$.[698] Anodic methoxylation of methylbenzenes and anisoles leads to both nuclear and side-chain methoxylated products; nuclear substitution occurs by nucleophilic attack of methoxide on the arene radical cation and/or coupling of the radical cation with methoxy-radicals.[699] Nuclear trifluoro-acetoxylation of chlorobenzene has been accomplished anodically,[700] whilst nuclear acetoxylation of benzene and anthracene has been brought about with cobalt(III) trifluoroacetate[701] and thallium(III) trifluoroacetate,[702] respectively; in each case the reaction proceeds *via* the arene radical cation. Oxidative acetoxylation of anisole occurs when CAN is used in acetic acid.[703] Substances, *e.g.* polycyclic aromatics, that can be adsorbed on the electrode surface decrease the *ortho*/*para*-ratio in anodic acetoxylation of anisole as a result of steric hindrance to attack at the *ortho*-position by the co-adsorbed substance.[704] Anodic cyanation of 4,4′-dialkoxybiphenyls gives 4-alkoxy-4′-cyanobiphenyls as a result of *ipso*-substitution on the intermediate radical cation.[705] 2-Cyano-5-methoxy-2,5-dimethylfuran is formed by the anodic cyanomethoxylation of 2,5-dimethylfuran.[706] Further studies of aromatic hydroxylation with peroxydisulphate have been reported;[707,708] the isomer ratio of phenols produced varies with the acidity of the medium.

Oxidation of cyclohexane with cerium(IV) salts proceeds *via* the radical cation to give cyclohex-2-en-1-ol and cyclohex-2-enone as the main products.[709] Labelling experiments indicate that hydroxylation of norbornane with liver cytochrome P-450 involves a carbon-radical intermediate and leads to a mixture of *exo*- and *endo*-2-norborneol.[710] Anodic oxidation of 1*H*-perfluoroalkanes gives perfluoroalkyl fluorosulphates by a radical process.[711] On photo-oxidation of alkenes in presence of iron(III) chloride,[712] chlorinated ketones are formed by the route formulated below.

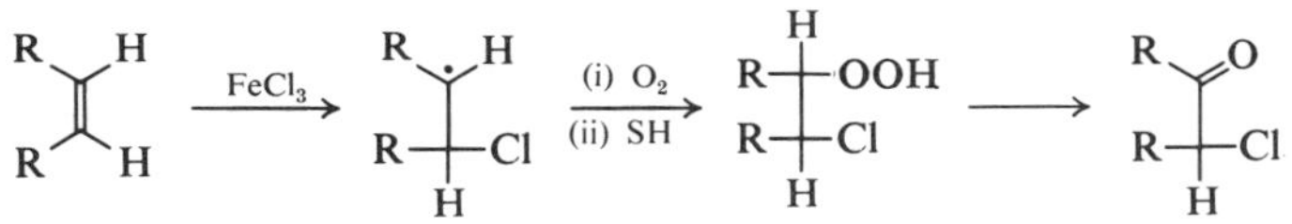

$$\mathrm{RCH{=}CHR} \xrightarrow{FeCl_3} \mathrm{R\dot{C}H{-}CHRCl} \xrightarrow[\text{(ii) SH}]{\text{(i) } O_2} \mathrm{RCH(OOH){-}CHRCl} \longrightarrow \mathrm{RC(=O){-}CHRCl}$$

3-Ethoxy-1,2,3-triphenylcyclopropene is converted by iron(III) chloride in ethanol into a mixture of 1,2,3-triphenylcyclopropene and 2,3-diphenylindenone;[713] the indenone is formed *via* a ring-opened vinylic radical. Oxidative dimerization of dienes at graphite anodes leads to dimers formed by addition of the initial radical cation to a second molecule of diene to give a 1,4-radical cation which, after oxidation, undergoes solvolysis or deprotonation.[714] Anodic cyanation of stilbene to 1,2-dicyano-1,2-phenylethane proceeds *via* consecutive one-electron transfer steps.[715]

Oxidation of alcohols with peroxydisulphate in presence of silver(I) leads to alkoxy-radicals, whilst in the absence of silver(I) carbon radicals are obtained:[716,717] the difference in behaviour is illustrated in the reactions of 2-methylbutan-2-ol with peroxydisulphate and copper(II) as an oxidant for the intermediate radicals: in the presence of silver(I) the main product is acetone and in its absence 2-methylbutane-2,3-diol (see Scheme 9).

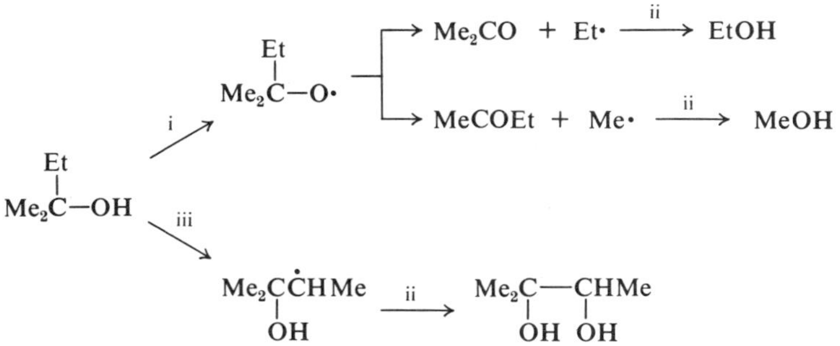

$$\mathrm{Me_2C(Et){-}OH} \xrightarrow{i} \mathrm{Me_2C(Et){-}O\cdot} \rightarrow \begin{cases} \mathrm{Me_2CO + Et\cdot} \xrightarrow{ii} \mathrm{EtOH} \\ \mathrm{MeCOEt + Me\cdot} \xrightarrow{ii} \mathrm{MeOH} \end{cases}$$

$$\mathrm{Me_2C(Et){-}OH} \xrightarrow{iii} \mathrm{Me_2C(OH)\dot{C}HMe} \xrightarrow{ii} \mathrm{Me_2C(OH){-}CH(OH)Me}$$

SCHEME 9

Reagents: i, $S_2O_8^{2-} - Ag^+$; ii, $Cu^{II} - H_2O$; iii, SO_4^-

Oxidation of pent-4-en-1-ol (**219**) with $S_2O_8^{2-}/Ag^+$ in the presence of protonated 4-methylquinoline leads to (**220**) and (**221**) as shown in Scheme 10.[718, 719]

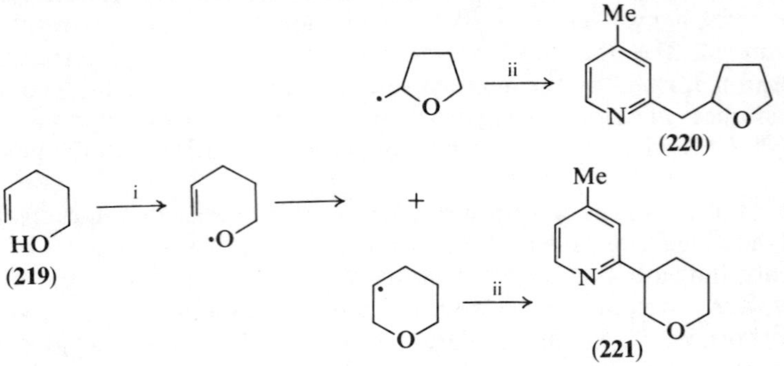

(**220**)

(**219**)

(**221**)

SCHEME 10

Reagents: i, $S_2O_8^{2-} - Ag^+$; ii, 4-$MeC_5H_5N^+$

Oxidations of aromatic compounds containing side-chain OH or COOH groups with $S_2O_8^{2-}$ give products derived from the aromatic radical cation; in the presence of Ag^+ products resulting from attack at the OH or COOH group are also obtained (see Scheme 11).[720, 721]

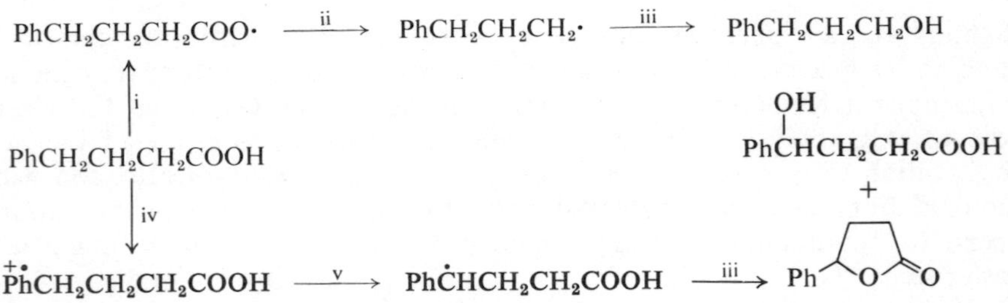

$PhCH_2CH_2CH_2COO\cdot \xrightarrow{ii} PhCH_2CH_2CH_2\cdot \xrightarrow{iii} PhCH_2CH_2CH_2OH$

$PhCH_2CH_2CH_2COOH \xrightarrow{i} PhCH_2CH_2CH_2COO\cdot$

$PhCH_2CH_2CH_2COOH \xrightarrow{iv} \overset{+\bullet}{Ph}CH_2CH_2CH_2COOH \xrightarrow{v} Ph\dot{C}HCH_2CH_2COOH \xrightarrow{iii} PhCH(OH)CH_2CH_2COOH$ + γ-phenylbutyrolactone

SCHEME 11

Reagents: i, $S_2O_8^{2-} - Ag^+$; ii, $-CO_2$; iii, $Cu^{II} - H_2O$; iv, $S_2O_8^{2-}$; v, $-H^+$

Oxidation of 1-methyl- and 1-ethylcyclohexanols with CAN gives rise to product derived from cleavage of the initial 1-alkylcyclohexyloxy-radical.[722] Anodic oxidation of *tert*-alkylbenzylcarbinols gives products formed from cleavage of the initial arene radical cation (*cf.* Scheme 12).[723]

$$ArCH_2CR^1R^2OH \xrightarrow{-e} \overset{\bullet}{Ar}CH_2CR^1R^2OH \longrightarrow ArCH_2\cdot + R^1COR^2 + H^+$$

SCHEME 12

p-Methoxybenzyl ether protecting groups can be removed under very mild conditions through oxidation with *tris*-(*p*-bromophenyl)aminium radical cations[724] (*cf.* Scheme 13).

$$(BrC_6H_4)_3N^{+\bullet} + R^1OCH_2R^2 \rightleftharpoons (BrC_6H_4)_3N + R^1\overset{+\bullet}{O}CH_2R^2$$

$$R^1\overset{+\bullet}{O}CH_2R^2 \xrightarrow{H_2O} R^1O\dot{C}HR^2 \xrightarrow{Ar_3N^{+\bullet}} R^1O\overset{+}{C}HR^2 \xrightarrow{H_2O} R^1OH + R^2CHO$$

SCHEME 13

The osmium(VIII)-catalysed oxidation of benzaldehydes with chloramine-T proceeds by a radical mechanism.[725] The anodic oxidation of 2,3,4,4-tetraphenylpyrrolin-5-one proceeds by two consecutive one-electron transfer steps.[726] The electrochemical trifluoromethylation of ethylene and of benzene has been accomplished by using the CF_3COOH/MeCN/substrate system.[727,728]

One-electron anodic oxidation of *cis*-4-phenylcyclohex-2-ene-1-carboxylate leads to a 1 : 2 : 0.4 mixture of the 4,4′-, 2,4′-, and 2,2′-diphenylbicyclohex-2-enyls.[729] The low proportion of the 2,2′-product is attributed to stabilization of the intermediate allylic radicals by adsorption; this increases the activation energy for combination. In the transition state C—C bond formation is only advanced sufficiently to be influenced by the most severe steric interactions. Voltammetric experiments and controlled potential analysis indicate that in the anodic oxidation of 1,3,4-trimethylcyclohex-3-ene-1-carboxylic acid the C=C group is more readily oxidized than the carboxylate function; electro-oxidation leads to a mixture of stereoisomeric lactones.[730] The rate-determining step in the ferricyanide oxidation of tartrate involves formation of the tartrate radical anion.[731] Aromatic radical zwitterions $\overset{+\cdot}{Ar}COO^-$ and acyloxy-radicals ArCOO· have been shown to be distinct species, by using spin-trapping studies with $CH_2NO_2^-$, as illustrated by the different behaviour of the 4-methyl-substituted species:[732] the acyloxy-radical leads to **(222)** and **(223)** whilst the radical zwitterion gives mainly **(224)** with some **(222)**; the results are explicable on the basis that acyloxy-radicals are σ-radicals (*cf.* p. 112) whilst the radical zwitterions are π-species.

$$ArCO_3H \xrightarrow{Ti^{III}} ArCOO\cdot \xrightarrow{-CO_2} Ar\cdot$$

$$ArCOO\cdot \xrightarrow{CH_2NO_2^-} ArCOOCH_2NO_2^{\overline{\cdot}}\ \textbf{(222)} \qquad Ar\cdot \xrightarrow{CH_2NO_2^-} ArCH_2NO_2^{\overline{\cdot}}\ \textbf{(223)}$$

$$ArCOOH \xrightarrow{S_2O_8^{2-}} \overset{+\cdot}{Ar}COO^- \longrightarrow ArCOO\cdot \xrightarrow{CH_2NO_2^-} \textbf{(222)}$$

$$\overset{+\cdot}{Ar}COO^- \xrightarrow{Ar\,=\,p\text{-}MeC_6H_4} \cdot CH_2\text{-}C_6H_4\text{-}COO^- \xrightarrow{CH_2NO_2^-} {}^{\overline{\cdot}}O_2NCH_2CH_2\text{-}C_6H_4\text{-}COO^-\ \textbf{(224)}$$

Anodic oxidation of *p*-(dimethylamino)dithiobenzoate in acetonitrile gives a tetrathian dication as a result of successive one-electron transfers.[733]

Anodic oxidations of phenols give dimeric products as a result of radical coupling.[734] Anodic oxidation of 4,4′-dimethoxy-2-methylbibenzyl gives a dihydrophenanthrene *via* intramolecular coupling of the diradical cation.[735] Coupling of 2,4,6-tris-(4-*tert*-butylphenyl)phenoxy-radicals leads to the dimer **(225)**, as a result of C—O coupling, which on further oxidation gives rise to a persistent radical.[736] A persistent biradical is formed by oxidation of 1-(3,5-di-*tert*-butyl-4-hydroxy-4′-biphenylyl)-2-(3,5-di-*tert*-butyl-4-hydroxyphenyl)-3-(3,5-di-*tert*-butyl-4-oxocyclohexa-2,5-dien-1-ylidene)cyclopropene.[737] Anodic oxidation of 2,6-dimethylphenol above pH 5 gives a polymeric product resulting from C–O coupling;

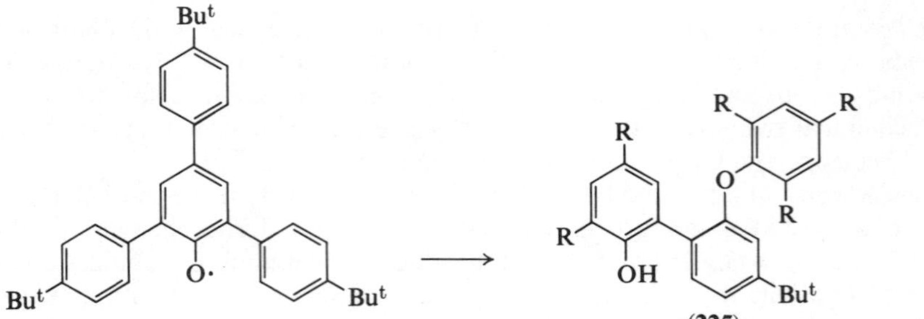

(225)

($R = p\text{-}Bu^tC_6H_4$)

in more acid solution phenoxonium ion intermediates are formed and give predominantly C–C coupled products.[738]

The proportion of C–O coupling in copper(II)-catalysed oxidations of 2,6-dimethylphenol is increased by the addition of triethylamine or piperidine.[739] Oxidation of 2-*tert*-butyl-4-methylphenol with potassium ferricyanide in alkaline solution gives mainly a product of Pummerer's ketone type, whereas oxidation with iron(II) chloride in acidic solution gives mainly 3,5′-di-*tert*-butyl-2,2′-dihydroxy-5,5′-dimethylphenol: the ferricyanide oxidation is believed to proceed *via* the free phenoxy-radical whilst the iron(III) chloride oxidations proceed by an inner-sphere mechanism not involving a *free* radical;[740] mixtures of *o–o*-coupled products were obtained from mixtures of 4-substituted 2-*tert*-butylphenols, which are consistent with a radical-pairing mechanism.[741, 742] Several new tetrahydrodibenzofuranones have been obtained by the *ortho–para*-cross-coupling of two phenols with similar oxidation potentials, *e.g.* oxidations of a mixture of 2-naphthol and 6-hydroxytetralin gives (**226**): the isomeric product (**227**) is not formed, presumably because

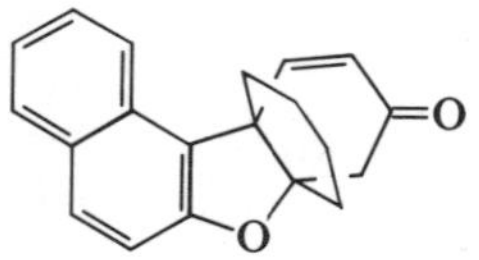

(226) 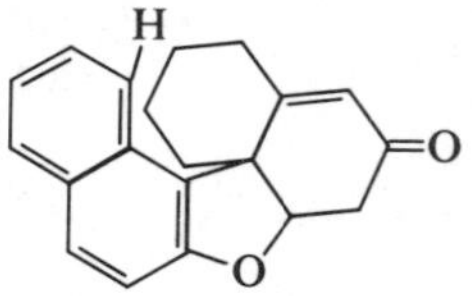 **(227)**

of steric interaction of the *peri*-hydrogen atoms with the tetramethylene chain.[743] Both C–O and C–C-cross-coupling occur in oxidations of binary mixtures of 3-*tert*-butyl-4-methoxyphenol, 3-isopropyl-4-methoxyphenol, and 4-*tert*-butyl-2-methylphenol; the products obtained are trimeric spiro-acetals.[744] Several studies on the oxidation of 2,6-di-*tert*-butyl-4-methylphenol have been reported.[745–747] Dimeric products are obtained from peroxydisulphate/silver(I) oxidations of nitrophenols.[748]

In contrast to the behaviour of 4-allyl-2,6-dimethoxyphenol, which undergoes β-O-coupling,[749] the only dimeric product obtained from the anodic oxidation of 4-allyl-2-methoxyphenol was the *o–o*-coupled diphenol.[750] Products derived from coupling at the β-position predominate in oxidations of *E*- and *Z*-eugenols.[751] Side-chain coupling also occurs in the oxidation of octadecyl 3-(3,5-di-*tert*-butyl-4-hydroxyphenyl)propanoate.[752] Oxidation of the diphenol (**228**; $X = CH_2$) fails to give any of the corresponding intramolecularly coupled product (**230**): this has been attributed to unfavourable 1,3-diaxial interactions in the intermediate

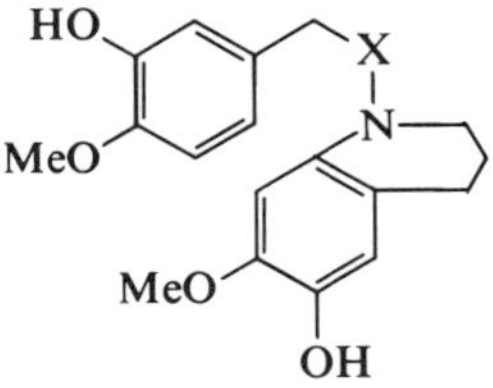

(**229**).[753] Two of the three unfavourable interactions are removed when the diphenol (**228**; X = CO) is oxidized: this gives a good yield of the intramolecularly coupled product. Other examples of intramolecular oxidative coupling have also been reported.[754,755]

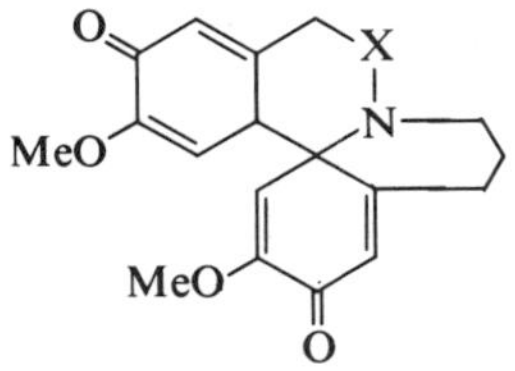

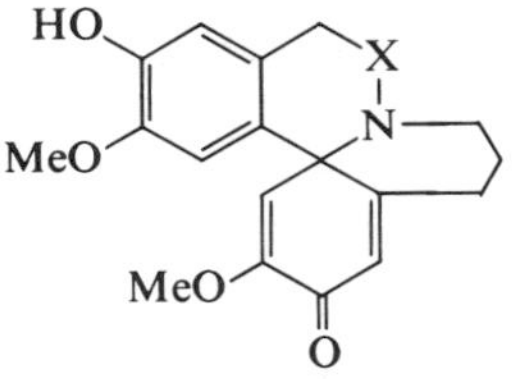

Oxidation of anions of primary amines leads to the corresponding radicals which couple to give hydrazines, subsequent oxidation of which leads to azoalkanes.[756] The horseradish peroxidase-catalysed oxidation of *N,N*-dialkylanilines gives products derived from α-aminoalkyl radical intermediates.[757] Anodic dimerizations of 9,9-disubstituted and 9,9,10-trisubstituted acridans,[758] and of 2,3-diphenylindole,[759] have been reported. Oxidative dimerizations of piperidines[760] and diphenylamine[761] have been effected using peroxydisulphate as the oxidant. The kinetics of oxidation of diphenylamine with dibenzoyl peroxide have been studied.[762] A cation-radical mechanism has been proposed for the cobalt(III) fluoride fluorination of 4-methylmorpholine.[763] Radical intermediates are involved in oxidations of 4,7-dihydroxybenzothiazole,[764] pyrazolines,[765,766] and octacyanoquinodimethane.[767] Aminyl radicals, generated by electrolysis of hydroxylamine in presence of titanium(IV), add to maleic acid to give aspartic acid.[768] Anodic oxidation of arylnickel(II) phosphine complexes gives rise to aryl radicals.[769]

β-Hydroxyalkyl radicals are oxidized to alkenes with copper(II) *via* alkylcopper intermediates;[770] both α- and β-hydroxyalkyl radicals oxidize copper(I). α-Hydroxyalkyl radicals are oxidized by iron(III) by an outer-sphere mechanism.[771] The relative redox reactivity of radicals derived from substituted carboxylic acids and glycols towards heteropoly electrolytes of molybdenum and tungsten have been determined.[772]

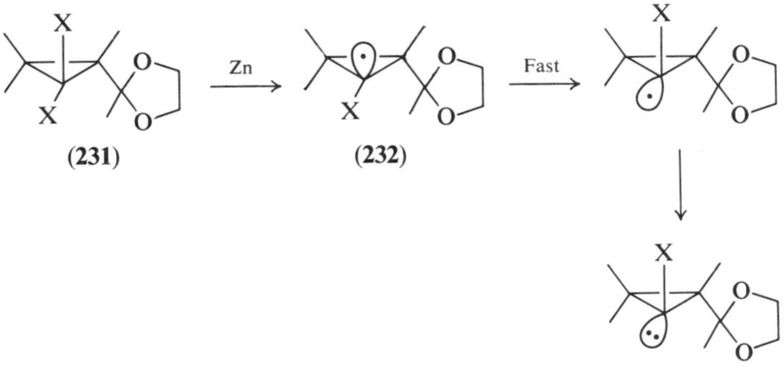

The monodehalogenation of the dioxalanes of *gem*-dihalocyclopropyl methyl ketones (**231**) with zinc occurs stereoselectively; the reaction proceeds *via* initial attack of zinc at the less hindered C—X bond, followed by inversion of the resultant α-halocyclopropyl radical (**232**) and rapid reduction of the radical to the anion;[773] the parent ketones, in which steric factors are less important, react without general stereoselectivity. Radical intermediates are invoked in the reduction of 1,4-dihalonorbornanes to [2.2.1]propellane.[774]

The reductive methylation of 4-acetylbiphenyl proceeds *via* the ketyl radical anion.[775] Acyclic products are formed in the controlled-potential reduction of 1-benzoyl-2-phenylcyclopropane as a result of ring-opening of the intermediate cyclopropylcarbinyl radical (**233**); no acyclic product is obtained from benzoylcyclopropane as there is less driving force for ring-opening of the intermediate

$$\text{PhCO}\triangledown\text{Ph} \xrightarrow{\text{e, H}^+} \text{Ph}-\dot{\text{C}}(\text{OH})\triangledown\text{Ph} \longrightarrow \text{Ph}-\text{C(OH)}=\text{CHCH}_2\dot{\text{C}}\text{HPh} \xrightarrow{\text{e, H}^+} \text{PhCOCH}_2\text{CH}_2\text{CH}_2\text{Ph}$$

(**233**)

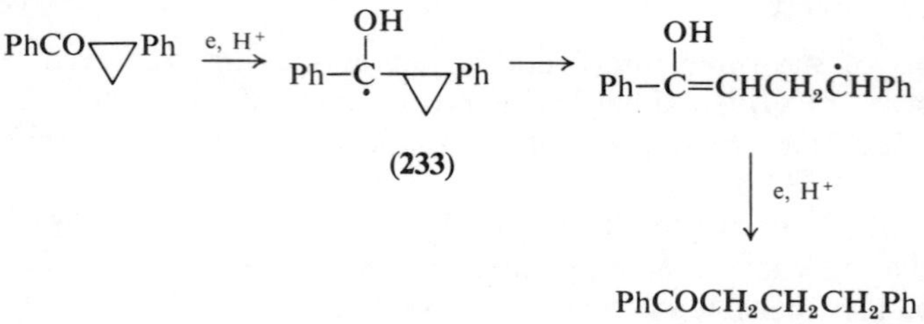

radical.[776] Asymmetric induction is achieved in the cathodic reduction of acetophenone in presence of a chiral quaternary ammonium salt in an achiral solvent; this is due to recombination of ketyl radical anions as intimate ion pairs with the chiral ammonium ion.[777] Cathodic reduction of 2-alkyl-2-cyanocycloalkanones gives mixtures of the corresponding cycloalkanol and 2-alkylcycloalkanone; the latter product is formed by loss of cyanide from the initial ketyl radical anion, followed by reduction of the resultant radical.[778] The reversibility of electrolytic reduction of aryl ketones can be prevented by addition of certain metal salts.[779] The electrochemical reduction of (benzylideneacetone)tricarbonyliron(0) gives the parent ketone in a one-electron transfer process.[780] Oligomerization of chalcone with lithium proceeds *via* the ketyl radical anion.[781]

The cathodic reduction of diphenylmethyl *p*-nitrophenyl sulphide involves fragmentation of the initial radical anion,[782] as shown in Scheme 14. The one-electron reduction of other alkyl aryl sulphides by lithium naphthalenide proceeds by a similar mechanism.[783]

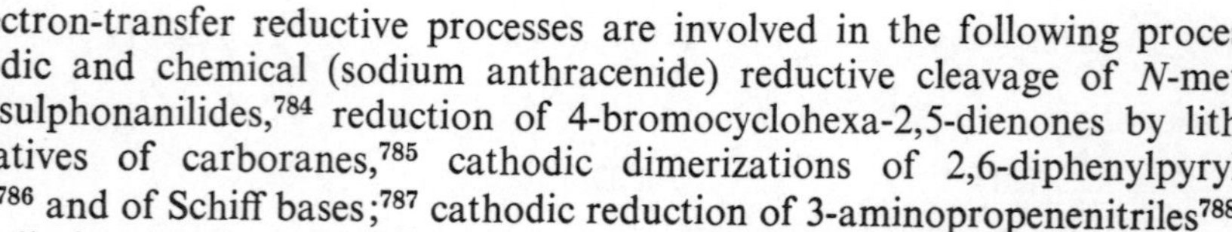

$$\text{Ph}_2\text{CHSC}_6\text{H}_4\text{NO}_2\text{-}p \xrightarrow{\text{e}} \text{Ph}_2\text{CHSC}_6\text{H}_4\text{NO}_2^{\dot{-}} \longrightarrow \text{Ph}_2\dot{\text{C}}\text{H} + \text{O}_2\text{NC}_6\text{H}_4\text{S}^-$$

$$\text{Ph}_2\dot{\text{C}}\text{H} \xrightarrow{\text{e, H}^+} \text{Ph}_2\text{CH}_2 \qquad \text{Ph}_2\dot{\text{C}}\text{H} \longrightarrow \text{Ph}_2\text{CHCHPh}_2$$

SCHEME 14

Electron-transfer reductive processes are involved in the following processes: cathodic and chemical (sodium anthracenide) reductive cleavage of *N*-methylarenesulphonanilides,[784] reduction of 4-bromocyclohexa-2,5-dienones by lithium derivatives of carboranes,[785] cathodic dimerizations of 2,6-diphenylpyrylium salts,[786] and of Schiff bases;[787] cathodic reduction of 3-aminopropenenitriles[788] and of cyclic immonium salts;[789] reduction of *tert*-alkyl nitriles with lithium;[790] and reductive silylation of esters.[791] Preferential cleavage of the allylic benzenesulphonyl group occurs in the reductive cleavage of 2-methyl-1,5-bis(benzenesulphonyl)pent-2-ene.[792] Cathodic reduction of 9-diazofluorene gives initially its radical anion and subsequently the fluorene radical anion.[793]

Electron-transfer Reactions

Triphenylaminium radical cations are reduced by DPPH, phenoxy-radicals, and triphenylverdazyl in one-electron processes.[794] There is a linear correlation between the rate of hydrogen-abstraction by DPPH from electron-donor molecules, *e.g.* amines, and their ionization potentials; this is consistent with a rate-determining electron-transfer step.[795] Bisbenzenechromium(0) is oxidized to the corresponding cation by 2,4,6-tri-*tert*-butylphenoxy-radicals.[796] Triarylmethyl radicals are oxidized to the corresponding cation by sulphur dioxide, while anthracene *endo*-peroxides lose oxygen and give the corresponding anthracene radical cation.[797] The perchlorotriphenylmethyl radical is reduced to the carbanion by ascorbic acid.[798]

Nitration of mesitylene does not appear to proceed by a radical-pair mechanism as the product composition from reactions of the mesitylene radical cation with nitrogen dioxide is distinct from that found in direct nitration.[799] The radical-pair mechanism has been postulated for electrophilic substitutions of [2.2]metacyclophane.[800]

The initiation step in the copper(I) chloride-catalysed addition of sulphonyl halides to alkenes involves the formation of alkylsulphonyl radicals in an electron-transfer step.[801] The electron-transfer-initiated addition of 2-phenyl-1-pyrrolinium perchlorate (**234**) to alkenes proceeds by a radical-pair mechanism, as shown.[802]

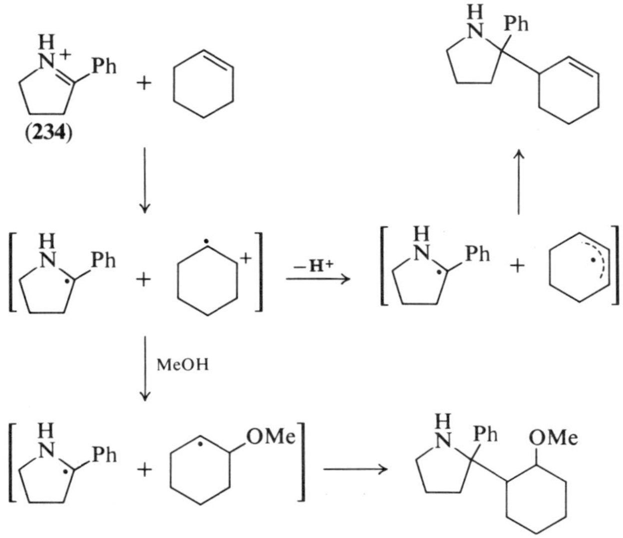

(**234**)

$-H^+$

MeOH

In the Lewis acid-catalysed oxygenations of 1,1′-bicyclohexenyl, α-terpinene,[803] ergosteryl acetate,[804] and 1,1-diarylethanes[805] radical-cation intermediates (generated by electron-transfer to the Lewis acid) subsequently react with oxygen and finally yield cyclic peroxides. Radical cations of tetrathiafulvalene are obtained by oxidation of the parent compound with DDQ.[806]

Nucleophilic substitution reactions involving one-electron transfer processes have been reviewed;[807] a specific review of the S_{RN} mechanism of nucleophilic aromatic substitution has also been published.[808] New examples of reactions proceeding by the S_{RN} mechanism include the phenylation of the dimsyi anion[809] and the synthesis of aryl phenyl selenides.[810] Reactions of *m*-haloiodobenzenes (*m*-XC_6H_4I) with diethyl phosphite [$Y^- = (EtO)_2PO^-$] lead to mixtures of the mono- and di-substituted products:[811–813] the monosubstituted product does not appear to be an intermediate in the formation of the disubstituted product; these results are explicable on the basis of the mechanism shown (reactions 2–7).

$$m\text{-}XC_6H_4I^{\bar{\cdot}} \longrightarrow m\text{-}XC_6H_4\cdot + I^{\bar{\cdot}} \quad (2)$$

$$m\text{-}XC_6H_4\cdot + Y^- \longrightarrow m\text{-}XC_6H_4Y^{\bar{\cdot}} \quad (3)$$

$$m\text{-}XC_6H_4Y^{\bar{\cdot}} + m\text{-}XC_6H_4I \longrightarrow m\text{-}XC_6H_4Y + m\text{-}XC_6H_4I^{\bar{\cdot}} \quad (4)$$

$$m\text{-}XC_6H_4Y^{\bar{\cdot}} \longrightarrow m\text{-}YC_6H_4\cdot + X^- \quad (5)$$

$$m\text{-}YC_6H_4\cdot + Y^- \longrightarrow m\text{-}C_6H_4Y_2^{\bar{\cdot}} \quad (6)$$

$$m\text{-}C_6H_4Y_2^{\bar{\cdot}} + m\text{-}XC_6H_4I \longrightarrow m\text{-}C_6H_4Y_2 + m\text{-}XC_6H_4I^{\bar{\cdot}} \quad (7)$$

The relative rates of reactions 4 and 5 control whether mono- or di-substitution occurs; disubstitution is preferred in reactions with diethyl phosphite when X is a good leaving group and occurs when X = Br, but not when X = Cl;[813] monosubstitution is favoured by strongly electron-attracting groups which stabilize the intermediate *m*-$XC_6H_4Y^-$ (*e.g.* monosubstitution occurs in the reaction of *m*-chloroiodobenzene with diethyl phosphite, but not with thiophenoxide).[811] A competing reaction in photostimulated S_{RN} reactions of iodobenzene with enolates, particularly tertiary enolates, is hydrogen-transfer.[814]

In a reaction closely related to S_{RN} reactions, *o*-dichlorobenzene reacts with sodium or potassium in liquid ammonia to give benzene in good yield, but no chlorobenzene; the absence of chlorobenzene has been attributed to the further reaction of chlorobenzene in an electron-rich zone;[815] as illustrated in Scheme 15.

$$C_6H_4Cl_2 \xrightarrow{e} C_6H_4Cl_2^{\bar{\cdot}} \xrightarrow{-Cl^-} C_6H_4Cl\cdot$$

$$C_6H_4Cl\cdot \xrightarrow{e} C_6H_4Cl^-$$

$$C_6H_6 \longleftarrow \longleftarrow C_6H_5Cl \xleftarrow{H^+} C_6H_4Cl^-$$

SCHEME 15

Nucleophilic aromatic substitution of halide by thiolate and cyanide has been induced by electrochemical generation of the aryl radical cation which subsequently reacts as in other S_{RN} processes.[816]

α,*p*-Dinitrocumene, *p*-$O_2NC_6H_4CMe_2NO_2$, undergoes electron-transfer nucleophilic substitution with nitroalkane anions or with sodium sulphide to give excellent yields of highly hindered nitro-compounds[817] and thiols,[818] respectively. Replacement of the nitro-groups in tertiary nitro-compounds by hydrogen occurs in the electron-transfer reaction with CH_3S^-[819] formulated below.

$$R_3CNO_2 + CH_3S^- \longrightarrow R_3CNO_2^{\bar{\cdot}} + CH_3S\cdot$$

$$R_3CNO_2^{\bar{\cdot}} \longrightarrow R_3C\cdot + NO_2^-$$

$$R_3C\cdot + CH_3S^- \longrightarrow R_3CH + H_2\dot{C}S^-$$

Replacement of halogen occurs in the reactions of 5-halo-2*H*,3*H*-benzo[*b*]-thiophene-2,3-diones with thiophenoxide, while reductive dehalogenation occurs in reactions with Bu^tO^-, EtO^-, HO^-, and PhO^-; both reaction types involve initial electron-transfer from the nucleophile.[820] Diorganophosphides react with allyl and benzyl halides to form trialkylphosphines; ^{31}P-CIDNP studies indicate that the product-forming radical-combination step is preceded by an electron-transfer step,[821] thus:

$$PR^1_2{}^- + R^2X \longrightarrow [R^1_2P\cdot \quad \cdot R^2] + X^-$$
$$[R^1_2P\cdot \quad \cdot R^2] \longrightarrow R^1_2PR^2$$

A similar process occurs in the reaction of sodium thiolates with benzyl halides.[822] Alkylperoxy-radicals are generated in the reactions of alkyl halides with superoxide.[823] Electron-transfer processes are also involved in the reactions of *p*-nitrobenzylidene chloride[824] and *p*-nitrobenzene[825] with hydroxide, of *p*-nitrobenzyl and 4-nitrobenzhydryl halides with *tert*-butoxide,[826] also in reductive elimination of β-nitrosulphones,[827] and in the reactions of 2-bromo-*N*,*N*-dimethyl-2,2-diphenylacetamide with methoxide,[828] and of diphenyliodinium salts with cyanide.[829]

Radical-anion reactions with organic halides have been reviewed.[830] Consideration of the nature of electron capture by alkyl halides RX indicates that this leads directly to R$\cdot$----X$^-$ and not to RX$^{\overline{\cdot}}$ in which the unpaired electron would lie in a C–X σ^* orbital.[831] Alkylation of lithium anthracenide with 1-methylheptyl halides occurs with partial inversion due to competition between electron-transfer and S_N2 processes.[832] Electrochemical reductive *tert*-butylation of pyridinecarboxamide (A) involves coupling of the *tert*-butyl radical with the carboxamide radical anion, as formulated.[833]

$$A^{\overline{\cdot}} + Bu^tX \longrightarrow A + Bu^t\cdot + X^-$$
$$Bu^t\cdot + A^- \longrightarrow ABu^{t-}$$
$$ABu^{t-} + H^+ \longrightarrow ABu^tH$$

Electron-transfer processes are involved in reduction of benzophenone with lithium di-isopropylamide,[834] halogenation of ester anions with carbon tetrahalides,[835] and in reductive cleavage of benzhydryl benzoates with lithium naphthalenide.[836] Decarbonylation of diphenylcyclopropene[837] and decarboxylation of *o*-phthaloyl peroxide[838] (to give *trans*-stilbene and biphenylene radical anions, respectively), with lithium in HMPA, occur by a series of electron-transfer processes.

α-Nitroalkyl radicals have been detected by spin-trapping in the reaction of tetranitromethane and α-nitro-carbanions:[839]

$$R^1R^2CNO_2{}^- + C(NO_2)_4 \longrightarrow R^1R^2\dot{C}NO_2 + C(NO_2)_4{}^{\overline{\cdot}}$$
$$C(NO_2)_4{}^{\overline{\cdot}} \longrightarrow NO_2\cdot + C(NO_2)_3{}^-$$

A similar electron-transfer process is involved in the reaction of tetranitromethane with ascorbic acid.[840] An electron-transfer process is likewise postulated for the oxidation of estradiol **(235)** and estradiol 3-methyl ether to the corresponding estrones **(236)**.[841]

Electron-transfer processes are involved in the reaction of 2,4,6-trimethylpyrylium salts with superoxide,[842] in reactions between arylphosphines and arylnitroso-compounds,[843] and of dehydroascorbic acid with aromatic amines.[844]

HO, H, H **(235)** → O$^-$, H, H → [O]$^{\overline{\cdot}}$, H + HOO$\cdot$ $\xrightarrow{CCl_4}$ O, H **(236)** + $\cdot CCl_3$ + Cl^-

Radical anions are formed in the oxidation of polyvinyl acetals with TCNQ.[845] ESR studies indicate that at least partial electron-transfer occurs in the complexes of electron-rich aromatic hydrocarbons with 4,5-dinitrophthalic anhydride,[846] and intramolecular charge-transfer has been detected in certain derivatives of 1,4-naphthoquinone[847] and cyclophanes.[848]

The absence of a ^{13}C-isotope effect in the formation of methylmagnesium iodide from $^{13}CH_3I$ indicates that the rate-determining step involves outer-sphere electron-transfer from magnesium to the alkyl halide, and not cleavage of the C—I bond.[849] The presence of methyl radicals in the preparation of methylmagnesium iodide in allyl phenyl ether is indicated by results of CIDNP studies.[850] Long-lived radicals are obtained on the surface of the magnesium in the attempted preparation of Grignard reagents from 2-, 3-, and 4-haloalkoxysilanes.[851] 1,2-Migration of phenyl groups occurs during reaction of magnesium with 2,2,2-triphenylethyl chloride in ether; this has been attributed to rearrangement of intermediate neophyl radicals.[852] Neophyl radicals are also in the reaction of neophylmagnesium chloride with 1,2-dibromoethane.[853] 3,3-Dimethylallenyl radicals are implicated as intermediates in the reactions of 3,3-dimethylallenyllithium (**237**) with benzyl

$$\underset{(\mathbf{237})}{Me_2C{=}C{=}CHLi + PhCH_2X} \longrightarrow \left[\begin{matrix} Me_2C{=}C{=}\dot{C}H \\ \updownarrow \\ Me_2\dot{C}{-}C{\equiv}CH \end{matrix} \quad PhCH_2\cdot\right] + LiX$$

$$\downarrow$$

$$\underset{(\mathbf{238})}{\begin{matrix} Me_2C{-}C{\equiv}CH \\ | \\ CH_2Ph \end{matrix}} + Me_2C{=}C{=}CHCH_2Ph$$

halides; coupling occurs preferentially at the propargyl position, to give (**238**), as that position has a higher spin density than the 1-position.[854] Radical processes can be detected in the reactions of Me_3Sn^- with alkyl halides;[855] reaction with *cis*- or *trans*-4-*tert*-butylcyclohexyl halides gives mixtures of *cis*- and *trans*-products, and reaction with cyclopropylmethyl bromide gives ring-opened products, indicating the formation of non-caged radicals; metallation of aryl halides by Bu_3Sn^- proceeds to only a minor extent by a radical route.[856]

cis-2,2,6,6-Tetramethylhept-4-en-3-one (**239**) has been used as a probe for detecting SET (single electron transfer) in reactions of Grignard and other organometallic reagents;[857] if SET occurs, some isomerization of (**239**) to the *trans*-isomer (**240**) occurs whilst polar addition yields only the *cis*-addition product. The results indicate that Bu^tMgCl, and probably $LiCuMe_2$ and Li_2CuMe_3, react by SET whereas allylGrignard reagents and MeLi react by a polar mechanism. *tert*-Butylmagnesium chloride reacts with benzophenone to give benzpinacol and the products of 1,2- and 1,6-addition: addition of small amounts of *m*-dinitrobenzene inhibits the formation of the pinacol by trapping free ketyl radical but has no effect on the ratio of the other products, indicating that they arise by collapse of the radical cation–radical anion pair.[858] Reaction of benzophenone with 1,1-dimethylhex-5-enylmagnesium bromide gives products of 1,2- and 1,6-addition: the products of 1,6-addition are 74% cyclized and this has been attributed to

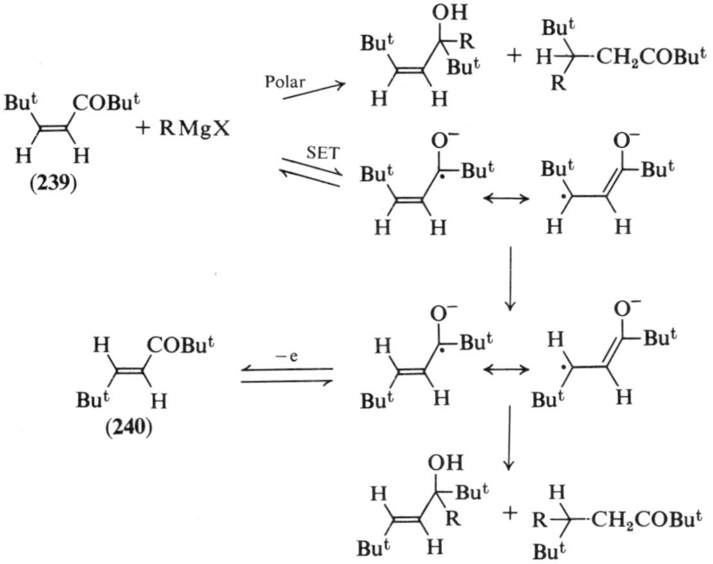

(239)

(240)

isomerization of an intermediate hexenyl radical.[859] Scheme 16 summarizes the mechanisms for these processes.

Radical intermediates formed in electron-transfer processes are involved in the following processes: homocoupling of phenylmagnesium bromide with benzyl halides, catalysed by iron complexes;[860] reaction of alkyl-lithiums with titanium tetrachloride;[861] reactions of bis(trialkylgermyl)mercurys with 4-bromo-2,4,6-tri-*tert*-butylcyclohexa-2,5-dienone;[862] reaction of diphenylmercury with phenanthraquinone;[863] and reaction of copper(II) halides with organopentafluorosilicates.[864] The kinetics of the electron-transfer reaction between 2,5-di-*tert*-butylbenzoquinone radical anion and metal porphyrins have been studied;[865] initial electron-transfer to the metal occurs with the iron(II) porphyrin.

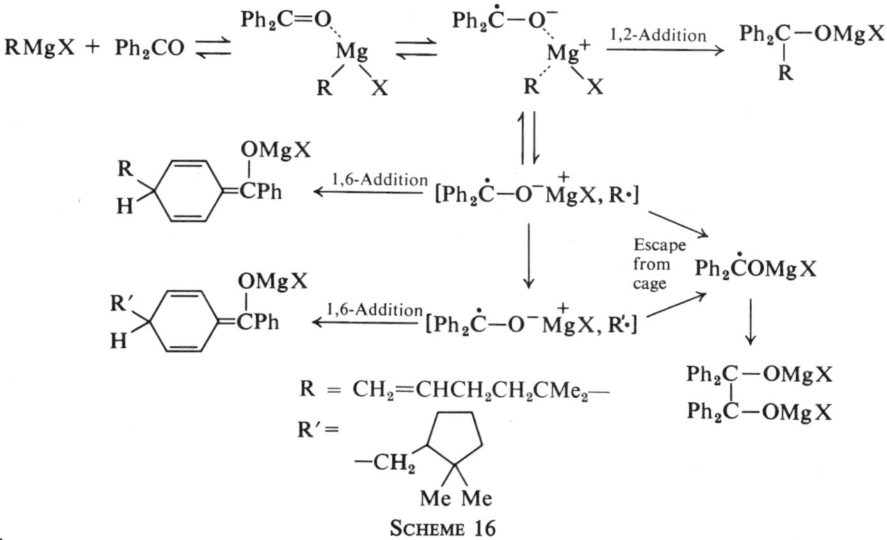

$R = CH_2{=}CHCH_2CH_2CMe_2{-}$

SCHEME 16

Chemiluminescent reactions of 1,2-dioxetanes have been reviewed[866] and further details have been reported of the origin of chemiluminescence in the decomposition of diphenyl peroxide[867] and *o*-xylylene peroxide,[868,869] in the presence of electron-rich aromatic hydrocarbons.

The photochemical reactions between aromatic amines and carbon tetrachloride or chloroform involve initial formation of a radical cation–radical anion pair which then undergoes proton transfer.[870–873] A similar process is involved in the photolysis of mixtures of 1,1-diarylethylenes (Ar = *p*-NCC_6H_4 or *p*-$MeOOCC_6H_4$) and tetramethylurea; in this case the reaction gives the 1 : 1 adducts (**241**).[874] The photochemical addition of amines to stilbenes proceeds similarly.[875]

$$Ar_2C{=}CH_2 + Me_2NCONMe_2 \xrightarrow{h\nu} [Ar_2C{=}CH_2^{\overline{\cdot}},\ Me_2NCONMe_2^{+\cdot}]$$

$$\downarrow$$

$$\underset{(\mathbf{241})}{Ar_2CHCH_2CH_2NMeCONMe_2} \longleftarrow [Ar_2CHCH_2\cdot \quad \cdot CH_2NMeCONMe_2]$$

The photo-induced solvent-incorporated addition of *N*-methylphthalimide to styrene proceeds by initial electron-transfer; the styrene radical-cation is then captured by methanol to give the $Ph\dot{C}HCH_2OMe$ radical which adds to the *N*-methylphthalimide radical-anion.[876] Electron-transfer occurs in the photolysis of a mixture of 9,10-dicyanoanthracene and methyl 1,2-diphenylcyclopropane-3-carboxylate; recombination of the ion pair gives the triplet-substituted cyclopropane which reacts further with more of the substituted cyclopropane to give a dimeric product.[877] The photochemically induced hydrogen exchange between anthracene and aliphatic amines occurs by means of an electron-transfer process.[878]

Photolysis of 1,1-diphenylethylene, 2-phenylnorbornene, or 1-phenylcyclohexene in presence of a good electron-acceptor (*e.g.* 1-cyanonaphthalene or methyl *p*-cyanobenzoate) leads to anti-Markovnikov addition of cyanide;[879] the same mechanism is operative for the electron-acceptor photosensitized anti-Markovnikov addition of methanol to 2-methylene-1,2,3,4-tetrahydro-1,4-ethanonaphthalene,[880]

$$Ph_2C{=}CH_2 + A \xrightarrow{h\nu} Ph_2C{=}CH_2^{+\cdot} + A^{\overline{\cdot}}$$

$$\downarrow CN^-$$

$$Ph_2CHCH_2CN \xleftarrow{H^+} Ph_2\bar{C}CH_2CN \xleftarrow{A^{\overline{\cdot}}} Ph_2\dot{C}CH_2CN$$

and for additions of furans to indene.[881] In contrast, the photosensitized addition of methanol to 1,1-diphenylethylene by means of an electron-donor (*e.g.* 1-methoxynaphthalene) as the photosensitizer leads to the Markovnikov addition product.[882]

$$Ph_2C{=}CH_2 + D \xrightarrow{h\nu} Ph_2C{=}CH_2^{\overline{\cdot}} + D^{+\cdot}$$

$$Ph_2C{=}CH_2^{\overline{\cdot}} \xrightarrow{H^+} Ph_2\dot{C}CH_3 \xrightarrow{D^{+\cdot}} Ph_2\overset{+}{C}CH_3$$

$$Ph_2\overset{+}{C}CH_3 \xrightarrow{MeOH} Ph_2C(OMe)CH_3$$

Photolysis of either *cis*- or *trans*-2,3-diphenyloxirane (**242**) in presence of an electron-acceptor as photosensitizer results in isomerization of the oxirane and cleavage to the isomeric carbonyl ylides (**243** and **244**).[883]

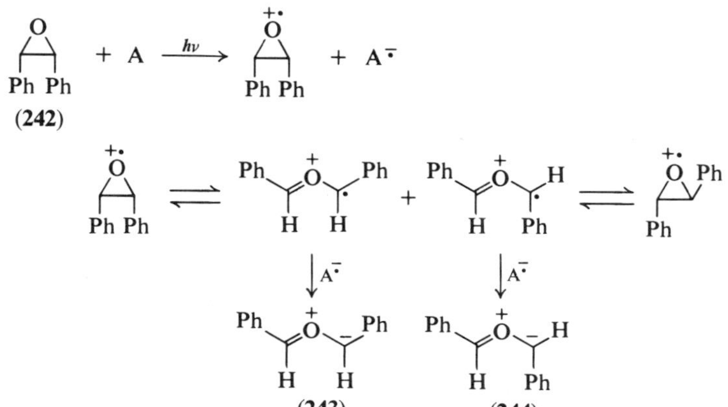

(242) + A $\xrightarrow{h\nu}$ $[\text{Ph Ph epoxide}]^{+\cdot}$ + $A^{\overline{\cdot}}$

(243) (244)

Indene dimers are cleaved by a redox photosensitized process with phenanthrene as an electron-donor and *p*-dicyanobenzene as an electron-acceptor;[884] a similar process had previously been used to effect the cyclodimerization of *N*-vinylcarbazole.[885]

The photo-induced dimerizations of 1-naphthyl oxide[886] and of allylphenoxide ions[887] have been reported; the reactions proceed *via* electron-transfer to give phenoxy-radicals.

Superoxide has been generated by irradiation of an oxygen-saturated solution of Rose Bengal,[888, 889] and shown to react with biadamantylidene to give the corresponding epoxide. The initial step in the reaction of oxygen with 1,5-dihydroflavin involves a one-electron transfer to give a flavin radical–superoxide ion pair which then collapses rapidly to the hydroperoxyflavin.[890] The reaction of nitroxides with

$$FlH^{-} + O_2 \longrightarrow [FlH\cdot\ O_2^{\overline{\cdot}}] \xrightarrow{H^+} FlH\cdot OOH$$

FlH = Dihydroflavin

1,5-dihydroflavins to give flavin radicals proceeds by an electron-transfer mechanism.[891] One-electron oxidations of several 1,5-dideazaisoalloxazines have been investigated.[892]

There is considerable confusion as to whether or not one-electron transfer processes are involved in reductions by NADH-model compounds: isotope studies using *N*-benzyldihydronicotinamides are more consistent with a mechanism involving electron-transfer and subsequent proton-transfer with radical intermediates rather than with hydride-transfer;[893] however, spin-trapping studies on the reduction of pyridine-2-aldehyde indicate that radical intermediates are not involved.[894] The calculated rates of one-electron transfer from *N*-benzyldihydronicotinamide to various electron-acceptors are several orders of magnitude less than the experimental rates for the overall reduction process and hence seem to exclude an outer-sphere electron-transfer process.[895] The failure to detect CIDNP in reactions of *N*-methyldihydronicotinamide with triphenylmethyl chloride has been attributed to insufficient lifetime of the radical pair intermediates.[896]

1-Alkyl-4-pyridinyl radicals are formed by electron-transfer oxidations of the corresponding 1,2- or 1,4-dihydropyridines[897] and by reduction of the corresponding pyridinium salts;[898] with magnesium and other metal halides these radicals form complexes which decompose in the presence of alcohols by intramolecular electron-transfer.[899]

Radical Ions

The anion and cation radicals,

$$R_3\overset{+}{P}CH{=}CH\dot{P}R_3 \quad \text{and} \quad PhP(O)(OEt)CH{=}CHP(O)(OEt)^{\overset{\cdot}{-}},$$

are both π-radicals in which the unpaired electron is delocalized over both phosphorus atoms and the unsaturated system.[900] A thiomethoxy-group stabilizes both radical anions and radical cations of naphthalenes.[901]

Arene radical cations are relatively long-lived in superacid media, *e.g.* HF;[902, 903] strong acid also increases the lifetime of tertiary aminium radicals, as proton loss to give a 1-aminoalkyl radical is prevented.[904] Nitromethane has a much greater stabilizing influence on aromatic radical cations than has DMF or acetonitrile.[905] The naphthalene radical cation salt $(NpH^{+\bullet})_2PF_6^{2-}$ is relatively stable.[906]

Theoretical assignments of coupling constants in the ethylene radical cation have been made.[907] The [3]radialene radical-cation has the structure **(245)**.[908]

(245) 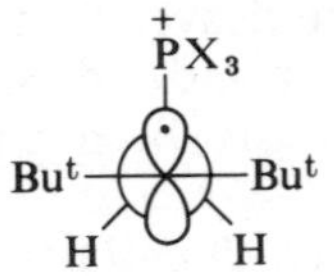 **(246)**

Trialkylphosphine radical cations have been trapped with 1,1-di-*tert*-butylethylene to give the radical cations $Bu^t{}_2\dot{C}CH_2\overset{+}{P}X_3$ which owing to steric and electronic factors exist in the eclipsed conformation **(246)**.[909] All the methyl groups in the 2,2,3,3-tetramethylbutane radical cation are equivalent.[910] A long-lived radical cation is obtained by oxidation of 1,5-dithiacyclo-octane; its ESR spectrum is consistent with a structure having a long S—S bond with the unpaired electron localized on one sulphur atom only.[911] MO-studies on the stability of substituted benzene radical cations have been carried out.[912] Trimethylsilyl substituents exert a stabilizing influence on aromatic radical cations as a result of π-hyperconjugation with the C—Si bonds.[913–915] The preferred orientations of the methoxy-groups in *O*-methyl ethers of phloroglucinol[916] and 4-propylveratrole have been studied.[917, 918] The radical cation derived from *p*-$Et_3\overset{+}{P}C_6H_4PEt_3$ exhibits hyperfine coupling with the peripheral CH_2 groups attached to phosphorus.[919] The spin density distribution in $(Me_3Si)_2N{-}N{=}N{-}N(SiMe_3)_2{}^{+\bullet}$ is comparable to that in other 5-electron π-systems such as butadiene radical anions.[920] ESR-studies of triarylaminium radical cations indicate that the hyperfine splitting constant a_N correlate with Hammett σ^+ substituents for electron-donating but not electron-withdrawing substituents.[921] The β-methylene proton couplings are much larger for mono-methylene-bridged bipyridyl radical cations than for the fluorene radical cation owing to stronger hyperconjugative interactions in the heterocyclic system.[922] A variety of cyclic hydrazine radical-cations with boron, carbon, silicon, and phosphorus substituents are more planar than the parent hydrazines: the radical cations are also relatively stable.[923] ESR-studies of *N,N'*-dialkyl-*N,N'*-diarylhydrazine radical cations also indicate that the nitrogen atoms are planar or nearly planar, though the hydrazinium system can be somewhat twisted about the N—N bond owing to steric interaction of the substituents.[924] The wide variation in oxidation potentials of tetra-alkylhydrazines arises from differences in strain of the neutral and the radical cation forms.[925] The conformations of 1,2-dialkyl-1,2-diazetidine radical cations have been studied and compared with their acyclic analogues.[926] The 1,3,6,8-tetra-azatricyclo[4.4.1.1^{3,8}]dodecane radical cation is relatively long lived: this is attributed to a three-electron bond between two nitrogen atoms.[927] The stability of phenothiazine radical cations is decreased by substituents which exert $-I$ effects.[928] The first persistent phosphine radical cation to be described is that of 2,3,5,6,7,8-hexamethyl-2,3,5,6,7,8-hexa-aza-1,4-diphosphabicyclo[2.2.2]octane.[929]

Reviews on the mechanisms of radical cation reactions have appeared.[930, 931] A theoretical study based on thermochemical analysis indicates that the reactivity of radical cations with nucleophiles is determined primarily by the oxidation potentials of the reactants.[932] When the standard free energy for the reaction $RH^{+\bullet} + Nu^- \rightarrow RH + Nu\cdot$ is negative, a diffusion-controlled electron-transfer reaction should occur, as in the reaction of the perylene radical cation with iodide.[933] As ΔG^0 for the reaction becomes increasingly positive, electron-transfer becomes less probable and direct bond-formation will occur preferentially. The importance of the oxidation potential of the nucleophile (and also its S_N2 nucleophilicity parameter) are seen in the reactions of 9,10-diphenylanthracene radical cation with nucleophile; I^-, Br^-, CN^-, SCN^-, and H_2S react by electron-transfer whereas water, pyridines, and piperidine react by addition.[934] Application of the Dewar–Zimmerman rules to reactions of radical cations shows that the direct suprafacial reaction corresponds to an unfavourable antiaromatic transition state, thereby accounting for the low reactivity of the perylene radical cation with halides; any reaction that occurs is of the electron-transfer type.[935] The chloropromazine radical cation undergoes direct reaction with water.[936] Thioanthrene and phenoxanthiin radical cations undergo *S*-alkylation and *S*-arylation on reaction with, respectively, a dialkyl- and a diaryl-mercury.[937] The 4,4′-dimethoxystilbene radical cation dimerizes to a tetralin derivative, 2,3,4,5-tetrakis-*p*-(methoxyphenyl)tetrahydrofuran or 1,4-dimethoxy-1,2,3,4-tetrakis-*p*-(methoxyphenyl)butane, depending on the solvent; in the presence of nucleophiles, *e.g.* MeOH, the radical cation also reacts with the nucleophile.[938] The dimerization of *N,N′*-disubstituted 4,4′-dipyridylium derivatives has also been studied.[939] A thermochemical analysis of proton abstraction from alkylaromatic radical cations, *e.g.*:

$$ArR_2CH^{+\bullet} + B^- \rightleftharpoons [ArR_2C\cdots H\cdots B] \longrightarrow ArR_2C\cdot + BH$$

indicates that this type of reaction should be extremely fast and seems to invalidate some previous mechanisms in which a process of rapid reversible electron-transfer followed by slow proton-loss has been postulated for the oxidative α-substitution of alkylaromatics by high-valence metal ions.[940] The rate of proton-loss from polymethylbenzene radical cations decreases by a factor of 10^3 as the number of methyl groups increases from 1 to 5; the rate constants correlate with the ionization potentials of the neutral compounds.[941] The pK_a values of some phenol radical cations have been reported.[942] The *N*-chloro-2,2,6,6-tetramethylpiperidinium radical cation abstracts hydrogen from methanol and ethanol and adds to cyclohexene, though much less rapidly than does the piperidinium radical cation;[943] the piperidinium radical cation reacts with NO to give *N*-nitrosopiperidine.[944] Pyridinium bis(alkoxycarbonyl)methylide radical cations have been trapped with diphenylcyclopropenone.[945]

Organometallic radical anions have been reviewed.[946] Radical anions of naphthalene and benzophenone can be efficiently generated by electron-transfer from trimethylsilyl potassium.[947] Perylene and anthracene radical anions are relatively long-lived in DMF solutions containing tetra-alkylammonium fluoroborates.[948] Stable ascorbic acid radical anions are generated by oxidation of ascorbic acid in presence of diarylthallium hydroxide.[949] The formation of radical anions, derived from 3,4-dicyanocyclobutene-1,2-dione,[950] cyclobutene-1,2-diones,[951] nitrodiphenylethylenes,[952] and 1,4-bis(dimethylphosphino)benzene,[953] has been reported.

The formation of the $(F_3C)_2C{=}C(CF_3)_2^{\bar{\cdot}}$ radical anion is the first example of a reversible one-electron reduction of an isolated double bond; the radical anion is stabilized by the electron-withdrawing influence of the CF_3 groups and has a planar or near-planar structure.[954] The unpaired electron is more extensively delocalized in $(F_3C)(NC)C{=}C(CN)(CF_3)^{\bar{\cdot}}$ and $(NC)_2C{=}C(CF_3)_2^{\bar{\cdot}}$ radical anions than in $(F_3C)_2C{=}C(CF_3)_2^{\bar{\cdot}}$ but less than in $TCNE^{\bar{\cdot}}$.[955] There is a weak bonding interaction between the triple bonds in (**247**).[956] *Ab initio* calculations indicate that

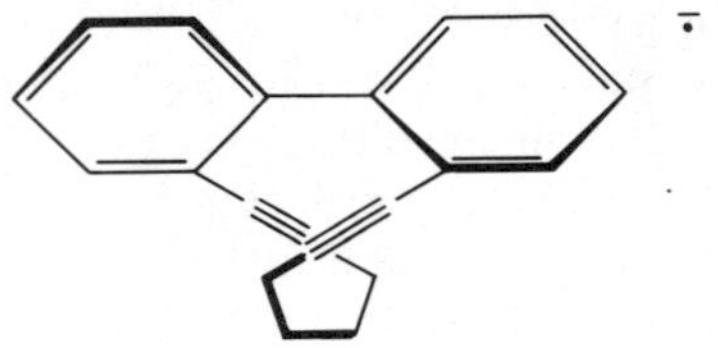

(**247**)

the benzene radical anion is distorted from regular hexagonal (D_{6h}) symmetry.[957] The spin density is greatest at the 2- and 6-positions in indene radical anions.[958] Spin-density distributions have been obtained also for the radical anions of 4,4′-dinitrodiphenylmethane,[959] *cis*-stilbene[960] (the results show that there is some distortion from planarity in the C(1), C(7), C(7′), C(1′) fragment), and aza-aromatics.[961] Electron-donor substituents increase the spin density at the nitro-group in substituted 3-nitropyridine radical anions.[962] The stability of 2-X-5-nitrofuran radical anions increases in the order: $X = I < Br < Cl$.[963] The conformational rigidity of radical anions of several bi- and ter-aryls containing the thiophene ring is greatly increased relative to the neutral compounds as the unpaired electron enters the LUMO which has strong bonding character in the interannular regions.[964] The extent of delocalization on to the nitro-group in alkylnitrothiazole radical anions increases in the order: $4\text{-}NO_2 < 5\text{-}NO_2 < 3\text{-}NO_2$.[965] The S—O group in both *cis*- and *trans*-9-methylthioxanthen *S*-oxide radical anions is in a pseudo-axial conformation whereas in the neutral compounds it is in a pseudoequatorial conformation.[966] The unpaired electron in the 1,2-diphenylcyclopropene radical anion is largely confined to the π-system,[967] while in the bromodifluoroacetamide radical anion the unpaired electron occupies the σ^* antibonding C–Br orbital.[968] The spin density in 2-thioxo-2-phenylacetamides is highest at the carbon of the thiocarbonyl group; the phenyl group is less twisted out of planarity in these radical anions than in monothiobenzil radical anions.[969] The coupling constant and hence the spin density is highest at the 2- and 7-positions in tropone radical anions;[970] long-range couplings have been observed for radical anions of polycyclic tropones.[971] Conformational studies have been carried out on several ketyl radical anions.[972–974] ESR-studies have been carried out on alkali-metal radical ion pairs of di-*tert*-butyl ketone,[975] biacetyl,[976] 3,5-dinitropyridine,[977] furil, and di-*tert*-butyl azodicarboxylate;[978] there is rapid intramolecular migration of the metal ion between the oxygen atoms of the furil system but not in the azodicarboxylate in which a tight ion-pair is formed. The dimeric radical anion $[(MeO)_3B{-}B(OMe)_3]^{\cdot-}$ is a σ-species with a one-electron bond.[979] The structures of radical anions of a quiniminocyclopropene,[980] a series of phosphorus-containing nitroso- and nitro-compounds[981] and alloxan have been studied.[982]

The diphenylcarbene radical anion $Ph_2C^{\cdot-}$ abstracts a proton from protic solvents to give the benzhydryl radical but in aprotic solvents hydrogen-abstraction to give the diphenylmethyl carbanion also occurs.[983] The Dewar–Zimmerman rules indicate that the rate of protonation of $(4n+2)$ aromatic radical anions is slow compared to that of the analogous carbanions.[985] The rates of protonation of aromatic hydrocarbon radical anions have also been correlated with the singlet energies of the parent hydrocarbon.[984] The cyclohexa-1,3-diene radical anion reacts with protons to give cyclohexenyl radicals, then dimerizes, or reacts with the substrate;[985] the radical anion of cyclo-octa-1,3-diene is significantly more basic, and the main reaction pathway is protonation, since the greater angle of twist in this system leads to loss of π-overlap and reduced electron affinity. Radical anions from polycyclic olefins readily undergo [2+2] cycloadditions.[986] The [2+2] cyclo-addition of tetrafluoroethylene to its radical anion has been reported.[987] The valence isomer bicyclo[4.2.0]octa-3,7-diene-2,5-semidione is more stable than the monocyclic cyclo-octa-2,5,7-triene-1,4-semidione.[988] The radical anion derived by reduction of 2,4,6-tri-*tert*-butyl-4-hydroxycyclohexa-2,5-dienone rearranges to the 2,4,6-tri-*tert*-butylcyclohex-4-ene-1,3-dione radical anion.[989] The rate of loss of phosphate from $\cdot CH(OMe)CH_2OPO_3^{2-}$ is less than that from $\cdot CH(OMe)CH_2\text{-}OPO_3H^-$.[990]

The equilibrium constant for the disproportionation of *cis*-stilbene radical anions is much greater than for the *trans*-radical anions; this implies that the geometry of the *cis*-radical anion resembles *cis*-stilbene and the resulting strain is released on formation of the dianion.[991] This idea was confirmed by the observation that the equilibrium constant for disproportionation of the [1,2 : 5,6]dibenzo-cycloheptatriene is low because in this case conversion to the dianion would not result in release of strain. The disproportionations of radical anions of aromatic hydrocarbons,[992] 2,2′-azonaphthalene,[993] and tetracyanodiphenoquinodimethane[994] have also been studied. Work on the dimerization of the radical anions of alkenyne[995] and 1,1-diphenylethylenes[996] has been reported.

Photolysis

Fragmentations

Investigation of the photochemistry of nitrosyl cyanide has proved hazardous; products arise from the intermediate radicals $(CN)_2NO\cdot$ and $(NC)N(NO)O\cdot$, formed by addition of $\cdot CN$ and $NO\cdot$ to the parent compound.[997] The photo-dissociation of CF_3NO in the gas phase[998] and of geminal chloro-*C*-nitroso-compounds in solution[999] have been investigated. Flash photolysis of arenesul-phonyl iodides or arenesulphinate anions leads to arenesulphonyl radicals, $ArSO_2\cdot$, in which the unpaired spin density is localized on the SO_2 portion.[1000] Photofission of $PhCH_2SO_2CH_2SO_2Ph$ and its derivatives leads to sulphones and sulphinic acids *via* $PhCH_2\cdot$ and $PhSO_2CH_2SO_2\cdot$ radicals; the latter do not extrude SO_2.[1001] In the series $PhCH_2SO_2CH_2Ar$ (where Ar = Ph, 1- or 2-naphthyl) direct irradiation leads to 55% cage recombination of arylmethyl radicals while triplet sensitization leads to <15% cage recombination, with recombination being much slower than the loss of SO_2 from one of the primary radical pairs.[1002] Cyclophanes can be prepared by the SO_2-photoextrusion reaction of the sulphones (**248**), prepared by *m*-chloroperbenzoic acid oxidation of the known bis-sulphides.[1003]

254 nm + SO_2
90–95% yield

(**248**) (**249**)

hν $-SO_2$

(**250**)

(**251**)

γ-Sultines, *e.g.* (**249**) form cyclopropanes upon photoextrusion of SO_2 (*cf.* decarboxylation of analogous γ-lactones). 3-Phenylbenzo-2,1-oxathiolane-1-one (**250**) extrudes SO_2 on irradiation to yield fluorene (**251**).[1004] ArSO· radicals have been detected by ESR during the photolysis and thermolysis of *S*-arylarenethiosulphonates; they are derived *via* the initial production of ArSO· and ArS· radicals.[1005]

Irradiation of 2-iodoadamantane in Et_2O or MeOH yields 2-alkoxyadamantane; in the presence of Et_3N acting as a remover of H^+, 2,4-dehydroadamantane and protoadamantane are formed from the intermediate 2-adamantanyl cation.[1006] *gem*-Di-iodides, if structurally incapable of undergoing a competing elimination to give a vinyl iodide, yield a carbenoid precursor on irradiation; thus CH_2I_2 with olefins gives cyclopropanes in >80% yield while 1,1-di-(iodomethyl)cyclohexane yields methylenecyclohexane and some 1-methylcyclohexene.[1007] Photochemical dechlorination of chlorinated *p*-terphenyls occurs mainly by homolytic dissociations.[1008] Chain dehalogenation of CCl_4 and C_2Cl_6 when photolysed in the presence of acetone and an alcohol appears to be efficient ($\phi \sim 100$) and to arise from the reducing action of ketyl radicals which are generated from the alcohol upon hydrogen-abstraction by $\cdot CCl_3$ or $\cdot C_2Cl_5$.[1009] Aryl chloroiodonium chlorides, $ArICl_2$, photolytically decompose to give aryl iodides and chlorinated products; the latter formed, in part, by replacement of ICl_2 with Cl.[1010] Generation of the oxyl-radical of the steroidal alcohol system (**252**) by photolysis of the corresponding hypoiodite leads to a double β-scission yielding (**253**) when a vicinal acylamino-function is present;[1011] such does not occur with a vicinal acyloxy-group. The labelling patterns in the products (**254**) and (**255**) from photolysis of (**256**) are consistent with formation of a diradical intermediate, by scission of the N(1)—C(6) bond, and subsequent bond formation between C(4) or C(6) and C(7).[1012]

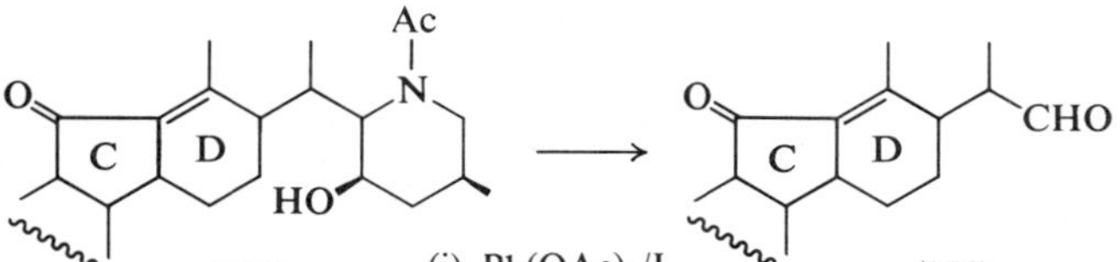

(252) $\xrightarrow{\text{(i) Pb(OAc)}_4/\text{I}_2 \quad \text{(ii) } h\nu}$ (253)

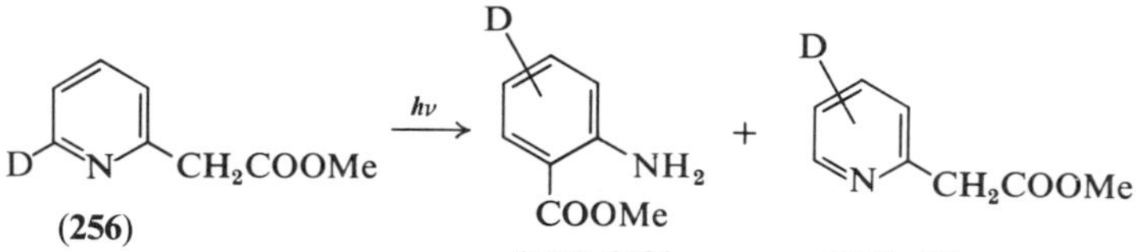

(256) $\xrightarrow{h\nu}$ (254) + (255)

(254) 20%
4-^{2}H 40–45%
6-^{2}H 35–40%

(255) 6%
3-^{2}H ~10%
6-^{2}H ~90%

The phosphate anion radical, $PO_4^{\cdot-}$, formed from $P_2O_8^{4-}$, behaves similarly to $SO_4^{\cdot-}$ in addition to olefins and H-abstraction from alcohols (cf. pp. 141, 142); it differs from $SO_4^{\cdot-}$ in reactions with aliphatic and aromatic carboxylic acids, in that direct oxidative electron-transfer (normally leading to decarboxylation) is not the dominant process.[1013]

A bimolecular redistribution, "metathesis", occurs in 30–60% yields in the Hg-sensitized photolysis of disilanes (Scheme 17), thus providing a feasible route to *Si*-fluoromethylsilanes.[1014]

$$X_3SiH + Hg(^3P) \longrightarrow X_3Si\cdot + H\cdot + Hg(^1S)$$

$$X_3Si\cdot \xrightarrow{Si_2Y_6} X_3Si{-}SiY_3 + Y_3Si\cdot \qquad H\cdot \xrightarrow{Si_2Y_6} Y_3SiH + Y_3Si\cdot$$

$$Y_3Si\cdot + X_3Si\cdot \underset{Hg(^3P)}{\rightleftharpoons} Y_3SiSiX_3$$

SCHEME 17

Photolysis of trialkyl-η^1-cyclopentadienyltin(IV) compounds provides a useful source of $R_3Sn\cdot$ radicals.[1015] In aprotic media photolysis of certain alkylcobalt(III) compounds (*e.g.* alkylcobaloximes) results in loss of H· from an equatorial ligand which is followed by a homolytic cleavage of the cobalt–alkyl bond.[1016] Polymerization of methyl methacrylate is initiated by a labile photolysis product obtained from di-η^5-cyclopentadienyldimethyltitanium(IV) by loss of CH_4 inside a tight radical–ion pair, and not by methyl radicals.[1017]

Hydrogen-atom Transfer

A theoretical analysis utilizing a dynamic model of reactivity has been presented for H abstraction by the lowest $^3(n,\pi^*)$ state of ketones.[1018] Intramolecular transfer of γ-hydrogen can occur in substituted cyclopropenes by a mechanism analogous to the Norrish type II process for carbonyl compounds.[1019] Triplet biacetyl abstracts H regiospecifically from the bridgehead position of adamantanes; with a large ρ^* value of -0.71, it leads to the exclusive formation of bridgehead acetyladamantanes.[1020] The elements of acetone $(H)(CH_2COCH_3)$ are added to the double bond of the monoepoxide of endodicyclopentadiene (**257**) on photolysis in acetone.[1021]

Benzhydryl radicals are found only to dehydrogenate the 2-pyrrolidino-radical formed on photolysis of (**258**) in presence of benzophenone; no coupling occurs and (**259**) is the sole product.[1022]

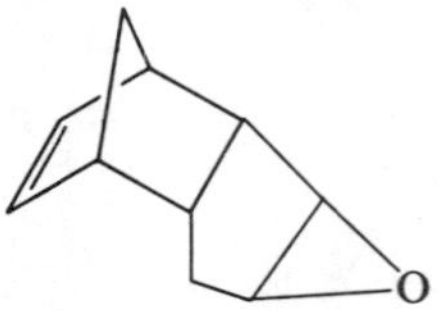

(**257**) 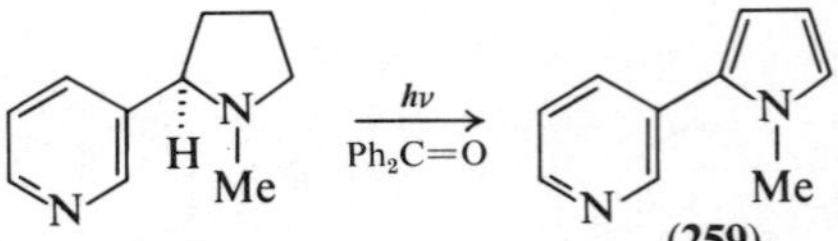 (**258**) $\xrightarrow[Ph_2C=O]{h\nu}$ (**259**)

Tos $\dot{N}$CH $(CH_2)_{n-3}$ ⟶ Tos N=CH$(CH_2)_{n-2}CH_2$·

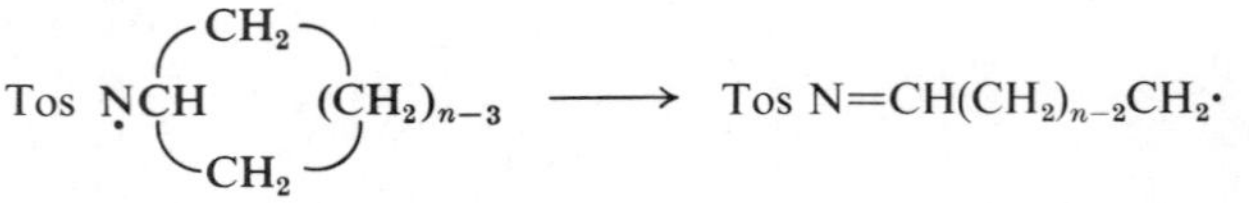

(**260**)

Loss of halogen (or NO) on photolysis of *N*-cycloalkyl-*N*-halo(or *N*-nitroso)-sulphonamides leads to the aminyl radicals (**260**) which subsequently undergo ring-opening (except for $n = 5$).[1023]

Photolysis of potassio-2,4-dimethyl-3-pentenone with iodobenzene leads, *via* electron transfer, to an enolate radical and a phenyl radical; the latter abstracts a β-H from the parent enolate anion and the resultant enolate anion radical reacts with enolate radical to yield, on protonation, 2,4,4,6,8-pentamethyl-3,7-nonadione.[1024] H-abstraction by maleic anhydride from THF and 1,4-dioxan[1025] and from alcohols[1026] has been studied. Maleimides form radical anions and alcohol-detived adduct radicals.[1026] Photoreaction of 1,2-naphthoquinones (**261**) with aldehydes leads to *C*- (**262**) and *O*-acylated (**263**) reduction products by coupling of the semiquinone and acyl radicals following H-abstraction; the ratio of (**262**) to (**263**) depends on the polarity of acyl radical and is increased by electron-withdrawing substituents in (**261**).[1027]

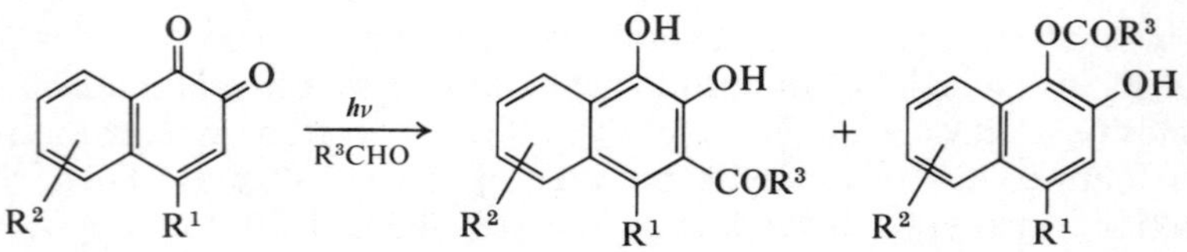

(**261**) $\xrightarrow[R^3CHO]{h\nu}$ (**262**) + (**263**)

The variety of products obtained from the reaction of $^3SO_2$ with alkanes has been shown to arise from further reactions of the sulphinic acids initially formed.[1028]

Displacements of two hydrogen atoms from alkanes to give alkenes are seldom observed in non-concerted processes (see Scheme 18, path a) because the radicals initially formed either recombine (path b) or undergo back reaction (path c). However, such has been observed in the photo-induced H-transfers in chlorinated tetrahydromethanoindanes since path c, requiring an impossible triplet energy transfer, is blocked.[1029] Primary and tertiary amines assist in the formation of the ion-pair (**264**) from the diradical (**265**) obtained by photolysis of (**266**) and then

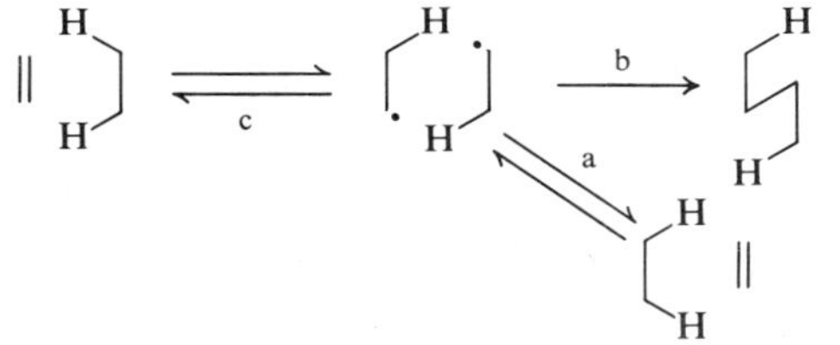

SCHEME 18

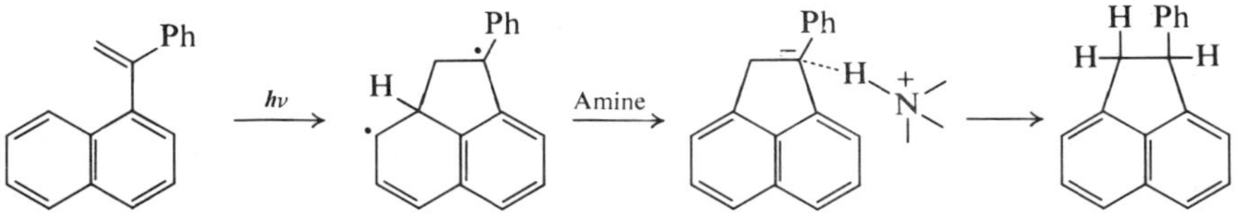

(266) (265) (264) (267)

to the cyclized photoproduct (**267**).[1030] The photoconversion of α-phenylcinnamic acid esters to 9,10-dihydrophenanthrenes is now thought to occur by a radical mechanism and not by a prototropic shift mechanism.[1031]

Carbonyl Compounds

H-abstraction and type I fission of triplet ketones have been considered by a new theoretical approach which involves the prediction of molecular deformations leading to a minimization of the separation between excited- and ground-state potential energy surfaces; this depends on a consideration of bond-order changes.[1032] A MINDO/3 study has been made of the Norrish type II reaction of butanal.[1033]

Quantum yields for the decomposition of formaldehyde alone,[1034] and in formaldehyde–butene mixtures,[1035] have been determined. The type I, α-bond, cleavage of aliphatic ketones is determined essentially by the α-substituent(s) while the type II, γ-H-abstraction, process is affected mainly by the substituent at the γ-carbon.[1036] The initial products from the photolysis of cyclopropyl methyl ketone are acetyl and cyclopropyl radicals, although on irradiation at shorter wavelengths decarbonylation of $CH_3\dot{C}O$ radical and ring-opening of cyclopropyl radical are in evidence.[1037] Bromine and iodine in δ-halovalerophenones, $PhC(=O)CH_2CH_2CH_2X$ ($X = Br$ or I), enhance the rate of γ-H-abstraction in the ketone triplet, *i.e.* anchimeric assistance occurs.[1038] Neighbouring-group participation in intramolecular reactions, including photo-cleavages of ketones, has been reviewed.[1039]

Decarboxylation is the main photoreaction of mercury trifluoroacetates; at low temperatures, *e.g.* −196 °C, cage reaction yields $(CF_3)_2Hg$ while at high temperatures, *e.g.* 20 °C, C_2F_6 is formed by dimerization of the mobile $CF_3\cdot$ radicals.[1040] Methyl and trityl radicals have been spin-trapped in the photo-Kolbe reaction of acetic and triphenylacetic acids catalysed by TiO_2.[1041] A comprehensive survey, by an ESR flow technique, of the radicals produced on irradiation of aliphatic acids and esters in solution, indicates that the carboxylic group is similar to the carbonyl group in its reaction modes (α-cleavage, photoreduction); for α-monohalogenated compounds β-dehalogenation was found to be the dominant process.[1042]

o-Phenylene oxalate (**268**) does not undergo the normal photo-Fries reaction but decarboxylates to give (**269**) in 94% yield.[1043] By a method designed to avoid polar kinetic effects in the perester method, the relative rate constants for formation of

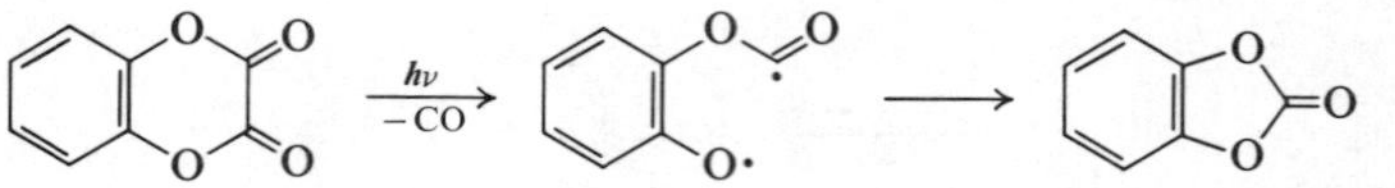

(268) (269) 94% yield

cycloalkyl radicals (k_c) and 1-methylcycloalkyl radicals (k_{mc}) by decarbonylation of acyl radicals have been obtained at 135 °C (see Table):[1044]

Ring size	3	4	5	6
$k_c(s^{-1})$	0.0121	0.0795	0.316	0.268
$k_{mc}(s^{-1})$	0.542	1.18	11.1	14.1

N,N-Disubstituted α-oxo-amides (**270**) undergo a type II photo-process to give the biradical (**271**) which can undergo a 1,4-H-migration (O to O) before cyclization to the enol of the oxazolidin-4-one product (**272**).[1045] The main products in the photolysis of *N*-acetylhydrazobenzene arise from (O=)C—N rather than N—N bond cleavage; the converse is true of *N,N'*-diacetylhydrazobenzene.[1046] The type I

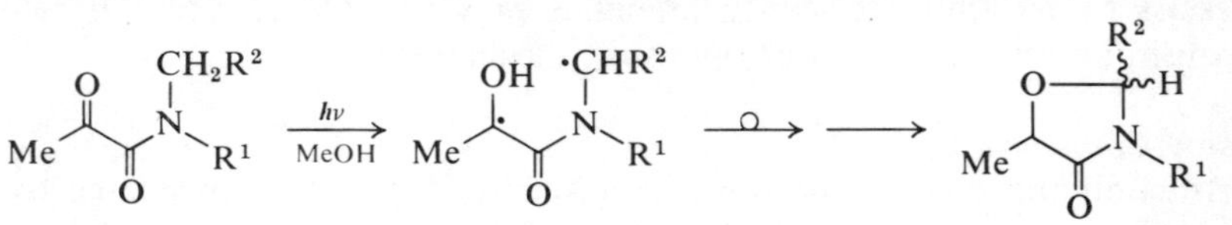

(270) (271) (272)

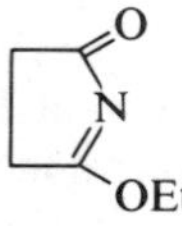

(273) 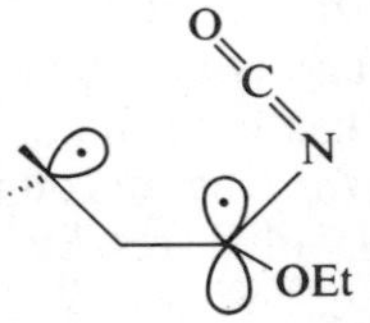 (274)

rearrangement of 2-ethoxypyrrolin-5-one (**273**) to 2,3-dimethylethoxycyclopropyl isocyanate, with 98% retention of stereochemistry at C(4) in (**273**), involves the diradical (**274**) formed by α cleavage of a $^1(n,\pi^*)$ state.[1047] Photolysis of dihydrothiophen-3(2*H*)-ones in methanol results in β-cleavage [S—C(2) bond] due to the relative stability of the thiyl radical centre (unlike the oxygen analogue), 5,6-dithiodecane-2,9-diones being the final products.[1048]

Chemical evidence has been obtained that $^3Ph_2C{=}O^*$ can deactivate radiationlessly by addition to the *ipso*-carbon of diphenyl ether.[1049] Biradical intermediates have been proposed in the photolyses of 2,3-dihydro-2,3-methanonaphthoquinones[1050] and of 2-alkyl-2,3-epoxy-2,3-dihydro-1,4-naphthoquinones.[1051] The latter compounds undergo photoaddition to olefins in two modes: (1) without ring opening of the oxirane, (2) 1,3-dipolar cycloaddition after fission of the C(2)—C(3) bond in the epoxyquinone component.[1052] Conversion of 1,4-naphthoquinones into 2,2'-bis(1,4-naphthoquinonyl) derivatives occurs by photo-dimerization of biradicals and subsequent photo-oxidation.[1053]

Miscellaneous Compounds

The photo-oxidations of isobutane[1054] and of toluene–air mixtures by nitrogen dioxide,[1055] and of DMSO by ·OH radicals during flash photolysis[1056] have been investigated. Photolytic oxidation of ethylene glycol dimethyl ether in aqueous hypochlorite has been studied, the ether being used as a model for polyethylene glycols; α/β-scissions follow initial H-abstraction and there is some evidence for insertion of an oxygen atom in a C—H bond.[1057] The simple assumption that the slow and fast eliminations of $HO_2\cdot$ from peroxy-radicals derived from glucose and other polyhydric alcohols are due to eliminations from α, ω, and other positions is inadequate to explain the kinetics; neighbouring hydroxyl groups may influence the elimination rates. The very fast component in D-glucose, observed conductimetrically after flash photolysis, has been ascribed to elimination from the peroxy-radical at C(1), but peroxyl radicals at C(2), C(3), and C(4) appear to eliminate at both slow and fast rates.[1058]

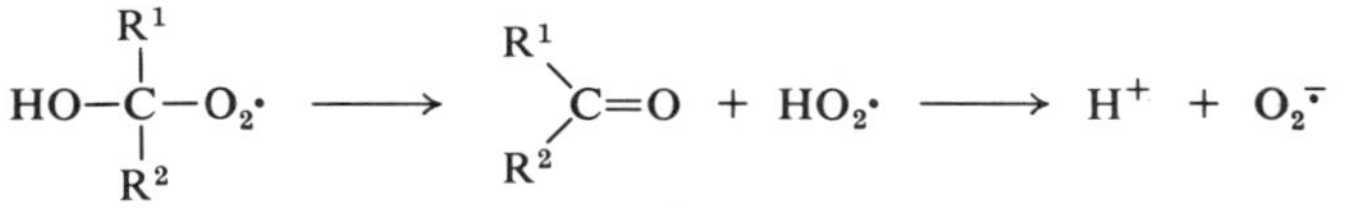

$$HO{-}C(R^1)(R^2){-}O_2\cdot \longrightarrow R^1R^2C{=}O + HO_2\cdot \longrightarrow H^+ + O_2^{\cdot-}$$

SCHEME 19

Vacuum-UV photolysis of liquid 1,3-dioxolan and 2,2-dimethyl-1,3-dioxolan involves ring cleavage to the biradical $\cdot CH_2CH_2OCR_2O\cdot$; dioxolan-2-yl radicals rearrange to β-acyloxyethyl radicals.[1059] A dipolar biradical is involved in the photoaddition of dienes to *N*-alkylphthalimides.[1060] Charge-transfer on photo-excitation of 1,1-diphenylethylene, complexed with $SbCl_5$, followed by addition of molecular oxygen leads ultimately to 3,3,6,6-tetraphenyl-1,2-dioxan.[1061] Photolysis of aromatic dipeptides in neutral (8M-$NaClO_4$ solution) and basic (8M-NaOH solution) glasses leads to phenoxyl radicals from tyrosine units and to indole cation radicals from tryptophan units, as determined by ESR methods.[1062] Arenethiyl radicals, produced by photofission of diaryl sulphides, add to the *para*-position of 2,6-di-*tert*-butylphenol, the *para*-H being subsequently removed to give the sulphenylated product.[1063] Two quenchable intermediates, including (**275**), are thought to exist in the photorearrangement of α- and γ-methylallyl chlorides to *cis*- and *trans*-2-chloro-1-methylcyclopropane.[1064]

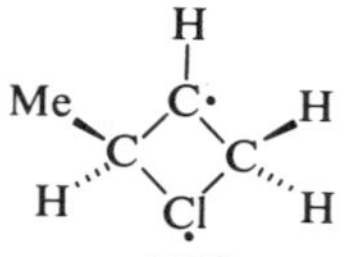

(**275**)

Pyrolysis

Hydrocarbons

Shock-tube studies of the thermal decomposition of propene in the presence of deuterium[1065] and of cyclopentane, cyclopropane, and pent-1-ene[1066] have been reported. *cis*- and *trans*-Pent-2-ene decompose at 500 °C to give mainly methane and butadiene by both molecular and free-radical processes; pent-1-ene formation occurs by abstraction of allylic H from pent-2-ene by a resonance-stabilized C_5H_9 radical.[1067] The heat of formation of Ph· radical has been determined from a shock-tube study of pyrolysis of benzene at 1200–1900 K,[1068] and a complete quantitative kinetic scheme has been given for the low-temperature pyrolysis of ethylbenzene.[1069] Pyrolysis of *n*-propylbenzene, at 500–600 °C, involves dissociation to $PhCH_2\cdot$ and Et· radicals; $PhCH_2CH_2CH_2\cdot$ radical, formed in the propagation step, dissociates to styrene and methyl radical.[1070] 3-Methylpent-1-ene is formed from ethylene and but-2-ene by an ene process involving a cyclic six-membered, non-biradical, transition state.[1071]

Other compounds

*iso*Butyl radicals from 1,1'-azoisobutane combine to yield 2,5-dimethylhexane and cleave to Me· and C_3H_6; the cross-combination ratio for Me· and Bu^i· radicals was found to be only 0.25 indicating that the geometric mean rule does not apply to these radicals.[1072]

Ethyl cyanide decomposes by a homogeneous free radical chain mechanism involving Me· and $\cdot CH_2CN$ radicals as determined by scavenging techniques;[1073] the stabilization energy of an α-cyano-group in an alkyl radical was confirmed as ~21 kJ mol^{-1}, and this is also supported by a kinetic study of the decomposition of *trans*-1,2-dicyanocyclobutane.[1074] Rates of methyl radical recombination in the pyrolysis of $(CH_3)_2O$ favour adiabatic open-channel or minimum state density theories for recombination reactions, rather than RRKM theory.[1075] Hydrogen-abstraction by Me· or Et· from acetaldehyde is *ca.* three-fold faster than from Et_2O.[1076] A mechanism proposed earlier has been modified by inclusion of further reactions, in order to account for ^{14}C-incorporation into the products (MeCHO, $CH_2{=}CH_2$, CO, and C_2H_6) of pyrolysis of Et_2O in the presence of ^{14}C-labelled ethanol.[1077] Above 200 °C, 2-fluoro-2,2-dinitroethyl methyl ethers, $FC(NO_2)_2CH_2$-OCH_2X (X = H, Cl or NO_2), decompose by radical cleavage of a NO_2 group. At lower temperatures, when X = N_2 or ONO_2, decomposition involves loss of the OCH_2X fragment; the dissociation energy, relative to the corresponding alkyl compound RX, is lowered ~5 kcal mol^{-1} by the α-oxygen.[1078] Re-combination of amino-radicals in the pulse shock-tube decomposition of *tert*-pentylamine appears to be analogous to that of alkyl radicals.[1079]

No evidence has been found for a concerted 1,5 H-shift in the pyrolytic de-carboxylation of arylacetic acids.[1080] Pyrolysis of α-phenylacetanilide involves homolysis of the amide C—N bond followed by interaction of the primary and secondary formed radicals with the rearrangement products and the solvent.[1081] A cross-over experiment has proved that the novel, synthetically useful, pyrolytic formation of 10-aryl-9-arsa-anthracenes from 9-aryl-10-benzyl-9,10-dihydro-9-arsa-anthracenes proceeds *via* loss of $PhCH_2$· from radical (**276**) with an intra-molecular 1,4 migration of the aryl group.[1082]

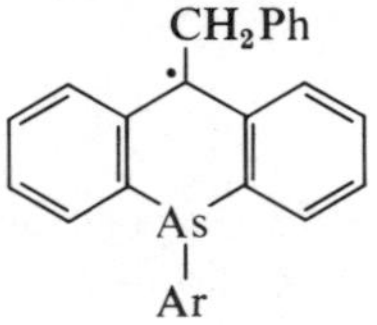

CH_2Ph

As

Ar

(276)

From the magnitude of the Arrhenius *A* factor ($\log_{10} A = 16.0$, $E_a = 38.5$ kcal mol^{-1}) decomposition of dibenzylmercury probably involves $R_2Hg \rightarrow R\cdot + Hg{-}R$ rather than $2R\cdot + Hg$.[1083] C—H and Si—C bond cleavages are considered to be unimportant in the pyrolysis of $MeSiH_3$ and Me_2SiH_2 which involves some contri-bution from radical chains initiated at the walls of the reaction vessel.[1084] The tetramethyl derivatives of silicon, germanium, and tin all decompose by first-order kinetics; methyl radicals act as chain carriers.[1085]

Radiolysis

Pulse radiolysis, [1086] reactions of organic free radicals in irradiated aqueous media,[1087] and radiation-induced organic hydrogen isotope-exchange reactions[1088] have been reviewed recently.

Compounds of Biological Interest

The alkoxy-radical generated by γ-irradiation of deoxycytidine 5′-phosphate at 77 K has been characterized by ESR-spectroscopy.[1089] The kinetics of electron-transfer processes in proteins such as ribonuclease, lysozyme, α-chymotrypsin, and $\alpha_{s,1}$-casein following radiolysis have been studied.[1090] Viscosity effects are important in the first-order decay of cystamine anion radical, $RSSR^{\dot{-}} \rightarrow RS\cdot + RS^-$, and solvent polarity influences its electron-transfer reaction with O_2.[1091] Irradiation of 6-mercaptopurine at 77 K by 4 MeV electrons leads to a thiyl radical following H-abstraction from N(7) of the purine.[1092]

Addition of hydrogen atom to the 5,6-double bond in 5-bromodeoxyuridine occurs on radiolysis,[1093] while σ^*-radicals as well as π^*-anion radicals are produced by electron-addition to 5-halouracils when the halogen is Br or I (but not for Cl or F, where halide ion displacement is more difficult).[1094] A cation such as (**277**) is

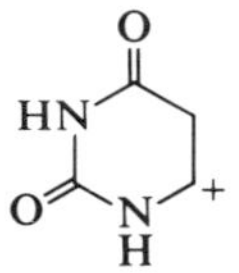

(**277**)

formed by oxidation of the corresponding hydrouracil radical, as well as by its disproportionation, following γ-irradiation of aqueous solutions of pyrimidines.[1095] An extensive survey of the radiation-induced oxidation of D-glucose in oxygenated aqueous solution has shown that the major products arise from initial H-abstraction by $\cdot$OH radical followed by addition of O_2 and subsequent elimination of $\cdot O_2H$; reaction of the glucose peroxyl radicals with $HO_2\cdot/O_2^{\dot{-}}$ leads to fragmentation of C—C bonds, the major product L-*threo*-tetradialdose having a C(5)-peroxy-radical precursor.[1096] Attack of $\cdot$OH radicals on D-ribose involves C(1), ~20%; C(2) and C(4), ~35%; C(3), ~20%; and (C)5, ~25%. Products of reaction in deoxygenated solution arise from disproportionation (and dimerization) of primary ribosyl radicals either directly or after transformations such as elimination of water or CO; in oxygenated solution, reactions analogous to those reported for D-glucose are found.[1097]

The deamination of 2-amino-2-deoxy-D-glucose on γ-irradiation in aqueous solution has also been studied.[1098]

ESR methods have been applied to determination of the structures of alkoxy-radicals derived from the sugar component of nucleosides and nucleotides on X-ray irradiation.[1099]

Other Compounds

Loss of OH^- rather than F^- on electron-attachment to fluorinated aliphatic alcohols is assisted by hydrogen-bonding and by H^+-transfer.[1100] Adducts of e^- and H$\cdot$ with many organic compounds have been stabilized by trapping in the torus-shaped central void of a cycloamylose host: such matrices, having the added advantage of control of void dimensions, compare very favourably with the extensively used adamantane matrices.[1101] Radiolysis of thiols at 77 K leads to thiyl radicals having a g_z factor of 2.158; dissociative electron-capture by thiols was observed only in aqueous and methanolic glasses.[1102] Species previously believed to be $R_2S^{+\bullet}$ radicals formed upon γ-irradiation of some organic sulphides and disulphides are better identified with $\cdot SR_2-SR_2^+$ radicals.[1103]

Pulse radiolysis of aqueous ammonia solutions leads to the amino-radical but attempts to generate $NH_3^{+\bullet}$ have been unsuccessful;[1104] $SO_4^{\overline{\bullet}}$ and $PO_4^{2\overline{\bullet}}$ radicals react with NH_3 by H-abstraction rather than by electron-transfer oxidation. Amino-radicals are unreactive towards benzene but react by electron-transfer with substituted phenoxide ions ($\rho = -3.3$).[1104] Reactions of hydroxyl radical with anedisulphonic acids,[1105] EDTA-complexed bivalent first-row transition metals,[1106] and anionic surfactant systems[1107, 1108] have been examined using pulse-radiolysis techniques.

The radiation-sensitized chain reaction of N_2O with methanol includes the O atom transfer reaction:[1109] $N_2O + \cdot CH_2OH \rightarrow \cdot OCH_2OH + N_2$. α-Hydroxyalkyl radicals are also chain carriers in the chain dechlorination of hexachloroethane by alcohols.[1110]

Addition of H· to thiophene is fast ($k = 9 \times 10^9\ M^{-1}\ s^{-1}$) in solution but addition of e^- is slow ($k = 4 \times 10^7\ M^{-1}\ s^{-1}$). The 2-hydrothienyl radical is observed, and decays at the same rate, under both conditions; however, the products obtained are not the same and addition of H· leads to eventual destruction of the thiophene ring while the electron addition mode leads to 2,2′-bithiophene.[1111]

The hydroxyl radical adduct of *N,N*-dimethylaniline decays to the amine cation radical and OH^- ($k = 7 \times 10^6\ s^{-1}$) and also to give the $PhN(Me)CH_2\cdot$ radical.[1112] Addition of HO· to nitroaniline in aqueous solutions leads to elimination of HNO_2.[1113] Reaction of some substituted nitroimidazoles and nitroaromatic compounds with reducing species such as e^-_{aq}, $CO_2^{\overline{\bullet}}$, and $Me_2C(OH)\cdot$ led only to the corresponding radical anions.[1114] Pulse radiolysis of $PhCH_2Cl$ in methanol resulted in dissociative electron capture ($PhCH_2\cdot + Cl^-$ products) while some evidence was found for H· atom addition to $PhCH_2Cl$.[1115] The hydroxy radical adduct with naphthalene at pH 0.5 decomposes to the naphthalene cation radical; this cation radical is also formed at pH 3 upon reaction of naphthalene with $SO_4^{\overline{\bullet}}$ radicals.[1116]

Paraquat cation radicals $PQ^{+\bullet}$, and their analogues $RQ^{+\bullet}$, react rapidly with O_2 (k's $4\text{–}9 \times 10^8\ M^{-1}\ s^{-1}$) and even faster with $O_2^{\overline{\bullet}}$ radicals; the rate constants correlate with corresponding redox potentials. The back-reaction $O_2^{\overline{\bullet}} + RQ^{+\bullet} \rightarrow O_2 + RQ$ was investigated when feasible from the redox potentials.[1117] Both cation and anion radicals are generated on pulse radiolysis of *trans*-stilbene in benzene.[1118] Reduction of *syn*- and *anti*-azobenzene, A, by e^-_{aq} or aliphatic ketyl radicals leads to the same radical transient, whose UV-absorption spectrum is pH-dependent owing to the equilibria $\cdot AH_2^+ \rightleftharpoons AH\cdot \rightleftharpoons A^{\overline{\bullet}}$ (pK_a values of 2.9 and 13.7, respectively).[1119]

Nanosecond pulse radiolysis of 1-methylstyrene and 1,1-diphenylethylene in aqueous solution leads to H· and ·OH adducts at the unsubstituted olefinic carbon atom.[1120] Since the rates of phenyl radical addition to aromatic systems should be of the same order of magnitude as for addition of *p*-carboxyphenyl radicals to *p*-bromobenzoate anion in water ($k = 7.6 \times 10^6\ M^{-1}\ s^{-1}$) and since H-abstraction by *p*-carboxyphenyl radicals from alcohols occurs at a comparable rate, it has been concluded that bimolecular (self-)reaction of phenyl radicals should be unimportant in the presence of many organic compounds.[1121]

γ-Radiolysis of phenanthrene (also indene) in liquid CO_2 leads to addition of $O(^3P)$ atoms to form the biradical (**278**) which can (i) cyclize to the dibenz[*b,d*]-oxepin (**279**), (ii) cleave and oxidize to (**280**), and (iii) add CO_2 to give the cyclic carbonate (**281**).[1122]

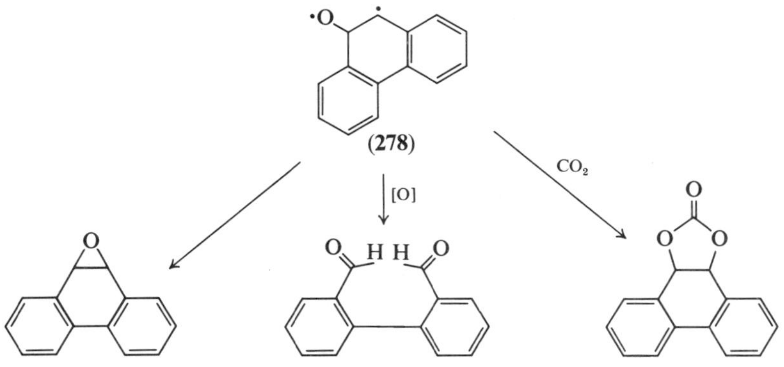

(**278**)

[O]

CO_2

(**279**) (**280**) (**281**)

CIDNP and CIDEP Methods

Review papers have appeared on pair substitution effects[1123] and the problem of parallel radical and non-radical mechanisms.[1124] Anomalous CIDNP enhancements in biradicals have been noted[1125] and a discussion of the field-dependence (at $B < 1000$ G) of the CIDNP effect in biradicals has commented on the deficiencies of the simple model based on a perturbational treatment in radical pair theory.[1126]

CIDEP of radiolytically produced radicals has been reviewed.[1127] A time-resolved ESR and flow FT NMR study of the reaction sequence following electron addition to chloroacetate and other anions has provided an example of Overhauser CIDNP.[1128] Radiolysis of carboxylate anions has also revealed $S–T_{\pm}$ polarization pathways in a CIDEP study.[1129] The decomposition characteristics of ω,ω-dialkoxy-ω-phenylacetophenones as initiators of the photopolymerization of methyl methacrylate have been examined by CIDNP using a slow square-wave light modulation technique coupled with ^{13}C-FT NMR spectroscopy to obtain absolute intensities.[1130]

The rate constant for H-abstraction from 2-methylpropane-2-thiol by the biradical derived from photolysis of cyclohexanone has been obtained from the magnetic field-dependent CIDNP.[1131] The formation of vinylic derivatives and adducts in the photoreduction of $Ph_2C{=}O$ by thioethers, $R^1CH_2SR^2$, occurs by way of disproportionation and combination of the triplet radical pair (**282**); the corresponding selenides show no CIDNP effects, possibly because the *g* values of the radicals in the pair are the same.[1132] Photoreductions of furil by phenols have been examined in CIDEP[1133] and zero-field CIDNP[1134] studies. Various geminate combination and disproportionation reactions, including the formation of pyruvic acid enol, have been deduced to occur in the photoreduction of pyruvic acid by alcohols and acetaldehyde; escape reactions involving H-exchange between ketyl radicals and pyruvic acid are significant. Photodecarboxylation of pyruvic acid is initiated by scission of the carbonyl–carboxy bond. Rapid reduction of ground-state pyruvic acid by carboxyl radicals followed by radical coupling yields 2-hydroxy-2-methylacetoacetic acid, which can isomerize to acetoin, as the primary photoproduct (Scheme 20).[1135]

$\overline{R^1\dot{C}HSR^2 \quad Ph_2\dot{C}OH}^{T}$

(282)

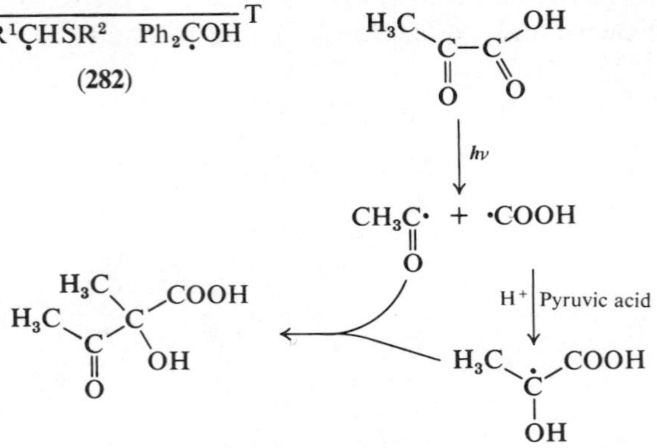

$H_3C{-}CO{-}COOH \xrightarrow{h\nu} CH_3\dot{C}{=}O + {\cdot}COOH$

${\cdot}COOH \xrightarrow{H^+,\ \text{Pyruvic acid}} H_3C\dot{C}(OH)COOH$

$CH_3\dot{C}{=}O + H_3C\dot{C}(OH)COOH \rightarrow H_3C{-}CO{-}C(CH_3)(OH)COOH$

SCHEME 20

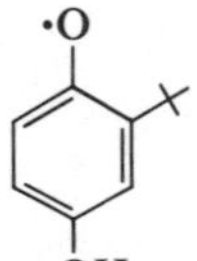

(283) 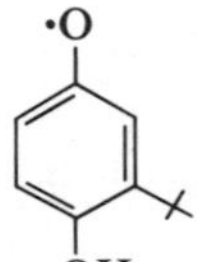 **(284)**

From CIDEP studies the quenching of quinone triplets is ~100 times less for propan-2-ol than for phenols and involves H-abstraction,[1136] while for Et_3N the quenching process involves electron-transfer.[1137] The photoreduction of *tert*-butyl-*p*-benzoquinone in Pr^iOH–toluene and acetic acid–phenol mixtures leads to the formation of radical (**284**) *ca.* 6 times faster than radical (**283**) in the primary step, but this is followed by chemical equilibration of the radicals (this could not be distinguished by ESR techniques alone but required CIDEP studies).[1138] Photo-CIDNP of the ring and methylene protons of tyrosyl units in peptides and other compounds is induced by reversible hydrogen-abstraction by excited triplet xanthene dyes and provides a tool in the NMR study of peptides.[1139] CIDNP effects are found in the photoreaction of acridans in alcohols *via* the triplet radical pair (**285**) when X=H and the singlet radical pair (**285**) when X = $CHCl_2$ or CCl_3.[1140]

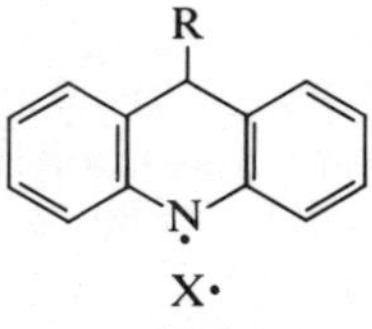

R

N

X·

(285)

The theory of the CIDNP effects observable in an electron donor D following an initial photo-induced electron-transfer to an acceptor A has been extended to cover the case where the free energy ΔG of the geminate recombination of cation–anion radical pairs exceeds the triplet energy of A; in this situation CIDNP effects stem predominantly from triplet pairs and reach a maximum when $\Delta G = E_T(A)$.[1141] Polarization of ^{19}F-nuclei following electron-transfer quenching of $PhCOCF_3$ and its derivatives by 1,4-dimethoxybenzene occurs *via* the triplet mechanism.[1142] The different concentration-dependences of nuclear-spin polarization and line broadening in the system $h\nu/m\text{-}FC_6H_4COCF_3/p\text{-}(MeO)_2C_6H_4$ have been interpreted as evidence for a normal triplet precursor ^3K to the radical ion pair $K^{\bar{\cdot}},Q^{+\cdot}$ and of a shorter-lived electron-spin polarized precursor $^3K_{\pm}$ generated by selective intersystem crossing from singlet excited ketone, ^1K.[1143] Pyrene has been shown to act as a one-electron donor in the singlet-state decomposition of arenediazonium salts.[1144]

Side-chain hydroxylation of alkyl-substituted methoxybenzenes on photolysis with nitrobenzenes involves a radical ion pair.[1145] CIDNP studies of the singlet photoreaction of anthracene and perhalomethanes have demonstrated initiation by the 9-chloro-10-anthryl, $CX_3\cdot$ radical pair.[1146]

High-field polarization in ethylene glycol formed on radiolysis of MeOH implies encounter of $\cdot CH_2OH$ radicals, before their dimerization, with species of higher *g* factor – probably MeO· radicals.[1147] Variable-field CIDNP studies have enabled various reaction pathways in the pulse radiolysis of $Me_2S{=}O$ to be delineated: H· atoms prefer to abstract H rather than add it; dissociative electron-capture occurs, giving $CH_3\cdot$ and $CH_3SO^{\bar{\cdot}}$ radicals.[1148]

β,γ-Unsaturated ketones undergo photolytic α-cleavage from a $^1(n,\pi^*)$ state if a γ-phenyl substituent is present but from a $^3(n,\pi^*)$ state when there is a γ-methyl substituent.[1149] CIDNP signals observed in the thermal decomposition of peracids RCO_3H in $(CCl_3)_2CO$ arise from an $R\cdot,\cdot CCl_3$ and not an $R\cdot,\cdot OH$ radical pair.[1150] Unsymmetric diazenes, $R^1{-}N{=}N{-}R^2$ (R^1 = Ph or cumyl; R^2 = trityl, cumyl, isobutyl or 1-norbornyl) decompose by one-bond cleavage to give $R{-}N{=}N\cdot$ as shown by ^{15}N-CIDNP and radical-trapping studies.[1151] The formation of nitriles in similar yields from the reaction of both *E*- and *Z*-isomers of benzaldoxime with thiobenzoyl chloride and *N,N*-dimethylthiocarbamoyl chloride occurs mainly *via* the radical pair (**286**) as shown by ESR detection of the phenyliminyl radical and the strong polarization of the carbon nuclei of the nitrile detected in the ^{13}C-NMR spectra.[1152]

Ph(H)C=N· O=C(·S)R

(**286**) R = Ph or NMe_2

^{19}F-CIDNP has been observed in the highly useful anodic functionalization of 1-*H*-perfluoroalkanes in fluorosulphuric acid, which involves H abstraction by $FSO_3\cdot$ radicals.[1153] ^{31}P-CIDNP evidence exists for the partial involvement of a one-electron transfer path in the coupling reactions of diorganophosphides with alkyl, allyl, and benzyl bromides and iodides.[1154]

Decomposition of bis(allyl)zinc compounds leads to randomly diffusing allyl radicals.[1155] CIDNP studies of bimolecular reactions of organometallic compounds have been reviewed.[1156]

Miscellaneous

The sulphuranes obtained by nucleophilic addition to 1-aryl-1-cyano-1-methoxy-carbonylmethyl sulphonium salts (**287**) undergo homolytic fragmentation[1157] as formulated in Scheme 21.

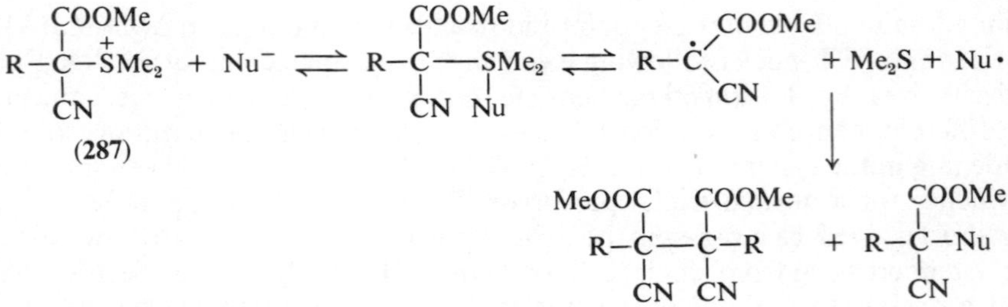

$$\underset{\textbf{(287)}}{R-\overset{COOMe}{\underset{CN}{C}}-\overset{+}{S}Me_2} + Nu^- \rightleftharpoons R-\overset{COOMe}{\underset{CN}{C}}-\underset{Nu}{S}Me_2 \rightleftharpoons R-\dot{C}(COOMe)(CN) + Me_2S + Nu\cdot$$

$$\downarrow$$

$$R-\overset{MeOOC}{\underset{CN}{C}}-\overset{COOMe}{\underset{CN}{C}}-R + R-\overset{COOMe}{\underset{CN}{C}}-Nu$$

SCHEME 21

The Darzens reaction between 2,2-dimethyltetrahydropyran-4-one and ethyl chloroacetate proceeds by a radical mechanism.[1158] Radical intermediates are involved in the reaction of ninhydrin with amines.[1159]

Alkyl radicals are generated from the alkylthallium(III) compound, PhCH(OMe)-$CH_2Tl(OAc)_2$, by reaction with ascorbic acid[1160] or with copper(I) bromide[1161] in acetonitrile: in both instances the intermediate radicals can be spin-trapped; in the former reaction they give rise (after reaction with oxygen and reduction of the resultant hydroperoxide) to the alcohol, and in the latter to the alkyl bromide.

Radicals react with the cobalt(II) complex of 5,7,7,12,14,14-hexamethyl-1,4,8,11-tetra-azocyclotetradeca-4,11-diene to give cobalt(II) adducts:[1162]

$$Co^{II}L + R\dot{C}HOH \longrightarrow Co^{III}L-CHR(OH)$$

The $Co^{III}L-CH(OH)CH_2OH$ complex breaks down in acid solution to give acetaldehyde: this is the first direct observation of the rearrangement of ethylene glycol to acetaldehyde while it is bound to a cobalt complex; thus, the reaction is a model for the behaviour of the diol dehydrase enzyme and shows that it can be initiated by a radical mechanism though the rearrangement proceeds by a different route. Photolysis (and also thermolysis) of 2-hydroxy-2,2-diphenylethylcobalt(III) complexes gives both rearranged and reduced products: the initial step involves Co—C-bond homolysis;[1163] for detail see Scheme 22.

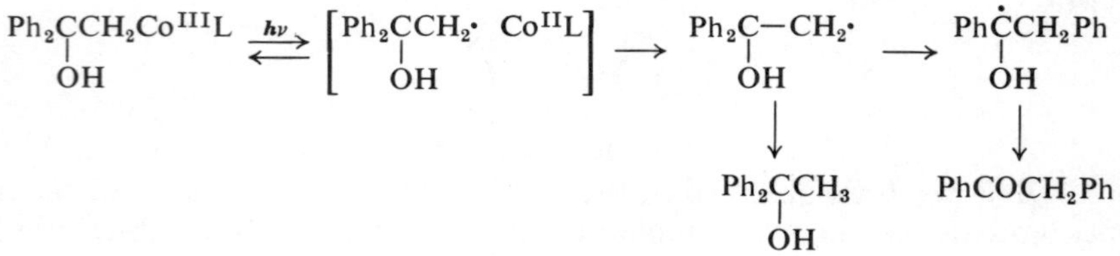

$$\underset{OH}{Ph_2\underset{|}{C}CH_2Co^{III}L} \underset{}{\overset{h\nu}{\rightleftharpoons}} \left[\underset{OH}{Ph_2\underset{|}{C}CH_2\cdot}\ Co^{II}L\right] \longrightarrow \underset{OH}{Ph_2\underset{|}{C}-CH_2\cdot} \longrightarrow \underset{OH}{Ph\underset{|}{\dot{C}}CH_2Ph}$$

$$\downarrow \qquad\qquad \downarrow$$

$$\underset{OH}{Ph_2\underset{|}{C}CH_3} \qquad PhCOCH_2Ph$$

SCHEME 22

The oxidative addition of bromobenzene to SnR_2 or $Sn(NR_2)_2$ proceeds by a radical-chain process: the reaction is initiated by ethyl bromide:[1164]

$$\begin{aligned} SnR_2 + EtBr &\longrightarrow \cdot SnR_2Br + Et\cdot \\ \cdot SnR_2Br + PhBr &\longrightarrow SnR_2Br_2 + Ph\cdot \\ Ph\cdot + SnR_2 &\longrightarrow \cdot SnR_2Ph \\ \cdot SnR_2Ph + PhBr &\longrightarrow SnR_2(Br)Ph + Ph\cdot \end{aligned}$$

The principal dehydration–decarbonylation reaction in the oscillating decomposition of formic acid in concentrated sulphuric acid is now attributed to ·OH radical catalysis, rather than to acid catalysis.[1165]

After light-modulated ESR and quantum-mechanical studies, the three radicals observed on irradiation of glyoxal in toluene have been assigned to the *cis*- and *trans*-conformers of 2-hydroxyethenyloxy radical (in which chemical exchange is controlled by the internal rotation kinetics of the OH group) and the adduct of formyl radical to glyoxal.[1166]

New syntheses and ESR/ENDOR studies of partially deuteriated and chlorinated perinaphthyl radicals have been reported.[1167]

References

1 Pryor, W. A. (Ed.), "*Free Radicals in Biology*," Vol. III, Academic Press, New York, 1977.
2 Singh, A. and Petkan, A. (Ed.), *Photochem. Photobiol.*, **28,** 429 (1978).
3 Pryor, W. A., *Am. Chem. Soc., Symp. Series*, No. 69, "Organic Free Radicals," 1978. (W. A. Pryor, Ed.).
4 "Radicaux Libres Organiques," Editorial du C.R.N.S., Paris, 1978.
5 *Ber. Bunsenges. Phys. Chem.*, **82,** 2 (1978).
6 *J. Phys. Chem.*, **81,** 2309 (1977).
7 *Yuki Gosei Kagaku Kyokaishi*, **32,** lf. (1978); *Chem. Abs.*, **89,** 41580 (1978); **89,** 89774–89778 (1978); **89,** 106951 (1978).
8 Mackiewicz, P., and Furstoss, R., *Tetrahedron*, **34,** 3241 (1978).
9 Chow, Y. L., Danen, W. C., Nelsen, S. F., and Rosenblatt, D. H., *Chem. Rev.*, **78,** 243 (1978).
10 Block, E., "*Reactions of Organosulphur Compound*," Academic Press, New York, 1978.
11 Benson, S. W., *Chem. Rev.*, **78,** 23 (1978).
12 See *NATO Ad., Study Inst. Ser., Ser. C*, C34, 39f (1977); *Chem. Abs.*, **88,** 10411–4 (1978).
13 Sagdeev, R. Z., (Ed.) *Tezisy Dokl. Vses. Konf.* "*Polyariz Yader Elektronov Eff. Magn. Polya Khim. Reakts.*" **1975,** 1; *Chem. Abs.*, **89,** 23514–17 (1978).
14 Fischer, M., *Angew. Chem. Int. Edn.*, **17,** 16, (1978).
15 Asano, T., and le Noble, W. J., *Chem. Rev.*, **78,** 407 (1978).
16 Hester, R. E., Grossman, W. E. L., and Ernstbrunner, E., *Proc. Int. Conf. Raman Spectrosc. 5th*, **1976,** 13; *Chem. Abs.*, **88,** 21449 (1978).
17 Gilbert, B. C., *Electron Spin Reson.*, **4,** 111 (1977); *Chem. Abs.*, **88,** 151585 (1978).
18 Mendenhall, G. B., *Sci. Prog. (Oxford)*, **65,** 1 (1978); *Chem. Abs.*, **88,** 104198 (1978).
19 Giese, B., and Beckhaus, H.-D., *Angew. Chem. Int. Edn.*, **17,** 594 (1978).
20 Beckhaus, H.-D., *Angew. Chem. Int. Edn.*, **17,** 594 (1978).
21 Griller, D., Ingold, K. U., Krusic, P. J., and Fischer, H., *J. Am. Chem. Soc.*, **100,** 6750 (1978).
22 Bonazzola, L., Leray, N., and Roncin, J., *J. Am. Chem. Soc.*, **99,** 8348 (1978).
23 Dyke, J., Jonathan, N., Lee, E., and Winter, M., *Phys. Ser.*, **16,** 197 (1977).
24 Cf. *Org. Reaction Mech.*, **1976,** 98.
25 Pacansky, J., Horne, D. E., Gardini, G. P., and Bargon, J. *J. Phys. Chem.*, **81,** 2149 (1977).
26 Pacansky, J., and Dupuis, M., *J. Chem. Phys.*, **68,** 4276 (1978).
27 Bernardi, F., Cherry, W., Shaik, S., and Epiotis, N. D., *J. Am. Chem. Soc.*, **100,** 1352 (1978).
28 Bernardi, F., and Epiotis, N. D., *Prog. Theor. Org. Chem.*, **2,** 1 (1977).
29 Malatesta, V., Forrest, D., and Ingold, K. U., *J. Phys. Chem.*, **82,** 2370 (1978).
30 Smolenskii, E. A., and Kocharova, L. V., *Dokl. Akad. Nauk SSSR*, **239,** 135 (1978); *Chem. Abs.*, **88,** 189825 (1978).
31 Gaze, C., Gilbert, B. C., and Symons, M. R. C., *J. Chem. Soc., Perkin Trans.* 2, **1978,** 235.
32 Forrest, D., and Ingold, K. U., *J. Am. Chem. Soc.*, **100,** 3868 (1978).
33 Kawamura, T., Tsumura, M., Yokomichi, Y., and Yonezawa, T., *J. Am. Chem. Soc.*, **99,** 8251 (1977).
34 Boche, G., and Schneider, D. R., *Tetrahedron Lett.*, **1978,** 2327.
35 Malatesta, V., Forrest, D., and Ingold, K. U., *J. Am. Chem. Soc.*, **100,** 7073 (1978).
36 Blum, P. M., Davies, A. G., and Henderson, R. A., *J. Chem. Soc., Chem. Commun.*, **1978,** 569.
37 Lloyd, R. V., and Wood, D. E., *J. Am. Chem. Soc.*, **99,** 8269 (1977).
38 Gaze, C., and Gilbert, B. C., *J. Chem. Soc., Perkin Trans.* 2, **1978,** 503.

[39] Kawamura, T., Kojima, T., and Yonezawa, T., *Tetrahedron Lett.*, **1978**, 2421.
[40] Fort, R. C., and Hiti, J., *J. Org. Chem.*, **42,** 3968 (1977).
[41] Kawamura, T., Matsunaga, M., and Yonezawa, T., *J. Am. Chem. Soc.*, **100,** 92 (1978).
[42] Ol'dekop, Yu. A., Maier, N. A., Erdman, M. A., Rubakha, T. A., and Shingel', I. A., *Dokl. Chem.*, **235,** 445 (1977).
[43] Stella, L., Janousek, Z., Merényi, R., and Viehe, H. G., *Angew. Chem. Int. Edn.*, **17,** 691, (1978).
[44] Heimer, N. E., *J. Org. Chem.*, **42,** 3767 (1977).
[45] de Vries, L., *J. Am. Chem. Soc.*, **100,** 926 (1978).
[46] Aurich, H. G., Deuschle, E., and Weiss, W., *J. Chem. Res.*, **1977,** (S) 301.
[47] Houle, F. A., and Beauchamp, J. L., *J. Am. Chem. Soc.*, **100,** 3290 (1978).
[48] Boche, G., and Schneider, D. R., *Angew. Chem. Int. Edn.*, **16,** 869 (1977).
[49] King, K. D., *Int. J. Chem. Kinet.* **10,** 545 (1978).
[50] Tsang, W., *Int. J. Chem. Kinet.*, **10,** 687 (1978).
[51] *Org. Reaction Mech.*, **1975,** 86.
[52] King, K. D., *Int. J. Chem. Kinet.*, **9,** 907 (1978).
[53] Zabel, F., Benson, S. W., and Golden, D. M., *Int. J. Chem. Kinet.*, **10,** 295 (1978).
[54] De Tannoux, N. M., and Pratt, D. W., *J. Chem. Soc., Chem. Commun.*, **1978,** 394.
[55] Sustmann, R., and Gellert, R. W., *Chem. Ber.*, **111,** 42 (1978).
[56] Koenig, T., and Chang, J. C., *J. Am. Chem. Soc.*, **100,** 2240 (1978).
[57] Poppinger, D., Radom, L., and Vincent, M. A., *Chem. Phys.*, **23,** 437 (1977).
[58] Kira, M., Watanabe, M., and Sakurai, H., *J. Am. Chem. Soc.*, **99,** 7780 (1977).
[59] Sugiyama, Y., Kawamura, T., and Yonezawa, T., *J. Chem. Soc., Chem. Commun.*, 1978, 804.
[60] Sugiyama, Y., Kawamura, T., and Yonezawa, T., *Chem. Lett.*, **1978,** 639.
[61] Fischer, T. H., and Meierhoefer, A. W., *J. Org. Chem.*, **43,** 224 (1978).
[62] Polishchuk, V. R., Bubnov, N. N., German, L. S., Liv'e, E. P., Solodovnikov, S. P., Tumanskii, B. L., and Knunyants, I. L., *Dokl. Akad. Nauk SSSR*, **241,** 373 (1978); *Chem. Abs.*, **89,** 162755 (1978).
[63] Gey, E., and Tino, J., *Croat. Chem. Acta*, **51,** 11 (1978); *Chem. Abs.*, **89,** 90002 (1978).
[64] Hinchcliffe, A., *J. Mol. Struct.*, **43,** 273 (1978).
[65] Ziebarth, M., and Neumann, N. P., *Justus Liebigs Ann. Chim.*, **1978,** 1765.
[66] Cf. *Org. Reaction Mech.*, **1975,** 85; **1976,** 101.
[67] Neugebauer, F., Hellwinkel, D., and Aulmich, G., *Tetrahedron Lett.*, **1978,** 4871.
[68] Haddon, R. C., Wudl., F., Kaplan, M. L., Marshall, J. H., and Bramwell, F. B., *J. Chem. Soc., Chem. Commun.*, **1978,** 429.
[69] Stein, M., Winter, W., and Rieker, A., *Angew. Chem. Int. Edn.*, **17,** 692 (1978).
[70] Kosower, E. M., Waits, H. P., Teuerstein, A., and Butler, L. C., *J. Org. Chem.*, **43,** 800 (1978).
[71] Alberti, A., and Pedulli, G. F., *Tetrahedron Lett.*, **1978,** 3283.
[72] Sakaguchi, K., and Inuzuka, K., *Res. Rep. Tokyo Electr. Eng. Coll.*, **23,** 13 (1975); *Chem. Abs.*, **89,** 128926 (1978).
[73] Roberts, B. P., and Winter, J. N., *J. Chem. Soc., Chem. Commun.*, **1978,** 960.
[74] Clark, D. T., and Beer, H. F., *Tetrahedron*, **34,** 491 (1978).
[75] Baban, J. A., and Roberts, B. P., *J. Chem. Soc., Perkin Trans. 2*, **1978,** 678.
[76] Nelsen, S. F., Landis, R. T., and Calabrese, J. C., *J. Org. Chem.*, **42,** 4192 (1977).
[77] Colussi, A. J., and Benson, S. W., *Int. J. Chem. Kinet.*, **10,** 1139 (1978).
[78] Miura, Y., Asada, H., and Kinoshita, M., *Chem. Lett.*, **1978,** 1085.
[79] Mayer, R., Domschke, G., Bleisch, S., and Bartl, A., *Tetrahedron Lett.*, **1978,** 4003.
[80] Miura, Y., Nakamura, Y., and Kinoshita, M., *Chem. Lett.*, **1978,** 521.
[81] Miura, Y., Nakamura, Y., and Kinoshita, M., *Bull. Chem. Soc. Jpn.* **51,** 947 (1978).
[82] Pirkle, W. H., and Gravel, P. L., *J. Org. Chem.*, **43,** 808 (1978).
[83] Bernardi, F., Guerra, M., Lunazzi, L., Panciera, G., and Placucci, G., *J. Am. Chem. Soc.*, **100,** 1607 (1978).
[84] Skell, P. S., Day, J. C., and Slanga, J. P., *Angew. Chem. Int. Edn.*, **17,** 515 (1978).
[85] Skell, P. S., and Day, J. C., *J. Am. Chem. Soc.*, **100,** 1951 (1978).
[86] Yim, M. B., Kikuchi, O., and Wood, D. E., *J. Am. Chem. Soc.*, **100,** 1869 (1978).
[87] Janousek, B. K., Zimmerman, A. H., Reed, K. J., and Brauman, J. I., *J. Am. Chem. Soc.*, **100,** 6142 (1978).
[88] Shinagawa, Y., and Shinagawa, Y., *J. Am. Chem. Soc.*, **100,** 67 (1978).

[89] Kirste, B., Kurreck, H., and Schubert, K., *Tetrahedron, Lett.*, **1978,** 777.
[90] Prokof'ev, A. I., Khodak, A. A., Malysheva, N. A., Petrovskii, P. V., Bubnov, N. N., Solodovnikov, S. P., and Kabachnik, M. I., *Dokl. Akad. Nauk SSSR*, **240,** 92 (1978).
[91] Mackor, A., *J. Org. Chem.*, **43,** 3241 (1978).
[92] Il'yasov, A. V., Liober, B. G., Barlev, A. A., Mukhtarov, A., Sh., Sokolov, M. P., Pavlov, V. A., and Razumov, A. I., *Teor. Eksp. Khim.*, **13,** 693 (1977); *Chem. Abs.*, **88,** 21583 (1978).
[93] Barlev, A. A., Il'yasov, A. V., Mukhtarov, A. Sh., Shermergorn, I. M., and Vafina, N. M., *Izv. Akad. Nauk SSSR, Ser. Khim.*, **5,** 1187 (1978); *Chem. Abs.*, **89,** 107096 (1978).
[94] Howard, J. A., and Tait, J. C., *J. Am. Chem. Soc.*, **99,** 8349 (1977).
[95] Prokof'ev, A. I., Prokof'eva, T. I., Bubnov, N. N., Solodovnikov, S. P., Belostotskaya, I. S., Ershov, V. V., and Kabachnik, M. I., *Dokl. Akad, Nauk SSSR*, **236,** 897 (1977); *Chem. Abs.*, **88,** 21697 (1978).
[96] Berclaz, T., and Geoffroy, M., *Helv. Chim. Acta*, **91,** 684 (1978).
[97] Berclaz, T., and Geoffroy, M., *Chem. Phys. Lett.*, **52,** 606 (1977).
[98] Litinskii, A. O., Rakhimov, A. I., Balexicius, L., and Bolotin, A. B., *Fiz. Mol.* (*Kiev*), **4,** 58 (1977); *Chem. Abs.*, **88,** 151596 (1978).
[99] Tang, F., and Huyser, E. S., *J. Org. Chem.*, **43,** 1061 (1978).
[100] Hiatt, R., and Rahimi, P. M., *Int. J. Chem. Kinet.*, **10,** 185 (1978).
[101] Yablokov, V. A., Ganyushkin, A. V., Yablokova, N. V., Mikhlin, V. E., and Antonova, T. M., *Zh. Obshch. Khim.*, **47,** 1786 (1977); *Chem. Abs.*, **87,** 167277 (1977).
[102] Turovskii, A. A., Petrenko, V. V., Torovskii, A. A., and Litkovets, A. K., *Zh. Obshch. Khim.*, **47,** 2556 (1977); *Chem. Abs.*, **88,** 61817 (1978).
[103] Serdyuk, A. I., Kucher, R. V., Turovskii, A. A., and Batog, A. E., *Zh. Org. Khim.*, **13,** 1645 (1977); *Chem. Abs.*, **87,** 183767 (1977).
[104] Lind, H., Loeliger, H., and Winkler, T., *Tetrahedron Lett.*, **1978,** 1571.
[105] Cf. *Org. Reaction Mech.*, **1976,** 171.
[106] Sosnovsky, G., and Karas, G. A., *Z. Naturforsch.*, *B*, **33**, 1165 (1978).
[107] Sosnovsky, G., and Karas, G. A., *Z. Naturforsch.*, *B*, **33**, 1177 (1978).
[108] Lee, M., and Minoura, Y., *J. Chem. Soc.*, *Faraday Trans.* 1, **74,** 1726 (1978).
[109] Lee, M., Morigami, T., and Minoura, Y., *J. Chem. Soc.*, *Faraday Trans.* 1, **74,** 1738 (1978).
[110] Horn, K. A., Schmidt, S. P., Koo, J.-Y., and Schuster, G. B., *U.S. NTIS, AD Rep.*, 1978, AD–A053290; *Chem. Abs.*, **89,** 162687 (1978).
[111] Richardson, W. H., Anderegg, J. H., Price, M. E., Tappen, W. A., and O'Neal, H. E., *J. Org. Chem.*, **43,** 2236 (1978).
[112] Richardson, W. A., Anderegg, J. H., Price, M. E., and Crawford, W., *J. Org. Chem.*, **43,** 4045 (1978).
[113] Neill, C., and Stauff, J., *Z. Naturforsch.*, *B*, **33,** 763 (1978).
[114] Pedersen, C. L., and Lohse, C., *Tetrahedron Lett.*, **1978,** 3141.
[115] Lechtken, P., *Chem. Ber.*, **111,** 1413 (1978).
[116] McCapra, F., *J. Chem. Soc.*, *Chem. Commun.*, **1977,** 946.
[117] McCapra, F., Beheshti, I., Burford, A., Hann, R. A., and Zaklika, K. A., *J. Chem. Soc.*, *Chem. Commun.*, **1977,** 944.
[118] Zaklika, K. A., Thayer, A. L., and Schaap, A. P., *J. Am. Chem. Soc.*, **100,** 4916 (1978).
[119] Pryor, W. A., and Graham, W. D., *J. Org. Chem.*, **43,** 770 (1978).
[120] Camacho Rubio, F., Diaz Rodriguez, F., Delgado Diaz, S., and Moreno Jimenez, V., *An. Quim.*, **74,** 117 (1978); *Chem. Abs.*, **88,** 162747 (1978).
[121] Bridgewater, A. J., and Sexton, M. D., *J. Chem. Soc.*, *Perkin Trans* 2, **1978,** 530.
[122] Solyanikov, V. M., Pustarnakova, G. F., and Solov'eva, T. Yu., *Bull. Acad, Sci. USSR*, **26,** 2061 (1977).
[123] Matisova-Rychla, L., Rychly, J., Chodak, I., and Lazar, M., *Chem. Zvesti*, **31,** 271 (1977); *Chem. Abs.*, **88,** 104445 (1978).
[124] Grigoryan, S. K., and Beileryan, N. M., *Arm. Khim. Zh.*, **30,** 634 (1977); *Chem. Abs.*, **88,** 49974 (1978).
[125] Grigoryan, S. K., *Arm. Khim. Zh.*, **31,** 357 (1978); *Chem. Abs.*, **89,** 17406 (1978).
[126] Kenigsberg, T. P., Ariko, N. G., and Mitskevich, N. I., *Kinet. Katal.*, **18,** 1067 (1977); *Chem. Abs.*, **87,** 183770 (1977).
[127] Drozdova, T. I., Aleksandrov, A. L., and Denisov, E. T., *Izv. Akad, Nauk SSSR, Ser. Khim.*, **1978,** 965, 970; Chem. Abs., **89,** 42063, 42064 (1978).
[128] Pavlinec, J., and Lazar, M., *Collect. Czech. Chem. Comm.*, **1978,** 122.

[129] Kovač, F., and Plesničar, B., *J. Chem. Soc., Chem. Commun.*, **1978,** 122.
[130] Dmitruk, A. F., *Fiz. Mol. (Kiev)*, **4,** 47 (1977); *Chem. Abs.*, **88,** 151596 (1978).
[131] Garibyan, E. G., Muradyan, A. A., and Garibyan, T. A., *Arm. Khim. Zh.*, **31,** 67 (1978); *Chem. Abs.*, **89,** 128638 (1978).
[132] Minato, T., Yamabe, S., Fujimoto, H., and Fukui, K., *Bull. Chem. Soc. Jpn.*, **51,** 682 (1978).
[133] *Org. Reaction Mech.*, **1976,** 110.
[134] Kirsch, L. J., Parkes, D. A., Waddington, D. J., and Wooley, A., *J. Chem. Soc., Faraday Trans.* 1, **74,** 2293 (1978).
[135] Bothe, E., and Schulte-Frohlinde, D., *Z. Naturforsch., B,* **33,** 786 (1978).
[136] Opeida, I. A., Timokhin, V. I., and Galet, V. F., *Teor. Eksp. Khim.*, **14,** 554 (1978); *Chem. Abs.*, **89,** 162827 (1978).
[137] Kikuchi, O., Hiyama, A., Yoshida, H., and Suzuki, K., *Bull. Chem. Soc. Jpn.*, **51,** 11 (1978).
[138] Pacansky, J., Gardini, G. P., and Bargon, J., *Ber. Bunsenges., Phys. Chem.*, **82,** 19 (1978).
[139] Lawler, R. G., Barbara, P. F., and Jacobs, D., *J. Am. Chem. Soc.*, **100,** 4912 (1978).
[140] Rykov, S. V., Ignatenko, A. V., and Kessenikh, V. A., *Tezisy Dokl.-Vses. Konf. "Polyariz. Yader Elektronov. Eff. Magn. Polya Khim, Reakts.,"* **1975,** 7; *Chem. Abs.*, **89,** 23514 (1978).
[141] Shatkina, T. N., Lovtsova, A. N., Mazel', K. S., Pekhk, T. I., Lippmaa, E. T., and Reutov, O. A., *Dokl. Chem.*, **235,** 675 (1977).
[142] Sukin, A. V., Bulatov, M. A., Spasskii, S. S., and Suvorov, A. L., *Bull. Acad. Sci. USSR,* **27,** 619 (1978).
[143] Albeck, M., Levinger, A., and Schiff, K., *J. Chem. Soc., Faraday Trans.* 1, **74,** 1488 (1978).
[144] Danen, W. C., and Arudi, R. L., *J. Am. Chem. Soc.*, **100,** 3944 (1978).
[145] Suehiro, T., Matsui, Y, and Inoguchi, Y., *Bull. Chem. Soc., Jpn.*, **51,** 2609 (1978).
[146] Cf. *Org. Reaction Mech.*, **1975,** 118.
[147] Luh, T.-Y., and Stock, L. M., *J. Org. Chem.*, **43,** 3271 (1978).
[148] Tabushi, I., Kojo, S., and Fukunishi, K., *J. Org. Chem.*, **43,** 2370 (1978).
[149] Bourgeois, M. J., Filliatre, C., Lalande, R., Maillard, B., and Villenave, J. J., *Tetrahedron Lett.*, **1978,** 3355.
[150] Razuvaev, G. A., Brevnova, T. N., Chesnokova, T. A., Semenov, V. V., and Cherepennikova, N. F., *Zh. Obshch. Khim.*, **47,** 1760 (1977); *Chem. Abs.*, **87,** 183768 (1977).
[151] Adam, W., Cueto, O., Guedes, L. N., and Rodriguez, L. O., *J. Org. Chem.*, **43,** 1466 (1978).
[152] Gibson, D. H., and Joseph, J. T., *J. Am. Chem. Soc.*, **100,** 3641 (1978).
[153] Cf. *Org. Reaction Mech.*, **1974,** 86.
[154] van Swieten, A. P., Louw, R., and Arnoldy, P., *Recl. Trav. Chim. Pays-Bas*, **97,** 61 (1978).
[155] Niki, H., Maker, P. D., Savage, C. M., and Breitenbach, L. P., *Chem. Phys. Lett.*, **55,** 289 (1978).
[156] Ogata, Y., and Tomizawa, K., *J. Org. Chem.*, **43,** 1920 (1978).
[157] Ogata, Y., and Tomizawa, K., *J. Org. Chem.*, **43,** 261 (1978).
[158] Harding, L. B., and Goddard, W. A., *J. Am. Chem., Soc.*, **100,** 7180 (1978).
[159] Herron, J. T., and Huie, R. E., *Int. J. Chem. Kinet.*, **10,** 1019 (1978).
[160] Hiberty, P. C., and Leforestier, C., *J. Am. Chem. Soc.*, **100,** 2012 (1978).
[161] Pryor, W. A., and Kurz, M. E., *Tetrahedron Lett.*, **1978,** 697.
[162] Komissarov, V. D., Galimova, L. G., Komissarova, I. N., Shchereshovets, V. V., and Denisov, E. T., *Dokl. Akad. Nauk SSSR*, **235,** 1350 (1977); *Chem. Abs.*, **87,** 200499 (1977),
[163] Howard, J. A., Chenier, J. H. B., and Holden, D. A., *Can. J. Chem.*, **56,** 170 (1978).
[164] Timokhin, V. I., Opeida, I. A., and Kucher, R. V., *Neftekhimiya*, **17,** 555 (1977); *Chem. Abs.* **87,** 555 (1977).
[165] Camacho, F., Rodriguez, D., Delgado, S., and Quintana, J., *An. Quim.*, **73,** 898 (1977); *Chem. Abs.*, **87,** 167229 (1977).
[166] Camacho, F., Diaz, F., Delgado, D., and Moreno, V., *An. Quim.*, **73,** 903 (1977); *Chem. Abs.*, **87,** 167230 (1977).
[167] Vidóczy, T., and Gal. D., *Z. Phys. Chem. (Wiesbaden)*, **106,** 269 (1977).
[168] Ohkatsu, Y., and Tsuruta, T., *Bull. Chem. Soc., Jpn.*, **51,** 188 (1978).
[169] Mizukami, F., and Imamura, J., *Bull. Chem. Soc., Jpn.*, **51,** 1404 (1978).
[170] Hasegawa, R., and Kamiya, Y., *Bull. Chem. Soc. Jpn.*, **51,** 1490 (1978).
[171] Ando, W. and Kohmoto, S., *J. Chem. Soc., Chem. Commun.*, **1978,** 120.
[172] Bartlett, P. D., and Becherer, J., *Tetrahedron Lett.*, **1978,** 2983.
[173] Agisheva, S. A., Pastushenko, E. V., Klyavlin, M. S., Martem'yanov, V. S., Zlotskii, S. S., and Rakhmankulov, D. L., *Zh. Prikl. Khim. (Leningrad)*, **50,** 2732 (1977); *Chem. Abs.*, **89,** 151758 (1978).

[174] Agisheva, S. A., Estrina, G. Ya., Imashev, U. B., Aleksandrov, A. L., Zlotskii, S. S., and Rakhmankulov, D. L., *Zh. Prikl. Khim.* (*Leningrad*), **51,** 1809 (1978); *Chem. Abs.*, **89,** 162798 (1978).
[175] Hendriks, C. F., Van Beek, H. C. A., and Heertjes, P. M., *Ind. Eng. Chem. Prod. Res. Dev.*, **16,** 270 (1977); *Chem. Abs.*, **87,** 183743 (1977).
[176] Sakamoto, H., Funabiki, T., and Tarama, K., *J. Catal.*, **48,** 427 (1977); *Chem. Abs.*, **87,** 167200 (1977).
[177] Druliner, J. D., *J. Org. Chem.*, **43,** 2069 (1978).
[178] Ishii, Y., Adachi, A., Imai, R., and Ogawa, M., *Chem. Lett.*, **1978,** 611.
[179] Turro, N. J., Chow, M.-F., and Yashikatsu, I., *J. Am. Chem. Soc.*, **100,** 5580 (1978).
[180] Jefford, C. W., Knöpfel, W., and Cadby, P. A., *J. Am. Chem. Soc.*, **100,** 6432 (1978).
[181] Lukacs, J., *Magy. Kem. Lapja*, **32,** 570 (1977); *Chem. Abs.*, **89,** 107364 (1978).
[182] Cáceres, T., Lissi, E. A., and Sanhueza, E., *Int. J. Chem. Kinet.*, **10,** 1167 (1978).
[183] Plis, E. M., Aleksandrov, A. L., and Mogilevich, M. M., *Bull. Acad. Sci. USSR*, **26,** 1327 (1977).
[184] Pliss, E. M., and Aleksandrov, A. L., *Izv. Akad. Nauk SSSR, Ser. Khim.*, **1978,** 214; *Chem. Abs.*, **88,** 151756 (1978).
[185] Häusler, J., Jahn, R., and Schmidt, U., *Chem. Ber.*, **111,** 361 (1978).
[186] Chenier, J. H. B., Howard, J. A., and Tait, J. C., *Can. J. Chem.*, **56,** 157 (1978).
[187] Howard, J. A., and Tait, J. C., *Can. J. Chem.*, **56,** 164 (1978).
[188] Zakhorova, N. A., Kuz'min, V. I., Kruglyakova, K. E., Smirnov, L. D., Dyumaev, K. M., and Emanuel, N. M., *Bull. Acad. Sci. USSR*, **26,** 929 (1977).
[189] Holčík, J., Maňásek, Z., and Pospíšil, J., *Collect. Czech. Chem. Commun.*, **43,** 142 (1978).
[190] Kovářová-Lerchová, J., and Pospíšil, J., *Chem. Ind.* (*London*), **1978,** 91.
[191] Vetchinka, V. N., Skibida, I. P., and Maizus, Z. K., *Bull. Acad. Sci. USSR*, **26,** 925 (1977).
[192] Porter, N. A., Dubay, G. R., and Green, J. G., *J. Am. Chem. Soc.*, **100,** 920 (1978).
[193] Martin, G., and Maccoll, A., *J. Chem. Soc., Perkin Trans. 2*, **1977,** 1887.
[194] Marshall, R. M., and Page, N. D., *J. Chem. Soc., Faraday Trans. 1*, **74,** 2121 (1978).
[195] Engel, P. S., Bishop, D. J., and Page, M. A., *J. Am. Chem. Soc.*, **100,** 7009 (1978).
[196] Evanochko, W. T., and Shevlin, P. B., *J. Am. Chem. Soc.*, **100,** 6428 (1978).
[197] Bigot, B., Sevin, A., and Devaquet, A., *J. Am. Chem. Soc.*, **100,** 2639 (1978).
[198] Meier, H., and Zeller, K.-P., *Angew. Chem. Int. Edn.*, **16,** 835 (1977).
[199] Mirbach, M. J., Liu, K.-C., Mirbach, M. F., Cherry, W. R., Turro, N. J., and Engel, P. S. *J. Am. Chem. Soc.*, **100,** 5122 (1978).
[200] Duisman, W., and Rüchardt, C., *Chem. Ber.*, **111,** 596 (1978).
[201] Engel, P. S., Hayes, R. A., Keifer, L., Szilagyi, S., and Timberlake, J. W., *J. Am. Chem. Soc.*, **100,** 1876 (1978).
[202] White, D. K., and Greene, F. D., *J. Am. Chem. Soc.*, **100,** 6760 (1978).
[203] Yeung, D. W. K., McAlpine, G. A., and Warkentin, J., *J. Am. Chem. Soc.*, **100,** 1962 (1978).
[204] Tsuchiya, T., Enkaku, M., and Sawanishi, H., *J. Chem. Soc., Chem. Commun.*, **1978,** 568.
[205] Turro, N. J., Cherry, W. R., Mirbach, M. F., and Mirbach, M. J., *J. Am. Chem. Soc.*, **99,** 7388 (1977).
[206] Turro, N. J., Liu, K.-C., Cherry, W., Liu, J.-M., and Jacobson, B., *Tetrahedron Lett.*, **1978,** 555.
[207] Olsen, H., and Synder, J. P., *J. Am. Chem. Soc.*, **100,** 285 (1978).
[208] Barton, J. W., and Walker, R. B., *Tetrahedron Lett.*, **1978,** 1005.
[209] Cichra, D., Platz, M. S., and Berson, J. A., *J. Am. Chem. Soc.*, **99,** 8507 (1977).
[210] Crawford, R. J., Tokunaga, H., Schrijver, L. M. H. C., and Godard, J. C., *Can. J. Chem.*, **56,** 998 (1978).
[211] Bushby, R. J., and Pollard, M. D., *Tetrahedron Lett.*, **1978,** 3855, 3859.
[212] Begley, M. J., Dean, F. M., Houghton, L. E., Johnson, R. S., and Park, B. K., *J. Chem. Soc., Chem. Commun.*, **1978,** 461.
[213] Dushkin, A. V., Leshina, T. V., Sycheva, I. M., and Sagdeev, R. Z., *Tezisy Dokl.-Vses. Konf. "Polyariz. Yader Elektronov. Eff. Magn. Polya Khim. Reakts."*, **1975,** 22; *Chem. Abs.*, **89.** 23516 (1978).
[214] Vernin, G., Siv, C., Metzger, J., Chanon, M., and Archavalis, A., *Nouveau J. de Chimie*, **2,** 625 (1978).
[215] Oda, T., Maeshima, T., and Sugiyama, K., *Makromol. Chem.*, **179,** 2331 (1978).
[216] Sugiyama, K., Tabuchi, T., and Maeshima, T., *Nippon Kagaku Kaishi*, **1977,** 1158; *Chem. Abs.*, **87** 200550 (1977).

[217] Sugiyama, K., Tabuchi, T., and Maeshima, T., *Nippon Kagaku Kaishi*, **1978,** 150; *Chem. Abs.*, **88** 104367 (1978).
[218] Sugiyama, K., Tabuchi, T., and Maeshima, T., *J. Chem. Soc. Jpn. Chem. Ind. Chem.*, **1978,** 150.
[219] Sugiyama, K., and Tabuchi, T., *Makromol. Chem.*, **179,** 259 (1978).
[220] Madgzinski, L. J., Pillay, K. S., Richard, H., and Chow, Y. L., *Can. J. Chem.*, **56,** 1657 (1978).
[221] Szele, I., and Zollinger, H., *Helv. Chim. Acta*, **61,** 1721 (1978).
[222] Zollinger, H., *Angew. Chem. Int. Edn.*, **17,** 141 (1978).
[223] Doyle, M. P., Dellaria, J. P., Siegreid, B., and Bishop, S. W., *J. Org. Chem.*, **42,** 3494 (1977).
[224] Rykova, L. A., Kiprianova, L. A., Levit, A. F., and Gragerov, I. P., *Teor, Eksp. Khim.*, **14,** 256 (1978); *Chem. Abs.*, **89,** 59445 (1978).
[225] Becker, H. G. O., Pfeifer, D., and Radeglia, R., *Z. Chem.*, **17,** 439 (1977).
[226] Sterk, H., and Lechner, G., *Org. Magn. Reson.*, **11,** 453 (1978).
[227] Minato, T., Yamabe, S., Fujimoto, H., and Fukui, K., *Bull. Chem. Soc., Jpn.*, **51,** 1 (1978).
[228] Schuh, H.-H., and Fischer, H., *Helv. Chim. Acta*, **61,** 2130 (1978).
[229] Schuh, H.-H., and Fischer, H., *Helv. Chim. Acta*, **61,** 2463 (1978).
[230] Tilquin, B., Allaert, J., and Claes, P., *J. Phys. Chem.*, **82, 277** (1978).
[231] Beckwith, A. L. J., and Easton, C., *J. Am. Chem. Soc.*, **100,** 2913 (1978).
[232] Beckhaus, H.-D., Hellmann, G., and Rüchardt, C., *Chem. Ber.*, **111,** 72 (1978).
[233] Langhals, H., and Fischer, H., *Chem. Ber.*, **111,** 543 (1978).
[234] Colle, T. H., Glaspie, P. S., and Lewis, E. S., *J. Org. Chem.*, **43,** 2722 (1978).
[235] Buschhaus, H. U., and Neumann, W. P., *Angew. Chem. Int. Edn.*, **17,** 59 (1978).
[236] Polishchuk, V. R., Bubnov, N. N., German, L. S., Lur'e, E. P., Solodovnikov, S. P., Tumanskii, B. L., and Knunyants, I. L., *Dokl. Akad. Nauk SSSR*, **241,** 373 (1978); *Chem. Abs.*, **89,** 162755 (1978).
[237] Ayscough, P. B., and Lambert, G., *J. Chem. Soc., Faraday Trans.* 1, **74**, 2481 (1978).
[238] Pfau, M., Gobert, F., Gramain, J.-F., and Lhomme, M.-F., *J. Chem. Soc., Perkin Trans.* 1, **1978,** 509.
[239] Adachi, H., Basco, N., and James, D. G. L., *Chem. Phys. Lett.*, **59,** 502 (1978).
[240] Khudyakov, I. V., Kokrashvili, T. A., Levin, P. P., and Kuz'min, V. A., *Izv. Akad. Nauk SSSR, Ser. Khim.*, **1977,** 2364; *Chem. Abs.*, **88,** 36911 (1978).
[241] Khudyakov, I. V., Kuz'min, V. A., and Emanuel, N. M., *Int. J. Chem. Kinetics*, **10,** 1015 (1978).
[242] Pastushenko, E. V., Zlotskii, S. S., and Rakhmankulov, D. L., *React. Kinet. Catal. Lett.*, **8,** 209 (1978); *Chem. Abs.*, **89,** 128799 (1978).
[243] Kubarev, S. I., Pschenichnov, E. A., and Shustov, A. S., *Tezisy Dokl.-Vses. Konf. "Polyariz. Yader Elektronov. Eff. Magn. Polya Khim. Reakts."*, **1975,** 36; *Chem. Abs.*, **89,** 23479 (1978).
[244] Broomhead, E. J., and McLauchlan, K. A., *J. Chem. Soc., Faraday Trans.* 2, **74,** 775 (1978).
[245] Minato, T., Yamabe, S., Fujimoto, H., and Fukui, K., *Bull. Chem. Soc. Jpn.*, **51,** 682 (1978).
[246] Kirsch, L. J., Parkes, D. A., Waddington, D. J., and Woolley, A., *J. Chem. Soc., Faraday Trans.* 1, **74,** 2293 (1978).
[247] Kosower, E. M., Teuerstein, A., Burrows, H. D., and Swallow, A. J., *J. Am. Chem. Soc.*, **100,** 5185 (1978).
[248] Domány, G., Nyitrai, J., Lempert, K., Voelter, W., and Horn, H., *Chem. Ber.*, **111,** 1464 (1978).
[249] Tsang, W., *Int. J. Chem. Kinet.*, **10,** 821 (1978).
[250] Glaenzer, K., Quack, M., and Troe, J., *Symp. (Int.) Combust.* (Proc.), 1976 (pub. 1977), **16,** 949; *Chem. Abs.*, **87,** 200442 (1977).
[251] Tsuboi, T., *Jpn. J. Appl. Phys.*, **17,** 709 (1978); *Chem. Abs.*, **88,** 189633 (1978).
[252] Kao, W., and Yeh, C., *J. Phys. Chem.*, **81,** 2304 (1977).
[253] Skorobogatov, G. A., Slesar, O. N., and Seleznev, V. G., *Dokl. Akad. Nauk SSSR*, **237,** 889 (1977); *Chem. Abs.*, **88,** 49978 (1978).
[254] Lesclaux, R., and Demissy, M., *J. Photochem.*, **9,** 110 (1978); *Chem. Abs.*, **89,** 162760 (1978).
[255] Tezuka, M., and Miller, L. L., *J. Am. Chem. Soc.*, **100,** 4201 (1978).
[256] Spangenberg, H.-J., Börge, I., Hoffmann, H., and Mögel, G., *Z. Phys. Chem. (Leipzig)*, **259,** 531 (1978).
[257] Schuchmann, H.-P., Ritter, A., and von Sonntag, C., *J. Organometal. Chem.*, **148,** 213 (1978).

[258] Lewis, E. S., *Top. Curr. Chem.*, **74,** 31 (1978).
[259] Bell, T. N., and Perkins, P. G., *J. Phys. Chem.*, **81,** 2012 (1977).
[260] Rayez-Meaume, M. T., Dannenberg, J. J., and Whitten, J. L., *J. Am. Chem. Soc.*, **100,** 747 (1978).
[261] Held, A. M., Manthorne, K. C., Pacey, P. D., and Reinholdt, H. P., *Can. J. Chem.*, **55,** 4128 (1977).
[262] Manthorne, K. C., and Pacey, P. D., *Can. J. Chem.*, **56,** 1307 (1978).
[263] Ballod, A. P., Fedorova, T. V., Kumova, A. A., Titarchuk, T. A., Chikvaidze, N., and Yanyukova, A. M., *Kinet. Katal.* **19,** 814 (1978); *Chem. Abs.*, **89,** 128792 (1978).
[264] Ballod, A. P., and Titarchuk, T. A., *Kinet. Katal.*, **18,** 1359 (1977); *Chem. Abs.*, **88,** 36921 (1978).
[265] Baldwin, A. C., and Golden, D. M., *Chem. Phys. Lett.*, **55,** 350 (1978).
[266] Budeiko, N. L., Agabekov, V. E., Mitskevich, N. I., and Denisov, E. T., *Kinet. Catal. Lett.*, **8,** 71 (1978); *Chem. Abs.*, **89,** 5663 (1978).
[267] Agabekov, V. E., Budeiko, N. L., Denisov, E. T., and Mitskevich, N. I., *React. Kinet. Catal. Lett.*, **7,** 437 (1977); *Chem. Abs.*, **88,** 88793 (1978).
[268] Kardos, J., Fitos, I., Kovács, I., Szammer, J., and Simonyi, M., *J. Chem. Soc., Perkin Trans.* 2, **1978,** 405.
[269] Myshkin, V. E., Shostenko, A. G., Zagorets, P. A., and Sharapova, N. M., *Izv. Vyssh. Uchebn. Zaved., Khim. Khim. Tekhnol.*, **20,** 1446 (1977); *Chem. Abs.*, **88,** 49964 (1978).
[270] Grinevich, I. A., Zagorets, P. A., Shostenko, A. G., Dodonov, A. M., Myshkin, V. E., and Tarasova, N. P., *Izv. Vyssh. Uchebn. Zaved., Khim. Khim. Tekhnol.*, **20,** 1575 (1977); *Chem. Abs.*, **88,** 49965 (1978).
[271] Onishchenko, T. A., Zhuk, B. L., and Friedlina, R. Kh., *Izv. Akad. Nauk SSSR, Ser. Khim.*, **1977,** 2297.
[272] Terent'ev, A. B., *Izv. Akad. Nauk SSSR, Ser. Khim.*, **1977,** 2614.
[273] Terent'ev, A. B., Ikonnikov, N. S., and Freidlina, R. Kh., *Dokl. Akad. Nauk SSSR*, **238,** 870 (1978).
[274] Moskalenko, M. A., and Terent'ev, A. B., *Izv. Akad. Nauk SSSR, Ser. Khim.*, **1977,** 2150.
[275] Churilova, M. A., and Terent'ev, A. B., *Bull. Acad. Sci. USSR*, **26,** 1073, (1977).
[276] Belova, L. I., and Karpukhina, G. V., *Bull. Acad. Sci. USSR*, **26,** 1605 (1977).
[277] Kryger, R. G., Lorand, J. P., Stevens, N. R., and Herron, N. R., *J. Am. Chem. Soc.*, **99,** 7589 (1977).
[278] Suehiro, T., Suzuki, A., Tsuchida, Y., and Yamazaki, J., *Bull. Chem. Soc., Jpn.*, **50,** 3324 (1977).
[279] Levy, L. A., *J. Chem. Soc., Chem. Commun.*, **1978,** 574.
[280] Sanders, D. G., Tedder, J. M., and Walton, J. C., *J. Chem. Soc., Perkin Trans.* 2, **1978,** 580.
[281] Majer, J. R., and Al-Saigh, Z. Y., *J. Fluorine Chem.*, **10,** 565 (1977).
[282] Katz, M. G., Baruch, G., and Rajbenbach, L. A., *Int. J. Chem. Kinet.*, **10,** 905 (1978).
[283] Bell, T. N., Perkins, K. A., and Perkins, P. G., *J. Phys. Chem.*, **81,** 2610 (1977).
[284] Aver'yanov, V. A., Lebedev, N. N., and Lebedeva, G. F., *Zh. Fiz. Khim.*, **51,** 2834 (1977); *Chem. Abs.*, **88,** 49966 (1976).
[285] Aver'yanov, V. A., Somov, G. V., and Gorlova, S. I., *Zh. Fiz. Khim.*, **51,** 2693 (1977); *Chem. Abs.*, **88,** 36909 (1978).
[286] Everly, C. R., Schweinsberg, F., and Traynham, J. G., *J. Am. Chem, Soc.*, **100,** 1200 (1978).
[287] Perkins, R. R., and Pincock, R. E., *Can. J. Chem.*, **56,** 1269 (1978).
[288] Mosher, M. W., and Estes, G. W., *J. Am. Chem. Soc.*, **99,** 6928 (1977).
[289] Roberts, B. P., and Wilson, C., *J. Chem. Soc., Chem. Commun.*, **1978**, 752.
[290] Pews, R. G., and Davis, R. A., *J. Chem. Soc., Chem. Commun.*, **1978,** 105.
[291] Zhulin, V. M., and Rubinshtein, B. I., *Izv. Akad. Nauk SSSR, Ser. Khim.*, **1977,** 2248; *Chem. Abs.*, **88,** 36847 (1978).
[292] Bell, T. N., and Davidson, I. M. T., *J. Organometal. Chem.*, **162,** 347 (1978).
[293] Nagase, S., and Morokuma, K., *J. Am. Chem. Soc.*, **100,** 1666 (1978).
[294] Gilliom, R. D., *J. Am. Chem. Soc.*, **99,** 8399 (1977).
[295] Krech, R. H., and McFadden, D. L., *J. Am. Chem. Soc.*, **99,** 8402 (1977).
[296] Oldershaw, G. A., and Robinson, E. A., *Chem. Phys. Lett.*, **54,** 527 (1978); *Chem. Abs.*, **88,** 189571 (1978).
[297] Volpe, P., and Castigliani, M., *Gazz. Chim. Ital.*, **107,** 279 (1977); *Chem. Abs.*, **88,** 36912 (1978).

[298] Castigliani, M., and Volpe, P., *Radiochem. Radioanal. Lett.*, **33,** 211 (1978); *Chem. Abs.*, **89,** 59448 (1978).
[299] Pryor, W. A., and Olsen, E. G., *J. Am. Chem. Soc.*, **100,** 2852 (1978).
[300] Horie, O., Nishino, J., and Amano, A., *Int. J. Chem. Kinet.*, **10,** 1043 (1978).
[301] Bogan, D. J., and Hand, C. W., *J. Phys, Chem.*, **82,** 2067 (1978).
[302] Clynne, M. A. A., and Nip, W. S., *Int. J. Chem. Kinet.*, **10,** 367 (1978).
[303] Zahniser, M. S., Berquist, B. M., and Kaufman, F., *Int. J. Chem. Kinet.*, **10,** 15 (1978).
[304] Keyser, L. F., *J. Chem. Phys.*, **69,** 214 (1978).
[305] Lin, C. L., Leu, M. T., and DeMore, W. B., *J. Phys. Chem.*, **82,** 1772 (1978).
[306] Baghal-Vayjooee, M. H., Colussi, A. J., and Benson, S. W., *J. Am. Chem. Soc.*, **100,** 3214 (1978).
[307] Motell, E. L., Boone, A. W., and Fink, W. H., *Tetrahedron*, **34,** 1619 (1978).
[308] Fischer, T. H., and Meierhoefer, A. W., *J. Org. Chem.*, **43,** 220 (1978).
[309] Fischer, T. H., and Meierhoefer, A. W., *J. Org. Chem.*, **43,** 224 (1978).
[310] Dneprovskii, A. S., and Belyakov, N. V., *Dokl. Akad. Nauk SSSR*, **237,** 340 (1977); *Chem. Abs.*, **88,** 73804 (1978).
[311] Doncaster, A. M., and Walsh, R., *Int. J. Chem. Kinet.*, **10,** 101 (1978).
[312] Shinohara, H., Imamura, A., Masuda, J., and Kondo, M., *Bull. Chem. Soc. Jpn.*, **51,** 1917 (1978).
[313] Shinohara, H., Imamura, A., Masuda, T., and Kondo, M., *Bull. Chem. Soc., Jpn.* **51,** 98 (1978).
[314] Darnall, K. R., Atkinson, R., and Pitts, J. N., Jr., *J. Phys. Chem.*, **82,** 1581 (1978).
[315] Smith, R. H., *Int. J. Chem. Kinet.*, **10,** 519 (1978).
[316] Atkinson, R., and Pitts, J. N., Jr., *J. Chem. Phys.*, **68,** 3581 (1978).
[317] Niki, H., Maker, P. D., Savage, C. M., and Breitenbach, L. P., *J. Phys. Chem.*, **82,** 132 (1978).
[318] Overend, R., and Paraskevopoulos, G., *J. Phys. Chem.*, **82,** 1329 (1978).
[319] Ogata, Y., Tomizawa, K., and Fujii, K., *Bull. Chem. Soc. Jpn.*, **51,** 2628 (1978).
[320] Zorin, V. V., Zlotskii, S. S., Shuvalov, V. F., Moravskii, A. P., Rakhmankulov, D. L., and Paushkin, Ya. M., *Dokl. Akad. Nauk SSSR*, **236,** 106 (1977); *Chem. Abs.*, **87,** 200462 (1977).
[321] Campbell, I. M., and Parkinson, P. E., *Chem. Phys. Lett.*, **53,** 385 (1978); *Chem. Abs.*, **88,** 104377 (1978).
[322] Atkinson, R., Perry, R. A., and Pitts, J. N., Jr., *J. Chem. Phys.*, **68,** 1850 (1978).
[323] Kwart, H., Benko, D. A., and Bromberg, M. E., *J. Am. Chem. Soc.*, **100,** 7093 (1978).
[324] Detty, M. R., and Paquette, L. A., *J. Org. Chem.*, **43,** 1118 (1978).
[325] Chodák, I., and Bakoš, D., *Collect. Czech. Chem. Commun.*, **43,** 2574 (1978).
[326] Hargis, J. H. and Hsu, H.-H., *J. Am. Chem. Soc.*, **99,** 8114 (1977).
[327] Davies, A. G., and Quintard, J.-P., *J. Chem. Soc., Perkin Trans.* 2, **1978,** 1163.
[328] Small, R. D., Jr., and Scaiano, J. C., *J. Am. Chem. Soc.*, **100,** 296 (1978).
[329] Paul, H., Small, R. D., Jr., and Scaiano, J. C., *J. Am. Chem. Soc.*, **100,** 4520 (1978).
[330] Kelly, N., and Heicklen, J., *J. Photochem.*, **8,** 83 (1978); *Chem. Abs.*, **88,** 120307 (1978).
[331] Zorin, V. V., Zlotskii, S. S., Il'yasov, A. V. and Rakhmankulov, D. L., *Zh. Org. Khim.*, **13,** 2430 (1977); *Chem. Abs.*, **88,** 61775 (1978).
[332] Encina, M. V., and Lissi, E. A., *Int. J. Chem. Kinet.*, **10,** 653 (1978).
[333] Abuin, E., Encina, M. V., Diaz, S., and Lissi, E. A., *Int. J. Chem. Kinet.*, **10,** 677 (1978).
[334] Choo, K. Y., and Gaspar, P. P., *Taehan Hwahak Hoechi*, **21,** 270 (1977); *Chem. Abs.*, **87,** 200475 (1977).
[335] Davies, A. G., Roberts, B. P., and Tse, M.-W., *J. Chem. Soc., Perkin Trans.* 2, **1978,** 145.
[336] Nishimura, N., Okahashi, K., Yukutomi, T., Fujiwara, A., and Kubo, S., *Aust. J. Chem.*, **31,** 1201 (1978.
[337] Kornilova, N. N., Arko, N. G., Agabekov, V. E., and Mitskevich, N. I., *React. Kinet. Catal. Lett.*, **7,** 241 (1977); *Chem. Abs.*, **88,** 21715 (1978).
[338] Howard, J. A., Chenier, J. H. B., and Holden, D. A., *Can. J. Chem.*, **56,** 170 (1978).
[339] Golding, B. T., Kemp, T. J., Sell, C. S., Sellars, P. J., and Watson, W. P. *J. Chem. Soc., Perkin Trans.* 2, **1978,** 839.
[340] Rudakova, I. P., Ershova, T. E., Belikov, A. B., and Yurkevich, A. M., *J. Chem. Soc., Chem. Commun.*, **1978,** 592.
[341] Pryor, W. A., Gojon, G., and Church, D. F., *J. Org. Chem.*, **43,** 793 (1978).
[342] Harris, J. F., Jr., *J. Org. Chem.*, **43,** 1319 (1978).

[343] Skell, P. S., and Day, J. C., *Acc. Chem. Res.*, **11,** 381 (1978).
[344] Skell, P. S., and Day, J. C., *J. Am. Chem. Soc.*, **100,** 1951 (1978).
[345] Skell, P. S., Day, J. C., and Slanga, J. P., *Angew. Chem. Int. Edn.* **17,** 515 (1978).
[346] Roberts, B. P., and Winter, J. N., *J. Chem. Soc., Chem. Commun,*, **1978,** 545.
[347] Yip, R. W., Vidoczy, T., Snyder, R. W., and Chow, Y. L., *J. Phys. Chem.*, **82** 1194 (1978).
[348] Pokhodenko, V. D., Koshechko, V. G., and Atamanyuk, V. Yu., *Teor. Eksp. Khim.*, **14,** 240 (1978); *Chem. Abs.*, **89,** 59472 (1978).
[349] Beileryan, N. M., *Arm. Khim. Zh.*, **30,** 540 (1977); *Chem. Abs.*, **88,** 36905 (1978).
[350] Arzamanova, I. G., Gurvich, Ya. A., and Rybak, A. I., *Zh. Fiz. Khim.*, **52,** 1349 (1978); *Chem. Abs.*, **89,** 59450 (1978).
[351] Wilt., J. W., and Aznavoorian, P. M., *J. Org. Chem.*, **43,** 1285 (1978).
[352] Baldwin, A. C., Davidson, I. M. T., and Reed, M. D., *J. Chem. Soc., Faraday Trans.* 1, **74,** 2171 (1978).
[353] Wolff, W. D., Goure, W. F., and Barton, T. J., *J. Am. Chem. Soc.*, **100,** 6236 (1978).
[354] Baldwin, A. C., and Golden, D. M., *Chem. Phys. Lett.*, **60,** 108 (1978).
[355] Gasanov, R. G., Terent'ev, A. B., and Freidlina, R. L., *Bull. Acad. Sci. USSR*, **26,** 2433 (1977).
[356] Gouverneur, P. J., and Mulangala, J. M., *Bull. Soc., Chim. Belg.*, **86,** 699 (1977).
[357] Machida, M., Takechi, H., and Kanaoka, Y., *Heterocycles*, **7,** 273 (1977).
[358] Breslow, R., Rothbard, J., Herman, F., and Rodriguez, M. L., *J. Am. Chem. Soc.*, **100,** 1213 (1978).
[359] Masalimov, A. S., Prokof'ev, A. I., Bubnov, N. N., Solodovnikov, S. P., and Kabachnik, M. I., *Dokl. Akad. Nauk SSSR*, **236,** 116 (1977); *Chem. Abs.*, **87,** 200463 (1977).
[360] Pines, S. H., Purick, R. M., Reamer, R. A., and Gal, G., *J. Org. Chem.*, **43,** 1337 (1978).
[361] Evans, B. S., and Whittle, E., *Int. J. Chem. Kinet.*, **10,** 745 (1978).
[362] Luh, T.-Y., and Stock, L. M., *J. Org. Chem.*, **43,** 3271 (1978).
[363] Katz, M. G., and Rajbenbach, L. A., *Int. J. Chem. Kinet.*, **10,** 955 (1978).
[364] Bamford, C. H., and Basahel, S. N., *J. Chem. Soc., Faraday Trans.* 1, **74,** 1020 (1978).
[365] Itoh, K., *Nippon Kagaku Kaishi*, **1977,** 1359; *Chem. Abs.*, **88,** 104291 (1978).
[366] Malysheva, N. A., Prokof'ev, A. I., Bubnov, N. N., Solodovnikov, S. P., Prokof'eva, T. I., Volod'kin, A. A., and Ershov, V. V., *Bull. Acad. Sci. USSR*, **26,** 1398 (1977)
[367] Sakurai, H., and Murakami, M., *Bull. Chem. Soc. Jpn.*, **50,** 3384 (1977).
[368] Coates, D. A., and Tedder, J. M., *J. Chem. Soc., Perkin Trans.* 2, **1978,** 725.
[369] Sydnes, L. Kr., *Acta Chem. Scand. Ser. B*, **32,** 47 (1978).
[370] Okafo, E. N., and Whittle, E., *Int. J. Chem. Kinet.*, **10,** 591 (1978).
[371] Pickard, J. M., and Rodgers, A. S., *Int. J. Chem. Kinet.*, **9,** 759 (1977).
[372] Cross, P. J., and Husain, D., *J. Photochem.*, **8,** 183 (1978); *Chem. Abs.*, **88,** 120315 (1978).
[373] Mintz, K. J., and Le Roy, D. J., *Can. J. Chem.*, **56,** 941 (1978).
[374] Knoll, H., *Z. Chem.*, **18,** 302 (1978).
[375] Dewar, M. J. S., and Olivella, S., *J. Am. Chem. Soc.*, **100,** 5290 (1978).
[376] Giese, B., and Zwick, W., *Angew. Chem. Int. Edn.*, **17,** 66 (1978).
[377] Marshall, R. M., and Rahman, L., *Int. J. Chem. Kinet.*, **9,** 705 (1977).
[378] Myshkin, V. E., Shostenko, A. G., and Zagorets, P. A., *Kinet. Katal.*, **18,** 1076 (1977); *Chem. Abs.*, **87,** 167189 (1977).
[379] Shostenko, A. G., Myshkin, V. E., and Shapovalov, N. A., *Kinet. Katal.*, **18,** 1358 (1977); *Chem. Abs.*, **88,** 36920 (1978).
[380] Myshkin, V. E., Shostenko, A. G., and Zagorets, P. A., *React. Kinet. Catal. Lett.*, **8,** 41 (1978); *Chem. Abs.*, **89,** 41950 (1978).
[381] Citterio, A., Miniki, F., Arnoldi, A., Pagano, R., Parravicini, A., and Porta, O., *J. Chem. Soc., Perkin Trans.* 2, **1978,** 519.
[382] (a) Kopecky, K. R., and Lau, M.-P., *J. Org. Chem.*, **43,** 525 (1978). (b) Pryor, W. A., Graham, W. D., and Green, J. G., *J. Org. Chem.*, **43,** 526 (1978).
[383] Edwards, J. H., McQuillin, F. J., and Wood, M., *J. Chem. Soc., Chem. Commun.*, **1978,** 438.
[384] Montaudon, E., Thépenier, J., and Lalande, R., *Tetrahedron*, **34,** 897 (1978).
[385] Sorba, J., Fossey, J., and Lefort, D., *Bull. Soc. Chim. Fr. II*, **1977,** 967.
[386] Stella, L., Janousek, Z., Merényi, R., and Viehe, G., *Angew. Chem. Int. Edn.*, **17,** 691 (1978).
[387] Citterio, A., *Tetrahedron Lett.*, **1978,** 2701.
[388] Amriev, R. A., and Velichko, F. K., *Izv. Akad. Nauk SSSR, Ser. Khim.*, **1977,** 2360.
[389] James, F. C., Choi, J., Strausz, O. P., and Bell, T. N., *Chem. Phys. Lett.*, **53,** 206 (1978); *Chem. Abs.*, **88,** 104376 (1978).

[390] Chkhubianishvili, N. G., Dzheiranishvili, M. S., Levinskii, M. B., and Afanas'ev, I. B., *Zh. Org. Khim.*, **13,** 1578 (1977); *Chem. Abs.*, **87,** 167172 (1977).
[391] Inoue, Y., Ohno, S., and Hashimoto, H., *Chem. Lett.*, **1978,** 367.
[392] Matsumoto, H., Nakano, T., Takasu, K., and Nagai, Y., *J. Org. Chem.*, **43,** 1734 (1978).
[393] Paterson, I., Tedder, J. M., and Walton, J. C., *J. Chem. Soc., Perkin Trans.* 2, **1978,** 884.
[394] Sass, V. P., Serov, S. I., and Sokolov, S. V., *Zh. Org. Khim.*, **13,** 2298 (1977); *Chem. Abs.*, **88,** 73802 (1978).
[395] *Shanghai Institute of Organic Chemistry, Sci. Sin.* (*Engl. Ed.*), **20,** 353 (1977); *Chem. Abs.*, **87,** 167441 (1977).
[396] Tedder, J. M., and Walton, J. C., *ACS Symp. Ser.* 1978, **66** (Fluorine-containing Free Radicals), 107; *Chem. Abs.*, **88,** 120118 (1978).
[397] El Soueni, A., Tedder, J. M., and Walton, J. C., *J. Fluorine Chem.*, **11,** 407 (1978); *Chem. Abs.*, **89,** 5657 (1978).
[398] Brookes, C. J., Coe, P. L., Pedler, A. E., and Tatlow, J. C., *J. Chem. Soc., Perkin Trans.* 1, **1978,** 202.
[399] Low, H. C., Tedder, J. M., and Walton, J. C., *Int. J. Chem. Kinet.*, **10,** 325 (1978).
[400] Kandror, I. I., Kopylova, B. V., and Freidlina, R. Kh., *Tetrahedron Lett.*, **1978,** 3087.
[401] Roberts, B. P., and Winter, J. N., *J. Chem. Soc., Chem. Commun.*, **1978,** 960.
[402] Nadervel, T. A., Sass, V. P. Serov, S. I., and Sokolov, S. V., *Zh. Org. Khim.*, **13,** 2294 (1977); *Chem. Abs.*, **88,** 73842 (1978).
[403] Rowland, F. S., Rust, F., and Frank, J. P., *ACS Symp. Ser. 1978*, **66** (*Fluorine-containing Free Radicals*), 26; *Chem. Abs.*, **88,** 120117 (1978).
[404] Stevens, D. J., and Spicer, L. D., *J. Am. Chem. Soc.*, **100,** 3295 (1978).
[405] Clark, D. T., Scanlan, I. W., and Walton, J. C., *Chem. Phys. Lett.*, **55,** 102 (1978); *Chem. Abs.*, **89,** 75174 (1978).
[406] Shellhamer, D. F., and Oakes, M. L., *J. Org. Chem.*, **43,** 1316 (1978).
[407] Vigalok, I. V., Ostrosvskaya, A. V., Petrova, G. G., and Levin, Ya. A., *Zh. Org. Khim.*, **14,** 1255 (1978); *Chem. Abs.*, **89,** 107402 (1978).
[408] Sergeev, G. B., Smirnov, V. V., and Popov, E. A., *Dokl. Akad. Nauk SSSR*, **240,** 1161 (1978); *Chem. Abs.*, **89,** 146108 (1978).
[409] Bottini, A. T., and Cabral, L. J., *Tetrahedron*, **34,** 3195 (1978).
[410] Giacomelli, G., Lardicci, L., and Saba, A., *J. Chem. Soc., Perkin Trans.* 1, **1978,** 314.
[411] Faubel, C., and Wagner, H. G., *Ber. Bunsenges. Phys. Chem.*, **81,** 694 (1977); *Chem. Abs.*, **87,** 183717 (1977).
[412] Ishikawa, Y., Yamabe, M., Noda, A., and Sato, S., *Bull. Chem. Soc. Jpn.*, **51,** 2488 (1978).
[413] Koda, S., Nakamura, K., Hoshino, T., and Hikita, T., *Bull. Soc. Chem. Jpn.*, **51,** 957 (1978).
[414] Havel, J. J., *J. Org. Chem.*, **43,** 762 (1978).
[415] Atkinson, R., and Pitts, J. N., Jr., *J. Chem. Phys.*, **68,** 2992 (1978); *Chem. Abs.*, **88,** 189635 (1978).
[416] Perry, R. A., Atkinson, R., and Pitts, J. N., Jr., *J. Chem. Phys.*, **67,** 5577 (1977).
[417] Atkinson, R., Perry, R. A., and Pitts, J. N., Jr., *J. Chem. Phys.*, **67,** 3170 (1977); *Chem. Abs.*, **88,** 21699 (1978).
[418] Nip, W. S., and Paraskevopoulos, G., *J. Photochem.*, **9,** 119 (1978); *Chem. Abs.*, **89,** 162761 (1978).
[419] Niki, H., Maker, P. D., Savage, C. M., and Breitenbach, L. P., *J. Phys. Chem.*, **82,** 135 (1978).
[420] Anastasi, C., and Smith, I. W. M., *J. Chem. Soc., Faraday Trans.* 2, **74,** 1056 (1978).
[421] Hoshino, M., Akimoto, H., and Okuda, M., *Bull. Chem. Soc. Jpn.*, **51,** 718 (1978).
[422] Ravishankara, A. R., Wagner, S., Fischer, S., Smith, G., Schiff, R., Watson, R. T., Tesi, G., and Davis, D. D., *Int. J. Chem. Kinet.*, **10,** 783 (1978).
[423] Castle, L., and Lindsay Smith, J. R., *J. Chem. Soc., Chem. Commun.*, **1978,** 704.
[424] Anderson, R. F., and Schulte-Frohlinde, D., *J. Phys. Chem.*, **82,** 22 (1978).
[425] Melhuish, W. H., and Sutton, H. C., *J. Chem. Soc., Chem. Commun.*, **1978,** 970.
[426] Saunders, B. B., Kaufman, P. C., and Matheson, M. S., *J. Phys. Chem.*, **82,** 142 (1978).
[427] Shellhamer, D. F., Heasley, V. L., Foster, J. E., Luttrull, J. K., and Heasley, G. E., *J. Org. Chem.*, **43,** 2652 (1978).
[428] Heasley, G., E., Emery, W. E., III, Hinton, R., Shellhamer, D. F., Heasley, V. L., and Rodgers, S. L., *J. Org. Chem.*, **43,** 361 (1978).
[429] Alexandrov, Yu. A., Garbatov, V. V., Yablokova, N. V., and Tsvetkov, V. G., *J. Organometal. Chem.*, **157,** 267 (1978).

[430] van Roodselaar, A., Safarik, I., Strausz, O. P., and Gunning, H. E., *J. Am. Chem. Soc.*, **100,** 4068 (1978).
[431] Miyata, T., Sakumoto, A., and Washino, M., *Bull. Chem. Soc. Jpn.*, **50,** 2950 (1977).
[432] Ito, O., and Matsuda, M., *Bull. Chem. Soc. Jpn.*, **51,** 427 (1978).
[433] Gasanov, R. G., Petrova, R. G., and Bragina, I. O., *Izv. Akad. Nauk SSSR, Ser. Khim.* **1977,** 2464; *Chem. Abs.*, **88,** 36931 (1978).
[434] Szeimies, G., Schlosser, A., Philipp, F., Dietz, P., and Mickler, W., *Chem. Ber.*, **111,** 1922 (1978).
[435] Guy, R. G., and Thompson, J. J., *Tetrahedron*, **34,** 541 (1978).
[436] Berry, A. D., and Fox, W. B., *J. Org. Chem.*, **43,** 365 (1978).
[437] Wozny, J. C., Miles, M. L., Purrington, S. T., and Bumgardner, C. L., *J. Fluorine Chem.*, **11,** 175 (1978); *Chem. Abs.*, **89,** 5649 (1978).
[438] Magdzinski, L. J., and Chow, Y. L., *J. Am. Chem. Soc.*, **100,** 2444 (1978).
[439] Day, J. C., Katsaros, M. G., Kocher, W. D., Scott, A. E., and Skell, P. S., *J. Am. Chem. Soc.*, **100,** 1950 (1978).
[440] Driguez, H., Vermes, J.-P., and Lessard, J., *Can. J. Chem.*, **56,** 119 (1978).
[441] Olejniczak, B., Osowska, K., and Zwierzak, A., *Tetrahedron*, **34,** 2051 (1978).
[442] Alberti, A., and Pedulli, G. F., *Tetrahedron Lett.*, **1978,** 3283.
[443] Rivière, P., Rivière-Baudet, M., Richelme, S., and Satgé, J., *Bull. Soc. Chim. Fr., II,* **1978,** 193.
[444] Howard, J. A., and Tait, J. C., *J. Am. Chem. Soc.*, **99,** 8349 (1977).
[445] Alberti, A., and Hudson, A., *J. Chem. Soc. Perkin Trans.* 2, **1978,** 1098.
[446] Alberti, A., and Camaggi, C. M., *J. Organometal. Chem.*, **161,** C63 (1978).
[447] Sweany, R. L., and Halpern, J., *J. Am. Chem. Soc.*, **99,** 8335 (1977).
[448] Clark, H. C., and Wong, C. S., *J. Am. Chem. Soc.*, **99,** 7073 (1977).
[449] Smith, T. W., and Butler, G. B., *J. Org. Chem.*, **43,** 6 (1978).
[450] Monahan, A., Campbell, P., Cheh, S., Forg, J., Grossman, S., Miller, J., Rankin, P., and Vallee, J., *Syn. Commun.*, **7,** 553 (1977).
[451] Lopez Aparicio, F. J., Fernandez Sanchez, J., and Plaza Lópe-Espinosa, M. T., *An. Quim.*, **74,** 294 (1978); *Chem. Abs.*, **89,** 162752 (1978).
[452] Machiewicz, P., Furstoss, R., and Waegell, B., *J. Org. Chem.*, **43,** 3746 (1978).
[453] Lessard, J., Cote, R., Mackiewicz, P., Furstoss, R., and Waegell, B., *J. Org. Chem.*, **43,** 3750 (1978).
[454] Mondon, M., and Lessard, J., *Can. J. Chem.*, **56,** 2590 (1978).
[455] Lockhart, R. W., Kitadani, M., Einstein, F. W. B., and Chow, Y. L., *Can. J. Chem.*, **56,** 2897 (1978).
[456] Perry, R. A., Lockhart, R. W., Kitadani, M., and Chow, Y. L., *Can. J. Chem.*, **56,** 2906 (1978).
[457] Benati, L., Montevecchi, P. C., and Spagnolo, P., *Tetrahedron Lett.*, **1978,** 815
[458] Applequist, D. E., and Klug, J. H., *J. Org. Chem.*, **43,** 1729 (1978).
[459] Batt, L., Islam, T. S. A., and Rattray, G. N., *Int. J. Chem. Kinet.*, **10,** 567 (1978).
[460] Imashev, U. B., Glukhova, S. S., Kalashnikov, S. M., Zlotskii, S. S., Rakhmankulov, D. L., and Paushkin, Ya. M., *Dokl. Chem.*, **235,** 683 (1977).
[461] Pastushenko, E. V., Zlotskii, S. S., and Rakhmankulov, D. L., *React. Kinet. Catal. Lett.*, **8,** 209 (1978); *Chem. Abs.*, **89,** 128799 (1978).
[462] Pastushenko, E. V., and Rakhmankulov, D. L., *Neftekhim. Sint. Tekh. Prog.*, **1976,** 29; *Chem. Abs.*, **89,** 5716 (1978).
[463] Boothe, T. E., Greene, J. L., Shevlin, P. B. Willcott, M. R., Inners, R. R., and Cornelis, A., *J. Am. Chem. Soc.*, **100,** 3874 (1978).
[464] Imashev, U. B., Kravets, E. Kh., Zlotskii, S. S., and Rakhmankulov, D. L., *Zh. Prikl. Khim. (Leningrad)*, **51,** 1431 (1978); *Chem. Abs.*, **89,** 146137 (1978).
[465] Nakipelova, T. D., Gagarina, A. B., and Emanuel, N. M., *Dokl. Akad. Naukl SSSR.*, **238,** 392 (1978); *Chem. Abs.*, **88,** 104405 (1978).
[466] Hudson, R. L., and Williams, F., *J. Phys. Chem.*, **82,** 967 (1978).
[467] Deshayes, H., and Pete, J.-P., *J. Chem. Soc., Chem. Commun.*, **1978,** 567.
[468] Bolar, R. B., Joukhadar, L., McGhie, J. F., Misra, S. C., Barrett, A. G. M., Barton, D. H. R., and Prokopiou, P. A., *J. Chem. Soc., Chem. Commun.*, **1978,** 68.
[469] Behrens, G., Bothe, E., Eibenberger, J., Koltzenburg, G., and Schulte-Frohlinde, D., *Angew. Chem. Int. Edn.*, **17,** 604 (1978).
[470] Challand, S. R., *Aromat. Heteroaromat. Chem.*, **5,** 339 (1977): *Chem. Abs.*, **88,** 21471 (1978).

[471] Kryger, R. G., Lorand, J. P., Stevens, N. R., and Herron, N. R., *J. Am. Chem. Soc.*, **99,** 7589 (1977).
[472] Madhavan, V., Schuler, R. H., and Fessenden, R. W., *J. Am. Chem. Soc.*, **100,** 888 (1978).
[473] Hasegawa, K., and Neta, P., *J. Phys. Chem.*, **82,** 854 (1978).
[474] Citterio, A., Minisci, F., Porta, O., and Sesana, G., *J. Am. Chem. Soc.*, **99,** 7960 (1977).
[475] See *Org. Reaction. Mech.*, **1974,** 112.
[476] El Zein, M., Gardrat, C., and Maillard, B., *Bull. Soc., Chim. Belg.*, **87,** 67 (1978).
[477] Kurz, M. E., and Chen, T. R., *J. Org. Chem.*, **43,** 239 (1978).
[478] Mitsuhashi, T., and Ōki, M., *Chem. Lett.*, **1978,** 1089.
[479] Porta, O., and Sesana, G., *Tetrahedron Lett.*, **1978,** 3571.
[480] Aversa, M. C., Cum, G., Gianetto, P., and Romeo, G., *J. Chem. Res.*, **1978,** (S) 292.
[481] Dorrepaal, W., and Louw, R., *Int. J. Chem. Kinet.*, **10,** 249 (1978).
[482] Olah, G. A., Lin, H. C., Olah, J. C., and Narang, S. C., *Proc. Natl. Acad. Sci. USA*, **75,** 1045 (1978).
[483] Hori, A., Matsumoto, H., Takamuku, S., and Sakurai, H., *J. Chem. Soc., Chem. Commun.*, **1978,** 16.
[484] Hori, A., Matsumoto, H., Takamuku, S., and Sakarai, H., *Chem. Lett.*, **1978,** 467.
[485] Castle, L., and Lindsay Smith, J. R., *J. Chem. Soc., Chem. Commun.*, **1978,** 704.
[486] Steenken, S., and O'Neill, P., *J. Phys. Chem.*, **82,** 372 (1978).
[487] Ishikawa, H., Watanabe, K., and Ando, W., *Bull. Chem., Soc. Japan.*, **51,** 2173 (1978).
[488] Hoshino, M., Akimoto, H., and Okuda, M., *Bull. Chem. Soc. Jpn.*, **51,** 718 (1978).
[489] Watanabe, K., Ishikawa, H., and Anpo, W., *Bull. Chem. Soc., Jpn.*, **51,** 1253 (1978).
[490] Grimshaw, J., Haslett, R. J., and Trocha-Grimshaw, J., *J. Chem. Soc., Perkin Trans.* **1, 1977,** 2448.
[491] Grimshaw, J., and Mannus, D., *J. Chem. Soc., Perkin Trans.* 1, **1977,** 2456.
[492] Cf. *Org. Reaction Mech.*, **1977,** 113.
[493] Testaferri, L., Tiecco, M., Tingoli, M., Fiorentino, M., and Troisi, L., *J. Chem. Soc., Chem. Commun.*, **1978,** 93.
[494] Bolton, R., Sandall, J. P. B., and Williams, G. H., *J. Fluorine Chem.*, **11,** 591 (1978).
[495] Zupan, M., Sket, B., and Pahor, B., *J. Org. Chem.*, **43,** 2297 (1978).
[496] Everly, C. R., and Traynham, J. G., *J. Am. Chem. Soc.*, **100,** 4316 (1978).
[497] Okafo, E. N., and Whittle, E., *Int. J. Chem. Kinet.*, **10,** 591 (1978).
[498] Benati, L., Montevecchi, P. C., and Tundo, A., *J. Chem. Soc., Chem. Commun.*, **1978,** 530.
[499] Speckamp, W. N., and Kohler, J. J., *J. Chem. Soc., Chem. Commun.*, **1978,** 166.
[500] See *Org. Reaction. Mech.*, **1977,** 115.
[501] Nowada, K., Hisoaka, M., Sakuragi, H., Tokumaru, K., and Yoshida, M., *Tetrahedron Lett.*, **1978,** 137.
[502] See *Org. Reaction Mech.*, **1977,** 113.
[503] Gorbachev, V. M., Kalyazin, E. P., Petryaev, E. P., and Shadyro, O. I., *Khim. Vys. Energ.*, **11,** 462, (1977); *Chem. Abs.*, **88,** 36984 (1978).
[504] Iwasaki, M., and Toriyama, K., *J. Am. Chem. Soc.*, **100,** 1964 (1978).
[505] Brunton, G., Gray, J. A., Griller, D., Barclay, L. R. C., and Ingold, K. U., *J. Am. Chem. Soc.*, **100,** 4197 (1978).
[506] Bischof, P., *J. Am. Chem. Soc.*, **99,** 8145 (1977).
[507] Biehl, R., Hass, Ch., Kurreck, H., Lubitz, W., and Oestreich, S., *Tetrahedron*, **34,** 419 (1978).
[508] Alberti, A., and Hudson, A., *J. Chem. Soc., Perkin Trans.* 2, **1978,** 1098.
[509] Prokof'ev, A. I., Prokof'eva, T. I., Bubnov, N. N., Solodovnikov, S. P., Belostotskaya, I. S., Ershov, V. V., and Kabachnik, M. I., *Dokl. Akad. Nauk SSSR*, **239,** 1367 (1978); *Chem. Abs.*, **89,** 23443 (1978).
[510] Barany, F., Wolff, S., and Agosta, W. C., *J. Am. Chem. Soc.*, **100,** 1946 (1978).
[511] Wilt, J. W., Kolewe, O., and Kramer, J. F., *J. Am. Chem. Soc.*, **91,** 2624 (1969).
[512] Wilt, J. W., Chwang, W. K., Dockus, C. F., and Tomuik, N. M., *J. Am. Chem. Soc.*, **100,** 5534 (1978).
[513] Domány, G., Nyitrai, J., Lempert, K., Voelter, W., and Horn, H., *Chem. Ber.*, **111,** 1464 (1978).
[514] Weustink, R. J. M., Lourens, R., and Bickelhaupt, F., *Justus Liebigs Ann. Chem.*, **1978,** 214.
[515] Shawali, A. S., Hassaneen, H. M., and Almousawi, S., *Bull. Chem. Soc. Jpn.*, **51,** 512 (1978).
[516] Davies, A. G., and Tse, M.-W., *J. Organometal. Chem.*, **155,** 25 (1978).
[517] Blum, P., Davies, A. G., and Henderson, R. A., *J. Chem. Soc., Chem. Commun.*, **1978,** 569.

[518] Shellhamer, D. F., and Oakes, M. L., *J. Org. Chem.*, **43,** 2616 (1978).
[519] Sugiyama, Y., Kawamura, T., and Yonezawa, T., *J. Am. Chem. Soc.*, **100,** 6525 (1978).
[520] Sustmann, R., and Gellert, R. W., *Chem. Ber.*, **111,** 388 (1978).
[521] Washburn, W. N., *J. Am. Chem. Soc.*, **100,** 6235 (1978).
[522] Lambert, J. B., Kobayashi, K., and Mueller, P. H., *Tetrahedron Lett.*, **1978,** 4253.
[523] Gasanov, R. G., and Freidlina, R. Kh., *Izv. Akad. Nauk SSSR. Ser. Khim.*, **1977,** 2242; *Chem. Abs.*, **88,** 36910 (1978).
[524] Rakhmankulov, D. L., Zlotskii, S. S., Martem'yanov, V. S., Karakhanov, R. A., and Bartok, M., *Acta Chim. Acad. Sci. Hung.*, **95,** 267 (1977); *Chem. Abs.*, **89,** 762 (1978).
[525] Zorin, V. V., Zlotskii, S. S., and Rakhmankulov, D. L., *Zh. Org. Khim.*, **13,** 2614 (1977); *Chem. Abs.*, **88,** 88845 (1978).
[526] Zlotskii, S. S., Martem'yanov, V. S., Kravets, E. Kh., Samirkhanov, Sh. M., and Rakhmankulov, D. L., *Neftekhim. Sint. Tekh. Prog.*, **1976,** 15; *Chem. Abs.*, **89,** 42034 (1978).
[527] Samorkhanov, Sh. M., Zlotskii, S. S., and Rakhmankulov, D. L., *Dokl. Akad. Nauk SSSR*, **236,** 935 (1977); *Chem. Abs.*, **88,** 21698 (1978).
[528] Nishinaga, A., Nakamura, K., and Matsuura, T., *Tetrahedron Lett.*, **1978,** 3557.
[529] Kalvoda, J., and Grob, J., *Helv. Chim. Acta*, **61,** 1996 (1978).
[530] Suginome, H., Sugiura, S., Yonekura, N., Masamune, T., and Osowa, E., *J. Chem. Soc., Perkin* 1, **1978,** 612.
[531] Freidlina, R. Kh., Gasanov, R. G., Petrova, R. G., and Kandror, I. I., *Tezisy Dokl. Nauchn. Sess. Khim. Tekhnol. Org. Soedin. Sery Sernistykh Neftei, 14th*, **1975** (publ. 1976) 124; *Chem. Abs.*, **89,** 23498 (1978).
[532] Gerber, U., Widmer, U., Schmid, R., and Schmid, H., *Helv. Chim. Acta*, **61,** 83 (1978).
[533] Boriack, C., Laganis, E. D., and Lemal, D. M., *Tetrahedron Lett.*, **1978,** 1015.
[534] Chan. H. W.-S., Levett, G., and Matthew, J. A., *J. Chem. Soc., Chem. Commun.*, **1978,** 756.
[535] Evans, D. A., and Baillargeon, D. J., *Tetrahedron Lett.*, **1978,** 3315, 3319.
[536] Villarreal, J. A., and Grubbs, E. J., *J. Org. Chem.*, **43,** 1896 (1978).
[537] Villarreal, J. A., Dobashi, T, S., and Grubbs, E. J., *J. Org. Chem.*, **43,** 1890 (1978).
[538] Ankers, W. B., Brown, C., Hudson, R. F., and Lawson, A. J., *J. Chem. Soc., Perkin Trans.* 2, **1978,** 127.
[539] Hochstein, W., and Schöllkopf, U., *Justus Liebigs Ann. Chem.*, **1978,** 1823.
[540] Reetz, M. T., Kliment, M., and Greif, N., *Chem. Ber.*, **111,** 1083 (1978).
[541] Reetz, M. T., Greif, N., and Kliment, M., *Chem. Ber.*, **111,** 1095 (1978).
[542] Waespe, H.-R., Heimgartner, H., Schmid, H., Hansen, H.-J., Paul, H., and Fischer, H., *Helv. Chim. Acta*, **61,** 401 (1978).
[543] Barnes, C. E., and Myhra, P. C., *J. Am. Chem. Soc.*, **100,** 973, 975 (1978).
[544] Baldwin, J. E., Branz, S. E., and Walker, J. A., *J. Org. Chem.*, **42,** 4142 (1977).
[545] Chantrapromma, K., Ollis, W. D., and Sutherland, I. O., *J. Chem. Soc., Chem. Commun.*, **1978,** 670.
[546] Chantrapromma, K., Ollis, W. D., and Sutherland, I. O., *J. Chem. Soc., Chem. Commun.*, **1978,** 672.
[547] Chantrapromma, K., Ollis, W. D., and Sutherland, I. O., *J. Chem. Soc., Chem. Commun.*, **1978,** 673.
[548] Ollis, W. D., Rey, M., and Sutherland, I. O., *J. Chem. Soc., Chem. Commun.*, **1978**, 675.
[549] Hori, M., Katoaka, T., Shimuzu, H., and Ohno, S., *Tetrahedron Lett.*, **1978**, 255.
[550] Hudson, A., *Electron Spin Reson.*, **4,** 30 (1977); *Chem. Abs.*, **88,** 135767 (1978).
[551] Schoeller, W. W., *J. Chem. Soc., Perkin Trans.* 2, **1978,** 525.
[552] Yamaguchi, K., Nishio, A., Yabushita, S., and Fueno, T., *Chem. Phys. Lett.*, **53,** 109 (1978); *Chem. Abs.*, **88,** 120428 (1978).
[553] Yamaguchi, K., Nishio, A., Yabushita, S., and Fueno, T., *Chem. Lett.*, **1977,** 1479.
[554] Dixon, D. A., Foster, R., Halgren, T. A., and Lipscomb, W. N., *J. Am. Chem. Soc.*, **100,** 1359 (1978).
[555] Japenga, J., Klumpp, G. W., and Schakel, M., *Tetrahedron Lett.*, **1978,** 1869.
[556] Zimmerman, H. E., and Cutler, T. P., *J. Chem. Soc., Chem. Commun.*, **1978,** 232.
[557] Yamaguchi, K., Ohta, K., Yabushita, S., and Fueno, T., *J. Chem. Phys.*, **68,** 4323 (1978); *Chem. Abs.*, **89,** 107642 (1978).
[558] Böhm, M. C., and Gleiter, R., *Tetrahedron Lett.*, **1978,** 1179.
[559] Montague, D. C., *J. Chem. Soc., Faraday Trans.* 1, **74,** 262 (1978).
[560] Montague, D. C., *J. Chem. Soc., Faraday Trans.* 1, **74,** 277 (1978).
[561] Muller, J.-F., Muller, D., Dewey, W. J., and Michl, J., *J. Am. Chem. Soc.*, **100, 1629 (1978).**

[562] Rule, M., and Berson, J. A., *Tetrahedron Lett.*, **1978**, 3191.
[563] Padwa, A., Blacklock, T. J., Getman, D., Hatanaka, N., and Loza, R., *J. Org. Chem.*, **43**, 1481 (1978).
[564] Zimmermann, H. E., and Aasen, S. M., *J. Org. Chem.*, **43**, 1493 (1978).
[565] Parlar, H., *Tetrahedron Lett.*, **1978**, 3885.
[566] Cheng, Y. S. P., Dominguez, E., Garrett, P. J., and Neoh, S. B., *Tetrahedron Lett.*, **1978**, 691.
[567] Schneider, M. P., and Csacsko, B., *J. Chem. Soc., Chem. Commun.*, **1978**, 964.
[568] Gerson, F., Huber, W., and Müllen, K., *Angew. Chem. Int. Edn.*, **17**, 208 (1978).
[569] Ollis, W. D., Rey, M., and Sutherland, I. O., *J. Chem. Soc., Chem. Commun.*, **1978**, 675.
[570] Schaden, G., *Z. Naturforsch., B: Anorg. Chem., Org. Chem.*, **33B**, 663, (1978); *Chem. Abs.*, **89**, 107561 (1978).
[571] Wolff, S., and Agosta, W. C., *J. Org. Chem.*, **43**, 3627 (1978).
[572] Wilson, R. M., and Geiser, F., *J. Am. Chem. Soc.*, **100**, 2225 (1978).
[573] Siemionko, R., Shaw, A., O'Connell, G., Little, L. D., Carpenter, B. K., Shen, L., and Berson, J. A., *Tetrahedron Lett.*, **1978**, 3529.
[574] Weigert, F. J., *J. Org. Chem.*, **42**, 3859 (1977).
[575] Huybrechts, G., and van Mele, B., *Int. J. Chem. Kinetics*, **10**, 1183 (1978).
[576] Harcourt, R. D., *Tetrahedron*, **34**, 3125 (1978).
[577] Caldwell, R. A., and Creed, D., *J. Am. Chem. Soc.*, **99**, 8360 (1977).
[578] Bergmark, W. R., Jones, G., II, Reinhardt, T. E., and Halpern, A. M., *J. Am. Chem. Soc.*, **100**, 6665 (1978).
[579] Cichra, D., Platz, M. S., and Berson, J. A., *J. Am. Chem. Soc.*, **99**, 8507 (1977).
[580] Palladino, G. F., and McEwen, W. E., *J. Org. Chem.*, **43**, 2420 (1978).
[581] Kosower, E. M., and Teuerstein, A., *J. Am. Chem. Soc.*, **100**, 1182 (1978).
[582] Kosower, E. M., Hajdu, J., and Nagy, J. B., *J. Am. Chem. Soc.*, **100**, 1186 (1978).
[583] Rousseau, G., Le Perchec, P., and Conia, J. M., *Tetrahedron Lett.*, **1977**, 2517.
[584] Harding, L. R., and Goddard, W. A., III, *Tetrahedron Lett.*, **1978**, 747.
[585] Dice, D. R., and Steer, R. P., *Can. J. Chem.*, **56**, 114 (1978).
[586] Small, R. D., Jr., and Scaiano, J. C., *J. Phys. Chem.*, **81**, 2126 (1977).
[587] Small, R. D., Jr., and Scaiano, J. C., *J. Am. Chem. Soc.*, **99**, 7713 (1977).
[588] Small, R. D., Jr., and Scaiano, J. C., *J. Am. Chem. Soc.*, **100**, 4512 (1978).
[589] Wagner, P. J., Sedon, J. H., and Lindstrom, M. J., *J. Am. Chem. Soc.*, **100**, 2579 (1978).
[590] Scaiano, J. C., *J. Org. Chem.*, **43**, 568 (1978).
[591] Neuman, R. C., Jr., and Berge, C. T., *Tetrahedron Lett.*, **1978**, 1709.
[592] Encinas, M. V., and Scaiano, J. C., *J. Am. Chem. Soc.*, **100**, 7108 (1978).
[593] Marthamuthu, P., and Scaiano, J. C., *J. Phys. Chem.*, **82**, 1588 (1978).
[594] Weiss, D. S., *Tetrahedron Lett.*, **1978**, 1039.
[595] Enger, A., Feigenbaum, A., Pete, J.-P., and Wolfhugel, J.-L., *Tetrahedron*, **34**, 1509 (1978).
[596] Ghisalberti, E. L., Jefferies, P. R., and Sefton, M. A., *Tetrahedron*, **34**, 3337 (1978).
[597] Murray, R. K., and Andruskiewicz, C. A., *J. Org. Chem.*, **42**, 3994 (1977).
[598] Barany, F., Wolff, F., and Agosta, W. C., *J. Am. Chem. Soc.*, **100**, 1946 (1978).
[599] Takakis, I. M., and Agosta, W. C., *Tetrahedron Lett.*, **1978**, 2387.
[600] Keana, J. F. W., *Chem. Rev.*, **78**, 37 (1978).
[601] Alewood, P. F., Hussain, S. A., Jenkins, T. C., Perkins, M. J., Sharma, A. H., Siew, N. P. Y., and Ward, P., *J. Chem. Soc., Perkin Trans. 2*, **1978**, 1066.
[602] Espie, J.-C., Ramasseul, R., and Rassat, A., *Tetrahedron Lett.*, **1978**, 795.
[603] Volodarskii, L. B., and Tikhonov, A., Ya., *Bull. Acad. Sci. USSR*, **26**, 2425 (1977).
[604] Hewgill, F. R., and Webb, R. J., *Aust. J. Chem.*, **30**, 2565 (1977).
[605] Yamaguchi, T., Yamauchi, A., and Kimizuka, H., *Chem. Lett.*, **1978**, 941.
[606] Kwan, C. L., Atik, S., and Singer, L. A., *J. Am. Chem. Soc.*, **100**, 4783 (1978).
[607] Braun, D., and Törmälä, P., *Makromol. Chem.*, **179**, 1025 (1978).
[608] Lee, T. D., Birrell, G. B., and Keana, J. F. W., *J. Am. Chem. Soc.*, **100**, 1618 (1978).
[609] Lee, T. D., and Keana, J. F. W., *J. Org. Chem.*, **43**, 4226 (1978).
[610] Keana, J. F. W., Norton, R. S., Morello, M., Van Engen, D., and Clardy, J., *J. Am. Chem. Soc.*, **100**, 934 (1978).
[611] Bartolini, G., and Piette, L., *J. Chem. Soc., Chem. Commun.*, **1978**, 215.
[612] Storzer, W., and Röschenthaler, G.-V., *Z. Naturforsch., B*, **33**, 305 (1978).
[613] Sukhanov, V. A., Shvets, V. I., Zhanov, R. I., and Evstigneeva, R. P., *Dokl. Chem.*, **235**, 616 (1977).

[614] Sosnoksky, G., and Konieczny, M., *Z. Naturforsch.*, *B*, **33**, 797 (1978).
[615] Chen, S.-C., and Gaffney, B. J., *J. Magn. Reson.*, **29**, 341 (1978).
[616] Hall, L. D., and Aplin, J. D., *J. Am. Chem. Soc.*, **100**, 1934 (1978).
[617] Ballester, M., Castañer, J., Riera, J., and Veciana, J., *Tetrahedron Lett.*, **1978**, 479.
[618] Aurich, H. G., and Czepluch, H., *Tetrahedron Lett.*, **1978**, 1187.
[619] Dupeyre, R. M., and Rassat, A., *Tetrahedron*, **34**, 1501 (1978).
[620] Dupeyre, R. M., and Capiomont, A., *J. Mol. Struct.*, **42**, 243 (1977).
[621] Forrester, A. R., Neugebauer, F. A., and Fischer, H., *J. Chem. Soc., Perkin Trans.* 2, **1978**, 1014.
[622] Flesia, E., Surzur, J.-M., and Tordo, P., *Org. Magn. Reson.*, **11**, 123 (1978).
[623] Jenkins, T. C., Perkins, M. J., and Terem, B., *Tetrahedron Lett.*, **1978**, 2925.
[624] Forrester, A. R., Henderson, J., Johanson, E. M., and Thomson, R. H., *Tetrahedron Lett.*, **1978**, 5139.
[625] Tel, R. M., and Egberts, J. B. F. N., *J. Org. Chem.*, **42**, 3542 (1977).
[626] Nickel, B., and Rassat, A., *Tetrahedron Lett.*, **1978**, 633.
[627] Kokorin, A. I., and Parmon, V. N., *Bull. Acad. Sci. USSR*, **26**, 496 (1977).
[628] Michon, J., Michon, P., and Rassat, A., *Nouveau J. Chimie.*, **2**, 619 (1978).
[629] Putirszkaja, G. V., Matus, L., Kroo, E., Kovacs, L., and Rockenbauer, A., *Magy. Kem. Lapja*, **32**, 230 (1977); *Chem. Abs.*, **88**, 49951 (1978).
[630] Putirszkaya, G. V., Matus, L., and Rockenbauer, A., *Acta Chim. Acad. Sci. Hung.*, **95**, 133 (1977); *Chem. Abs.*, **89**, 107400 (1978).
[631] Putirszkaya, G. V., Matus, L., Maratbekov, M. B., and Rockenbauer, A., *Magy. Kem. Foly.*, **84**, 229 (1978); *Chem. Abs.*, **89**, 107370 (1978).
[632] Putirszkaja, G., Matus, L., and Rockenbauer, A., *Magy. Kem. Foly.*, **83**, 178 (1977); *Chem. Abs.*, **88**, 49950 (1978).
[633] Daveloose, D., and Leterrier, F., *Photochem. Photobiol.*, **28**, 23 (1978).
[634] Hayashi, T., and Hegedus, L. S., *J. Am. Chem. Soc.*, **99**, 7091 (1977).
[635] Atik, S., and Singer, L. A., *J. Am. Chem. Soc.*, **100**, 3234 (1978).
[636] Zhulin, V. M., Stashina, G. A., and Rozantsev, E. G., *Bull. Acad. Sci. USSR*, **26**, 1387 (1977).
[637] Zhulin, V. M., and Stashina, G. A., *Bull. Acad. Sci. USSR*, **26**, 2396 (1977).
[638] Podedimskii, D. G., Rzhevskaya, N. N., Nasybullin, Sh. A., and Yablonskii, O. P., *Kinet. Katal.*, **18**, 830 (1977); *Chem. Abs.*, **88**, 5926 (1978).
[639] Golubev, V. A., Salmina, N. N., Korsunskii, B. L., Aleinikov, N. N., and Dubovitskii, F. I., *Izv. Akad. Nauk SSSR, Ser. Khim.*, **1978**, 778; *Chem. Abs.*, **89**, 42006 (1978).
[640] Vorob'eva, T. P., Kozlov, Yu. N., and Purmalis, A., *Zh. Fiz. Khim.*, **51**, 2686 (1977); *Chem. Abs.*, **88**, 21778 (1978).
[641] Polumbrik, O. M., and Ryabokon, I. G., *Zh. Org. Khim.*, **14**, 1332 (1978); *Chem. Abs.*, **89**, 107689 (1978).
[642] Benitez, F. M., and Grunwell, J. R., *J. Org. Chem.*, **43**, 2914 (1978).
[643] Janzen, E. G., in "Creation and Detection of the Excited State", (1976) Vol. 4, p. 83 (W. R. Ware, Ed.,) Marcel Dekker Inc., New York.
[644] Janzen, E. G., Evans, C. A., and Davis, E. R., *Am. Chem. Soc., Symp. Series*, No. 69, "Organic Free Radicals" (W. A. Pryor, Ed.) 1978.
[645] Janzen, E. G., and Zawalski, R. C., *J. Org. Chem.*, **43**, 1900 (1978).
[646] Rockenbauer, A., Sümegi, L., Möger, G., Simon, P., Azori, M., and Györ, M., *Tetrahedron Lett.*, **1978**, 5057.
[647] Doba, T., Ichikawa, T., and Yoshida, H., *Bull. Chem. Soc. Jpn.*, **50**, 3124 (1978).
[648] Maeda, Y., Schmid, P., Griller, D., and Ingold, K. U., *J. Chem. Soc., Chem. Commun.*, **1978**, 525.
[649] Schmid, P., and Ingold, K. U., *J. Am. Chem. Soc.*, **100**, 2493 (1978).
[650] Doba, T., Ichikawa, T., and Yoshida, H., *Bull. Chem. Soc. Jpn.*, **50**, 3158 (1977).
[651] Janzen, E. G., Nutter, D. E., Jr., Davis, E. R., Blackburn, B. J., Poyer, J. L., and McCay, P. B., *Can. J.Chem.*, **56**, 2237 (1978).
[652] Janzen, E. G., Wang, Y. Y., and Shetty, R. V., *J. Am. Chem. Soc.*, **100**, 2923 (1978).
[653] Pfab, J., *Tetrahedron Lett.*, **1978**, 843.
[654] Roginskii, V. A., and Belyakov, V. A., *Dokl. Akad. Nauk SSSR*, **237**, 1404 (1977); *Chem. Abs.*, **88**, 120302 (1978).
[655] Steenken, S., and Lyda, M., *J. Phys. Chem.*, **81**, 2201 (1977).

656 Janzen, E. G., Dudley, R. L., Davis, E. R., and Evans, C. A., *J. Phys. Chem.*, **82**, 2444 (1978).
657 Harbour, J. R., and Hair, M. L., *J. Phys. Chem.*, **82**, 1397 (1978).
658 Kraeutler, B., Jaeger, C. D., and Bard, A. J., *J. Am. Chem. Soc.*, **100**, 4903 (1978).
659 Griffin, B. W., and Ting, P. L., *FEBS Lett.*, **89**, 196 (1978).
660 Yarkov, S. P., Belevskii, V. N., Zubarev, V. E., and Bugaenko, L. T., *Khim. Vys. Energ.*, **12**, 131 (1978); *Chem. Abs.*, **89**, 41935 (1978).
661 Howard, J. A., and Tait, J. C., *Can. J. Chem.*, **56**, 176 (1978).
662 Sargent, F. P., and Gardy, E. M., *J. Chem. Phys.*, **67**, 1793 (1977).
663 Gasanov, R. G., Vasil'eva, T. T., and Freidlina, R. Kh., *Izv. Akad. Nauk SSSR, Ser. Khim.*, **1978**, 817; *Chem. Abs.*, **89**, 107363 (1978).
664 Gasanov, R. G., and Freidlina, R. Kh., *Dokl. Chem.*, **236**, 96 (1978).
665 Poyer, J. L., Floyd, R. A., McCay, P. B., Janzen, E. G., and Davis, E. R., *Biochim. Biophys. Acta*, **539**, 402 (1978).
666 Freidlina, R. Kh., Gasanov, R. G., Petrova, R. G., and Kandror, I. I., *Tezisy Dokl. Nauchn. Sess. Khim. Tekhnol. Org. Soedin. Sery Sernistykh Neftei, 14th* **1975**, (publ. 1976), 124; *Chem. Abs.*, **89**, 23498 (1978).
667 Janzen, E. G., Davis, E. R., and Nutter, D. E., *Tetrahedron Lett.*, **1978**, 3309.
668 Perry, R. A., Lockhart, R. W., Kitadani, M., and Chow, Y. L., *Can. J. Chem.*, **56**, 2906 (1978).
669 Okhlobystina, L. V. and Tyurikov, V. A., *Izv. Akad. Nauk SSSR, Ser. Khim.* **1977**, 2035; *Chem. Abs.*, **88**, 21694 (1978).
670 Freidlina, R. Kh., Kandror, I. I., and Gasanov, R. G., *Usp. Khim.*, **47**, 508 (1978); *Chem. Abs.*, **88**, 169178 (1978).
671 Peyronel, J. F., and Kagran, H. B., *Nouveau J. de Chimie*, **2**, 211 (1978).
672 Rivière, P., Richelme, S., Rivière-Baudet, M., Satgé, J., Gynane, M. J. S., and Lappert, M. F., *J. Chem. Res.*, **1978**, (S) 218; (M) 2801.
673 Wardell, J. L., and Grant, D. W., *J. Organomet. Chem.*, **149**, C13 (1978).
674 Mulvey, D., and Waters, W. A., *J. Chem. Soc.*, *Perkin Trans.* 2, **1978**, 1059.
675 Tordo, P., Beyer, M., Cerri, V., and Vila, F., *Phosphorus Sulphur*, **3**, 373 (1977); *Chem. Abs.*, **88**, 189605 (1978).
676 Aurich, H. G., Lotz, I., and Weiss, W., *Tetrahedron*, **34**, 879 (1978).
677 Shibuya, K., Ebata, T., Obi, K., and Tanaka, I., *J. Phys. Chem.*, **81**, 2292 (1977).
678 Dawson, J. H. J., and Nibberung, N. M. M., *J. Am. Chem. Soc.*, **100**, 1928 (1978).
679 Cadman, P., Inel, Y., and Trotman-Dickenson, A. F., *J. Chem. Soc.*, *Faraday Trans.* 1, **74**, 2301 (1978).
680 Porter, N. A., and Nixon, J. R., *J. Am. Chem. Soc.*, **100**, 7116 (1978).
681 Kandror, I. I., Kopylova, B. U., Yashkina, L. V., and Freidlina, R. Kh., *Bull. Acad. Sci. USSR*, **26**, 629 (1977).
682 Baban, J. A., and Roberts, B. P., *J. Chem. Soc.*, *Perkin Trans.* 2, **1978**, 678.
683 Forrest, D., and Ingold, K. U., *J. Am. Chem. Soc.*, **100**, 3868 (1978).
684 Lehnig, M., Neumann, W. P., and Seifert, P., *J. Organometal. Chem.*, **162**, 145 (1978).
685 Hay, R. S., and Roberts, B. P., *J. Chem. Soc.*, *Perkin Trans.* 2, **1978**, 770.
686 Nakanishi, A., Nishikida, K., and Bentrude, W. G., *J. Am. Chem. Soc.*, **100**, 6398 (1978).
687 Nakanishi, A., Nishikida, K., and Bentrude, W. G., *J. Am. Chem. Soc.*, **100**, 6403 (1978).
688 Nakanishi, A., and Bentrude, W. G., *J. Am. Chem. Soc.*, **100**, 6271 (1978).
689 Howard, J. A., and Tait, J. C., *Can. J. Chem.*, **56**, 2163 (1978).
690 Tumanskii, B. L., Solodovnikov, S. P., Leont'eva, I. V., Bubnov, N. N., and Kabachnik, M. I., *Izv. Akad. Nauk SSSR, Ser. Khim.*, **9**, 2128 (1977); *Chem. Abs.*, **88**, 21695 (1978).
691 Roberts, B. P., and Singh, K., *J. Organometal. Chem.*, **159**, 31 (1978).
692 Giles, J. R. M., and Roberts, B. P., *J. Chem. Soc.*, *Chem. Commun.*, **1978**, 623.
693 Boothe, T. E., Greene, J. L., Jr., Shevlin, P. B., Willcott, M. R., III, Inners, R. R., and Cornelis, A., *J. Am. Chem. Soc.*, **100**, 3874 (1978).
694 Wen. L.-S., and Kovacic, P., *Tetrahedron*, **34**, 2723 (1978).
695 Wen. L.-S. Zawalski, R. C., and Kovacic, P., *J. Org. Chem.*, **43**, 2435 (1978).
696 Tanaka, S., Uemura, S., and Okano, M., *J. Chem. Soc.*, *Perkin Trans.* 1, **1978**, 431,
697 Baciocchi, E., Rol. C., and Mandolini, L., *J. Org. Chem.*, **42**, 3682 (1977).
698 Zhumadylov, T., Levanevskii, O. E., Ikonnikova, N. N., Fedichkina, N. G., and Mariko, K. M., *Izv. Akad. Nauk Kirg. SSR*, **1977**, 51; *Chem. Abs.*, **87**, 167227 (1977).

699 Nilsson, A., Palmquist, U., Pettersson, T., and Ronlán, A., *J. Chem. Soc., Perkin Trans.* 1, **1978,** 708.
700 Fritz, H. P., and Gebauer, *Z. Naturforsch., B.*, **33,** 702 (1978).
701 Kurz, M. E., and Hage, G. W., *J. Org. Chem.*, **42,** 4080 (1977).
702 Sullivan, P. D., Menger, E. M., Reddoch, A. H., and Paskovich, D. H., *J. Phys. Chem.*, **82,** 1158 (1978).
703 Baciocchi, E., Mei, S., Rol. C., and Mandolini, L., *J. Org. Chem.*, **43,** 2919 (1978).
704 van Tilborg, W. J. M., Scheele, J. J., and Eberson, L., *Acta Chem. Scand., Ser. B*, **32,** 36 (1978).
705 Eberson, L., and Helgée, B., *Acta Chem. Scand., Ser. B*, **31,** 813 (1977).
706 Yoshida, K., *J. Chem. Soc., Chem. Commun.*, **1978,** 20.
707 Walling, C., Camaioni, D. M., and Kim, S. S., *J. Am. Chem. Soc.*, **100,** 4814 (1978).
708 Cf. *Org. Reaction Mech.*, **1977,** 124.
709 Nikishin, G. I., Spektor, S. S., Kapustina, N. I., and Kaplan, E. P., *Bull. Acad. Sci. USSR*, **26,** 1534 (1976).
710 Groves, J. T., McClusky, G. A., White, R. E., and Coon, M. J., *Biochem. Biophys. Acta*, **81,** 154 (1978).
711 Germain, A., and Commeyras, A., *J. Chem. Soc., Chem. Commun.*, **1978,** 118.
712 Murayama, E., Kohda, A., and Sato, T., *Chem. Lett.*, **1978,** 161.
713 Monahan, A. S., Frêilich, J. D., Fong, J.-J., and Kronenthal, D., *J. Org. Chem.*, **43,** 232 (1978).
714 Baltes, H., Steckhan, E., and Schäfer, H. J., *Chem. Ber.*, **111,** 1294 (1978).
715 Eberson, L., and Helgée, B., *Acta Chem. Scand., Ser. B*, **32,** 313 (1978).
716 Walling, C., and Camaioni, D. M., *J. Org. Chem.*, **43,** 3266 (1978).
717 *Org. Reaction Mech.*, **1976,** 148.
718 Clerici, A., Minisci, F., Ogawa, K., and Surzur, J.-M., *Tetrahedron Lett.*, **1978,** 1149.
719 *Org. Reaction Mech.*, **1976,** 150.
720 Ogibin, Yu. N., Troyanskii, E. I., Rakhamtullina, L. Kh., and Nikishin, G. I., *Ref. Dokl. Soobshch.-Mendeleevrk. S'ezd Obshch, Prikl. Khim., 11th*, **2,** 56 (1975); *Chem. Abs.*, **88,** 135902 (1978).
721 Radhakrishnamurti, P. S., and Swamy, B. R. K., *Indian J. Chem. Sect. A*, **15A,** 1115 (1977)
722 Nikishin, G. I., Kapustina, N. I., and Kaplan, E. P., *Bull. Acad. Sci. USSR*, **26,** 1420 (1977).
723 Khalifa, M. H., and Rieker, A., *J. Chem. Res.*, **1977,** (S) 316.
724 Schmidt. W., and Steckhan, E., *Angew. Chem. Int. Edn.*, **17,** 673 (1978).
725 Radhakrishnamurti, P. S., and Sahu, B., *Indian J. Chem., Sect. A*, **15A,** 700 (1977).
726 Guyon, J.-Y., Libert, M., and Caullet, C., *J. Chem. Res.*, **1978,** (S) 202.
727 Grinberg, V. A., Polishchuk, V. R., Mysov, E. I., German, L. S., Kanevskii, L. S., Tsodikov, V. V., and Vasil'ev, Yu. B., *Bull. Acad. Sci. USSR*, **26,** 2615 (1977).
728 Grinberg, V. A., Polishchuk, V. R., German, L. S., Kanevskii, L. S., and Vasil'ev, Yu. B., *Bull. Acad. Sci. USSR*, **26,** 580 (1977).
728 Utley, J. H. P., and Yates, G. B., *J. Chem. Soc., Perkin Trans* 2, **1978,** 395.
730 Adams, C., Jacobsen, N., and Utley, J. H. P., *J. Chem. Soc., Perkin Trans* 2, **1978,** 1071.
731 Singh, M. P., Singh, M. S., Arya, B. S., Singh, A. K., Tripathi, V., and Singh, A. K., *Indian J. Chem., Sect. A*, **15,** 716 (1977).
732 Ashworth, B., Gilbert, B. C., Holmes, R. G. G., and Norman, R. O. C., *J. Chem. Soc., Perkin Trans.* 2, **1978,** 951.
733 Caquis, G., and Deronzier, A., *J. Chem. Soc., Chem. Commun.*, **1978,** 809.
734 Nilsson, A., Palmquist, U., Pettersson, T., and Ronlán, A., *J. Chem. Soc., Perkin Trans.* 1, **1978,** 696.
735 Miller, L. L., Stewart, R. F., Gillespie, J. P., Ramachandran, V., So, Y. H., and Stermitz, F. R., *J. Org. Chem.*, **43,** 1580 (1978).
736 Dimroth, K., Tüncher, W., and Kaletsch, H., *Chem. Ber.*, **111,** 264 (1978).
737 Wendling, L. A., and West, R., *J. Org. Chem.*, **43,** 1573 (1978).
738 Roubaty, J.-L., Bréant, M., Lavergne, M., and Revillon, A., *Makromol. Chem.*, **179,** 1151 (1978).
739 Tsuruya, S., Kishikawa, Y., Tanaka, R., and Kuse, T., *J. Catal.*, **49,** 259 (1977); *Chem. Abs.*, **87,** 200495 (1977).
740 Hewgill, F. R., and Howie, G. B., *Aust. J. Chem.*, **31,** 907 (1978).
741 Hewgill, F. R., and Howie, G. B., *Aust. J. Chem.*, **31,** 1061 (1978).

[742] Cf. *Org. Reaction Mech.*, **1977**, 126
[743] Bird, C. W., and Chauhan, Y. S., *Tetrahedron Lett.*, **1978**, 2133.
[744] Hewgill, F. R., and Howie, G. B., *Aust. J. Chem.*, **31**, 1069 (1978).
[745] Benjamin, B. M., Raaen, V. F., Hagaman, E. W., and Brown, L. L., *J. Org. Chem.*, **43**, 2986 (1978).
[746] Volod'kin, A. A., Ershov, V. V., and Kudinova, L. I., *Bull. Acad. Sci. USSR*, **26**, 448 (1977).
[747] Kowal, J., and Waligóra, B., *Makromol. Chem.*, **179**, 707 (1978).
[748] Srivastava, S. P., and Dua, V. K., *Indian J. Chem., Sect. B.*, **15**, 966 (1977).
[749] *Org. Reaction Mech.*, **1977**, 126.
[750] Iguchi, M., Nishiyama, A., Terada, Y., and Yamamura, S., *Chem. Lett.*, **1978**, 451.
[751] Iguchi, M., Nishiyama, A., Hara, M., Terada, Y., and Yamamura, S., *Chem. Lett.*, **1978**, 1015.
[752] Bartlelink, H. J. M., Beulen, J., Bielders, G., Cremers, L., Duynstee, E. F. L., and Konijnenberg, E., *Chem. Ind. (London)*, **1978**, 586.
[753] McDonald, E., and Susksamrarn, A., *J. Chem. Soc., Perkin Trans.* 1, **1978**, 440.
[754] Kupchan, S. M., Dhingra, O. P., and Kim, C.-K., *J. Org. Chem.*, **43**, 4076 (1978).
[755] Whiting, D. A., and Wood, A. F., *Tetrahedron Lett.*, **1978**, 2335.
[756] Bauer, R., and Wendt, H., *Angew. Chem. Int. Edn.*, **17**, 202 (1978).
[757] Galliani, G., Rindone, B., and Marchesini, A., *J. Chem. Soc., Perkin Trans.* 1, **1978**, 456.
[758] Sturm, E., Kiesele, H., and Daltrozzo, E., *Chem. Ber.*, **111**, 227 (1978).
[759] Cheek, G. T., and Nelson, R. F., *J. Org. Chem.*, **43**, 1230 (1978).
[760] Nomura, Y., Ogawa, K., Takeuchi, Y., and Tomoda, S., *Chem. Lett.*, **1978**, 271.
[761] Ram, N., and Sidhu, K. S., *Indian J. Chem. Sect. A*, **16**, 195 (1978).
[762] Sarukhanyan, E. R., and Beileryan, N. M., *Arm. Khim. Zh.*, **31**, 477 (1978); *Chem. Abs.*, **89**, 162910 (1978).
[763] Rendell, R. W., and Wright, B., *Tetrahedron*, **34**, 197 (1978).
[764] Pelizetti, E., Mentasti, E., and Barni, E., *J. Chem. Soc., Perkin Trans.* 2, **1978**, 623.
[765] Shah, J. N., and Shah, C. K., *J. Org. Chem.*, **43**, 1286 (1978).
[766] Sharma, T. C., Saksena, V., and Reddy, N. J., *Acta Chim. Acad. Sci. Hung.*, **93**, 415 (1977); *Chem. Abs.*, **88**, 104386 (1978).
[767] Bucsis, L., and Kiesele, H., *Acta Chim. Acad. Sci. Hung.*, **93**, 141 (1977); *Chem. Abs.*, **88**, 88806 (1978).
[768] Farnia, G., Sandona, G., and Vianello, E., *J. Electroanal. Chem. Interfacial Electrochem.*, **88**, 147 (1978); *Chem. Abs.*, **88**, 151734 (1978).
[768] Almemark, M., and Åkermark, B., *J. Chem. Soc., Chem. Commun.*, **1978**, 66.
[770] Buxton, G. V., and Green, J. C., *J. Chem. Soc., Faraday Trans.* 1, **74**, 697 (1978).
[771] Berdnikov, V. M., Zhuravleva, O. S., and Terent'eva, L. A., *Izv. Akad. Nauk SSSR, Ser. Khim.*, **1977**, 2214; *Chem. Abs.*, **88**, 36953 (1978).
[772] Papaconstantinou, E., *Z. Phys. Chem. (Wiesbaden)*, **106**, 283 (1977).
[773] Barlet, R., *J. Org. Chem.*, **43**, 3500 (1978).
[774] Carroll, W. T., and Peters, D. G., *Tetrahedron Lett.*, **1978**, 3543.
[775] Mejer, S., and Marcinow, Z., *Rocz. Chem.*, **51**, 1281 (1977); *Chem. Abs.*, **88**, 73856 (1978).
[776] Mandell, L., Johnston, J. C., and Day, R. A., *J. Org. Chem.*, **43**, 1616 (1978).
[777] van Tilborg, W. J. M., and Smit, C. J., *Recl. Trav. Chim. Pays-Bas*, **100**, 89 (1978).
[778] Le Guillanton, G., and Lamant, M., *Nouveau J. de Chimie*, **2**, 157 (1978).
[779] Diaz, A., Parra, M., Banuelos, R., and Contreras, R., *J. Org. Chem.*, **43**, 4461 (1978).
[780] El Murr, N., Riveccie, M., and Dixneuf, P., *J. Chem. Soc., Chem. Commun.*, **1978**, 552.
[781] Honzl, J., Doskočilová, D., Pokorný, S., and Uchytil, B., *Collect. Czech. Chem. Commun.*, **47**, 1895 (1978).
[782] Farnia, G., Severin, M. G., Capobianco, G., and Vianello, E., *J. Chem. Soc., Perkin Trans.* 2, **1978**, 1.
[783] Screttas, C. G., and Micha-Screttas, M., *J. Org. Chem.*, **43**, 1064 (1978).
[784] Quaal, K. S., Ji, S., Kim, Y. M., Closson, W. D., and Zubieta, J. A., *J. Org. Chem.*, 43, 1311 (1978).
[785] Milaeva, E. R., Panov, V. B., and Okhlobystin, O. Yu., *Teor. Eksp. Khim.*, **13**, 543 (1977); *Chem. Abs.*, **87**, 200498 (1977).
[786] Pragst, F., and Seydewitz, *J. Prakt. Chem.*, **319**, 952 (1977).
[787] Smith, J. G., and Chun, Y.-I., *Tetrahedron Lett.*, **1978**, 413.
[788] Mabon, G., and Le Guillanton, G., *C. R. Hebd. Séances Acad. Sci., Sér C*, **286**, 245 (1978),
[789] Kerr, J. B., and Iversen, I. E., *Acta Chem. Scand., Ser. B*, **32**, 405 (1978).
[790] Mazaleyrat, J.-P., *Can. J. Chem.*, **56**, 2731 (1978).

[791] Kuwajima, I., Minami, N., Abe, T., and Sato, T., *Bull. Chem. Soc. Jpn*, **51,** 2391 (1978).
[792] Simonet, J., and Lund, H., *Acta Chem. Scand., Ser. B,* **31,** 909 (1977).
[793] McDonald, R. N., Borhani, K. J., and Hawley, M. D., *J. Am. Chem. Soc.*, **100,** 995 (1978).
[794] Koshechki, V. G., and Atamanyuk, V. Yu., *Tezisy Dokl.-Resp. Konf. Molodykh Uch.-Khim., 2nd,* **1,** 87 (1977); *Chem. Abs.*, **89,** 89891 (1978).
[795] Beileryan, N. M., *Arm. Khim. Zh,,* **30,** 540 (1977); *Chem. Abs.*, **88,** 36905 (1978).
[796] Sorokin, A. Yu., Shevelev, Yu. A., and Domrachev, G. A., *Bull. Acad. Sci. USSR,* **27,** 618 (1978).
[797] Nojima, M., Takagi, M., Morinaga, M., Nagao, G., and Tokura, N., *J. Chem. Soc., Perkin Trans.* 1, **1978,** 488.
[798] Ballester, M., Riera, J., Castañer, J., and Easulleras, M., *Tetrahedron Lett.*, **1978,** 643.
[788] Draper, M. R., and Ridd, J. H., *J. Chem. Soc., Chem. Commun.*, **1978,** 445.
[800] Sato, T., and Torizuka, K., *J. Chem. Soc., Perkin Trans.* 2, **1978,** 1199.
[801] Tanaskov, M. M., and Stadnichuk, M. D., *Zh. Obshch. Khim.*, **48,** 1140 (1978); *Chem. Abs.*, **89,** 41911 (1978).
[802] Mariano, P. S., Stavinoha, J. L., Pèpe, G., and Meyer, E. F., *J. Am. Chem. Soc.*, **100,** 7114 (1978).
[803] Haynes, R. K., *Aust. J. Chem.*, **31,** 131 (1978).
[804] Haynes, R. K., *Aust. J. Chem.*, **31,** 121 (1978).
[805] Haynes, R. K., Probert, M. K. S., and Wilmot, I. D., *Aust. J. Chem.*, **31,** 1737 (1978).
[806] Ueno, Y., Nakayama, A., and Okawara, M., *J. Chem. Soc., Chem. Commun.*, **1978,** 74.
[807] Todres, Z. V., *Usp. Khim.*, **47,** 260 (1978); *Chem. Abs.*, **88,** 151599 (1978).
[808] Bunnett, J. F., *Acc. Chem. Res.*, **11,** 413 (1978).
[809] Rajan, S., and Muralimohan, K., *Tetrahedron Lett.*, **1978,** 483.
[810] Pierini, A. B., and Rossi, R. A., *J. Organometal. Chem.*, **144,** C12 (1978).
[811] Bunnett, J. F., and Traber, R. P., *J. Org. Chem.*, **43,** 1867 (1978).
[812] Bunnett, J. F., and Shafer, S. J., *J. Org. Chem.*, **43,** 1873 (1978).
[813] Bunnett, J. F., and Shafer, S. J., *J. Org. Chem.*, **43,** 1877 (1978).
[814] Wolfe, J. F., Moon, M. P., Sleevi, M. C., Bunnett, J. F., and Bard, R. R., *J. Org. Chem.*, **43,** 1028 (1978).
[815] Rossi, R. A., Pierini, A., and de Rossi, R. H., *J. Org. Chem.*, **43,** 1276 (1978).
[816] Pinson, J., and Saveant, J.-M., *J. Am. Chem. Soc.*, **100,** 1506 (1978).
[817] Kornblum, N., Carlson, S. C., Widmer, J., Fifolt, M. J., Newton, B. N., and Smith, R. G., *J. Org. Chem.*, **43,** 1394 (1978).
[818] Kornblum, N., and Widmer, J., *J. Am. Chem. Soc.*, **100,** 7086 (1978).
[819] Kornblum, N., Carlson, S. C., and Smith, R. G., *J. Am. Chem. Soc.*, **100,** 289 (1978).
[820] Cimanele, F., Bruno, G., Testaferri, L., Tiecco, M., and Martelli, G., *J. Org. Chem.*, **43,** 4509 (1978).
[821] Bangerter, B. W., Beatty, R. P., Kouba, J. K., and Wreford, S. S., *J. Org. Chem.*, **42,** 3247 (1977).
[822] Flesia, E., Crozet, M. P., Surzur, J.-M., Jauffred, R., and Ghighione, C., *Tetrahedron*, **34,** 1699 (1978).
[823] Chern, C. I., DiCosimo, R., De Jesus, R., and San Filippo, J., *J. Am. Chem. Soc.*, **100,** 7317 (1978).
[824] Goh, S. H., and Kam, T. S., *J. Chem. Soc., Perkin Trans.* 2, **1978,** 648.
[825] Abe, J., and Ikegami, Y., *Bull. Chem. Soc. Jpn.*, **51,** 196 (1978).
[826] Bethell, D., and Bird, R., *J. Chem. Soc., Perkin Trans.* 2, **1977,** 1856.
[827] Ono, N., Tamura, R., Hayami, J., and Kaji, A., *Tetrahedron Lett.*, **1978,** 763.
[828] Simig, Gy., Lempert, K., Váli, Z., Tóth, G., and Tamás, J., *Tetrahedron*, **34,** 2371 (1978).
[828] Lubinkowski, J. J., Gomez, M., Calderon, J. L., and McEwen, W. E., *J. Org. Chem.*, **43,** 2432 (1978).
[830] Bank, S., and Bank, J. F., in ref. 3, p. 343.
[831] Symons, M. C. R., *J. Chem. Res.*, **1978** (S) 360.
[832] Mazaleyrat, J.-P., and Welvart, Z., *C. R., Hebd. Séances Acad. Sci.. Sér. C,* **287,** 379 (1978)
[833] Malissard, M., Mazaleyrat, J.-P., and Welvart, Z., *J. Am. Chem. Soc.*, **99,** 6933 (1977).
[834] Degrand, C., Jacquin, D., and Compagnon, P.-L., *J. Chem. Res.*, **1978,** (S) 246.
[835] Scott, L. T., Carlin, K. J., and Schultz, T. H., *Tetrahedron Lett.*, **1978,** 4637.
[836] Tochtermann, W., and Kirrstetter, R. G. H., *Chem. Ber.*, **111,** 1228 (1978).
[837] Arnold, R. T., and Kulenovic, S. T., *J. Org. Chem.*, **43,** 3687 (1978).
[838] Russell, G. A., Malatesta, V., Lawson, D. E., and Steg, R., *J. Org. Chem.*, **43,** 2242 (1978).

[839] Okhlobystina, L. V., and Tyurikov, V. A., *Izv. Akad. Nauk SSSR, Ser. Khim.*, **1977,** 2035; *Chem. Abs.*, **88,** 21694 (1978).
[840] Nadezhdin, A. D., *Bull. Acad. Sci. USSR*, **26,** 932 (1977).
[841] Meyers, C. Y., and Kolb, V. M., *J. Org. Chem.*, **43,** 1985 (1978).
[842] Kobayashi, S., and Ando, W., *Chem. Lett.*, **1978,** 1159.
[843] Tordo, P., Boyer, M., Vila, F., and Pujol, L., *Phosphorus Sulphur*, **3,** 43 (1977); *Chem. Abs.*, **89,** 59438 (1978).
[844] Yano, M., Hayashi, T., and Namiki, M., *Agric. Biol. Chem.*, **42,** 809 (1978); *Chem. Abs.*, **89,** 107365 (1978).
[845] Löliger, H., and Braun, A. M., *Makromol. Chem.*, **179,** 1369 (1978).
[846] Kondratov, V. K., Zhdanov, V. S., and Popov, V. K., *Zh. Fiz. Khim.*, **51,** 2739 (1977); *Chem. Abs.*, **88,** 36804 (1978).
[847] Turovskis, I., Stradinš, J., Volke, J., Freimanis, J., and Glezer, V., *Collect. Czech. Chem. Commun.*, **43,** 909 (1978).
[848] Reimann, R., and Staab, H. A., *Angew. Chem. Int. Edn.*, **17,** 374 (1978).
[849] Vogler, E. A., Stein, R. L., and Hayes, J. M., *J. Am. Chem. Soc.*, **100,** 3163 (1978).
[850] Bodewitz, H. W. J. J., Schaart, B. J., van der Niet, J. D., Blomberg, C., Bickelhaupt, F., and den Hollander, J. A., *Tetrahedron*, **34,** 2523 (1978).
[851] Fostein, P., and Pommier, J.-C., *J. Organometal. Chem.*, **150,** 187 (1978).
[852] Grovenstein, E., Cottingham, A. B., and Gelbaum, L. T., *J. Org. Chem.*, **43,** 3332 (1978).
[853] Ljungquist, A., *J. Organometal. Chem.*, **159,** 1 (1978).
[854] Creary, X., *J. Am. Chem. Soc.*, **99,** 7632 (1977).
[855] San Filippo, J., Silbermann, J., and Fagan, P. J., *J. Am. Chem. Soc.*, **100,** 4834 (1978).
[856] Quintard, J.-P., Hauvette-Frey, S., and Pereyre, M., *Bull. Soc., Chim. Belg.*, **7,** 505 (1978).
[857] Ashby, E. C., and Wiesemann, T. L., *J. Am. Chem. Soc.*, **100,** 3101 (1978).
[858] Ashby, E. C., and Wiesemann, T. L., *J. Am. Chem. Soc.*, **100,** 189 (1978).
[859] Ashby, E. C., and Bowers, J. C., *J. Am. Chem. Soc.*, **99,** 8504 (1977).
[860] Felkin, H., and Meunier, B., *J. Organometal. Chem.*, **146,** 140, 169 (1978).
[861] Russell, G. A., and Lamson, D. W., *J. Organometal. Chem.*, **156,** 17 (1978).
[862] Gladyshev, E. N., Bayushkin, P. Ya., and Sokolov, V. S., *Bull. Acad. Sci. USSR*, **26,** 592 (1977).
[863] Chen, K. S., Smith, R. T., and Wan, J. K. S., *Can. J. Chem.*, **56,** 2503 (1978).
[864] Yoshida, J., Tamao, K., Kurita, A., and Kumada, M., *Tetrahedron Lett.*, **1978,** 1809.
[865] Yamagishi, A., Watanabe, F., and Masui, T., *J. Chem. Soc., Chem. Commun.*, **1978,** 361.
[866] Horn, K. A., Schmidt, S. P., Koo, J.-Y., and Schuster, G. B., *U.S. NTIS, AD Rep.*, **1978,** AD–AO53290; *Chem. Abs.*, **89,** 162687 (1978).
[867] Smith, J. P., and Schuster, G. B., *J. Am. Chem. Soc.*, **100,** 2564 (1978).
[868] Koo, J.-Y., and Schuster, G. B., *J. Am. Chem. Soc.*, **100,** 4496 (1978).
[869] Cf. *Org. Reaction Mech.*, **1977,** 138.
[870] Iwasaki, T., Sawada, T., Okuyama, M., and Kamada, H., *J. Phys. Chem.*, **82,** 371 (1978).
[871] Wyrzykowska, K., Grodowski, M., Weiss, K., and Latowski, T., *Photochem. Photobiol.*, **28,** 311 (1978).
[872] Latowski, T., and Zelent, B., *Rocz. Chem.*, **51,** 1883 (1977); *Chem. Abs.*, **88,** 104348 (1978).
[873] Latowski, T., and Zelent, B., *Rocz. Chem.*, **51,** 1709 (1977); *Chem. Abs.*, **88,** 49958 (1978).
[874] Miyamoto, T., Tsujimoto, Y., Tsuchinaga, T., Nishimura, Y., and Odairo, Y., *Tetrahedron Lett.*, **1978,** 2155.
[875] Maruyama, K., and Kubo, Y., *Chem. Lett.*, **1978,** 851.
[876] Lewis, F. D., and Ho, T.-I., *J. Am. Chem. Soc.*, **99,** 7991 (1977).
[877] Brown-Wensley, K. A., Mattes, S. A., and Farid, S., *J. Am. Chem. Soc.* **100,** 4162 (1978).
[878] Gebicki, J., Reimschüssel, W., and Nowicki, T., *Chem. Phys. Lett.*, **59,** 197 (1978).
[879] Maroulis, A. J., Shigemitsu, Y., and Arnold, D. R., *J. Am. Chem. Soc.*, **100,** 535 (1978).
[880] Neidigk, D. D., and Morrison, H., *J. Chem. Soc., Chem. Commun.*, **1978,** 600.
[881] Mizuno, K., Kaji, R., Okada, H., and Otsuji, Y., *J. Chem. Soc., Chem. Commun.*, **1978,** 594.
[882] Arnold, D. R., and Maroulis, A. J., *J. Am. Chem. Soc.*, **100,** 7355 (1977).
[883] Albini, A., and Arnold, D. R., *Can. J. Chem.*, **56,** 2985 (1978).
[884] Majima, T., Pac, C., Nakasona, A., and Sakurai, H., *J. Chem. Soc., Chem. Commun.*, **1978,** 490.
[885] Cf. *Org. Reaction Mech.*, **1977,** 139.
[886] Kitamura, T., Imagawa, T., and Kawanisi, M., *Tetrahedron*, **34,** 3261 (1978).

[887] Kitamura, T., Imagawa, T., and Kawanisi, M., *Tetrahedron.* **34,** 3451 (1978).
[888] Srinivasan, V. S., Podolski, D., Westrick, N. J., and Neckers, D. C., *J. Am. Chem. Soc.*, **100,** 6513 (1978).
[889] Jefford, C. W., and Boschung, A., *Helv. Chim. Acta*, **60,** 2673 (1977).
[890] Kemal, C., Chan, T. W., and Bruice, T. C., *J. Am. Chem. Soc.*, **99,** 7272 (1977).
[891] Chan, T. W., and Bruice, T. C., *J. Am. Chem. Soc.*, **99,** 7287 (1977).
[892] Chan. R. L., and Bruice, T. C., *J. Am. Chem. Soc.*, **100,** 7375 (1978).
[893] Von Eikeren, P., and Grier, D. L., *J. Am. Chem. Soc.*, **99,** 8057 (1977).
[894] Hood, R. A., Prince, R. H., and Rubinson, K. A., *J. Chem. Soc., Chem. Commun.*, **1978,** 300.
[895] Martens, F. M., Verhoeven, J. W., Gase, R. A., Pandit, U. K., and de Boer, Th. J., *Tetrahedron*, **34,** 443 (1978).
[896] Ioffe, N. T., Yuzefovich, L. Yu., and Mairanovskii, V. G., *Teor. Eksp. Khim.*, **14,** 378 (1978); *Chem. Abs.*, **89,** 89914 (1978).
[897] Eisner, U., Sadegahi, M. M., and Hambright, W. P., *Tetrahedron Lett.*, **1978,** 303.
[898] Moracci, F. M., Liberatore, F., Carelli, V., Arnone, A., Carelli, I., and Cardinali, M. E., *J. Org. Chem.*, **43,** 3420 (1978).
[899] Kosower, E. M., Hajdu, J., and Nagy, J. B., *J. Am. Chem. Soc.*, **100,** 1186 (1978).
[900] Vafina, A. A., Il'yasov, A. V., Berdnikov, E. A., Morozova, I. D., Tantasheva, F. R., and Morozov, V. I., *Izv. Akad. Nauk SSSR, Ser. Khim.*, **1977,** 2487; *Chem. Abs.*, **88,** 88660 (1978).
[901] Block, H., and Brähler, G., *Angew. Chem. Int. Edn.*, **16,** 855 (1977).
[902] Thiebault, A., Mathieu, C., and Oliva, P., *C. R. Hebd. Séances Acad. Sci., Sér. C*, **286,** 417 (1978).
[903] Devynck, J., and Ben Hadid, A., *C. R. Hebd. Séances Acad. Sci. Sér. C*, **286,** 389 (1978).
[904] Kelly, R. P., and Lindsay Smith, J. R., *J. Chem. Soc., Chem. Commun.*, **1978,** 329.
[905] Kitani, A., and Sasaki, K., *Denki Kagaku Oyobi Kogyo Butsuri Kagaku*, **45,** 484 (1977); *Chem. Abs.*, **88,** 49981 (1978).
[906] Fritz, H. P., Gebauer, H., Friedrich, P., Ecker, P., Artes, R., and Schubert, U., *Z. Naturforsch.* B, **33,** 498 (1978).
[907] Clark, D. T., Shuttleworth, D., Rodwell, W. R., and Guest, M. F., *Prog. Theor. Org. Chem.*, **2,** 415 (1977); *Chem. Abs.*, **88,** 151867 (1978).
[908] Bally, T., Haselbach, E., Lanyiova, Z., and Baertschi, P., *Helv. Chim. Acta*, **61,** 2488 (1978).
[909] Gara, W. B., and Roberts, B. P., *J. Chem. Soc., Perkin Trans.* 2, **1978,** 150.
[910] Symons, M. C. R., *J. Chem. Soc., Chem. Commun.*, **1978,** 686.
[911] Musker, W. K., Wolford, W. L., and Roush, P. B., *J. Am. Chem. Soc.*, **100,** 6416 (1978).
[912] Bernardi, F., Guerra, M., and Pedulli, G. F., *Tetrahedron*, **34,** 2141 (1978).
[913] Bock, H., and Kaim, W., *Chem. Ber.*, **111,** 3552 (1978).
[914] Bock, H., Kaim, W., and Rohwer, H. E., *Chem. Ber.*, **111,** 3573 (1978).
[915] Kaim, W., and Bock, H., *Chem. Ber.*, **111,** 3585 (1978).
[916] Dixon, W. T., and Murphy, D., *J. Chem. Soc., Faraday Trans.* 2, **73,** 1475 (1977).
[917] Sundholm, F., *Finn. Chem. Lett.*, **1977,** 200.
[918] Sundholm, F., and Sundholm, G., *Finn. Chem. Lett.*, **1978,** 216.
[919] Rieke, R. D., and White, C. K., *J. Org. Chem.*, **42,** 3759 (1977).
[920] Bock, H., Kaim, W., Wiberg, N., and Ziegleder, G., *Chem. Ber.*, **111,** 3150 (1978).
[921] Pearson, G. A., Rocek, M., and Walter, R. T., *J. Phys. Chem.*, **82,** 1185 (1978).
[922] Sullivan, P. D., Fong, J. Y., Williams, M. L., and Parker, V. D., *J. Phys. Chem.*, **82,** 1181 (1978).
[923] Bock, H., Kaim, W., Semkow, A., and Noth, H., *Angew. Chem. Int. Edn.*, **17,** 286 (1978).
[924] Neugebauer, F. A., and Weger, H., *J. Phys. Chem.*, **82,** 1152 (1978).
[925] Nelsen, R. F., Peacock, V. E., and Kessel, C. R., *J. Am. Chem. Soc.*, **100,** 7017 (1978).
[926] Nelsen, R. F., Peacock, V. E., Weisman, G. R., Landis, M. E., and Spencer, J. A., *J. Am. Chem. Soc.*, **100,** 2806 (1978).
[927] Nelsen, R. F., Haselbach, E., Gschwind, R., Klemm, U., and Lanyova, S., *J. Am. Chem. Soc.*, **100,** 4367 (1978).
[928] Clarke, D., Gilbert, B. C., Hanson, P., and Kirk, C. M., *J. Chem. Soc., Perkin Trans.* 2, **1978,** 1103.
[929] Kaim, W., Bock, H., and Nöth, H., *Chem. Ber.*, **111,** 3276 (1978).
[930] Jawdosiuk, M., and Golinski, J., *Wiad. Chem.*, **32,** 421 (1978); *Chem. Abs.*, **89,** 162586 (1978).

931 Blount, H. N., and Evans, J. T., *Proc. Symp. Spectrosc. Electrochem., Charact. Solute Species Nonaqueous Solvents*, **1978,** 105, Plenum, New York, N.Y.; *Chem. Abs.*, **89,** 59324 (1978).
932 Eberson, L., and Nyberg, K., *Acta Chem. Scand., Ser. B.*, **32,** 235 (1978).
933 Shine, H. J., Bandlish, B. K., and Stephenson, M. T., *Tetrahedron Lett.*, **1978,** 733.
934 Evans, J. F., and Blount, H. N., *J. Am. Chem. Soc.*, **100,** 4191 (1978).
935 Eberson, L., Blum, Z., Helgée, B., and Nyberg, K., *Tetrahedron* **34,** 731 (1978).
936 Cheng, H. Y., Sackett, P. H., and McCreery, R. L., *J. Am. Chem. Soc.*, **100,** 962 (1978).
937 Bandlish, B. K., Porter, W. R., and Shine, H. J., *J. Phys. Chem.*, **82,** 1168 (1978).
938 Steckhan, E., *J. Am. Chem. Soc.*, **100,** 3526 (1978).
938 Ivanov, V. F., and Grishina, A. D., *Bull. Acad. Sci. USSR*, **26,** 1735 (1977).
940 Eberson, L., Jönnsson, L., and Wistrand, L.-G., *Acta Chem. Scand., Ser. B*, **32,** 520 (1978).
941 Sehested, K., and Holcman, J., *J. Phys. Chem.*, **82,** 651 (1978).
942 Dixon, W. T., and Murphy, D., *J. Chem. Soc., Faraday Trans.* 2, **74,** 432 (1978).
943 Yip, R. W., Vidoczy, T., Synder, R. W., and Chow, Y. L., *J. Phys. Chem.* **82,** 1194 (1978).
944 Challis, B. C., and Outram, J. R., *J. Chem. Soc., Chem. Commun.*, **1978,** 707.
945 Matsumoto, K., Fujita, H., and Deguchi, Y., *J. Chem. Soc., Chem. Commun.*, **1978,** 817.
946 Jones, P. R., *Adv. Organometal. Chem.*, **15,** 273 (1977).
947 Sakurai, H., Kira, M., and Umino, H., *Chem. Lett.*, **1977,** 1265.
948 Lines, R., Jensen, B. S., and Parker, V. D., *Acta Chem. Scand., Ser. B*, **32,** 513 (1978).
949 Stegmann, H. B., Scheffler, K., and Schuler, P., *Angew. Chem. Int. Edn.*, **17,** 365 (1978).
950 Farnia, G., Capobianco, G., and Lunelli, B., *Chim. Ind. (Milan)*, **59,** 215 (1977); *Chem. Abs.*, **88,** 5915 (1978).
951 Russell, G. A., Malatesta, V., and Blankespoor, R. A., *J. Org. Chem.*, **43,** 1837 (1978).
952 Barzaghi, M., Beltrame, P. L., Gamba, A., and Simonetta, M., *J. Am. Chem. Soc.*, **100,** 251 (1978).
953 Kaim, W., and Bock, H., *J. Am. Chem. Soc.*, **100,** 6504 (1978).
954 Polenov, E. A., Minin, V. V., Sterlin, S. R., and Yagupol'skii, L. M., *Bull. Acad. Sci. USSR*, **26,** 419 (1977).
955 Polenov, E. A., and Sterlin, S. R., *Bull. Acad. Sci. USSR*, **26,** 609 (1977).
956 Elschenbroich, C., Gerson, F., Ohya-Nishiguchi, H., and Wydler, Ch., *Helv. Chim. Acta*, **60,** 2530 (1977).
957 Hinde, A. L., Poppinger, D., and Radom, L., *J. Am. Chem. Soc.*, **100,** 4681 (1978).
958 Kiesele, H., *Chem. Ber.*, **111,** 1908 (1978).
959 Anikolenko, V. A., Kazakova, V. M., and Mikhailov, A. I., *Bull. Acad. Sci. USSR*, **26,** 1061 (1977).
960 Szwarc, M., and Levin, G., *Chem. Phys. Lett.*, **52,** 587 (1977).
961 Sharma, K. K., *Indian J. Chem.*, **15A,** 371 (1977). *Chem. Abs.*, **87,** 167403 (1977).
962 Barzaghi, M., Gamba, A., Morosi, G., and Simonetta, M., *J. Phys. Chem.*, **82,** 2105 (1978).
963 Sosonkin, I. M., Strogov, G. N., Novikov, V. N., and Ponomareva, T. K., *Nauchn. Tr., Kuban. Gos. Univ.*, **256,** 65 (1977) *Chem. Abs.*, **89,** 41874 (1978).
964 Pedulli, G. F., Tiecco, M., Guerra, M., Martelli, G., and Zanirato, P., *J. Chem. Soc., Perkin Trans.* 2, **1978,** 212.
965 Tordo, P., Pouzard, G., Babadjamian, A., Dou, H. J.-M., and Metzger, J., *Nouveau J. de Chimie*, **1,** 493 (1977).
966 Lambelet, P., and Lucken, E. A. C., *J. Chem. Soc., Perkin Trans.* 2, **1978,** 617.
967 Konishi, S., and Reddoch, A. H., *J. Magn. Reson.*, **29,** 113 (1978).
968 Kispert, L. D., Reeves, R., and Chen. T. C. S., *J. Chem. Soc., Faraday Trans.* 2, **74,** 871 (1978).
969 Schmüser, W., and Voss, J., *J. Chem. Res.*, **1978,** (S) 149.
970 Fürderer, P., and Gerson, F., *J. Phys. Chem.*, **82,** 1125 (1978).
971 Fürderer, P., Gerson, F., Murata, I., and Nakazawa, T., *J. Phys. Chem.*, **82,** 1129 (1978).
972 Kazama, S., Sato, E., Kamiya, M., and Akahori, Y., *Yakugaku Zasshi*, **87,** 1279 (1977); *Chem. Abs.*, **88,** 88971 (1978).
973 Pasimeni, L., Brustolon, M., and Corvaja, C., *J. Chem. Soc., Perkin Trans.* 2, **1978,** 420.
974 Samokhvalova, A. I., Sokol, O. G., and Kazakova, V. M., *Zh. Strukt. Khim.*, **18,** 964 (1977); *Chem. Abs.*, **88,** 21931 (1978).
975 Brustolon, M., Pasimeni, L., and Corvaja, C., *J. Chem. Soc., Faraday Trans.* 2, **73,** 1729 (1977).
976 Russell, G. A., Wallraff, G., and Gerlock, J. L., *J. Phys. Chem.*, **82,** 1161 (1978).

[977] Barzaghi, M., Cremaschi, P., Gamba A., Morosi, G., Oliva, C., and Simonetta, M., *J. Am. Chem. Soc.*, **100,** 3132 (1978).
[978] Chen, K. S., and Wen, J. K. S., *J. Am. Chem. Soc.*, **100,** 6051 (1978).
[979] Hudson, R. L., and Williams, F., *J. Am. Chem. Soc.*, **99,** 7714 (1977).
[980] Wendling, L. A., and West, R., *J. Org. Chem.*, **43,** 1577 (1978).
[981] Il'yasov, A. V., Levin, Ya. A., Mukhtarov, A. Sh., Skorobogotova, M. S., and Barlev, A. A., *Zh. Strukt. Khim.*, **19,** 69 (1978); *Chem. Abs.*, **88,** 189618 (1978).
[982] Endo, T., and Okawara, M., *Chem. Lett.*, **1977,** 1487.
[983] Wright, R. E., and Vogler, A., *J. Organometal. Chem.*, **160,** 197 (1978).
[984] Levanon, H., Neta, P., and Trozzolo, A. M., *Chem. Phys. Lett.*, **54,** 181 (1978).
[985] Stroebel, G. G., Myers, D. Y., Grabbe, R. R., and Gardner, P. D., *J. Phys. Chem.*, **82,** 1121 (1978).
[986] Müllen, K., and Huber, W., *Helv. Chim. Acta.* **61,** 1310 (1978).
[987] McNeil, R. I., Shiotani, M., Williams, F., and Yim, M. B., *Chem. Phys. Lett.*, **51,** 438 (1977).
[988] Russell, G. A., Tanger, C. M., and Kosugi, Y., *J. Org. Chem.*, **43,** 3278 (1978).
[989] Volod'kin, A. A., Ershov, V. V., and Prokof'eva, T. I., *Bull. Acad. Sci. USSR*, **26,** 1108 (1977).
[990] Behrens, G., Koltzenburg, G., Ritter, A., and Schulte-Frohlinde, D., *Int. J. Radiat. Biol. Relat. Stud. Phy. Chem. Med.*, **33,** 163 (1978); *Chem. Abs.*, **89,** 107351 (1978).
[991] Jachimowicz, F., Levin, G., and Szwarc, M., *J. Am. Chem. Soc.*, **100,** 5426 (1978).
[992] Jachimowicz, F., Wang. H. C., Levin, G., and Szwarc, M., *J. Phys. Chem.*, **82,** 1371 (1978).
[993] Kapturkiewicz, A., and Kalinowski, M. K., *Rocz. Chem.*, **51,** 1483 (1977); *Chem. Abs.*, **88,** 36832 (1978).
[994] Addison, A. W., Dalal, N. S., Hoyano, Y., Huizinga, S., and Weiler, L., *Can. J. Chem.*, **55,** 4191 (1977).
[995] Zubritskii, L. M., Romanov, O. E., and Bal'yan, Kh. V., *Tezisy Dokl.-Vses. Konf. Khim, Atsetilena, 5th*, **1975,** 88; *Chem. Abs.*, **88,** 151735 (1978).
[996] Wang. H. C., Levin, G., and Szwarc, M., *J. Am. Chem. Soc.*, **100,** 6137 (1978).
[997] Keary, C. M., Gowenlock, B. G., and Pfab, J., *J. Chem. Soc.*, *Perkin Trans.* 2, **1978,** 242.
[998] Roellig, M. P., and Houston, P. L., *Chem. Phys. Lett.*, **57,** 75 (1978).
[999] Forrest, D., Gowenlock, B. G., and Pfab, J., *J. Chem. Soc.*, *Perkin Trans.* 2, **1978,** 12.
[1000] Thoi, H. H., Ito, O., Iino, M., and Matsuda, M., *J. Phys. Chem.*, **82,** 314 (1978).
[1001] Langler, R. F., Marini, Z. A., and Pincock, J. A., *Can. J. Chem.* **56,** 903 (1978).
[1002] Givens, R. S., and Matuszewski, B., *Tetrahedron Lett.*, **1978,** 861.
[1003] Givens, R. S., and Wylie, P. L., *Tetrahedron Lett.*, **1978,** 865.
[1004] Durst, T., Huang, J. C., Sharma, N. K., and Smith, D. J. H., *Can. J. Chem.*, **56,** 512 (1978).
[1005] Gilbert, B. C., Gill, B., and Sexton, M. D., *J. Chem. Soc., Chem. Commun.*, **1978,** 78.
[1006] Kropp, P. J., Gibson, J. R., Snyder, J. J., and Poindexter, G. S., *Tetrahedron Lett.*, **1978,** 207.
[1007] Pienta, N. J., and Kropp, P. J., *J. Am. Chem. Soc.*, **100,** 655 (1978).
[1008] Chittim, B., Safe, S., Bunce, N. J., Ruzo, L. O., Olie, K., and Hutzinger, O., *Can. J. Chem.*, **56,** 1253 (1978).
[1009] van Beek, H. C. A., and van der Stoep, H. J., *Recl. Trav. Chim. Pays-Bas*, **97,** 279 (1978).
[1010] Dneprovskii, A. S., Shkurov, V. A., and Temnikova, T. I., *Zh. Org. Khim.*, **13,** 2587 (1977).
[1011] Suginome, H., Umeda, H., Sugiura, S., and Masamune, T., *J. Chem. Res.*, **1978,** (S) 380; (M) 4523.
[1012] Takagi, K., and Ogata, Y., *J. Chem. Soc., Perkin Trans.* 2, **1977,** 1980.
[1013] Maruthamuthu, P., and Taniguchi, H., *J. Phys. Chem.*, **81,** 1944 (1977).
[1014] Sharp, K. G., and Sutor, P. A., *J. Am. Chem. Soc.*, **99,** 8046 (1977).
[1015] Davies, A. G., and Tse, M.-W., *J. Chem. Soc., Chem. Commun.*, **1978,** 353.
[1016] Maillard, Ph., Massot, J. C., and Giannotti, C., *J. Organometal. Chem.*, **159,** 219 (1978).
[1017] Bamford, C. H., Puddephatt, R. J., and Slater, D. M., *J. Organometal. Chem.*, **159,** C31 (1978).
[1018] Cardy, H., Poquet, E., Chaillet, M., and Dargelos, A., *Nouveau J. de Chimie* **2,** 603 (1978). (1978).
[1019] Padwa, A., Chiacchio, V., and Hatanaka, N., *J. Am. Chem. Soc.*, **100,** 3928 (1978).
[1020] Tabushi, I., Kojo, S., and Fukunishi, K., *J. Org. Chem.*, **43,** 2370 (1978).
[1021] Paulson, D. R., Murray, A. S., and Fornoret, E. J., *J. Org. Chem.*, **43,** 2010 (1978).
[1022] Cossy, J., and Pete, J.-P., *Tetrahedron Lett.*, **1978,** 4941.

1023 Dekker, E. J. J., Engberts, J. B. F. N., and de Boer, Th. J., *Recl. Trav. Chim. Pays-Bas.*, **97,** 39 (1978).
1024 Wolfe, J. F., Moon, M. P., Sleevi, M. C., Bunnett, J. F., and Bard, R. R., *J. Org. Chem.*, **43,** 1019 (1978).
1025 Kasperski, H., Hellebrand, J., Roth, H.-K., and Rätzsch, M., *Z. Phys. Chem.* (*Leipzig*), **259,** 167 (1978).
1026 Zott, H., and Heusinger, H., *Radiat. Phys. Chem.*, **10,** 297 (1977); *Chem. Abs.*, **89,** 41931 (1978).
1027 Takuwa, A., *Bull. Chem. Soc. Jpn.*, **50,** 2973 (1977).
1028 Sherwell, J., and Tedder, J. M., *J. Chem. Soc.*, *Perkin Trans.* 2, **1978,** 1076.
1029 Parlar, H., Mansour, M., and Gäb, S., *Tetrahedron Lett.*, **1978,** 1597.
1030 Lapouyade, R., Koussini, R., and Bouas-Laurent, H., *J. Am. Chem. Soc.*, **99,** 7374 (1977).
1031 op het Veld, P. H. G., and Laarhoven, W. H., *J. Am. Chem. Soc.*, **99,** 7221 (1977).
1032 Zimmerman, H. E., and Steinmetz, M. G., *J. Chem. Soc.*, *Chem. Commun.*, **1978,** 230.
1033 Dewar, M. J. S., and Doubleday, C., *J. Am. Chem. Soc.*, **100,** 4935 (1978).
1034 Clark, J. H., Moore, C. B., and Nogar, N. S., *J. Chem. Phys.*, **68,** 1264 (1978); *Chem. Abs.*, **88,** 189603 (1978).
1035 Lewis, R. S., and Lee, E. K. C., *J. Phys. Chem.*, **82,** 249 (1978).
1036 Encina, M. V., and Lissi, E. A., *J. Photochem.*, **8,** 131 (1978); *Chem. Abs.*, **88,** 104369 (1978).
1037 Koskikallio, J., *Finn. Chem. Lett.*, **1977,** 231.
1038 Wagner, P. J., and Sedon, J. H., *Tetrahedron Lett.*, **1978,** 1927.
1039 Tanida, H., *Yuki Gosei Kagaku Kyokaishi*, **35,** 439 (1977); *Chem. Abs.*, **87,** 166928 (1977).
1040 Kagramanov, N. D., Mal'tsev, A. K., and Nefedov, O. M., *Izv. Akad. Nauk SSSR, Ser. Khim.*, **1977,** 1835; *Chem. Abs.*, **87,** 167173 (1977).
1041 Kraeutler, B., Jaeger, C. D., and Bard, A. J., *J. Am. Chem. Soc.*, **100,** 4903 (1978).
1042 Kaiser, T., Grossi, L., and Fischer, H., *Helv. Chim. Acta*, **61,** 223 (1978).
1043 Schmidt, S. P., and Schuster, G. B., *J. Org. Chem.*, **43,** 1823 (1978).
1044 Applequist, D. E., and Klug, J. H., *J. Org. Chem.*, **43,** 1729 (1978).
1045 Aoyama, H., Hasegawa, T., Watabe, M., Shiraishi, H., and Omote, Y., *J. Org. Chem.*, **43,** 419 (1978).
1046 Shibata, K., Fadaka, H., Matsui, M., and Takase, Y., *J. Chem. Soc. Jap. Chem. Ind. Chem.*, **1978,** 75.
1047 Burns, J. M., Ashley, M. E., Crockett, G. C., and Koch, T. H., *J. Am. Chem. Soc.*, **99,** 6924 (1977).
1048 Yates, P., and Toong, Y. C., *J. Chem. Soc.*, *Chem. Commun.*, **1978,** 205.
1049 Nowada, K., Hisaoka, M., Sakuragi, H., Tokumaru, K., and Yoshida, M., *Tetrahedron Lett.*, **1978,** 137.
1050 Maruyama, K., and Tanioka, S., *J. Org. Chem.*, **43,** 310 (1978).
1051 Maruyama, K., and Arakawa, S., *J. Org. Chem.*, **42,** 3793 (1977).
1052 Arakawa, S., *J. Org. Chem.*, **42,** 3800 (1977).
1053 Martins, F. J. C., Dekker, J., Viljoen, A. M., and Venter, D. P., *S. Afr. J. Chem.*, **30,** 89 (1977); *Chem. Abs.*, **88,** 5871 (1978).
1054 Paraskevopoulos, G., and Cvetanovic, R. J., *J. Phys. Chem.*, **81,** 2598 (1977).
1055 Akimoto, H., Hoshino, M., Inoue, G., Okuda, M., and Washida, N., *Bull. Chem. Soc. Jpn.*, **51,** 2496 (1978).
1056 Kurylo, M. J., *Chem. Phys. Lett.*, **58,** 233 (1978).
1057 Ogata, Y., Takagi, K., and Suzuki, T., *J. Chem. Soc.*, *Perkin Trans.* 2, **1978,** 562.
1058 Bothe, E., Schulte-Frohlinde, D., and von Sonntag, C., *J. Chem. Soc.*, *Perkin Trans.* 2, **1978,** 416.
1059 Cetinkaya, E., Schuchmann, H.-P., and von Sonntag, C., *J. Chem. Soc.*, *Perkin Trans.* 2, **1978,** 985.
1060 Mazzocchi, P. H., Bowen, M. J., and Narain, N. K., *J. Am. Chem. Soc.*, **99,** 7063 (1977).
1061 Haynes, R. K., Probert, M. K. S., and Wilmot, I. D., *Aust. J. Chem.*, **31,** 1737 (1978).
1062 Sevilla, M. D., and D'Arcy, J. B., *J. Phys. Chem.*, **82,** 338 (1978).
1063 Fujisawa, T., and Kojima, T., *Bull. Chem. Soc. Jpn.*, **50,** 3061 (1977).
1064 Cristol, S. J., and Micheli, R. P., *J. Am. Chem. Soc.*, **100,** 850 (1978).
1065 Yano, T., *Int. J. Chem. Kinet.*, **9,** 725 (1977).
1066 Tsang, W., *Int. J. Chem. Kinet.*, **10,** 599 (1978).
1067 Perrin, D., Richard, C., and Martin R., *Int. J. Chem. Kinet.*, **10,** 529 (1978).

[1068] Fujii, N., and Asaba, T., *J. Fac. Eng., Univ. Tokyo, Ser. B*, **34,** 189 (1977); *Chem. Abs.*, **87,** 167164 (1977).

[1069] Ebert, K. H., Ederer, H. J., and Schmidt., P., *Am. Chem. Soc., Symp. Series*, No. 65, "Chem. React. Eng.—Houston", 313; *Chem. Abs.*, **88,** 104438 (1978).

[1070] Ishiwatari, M., Yahagi, N., and Tominaga, H., *Sekiyu Gakkai Shi*, **20,** 237 (1977); *Chem. Abs.*, **87,** 167272 (1977).

[1071] Richard, C., Scacchi, G., and Back, M. H., *Int. J. Chem., Kinet.*, **10,** 307 (1978).

[1072] McKay, G., and Turner, J. M. C., *Int. J. Chem. Kinet.*, **10,** 89 (1978).

[1073] King, K. D., and Goddard, R. D., *J. Phys. Chem.*, **82,** 1675 (1978).

[1074] King, K. D., and Goddard, R. D., *Int. J. Chem. Kinet.*, **10,** 453 (1978).

[1075] Held, A. M., Manthorne, K. C., Pacey, P. D., and Reinholdt, H. P., *Can. J. Chem.*, **55,** 4128 (1977).

[1076] Seres, I., Labadi, I., and Huhn, P., *Magy. Kem. Foly.*, **83,** 151 (1977); *Chem. Abs.*, **88,** 36997 (1978).

[1077] Seres, I., and Huhn, P., *Magy, Kem. Foly.*, **83,** 158 (1977); *Chem. Abs.*, **88,** 36998 (1978).

[1078] Grebennikov, V. N., and Nazin, G. M., *Izv. Akad. Nauk SSSR, Ser. Khim.*, **1978,** 1635; *Chem. Abs.*, **89,** 146207 (1978).

[1079] Tsang, W., *Int. J. Chem. Kinet.*, **10,** 41 (1978).

[1080] Bigley, D. B., and Weatherhead, R. H., *Can. J. Chem.*, **56,** 628 (1978).

[1081] Badr, M. Z. A., Aly, M. M., and Salem, S. S., *Tetrahedron*, **33,** 3155 (1977).

[1082] Ruud, J. M., Weustink, R. L., and Bickelhaupt, F., *Justus Liebigs Ann. Chem.*, **1978,** 214.

[1083] Jackson, R. A., and O'Neill, D. W., *J. Chem. Soc., Perkin Trans. 2*, **1978,** 509.

[1084] Neudorfl, P. S., and Strausz, O. P., *J. Phys. Chem.*, **82,** 241 (1978).

[1085] Taylor, J. E., and Milazzo, T. S., *J. Phys. Chem.*, **82,** 847 (1978).

[1086] Baxendale, J. H., and Rodgers, N. A. J., *Chem. Soc., Rev.*, **7,** 235 (1978).

[1087] Swallow, A. J., *Progr. Reaction Kinetics*, **9,** 197 (1978).

[1088] Gold, V., and McAdam, M. E., *Acc. Chem. Res.*, **11,** 36 (1978).

[1089] Krilov, D., Velenik, A., and Herak, J. N., *J. Chem. Phys.*, **69,** 2420 (1978).

[1090] Faraggi, M., Klapper, M. H., and Dorfman, L. M., *J. Phys. Chem.*, **82,** 508 (1978).

[1091] Mićić, O. I., Nenadović, M. T., and Carapellucci, P. A., *J. Am. Chem. Soc.*, **100,** 2209 (1978).

[1092] Sagstuen, E., *J. Chem. Phys.*, **69,** 3206 (1978).

[1093] Haindl, E., and Hüttermann, J., *J. Magn. Reson.*, **30,** 13 (1978).

[1094] Riederer, H., Hüttermann, J., and Symons, M. C. R., *J. Chem. Soc., Chem. Commun.*, **1978,** 313.

[1095] Al-Yamoor, K. Y., Garner, A., Ali, K. M. I., and Scholes, G., *Proc. Tihany Symp. Radiat. Chem.* 1976 (pub. 1977), **4,** 845; *Chem. Abs.*, **88,** 104356 (1978).

[1096] von Sonntag, C., and Schuchmann, M. N., *J. Chem. Soc., Perkin Trans. 2*, **1977,** 1958.

[1097] von Sonntag, C., and Dizdaroglu, M., *Carbohydr. Res.*, **58,** 21 (1977).

[1098] Bradbury, A. G. W., and von Sonntag, C., *Carbohydr. Res.*, **62,** 223 (1978).

[1099] Box, H. C., Budzinski, E. E., and Potienko, G., *J. Chem. Phys.*, **69,** 1966 (1978).

[1100] Symons, M. C. R., *J. Chem. Res.*, **1978,** (S) 288; (M) 3565.

[1101] Baugh, P. J., Goodall, J. I., and Bardsley, J., *J. Chem. Soc., Perkin Trans. 2*, **1978,** 700.

[1102] Nelson, D. J., Petersen, R. L., and Symons, M. C. R., *J. Chem. Soc., Perkin Trans. 2*, **1977,** 2005.

[1103] Petersen, R. L., Nelson, D. J., and Symons, M. C. R., *J. Chem. Soc., Perkin Trans. 2*, **1978,** 225.

[1104] Neta, P., Maruthamuthu, P., Carton, P. M., and Fessenden, R. W., *J. Phys. Chem.*, **82,** 1875 (1978).

[1105] Lind, J., and Eriksen, T. E., *Radiat. Phys. Chem.*, **10,** 349 (1977); *Chem. Abs.*, **88,** 189614 (1978).

[1106] Lati, J., and Meyerstein, D., *J. Chem. Soc., Dalton Trans.*, **1978,** 1105.

[1107] Henglein, A., and Proske, Th., *J. Am. Chem. Soc.*, **100,** 3706 (1978).

[1108] Bakalik, D. P., and Thomas, J. K., *J. Phys. Chem.*, **81,** 1905 (1977).

[1109] Ryan, T. G., Sambrook, T. E. M., and Freeman, G. R., *J. Phys. Chem.*, **82,** 26 (1978).

[1110] Sawai, T., Ohara, N., and Shimokawa, T., *Bull. Chem. Soc. Jpn.*, **51,** 1300 (1978).

[1111] Saunders, B. B., *J. Phys. Chem.*, **82,** 151 (1978).

[1112] Holcman, J., and Sehested, K., *J. Phys. Chem.*, **81,** 1963 (1977).

[1113] van der Linde, H. J., *Radiat. Phys. Chem.*, **10,** 199 (1977); *Chem. Abs.*, **88,** 36894 (1978).

[1114] Ayscough, P. B., Elliot, A. J., and Salmon, G. A., *J. Chem. Soc., Faraday Trans.*, 1, **74,** 511 (1978).
[1115] Johnson, D. W., and Salmon, G. A., *J. Chem., Soc., Faraday Trans.* 1, **74,** 964 (1978).
[1116] Zevos, N., and Sehested, K., *J. Phys. Chem.*, **82,** 138 (1978).
[1117] Farrington, J. A., Ebert, M., and Land, E. J., *J. Chem., Soc., Faraday Trans.* 1, **74,** 665 (1978).
[1118] Robinson, E. A., and Salmon, G. A., *J. Phys. Chem.*, **82,** 382 (1978).
[1119] Neta, P., and Levanon, H., *J. Phys. Chem.*, **81,** 2288 (1977).
[1120] Brede, O., Bös, J., Helmstreit, W., and Mehnert, R., *Z. Chem.*, **17,** 447 (1977).
[1121] Madhavan, V., Schuler, R. H., and Fessenden, R. W., *J. Am. Chem. Soc.*, **100,** 888 (1978).
[1122] Goto, S., Hori, A., Takamuku, S., and Sakurai, H., *Bull. Soc. Chem. Jpn.*, **51,** 1569 (1978).
[1123] Kaptein, R., *NATO Adv. Study Inst. Ser., Ser. C* 1977, **C34** (*Chem. Induced Magn. Polariz.*), 257; *Chem. Abs.*, **88,** 104195 (1978).
[1124] Lawler, R. G., *NATO Adv. Study Inst. Ser., Ser. C* 1977, **C34** (*Chem. Induced Magn. Polariz.*), 275; *Chem. Abs.*, **88,** 104197 (1978).
[1125] Dushkin, A. V., Grishin, Yu. A., and Sagdeev, R. Z., *Chem. Phys. Lett.*, **55,** 174 (1978).
[1126] Closs, G. L., *NATO Adv. Study Inst. Ser., Ser. C* 1977, **C34,** (*Chem. Induced Magn. Polariz*)., 225; *Chem. Abs.*, **88,** 120433 (1978).
[1127] Fessenden, R. W., *NATO Adv. Study Inst. Ser., Ser. C* 1977, **C34** (*Chem. Induced Magn. Polariz.*), 119; *Chem. Abs.*, **88,** 104193 (1978).
[1128] Trifunac, A. D., and Nelson, D. J., *J. Am. Chem. Soc.*, **100,** 5244 (1978).
[1129] Trifunac, A. D., Nelson, D. J., and Mottley, C., *J. Magn. Reson.*, **30,** 263 (1978).
[1130] Borer, A., Kirchmayr, R., and Rist, G., *Helv. Chim. Acta*, **61,** 305 (1978).
[1131] de Kanter, F. J. J., and Kaptein, R., *Chem. Phys. Lett.*, **58,** 340 (1978).
[1132] Vermeersch, G., Marko, J., Febvay-Garot, N., Caplain, S., Couture, A., and Lablache-Combier, A., *Tetrahedron*, **34,** 2453 (1978).
[1133] Elliot, A. J., and Wan., J. K. S., *Can. J. Chem.*, **56,** 2499 (1978).
[1134] Vyas, H. M., and Wan, J. K. S., *Mol. Phys.*, **34,** 887 (1977); *Chem. Abs.*, **88,** 120325 (1978).
[1135] Closs, G. L., and Miller, R. J., *J. Am. Chem. Soc.*, **100,** 3483 (1978).
[1136] Elliot, A. J., and Wan, J. K. S., *J. Phys. Chem.*, **82,** 444 (1978).
[1137] Wong, S. K., *J. Am. Chem. Soc.*, **100,** 5488 (1978).
[1138] Foster, T., Elliot, A. J., Adeleke, B. B., and Wan, J. K. S., *Can. J. Chem.*, **56,** 869 (1978).
[1139] Muszkat, K. A., *J. Chem. Soc., Chem. Commun.*, **1977,** 872.
[1140] Schwarz, W., Hesse, P., and Doerr, F., *Ber. Bunsenges. Phys. Chem.*, **81,** 1231 (1977); *Chem. Abs.*, **88,** 120294 (1978).
[1141] Bargon, J., *J. Am. Chem. Soc.*, **99,** 8350 (1977).
[1142] Schilling, M. L. M., Hutton, R. S., and Roth, H. D., *J. Am. Chem. Soc.*, **99,** 7792 (1977).
[1143] Schilling, M. L. M., and Roth, H. D., *J. Am. Chem. Soc.*, **100,** 6533 (1978).
[1144] Becker, H. G. O., Pfeifer, D., and Radeglia, R., *Z. Chem.*, **17,** 439 (1977).
[1145] Libman, J., *J. Chem. Soc., Chem. Commun.*, **1977,** 868.
[1146] Vermeersch, G., Marko, J., Febvay-Garot, N., Caplain, S., and Lablache-Combier, A., *Tetrahedron*, **34,** 1493 (1978).
[1147] Nelson, D. J., Mottley, C., and Trifunac, A. D., *Chem. Phys. Lett.*, **55,** 323 (1978).
[1148] Nelson, D. J., *J. Phys. Chem.*, **82,** 1400 (1978).
[1149] van der Weerdt, A. J. A., Cerfontain, H., van der Ploeg, J. P. M., and den Hollander, J. A., *J. Chem. Soc., Perkin Trans.* 2, **1978,** 155.
[1150] Gruselle, M., and Nedelec, J. Y., *Tetrahedron*, **34,** 1813 (1978).
[1151] Porter, N. A., Dubay, G. R., and Green, J. G., *J. Am. Chem. Soc.*, **100,** 920 (1978).
[1152] Hudson, R. F., and Record, K. A. F., *J. Chem. Soc., Perkin Trans.* 2, **1978,** 1167.
[1153] Germain, A., and Commeyras, A., *J. Chem. Soc., Chem. Commun.*, **1978,** 118.
[1154] Bangerter, B. W., Beatty, R. P., Kouba, J. K., and Wreford, S. S., *J. Org. Chem.*, **42,** 3247 (1977).
[1155] Benn, R., Hoffmann, E. G., Lehmkuhl, H., and Nehl, H., *J. Organometal. Chem.*, **146,** 103 (1978).
[1156] Lawler, R. G., *NATO Adv. Study Inst. Ser., Ser. C* 1977, **C34** (*Chem. Induced Magn. Polariz.*), 267; *Chem. Abs.*, **88,** 104196 (1978).
[1157] Morel, G., Marchand, E., and Foucaud, A., *Tetrahedron Lett.*, **1978,** 3719.
[1158] Kuroyan, R. A., Minasyan, S. A., Akopyan, R. A., and Vartanyan, S. A., *Arm. Khim. Zh.* **30,** 705 (1977); *Chem. Abs.*, **88,** 151674 (1978).

1159 Hayashi, T., Hirata, T., Yano, M., and Namiki, M., *Agric. Biol. Chem.*, **42,** 83 (1978); *Chem. Abs.*, **88,** 189597 (1978).
1160 Kurosawa, H., and Yasuda, M., *J. Chem. Soc., Chem. Commun.*, **1978,** 716.
1161 Uemura, S., Toshimitsu, A., Okano, M., Kawamura, T., Yonezawa, T., and Ichikawa, K., *J. Chem. Soc., Chem. Commun.*, **1978,** 65.
1162 Elroi, H., and Meyerstein, D., *J. Am. Chem. Soc.*, **100,** 5540 (1978).
1163 Tada, M., Okabe, M., and Miura, K., *Chem. Lett.*, **1978,** 1135.
1164 Gynane, M. J. D., Lappert, M. F., Miles, S. J., and Power, P. P., *J. Chem. Soc., Chem. Commun.*, **1978,** 192.
1165 Showalter, K., and Noyes, R. M., *J. Am. Chem. Soc.*, **100,** 1042 (1978).
1166 Graf, F., Loth, K., Rudin, M., Forster, M., and Ha, T. K., *Chem. Phys.*, **23,** 327 (1977); *Chem. Abs.*, **87,** 167175 (1977).
1167 Biehl, R., Hass, Ch., Kurreck, H., Lubitz, W., and Oestreich, S., *Tetrahedron*, **34,** 419 (1978).

CHAPTER 4

Oxidation and Reduction

G. W. J. FLEET

Dyson Perrins Laboratory, South Parks Road, Oxford

Oxidation by Metal Ions

Chromium

Picolinic acid catalyses the chromium(VI) oxidation of alcohols to carbonyl compounds by forming intermediate (**1**); picolinic acids with substituents at C(6) showed greatly reduced catalytic activity.[1] Studies on the oxidation of a series of dihydroxy-acids, $HO(CH_2)_nCHOHCOOH$, show that 2,7-dihydroxyheptanoic acid behaves anomalously since the reaction proceeds *via* a three-electron oxidation within the chromium(VI)–dihydroxy-acid complex. It is concluded that (*a*) carboxylate is a ligand to chromium, (*b*) Cr(VI) can easily co-ordinate with more than four oxygen atoms, (*c*) hydrogen is transferred intramolecularly from the C—H bond of the hydroxy-carbon to the oxygen ligand of chromium, and (*d*) the hydrogen is transferred as an atom or hydride, rather than as a proton; this transfer mechanism for hydrogen presumably holds for the general chromium(VI) oxidation of alcohols.[2] Chromium(V) plays a crucial rôle in organic oxidation; a water-soluble stable chromium(V) compound, potassium bis-(2-hydroxy-2-methylbutanoate)oxochromate(V) (**2**), has been shown to have similar geometry to the isoelectronic VO^{2+} complexes, intermediate between square-pyramidal and trigonal-bipyramidal.[3]

Pyridinium chlorochromate causes oxidative cyclizations, such as (**3**)→(**4**), with the chromium species acting as both oxidant and acid catalyst.[4] Studies have been

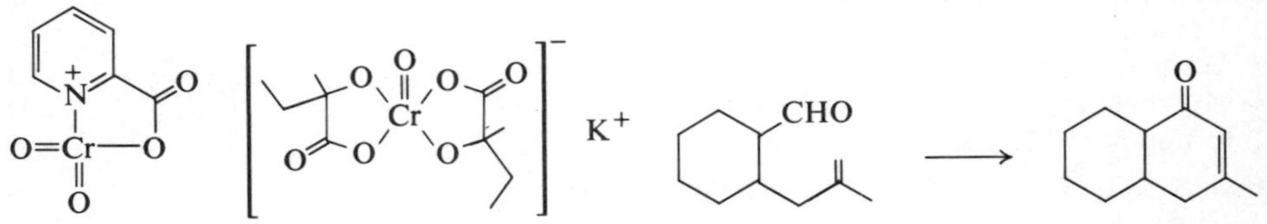

(**1**) (**2**) (**3**) (**4**)

made of the kinetics of the pyridinium chlorochromate oxidation of benzyl alcohol,[5] of substituted benzhydrols,[6] and of substituted mandelic acids.[7] Poly[vinyl-(pyridinium chlorochromate)] is an efficient recyclable oxidizing agent.[8] Allylic oxidation with 3,5-dimethylpyrazole–chromium trioxide (**5**) is remarkably swift, possibly owing to intramolecular assistance by the pyrazole (Scheme 1).[9] Chromyl

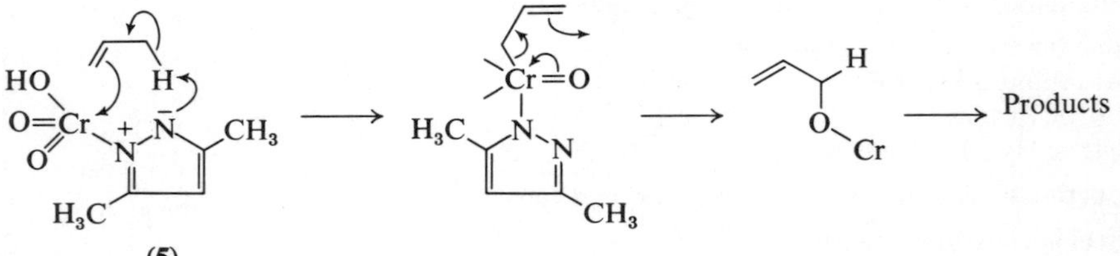

(**5**) → → Products

SCHEME 1

chloride oxidation of norbornadiene gives products mainly arising out of direct 1,2-addition and Wagner–Meerwein rearrangement; this indicates only partial carbonium-ion character in the transition state since stronger electrophiles give more fully developed cations which lead to tricyclic products.[10] Silver chromate–iodine oxidizes olefins to α-iodo-ketones by a mechanism involving nucleophilic attack by chromate on an intermediate iodonium ion.[11] A comparative study of the Jones, Collins, and Corey reagents in the oxidation of allylic alcohols has been made.[12]

Steric effects in the chromium(VI) oxidation of epimeric tetrahydro-1-naphthols[13] and the relative rates of oxidation of axial and equatorial piperidin-4-ols[14] have been investigated. Secondary deuterium kinetic isotope effects in the slow stage of Cr(VI) oxidation of alcohols[15] and the phase-transfer catalysis of dichromate oxidation of alcohols to aldehydes[16] have been studied.

Kinetic and mechanistic investigations of the oxidations of the following systems with chromium(VI) have been reported: cinnamic acids;[17] benzimidazole;[18] ethylene glycol;[19] isobutyraldehyde;[20] acenaphthene derivatives;[21] methoxy-oestratriene;[22] ethanolamines;[23] naphthalene;[24] thiolactic, thiomalic, and thioglycollic acids;[25] and heterocyclic alcohols.[26]

Manganese

The oxidizing power of permanganate rises markedly as the concentration of trifluoracetic acid (in aqueous trifluoracetic acid) rises; aliphatic hydrocarbons are oxidized by hydrogen abstraction by MnO_3^+. The oxidation of arenes C_6H_5Z has a ρ^+ value of -5.29 when appropriate σ^+ values of Z are used and there is no isotope effect when the reaction rates of $C_6H_5CH_3$, $C_6H_5CD_3$, and $C_6D_5CD_3$ are compared; the mechanism probably involves electrophilic attack of MnO_3^+ at the activated positions of the ring, followed by rapid degradation of the ring system.[27] Oxidation of benzaldehydes by acidic permanganate possibly involves rate-determining reaction of $ArCH(OH)OMnO_3$ with water[28] although oxidation of enolizable ketones may proceed by two-electron transfers from the enolates to MnO_4^- in the rate-determining step.[29] The pH/rate profile for the oxidation of butyl alcohol by aqueous permanganate shows that increased oxidation occurs at higher pH;[30] the product distribution of the permanganate oxidation of thymidine has been studied as a function of pH.[31]

gem-Disulphides are selectively oxidized to monosulphones by permanganate.[32] The oxidizing properties of barium manganate[33] and of tetrabutylammonium manganate in organic solvents[34] have been studied; dimethylpolyethylene glycol solubilizes potassium permangante in benzene or methylene chloride.[35] Kinetic studies have been reported on the permanganate oxidation of 4-hydroxypyrimidine-2-thiols,[36] tertiary amines,[37] and diphenylmethane.[38]

Oxidation of phenols containing substituent pyrazolyl groups by manganese dioxide involves a radical mechanism.[39] The kinetics of manganese dioxide oxidation of aryl-1,2-diaminoimidazoles have been investigated.[40] Haematoporphyrin manganese(VI) (HmMnIV) oxidizes many RCH_2X systems, including ethers, to aldehydes; the reaction may be made catalytic by regenerating HmMnIV by sodium chlorate(I) (Scheme 2).[41]

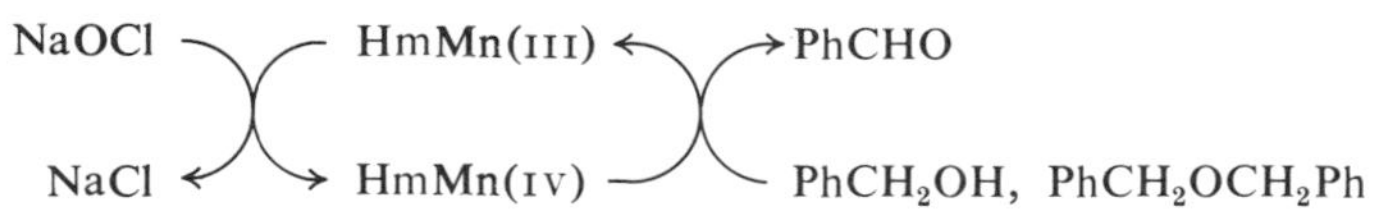

NaOCl ⟶ NaCl; HmMn(III) ⟶ HmMn(IV); HmMn(IV) ⟶ HmMn(III); $PhCH_2OH$, $PhCH_2OCH_2Ph$ ⟶ PhCHO

Scheme 2

Manganese(III) acetate with nitromethane gives nitromethylation of aromatic compounds by a radical oxidation involving $\cdot CH_2NO_2$ as an intermediate.[42] A cation radical is an intermediate in the oxidation of acetophenone by manganese triacetate.[43] Other mechanistic studies on manganese(III) include oxidations of *p*-xylene,[44] methyl mandelate,[45] propyl mandelate,[46] toluene,[47] and aromatic rings.[48]

Silver, Copper, Mercury, and Thallium

The kinetics of oxidation of aliphatic aldehydes by silver(II) in aqueous perchloric acid have been studied by a stop–flow technique; the rate-determining step involves both Ag^{2+} and $AgOH^{+}$, and the *gem*-diol form of the aldehyde is considered to be the reactive species.[49] Silver(II) complexes with nitrogen-containing ligands oxidize aromatic compounds in acetic acid to yield acetoxy-derivatives; the mechanism proceeds by one-electron transfer from the aromatic substrate to silver(II), forming a cation radical as an intermediate.[50] Oxidation of bis(unsaturated) α-diols (**6**) by

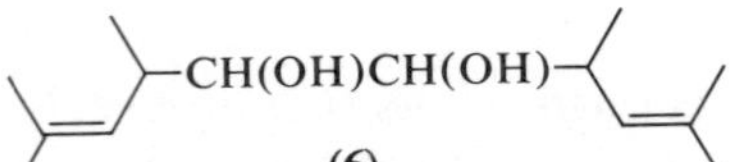

-CH(OH)CH(OH)-

(**6**)

silver(I) carbonate is highly selective; the *threo*-alcohols are totally cleaved whereas the *erythro*-alcohols lead to α-ketols. During the reaction, the geometry of the radical intermediate is conserved; the ketol obtained is the kinetic product, whilst the site of oxidation is determined solely by electronic factors.[51]

The monocopper enzyme, galactose oxidase, provides the first direct evidence for copper(III) in an enzymic reaction. The mechanism involves oxidation of the alcohol to the aldehyde by copper(III) with formation of copper(I)–enzyme which is reoxidized to copper(III) with a copper(II)–peroxide species; the probable rôle of superoxide in most oxidase and oxygenase reactions is discussed elsewhere.[52] The mechanisms of action of copper(II) in the cleavage of organofluorosilicates[53] and in the oxidation of fructose and sorbose[54] have been studied.

Unstable hydrazonomercury acetate intermediates have been isolated in the oxidation of benzil bis(toluene-*p*-sulphonyl)hydrazone to *N*-(4,5-diphenyl-1,2,3-triazol-1-yl) toluene-*p*-sulphonamide by mercury(II) and lead(IV).[55] Cyclopropanyl-mercurials are formed in the oxidation of longicycline by mercury(II) salts.[56] HgI_3^- has been claimed to be the reactive species in the oxidation of formaldehyde by HgI_4^{2-} in aqueous sodium hydroxide.[56a]

The oxidation of alkenes by thallium(III) salts has been reviewed,[57] and the effect of temperature on the oxidation of alkenes by thallium(III) sulphate has been studied.[58] Medium and substituent effects on partitioning of (7) during oxidation of

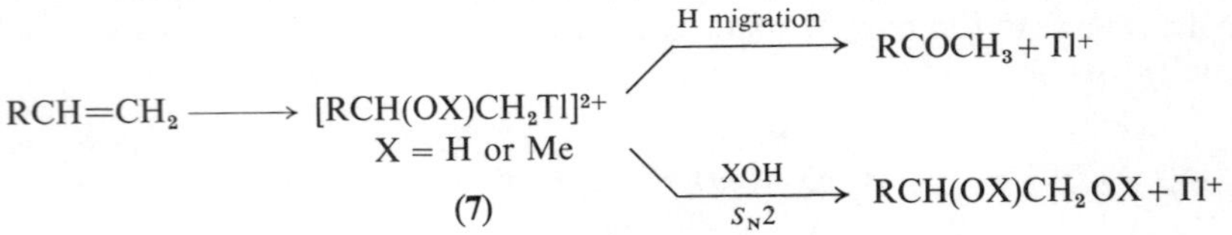

$$RCH{=}CH_2 \longrightarrow \underset{\substack{X = H \text{ or Me} \\ (\mathbf{7})}}{[RCH(OX)CH_2Tl]^{2+}} \begin{cases} \xrightarrow{\text{H migration}} RCOCH_3 + Tl^+ \\ \xrightarrow[S_N2]{XOH} RCH(OX)CH_2OX + Tl^+ \end{cases}$$

SCHEME 3

terminal alkenes by thallium(III) have been discussed; increase in dielectric constant of solvent increases ketonic products since the transition state for hydride shift is stabilized – also methanol is more nucleophilic so that more of the S_N2 pathway is observed in methanol than in water.[59] The oxidative dimerization of 4-alkoxy-cinnamic acids to (**8**) by thallium(III) trifluoracetate involves one-electron oxidation, followed by dimerization of the radical cations to (**9**) (Scheme 4).[60] A novel intramolecular benzyl-aryl coupling reaction by thallium(III) has been observed.[61] Oxidations of phenols by thallium(III) in the presence of acids may involve a two-electron transfer from the substrate, forming a dienone,[62] whereas similar oxidation of diarylpropenones, ArCH=CHCOAr′, proceeds by oxythallation followed by

1,2-aryl migration.[63] Good correlation has been found between the rate of oxythallation of alkenes and their ionization potentials.[64] Kinetic studies on thallium(III) oxidation of the following have been reported: isomeric hexenes,[65] aliphatic oximes,[66] and propan-2-ol.[67]

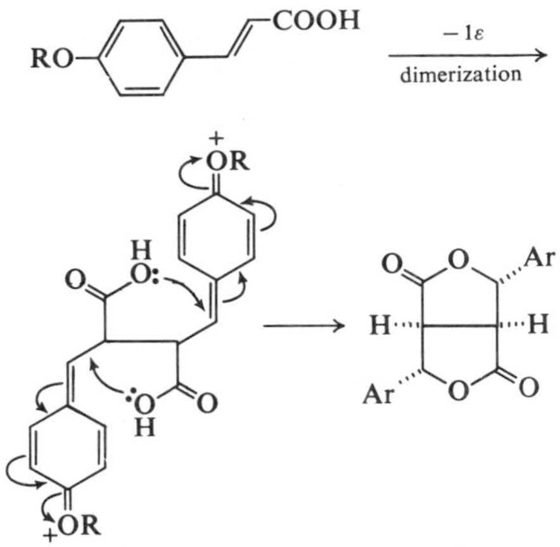

SCHEME 4

Vanadium

In the vanadium(V)-catalysed oxidation of sulphides by *tert*-butyl hydroperoxide, the most active catalyst is the vanadate ester, $VO(OR)_3$, formed by the reaction of alcohols, ROH, on $VO(acac)_2$.[68] Oxidation of phenacyl bromides by vanadium(V) is initiated by formation of radicals produced from C—C fission of a chelate ring complex of the phenacyl bromide and vanadium.[69] The rates of oxidation of cyclohexanone and of lactic acid by vanadium(V) are increased by addition of copper(II); a mechanism involving oxidation of a copper(II)-substrate to a copper(III)-substrate is proposed.[70] A linear free-energy correlation of the oxidation of xylenes by vanadium(V) and cerium(IV) with the ionization potential has been observed and is the evidence on which a mechanism involving rate-limiting electron abstraction has been suggested.[71] Other mechanistic studies include the vanadium(V) oxidation of butanone,[72] aldoses to formic acid and the next lower aldose,[73] substituted benzaldehydes,[74] aroylhydrazines,[75] glyoxylic and pyruvic acids,[76] and fluorenes.[77]

Cerium, Lead, and Tin

Studies[78] on the Belousov–Zhabotinskii oscillating reaction [bromate–cerium(IV)–malonate] have shown that iodide ion can induce and inhibit chemical oscillation.[79] Oxidations of catechol and *o*-aminophenol with cerium sulphate–potassium bromate in aqueous acid show chemical oscillations in the electrode potential of the mixture; *m*- and *p*-substituted analogues do not lead to chemical oscillations under similar conditions.[80]

Oxidative acetoxylation of anisole by ceric ammonium nitrate in acetic acid involves one-electron transfer processes, forming radical cations.[81] Cerium(IV) pyridinium chloride has been used in the oxidation of the side-chains of aromatic rings.[82] Kinetic studies have been reported of the cerium(IV) oxidation of carboxylic acids in the presence of silver(I) and copper(I),[83] *p*-methylmandelic acid,[84] propionic acid,[85] and *p*-nitrophenol.[86]

The mechanism of the tetrahydroisoquinoline oxidative rearrangement by lead tetra-acetate involves a retro-Mannich reaction (Scheme 5).[87]

A mechanism has been proposed for the oxidative decarboxylation of nitrosoproline by lead tetra-acetate (see Scheme 6).[88] Five types of heterocyclic product

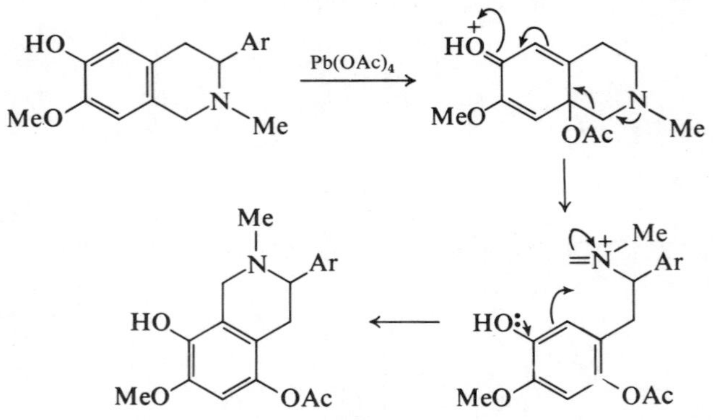

SCHEME 5

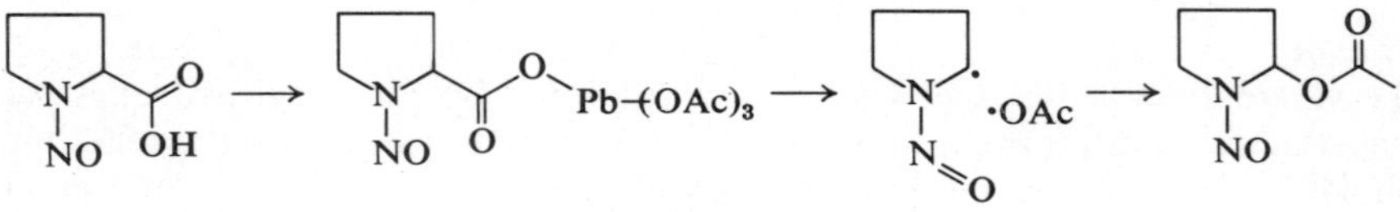

SCHEME 6

formed on lead(IV) oxidation of *N*-alkylaminofumarates have been identified and preliminary consideration has been given to the mechanism of their formation.[89] The oxidative cyclizations by lead(IV) of N^1-methylated acid hydrazides[90] and of 1,3,4,6-tetraketones, followed by rearrangement to dehydroacetic acid analogues,[91] have been studied. In the lead(IV) oxidation of benzyl alcohol,[92] decomposition of a lead(IV)-alkoxide, with the heterolysis of the Pb—O bond, is rate-limiting. Rate-determining enolization was observed in the oxidation of ketones to α-acetoxylated ketones by lead tetra-acetate.[93]

Kinetic studies have been conducted on the co-oxidation of benzene and tin(II) chloride.[94]

Molybdenum, Tungsten, and Uranium

Peroxomolybdenum complexes, stabilized by picolinate and pyridine-2,6-dicarboxylate ligands, catalyse the oxidation of cyclic ketones by hydrogen peroxide to lactones. This reaction is a catalytic analogue of the Baeyer–Villiger reaction and proceeds *via* a metalozonide (**10**).[95] In contrast, cyclohexanone is oxidized to adipic acid by H_2MoO_4 and H_2WO_4; 1-hydroxy-1-hydroperoxycyclohexane (**11**) is an intermediate.[96] Enolates of ketones, esters, and lactones are oxidized by MoO_5–pyridine–HMPA to α-hydroxy-carbonyl compounds; the hydroxylation of

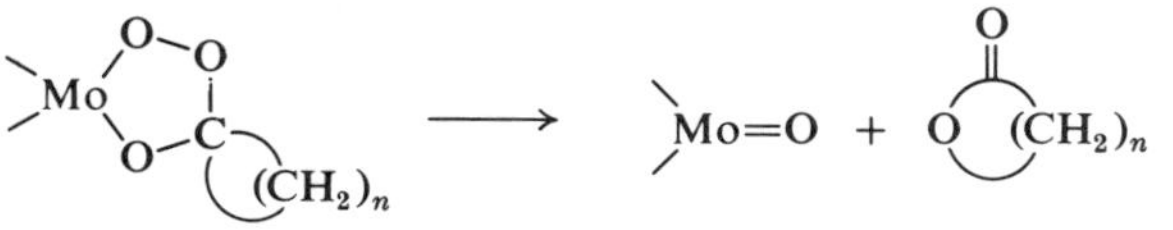

(10)

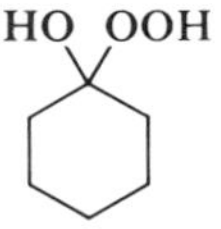

(11)

enolates of α,β-unsaturated carbonyl compounds leads to complicated side-reactions.[97] Metallo-oxaziridines of Group VI metals (**12**) have been isolated and studied in amination of olefins[98] and in reaction with cyclohexanone.[99] The use of a template to direct the peroxymolybdenum epoxidation of a steroidal diene at a particular C=C bond and with stereochemical control[100] has been used to discover

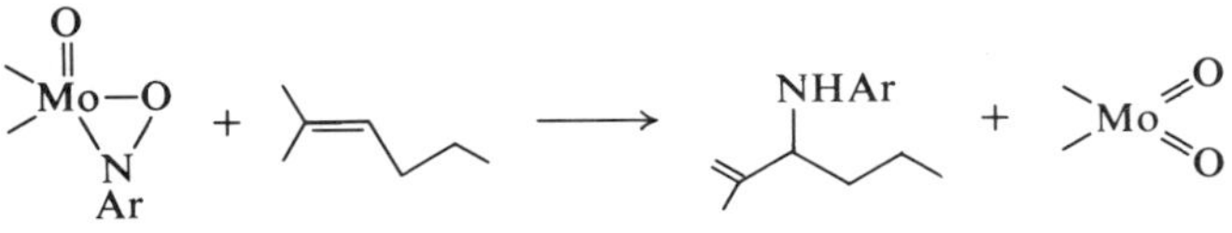

(12)

features of conformational preference in the flexible polyenes, *all-trans*-farnesol and *all-trans*-geranylgeraniol.[101] Peroxouranium oxide oxidizes alkenes through an oxyuranylation pathway (**13**) with resulting carbocationic rearrangement of the intermediates.[102] Uranium hexafluoride is an active, though selective, oxidizing agent for the conversion of methyl ethers, benzylic bromides and alcohols, oximes,

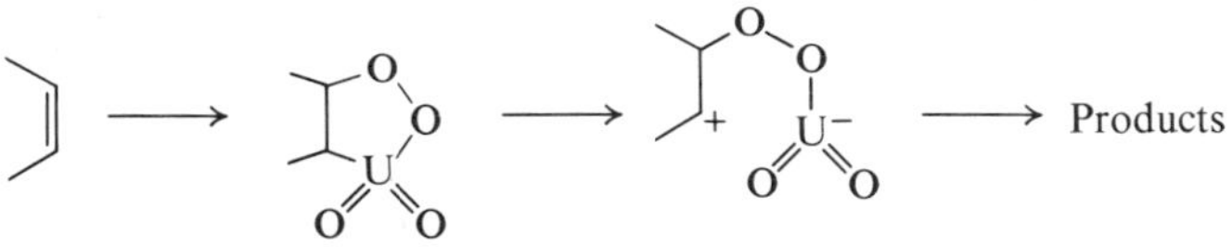

Products

(13)

and hydrazones to carbonyl compounds; the intermediate oxonium ions in the oxidation of methyl ethers were trapped with dithiols.[103]

Group VIII Metals

The kinetics of oxidation by cobalt(III) of hexanoic acid,[104] decyl alcohol,[105] decanone,[106] and iminodiacetic acid[107] have been reported.

Oxidation of deuteriated aromatic compounds by iron perchlorate and hydrogen peroxide show large NIH shifts; it is proposed that the active oxidant behaves as a radical – possibly a hydroxy radical or an iron derivative thereof, such as FeO^{3+} or FeO^{2+}, or a triplet oxenoid species – and that the initially formed radical adduct is rapidly oxidized to a cyclohexadienyl cation which leads to a phenol with a large NIH shift.[108] Kinetic investigations on the oxidation of the following systems with hexacyanferrate(III) have been reported: diphenylthiocarbazone as a function of pH;[109] tartrate ion;[110] diols;[111] chromans;[112] pinacol;[113] phenylhydroxylamine;[114] 2-*tert*-butyl-*p*-cresol;[115] cross-coupling of -4-substituted 2-*tert*-butylphenols;[116] and other cross-coupling of phenols.[117] The oxidation of heteroaromatic cations by hexacyanoferrate(III) involves rate-determining abstraction of hydride ion to give $HFe(CN)_6{}^{4-}$ which reacts with hexacyanoferrate(III) to form hexacyanoferrate(II).[118] Oxidation of phenols by iron(III)–1,10-phenanthroline complexes forms intermediate mixed-ligand iron(III) complexes of both the phenol and phenanthroline.[119]

The kinetics of oxidation of glycolic acid by osmium tetroxide have been studied.[120] The mechanism of the stereospecific vicinal oxyamination of olefins by trioxo-(*tert*-alkylimido)osmium(VIII) complexes[121] and by *N*-argento-*N*-chlorocarbamates[122] have been discussed.

Although the iron(III) oxidation of diols in the presence of catalytic amounts of osmium tetroxide is complicated,[123] oxidation of alcohols,[124] diols,[125] and cyclopentanol[126] by ruthenium(III) chloride catalysed by hexacyanoferrate(III) is of zero order in iron(III) and first-order in ruthenium(III).

Ruthenium tetroxide, generated *in situ* from the water-soluble ruthenium dioxide in a two-phase system, oxidizes sulphilimines to sulphoximines.[127] The selective cleavage of biphenyls by ruthenium tetroxide occurs with oxidation of rings that have activating substituents.[128] The more strongly basic alkylamines are protected in dilute acid from oxidation by ruthenium tetroxide, in contrast to aromatic amines which are degraded under similar conditions.[129]

Oxidative arylation by benzene in the presence of palladium(II) acetate proceeds by successive formation of π- and σ-complexes of benzene with palladium, in which formation of σ-complex is rate-determining.[130] Vicinal diamination of olefins by amines is catalysed by palladium(II) chloride.[131] The stereochemistry and mechanism of oxidative addition of benzyl-α-d chloride, C_6H_5CHDCl, to tris(triethyl)-palladium(0) have been studied with observation of optical activity in a carbon–palladium σ-bonded complex.[132,133] Oxidative addition of organic halides to Group VIII transition metal complexes has been reviewed.[134]

Oxidation by Compounds of Non-metallic Elements

Nitrogen

Thiols are oxidized to thionitrites (RSNO) by dinitrogen tetroxide; the thionitrite may be further oxidized to intermediates, such as RS(O)NO and $RS(O)NO_2$, which lead to thiosulphonate esters and sulphonic acids as products.[135] Oxidation of unsymmetrical disulphides and thiolsulphinates with an excess of dinitrogen tetroxide produces symmetrical thiolsulphonates as the main products by intermediate formation of thionitrites and sulphinyl nitrites.[136] The liquid-phase oxidation of butyl nitrate by nitrogen dioxide has also been studied.[137] Alkylhydrazines are smoothly oxidized to paraffins by toluene-*p*-sulphonyl azide under phase-transfer conditions.[138] Oxidation of alcohols by *N*-hydroxypiperidinium salts **(14)**

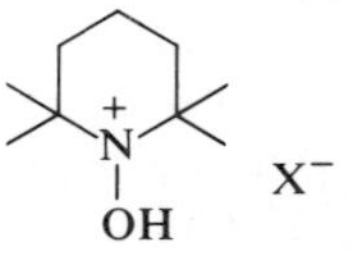

X^- N⁺–OH

(14)

$$PhCHO + ArNO_2 \xrightarrow{^-CN} PhC^-(OH)(CN) \quad \overset{+}{N}(O^-)(=O)Ar \longrightarrow Ph{-}C(OH)(CN){-}\overset{+}{N}(O^-)(O^-){-}Ar$$

$$\longrightarrow PhCOCN$$

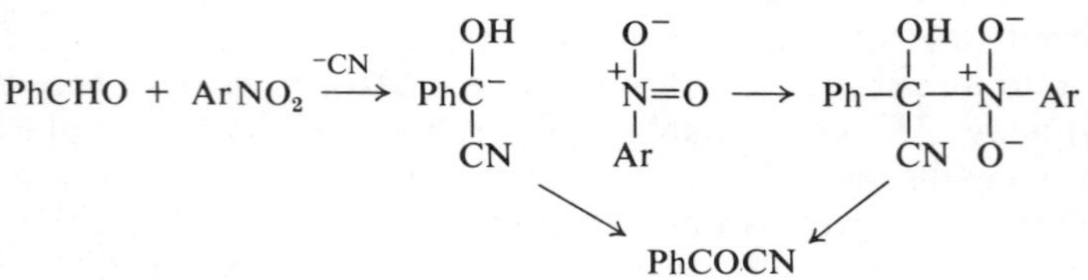

SCHEME 7

involves hydride transfer from the alcohol to (**14**); nitroxyl radicals are intermediates.[139] Cyanide ion catalyses the oxidation of aldehydes by nitro-compounds (Scheme 7).[140]

Sulphur

The mechanism of oxidation of tertiary alicyclic amines with sulphur has been investigated.[141] Dehydrogenation of cyclohexanones by sulphenylation with $(PhS)_2$ may involve a Favorskii-type reaction.[142] The oxidation of sulphides to sulphoxides by sulphuryl chloride in the presence of wet silica has been used for making ^{18}O-labelled sulphoxides.[143] A hydroxylamine-*O*-sulphonate (**15**) is an intermediate in the oxidation of amines with arylsulphonyl peroxides since carbon-to-nitrogen rearrangement is found in (**15**) where a good migrating group is present and there is no α-proton to allow elimination.[144]

$$RCH_2NHR' \xrightarrow{(ArSO_2O)_2} RCH_2N(R')OSO_2Ar \longrightarrow ArSO_3H + RCH{=}NR'$$

(**15**)

A thiosulphonium ion, $Me_2\overset{+}{S}{-}SAr$, has been isolated as an intermediate in the dimethyl sulphoxide oxidation of ArSH to the disulphide.[145] The mechanism of the oxidation of alcohols to carbonyl compounds by dimethyl sulphoxide, activated by a wide range of electrophiles (such as oxalyl chloride, thionyl chloride), has been discussed.[146] The reduction of dichlorocarbene under phase-transfer catalysis is much more efficient by aryl sulphoxides than alkyl sulphoxides.[147] Silver(I) catalyses the DMSO oxidation of β-bromoselenides, $RCHBr{-}CH_2SePh$, to α-phenylseleno-ketones; an alkoxydimethylsulphonium salt and a seleniranium ion have been postulated as intermediates.[148] The kinetics of the oxidation of phenothiazine by dimethyl sulphoxide in sulphuric acid have been determined.[149]

Kinetic studies on the peroxydisulphate oxidation of substituted phenols are consistent with rate-determining electrophilic attack on the phenol by peroxydisulphate;[150,151] however, a radical mechanism is probably involved when the reaction is catalysed by silver(I).[152] The peroxydisulphate oxidation of phloroglucinol has been described.[153] The kinetics of,[154] and the effect of ionic[155] and non-ionic[156] micelles in, the oxidation of diphenylamine by peroxydisulphate have been studied. Peroxydisulphate oxidation of anilines is accelerated by electron-withdrawing groups and slowed down by electron-donating groups;[157] the influence of *ortho*-substituents has been explained in terms of polar and steric factors.[158] Oxidation mechanisms for the oxidation of butyramide[159] and alcohols and olefins[160] by peroxydisulphate have been proposed.

In silver(I) catalysis of peroxydisulphate oxidations, both $SO_4^{\cdot-}$ and silver(II) have been shown to be present; silver(II) oxidizes alcohols to alkoxy-radicals.[161] Mechanistic studies have been reported on the silver(I)-catalysed peroxydisulphate oxidation of *tert*-pentyl alcohol,[162] glyoxal and glyoxylic acid,[163] propane-1,3-diol,[164] dicarboxylic acids,[165,166] aromatic azo-compounds,[167] and diols.[168]

Selenium

Modern organoselenium chemistry has been reviewed.[169] Benzeneseleninic anhydride oxidizes alcohols to carbonyl compounds by fragmentation of the initially formed seleninic ester (**16**);[170] oxidation of cyclic ketones to α,β-unsaturated ketones is accompanied by ring contraction which may be accounted for by a Pummerer-type of rearrangement (Scheme 8).[171] Benzeneseleninic anhydride converts thiocarbonyl compounds into their oxo-analogues,[172] hydrazines into

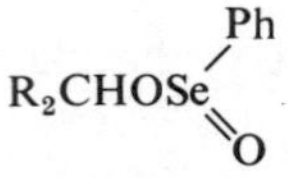

$R_2CHOSe(=O)Ph$

(**16**)

azo-compounds or unstable di-imide species,[173] and with aldehyde hydrazones, hydrazo-compounds and hydroxylamines it gives high yields of azo- and nitroso-products;[174] phenyl- and *p*-nitrophenyl-hydrazones lead to acylazo-derivatives (Scheme 9).

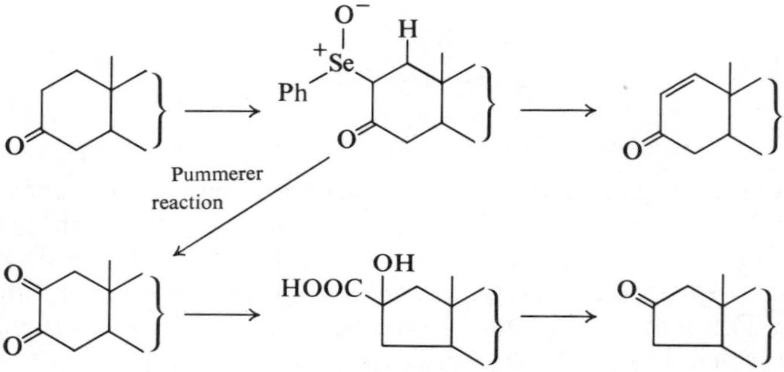

Pummerer reaction

HOOC OH

SCHEME 8

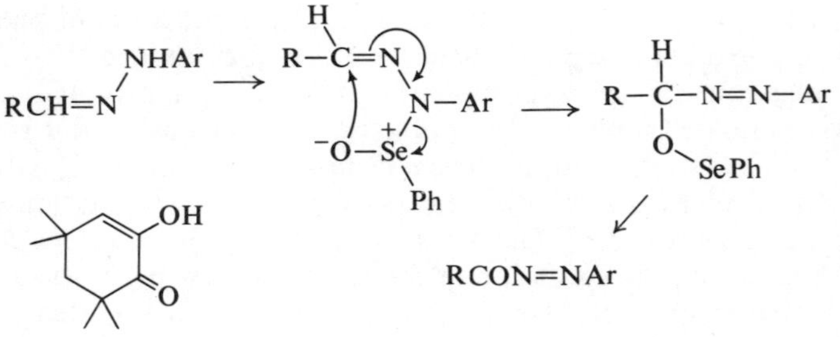

RCH=N–NHAr → R–CH=N–N(Ar)–Se⁺(Ph)–O⁻ → R–CH(OSePh)–N=N–Ar → RCON=NAr

(**17**)

SCHEME 9

Benzeneperoxyseleninic acid, generated *in situ*, is a new reagent for Baeyer–Villiger oxidation.[175] Selenium-containing by-products have been isolated and characterized in the selenium dioxide oxidation of (**17**).[176] The behaviour of aromatic systems with selenium dioxide has been contrasted with that of tellurium dioxide.[177] *tert*-Butyl hydroperoxide is an excellent oxidant for the oxidation of selenides to selenoxides to be used for the subsequent elimination reaction;[178] this reaction has been applied to the conversion of β-selenyl alcohols, $RCH_2CH(SeR')CH_2OH$, to allylic alcohols, $RCH{=}CHCH_2OH$.[179] Diphenyl selenide, bromine, and hexabutyl distannoxane combine to oxidize alkenes to α-phenylselenocarbonyl compounds.[180] Chiral acyclic phosphorus(III) compounds are oxidized by dimethyl selenoxide with inversion of configuration, whereas six-membered ring phosphite oxidation occurs with retention.[181] The mechanism of the reaction of seleninic acids with thiols has been investigated.[182]

Halogens, Including Periodate

A review has appeared on choramine T and related *N*-halo-*N*-metallo reagents.[183] Kinetic studies on the oxidation of the following systems with chloramine T have been described: alcohols[184] and aromatic aldehydes[185] in the presence of osmium-(VIII); acetophenone;[186] aliphatic ketones;[187] α-hydroxy-acids;[188] butanol in the presence of copper(II);[189] diethanolamine;[190] methyl phenyl sulphoxide;[191] dimethyl sulphoxide;[192] and thiols.[193]

Treatment of an alkyl iodide with a peracid oxidizes the halogen and leads to formation of an alcohol; oxidation of iodides with strongly electron-withdrawing groups, such as (**18**), produces alkenes, presumably *via* the iodoso-compound (**19**).[194] Kinetics of the oxidation of *sec*-butyl alcohol,[195] propionaldehyde,[196] and

$PhSO_2$ / I (**18**) ⟶ $PhSO_2$ / ^{+}I–O^{-} / H (**19**) ⟶ $=CH_2$ / SO_2Ph

alcohols[197] by *tert*-butyl hypochlorite have been determined. Arene oxides, such as phenanthrene 9,10-oxide, are formed directly by treating the arene with aqueous sodium hypochlorite in the presence of a phase-transfer catalyst.[198] Arylacetic acids are degraded to aromatic aldehydes and/or carboxylic acids by sodium hypochlorite; α-keto- and α-hydroxy-acids are intermediates in the oxidation.[199] It is claimed that aqueous chlorine oxidizes propan-2-ol *via* the hypochlorite, Me_2-CHOCl, from which a proton is abstracted in the rate-determining step.[200] The pH/rate profile of oxidation of malic acid has a maximum at pH 9.4 with HOBr a more reactive species than Br_2.[201] The mechanisms of the *N*-bromosuccinimide oxidations of cyclohexanone in the presence of mercury(II),[202] of alkyl trimethylsilyl ethers,[203] and of diethyl ketone[204] have been discussed. The kinetics of the iodine oxidation of aroylhydrazines[205] and the bromate oxidation of phenols[206] have been determined; also the oxidation of amines by polyhalomethanes has been reviewed.[207]

Oxidation of α-keto-γ-lactams by periodate results in β-lactam formation by oxidative ring contraction; factors that determine which of two pathways of enol oxidation predominates provide a unified mechanistic interpretation for periodate hydroxylation.[208] Periodate selectively oxidizes unsymmetrical thiolsulphinates to the corresponding thiolsulphonates with no cleavage of the S—S bond; unsymmetrical disulphides are oxidized to thiolsulphonates derived by cleavage of the S—S bond. Oxidation of the disulphide is probably initiated by electrophilic attack of periodate on sulphur whereas thiolsulphinate oxidation occurs by nucleophilic attack of periodate on the sulphinyl sulphur.[209]

Secondary amino-acids undergo oxidative decarboxylation on treatment with periodate.[210] Alumina-supported sodium metaperiodate has been used as an efficient reagent for oxidation of sulphides to sulphoxides.[211] Kinetic evidence has been obtained for the formation of hemiacetals during the oxidation of dextran by aqueous periodate;[212] the mechanisms of oxidation of phenols,[213] α,β-unsaturated acids,[214] and aromatic aldehydes[215] by periodate in the presence of osmium(VIII) have been studied.

Ozonolysis and Ozonation

The first in a series of books on ozonation has appeared.[216] The oxidation of hydrocarbons by ozone,[217] electrophilic oxygenation of alkanes by ozone in super-acid solutions,[218] the kinetics and mechanism of oxidation by ozone of organic compounds of Group Va and VIa,[219] and the reactions of ozone with organic compounds[220] have been reviewed. The first unambiguous crystal structure of an ozonide (**20**) has been obtained;[221] a stable ozonide (**21**) has been found to occur naturally.[222]

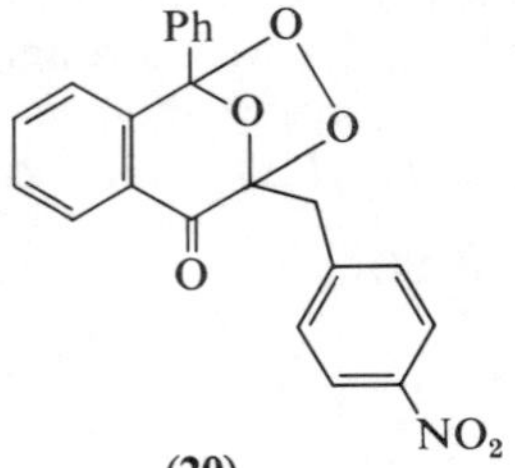

(**20**)

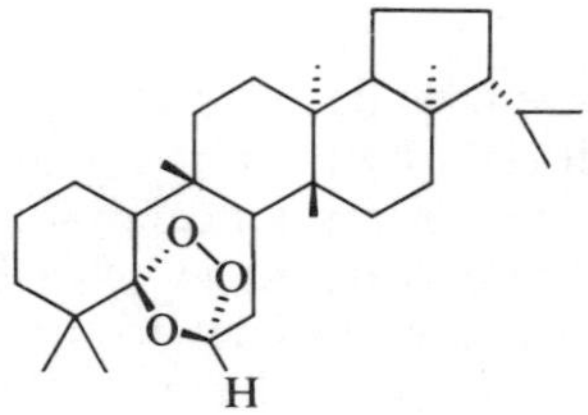

(**21**)

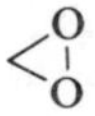

(**22**)

Gas-phase ozonolysis of ethylene has been studied by combining a linear reactor with infrared matrix or microwave spectroscopy; the products observed were $(HCO)_2O$, $HOCH_2CHO$, ethylene oxide and methanol, as well as the previously reported products. Ozonolysis of $^{12}CH_2{=}^{13}CH_2$ shows there are at least two different pathways; a Criegee-type mechanism and a further mechanism in which the C—C bond is not cleaved.[223, 224] The photoelectron spectra of the ozonides of ethylene, cyclopentene, and cyclohexene have been recorded.[225] Dioxirane (**22**) has been identified in the low-temperature reaction of ozone with ethylene.[226] *Ab initio* calculations have been applied to peroxymethylene, dioxirane, and dioxymethylene in studies of gas- and liquid-phase mechanisms of ozonolysis.[227] A set of MINDO/3 calculations support the Bailey modification of the Criegee ozonolysis mechanism.[228] The Bailey modification has been elaborated in order to provide greater flexibility in rationalizing sterochemical observations.[229] Evidence for complexation of the carbonyl oxide intermediate at low temperatures has been presented; either *syn–anti* interconversion is prevented, or the *anti* isomer is selectively formed.[230] The transient red colour observed when tetra-(*p*-anisyl)ethylene is treated with ozone adsorbed on silica gel may be due to the radical cation of the alkene and $O_3^{\cdot-}$.[231] Ozonation of 10,10′-dimethyl-9,9′-biacridylidene in the presence of tetracyanoethylene is an efficient chemiluminescent reaction.[232] Studies of the effect of temperature on the ozonolysis of *cis*- and *trans*-di-isopropylethylene show that the *syn*- and *anti*-carbonyl oxide are in equilibrium but have very different reactivities.[233] The formation of the ozonide intermediate in the ozonolysis of *trans*-di-*tert*-butylethylene is not concerted.[234] Hammett studies on the ozonlysis of cyanostilbenes[235] and stilbenes[236] show a change from a linear transition state with electron-donating groups to a cyclic one with electron-withdrawing substituents. Ozonolysis of vinylsilanes gives products that are derived from dioxetanes and silylperoxy-ketones.[237] The nature of intermediates in the ozonation of trimethylsilylketenes[238] and ketenes[239] has been investigated. Ozonolysis of the following alkenes has been reported: methyl linoleate;[240] trichloroethylene;[241] maleic and fumaric acids;[242] and 1-phenylaziridines.[243]

The reaction of ozone with C—H bonds, both in solution and in the dry ozonation technique, exhibits preferential attack at tertiary C—H; ozonation of cyclododecyl acetate on silica gel gives keto-acetates by regioselective oxidation similar to that observed in microbial oxidation.[244] Cyclopropanes are normally stable to treatment with ozone; however, highly strained bicyclo[*n*.1.0]alkanes do react; for example, bicyclo[2.1.0]pentane with ozone forms succindialdehyde and cyclopropylacetic acid.[245] Ozone on silica gel has been used for selective attack at the tertiary C—H bonds in cedrol (**23**; X = H) to give the diol (**23**; X = OH).[246] Oxidation of cyclohexane[247, 248] and of *p*-nitrotoluene[249, 250] by ozone–oxygen mixtures has been studied.

The reaction of ozone with acetals at low temperatures forms acetal hydrotrioxides (**24**) with two characteristic OOOH absorptions at low field, tenuously explained by the intramolecular hydrogen bonding of the OOOH group with an acetal oxygen; all of the hydrotrioxides decompose below 0 °C to produce singlet oxygen.[251] Kinetic data on a series of reactions of ozone with cyclic acetals (**25**) suggest that the transition state for the reaction is product-like,[252]

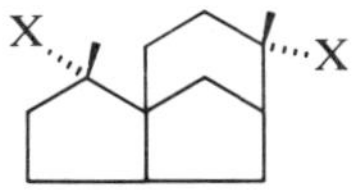

(23) 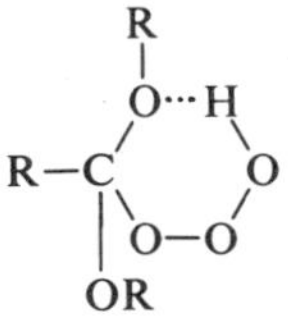 **(24)** 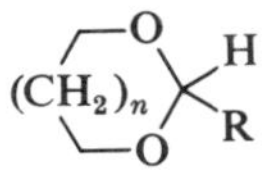 **(25)**

Aldehydes are directly oxidized to esters by treatment with ozone and alcoholic hydroxide; a mechanism similar to a Cannizzaro reaction has been suggested.[253] The reactions of ozone with *tert*-butyl hydroperoxide,[254] with alcohols under UV irradiation,[255] and with dibenzo[18]-crown-6-ether[256] have been investigated.

The reaction of ozone with dimethyl sulphide has been studied by stop–flow techniques; the rapid overall rate is due to a chain process initiated by a slow primary attack of ozone on dimethyl sulphide.[257] Ozone reacts with vinyl sulphides to give compounds with unmodified carbon chains if there is a hydrogen atom *cis* to the thioalkyl group [e.g. (**26**)→(**27**)]; if no *cis*-hydrogen is present, normal ozonolytic cleavage is observed.[258] Ozonolysis of dibenzyl sulphide[259] leads to oxidation of both sulphur and the side-chain, possibly by pathways derived from (**28**), giving products analogous to those found in the ozonation of amines.[260, 261] The reaction of thiophene with ozone has also been studied.[262]

(26) (27) 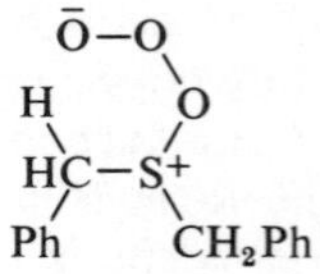 (28)

Peracids, Peroxides, and Superoxide

In normal Baeyer–Villiger oxidations, addition of peracid to the carbonyl group is fast and the subsequent concerted migration is slow; oxidation of strained-cage compounds, such as 1,3-bishomocubanones, proceeds *via* carbonium ions as intermediates.[263] Studies of the Baeyer–Villiger oxidation of 5-*endo*-7-*anti*-substituted bicyclo[2.2.1]heptan-2-ones (**29**) show that the outcome of the reaction

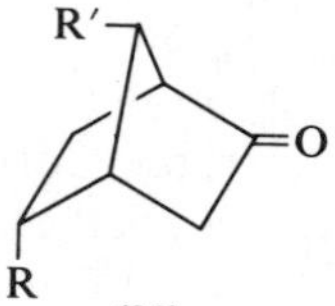

(29)

depends on the electronegativity of the *anti*-group (R′), the hydrogen bonding capability of the *endo*-group (R), and the peracid employed.[264] The reaction of acetophenone with monoperphosphoric acid (H_3PO_5) is catalysed by sulphuric acid and the rate is correlated with the acidity function, H_0.[265] No intermediate could be isolated in the oxidation of *o*-, *m*-, and *p*-hydroxyacetophenones with peracetic acid or *tert*-butyl hydroperoxide; however, on oxidation of the *p*-isomer with alkaline hydrogen peroxide (Dakin reaction), hydroquinone monoacetate was isolated.[266] A microbial analogy of the Baeyer–Villiger oxidation has been observed in the oxidation of chrysazin.[267]

Acetic anhydride has been demonstrated to be an intermediate in the oxidation of biacetyl by hydrogen peroxide.[268] Oxidative cleavage of α-ketols by alkaline hydrogen peroxide proceeds by rate-determining fragmentation of (**30**);[269] similarly, reaction of epoxides with hydrogen peroxide involves the β-hydroperoxy-alcohol (**31**) which is subject to base-catalysed decomposition.[270] The mechanisms of the

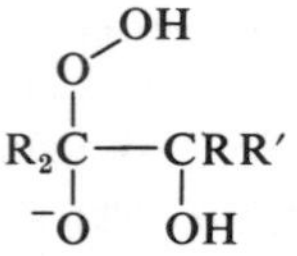

(30) 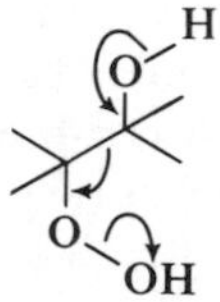 (31)

oxidations of butyraldehyde by peroxybutyric acid,[271] aliphatic aldehydes by peroxydiphosphate,[272] α-alkylacroleins,[273] and also of thiols, disulphides,[274] and thioureas[275] by hydrogen peroxide have been investigated.

Ab initio calculations indicate that unsymmetric, rather than symmetric, transition states are favoured for the epoxidation of ethylene by peroxyformic acid.[276] *trans*-Epoxides are the major products from the reaction of *cis*- and *trans*-alkenes with hydrogen peroxide–tris(acetylacetonato)iron(III); FeO^{3+} may take part in the reaction.[277] In general, peracid oxidation of methylenecyclopropanes produces cyclobutanones *via* oxaspiropentane intermediates; however, (**32**) forms the lactone (**33**).[278] Other epoxidation studies include: cholesterol by hydrogen peroxide in sodium stearate dispersions;[279] steroidal A-ring dihydropyrans;[280] 3-β-acetoxylanost-9(11)-en-7-one;[281] and enol silyl ethers.[282]

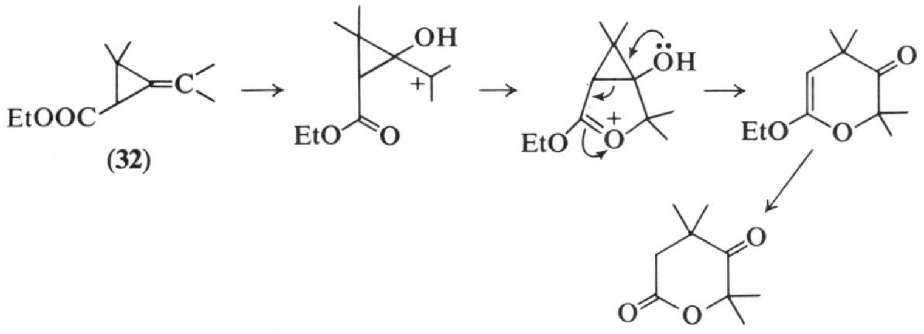

(**32**)

(**33**)

In a number of alkaline peroxide–chiral phase-transfer epoxidations of α,β-unsaturated ketones, the enantiomeric yield of products was found to be inversely proportional to the dielectric constant of the solvent.[283] The large negative entropy of activation of alkali-catalysed epoxidation of cyclohex-1-en-3-one by cumyl hydroperoxide is consistent with a highly ordered transition state such as a six-membered cyclic structure.[284] α-Unsubstituted α,β-epoxy-ketoximes (**34**) yield α-hydroxy-α-nitroalkenes (**35**) on treatment with trifluoroperoxyacetic acid.[285] Unsaturated peroxides cyclize when treated with mercuric salts, giving β-mercurated cyclic peroxides.[286] Peroxymercuration of cyclo-octa-1,5-diene, followed by borohydride treatment, produces (**36**) whereas cyclo-octa-1,3-diene with singlet oxygen followed by di-imide yields (**37**).[287] Benzene, alkylbenzenes, and halobenzenes are hydroxylated by hydrogen peroxide in super-acid media; the phenols

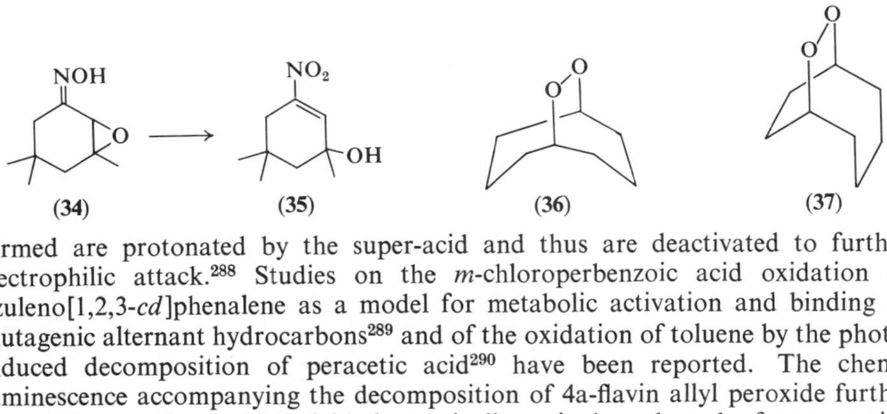

(**34**) (**35**) (**36**) (**37**)

formed are protonated by the super-acid and thus are deactivated to further electrophilic attack.[288] Studies on the *m*-chloroperbenzoic acid oxidation of azuleno[1,2,3-*cd*]phenalene as a model for metabolic activation and binding of mutagenic alternant hydrocarbons[289] and of the oxidation of toluene by the photo-induced decomposition of peracetic acid[290] have been reported. The chemiluminescence accompanying the decomposition of 4a-flavin allyl peroxide further exemplifies reactions which yield electronically excited products by fragmentation of a mixed peroxide formed by addition of R^1R^2CHOOH to the $R_2C{=}\overset{+}{N}R_2$ unit of suitable nitrogen heterocycles.[291]

N-Acyl-α-amino-acids are oxidatively decarboxylated by treatment of the *p*-nitrophenyl ester with peracid.[292] *N*-Alkyl-*N*′-toluene-*p*-sulphonylhydrazines

(**38**) with hydrogen peroxide form the corresponding hydroperoxides by a mechanism involving ion-pair intermediates; toluene-*p*-sulphonylhydrazones under similar conditions give dihydroperoxides.[293] Oxidation of benzylamines with

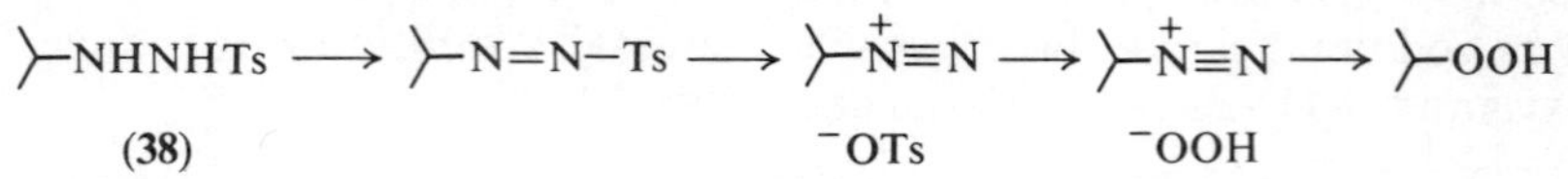

$$\succ\!\!-NHNHTs \longrightarrow \succ\!\!-N{=}N{-}Ts \longrightarrow \succ\!\!-\overset{+}{N}{\equiv}N \;\; {}^{-}OTs \longrightarrow \succ\!\!-\overset{+}{N}{\equiv}N \;\; {}^{-}OOH \longrightarrow \succ\!\!-OOH$$

(**38**)

sulphonyl peroxides, $ArSO_2{-}O{-}O{-}SO_2Ar$, proceeds by a two-electron pathway.[294] The stereochemistry of the bisoxaziridines from peracid oxidation of bis-*N*-alkylimines of glyoxal and terephthaldehyde,[295] and the oxidation of organic sulphoxides by peroxide in the presence of crown ethers[296] have been studied. The rôle of aromatic radical zwitterions and acyloxy-radicals in one-electron oxidation of aromatic carboxylates and reduction of aromatic peracids has been investigated by ESR spectroscopy; $ArCOO^{\bullet}$ and $\dot{+}ArCOO^{-}$ are distinct chemical entities rather than canonical forms of a single intermediate.[297] Investigations have been reported on the chemistry of anthracene endoperoxides,[298–300] and of naphthacene photoxide.[301] Thermal rearrangement of a model prostaglandin endoperoxide, *viz.* the highly strained 2,3-dioxabicyclo[2.2.1]heptane, involves a polar transition state.[302]

From a thermodynamic standpoint, superoxide ion is a moderate reducing agent, but very weak oxidizing agent; the normally observed oxidation arises in the presence of proton sources from disproportionation to peroxide ion and oxygen. The oxidative inertness of $O_2^{\dot{-}}$ is confirmed by its non-reactivity with a wide range of functional groups in rigorously aprotic solvents.[303] Superoxide ion is involved in initiating several autoxidation reactions.[304, 305] Superoxide is produced on near-UV photo-oxidation of trytophan and has at least two fates, one of which is the formation of hydrogen peroxide.[306].

Atomic Oxygen and Singlet Oxygen

The reactions of atomic oxygen with but-1- and -2-yne[307] and with methylenecycloalkenes[308] have been studied.

SCF calculations show that a stable perepoxide (**39**) is a possible intermediate in the formation of dioxetanes from alkenes and singlet oxygen;[309] however, stereochemical studies suggest that the reaction involves a biradical peroxyl intermediate and are inconsistent with an intermediate (**39**).[310] Earlier results[311] are consistent with the biradical mechanism.[310]

Reactions of $^1O_2$ with 2-methylnorbornadiene and 2-methylidenenorbornene show that hydroperoxidation of an olefin is a typical electrophilic reaction, in which the rate is controlled by the HOMO of the double bond, regardless of whether the activated complex resembles an ene reaction or an approach to a perepoxide.[312] Sensitized photo-oxygenation of norbornyl ethers gives zwitterionic intermediates, such as (**40**) and (**41**), which either give dioxetanes or, in the presence

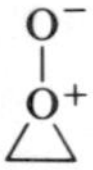

O^- / O^+ (**39**) O–O$^-$ / $\overset{+}{O}Me$ (**40**) O^+–O^- / OMe (**41**)

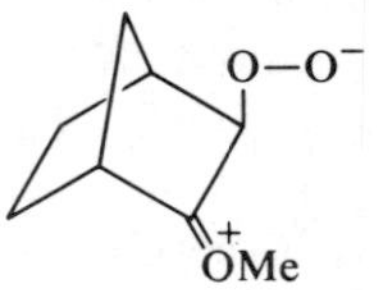

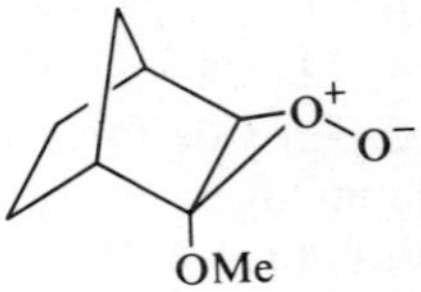

of solvent, are trapped by nucleophilic attack.[313–315] Electronic, rather than stereoselective, control is observed in the addition of $^1O_2$ with 7-isopropylidenenorbornene derivatives.[316] Chalcones with 2′-OH groups are photo-oxygenated to dioxetanes which decompose to benzaldehyde and acidic products.[317] Several studies on the chemistry of dioxetanes have been reported.[318–323]

Singlet oxygen is produced by the reaction of superoxide anion radical with diacyl peroxides[324] and by dehydration of hydrogen peroxide.[325] Photosensitized oxygenation[326] of α-keto-acids leads to decarboxylation *via* the intermediate (**42**)

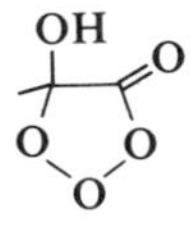

(42)

which is capable of oxygen transfer to unsaturated substrates;[327] this observation may explain the rôle of α-oxoglutarate in α-oxoglutarate-dependent oxygenases. The preference for *syn*-ene reactions of $^1O_2$ with trisubstituted acyclic olefins[328] and the effect of methoxy-groups on addition of singlet oxygen to styrene[329] have been discussed. The reaction of singlet oxgen with cyclohexa-1,4-diene has been used in a synthetic approach to benzene -1,4-endoperoxide.[330]

The reactions of singlet oxygen with the following substrates have been reported: bicyclobutane-bridged dienes;[331] sterically hindered phenols;[332] *m*-hydroxydiphenylamine;[333] alkenes;[334] phosphorus ylides;[335] 9-vinylphenanthrenes;[336] fluorenylidene derivatives;[337] unsaturated acids;[338] and nitronate salts.[339]

Autoxidation and Other Reactions of Triplet Oxygen

Reviews have appeared on: catalysis in oxygenation of organic compounds;[340] transition-metal complexes as catalysts for addition of oxygen to reactive organic substrates;[341] platinum(II) complexes in oxidation of saturated hydrocarbons;[342] autoxidation;[343] and metal catalysis of autoxidation.[344]

Metal ions change from inhibitors to potent inhibitors of autoxidation as their concentration rises above a sharply defined critical level. This inhibition has three characteristics: it is extremely abrupt, it is only observed in solvents of low polarity, and it occurs at a particular concentration of metal ion. All of these characteristics may be explained by a mechanism involving initiation through formation of a metal–hydroperoxide complex which takes place between multivalent metal ions and hydroperoxides in media of low polarity.[345]

The following studies of autoxidation of benzylic hydrogen have been reported: cumene[346–348] in the presence of acetic anhydride,[349] organic sulphide,[350] manganese dioxide,[351] cobalt(II) with increasing concentration of manganese(II),[352] iron(III) and manganese(III)[353] dihydroxybenzenes,[354] and cobalt complexes;[355] 4-isopropylpyridine under γ-irradiation[356] and in the presence of cobalt compounds;[357] *p*-cymene[358] in the presence of cobalt(III) and manganese(III);[359] tetralin[360] in the presence of phenols;[361] xylenes[362] in the presence of vanadium catalysts,[363] and ferric molybdate;[364] ethylbenzene[365] initiated by metal acetoacetonates[366] and inhibited by phenols;[367] toluene[368,369] and other alkyl-aromatics[370–372] in the presence of cobalt(II) salts.

The reactive species in the oxidation of phenol[373] and benzene[374] by xenon difluoride is either O_2 or XeO. The influence of copper catalysts on the oxidation of catechols[375] and *o*-benzoquinones,[376] phenol,[377] and xylenols,[378, 379] has been investigated. The oxygenations of *m*-tolyl acetate,[380] of pyrogallol derivatives,[381] and of highly hindered *tert*-butyl-substituted phenols[382–384] have been studied. Kinetic investigations have been conducted on the gas-phase oxidations of methane[385] and ethane[386] and on oxidation of alkanes in sulphuric acid in which removal of hydrogen as $H_3SO_4^+$ is the rate-determining step.[387] The different reactivities in the oxidation of C_5 to C_{12} cycloalkanes are ascribed to conformational effects.[388] The effects of cobalt complexes on the oxidation of alkanes[389] and cyclododecane,[390] and the mechanisms of oxidation of saturated hydrocarbons by palladium(II)[391] and other metal complexes[392] in highly acidic media have been studied. The inhibition of autoxidation of hydrocarbon by cupric complexes has been investigated.[393–395] The kinetics of decomposition of hydroperoxides by dodecamolybdophosphoric acid,[396] chromium(III) salts,[397] and sulphides[398, 399] have been determined. Other studies on hydroperoxide decomposition have also been reported.[400–403]

The autoxidation of linoleic acid is not regioselective although non-haem iron enzymes, ipoxygenases, catalyse regio- and enantio-selective oxygenation; the oxidation of linoleic acid catalysed by copper(II) gives mainly the C_{13} hydroperoxide.[404] Thermal rearrangement of hydroperoxides of linoleic acid is accompanied by exchange with atmospheric oxygen.[405] Mechanistic studies of the autoxidation of monoene[406] and nonconjugated polyene[407] fatty-acid methyl esters including methyl linoleate,[408] of methyl oleate,[409] of stearic acid,[410] and alkoxy–lipids[411, 412] have been carried out. The kinetics of autoxidation of the following aldehydes[413] have been studied: benzaldehyde[414, 415] in the presence of palladium(II);[416] *p*-methoxycarbonylbenzaldehyde;[417] formaldehyde;[418] acetaldehyde;[419] and crotonaldehyde.[420]

Autoxidation of ketenes by triplet oxygen proceeds by way of diradical and zwitterionic intermediates which may be trapped.[421] Oxygenation of cyclic dienes to endoperoxides in the presence of ammoniumyl radicals and certain Lewis acids is not due to the catalyst overcoming the spin barrier of addition of $^3O_2$, as has been suggested,[422, 423] but probably proceeds by a cation-radical chain mechanism.[424] The oxygenation of (**43**) to (**44**) provides a rare example of a spontaneous endoperoxidation by triplet oxygen.[425] Benzyl hydroperoxide is formed on autoxidation of 5-methylenecyclohexa-1,3-diene (**45**).[246]

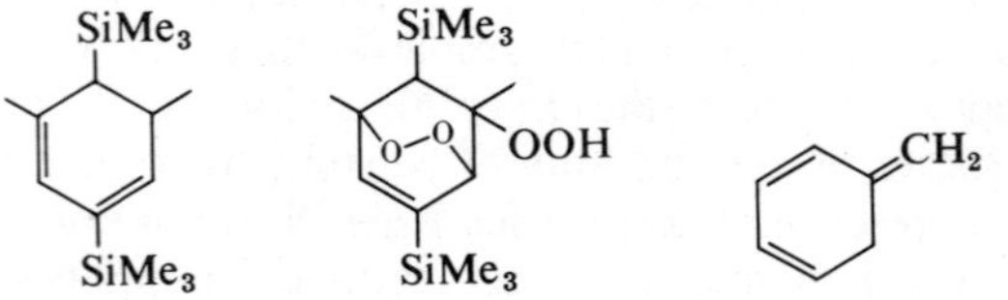

SiMe$_3$ SiMe$_3$ SiMe$_3$ SiMe$_3$ OOH CH_2

(**43**) (**44**) (**45**)

Rhodium-catalysed oxidation of terminal alkenes may occur by coupling an oxygen activation with a Wacker-type oxidation;[427] a possible mechanism (see

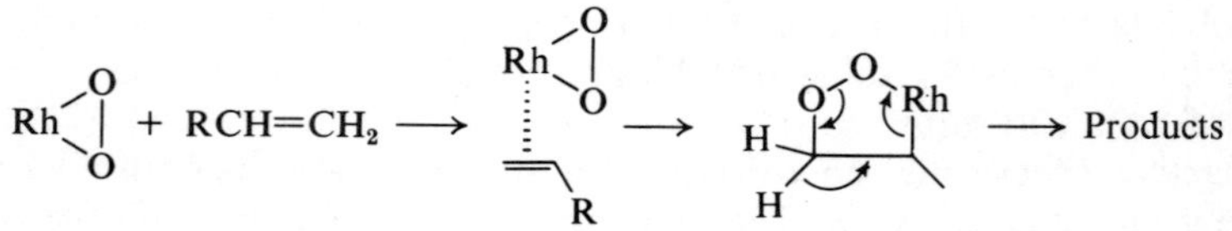

$$\mathrm{Rh}(\mathrm{O_2}) + \mathrm{RCH{=}CH_2} \longrightarrow \longrightarrow \longrightarrow \text{Products}$$

preceding reaction sequence) is indicated.[428] In the co-oxygenation of oct-1-ene and triphenylphosphine in the presence of rhodium(I), octan-2-ol arises from the reduction of octan-2-one by an Rh—H species.[429] The rhodium(I)-catalysed oxygenations of 3-(trimethylsiyl)cycloalkenes to β-silylcycloalk-2-enones[430] and of butadiene[431] have been studied. The rôles of palladium(II) in the oxidation of cyclohexa-1,3-diene,[432] thiophene,[433] higher alkenes,[434] and styrene[435] have been discussed.

The autoxidations of acetylenes,[436] oct-1-yne,[437] propargyl alcohol,[438] and acetylenic alcohols[439, 440] have been investigated. So have the reactions of molecular oxygen with acetals,[441] 2-alkoxytetrahydropyrans,[442] 1,3-dioxacyclanes,[443] and other 1,3-dihetero-analogues of cyclohexane.[444]

The oxidations of phosphines by platinum dioxygen[445] and iron[446] complexes, and of sulphides in the presence of copper(II)[447] and cobalt(II)[448] have been investigated.

Mechanistic studies on the reaction of oxygen with following systems have also been reported: benzene;[449–451] indoles[452, 453] and tryptophan;[454] nitrosobenzenes,[455] aniline,[456] Reissert compounds,[457] and ethoxyquin;[458] 4-hydroxy-[459] and 4-methoxyphenylpyruvate;[460] benzyl alcohol;[461] (1-arylvinyl)ferrocenes;[462] 1-methylcyclohexanol;[463] 1,2-dialkoxy-1-halogenoethanes;[464] α-keto-imine carbanions;[465] naphthyl ketones[466] and methylcyclohexenones[467] in the presence of base; valeric acid;[468] 6-aryl-3,6-dihydro-5-methyl-2-pyrones;[469] 2-methylbut-2-ene,[470] and styrene in the presence of phosphines.[471]

Reduction by Complex Hydrides

Aluminium Hydrides

The kinetics of the reduction of camphor by $LiAlH_4$ are of the first order in both camphor and the reducing agent; reduction of camphor and of 3,3,5-trimethylcyclohexanone by $LiAlH(OR)_3$ occurs with disproportionation of the alkoxyhydride, while $LiAlH_4$ is the predominant reducing species. The rate of reduction is depressed by addition of lithium bromide or of crown ether; reduction by sodium aluminium hydride is much slower, and non-linearly dependent on the concentration of $NaAlH_4$.[472] Interest in the stereochemical aspects of reductions of ketones by hydride continues; theoretical studies indicate that axial attack is favoured over equatorial attack on cyclohexanone.[473] The stereochemical aspects of the reduction of 2,3-substituted indan-1-ones,[474] aza-analogues of bicyclo[3.3.1]-nonan-9-ones,[475] and cyclopentanones[476] by lithium aluminium hydride, and of aryl alkyl ketones by sterically hindered chiral aluminium compounds[477] have been discussed. The complex derived from treatment of (**46**) with $LiAlH_4$ gives a high enantiomeric ratio of propargylic alcohols in reduction of acetylenic ketones.[478] Asymmetric reduction of α-amino-ketones, $RCOCH_2NMe_2$, by $LiAlH_4$ partially decomposed with (−)-menthol gives optically active alcohols *via* a transition state which involves participation of nitrogen.[479] The *cis*-diketone (**47**) gives only the *trans*-fused diol (**48**) with $LiAlH(OMe)_3$, whereas (**48**) is not obtained from the *trans*-diketone; thus, *cis*-to-*trans* epimerization must occur after the first reduction.[480]

The effect of crown ethers on the regioselectivity of reduction of cyclohexenone by $LiAlH_4$ has been investigated; in the absence of cryptand, attack at the carbonyl occurs exclusively whereas in the presence of crown ethers high yields of products

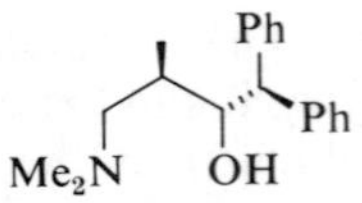

(46) 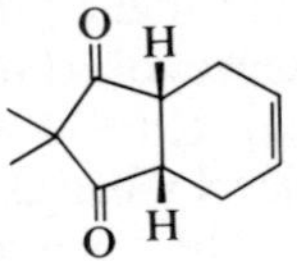 (47) 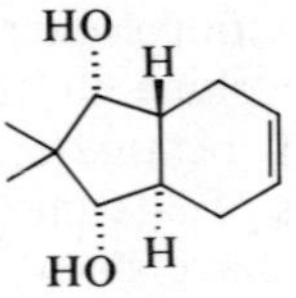 (48)

from conjugate attack by hydride are obtained. This behaviour is due to differential complexation of the carbonyl group by lithium.[481] The stereochemical features of conjugate addition by lithium aluminium hydride to (**49**)[482] and to norbornadiene-2,3-dicarboxylate esters (**50**)[483] have been studied. $LiAlH_4$ reduction of androsta-4,6-dien-3β-ol (**51**) involves intramolecular hydride transfer leading to an allyl-lithium intermediate.[484] Examples of *N*-participation in the reduction of isolated

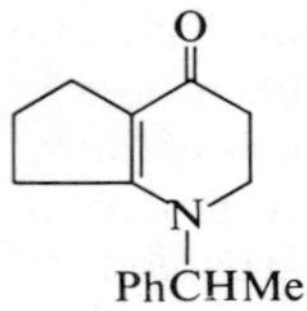

(49) 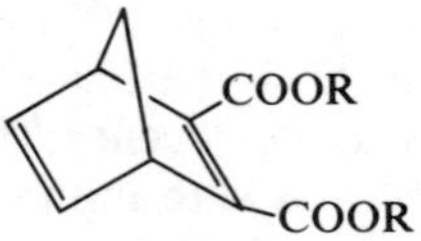 (50) 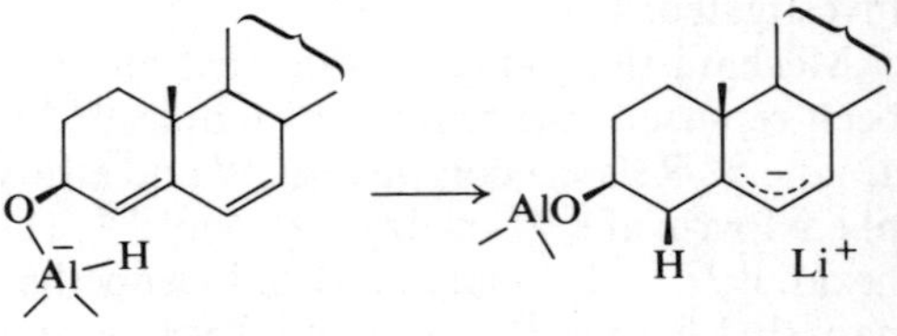 (51)

carbon–carbon double bonds are rare; however, (**52**) may be an intermediate in the reduction of (**53**) to (**54**).[485]

In the reduction of 5,6-dimethylidene-*exo*-2,3-epoxynorbornane (**55**) by $LiAlH_4$, electrophilic ring opening of the epoxide to a cationic species competes with direct attack of hydride on the methylidene double bond (homoallylic S_N2').[486] The ring openings of α-trimethylsilyl and α-trimethylgermyl epoxides[487] and of optically active styrene oxide[488] have been studied.

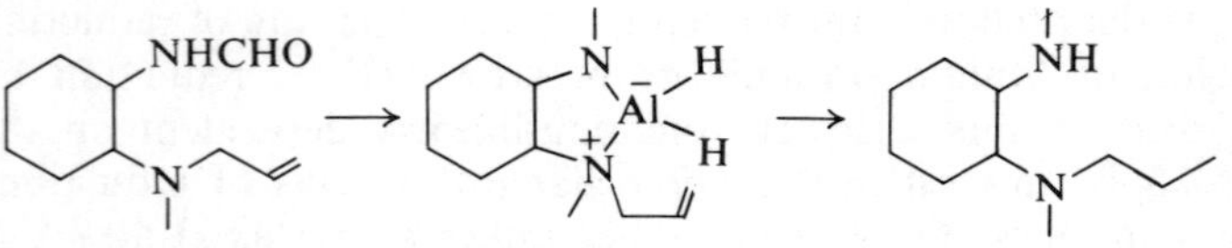

(53) (52) (54) 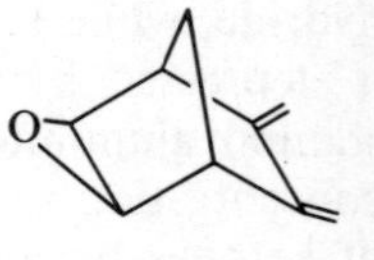 (55)

Selectivity in the reduction of monosubstituted succinic anhydrides,[489] tetramethyl-1,5-diazabicyclo[3.3.0]octane-2,4,6,8-tetraone,[490] and isoxazolines[491] by lithium aluminium hydride has been discussed. The selective reduction of imidoyl chlorides to imines by lithium aluminium tri-*tert*-butoxyhydride[492] has been used for the cleavage of the 7-amido-group in cephamycins.[493]

Stereoselective addition of the ketene dithioacetal (**56**) with $LiAlD_4$ occurs by intramolecular delivery of deuterium from an alkoxyaluminum deuteride salt.[494] Reduction of sulphones by $LiAlH_4$ results in cleavage of C—S bonds.[495] Ion-pair dissociation has been shown to be the controlling factor in the stereochemistry of attack by lithium aluminium hydride at silicon.[496]

The reactions of $LiAlH_4$ modified by first-row transition-metal halides with alkenes and alkynes,[497–502] with alkyl and aryl halides,[503] bromohydrins and epoxides,[504] and diketones[505] have been studied in detail. Lithium aluminium

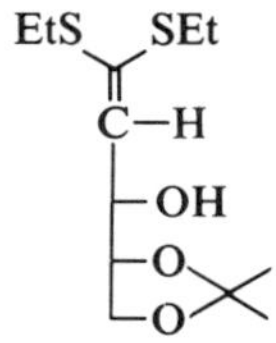

(56)

hydride with tungsten hexachloride, or zerovalent metal carbonyl complexes of Group VI metals causes reductive coupling of carbonyl compounds; a carbene–tungsten complex is proposed as an intermediate.[506]

Borohydrides

Hydroxyborohydride ion, BH_3OH^-, formed during the hydrolysis of borohydride ion, is more reactive than BH_4^-.[507] Phenyl esters are reduced by sodium borohydride whereas benzyl and alkyl esters are unaffected.[508] Regiospecific reduction by sodium borohydride of the most substituted carbonyl group in *gem*-disubstituted succinimides has been observed.[509] The isomer ratio of the possible phthalides in the reduction of substituted phthalic anhydrides by lithium borohydride has been explained by the electronic effect of the substituents.[510] Selective reductions with lithium borohydride have been reviewed.[511]

Reductions of prochiral ketones to secondary alcohols by electro-reduction and by sodium borohydride, using identical ephedrinium salts as optically active supporting electrolytes or as a chiral phase-transfer catalyst, give very different relative optical yields, showing that the transition states for the two reductions have very different structures.[512] Asymmetric reduction of ketones by borohydride in the presence of bovine serum albumin,[513] and chiral ammonium salts as phase-transfer catalysts,[514] and in micelles of lecithin,[515] sodium cholate,[516] and (+)-*N*-dodecyl-*N*,*N*-dimethyl-1-phenylethylammonium bromide[517] have been investigated. Examination of unconsumed borohydride after reduction of ketones by mixtures of $NaBH_4$ and $NaBD_4$ shows that no $NaBH_{4-n}D_n$ species are present, so that disproportionation of intermediate to sodium borohydride plays no significant rôle in the reduction of ketones.[518] In the reduction of substituted 1- and 2-acetonaphthones by $NaBH_4$, electronic effects of substituents are less pronounced in 1-acetonaphthones, presumably because the acetyl group is twisted out of the plane of the ring by *peri*-interactions.[518] The stereoselectivity of reduction by sodium borohydride of hydroxy-substituted cyclohexanones,[520] tricarbonyl indanonechromium complexes,[521] aza-[522] and oxa-cyclic ketones,[523] α,β-epoxyketones[524] and steroidal ketones[525] has been investigated. MNDO calculations on the reduction of formaldehyde and methyl formate by borohydride have been reported.[526]

Initial attack by borohydride on methyleneanthrones (as **57**) occurs at the methylene carbon.[527] Reduction of cyclic α,β-unsaturated ketones with bulky

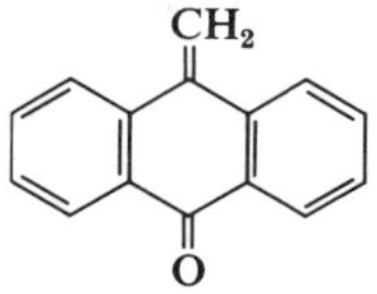

(57)

trialkyl borohydrides leads predominantly to the thermodynamically more stable pseudo-equatorial alcohols.[528] Reaction of α,β-unsaturated nitriles with sodium borohydride yields saturated nitriles.[529] Sodium borohydride reductions of 2-thiazoline-2-thiol esters of carboxylic acids,[530] 1,3-dimethylpyridinium iodide,[531] α-nitro-epoxides,[532] elemental sulphur,[533] carbonyl tosylhydrazones,[534] tertiary halides in dipolar aprotic solvents,[535] and azo-derivatives of β-keto-esters[536] have been reported.

Reduction of acetone by sodium cyanoborohydride in the presence of weakly basic amines gives predominantly reductive amination whereas more strongly basic amines act as general acid catalysts for reduction to propan-2-ol.[537] Reduction of quinolizinium salts to dihydropyridines[538] and of acetals and ketals to methyl ethers[539] by sodium cyanoborohydride have been described.

The first reports of the use of transition-metal borohydrides as organic reagents have appeared. Bis(triphenylphosphine)copper(I) tetrahydroborate (**58**) reduces

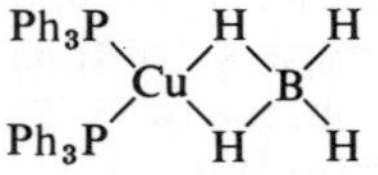

$(Ph_3P)_2Cu(\mu\text{-}H)_2BH_2$

(**58**)

acid halides to aldehydes;[540, 541] the reaction is subject to nucleophilic catalysis by triphenylphosphine.[540] Bis(cyclopentadienyl)chlorotetrahydroboratozirconium(IV) reduces aldehydes and ketones to alcohols with a surprising lack of either regio- or stereo-selectivity.[542] The effects of cerium(III)[543] and samarium(III)[544] on the reduction of α,β-unsaturated carbonyl compounds, and of cadmium chloride on the reduction of acid halides by sodium borohydride[545] have been reported.

Other Hydride Reductions

Tributyltin hydride reduces vicinal diol bisdithiocarbonates (**59**) to the corresponding alkenes by a radical mechanism.[546] Reduction of (**60**) by tributyltin hydride shows that chlorosilanes are more readily reduced than chloroalkanes,

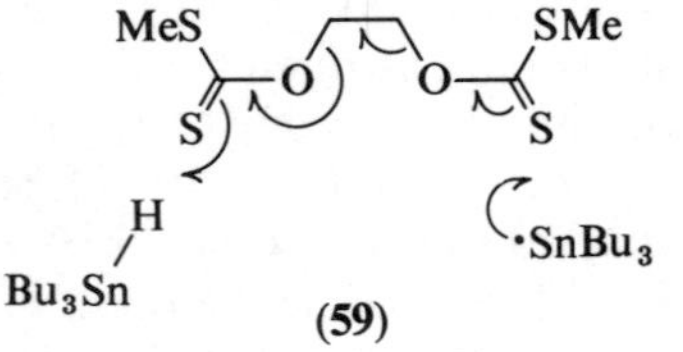

MeS, SMe, O, O, S, S, H, Bu_3Sn, $\cdot SnBu_3$

(**59**)

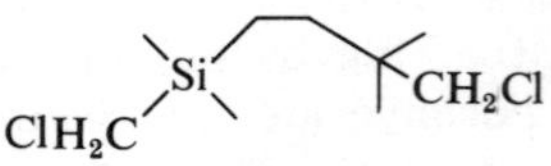

Si, ClH_2C, CH_2Cl

(**60**)

possibly because of the greater stability of the α-silyl radicals.[547] Aldehydes are selectively reduced in the presence of ketones by tributyltin hydride adsorbed on silica gel.[548] The reactions of tributyltin hydride with 4-chloroazetidinones,[549] phenyl selenides and seleno-acetals,[550] and a sugar dithiocarbonate[551] have been discussed.

A series of complex copper hydrides, Li_nCuH_{n+1} ($n = 1$–5), exhibit different regioselectivities with enones and different stereoselectivities towards cyclic ketones, so that the hydrides are discrete compounds rather than a mixture; Li_4CuH_5 is the most reactive.[552] The degree of molecular association of magnesium hydrides controls the stereoselectivity in the reduction of ketones.[553, 554] Other studies on magnesium hydride reductions of ketones[555] and alkynes in the presence of titanium[556] and copper[557] have been reported.

The binuclear cluster compound, $NaHFe_2(CO)_8$, reduces only the carbon–carbon double bond in α,β-unsaturated esters, ketones, nitriles, and lactones; isotopic labelling experiments show a reversible regiospecific hydride-addition step and a large inverse isotope effect.[558] Metal formyl complexes, such as Et_4N^+-$[(PhO)_3P(CO)_3FeCHO]^-$, act as hydride donors to ketones and alkyl halides.[559] The mechanisms of the reduction of organic compounds with hydrides of cobalt,[560] vanadium,[561] and zirconium[562–564] have been studied.

Hydride transfer from hydrocarbons to carbonium ions in super-acids has been studied under kinetic[565] and thermodynamic[566] control. Triphenylmethyl tetrafluoroborate oxidizes secondary alcohols to ketones in the presence of primary alcohols; triphenylmethyl ethers, $R_2CHOCPh_3$, are intermediates.[567] The rates of hydride transfer from cyclic ethers and acetals to the triphenylmethyl cation have been studied.[568]

Gas-phase reduction reactions have been investigated using the cyclohexadienyl anion (**61**) (from hydroxide and cyclohexa-1,4-diene) as a convenient source of hydride.[569] Reduction of benzophenone by lithium *sec*-butoxide[570] and of fluorenone by quinine[571] involve single-electron transfer processes, as well as hydride transfers. Competing one-electron and hydride-ion transfer pathways are also evident in the reduction of ketones by lithium di-isopropylamide.[572] However, (*S*)(+)-1-lithio-2-methylpiperidine (**62**) produces asymmetric reduction of diaryl ketones by concerted hydride transfer.[573] Hydride transfer from carbon is also involved in the

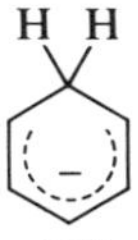

H H

(**61**)

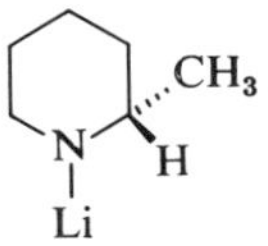

CH_3 H N Li

(**62**)

reduction of aldehydes by *B*-alkyl-9-borabicyclo[3.3.1]nonanes,[574] of pyridine *N*-oxide by Hantzsch esters,[575] and the disproportionation of several heterocyclic pseudobases.[576]

The ionic hydrogenation of cycloalkenes by hydrosilanes in an acidic medium proceeds by hydride transfer to a symmetrically solvated carbonium ion.[577, 578] Among other studies of hydrosilylation,[579, 580] it has been observed that ketones and aldehydes are selectively reduced to alkanes (Scheme 10) by hydrosilanes in the presence of boron trifluoride.[581]

Sodium formate in dimethylformamide in the presence of tetrakis(triphenylphosphine)palladium reduces aryl bromides by a mechanism involving initial oxidative addition, followed by displacement of halide and reductive elimination (Scheme 11).[582]

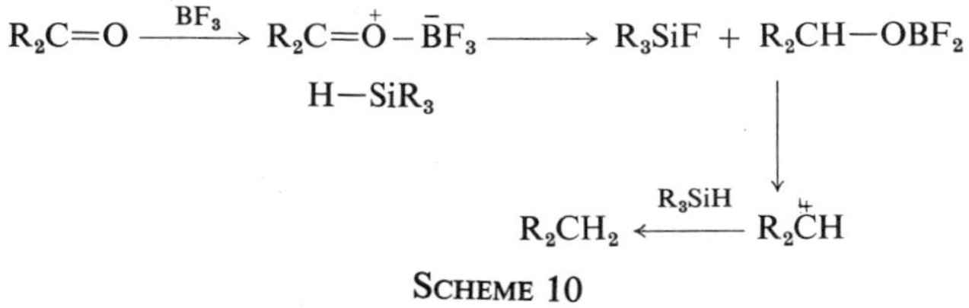

$$R_2C{=}O \xrightarrow{BF_3} R_2C{=}\overset{+}{O}-\overset{-}{B}F_3 \longrightarrow R_3SiF + R_2CH{-}OBF_2$$

$$H{-}SiR_3$$

$$R_2CH_2 \xleftarrow{R_3SiH} R_2\overset{+}{C}H$$

Scheme 10

$$ArBr \longrightarrow Pd(PPh_3)_2ArBr \longrightarrow Pd(PPh_3)_2Ar(HCO_2^-) \longrightarrow Pd(PPh_3)_2ArH \longrightarrow ArH$$

SCHEME 11

Reduction by Metals, Metal Complexes, and Ions

The stereochemistry of the dissolving-metal reductions of ketones has been discussed.[583–586] The lithium salts of 1,2-diols isomerize when heated, by a hydride transfer rather than a reverse pinacolic coupling pathway.[587] The reductive dimerizations by alkali metals of adamantanone in trimethylsilyl chloride[588] and Schiff bases in the presence of crown ethers[589] have been investigated. The double ring opening[590] of bridged dicyclopropyl ketones, *e.g.* (**63**) to (**64**), provides an example of the generation of pericyclic anions by lithium.[591] The mechanism of ring opening in the reduction of 1-methyltricyclo[4.4.0.0^{2,6}]decan–3–one by lithium in ammonia has also been studied.[592]

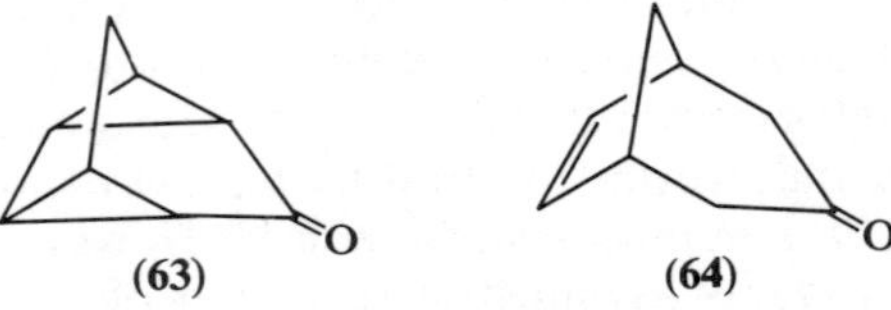

The behaviour of aromatic and polynuclear aromatic compounds in sodium–liquid ammonia reductions has been categorized according to the nature of the intermediates involved – radical anions, monoanions and dianions; the outcome of many reductions and reductive alkylations is a result of reactions occurring during the quenching process.[593] The novel cleavage of terphenyls by lithium in HMPA to 3-phenylcyclohexa-1,4-diene and cyclohexa-1,3-diene has been observed.[594] Significant yields of alcohol and hydrocarbon products are obtained in reduction of biphenyl-4-carboxylic acid by sodium in liquid ammonia.[595] Birch reduction of methoxy-*N*,*N*-dimethylanilines[596] and of a series of benzamides[597] has been reported. Reductions of isocyanides,[598] esters,[599] and furanosyl and pyranosyl halides[600] by dissolving alkali metals have been studied.

Irradiation of $(C_5H_5)_2WH_2$ in propylene oxide causes deoxygenation to propene;[601] the system (Scheme 12) may be catalytic and the intermediate (**65**) is analogous to that proposed for transfer of oxygen atoms between transition-metal

$$W(C_5H_5)_2 \xrightarrow{\text{propylene oxide}} \text{O-W}(C_5H_5)_2 \text{ metallacycle } (\mathbf{65}) \longrightarrow O{=}W(C_5H_5)_2 + \text{propene}; \quad O{=}W(C_5H_5)_2 \xrightarrow{Na/Hg} W(C_5H_5)_2$$

SCHEME 12

complexes and alkenes in epoxidation reactions.[602] Uranocenes react with nitro-compounds to produce azo-compounds; free nitroso-compounds are probable intermediates.[603]

Reduction of α,α′-dibromo-ketones by ultrasonically dispersed mercury in acetic acid gives excellent yields of α-acetoxy-ketones with highly hindered systems, and ketones with less sterically congenested substrates;[604] reduction in ketones as solvent produces 4-isopropylidene-1,3-dioxalans (**66**) (Scheme 13).[605] Reductive 1,3-dehalogenations of stereoisomeric 2,4-dibromo-3-methylpentanes with zinc,

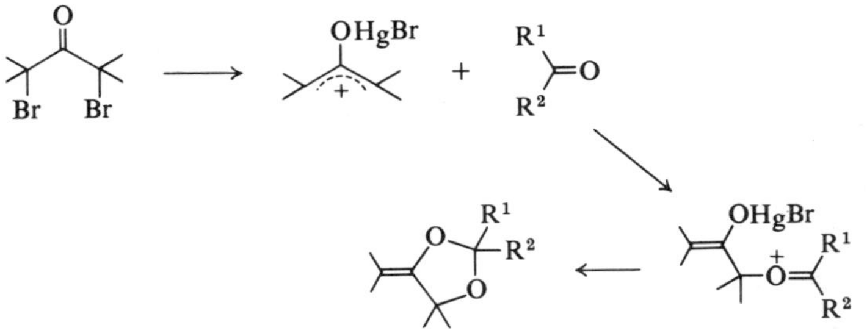

(**66**)

SCHEME 13

chromium(II), or sodium proceed by inversion at one carbon and a non-stereospecific reaction at the other.[606] The mechanism for acid-catalysed α-to-α'-halogen migration in reductive dehalogenation of α-halo-ketones has been elucidated.[607] The reductions of dihalomethanes,[608] β-halo-ketones,[609] and acetophenone[610] by zinc have been the subject of further investigations.

Titanium metal reductively couples carbonyl compounds to form olefins; intramolecular dicarbonyl coupling gives cycloalkenes of ring sizes 4 to 16; the coupling reaction occurs on the surface of the metal.[611] Titanium in tetrahydrofuran reduces enol phosphates to alkenes.[612] The reduction of compounds by electrochemically generated titanium(III) is similar to that by aqueous titanium(III) chloride.[613] The mechanism for reduction of $RCOCBr_2R'$ by titanium(III) chloride[614] and the gas-phase chemistry of chlorotitanium ions with organic compounds[615] have been discussed.

The selective reduction of the ketonic carbonyl group in pyruvic acid by chromous ion occurs only on the free carbonyl group and not on the hydrate; the product is a binuclear complex of chromium(III) with lactic acid which eventually undergoes hydrolysis.[616] The chromous ion reduction of substituted cyclopropenium cations has been studied.[617]

The reaction of methylmagnesium halide with benzophenone is much altered by the presence of traces of transition metals in the magnesium, whereas the *tert*-butyl-Grignard reagent is not affected; an iron(I) complex is the active species leading to iron-catalysed pinacol reductions: the radical anion, $Ph_2\dot{C}{-}O^-$, can be trapped with *p*-dinitrobenzene. In contrast, MeMgX is unaffected by acetone, whereas *tert*-BuMgBr with iron gives a high yield of reduction product *via* an Fe—H intermediate.[618] The reduction of allylic compounds with organocuprates involves one-electron transfer.[619]

Reduction of halophenols with Raney nickel alloys in $NaOD/D_2O$ gives all possible deuteriated phenols; bromophenols give largely the corresponding deuteriated phenol, but chlorophenols give extensive further exchange.[620] Stereospecific formation of β-lactams is observed in the Raney nickel desulphurization of mesoionic thiazol-4-ones; treatment of anhydro-4-oxy-2,3,5-triphenylthiazolium (**67**) with nickel gives (**68**) as a pure compound formed stereospecifically. The most

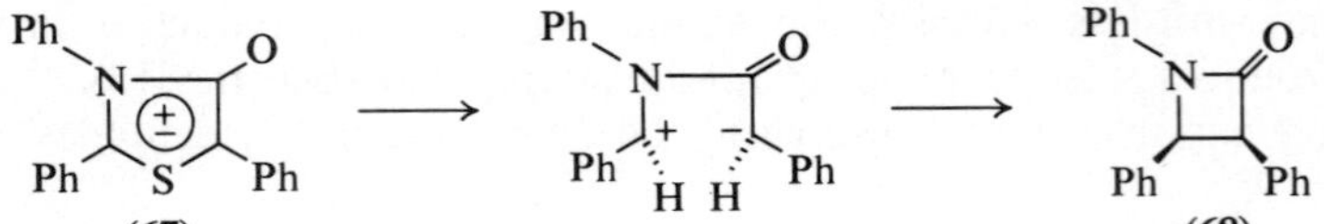

(67) (68)

likely mechanism requires conservation of dipolar character during the desulphurization.[621] Non-aqueous reductive lanthanide chemistry has been studied in the formation of complexes of lanthanide ions with butadienes[622] and the conversion of *cis,cis*-cyclo-octa-1,5-diene into cyclo-octateraene dianion by reduced praseodymium.[623]

Miscellaneous Reductions

Reviews have appeared on reduction,[624] stereoselectivity in reduction,[625] photoreduction,[626] and hydrazine reduction.[627] The mechanisms of reduction by hydrazine of *p*-nitrotoluene[628] and epoxystearates[629] have been studied. Primary amines undergo reductive deamination on treatment with an excess of hydroxylamine-sulphonic acid (Scheme 14).[630] An abnormal Wolff–Kishner reaction is observed

$$RNH_2 + NH_2OSO_3H \longrightarrow RNHNH_2 \longrightarrow R{-}N{=}N{-}H \longrightarrow RH$$

$$NH_2OSO_3H \xrightarrow{\bar{O}H} {:}NH$$

SCHEME 14

with 7-methylenebicyclo[3.3.1]nonan-3-one.[631] In the McFadyen–Stevens reduction, vacuum pyrolysis of lithium 1-acyl-2-toluene-*p*-sulphonylhydrazines produces aldehydes.[632] The mechanism of reduction of quinones by phenylhydrazine has been studied by the CIDNP method.[633] α-Amino-acids reduce 4,6-di-*tert*-butylquinone to benzoxazoles (**69**)[634] rather than undergoing the normal Strecker degradation to an aldehyde.[635]

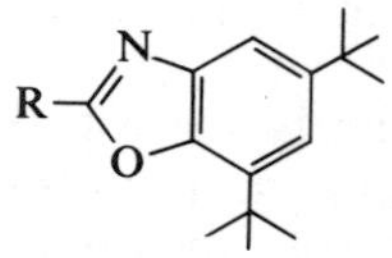

(69)

Phosphorus pentasulphide is a mild and effective reducing agent for the conversion of sulphoxides into sulphides;[636] thiosulphoxides, $R_2S{=}S$, have been shown to be intermediates in the reaction[637] and have been trapped in the intramolecular rearrangement of allyl sulphoxides into allyl disulphides (Scheme 15).[638] Allenyl phenyl sulphoxides are converted into the corresponding sulphides by phosphorus pentasulphide.[639] A thiosulphoxide is also an intermediate in the reduction of sulphoxides by hexamethyldisilathiane, $(Me_3Si)_2S$.[640]

The rate of reduction of azides to amines by propane-1,3-dithiol is very sensitive to the functional groups present.[641] The reduction of *N,N*-dimethylaniline *N*-oxide by sulphides involves protonation of the amine oxide followed by rate-determining

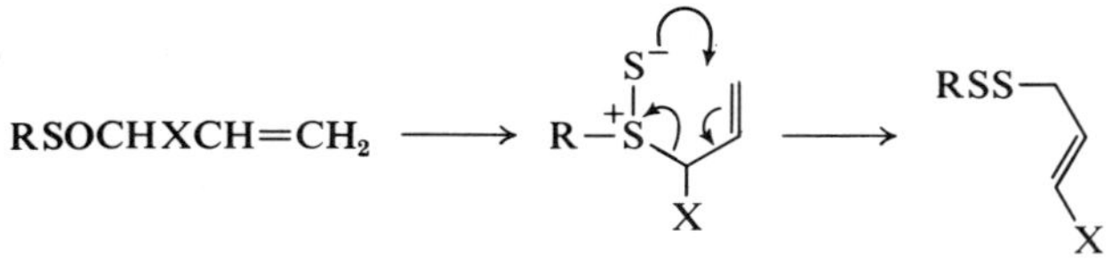

RSOCHXCH=CH_2 ⟶ R–$\overset{+}{S}$(–S^-)… ⟶ RSS–CH$_2$CH=CHX

SCHEME 15

nucleophilic attack by sulphide.[642] The rate-determining step in the reduction of dehydromethionine by thiols is general acid catalysis of an S_N2 displacement in which thiolate anion attacks the sulphur of dehydromethionine (**70**) in a reaction concerted with protonation of nitrogen.[643]

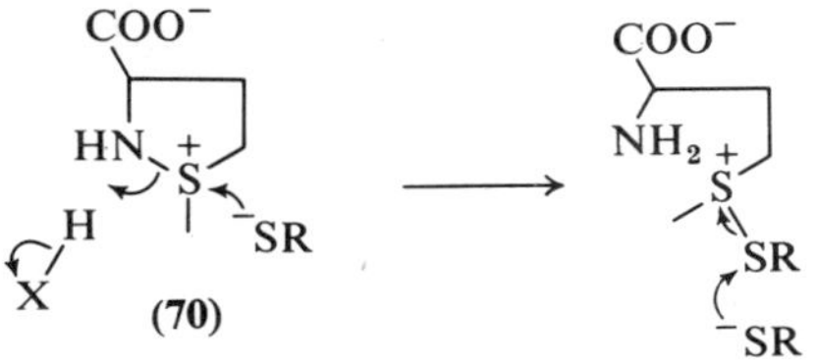

(**70**)

Replacement of nitro-groups in tertiary nitro-compounds by hydrogen is achieved directly by treatment with sodium methyl sulphide, by a mechanism involving radicals and radical anions.[644] β-Nitro-sulphones are reduced to alkenes by sodium sulphide by a radical-chain mechanism.[645] The sodium dithionite reduction of carbonyl compounds has been studied.[646] Reduction of the cyclic monosulphoxide (**71**) by iodide is faster than that of simple sulphoxides and probably proceeds *via* the dithioether dication (**72**) as an intermediate.[647]

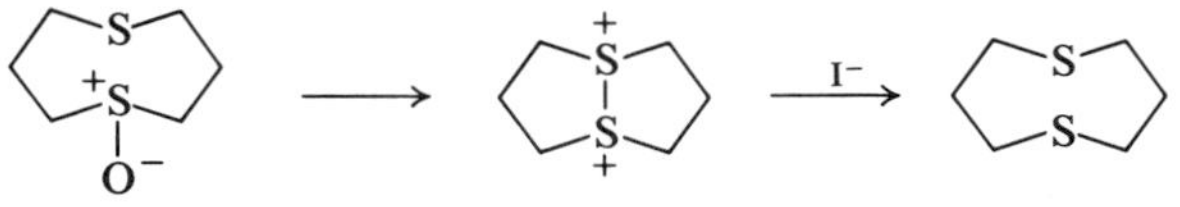

(**71**) ⟶ (**72**) $\xrightarrow{I^-}$

A CNDO/2 study of the ethylene–borane complex has been used as a model of the transition state in hydroboration.[648] The reduction of toluenesulphonyl-hydrazones of α,β-acetylenic ketones with catechol–borane occurs with migration of an acetylenic bond to form an allene.[649] Direct evidence has been obtained that hydroboration involves *cis*-addition of B—H to an alkene.[650] The high sensitivity of 9-borabicyclo[3.3.1]nonane to structure, observed with simple olefins, is also observed with conjugated dienes.[651] The first convenient hydroboration by a polyhedral borane has been reported.[652]

Hydrogenation, Hydrogenolysis, and Dehydrogenation

Reviews have appeared on catalytic asymmetric dehydrogenation,[653] reduction of nitrogen compounds in the presence of transition-metal complexes,[654] selective hydrogen-transfer reactions in alcohols, ketones and olefins by transition-metal catalysts,[655] and selective or complete hydrogenation of unsaturated ketones with retention of the carbonyl group in the products.[656]

There has been considerable interest in homogeneous asymmetric hydrogenation in the presence of rhodium complexes, including catalysts with low ratios of phosphine to rhodium.[657]

On the basis of kinetic studies of hydrogenation of cyclohexene by chlorotris-(triphenylphosphine)rhodium (**73**), a new mechanistic scheme has been proposed in which (**74**) is the active species in the hydrogenation; (**74**) is formed by oxidative addition of hydrogen to (**73**), followed by dissociation of triphenylphosphine.[658]

$(Ph_3P)_3RhCl$ (**73**)

$(Ph_3P)_2RhH_2Cl$ (**74**)

The kinetic studies indicate the complexation of a benzene solvent molecule in the catalytic sequence.[659] Homogeneous catalysis by rhodium complexed with amino-acids of hydrogenation of olefins is first-order both in substrate and catalyst.[660] The mechanism of asymmetric hydrogenation of olefins by complexes of rhodium with chiral phosphines[661] has been studied by ^{31}P-NMR spectroscopy; the stereochemistry of the reaction is defined in an olefin-binding complex whereby an intermediate square-planar chelate is formed; the subsequent addition of hydrogen has only a minor influence on the optical yield.[662] The enantioselective hydrogenation of α-(acetylamino)cinnamic acid by DIOP–rhodium complex involves hydrogen transfer within the co-ordination sphere of the metal as the rate-determining step.[663]

The co-ordination chemistry and catalytic properties of cationic 1,2-bis(diphenylphosphino)ethanerhodium(I) complexes are different from those of corresponding complexes containing monodentate phosphorus ligands. The striking stereoselectivity of complexes with bidentate ligands in asymmetric catalytic hydrogenation reflects the tendency of the functional groups typically present (such as Ph, COOH, and NHCOR) to co-ordinate to rhodium and thus exhibit a strong orienting influence.[664] The hydrogenation of itaconic acid with cationic chiral pyrrolidinophosphinerhodium(I) complexes gives a high yield of (*S*)(−)-methylsuccinic acid.[665–667] (*R*)-Prophos (**75**) has only a single methyl group to constrain the chirality of the chelate ring; however, soluble rhodium(I) complexes with this ligand act as efficient asymmetric hydrogenation catalysts for the formation of chiral amino-acids. Moreover, the complex is capable of breeding its own chirality, so that large quantities of (*R*)-prophos can be produced.[668]

Among many examples of the reduction of prochiral olefinic carboxylic acids by asymmetric hydrogenation catalysed by phosphinerhodium(I) complexes,[669–672] the following chiral ligands have been used: carbocyclic analogues of DIOP;[673]

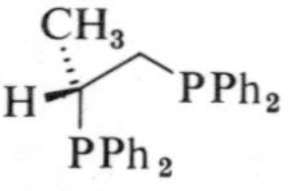

CH_3, H, PPh_2, PPh_2
(**75**)

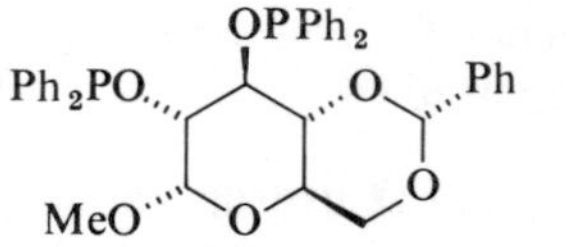

$OPPh_2$, Ph_2PO, O, Ph, MeO, O, O
(**76**)

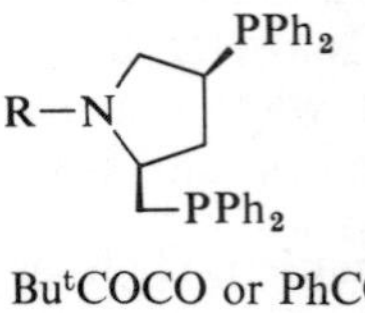

PPh_2, R—N, PPh_2
(R = Bu^{t}COCO or PhCO)
(**77**)

(1*R*, 2*R*)-*trans*-1,2-bis(diphenylphosphinomethyl)cyclobutane;[674] the chiral diphosphinite (**76**);[675] the pyrrolidine-substituted phosphines (**77**);[676] (1*S*, 2*S*)-*trans*-1,2-bis(diphenylphosphinomethyl)cyclohexane;[677] (1*R*, 2*R*)-bis-(*N*-diphenylphosphinomethylamino)cyclohexane,[678] and other chiral amino-phosphines.[679] Asymmetric hydrogenation of ketones has been achieved with a chiral amino-phosphine-rhodium catalyst.[680] There have been several examples of polymer-supported chiral phosphine-rhodium asymmetric hydrogenation catalysts;[681–683] a protein has been converted into a homogeneous hydrogenation catalyst by site-specific modification with a diphosphinerhodium system.[684]

An efficient polymer-bound hydrogenation catalyst **(78)**[685] is produced by anchoring bidentate anthranilic acid to chloromethylated polystyrene beads, followed by treatment with palladium chloride in refluxing solution; ESCA studies have established that **(78)** is the first example of an effective palladium(II) hydrogenation catalyst.[686] The small-angle X-ray scattering determination of particle size

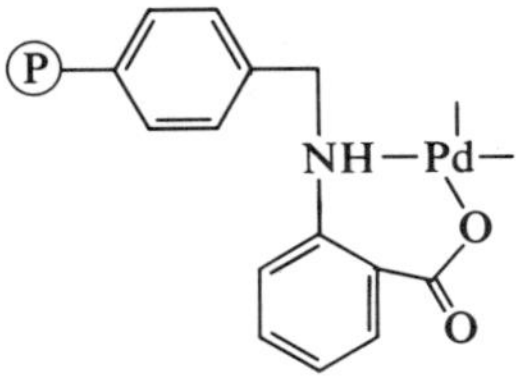

(78)

of palladium metal supported on vitreous materials shows a very high degree of dispersion of the metal, which is linearly correlated to the catalytic activity in hydrogenation reactions.[687] The kinetics of the hydrogenolysis of 1-phenylethane-1,2-diol to 2-phenylethanol[688] and of the hydrogenation of 1,3-dienes[689] in the presence of palladium catalysts have been studied. Trialkylammonium formate–palladium reduces α,β-unsaturated carbonyl compounds to saturated carbonyl compounds.[690] The solvent effects on the catalytic properties of phosphine–palladium complexes[691] and some hydrogenolytic asymmetric transamination reactions catalysed by palladium[692] have been investigated.

A chelating chiral sulphoxide ligand, (2*R*, 3*R*)-2,3-*O*-isopropylidene-2,3-dihydroxy-1,4-bis(methylsulphinyl)butane complexed to ruthenium(II) is an efficient asymmetric homogeneous catalyst for the hydrogenation of prochiral olefinic substrates.[693] Both $RuHCl(\eta\text{-}C_6Me_6)PPh_3$[694] and $(\eta^6\text{-}C_6Me_3)Ru(\eta^4\text{-}C_6Me_6)$[695] are active homogeneous catalysts for the hydrogenation of arenes. Reaction of tertiary phosphine complexes of ruthenium(0) with allylic olefins produces isolable hydrido(π-allyl)ruthenium(II) compounds.[696] The reduction of unsymmetrical cyclic anhydrides to γ-lactones by $RuH_2(PPh_3)_4$ is regioselective; oxidative addition to the anhydride produces a ruthenium(IV) complex which undergoes reductive elimination.[697] Solvent effects on homogeneous palladium hydrogenation catalysts have been studied;[698] the homogeneous hydrogenations by palladium complexes of olefins,[699] carbonyl compounds,[700] and sugars[701] have been investigated. The enantioselective dehydrogenation of racemic secondary alcohols by chiral phosphine complexes of ruthenium(II) is sterically, rather than electronically, controlled.[702]

The effect of temperature[703] and of other factors[704] on the optical yield in the reduction of ketones by modified Raney nickel have been studied. The kinetics of hydrogenation of substituted acetophenones by various nickel catalysts[705] and the selective hydrogenation of isocyanides and nitriles by a nickel–isocyanide cluster catalyst[706] have been investigated.

High selectivity is observed in the hydrogenation of cyclopentadiene to cyclopentene[707] and of cyclopentenone to cyclopentanone[708] with bis(dimethylglyoximato)pyridinecobalt(II) as a catalyst; σ-cobaloxime complexes are involved as intermediates. Radicals are formed as intermediates in the homogeneous hydrogenations of methacrylic acid[709] and of nitrosobenzene[710] with a pentacyanocobaltate(II) catalyst. Homogeneous hydrogenation of long-chain conjugated diene esters, catalysed by octacarbonyldicobalt shows that the affinity of $HCo(CO)_3$ is greater for *cis* than for *trans* double bonds.[711]

Whether or not π-allylic structures are intermediates in platinum metal-catalysed hydrogenation of alkenes is an open question;[712] the formation of *trans*-cyclononene from cyclonona-1,2-diene with supported platinum metal provides some evidence for π-allylic intermediates.[713] The rôles of platinum catalysts in the hydrogenolysis of cyclopropane,[714] the hydrogenation of benzofuran derivatives,[715] and in the simultaneous hydrogenation and hydroformylation of olefins[716] have been discussed. The ability of fluidized powders of SiO_2–Al_2O_3 to catalyse the hydrogenation of ethene is greatly increased when they receive hydrogen atoms spilling over from platinum–aluminium oxide pellets.[717]

Although catalytic hydrogenation of the olefinic double bond in α,β-unsaturated carbonyl is performed by a number of transition-metal complexes, the selective reduction of the carbonyl function is rare; α,β-unsaturated aldehydes are converted into the allylic alcohols by a hydridoiridium–sulphoxide complex, $HIrCl_2(Me_2SO)_3$, in propanol as solvent and source of hydrogen.[718] Hydrogenation of penta-1,3-diene in the presence of zinc oxide produces mainly pent-1-ene whereas other catalysts give predominantly pent-2-ene.[719] Copper(II) complexes of tartaric acid can act as asymmetric hydrogenation catalysts.[720] Other hydrogenation catalysts have included: bicyclopentadienylniobium species,[721] hydridopentacarbonylmanganese(I),[722] arenetricarbonyl chromium complexes,[723] and lanthanum cobalt hydride[724] and nickel hydride.[725] Carbon monoxide and water may replace hydrogen in the reduction of nitro-compounds to amines in the presence of pentacarbonyliron as catalyst;[726] the rôle of carbon monoxide in the carbon monoxide–water reduction of organic compounds has been rationalized.[727] The effect of alcohol on the hydrogenation of *p*-nitrotoluene,[728] the rapid metal-catalysed transfer reduction of nitro-compounds to *N*-substituted hydroxylamines,[729] and the competitive hydrogenation of olefins on multicomponent systems[730] have been studied. The heats of hydrogenation of *cis*- and *trans*-cyclooctene have been redetermined in hexane in order to avoid solvation problems; the value for the *cis*-isomer is comparable to earlier values but the value for the *trans*-isomer is 2 kcal mol^{-1} greater than was previously reported.[731]

Reviews on the dehydrogenation of polycyclic hydroaromatic compounds,[732] the oxidative dehydrogenation of hydrocarbons,[733] and the dehydrogenation of cycloalkanes by palladium(II) complexes[734] have appeared. Mechanisms of dehydrogenations of the following systems have been proposed: of tertiary amines[735] and enamines[736] by diacyldi-imides; of ethanol by atomic vapours of zinc;[737] of hydroaromatic compounds[738] and benzylic positions[739] by quinones; and of reducing sugars in the presence of transition metals.[740]

Reduction and Oxidation of Biological Interest

The NADH reduction of unsaturated linkages has been reviewed[741] and the one-electron transfer mechanism in reduction by NADH models has been discussed.[742] Several investigations on the rôle of metal ions in NADH–model reductions[743–746] and asymmetric reductions by chiral NADH models[747–750] have been reported. The stereochemistry of the reduction of iminium salts by NADH models has been studied; in the reduction of a series of iminium perchlorates of 4-*tert*-butylcyclohexanone by Hantzsch esters, it was found that the higher the energy level of the LUMO of the iminium salt, the more axial attack by Hantzsch ester was observed.[751]

Analysis of reactants and products in the reduction of pyridinecarboxaldehyde by the benzyl analogue (**79**) of NADH shows that radical intermediates are not involved in either the photochemical or the thermal reaction; photolysis leads to loss of (**79**) but not to the production of reduced substrates, and the radicals detected by ESR spectroscopy are only artefacts.[752] The NADH model (**80**) is

$CONH_2$ N CH_2Ph

(**79**)

$CONH_2$ N CH_2Ph

(**80**)

resistant to acid-catalysed decomposition; added acid greatly accelerates the reduction of carbonyl substrates in a general acid-catalysed process.[753] Several other studies on the chemistry of NADH have also been reported.[754–770]

Mechanistic work on the oxidation of vitamin C,[771–774] microbial reduction,[775–778] and many other oxidation–reduction systems of biological interest has continued.[782–800]

References

[1] Rocek, J., and Peng, T.-Y., *J. Am. Chem. Soc.*, **99,** 7622 (1977).
[2] Srinivasan, K. G., and Rocek, J., *J. Am. Chem. Soc.*, **100,** 2789 (1978).
[3] Krumpoic, M., DeBoer, B. G., and Rocek, J., *J. Am. Chem. Soc.*, **100,** 145 (1978).
[4] Corey, E. J., and Boger, D. L., *Tetrahedron Lett.*, **1978,** 2461.
[5] Banerji, K. K., *J. Chem. Soc., Perkin Trans.* 2, **1978,** 639.
[6] Venkataraman, K. S., Sundaram, S., and Venkatasubramanian, N., *Indian J. Chem.*, **15B,** 84 (1978); *Chem. Abs.*, **88,** 135915 (1978).
[7] Banerji, K. K., *J. Chem. Res.*, **1978,** (2) 193; (M) 2561.
[8] Frechet, J. M. J., Warnock, J., and Farrall, M. J., *J. Org. Chem.*, **43,** 2618 (1978).
[9] Salmond, W. G., Barta, M. A., and Havens, J. L., *J. Org. Chem.*, **43,** 2057 (1978).
[10] Chung, S. K., *Tetrahedron Lett.*, **1978,** 3211.
[11] Cardillo, G., and Shimizu, M., *J. Org. Chem.*, **42,** 4268 (1977).
[12] Warrener, R. N., Lee, T. S., Russell, R. A., and Paddon-Row, M. N., *Aust. J. Chem.*, **31,** 1113 (1978).
[13] Hanaya, K., Awano, S., and Kudo, H., *Bull. Chem. Soc. Jpn.*, **51,** 342 (1978).
[14] Baliah, V., and Chandrasekharan, J., *Indian J. Chem.*, **15B,** 1035 (1977); *Chem. Abs.*, **88,** 136020 (1978).
[15] Durand, R., Geneste, P., Lamaty, G., Moreau, C., Pomares, O., and Roque, J. P., *Recl. Trav. Chim. Pays-Bas*, **97,** 42 (1978).
[16] Pletcher, D., and Tait, S. J. D., *Tetrahedron Lett.*, **1978,** 1601.
[17] Vyas, D. N., Malkani, R. K., Mehta, S. B., and Mehta, S. H., *Chem. Era*, **13,** 33 (1977); *Chem. Abs.*, **88,** 49999 (1978).
[18] Tyupako, N. F., and Stepanyan, A. A., *Zh. Prikl. Khim. (Leningrad)*, **50,** 1876 (1977); *Chem. Abs.* **87,** 183728 (1977).
[19] Das, K. N., and Huque, M. M., *Dacca Univ. Stud.*, **25,** 1 (1977); *Chem. Abs.*, **89,** 50481 (1978).
[20] Bhalekar, A. A., Shanker, R., and Bakne, G. V., *Indian J. Chem.*, **15A,** 705 (1977); *Chem. Abs.*, **88,** 189640 (1978).

[21] Kozorez, L. A., Zernov, A. G., and Yakobi, V. A., *Zh. Prikl, Khim.* (*Leningrad*), **51,** 1826 (1978); *Chem. Abs.*, **89,** 146179 (1978).
[22] Aclinou, P., Gastambide, B., Deruaz, D., Desage, Y., and Guilluy, R., *Tetrahedron*, **34,** 393 (1978).
[23] Antelo, J. M., Cachaza, J. M., and Castro, A., *Acta Cient. Compostelana*, **14,** 375 (1977); *Chem. Abs.*, **87,** 167207 (1977).
[24] Ramesh, S., and Bhatt, M. V., *Curr. Sci.*, **46,** 809 (1977); *Chem Abs.*, **88,** 104387 (1978).
[25] Sen Gupta, K. K., Sen Gupta, S., and Sen, P. K., *Indian J. Chem.*, **15A,** 1082 (1977); *Chem. Abs.*, **88,** 169339 (1978).
[26] Baliah, V., and Chandrasekharan, J., *Indian J. Chem.*, **15B,** 558 (1977); *Chem. Abs.*, **88,** 5951 (1978).
[27] Stewart, R., and Spitzer, U. A., *Can. J. Chem.*, **56,** 1273, (1978).
[28] Rao, P. R., and Sundaram, E. V., *J. Indian Chem. Soc.*, **55,** 107 (1978); *Chem. Abs.*, **89,** 89920 (1978).
[29] Radhakrishnamurti, P. S., and Rao, M. D. P., *Indian J. Chem.*, **15A,** 524 (1977); *Chem. Abs.*, **88,** 5969 (1978).
[30] Garshina, S. I., Prokopenkova, L. V., Tararykov, G. M., and Perelyin, V. M., *Izv. Vyssh. Uchebn. Zaved., Khim. Khim. Tekhnol.*, **21,** 762 (1978); *Chem Abs.*, **89,** 89912 (1978).
[31] Cadet, J., and Teoule, R., *J. Chem. Res.*, **1977,** (S) 296; (M) 3215.
[32] Poje, M., and Balenovic, K., *Tetrahedron Lett.*, **1978,** 1231.
[33] Firouzabadi, H., and Ghaderi, E., *Tetrahedron Lett.*, **1978,** 839.
[34] Sala, T., and Sargent, M. V., *J. Chem. Soc., Chem. Commun.*, **1978,** 253.
[35] Lee, D. G., and Chang, V. S., *J. Org. Chem.*, **43,** 1532 (1978).
[36] Freeman, F., Bond, D. L., Chernow, S., Davidson, P. A., and Karchefski, E. M., *Int. J. Chem. Kinet.*, **10,** 911 (1978).
[37] Senent, S., Mata, F., De La Pena, M. A., *Rev. Roum. Chim.*, **23,** 841 (1978).
[38] Gopalan, R., and Sugumar, R. W., *Indian J. Chem.*, **15A,** 198 (1978); *Chem. Abs.*, **89,** 23481 (1978).
[39] Sharma, T. C., Saksena, V., and Reddy, N. J., *Acta Chim. Acad. Sci. Hung.*, **93,** 415 (1977); *Chem. Abs.*, **88,** 104386 (1978).
[40] Nakajima, M., Hisada, R., and Anselme, J.-P., *J. Org. Chem.*, **43,** 2693 (1978).
[41] Tabushi, I., and Koga, N., *Tetrahedron Lett.*, **1978,** 5017.
[42] Kurz, M. E., and Chen, T. R., *J. Org. Chem.*, **43,** 239 (1978).
[43] Digurov, N. G., Gavrilenko, N. D., Bukharkina, T. V., and Lebedev, N. N., *Deposited Doc.*, **1976,** Viniti 1348; *Chem. Abs.*, **88,** 50010 (1978).
[44] Bhatnagar, R. P., and Fadnis, A. G., *Monatsh. Chem.*, 109, **319,** (1978).
[45] Malkani, R. K., Suresh, K. S., and Bakore, G. V., *J. Indian Chem. Soc.*, **55,** 215 (1978); *Chem. Abs.*, **89,** 146158 (1978).
[46] Malkani, R. K., Suresh, K. S., and Bakore, G. V., *Chem. Era*, **12,** 390 (1976); *Chem. Abs.*, **87,** 200481 (1977).
[47] Bukharkina, T. V., Gavilenko, N. D., Digurov, N. G., and Knyazeva, N. A., *Kinet, Katal.*, **19,** 506 (1978); *Chem. Abs.*, **89,** 5708 (1978).
[48] Periasamy, M., and Bhatt, M. V., *Tetrahedron Lett.*, **1978,** 4561.
[49] Mentasti, E., Pelizetti, E., and Baiocchi, C., *J. Chem. Soc., Perkin Trans.* 2, **1978,** 77.
[50] Nyberg, K., and Wistrand, L.-G., *J. Org. Chem.*, **43,** 2613 (1978).
[51] Le Thuan, S., and Maitte, M. P., *Tetrahedron*, **34,** 1469 (1978).
[52] Hamilton, G. A., Adolf, P. K., de Jersey, J., DuBois, G. C., Dyrkacz, G. R., and Libby, R. D., *J. Am Chem. Soc.*, **100,** 1899 (1978).
[53] Whitesitt, C. A., and Herron, D. K., *Tetrahedron Lett.*, **1978,** 1737.
[54] Singh, M. P., Singh, A. K., and Tripathi, V., *J. Phys. Chem.*, **82,** 1222 (1978).
[55] Butler, R. N., Hanahoe, A. B., and King, W. B., *J. Chem. Soc., Perkin Trans.* 1, **1978,** 881.
[56] Singh, M. P., Singh, D. P., Singh, A. K., and Kumar, M., *Indian J. Chem.*, **15A,** 718 (1977); *Chem. Abs.*, **88,** 88809 (1978).
[56a] Suryawanshi, S. N., and Nayak, U. R., *Tetrahedron Lett.*, **1978,** 465.
[57] Strasak, K., *Chem. Listy*, **72,** 673 (1978); *Chem. Abs.*, **89,** 89790 (1978).
[58] Strasak, M., *React. Kinet, Catal. Lett.*, **7,** 387 (1977); *Chem. Abs.*, **88,** 88825 (1978).
[59] Strasak, M., and Majer, J., *Tetrahedron Lett.*, **1978,** 5037.
[60] Taylor, E. C., Andrade, J. G., Rall, G. J. H., and McKillop, A., *Tetrahedron Lett.*, **1978,** 3623.
[61] Kende, A. S., Liebeskind, L. S., Mills, J. E., Rutledge, P. S., and Curran, D. P., *J. Am. Chem. Soc.*, **99,** 7082 (1977).

[62] Radhakrishnamurti, P. S., and Pati, S. N., *Indian J. Chem.*, **16A**, 139 (1978); *Chem. Abs.*, **88**, 189675 (1978).
[63] Leon, M. A., and Cabaleiro, M. C., *An. Asoc. Quim. Argent.*, **64**, 331 (1976); *Chem. Abs.*, **89**, 107486 (1978)
[64] Strasak, M., *Z. Naturforsch.*, **33B**, 224 (1978); *Chem. Abs.*, **88**, 169455 (1978).
[65] Urbanec, J., Strasak, M., Repasova, I., and Hrusovsky, M., *Chem. Prum.*, **27**, 346 (1977); *Chem. Abs.*, **87**, 167198 (1977).
[66] Balakrishnan, T., and Santappa, M., *Int. J. Chem. Kinet.*, **10**, 833 (1978).
[67] Srinivasan, V. S., and Venkatasubramanian, N., *Indian J. Chem.* **15A**, 791 (1977); *Chem. Abs.*, **88**, 61877 (1978).
[68] Cenci, S., Di Furia, F., Modena, G., Curci, R., and Edwards, J. O. *J. Chem. Soc., Perkin Trans.* 2, **1978**, 979.
[69] Kishan, B. H., and Sundaram, E. V., *Indian J. Chem.*, **16A**, 322 (1978); *Chem. Abs.*, **89**, 146153 (1978).
[70] Murthy, G. S. S., Sethuram, B., and Rao, T. N., *Indian J. Chem.*, **15A**, 880 (1977); *Chem. Abs.*, **88**, 88815 (1978).
[71] Narasimhan, S., and Venkatasubramanian, N., *Indian J. Chem.*, **16A**, 349 (1978); *Chem. Abs.*, **89**, 146265 (1978).
[72] Pillai, G. C., Rajaram, J., and Kuriacose, J. C., *Indian J. Chem.*, **15A**, 608 (1977); *Chem. Abs.*, **88**, 104382 (1978).
[73] Haldorsen, K. M., *Carbohydr. Res.*, **63**, 61 (1978).
[74] Murty, G. S. S., Sethuram, B., and Rao, T. N., *Indian J. Chem.*, **15A**, 1073 (1977); *Chem. Abs.*, **88**, 189652 (1978).
[75] Rao, P. V. K., and Ramaiah, A. K., *Indian J. Chem.*, **16A**, 415 (1978); *Chem. Abs.*, **89**, 146167 (1978).
[76] Sen Gupta, K. K., and Chatterjee, H. R., *Inorg. Chem.*, **17**, 2429 (1978); *Chem. Abs.*, **89**, 146132 (1978).
[77] Narasimhan, S., and Venkatasubramanian, N., *Indian J. Chem.*, **15A**, 355 (1977); *Chem. Abs.*, **87**, 167237 (1977).
[78] Barkin, S., Bixon, M., Noyes, R. M., and Bar-Eli, K., *Int. J. Chem. Kinet.*, **10**, 619 (1978).
[79] Kaner, R. J., and Epstein, I. R., *J. Am. Chem. Soc.*, **100**, 4073 (1978).
[80] Vadava, K. D. S., *Indian J. Chem.*, **16A**, 347 (1978); *Chem. Abs.*, **89**, 146155 (1978).
[81] Baciocchi, E., Mei, S., and Rol, C., *J. Org. Chem.*, **43**, 2919 (1978).
[82] Maini, S., and Mandolini, L., *J. Org. Chem.*, **43**, 3236 (1978).
[83] Rao, M. A., Adinarayana, M., Sethuram, B., and Rao, T. N., *Indian J. Chem.*, **16A**, 440 (1978); *Chem. Abs.*, **89**, 146170 (1978).
[84] Arcoleo, G., Calvaruso, G., Cavasino, F. P., and Sbriziolo, C., *Inorg. Chim. Acta*, **23**, 227 (1977); *Chem. Abs.*, **87**, 167197 (1977).
[85] Singh, B., Saxena, P. K., Shukla, R. K., and Krishna, B., *J. Indian Chem. Soc.*, **54**, 378 (1977); *Chem. Abs.*, **88**, 5955 (1978).
[86] Ignaczak, M., and Deka, M., *Rocz. Chem.*, **51**, 1683 (1977); *Chem. Abs.*, **88**, 36945 (1978).
[87] Hara, H., Hosaka, M., Hoshino, O., and Umezawa, B., *Tetrahedron Lett.*, **1978**, 3809.
[88] Saavedra, J. E., *Tetrahedron Lett.*, **1978**, 1923.
[89] Carr, R. M., Norman, R. O. C., and Vernon, J. M., *J. Chem. Soc., Chem. Commun.*, **1977**, 854.
[90] Zinner, G., and Krause, T., *Arch. Pharm.* (*Weinheim*), **310**, 704 (1977).
[91] Poje, M., Gaspert, B., and Balenovic, K., *Recl. Trav. Chim. Pays-Bas*, **97**, 242 (1978).
[92] Banerji, K. K., Banerjee, S. K., and Shanker, R., *Indian J. Chem.*, **15A**, 702 (1977); *Chem. Abs.*, **88**, 189639 (1978).
[93] Radhakrishnamurti, P. S., and Pati, S. N., *Indian J. Chem.*, **16A**, 322 (1978); *Chem. Abs.*, **89**, 146152 (1978).
[94] Berentzveig, V. V., Fedorov, A. V., Pleskach, N. I., and Rudenko, A. P., *Vestn. Mosk. Univ., Ser.* **2**: *Khim.*, **19**, 75 (1978); *Chem. Abs.*, **89**, 5688 (1978).
[95] Jacobson, S. E., Tang, R., and Mares, F., *J. Chem. Soc., Chem. Commun.*, **1978**, 888.
[96] Ishii, Y., Adachi, A., Imai, R., and Ogawa, M., *Chem. Lett.*, **1978**, 611.
[97] Vedejs, E., Engler, D. A., and Telschow, J. E., *J. Org. Chem.*, **43**, 188 (1978).
[98] Liebeskind, L. S., Sharples, K. B., Wilson, R. D., and Ibers, J. A., *J. Am. Chem. Soc.*, **100**, 7061 (1978).
[99] Muccigrosso, D. A., Jacobson, S. E., Apgar, P. A., and Mares, F., *J. Am. Chem. Soc.*, **100**, 7063 (1978).
[100] Breslow, R., and Maresca, L. M., *Tetrahedron Lett.*, **1977**, 623.

[101] Breslow, R., and Maresca, L. M., *Tetrahedron Lett.*, **1978,** 887.
[102] Olah, G. A., and Welch, J., *J. Org. Chem.*, **43,** 2830 (1978).
[103] Olah, G. A., and Welch, J., *J. Am. Chem. Soc.*, **100,** 5396 (1978).
[104] Savel'yanova, R. T., and Manakov, M. N., *Izv. Vyssh. Uchebn. Zaved., Khim. Khim. Tekhnol.*, **20,** 1777 (1977); *Chem. Abs.*, **88,** 88865 (1978).
[105] Savel'yanova, R. T., and Manakov, M. N., *Izv. Vyssh. Uchebn. Zaved., Khim. Khim. Tekhnol.*, **20,** 1303 (1977); *Chem. Abs.*, **88,** 5985 (1978).
[106] Savel'yanova, R. T., and Manakov, M. N., *Izv. Vyssh. Uchebn. Zaved., Khim. Khim. Tekhnol.*, **20,** 1454 (1977); *Chem. Abs.*, **88,** 50001 (1978).
[107] Varadarajan, R., and Hossain, M. A., *Indian J. Chem.*, **16A,** 410 (1978); *Chem. Abs.*, **89,** 146166 (1978).
[108] Castle, L., and Smith, J. R. L., *J. Chem. Soc., Chem. Commun.*, **1978,** 704.
[109] Kassim, A. Y., *Inorg. Chim. Acta*, **27,** 243 (1978); *Chem. Abs.*, **89,** 23470 (1970).
[110] Singh, M. P., Singh, H. S., Arya, B. S., Singh, A. K., Tripathi, V., and Singh, A. K., *Indian J. Chem.*, **15A,** 716 (1977); *Chem. Abs.*, **88,** 88808 (1978).
[111] Singh, H. S., Singh, V. P., Singh, J. M., and Srivastava, P. N., *Indian J. Chem.*, **15A,** 517 (1977); *Chem. Abs.*, **88,** 5967 (1978).
[112] Igarashi, O., *J. Nutr. Sci. Vitaminol.*, **22,** 169 (1977); *Chem. Abs.*, **88,** 21749 (1978).
[113] Singh, H. S., Singh, V. P., Singh, J. M., and Srivastava, P. N., *Proc. Indian Natl. Sci. Acad.*, **43A,** 24 (1977); *Chem. Abs.*, **88,** 120342 (1978).
[114] Mulvey, D., and Waters, W. A., *J. Chem. Soc., Perkin Trans.* 2, **1977,** 1868.
[115] Hewgill, F. R., and Howie, G. B., *Aust. J. Chem.*, **31,** 907 (1978).
[116] Hewgill, F. R., and Howie, G. B., *Aust. J. Chem.*, **31,** 1061 (1978).
[117] Hewgill, F. R., and Howie, G. B., *Aust. J. Chem.*, **31,** 1069 (1978).
[118] Bunting, J. W., Lee-Young, P. A., and Norris, D. J., *J. Org. Chem.*, **43,** 1132 (1978).
[119] Rao, P. V. S., Murty, B. A. N., Murty, R. V. S., and Murty, K. S., *J. Indian Chem. Soc.*, **55,** 207 (1978); *Chem. Abs.*, **89,** 146157 (1978).
[120] Chandwani, B. L., and Menghani, G. D., *Indian J. Chem.*, **15A,** 560 (1977); *Chem. Abs.*, **88,** 5970 (1978).
[121] Patrick, D. W., Truesdale, L. K., Biller, S. A., and Sharpless, K. B., *J. Org. Chem.*, **43,** 2628 (1978)
[122] Herranz, E., Biller, S. A., and Sharpless, K. B., *J. Am. Chem. Soc.*, **100,** 3598 (1978).
[123] Singh, H. S., Singh, V. P., Singh, J. M., and Srivastava, P. N., *Indian J. Chem.*, **15A,** 520 (1977); *Chem. Abs.*, **88,** 5968 (1978).
[124] Singh, R. N., Singh, R. K., and Singh, H. S., *J. Chem. Res.*, **1978,** (S) 176; (M) 2182.
[125] Singh, R. N., Singh, R. K., and Singh, H. S., *J. Chem. Res.*, **1977,** (S) 249; (M) 2971.
[126] Singh, R. N., and Singh, H. S., *Indian J. Chem.*, **16A,** 145 (1978); *Chem. Abs.*, **88,** 189676 (1978).
[127] Veale, H. S., Levin, J., and Swern, D., *Tetrahedron Lett.*, **1978,** 503.
[128] Ayres, D. C., and Gopalan, R., *J. Chem. Soc., Perkin Trans.* 1, **1978,** 588.
[129] Ayres, D. C., *J. Chem. Soc., Perkin Trans.* 1, **1978,** 585.
[130] Yatsimirskii, A. K., Ryabov, A. D., and Berezin, I. V., *Dokl. Akad. Nauk SSSR*, **236,** 942 (1977); *Chem. Abs.*, **88,** 5973 (1978).
[131] Backvall, J. E., *Tetrahedron Lett.*, **1978,** 163.
[132] Becker, Y., and Stille, J. K., *J. Am. Chem. Soc.*, **100,** 838 (1978).
[133] Becker, Y., and Stille, J. K., *J. Am. Chem. Soc.*, **100,** 845 (1978).
[134] Stille, J. K., and Lau, K. S. Y., *Acc. Chem. Res.*, **10,** 434 (1977).
[135] Kim, Y. H., Shinhama, K., Fukushima, D., and Oae, S., *Tetrahedron Lett.*, **1978,** 1211.
[136] Oae, S., Fukushima, D., and Kim, Y. H., *Chem. Lett.*, **1978,** 279.
[137] Lur'e, B. A., Svetlov, B. S., and Smagin, V. V., *Izv. Vyssh. Uchebn. Zaved., Khim. Khim. Tekhnol.* **20,** 1259 (1977); *Chem. Abs.*, **88,** 5958 (1978).
[138] Stanovnik, B., Tisle, M., Kunaver, M., Gabrijelcic, D., and Kocevar, K., *Tetrahedron Lett.*, **1978,** 3059.
[139] Golubev, V. A., Borislavskii, V. N., and Aleskandrov, A. L., *Izv. Akad. Nauk SSSR, Ser. Khim.*, **1977,** 2025; *Chem. Abs.*, **88,** 21757 (1978).
[140] Castells, J., Moreno-Maras, M., and Pujol. F., *Tetrahedron Lett.*, **1978,** 385.
[141] Perregaard, J., Scheibye, S., Meyer, H. J., Thomsen, I., and Lawesson, S. P., *Bull. Soc. Chim. Belg.*, **86,** 679 (1977).
[142] Trost, B. M., and Rigby, J. H., *Tetrahedron Lett.*, **1978,** 1667.
[143] Hojo, M., Masuda, R., and Hakotani, K., *Tetrahedron Lett.*, **1978,** 1121.

[144] Hoffman, R. V., Cadena, R., and Peolker, D. J., *Tetrahedron Lett.*, **1978,** 203.
[145] Tanikaga, R., Tanaka, K., and Kaji, J., *J. Chem. Soc., Chem. Commun.*, **1978,** 865.
[146] Omura, K., and Swern, D., *Tetrahedron*, **34,** 1651 (1978).
[147] Soysa, H. S. D., and Weber, W. P., *Tetrahedron Lett.*, **1978,** 1969.
[148] Raucher, S., *Tetrahedron Lett.*, **1978,** 2261.
[149] Airinei, A., and Oita, N., *Farmacia (Bucharest)* **25,** 149 (1977); *Chem. Abs.*, 88, **5978** (1978).
[150] Srinivasan, C., and Rajagopal, S., *Curr. Sci.*, **46,** 669 (1977); *Chem. Abs.*, **88,** 5971 (1978).
[151] Panigrahi, G. P., and Panda, R., *Indian J. Chem.*, **15A,** 1070 (1977); *Chem. Abs.*, **88,** 169338 (1978).
[152] Srivastava, S. P., and Shukla, A. K., *Indian J. Chem.*, **15A,** 603 (1977); *Chem. Abs.*, **88,** 73863 (1978).
[153] Jensen, A., and Ragan, M. A., *Tetrahedron Lett.*, **1978,** 847.
[154] Ram, N., and Sidhu, K. S., *Indian J. Chem. Soc.*, **16A,** 195 (1978); *Chem. Abs.*, **89,** 23480 (1978).
[155] Gevorkyan, M. G., Beileryan, N. M., and Ashtyan, A. P., *Arm. Khim. Zh.*, **31,** 291 (1978); *Chem. Abs.*, **89,** 128742 (1978).
[156] Gevorkyan, M. G., Karupetyan, T. G., Mar'yanyan, A. I., Levonyan, A. V., and Melkonyan, L. G., *Arm. Khim. Zh.*, **31,** 296 (1978); *Chem. Abs.*, **89,** 128743 (1978).
[157] Srivastava, S. P., Gupta, R. C., and Shukla, A. K., *Indian J. Chem.*, **15A,** 605 (1977); *Chem. Abs.*, **88,** 73914 (1978).
[158] Singh, R. N., Singh, L. N., and Singh, H. S., *Indian J. Chem.*, **15A,** 1117 (1977); *Chem. Abs.*, **88,** 189656 (1978).
[159] Sankpal, S. G., Patil, J. B., and Jagdale, M. H., *Sci. J. Shivagi Univ.*, **15,** 157 (1977); *Chem. Abs.*, **88,** 5981 (1978).
[160] Clerici, A., Minisci, F., Ogawa, K., and Surzur, J.-M., *Tetrahedron Lett.*, **1978,** 1149.
[161] Walling, C., and Camaioni, D. M., *J. Org. Chem.*, **43,** 3266 (1978).
[162] Singh, R. N., *Acta Cienc. Indica*, **3,** 320 (1977); *Chem., Abs.* **89,** 5701 (1978).
[163] Singh, R. N., Singh, L. N., and Singh, H. S., *Indian J. Chem.*, **15A,** 1117 (1977).
[164] Srivastava, S. P., and Kumar, A., *Indian J. Chem.*, **15A,** 1114 (1977); *Chem. Abs.*, **88,** 189654 (1978).
[165] Radhakrishnamurti, P. S., and Swany, B. R. K., *Indian J. Chem.*, **15A,** 1115 (1977); *Chem. Abs.*, **88,** 189655 (1978).
[166] Mehrotra, R. N., Shukla, A. K., and Srivastava, S. P., *J. Indian Chem. Soc.*, **54,** 799 (1977); *Chem. Abs.*, **89,** 146178 (1978).
[167] Murty, P. S. N., and Rao, P. V. S., *J. Indian Chem. Soc.*, **54,** 1043 (1977); *Chem. Abs.*, **89,** 107454 (1978).
[168] Srivastava, S. P., and Kumar, A., *Indian J. Chem.*, **15B,** 967 (1977); *Chem. Abs.*, **88,** 104393 (1978).
[169] Clive, D. J., *Tetrahedron*, **34,** 1049 (1978).
[170] Barton, D. H. R., Brewster, A. G., Hui, R. A. H. F., Lester, D. J., Ley, S. V., and Buck, T. G., *J. Chem. Soc., Chem. Commun.*, **1978,** 952.
[171] Barton, D. H. R., Lester, D. J., and Ley, S. V., *J. Chem. Soc., Chem. Commun.*, **1978,** 130.
[172] Barton, D. H. R., Cussans, N. J., and Ley, S. V., *J. Chem. Soc., Chem. Commun.*, **1978,** 393,
[173] Back, T. G., *J. Chem. Soc., Chem. Commun.*, **1978,** 278.
[174] Barton, D. H. R., Lester, D. J., and Ley, S. V., *J. Chem. Soc., Chem. Commun.*, **1978,** 276.
[175] Grieco, P. A., Yokoyama, Y., Gilman, S., and Ohfune, Y., *J. Chem. Soc., Chem. Commun.*, **1977,** 870.
[176] Laitalainen, T., Simonen, T., and Kivekas, R., *Tetrahedron Lett.*, **1978,** 3079.
[177] Bergman, J., and Engman, L., *Tetrahedron Lett.*, **1978,** 3279.
[178] Labar, D., Hevesi, L., Dumont, W., and Krief, A., *Tetrahedron Lett.*, **1978,** 1141.
[179] Labar, D., Dumont, W., Hevesi, L., and Krief, A., *Tetrahedron Lett.*, **1978,** 1145.
[180] Kuwajima, I., and Shimizu, M., *Tetrahedron Lett.*, **1978,** 1277.
[181] Mikolajczyk, M., and Luczak, J., *J. Org. Chem.*, **43,** 2132 (1978).
[182] Kice, J. L., and Lee, T. S. W., *J. Am. Chem. Soc.*, **100,** 5094 (1978).
[183] Campbell, M. M., and Johnson, G., *Chem. Rev.*, 78, 65 (1978).
[184] Radhakrishnamurti, P. S., and Sahu, B., *Indian J. Chem.*, **16A,** 259 (1978); *Chem. Abs.*, **88,** 42011 (1978).
[185] Radhakrishnamurti, P. S., and Sahu, B., *Indian J. Chem.*, **15A,** 700 (1977); *Chem. Abs.*, **88,** 88807 (1978).
[186] Singh, K. J., and Raine, N., *J. Indian Chem. Soc.*, **54,** 482 (1977); *Chem. Abs.*, **88,** 5977 (1978).

[187] Naidu, H. M., and Mahadevappa, D. S., *Monatsh. Chem.*, **109**, 269 (1978).
[188] Banerji, K. K., *Indian J. Chem.*, **15A**, 615 (1977); *Chem. Abs.*, **88**, 104383 (1978).
[189] Sharma, S. L., and Bansal, O. P., *Chem. Era*, **13**, 178 (1977), *Chem. Abs.*, **88**, 135919 (1978).
[190] Chandra, M., Lal, S., and Bansal, O. P., *J. Indian Chem. Soc.*, **54**, 1040 (1977); *Chem. Abs.* **89**, 42021 (1978).
[191] Ganapathy, K., and Jayagandhi, P., *Indian J. Chem.*, **16A**, 421 (1978); *Chem. Abs.*, **89**, 146169 (1978).
[192] Radhakrishnamurti, P. S., and Padhi, S. C., *Curr. Sci.*, **46**, 517 (1977); *Chem. Abs.*, **87**, 183721 (1977).
[193] Shanmuganathan, S., and Vivekanandan, S., *Indian J. Chem.*, **15A**, 428 (1977); *Chem. Abs.*, **87**, 167236 (1977).
[194] Reich, H. J., and Peake, S. L., *J. Am. Chem. Soc.*, **100**, 4888 (1978).
[195] Kudesia, V. P., and Sharma, B. K., *Indian J. Chem.*, **16A**, 347 (1978); *Chem. Abs.*, **89**, 146154 (1978).
[196] Sharma, S. C., Sharma, B. K., and Kudesia, V. P., *Acta Cienc. Indica*, **3**, 193 (1977); *Chem. Abs.*, **88**, 135921 (1978).
[197] Kudesia, V. P., and Sharma, B. K., *Acta Cienc. Indica*, **3**, 116 (1977); *Chem. Abs.*, **88**, 135920 (1978).
[198] Krishnan, S., Kuhn, D. G., and Hamilton, G. A., *J. Am. Chem. Soc.*, **99**, 8121 (1977).
[199] Kaberia, F., and Vickery, B., *J. Chem. Soc., Chem. Commun.*, **1978**, 459.
[200] Kudesia, V. P., and Mukherjee, S. K., *Indian J. Chem.*, **15A**, 513 (1977); *Chem. Abs.*, **88**, 5966 (1978).
[201] Rathor, B. S., and Grover, K. C., *J. Indian Chem. Soc.*, **54**, 1032 (1977); *Chem. Abs.*, **89**, 42020 (1978).
[202] Mushran, S. P., Bose, A. K., and Tiwari, J. N., *J. Indian Chem. Soc.*, **55**, 41 (1978); *Chem. Abs.*, **89**, 107482 (1978).
[203] Pinnick, H. W., and Lagis, N. G., *J. Org. Chem.*, **43**, 371 (1978).
[204] Singh, K., Tiwari, J. N., and Mushran, S. P., *Int. J. Chem. Kinet.*, **10**, 995 (1978).
[205] Rao, P. V. K., Frank, M. S., and Ramaiah, A. K., *Indian J. Chem.*, **16A**, 418 (1978); *Chem. Abs.*, **89**, 146168 (1978).
[206] Vijayalaxmi, V., and Sundaram, E. V., *Indian J. Chem.*, **15A**, 612 (1977); *Chem. Abs.*, **88**, 73915 (1978).
[207] Malik, Z. A., *Arabian J. Sci. Eng.*, **2**, 89 (1977); *Chem. Abs.*, **87**, 183541 (1977).
[208] Bender, D. R., Brennan, J. and Rapoport, H., *J. Org. Chem.*, **43**, 3354 (1978).
[209] Kim, Y. H., Takata, T., and Oea, S., *Tetrahedron Lett.*, **1978**, 2305.
[210] Allan, R. D., *Tetrahedron Lett.*, **1978**, 2199.
[211] Liu, K.-T., and Tong, Y.-C., *J. Org. Chem.*, **43**, 2717 (1978).
[212] Ishak, M. F., and Painter, T. J., *Carbohydr. Res.*, **64**, 189 (1978).
[213] Pati, S. C., and Sriramulu, Y., *Indian J. Chem.*, **16A**, 74 (1978); *Chem. Abs.*, **88**, 189648 (1978).
[214] Panigrahi, G. P., and Misro, P. K., *Indian J. Chem.*, **16A**, 201 (1978); *Chem. Abs.*, **89**, 23482 (1978).
[215] Panigrahi, G. P., and Misro, P. K., *Indian J. Chem.*, **15A**, 1066 (1977); *Chem. Abs.* **88**, 169337 (1978).
[216] Bailey, P. S., *Ozonation in Organic Chemistry*, Vol. 1, Academic Press, New York, 1978.
[217] Miyazaki, S., *Tokoshi Nyusu, Kagaku Kogyo Shriyo*, **12**, 55 (1977); *Chem. Abs.*, **88**, 73734 (1978).
[218] Yoneda, N., *Yuki Gosei Kagaku Kyokaishi*, **35**, 711 (1977); *Chem. Abs.*, **88**, 5638 (1978).
[219] Aleksandrov, Y. A., and Tarunin, B. I., *Khim. Elementoorg. Soedin.*, **4**, 16, (1977); *Chem. Abs.*, **88**, 5648 (1978).
[220] Suharo, Y., *Yukagaku*, **26**, 117 (1977); *Chem. Abs.*, **88**, 36650 (1978).
[221] Karban, J., McAttee, J. L., Belew, J. S., Mullica, D. F., Milligan, W. O., and Korp, D., *J. Chem. Soc., Chem. Commun.*, **1978**, 729.
[222] Ageta, H., Shiojima, K., Kamaya, R., and Masuda, K., *Tetrahedron Lett.*, **1978**, 899.
[223] Kuehne, H., Vaccani, S., Bauder, A., and Guenthard, H. H., *Chem. Phys.*, **28**, 11 (1978); *Chem. Abs.*, **88**, 189561 (1978).
[224] Kuehne, H., *Ber. Bunsenges. Phys. Chem.*, **82**, 15 (1978); *Chem. Abs.*, **88**, 120346 (1978).
[225] Brown, R. S., and Marcinko, R. W., *J. Am. Chem. Soc.*, **100**, 5584 (1978).
[226] Suenram, R. D., and Lovas, F. J., *J. Am. Chem. Soc.*, **100**, 5117 (1978).
[227] Harding, L. B., and Goddard, W. A., *J. Am. Chem. Soc.*, **100**, 7180 (1978).

[228] Hull, L. A., *J. Org. Chem.*, **43,** 2780 (1978).
[229] Bailey, P. S., and Ferrell, T. M., *J. Am. Chem. Soc.*, **100,** 899 (1978).
[230] Bailey, P. S., Ferrell, T. M., Rustaiyan, A., Seyhan, S., and Unruh, L. E., *J. Am. Chem. Soc.*, **100,** 894 (1978).
[231] Desvergne, J. P., and Bouas-Laurent, H., *J. Catal.*, **51,** 126 (1978); *Chem. Abs.*, **88,** 104396 (1978).
[232] Nikokavouras, J., and Vassilopoulos, G., *Monatsh. Chem.*, **109,** 831 (1978).
[233] Mile, B., and Morris, G. M., *J. Chem. Soc., Chem. Commun.*, **1978,** 263.
[234] Ramachandran, V., and Murray, R. W., *J. Am. Chem. Soc.*, **100,** 2197 (1978).
[235] Cherneva, D., Rakovski, S., and Shopov, D., *Izv. Khim.*, **10,** 84 (1977); *Chem Abs..*, **87,** 167401 (1977).
[236] Rakovski, S., Cherneva, D., and Shopov, D., *Izv. Khim.*, **10,** 395 (1977); *Chem. Abs.*, **89,** 23464 (1978).
[237] Buchi, G., and Wuerst, H., *J. Am. Chem. Soc.*, **100,** 294 (1978).
[238] Brady, W. T., and Saidi, K., *Tetrahedron Lett.*, **1978,** 721.
[239] Moriarty, R. M., White, K. B., and Chin, A., *J. Am. Chem. Soc.*, **100,** 5582 (1978).
[240] Sebedio, J.-L., and Ackman, R. G., *Can. J. Chem.*, **56,** 2480 (1978).
[241] Tarunin, B. I., Aleksandrov, Y. A., Perepltechikova, V. N., and Baklanov, N. V., *Kinet. Katal.*, **18,** 1329 (1977); *Chem. Abs.*, **88,** 36966 (1978).
[242] Gilbert, E., *Z. Naturforsch.*, **32B,** 1308 (1977); *Chem. Abs.*, **88,** 49998 (1978).
[243] Ito, Y., Ida, H., and Matsuura, T., *Tetrahedron Lett.*, **1978,** 3119.
[244] Beckwith, A. L. J., and Duong, T., *J. Chem. Soc., Chem. Commun.*, **1978,** 413.
[245] Preuss, T., Proksch, E., and Meijere, A., *Tetrahedron Lett.*, **1978,** 833.
[246] Trifilieff, E., Bang, L., Narula, A. S., and Ourisson, G., *J. Chem. Res.*, **1978,** (S) 64; (M) 601.
[247] Komissarov, V. D., Galimova, L. G., Komissarova, I. N., Shchereshovets, V. V., and Denisov, E. T., *Dokl. Akad. Nauk SSR*, **235,** 1350 (1977): *Chem. Abs.*, **87,** 200499 (1977).
[248] Vikhorev, A. A. Syroezhko, A. M., Korotkova, N. P., Fedotova, V. M., and Proskuryakov V. A., *Mater.-Vses, Mezhvuz. Konf. Ozonu, 2nd*, **1977,** 58; *Chem. Abs.*, **89,** 146176 (1978).
[249] Galstyan, G. S., and Dvortsevoi, M. M., *Mater.-Vses. Mezhvuz. Konf. Ozonu, 2nd*, **1977,** 22; *Chem. Abs.*, **89,** 146175 (1978).
[250] Galstyan, G. A., Yakobi, V. A., Dvortsevoi, M. M., and Galstyan, T. M., *Zh. Prikl. Khim. (Leningrad)*, **51,** 133 (1978); *Chem. Abs.*, **88,** 151765 (1978).
[251] Kovac, F., and Plasnicar, B., *J. Chem. Soc., Chem. Commun.*, **1978,** 122.
[252] Brudnik, B. M., Zlotskii, S. S., Imashev, U. B., and Rakhmankulov, D. L., *Dokl. Akad. Nauk SSSR*, **241,** 129 (1978); *Chem. Abs.*, **89,** 107496 (1978).
[253] Sundararaman, P., Salker, E. C., and Djerassi, C., *Tetrahedron Lett.*, **1978,** 1627.
[254] Pryor, W. A., and Kurz, M. E., *Tetrahedron Lett.*, **1978,** 697.
[255] Nakayamu, S., and Esaki, K., *Mizu Shori Gijutsu*, **91,** 241 (1978); *Chem. Abs.*, **89,** 89906 (1978).
[256] Kogel, W., and Schröder, G., *Tetrahedron Lett.*, **1978,** 623.
[257] Martinez, R. I., and Herron, J. T., *J. Chem. Kinet.*, **10,** 433 (1978); *Chem. Abs.*, **89,** 23469 (1978).
[258] Chaussin, R., Leriverend, P., and Paquer, D., *J. Chem. Soc., Chem. Commun.*, **1978,** 1032.
[259] Bailey, P. S., and Khashab, A.-I. Y., *J. Org. Chem.*, **43,** 675 (1978).
[260] Bailey, P. S., Southwick, L. M., and Carter, T. P., *J. Org. Chem.*, **43,** 2657 (1978).
[261] Bailey, P. S., Lerdal, D. A., and Carter, T. P., *J. Org. Chem.*, **43,** 2662 (1978).
[262] Kaduk, B. A., and Toby, S., *Int. J. Chem. Kinet.*, **9,** 829 (1977).
[263] Miura, H., Hirao, K., and Yonemitsu, O., *Tetrahedron*, **34,** 1805 (1978).
[264] Grudzinski, Z., and Roberts, S. M., *J. Chem. Soc., Perkin Trans.* 1, **1978,** 1182.
[265] Ogata, Y., Tomizawa, K., and Ikeda, T., *J. Org. Chem.*, **43,** 2417 (1978).
[266] Hocking, M. B., Ko, M., and Smyth, T. A., *Can. J. Chem.*, **56,** 2646 (1978).
[267] Cudlin, J., Steinerova, N., Sedmera, P., and Vokoun, J., *Collect. Czech Chem. Commun.*, **43,** 1808 (1978).
[268] Frankvoort, W., *Thermochim. Acta*, **25,** 35 (1978); *Chem. Abs.*, **89,** 146145 (1978).
[269] Ogata, Y., Sawaki, Y., and Shiroyama, M., *J. Org. Chem.*, **42,** 4061 (1977).
[270] Ogata, Y., Sawaki, Y., and Shimizu, H., *J. Org. Chem.*, **43,** 1760 (1978).
[271] Litovka, A. P., Baluev, V. V., and Rakhimshanova, S. A., *Kinet. Katal.*, **19,** 567 (1978); *Chem. Abs.*, **89,** 146138 (1978).

[272] Maruthamuthu, P., and Santappa, M., *Indian J. Chem.*, **15A,** 420 (1977); *Chem. Abs.*, **87,** 167235 (1977).
[273] Fedevish, M. D., Yatchishin, I. I., and Pirig, Y. N., *Zh. Fiz. Khim.*, **52,** 242 (1978); *Chem. Abs.*, **89,** 23457 (1978).
[274] Oae, S., Kim, Y. H., Fukushima, D., and Takata, T., *Pure Appl. Chem.*, **49,** 153 (1977); *Chem. Abs.*, **87,** 183714 (1977).
[275] Hoffman, M., and Edwards, J. O., *Inorg. Chem.*, **16,** 3333 (1977); *Chem. Abs.*, **87,** 200491 (1977).
[276] Plesnicar, B., Tasevski, M., and Asman, A., *J. Am. Chem. Soc.*, **100,** 743 (1978).
[277] Yamamoto, T., and Kimura, M., *J. Chem. Soc., Chem. Commun.*, **1977,** 948.
[278] Crandall, J. K., and Conover, W. W., *J. Org. Chem.*, **43,** 3533 (1978).
[279] Smith, L. L., Kulig, M. J., Miller, D., and Ansari, G. A. S., *J. Am. Chem. Soc.*, **100,** 6206 (1978).
[280] Dehal, S. S., Marples, B. A., and Stretton, R. J., *Tetrahedron Lett.*, **1978,** 2183.
[281] Paryzek, Z., *J. Chem. Soc., Perkin Trans.* 1, **1978,** 329.
[282] Rubottom, G. M., Gruber, J. M., Boeckman, R. K., Ramaiah, M., and Medwid, J. B., *Tetrahedron Lett.*, **1978,** 4603.
[283] Wynberg, H., and Greijdanus, B., *J. Chem. Soc., Chem. Commun.*, **1978,** 427.
[284] Shibaeva, L. V., Agabekov, V. E., Mitskevich, N. I., and Nekrashevich, G. G., *Vestsi Akad, Navuk BSSR, Ser,. Khim. Navuk*, **1977,** 12; *Chem. Abs.*, **87,** 167224 (1977).
[285] Takamoto, T., Ikeda, Y., Seta, A., and Sudoh, R., *J. Chem. Soc., Chem. Commun.*, **1978,** 350.
[286] Nixon, J. R., Cudd, M. A., and Porter, N. A., *J. Org. Chem.*, **43,** 4048 (1978).
[287] Adam, W., Bloodworth, J. A., Eggelte, H. J., and Loveitt, M. E., *Angew. Chem. Int. Edn.*, **17,** 209 (1978).
[288] Olah, G. S., and Ohnishi, R., *J. Org. Chem.*, **43,** 865 (1978).
[289] Nakasuji, K., Nakamura, T., and Murata, I., *Tetrahedron Lett.*, **1978,** 1539.
[290] Ogata, Y., and Tomizawa, K., *J. Org. Chem.*, 43, **261,** (1978).
[291] Kemal, C., and Bruice, T. C., *J. Am. Chem. Soc.*, **99,** 7064 (1977).
[292] Lucente, G., Pinnen, F., and Zanotti, G., *Tetrahedron Lett.*, **1978,** 3155.
[293] Caglioti, L. Gasparrini, F., Misiti, D., and Palmieri, G., *Tetrahedron*, **34,** 135 (1978).
[294] Hoffman, R. V., and Cadena, R., *J. Am. Chem. Soc.*, **99,** 8226 (1977).
[295] Boyd, D. R., Waring, L. C., and Jennings, W. B., *J. Chem. Soc., Perkin Trans.* 1, **1978,** 243.
[296] Curci, R., DiFuria, F., and Modena, G., *J. Chem. Soc., Perkin Trans.* 2, **1978,** 603.
[297] Ashworth, B., Gilbert, B. C., Holmes, R. G. G., and Norman, R. O. C., *J. Chem. Soc., Perkin Trans.* 2, **1978,** 951.
[298] Rigaudy, J., Breliere, C., and Scribe, P., *Tetrahedron Lett.*, **1978,** 687.
[299] Rigaudy, J., Baronne-Lafont, J., Defoin, A., and Cuong, N. K., *Tetrahedron*, **34,** 73 (1978).
[300] Defoin, A., Baronne-Lafont, J., Rigaudy, J., and Cuilhem, J., *Tetrahedron*, **34,** 83 (1978).
[301] Rigaudy, J., and Sparfel, D., *Tetrahedron*, **34,** 113 (1978).
[302] Salomon, R. G., Salomon, M. F., and Coughlin, D. J., *J. Am. Chem. Soc.*, **100,** 646 (1978).
[303] Sawyer, D. T., Gibian, M. J. Morrison, M. M., and Seo, E. T., *J. Am. Chem. Soc.*, **100,** 627 (1978).
[304] Ohkatsu, Y., and Tsuruta, T., *Bull. Chem. Soc. Jpn.*, **51,** 188 (1978).
[305] Takabe, T., Miyakawa, M., and Nikai, S., *Bull. Chem. Soc. Jpn.*, **51,** 321 (1978).
[306] McCormick, J. P., and Thomason, T., *J. Am. Chem. Soc.*, **100,** 312 (1978).
[307] Umstead, M. E., and Lin, M. C., *Chem. Phys.*, **23,** 353 (1977); *Chem. Abs.*, **88,** 21783 (1978).
[308] Havel, J. J., *J. Org. Chem.*, **43,** 762 (1978).
[309] Politsev, P., and Daikev, K. C., *Jerusalem Symp. Quantum Chem., Biochem.*, **1977,** 331; *Chem. Abs.*, **88,** 120434 (1978).
[310] Harding, L. B., and Goddard, W. A., *Tetrahedron Lett.*, **1978,** 747.
[311] Rousseau, G., LePerchec, P., and Conia, J. M., *Tetrahedron Lett.*, **1977,** 2517.
[312] Jefford, C. W., and Rimbault, C. G., *J. Org. Chem.*, **43,** 1908 (1978).
[313] Jefford, C. W., and Rimbault, C. G., *J. Am. Chem. Soc.*, **100,** 295 (1978).
[314] Jefford, C. W., and Rimbault, C. G., *J. Am. Chem. Soc.*, **100,** 6437 (1978).
[315] Jefford, C. W., and Rimbault, C. G., *J. Am. Chem. Soc.*, **100,** 6515 (1978).
[316] Okada, K., and Mukai, T., *J. Am. Chem. Soc.*, **100,** 6509 (1978).
[317] Chawla, H. M., Chibber, S. S., and Saigal, R., *Indian J. Chem.*, **15B,** 975 (1977); *Chem. Abs.*, **88,** 104394 (1978).
[318] Takeshita, H., Hatsui, T., and Shimooda, I., *Tetrahedron Lett.*, **1978,** 2889.

[319] McCapra, F., *J. Chem. Soc., Chem. Commun.*, **1977,** 946.
[320] McCapra, F., Beheshti, I., Burford, A., Hann, R. A., and Zaklika, K. A., *J. Chem. Soc., Chem. Commun.*, **1977,** 944.
[321] McCapra, F., and Burford, A., *J. Chem. Soc., Chem. Commun.*, **1977,** 874.
[322] Pedersen, C. L., and Lohse, C., *Tetrahedron Lett.*, **1978,** 3141.
[323] Zaklika, K. A., Burns, P. A., and Schaap, A. P., *J. Am. Chem. Soc.*, **100,** 318 (1978).
[324] Danen, W. C., and Arudi, R. L., *J. Am. Chem. Soc.*, **100,** 3944 (1978).
[325] Rebek, J., Wolf, S., and Mossman, A., *J. Org. Chem.*, **43,** 180 (1978).
[326] Jefford, C. W., Exarchou, A., and Cadby, P. A., *Tetrahedron Lett.*, **1978,** 2053.
[327] Moriarty, R. M., Chin, A., and Tucker, M. P., *J. Am. Chem. Soc.*, **100,** 5578 (1978).
[328] Schulte-Elte, K. H., Muller, B. L., and Rautenstrauch, V., *Helv. Chim. Acta*, **61,** 2777 (1978).
[329] Lerdal, D., and Foote, C. S., *Tetrahedron Lett.*, **1978,** 3227.
[330] Saito, I., Tamoto, K., Katsumara, A., Sugiyama, H., and Matsuura, T., *Chem. Lett.*, **1978,** 127.
[331] Heldevey, R. F., Hogeveen, H., and Schudde, E. P., *J. Org. Chem.*, **43,** 1912 (1978).
[332] Snyakin, A. P., Samsonova, L. V., Shlyapintoch, V. Y., and Ershov, V. V., *Izv. Akad. Nauk SSSR, Ser. Khim.*, **1978,** 55; *Chem. Abs.*, **88,** 120352 (1978).
[333] Ram, N., Bansal, W. R., Pandav, B. V., Sidhu, K. S., *Indian J. Chem.*, **15A,** 820 (1977); *Chem Abs.*, **88,** 104390 (1978).
[334] Monroe, B. M., *J. Phys. Chem.*, **82,** 15 (1978).
[335] Jefford, C. W., and Barchietto, G., *Tetrahedron Lett.*, **1977,** 4531.
[336] Matsumoto, M., Dobashi, S., and Kondo, K., *Bull. Chem. Soc. Jpn.*, **51,** 185 (1978).
[337] Ando, W., and Kohmoto, S., *J. Chem. Soc., Chem. Commun.*, **1978,** 120.
[338] Kadnikov, O. G., *Pism'ma Zh. Tekh. Fiz.*, **4,** 32 (1978); *Chem. Abs.*, **88,** 151572 (1978).
[339] Williams, J. R., Unger, L. R., and Moore, R. H., *J. Org. Chem.*, **43** 1271 (1978).
[340] Golodets, G. I., *Katal. Svoistva Veshchestu*, **4,** 187 (1977); *Chem. Abs.*, **88,** 5617 (1978).
[341] Lyons, J. E., *Aspects Homogeneous Catal.*, **3,** 1 (1977); *Chem. Abs.*, **88,** 151564 (1978).
[342] Tret'yakov, V. P., and Rudakov, E. S., *Metallokompleksnyi Katal.*, **1977,** 63; *Chem. Abs.*, **89,** 23240 (1978).
[343] Oda, R., *Kagaku* (*Kyoto*), **32,** 328 (1977); *Chem. Abs.*, **88,** 5659 (1978).
[344] Tatsuno, Y., Otsuka, S., Saigo, K., Mukaiyama, T., Yasuda, A., Yamamoto, H., and Sonoda, N., *Kagaku No Roiki, Zohan*, **117,** 329 (1977); *Chem. Abs.*, **89,** 128540 (1978).
[345] Black, J. F., *J. Am. Chem. Soc.*, **100,** 527 (1978).
[346] Makalets, B. I., Kirichenko, G. S., Strygin, E. I., Kolkotina, S. N., and Kamneva, S. A., *Neftekhimiya*, **18,** 250 (1978); *Chem. Abs.*, **89,** 5698 (1978).
[347] Belakov, V. A., Vasil'ev, R. F., and Fedorova, G. F., *Dokl. Akad. Nauk SSR*, **239,** 344 (1978); *Chem. Abs.*, **89,** 5691 (1978).
[348] Tsepalov, V. F., Kharitonova, A. A., Gladyshev, G. P., and Emanuel, N. M., *Kinet, Katal.*, **18,** 1395 (1977); *Chem. Abs.*, **88,** 120185 (1978).
[349] Timokhin, V. I., Opeida, I. A., and Kucher, R. V., *Neftekhimiya*, **17,** 563 (1977): *Chem. Abs.*, **87,** 167217 (1977).
[350] Kulicki, Z., and Zawadiak, J., *Chem. Stosow.*, **21,** 471 (1977); *Chem. Abs.*, **88,** 120339 (1978).
[351] Krishna, L. V. G., Srivastava, R. D., and Rao, M. S., *J. Appl. Chem. Biotechnol.*, **27,** 522 (1977); *Chem. Abs.*, **88,** 120324 (1978).
[352] Nemirova, L. I., Evmenenko, N. P., and Matkovskii, K. I., *Kinet. Katal.*, **18,** 1289 (1977): *Chem. Abs.*, **88,** 36964 (1978).
[353] Tsyskovskii, V. K., and Monakhova, N. E., *Zh. Prikl. Khim.* (*Leningrad*), **51,** 129 (1978); *Chem. Abs.*, **88,** 151764 (1978).
[354] Kaloerova, V. G., Nikolaevskii, A. N., and Vinter, P., *Deposited Dock.*, **1976,** VINITI 2957; *Chem. Abs.*, **89,** 107497 (1978).
[355] Kovtun, G. A., Berenblyum, A. S., Pustarnakova, G. F., and Moiseev, I. I., *Dokl. Akad. Nauk SSSR*, **236,** 150 (1977); *Chem. Abs.*, **87,** 183742 (1977).
[356] Torres, L., Riba, M. L., and Mathieu, J., *J. Chim. Phys., Phys. Chim. Biol.*, **74,** 1119 (1977); *Chem. Abs.*, **88,** 135918 (1978).
[357] Liberto, T., Riba, M. L., and Mathieu, J., *J. Chim. Phys. Phys.-Chim. Biol.*, **75,** 341 (1978); *Chem. Abs.*, **88,** 189692 (1978).
[358] Aleksandrov, V. N., Ocvhinnikov, V. I., Simonova, T. A., and Minaeva, S. A., *Khim. Prom.-st* (*Moscow*), **1977,** 895; *Chem. Abs.*, **88,** 61802 (1978).
[359] Aleksandrov, V. N., *Zh. Org. Khim.*, **14,** 1517 (1978); *Chem. Abs.*, **89,** 107501 (1978).

[360] Gun, S. R., and Nandi, U. S., *Indian J. Chem.*, **15A,** 483 (1977); *Chem. Abs.*, **88,** 21756 (1978).
[361] Holcik, J., Manasek, Z., and Pospisil, J., *Collect. Czech. Chem. Commun.*, **43,** 142 (1978).
[362] Gasior, M., and Gryzybowska, B. J., *J. Catal.*, **52,** 534 (1978); *Chem. Abs.*, **89,** 23487 (1978).
[363] Il'inich, O. M., Boreskov, G. K., Ivaninv, A. A., Lyakhova, V. F., and Belyaeva, N. P., *Neftekhimiya*, **18,** 270 (1978); *Chem. Abs.*, **89,** 5700 (1978).
[364] Mathur, B. C., and Viswanath, D. S., *Can. J. Chem. Eng.*, **56,** 223 (1978); *Chem. Abs.*, **89,** 23485 (1978).
[365] Vidoczy, T., and Gal, D., *Z. Phys. Chem.* (*Frankfurt-am-Main*), **106,** 269 (1977).
[366] Morita, Y., *Asahikawa Kogyo Koto Semmon Gakko Kenkyu Hobun*, **15,** 147 (1978); *Chem. Abs.*, **89,** 128825 (1978).
[367] Vidoczy, T., and Gal, D., *Magy. Kem. Fdy.*, **83,** 457 (1977); *Chem. Abs.*, **88,** 104388 (1978).
[368] Sakharov, I. V., and Geletii, Y. V., *Neftekhimiya*, **18,** 261 (1978); *Chem. Abs.*, **89,** 5699 (1978).
[369] Ivanov, S., *Dokl. Bolg. Akad. Nauk*, **30,** 1141 (1977); *Chem. Abs.*, **88,** 3697 (1978).
[370] Khodzhaev, O. M., Ruvinskii, M. E., Shik, G. L., Baidov, V. N., and Shakhtakhtinskii, T. N., *Azerb. Khim. Zh.*, **1977,** 8; *Chem. Abs.*, **89,** 23466 (1978).
[371] Tmenov, D. N., Lysukho, T. V., and Scherbina, F. F., *Khim. Prom.-st.* (*Moscow*), **1978,** 259; *Chem. Abs.*, **89,** 23468 (1978).
[372] Morozov, A. N., Polmane, G., Kampars, V., Trusov, S. R., and Neilands, O., *Latv. PSR Zinat. Akad. Vestis, Kim. Ser.*, **1977,** 476; *Chem. Abs.*, **87,** 183738 (1977).
[373] Goncharov, A. A., and Kozlov, Y. N., *Zh. Fiz. Khim.*, **52,** 945 (1978); *Chem. Abs.*, **89,** 5703 (1978).
[374] Goncharov, A. A., Kozlov, Y. N., and Purmal, A. P., *Zh. Fiz. Khim.*, **51,** 2839 (1977); *Chem. Abs.*, **88,** 36967 (1978).
[375] Tsuji, J., and Takayanagi, H., *Tetrahedron*, **34,** 641 (1978).
[376] Rogic, M. M., and Demmin, T. R., *J. Am. Chem. Soc.*, **100,** 5472 (1978).
[377] Sadana, A. J., *U.S. NTIS PB Rep.*, **1975,** PB-267810; *Chem. Abs.*, **88,** 49985 (1978).
[378] Tsuruya, S., Kishikawa, Y., Tanaka, R., and Kuse, T., *J. Catal.*, **49,** 254 (1977); *Chem. Abs.*, **87,** 200495 (1977).
[379] Hudec, P., *J. Catal.*, **53,** 228 (1978); *Chem. Abs.*, **89,** 75147 (1978).
[380] Vygodskaya, I. V., Kugel, V. Y., Novak, F. I., and Bashkirov, A. N., *Neftekhimiya*, **18,** 409 (1978); *Chem. Abs.*, **89,** 89919 (1978).
[381] Abrash, H. I., *Carlsberg Res. Commun.*, 42, 11 (1977); *Chem. Abs.*, **87,** 200493 (1977).
[382] Nishinaga, A., Shimizu, T., and Matsuura, T., *Tetrahedron Lett.*, **1978,** 3747.
[383] Nishinaga, A., Itahara, T., Shimizu, T., and Matsuura, T., *J. Am. Chem. Soc.*, **100,** 1820 (1978).
[384] Nishinaga, A., Itahara, T., Matsuura, T., Rieker, A., Koch, D., Albert, K., and Hitchcock, P. B., *J. Am. Chem. Soc.*, **100,** 1826 (1978).
[385] Gutor, I. M., *Dopov. Akad. Nauk Ukr., RSR, Ser. B: Geol., Khim. Biol. Nauki*, **1977,** 910; *Chem. Abs.*, **88,** 36956 (1978).
[386] Mantashyan, A. A., Khachatryan, L. A., and Niazyan, O. M., *Arm. Khim. Zh.*, **31,** 49 (1978); *Chem. Abs.*, **89,** 5694 (1978).
[387] Lutsyk, A. I., Rudakov, E. S., Tret'yakov, V. P., Suikov, S. Y., and Galenin, A. A., *Dopov. Akad. Nauk Ukr. RSR, Ser. B: Geol., Khim. Biol. Nauki*, **1978,** 526; *Chem. Abs.*, **89,** 89922 (1978).
[388] Koshel, G. N., Farberov, M. I., Antonova, T. N., and Yablonskii, O. P., *Tezisy Dokl.-Vses. Konf. Stereokhim, Konform. Anal. Org. Neftekhim. Sint., 3rd*, **1976,** 59; *Chem. Abs.*, **88,** 151767 (1978).
[389] Drimus, I., and Papahaai, L., *Rev. Roum. Chim.*, **23,** 209 (1978); *Chem. Abs.*, **88,** 169478 (1978).
[390] Moskovich, Y. L., Pavlyuchenko, A. I., Rassadina, L. I., and Freidin, B. G., *Zh. Prikl. Khim.* (*Leningrad*), **51,** 1137 (1978); *Chem. Abs.*, **89,** 42025 (1978).
[391] Rudakov, E. S., and Lutsyk, A. I., *Metallokompleksyni Katal.*, **1977,** 116; *Chem. Abs.*, **89,** 23486 (1978).
[392] Rudakov, E. S., and Zamashchikov, V. V., *Metallokompleksyni Katal.*, **1977,** 129; *Chem. Abs.*, **89,** 23242 (1978).
[393] Chernier, J. H. B., Howard, J. A., and Tait, J. C., *Can. J. Chem.*, **56,** 157 (1978).
[394] Howard, J. A., Chernier, J. H. B., and Holden, D. A., *Can. J. Chem.*, **56,** 170 (1978).
[395] Howard, J. A., and Tait, J. C., *Can. J. Chem.*, **56,** 164 (1978).

396 Kozhevnikov, I. V., Mastikhin, V. M., Matveev, K. I., Kirichenko, V. N., Churkin, Y. V., and Glazunova, V. I., *React. Kinet. Catal. Lett.*, **7,** 291 (1977); *Chem. Abs.*, **88,** 21819 (1978).
397 Solyanikov, V. M., Pustarnakova, G. F., and Solov'eva, T. Y., *Izv. Akad. Nauk SSSR, Ser. Khim.*, **1977,** 2225; *Chem. Abs.*, **88,** 21817 (1978).
398 Luca, M., Farzaliev, V. M., Aliev, A. S., Guseinov, M. M., and Muganlinski, F. F., *Rev. Chim.* (*Bucharest*), **18,** 727 (1977); *Chem. Abs.*, **88,** 21813 (1978).
399 Bridgewater, A. J., and Sexton, M. D., *J. Chem. Soc., Perkin Trans.* 2, **1978,** 530.
400 Mizukami, F., Ando, M., Imamura, J., Furuhashi, S., and Tanaka, T., *Nippon Kagaku Kaishi*, **1977,** 1871; *Chem. Abs.*, **88,** 88811 (1978).
401 Ramos, R., and Valov, P. I., *Rev. CENIC, Cienc. Fis.*, **6,** 69 (1975); *Chem. Abs.*, **89,** 128865 (1978).
402 Ramos, R., and Valov, P. I., *Rev. CENIC, Cienc. Fis.*, **6,** 85 (1975); *Chem. Abs.*, **89,** 128866 (1978).
403 Agliullina, G. G., Marteniyanov, V. S., Ivanova, I. A., and Denisov, E. T., *Izv. Akad. Nauk SSSR, Ser. Khim.*, **1977,** 2221; *Chem. Abs.*, **88,** 21816 (1978).
404 Chan, H. W.-S., Newby, V. K., and Levett, G., *J. Chem. Soc., Chem. Commun.*, **1978,** 82.
405 Chan, H. W.-S., Levett, G., and Matthew, S. A., *J. Chem. Soc., Chem. Commun.*, **1978,** 756.
406 Ikeda, N., and Fukuzumi, K., *Yukagaku*, **27,** 21 (1978); *Chem. Abs.*, **88,** 104400 (1978).
407 Ikeda, N., and Fukuzumi, K., *Yukagaku*, **27,** 26 (1978); *Chem. Abs.*, **88,** 104401 (1978).
408 Ikeda, N., and Fukuzumi, K., *Yukagaku*, **27,** 33 (1978); *Chem. Abs.*, **88,** 104402 (1978).
409 Piretti, M. V., Cavani, C., and Zeli, F., *Rev. Fr. Corps Gras*, **25,** 73 (1978); *Chem. Abs.*, **89,** 59468 (1978).
410 Yanishlieva, N., *Riv. Ital. Sostanze Grasse*, **55,** 55 (1978); *Chem. Abs.*, **89,** 41995 (1978).
411 Yanishlieva, N., Beckev, H., and Mangold, H. K., *Actes Congr. Mond.-Soc. Int. Etude Corps Graz., 13th*, **1976,** 29; *Chem. Abs.*, **87,** 183737 (1977).
412 Yanishlieva, N., and Mongold, H. K., *Chem. Phys. Lipids*, **20,** 21 (1977); *Chem. Abs.*, **88,** 5956 (1978).
413 Hendriks, C. F., Van Beek, H. C. A., and Heertjes, P. M., *Ind. Eng. Chem. Prod. Res. Dev.*, **16,** 270 (1977); *Chem. Abs.*, **87,** 183743 (1977).
414 Tavadyan, L. A., Maslov, S. A., and Blyumberg, E. A., *Dokl. Akad. Nauk SSSR*, **237,** 657 (1977); *Chem. Abs.*, **88,** 50007 (1978).
415 Ivanov, A. M., *Zh. Prikl. Khim.* (*Leningrad*), 50, 2529 (1977); *Chem. Abs.*, **88,** 61793 (1978).
416 Sakamoto, H., Funabiki, T., and Tarama, K., *J. Catal.*, **48,** 427 (1977); *Chem. Abs.*, 87, 167200 (1977).
417 Petkevich, T. S., Mitskevich, N. I., and Bal'kov, B. G., *Vesti Akad. Navuk BSSR, Ser. Khim. Navuk*, **1977,** 31; *Chem. Abs.*, **88,** 88801 (1978).
418 Sarkisyan, E. G., Vardanyan, I. A., and Nalbandyan, A. B., *Arm. Khim. Zh.*, **30,** 619 (1977); *Chem. Abs.*, **88,** 50005 (1978).
419 Naegeli, D. W., and Glassman, I., *Symp.* (*Int.*) *Combust., Proc.*, **16,** 1023 (1976); *Chem. Abs.*, **87,** 167203 (1977).
420 Abdomaarani, M., Nikipanchuk, M. V., and Chernyak, B. I., *Kinet. Katal.*, **19,** 499 (1978); *Chem. Abs.*, **89,** 5707 (1978).
421 Turro, N. J., Chow., M.-F., and Ito, Y., *J. Am. Chem. Soc.*, **100,** 5580 (1978).
422 Tsng, R., Yue, H. J., Wolf, J. F., and Mares, F., *J. Am. Chem. Soc.*, **100,** 5248 (1978).
423 Haynes, R. K., *Aust. J. Chem.*, **31,** 121, 131 (1978).
424 Barton, D. H. R., Haynes, R. K., Magnus, P. D., and Menzies, I. D., *J. Chem. Soc., Chem. Commun.*, **1974,** 511.
425 Laguerre, M., Dunogues, J., and Calas, R., *J. Chem. Res.*, **1978,** (S) 295.
426 Pryor, W. A., and Graham, W. D., *J. Org. Chem.*, **43,** 770 (1978).
427 Mimoun, H., Machirant, M. M. P. M., and de Roch, I. S., *J. Am. Chem. Soc.*, **100,** 5437 (1978).
428 Igersheim, F., and Mimoun, H., *J. Chem. Soc., Chem. Commun.*, **1978,** 559.
429 Read, G., *J. Mol. Catal.*, **4,** 83 (1978); *Chem. Abs.*, **89,** 23489 (1978).
430 Reuter, J. M., Sinha, A., and Salomon, R. G., *J. Org., Chem.* **43,** 2438 (1978).
431 Mares, F., and Tang, R., *J. Org. Chem.*, **43,** 4631 (1978).
432 Paraskewas, S., *Chem.-Ing.-Tech.*, **49,** 581 (1977); *Chem. Abs.*, **87,** 183723 (1977).
433 Rudakov, E. S., and Ignatenko, V. M., *Dokl. Akad. Nauk SSSR*, **241,** 148 (1978); *Chem. Abs.*, **89,** 128754 (1978).
434 Kolb, M., Bratz, E., and Dialev, K., *J. Mol. Catal.*, **2,** 399 (1977); *Chem. Abs.*, **88,** 21760 (1978).

[435] Vojtko, J., Hrusovsky, M., and Cihova, M., *Zb. Pr. Chemichotecknol. Fak., SVST*, **1973,** 195; *Chem. Abs.*, **89,** 146285 (1978).
[436] Chirko, A. I., Tishchenko, I. G., and Kudrevatykh, M. V., *Vesti Akad. Navuk BSSR, Ser. Khim. Navuk*, **1977,** 78; *Chem. Abs.*, **87,** 167225 (1977).
[437] Chirko, A. I., Tishchenko, I. G., and Sosnovskii, G. M., *Vestsi Akad. Navuk BSSR, Ser. Khim. Navuk*, **1978,** 114; *Chem. Abs.*, **89,** 107442 (1978).
[438] Fedenok, L. G., Berdnikov, V. M., and Shvartsberg, M. S., *Zh. Org. Khim.*, **14,** 1429 (1978); *Chem. Abs.*, **89,** 107500 (1978).
[439] Chirko, A. I., Tichenko, I. G., Stepin, S. G., and Mikhailova, I. P., *Vestsi Akad. Navuk BSSR, Ser. Khim. Navuk*, **1977,** 87; *Chem. Abs.*, **88,** 36949 (1978).
[440] Chirko, A. T., Tishchenko, I. G., and Sosnovskii, G. M., *Tezisy Dokl.-Vses. Konf. Khim. Atsetilena, 5th*, **1975,** 411; *Chem. Abs.*, **88,** 189659 (1978).
[441] Agisheva, S. A., Estrina, G. Y., Imashev, U. B., Aleksandrov, A. L., Zlotskii, S. S., and Rakhmankulov, D. L., *Zh. Prikl. Khim. (Leningrad)*, **51,** 1809 (1978); *Chem. Abs.*, **89,** 162798 (1978).
[442] Agisheva, S. A., Pastushenko, E. V., Klyavlin, M. S., Martem'yanov, V. S., Zlotskii, S. S., and Rakhmankulov, D. L., *Zh. Prikl. Khim. (Leningrad)*, **50,** 2732 (1977); *Chem. Abs.*, **88,** 151758 (1978).
[443] Agisheva, S. A., and Rakhmankulov, D. L., *Neftekhim. Sint. Tekh. Prog.*, **1976,** 19; *Chem. Abs.*, **88,** 189674 (1978).
[444] Agisheva, S. A., Latypova, F. N., Zlotskii, S. S., Martem'yanov, V. S., Gren, A. I., and Rakhmankulov, D. L., *Tezisy Dokl. Nauchn. Sess. Khim. Teckhnol. Org. Soedin. Sery Sernistykh Neftei*, **14*th***, 1975, 176; *Chem. Abs.*, **88,** 189680 (1978).
[445] Sen., A., and Halpern, J., *J. Am. Chem. Soc.*, **99,** 8337 (1977).
[446] Ohkatsu, Y., Okuyama, T., and Osa, T., *Yukagaku*, **27,** 142 (1978); *Chem. Abs.*, **89,** 189689 (1978).
[447] Avdeeva, L. B., and Shklyaev, A. A., *Tezisy Dokl. Nauchn. Sess. Khim. Technol. Org. Soedin. Sery Sernistykh Neftei, 14th*, **1975,** 218; *Chem. Abs.*, **88,** 189682 (1978).
[448] Numata, T., Watanabe, Y., and Oae, S., *Tetrahedron Lett.*, **1978,** 4933.
[449] Bourceanu, G., Ababi, V., Popa, G., and Sanduloviciu, M., *Rev., Roum. Chim.*, **22,** 1391 (1977).
[450] Fujii, N., and Asaba, T., *Nippon Kagaku Kaishi*, **1977,** 1097; *Chem. Abs.*, **87,** 167211 (1977).
[451] Lyubarskii, A. G., Gorelik, A. G., Sal'nikova, O. N., Glukhova, R. G., Malygin, E. N., and Grabova, M. N., *Neftekhimya*, **18,** 397 (1978); *Chem. Abs.*, **89,** 89918 (1978).
[452] Kutney, J. P., Balsevich, J., Bokelman, G. H., Hibino, T., Honda, T., Itoh, I., Ratcliffe, A. H., and Worth, B. R., *Can. J. Chem.*, **56,** 62 (1978).
[453] Jawdosiuk, M., Kmiotek-Skarzynska, I., and Wilczynski, W., *Can. J. Chem.*, **56,** 218 (1978).
[454] Nakagawa, M., Watanabe, H., Kodato, S., Okajima, H., Hino, T., Flippen, J. L., and Witkop, B., *Proc. Natl. Acad. Sci. U.S.A.*, **75,** 4730 (1977).
[455] Rozwadowska, M. D., and Brozda, D., *Tetrahedron Lett.*, **1978,** 589.
[456] Bellobono, I. R., Beltrame, P. L., Marcandalli, B., Fumagalli, A., and Trinchierf, M., *J. Chem. Soc., Perkin Trans. 2*, **1977,** 1989.
[457] Hema, M. A., Ramakrishnan, V., and Kuriacose, J. C., *Indian J. Chem.*, **15B,** 947 (1977); *Chem. Abs.*, **88,** 120330 (1978).
[458] Nekipelova, T. D., Gagarina, A. B., and Emanuel, N. M., *Dokl. Akad. Nauk SSSR*, **238,** 392 (1978); *Chem. Abs.*, **88,** 104405 (1978).
[459] Jefford, C. W., Knopfel, W., and Cadby, P. A., *J. Am. Chem. Soc.*, **100,** 6432 (1978).
[460] Jefford, C. W., Knopfel, W., and Cadby, P. A., *Tetrahedron Lett.*, **1978,** 3585 .
[461] Shendrik, A. N., Mytsyk, N. P., and Opeida, I. A., *Kinet. Katal.*, **18,** 1077 (1977); *Chem, Abs.*, **87,** 167238 (1977).
[462] Hisatome, M., Koshikawa, S., Chimura, K., Hashimoto, H., and Yamakawa, K., *J. Organomet. Chem.*, **145,** 225 (1978).
[463] Sukarev, B. N., Potekhin, V. M., Ptroskuryakov, V. A., and Fedyanin, P. N., *Zh. Prikl. Khim. (Leningrad)*, **51,** 413 (1978); *Chem. Abs.*, **88,** 169333 (1978).
[464] Pericas, M. A., and Serratosa, F., *Tetrahedron Lett.*, **1978,** 4969.
[465] Cuvigny, T., Valette, G., Larcheveque, M., and Normant, H., *J. Organomet. Chem.*, **155,** 147 (1978).
[466] Chatterjee, A., Raychaudhuri, S. R., and Chatterjee, S. K., *Tetrahedron Lett.*, **1978,** 3487.
[467] Sydnes, L. K., *Acta Chem. Scand.*, **31B,** 903 (1977).

[468] Butovskaya, G. V., Agabekov, V. E., and Mitskevich, N. I., *Dokl. Akad. Nauk BSSR*, **22,** 155 (1978); *Chem. Abs.*, **88,** 151762 (1978).
[469] Rebuffat, S., Giraud, M., and Molho, D., *Bull. Mus. Natl. Hist. Nat., Sci. Phys.-Chim.*, **1976,** 7; *Chem. Abs.*, **87,** 200667 (1977).
[470] Yurzhenko, S. A., and Luong, B. P., *Neftepererab Neftekhim. (Kiev)*, **15,** 88 (1977); *Chem. Abs.*, **88,** 88830 (1978).
[471] Kurashov, V. I., Pobedimskii, D. G., and Kirpichhnikov, P. A., *Dokl. Akad. Nauk SSSR*, **238,** 1407 (1978); *Chem. Abs.*, **89,** 107434 (1978).
[472] Wiegers, K. E., and Smith, S. G., *J. Org. Chem.*, **43,** 1126 (1978).
[473] Royer, J., *Tetrahedron Lett.*, **1978,** 1343.
[474] Hanaya, K., Kitamoto, S., Kudo, H., and Mitsui, S., *Nippon Kagaku Kaishi*, **1978,** 238 (1978); *Chem. Abs.*, **88,** 151761 (1978).
[475] Baliah, V., and Usha, R., *Indian J. Chem.*, **15B,** 684 (1977); *Chem. Abs.*, **88,** 104541 (1978).
[476] Rei, M.-H., *J. Org. Chem.*, **43,** 2173 (1978).
[477] Giacomelli, G., Menicagli, R., Caporusso, A. M., and Lardicci, L., *J. Org. Chem.*, **43,** 1790 (1978).
[478] Brinkmeyer, R. S., and Kapoor, V. M., *J. Am. Chem. Soc.*, **99,** 8341 (1977).
[479] Angeloni, A. S., Marzocchi, S., and Scapini, G., *Gazz. Chim. Ital.*, **107,** 421 (1977); *Chem. Abs.*, **88,** 169482 (1978).
[480] Jefferson, P., McDonald, E., and Smith, P., *Tetrahedron Lett.*, **1978,** 585.
[481] Loupy, A., and Seyden-Penne, J., *Tetrahedron Lett.*, **1978,** 2571.
[482] Grishina, G. V., Potapov, V. M., and Liberchuk, T. A., *Khim. Geterotsikl. Soedin*, **1978,** 240; *Chem. Abs.*, **89,** 23631 (1978).
[483] Jackson, W. R., Norman, J. W., and Rae, I. P. D., *Tetrahedron Lett.*, **1978,** 2061.
[484] Fetizon, M., Huy, H. T., and Mourgues, P., *Tetrahedron*, **34,** 209 (1978).
[485] Szimuszkovicz, J., Musser, J. H., and Laurian, L. G., *Tetrahedron Lett.*, **1978,** 1411.
[486] Chollet, A., and Vogel, P., *Helv. Chim. Acta*, **61,** 732 (1978).
[487] Richer, J.-C., Poirier, J.-A., Maroni, Y., and Manuel, G., *Can. J. Chem.*, **56,** 2049 (1978).
[488] Sankawa, U., and Sato, T., *Tetrahedron Lett.*, **1978,** 981.
[489] Kayser, M. M., and Morand, P., *Can. J. Chem.*, **56,** 1524 (1978).
[490] Kemp, D. S., Chabala, J. C., and Marson, S. A., *Tetrahedron Lett.*, **1978,** 543.
[491] Jager, V., Buss, V., and Schwab, W., *Tetrahedron Lett.*, **1978,** 3133.
[492] Karady, S., Amato, J. S., Weinstock, L. M., and Sletzinger, M., *Tetrahedron Lett.*, **1978,** 403.
[493] Karady, S., Amato, J. S., Weinstock, L. M., and Sletzinger, M., *Tetrahedron Lett.*, **1978,** 407.
[494] Wong, M. Y. H., and Gray, G. R., *J. Am. Chem. Soc.*, **100,** 3548 (1978).
[495] Dufort, N., and Jodoin, B., *Can. J. Chem.*, **56,** 1779 (1978).
[496] Corriu, R. J. P., Fernandez, J. M., and Guerin, C., *Tetrahedron Lett.*, **1978,** 3391.
[497] Ashby, E. C., and Lin, J. J., *J. Org. Chem.*, **43,** 2567 (1978).
[498] Ashby, E. C., and Lin, J. J., *Tetrahedron Lett.*, **1977,** 4481.
[499] Ashby, E. C., and Noding, S. A., *Tetrahedron Lett.*, **1977,** 4579.
[500] Isagawa, K., Tatsumi, K., Kosugi, H., and Otsuji, Y., *Chem. Lett.*, **1977,** 1017.
[501] Isagawa, K., Tatsumi, K., and Otsuji, Y., *Chem. Lett.*, **1977,** 1117.
[502] Tsuji, J., Yamakawa, T., and Mandai, T., *Tetrahedron Lett.*, **1978,** 565.
[503] Ashby, E. C., and Lin, J. J., *J. Org. Chem.*, **43,** 1263 (1978).
[504] McMurry, J. E., Silvestri, M. G., Fleming, M. P., Hoz, T., and Grayston, M. W., *J. Org. Chem.*, **43,** 3249 (1978).
[505] Baumstark, A. L., McCloskey, C. J., and Witt, K. E., *J. Org. Chem.*, **43,** 3609 (1978).
[506] Fugiwara, Y., Ishikawa, R., Akiyama, F., and Teranishi, S., *J. Org. Chem.*, **43,** 2477 (1978).
[507] Reed, J. W., and Jolly, W. L., *J. Org. Chem.*, **42,** 3963 (1977).
[508] Quick, J., and Crelling, J. K., *J. Org. Chem.*, **43,** 155 (1978).
[509] Wijnberg, J. B. P. A., Schoemaker, H. E., and Speckamp, W. N., *Tetrahedron*, **34,** 179 (1978).
[510] Etogo Nzue, S., Bodo, B., and Molho, D., *Bull. Mus. Natl. Hist. Nat., Sci. Phys.-Chim.*, **14,** 65 (1977); *Chem. Abs.*, **89,** 128822 (1978).
[511] Yoon, N.-M., and Cha, J.-S., *Taehan Hwahak Hoechi*, **21,** 108 (1977); *Chem. Abs.*, **87,** 167312 (1977).
[512] Horner, L., and Brich, W., *Justus Liebigs Ann. Chem.*, **1978,** 710.

[513] Sugimoto, T., Matsumura, Y., Tanimoto, S., and Okano, M., *J. Chem. Soc., Chem. Commun.*, **1978,** 926.
[514] Colonna, S., and Fornasier, R., *J. Chem. Soc., Perkin Trans.* 1, **1978,** 371.
[515] Doiuchi, T., and Minoura, Y., *Isr. J. Chem.*, **15,** 84 (1977); *Chem. Abs.*, **87,** 200492 (1977).
[516] Sugimoto, T., Matsumura, Y., Imanishi, T., Tanimoto, S., and Okano, M., *Tetrahedron Lett.*, **1978,** 3431.
[517] Goldberg, S. I., Baba, N., Green, R. L., Pandian, R., Stowers, J., and Dunlap, R. B., *J. Am. Chem. Soc.*, **100,** 6768 (1978).
[518] Wigfield, D. C., and Gowland, F. W., *Can. J. Chem.*, **56,** 786 (1978).
[519] Nadar, P. A., Gnanasekaran, C., and Chandrasekharan, J., *J. Chem. Res.*, **1978,** (S) 424; (M) 4680.
[520] Brown, H. C., and Vogel, F. G. M., *Justus Liebigs Ann. Chem.*, **1978,** 695.
[521] Caro, B., Gentric, E., Grandjean, D., and Jaouen, G., *Tetrahedron Lett.*, **1978,** 3009.
[522] Baliah, V., and Jeyaraman, R., *Indian J. Chem.*, **15B,** 791 (1977); *Chem. Abs.*, **88,** 88972 (1978).
[523] Wigfield, D. C., and Feiner, S., *Can. J. Chem.*, **56,** 789 (1978).
[534] Kasal, A., *Collect. Czech. Chem. Commun.*, **43,** 498 (1978).
[525] Weissenberg, M., Krinsky, P., and Glotter, E., *J. Chem. Soc., Perkin Trans.* 1, **1978, 565**; Weissenberg, M., and Glotter, E., *J. Chem. Soc., Perkin Trans.* 1, **1978,** 568.
[526] Dewar, M. J. S., and McKee, M. L., *J. Am. Chem. Soc.*, **100,** 7499 (1978).
[527] Schmid, G., Luedemann, H. D., and Jaenicke, R., *Eur. J. Biochem.*, **86,** 219 (1978).
[528] Dauben, W. G., and Ashmore, J. W., *Tetrahedron Lett.*, **1978,** 4487.
[529] Pepin, Y., Nazemi, H., and Payette, D., *Can. J. Chem.*, **56,** 41 (1978).
[530] Nagao, Y., Kawabata, K., and Fujita, E., *J. Chem. Soc., Chem. Commun.*, **1978,** 330.
[531] Casini, A., di Rienzo, B., Moracci, F. M., Tortorella, S., Liberatore, F., and Arnone, A., *Tetrahedron Lett.*, **1978,** 2139.
[532] Baer, H. H., and Madumelu, C. B., *Can. J. Chem.*, **56,** 1177 (1978).
[533] Gladysz, J. A., Wong, V. K., and Jick, B. S., *J. Chem. Soc., Chem., Commun.*, **1978,** 838.
[534] Hutchins, R. O., and Natale, N. R., *J. Org. Chem.*, **43,** 2299 (1978).
[535] Hutchins, R. O., Kandasamy, D., Dux, F., Maryanoff, C. A., Rotstein, D., Goldsmith, B., Burgoyne, W., Cistone, F., Dalessandron, J., and Puglis, J., *J. Org. Chem.*, **43,** 2259 (1978).
[536] Kozikowski, A. P., and Floyd, W. C., *Tetrahedron Lett.*, **1978,** 19.
[537] Chow, M. C., *Tung Wu Shu Li Hsueh Pao*, **1,** 139 (1975); *Chem. Abs.*, **89,** 59467 (1978).
[538] Szychowski, J., Leniewski, A., and Wrobel, J. T., *Chem. Ind.* (*London*), **1978,** 273.
[539] Horne, D. A., and Jordan, A., *Tetrahedron Lett.*, **1978,** 1357.
[540] Fleet, G. W. J., Fuller, C. J., and Harding, P. J. C., *Tetrahedron Lett.*, **1978,** 1473.
[541] Sorrell, T. N., and Spillane, R. J., *Tetrahedron Lett.*, 1978, 2473.
[542] Sorrell, T. N., *Tetrahedron Lett.*, **1978,** 4985.
[543] Luche, J.-L., Rodriguez-Hahn, L., and Crabbe, P. J., *J. Chem. Soc., Chem. Commun.*, **1978,** 601.
[544] Luche, J.-L., *J. Am. Chem. Soc.*, **100,** 2226 (1978).
[545] Johnstone, R. A. W., and Telford, R. P., *J. Chem. Soc., Chem. Commun.*, 1978, **354.**
[546] Barrett, A. G. M., Barton, D. H. R., Bielski, R., and McCombie, S. W., *J. Chem. Soc., Chem. Commun.*, **1977,** 866.
[547] Wilt, J. W., and Aznavoorian, P. M., *J. Org. Chem.*, **43,** 1285 (1978).
[548] Fung, N. Y. M., de Mayo, P., Schauble, J. H., and Weedon, A. C., *J. Org. Chem.*, **43,** 3977 (1978).
[549] Whitesitt, C. A., and Herron, D. K., *Tetrahedron Lett.*, **1978,** 1737.
[550] Clive, D. L. J., Chittattu, G., add Wong, C. K., *J. Chem. Soc., Chem. Commun.*, **1978,** 41.
[551] Patroni, J. J., and Stick, R. V., *J. Chem. Soc., Chem. Commun.*, **1978,** 449.
[552] Ashby, E. C., Lin, J. J., and Goel, A. B., *J. Org. Chem.*, **43,** 183 (1978).
[553] Ashby, E. C., Lin, J. J., and Goel, A. B., *J. Org. Chem.*, **43,** 1557 (1978).
[554] Ashby, E. C., Lin, J. J., and Goel, A. B., *J. Org. Chem.*, **43,** 1560 (1978).
[555] Ashby, E. C., Lin, J. J., and Goel, A. B., *J. Org. Chem.*, **43,** 1564 (1978).
[556] Ashby, E. C., and Smith, T., *J. Chem. Soc., Chem. Commun.*, **1978,** 30.
[557] Ashby, E. C., Lin, J. J., and Goel, A. B., *J. Oıg. Chem.*, **43,** 757 (1978).
[558] Collman, J. P., Finke, R. G., Matlock, P. L., Wahren, R., Komoto, R. G., and Brauman, J. I., *J. Am. Chem. Soc.*, **100,** 1119 (1978).
[559] Casey, C. P., and Neumann, S. M., *J. Am. Chem. Soc.*, **100,** 2544 (1978).
[560] Funabiki, T., Yamazaki, Y., and Tarama, K., *J. Chem. Soc., Chem. Commun.*, **1978,** 63.

[561] Kinney, R. J., Jones, W. D., and Bergman, R. G., *J. Am. Chem. Soc.*, **100,** 635 (1978).
[562] Fachinetti, G., Floriani, C., Roselli, A., and Pucci, S., *J. Chem. Soc., Chem. Commun.*, **1978,** 269.
[563] Messing, A. W., Ross, F. P., Norman, A. W., and Okamura, W. H., *Tetrahedron Lett.*, **1978,** 3635.
[564] Gell, K. I., and Schwartz, J., *J. Am. Chem. Soc.*, **100,** 3246 (1978).
[565] Coustard, J.-M., Douteau, M.-H., Jacquesy, R., Longevialle, P., and Zimmerman, D., *J. Chem. Res.*, **1978,** (S) 16; (M) 337.
[566] Coustard, J.-M., Douteau, M.-H., and Jacquesy, R., *J. Chem. Res.*, **1978,** (S) 18, (M) 357.
[567] Jung, M. E., and Brown, R. W., *Tetrahedron Lett.*, **1978,** 2771.
[568] Kabir-ud-Din and Plesch, P. H., *J. Chem. Soc., Perkin Trans. 2*, **1978,** 937.
[569] DePuy, C. H., Bierbaum, V. M., Schmitt, R. J., and Shapiro, R. H., *J. Am. Chem. Soc.*, **100,** 2920 (1978).
[570] Screttas, C. G., and Cazianis, C. T., *Tetrahedron*, **34,** 933 (1978).
[571] Koenig, J.-J., de Rostolan, J., Bourbier, J.-C., and Jarreau, F.-X., *Tetrahedron Lett.*, **1978,** 2779.
[572] Scott, L. T., Carlin, K. J., and Schultz, T. H., *Tetrahedron Lett.*, **1978,** 4637.
[573] Cervinka, O., Dudek, V., and Scholzova, I., *Collect. Czech. Chem. Commun.*, **43,** 1091 (1978).
[574] Midland, M. M., Tramontano, A., and Zderic, S. A., *J. Organomet. Chem.*, **156,** 203 (1978).
[575] Kireev, G. V., Leont'ev, V. B., Kurbatov, Y. V., Kartavtsev, O. T., Otroshchenko, O. S., and Sadykov, A. S., *Izv. Akad. Nauk SSSR, Ser. Khim.*, **1978,** 807; *Chem. Abs.*, **89,** 23471 (1978).
[576] Bunting, J. W., and Kabir, S. H., *J. Org. Chem.*, **43,** 3662 (1978).
[577] Doyle, M. P., and McOsket, C. C., *J. Org. Chem.*, **43,** 693 (1978).
[578] Kursanov, D. B., and Parnes, Z. N., *Ref. Dokl. Soobshch.-Mendeleevsk S'ezd Obshch. Prikl. Khim., 11th*, **2,** 7 (1975); *Chem. Abs.*, **88,** 135758 (1978).
[579] Tamao, K., Yoshida, J., Takahashi, M., Yamamoto, H., Kakui, T., Matsumoto, H., Kurita, A., and Kumada, M., *J. Am. Chem. Soc.*, **100,** 290 (1978).
[580] Yamamoto, K., Tsuruoka, K., and Tsuji, J., *Chem. Lett.*, **1977,** 1115.
[581] Fry, J. L., Orfanopoulos, M., Adlington, M. G., Dittman, W. R., and Silverman, S. B., *J. Org. Chem.*, **43,** 375 (1978).
[582] Helquist, P., *Tetrahedron Lett.*, **1978,** 1913.
[583] Rassat, A., *Pure Appl. Chem.*, **49,** 1049 (1977).
[584] Caine, D., and Smith, T. L., *J. Org. Chem.*, **43,** 755 (1978).
[585] McMurry, J. E., Blaszczak, L. C., and Johnson, M. A., *Tetrahedron Lett.*, **1978,** 1633.
[586] Hanselaer, R., Samson, M., and Vandewalle, M., *Bull. Soc. Chim. Belg.*, **87,** 359 (1978).
[587] McMurry, J. E., and Choy, W., *J. Org. Chem.*, **43,** 1800 (1978).
[588] Ekouya, A., Dunogues, J., and Calas, R., *J. Chem. Res.*, **1978,** (S) 296.
[589] Smith, J. G., and Chun, Y.-I., *Tetrahedron Lett.*, **1978,** 413.
[590] Bos, H., and Klumpp, G. W., *Tetrahedron Lett.*, **1978,** 1863.
[591] Boz, H., and Klumpp, G. W., *Tetrahedron Lett.*, **1978,** 1865.
[592] Caine, D., Pennington, W. R., and Smith, T. L., *Tetrahedron Lett.*, **1978,** 2663.
[593] Rabideau, P. W., and Burkholder, E. G., *J. Org. Chem.*, **43,** 4283 (1978).
[594] Kotlarek, W., and Pacut, R., *J. Chem. Soc., Chem. Commun.*, **1978,** 153.
[595] Franks, D., Grossel, M. C., Hayward, R. C., and Knutsen, L. J. S., *J. Chem. Soc., Chem. Commun.*, **1978,** 941.
[596] Birch, A. J., and Dyke, S. F., *Aust. J. Chem.*, **31,** 1625 (1978).
[597] Applequist, D. E., and Pfohl, W. F., *J. Org. Chem.*, **43,** 867 (1978).
[598] Niznik, G. E., and Walborsky, H. M., *J. Org. Chem.*, **43,** 2396 (1978).
[599] Deshayes, H., and Pete, J.-P., *J. Chem. Soc., Chem. Commun.*, **1978,** 567.
[600] Ireland, R. E., Wilcox, C. S., and Thaisrivongs, S., J. *Org. Chem.*, **43,** 786 (1978).
[601] Berry, M., Davies, S. G., and Green, M. L. H., *J. Chem. Soc., Chem. Commun.*, **1978,** 99.
[602] Sharpless, K. B., Teranishi, A. Y., and Backvall, J.-E., *J. Am. Chem. Soc.*, **99,** 3120 (1977).
[603] Grant, C. B., and Streitwieser, A., *J. Am. Chem. Soc.*, **100,** 2433 (1978).
[604] Fry, A. J., and Herr, D., *Tetrahedron Lett.*, **1978,** 1721.
[605] Fry, A. J., Ginsburg, G. S., and Parente, R. A., *J. Chem. Soc., Chem. Commun.*, **1978,** 1040.
[606] Applequist, D. E., and Pfohl, W. F., *J. Org. Chem.*, **43,** 867 (1978).
[607] Warnhoff, E., Rampersad, M., Raman, P. S., and Yerhoff, F. W., *Tetrahedron Lett.*, **1978,** 1659.

[608] Takai, K., Hotta, Y., Oshimo, K., and Nozaki, H., *Tetrahedron Lett.*, **1978,** 2417.
[609] Cros, E., Elphimoff-Felkin, I., and Sarda, P., *C. R. Hebb. Séances Acad. Sci.*, **286C,** 261 (1978).
[610] Horner, L., and Schmitt, E., *Justus Liebigs Ann. Chem.*, **1978,** 1617.
[611] McMurry, J. E., Fleming, M. P., Kees, K. L., and Krepski, L. R., *J. Org. Chem.*, **43,** 3255 (1978).
[612] Welch, S. C., and Walter M. E., *J. Org. Chem.*, **43,** 2715 (1978).
[613] Christofis, O., Habeeb, J. J., Steevensz, R. S., and Tuck, D. G., *Can. J. Chem.*, **56,** 2269 (1978).
[614] Bright-Angrand, D., and Muckensturm, *J. Chem. Res.*, **1977,** (S) 274.
[615] Allison, J., and Ridge, D. P., *J. Am. Chem. Soc.*, **100,** 163 (1978)
[616] Konstantos, J., Katsaros, N., Vrachnou-Astra, E., and Katakis, D., *J. Am. Chem. Soc.*, **100,** 3128 (1978).
[617] Komatsu, K., Tomioka, I., and Okamoto, K., *Tetrahedron Lett.*, **1978,** 803.
[618] Ashby, E. C., and Wiesemann, T. L., *J. Am. Chem. Soc.*, **100,** 189 (1978).
[619] Claesson, A., and Sahlberg, C., *Tetrahedron Lett.*, **1978,** 5049.
[620] Tashiro, M., Iwasaki, A., and Fukawata, G., *J. Org. Chem.*, **43,** 196 (1978).
[621] Sheradsky, T., and Zbaida, D., *Tetrahedron Lett.*, **1978,** 2037.
[622] Evans, W. J., Engerer, S. C., and Neville, A. C., *J. Am. Chem. Soc.*, **100,** 331 (1978).
[623] Evans, W. J., Wayda, A. L., Chang, C.-W., and Cwirla, W. M., *J. Am. Chem. Soc.*, **100,** 333 (1978).
[624] Huenig, S., *Chim. Ind.* (*Milan*), **59,** 457 (1977); *Chem. Abs.*, **87,** 200263 (1977).
[625] Doyle, M. P., and West, C. T., *Stereoselective Reductions* (*Benchmark Papers in Organic Chemistry*, Vol. 6), Wiley/Halstead, New York, 1977.
[627] Carless, H. A. J., *Photochemistry*, **8,** 413 (1977); *Chem. Abs.*, **88,** 5655 (1978).
[627] Nelsen, S. F., *ACS Symp. Ser.*, **69,** 309 (1978); *Chem. Abs.*, **89,** 23231 (1978).
[628] Miyata, T., Endo, Y., and Hirashima, T., *Nippon Kagaku Kaishi*, **1978,** 858; *Chem. Abs.*, **89,** 89904 (1978).
[629] Adlakha, P. K., Rao, M. R. T., Tigare, D. V., Yede, A. O., and Bhakare, H. A., *J. Oil Technol. Assoc. India*, **9,** 5 (1977); *Chem. Abs.*, **88,** 21762 (1978).
[630] Doldourus, G. A., and Kollonitsch, J., *J. Am. Chem. Soc.*, **100,** 341 (1978).
[631] Momose, T., and Muraoka, O., *Tetrahedron Lett.*, **1978,** 1125.
[632] Nair, M., and Schechter, H., *J. Chem. Soc., Chem. Commun.*, **1978,** 793.
[633] Kiprianova, L. A., Levit, A. F., Sterleva, T. G., and Gragerov, I. P., *Tezisy Dokl.-Vses. Konf. Polyariz, Yader Elektranov Eff. Magn. Polya Khim. Reakts.*, **1975,** 14; *Chem. Abs.*, **89,** 23476 (1978).
[634] Zwan, M. C. V., Hartner, F. W., Reamer, R. A., and Tull, R., *J. Org. Chem.*, **43,** 509 (1978).
[635] Strecker, A., *Justus Liebigs Ann. Chem.*, **123,** 363 (1862).
[636] Still, I. W. J., Hasan, S. K., and Turnbull, K., *Cann. J. Chem.*, **56,** 1423 (1978).
[637] Baechler, R. D., and Daley, S. K., *Tetrahedron Lett.*, **1978,** 101.
[638] Baechler, R. D., Daley, S. K., Daley, B., and McGlynn, K., *Tetrahedron Lett.*, **1978,** 105.
[639] Cookson, R. C., and Parsons, P. J., *J. Chem. Soc., Chem. Commun.*, **1978,** 821.
[640] Soysa, H. S. D., and Weber, W. P., *Tetrahedron Lett.*, **1978,** 235.
[641] Bayley, H., Standring, D. N., and Knowles, J. R., *Tetrahedron Lett.*, **1978,** 3633.
[642] Tamagaki, S., Ichihara, R., *Mem. Fac. Eng. Osaka City Univ.*, **17,** 113 (1976); *Chem. Abs.*, **88,** 21779 (1978).
[643] Lambeth, D. O., *J. Am. Chem. Soc.*, **100,** 4808 (1978).
[644] Kornblum, N., Carlson, S. C., and Smith, R. G., *J. Am. Chem. Soc.*, **100,** 289 (1978).
[645] Ono, N., Tamura, R., Hayami, J., and Kaji, A., *Tetrahedron Lett.*, **1978,** 763.
[646] Minato, H., Fujie, S., Okuma, K., and Kobayashi, M., *Chem. Lett.*, **1977,** 1091.
[647] Doi, J. T., and Musher, W. K., *J. Am. Chem. Soc.*, **100,** 3533 (1978).
[648] Dasgupta, S., Datta, M. K., and Datta, R., *Tetrahedron Lett.*, **1978,** 1309.
[649] Kabalka, G. W., Newton, R. J., Chandler, J. H., and Yang, D. T. C., *J. Chem. Soc., Chem. Commun.*, **1978,** 726.
[650] Kabalka, G. W., Newton, R. J., and Jacobus, J., *J. Org. Chem.*, **43,** 1567 (1978).
[651] Brown, H. C., Liotta, R., and Kramer, G. W., *J. Org. Chem.*, **43,** 1058 (1978).
[652] Meneghelli, B. J., and Rudolph, R. W., *J. Am. Chem. Soc.*, **100,** 4626 (1978).
[653] Valentine, J. T., and Scott, J. W., *Synthesis*, **1978,** 329.
[654] Avar, G., and Kisch, H., *Monatsh. Chem.*, **109,** 89 (1978).

[655] Blum, J., *J. Mol. Catal.*, 3, 33 (1977); *Chem. Abs.*, **88,** 36845 (1978).
[656] Bayer, O., and Haefner, M., *Methoden Org. Chem.* (*Houben-Weyl*), *4th Edn.*, **7,** 2079 (1977); *Chem. Abs.*, **88,** 151565 (1978).
[657] Nagy-Mages, Z., Vastag, S., Heil, B., and Murko, L., *Transition Met. Chem.* (*Weinheim*), **3,** 123 (1978); *Chem. Abs.*, **89,** 146134 (1978).
[658] Rousseau, C., Evrard, M., and Petit, F., *J. Mol. Catal.*, **3,** 309 (1978); *Chem. Abs.*, **88,** 189644 (1978).
[659] de Croon, M. H. J. M., Van Nissellrooij, P. F. M. T., Kuipers, H. J. A. M., and Coenen, J. W., *J. Mol. Catal.*, **4,** 325 (1978).
[660] Rajca, I., Borowski, A., and Marzec, A., *Neftekhimiya*, **17,** 672 (1977); *Chem. Abs.*, **88,** 36672 (1978).
[661] Brown, J. M., and Chaloner, P. A., *Tetrahedron Lett.*, **1978,** 1877.
[662] Brown, J. M., and Chaloner, P. A., *J. Chem. Soc.*, *Chem. Commun.*, **1978,** 321.
[663] Vilivn, J., and Hetflejs, J., *Collect. Czech. Chem. Commun.*, **43,** 122 (1978).
[664] van Eikeren, P., and Grier, D. L., *J. Am. Chem. Soc.*, **99,** 8057 (1977).
[665] Achiwa, K., *Chem. Lett.*, **1978,** 561.
[666] Ojima, I., Kogure, T., and Achiwa, K., *Chem. Lett.*, **1978,** 567.
[667] Achiwa, K., *Tetrahedron Lett.*, **1978,** 1475.
[668] Fryzuk, M. D., and Bosnich, B. J., *J. Am. Chem. Soc.*, **100,** 5491 (1978).
[669] James, B. R., and Mahajan, D., *Isr. J. Chem.*, **15,** 214 (1977); *Chem. Abs.*, **88,** 61791 (1978).
[670] James, B. R., and Mahajan, D., *Isr. J. Chem.*, **15,** 214 (1977).
[671] Horner, L., and Schlotthauer, B., *Phosphorus Sulfur*, **4,** 155 (1978); *Chem. Abs.*, **89,** 23645 (1978).
[672] Achiwa, K., *Tetrahedron Lett.*, **1978,** 2583.
[673] Glaser, R., Twaik, M., Gerush, S., and Blumenfeld, J., *Tetrahedron Lett.*, **1977,** 4635.
[674] Glaser, R., Blumenfeld, J., and Twaik, M., *Tetrahedron Lett.*, **1977,** 4639.
[675] Cullen, W. R., and Sugi, Y., *Tetrahedron Lett.*, **1978,** 1635.
[676] Achiwa, K., and Soga, T., *Tetrahedron Lett.*, **1978,** 1119.
[677] Glaser, R., Geresh, S., Blumenfeld, J., and Twaik, M., *Tetrahedron*, **34,** 2405 (1978).
[678] Hanaki, K., Kashiwabara, K., and Fujita, J., *Chem. Lett.*, **1978,** 489.
[679] Fiorini, M., Marcati, F., and Giongo, G. M., *J. Mol. Catal.*, **4,** 125 (1978); *Chem. Abs.*, **89,** 162773 (1978).
[680] Fiorini, M., Marcati, F., and Giongo, G. M., *J. Mol. Catal.*, **3,** 385 (1978); *Chem. Abs.*, **88,** 135751 (1978).
[681] Achiwa, K., *Chem. Lett.*, **1978,** 905.
[682] Takaishi, N., Imai, H., Bertelo, C. A., and Stille, J. K., *J. Am. Chem. Soc.*, **100,** 264 (1978).
[683] Masuda, T., and Stille, J. K., *J. Am. Chem. Soc.*, **100,** 268 (1978).
[684] Wilson, M. E., and Whitesides, G., *J. Am. Chem. Soc.*, **100,** 306 (1978).
[685] Holy, N. L., *J. Org. Chem.*, **43,** 4686 (1978).
[686] Holy, N. L., *J. Chem. Soc.*, *Chem. Commun.*, **1978,** 1074.
[687] Cocco, G., Fagherazzi, G., Carturan, G., and Gottardi, V., *J. Chem. Soc.*, *Chem. Commun.*, **1978,** 979.
[688] Kripylo, P., Muench, P., Borchert, T., and Klose, D., *Chem. Tech.* (*Leipzig*), **29,** 500 (1977); *Chem. Abs.*, **88,** 5972 (1978).
[689] Cerveny, C., Marhoul, A., Berka, Z., and Ruzicka, V., *Chem. Prum.*, **27,** 560 (1977); *Chem. Abs.*, **88,** 73857 (1978).
[690] Cortese, N. A., and Heck, R. F., *J. Org. Chem.*, **43,** 3985 (1978).
[691] Berenblyum, A. S., Aseeva, A. P., Lakhman, L. I., and Moiseev, I. I., *Izv. Akad. Nauk SSR*, *Ser. Khim.* **1977,** 2163; *Chem. Abs.*, **88,** 21758 (1978).
[692] Harada, K., and Kataoka, Y., *Tetrahedron Lett.*, **1978,** 2103.
[693] James, B. R., and McMillan, R. S., *Can. J. Chem.*, **56,** 3927 (1978).
[694] Bennett, M. A., Huang, T.-N., Smith, A. K., and Turney, T. W., *J. Chem. Soc.*, *Chem. Commun.*, **1978,** 582.
[695] Johnson, J. W., and Muetterties, E. L., *J. Am. Chem. Soc.*, **99,** 9375 (1977).
[696] Sherman, E. O., and Shreiner, P. R., *J. Chem. Soc.*, *Chem. Commun.*, **1978,** 223.
[697] Ikariya, T., Osakada, K., and Yoshikawa, S., *Tetrahedron Lett.*, **1978,** 3749.
[698] Vancheesan, S., Sethi, S. P., Rajaram, J., and Kuriacose, J. C., *Indian J. Chem.*, **16A,** 399 (1978); *Chem. Abs.*, **89,** 162791 (1978).
[699] Tsuji, J., and Suzuki, H., *Chem. Lett.*, **1977,** 1083.
[700] Tsuji, J., and Suzuki, H., *Chem. Lett.*, **1977,** 1085.

[701] Kruse, M. W., and Wright, L. W., *Carbohydr. Res.*, **64,** 293 (1978).
[702] Ohkubo, K., Shoji, T., Terada, I., and Yoshinaga, K., *Inorg. Nucl. Chem. Lett.*, **13,** 433 (1977); *Chem. Abs.*, **87,** 167228 (1977).
[703] Ozaki, H., *Bull. Chem. Soc. Jpn.*, **51,** 257 (1978).
[704] Tai, A., Imaida, M., Oda, T., and Watanabe, H., *Chem. Lett.*, **1978,** 61.
[705] Nitta, Y., Okamoto, Y., Imanaka, T., and Teranishi, S., *Nippon Kagaku Kaishi*, **1978,** 634; *Chem. Abs.*, **88,** 189698 (1978).
[706] Band, E., Pretzer, W. R., Thomas, M. G., and Muetterties, E. L., *J. Am. Chem. Soc.*, **99,** 7380 (1977).
[707] Miyagawa, R., and Yamaguchi, T., *Nippon Kagaku Kaishi*, **1978,** 160; *Chem. Abs.*, **89,** 59456 (1978).
[708] Miyagawa, R., and Yamaguchi, T., *Chiba Kogyo Daigaku Kenkyu Hokoku, Riko Hen*, **22,** 163 (1977); *Chem. Abs.*, **88,** 135916 (1978).
[709] Zakhariev, A., Petrov, L., and Shopov, D., *React. Kinet. Catal. Lett.*, **7,** 253 (1977); *Chem. Abs.*, **88,** 21782 (1978).
[710] Zakhariev, A., and Ivanova, V., *Izv. Khim.*, **10,** 58 (1977); *Chem. Abs.*, **87,** 200500 (1977).
[711] Ucciani, E., Pelloquin, A., and Cecchi, G., *J. Mol. Catal.*, **3,** 363 (1978); *Chem. Abs.*, **88,** 120507 (1978).
[712] Olivier, R. G., and Wells, P. B., *J. Catal.*, **47,** 364 (1977).
[713] Siegel, S., and Perot, G., *J. Chem. Soc., Chem. Commun.*, **1978,** 114.
[714] Hattori, T., and Burwell, R. L., *J. Chem. Soc., Chem. Commun.*, **1978,** 127.
[715] Karakhanov, E. A., Dedov, A. G., Pakhomova, I. I., Saginova, L. G., and Viktorova, E. A., *Vestn. Mosk. Univ., Ser.* **2***:* Khim., **18,** 334 (1977); *Chem. Abs.*, **88,** 21748 (1978).
[716] Consiglio, G., and Pino, P., *Isr. J. Chem.*, **15,** 221 (1977).
[717] Lau, M. S. W., and Sermon, P. A., *J. Chem. Soc., Chem. Commun.*, **1978,** 891.
[718] James, B. R., and Morris, R. H., *J. Chem. Soc., Chem. Commun.*, **1978,** 929.
[719] Okuhara, T., and Tanaka, K., *J. Chem. Soc., Chem. Commun.*, **1978,** 53.
[720] Vedenyapin, A. A., Klabunovskii, E. I., Leonova, E. V., Areshidze, G.K., and Barannikova, N. E., *Izv. Akad. Nauk SSSR, Ser. Khim.*, **1978,** 206; *Chem. Abs.*, **88,** 135922 (1978).
[721] Federicks, S., and Thomas, J. L., *J. Am. Chem. Soc.*, **100,** 350 (1978).
[722] Sweany, R. L., and Halpern, J., *J. Am. Chem. Soc.*, **99,** 8335 (1977).
[723] Eden, Y., Fraenkel, D., Cais, M., and Halevi, E. A., *Isr. J. Chem.*, **15,** 223 (1977).
[724] Soga, K., Imamura, H., and Ikeda, S., *Nippon Kagaku Kaishi*, **1978,** 923; *Chem. Abs.*, **89,** 128823 (1978).
[725] Soga, K., Sano, T., Imamura, H., Sato, M., and Ikeda, S., *Nippon Kagaku Kaishi*, **1978,** 930; *Chem. Abs.*, **89,** 128824 (1978).
[726] Cann, K., Cole, T., Slegeir, W. and Pettit, R., *J. Am. Chem. Soc.*, **100,** 3969 (1978).
[727] Farrell, N., Dolphin, D. H., and James, B. R., *J. Am. Chem. Soc.*, **100,** 324 (1978).
[728] Reutov, G. A., and Finkel-stein, A. V., *Zh. Fiz. Khim.*, **52,** 1696 (1978); *Chem. Abs.*, **89,** 128951 (1978).
[729] Entwistle, I. D., Gilkerson, T., and Johnstone, R. A. W., *Tetrahedron*, **34,** 213 (1978).
[730] Cerveny, L., Plechacova, D., and Ruzicka, V., *Collect. Czech. Chem., Commun.* **43,** 2387 (1978).
[731] Rogers, D. W., von Voithenberg, H., and Allinger, N. L., *J. Org. Chem.*, **43,** 360 (1978).
[732] Fu, P. P., and Harvey, R. G., *Chem. Rev.*, **78,** 317 (1978).
[733] Skarchenko, V. K., *Usp. Khim.*, **46,** 1411 (1977).
[734] Rudakov, E. S., and Rudakova, R. I., *Metallokompleksnyi Katal.*, **1977,** 85; *Chem. Abs.*, **89,** 23241 (1978).
[735] Marchetti, L., *J. Chem. Soc., Perkin Trans.* 2, **1977,** 1977.
[736] Marchetti, L., *J. Chem. Soc., Perkin Trans.* 2, **1978,** 282.
[737] Danchevskaya, M. N., and Torkin, S. N., *Zh. Fiz. Khim.*, **52,** 454 (1978); *Chem. Abs.*, **88,** 169342 (1978).
[738] Fu, P. P., Lee, H. M., and Harvey, R. G., *Tetrahedron Lett.*, **1978,** 551.
[739] Brown, D. R., and Turner, A. B., *J. Chem. Soc., Perkin Trans.* 2, **1978,** 165.
[740] de Wit, G., de Vlieger, J. J., Kock-van Dalen, A. C., Kieboom, A. P. G., and van Bekkum, H., *Tetrahedron Lett.*, **1978,** 1327.
[741] Akhtar, M., and Jones, C., *Tetrahedron*, **34,** 813 (1978).
[742] Martens, F. M., Verhoeven, J. W., Gase, R. A., Pandit, U. K., and de Boer, T. J., *Tetrahedron*, **34,** 443 (1978).
[743] Watanabe, K., Kawaguchi, R., and Lato, H., *Chem. Lett.*, **1978,** 255.

[744] Ohno, A., Yasui, S., Nakamura, K., and Oka, S., *Bull. Chem. Soc. Jpn.*, **51,** 290 (1978).
[745] Ohno, A., Kimura, T., Yamamoto, H., Kim, S. G., Oka, S., and Ohnishi, Y., *Bull. Chem. Soc. Jpn.*, **50,** 1535 (1977).
[746] Ohno, A., Yasui, S., Yamamoto, H., Oka, S., and Ohnishi, Y., *Bull. Chem. Soc. Jpn.*, **51,** 294 (1978).
[747] Van Ramesdonk, H. J., Verhoeven, J. W., Pandit, U. K., and De Boer, T. J., *Recl. Trav. Chim. Pays-Bas*, **97,** 195 (1978).
[748] Ohno, A., Kimura, T., Oka, S., and Ohnishi, Y., *Tetrahedron Lett.*, **1978,** 757.
[749] Ohno, A., Ikeguchi, M., Kimura, T., and Oka, S., *J. Chem. Soc., Chem. Commun.*, **1978,** 328.
[750] Baba, N., Matsumura, Y., and Sugimoto, T., *Tetrahedron Lett.*, **1978,** 4281.
[751] de Nie-Sarink, M. J., and Pandit, U. K., *Tetrahedron Lett.*, **1978,** 1335.
[752] Hood, R. A., Prince, R. H., and Rubinson, K. A., *J. Chem. Soc., Chem. Commun.*, **1978,** 300.
[753] Shinkai, S., Hamada, H., Ide, T., and Manabe, O., *Chem. Lett.*, **1978,** 685.
[754] Shinkai, S., Ide, T., Hamada, H., Manabe, O., and Kinitahe, T., *J. Chem. Soc., Chem. Commun.*, **1977,** 848.
[755] Makamura, K., Ohno, A., Yasui, S., and Oka, S., *Tetrahedron Lett.*, **1978,** 4815.
[756] Behr, J.-P., and Lehn, J.-M., *J. Chem. Soc., Chem. Commun.*, **1978,** 143.
[757] Stewart, R., and Norris, D. J., *J. Chem. Soc., Perkin Trans.* 2, **1978,** 246.
[758] Yoneda, F., Sakuma, Y., and Matsushita, Y., *J. Chem. Soc., Chem. Commun.*, **1978,** 398.
[759] Ohno, A., Yamamoto, H., Okamoto, T., Oka, S., and Ohnishi, Y., *Chem. Lett.*, **1978,** 65.
[760] Hedstrand, D. M., Kruizinga, W. H., and Kellogg, R. M., *Tetrahedron Lett.*, **1978,** 1255.
[761] Ohnishi, Y., and Kitami, M., *Tetrahedron Lett.*, **1978,** 4033.
[762] Ohnishi, Y., and Kitami, M., *Tetrahedron Lett.*, **1978,** 4035.
[763] Foos, J. S., Killian, W., Rizvi, S. Q. A., Unger, M., and Fraenkel, G., *Tetrahedron Lett.*, **1978,** 1407.
[764] Eisner, U., Sadeghi, M. M., and Hambright, W. P., *Tetrahedron Lett.*, **1978,** 303.
[765] Biellmann, J.-F., and Lapinte, C., *Tetrahedron Lett.*, **1978,** 683.
[766] van Eikeren, P., and Grier, D. L., *J. Am. Chem. Soc.*, **99,** 8057 (1977).
[767] Nakamura, K., Ohno, A., and Oka, S., *Tetrahedron Lett.*, **1977,** 4593.
[768] Singh, S., Trehan, A. K., and Sharma, V. K., *Tetrahedron Lett.*, **1978,** 5029.
[769] Lluis, C., and Bozal, J., *Biochim. Biophys. Acta*, **523,** 273 (1978).
[770] Van Osselaer, T. A., Lemiere, G. L., Lepoivre, J. A., and Alderweireldt, F. C., *J. Chem. Soc., Perkin Trans.* 2, **1978,** 1181.
[771] Mushran, S. P., and Agrawal, M. C., *J. Sci. Ind. Res.*, **36,** 274 (1977); *Chem. Abs.*, **89,** 5532 (1978).
[772] Pelizetti, E., Mentasti, E., and Pramauro, E., *Inorg. Chem.*, **17,** 1181 (1978); *Chem. Abs.*, **89,** 23451 (1978).
[773] Yatsimirskii, K. B., Bratushko, Y. I., Zatsny, I. L., Kriss, E. E., and Kurbatova, G. T., *Zh. Neorg. Khim.*, **22,** 2441 (1977); *Chem. Abs.*, **87,** 183733 (1977).
[774] Sychev, A. Y., Isak, V. G., Hiem, M. H., and Shemyakova, L. D., *Zh. Fiz. Khim.*, **51,** 2138 (1977); *Chem. Abs.*, **87,** 183726 (1977).
[775] Imuta, M., and Ziffer, H., *J. Org. Chem.*, **43,** 3319 (1978).
[776] Nakazaki, M., Chikamatsu, H., Naemura, K., Hirose, Y., Shimizu, T., and Asao, M., *J. Chem. Soc., Chem. Commun.*, **1978,** 668.
[777] Bucciarelli, M. Forni, A., Moretti, I., and Torre, G., *J. Chem. Soc., Chem. Commun.*, **1978,** 456.
[778] Nakazaki, M., Chikamatsu, H., Naemura, K., Nishino, M., Murakami, H., and Asao, M., *J. Chem. Soc., Chem. Commun.*, **1978,** 667.
[779] Ohta, H., and Tetsukawa, H., *J. Chem. Soc., Chem. Commun.*, **1978,** 849.
[780] Schnarr, G. W., and Szarek, W. A., *Can. J. Chem.*, **56,** 1752 (1978).
[781] Abushanab, E., Reed, D., Suzuki, F., and Sih, C. J., *Tetrahedron Lett.*, **1978,** 3415.
[782] Leete, E., *J. Chem. Soc., Chem. Commun.*, **1978,** 610.
[783] Rosenfeld, H. J., Watanabe, K. A., and Roberts, J., *J. Biol. Chem.*, **252,** 6970 (1977).
[784] Dallocchio, F., Signorini, M., and Rippa, M., *Arch. Biochem. Biophys.*, **185,** 57 (1978).
[785] Schmid, G., Luedemann, H. D., Jaenicke, R., *Eur. J. Biochem.*, **86,** 219 (1978).
[786] Rotenberg, S. L., and Sprinson, D. B., *J. Biol. Chem.*, **253,** 2210 (1978).
[787] Shiakai, S., Ide, T., and Manabe, O., *Chem. Lett.*, **1978,** 583.
[788] Rich, P. R., Wiegand, N. K., Blum, H., Moore, A. L., and Bonner, W. D., *Biochim. Biophys. Acta*, **252,** 325 (1978).
[789] Suva, R. H., and Abeles, R. H., *Biochemistry*, **17,** 3538 (1978).

[790] Griffin, B. W., and Ting, P. L., *Biochemistry*, **17,** 2206 (1978).
[791] Roseboom, H., and Perrin, J. H., *J. Pharm. Sci.*, **66,** 1392 (1977).
[792] Roseboom, H., and Perrin, J. H., *J. Pharm. Sci.*, **66,** 1395 (1977).
[793] Scott, A. I., Irwin, A. J., Siegel, L. M., and Shoolery, J. N., *J. Am. Chem. Soc.*, **100,** 316 (1978).
[794] Kaufmann, E. J., and Espenson, J. H., *J. Am. Chem. Soc.*, **99,** 7051 (1977).
[795] Kemla, C., Chan, T. N., and Bruice, T. C., *J. Am. Chem. Soc.*, **99,** 7272 (1977).
[796] Gustafson, M. E., Miller, D., Davis, P. J., Rosazza, J. P., Chang, C., and Floss, H. G., *J. Chem. Soc., Chem. Commun.*, **1977,** 842.
[797] Moad, G., Luthy, C. L., and Benkovic, S. J., *Tetrahedron Lett.*, **1978,** 2271.
[798] Awruch, J., and Frydman, B., *Tetrahedron Lett.*, **1978,** 2637.
[799] Galliani, G., Rindone, B., and Marchesini, A., *J. Chem. Soc., Perkin Trans.* 1, **1978,** 456.
[800] Young, P. R., and Hsieh, L.-S., *J. Am. Chem. Soc.*, **100,** 7121 (1978).

CHAPTER 5

Carbenes and Nitrenes

M. S. Baird

School of Chemistry, The University, Newcastle-upon-Tyne

Reviews of fluorine-containing[1,2] and unsaturated carbenes,[3] of the reactions of halocarbenes in two-phase systems,[4] of the generation, spectra, and reactions of triplet carbenes,[5] of carbene analogues,[6a] and of the synthesis of heterocycles using nitrenes[6b] have appeared, as well as a general review of carbenes and nitrenes.[7]

Structure and Reactivity

Ab initio calculations are reported for methylene[8] and a range of substituted methylenes[9,10] and trends in singlet–triplet energy separation are discussed; an S–T separation of 10.5 kcal mol^{-1} for methylene itself is reported. MINDO/3 calculations on methylene and silyl-substituted derivatives suggest that interaction of the silyl group to some extent increases the electron affinity of the carbene carbon; a bent geometry is suggested for triplet $Me_3SiCH\colon$ in conflict with previous postulates.[11]

A theoretical study of the intersystem crossing of singlet and triplet methylenes has been reported.[12]

Calculations, including Hartree–Fock and spin-unrestricted many-body perturbations, show singlet ground states for $CH_2{=}C$: and $CH_2{=}C{=}C$: with S–T separations of 51.1 and 48.7 kcal mol^{-1}, respectively (a separation of −13.1 kcal mol^{-1} is found for methylene). Singlet ground states are predicted for carbenes containing extended unsaturation, and d-type carbon polarization functions are found to have considerable effect on structure.[13] *Ab initio* calculations on the singlet states of a range of substituted alkylidene-carbenes $R_2C{=}C$: show that electropositive substituents stabilize the carbene, principally by σ-inductive electron donation and hyperconjugation, and that electronegative substituents are destabilizing.[14]

Calculations on C_2H_3N isomers including vinylnitrene, iminoethylidene, and aziridinylidene (**1**)[15], on HCNO isomers including formylnitrene and oxaziridinylidene (**2**),[16] on oxirene isomers including oxiranylidene (**3**; X = O) and formylmethylene,[17] and on C_2H_2S isomers including (**3**; X = S)[18] are also reported.

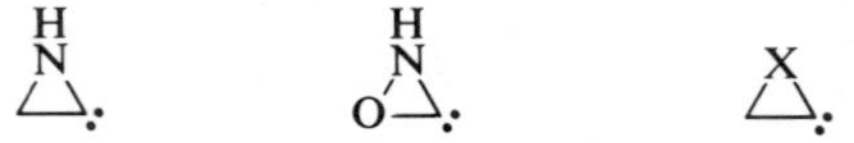

(**1**) (**2**) (**3**)

The heat of formation of dichlorocarbene has been found to be 53–54 ± 2 kcal mol^{-1} from the enthalpy of deprotonation of the dichloromethyl cation.[19] Further measures of the proton affinity of the carbene have been obtained by reaction of Cl_2DC^+ with a variety of bases.[20]

The spectroscopy and photophysics of laser-formed difluorocarbene,[21] its initial energy distribution,[22] and formation from fluorocarbons by lasers[23] and in the pyrolysis in reflected-shock waves[24] are also reported.

ESR studies of unsymmetrically substituted diphenylmethylenes show a triplet ground state stabilized by mero-stabilization,[25] while alkoxycarbonylcarbenes have two ground-state triplets that are assigned as rotamers.[26]

There has been some questioning of the characteristic reactions of singlet and triplet carbenes. Unsensitized photolysis of dimethyl diazomalonate in cyclobutene leads to dimethylbicyclo[2.1.0]pentane-5:5-dicarboxylate, but sensitized photolysis leads in addition to (**4**; X = COOMe, Y = H) and (**4**; X = H, Y = COOMe). The last product (**4**) is formed by direct reaction of the triplet carbene. The products from 1-diazopent-4-ene can then be assigned as penta-1,4-diene from the singlet, and bicyclo[2.1.0]pentane from the triplet carbene.[27] Rearranged products are also obtained from the reaction of 1,2-dihaloalkenes with dibromocarbene; thus, *trans*-1,2-dichloroethylene gives (**5**). The latter is not a cyclopropane rearrangement product and it is suggested that the carbene can undergo S–T equilibration and that the rearranged product is derived from the triplet *via* the diradical (**6**).[28] Support for this is found in calculations which suggest that S–T energy

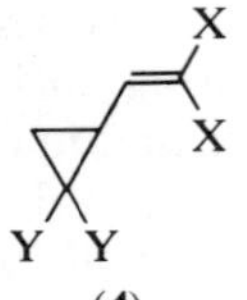

(**4**) 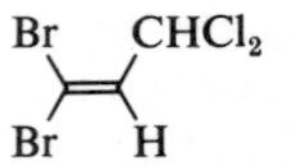 (**5**) 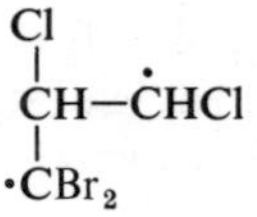 (**6**) 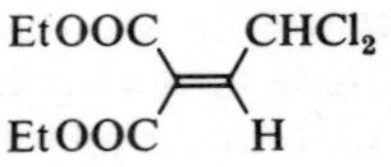 (**7**)

separations for some halo-carbenes are indeed very small, when only a small triplet reactivity preference would lead to the observation of products from both spin states. Furthermore, triplet di(ethoxycarbonyl)carbene does indeed give a rearranged product (**7**) with *cis*-dichloroethylene, whereas the singlet leads to a cyclopropane.

The ground and excited states of aminonitrene have been studied by means of generalized valence-bond and configuration-interaction wave functions and the ground state is found to be a singlet ($^1A_1$) some 15 kcal mol^{-1} below the triplet ($^3A_2$). The stabilization is due to a high double-bond character.[30] ESR spectra of a wide range of triplet 4-substituted phenylnitrenes have been measured in methylcyclohexane glasses at 77 K, and INDO calculations have been carried out.[31] Spectroscopic data are also reported for *N*-(2,2,6,6-tetramethylpiperidyl)nitrene.[32]

Generation

Carbenes

Reviews have appeared of the generation of carbenes from organometallic precursors,[33] and in particular organomercurials,[34] and of carbenes from phase-transfer reactions.[4] In addition, a review of thermal and photochemical nitrogen extrusion includes many carbene examples.[35]

(Chloromethyl)trimethylsilane reacts with lithium 2,2,6,6-tetramethylpiperidide to produce singlet Me_3SiCH:, which adds stereospecifically to alkenes. Similarly, Me_3SnCH_2Cl gives the previously unreported Me_3SnCH:.[36] Crown-ether-catalysed dehydrobromination of *ω*-bromomethyleneadamantane produces another new carbene, adamantylidene-carbene, which undergoes a range of intermolecular additions.[37]

Photolysis of methylene iodide in the presence of alkenes leads to good yields of cyclopropanes in a reaction that is relatively selective but shows little steric effect.[38] The selectivity and a lack of C—H insertion suggest that the reaction does not involve a free carbene, and an intermediate involving methylene stabilized by co-ordination to iodine is proposed.

Sodium salts of hydrazones derived from 2,4,6-tri-isopropylbenzenesulphonylhydrazine decompose to carbenes under milder conditions than suffice for toluene-*p*-sulphonyl- or trifluoromethanesulphonyl-hydrazones.[39]

Photolysis of (**8**) leads to nitrogen-free products. These have been shown to arise by initial isomerization to (**9**), carbene formation and insertion, and addition or rearrangement reactions of the carbene.[40] Photolysis of 3,3-dimethyl-3H-pyrazoles

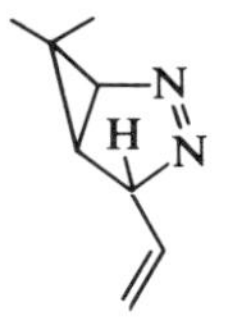

H N N N H N$_2$ R N N R′ Me Me

(**8**) 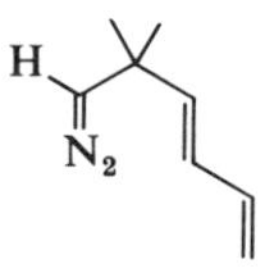 (**9**) 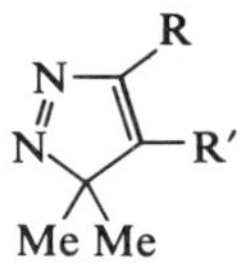 (**10**)

(**10**) leads to vinylcarbenes which add in good yield and with high stereoselectivity to alkenes. Cyano-substituted vinylcarbenes are the most efficiently trapped but alkoxycarbonyl and even ketovinyl derivatives undergo cyclopropanation with dienes, whilst there is little competition from Wolff rearrangement in the latter case.[41] Thermal decomposition of 1-methyl-1-phenyldiazirene leads to $PhC(Me)N_2$ which at higher temperatures in polar solvents leads to PhC(Me)=N—N=C(Me)Ph but in non-polar solvents produces styrene-derived products.[42]

Alkylchlorocarbenes have been generated by photolysis of diazirines (**11**), and added to alkenes. The partition between inter- and intra-molecular interception (generally a hydrogen shift) depends mainly on the nature of R.[43] Photolysis of (**11**; R = cyclopropyl) gives cyclopropylchlorocarbene; this can be relatively efficiently trapped by alkenes but, in the absence of a trap, the main intramolecular product is 1-chlorocyclobutene and not, as earlier reported, chloromethylenecyclopropane. The efficiency of addition of this carbene to alkenes is ascribed to a bisected geometry (**12**) in which hydrogen migration is unfavourable but the substituents best stabilize the vacant *p*-orbital. Even cyclopropylcarbene itself is found to add to alkenes, albeit in low yield.[44]

(**11**) (**12**)

The diazirine (**11**; R = OMe) is explosive in the neat form but can be handled readily in solution or as a gas. At 25 °C, it loses nitrogen to produce methoxychlorocarbene which adds to olefins, probably stereospecifically, and forms cyclopropenones with acetylenes. As predicted, the carbene exhibits ambiphilic behaviour, and it shows no products of intramolecular rearrangement or insertion, in contrast to the behaviour of other reported alkoxyhalocarbenes.[45,46]

Optimum conditions for generation of dichlorocarbene in a phase-transfer system have been investigated, and side-reactions and catalyst differences are reported.[47] Anhydrous potassium carbonate is found to be a good base for the generation of dibromocarbene from bromoform in the presence of 18-crown-6 ether.[48] Me_3SiCCl_3 and Me_3SiCBr_3 react with anhydrous potassium fluoride in diglyme to give good yields of dihalocarbene-transfer products with alkenes, in a reaction which is catalysed by 18-crown-6 ether;[49] galvanostatic reduction of carbon tetrachloride or chloroform at lead cathodes in the presence of alkenes and tetrabutylammonium bromide also leads efficiently to dihalocyclopropanes.[50] Thallium(I) butanethiolate and chloroform do give the carbene, but in this case yields of adducts are low.[51] The effect of association in organolithium compounds on the reactivity of carbenes produced by their decomposition has been discussed, and dichlorocarbene from the thermolysis of trichloromethyl-lithium has been found to insert into the carbon–lithium bond of phenyl-lithium in a cage;[52] CIDNP and product analysis shows that the carbene is produced in radical pairs from the reaction of *tert*-butylmagnesium chloride and $CXCl_3$.[53]

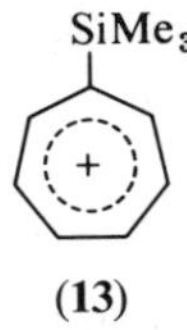

(**13**)

Reaction of (**13**) with tetrabutylammonium fluoride leads to cycloheptatrienylidene which undergoes dimerization or cycloaddition to electron-deficient alkenes.[54]

Photolysis of diazoxanthene leads to xanthenylidene-carbene which reacts as an electrophilic σ^2-singlet that does not insert into the C—H bonds of alkanes; insertion does, however, occur into the C—H bonds of nitromethane, into the *para*-C—H bond of *N,N*-dimethylaniline, and into water.[55] The S and T states do not react with alkylated olefins.[56] In a second study it was found, in contrast with earlier results, that the carbene adds stereospecifically to styrenes with a ρ-value of $+0.97$, inserts into saturated C—H bonds, and adds to alkylalkenes.[57] Insertion into allylic C—H bonds occurs without double-bond rearrangement.

Thioxanthenylidene, obtained similarly, undergoes dimerization but can be trapped by dimethyl fumarate. With dimethyl maleate, an open-chain product (**14**) is obtained.[58]

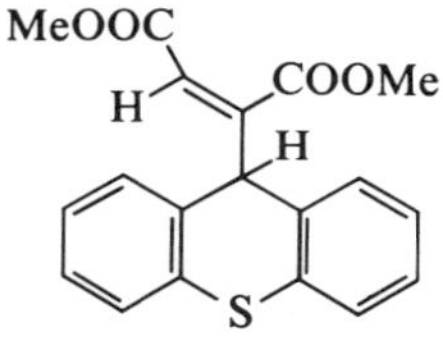

(**14**) 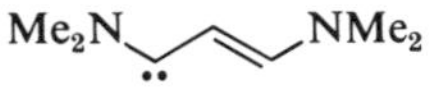 (**15**) 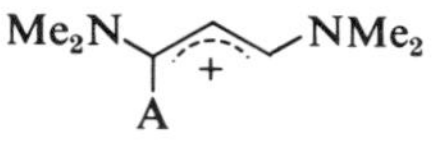 (**16**) 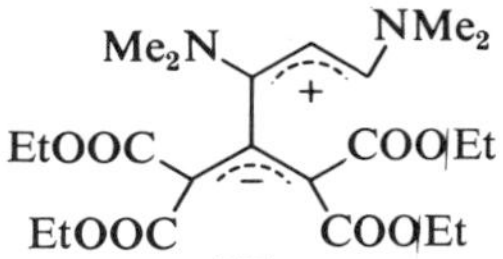 (**17**) 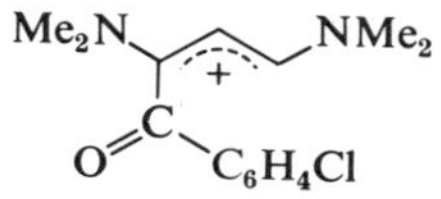 (**18**)

The nucleophilic carbene (**15**) can be obtained from vinamidinium salts (**16**; A = COPh or COOEt) and trapped by formation of (**17**) with tetraethyl allenetetracarboxylate; on its reaction with 4-chlorobenzoyl chloride followed by sodium perchlorate, the ion (**18**) is obtained.[59]

Thermal decomposition of (**19**), catalysed by $Cu(acac)_2$, in the presence of alkenes leads to good yields of bis(methoxycarbonyl)carbene adducts, though a metal-carbenoid is probably involved in the reaction.[60] Irradiation of (**20**) in alkenes leads to cyclopropanes stereospecifically and in high yield, and C—H insertion is seldom observed.[61] A high degree of *trans*-stereoselectivity is observed in the addition to unsymmetrical alkenes; Lewis acids affect both the rate of decomposition and the efficiency of the addition. Photolysis of (**21**) leads to a carbene which forms benzoylphosphorane PhP(O)=C(Ph)COPh.[62] The related species (**22**) undergo intramolecular cyclization to produce cyclopropenes, e.g. (**23**), though when R^1 = Ph some indene is observed.[63] Carbenes (**22**; X = O) or related carbenoids are also intermediates in the copper-catalysed transformation of

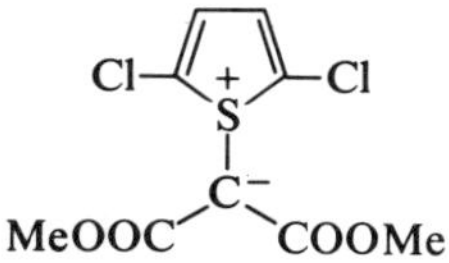

(**19**) 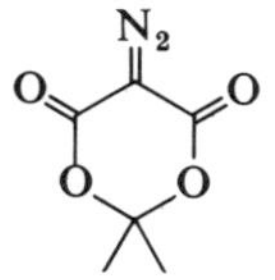 (**20**) 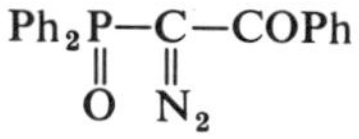 (**21**)

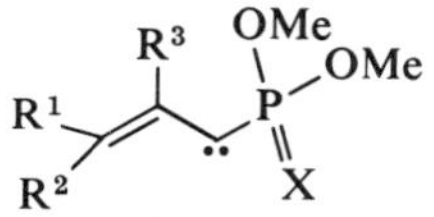

(**22**) 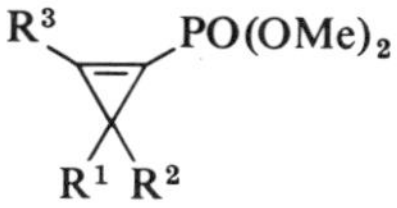 (**23**)

(**23**; R^1 = Me) to (**24**), either by a 1,4-hydrogen shift or by cyclization to a cyclobutene followed by ring-opening.

Photolysis of phosphorus-substituted oxiranes of type (**25**) provides an alternative source of phosphonocarbenes $Ph\ddot{C}P(O)(OR)_2$; cleavage to diphenylmethylene produces a complicating alternative.[64]

Gas-phase photolysis of carbon suboxide leads to carbonylcarbene (:C=C=O) which reacts with tetrahydrofuran either by deoxygenation to the tetramethylene biradical or by C—O insertion and photodecarbonylation to (**26**) and thence to cyclopentanone.[65]

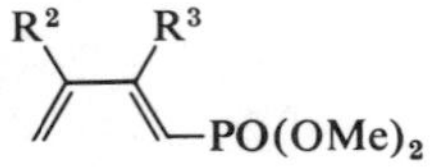

(**24**) 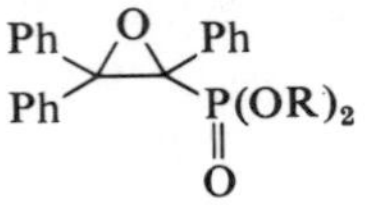 (**25**) 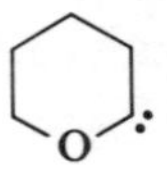 (**26**)

Nitrenes

Electrochemical reduction of Cl_2NCOOR (R = Me or Et) in 1,4-dioxan leads to the carbamate (**27**) apparently by insertion of alkoxycarbonylnitrene;[66] decomposition

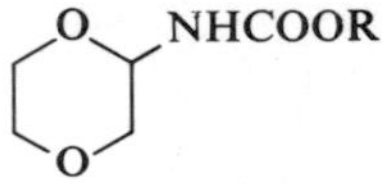

(**27**)

of ethyl *p*-nitrobenzenesulphonyloxycarbamate by base in a two-phase system also leads to products consistent with formation of :NCOOEt, as in the generation under homogeneous conditions.[67] Treatment of n-$C_6F_{13}S(O)Cl$ with tetraethylammonium azide leads to products apparently derived from n-$C_6F_{13}S(O)\ddot{N}$, which does not undergo a Curtius rearrangement or add to cyclohexene.[68] Aminonitrenes are reported in the reaction of $NaONNaNO_2$ with secondary amines.[69]

Reaction of *N*,*N*-dichlorourethane with Cu or Cu(I) chloride in the presence of various substrates has been shown to proceed by a metal-radical mechanism, and not by α-elimination to a nitrene, by comparison of products with those from the nitrene derived by photolysis of ethyl azidoformate.[70]

Reactions thought to involve Carbenes or Nitrenes

Photolysis of tri- and tetra-substituted alkenes with high-energy photons is reported. Norbornene leads to 5-methylenebicyclo[2.1.1]hexane and tricyclo[2.2.1.0^{2,6}]-heptane, apparently through two different carbenes (**28**) and (**29**) derived by either initial carbon–carbon or carbon–hydrogen migration; similar results have been obtained for other alkenes.[71] 2,2,6,6-Tetradeuteriocyclohexene leads to methylenecyclopentane with 92% of the label at C-2 and C-5 of the ring, supporting an intermediate cyclopentylcarbene rather than cyclohexylidene. 4-Vinylcyclohexene provides products which can be explained by a 1,2-hydrogen shift or C—H insertion at C-2 or C-5 in an intermediate (**30**).[72]

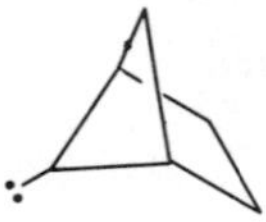

(**28**) 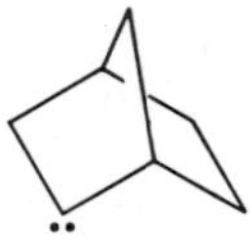 (**29**) 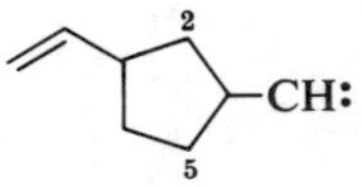 (**30**) 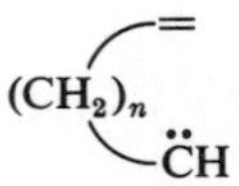 (**31**)

Photolysis of bicyclo[*n*.1.0]alkanes leads to dienes, apparently through fragmentation to (**31**).[73]

Photolysis of (**32**) leads to (**33**), and D-labelling studies support a mechanism involving photoelimination of carbene to produce (**34**), followed by a 1,2-hydrogen shift.[74]

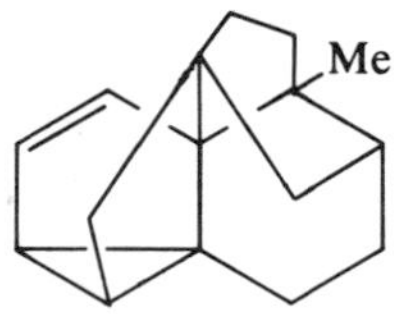

(**32**) 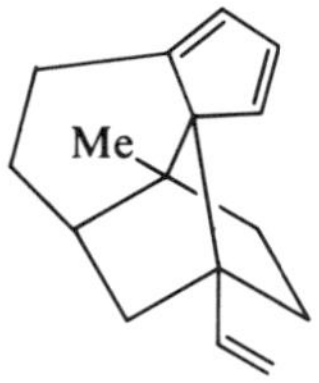 (**33**) 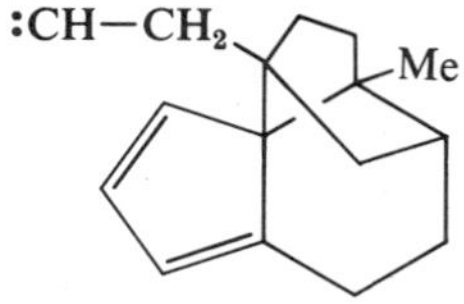 (**34**)

Hexadienylcarbenes may be involved in the intramolecular cyclization of 2-alkylbiphenyl-2′-yldiazonium salts.[75]

Reaction of alkyl Grignard reagents, free from transition-metal impurities, with terminal propargylic chlorides leads to an allene-carbene zwitterionic intermediate which reacts with a second mole of Grignard reagent to produce a mixture of propargyl and allenyl Grignard reagents.[76] However, 3-chloro-3-methylbut-1-yne reacts with triethylamine and copper powder to produce the octamethylcyclododeca-1,3,8,10-tetraene (**35**), apparently by dimerization of $Me_2C{=}C{=}C{:}$ to 2,7-dimethylocta-2,3,4,5,6-pentaene followed by a second dimerization.[77]

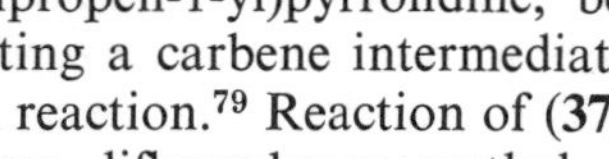

(**35**) 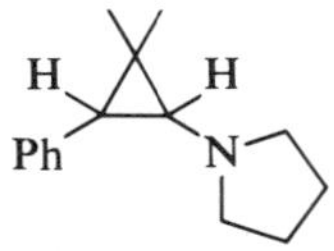 (**36**) 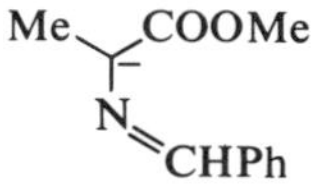 (**37**)

Tungsten hexacarbonyl and WCl_6–$LiAlH_4$ cause coupling of *gem*-dihalides and benzyl halides to alkenes and dibenzyls, respectively; in the presence of 1-(2-methylpropen-1-yl)pyrrolidine, benzylidene chloride forms the compound (**36**), suggesting a carbene intermediate.[78] Carbenes are also implicated in the WCl_6–CD_3Li reaction.[79] Reaction of (**37**) with difluorobromomethane leads to the corresponding difluorobromomethyl and difluoromethyl derivatives and a carbene mechanism is again suggested.[80]

Decomposition of (dimethylamino)malononitrile is thought to proceed by an allowed $[\sigma^{2s}+\sigma^{2a}]$ non-linear chelotropic fragmentation which eliminates HCN to produce (dimethylamino)cyanocarbene; the latter abstracts a proton from the starting material.[81] Kinetic measurements show that the decomposition of $(PhS)_3CLi$ proceeds through a pre-equilibrium with $(PhS)_2C{:}$ and PhSLi, followed by reaction of the carbene with starting material to produce, eventually, $(PhS)_2C{=}C(SPh)_2$.[82]

Gas-phase thermolysis of 4- or 5-substituted isothiazoles (**38**) leads to thioketenes, possibly *via* a thioxocarbene $R^2C(S)\ddot{C}R^1$;[83] the outcome of photolysis of $Hg(N_2CCOOEt)_2$ indicates the formation of products derived from ethoxycarbonylcarbyne, and the formation of mercury-containing products from the decomposition in pyrrolidine provides evidence for major involvement of a mercury carbene, $EtO_2C\ddot{C}HgCN_2COOEt$.[84]

Reaction of primary amines (RNH_2) with hydroxylamine-*O*-sulphonic acid and hydroxide ion leads to RH, nitrogen, and sulphate ion. The reaction is thought to proceed by formation of $RNHNH_2$, for which there is precedent, and reaction of this with :NH from the reaction of the sulphonic acid with base, to produce RN=NH which loses nitrogen.[85]

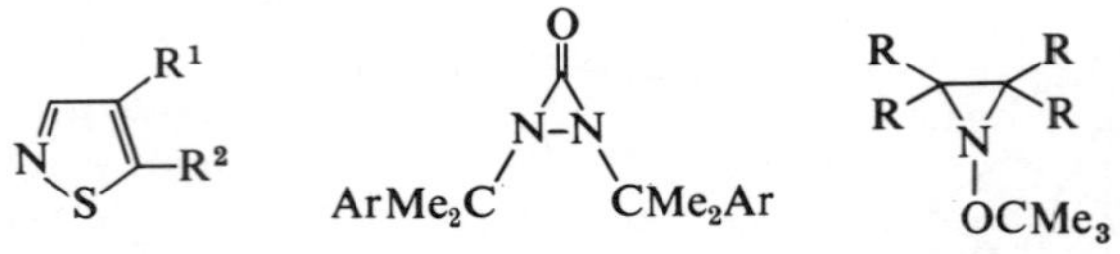

(38) (39) (40)

Thermal decomposition of the diaziridinone (**39**) leads to a complex product mixture but the major decomposition pathway is fragmentation to an isocyanate and $ArC(Me_2)N$: which rearranges to $ArN{=}CMe_2$ by aryl migration.[86]

Acyl-nitrenes are proposed as intermediates in the thermolytic conversion of $RCON(OSiMe_3)SiMe_3$ (*i.e.* silylated hydroxamic acids) to isocyanates,[87] and in the formation of (**40**) from the reaction of *O*-*tert*-butylhydroxylamine with lead tetra-acetate in the presence of alkenes.[88] *N*-Dibenzylaminonitrene is also apparently formed in the reaction of *N*-nitrosobenzylamine with phenacyl bromides and silver hexafluoroantimonate.[89]

An increase in the photochemical reactivity of (**41**) is observed at temperatures below 120 K, and the products observed are derived from (**42**), presumably obtained in turn from the corresponding nitrene.[90]

(41) (42)

Addition

Intermolecular

Carbenes. Reviews have appeared of enantioselective cyclopropanation through chiral metal-carbene intermediates,[91] and the synthesis of carbocyclic spiro-compounds by cycloadditions including carbene additions.[92]

Ab initio calculations on the addition of $^1A_1$ methylene to ethylene to produce cyclopropane show no barrier and indicate an unsymmetric approach which is initiated by an electrophilic phase and then followed by a nucleophilic phase.[93]

The selectivity of addition of $:CX_2$ (X = halogen) to alkenes varies with temperature. At temperatures below 330 K, the order is $:CF_2 > :CCl_2 > :CBr_2$, but this is reversed at temperatures above 330 K; the latter is an isoselective temperature.[94] Addition of $:CCl_2$ to alkenes does not show a pronounced rate-enhancement due to the presence of an adjacent cyclopropyl group,[95] but reaction of $\alpha\beta$-unsaturated ketones and esters with CHX_3(X = Cl or Br)–aq. NaOH in a phase-transfer system frequently leads to high yields of dibromocyclopropanes.[96] The last-mentioned reaction may well proceed by Michael addition of $^-CX_3$ rather than by cyclopropanation by $:CX_2$. Addition of $:CBr_2$ (generated under phase-transfer conditions) to allyl bromide leads to a cyclopropane, but 4,4,4-tribromobut-1-ene is obtained as a side-product.[97]

Relative rates of addition of :CXY (X or Y = halogen) to substituted styrenes lead to linear $\rho\sigma$-plots for each carbene;[98,99] addition to 1,4-dioxins and 1,4-dithins is also discussed.[100] Reaction of $:CCl_2$, from the thermolysis of Cl_3CSiX_3, with alkenes leads to chemically-activated cyclopropanes, for which the kinetics of rearrangement have been determined.[101]

Addition of $:CCl_2$ to the alkene bond of enynes,[102] to polyunsaturated acetals,[103] and to allenes,[104,105] vinylallenes,[106] and cumulenes[107] is discussed; in one report, 1,1-dicyclopropylallene is found to react with $:CCl_2$ only at the alkene 1,2-double bond.[104,105] Addition of dihalocarbenes to benzvalene[108] and of several carbenes to a cyclo-octatetraene dimer[109] is also recorded.

Relative rates and stereochemistry of addition of halocarbenes to compounds $RC_6H_4N{=}CR'C_6H_4R''$ lead to a ρ-value of less than 0.1; when the halogen atoms of a dihalocarbene are different, there is a tendency for the halogen *syn*- to the N-lone pair in the product to be in the order F > Cl > Br. MO calculations indicate that a π-approach of the carbene to the imine bond is preferred.[110]

An improved synthesis of *gem*-dichloroaziridines from imines and $:CCl_2$ in a phase-transfer system is reported,[111] while in other cases ring-opened products are observed.[112] Addition of diphenylcarbene, from the copper-chelate-catalysed decomposition of diphenyldiazomethane, to the C=N bond of diarylazomethines gives similar yields of cyclopropane independently of *para*-substitution, apparently *via* a copper carbenoid.[113]

Dichlorocarbene adds smoothly to the thioketone (**43**), but the reaction leads to (**44**) as well as the product resulting from addition only at the double bond.[114]

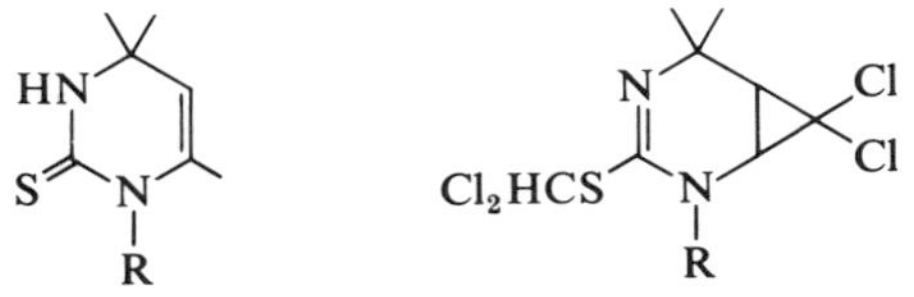

(**43**) (**44**)

The nature of the catalytic effect in the carbenoid decomposition of ethyl diazoacetate by various Pd complexes has been investigated. Chiral ligands did not induce appreciable enantioselectivity into the cyclopropanation of styrene.[115] However, cyclopropanation by diazoalkanes catalysed by [(−)camphorquinone-α-dioximato]cobalt(II) led to high enantioselectivity (up to 88%) and to high chemical yield (90–95%) with diazo-compounds bearing electron-withdrawing groups.[116] Kinetic and stereochemical effects in the enantioselective cyclopropanation of styrene with three cobalt-complex catalysts have also been analysed.[117] Copper(I) chloride-catalysed decomposition of diazo-esters of chiral alcohols in the presence of styrene leads to chiral cyclopropanes.[118]

Rhodium carboxylate-catalysed decomposition of methyl diazoacetate in the presence of alkynes leads to good yields of cyclopropenes, though with propargylic alcohols competitive addition and "insertion" into the O—H bond are observed.[119] In the gas phase, ethoxycarbonylcarbene adds to propyne and also forms an apparent insertion product, $MeC{\equiv}CCH_2COOEt$.[120] The carbene, generated by copper-catalysed diazo-decomposition, adds to both bonds of 1,1-dimethylallene and, when generated photochemically, gives also products derived by insertion into both C—H bonds. Photosensitized decomposition leads to differing product ratios, and a perpendicular trimethylenemethane is discussed as an intermediate.[121]

Cycloaddition of methoxycarbonylcarbene to aryl cyanides leads to oxazoles; the kinetics of the reaction and a theoretical analysis are reported for singlet and triplet carbenes.[122]

Addition of keto-carbenes to silylacetylenes to produce cyclopropenes is reported.[123] The carbene (**45**) reacts with phenylacetylene followed by perchloric acid to produce the salt (**46**).[124]

Addition of $(MeO)_2P(O)CH$: to dienes is discussed.[125]

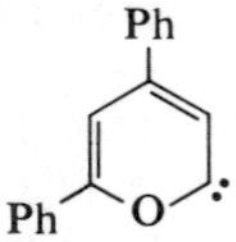

(**45**) 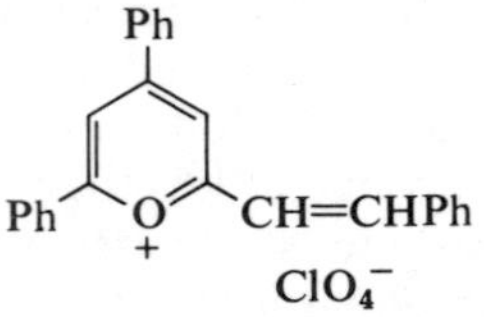 (**46**) 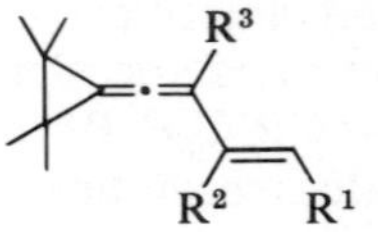 (**47**)

The olefinic selectivity of "free" dimethylvinylidenecarbene, generated by reaction of $Me_2C{=}C{=}CHBr$ with potassium *tert*-butoxide in the presence of 18-crown-6 ether, has been determined. In the absence of the crown ether, the reaction appears to proceed *via* a carbenoid; however, when $KOBu^t$–Bu^tOH is used, or when the carbene is generated in a two-phase system, the results are close to those found for the free carbene.[126]

Addition of vinylallenic carbenes to alkenes to produce (**47**) is also discussed.[127]

Relative rates of addition of phenylsulphinyl-carbene to alkenes reveal a large steric effect but, when account is taken for this, an *m*-value of 0.51 is obtained.[128] The carbene is about as selective as methylchlorocarbene and nearly as selective as dibromocarbene; σ_R^+ for the phenylsulphinyl substituent can be estimated as −0.50. The carbene adds to acetylenes to give low yields of cyclopropenes (**48**).[129]

Nitrenes. Addition of photochemically-generated ethoxycarbonylnitrene to *cis*-4-methylpent-2-ene in 1,4-dioxan-cyclohexane shows that both the stereospecificity and total singlet-nitrene-derived products increase with increasing dioxan concentration. This is explained by the formation of a singlet nitrene–dioxan complex rather than a free nitrene. For thermal generation, the stereochemistry is practically independent of the dioxan.[130] Addition of the nitrene occurs at all double bonds of *syn*-9-chloro-*cis*-bicyclo[6.1.0]nona-2,4,6-triene;[131] adducts of a cyclo-octatetraene dimer are also reported.[132]

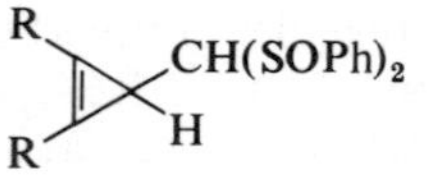

(**48**) 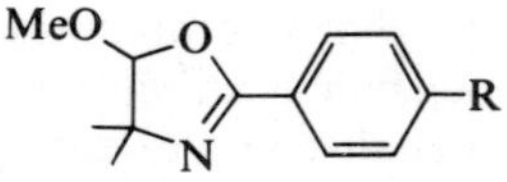 (**49**)

Photolysis of acyl azides p-$RC_6H_4CON_3$ in $Me_2C{=}CHOMe$ leads to about equal amounts of aryl isocyanate and singlet nitrene which is trapped by the enol ether as (**49**).[133]

Intramolecular

Copper-catalysed generation of the keto-carbene from (**50**) and intramolecular addition at the arrowed alkene bond is a key step in a synthesis of (±)-prostaglandin $F_{2\alpha}$.[134]

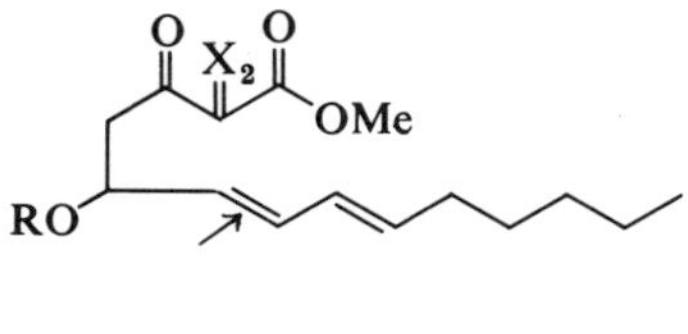

(50) 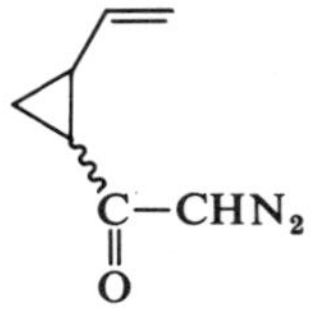 (51)

A similar addition to an alkene double bond 5,6-related to the carbene centre is reported in the decomposition of (**51**),[135] while the conversion of (**52**) into (**54**) can be explained in terms of initial carbene addition to the 6,7-related carbonyl group to produce (**53**), followed by a rearrangement.[136]

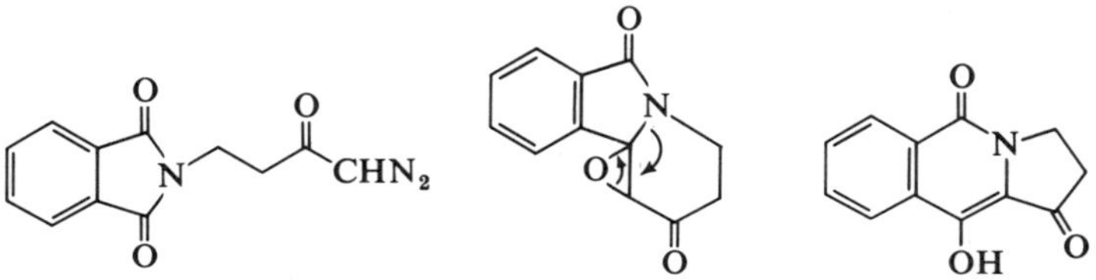

(52) (53) (54)

Reaction of (**55**; X = H, Y = Cl, R = Me) with BuLi leads to dimers of the trimethylene methane (**56**), together with the previously reported dimers of the cyclopropane form (**57**). However, (**55**; X, Y = halogen, R = H) leads only to dimers of methylenecyclopropane corresponding to (**57**). The results are taken to show a clear difference between the trimethylenemethane and methylenecyclopropane forms in these systems.[137]

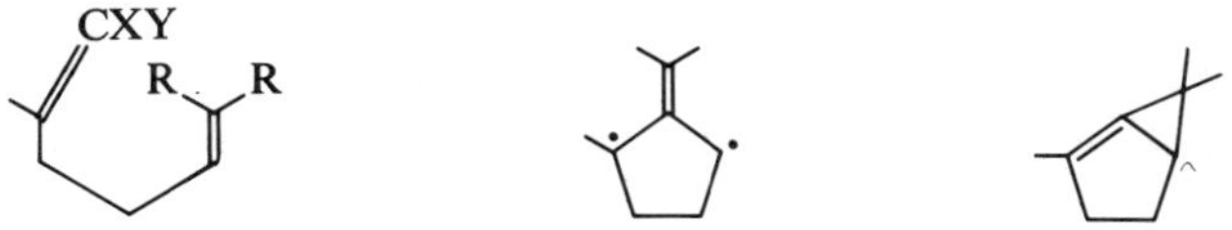

(55) (56) (57)

The nitrenes (**58**) and (**59**), obtained by oxidation of the corresponding *N*-amino-compound at −70 °C, undergo efficient intramolecular addition. The nitrene (**60**), however, does not react in this way, and dimers and cyclo-octa-1,5-diene are obtained.[138]

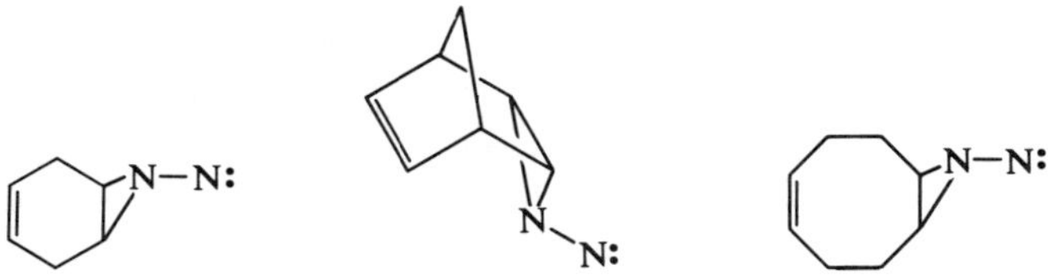

(58) (59) (60)

Insertion and Abstraction

Intermolecular

Various theoretical studies of the reaction of singlet and triplet methylene with hydrogen, to produce either methane[139,140] or methyl and hydrogen radicals,[141] are reported. Modified CNDO calculations of the reaction of hydrogen with cyclopropenylidene have been used as a model for the insertion and addition reactions of nucleophilic carbenes and predict a high activation energy (41 kcal mol^{-1}), in contrast to the methylene–hydrogen reaction, with a rate-determining electrophilic phase.[142]

The reaction of diphenylcarbene with 2-methylpropene at -196 °C leads almost exclusively to (**61**) rather than to 1,1-dimethyl-2,2-diphenylcyclopropane, which is the major product at 0–25 °C. Compound (**61**) is a true abstraction–recombination product, and a correlation of product ratio with temperature reveals no enhancement of alkene formation when the reaction is performed in the solid phase.[143] Low temperature favours the reactions of triplet rather than singlet fluorenylidene; the energetically favoured reaction with butenes is abstraction of an allylic hydrogen rather than addition.[144]

$Ph_2CH{-}CH_2{-}C({=}CH_2){-}CH_3$

(**61**)

R, Cl_2CH–C(–O–CH_2–$(CH_2)_n$–O–)

(**62**)

$CH_2{=}C{=}C(CH_2Cl)(CHClCOOEt)$

(**63**)

Dichlorocarbene, generated in a phase-transfer system, inserts into acetal C—H bonds to produce (**62**).[145] Competition studies lead to a ρ-value of -0.63 when R is aromatic, and -0.73 with a steric factor of about zero when R is aliphatic.[146]

Ethoxycarbonylcarbene, generated by copper salt-catalysed decomposition of the diazo-compound, reacts with 3,6-dibromocyclohexene by apparent insertion into the C—Br bond (4%);[147] with 1,4-dichlorobut-2-yne there is formed (**63**), which is considered to arise by insertion of the carbene into the C—Cl bond, followed by a sigmatropic shift.[148] It does not seem unlikely, however, that both processes could proceed by radical abstraction–recombination processes.

Photolysis of (**64**) leads to singlet 4*H*-imidazolylidenes which insert into the C—H bonds of alcohols to produce, after hydrogen shifts, (**65**; R = CR^1R^2OH) or (**65**; R = $CH_2CR^1R^2OH$).[149] Thermal generation leads to carbenes which insert into both C—H and O—H bonds of *tert*-butyl alcohol.

H_2N–C(=O)–[4-diazo-4*H*-imidazol-5-yl] (N_2)

(**64**)

H_2N–C(=O)–[5-R-imidazol-4-yl] (NH)

(**65**)

Ethoxycarbonylnitrene, from the photolysis of *S*,*S*-dimethyl-*N*-(ethoxycarbonyl)-sulphimides ($Me_2\overset{+}{S}{-}\overset{-}{N}COOEt$) inserts preferentially into the C—H bonds α to oxygen in cyclic ethers. The relative reactivity of C—H bonds in 2,4-dimethylpentane is in the order primary : secondary : tertiary = 1 : 9 : 35, and stereospecific insertion occurs into the tertiary C—H bonds of 1,4-dimethylcyclohexane. Insertion also occurs into the O—H bond of alcohols. Comparison with the results for the nitrene from ethyl azidoformate suggests that the insertions occur *via* a singlet nitrene.[150] The selectivity for ethoxycarbonyl-nitrene insertion into C—H bonds has been determined at various dioxan concentrations; the relative selectivity for primary and secondary bonds is almost independent of dioxan, but the tertiary/primary ratio decreases as the concentration of dioxan increases. This has been explained in terms of steric hindrance in the transition state for reaction between a nitrene–dioxan complex and the tertiary bond.[151] Evidence is also presented for a solvent–singlet nitrene complex and for involvement of the triplet nitrene in insertion into benzylic C—H bonds.[152]

Intramolecular

The reaction of dibromocyclopropanes (**66**) with methyl-lithium is reported to lead to cyclopropyl aldehydes (**67**) and derived alcohols, apparently by insertion of a carbenoid into the C—H bonds adjacent to oxygen. This is in contrast to earlier results in similar systems when cyclopropylidene–allene rearrangement was observed.[153]

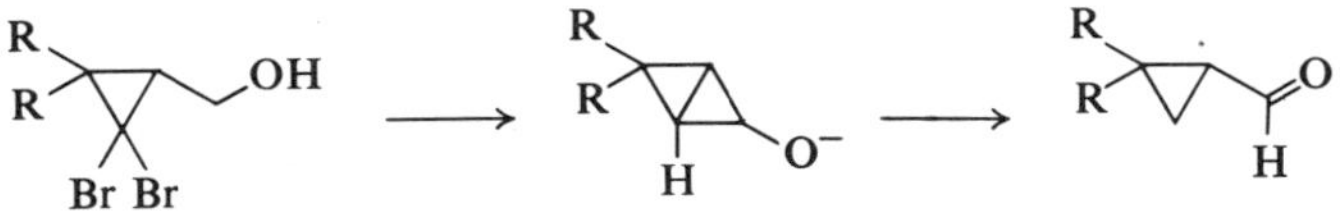

(**66**) (**67**)

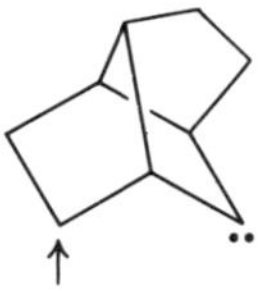

(**68**)  (**69**)

Insertion into 3,4-related bonds has also been used in the synthesis of brexanes and brendanes from, *e.g.*, (**68**),[154] while insertion into a 5,6-bond (arrowed) occurs for (**69**).[155]

Aromatics

Aromatic substitution by carbenes and nitrenes has been reviewed.[156]

The carbene (**70**), generated by pyrolysis of a toluene-*p*-sulphonylhydrazone sodium salt, adds to benzene to produce (**71**) by means of addition to form a norcaradiene, valence tautomerism to the cycloheptatriene, and a sigmatropic shift.[157]

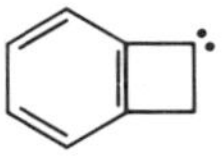

(**70**) 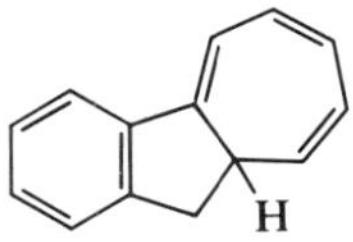 (**71**)

Reaction of 2-*tert*-butyl-5-methylpyrrole with $PhHgCCl_3$ leads to the mercury-containing intermediate (**72**; X = HgPh), which reacts with acids and bases to give 2-*tert*-butyl-5-chloro-6-methylpyridine and (**72**; X = H). Since these are the normal products from "dichlorocarbene" and the pyrrole, it is suggested that the last reaction proceeds *via* the formation of a common intermediate related to (**72**; X = HgPh).[158]

Reaction of 2-methylisoquinolone with ethoxycarbonyl- and dichloro-carbenes is reported and the reactions of uracil and 3-methyluridine with dichlorocarbene are reviewed.[159]

Intramolecular reaction of the keto-carbene derived by copper-halide-catalysed decomposition of (**73**) leads eventually to (**74**), presumably *via* addition to the aromatic system followed by a rearrangement; in some cases valence tautomerism of the initial adduct leads, after a sigmatropic shift, to (**75**).[160]

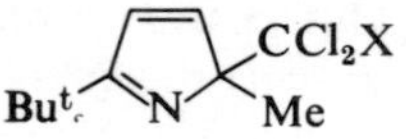

(**72**) 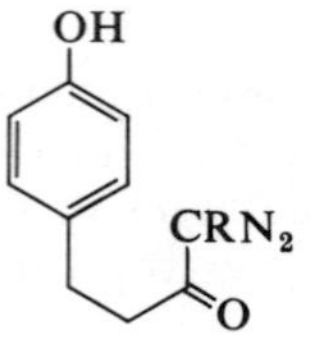 (**73**) 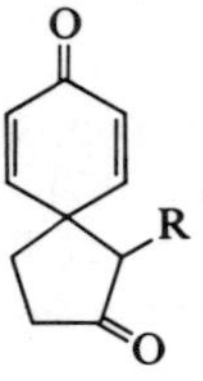 (**74**) 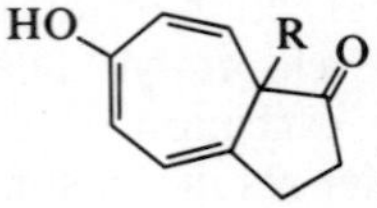 (**75**)

Flash pyrolysis of tetrazoles (**76**) at 400–500 °C leads to (**77**); at 800 °C, hetero-substituted fluorenes (**78**) are formed. Similarly, thermolysis of (**79**) leads to 3-phenylindazole at 450 °C and to a fluorene at 750 °C. The results can be explained by loss of N_2 or CO_2, respectively, to produce nitrile imines, which are a resonace hybrid of a bent dipolar form, and the carbene (**80**) which cyclizes into the remote aromatic ring.[161]

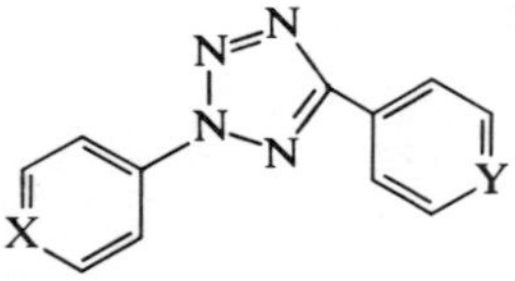

(**76**) 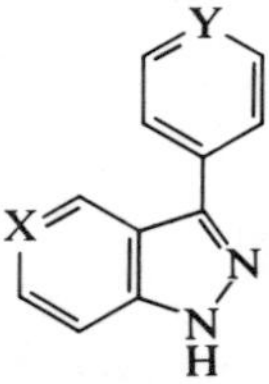 (**77**) 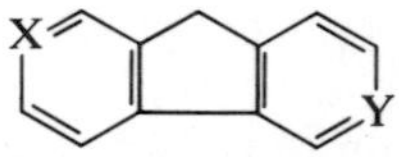 (**78**)

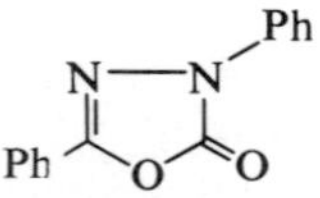

(**79**) 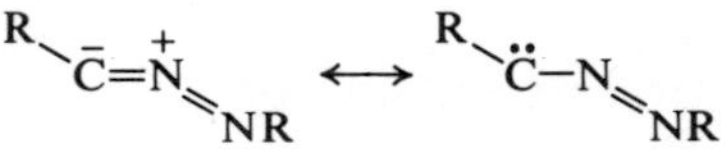 (**80**) 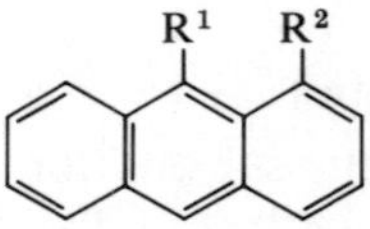 (**81**)

Reaction of benzenesulphonylnitrene with anthracene leads to (**81**; $R^1 = NHSO_2Ph$, $R^2 = H$) and (**81**; $R^1 = H$, $R^2 = NHSO_2Ph$) and the relative amounts of 9-substituted product are found by concentration variation to be a measure of the triplet character of the nitrene. Aziridines are thought to be intermediates in these reactions.[162]

Nitrenes derived from azides (**82**) generally undergo intramolecular attack to produce eventually tricyclo-compounds (**83**); when R is alkyl, there is no evidence of attack at the side-chain. A complex mechanism involving extrusion of sulphur is suggested, as shown in Scheme 1.[163]

Thermolysis of the azides (**84**) leads to 9-phenylacridans, 9-phenylacridines, and 11-phenyl-10*H*-azepino[1,2-a]indoles together with tetracyclic derivatives such as (**85**) (from **84**; Ar = 4-$MeOC_6H_4$). Comparison of this with photolytic generation of the nitrene in various solvents suggests that acridans and acridines are triplet-derived whereas indoles and tetracyclic compounds are singlet-derived.[164]

Thermolysis of 2-phenoxybenzenesulphonyl azide leads to (**86**) (the first seven-membered ring formed by intramolecular sulphonylnitrene cyclization) and other products. 2-(Phenylthio)benzenesulphonyl azide gives some of the heterocycle (**87**) but no product containing a seven-membered ring.[165]

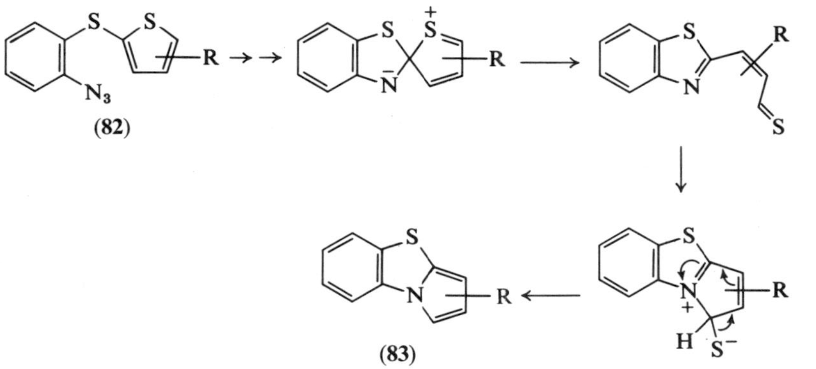

(82)

(83)

SCHEME 1

Intramolecular cyclization of nitrenes derived by oxidation of 6-nitro-7-(substituted amino)quinazoline-4-ones is also reported.[166]

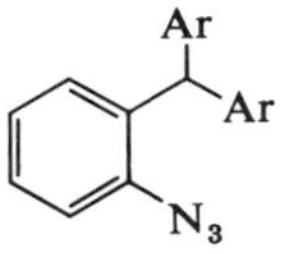

(84)

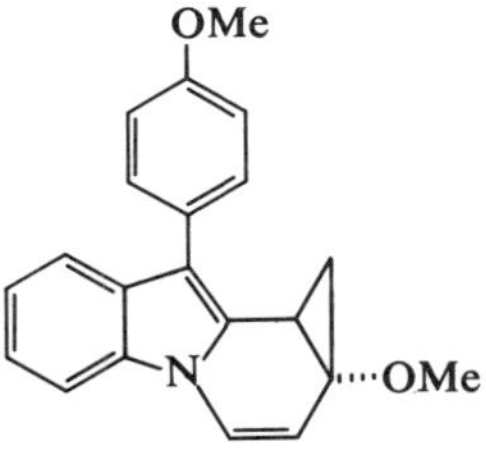

(85)

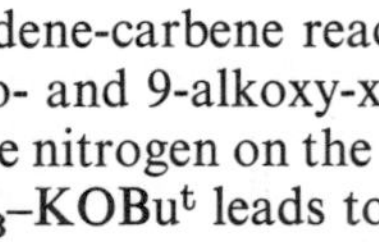

(86)

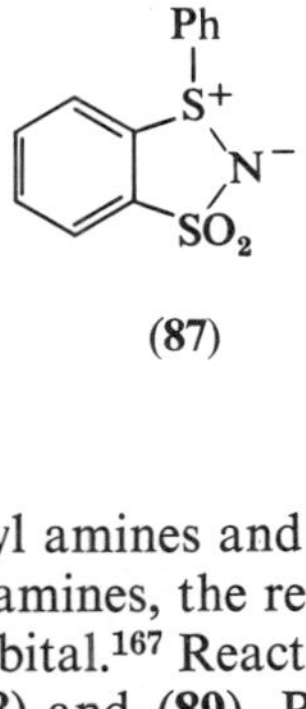

(87)

Reaction with Nucleophiles

Xanthenylidene-carbene reacts with primary aryl amines and alcohols to produce 9-arylamino- and 9-alkoxy-xanthenes; with the amines, the reaction is initiated by attack of the nitrogen on the vacant carbene *p*-orbital.[167] Reaction of benzimidazole with $CHCl_3$–$KOBu^t$ leads to the derivatives (**88**) and (**89**). Related products are obtained from reaction with $CH_2Cl_2/KOBu^t$ and a non-carbenic mechanism may be involved.[168]

Alkali-metal perfluorodiphenylamides react with difluorocarbene to give (difluoromethyl)diphenyl amines, whereas dibromo- and dichloro-carbenes are known to give *N*-formyl derivatives.[169]

The carbene $Me_2C{=}C{=}C$:, generated in a phase-transfer system, reacts with thiophenol to produce (**90**) and with RSSR leads to (**91**). The latter product may arise by a Stevens rearrangement of an initially formed sulphonium ylid.[170] Reaction of methoxycarbonylcarbene with vinyl sulphides leads largely to methyl vinylthioacetate and some cyclopropane, of which the former apparently arises *via* (**92**).

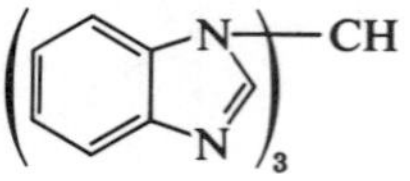

(88) 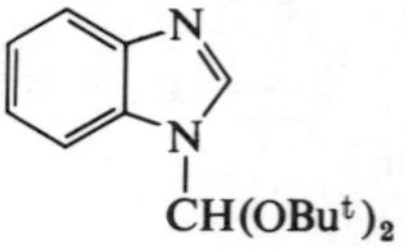 (89)

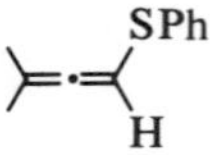

(90) 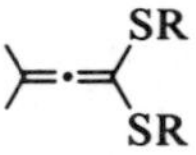 (91) 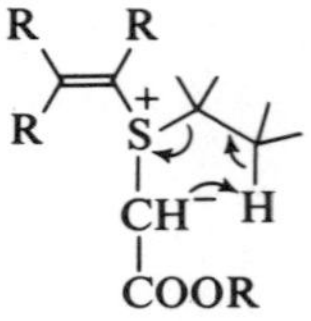 (92)

With bis(methoxycarbonyl)carbene, a stable sulphonium ylid is obtained, the thermal rearrangement of which can then be studied. Further examples of the reaction of allyl sulphides and dichlorocarbene, leading to mercaptobutadienes, are also discussed.[171] The unsaturated carbene (**93**), generated by photolysis of (**94**), leads to dimers or $CH_2{=}CMeCH{=}CHSR$, but in the presence of $Me_2C{=}CH$-CH_2SEt is trapped as either (**95**) or (**96**); these may arise by rearrangement of an initially formed ylid (**97**).[172]

α-Hydroxy-ketoximes are cleaved by reaction with dichlorocarbene in a two-phase system, yielding aldehyde or ketone and nitrile; the reaction possibly proceeds by initial formation of (**98**).[173]

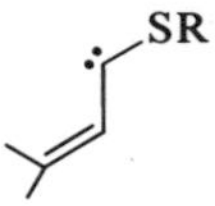

(93) 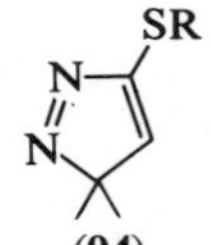 (94)

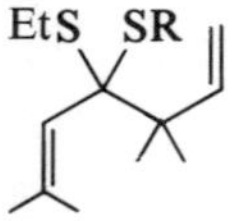

(95) 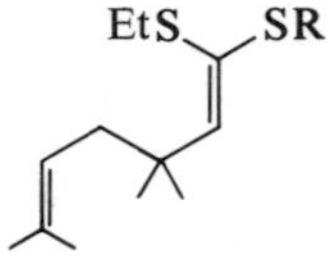 (96) 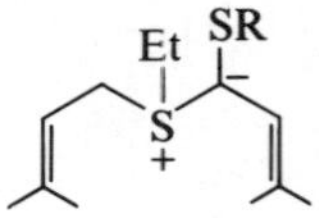 (97)

The imine (**99**) reacts with dichlorocarbene in a phase-transfer system to produce the formylamino compound (**100**), apparently by initial ylid formation, proton loss, and hydrolysis of the dichloride.[174] Copper-catalysed decomposition of diaryldiazomethane in the presence of 1,1-diphenylmethyleneimine leads to $Ph_2CHN{=}CAr_2$ and $Ph_2C{=}NCHAr_2$, presumably once again *via* an initial ylid $Ph_2C{=}\overset{+}{N}H{-}\overset{-}{C}Ar_2$.[175]

The formation of enol ethers by reaction of ketones with ethyl diazoacetate is effectively catalysed by copper(I) chloride, presumably *via* a carbonyl-ylid (**101**).[176] Complexation between $PhHgCCl_2Br$ and aromatic aldehydes enhances the rate of dichlorocarbene formation; the selectivity of the latter is the same as in the absence of aldehyde and the evidence is interpreted in terms of dichlorocarbonyl-ylids.[177]

Phase-transfer-generated dichlorocarbene deoxygenates alkyl and aryl sulphoxides to sulphides.[178] Triazinylnitrenes react with acetone and acetonitrile to give

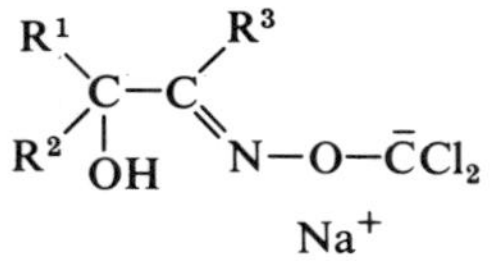

(98) 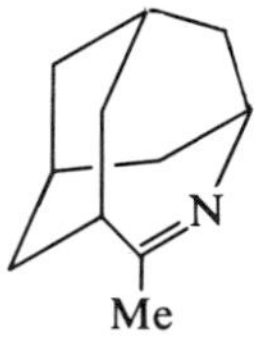 (99) 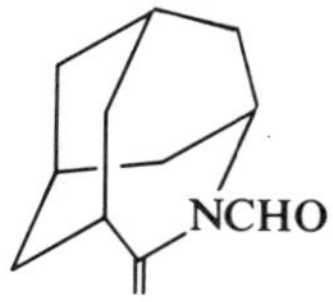 (100)

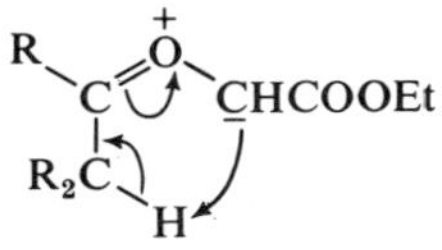

(101)

cycloadduct and polymer, but with dimethyl sulphoxide apparently form an ylid, polymer, and (from triplet nitrene) an aminotriazine.[179] With Ar′N=S=NAr′, aryl azides (ArN_3) afford a complex mixture of products including ArN=NAr, ArN=NAr′, Ar′N=NAr′, $ArNH_2$, and $Ar'NH_2$, apparently through an initial diradical (**102**) derived from triplet nitrene.[180]

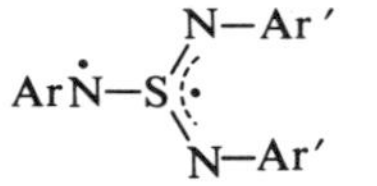

(102) 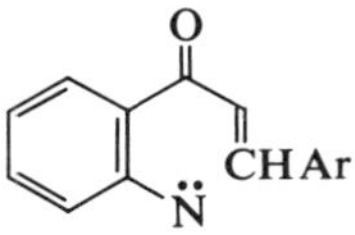 (103)

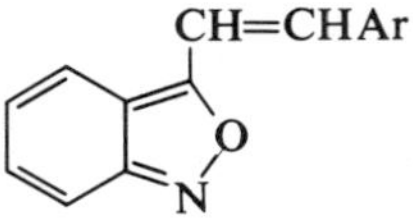

(104) 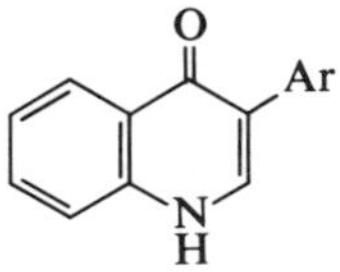 (105) 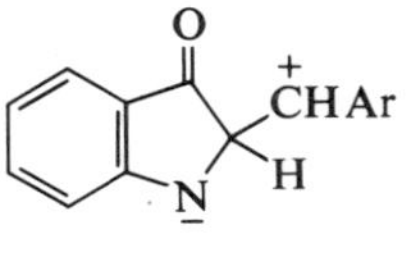 (106)

The nitrene (**103**), derived by azide thermolysis, leads to (**104**) and a small amount of (**105**), apparently through (**106**).[181]

Rearrangement

A wide range of carbene rearrangements has been reviewed.[182] Pyrolysis and photolysis of lithium salts of toluenesulphonylhydrazones of 3-deuteriated 6,6-dimethylnorbornan-2-ones reveal a preference for *exo*-C-3→*exo*-C-2 rather than *endo*-C-3→*endo*-C-2 hydride migration by a factor of about twenty to one. This is interpreted in terms of torsional effects in the transition state for migration to a classical singlet carbene.[183]

A model of the conformationally rigid carbene (**107**) shows that the axial H_a is about 10° away from alignment with the vacant carbene *p*-orbital and that H_e is about 10° away from alignment with the sp^2 orbital. In this case labelling studies show that the relative migratory aptitude H_a/H_e is only 1.5. These results agree with earlier ones for 1,2-phenyl migration in a cyclohexylidene but not with the predictions of a least-motion calculation which suggest overwhelming axial preference.[184] However, recalculations using MINDO/3 and MNDO show that, when these are carried out with no geometric restraints, the molecules have sufficient flexibility to rearrange by migration of the hydrogen aligned with the empty carbene *p*-orbital whatever the stereochemical origin of this hydrogen.[185]

Deuterium isotope effects have also been measured for the rearrangement of carbenes derived from thermal decomposition of 1-aryl-2-diazopropanes-1-d by H(D) migration to produce *cis*- and *trans*-β-methylstyrenes. Isotope effects vary from 1.2 to 1.5 and a detailed analysis indicates a "push–pull" mechanism pictured as electrophilic attack of a "phantom" *p*-orbital of the carbene on the C—H bond and backside nucleophilic attack to push the hydrogen away and form the π-bond.[186]

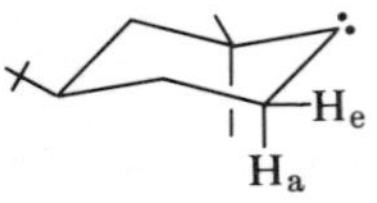

H_e H_a CH:

(107) 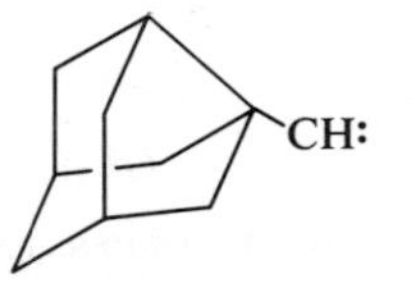 **(108)** 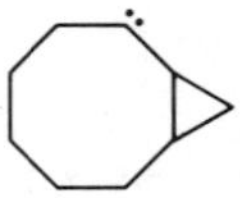 **(109)**

Carbene (**108**) (from pyrolysis of the lithium salt of the corresponding toluene-*p*-sulphonylhydrazone) leads to biadamantyls and dimers of adamantene. In the presence of butadiene, the latter, which is formed by a 1,2-shift, is trapped as a Diels–Alder adduct. The biadamantyls have not been observed in other claimed syntheses of adamantene.[187] The carbene (**109**) rearranges in very high yield to bicyclo[5.2.0]non-1(9)-ene.[188]

Ab initio calculations for rearrangements of singlet and triplet cyclopropylidene to allene show some significant differences from earlier calculations:[189] the triplet is found to be 8.4 kcal mol^{-1} lower in energy than the singlet, and both species open in a disrotatory manner though the processes differ considerably in detail. MINDO/3 calculations of the singlet vinylcyclopropylidene-to-cyclopentenylidene reaction show a very early transition state with a barrier of less than 13.8 kcal mol^{-1}; the reaction is initiated by π-complex formation between the alkene and the empty carbene orbital followed by formation of a non-classical carbene (**110**) in which the electron density is shifted towards the carbene site; electron-donating substituents on the alkene should enhance the rate but hetero-substituted carbenes such as (**111**) should be less reactive towards nucleophiles.[190]

X Li Br MeO

(110) 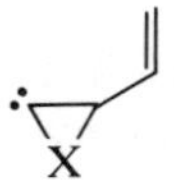 **(111)** 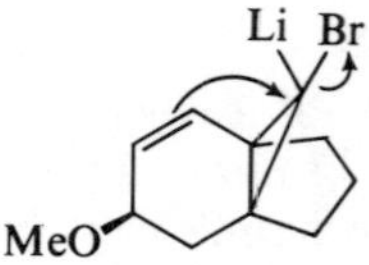 **(112)**

It has been suggested that a number of vinylcyclopropylidene-to-cyclopentenylidene rearrangements, initiated by reaction of *gem*-dibromocyclopropanes with MeLi, may not proceed through a carbene but rather by displacement of bromide from a lithiobromide such as (**112**) and trapping of a non-classical carbonium ion.[191]

Carbene–carbene rearrangements have, however, been observed in various reactions of bridged diazonium salts,[192] and a full account has appeared of rearrangements of (**113**) and (**114**).[193]

(113) (114)

The self-consistent electron-pairs method has been used to study the unimolecular rearrangement of vinylidene to acetylene, and the effect of electrostatic interaction with a positive centre, Li^+, during the reaction has been estimated.[194] Flash-vacuum pyrolysis of (**115**) leads to acetylenes R—C≡CH presumably *via* the carbene RCH=C:;[195] similar treatment of (**116**) leads to cycloalkylidenecarbenes which rearrange to acetylenes, bicyclic cyclopropenes, or products derived from these; thus (**116**; $n = 8$) leads to (**117**) from a secondary carbene, cyclononen-3-ylidene, derived probably from the corresponding cyclopropene.[196] Reaction of either *E*- or *Z*-(**118**) with $KOBu^t$ leads to $Me^{14}C{\equiv}CPh$, suggesting a non-stereoselective process *via* a common intermediate.[197]

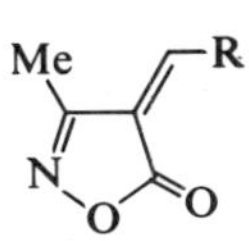

(**115**) 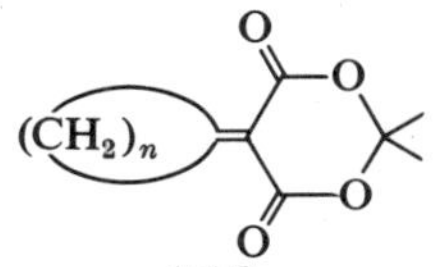 (**116**)

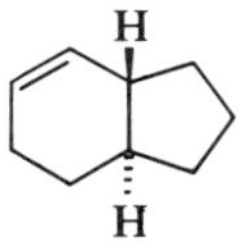

(**117**) 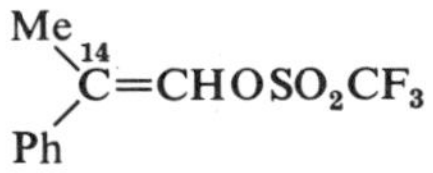 (**118**) $\mathrm{Me(Ph)}{}^{14}\mathrm{C{=}CHOSO_2CF_3}$

The equilibration of (**119**) and (**120**) at temperatures above 700 °C probably occurs *via* a methylene-carbene intermediate.[198]

$$\underset{(\mathbf{119})}{\mathrm{H{-}C{\equiv}{}^{13}C{-}D}} \rightleftharpoons \underset{(\mathbf{120})}{\mathrm{H{-}{}^{13}C{\equiv}C{-}D}}$$

Cyclobutenylidene, obtained by deoxygenation of cyclobutenone by atomic carbon, has been reported to rearrange in low yield to vinylacetylene; the authors examine the intermediacy of bicyclo[1.1.0]but-1(2)-ene but suggest that calculations of the equilibrium geometry and rearrangement of the carbene support a late bicyclobutene-like transition state.[199]

Ab initio calculations on silabenzene and (**121**; R = H) have been carried out[200] in an attempt to explain why the derivative (**121**; R = Me) does not aromatize.[201]

A wide range of papers has appeared concerning other unsaturated carbenes and nitrenes. The vinylcarbene (**122**) can be trapped as a cycloadduct with diethyl maleate or fumarate but in the absence of a trap is converted into the cyclopropenes (**123**). The related carbene (**124**) rearranges to (**125**) and (**126**), but this carbene also can be trapped by diethyl fumarate.[202] Photo-fragmentation of (**127**) also leads to a vinylcarbene which rearranges to a cyclopropene (**128**), and products derived by attack of the carbene on the aromatic ring are sometimes observed.[203] Evidence for the formation of vinyl-carbenes in cyclopropene photolyses is also presented.[204]

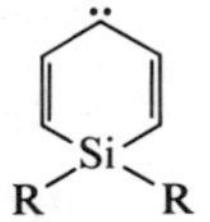

(121) 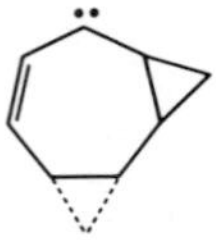 (122) 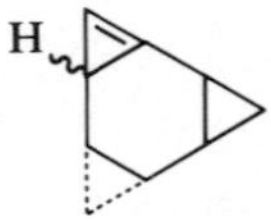 (123)

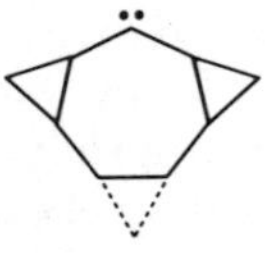

(124) 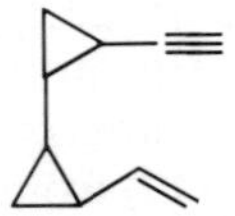 (125) 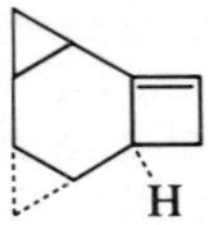 (126)

Though dilution and high temperature are not sufficient to allow the rearrangement of cycloheptatrienylidene to phenylcarbene to compete with dimerization, 2,7-disubstitution is apparently effective. Thus, 2,7-diphenylcycloheptatrienylidene leads to 9-phenylfluorene, the final product expected from the carbene–carbene rearrangement. In the case of the 2,7-dimethyl derivative, rearrangement competes with dimerization at high temperature.[205] Gas-phase generation of phenyl(trimethylsilyl)carbene leads to (**129**) as a major product and this has been shown to arise through a carbene–carbene rearrangement to (2-trimethylsilyl)phenylcarbene followed by intramolecular trapping.[206,207]

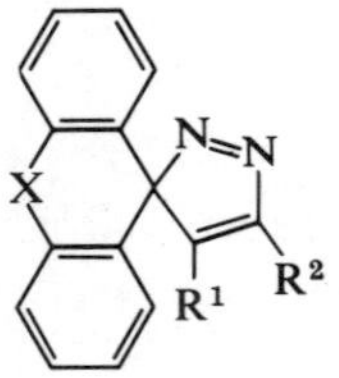

(127) 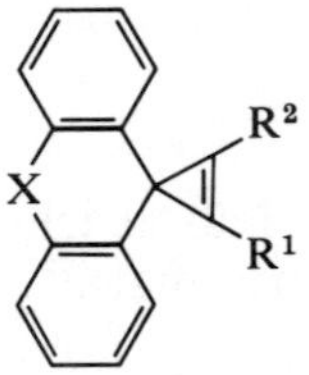 (128) 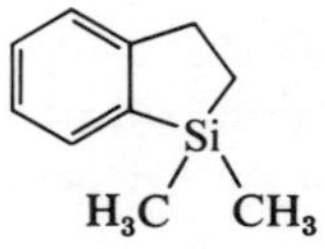 (129)

The initial carbene also undergoes an alkyl shift, to lead to products derived from 1,2,2-trimethyl-1-phenylsilene. Carbene (**130**) may well be better described as a twisted allene; trapping supports an allenic form.[208]

Aromatic carbene–carbene rearrangements have also been observed in photolysis of aryl azides. ESR studies of phenyl azide photolysis in an argon matrix show the formation of phenylnitrene which equilibrates with 2-pyridylmethylene. Photolysis of 4-diazomethylpyridine leads to 4-pyridylmethylene which, on continued photolysis, produces 3- and 2-substituted isomers, and phenylnitrene. When the irradiation of either 4- or 3-(diazomethyl)pyridine is followed by IR spectroscopy, a common primary photo-product, characterized as (**131**), is observed.[209,210]

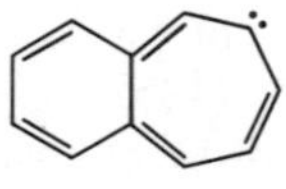

(130) 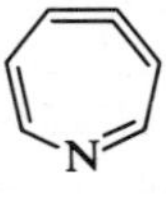 (131) 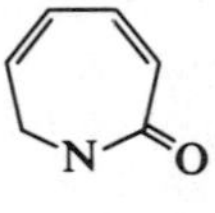 (132)

Photolysis of phenyl azide in methanol does not lead to (**132**), but in KOMe/MeOH/dioxan the azepinone is formed; the yield is increased by addition of 18-crown-6 ether. Co-ordination of singlet phenylnitrene to dioxan and more efficient trapping by methoxide ion is thought to be reponsible. No evidence of ring expansion was seen in the case of 3-azidopyridine photolysis.[211] Thermolysis of methyl *o*-azidobenzoates in alcohol leads to mixtures of 3*H*-azepines and triplet nitrene-derived products,[212] and photolysis of (**133**) in the presence of amines leads to (**134**) by nitrene–azirine equilibration and nucleophilic opening.[213]

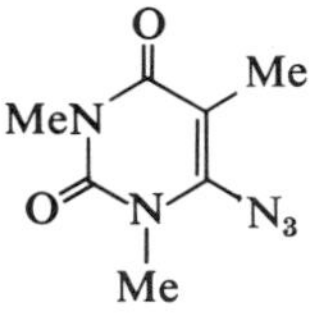

(**133**) 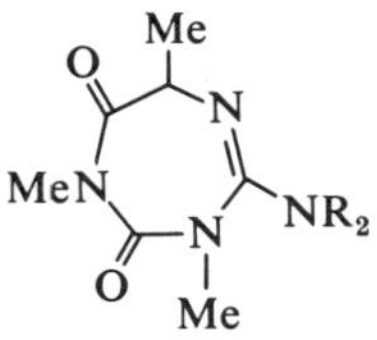 (**134**)

Ab initio calculations have been performed on various azirine cleavages, including loss of methylene to produce HCN.[214] The thermal rearrangement of 2-allyl-2*H*-azirines to 3-azabicyclo[3.1.0]hex-3-enes is best rationalized in terms of equilibration of the azirine with a transient vinyl-nitrene which adds to the π-bond.[215] 2-Allyl-3-methyl-2-phenyl-2*H*-azirine leads to 2-methyl-1-phenyl-3-azabicyclo[3.1.0]hex-2-ene and (**135**), the former apparently arising by cyclization of an intermediate zwitterion (**136**) and the latter by electrocyclic closure of the nitrene. Azirine (**137**; R^1 = Ph, R^2 = Me) leads to products apparently derived from a similar species, while those from (**137**; R^1 = Me, R^2 = Ph) are accounted for in terms of an intermediate of the form (**138**).[216]

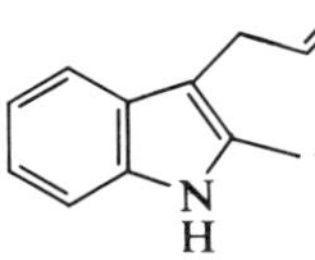

(**135**) 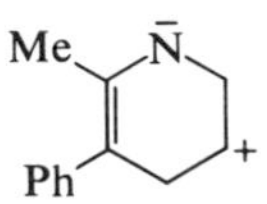 (**136**) 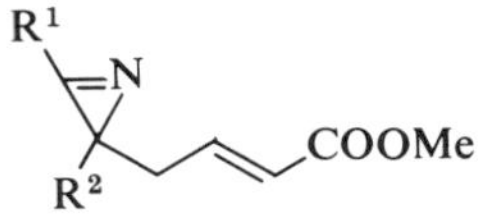 (**137**)

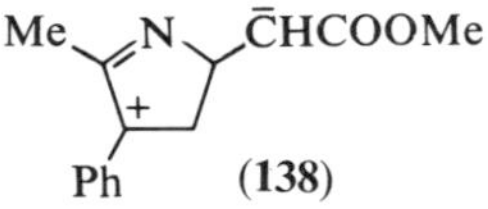

(**138**)

Reaction of (**139**) with Et_3PO leads to 1-aryl-3-cyanopyrroles, possibly by equilibration of an initially formed nitrene with the azirine (**140**) which can open as shown and recyclize with loss of sulphur.[217] Oxidation of 1,2-diaminoimidazoles leads to 1,2,3-triazoles and 3-amino-1,2,4-triazoles; this apparently occurs by opening of an initially formed C-nitrene or nitrenoid to (**141**) followed by cyclization; the formation of some acetophenone is explained in terms of the formation of an *N*-nitrene.[218]

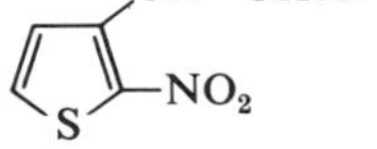

(**139**) 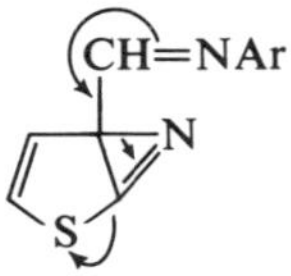 (**140**) 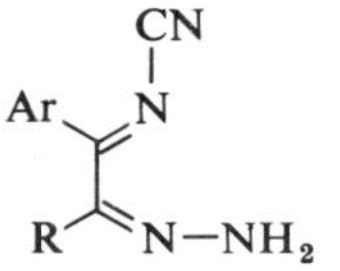 (**141**) 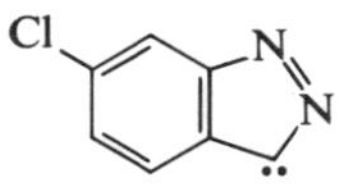 (**142**)

Carbene (**142**), the apparent initial product of photolysis of the corresponding azo-compound, rearranges to 5-chloro-2-cyanophenylnitrene; the latter leads to a dimer.[219]

Alkynes of the form $R^3SC{\equiv}CNR^1R^2$ and sulphonylazides $R^4SO_2N_3$ react with evolution of nitrogen and an oxygen shift to form *S*-alkyl and *S*-aryl amidinothioformates; the reaction is thought to proceed through (**143**), derived from the corresponding diazo-compound, and (**144**).[220]

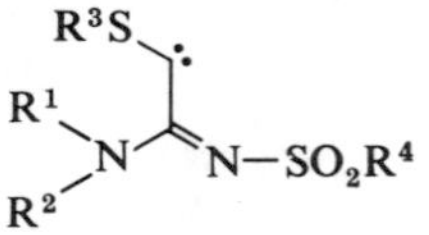

(**143**) 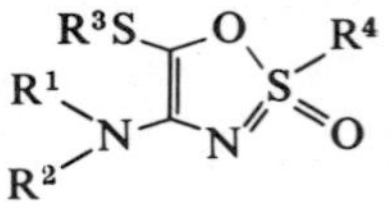 (**144**)

The ketocarbene-oxirene rearrangement has also been further investigated. Carbene (**145**; $R^1 = Me$, $R^2 = H$), generated thermally and photochemically, undergoes a 1,2-oxygen shift to produce (**145**; $R^1 = H$, $R^2 = Me$) *via* the oxirene; there is no oxirene participation, however, in silver oxide-catalysed generation of the carbene.[221] *Ab initio* calculations have been carried out on the ketocarbene–ketene rearrangement.[222] The carbene derived by silver perchlorate-catalysed decomposition of (**146**) reacts with vinyl acetate to produce (**147**); this is thought to occur either by 1,3-dipolar addition followed by rearrangement or by prior rearrangement of the carbene to (**148**), followed by 1,3-dipolar addition.[223] While the carbene $Et_3Ge\ddot{C}COR$ undergoes Wolff rearrangement, the silicon analogue, $Et_3Si\ddot{C}COCH_3$ leads to (**149**), which may well arise by rearrangement of an initial silacyclopropane derived by carbene insertion into a C—H bond α to silicon.[224]

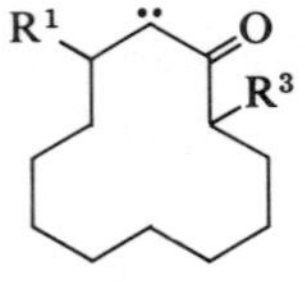

(**145**) 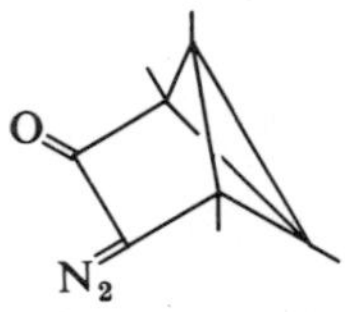 (**146**) 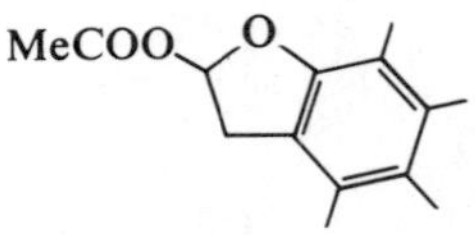 (**147**)

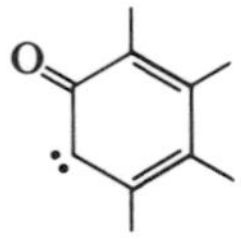

(**148**) 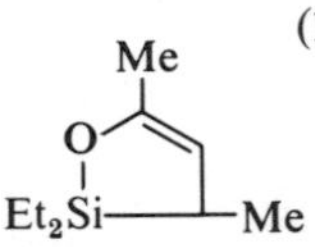 (**149**)

Several theoretical studies of oxacarbene reactions have appeared.[225–227] Calculations on the :CHOH to formaldehyde energy surface using configurational interaction show that the two geometrical forms of the singlet carbene and the triplet all lie below the formaldehyde S_1 state. All are therefore accessible during the photo-dissociation of formaldehyde, and the S_0 *trans*-form is a particularly attractive intermediate.[226] Non-empirical LCAO-SCF-MO calculations on the effect of basis set size and alkyl substitution on the rearrangement of triplet carbonyl compounds to oxacarbenes, as modelled by formaldehyde, lead to a concerted pathway, switching to a diradical pathway when H becomes alkyl.[227]

Photolysis of (**150**) in methanol leads to (**151**), a known prostaglandin precursor, in reasonable yield, but other alcohols are less effective traps for the intermediate carbene.[228] Carbenes (**152**), obtained by photolysis of 1-alkoxytriptycenes, rearrange in inert solvents to the corresponding ketones and insert to form (**153**);

migratory aptitudes are very similar to those in the Wittig rearrangement of α-metallated ethers.[229]

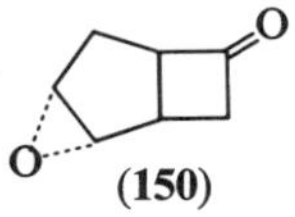

(150) (151) (152)

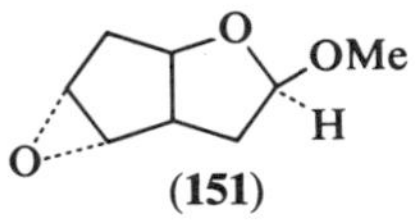

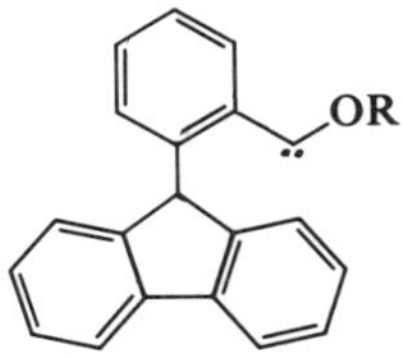

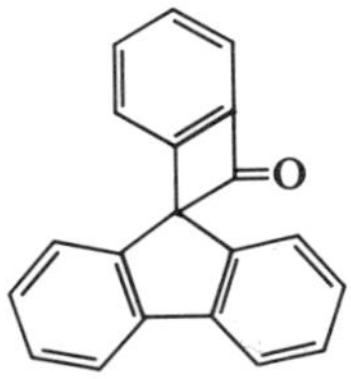

(153)

Pyrolysis of (**154**) is best rationalized in terms of formation of a carbenadioxalane which does not undergo concerted loss of CO_2 but rather undergoes one-bond fission to produce diradicals of the type (**155**).[230]

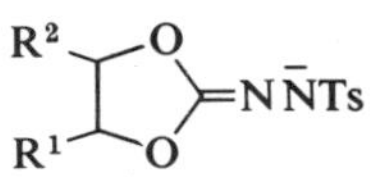

(154) (155)

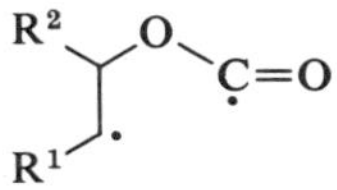

The photolysis of azides such as (**156**) apparently leads to a nitrene that undergoes either methyl or cyclopentyl migration; when acetone-sensitized photolysis is used the products include the cyano-ketone (**157**); this is apparently derived by hydrogen abstraction from the methyl group by the nitrene, cyclization to (**158**), and transformation to (**157**) by ring-opening and hydrolysis of the imine.[231]

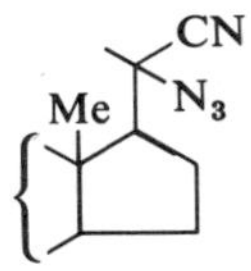

(156) (157) (158)

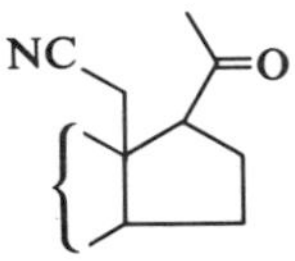

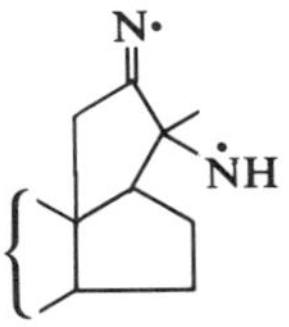

No PhMeC=$GeMe_2$ is observed in the generation of Ph$\ddot{C}$$GeMe_3$ from the tosylhydrazone or diazo-compound; one unexpected product, however, is propenylbenzene.[232]

Silylenes

Silylenes have been discussed in a review,[233] and calculations have been performed on the methylsilylene (MeSiH:)–silaethene ($H_2Si=CH_2$)–silylcarbene (H_3SiCH:) rearrangement which show the silylene to be of lower energy than the carbene for the closed-shell states.[234]

Photolysis of $Me_3SiSiMePhSiMe_3$ leads to methylphenylsilylene which can be trapped by alkenes as rearrangement products of silacyclopropanes.[235] The silylene is also reported to insert into Si—O bonds of cyclic siloxanes but to be less reactive than dimethylsilylene.[236]

Evidence is presented that reaction of difluorosilylene with alkenes involves attack of $(SiF_2)_n$ followed by rearrangement.[237] Thermolysis of 2-chloroheptamethyltrisilane produces triethylsilyl chloride and methyl(trimethylsilyl)silylene which lead to (**159**) and (**160**). Trapping studies show that only a minor portion is formed *via* $Me_2Si{=}SiMe_2$.[238]

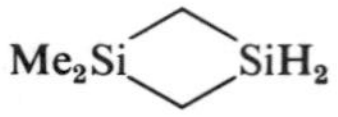

Me_2Si SiH_2

(**159**)

Me H Si Si H Me

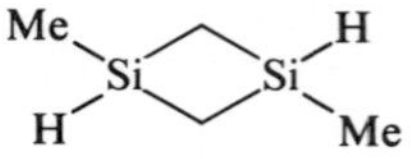

(**160**)

Transition-metal Complexes

Selectivity and specificity in the reactions of carbene and carbyne complexes have been reviewed,[239] as has carbene chemistry of Pd(II) and Pt(II).[240] Calculations on a series of carbene ligands :CXY and complexes $(OC)_5CrCXY$ show that methoxycarbenes are better π-acceptors than aminocarbenes, but that all accept less charge from Cr than does carbon monoxide.[241] A theoretical structure is derived for a model complex $MnCH_2$.[242]

A mechanism for Ziegler–Natta catalysis has been discussed which involves a metal carbene intermediate,[243] and the involvement of carbenes in metathesis has been further analysed.[244]

Reaction of alkylnickel complexes [R_2Ni(2,2′-bipyridine)] with methylene halides leads to RCH_2CH_2R, RCH_2R, RMe, and alkenes. These products are considered to be formed by insertion of methylene into the Ni—R bond.[245] Diazoalkanes react with Co(III) porphyrins to give vinyl or halomethyl derivatives and here the reaction pathway involves insertion into the Co—N bond.[246] An X-ray study of an iron porphyrin with dichlorocarbene as a ligand is reported.[247]

Complexes of the widely studied carbene cycloheptatrienylidene are reported,[248] as are alkoxycarbonylcarbene complexes.[249] Reaction of $(C_5H_5)Mn(CO)_2(THF)$ with diazocyclopentadiene leads to dimers (**161**) of the expected carbene derivative; these dimers show high thermal stability.[250]

Reaction of the complex (Fe^{II})(tetraphenylporphyrin)(:CCl_2) with primary amines leads to isocyanides.[251] Conversely, various cationic and neutral isocyanide complexes react with primary amines to produce diamino-carbene complexes.[252] Cyclic enol ethers are cleaved by $(OC)_5W{=}CPh_2$; thus 2-ethoxynorbornene leads to (**162**).[253] Various transformations of carbene complexes with alkynes are also discussed.[254,255] Thus diphenylacetylene and $(OC)_5CrC(Me)OMe$ produce (**163**).[254]

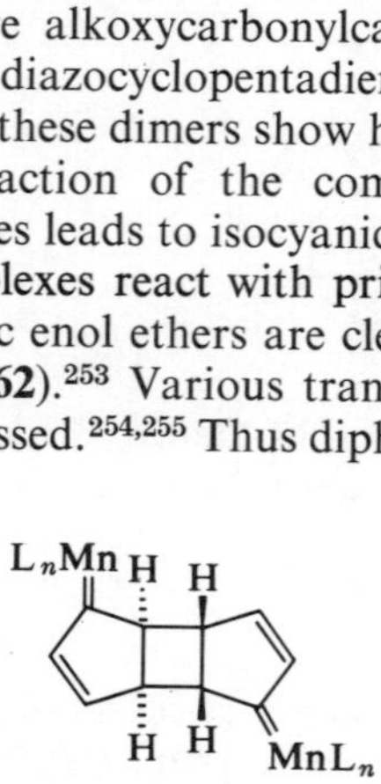

L_nMn H H H H MnL_n

(**161**)

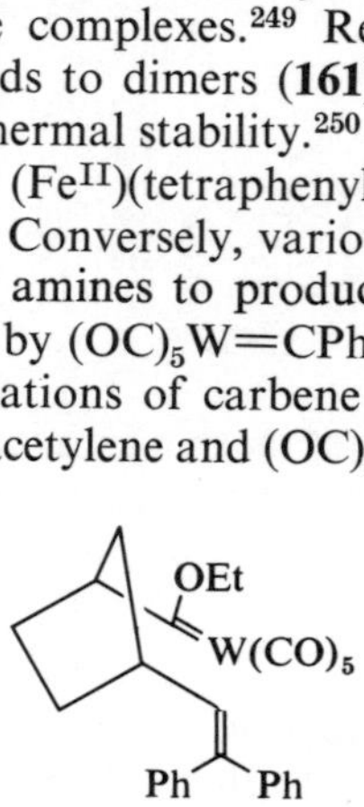

OEt $W(CO)_5$ Ph Ph

(**162**)

OMe O Me Ph Ph $[Cr(CO)_3]$

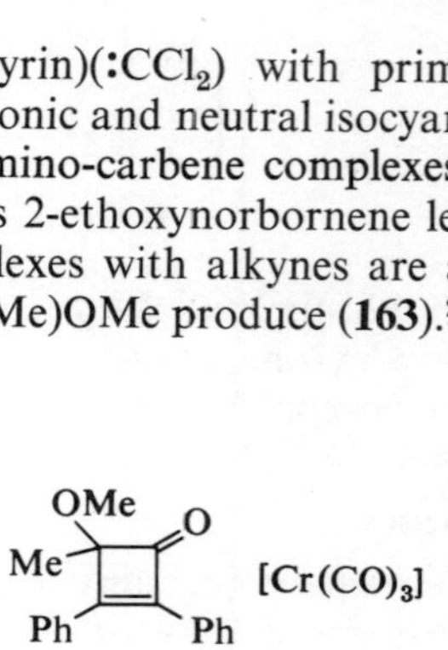

(**163**)

Reaction of silyl ylids such as $Me_3P{=}CHSiMe_3$ with, *e.g.* $W(CO)_6$ leads to direct transformation of complexed CO into an ylid carbene ligand, producing $(CO)_5W{=}C(OSiMe_3)CH{=}PMe_3$.[256] There are several reports of more conventional alkoxy-carbene complexes.[257] There are also reports of thioalkoxycarbene,[258] amino-carbene,[259] vinylidene[260] and alkylidene,[261] bisneopentylidene,[262] and perfluoro-carbene complexes.[263] Carbene complexes of iron[264] and osmium[265] and a nitrene complex of tungsten[266] are also reported, as are various carbyne complexes.[267]

References

1 Burton, D. J., and Hahnfeld, J. L., *Fluorine Chem. Rev.*, **8**, 119 (1977).
2 Hsu, D. S. Y., Umstead, M. E., and Lin, M. C., *ACS Symp. Series*, **66**, 128 (1978).
3 Stang, P. J., *Chem. Rev.*, **78**, 383 (1978); *Acc. Chem. Res.*, **11**, 107 (1978).
4 Makosza, M., *Usp. Khim.*, **46**, 2174 (1977); *Chem. Abs.*, **88**, 61659 (1978); *Russ. Chem. Rev.*, **46**, 1151 (1977).
5 Tomioka, H., *Kagaku No Ryoiki*, **32**, 508 (1978); *Chem. Abs.*, **89**, 146326 (1978).
6 (a) Nefedov, O. M., Kolesnikov, S. P., and Ioffe, A. I., *J. Organomet. Chem. Libr.*, **5**, 181 (1977); *Chem. Abs.*, **89**, 5511 (1978); (b) Semenov, V. P., Studenikov, A. N., and Potekhin, A. A., *Khim. Geterotsikl. Soedin.*, 1978, 291; *Chem. Abs.*, **89**, 6249 (1978).
7 Matlin, S. A., *Ann. Rep. Prog. Chem.*, *Sect. B*, 1977, 105.
8 Bauschlicher, C. W., and Shavitt, I., *J. Am. Chem. Soc.*, **100**, 739 (1978); Roos, B. O., and Siegbahn, P. M., *ibid.*, **99**, 7716 (1977).
9 Baird, N. C., and Taylor, K. F., *J. Am. Chem. Soc.*, **100**, 1333 (1978).
10 Bauschlicher, C. W., Schaefer, H. F., and Bagus, P. S., *J. Am. Chem. Soc.*, **99**, 7106 (1977).
11 Noyori, R., Yamakawa, M., and Ando, W., *Bull. Chem. Soc. Jpn.*, **51**, 811 (1978).
12 Halevi, E. A., and Trindle, C., *Isr. J. Chem.*, **16**, 283 (1977).
13 Kenney, J. W., Simons, J., Purvis, G. D., and Bartlett, R. J., *J. Am. Chem. Soc.*, **100**, 6930 (1978).
14 Apeloig, Y., and Schreiber, R., *Tetrahedron Lett.*, **1978**, 4555.
15 Hopkinson, A. C., Lien, M. H., Yates, K., and Csizmadia, I. G., *Int. J. Quantum Chem.*, **12**, 355 (1977); *Chem. Abs.*, **88**, 21874 (1978).
16 Poppinger, D., Radom, L., and Pople, J. A., *J. Am. Chem. Soc.*, **99**, 7806 (1977).
17 Hopkinson, A. C., Lien, M., Yates, K., and Csizmadia, I. G., *Prog. Theor. Org. Chem.*, **2**, 230 (1977); *Chem. Abs.*, **88**, 189771 (1978).
18 Strausz, O. P., Gosavi, R. K., Bernardi, F., Mezey, P. G., Goddard, J. D., and Csizmadia, I. G., *Chem. Phys. Lett.*, **53**, 211 (1978); *Chem. Abs.*, **88**, 135974 (1978).
19 Levi, B. A., Taft, R. W., and Hehre, W. J., *J. Am. Chem. Soc.*, **99**, 8454 (1977).
20 Ausloos, P., and Lias, S. G., *J. Am. Chem. Soc.*, **100**, 4594 (1978).
21 King, D. S., Schenck, P. K., and Stephenson, J. C., *Lasers Chem. Proc. Conf.*, **1977**, 340; *Chem. Abs.*, **88**, 88788 (1978).
22 King, D. S., and Stephenson, J. C., *Chem. Phys. Lett.*, **51**, 48 (1977); *Chem. Abs.*, **88**, 36720 (1978).
23 Ritter, J. J., *J. Am. Chem. Soc.*, **100**, 2441 (1978).
24 Foon, R., and Schug, K. P., *Mol. Rate Processes*, *Pap. Symp.*, **1975**, E3; *Chem. Abs.*, **88**, 104440 (1978).
25 Arnold, D. R., and Humphreys, R. W. R., *J. Chem. Soc.*, *Chem. Commun.*, **1978**, 181.
26 Hutton, R. S., and Roth, H. D., *J. Am. Chem. Soc.*, **100**, 4324 (1978).
27 Hendrick, M. E., and Jones, M., *Tetrahedron Lett.*, **1978**, 4249.
28 Lambert, J. B., Kobayashi, K., and Mueller, P. H., *Tetrahedron Lett.*, **1978**, 4253.
29 Jones, M., Gaspar, P. P., and Lambert, J. B., *Tetrahedron Lett.*, **1978**, 4257.
30 Davis, J. H., and Goddard, W. A., *J. Am. Chem. Soc.*, **99**, 7111 (1977).
31 Hall, J. H., Fargher, J. M., and Gisler, M. R., *J. Am. Chem. Soc.*, **100**, 2029 (1978).
32 Hinsberg, W. D., and Dervan, P. B., *J. Am. Chem. Soc.*, **100**, 1608 (1978).
33 Nefedov, O. M., D'yachenko, A. I., Prokof'ev, A. K., *Russ. Chem. Rev.*, **46**, 941 (1977); *Usp. Khim.*, **46**, 1787 (1977); *Chem. Abs.*, **88**, 36703 (1978).
34 Larock, R. C., *Angew. Chem. Int. Edn.*, **17**, 27 (1978).

[35] Meier, H., and Zeller, K.-P., *Angew. Chem. Int. Edn.*, **16**, 835 (1977).
[36] Sasaki, T., Eguchi, S., and Nakata, F., *Tetrahedron Lett.*, **1978**, 1999.
[37] Pienta, N. J., and Kropp, P. J., *J. Am. Chem. Soc.*, **100**, 655 (1978).
[38] Olofson, R. A., Hoskin, D. H., and Lotts, K. D., *Tetrahedron Lett.*, **1978**, 1677.
[39] Chamberlin, A. R., and Bond, F. T., *J. Org. Chem.*, **43**, 154 (1978).
[40] Schneider, M., and Csacsko, B., *Angew. Chem. Int. Edn.*, **16**, 867 (1977).
[41] Franck-Neumann, M., and Dietrich-Buchecker, C., *Tetrahedron*, **34**, 2797 (1978).
[42] Liu, M. T. H., and Ramakrishnan, K., *J. Org. Chem.*, **42**, 3450 (1977).
[43] Moss, R. A., and Munjal, R. C., *J. Chem. Soc., Chem. Commun.*, **1978**, 775.
[44] Moss, R. A., and Fantina, M. E., *J. Am. Chem. Soc.*, **100**, 6788 (1978).
[45] Smith, N. P., and Stevens, I. D. R., *Tetrahedron Lett.*, **1978**, 1931.
[46] Moss, R. A., and Shieh, W.-C., *Tetrahedron Lett.*, **1978**, 1935.
[47] Dehmlow, E. V., and Lissel, M., *J. Chem. Res.*, (*S*) **1978**, 310.
[48] Fedorynski, M., Wojciechowski, K., Matacz, Z., and Makosza, M., *J. Org. Chem.*, **43**, 4683 (1978).
[49] Cunico, R. F., and Chou, B. B., *J. Organomet. Chem.*, **154**, C45 (1978).
[50] Fritz, H. P., and Kornrumpf, W., *Justus Liebigs Ann. Chem.*, **1978**, 1416.
[51] Uemura, S., Tanaka, S., and Okano, M., *Bull. Inst. Chem. Res., Kyoto Univ.*, **55**, 273 (1977); *Chem. Abs.*, **88**, 6262 (1978).
[52] Nefedov, O. M., and D'yachenko, A. I., *Izv. Akad. Nauk SSSR, Ser. Khim.*, **1978**, 1085; *Chem. Abs.*, **89**, 107556 (1978).
[53] Savin, V. I., and Kitaev, Y. P., *Tezisy Dokl.-Vses. Konf. "Polyariz. Yader Elektronov Eff. Magn. Polya Khim. Reacts."*, **1975**, 19; *Chem. Abs.*, **89**, 23436 (1978).
[54] Reiffen, M., and Hoffmann, R. W., *Tetrahedron Lett.*, **1978**, 1107.
[55] Reverdy, G., *Bull. Soc. Chim. Fr. II*, **1976**, 1131.
[56] Reverdy, G., *Bull. Soc. Chim. Fr. II.*, **1976**, 1141.
[57] Jones, G. W., Chang, K. T., Munjal, R., and Shechter, H., *J. Am. Chem. Soc.*, **100**, 2922 (1978).
[58] Patrick, T. B., Dorton, M. A., and Dolan, J. G., *J. Org. Chem.*, **43**, 3303 (1978).
[59] Gompper, R., and Sobotta, R., *Angew. Chem. Int. Edn.*, **17**, 762 (1978).
[60] Cuffe, J., Gillespie, R. J., and Porter, A. E. A., *J. Chem. Soc., Chem. Commun.*, **1978**, 641.
[61] Livinghouse, T., and Stevens, R. V., *J. Am. Chem. Soc.*, **100**, 6479 (1978).
[62] Regitz, M., Illger, W., and Maas, G., *Chem. Ber.*, **111**, 705 (1978).
[63] Welter, W., Hartmann, A., and Regitz, M., *Chem. Ber.*, **111**, 3068 (1978); Heydt, H., and Regitz, M., *J. Chem. Res.*, (*S*) **1978**, 326.
[64] Griffin, C. E., Kraas, E., Terasawa, H., Griffin, G. W., and Lankin, D. C., *J. Heterocycl. Chem.*, **15**, 523 (1978).
[65] Forbus, T. R., Birdsong, P. A., and Shevlin, P. B., *J. Am. Chem. Soc.*, **100**, 6425 (1978).
[66] Fuchigama, T., and Nonaka, T., *Chem. Lett.*, **1977**, 1087; *Chem. Abs.*, **88**, 22111 (1978).
[67] Seno, M., Namba, T., and Kise, H., *J. Org. Chem.*, **43**, 3345 (1978).
[68] Von Werner, K., *J. Fluorine Chem.*, **10**, 163 (1977); *Chem. Abs.*, **88**, 36811 (1978).
[69] Ioffe, B. V., *Ref. Dokl. Soobshch.-Mendeleevsk. S'ezd. Obshch. Prikl. Khim., 11th*, **3**, 184 (1975); *Chem. Abs.*, **88**, 189976 (1978).
[70] Torimoto, N., Shingaki, T., and Nagai, T., *Bull. Chem. Soc. Jpn.*, **51**, 2983 (1978).
[71] Srinivasan, R., and Brown, K. H., *J. Am. Chem. Soc.*, **100**, 4602 (1978).
[72] Srinivasan, R., and Brown, K. H., *Tetrahedron Lett.*, **1978**, 3645.
[73] Srinivasan, R., and Ors, J. A., *J. Am. Chem. Soc.*, **100**, 7089 (1978).
[74] Pallmer, M., and Morrison, H., *J. Chem. Soc., Chem. Commun.*, **1978**, 558.
[75] Daudpota, A. S., and Heaney, H., *Tetrahedron Lett.*, **1978**, 3471.
[76] Pasto, D. J., Shults, R. H., McGrath, J. A., and Waterhouse, A., *J. Org. Chem.*, **43**, 1382 (1978).
[77] Buchi, G., and Luk, K.-C. T., *J. Org. Chem.*, **43**, 169 (1978).
[78] Fujiwara, Y., Ishikawa, R., and Teranishi, S., *Bull. Chem. Soc. Jpn.*, **51**, 589 (1978).
[79] Smirnov, S. A., Oreshkin, I. A., and Dolgoplosk, B. A., *Dokl. Akad. Nauk SSSR*, **239**, 1375 (1978); *Chem. Abs.*, **89**, 24417 (1978).
[80] Bey, P., and Vevert, J. P., *Tetrahedron Lett.*, **1978**, 1215.
[81] De Vries, L., *J. Am. Chem. Soc.*, **100**, 926 (1978).
[82] Nitsche, M., Seebach, D., and Beck, A. K., *Chem. Ber.*, **111**, 3644 (1978).
[83] Castillo, G. E., and Bertorello, H. E., *J. Chem. Soc., Perkin Trans.* 1, **1978**, 325.
[84] Patrick, T. B., and Wu, T.-T., *J. Org. Chem.*, **43**, 1506 (1978).

[85] Doldouras, G. A., and Kollonitsch, J., *J. Am. Chem. Soc.*, **100**, 341 (1978).
[86] McGann, P. E., Groves, J. T., Greene, F. D., Stack, G. M., Majeste, R. J., Trefonas, L. M., *J. Org. Chem.*, **43**, 922 (1978).
[87] King, F. D., Pike, S., and Walton, D. R. M., *J. Chem. Soc., Chem. Commun.*, **1978**, 351.
[88] Artsybasheva, Y. P., and Ioffe, B. V., *Zh. Org. Khim.*, **14**, 445 (1978); *Chem. Abs.*, **88**. 189978 (1978).
[89] Nishiyama, K., and Anselme, J.-P., *J. Org. Chem.*, **43**, 2045 (1978).
[90] Heinzelmann, W., *Helv. Chim. Acta*, **61**, 618 (1978).
[91] Nakamura, A., *Pure Appl. Chem.*, **50**, 37 (1978).
[92] Krapcho, A. P., *Synthesis*, **1978**, 77.
[93] Zurawski, B., and Kutzelnigg, W., *J. Am. Chem. Soc.*, **100**, 2654 (1978).
[94] Giese, B., and Meister, J., *Angew. Chem. Int. Edn.*, **17**, 595 (1978).
[95] Dehmlow, E. V., and Eulenberger, A., *Angew. Chem. Int. Edn.*, **17**, 674 (1978).
[96] Sydnes, L. K., *Acta Chem. Scand.*, **31B**, 823 (1977).
[97] Labeish, N. N., Kharicheva, E. M., Mandel'shtam, T. V., and Kostikov, R. R., *Zh. Org. Khim.*, **14**, 878 (1978); *Chem. Abs.*, **89**, 23809 (1978).
[98] Kostikov, R. R., Molchanov, A. P., Golovanova, G. V., and Zenkevich, I. G., *Zh. Org. Khim.*, **13**, 1846 (1977); *Chem. Abs.*, **88**, 37099 (1978); Kostikov, R. R., and Molchanov, A. P., *Dokl. Akad. Nauk SSSR*, **236**, 110 (1977); *Chem. Abs.*, **87**, 167190 (1977).
[99] Nefedov, O. M., Shafran, R. N., and Ioffe, A. I., *Izv. Akad. Nauk SSSR, Ser. Khim.*, **1977**, 2292; *Chem. Abs.*, **88**, 49914 (1978).
[100] Schroth, W., and Kaufmann, W., *Z. Chem.*, **18**, 15 (1978); *Chem. Abs.*, **88**, 136478 (1978).
[101] Heydtmann, H., Bardorff, W., and Rullmann, H., *Ber. Bunsenges. Phys. Chem.*, **80**, 311 (1976); *Chem. Abs.*, **88**, 21590 (1978).
[102] Tishchenko, I. G., and Kulinkovich, O. G., *Tezisy Dokl.-Vses. Konf. Khim. Atsetilena, 5th*, **1975**, 79; *Chem. Abs.*, **88**, 169560 (1978).
[103] Khusid, A. K., Kryshtal, G. V., and Kucherov, V. F., *Izv. Akad. Nauk SSSR, Ser. Khim.*, **1977**, 2135; *Chem. Abs.*, **88**, 22169 (1978).
[104] Nefedov, O. M., Dolgii, I. E., and Bulusheva, E. V., *Izv. Akad. Nauk SSSR, Ser. Khim.*, **1978**, 1454; *Chem. Abs.*, **89**, 129077 (1978).
[105] Slobodin, Y. M., and Vasil'eva, I. A., *Zh. Org. Khim.*, **13**, 2450 (1977); *Chem. Abs.*, **88**, 62014 (1978).
[106] Slobodin, Y. M., Egenburg, I. Z., and Khachaturov, A. S., *Zh. Org. Khim.*, **14**, 871 (1978); *Chem. Abs.*, **89**, 59681 (1978).
[107] Vasil'eva, I. A., *Tezisy Dokl.-Resp. Konf. Molodykh Uch.-Khim., 2nd*, **1**, 48 (1977); *Chem. Abs.*, **89**, 108234 (1978).
[108] Christl, M., Freitag, G., and Brüntrup, G., *Chem. Ber.*, **111**, 2307 (1978).
[109] Hildenbrand, P., Plinke, G., Oth, J. F. M., and Schröder, G., *Chem. Ber.*, **111**, 107 (1978).
[110] Kostikov, R. R., Khlebnikov, A. F., and Ogloblin, K. A., *Zh. Org. Khim.*, **13**, 1857 (1977); *Chem. Abs.*, **88**, 6061 (1978).
[111] Meilahn, M. K., Olsen, D. K., Brittain, W. J., and Anders, R. T., *J. Org. Chem.*, **43**, 1346 (1978).
[112] Lai, C. S., and Zee, S. H., *J. Chin. Chem. Soc. (Taipei)*, **24**, 163 (1977); *Chem. Abs.*, **88**, 190259 (1978).
[113] Sano, H., and Takebayashi, M., *Kinki Daigaku Rikogakubu Kenkyu Hokuku*, **11**, 49 (1976); *Chem. Abs.*, **88**, 36836 (1978).
[114] Singh, H., and Singh, P., *Chem. Ind. (London)*, **1978**, 807.
[115] Nakamura, A., Koyama, T., and Otsuka, S., *Bull. Chem. Soc. Jpn.*, **51**, 593 (1978).
[116] Nakamura, A., Konishi, A., Tatsuno, Y., and Otsuka, S., *J. Am. Chem. Soc.*, **100**, 3443 (1978).
[117] Nakamura, A., Konishi, A., Tsujitani, R., Kudo, M., and Otsuka, S., *J. Am. Chem. Soc.*, **100**, 3449 (1978).
[118] Krieger, P. E., and Landgrebe, J. A., *J. Org. Chem.*, **43**, 4447 (1978).
[119] Petiniot, N., Anciaux, A. J., Noels, A. F., Hubert, A. J., and Teyssie, P., *Tetrahedron Lett.*, **1978**, 1239.
[120] Nefedov, O. M., Dolgii, I. E., and Shapiro, E. A., *Tezisy Dokl.-Vses. Konf. Khim. Atsetilena, 5th*, **1975**, 68; *Chem. Abs.*, **88**, 190136 (1978).
[121] Creary, Z., *J. Org. Chem.*, **43**, 1777 (1978).
[122] Komendantov, M. I., Bekmukhametov, R. R., and Kostikov, R. R., *Zh. Org. Khim.*, **14**, 1448 (1978); *Chem. Abs.*, **89**, 107341 (1978).

[123] Dolgii, I. E., Okonnishnikova, G. P., Baidzhigitova, E. A., and Nefedov, O. M., *Tezisy Dokl.-Vses. Konf. Khim. Atsetilena, 5th*, **1975**, 183; *Chem. Abs.*, **88**, 190957 (1978).
[124] Andreichikov, Y. P., Kholodova, N. V., and Dorofeenko, G. N., *Zh. Org. Khim.*, **13**, 1565 (1977); *Chem. Abs.*, **87**, 184316 (1977).
[125] Reid, J. R., and Marmor, R. S., *J. Org. Chem.*, **43**, 999 (1978).
[126] Patrick, T. B., and Schmidt, D. J., *J. Org. Chem.*, **42**, 3354 (1977).
[127] Doutheau, A., and Gore, J., *Bull. Soc. Chim. Fr. II*, **1976**, 1189.
[128] Venier, C. G., and Ward, M. A., *Tetrahedron Lett.*, **1978**, 3215.
[129] Venier, C. G., and Beckhaus, H., *Tetrahedron Lett.*, **1978**, 109.
[130] Takeuchi, H., Igura, T., Mitani, M., Tsuchida, T., and Koyama, K., *J. Chem. Soc., Perkin Trans. 2*, **1978**, 783.
[131] Anastassiou, A. G., and Mahaffey, R. L., *J. Chem. Soc., Chem. Commun.*, **1978**, 915.
[132] Röttele, H., Heil, G., and Schröder, G., *Chem. Ber.*, **111**, 84 (1978).
[133] Semenov, V. P., Studenikov, A. N., Bespalov, A. D., and Ogloblin, K. A., *Zh. Org. Khim.*, **13**, 2202 (1977); *Chem. Abs.*, **88**, 49962 (1978).
[134] Kondo, K., Umemoto, T., Yako, K., and Tunemoto, D., *Tetrahedron Lett.*, **1978**, 3927.
[135] Dolbier, W. R., and Garza, O. T., *J. Org. Chem.*, **43**, 3848 (1978).
[136] Krauser, S. F., and Watterson, A. C., *J. Org. Chem.*, **43**, 2026 (1978).
[137] Rule, M., and Berson, J. A., *Tetrahedron Lett.*, **1978**, 3191.
[138] Hoesch, L., Egger, N., and Dreiding, A. S., *Helv. Chim. Acta*, **61**, 795 (1978).
[139] Janoschek, R., *Jerusalem Symp. Quantum Chem. Biochem.*, **10**, 419 (1977); *Chem. Abs.*, **88**, 135972 (1978).
[140] Khalil, S. M., and Shansham, *Z. Naturforsch.*, **33A**, 722 (1978); *Chem. Abs.*, **89**, 75189 (1978).
[141] Tan, L. P., *Chem. Phys. Lett.*, **57**, 239 (1978); *Chem. Abs.*, **89**, 146283 (1978).
[142] Kollmar, H., *J. Am. Chem. Soc.*, **100**, 2660 (1978).
[143] Moss, R. A., and Huselton, J. K., *J. Am. Chem. Soc.*, **100**, 1314 (1978).
[144] Moss, R. A., and Joyce, M. A., *J. Am. Chem. Soc.*, **100**, 4475 (1978).
[145] Steinbeck, K., *Tetrahedron Lett.*, **1978**, 1103.
[146] Steinbeck, K., and Klein, J., *J. Chem. Res.*, (*S*) **1978**, 396.
[147] Kharicheva, E. M., and Mandel'shtam, T. V., *Zh. Org. Khim.*, **14**, 888 (1978); *Chem. Abs.*, **89**, 59691 (1978).
[148] Ivanova, T. V., Mandel'shtam, T. V., and Kharicheva, E. M., *Tezisy Dokl.-Vses. Konf. Khim. Atsetilena, 5th*, **1975**, 70; *Chem. Abs.*, **88**, 169559 (1978).
[149] Kang, U. G., and Shechter, H., *J. Am. Chem. Soc.*, **100**, 651 (1978).
[150] Torimoto, N., Shingaki, T., and Nagai, T., *Bull. Chem. Soc. Jpn.*, **51**, 1200 (1978).
[151] Takeuchi, H., Kasamatsu, Y., Mitani, M., Tsuchida, T., and Koyama, K., *J. Chem. Soc., Perkin Trans. 2*, **1978**, 780.
[152] Casagrande, P., Pellacani, L., and Tardella, P. A., *J. Org. Chem.*, **43**, 2725 (1978).
[153] Nilsen, N. O., Sydnes, L. K., and Skattebøl, L., *J. Chem. Soc., Chem. Commun.*, **1978**, 128.
[154] Nickon, A., Kwasnik, H. R., Mathew, C. T., Swartz, T. D., Williams, R. O., and Di Giorgio, J. B., *J. Org. Chem.*, **43**, 3904 (1978).
[155] Sasaki, T., Eguchi, S., and Hioki, T., *J. Org. Chem.*, **43**, 3808, (1978).
[156] Challand, S. R., *Aromat. Heteroaromat. Chem.*, **5**, 339 (1977); *Chem. Abs.*, **88**, 21471 (1978).
[157] O'Leary, M. A., and Wege, D., *Tetrahedron Lett.*, **1978**, 2811.
[158] Gambacorta, A., Nicoletti, R., Cerrini, S., Fedeli, W., and Gavuzzo, E., *Tetrahedron Lett.* **1978**, 2439.
[159] Pandit, U. K., *Heterocycles*, **6**, 1520 (1977); *Chem. Abs.*, **88**, 7139 (1978).
[160] Iwata, C., Yamada, M., Shinoo, Y., Kobayashi, K., and Okada, H., *J. Chem. Soc., Chem. Commun.*, **1977**, 888.
[161] Wentrup, C., Damerius, A., and Reichen, W., *J. Org. Chem.*, **43**, 2037 (1978).
[162] Krause, J. G., *Chem. Ind.* (*London*), **1978**, 271.
[163] Lindley, J. M., Meth-Cohn, O., and Suschitzky, H., *J. Chem. Soc., Perkin Trans. 1*, **1978**, 1198.
[164] Carde, R. N., Jones, G., McKinley, W. H., and Price, C., *J. Chem. Soc., Perkin Trans. 1*, **1978**, 1211.
[165] Abramovitch, R. A., Azogu, C. I., McMaster, I. T., and Vanderpool, D. P., *J. Org. Chem.*, **43**, 1218 (1978).
[166] Roy, J., Seth, M., and Bhaduri, A. P., *Indian J. Chem.*, **16B**, 41 (1978); *Chem. Abs.*, **88**, 152543 (1978).
[167] Reverdy, G., *Bull. Soc. Chim. Fr. II*, **1978**, 1136.

[168] Singh, H., and Singh, P., *Chem. Ind.* (*London*), **1978**, 126.
[169] Koppang, R., *J. Fluorine Chem.*, **11**, 19 (1978).
[170] Clinet, J.-C., and Julia, S., *J. Chem. Res.*, (*S*) **1978**, 125.
[171] Ando, W., Higuchi, H., and Migita, T., *J. Org. Chem.*, **42**, 3365 (1977).
[172] Franck-Neumann, M., and Lohmann, J. J., *Tetrahedron Lett.*, **1978**, 3729.
[173] Shah, J. N., Mehta, Y. P., and Shah, G. M., *J. Org. Chem.*, **43**, 2078 (1978).
[174] Sasaki, T., Eguchi, S., and Toi, N., *J. Org. Chem.*, **43**, 3810 (1978).
[175] Mehrotra, K. N., and Prasad, G., *Tetrahedron Lett.*, **1978**, 4179.
[176] Landgrebe, J. A., and Iranmanesh, H., *J. Org. Chem.*, **43**, 1244 (1978).
[177] Martin, C. W., Lund, P. R., Rapp, E., and Landgrebe, J. A., *J. Org. Chem.*, **43**, 1071 (1978).
[178] Soysa, H. S. D., and Weber, W. P., *Tetrahedron Lett.*, **1978**, 1969.
[179] Goka, T., Shizuka, H., and Matsui, K., *J. Org. Chem.*, **43**, 1361 (1978).
[180] De Luca, G., Renzi, G., Cipollini, R., and Pizzabiocca, A., *Chem. Ind.* (*London*), **1978**, 803.
[181] Smalley, R. K., Smith, R. H., and Suschitzky, H., *Tetrahedron Lett.*, **1978**, 2309.
[182] Jones, W. M., and Brinker, U. H., "Some Pericyclic Reactions of Carbenes", Pericyclic Reactions, Vol. 1, Academic Press, 1977.
[183] Freeman, P. K., Hardy, T. A., Balyeat, J. R., and Wescott, L. D., *J. Org. Chem.*, **42**, 3356 (1977).
[184] Kyba, E. P., and John, A. M., *J. Am. Chem. Soc.*, **99**, 8329 (1977).
[185] Kyba, E. P., *J. Am. Chem. Soc.*, **99**, 8330 (1977).
[186] Su, D. T. T., and Thornton, E. R., *J. Am. Chem. Soc.*, **100**, 1872 (1978).
[187] Martella, D. J., Jones, M., and Schleyer, P. von R., *J. Am. Chem. Soc.*, **100**, 2896 (1978).
[188] Levashova, T. V., Semeikin, O. V., Balenkova, E. S., and Luzikov, Y. N., *Zh. Org. Khim.*, **14**, 885 (1978); *Chem. Abs.*, **89**, 42556 (1978).
[189] Pasto, D. J., Haley, M., and Chipman, D. M., *J. Am. Chem. Soc.*, **100**, 5272 (1978).
[190] Schoeller, W. W., and Brinker, U. H., *J. Am. Chem. Soc.*, **100**, 6012 (1978).
[191] Warner, P., and Chang, S.-C., *Tetrahedron Lett.*, **1978**, 3981.
[192] Kirmse, W., and Jendralla, H., *Chem. Ber.*, **111**, 1857, 1873 (1978), and references therein.
[193] Freeman, P. K., Hardy, T. A., Raghavan, R. S., and Kuper, D. G., *J. Org. Chem.*, **42**, 3882 (1977).
[194] Dykstra, C. E., and Schaefer, H. F., *J. Am. Chem. Soc.*, **100**, 1378 (1978).
[195] Wentrup, C., and Winter, H.-W., *Angew. Chem. Int. Edn.*, **17**, 609 (1978).
[196] Baxter, G. J., and Brown, R. F. C., *Aust. J. Chem.*, **31**, 327 (1978).
[197] Stang, P. J., Fox, D. P., Collins, C. J., and Watson, C. R., *J. Org. Chem.*, **43**, 364 (1978).
[198] Brown, R. F. C., Eastwood, F. W., and Jackman, G. P., *Aust. J. Chem.*, **31**, 579 (1978).
[199] Dyer, S. F., Kammula, S., and Shevlin, P. B., *J. Am. Chem. Soc.*, **99**, 8104 (1977).
[200] Schlegel, H. B., Coleman, B., and Jones, M., *J. Am. Chem. Soc.*, **100**, 6499 (1978).
[201] Barton, T. J., and Banasiak, D. S., *J. Organomet. Chem.*, **157**, 255 (1978).
[202] Oda, M., Ito, Y., and Kitahara, Y., *Tetrahedron Lett.*, **1978**, 977.
[203] Dürr, H., Fröhlich, S., Schley, B., and Weisgerber, H., *J. Chem. Soc., Chem. Commun.*, **1977**, 843.
[204] Padwa, A., Blacklock, T. J., Getman, D., Hatanaka, N., and Loza, R., *J. Org. Chem.*, **43**, 1481 (1978).
[205] Mayor, C., and Jones, W. M., *J. Org. Chem.*, **43**, 4498 (1978).
[206] Ando, W., Sekiguchi, A., Rothschild, A. J., Gallucci, R. R., Jones, M., Barton, T. J., and Kilgour, J. A., *J. Am. Chem. Soc.*, **99**, 6995 (1977).
[207] Sekiguchi, A., and Ando, W., *Bull. Chem. Soc. Jpn.*, **50**, 3067 (1977).
[208] Waali, E. E., Lewis, J. M., Lee, D. E., Allen, E. W., and Chappell, A. K., *J. Org. Chem.*, **42**, 3460 (1977).
[209] Chapman, O. L., and Le Roux, J.-P., *J. Am. Chem. Soc.*, **100**, 282 (1978).
[210] Chapman, O. L., Sheridan, R. S., and Le Roux, J.-P., *J. Am. Chem. Soc.*, **100**, 6245 (1978).
[211] Scriven, E. F. V., and Thomas, D. R., *Chem. Ind.* (*London*), **1978**, 385.
[212] Purvis, R., Smalley, R. K., Strachan, W. A., and Suschitzky, H., *J. Chem. Soc. Perkin Trans. 1*, **1978**, 191.
[213] Senda, S., Hirota, K., and Asao, T., *Tetrahedron Lett.*, **1978**, 1531.
[214] Bigot, B., Sevin, A., and Devaquet, A., *J. Am. Chem. Soc.*, **100**, 6924 (1978).
[215] Padwa, A., and Carlsen, P. H. J., *J. Org. Chem.*, **43**, 2029 (1978).
[216] Padwa, A., and Carlsen, P. H. J., *Tetrahedron Lett.*, **1978**, 433.
[217] Colburn, V. M., Iddon, B., Suschitzky, H., and Gallagher, P. T., *J. Chem. Soc., Chem. Commun.*, **1978**, 453.

[218] Nakajima, M., Hisada, R., and Anselme, J.-P., *J. Org. Chem.*, **43**, 2693 (1978).
[219] Dürr, H., and Schmitz, H., *Chem. Ber.*, **111**, 2258 (1978).
[220] Frank, D., Himbert, G., and Regitz, M., *Chem. Ber.*, **111**, 183 (1978).
[221] Timm, U., Zeller, K.-P., and Meier, H., *Chem. Ber.*, **111**, 1549 (1978).
[222] Strausz, O. P., Gosavi, R. K., and Gunning, H. E., *J. Chem. Phys.*, **67**, 3057 (1977); *Chem. Abs.*, **88**, 6121 (1978).
[223] Elzinga, J., Heldeweg, R. F., Hogeveen, H., and Schudde, E. P., *Tetrahedron Lett.*, **1978**, 2107.
[224] Kruglaya, D. A., Fedot'eva, I. B., Fedot'ev, B. V., Kalikhman, I. D., Brodskaya, E. I., and Vyazankin, N. S., *J. Organomet. Chem.*, **142**, 155 (1978).
[225] Altmann, J. A., Yates, K., and Yates, P., *Prog. Theor. Org. Chem.*, **2**, 378 (1977); *Chem. Abs.*, **88**, 151727 (1978).
[226] Lucchese, R. R., and Schaefer, H. F., *J. Am. Chem. Soc.*, **100**, 298 (1978).
[227] Altmann, J. A., Csizmadia, I. G., Robb, M. A., Yates, K., and Yates, P., *J. Am. Chem. Soc.*, **100**, 1653 (1978).
[228] Crossland, N. M., Roberts, S. M., and Newton, R. F., *J. Chem. Soc., Chem. Commun.*, **1978**, 661.
[229] Iwamura, H., and Tukada, H., *Tetrahedron Lett.*, **1978**, 3451.
[230] Borden, W. T., Hoo, L. H., *J. Am. Chem. Soc.*, **100**, 6274 (1978).
[231] Barone, A. D., and Watt, D. S., *Tetrahedron Lett.*, **1978**, 3673.
[232] Norsoph, E. B., Coleman, B., and Jones, M., *J. Am. Chem. Soc.*, **100**, 994 (1978).
[233] Sekiguchi, A., and Watura, A., *Yuki Gosei Kagaku Kyokaishi*, **35**, 897 (1977); *Chem. Abs.*, **88**, 121259 (1978).
[234] Gordon, M. S., *Chem. Phys. Lett.*, **54**, 9 (1978); *Chem. Abs.*, **89**, 75167 (1978).
[235] Ishikawa, M., Nakagawa, K.-I., Ishiguro, M., Ohi, F., and Kumada, M., *J. Organomet. Chem.*, **152**, 155 (1978).
[236] Okinoshima, H., and Weber, W. P., *J. Organomet. Chem.*, **150**, C25 (1978).
[237] Liu, C., and Hwang, T., *J. Am. Chem. Soc.*, **100**, 2577 (1978).
[238] Wulff, W. D., Goure, W. F., and Barton, T. J., *J. Am. Chem. Soc.*, **100**, 6236 (1978).
[239] Fischer, E. O., Schubert, U., and Fischer, H., *Pure Appl. Chem.*, **50**, 857 (1978).
[240] Clark, H. C., *Pure Appl. Chem.*, **50**, 43 (1978).
[241] Block, T. F., and Fenske, R. F., *J. Organometal. Chem.*, **139**, 235 (1977).
[242] Brooks, B. R., and Schaefer, H. F., *Mol. Phys.*, **34**, 193 (1977); *Chem. Abs.*, **88**, 21843 (1978).
[243] Ivin, K. J., Rooney, J. J., Stewart, C. D., Green, M. L. H., and Mahtab, R., *J. Chem. Soc., Chem. Commun.*, **1978**, 604.
[244] Leconte, M., Bilhou, J. L., Reimann, W., and Basset, J. M., *J. Chem. Soc., Chem. Commun.*, **1978**, 341.
[245] Yamamoto, T., *J. Chem. Soc., Chem. Commun.*, **1978**, 1003.
[246] Callot, H. J., and Schaeffer, E., *J. Organomet. Chem.*, **145**, 91 (1978).
[247] Mansuy, D., Lange, M., Chottard, J. C., Bartoli, J. F., Chevrier, B., and Weiss, R., *Angew. Chem. Int. Edn.*, **17**, 781 (1978).
[248] Allison, N. T., Kawada, Y., and Jones, W. M., *J. Am. Chem. Soc.*, **100**, 5224 (1978).
[249] Herrmann, W. A., *Chem. Ber.*, **111**, 1077 (1978).
[250] Herrmann, W. A., Plank, J., Ziegler, M. L., and Weidenhammer, K., *Angew. Chem. Int. Edn.*, **17**, 777 (1978).
[251] Mansuy, D., Lange, M., Chottard, J. C., and Bartoli, J. F., *Tetrahedron Lett.*, **1978**, 3027.
[252] Johnson, B. V., Sturtzel, D. P., and Shade, J. E., *J. Organomet. Chem.*, **154**, 89 (1978).
[253] Levisalles, J., Rudler, H., and Villemin, D., *J. Organomet. Chem.*, **146**, 259 (1978); Levisalles, J., Rudler, H., Villemin, D., Daran, J., Jeannin, Y., and Martin, L., *ibid.*, **155**, C1 (1978).
[254] Dietz, R., Doetz, K.-H., and Neugebauer, D., *Nouveau J. Chim.*, **2**, 59 (1978); Doetz, K.-H., and Dietz, R., *J. Organomet. Chem.*, **157**, C55 (1978).
[255] Bruce, M. I., and Wallis, R. C., *J. Organomet. Chem.*, **161**, C1 (1978).
[256] Malisch, W., Blau, H., and Voran, S., *Angew. Chem. Int. Edn.*, **17**, 780 (1978).
[257] Oguro, K., Wada, M., and Okawara, R., *J. Organomet. Chem.*, **159**, 417 (1978); Fischer, E. O., and Besl, G., *ibid.*, **157**, C33 (1978); Fontana, S., Schubert, U., and Fischer, E. O., *ibid.*, **146**, 39 (1978); Klemarczyk, P., Price, T., Priester, W., and Rosenblum, M., *ibid.*, **159**, C25 (1978); Raubenheimer, H. G., Lotz, S., Viljoen, H. W., and Chalmers, A. A., *ibid.*, **152**, 73 (1978); Priester, W., and Rosenblum, M., *J. Chem. Soc., Chem. Commun.*, **1978**, 26; Fischer, E. O., Selmayr, T., and Kreissl, F. R., *Monatsh. Chem.*, **108**, 759 (1977); Pfiz, R., and Daub, B., *J. Organomet. Chem.*, **152**, C32 (1978).

[258] Le Bozec, H., Gorgues, A., and Dixneuf, P. H., *J. Am. Chem. Soc.*, **100**, 3946 (1978).
[259] Ito, Y., Hirao, T., Tsubata, K., and Saegusa, T., *Tetrahedron Lett.*, **1978**, 1535; Kreiter, C. G., and Aumann, R., *Chem. Ber.*, **111**, 1223 (1978); Hiraki, K., and Fuchita, Y., *Chem. Lett.*, **1978**, 841; Hiraki, K., Fuchita, Y., and Morinaga, S., *ibid.*, p. 1.
[260] Kolobova, N. E., Antonova, A. B., and Khitrova, O. M., *J. Organomet. Chem.*, **146**, C17 (1978); Mansuy, D., Lange, M., and Chottard, C., *J. Am. Chem. Soc.*, **100**, 3213 (1978).
[261] Schrock, R. R., Messerle, L. W., Wood, C. D., and Guggenberger, L. J., *J. Am. Chem. Soc.*, **100**, 3793 (1978).
[262] Fellmann, J. D., Rupprecht, G. A., Wood, C. D., and Schrock, R. R., *J. Am. Chem. Soc.*, **100**, 5964 (1978).
[263] Reger, D. L., and Dukes, M. D., *J. Organomet. Chem.*, **153**, 67 (1978).
[264] Collins, T. J., and Roper, W. R., *J. Organomet. Chem.*, **159**, 73 (1978); Felkin, H., Mennier, B., Pascard, C., and Prange, T., *ibid.*, **135**, 361 (1977).
[265] Dean, W. K., and Vanderveer, D. G., *J. Organomet. Chem.*, **145**, 49 (1978).
[266] Hillhouse, G. L., and Haymore, B. L., *J. Organomet. Chem.*, **162**, C23 (1978).
[267] Clark, D. N., and Schrock, R. R., *J. Am. Chem. Soc.*, **100**, 6774 (1978); Fischer, E. O., Dao, N. Q., and Wagner, W. R., *Angew. Chem. Int. Edn.*, **17**, 50 (1978); Fischer, H., Matsch, A., and Kleine, W., *ibid.*, p. 842; Kreissl, F. R., Vedelhoven, W., and Eberl, K., *ibid.*, p. 859; Fischer, E. O., and Ruhs, A., *Chem. Ber.*, **111**, 2774 (1978); Fischer, E. O., Kleine, W., Schubert, U., and Neugebauer, D., *J. Organomet. Chem.*, **149**, C40 (1978); Fontana, S., Orama, O., Fischer, E. O., Schubert, U., and Kreissl, F. R., *ibid.*, **149**, C57 (1978); Neugebauer, D., Fischer, E. O., Dao, N. Q., and Schubert, U., *ibid.*, **153**, C41 (1978); Frank, A., Fischer, E. O., and Huttner, G., *ibid.*, **161**, C27 (1978).

CHAPTER 6

Nucleophilic Aromatic Substitution

MICHAEL R. CRAMPTON

Department of Chemistry, Durham University

General

The chemistry of diazonium and diazo-groups has been summarised.[1] Dediazoniation may occur by homolytic or heterolytic pathways:[2] in solvents of high nucleophilicity homolysis, giving aryl radicals, is favoured and is catalysed by the addition of nucleophiles which form relatively stable radicals by electron-transfer; however, in solvents of low nucleophilicity, such as trifluoroethanol, benzenediazonium ions decompose heterolytically. Detailed investigations have been reported[3,4] of the dediazoniation rate, the isotopic N_α–N_β rearrangement and the exchange reaction of the diazonio-group with external nitrogen. They provide evidence for two reaction intermediates, a tight molecule–ion pair and the free aryl cation. *Ab initio* MO calculations[5] of the N_α–N_β rearrangement favour an asymmetric transition state in which one of the nitrogens only is weakly bound to a ring-carbon atom.

There has been a further study by CIDNP of the radicals produced during reaction of aryldiazonium fluoroborates with oxygen- and nitrogen-containing nucleophiles.[6] Kinetic data have been reported[7] for the diazo-coupling of substituted benzenediazonium ions with *N,N*-dimethylaniline in non-aqueous media, and reaction of diazonium salts with phosphorus-containing nucleophiles has been examined.[8] There have been reports of the effects of isopolymolybdates on the thermal stability of diazonium ions in acidic media[9] and of the synthesis of *p*-fluoroarylamines from diazonium salts.[10]

The role of single-electron transfers in substitution has been reviewed.[11] The radical chain $S_{RN}1$ mechanism provides a means for nucleophilic substitution in aryl halides which do not contain strongly electron-withdrawing groups. Thus under photostimulation aryl iodides react with diethyl phosphite ion in liquid ammonia to give diethyl arylphosphonates. The reactions of dihalobenzenes with nucleophiles may result in replacement of one or two halogen atoms depending on the nucleophile and the orientation of halogen atoms in the starting material.[12]

Detailed consideration of the reaction of diethyl phosphite with *m*-bromoiodobenzene which gives mainly the disubstituted product, and with *m*-chloroiodobenzene, which gives mainly the iodo-substituted product, provides further strong support for the $S_{RN}1$ mechanism and excludes other mechanisms.[13,14] Further study of the reaction of iodobenzene with the enolate derived from 2,4-dimethylpentan-3-one shows that, in addition to the phenylated ketone, the expected product of $S_{RN}1$ reaction, benzene and a dimeric product are formed in a process originating in β-hydrogen abstraction by phenyl radicals.[15] Dimeric and reduced products are also formed on the reaction of an α-haloamide with sodium methoxide.[16] The $S_{RN}1$ phenylation of the dimsyl anion under stimulation by sunlight has been reported.[17] Electrochemistry can be used to induce substitutions of the $S_{RN}1$ type by setting the electrode potential at the reduction level of the substrate in the presence of a nucleophile; substitutions of halo-derivatives of benzophenone, benzonitrile, and napthalene by thiolates and cyanide ions have been achieved.[18]

Radical alkyldenitration provides a synthetically useful method of aromatic *ipso*-substitution in nitro-aromatics containing electron-withdrawing substituents in the *para*-position.[19] Examples have been reported also of intramolecular *ipso*-substitution in halobenzenes by aryl radicals[20] and in sulphonylbenzenes by substituted methyl radicals.[21] The reaction of 4-substituted 1-halo-2-nitrobenzenes with caesium fluoride in dimethylformamide may result in halogen exchange, or nucleophilic substitution by dimethylamine, or dehalogenation;[22] the last of these processes involves fragmentation of a radical anion derived from the substrate. The radical anion of *p*-dinitrobenzene has been identified spectroscopically in the reaction with hydroxide ions in dimethyl sulphoxide and is thought to be a precursor in the substitution reaction.[23]

The S_NAr Mechanism

Nucleophilic substitution reactions have been reviewed.[24] It has been recognized for some time that the nucleophilicity of a reagent depends both on basicity and polarizability factors. A new relation linking reactivity to substrate polarizability has been devised[25] and applied to the reactions of the four halo-2,4-dinitrobenzenes with seven nucleophiles in methanol. A theoretical study[26] of the qualitative potential energy surfaces for aromatic substitution predicts that the orientation of nucleophilic substitution will be controlled by the electron density of the lowest unoccupied MO of the substrate whereas in photonucleophilic substitutions the orientation of attack depends on the electron density of the highest occupied MO of the substrate.

Substitutions by amines proceed[27] by the mechanism of Scheme 1. Base catalysis is observed when the conversion of (**1**) into products is slower than the k_{-1} step. If deprotonation of (**1**) is rate-limiting [$k_4 \gg k_{-3}$ (BH)], general base catalysis is observed, while rate-limiting leaving group departure [$k_4 \ll k_{-3}$ (BH)] leads to specific base catalysis. In a recent paper[28] the possibility of observing base catalysis in the addition step has been examined: the examples chosen were the reactions of aniline and imidazole with 1-fluoro-2,4-dinitrobenzene where the k_1 step is rate-determining. Although the rate of the imidazole reaction increases with increasing pH it was shown that this is due to reaction of the substrate with imidazole anion rather than reaction by a concerted mechanism (k_1' step in Scheme 1). There has been further kinetic work[29,30] on the observation of base catalysis in the reactions of activated nitrophenyl ethers with aniline in benzene.

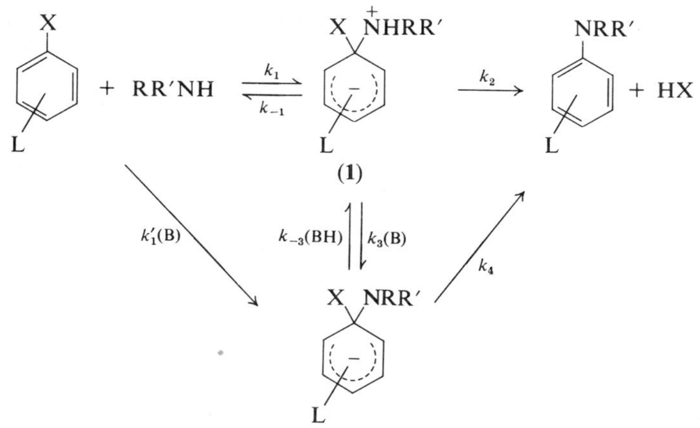

X, L substituted benzene + RR′NH $\underset{k_{-1}}{\overset{k_1}{\rightleftharpoons}}$ (1) $\xrightarrow{k_2}$ NRR′ + HX

k'_1(B); k_{-3}(BH); k_3(B); k_4

SCHEME 1

Nucleophilic substitution reactions of azide, hydroxide or methoxide ions with a series of 4-chloro-3-nitro-X-benzenes have been used[31] to obtain Hammett σ^- values for the following X-groups; CH=NPh, CH=$NC_6H_4NO_2$, CH=NOH, and CH=$N^+(O^-)$Ph. Substituent effects in 1-naphthoate and 2-naphthoate ions have been examined in the reaction with picryl bromide.[32] A methoxydechlorination reaction has been used to show that there is no transmission of electronic effects between benzene rings separated by a cyclopropyl ring,[33] and there has been an analysis of the reactivity of polynuclear dinitro- and diamino-compounds.[34] The reaction of benzenethiolate ions with 2,4-dinitrophenyl benzoate and acetate can result in either carbonyl carbon–oxygen or aryl carbon–oxygen scission (Scheme 2); the latter process is the more susceptible to polar substituent effects

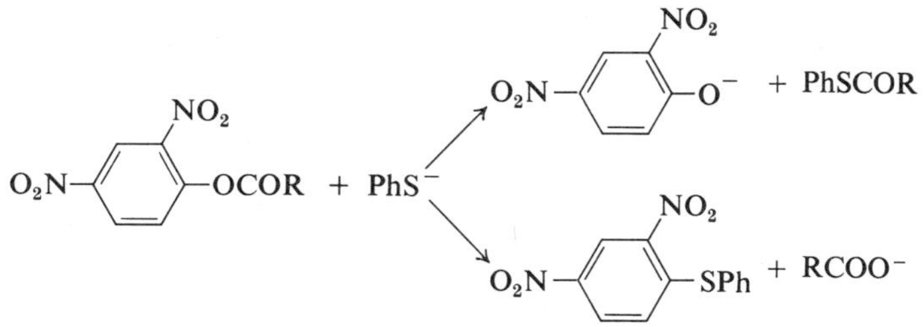

O_2N–$C_6H_3(NO_2)$–OCOR + PhS^- → O_2N–$C_6H_3(NO_2)$–O^- + PhSCOR

→ O_2N–$C_6H_3(NO_2)$–SPh + $RCOO^-$

SCHEME 2

in the nucleophile.[35] There has also been study, in this reaction,[36] of the steric effects of *ortho*-substituents in the benzenethiolate: it is found that increasing steric bulk has a decelerating effect on substitution at aromatic-carbon but an accelerating effect on substitution at carbonyl-carbon; steric inhibition of solvation of the sulphur nucleophile is invoked to account for the latter effect. Competition between attack at aryl-carbon or the sulphonyl group is observed in the reaction of fluoride ion with 2,4-dinitrophenyl sulphonate.[37]

The effects of micelles on chemical reactivity, including nucleophilic substitutions, have been reviewed.[38,39] It has been reported that the micellar catalysis by cetyltrimethylammonium bromide of the reaction of 1-chloro-2,4-dinitrobenzene with hydroxide ions is enhanced by added sugars. It has now been shown[40] that this rate acceleration is due to formation of an intermediate ether which subsequently decomposes to 2,4-dinitrophenoxide ion rather than to a physical effect of the sugar on the structure of the micelle. The rate of reaction of 1-chloro-2,4-dinitrobenzene with potassium methoxide in benzene is greatly enhanced[41] in the presence of crown-ethers, which may have the function of dissociating the base. The increased reactivity of nucleophiles in dipolar aprotic solvents has been put to good effect in the synthesis[42] of substituted benzothiazoles involving an intramolecular substitution by thioamide anion. There have also been reports of solvent effects on the reaction of 1-chloro-2,4-dinitrobenzene with substituted anilines.[43,44]

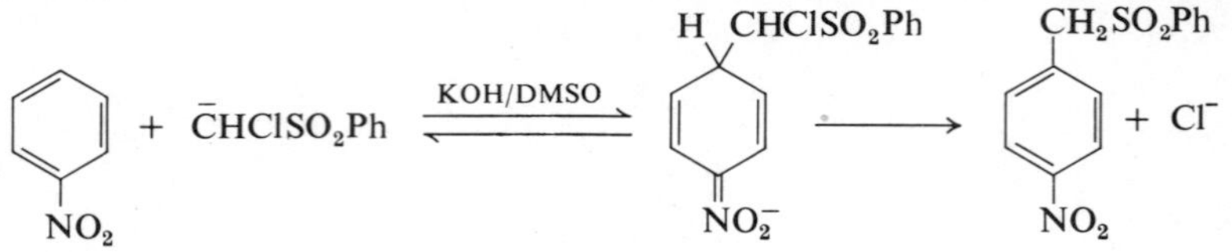

H CHClSO$_2$Ph   CH$_2$SO$_2$Ph

NO_2 + $\bar{C}$HClSO$_2$Ph $\xrightleftharpoons{KOH/DMSO}$ NO_2^- ⟶ NO_2 + Cl$^-$

SCHEME 3

The unusual substitution of hydrogen by a carbanion has been proposed[45] in a reaction (Scheme 3) in which the nucleophile contains a substituent able to leave as an anion. Nucleophilic substitution in chloronitrobenzenes by carbanions derived from *N*-methyloxindoles has been described.[46] There have also been studies of substitutions by anions derived from *N*-heterocycles. Thus the scope of the nucleophilic displacement of aromatic halogens on 1,4-benzodiazepine precursors, such as (**2**), by the anions of pyrroles, pyrazoles, and imidazoles has been examined;[47]

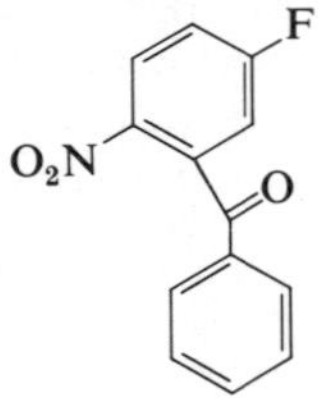

F, O_2N, O

(**2**)

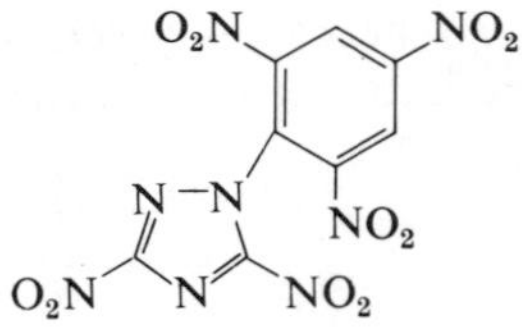

O_2N, NO_2, N—N, NO_2, O_2N, N, NO_2

(**3**)

some of the products are useful intermediates in the synthesis of fused 1,4-benzodiazepines. The reaction of the anion of 3,5-dinitrotriazole with picryl chloride in acetonitrile is thought[48] to yield (**3**) as an intermediate although a mixture of products is ultimately formed, and there has been kinetic study[49] of the reaction of nitrodiphenylamine anions with 1-fluoro-2,4-dinitrobenzene in dimethylformamide. There have been reports of the reactions of substituted halobenzenes with tertiary amines to give the corresponding dialkylamine and the trialkylammonium salt of the corresponding phenol[50,51] and of the reaction of picryl chloride with triethyl phosphite.[52]

The Smiles rearrangement of (**4**) to (**5**) proceeds under stimulation by ultraviolet radiation.[53] A free-radical mechanism has been proposed for the thermal isomerization of aryl *N*-(*p*-nitrophenyl)alkanehydrazonates into *N*-acyl-*N'*,*N'*-diarylhydrazines.[54]

Nucleophilic displacement reactions involving NO_2 as a leaving group have been reviewed[55] and it is reported that bis- and tris-(methylthio)benzenes can be synthesized by replacement of nitro-groups by methanethiol anion in aqueous alcohol.[56] A series of papers reports the reactions of nucleophiles with phthalimides (**6**),

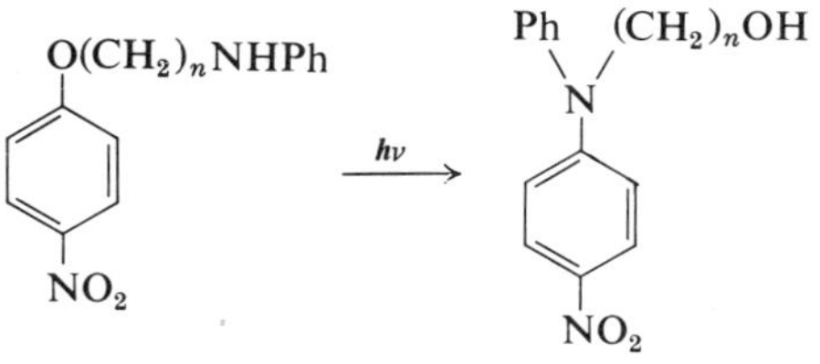

(4) $\xrightarrow{h\nu}$ (5)

phthalate esters (**7**) or phthalic anhydrides (**8**), each carrying nitro- or halogen-substituents at the 3- or 4-position. With phenoxide ions as nucleophiles[57–59] fairly drastic conditions are required for substitution, which is accompanied in (**7**) by attack at the alkyl group of the ester and in (**8**) by opening of the anhydride ring.

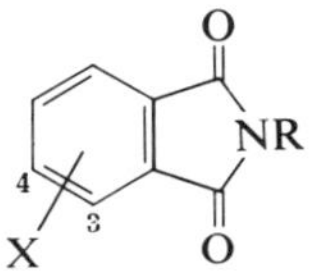

(**6**) 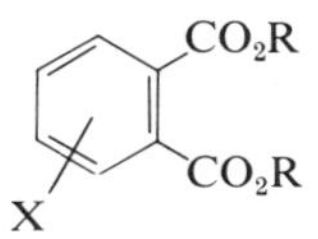 (**7**) 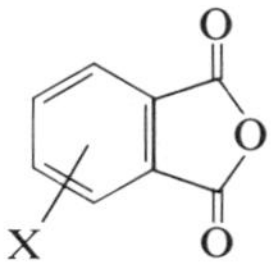 (**8**)

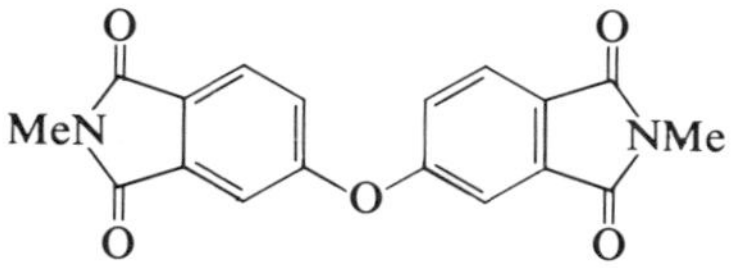

(**9**)

Reaction with thiophenoxide ion, a better nucleophile, proceeds under milder conditions[60,61], as expected and there is little competitive attack at the carbonyl group of (**8**). The reaction of *N*-methyl-4-nitrophthalimide with either fluoride ions or nitrite ions in dipolar aprotic solvents gives[62] the diaryl ether (**9**) *via* 4-hydroxy-*N*-methylphthalimide which is formed by oxygen attack of the ambident nitrite ion. Ring-opening of the anhydride, again involving attack by the ambident nitrite ion, is a side-reaction in the fluorodenitration of 4-nitrophthalic anhydride.[63]

The reaction of nitrite ion with diarylchloronium salts results in *ipso*-attack to give nitro- and chloro-benzene derivatives.[64] The product distributions from unsymmetrically substituted compounds are in accord with an S_N2-like mechanism controlling the collapse of ionic diarylchloronium nitrites.

Further study[65] of the reaction between water and the carcinogen *N*-acetoxy 2-acetamidofluorene shows that the product is 2-acetamidofluoren-4-ol (**10**) rather

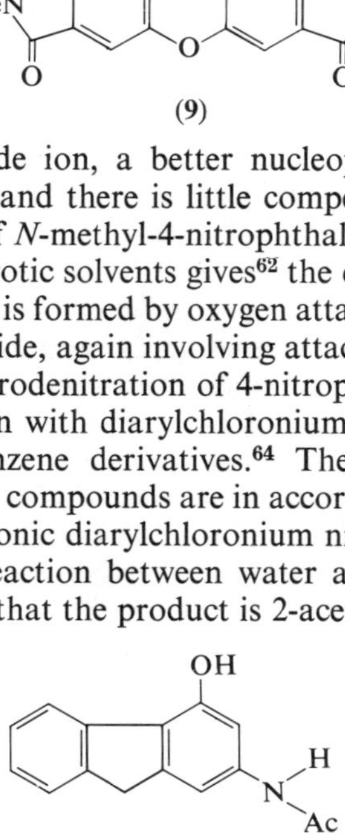

(**10**)

than 2-acetamido-1-acetoxy-fluorene. There have been miscellaneous studies of alkoxy-exchange reactions and isomerization in α- and β-narcotine,[66] of parallel substitution and reduction in reactions of 4-halonitrobenzenes with methoxide ions[67] and of the effects of uv radiation on aqueous solutions of nitrophenols.[68]

There have been several theoretical and experimental studies of substitution in polyhalobenzenes. A recent attempt to apply frontier-orbital formalism to these reactions has been criticized[69] since the method fails when applied to a number of substituted benzenes and perfluoropolycyclic aromatics; it is argued that the older I_π-repulsive method can more readily rationalize these substitutions. However, it is clear that factors other than purely electronic effects can influence the position of substitution. Thus steric factors have been invoked[70] to account for the substitution of 1,2,3,4-tetrachloro-5,6-dinitrobenzene with amines, which may result in displacement of either a nitro-group or a chlorine atom. There has also been support[71] for the hypothesis that the reaction of ammonia with fluoronitrobenzenes, which results mainly in *o*-fluorine displacement, is influenced by hydrogen-bonding between the attacking amine and the nitro-group. Kinetic studies of methoxydefluorination of polybromofluorobenzenes[72] and polyfluoronitrobenzenes[73] show that, owing to the mutual interaction of substituents, the additivity of substituent effects is poorer than in compounds containing only fluorine and chlorine substituents. Further study[74] of methoxydefluorination of some 2-substituted heptafluoronaphthalenes confirms that displacement occurs mainly at C-6. There is evidence[75] that the gas-phase bromodeiodination of pentafluoroiodobenzene involves the radical (**11**), and that the products of the photoreaction of hexafluorobenzene with cyclohexane[76] are derived from the intermediate (**12**). It has

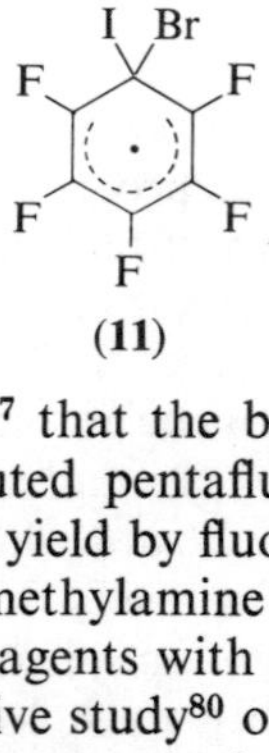

(**11**) 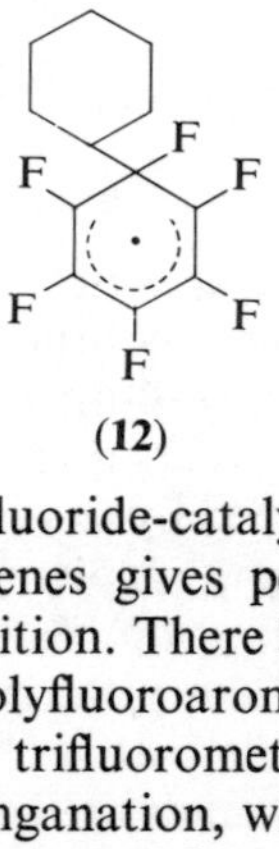 (**12**) 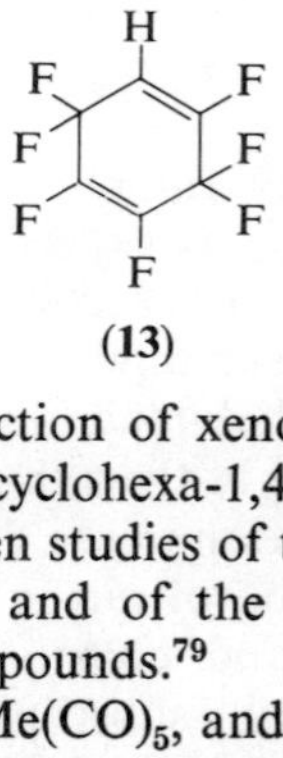 (**13**)

been reported[77] that the boron trifluoride-catalysed reaction of xenon difluoride with 1-substituted pentafluorobenzenes gives perfluorocyclohexa-1,4-dienes such as (**13**) in high yield by fluorine addition. There have been studies of the gas–solid reactions of methylamine with polyfluoroaromatics,[78] and of the reactions of nucleophilic reagents with aromatic trifluoromethyl compounds.[79]

A comparative study[80] of the manganation, with $MnMe(CO)_5$, and palladation, with $PdCl_2$, of some substituted azobenzenes has shown that these complexes behave as nucleophilic and electrophilic reagents, respectively. Metallation is preceded by initial co-ordination of one of the azo-nitrogen atoms. Detailed study[81] of the reaction of (trimethylstannyl)sodium with halobenzenes in tetraglyme indicates a mechanism in which the initial step is halogen–metal exchange; the resulting halo(trimethyl)stannane and arylsodium can react within a solvent cage or may diffuse out of the cage to react with other species present in the bulk of the medium. The stannylation of halotoluenes by (tributylstannyl)lithium in hexamethylphosphoramide has been interpreted in terms of competing anionic and radical mechanisms.[82]

Co-ordination of aromatics to transition-metal cations greatly enhances their susceptibility to nucleophilic attack: these reactions have been surveyed and interpreted[83] in terms of three rules which allow prediction of the most favourable position of attack. Further examples have appeared[84] of the ability of an oxazoline

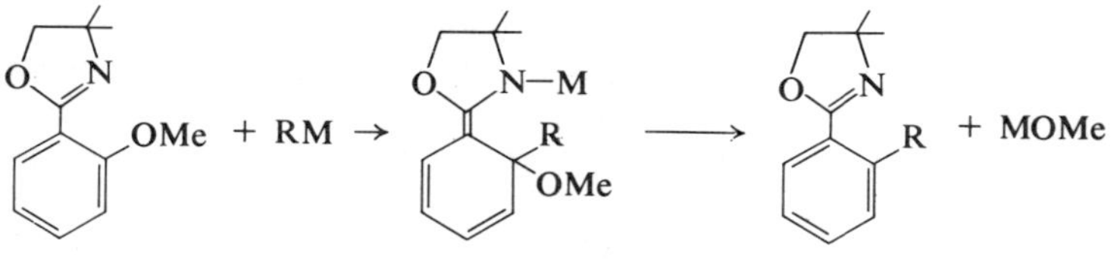

SCHEME 4

group to activate an aromatic ring towards substitution by organometallic reagents (RM). As indicated in Scheme 4, the oxazoline group plays a crucial role in complexing the organometallic reagent. In a similar way the oxazoline function can activate *o*-fluoro-substituents towards substitution by organometallics and lithioamides.[85] It has been reported that aromatic nitro-compounds may be alkylated by reaction with alkyl-lithium or alkyl-Grignard reagents.[86] Palladium in the form of $Pd(PPh_3)_4$ has found use as a catalyst in the reactions of aryl halides with thiolate ions[87] to give aryl sulphides, with alkynylzinc reagents to give arylalkynes,[88] and with sodium methoxide to give arenes.[89] The use of (tri-*o*-tolylphosphine)palladium as a catalyst in the reactions of ethylene with aryl bromides to give styrenes[90] has also been reported. Cobalt(III) trifluoroacetate has been used to promote the halogenation of benzene by a mechanism that probably involves attack by halide ion on aromatic radical cations.[91] The Ullman condensation of 1-bromoanthraquinone with 2-aminoethanol in the presence of copper(I) salts in aprotic solvents has been studied;[92] the ESR spectra indicate formation of the radical anion from the aromatic and also copper(II) species.[93] The aromatic thiocyanation of 1-alkoxynaphthalene by copper(II) thiocyanate has been reported.[94]

Heterocyclic Systems

Measurement of the rates of *p*-tolylthiodenitration of 2,5-dinitrofuran, 2,5-dinitrothiophene, 1-methyl-2,5-dinitropyrrole, and 1,4-dinitrobenzene shows that reactivity decreases in the given order. This order differs from that of piperidinodenitration of the same compounds, a fact attributed to the differing polarizabilities of the substrates.[95] There has been study of base catalysis in the reactions of 2-L-3-nitro- and 2-L-5-nitro-thiophenes (L = leaving group) with piperidine:[96] in general the latter isomers are more susceptible to catalysis, and the observation of specific methoxide ion catalysis in the reaction of 2-methoxy-3-nitrothiophene with piperidine is thought to be a consequence of the strong conjugative interaction of substituents linked to the 2- and 3-positions of the thiophene ring (hyper-*ortho*-effect).[97] There is evidence, including the isolation of intermediates, that the *cine*-substitution of 3,4-dinitrothiophene with arenethiolates to give 2-arylthio-4-nitrothiophenes proceeds by a set of consective addition–elimination steps.[98] *cine*-Substitution *via* the intermediate **(14)** is also observed in the methoxydenitration of 1-methyl-3,4-dinitropyrrole.[99] The reaction of 2-methyl-1,4-dinitropyrrole with base has been examined.[100]

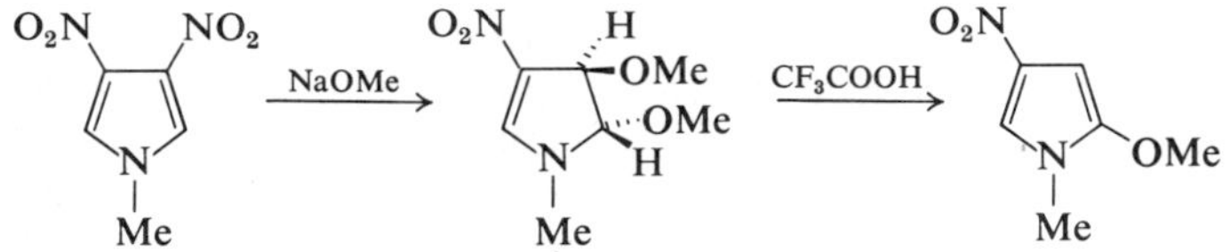

(14)

The relative reactivities of 2- and 4-halothiazoles are sensitive to the nature of the nucleophile and of its counter-ion, and to the solvent; it has been argued that steric effects are more important for the 2-halo-isomers.[101] The reaction of 2-nitrobenzothiazole with aliphatic amines is subject to base catalysis in the presence of DABCO, indicating that decomposition of the intermediate is rate-determining.[102]

There has been further study[103] of the effects of micellar catalysis on the hydrolysis of *N*-alkyl-4-cyanopyridinium ions. It has been reported[104] that di-1,3,5-triazinyl ethers may be prepared by condensation of 2-chloro-1,3,5-triazine with potassium 1,3,5-triazin-2-olate in acetonitrile containing 18-crown-6-polyether. Cyclic polyethers incorporating 2,6-pyrazino- and 2,4-pyrimidino-units have been prepared by nucleophilic substitutions on 2,6-dichloropyrazine[105] and 2,4-dichloropyrimidine,[106] respectively. However, attempts to prepare the related thio-crown ethers from 2,6-dihalopyridines met with limited success.[107]

The preparation of 6-chloropterin (**15**) has been reported[108]. Nucleophilic displacement of the 6-chloro-group by sulphur bases proceeds smoothly although reaction with amines leads to decomposition. Replacement of the chlorine atom in

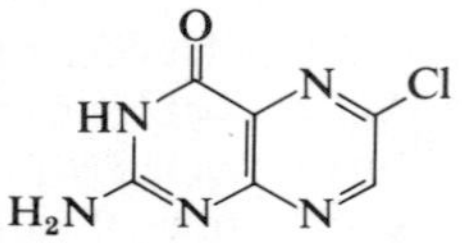

(**15**) 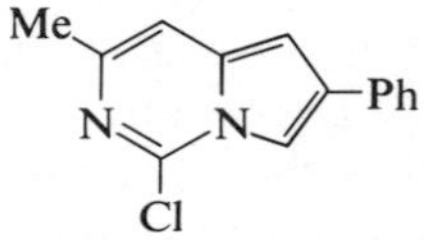 (**16**) 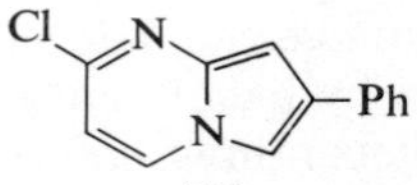 (**17**)

5-chloro-6-azaindolizine (**16**) by hydroxide, methoxide, or ammonia has been accomplished,[109] whereas only methoxylation occurs with 7-chloro-8-azaindolizine (**17**). Kinetic and theoretical studies[110] have been carried out on the substitution of isomeric halo-*N*-methyl-1,2,4-triazoles by piperidine or methoxide ions: it is found that the 5-halo-1-methyl isomer is more reactive than the 3-halo-4-methyl-isomer. Some 3-substituted 1,2,4-triazine-2-oxides have been found[111] to undergo deoxygenative 6-alkoxylation with alcohols to yield the corresponding 6-alkoxy-1,2,4-triazines.

The reactions of heterocyclic compounds with nucleophiles are often complicated by processes other than straightforward substitution. Thus reaction of the dinitroquinoxaline (**18**) with piperidine in ethanol leads to normal and *cine*-

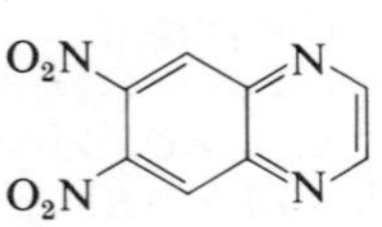

(**18**) 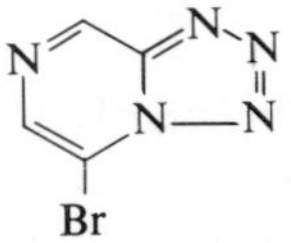 (**19**) 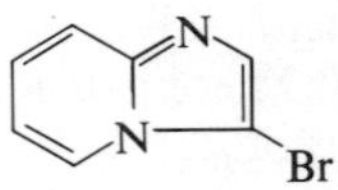 (**20**)

substituted products.[112] Kinetic study shows that the latter process has a squared dependence on base concentration. 5-Bromo-*s*-triazolo[4,3-*a*]pyrazine (**19**) reacts with nucleophiles to give the expected 5-substituted derivative or at the 8-position to give *tele*-substituted products.[113] *tele*-Amination has also been reported for the reaction of 3-bromoimidazo[1,2-*a*]pyridine (**20**) with metal amides.[114] The kinetic effects of *N*-alkyl substituents on nucleophilic substitution in benzimidazoles have been studied.[115] Intramolecular substitution of the nitro-group in 9-[(2-methylamino)ethylamino]-1-nitroacridine has also been reported.[116]

The pyridine ring is activated by the presence of an oxazoline group. Thus, metallation at the 3-position of 4-pyridyloxazoline can be accomplished with methyllithium,[117] and addition of organolithium reagents to 3-(4,4-dimethyloxazolin-2-yl)-pyridine leads to stable 1,4-dihydropyridines.[118] There has been study of solvent and substituent effects on the regioselective metallation of methylpyridines and on the reaction of the resulting pyridylmethanide anions with substituted benzonitriles. The latter reaction may involve nucleophilic addition to the cyano-group or, when the nitrile carries an acceptor group, nucleophilic aromatic substitution.[119] The reactions of *N*-alkoxypyridinium salts with cyanide ions may result in the formation of cyanopyridines or decomposition to pyridine.[120] The reaction of 3,6-dihalopyridazine with pyridazinethiones yields the double substitution product that is thought to be formed by successive nucleophilic displacements upon the protonated reactant.[121]

There have been kinetic studies of the relative reactivities of the chlorine atoms in 2,4-dichloropyrimidine upon reaction with ammonia and amines in "*iso*octane" and ethanol,[122] and of the reactions of some chloropyrimidines with ethoxide ions.[123] Nucleophilic substitution in pyrimidines often occurs by the S_N (ANRORC) mechanism (Addition of Nucleophile, Ring Opening and Ring Closure) and a ^{15}N-labelling study indicates that this mechanism operates in the aminodemethoxylation of 4,6-dimethoxypyrimidine.[124] However, an S_N (ANRORC) mechanism does not operate in the amination of 4-*tert*-butyl-5-chloropyrimidine by potassium amide in ammonia; amide addition at the 6-position gives (**21**) which may give 6-amino-4-*tert*-butyl-5-chloropyrimidine by loss of a hydride ion, or the *cine*-substitution product 6-amino-4-*tert*-butylpyrimidine by protonation and loss of

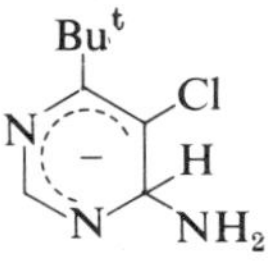

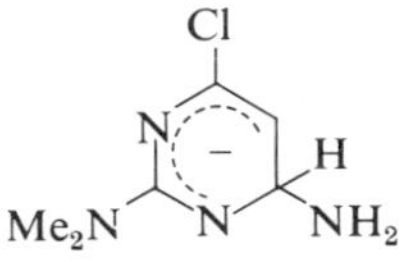

hydrogen chloride.[125] The reaction of 4-chloro-2-dimethylaminopyrimidine with potassium amide yields 2-dimethylamino-4-methyl-*s*-triazine via the intermediate (**22**) which has been identified by ^{13}C-NMR.[126] The first conversion of a pyrimidine ring into a benzene ring has been achieved by reaction of 5-nitropyrimidine with ketones in the presence of base,[127] and it has been reported that nitropyridinium salts react with amines to form anilines *via* ring-opened intermediates.[128] There is strong ^{1}H-NMR evidence that (**23**) is an intermediate in the ring contraction of *N*-methylpyrimidinium salts into pyrazoles[129] brought about by deuteriohydrazine. A novel ring-opening has been reported in the acylation of 6-amino-1,3-dimethyluracil.[130] Incorporation of ^{15}N into the tetrazine ring of the 3-hydrazino-6-methyl-1,2,4,5-tetrazine formed upon hydrazinolysis (with ^{15}N-labelled hydrazine) of 3-amino- and 3-bromo-6-methyl-1,2,4,5-tetrazine indicates that the reaction proceeds by an S_N (ANRORC) mechanism.[131] There has been further study of the reaction of 3,6-disubstituted 4-nitropyridazine 1-oxides with ammonia.[132]

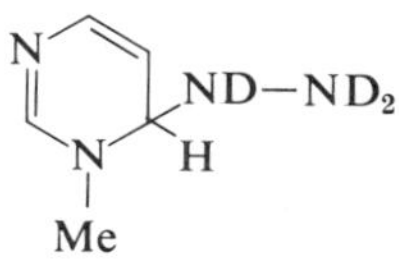

Several studies have biochemical significance. Cytidine is readily deaminated to give uridine in the presence of bisulphite and a scheme has been proposed which includes both the protonated and the non-protonated cytidine–bisulphite adducts.[133] By use of ^{13}C-NMR it has been shown[134] that the two diastereomeric bisulphite adducts (**24**) and (**25**) of cytidine 5′-monophosphate have significantly different pK_a values, which may lead to mechanistic complications. Benzofuroxan has been

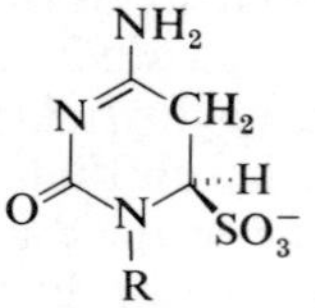

(**24**) 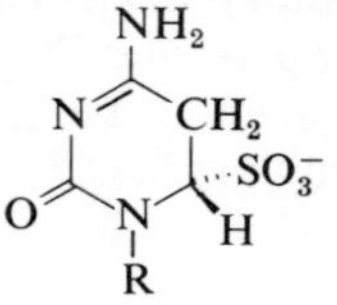 (**25**) 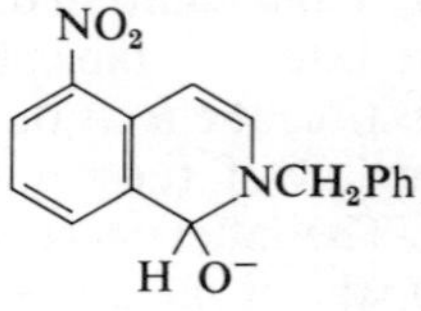 (**26**)

used as a thiol-specific probe for biological molecules, and a kinetic study[135] of its reaction with 2-mercaptoethanol indicates rate-determining attack of thiolate ion to give an adduct which reacts rapidly with a second molecule of thiol yielding disulphide and *o*-benzoquinone dioxime. In alkaline solution *N*-benzyl-5-nitro-isoquinolinium cations form hydroxide adducts (pseudo-bases) which undergo deprotonation to give (**26**). Hydride transfer from (**26**) to ferricyanide ions[136] and 2-methylisoquinolinium cations[137] as well as from 1,4-dihydropyridines to pyridinium salts[138] have been studied.

Meisenheimer and Related Complexes

The kinetic behaviour of anionic σ-complexes has been summarized[139] in an account that shows how study of these complexes provides information on the following areas: the mechanism of base catalysis in S_NAr reactions; solvent and steric effects on proton-transfer reactions; the transition state of concerted acid–base catalysed reactions; leaving-group abilities of amines, alkoxide, and aryloxide ions. The complex (**27**) has been observed as a transient species in the reaction of 2,4,6-trinitroanisole with phenoxide ions in aqueous dimethyl sulphoxide.[140] Departure of phenoxide ion from this complex is found to be more than 10^6 times faster than departure of methoxide ion from the corresponding 1,1-dimethoxy-complex although the rates of phenoxide and methoxide attack on 2,4,6-trinitro-anisole are similar. Study of the reaction of 2,4,6-trinitrobenzenesulphonate ions

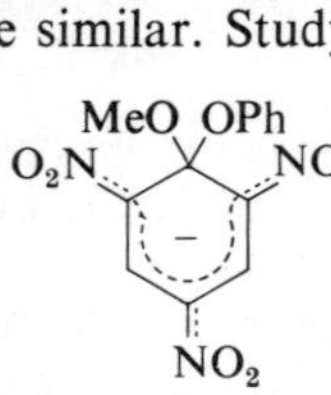

(**27**) 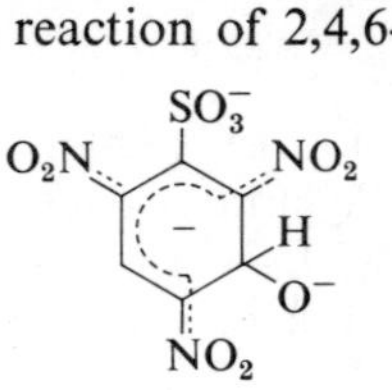 (**28**) 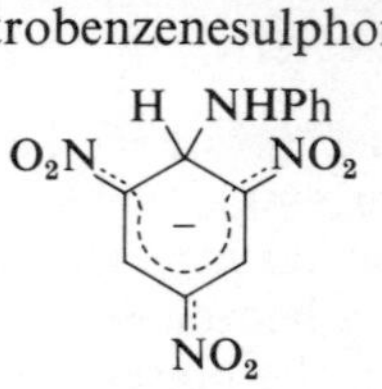 (**29**)

with aqueous sodium hydroxide shows that initial hydroxide attack occurs at the 3-position and that ionization of the added hydroxyl group occurs to give (**28**); slower addition of hydroxide at the 1-position leads to nucleophilic substitution, giving picrate ions.[141] Similarly hydroxide attack on methyl 2,4,6-trinitrobenzoate occurs initially at the 3-position.[142]

Equilibrium and kinetic data have been reported for the reaction of 1,3,5-trinitrobenzene with aniline in the presence of DABCO in dimethyl sulphoxide.[143] The reaction yields the anilide complex (**29**) and protonated DABCO. A large increase in equilibrium constant was found in the presence of tetraethylammonium chloride which was attributed to association of Cl^- with protonated DABCO to yield the DABCO,H^+...Cl^- complex with a consequent decrease in rate of the reverse reaction.

In the heterocyclic series there have been rate and equilibrium measurements of methoxide addition to 2-nitrofuran and 4-cyano-2-nitrofuran[144] in methanol. Comparison with data for the corresponding thiophene adducts shows the accelerating and stabilizing effect of the furan ring. Data for methoxide addition to 2-methoxy-3-nitrothiophene to give (**30**) have been presented[145] and it has been argued that comparison of the benzene and thiophene ring systems is complicated

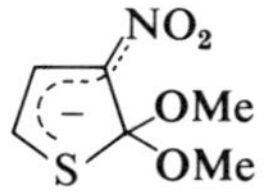

NO_2 OMe OMe S

(**30**)

NO_2^- S $\overset{+}{O}Me$

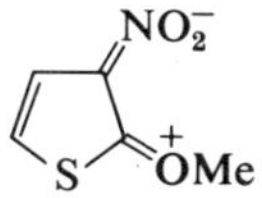

(**31**)

because of the importance of hyper-*ortho* interaction (**31**) in the ground state of the thiophene. Reaction of methoxide with 2,6-diphenyl- and 4-methoxy-2,6-diphenylpyrilium cations can lead to addition at the 2- or 4-position.[146]

Interest in spiro-complexes continues. The stabilities of spiro-complexes derived from nitro-aromatics decrease as the size of the dioxolan ring increases; the factors involved have been fully discussed.[147] Kinetic and equilibrium data relating to the formation of (**32**) and (**33**) have been compared:[148] (**32**) is only 2.5 times more

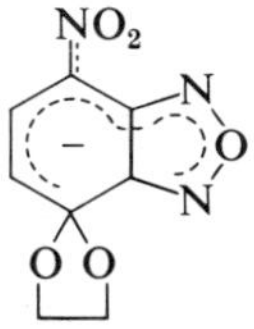

(**32**)

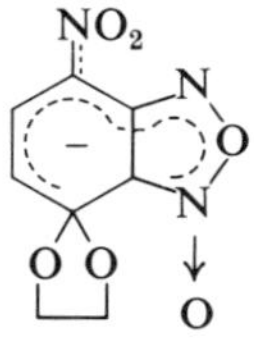

(**33**)

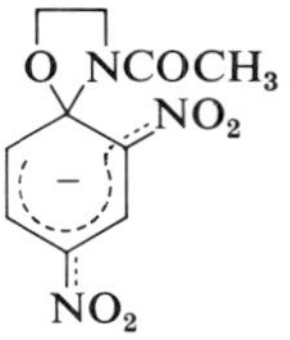

(**34**)

stable than (**33**), yet it is formed and decomposes much more rapidly; this has been attributed to the electrostatic effect of the *N*-oxide group in the furoxan. The complex (**34**) has been detected spectrometrically during the Smiles' rearrangement of 2-(acetylamino)ethyl-2,4-dinitrophenyl ether in dimethyl sulphoxide;[149] the rate of the rearrangement depends critically on the rate of decomposition of the intermediate.[150] The role of Meisenheimer spirocyclic complexes in intramolecular aromatic substitution reactions has been reviewed;[151] and there have been studies of the intramolecular cyclizations of 1-[*o*-(*N*-methylamino)phenylthio]-2,4,6- trinitrobenzene,[152] and 1-(3-hydroxypropylthio)-2,4,6-trinitrobenzene.[153] Also, in the 2,4,6-trinitrobenzene series, there have been reports of the preparation of compounds containing alkylated spiro-dioxolan rings[153,154] and of a compound containing two spiro-dioxolan rings – at the 1- and the 3-position.[155] The oxygen of the nitro group of a silver salt of a spiro-complex has been alkylated.[156]

It has previously been reported that the nitrogen and α-carbon centres of α-phenylacetamidines can act as nucleophilic centres in the *meta*-bridging of polynitroaromatics. However, reaction of *N*,*N*-dimethylphenylacetamidine with 3,5-dinitrobenzonitrile yields (**35**) and not a bicyclic adduct;[157] the electron-withdrawing ability of the ring apparently plays a major role in directing the course of the reaction. The reactions of α-phenylacetamidines with *o*-halonitro-aromatics yield displacement–addition products in which the amidine is annelated across the ring-carbon and the nitrogen of an adjacent nitro-group.[158] A new tricyclic *meta*-bridged adduct has been reported from reaction of 1,3,5-trinitrobenzene with ammonia and acetone in water,[159] and the kinetics of the Janovsky reaction of

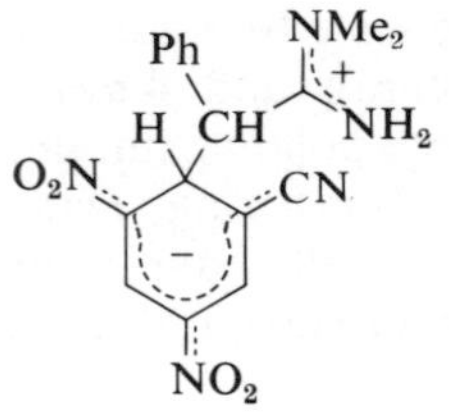

(35)

1,3-dinitrobenzene with acetone in basic solution have been studied.[160,161] The reactions of 2,4,6-trinitroanisole[162] and 1-chloro-2,4,6-trinitrobenzene[163] with dimethyl malonate ions in methanol initially yield mono- and di-adducts through addition at unsubstituted ring positions; the slower attack of methoxide or dimethyl malonate ions at the 1-position yields substitution products.

The use of NMR spectroscopy to characterize σ-adducts formed in flowing systems has been reviewed.[164] Cyclohexadienyl anions have been prepared by proton abstraction from cyclohexa-1,3- and -1,4-dienes and their ^{1}H- and ^{13}C-spectra have been examined;[165] the results, in agreement with MINDO/3 calculations, show that these cyclohexadienyl anions are planar rather than non-planar homoaromatic species. ^{13}C-Spectra of 2,6-dinitro-4-X-anisoles show deviations from additivity of substituent effects consistent with inhibition of resonance of the *o*-nitro-groups;[166] shifts for the corresponding 1,1-dimethoxy-adducts show an increase in negative charge at the 2-, 4-, and 6-positions but a decrease at the 3- and 5-positions, in agreement with SCF MO calculations. ^{13}C-NMR spectra of the acetonate adducts of 1,3-dinitrobenzene and 1,3,5-trinitrobenzene, and the related Zimmermann products, have been reported.[167] ^{1}H- and ^{13}C-studies[168] of the attack of amide ions or ammonia on 1,2,4-triazines show that addition occurs at the 5-position to give **(36)**. 1,5-, 1,6-, and 1,8-Naphthyridines undergo amide attack at the 2-position to give, **(37)**, for example, although 1,7-naphthyridine showed a

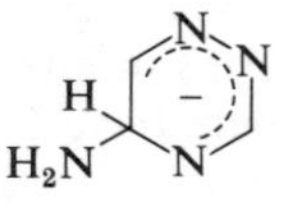

(36) 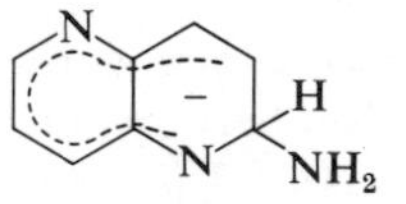 **(37)** 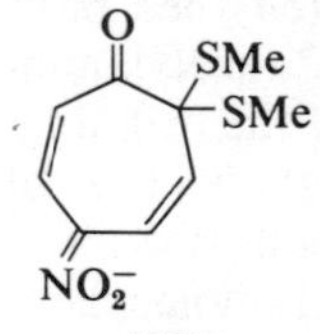 **(38)**

more complex reactivity pattern.[169] ^{13}C-NMR has been used to determine the position of attack of phenyl-lithium on pyrimidine, pyridazine, and pyrazine,[170] and of methoxide attack on flavinium salts.[171]

Attack of methanethiolate ions at the 2-position of 2-methylthio-5-nitrotropane gives[172] the *gem*-di(alkylthio)-adduct **(38)**; the spiro-analogue of this complex has also been observed. It has been reported that addition of thiolate ions to methoxytropenylium cation occurs exclusively at the 7-position to give 1-methoxy-7-(alkyl- or aryl-thio)cycloheptatrienes,[173] and that reaction of chlorotropylium cation with sodium fluoride in the presence of crown ether yields 7,7-difluorocycloheptatriene.[174]

There have been reports of σ-complexes from the reaction of 2,4-bis(trifluoromethylsulphonyl)anisole with sodium methoxide,[175] from that of *N*-phenylpicramide with hydroxylamine[176] and of 1,3,5-trinitrobenzene with phosphorus(III) compounds.[177] Formation of anionic hydride σ-complexes of

1,3,5-trinitrobenzene on reaction with the coenzyme NAD(H) and dihydro-derivatives of other nitrogen heterocycles[178] has been reported.[178] Also reactions of σ-complexes with carbon, nitrogen, and oxygen acids,[179] and of σ-complexes derived from 1-halo-2,4-dinitrobenzenes,[180] have been studied.

Benzyne and Related Intermediates

Arynes have been reviewed,[181] and there has been a summary[182] of the effects of using "complex" bases, such as mixtures of sodium amide and sodium butoxide, on the formation and reactivity of benzyne.

The addition of radioactive lithium bromide to perhaloarynes generated from perhalophenyl-lithium derivatives has been used as a gauge of aryne formation.[183] The decomposition of *o*-fluorophenyl-lithium tetrahydrofuranate in the presence of furan has been studied.[184] Evidence has been presented in favour of intermediate formation of 3,4-didehydro-1,8-naphthyridine during reaction of 3-halo- and 4-halo-1,8-naphthyridines with potassium amide in liquid ammonia.[185] Benzyne intermediates are implicated in the intramolecular cyclizations of enaminones to give γ-lycorane[186] and of 4- and 2-[2-(*o*-chlorophenyl)ethyl]pyridines to give benzisoquinolines and 1-pyridylbenzocyclobutenes.[187]

The Diels–Alder reaction of unsymmetrically substituted benzynes with hexamethylcyclohexa-2,4-dienone has been studied; the relative proportions of the two isomeric products are thought to be controlled by steric interactions between the substitutents on the benzynes and the *gem*-dimethyl groups of the dienone.[188] The effects of ring-size on the reactions of cyclic olefins with benzyne have been discussed in terms of the stereochemical requirements along the reaction pathway.[189]

References

1 Patai, S. (Ed.) "The Chemistry of Diazonium and Diazo Groups", Wiley, London–New York, 1978.
2 Szele, I., and Zollinger, H., *Helv. Chim. Acta*, **61**, 1721 (1978).
3 Szele, I., and Zollinger, H., *J. Am. Chem. Soc.*, **100**, 2811 (1978).
4 Hashida, Y., Landells, R. G. M., Lewis, G. E., Szele, I., and Zollinger, H., *J. Am. Chem. Soc.*, **100**, 2816 (1978).
5 Vincent, M. A., and Radom, L., *J. Am. Chem. Soc.*, **100**, 3306 (1978).
6 Leshina, T. V., Dushkin, A. V., Shuvaeva, O. V., and Sagdeev, R. Z., *Tezisy Dokl.-Vses. Kont.*, "*Polyariz. Yader Electronov Eff. Magn. Polya Khim. Reakts.*", **1975**, 34; *Chem. Abs.*, **89**, 23438 (1978).
7 Bagal, I. L., Skvortsov, S. A., and El'tsov, A. V., *Zh. Org. Khim.*, **14**, 361 (1978); *Chem Abs.*, **88**, 151696 (1978).
8 Abramov, V. N., Zyk, N. V. and Kazitsyna, L. A., *Vestn. Mosk. Univ., Ser. 2; Khim.*, **1978**, 190; *Chem. Abs.*, **89**, 162856 (1978).
9 Kozlov, V. V., Smirnova, V. G., and Sagalovich, V. P., *Zh. Obshch. Khim.*, **47**, 2244 (1977); *Chem. Abs.*, **88**, 61814 (1978).
10 Mulvey, D. M., DeMarco, A. M., and Weinstock, L. M., *Tetrahedron Lett.*, **1978**, 1419.
11 Todres, Z. V., *Usp. Khim.*, **47**, 260 (1978).
12 Bunnett, J. F., and Traber, R. P., *J. Org. Chem.*, **43**, 1867 (1978).
13 Bunnett, J. F., and Shafer, S. J., *J. Org. Chem.*, **43**, 1873 (1978).
14 Bunnett, J. F., and Shafer, S. J., *J. Org. Chem.*, **43**, 1877 (1978).
15 Wolfe, J. F., Moon, M. P., Sleevi, M. C., Bunnett, J. F., and Bard, R. R., *J. Org. Chem.*, **43**, 1019 (1978).
16 Simig, Gy., Lempert, K., Váli, Zs., Tóth, G., and Tamás, J., *Tetrahedron*, **34**, 2371 (1978).
17 Rajan, S., and Muralimohan, K., *Tetrahedron Lett.*, **1978**, 483.
18 Pinson, J., and Saveant, J.-M., *J. Am. Chem. Soc.*, **100**, 1506 (1978).

[19] Testaferri, L., Tiecco, M., Tingoli, M., Fiorentino, M., and Troisi, L., *J. Chem. Soc., Chem. Commun.*, **1978**, 93.
[20] Benati, L., Montevecchi, P. C., and Tundo, A., *J. Chem. Soc., Chem. Commun.*, **1978**, 530.
[21] Speckamp, W. N., and Köhler, J. J., *J. Chem. Soc., Chem. Commun.*, **1978**, 166.
[22] Morgan, S. F., Rackham, D. M., Swann, B. P., and Turner, S. P., *Tetrahedron Lett.*, **1978**, 4837.
[23] Abe, T., and Ikegami, Y., *Bull. Chem. Soc. Jpn.*, **51**, 196 (1978).
[24] Brooke, G. M., *Aromat. Heteroaromat. Chem.*, **5**, 314 (1977).
[25] Bartoli, G., Todesco, P. E., and Fiorentino, M., *J. Am. Chem. Soc.*, **99**, 6874 (1977).
[26] Epiotis, N. D., and Shaik, S., *J. Am. Chem. Soc.*, **100**, 29 (1978).
[27] Bernasconi, C. F., de Rossi, R. H., and Schmid, P., *J. Am. Chem. Soc.*, **99**, 4090 (1977).
[28] de Rossi, R. H., Pierini, A. B., and Rossi, R. A., *J. Org. Chem.*, **43**, 2982 (1978).
[29] Titsky, G. D., and Shumeiko, A. E., *Org. React. (Tartu)*, **14**, 285 (1977).
[30] Shumeiko, A. E., and Titsky, G. D., *Zh. Org. Khim.*, **14**, 1273 (1978); *Chem. Abs.*, **89**, 107304 (1978).
[31] Freire, H. R., and Miller, J., *J. Chem. Soc., Perkin Trans.* 2, **1978**, 108.
[32] Nadar, P. A., and Gnanasekaran, C., *J. Chem. Soc., Perkin Trans.* 2, **1978**, 671.
[33] Mancini, V., Morelli, G., and Standoli, L., *Gazz. Chim. Ital.*, **107**, 47 (1977); *Chem. Abs.*, **87**, 183775 (1977).
[34] Glaz, A. M., Ivanov, A. V., and Gitis, S. S., *Sintez Analiz i Struktura Organ. Soedin.*, **1977**, 15; *Chem. Abs.*, **89**, 42019 (1978).
[35] Guanti, G., Dell'Erba, C., Pero, F., and Cevasco, G., *J. Chem. Soc., Perkin Trans.* 2, **1978**, 422.
[36] Guanti, G., Dell'Erba, C., Cevasco, G., and Narisano, E., *J. Chem. Soc., Chem. Commun.*, **1978**, 613.
[37] de Oliveira Baptista, M. J. V., and Widdowson, D. A., *J. Chem. Soc., Perkin Trans.* 1, **1978**, 295.
[38] Bunton, C. A., *Pure Appl. Chem.*, **49**, 969 (1977).
[39] Negoro, K., *Sen'i Kako*, **29**, 427 (1977); *Chem. Abs.*, **88**, 49924 (1978).
[40] Bunton, C. A., Savelli, G., and Sepulveda, L., *J. Org. Chem.*, **43**, 1925 (1978).
[41] Mariani, C., Modena, G., and Scorrano, G., *J. Chem. Res.*, **1978**, (S) 392; (M) 4601.
[42] Spitulnik, M. J., *J. Heterocyclic Chem.*, **14**, 1073 (1977).
[43] Balundgi, R. H., and Raju, J. R., *J. Karnatak Univ., Sci.*, **21**, 188 (1976); *Chem. Abs.*, **88**, 104338 (1978).
[44] Lee, H.-W., and Lee, I., *Taehan, Hwahak Hoechi*, **21**, 83 (1977); *Chem. Abs.*, **87**, 167080 (1977).
[45] Golinski, J., and Makosza, M., *Tetrahedron Lett.*, **1978**, 3495.
[46] Makosza, M., Wojciechowski, K., and Jawdosink, M., *Pol. J. Chem.*, **52**, 1173 (1978); *Chem. Abs.*, **89**, 146079 (1978).
[47] Gilman, N. N., Holland, B. C., Walsh, G. R., and Fryer, R. I., *J. Heterocycl. Chem.*, **14**, 1157 (1977).
[48] Sitzmann, M. E., *J. Org. Chem.*, **43**, 3389 (1978).
[49] Sharnin, G. P., Levinson, F. S., and Shapshin, M. I., *Zh. Org. Khim.*, **13**, 1653 (1977); *Chem. Abs.*, **87**, 183672 (1977).
[50] Ivanova, T. M., and Shein, S. M., *Zh. Org. Khim.*, **14**, 578, 1978; *Chem. Abs.*, **89**, 108396 (1978).
[51] Shein, S. M., and Ivanova, T. M., *Teor. Eksp. Khim.*, **13**, 565 (1977); *Chem. Abs.*, **87**, 167137 (1977).
[52] Gololobov, Yu. G., Onys'ko, P. P., and Prokopenko, V. P., *Zh. Obshch. Khim.*, **47**, 2832 (1977); *Chem. Abs.*, **88**, 61753 (1978).
[53] Mutai, K., Kanno, S., and Kobayashi, K., *Tetrahedron Lett.*, **1978**, 1273.
[54] Shawali, A. S., Hassaneen, H. M., and Almousawi, S., *Bull. Chem. Soc. Jpn.*, **51**, 512 (1978).
[55] Beck, J. R., *Tetrahedron*, **34**, 2057 (1978).
[56] Beck, J. R., and Yahner, J. A., *J. Org. Chem.*, **43**, 2048, 2052 (1978).
[57] Williams, F. J., and Donahue, P. E., *J. Org. Chem.*, **32**, 3414 (1977).
[58] Williams, F. J., Relles, H. M., Manello, J. S., and Donahue, P. E., *J. Org. Chem.*, **42**, 3419 (1977).
[59] Williams, F. J., Relles, H. M., Donahue, P. E., and Manello, J. S., *J. Org. Chem.*, **42**, 3425 (1977).

[60] Williams, F. J., and Donahue, P. E., *J. Org. Chem.*, **43**, 250 (1978).
[61] Williams, F. J., and Donahue, P. E., *J. Org. Chem.*, **43**, 255 (1978).
[62] Markezich, R. L., and Zamek, O. S., *J. Org. Chem.*, **42**, 3431 (1977).
[63] Markezich, R. L., Zamek, O. S., and Donahue, P. E., *J. Org. Chem.*, **42**, 3435 (1977).
[64] Olah, G. A., Sakakibara, T., and Asensio, G., *J. Org. Chem.*, **43**, 463 (1978).
[65] Scribner, J. D., *J. Am. Chem. Soc.*, **99**, 7383 (1977).
[66] Schmidhammer, H., and Klötzer, W., *Arch. Pharm.* (*Weinheim, Ger.*), **311**, 664 (1978).
[67] Solodovnikov, S. P., and Tumanskii, B. L., *Izv. Akad. Nauk SSSR, Ser. Khim.*, **1978**, 1678; *Chem. Abs.*, **89**, 162783 (1978).
[68] Ishag, M. I. O., and Moseley, P. G. N., *Tetrahedron*, **33**, 3141 (1977).
[69] Burdon, J., and Parsons, I. W., *J. Am. Chem. Soc.*, **99**, 7445 (1977).
[70] Heaton, A., and Hunt, M., *J. Chem. Soc., Perkin Trans* 1, **1978**, 1204.
[71] Sitzmann, M. E., *J. Org. Chem.*, **43**, 1241 (1978).
[72] Bolton, R., and Sandall, J. P. B., *J. Chem. Soc., Perkin Trans.* 2, **1978**, 137.
[73] Bolton, R., and Sandall, J. P. B., *J. Chem. Soc., Perkin Trans.* 2, **1978**, 141.
[74] Bolton, R., and Sandall, J. P. B., *J. Chem. Soc., Perkin Trans.* 2, **1978**, 746.
[75] Okafo, E. N., and Whittle, E., *Int. J. Chem. Kinet.*, **10**, 591 (1978).
[76] Zupan, M., Šket, B., and Pahor, B., *J. Org. Chem.*, **43**, 2297 (1978).
[77] Stavber, S., and Zupan, M., *J. Chem. Soc., Chem. Commun.*, **1978**, 969.
[78] Lin, T., and Naae, D. G., *Tetrahedron Lett.*, **1978**, 1653.
[79] Kobayashi, Y., and Kumadaki, I., *Acc. Chem. Res.*, **11**, 197 (1978).
[80] Bruce, M. I., Goodall, B. L., and Stone, F. G. A., *J. Chem. Soc., Dalton Trans.*, **1978**, 687.
[81] Wursthorn, K. R., Kuivila, H. G., and Smith, G. F., *J. Am. Chem. Soc.*, **100**, 2779 (1978).
[82] Quintard, J.-P., Hauvette-Frey, S., and Pereyre, M., *Bull. Soc. Chim. Belg.*, **87**, 505 (1978).
[83] Davies, S. G., Green, M. L. H., and Mingos, D. M. P., *Tetrahedron*, **34**, 3047 (1978).
[84] Meyers, A. I., Gabel, R., and Mihelich, E. D., *J. Org. Chem.*, **43**, 1372 (1978).
[85] Meyers, A. I., and Williams, B. E., *Tetrahedron Lett.*, **1978**, 223.
[86] Kienzle, F., *Helv. Chim. Acta*, **61**, 449 (1978).
[87] Kosugi, M., Shimizu, T., and Migita, T., *Chem. Lett.*, **1978**, 13.
[88] King, A. O., Negishi, E., Villani, F. J., and Silveira, A., *J. Org. Chem.*, **43**, 358 (1978).
[89] Zask, A., and Helquist, P., *J. Org. Chem.*, **43**, 1619 (1978).
[90] Plevyak, J. E., and Heck, R. F., *J. Org. Chem.*, **43**, 2454 (1978).
[91] Kurz, M. E., and Hage, G. N., *J. Org. Chem.*, **42**, 4080 (1977).
[92] Arai, S., Hida, M., Yamagishi, T., and Ototake, S., *Bull. Chem. Soc. Jpn.*, **50**, 2982 (1977).
[93] Arai, S., Hida, M., and Yamagishi, T., *Bull. Chem. Soc. Jpn.*, **51**, 277 (1978).
[94] Fujiki, K., *Bull. Chem. Soc. Jpn.*, **50**, 3065 (1977).
[95] Mencarelli, P., and Stegel, F., *J. Org. Chem.*, **42**, 3550 (1977).
[96] Spinelli, D., Consiglio, G., and Noto, R., *J. Heterocycl. Chem.*, **14**, 1325 (1977).
[97] Spinelli, D., Consiglio, G., and Noto, R., *J. Org. Chem.*, **43**, 4038 (1978).
[98] Novi, M., Guanti, G., Sancassan, F., and Dell'Erba, C., *J. Chem. Soc., Perkin Trans* 1, **1978**, 1140.
[99] Mencarelli, P., and Stegel, F., *J. Chem. Soc., Chem. Commun.*, **1978**, 564.
[100] Kito, Y., Namiki, M., and Tsuji, K., *Tetrahedron*, **34**, 505 (1978).
[101] Forlani, L., Todesco, P. E., and Troisi, L., *J. Chem. Soc., Perkin Trans.* 2, **1978**, 1016.
[102] Di Pietro, S., Forlani, L., and Todesco, P. E., *Gazz. Chim. Ital.*, **107**, 135 (1978); *Chem. Abs.* **88**, 61728 (1978).
[103] Politi, M., Cuccovia, I. M., Chaimovich, H., de Almeida, M. L. C., Bonilha, J. B. S., and Quina, F. H., *Tetrahedron Lett.*, **1978**, 115.
[104] Sakurai, R., Hashido, Y., and Matsui, K., *Bull. Chem. Soc. Jpn.*, **51**, 2175 (1978).
[105] Newkome, G. R., and Nayak, A., *J. Org. Chem.*, **43**, 409 (1978).
[106] Newkome, G. R., Nayak, A., Otemaa, J., Van, D. A., and Benton, W. H., *J. Org. Chem.*, **43**, 3362 (1978).
[107] Newkome, G. R., Danesh-Khoshboo, F., Nayak, A., and Benton, W. H., *J. Org. Chem.*, **43**, 2685 (1978).
[108] Taylor, E. C., and Kobylecki, R., *J. Org. Chem.*, **43**, 680 (1978).
[109] Buchan, R., Fraser, M., and Shand, C., *J. Org. Chem.*, **43**, 3544 (1978).
[110] Maury, G., Fkih-Tétouani, S., Arriau, J., and Sauvaître, H., *J. Heterocycl. Chem.*, **14**, 1311 (1977).
[111] Keen, B. T., Radel, R. J., and Paudler, W. W., *J. Org. Chem.*, **42**, 3498 (1977).

[112] Nasielski-Hinkens, R., Pauwels, D., and Nasielski, J., *Tetrahedron Lett.*, **1978**, 2125.
[113] Bradač, J., Furek, Z., Janežič, D., Molan, S., Smerkolj, I., Stanovnik, B., Tišler, M., and Verček, B., *J. Org. Chem.*, **42**, 4197 (1977).
[114] Hand, E. S., and Paudler, W. W., *J. Org. Chem.*, **43**, 2900 (1978).
[115] Medvedeva, M. M., Doron'kin, V. N., Pozharskii, A. F., and Novikov, V. N., *Khim. Geterotsikl. Soedin*, **1977**, 1120; *Chem. Abs.*, **88**, 49900 (1978).
[116] Skonieczny, S., and Ledochowski, A., *Rocz. Chem.*, **51**, 2279 (1977); *Chem. Abs.*, **88**, 120242 (1978).
[117] Meyers, A. I., and Gabel, R. A., *Tetrahedron Lett.*, **1978**, 227.
[118] Giam, C. S., and Hauck, A. E., *J. Chem. Soc., Chem. Commun.*, **1978**, 615.
[119] Compagnon, O., Compagnon, P.-L., and Moise, C., *J. Chem. Res.*, **1978**, (S) 304; (M) 3801.
[120] Sliwa, H., and Tartav, A., *J. Heterocycl. Chem.*, **15**, 145 (1978).
[121] Phillips, R. B., and Wamser, C. C., *J. Org. Chem.*, **43**, 1190 (1978).
[122] Zagulyaeva, O. A., Bukhatkina, N. V., and Mamaev, V. P., *Zh. Org. Khim.*, **14**, 409 (1978); *Chem. Abs.*, **88**, 151697 (1978).
[123] Salickaite, L., *Tezisy Dokl. – Resp. Kont. Molodykh Uch.-Khim., 2nd*, **1977**, 37; *Chem. Abs.*, **89**, 107311 (1978).
[124] Rasmussen, C. A. H., and van der Plas, H. C., *Tetrahedron Lett.*, **1978**, 3841.
[125] Rasmussen, C. A. H., and van der Plas, H. C., *Recl. Trav. Chim. Pays-Bas*, **97**, 288 (1978).
[126] Geerts, J. P., and van der Plas, H. C., *J. Org. Chem.*, **43**, 2682 (1978).
[127] Barczyński, P., and van der Plas, H. C., *Recl. Trav. Chim. Pays-Bas*, **97**, 256 (1978).
[128] Sagitullin, R. S., Gromov, S. P., and Kost, A. N., *Tetrahedron*, **34**, 2213 (1978).
[129] Brouwer, M. S., van der Plas, H. C., and van Veldhuizen, A., *Recl. Trav. Chim. Pays-Bas*, **97**, 110 (1978).
[130] Anderson, G. L., and Broom, A. D., *J. Org. Chem.*, **42**, 4159 (1977).
[131] Counotte-Potman, A. D., and van der Plas, H. C., *J. Heterocycl. Chem.*, **15**, 445 (1978).
[132] Sakamoto, T., and van der Plas, H. C., *J. Heterocycl. Chem.*, **14**, 789 (1977).
[133] Slae, S., and Shapiro, R., *J. Org. Chem.*, **43**, 4197 (1978).
[134] Chow, N. H., Triplett, J. W., Smith, S. C., and Digenis, G. A., *J. Org. Chem.*, **43**, 3411 (1978).
[135] Shipton, M., and Brocklehurst, K., *Biochem. J.*, **167**, 799 (1977).
[136] Bunting, J. N., Lee-Young, P. A., and Norris, D. J., *J. Org. Chem.*, **43**, 1132 (1978).
[137] Bunting, J. N., and Kabir, S. H., *J. Org. Chem.*, **43**, 3662 (1978).
[138] van Bergen, T. J., Mulder, T., van der Veen, R. A., and Kellogg, R. M., *Tetrahedron*, **34**, 2377 (1978).
[139] Bernasconi, C. F., *Acc. Chem. Res.*, **11**, 147 (1978).
[140] Bernasconi, C. F., and Muller, M. C., *J. Am. Chem. Soc.*, **100**, 5530 (1978).
[141] Crampton, M. R., *J. Chem. Soc., Perkin Trans. 2*, **1978**, 343.
[142] Kaminskaya, E. G., Gitis, S. S., Kaminskii, A. Ya., and Grudtsyn, Yu. D., *Zh. Org. Khim.*, **12**, 917 (1976); *Chem. Abs.*, **85**, 20730 (1976).
[143] Buncel, E., and Eggimann, W., *J. Chem. Soc., Perkin Trans. 2*, **1978**, 673.
[144] Doddi, G., Stegel, F., and Tanasi, M. T., *J. Org. Chem.*, **43**, 4303 (1978).
[145] Spinelli, D., Consiglio, G., and Noto, R., *J. Chem. Res.*, **1978**, (S), 242; (M), 2984.
[146] Bersani, S., Doddi, G., Fornarini, S., and Stegel, F., *J. Org. Chem.*, **43**, 4112 (1978).
[147] Bernasconi, C. F., and Gandler, J. R., *J. Org. Chem.*, **42**, 3387 (1977).
[148] Ah-Kow, G., Terrier, F., and Lessard, F., *J. Org. Chem.*, **43**, 3578 (1978).
[149] Okada, K., Matsui, K., and Sekiguchi, S., *Bull. Chem. Soc. Jpn.*, **51**, 2601 (1978).
[150] Okada, K., and Sekiguchi, S., *J. Org. Chem.*, **43**, 441 (1978).
[151] Knyazev, V. N., and Drozd, V. N., *Izv. Timiryazevsk S-kh. Akad.*, **1977**, 203; *Chem. Abs.*, **88**, 36715 (1978).
[152] Knyazev, V. N., and Drozd, V. N., *Zh. Org. Khim.*, **14**, 893 (1978); *Chem. Abs.*, **89**, 109309 (1978).
[153] Knyazev, V. N., Drozd, V. N., and Minov, V. M., *Zh. Org. Khim.*, **14**, 105 (1978); *Chem. Abs.*, **89**, 43360 (1978).
[154] Morozova, T. I., Gitis, S. S., Kaminskii, A. Ya., Glaz, A. I., Mel'nikov, A. I., and Grudtsyn, Yu. D., *Zh. Org. Khim.*, **13**, 1917 (1977); *Chem. Abs.*, **88**, 22716 (1978).
[155] Gitis, S. S., Kaminskii, A. Ya., Mel'nikov, A. I., and Nikitin, N. R., *Zh. Org. Khim.*, **14**, 1343 (1978); *Chem. Abs.*, **89**, 109193 (1978).
[156] Drozd, V. N., and Grandberg, N. V., *Zh. Org. Khim.*, **14**, 1116 (1978); *Chem. Abs.*, **89**, 109188 (1978).

[157] Strauss, M. J., and Bard, R. R., *J. Org. Chem.*, **43**, 3600 (1978).
[158] Strauss, M. J., Palmer, D. C., and Bard, R. R., *J. Org. Chem.*, **43**, 2041 (1978).
[159] Kaminskii, A. Ya., and Gershkovich, I. M., *Zh. Org. Khim.*, **14**, 1341 (1978); *Chem Abs.*, **89**, 108990 (1978).
[160] Savinova, L. N., Gitis, S. S., and Kaminskii, A. Ya., *Zh. Strukt. Khim.*, **18**, 778 (1977); *Chem. Abs.*, **88**, 21613 (1978).
[161] Savinova, L. N., Gitis, S. S., Glaz, A. I. and Gordeeva L. V. *Sintez Analiz i Structura Organ. Soedin.* **1977** 3; *Chem. Abs.*, **89**, 107331 (1978).
[162] Kaválek, J., Macháček, V., Pastrnek, M., and Štěrba, V., *Collect. Czech. Chem. Commun.*, **42**, 2928 (1977).
[163] Kaválek, J., Pastrnek, M., and Štěrba, V., *Collect. Czech. Chem. Commun.*, **43**, 1401 (1978).
[164] Fyfe, C. A., Cocivera, M., and Damji, S. W. H., *Acc. Chem. Res.*, **11**, 277 (1978).
[165] Olah, G., Asensio, G., Mayr, H., and Schleyer, P. v. R., *J. Am. Chem. Soc.*, **100**, 4347 (1978).
[166] Simonnin, M.-P., Pouet, M.-J., and Terrier, F., *J. Org. Chem.*, **43**, 855 (1978).
[167] Kovar, K. A., and Breitmaier, E., *Chem. Ber.*, **111**, 1646 (1978).
[168] Rykowski, A., van der Plas, H. C., and van Veldhuizen, A., *Recl. Trav. Chim. Pays-Bas*, **97**, 273 (1978).
[169] van der Plas, H. C., van Veldhuizen, A., Woźniak, M., and Smit, P., *J. Org. Chem.*, **43**, 1673 (1978).
[170] van der Stoel, R. E., and van der Plas, H. C., *Recl. Trav. Chim. Pays-Bas*, **97**, 116 (1978).
[171] van Schagen, C. G., Grande, H. J., and Müller, F., *Recl. Trav. Chim. Pays-Bas*, **97**, 179 (1978).
[172] Cavazza, M., Veracini, C. A., Morganti, G., and Pietra, F., *J. Chem. Soc., Chem. Commun.*, **1978**, 167.
[173] Cavazza, M., Morganti, G., and Pietra, F., *J. Chem. Soc., Chem. Commun.*, **1978**, 710.
[174] Föhlisch, B., and Welt, G., *Tetrahedron Lett.*, **1978**, 4019.
[175] Boiko, V. N., Shchupak, G. M., and Yagupol'skii, L. M., *Zh. Org. Khim.*, **14**, 670 (1978); *Chem. Abs.*, **89**, 108469 (1978); Yagupol'skii, L. M., Kondratenko, N. V., Alekseeva, L. A., Belous, V. M., Boiko, V. N., Lukmanov, V. G., Popov, V. I., Sambur, V. P., and Shchupak, G. M., *Ref. Dokl. Soobshch. – Mendeleevsk. S'ezd Obshch, Prikl. Khim.*, *11th*, **1975**, 73; *Chem. Abs.*, **88**, 190255 (1978).
[176] Grudtsyn, Yu. D., and Gitis, S. S., *Sint. Anal. Strukt. Org. Soedin.*, **7**, 3 (1976); *Chem. Abs.*, **88**, 135979 (1978).
[177] Onys'ko, P. P., and Gololobov, Yu. G., *Zh. Obshch. Khim.*, **47**, 2480 (1977); *Chem. Abs.*, **88**, 61752 (1978): Onys'ko, P. P., Kasukhin, L. F., and Gololobov, Yu. G., *Zh. Obshch. Khim.*, **48**, 342 (1978); *Chem. Abs.*, **88**, 189536 (1978).
[178] Pozharskii, A. F., Suslov, A. N., and Kataev, V. V., *Dokl. Akad. Nauk SSSR*, **234**, 841 (1977); *Chem. Abs.*, **87**, 183573 (1977).
[179] Artamkina, G. A., Egorov, M. P., Beletskaya, I. P., and Reutov, O. A., *Izv. Akad. Nauk SSSR, Ser. Khim.*, **1978**; 1694; *Chem. Abs.*, **89**, 162872 (1978).
[180] Myakisheva, Z. A., Illarionova, L. V., Gitis, S. S., Kaminskii, A. Ya., and Glaz, A. I., *Sintez, Analiz i Struktura Organ. Soedin.*, **1977**, 11; *Chem. Abs.*, **89**, 128870 (1978).
[181] Moss, R. A., and Jones, M., *Reactive Intermediates*, Wiley, London–Chichester, 1978.
[182] Caubère, P., *Top. Curr. Chem.*, **73**, 49 (1978).
[183] Malcolme-Lawes, D. J., Massey, A. G., and Wickens, D., *J. Chem. Soc., Chem. Commun.*, **1977**, 933.
[184] Netedov, O. M., and D'yachenko, A. I., *Izv. Akad. Nauk SSSR, Ser. Khim.*, **1978**, 1089; *Chem. Abs.*, **89**, 107265 (1978).
[185] van der Plas, H. C., Woźniak, M., and van Veldhuizen, A., *Recl. Trav. Chim. Pays-Bas*, **97**, 130 (1978).
[186] Iida, H., Yuasa, Y., and Kibayashi, C., *J. Am. Chem. Soc.*, **100**, 3598, (1978); *Tetrahedron Lett.*, **1978**, 3817.
[187] Kessar, S. V., Nadir, U. K., Singh, P., and Gupta, Y. P., *Tetrahedron*, **34**, 449 (1978).
[188] Oku, A., and Matsui, A., *Bull. Chem. Soc. Jpn.*, **50**, 3338 (1977).
[189] Kato, M., Okamoto, Y., Chikamoto, T., and Miwa, T., *Bull. Chem. Soc. Jpn.*, **51**, 1163 (1978).

CHAPTER 7

Electrophilic Aromatic Substitution

A. R. Butler

Department of Chemistry, The Purdie Building,
The University, St. Andrews KY16 8ST

General

Electrophilic aromatic substitution has been reviewed.[1,2] Qualitative potential energy surfaces in this type of substitution have been calculated[3] and the reactivity–selectivity principle has been considered in relation to electrophilic substitution in heteroaromatic compounds.[4] The effect of polysubstitution in benzene compounds is not additive[5] and an examination has shown that there are considerable differences in the effect of substituents upon ring substitution and sidechain solvolysis for both benzene compounds and a number of aromatic heterocycles.[6] With bromine hexahelicene (**1**) gives addition products at positions 5/6 and 11/12, but

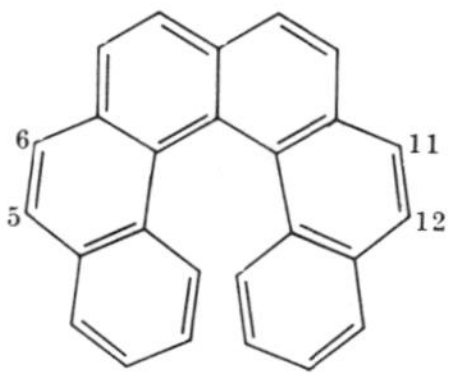

(**1**)

nitration and acetylation result in substitution at positions 5 and 12; the correlation with various reactivity parameters is not very successful.[7] With 2-(2-thienyl)-pyrimidine and 2-(3-thienyl)pyrimidine substitution occurs at all positions in the thiophen ring, indicating unusual directive effects of the pyrimidine group.[8] By using the well-established pyrolysis technique, the positional reactivities of the positions 2 and 3 of benzo[*b*]furan and of all positions of benzo[*b*]thiophen have been determined: for benzo[*b*]furan the order is $3 > 2$, based on rates of eliminative fragmentation of the corresponding 1-arylethyl esters, but this order is reversed in the case of intermolecular electrophilic substitutions for which the transition state more closely resembles the Wheland intermediate; for pyrolysis of each system the benzo-group activates position 3 but deactivates position 2, relative to the parent furan and thiophen; the relevance of the extended selectivity relationship to π-excessive heterocycles has been discussed.[9] The reactions of heteroaromatic cations[10] and 1,2-dialkoxybenzenes have been studied.[11]

Halogenation

A mixture of sulphur monochloride, anhydrous aluminium chloride and sulphuryl chloride has proved to be an effective perchlorinating agent under very mild conditions. With aryl fluorides some replacement of fluorine occurs, and in general, the effect of substituents suggests that the chlorinating agent is electrophilic (SCl_3^+ or SO_2Cl^+).[12] With alkanes and alkenes there is aromatization as well as chlorination; elucidation of the mechanism of this unexpected reaction is awaited with interest.[13] It appears that pyridine is not a significant catalyst in chlorination by molecular chlorine, although its use has been recommended.[14] The presence of micelles in aqueous solution can dramatically affect the position of aromatic halogenation: in general, the *p*-position of phenyl ethers is favoured and, this is readily understood if it is assumed that the substrate is incorporated into the micelle.[15]

From a study of the chlorination of *o*-xylene, indane (**2**), and tetralin it has been concluded that the Mills–Nixon effect is not the only one affecting reactivity of the last-mentioned two compounds.[16] On the other hand, monofluorination of indan by XeF_2/HF occurs at only the β-position.[17] A study of the kinetics of aqueous bromination of benzimidazole (**3**) has shown that attack occurs on the neutral species; the effect of annelation is to reduce the reactivity of the 2-position

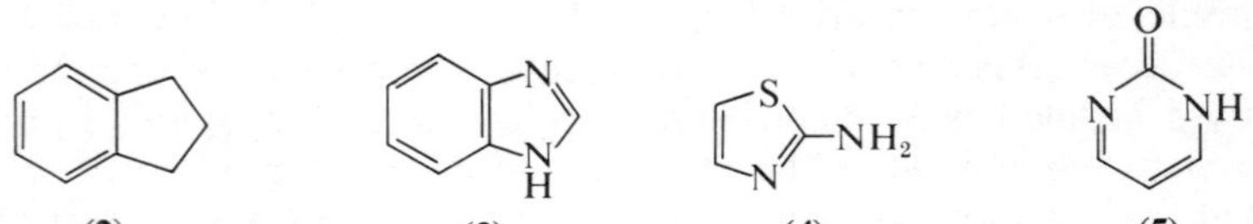

(**2**) (**3**) (**4**) (**5**)

by a factor of 5000.[18] Halogenation of 2-aminothiazole (**4**) occurs at the 5-position, involving an addition–elimination mechanism.[19] Haloarenes may be dehalogenated in the gas phase by reaction with H_3^+; the mechanism involves initial formation of a phenyl cation followed by reaction with H_2, rather than direct reaction with H_3^+. A similar reaction occurs with CH_5^+, but here two pathways, analogous to those described above, are operative.[20] Radiolytically formed D_2T^+ may attack the ring or the halogen of halobenzenes and cause consequent halogen migration.[21]

The nature of the brominating agent in aqueous HOBr has been investigated. Although the data obtained for the catalytic effect of H^+ suggest that H_2OBr^+ is an unlikely intermediate, the alternative mechanism given by the authors is unconvincing because of a lack of direct evidence.[22] Bromophenols undergo intermolecular bromination in the presence of catalytic amounts of aluminium phenoxide; the proposed mechanism is removal of Br^+ from the σ-complex formed from bromophenol and the catalyst.[23] From a kinetic study it has been shown that bromination of pyrimidinones (**5**) results from attack of molecular bromine on the hydrate.[24] Aniline, as the free base, is brominated by both Br_2 and Br_3^- in dilute sulphuric acid.[25]

The halogenations of aromatic amines,[26-28] phenols,[29,30] and phenyltrichlorosilane[31] have been examined.

Reaction of elemental fluorine with aromatic compounds is a violent reaction and leads to polymerization, addition, and sidechain attack; however, when the ratio $[F_2]/[substrate]$ is very low, ring substitution predominates. Fluorine behaves as a regular, but unselective, electrophile and ρ for the reaction is only -2.4; no radical pathway was detected.[32] The kinetics of the isomerization of fluorobenzenes in the presence of $HF–SbF_5$ have been examined.[33]

The nature of the transition state in the reaction of ICl with aza-aromatics has also been discussed.[34]

Hydrogen Exchange

The reagents C_6D_6/BBr_3 and HOT/BBr_3 may be used to label aromatic compounds selectively with D or T; the reactions follow the normal pattern of electrophilic substitution and are of synthetic utility.[35] The formation of spiroenones (*e.g.*(**6**)) from bisphenylpropanes[36] and methoxydiarylalkanes[37] in superacid media (SbF_5/HF) involves diprotonation of the phenyl ring.

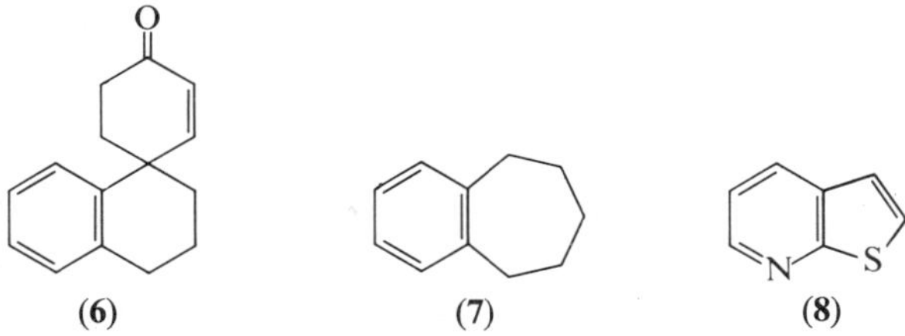

(**6**) (**7**) (**8**)

Calculations show that, as suggested previously,[38] 1,2-hydride shifts in protonated benzene and toluene are possible.[39] In superacid ($HSO_3F/SbF_5/SO_2ClF$) biphenylene undergoes degenerative rearrangement, a reaction also involving hydride shifts.[40]

As in the past, the study of hydrogen exchange has proved of value in determining the relative reactivities of different positions in aromatic compounds. Both aromatic positions in 6,7,8,9-tetrahydrobenzocycloheptene (**7**) are less reactive than the corresponding position in indane (**2**), but only by a small amount.[41] Studies with tetraphenylmethanes show that there is definitely steric hindrance to hydrogen exchange at the *o*-positions, while reaction at the *m*- and *p*-positions is anomalously fast. The Arrhenius plot for these reactions is curved, but this appears to be due to changes in acidity of the reaction medium with temperature.[42] The effect of additional methyl groups on hydrogen exchange in methylthiophens is not additive.[43] Exchange on thieno[2,3-*b*]pyridine (**8**), and related compounds, in acid solution occurs *via* the protonated species.[44] For hydrogen exchange on uracil, 1,3-dimethyluracil, 2,6-dimethoxyuracil, 2,6-dimethoxypyridine, 4-dimethylaminopyridine, and cinnoline there is no unique linear relation to the rates of nitration of the same compounds.[45] Hydrogen exchange on a series of compounds of general formula $Cl_3Si(CH_2)_nC_6H_5$ has been examined: the effect of the $SiCl_3$ group decreases as n increases.[46]

Nitrosation

The mechanism of nucleophilic denitrosation of *N*-nitrosodiphenylamine is attack by the nucleophile (*e.g.* halide or azide) on the conjugate acid of the substrate. However, at high nucleophile concentrations the rate-determining step changes to proton-transfer to the substrate. When the nucleophile is aniline, NO^+ is transferred to the aromatic ring before intramolecular migration to the amino-nitrogen; this report[47] also includes valuable information on the efficacy of traps for free nitrosating agents, the efficacy being $HN_3 > NH_2SO_3H > {}^+NH_3OH$.[47] Products obtained from the competitive nitrosation of pyrrolidine, cresol, and cysteine depend upon the pH and the temperature.[48]

Nitration

Two new nitrating agents of some interest have been reported: activated aromatic compounds may be nitrated by acetone cyanohydrin nitrate with BF_3–ether as catalyst;[49] sodium nitrite in TFA is also effective and the reaction (not nitrosation followed by oxidation) is thought to involve formation of dinitrogen tetroxide which, in acid conditions, forms NO_2^+ and nitrous acid. This reagent will also effect *nitroso*demetallation and this result is further evidence that the occurrence of nitration or nitrosation depends upon the species displaced. Rather surprisingly it has been found that sodium nitrate in TFA is also a nitrating agent.[50] In the nitration of 4-iodo-*o*-xylene the main reaction is nitrodeprotonation, with nitrodeiodination only as a minor pathway, even in the presence of nitrite ions.[51] There are two pathways for the replacement of iodine by a nitro-group: direct nitrodeiodination or nitrosodeiodination followed by oxidation. The predominant pathway depends upon the substrate as both NO^+ and NO_2^+ are present in the reaction mixture. Another factor influencing the form of the reaction is that NO^+ is present at relatively high concentrations, while NO_2^+, which is a much stronger electrophile, is formed at low concentrations in a slow process.[52]

Nitration (as well as deuteriation) of octaethylbilindione (**9**) occurs preferentially at the bridge positions,[53] and some nitramine is formed during the *C*-nitration of

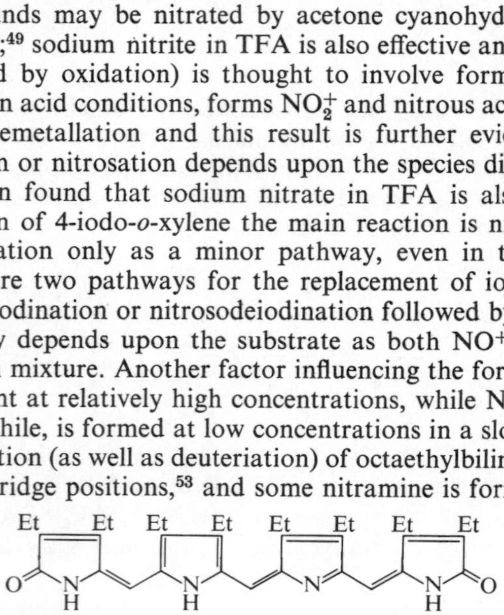

Et Et Et Et Et Et Et Et

(**9**)

Ph—NH—C(=NNO₂)—NH₂ → $Ph{-}NH{-}C(=NNO_2){-}NH_2$

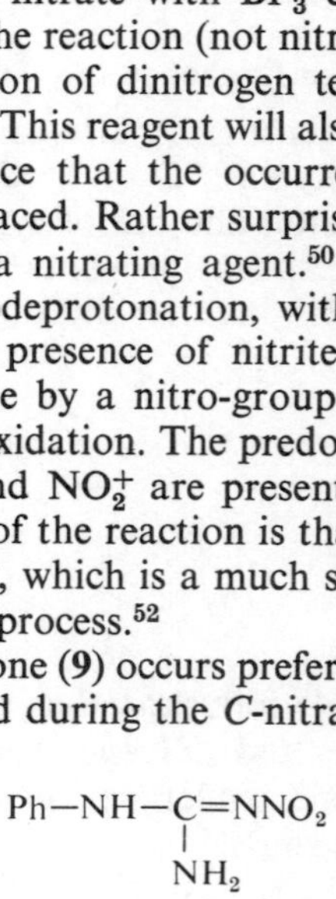

(**10**)

2,4-dinitroaniline.[54] The rearrangement of *N*-nitro-*N*′-phenylguanidine (**10**) to the *o*- and *p*-nitro-compounds is an intermolecular process which involves attack by NO_2^+ on the guanidinium cation.[55] The reactions of substituted 1-phenylbicyclo-[2,2,2]octanes (**11**) are consistent with a π-inductive effect.[56]

The rate of protonation of HNO_3 by *ca.* 78% H_2SO_4 to form nitronium ion and water has been determined, as have the activation parameters for formation of the encounter complex between NO_2^+ and toluene.[57] Nitration of indazole (**12**) in

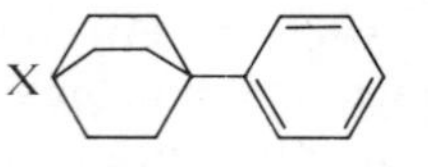

(**11**) 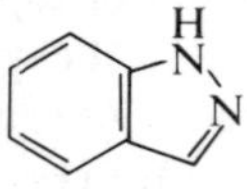 (**12**)

H_2SO_4 of concentration below 90% involves reaction of the conjugate acid.[58, 59] The kinetics of nitration of 2,4-dinitroaniline,[60] benzo[*h*]naphthyridines,[61] and various alkylbenzenes[62] have been examined. It has been suggested that the transition state in aromatic nitration resembles reactants or products according to the degree of substitution in the substrate.[63]

Among the most interesting developments of recent years in the, apparently, never ending study of aromatic nitration is the detection of *ipso*-attack in polysubstituted substrates and the isolation of the resulting adducts. For example, nitration of 4-methylphenol by HNO_3–H_2SO_4 results in 40% *ipso*-attack and the

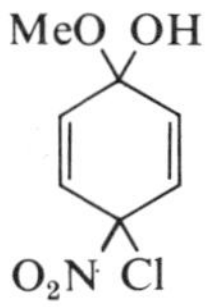

MeO OH / O_2N Cl **(13)** Et NO_2 / Me OAc **(14)** Me NO_2 / Et OAc **(15)**

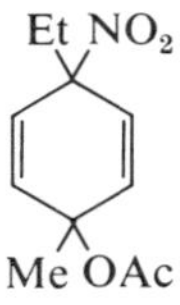

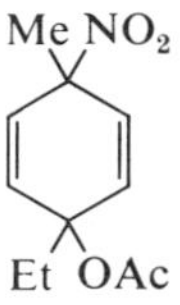

species formed rearranges, by an ionic mechanism, to 4-methyl-2-nitrophenol.[64] The adduct (**13**) is formed on nitration by HNO_3–H_2SO_4 of 4-chloroanisole. Loss of methoxide and migration of NO_2 within a solvent cage lead to the final product, 4-chloro-2-nitrophenol.[65] With acetic anhydride as solvent such adducts are formed very readily. Two diastereoisomers, (**14**) and (**15**), are formed from 4-ethyltoluene.[66] The main adduct formed from 5-*tert*-butyl-1,2,3-trimethylbenzene is (**16**),[67] while 2-chloro-1,3,5-trimethylbenzene gives mainly (**17**), with a little (**18**).[68]

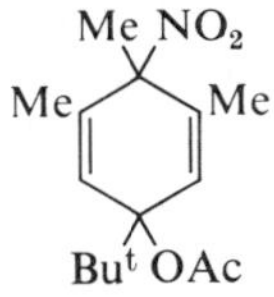

Me NO_2 Me Me / Bu^t OAc **(16)** Cl NO_2 Me Me / Me OAc **(17)** Me O_2N Cl Me H OAc Me **(18)**

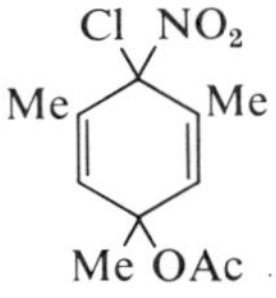

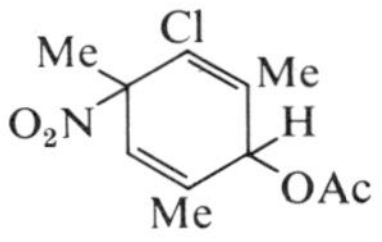

1,4-Adducts appear to form much more readily than 1,2-adducts and even the *tert*-butyl group in (**16**) does not change this. Formation of (**19**) from 1-chloro-2,3-dimethylbenzene indicates the directing effect of the chlorine; rearomatization of (**19**) occurs by loss of HNO_2 or AcOH, depending upon the conditions, and migration of the nitro-group.[69] Nitration of 4-*tert*-butyl-1,2-dimethylbenzene gives (**20**);

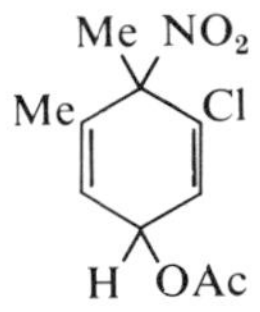

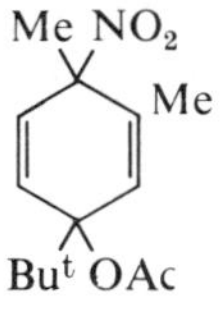

Me NO_2 Me Cl / H OAc **(19)** Me NO_2 Me / Bu^t OAc **(20)** Me NO_2 Me + **(21)**

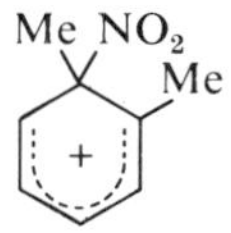

in moist AcOH this loses nitrite to give a cation, but in stronger acids a cation is generated by loss of acetate; basic solvents deprotonate (**20**) to give a triene and there is also nitro-group migration.[70] The rearrangement of (**22**) (obtained by nitration of 4-methylphenyl acetate) to 4-methyl-2-nitrophenol is thought to proceed by a radical dissociation–recombination mechanism (Scheme 1).[71] Rearrangement of (21), to give 1,2-dimethyl-3-nitrobenzene, involves a 1,2-shift of a nitro-group; isotopic labelling showed that a shift to the adjacent substituted position occurs more readily than that to the adjacent unsubstituted position (the latter is required for product formation).[72] The radical cation–radical pair mechanism suggested previously for such rearrangements[73] has been discussed and rejected.

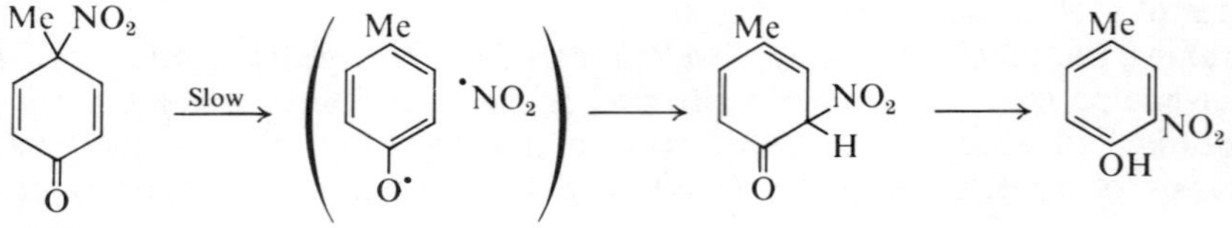

(22)

SCHEME 1

The nitrating mixture HNO_3–$HClO_4$ reacts at the encounter rate with a number of activated substrates. Conversion of HNO_3 into NO_2^+, even in 72% $HClO_4$, is slight. With some very activated substrates ring protonation occurs and this intermediate species is not nitrated.[74] Aromatic nitration is characterized by predominant *o*/*p*-substitution and it has been suggested that this selectivity is retained even when the reaction is diffusion-controlled.[75] In *o*-xylene all the positions are activated but the *o*/*p*-ratio varies with the nitrating agent even when reaction is under apparently diffusion-controlled conditions where, it is claimed, the two positions should be equally substituted. It has also been found that, in the nitration of anisole, there is no increase in the proportion of *m*-substituted product as the reactivity of the nitrating agent is increased. Thes results have led the authors to restate their belief that the primary step in aromatic nitration is formation of a π-complex.[76] They have presented further evidence to support this belief; thus, it has been found that the nitration of benzene and toluene under various conditions always proceeds with predominant attack at the *o*/*p*-positions; they state, "If all nitrations had involved σ-complex type transition states of highest energy then the decrease in substrate selectivity must have been accompanied by lowering *o*/*p* ratios and an increase in *m*-substitution. However, this is not observed." They believe that formation of the π-complex fixes the relative reactivity of substrates and that the *o*/*p*-ratio is fixed by rearrangement of the π-complex to a σ-complex; one would not expect, then, a simple reactivity–selectivity relationship.[77] The interpretation of the experimental results is made difficult by the problem of rapid mixing and in deciding the difference between diffusion-controlled formation of a π-complex and formation of an encounter pair. For a full discussion the reader is referred to two articles in *Accounts of Chemical Research*.[78, 79] The effect of mixing rates upon selectivity in nitration reactions involving nitronium salts in nitromethane has been examined; the results have been analysed in terms of a previously proposed model for mixing.[80]

Azo-coupling Reactions

The effect of the solvent on the rate of reaction of benzenediazonium ions with aromatic amines has been examined.[81] Azopyrroles (**23**) are difficult to prepare in an analytically pure state; however, a heterogeneous reaction between arenediazonium

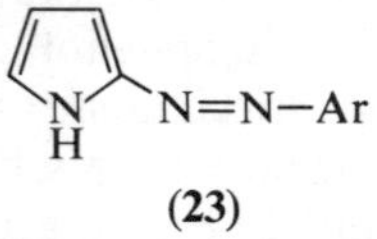

(**23**) 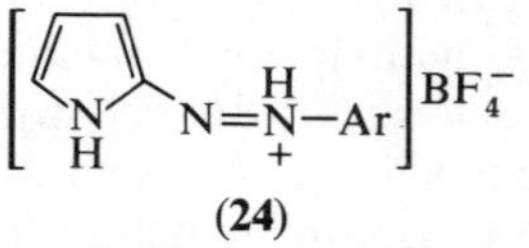 (**24**)

tetrafluoroborate and pyrrole dissolved in chloroform, in the presence of 18-crown-6 (which is known to solubilize diazonium salts[82]), gives quantitative yields of the azopyrrole salts (**24**); from a study of the kinetics of reaction under homogeneous conditions it was deduced that the crown acts as a solid–liquid phase-transfer catalyst.[83] The effect of mixing rates on the selectivity of azo-coupling reactions has also been examined.[84]

Sulphonation

Sulphonation of a number of biphenyl compounds containing deactivating substitutents has been examined and there is evidence for steric hindrance at the *o*-positions. A number of new σ^+ values have been determined.[85] Biphenyldisulphonic acids isomerize extensively in solution in sulphuric acid.[86] Sulphonation of *m*-xylene gives mainly the 5-substituted isomer, for steric reasons;[87] the kinetics of the sulphonation of indole derivatives show that the sulphonating species varies with the acid concentration.[88] Sulphonation of naphthalene derivatives,[89] phenanthrene, fluoroanthene,[90] and hydroxybenzoic acids[91] has been investigated.

An unusual pathway has been demonstrated for the sulphonation of *m*-aminobenzenesulphonic acid; formation of a sulphamic acid occurs and this is sulphonated in its protonated form; aminobenzenepolysulphonic acids can also be obtained.[92] The isomeric composition of the product obtained by the sulphonation of *N,N*-diethylaniline depends upon the amount of SO_3 in the H_2SO_4.[93]

Sulphonation of 4-iodoanisole is a reaction analogous to the Reverdin rearrangement (*i.e.* the product, 2-iodoanisole-4-sulphonic acid, is formed with apparent migration of the iodine from the 4- to the 2-position;[94] the mechanism of the Reverdin rearrangement has been investigated [95] but it is not known if the mechanism here is exactly parallel.

In H_2SO_4 phenols are immediately *O*-protonated but are sulphonated in a much slower reaction; the colours formed on addition of various metalloids are formed only from *O*-protonated phenol.[96]

Friedel–Crafts and Related Reactions

Different aspects of Friedel–Crafts reactions have been reviewed.[97-99] The long-held view that Friedel–Crafts acylations are irreversible has been challenged; in polyphosphoric acid, the fluoro-ketone (**25**) isomerizes to (**26**), probably *via* the

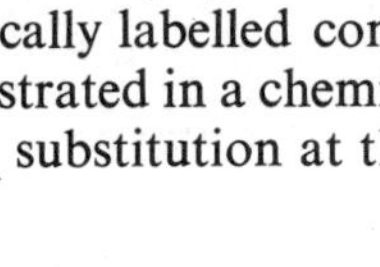

(**25**) 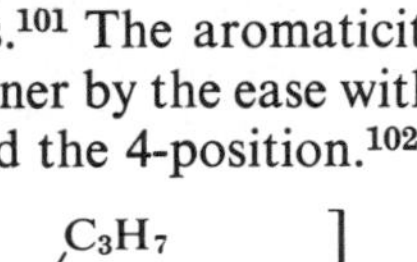 (**26**)  (**27**)

carbocation (**27**),[100] and there is also evidence of reversibility from a study of isotopically labelled compounds.[101] The aromaticity of arsabenzene (**28**) has been demonstrated in a chemical manner by the ease with which it undergoes acetylation to give substitution at the 2- and the 4-position.[102]

(**28**) 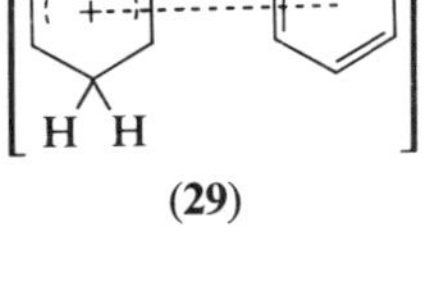 (**29**) 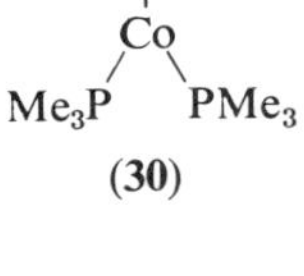 (**30**)

Some less common Friedel–Crafts catalysts have been examined: with BF_3 hydrate both benzylation[103] and ethylation[104] occur with intermediate formation of a free carbocation, but with BF_3 alone the electrophile is a donor–acceptor complex; palladium salts are effective, if expensive;[105] a mixture of $TiCl_4$ and carboxylic acid is milder than $AlCl_3$ and isomerization of alkyl groups does not occur.[106] An interesting development is the use of solid superacids (*eg*.. graphite-intercalated Lewis acids and perfluorinated sulphonic acid resin) in heterogeneous gas-phase reactions.[107]

Vinyl cations are intermediates in the reaction of vinyl triflates with substituted benzenes; the ρ-value for the reaction is unusually low for electrophilic aromatic substitution (−2.57).[108] In the radiolytically induced gas-phase isopropylation of benzene there is extensive ring-proton exchange between the σ-complex and benzene; exchange may occur within a π-complex (**29**).[109] The CF_3^+ ion has been generated radiolytically and is moderately electrophilic;[110] its reactions with benzene and toluene in the gas and the liquid phase have been compared.[111] The free acetylium ion may also be generated in the gas phase and, in reaction with benzene compounds, it shows great preference for sidechain attack (*e.g.* with phenol, phenyl acetate is formed); when ring substitution does occur (with anisole) there is a high *o*/*p* ratio, in sharp contrast to reaction in solution.[112]

There has been an important study of the effect of water upon the kinetics of Friedel–Crafts benzylation of benzene and toluene. The rate depends critically upon the water content, even when it is as low as 0.0006 wt %, but the ratio k_T/k_B remains relatively unaffected; these results throw a question mark over all previous work on these reactions. The situation was found to be simpler with nitromethane as solvent, although only at very low moisture contents. Under these conditions the reaction is of zero order in aromatic substrate, so the rate-determining step must be formation of the electrophile (*i.e.* the benzyl cation); consequently, k_T/k_B ratios are only meaningful if determined competitively.[113] Ring alkylation of the half-sandwich compound (**30**) occurs by initial alkylation of the metal followed by intramolecular migration.[114] Alkylation of benzene may be effected by a cyclo-alkane–alkyl chloride–$AlCl_3$ mixture; the alkyl cation, from the chloride, abstracts a hydride ion from the cycloalkane to generate a carbocation which reacts with benzene.[115] Other intramolecular hydride transfers have been suggested[116] and there have been many other studies of the mechanism of Friedel–Crafts and related reactions.[117–131]

Dibenzofuran may be benzoylated by benzoyl chloride–$AlCl_3$,[132] (3-oxocyclo-hexyl)acetyl chloride acylates anisole,[133] and pyrroles may be aroylated by a Vilsmeier–Haack reaction.[134] The cyclization of 2-phenylethyl isocyanate and related compounds, using Friedel–Crafts catalysts, has been studied.[135] In the conversion of β-allenic bromides into tetralins and related compounds the initial step is addition of H^+ to the allenic chain, followed by reaction with benzene.[136] In the reaction of cinnamoyl chloride–$AlCl_3$ with benzene, alkylation to give (**31**) is favoured over acylation to give (**32**).[137]

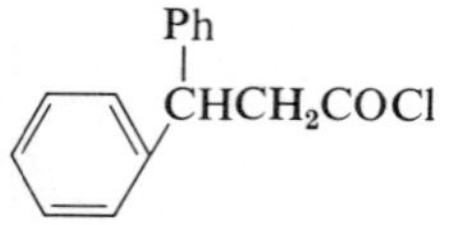

C_6H_5–CH(Ph)CH_2COCl

(**31**)

PhCH=CHCO–C_6H_5

(**32**)

The important carcinogen cyclopenta[*cd*]pyrene (**33**) has been synthesized by intramolecular Friedel–Crafts cyclization by means of $SOCl_2$–pyridine[138] or HF.[139] A major product from the reaction of thiophen-2,3-dicarbonyl chloride with

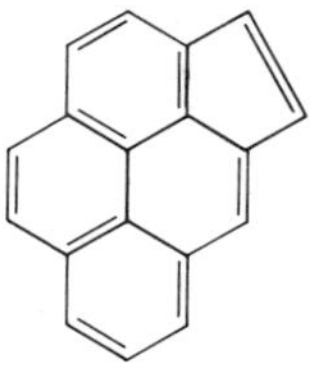

(33) 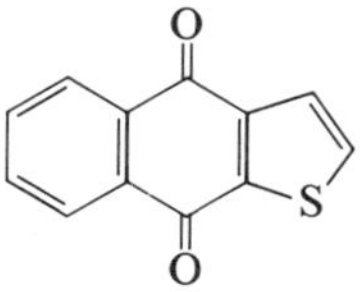 (34)

$AlCl_3$ and benzene is (**34**).[140] Cyclization of 4-(α-naphthyl)butanol to 1,2,3,4-tetrahydrophenanthrene involves initial attack at the *ipso*-position and subsequent migration.[141]

Substitution at the 4-position is the main reaction in the acetylation of 1-methylnaphthalene; there is some methyl migration since ketones derived from 2-methylnaphalene are formed as minor products.[142] There is an unexpectedly large amount of *meta*-attack in the benzylation[143] and allylation[144] of 2,6-dimethylphenol and related compounds; apparently the mechanism does not involve *ipso*-attack and subsequent migration although this sequence has been proposed for the methylation of 2,6-dimethylphenol.[145]

Electrophilic substitution at the 2-position of 3-alkylindoles is thought to occur by initial attack at the 3-position and subsequent migration; further evidence for this comes from a study of the relative migratory aptitudes of alkyl and acyl groups[146] and from the formation of tetrahydrocarbazoles;[147] there is evidence also for a similar mechanism in the acetylation of alkylbenzofuran.[148]

Metallation

By their reactions with azobenzenes it has been established that $MnMe(CO)_5$ and $PdCl_2$ are nucleophilic and electrophilic metalling reagents respectively.[149] Extensive isomerization, of the intermediate σ-complex and of the products formed during mercuration and thallation of arenes render isomer ratios determined at equilibrium meaningless; under kinetically controlled conditions there is predominant *para*-attack and the amount of *meta*-isomer formed is not anomalous.[150] The kinetics of thallation of fluorene by thallium trifluoroacetate in AcOH–H_2SO_4 have been examined.[151]

Reaction between phenyltriethyltin and mercury(II) salts occurs with a low activation energy and *via* a σ-complex.[152] Mercury(II) oxide reacts with 4-bromo- and 4-nitro-phenylhydrazine to give mercuration at the 2-position; the mechanism involves electrophilic attack on the nitrogen of the hydrazine, followed by intramolecular migration.[153]

Miscellaneous Reactions

Alkyl and halo-benzenes may be hydroxylated by reaction with hydrogen peroxide in superacid media at low temperatures; phenols are protonated in superacid and do not react, so hydroxylation ceases once the compound is monohydroxylated.[154] Aminohaloboranes may be used to effect specific *ortho*-hydroxybenzylation and *ortho*-hydroxyalkylation of secondary anilines.[155] Benzene reacts with *O*-phenyl-*N*-tosylhydroxylamine in the presence of TFA to give hydroxybiphenyls; the intermediacy of PhO^+ has been postulated.[156]

The C-atom of (**35**) is susceptible to electrophilic attack and hydrogen exchange must occur *via* the dication (**36**);[157] the chloro-compound (**37**) even undergoes

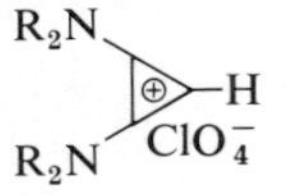

(**35**) 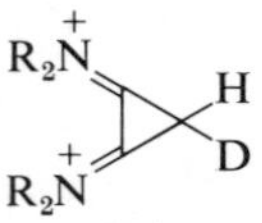 (**36**) 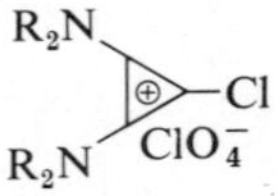 (**37**)

electrophilic substitution of chorine by lithium.[158] The coupling of aryl iodosoacetates with arenes to give diaryliodonium cations has been examined.[159]

The existence of monomeric methyl metaphosphate, which has been postulated as an intermediate in a number of important reactions, has been confirmed by isolation of the expected product (**38**) of electrophilic substitution by methyl

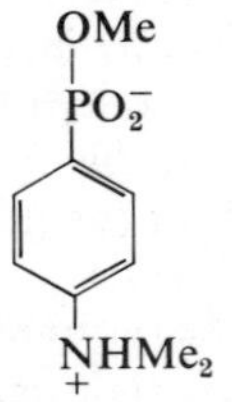

(**38**) 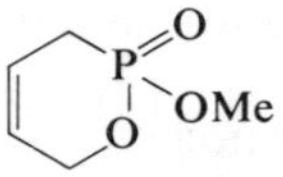 (**39**) 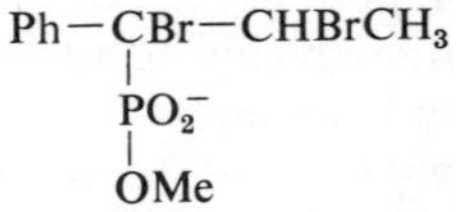 (**40**)

metaphosphate on *N,N*-dimethylaniline; methyl metaphosphate was generated by pyrolysis of methyl 2-butenylphostonate (**39**) in the gas phase, and by warming methyl hydrogen-1-phenyl-1,2-dibromopropylphosphonate (**40**) with triethylamine.[160]

Image dyes of many colour photography systems are formed by reactions involving quinone di-imines. The mechanism of such reactions has been studied by using, as substrates, naphthols modified so as to form micelles; the kinetic study indicated that the quinone di-imine rapidly distributes itself between the micelles and the aqueous solution and reacts at different rates in the two environments.[161] Reaction of the hydroxyguanidine-*O*-sulphonic acid (**41**), and related compounds, with arenes in the presence of $AlCl_3$ gives a substitution product (**42**); the mechanism does not involve formation of a nitrenium ion but proceeds by reaction of the arene with a complex of $AlCl_3$ and (**43**).[162]

$C_4H_9NH{-}C({=}NH){-}NHOSO_3H$ (**41**)

$C_4H_9NH{-}C({=}{}^+NH_2){-}NH{-}Ar$ (**42**)

$C_4H_9{-}NH{-}C^+(NH)(NH)$ (**43**)

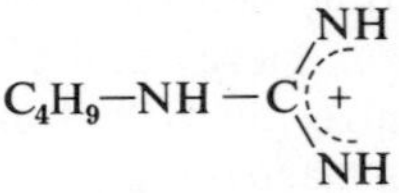

References

[1] Ridd, J. H., *Aromat. Heteroaromatic Chem.*, **5**, 260 (1977).
[2] Lister, J. H., *Aromat. Heteroaromatic Chem.*, **5**, 299 (1977).
[3] Epiotis, N. D., and Shaik, S., *J. Am. Chem. Soc.*, **100**, 29 (1978).
[4] Clementi, S., and Marino, G., *Chem. Scr.*, **11**, 87 (1977).
[5] Katritzky, A. R., *Cron. Chim.*, **53**, 3 (1977); *Chem. Abs.*, **88**, 189535 (1978).

[6] Clementi, S., Fringuelli, F., Linda, P., Marino, G., Savelli, G., Taticchi, A., and Piette, J. L., *Gazz. Chim. Ital.*, **107**, 339 (1977); *Chem. Abs.*, **88**, 36849 (1978).
[7] op den Brouw, P. M., and Laarhoven, W. H., *Recl. Trav. Chim. Pays-Bas*, **97**, 265 (1978).
[8] Gronowitz, S., and Liljefors, S., *Acta Chem. Scand.*, *Ser. B*, **31**, 771 (1977).
[9] Amin, H. B., and Taylor, R., *J. Chem. Soc.*, *Perkin Trans.* 2, **1978**, 1053.
[10] Kost, A. N., *Ref. Dokl. Soobshch. – Mendeleevsk. S'ezd Obshch. Prikl. Khim. 11th*, **2**, 4 (1975); *Chem. Abs.*, **88**, 135879 (1978).
[11] Udrenaite, E., Gineitite, V., Stelbiene, V., and Ruksenas, V., *Tezisy Dokl. – Resp. Konf. Molodykh Uch. – Khim.*, *2nd*, **1**, 36 (1977); *Chem. Abs.*, **89**, 107312 (1978).
[12] Andrews, A. F., Glidewell, C., and Walton, J. C., *J. Chem. Res.*, **1978**, (S) 294, (M) 3683.
[13] Glidewell, C., and Walton, J. C., *J. Chem. Soc.*, *Chem. Commun.*, **1977**, 915.
[14] Dunn, G. E., and Pincock, J. A., *Can. J. Chem.*, **55**, 3726 (1977).
[15] Jaeger, D. A., and Robertson, R. E., *J. Org. Chem.*, **42**, 3298 (1977).
[16] Novrocík, J., Poskočil, J., and Čepčianský, I., *Collect. Czech. Chem. Commun.*, **43**, 1488 (1978).
[17] Šket, B., and Zupan, M., *J. Org. Chem.*, **43**, 835 (1978).
[18] Evans, D. J., Thimm, H. F., and Coller, B. A. W., *J. Chem. Soc.*, *Perkin Trans.* 2, **1978**, 865.
[19] Forlani, L., and Medici, A., *J. Chem. Soc.*, *Perkin Trans* 1, **1978**, 1169.
[20] Leung, H. W., Ichikawa, H., Li, Y.-H., and Harrison, A. G., *J. Am. Chem. Soc.*, **100**, 2479 (1978).
[21] Speranza, M., and Cacace, F., *Proc. Tihany Symp. Radiat. Chem.*, **4**, 209 (1976); *Chem. Abs.*, **88**, 135886 (1978).
[22] Rao, T. S., Mali, S. I., and Dangat, V. T., *Tetrahedron*, **34**, 205 (1978).
[23] Jonkher, P. C., Moen, J., Wolters, J., and Kooyman, E. C., *Recl. Trav. Chim. Pays-Bas*, **97**, 223 (1978).
[24] Banerjee, S., Tee, O. S., and Wood, K. D., *J. Org. Chem.*, **42**, 3670 (1977).
[25] Park, B. B., Park, I. H., Kong, Y. K., and Choi, Q. W., *Taehan Hwahak Hoechi*, **21**, 227 (1977); *Chem. Abs.*, **87**, 200426 (1977).
[26] Radhakrishnamurti, P. S., and Rao; M. D. P., *Indian J. Chem. Sect. B* **15B**, 480 (1977); *Chem. Abs.*, **87**, 167106 (1977).
[27] Radhakrishnamurti, P. S., and Sahu, S. N., *Indian J. Chem.*, *Sect. A*, **15A**, 785 (1977); *Chem. Abs.*, **88**, 61762 (1978).
[28] Radhakrishnamurti, P. S., and Rao, M. D. P., *J. Indian Chem. Soc.*, **54**, 1048 (1977); *Chem. Abs.*, **89**, 59553 (1978).
[29] Radhakrishnamurti, P. S., and Janardhana, C., *Indian J. Chem.*, *Sect.* A, **16A**, 142 (1978): *Chem. Abs.*, **88**, 189559 (1978).
[30] Rajendran, G. P., and Nanjan, M. J., *Indian J. Chem.*, *Sect. B*, **16B**, 248 (1978); *Chem. Abs.*, **89**, 42176 (1978).
[31] Alikhanov, P. P., Motsarev, G. V., and Sakodynskii, K. I., *Dokl. Akad. Nauk SSSR*, **238**, 1351 (1978); *Chem. Abs.*, **89**, 5612 (1978).
[32] Cacace, F., and Wolf, A. P., *J. Am. Chem. Soc.*, **100**, 3639 (1978).
[33] Koptyug, V. A., Buraev, V. I., Isaev, I. S., and Perevyazkina, O. N., *Zh. Org. Khim.*, **14**, 328 (1978); *Chem. Abs.*, **88**, 151786 (1978).
[34] Dasgupta, A., and Rama-Basu, *J. Chim. Phys. Phys. Chim. Biol.*, **74**, 1174 (1977); *Chem. Abs.*, **88**, 120269 (1978).
[35] Long, M. A., Garrett, J. L., and West, J. C., *Tetrahedron Lett.*, **1978**, 4171.
[36] Jacquesy, J.-C., and Jouannetaud, M.-P., *Bull. Soc. Chim. Fr. II*, **1978**, 202.
[37] Geeson, J.-P., Jacquesy, J.-.C, and Jacquesy, R., *Nouveau J. de Chimie*, **1**, 511 (1977).
[38] Cacace, F., *Adv. Phys. Org. Chem.*, **8**, 79 (1970).
[39] Gleghorn, J. T., McConkey, F. W., and Lundy, K., *J. Chem. Res.*, **1978** (S) 418, (M) 4734.
[40] Bodoev, N. V., Mamatyuk, V. I., Krysin, A. P., and Koptyug, V. A., *Izv. Akad. Nauk SSSR*, *Ser. Khim.*, **1978**, 1199; *Chem. Abs.*, **89**, 107534 (1978).
[41] Stroud, M. M. A., and Taylor, R., *J. Chem. Res.*, **1978**, (S) 425.
[42] Ansell, H. V., and Taylor, R., *J. Chem. Soc.*, *Perkin Trans.* 2, **1978**, 751.
[43] Alexander, R. S., and Butler, A. R., *J. Chem. Soc.*, *Perkin Trans.* 2, **1977**, 1998.
[44] Clementi, S., Lepri, S., Sebastiani, G. V., Gronowitz, S., Westerlund, C., and Hörnfeldt, A.-B., *J. Chem. Soc.*, *Perkin Trans.* 2, **1978**, 861.
[45] Katritzky, A. R., Clementi, S., Milletti, G., and Sebastiani, G. V., *J. Chem. Soc.*, *Perkin Trans.* 2, **1978**, 613.

[46] Alikhanov, P. P., Bogatskaya, T. G., Kalinachenko, V. P., Motsarev, G. V., and Yakimenko, L. M., *Zh. Obshch. Khim.*, **48**, 603 (1978); *Chem. Abs.*, **89**, 41836 (1978).
[47] Thompson, J. T., and Williams, D. L. H., *J. Chem. Soc.*, *Perkin Trans* 2, **1977**, 1932.
[48] Davies, R., Massey, R. C., and McWeeny, D. J., *J. Sci. Food Agric.*, **29**, 62 (1978); *Chem. Abs.*, **88**, 189514 (1978).
[49] Narang, S. C., and Thompson, M. J., *Aust. J. Chem.*, **31**, 1839 (1978).
[50] Uemura, S., Toshimitsu, A., and Okano, M., *J. Chem. Soc.*, *Perkin Trans.* 1, **1978**, 1076.
[51] Zweig, A., Huffman, K. R., and Nachtigall, G. W., *J. Org. Chem.*, **42**, 4049 (1977).
[52] Johansson, I., and Olsson, K., *Acta Chem. Scand.*, *Ser. B*, **32**, 297 (1978).
[53] Bonfiglio, J. V., Bonnett, R., Husthouse, M. B., Abdu Malik, K. M., and Naithani, S. C., *J. Chem. Soc.*, *Chem. Commun.*, **1977**, 829.
[54] Bunya, G. F., Abramovich, L. D., Andreeva, L. R., Eremin, A. K., and Vinnik, M. I., *Izv. Akad. Nauk SSSR*, *Ser. Khim.*, **1978**, 304; *Chem. Abs.*, **89**, 128715 (1978).
[55] Heesing, A., and Schmaldt, W., *Chem. Ber.*, **111**, 320 (1978).
[56] Sotheeswaran, S., and Toyne, K. J., *J. Chem. Soc.*, *Perkin Trans.* 2, **1977**, 2042.
[57] Sheats, G. F., and Strachan, A. N., *Can. J. Chem.*, **56**, 1280 (1978).
[58] Austin, M. W., *Chem. Ind.* (*London*), **1978**, 40.
[59] Austin, W. A., *J. Chem. Soc.*, *Perkin Trans.* 2, **1978**, 632.
[60] Burya, G. F., Abramovich, L. D., Andreeva, L. R., Eremin, A. K., and Vinnik, M. I., *Izv. Akad. Nauk SSSR*, *Ser. Khim.*, **1978**, 556; *Chem. Abs.*, **89**, 5624 (1978).
[61] Sliwa, W., *Pol. J. Chem.*, **52**, 271 (1978); *Chem. Abs.*, **89**, 41864 (1978).
[62] Fujishima, S., and Dozen, Y., *Osaka Kogyo Gijutsu Shikensho Kiho*, **28**, 257 (1977); *Chem. Abs.*, **89**, 23394 (1978).
[63] Sohrabi, M., and Kaghazchi, T., *Pak. J. Sci. Ind. Res.*, **19**, 133 (1976); *Chem. Abs.*, **88**, 120436 (1978).
[64] Coombes, R. G., and Golding, J. G., *Tetrahedron Lett.*, **1978**, 3583.
[65] Moodie, R. B., Schofield, K., and Tobin, G. D., *J. Chem. Soc.*, *Chem. Commun.*, **1978**, 180.
[66] Fischer, A., Henderson, G. N., and Thompson, R. J., *Aust. J. Chem.*, **31**, 1241 (1978).
[67] Fischer, A., and Teo, K. C., *Can. J. Chem.*, **56**, 1758 (1978).
[68] Fischer, A., and Seyan, S. S., *Can. J. Chem.*, **56**, 1348 (1978).
[69] Fischer, A., and Greig., C. C., *Can. J. Chem.*, **56**, 1063 (1978).
[70] Fischer, A., and Teo, K. C., *Can. J. Chem.*, **56**, 258 (1978).
[71] Barnes, C. E., and Myhre, P. C., *J. Am. Chem. Soc.*, **100**, 973 (1978).
[72] Barnes, C. E., and Myhre, P. C., *J. Am. Chem. Soc.*, **100**, 975 (1978).
[73] Perrin, C. L., *J. Am. Chem. Soc.*, **99**, 5516 (1977).
[74] Moodie, R. B., Schofield, K., and Thomas, P. N., *J. Chem. Soc.*, *Perkin Trans.* 2, **1978**, 318.
[75] Moodie, R. B., Schofield, K., and Weston, J. B., *J. Chem. Soc.*, *Chem. Commun.*, **1974**, 382.
[76] Olah, G. A., Lin, H. C., Olah, J. A., and Narang, S. C., *Proc. Natl. Acad. Sci. U.S.A.*, **75**, 545 (1978).
[77] Olah, G. A., Lin, H. C., Olah, J. A., and Narang, S. C., *Proc. Natl. Acad. Sci.*, *U.S.A.*, **75**, 1045 (1978).
[78] Olah, G. A., *Accounts Chemical Research*, **4**, 240 (1971).
[79] Ridd, J. H., *Accounts Chemical Research*, **4**, 248 (1971).
[80] Nabholz, F., and Rys, P., *Helv. Chim. Acta*, **60**, 2937 (1977).
[81] Bagal, I. L., Skvortsov, S. A., and El'tsov, A. V., *Zh. Org. Khim.*, **14**, 1244 (1978); *Chem. Abs.*, **89**, 107685 (1978).
[82] Kyba, E. P., Helgeson, R. C., Madan, K., Gokel, G. W., Tarnowski, T. L., Moore, S. S., and Cram, D. J., *J. Am. Chem. Soc.*, **99**, 2564 (1977).
[83] Butler, A. R., and Shepherd, P. T., *J. Chem. Res.*, **1978**, (S) 339; (M), 4471.
[84] Bourne, J. R., Crivelli, E., and Rys, P., *Helv. Chim. Acta.*, **60**, 2944 (1977).
[85] Kortekaas, T. A., Cerfontain, H., and Gall, J. M., *J. Chem. Soc.*, *Perkin Trans.* 2, **1978**, 445.
[86] Kortekaas, T. A., and Cerfontain, H., *J. Chem. Soc.*, *Perkin Trans.* 2, **1978**, 742.
[87] Gnedin, B. G., Baranova, T. A., and Ocheretovji, A. S., *Izv. Vyssh. Uchebn. Zaved.*, *Khim. Khim. Tekhnol.*, **20**, 1800 (1977); *Chem. Abs.*, **88**, 88753 (1978).
[88] Librovich, N. B., Zhiguilin, A. G., Yudin, L. G., Yamashkin, S. A., Maiorov, V. D., Kost, A. N., and Vinnik, M. I., *Zh. Org. Khim.*, **13**, 2032 (1977); *Chem. Abs.*, **88**, 61740 (1978).
[89] Lammertsma, K., Verlaan, C. J., and Cerfontain, H., *J. Chem. Soc.*, *Perkin Trans.* 2, **1978**, 719.
[90] Velichko, L., and Kachurin, O., *Org. React.* (*Tartu*), **14**, 429 (1977).

[91] Mel'nikova, L. P., *Uch. Zap., Yarosl. Gos. Pedagog. Inst.*, **154**, 91 (1976); *Chem. Abs.*, **87**, 200392 (1977).
[92] Maarsen, P. K., Bregman, R., and Cerfontain, H., *J. Chem. Soc., Perkin Trans.* 2, **1977**, 1863.
[93] Khelevin, R. N., *Zh. Prikl. Khim. (Leningrad)*, **50**, 2035 (1977); *Chem. Abs.*, **87**, 200421 (1977).
[94] Muramoto, Y., Asakura, H., and Suzuki, H., *Nippon Kagaku Kaishi*, **1978**, 259; *Chem. Abs.*, **89**, 59387 (1978).
[95] Butler, A. R., and Sanderson, A. P., *J. Chem. Soc., Perkin Trans.* 2, **1971**, 989.
[96] Veselinovic, D. S., Markovic, D. A., and Jovanovic, D. M., *Glas. Hem. Drus, Beograd.*, **43**, 225 (1978); *Chem. Abs.*, **89**, 162722 (1978).
[97] Nakane, R., *Yuki Gosei Kagaku Kyokaishi*, **36**, 440 (1978); *Chem. Abs.*, **89**, 162581 (1978).
[98] Muneyuki, R., and Tanida, H., *Yuki Gosei Kagaku Kyokaishi*, **36**, 454 (1978); *Chem. Abs.*, **89** 162582 (1978).
[99] Oda, R., *Yuki Gosei Kagaku Kyokaishi*, **36**, 467 (1978); *Chem. Abs.*, **89**, 162583 (1978).
[100] Agranat, I., Bentor, Y., and Shih, Y.-S., *J. Am. Chem. Soc.*, **99**, 7068 (1977).
[101] Andreou, A. D., Gore, P. H., and Morris, D. F. C., *J. Chem. Soc., Chem. Commun.*, **1978**, 271.
[102] Ashe, A. J., Chan, T.-W., and Smith, T. W., *Tetrahedron Lett.*, **1978**, 2537.
[103] Takematsu, A., Sugita, K., and Nakane, R., *Bull. Chem. Soc. Jpn.*, **51**, 2082 (1978).
[104] Oyama, T., Hamano, T., Nagumo, K., and Nakane, R., *Bull. Chem. Soc. Jpn.*, **51**, 1441 (1978).
[105] Fujiwara, Y., Asamo, R., and Teranishi, S., *Isr. J. Chem.*, **15**, 252 (1977).
[106] Mach, K., Patzelová, V., Drahorádová, N., and Zelinka, J., *Collect. Czech. Chem. Commun.*, **42**, 2832 (1977).
[107] Olah, G. A., Kaspi, J., and Bukala, J., *J. Org. Chem.*, **42**, 4187 (1977).
[108] Strang, P. J., and Anderson, A. G., *J. Am. Chem. Soc.*, **100**, 1520 (1978).
[109] Yamamoto, Y., Takamuku, S., and Sakurai, H., *J. Am. Chem. Soc.*, **100**, 2474 (1978).
[110] Cipollini, R., Lilla, G., Pep, N., and Speranza, M., *J. Phys. Chem.*, **82**, 1207 (1978).
[111] Cacace, F., and Giacomello, P., *J. Chem. Soc. Perkin Trans.* 2, **1978**, 652.
[112] Giacomello, P., and Speranza, M., *J. Am. Chem. Soc.*, **99**, 7918 (1977).
[113] DeHaan, F. P., Covey, W. D., Anisman, M. S., Ezelle, R. L., Margetan, J. E., Miller, K. D., Pace, S. A., Pilmer, S. L., Sollenberger, M. J., and Wolf, D. S., *J. Am. Chem. Soc.*, **100**, 5944 (1978).
[114] Werner, H., and Hofmann, W., *Angew. Chem. Int. Edn.*, **16**, 794 (1977).
[115] Ndandji, C., Desbois, M., Gallo, R., and Metzger, J., *C. R. Hebd. Séances Acad. Sci., Ser. C*, **285**, 591 (1977).
[116] Inoue, M., Umaki, N., Sugita, T., and Ichikawa, K., *Nippon Kagaku Kaishi*, **1978**, 775; *Chem. Abs.*, **89**, 197528 (1978).
[117] Inoue, M., Sugita, T., and Ichikawa, K., *Bull. Chem. Soc. Jpn.*, **51**, 174 (1978).
[118] Zerkalenkov, A. A., and Kachurin, O. I., *Kinet. Katal.*, **18**, 1043 (1977); *Chem. Abs.*, **87**, 183679 (1977).
[119] Babin, E. P., Nasirova, K. Y., and Pisanenko, D. A., *Visn. Kiiv. Politekh. Inst. Ser. Khim., Mashinobuduv. Tekhnol.*, **14**, 22 (1977); *Chem. Abs.*, **89**, 59432 (1978).
[120] Răzuş, A. C., Arvay, Z., and Glatz, A. M., *Rev. Roum. Chim.*, **23**, 753 (1978).
[121] Alberola, A., Blanco, M., Esteban, S., and Marinas, J. M., *An. Quim.*, **73**, 736 (1977); *Chem. Abs.*, **87**, 200405 (1977).
[122] Gutierrez Jodra, L., Romero Salvador, A., and Munoz Andres, V., *An. Quim.*, **73**, 1485 (1977); *Chem. Abs.*, **89**, 128764 (1978).
[123] Gutierrez Jodra, L., Romero Salvador, A., and Munoz Andres, V., *An. Quim.*, **74**, 121 (1978); *Chem. Abs.*, **89**, 128762 (1978).
[124] Gutierrez Jodra, L., Romero Salvador, A., and Munoz Andres, V., *An. Quim.*, **74**, 128 (1978); *Chem. Abs.*, **89**, 128763 (1978).
[125] Kachurin, O., and Dereza, N., *Org. React. (Tartu)*, **15**, 5 (1978).
[126] Tiwari, R. K., and Sharma, M. M., *Chem. Eng. Sci.*, **32**, 1253 (1977); *Chem. Abs.*, **88**, 120235 (1978).
[127] Plotkina, N. I., Gein, N. V., Kachalkova, M. I., Shevchenko, N. A., and Kolenko, I. P., *Deposited Doc.*, **1976**, 92; *Chem. Abs.*, **88**, 61760 (1978).
[128] Okhrimenko, Z., Katchurin, O., and Chekhuta, V., *Org. React. (Tartu)*, **14**, 311 (1977).
[129] Plotkina, N. I., Gein, N. V., Kachalkova, M. I., and Kolenko, I. P., *Deposited Doc.*, **1976**, 4380; *Chem. Abs.*, **89**, 107335 (1978).

[130] Chavchanidze, D. G., Loladze, N. R., Vashakidze, M. S., Dvalishvili, A. I., Samsoniya, G. G., and Lagidze, R. M., *Tezisy Dokl. – Vses Konf. Khim. Atsetilena, 5th*, **1975**, 382; *Chem. Abs.*, **89**, 59402 (1978).
[131] Urbel, H., Joers, J., and Faingol'd, S. I., *Simp. Sint. Primen. Poverkhn.-Akt. Veshchestev Prom-sti. Kinofotomater*, **1977**, 31; *Chem. Abs.*, **89**, 146046 (1978).
[132] Keumi, T., Shimakawa, S., and Oshima, Y., *Nippon Kagaku Kaishi*, **1977**, 1518; *Chem. Abs.*, **88**, 36840 (1978).
[133] Dabral, V., Ila, H., and Nanda, N., *Chem. Ind. (London)*, **1977**, 952.
[134] White, J., and McGillivray, G., *J. Org. Chem.*, **42**, 4248 (1977).
[135] Davies, R. V., Iddon, B., Suschitzky, H., and Gittos, M. W., *J. Chem. Soc., Perkin Trans. 1*, **1978**, 180.
[136] Santelli, C., and Pèlerin, G., *J. Chem. Res.*, **1978**, (S) 204, (M) 2781.
[137] Shotter, R. G., Johnston, K. M., and Jones, J. F., *Tetrahedron*, **34**, 741 (1978).
[138] Gold, A., Schultz, J., and Eisenstadt, E., *Tetrahedron Lett.*, **1978**, 4491.
[139] Ittah, Y., and Jerina, D. M., *Tetrahedron Lett.*, **1978**, 4495.
[140] MacDowell, D. W. H., and Ballas, F. L., *J. Org. Chem.*, **42**, 3717 (1977).
[141] Jackson, A. H., Shannon, P. V. R., and Taylor, P. W., *J. Chem. Soc., Chem. Commun.*, **1978**, 734.
[142] Gore, P. H., Miri, A. Y., Rinaudo, J., and Bonnier, J.-M., *Bull. Soc. Chim. Fr. II*, **1978**, 104.
[143] McLaughlin, M. P., Creedon, V., and Miller, B., *Tetrahedron Lett.*, **1978**, 3537.
[144] Miller, B., and McLaughlin, M. P., *Tetrahedron Lett.*, **1978**, 3541.
[145] Leach, B. E., *J. Org. Chem.*, **43**, 1794 (1978).
[146] Jackson, A. H., Naidoo, B., Smith, A. E., Bailey, A. S., and Vandrevala, M. H., *J. Chem. Soc., Chem. Commun.*, **1978**, 779.
[147] Ibaceta-Lizana, J. S. L., Iyer, R., Jackson, A. H., and Shannon, P. V. R., *J. Chem. Soc., Perkin Trans. 2*, **1978**, 733.
[158] Baciocchi, E., Cipiciani, A., Clementi, S., and Sebastiani, G. V., *J. Chem. Soc., Chem. Commun.*, **1978**, 597.
[149] Bruce, M. I., Goodall, B. L., and Stone, F. G. A., *J. Chem. Soc., Dalton Trans.*, **1978**, 687.
[150] Olah, G. A., Hashimoto, I., and Lin, H. C., *Proc. Natl. Acad. Sci. U.S.A.*, **74**, 4121 (1977).
[151] Narasimhan, S., Ramani, P. V., and Venkatasubramanian, N., *Int. J. Chem. Kinet.*, **10**, 581 (1978).
[152] Abraham, M. H., and Sedaghat-Herati, M. R., *J. Chem. Soc., Perkin Trans. 2*, **1978**, 729.
[153] Butler, R. N., and O'Donohue, A. M., *Tetrahedron Lett.*, **1978**, 275.
[154] Olah, G. A., and Ohnishi, R., *J. Org. Chem.*, **43**, 865 (1978).
[155] Sugasawa, T., Toyoda, T., Adachi, M., and Sasakura, K., *J. Am. Chem. Soc.*, **100**, 4842 (1978).
[156] Endo, Y., Shudo, K., and Okamoto, T., *J. Am. Chem. Soc.*, **99**, 7721 (1977).
[157] Weiss, R., and Priesner, C., *Angew. Chem. Int. Edn.*, **17**, 445 (1978).
[158] Weiss, R., Priesner, C., and Wolf, H., *Angew. Chem. Int. Edn.*, **17**, 446 (1978).
[159] Hoffelner, H., Schneider, H., and Wendt, H., *Chem.-Ztg.*, **102**, 53 (1978); *Chem. Abs.*, **88**, 189534 (1978).
[160] Satterthwait, A. C., and Westheimer, F. H., *J. Am. Chem. Soc.*, **100**, 3197 (1978).
[161] Tong, L. K. J., and Glesman, M. C., *J. Am. Chem. Soc.*, **99**, 6991 (1977).
[162] Heesing, A., and Steinkamp, H., *Chem. Ber.*, **110**, 3862 (1977).

CHAPTER 8

Carbonium Ions

M. C. GROSSEL

Christ Church, Oxford

Bicyclic and Polycyclic Systems

Derivatives of Norbornane and Related Compounds

It has been shown that the ^{13}C chemical shift of the carbonium-ion centre in 1-arylethyl and 1-methyl-1-arylethyl cations (**1**) does not correlate linearly with σ^{+}.[1] This observation throws doubt on previous conclusions[2] relating to σ-delocalization in 2-aryl-2-norbornyl cations (**2**). An enhanced substituent constant σ^{++}, which

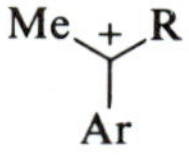

(**1**) R = H or Me 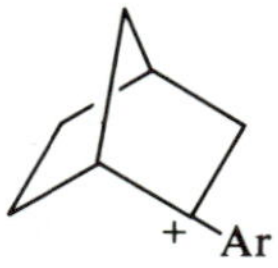 (**2**) 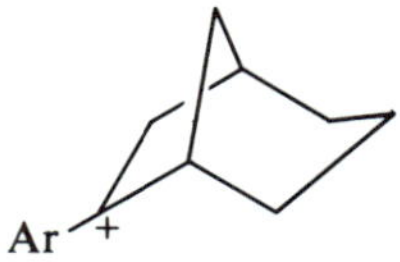 (**3**)

reflects the increased electron demand in stable carbonium ions when compared with solvolysis transition states, can, however, be derived from the data for (**1**). A plot of the C(2) chemical shift in 2-aryl-2-norbornyl cations (**2**) against such substituent parameters (see Figure 1) again shows a marked non-linearity in the chemical-shift variation for the relatively unstable cations, particularly in comparison with the results for a series of 6-arylbicyclo[3.2.1]oct-6-yl cations (**3**).[3]

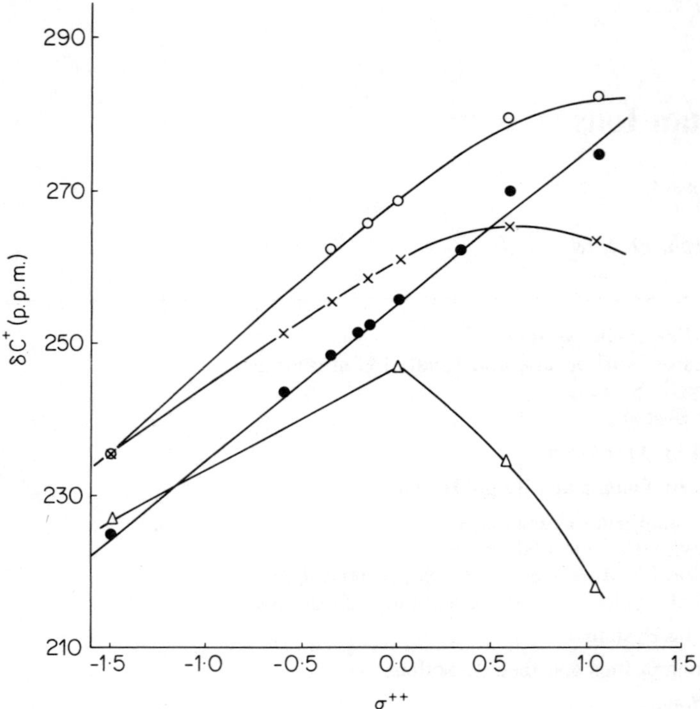

Figure 1. Studies of varying electron demand of arylbicycloalkyl cations: ●(**1**); × (**2**); ○ (**3**); △ 2-arylbicyclo[2.2.1]hept-5-en-2-yl

There is also evidence for the onset of σ-delocalization in 2-aryl-5,6-*exo*-trimethylene-2-norbornyl cations (**4**) but to a somewhat smaller extent, probably because of steric constraints introduced by the trimethylene bridge.[3] In contrast, there is no evidence for σ-participation in the solvolysis of 3-substituted 2-norbornyl tosylates (**5**) despite the fact that an electron-withdrawing C(3) substituent should increase the electron demand of the solvolysis transition state.[4] A similar conclusion is drawn

+Ar
(**4**)

X
OTs
(**5**)

from data relating to the solvolysis of various 2-norbornyl esters in a wide range of solvents: both *exo*- and *endo*-derivatives appear to react by simple unimolecular dissociation.[5] There is also considerable evidence suggesting that there is negligible solvent assistance in the reactions of the *endo*-esters.[5,6] In contrast, the isotopic label in 2-norbornyl trifluoroacetate formed by the addition of CF_3COOD to nortricyclene below 0 °C is thought to indicate an intermediate edge-protonated nortricyclene.[7]

Ab initio calculations suggest that classical and bridged 2-norbornyl cations are of very similar energy in the gas phase.[8] From the geometries predicted here, it has proved possible to simulate ESCA spectra for both structures: the results strongly support the analysis of the experimental spectra in terms of a σ-bridged ion.[9] However, Olah *et al.* have now stated that "σ-delocalized non-classical ions (like the 2-norbornyl cation) are not necessarily static symmetrically bridged species".[10] The enthalpy of formation of the 2-norbornyl cation in solution reveals that it is significantly more stable than simple secondary analogues.[11]

There is evidence for increased charge delocalization to C(1) in 2-aryl-1,7,7-trimethyl-2-norbornyl cations (**6**) arising from steric hindrance to benzylic conjugation. Indeed the phenyl cation (**6**; Ar = Ph) is unstable and readily rearranges

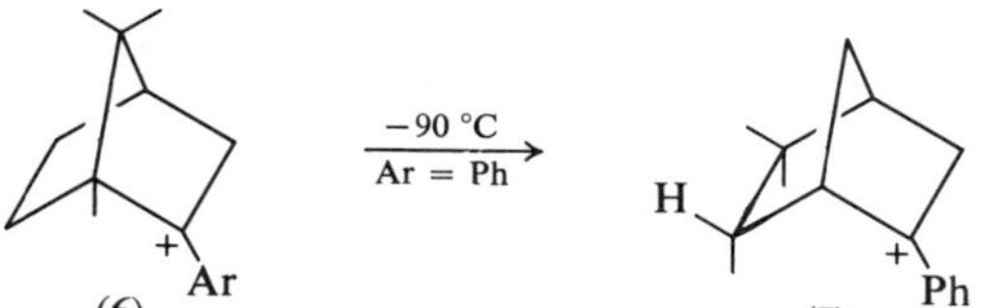

−90 °C, Ar = Ph

(6) (7)

to the 5,5,6-*endo*-trimethyl-2-norbornyl cation (**7**). Varying the electron demand of the carbonium-ion centre reveals that the chemical shift of C(2) in (**6**) is much more sensitive to charge density than that of C(3), when compared with simple 2-aryl-2-norbornyl cations.[12] In contrast, methylation of C(3) as in (**8**) has little effect on the relative charge distribution.[13] The *p*-tolyl cation (**8**; Ar = *p*-MeC_6H_4) rearranges into the fluorosulphonate (**9**) in fluorosulphonic acid at room temperature.[14]

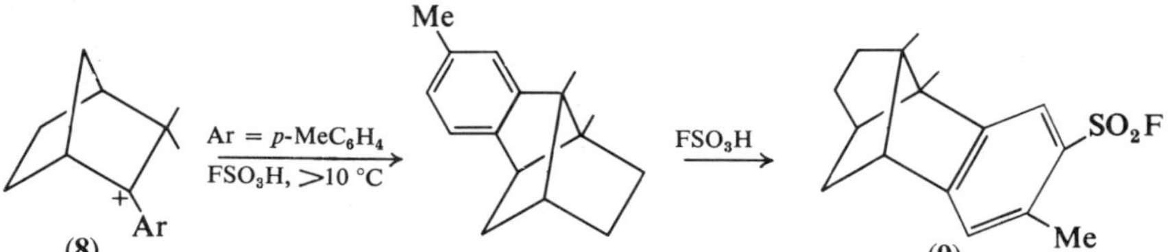

Ar = *p*-MeC_6H_4, FSO_3H, >10 °C; FSO_3H

(8) (9)

2-*endo*-(Dimethylamino)norbornane reacts with methyl iodide 51 times slower than its *exo*-isomer as a result of steric crowding in the reaction transition state involving the former compound.[15] The products arising from the $AgSbF_6$-catalysed dehalogenation of *endo*- and *exo*-3-bromocamphor, and the $HSbF_6$-catalysed decomposition of α-diazocamphor, are determined by the orientation of the leaving group, indicating that the reaction is anchimerically assisted and does not involve an α-keto-carbonium ion: the similarity of the two processes does not support the suggestion that a "hot" carbonium ion results from decomposition of the diazo-compound.[16]

The acid-catalysed hydrolysis of norbornane-2,3-*exo*- and -*endo*-epoxides leads to the formation of classical carbonium-ion intermediates which differ from those formed by similar treatment of nortricyclanol.[17]

A similar distribution of isotopic label is found in the products arising from acetolysis and trifluoroethanolysis of 2-^2H-5-norbornen-2-*exo*-yl *p*-bromobenzene-sulphonate: the reaction proceeds through a nortricyclyl cation and does not require the intervention of equilibrating homoallyl cations. Similar results are found for the less reactive *endo*-ester.[18] Solvolysis of 6,7-dimethoxy-1,2-dimethyl-benzonorbornen-2-*exo*-yl chloride and *p*-nitrobenzoate is anchimerically assisted, leading to an unsymmetrically bridged homobenzylic cation.[19]

The *exo*-ester (**10**) acetolyses at a rate similar to that of the corresponding norbornenyl ester but *ca.* 50 000 times faster than that of its *endo*-epimer (**11**). Both isomers produce a similar range of products, and isotopic labelling studies indicated that the reaction involves a symmetrical ion (**12**) or the corresponding ion pairs.[20]

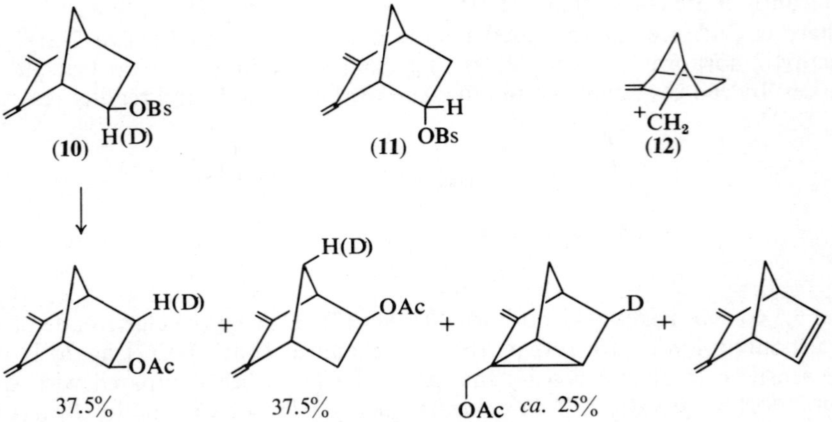

(**10**) (**11**) (**12**)

37.5% 37.5% *ca.* 25%

C(7)-Functionalized norbornenes are formed from the Lewis acid-catalysed rearrangement of methyl 7-(trimethylsilyl)bicyclo[2.2.1]hept-5-ene-2-carboxylate (**13**), or its epoxide (**14**), the reaction being encouraged by β-silyl-stabilization of the product-forming carbonium ion.[21] The results of MINDO/3 calculations on the 7-norbornyl, 2-norbornen-7-yl and 2,5-norbornadien-7-yl cations have been reported.[22]

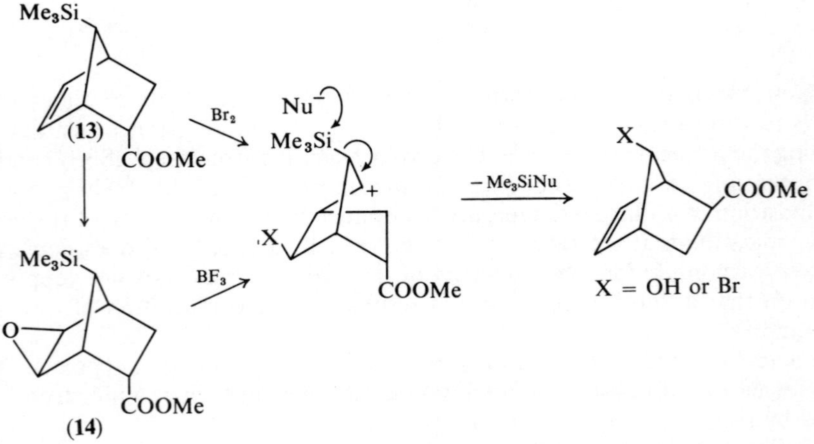

(**13**) (**14**)

X = OH or Br

There is a slight preference for *exo*-attack in the halofluorination of norbornadiene by, for example, *N*-halosuccinimides in the presence of polyhydrofluoric acid and pyridine.[23]

Wagner–Meerwein rearrangement of the 7-oxabicyclo[2.2.1]hept-2-yl cation (**15**) into the 2-oxabicyclo[2.2.1]hept-3-yl cation (**16**) provides a potentially useful stereoselective synthesis of 1,2,3-trisubstituted cyclopentanes.[24]

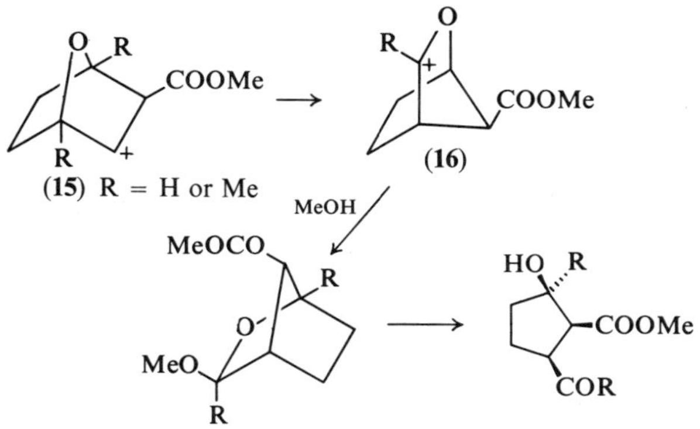

(15) R = H or Me

(16)

Other Bicyclic Systems[25, 26]

The bicyclic phosphetan (**17**) is formed from hydrolysis of the product of the reaction of α-pinene with $MePCl_2/AlCl_3$.[27]

There is evidence to suggest that the solvolysis of *endo*- and *exo*-bicyclo[3.2.1]-oct-3-yl *p*-toluenesulphonates is close to a limiting S_N1 process in all solvents.[28]

9-Chloro-9-methoxybicyclo[4.2.1]nona-2,4,7-triene (**18**) solvolyses in pyridine with high stereospecificity and without skeletal rearrangement at a rate considerably lower than those of its saturated analogues (**19**) and (**20**). It is thought that the

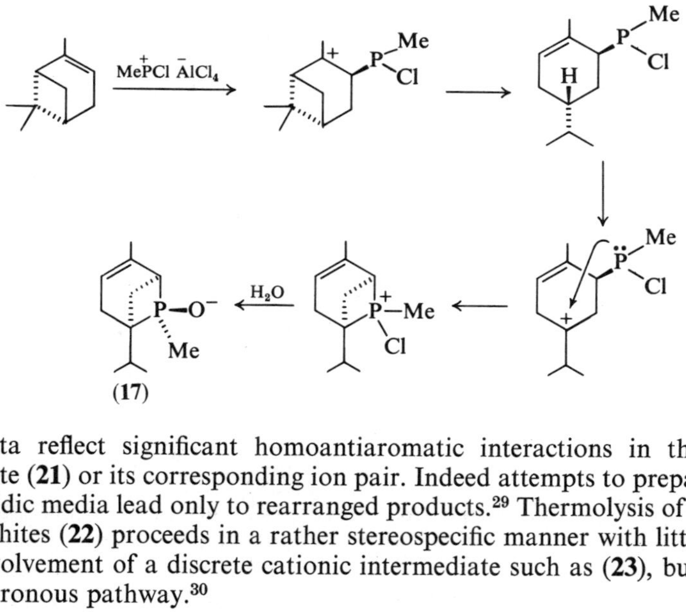

(17)

kinetic data reflect significant homoantiaromatic interactions in the reaction intermediate (**21**) or its corresponding ion pair. Indeed attempts to prepare this ion in superacidic media lead only to rearranged products.[29] Thermolysis of the related chlorosulphites (**22**) proceeds in a rather stereospecific manner with little evidence for the involvement of a discrete cationic intermediate such as (**23**), but probably by a synchronous pathway.[30]

Polycyclic Systems

The tricyclo[4.1.0.0^{2,7}]heptenyl cation (**24**) is too reactive to be observed directly in superacidic media unless stabilized by a hydroxyl substituent (**24**; X = OH); the presence of an aryl substituent results in exclusive formation of the bicyclo-[3.2.0]heptadienyl cation (**25**).[31] Buffered acetolysis of *syn*-tricyclo[4.2.1.1^{2,5}]dec-2-yl

Cl OMe (18) Cl OMe (19) Cl OMe (20) OSO_2Cl (22)

$k_{rel.}$ 4.5×10^{-7} 7.6×10^{-4}

(pyridene/CD_2Cl_2, −80 °C)

SO_2 FSO_3H

(21) R = OMe

(23) R = H

R = OMe → OMe → $-MeOH_2^+$ → + → X^- → X

(24) → X → > −70 °C, X = Me → X

X = Ph → (25) X; X = Me

H OTs (26) → 30% + H OAc 70%

p-toluenesulphonate (**26**) leads to products with the isomeric *anti*-skeleton, the driving force here being relief of strain in going from a boat to a chair cyclohexane.[32]

2,8-*endo*-Trimethylenebicyclo[3.3.0]octane and 2,4-bishomobrexane have been identified as intermediates in the trifluoromethanesulphonic acid-catalysed isomerization of *cis*-2,3-trimethylenebicyclo[3.3.0]octane into 4-homotwistane.[33] There is evidence for a [3,3] sigmatropic shift during the solvolysis of *exo*-tricyclo-[4.4.1.1^{2,5}]dodeca-3,7-dien-11-*anti*-yl tosylate (**27**). A similar process is found for the trienyl analogue (**28**), though the latter ester undergoes a competing fragmentation reaction leading to (**29**).[34]

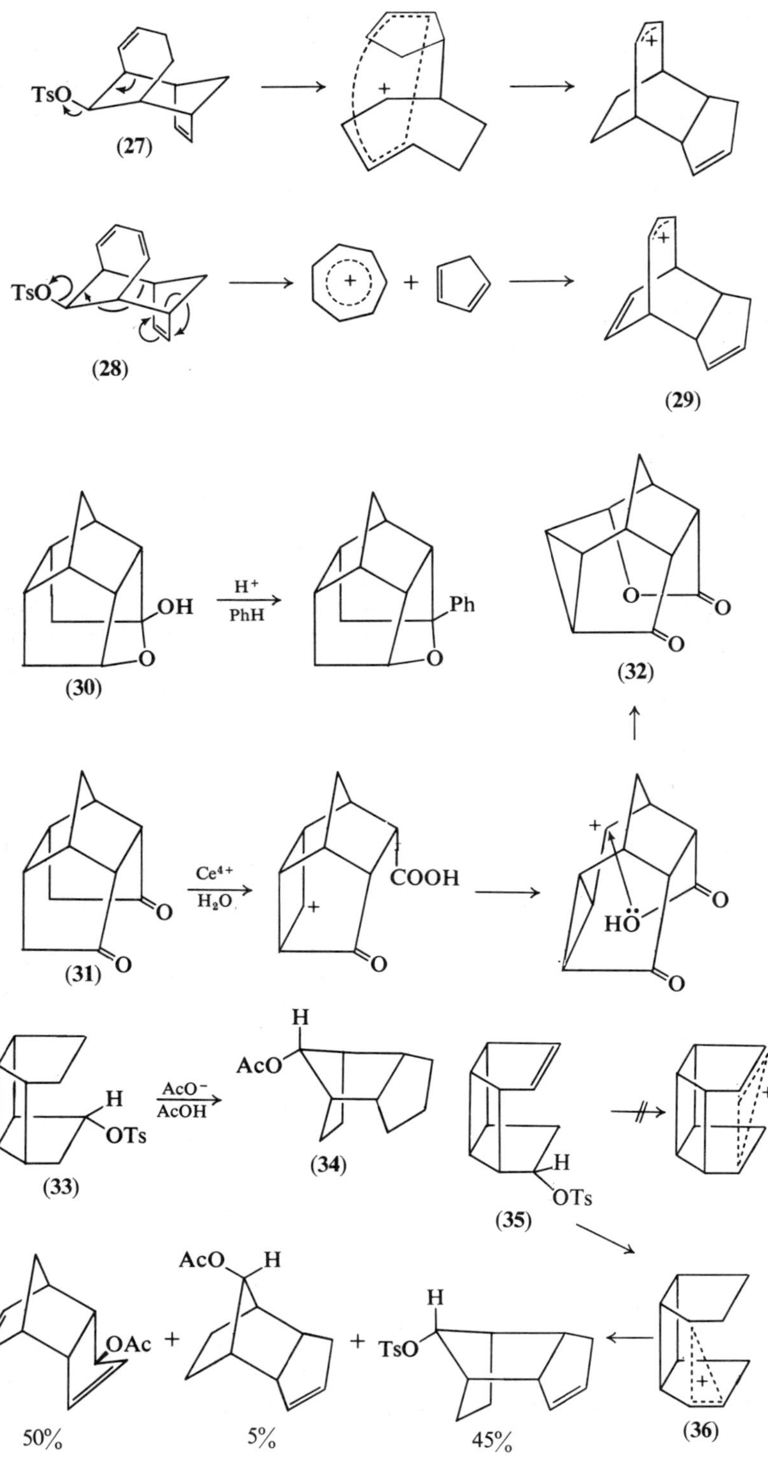
TsO
(27)
TsO
(28)
+
(29)
OH
H+
PhH
Ph
O
(30)
O
(32)
Ce4+
H2O
COOH
HÖ
(31)
AcO
H
AcO−
AcOH
OTs
(33)
(34)
(35)
(36)
AcO
OAc
TsO
50%
5%
45%

The attempted acid-catalysed dehydration of the hemiketal (**30**), a precursor of homohypostrophene, in benzene results instead in alkylation of the solvent.[35] Ceric ion-catalysed oxidation of the related pentacyclic dione (**31**) proceeds *via* a cyclobutyl-cyclopropylmethyl rearrangement leading to the keto-lactone (**32**).[26] Acetolysis of tetracyclo[5.3.0.0^{2,6}.0^{3,10}]dec-4-*exo*-yl tosylate (**33**) occurs 170 times faster (at 71 °C) than that of its *endo*-epimer, affording the rearranged acetate (**34**). Under comparable conditions, the unsaturated *exo*-tosylate (**35**) reacts *ca.* 50 times faster than its *endo*-isomer; in this case, the reaction is assisted by lateral σ-bond delocalization (**36**) which overrides proximate π-electron participation.[37]

Participation by Aryl Groups

A detailed study has been carried out on the effect of solvent on the solvolysis of 2-arylethyl and 2-aryl-1-methylethyl tosylates. In both cases, there is competition between the aryl-assisted and direct solvent-displacement pathways and no cross-over occurs between these two processes.[38] A ^{13}C-NMR tracer method has also been used to investigate the effect of substituents, solvents and added salts, or nucleophiles, on the acetolysis and formolysis of (2-arylethyl) arenesulphonates.[39] Complete inversion of configuration observed in the rearranged products isolated from hydrolysis of (−)-(*R*)-3-methyl-2-phenylbutyl tosylate in aqueous dioxan provides evidence for the intermediacy of a phenonium ion. Similar results are found in the deamination of the corresponding amine, though in this case phenyl, alkyl, and hydride migrations compete, reflecting the lower demand for anchimeric assistance here in comparison with the solvolysis transition state.[40] An intermediate arenium ion is also formed during the halogenation of some 3-arylpropenes.[41]

Participation by Double and Triple Bonds

Whereas studies of the varying electron demand in the solvolysis of 1-aryl-1-(cyclohex-1-enyl)ethyl *p*-nitrobenzoates (**37**) indicate major π-electron participation through an intermediate allyl cation, anchimeric assistance in the reaction of the homoallyl isomers (**38**) is only evident when the aryl group is very powerfully electron-withdrawing; this may indicate rearrangement of the initially formed

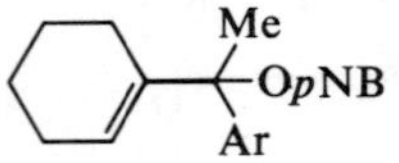

Me OpNB Ar (**37**) Me OpNB Ar (**38**) Me OpNB Ar (**40**)

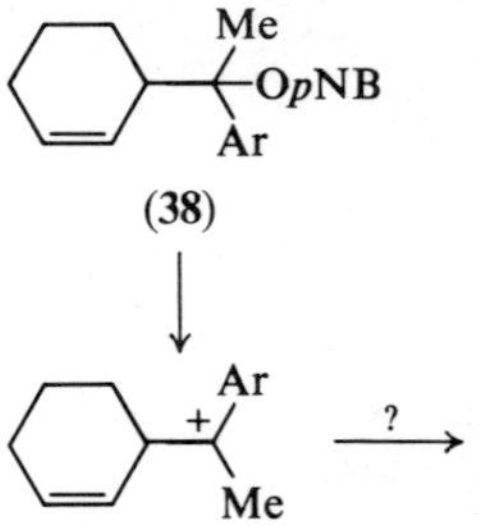

Ar + Me → ? → + Me H Ar (**39**)

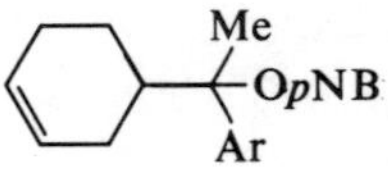

benzylic cation to the more stable allylic action (**39**). There is no evidence for π-electron participation in the cyclohex-3-enyl isomer (**40**).[42] The results from a detailed study of the hydrolysis of *exo*- and *endo*-bicyclo[3.2.1]oct-3-en-2-yl *p*-nitrobenzoate lend support to the proposal that stereospecific return in allyl ion pairs is the result of conformational factors.[43] The acetalation of 2-formyl-3-methoxypropionitrile in methanolic HCl also proceeds through an intermediate allyl cation.[44] A homoallyl cation, generated when $PhPCl_2/AlCl_3$ reacts with 3,3-

dimethylpenta-1,4-diene, is trapped intramolecularly by the phosphorus nucleophile to give a mixture of 6,6-dimethyl-3-oxo-3-phenyl-3-phosphabicyclo[3.1.0]-hexane and 2,2-dimethyl-1-oxo-1-phenyl-3-vinylphosphetane.[45]

The use of pyridinium chlorochromate to initiate oxidative cationic cyclizations of unsaturated alcohols and aldehydes provides a milder and more efficient route to α,β-unsaturated ketones than any previously available.[46] Methyl 14-oxo-labda-8(17),13-diene-4β-carboxylate cyclizes in formic acid to a mixture of pimaric and isopimaric aldehydes.[47] The reactions of the α-terpinyl cation in 85% polyphosphoric acid have also been investigated.[48]

Cyclization of γ-allenyl cations, *e.g.* those form (**41**), occurs exclusively through attack on the allenic CH_2 terminus, *via* a vinyl cation, to afford six-membered ring products.[49]

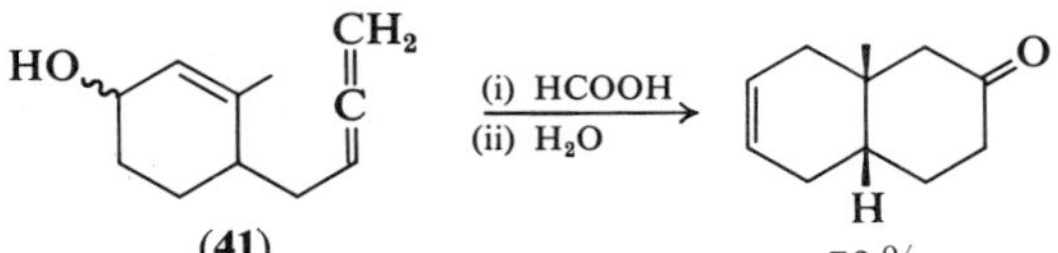

HO ... CH_2=C= ... (i) HCOOH, (ii) H_2O → ... H

(**41**) 70%

Treatment of the dithioketanal function with acid appears to offer a promising initiation of electrophilic polyene cyclizations.[50] 5β-D-Homoandrost-1-en-17-one has been synthesized by cyclization of (**42**), the trimethylsilyl-acetylene substituent favouring six-membered ring formation through silyl-stabilization of a β-carbonium ion.[51] Isotopic labelling studies of the biosynthetic route to fomannosin indicate that cyclization of *trans,trans*-farnesyl pyrophosphate leads to humulene and this, after protonation, cyclizes to a protoilludyl cation from which the product is derived by oxidative cleavage of the appropriate bond.[52]

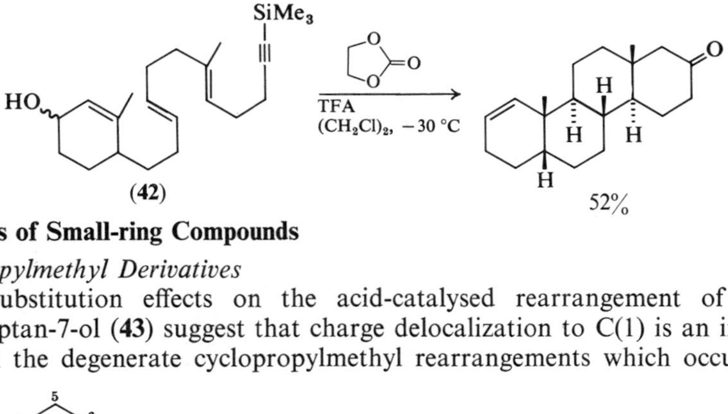

$SiMe_3$ ... HO ... TFA, $(CH_2Cl)_2$, −30 °C → ... H H H H

(**42**) 52%

Reactions of Small-ring Compounds

Cyclopropylmethyl Derivatives

Methyl-substitution effects on the acid-catalysed rearrangement of bicyclo-[4.1.0]heptan-7-ol (**43**) suggest that charge delocalization to C(1) is an important factor in the degenerate cyclopropylmethyl rearrangements which occur in this

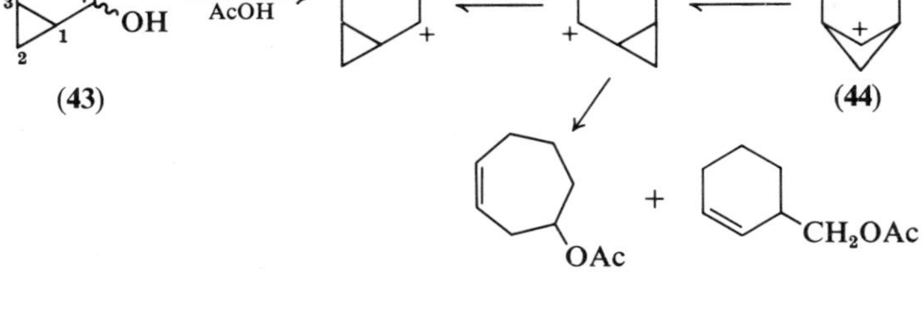

OH ... $HClO_4$, AcOH → ... + ... (**44**)

(**43**) ... OAc + ... CH_2OAc

system and that the latter processes probably involve a puckered cyclobutyl (**44**) or related delocalized cation.[53] The nortriquinacenyl cation (**45**), derived from deamination of the corresponding amine, readily rearranges into the considerably more stable tetracyclo[4.2.1.0^{3,5}.0^{4,9}]nonenyl cation (**46**). This species is also formed during the methanolysis of the corresponding *p*-nitrobenzoate (**47a**) and is preferentially captured from the *exo*-face affording a mixture of methyl ethers (**47b**) and (**48**) (7 : 1).[54]

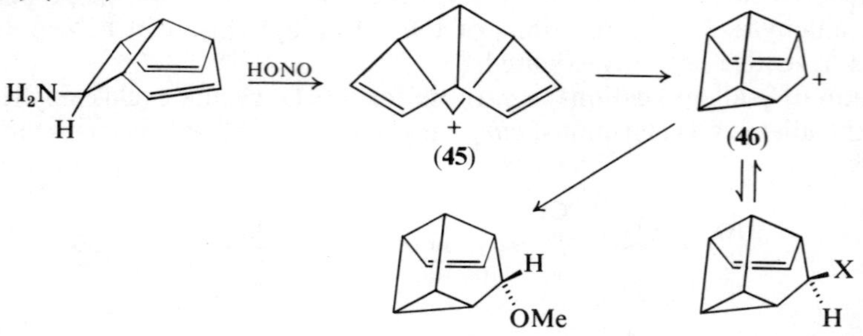

H_2N HONO (**45**) (**46**) (**48**) OMe

(**47**) **a**; X = O*p*NB
b; X = OMe

The hydrolysis of bicyclo[4.3.2]undecatetraen-9-*syn*-yl *p*-nitrobenzoate (**49**) proceeds initially by thermal rearrangement into the tetracyclic ester (**50a**) which then ionizes; there is no evidence for the involvement of the bicyclo[4.3.2]undecatetraenyl cation. The *anti*-tetracyclic and -pentacyclic esters (**50a**) and (**51a**) are both very reactive, ionizing faster, for example, than norbornadien-7-yl *p*-nitrobenzoate, through two distinct homoallyl cations (or ion pairs) which can equilibrate, and which are captured to give a mixture of the corresponding alcohols (**50b**) and (**51b**).[55] An intermediate cyclopropylmethyl cation is also implicated in the conversion of oxaspiropentanes into cyclobutanones *via* (**52**).[56]

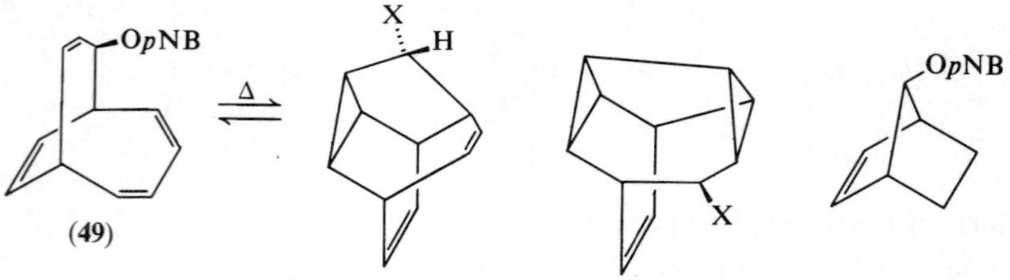

O*p*NB Δ (**49**)

(**50**) **a**; X = O*p*NB (**51**) **a**; X = O*p*NB
b; X = OH **b**; X = OH

$10^6 k^{OpNB}$/s^{-1} 426 1210 6.3

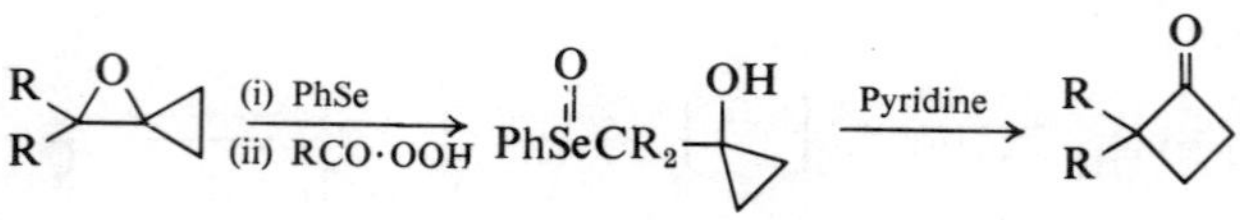

(i) PhSe (ii) RCO·OOH $PhSeCR_2$ OH Pyridine

(**52**)

Participation by More Remote Cyclopropane Rings

There is evidence for some cyclopropyl participation in the trifluoroethanolysis of 2-cyclopropylethyl tosylate (**53a**)[57] and its 1-methylated analogue (**53b**).[58] Products from the silver ion-catalysed hydrolysis of 4,4-dichlorotetracyclo[3.3.0.0^{2,8}.0^{3,6}]-

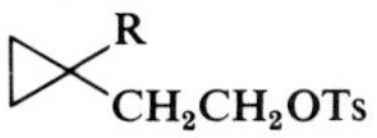

R
CH_2CH_2OTs
(**53**) **a**; R = H
b; R = Me

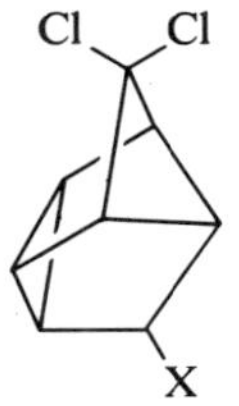

Cl Cl
X
(**54**) X = H or Bu^tO

octanes (**54**) arise from a cyclobutyl-to-cyclopropylcarbinyl rearrangement of the first-formed carbonium ion rather than *via* a trishomocyclopropenium or bishomo-square-pyramidal cation.[59]

Reactions of Cyclopropyl and Cyclobutyl Derivatives

There is evidence that the solvolysis of the propellane triflates (**55**) and (**56**) involves intermediate, essentially classical, bent cyclopropyl cations, rather than the partially ring-opened ions formed from reaction of the corresponding bromides.[60] Methanolysis of the chlorides of *syn*- and *anti*-1,5-bishomocycloheptatriene, (**57**) and (**58**), respectively, involves electrocyclic cyclopropyl ring-opening in concert with ionization with little evidence for π-electron participation. Whilst the *anti*-chloride (**58**) is the more reactive, a similar product mixture is formed in each case, probably from the allyl cation (**59**).[61]

The deamination of 4-aminospirohexane (**60**) leads to a product mixture similar to that obtained from solvolysis of the corresponding chloride or 3,5-dinitrobenzoate, demonstrating that, whilst the reactions of the spirocyclopentyl cation

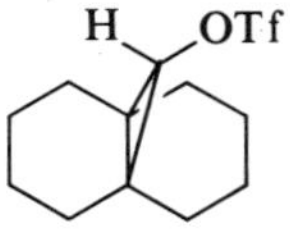

H OTf (**55**) Tf = $CF_3\overset{\prime}{C}O$ H OTf (**56**)

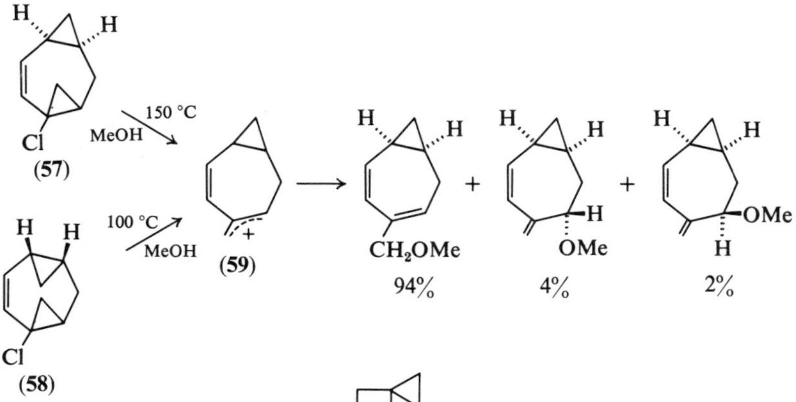

(**57**) Cl, H, H — 150 °C, MeOH → (**59**) ← 100 °C, MeOH — (**58**) Cl, H, H

(**59**) → CH_2OMe 94% + OMe, H 4% + OMe, H 2%

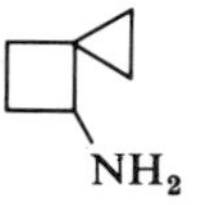

NH_2
(**60**)

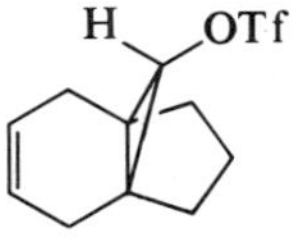

show striking sensitivity to the leaving group, those of the spirocyclohexyl cation do not.[62] The acetolysis of tricyclo[4.2.1.0^{2,5}]non-3-*exo*- and -*endo*-yl *p*-toluenesulphonates, (**61**)–(**64**), appears to involve disrotatory opening of the cyclobutane ring during ionization, as shown by large rate differences between the *inside* and *outside* isomers, and stereospecific product formation. The presence of a potentially interacting π-bond has little effect on the ionization step, apart from inductive retardation, except for the *outside*,*endo*-isomer (**64**) which reacts to give a single rearranged acetate (**65**).[63]

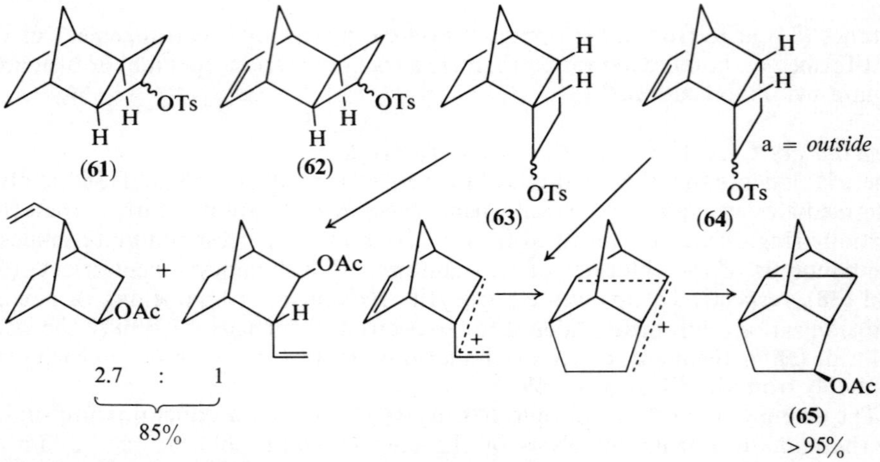

(**61**) (**62**) (**63**) (**64**) a = *outside*

2.7 : 1
85%

(**65**)
>95%

The *syn*- and *anti*-tricyclo[4.2.0.0^{2,5}]oct-7-en-3-*exo*-yl *p*-toluenesulphonates and their dihydro-derivatives are acetolysed some 50 times more slowly than cyclobutyl tosylate, with no evidence for neighbouring-group participation involving proximate π-bonds [in (**66**)] or peripheral cyclobutane bonds in the solvolysis transition state.[64] The esters (**68**) are among the least reactive cyclobutyl derivatives known.[65] A comparison of the rates of acetolysis of *anti*-tricyclo[5.2.0.0^{2,5}]non-3-en-6-*endo*-yl *p*-toluenesulphonate (**69**), its 6-*exo*-epimer, and their saturated analogues shows that homoallyl participation is much more important than σ-cyclobutyl

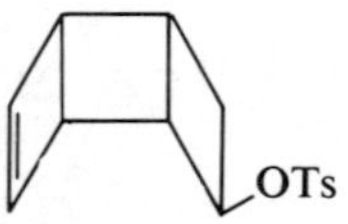

(**66**) 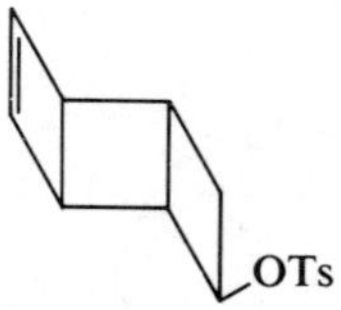 (**67**) 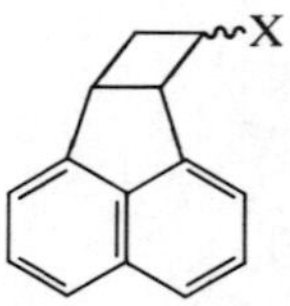 (**68**) X = OSO_2Me or OSO_2CF_3

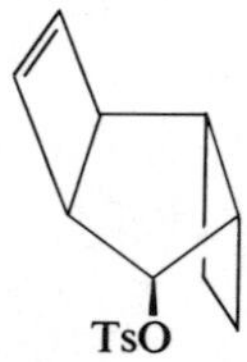

(**69**) *endo*

participation here, though the former is apparently much more sensitive to conformational factors.[66] Such effects are also thought to be important in determining the solvolytic reactivity of the *p*-toluenesulphonates (**70**) and (**71**) relative to (**72**).[67] Ring expansion of a cyclobutylmethyl cation has also been observed in the BF_3-etherate-catalysed rearrangement of a mixture of α-7,8-protoilludene oxides.[68]

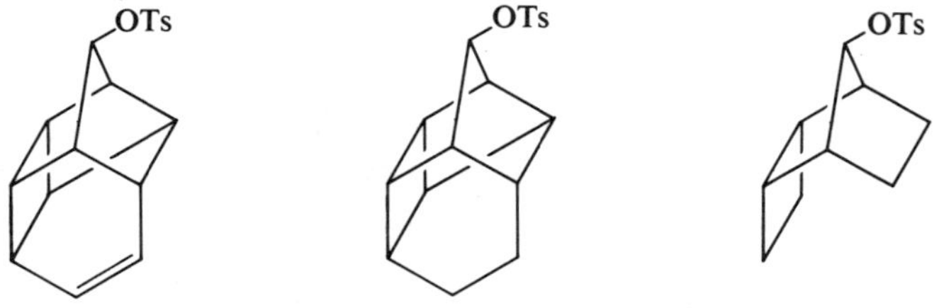
OTs
OTs
OTs

	(**70**)	(**71**)	(**72**)
$k_{rel.}^{100°}$ (AcO⁻/AcOH)	55	106	1

Organometallic Systems

Rules have been proposed which allow prediction of the most favourable position for nucleophilic attack on 18-electron organotransition-metal cations containing unsaturated hydrocarbon ligands.[69]

(Allyloxy)Fe(II) cations, *e.g.* (**73**), undergo a clean cationic [3+2]-cycloaddition to aryl-substituted olefins. This process provides a sterospecific route to 3- and 4-substituted cyclopentanones.[70] Two of the products obtained from the reaction of allyloxy-Fe(II) cations (**74**) with 2-methylpropene are derived from an ene reaction.[71] The kinetics of addition of tertiary phosphines to dicarbonyl(cyclobutadiene)-nitrosyl iron cations and tricarbonyl(cycloheptatriene)manganese cations have been studied.[72]

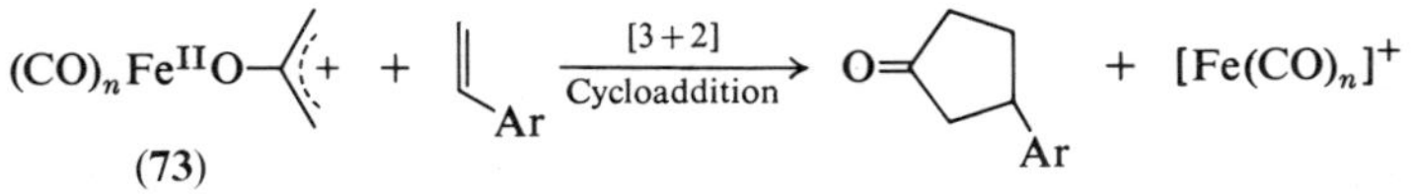
(CO)ₙFeᴵᴵO
Ar
[3+2]
Cycloaddition
O
Ar
+ [Fe(CO)ₙ]⁺

(**73**)

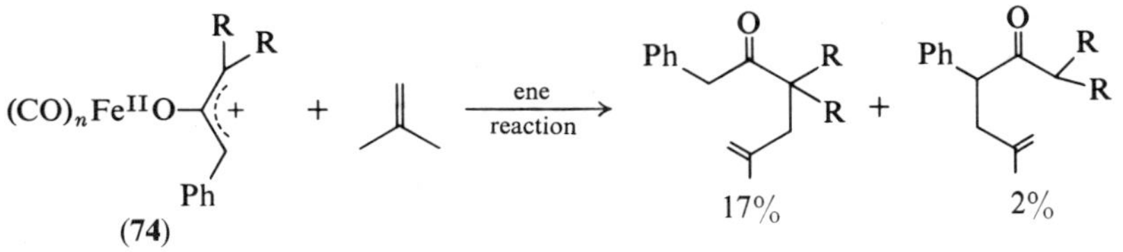
R
R
(CO)ₙFeᴵᴵO
Ph
ene
reaction
Ph
O
R
R
Ph
O
R
R
17%
2%

(**74**)

syn,syn- and *syn,anti*-Dienyltricarbonyliron cations are formed stereospecifically when the corresponding *ψ-exo,trans*- and *ψ-endo,trans*-dienoltricarbonyliron complexes react with FSO_3H at −60 °C.[73] A tricarbonylpentadienyliron cation is also formed from tricarbonylcyclooctatetraeneiron in FSO_3H, rather than a tricarbonyl(homotropylium)iron species.[74] Whilst *exo*-methyl groups [at C(7) and C(10)] suppress the addition of nucleophiles to the angular terminus, C(1), of tricarbonyl{1,3-6-η-4-methoxybicyclo[4.4.0]deca-3,5-dienylium}iron hexafluorophosphate, *endo*-substituents cause the formation of mixtures of angularly and non-angularly substituted products.[75] Only three of the tricarbonyliron complexes formed from the methyl esters of the various methoxy-1,4-dihydrobenzoic acids undergo hydride abstraction when treated with trityl tetrafluoroborate: all three pentadienyliron cations thus prepared react with sodiodimethylmalonate at the unsubstituted dienyl terminus.[76] A tricarbonylpentadienyliron cation, substituted at the termini, is attacked by water preferentially at the site bearing the smallest substituent.[77]

The formation of the vinyl cation (**75**) is the rate-determining step in the acid-catalysed hydration of ethynylferrocenes.[78] The crystal structure of diferrocenyl-methylium tetrafluoroborate reveals a *transoid*-conformation in which there is evidence for a metal-exocyclic carbon interaction.[79]

The reaction of iodide ion with $[M(C_7H_7)(CO)_3]^+BF_4^-$ salts, where M = Mo or W, involves initial formation of a metal–iodide bond after which iodide is transferred to the tropylium ring.[80] The reactions of tricarbonylchromium-complexed benzyl cations[81] and of bis(tricarbonylchromium)benzhydryl cations[82] with alcohols and amines have also been studied. INDO calculations suggest that the remarkable stability of benzyltricarbonylchromium cations arises from backbonding from the $d_{x^2-y^2}$ orbital of chromium into the non-bonding π-orbital of the benzyl ligand.[83]

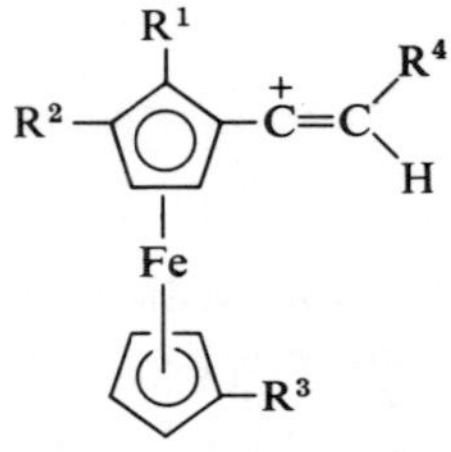

(**75**) R = H, Me, or Bu^t

Extended Hückel calculations favour a non-centred structure for cations of the type $(CO)_9Co_3CCH_2^+$.[84] The configuration of $Co_2(CO)_6$ complexes with α-ethynyl-carbonium ions has been studied.[85]

σ-Cyclobutadiene–aluminium complexes, *e.g.* (**76**), are readily formed when dialkylacetylenes react with $AlCl_3$. There is evidence for degenerate migration of $AlCl_3$ around the ring, here mainly through a series of 1,2-shifts.[86] Initiation of polymerization of cationically polymerizable monomers by organoaluminium complexes has also been investigated.[87]

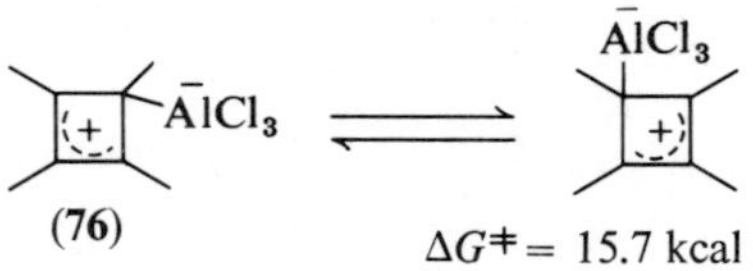

(**76**) $\Delta G^{\ddagger} = 15.7$ kcal mol^{-1}

Stable Carbonium Ions and their Reactions[88–91]

A calorimetric method has been described for measuring the enthalpies of reaction of alkyl halides with SbF_5 in a variety of solvents.[92] The reaction with tetrabutylammonium chloride under these conditions has also been studied. There is good correspondence between results of this method and those previously reported for the corresponding gas-phase reactions, though the 2-norbornyl cation is found to be rather less sensitive to substituent effects than are other related ions.[93] The effect of solvent nucleophilicity on the complex formation by and ionization of C_1–C_5 alkyl fluorides, when these are treated with SbF_5 and AsF_5, has been studied. As expected, whilst, for example, CH_3F/SbF_5 alkylates both SO_2 and SO_2ClF, only complex formation is observed in SO_2F_2.[94]

The rates of degenerate 1,2-hydride and methyl shifts ($\Delta G^{\ddagger} = ca.$ 3–4.5 kcal mol^{-1}) in several substituted 1-methylpropyl cations have been measured by ^{13}C-NMR spectroscopy.[95]

Methyl and ethyl fluoride react with HF/TaF_5 to form the corresponding alkyl cations: these abstract a hydride ion to form the corresponding alkane which then undergoes condensation and polymerization reactions.[96] Cycloalkanes undergo very selective acid-catalysed hydrogenolysis and isomerization reactions in HF/TaF_5 in the presence of molecular hydrogen.[97] Carbonium ions generated by protonation of several cyclohexenones in HF/SbF_5 are reduced by reaction with methylcyclopentane or cyclohexane.[98] If enone and hydrogen donor are added simultaneously, kinetically-controlled reduction is observed.[99]

There is evidence that addition of the *tert*-butyl cation to 2-methylpropene at low temperatures is reversible.[100] The 1,1,3,3,-tetramethylbutyl cation formed in this addition reaction is also an intermediate in the reaction of "iso-octane" with carbon monoxide in HF/SbF_5 at room temperature: once again products derived from the fragmentation of this species are observed.[101]

^1H- and ^{13}C-NMR spectra of the cyclodecyl cation[102] and its 1,6-dimethyl derivative[103] suggest that these ions have a symmetrical μ-hydrido-bridged structure (**77**). The ^{13}C—^1H apical coupling constant in (**78**) approximates to the value expected for an *sp*-hybridized carbon atom and is thought to be strong evidence in favour of a pyramidal bridged structure in this species.[104]

The ring contractions and expansions of a series of tertiary cycloalyl cations (**79**; n = 3–20) have been investigated. The carbonium-ion centre functions as a slip knot here and the maximum ring size that can be formed is controlled by the length of the alkyl side-chain. For smaller rings ($n \leqslant 10$) the alkylcyclohexyl cation is the

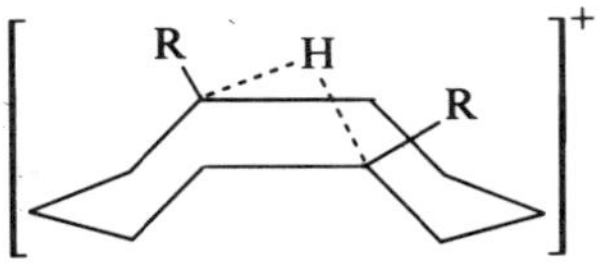

(**77**) R = H or Me

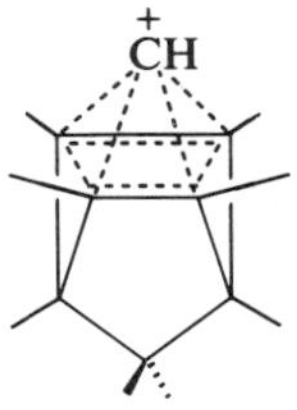

(**78**)

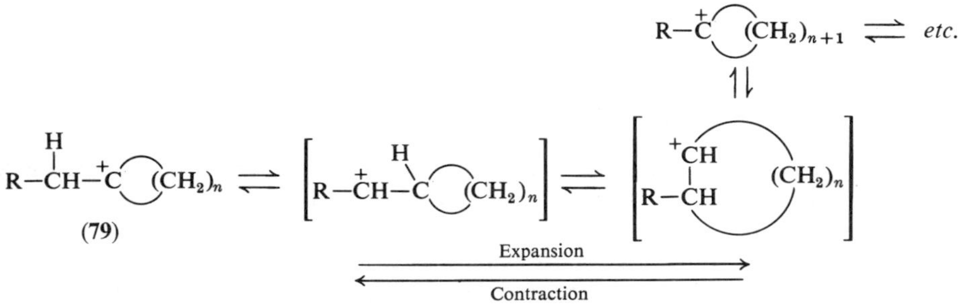

$$R\text{–}\overset{+}{C}(CH_2)_{n+1} \rightleftharpoons etc.$$

$$R\text{–}CH(H)\text{–}\overset{+}{C}(CH_2)_n \rightleftharpoons \left[R\text{–}\overset{+}{C}H\text{–}C(H)(CH_2)_n\right] \rightleftharpoons \left[\overset{+}{C}H(R\text{–}CH)(CH_2)_n\right]$$

(**79**)

Expansion

Contraction

$12.3 \leqslant \Delta G^{\ddagger} \leqslant 15.4$ kcal mol^{-1}

thermodynamically favoured structure, but for larger rings ($n \geqslant 11$) rearrangement to the largest tertiary cycloalkyl cation occurs.[105] The 1-methylcyclohexyl cation adopts a twist boat rather than a chair conformation: the methylcycloheptyl cation is fluxionally mobile, but the methylcyclo-octyl cation favours an unsymmetrical chair-twist boat conformation.[106] Spectroscopic data for the 1-methylcyclobutyl cation are now thought to reflect a σ-delocalized 1-methylcyclopropylmethyl structure (**80**).[10]

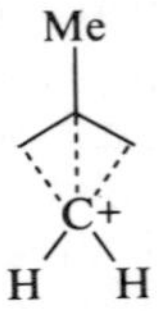

Me C+ H H

(**80**)

The cyclohex-4-enyl cation is formed as a minor product of the ionization of 4-substituted cyclohexenes in superacid media, but a rapid hydride shift converts it into its cyclohex-2-enyl isomer.[107]

Several 8,9-dehydroadamant-2-yl cations (**81**) have been prepared. The parent ion (**81**; R = H) undergoes a rapid degenerate cyclopropylmethyl rearrangement which is fast even at -120 °C. A fast three-fold degenerate rearrangement is also observed for the 4-methyl-2,5-dehydroprotoadamant-4-yl cation (**82**), probably through an intermediate 1-methyl-8,9-dehydroadamantyl cation (**83**).[108] The ground-state rotamer conformations of a series of "bisected" dicyclopropylmethyl cations, cyclopropylallyl cations, and phenylcyclopropylmethyl cations have been studied by NMR spectroscopy.[109]

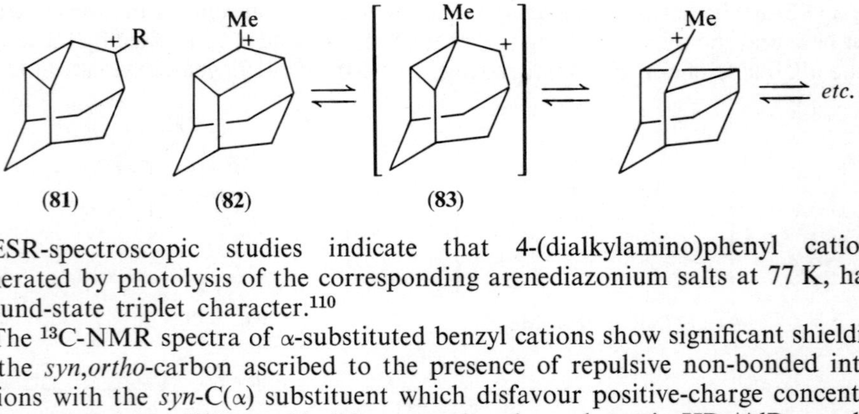

(**81**) (**82**) ⇌ [(**83**)] ⇌ ⇌ *etc.*

ESR-spectroscopic studies indicate that 4-(dialkylamino)phenyl cations, generated by photolysis of the corresponding arenediazonium salts at 77 K, have ground-state triplet character.[110]

The ^{13}C-NMR spectra of α-substituted benzyl cations show significant shielding of the *syn,ortho*-carbon ascribed to the presence of repulsive non-bonded interactions with the *syn*-C(α) substituent which disfavour positive-charge concentration at that carbon.[111] Benzyl halides are reduced to toluene in $HBr/AlBr_3$, and to a lesser extent in HF/TaF_5, even in the absence of an alkane hydride source: no reduction is observed in HF alone.[112]

The *trans*-phenylallyl cation is more stable than its *cis*-isomer, there being more effective charge stabilization in the former.[113] The 4-phenylpent-3-enyl cation, generated in "magic" acid from the corresponding alcohol, cyclizes above -80 °C, ultimately forming the 1,3-dimethylindanyl cation.[114] Despite potential anti-aromatic destabilization, the perfluoroindenyl cation (**84**) is stable for two years at room temperature.[115]

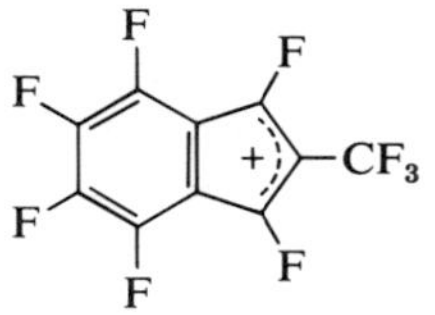

(84)

Full details of spectroscopic studies of the benzenium, naphthalenium and anthracenium cations have now appeared.[116] Substituent effects on the stability of 4-substituted polyfluorobenzenium ions have been investigated.[117,118] The polymethylbenzenium ion (**85**) undergoes a degenerate 1,2-benzyl shift ($\Delta G^{\ddagger}$= 10.9 kcal mol^{-1}).[119] The cation (**86**), which is formed when biphenylene is dissolved in "magic" acid, shows a degenerate hydride shift between C(2) and C(3).[120] A similar rearrangement has been observed with the 9,10-dimethylphenanthrenium ion (**87a**), and the hexamethylbenzenium ion.[121] Donor–acceptor interactions observed between the phenyl substituent and the carbonium-ion centre,[122] and the rate of the degenerate 1,2-phenyl shift,[123] in the ions (**87b**) and (**88**) correlate well with the electrophilicity of the carbonium-ion centre.

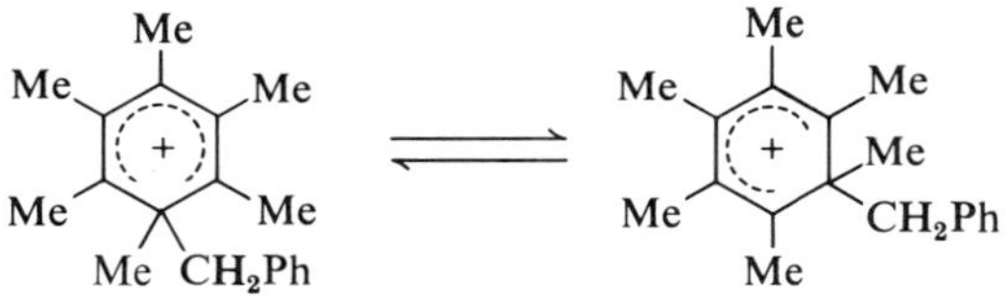

(85)

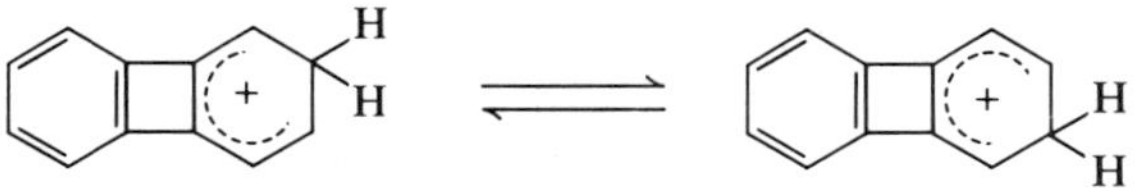

(86)

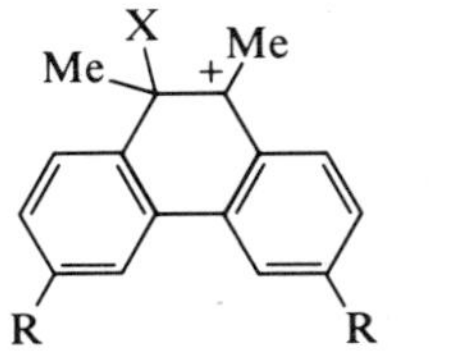

(87) **a**; X = H
b; X = Ph
R = H, Br, Me, or CF_3

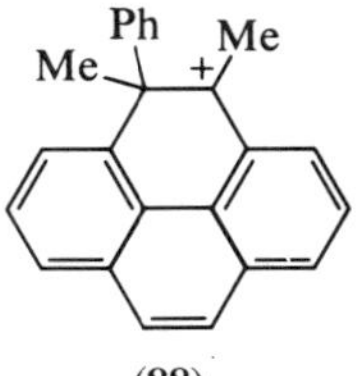

(88)

The trithienylcyclopropenium cation, which has been prepared by reaction of trichlorocyclopropenium tetrachloroaluminate with an excess of thiophene, is considerably more stable than its triphenyl analogue.[124] Ninhydrin is tri-protonated in "magic" acid to give (**89**) which is the first example of a Hückel trication: there is evidence for charge delocalization here, and it is suggested that (**89**) is stabilized by aromatic delocalization.[125] There is evidence for a diamagnetic ring current in the 1,9-dimethyldibenzo[*b*,*f*]pentalene dication (**90**), which is the first example of a pentalene dication.[126] The polyfluorofluorenyl cations (**91**) have also been characterized.[127]

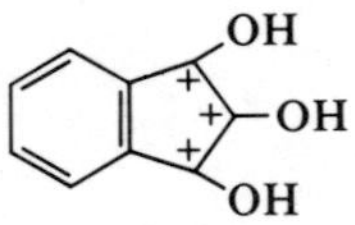

(89) 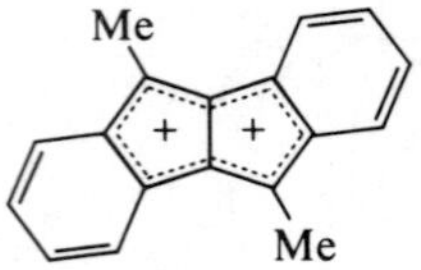 (90) 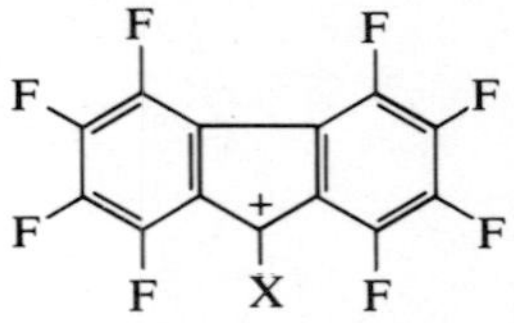 (91) X = F or C_6F_5

The equilibrium between (covalent) cycloheptatrienyl isothiocyanate and the corresponding tropylium salt has been observed by low-temperature NMR-spectroscopy.[128] The fluorotropylium cation has been prepared and found to be stabilized relative to the unsubstituted ion.[129] The methoxytropylium cation is captured regiospecifically by thiolate ions with exclusive formation of stable 1-methoxy-7-(alkyl- or arylthio-)cycloheptatrienes.[130] Conjugation which occurs between the nitrogen lone pair and the tropylium ring in aminotropylium cations is subject to steric hindrance for *N*-aryl derivatives: however, there is still evidence for some interaction even in the severely crowded dihydroindolyltropylium salt.[131] The hydride-exchange equilibrium between *N,N*-dimethylaminotropylium cation and methoxycycloheptatriene favours the nitrogen-substituted carbonium ion.[132] A simple synthesis of the triptopylium cation (**92**) has been reported.[133] Hydroxytropylium cations (**93**) are generated by dissolution of the corresponding tropones

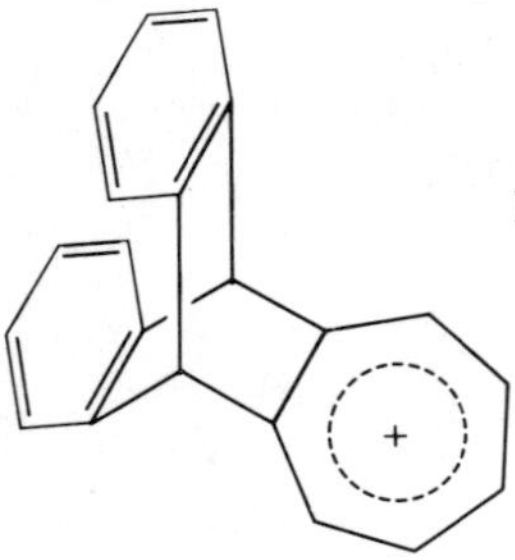

(**92**) 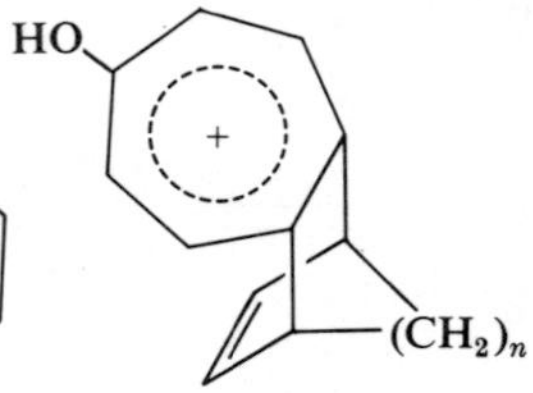 (**93**) n = 1 or 2 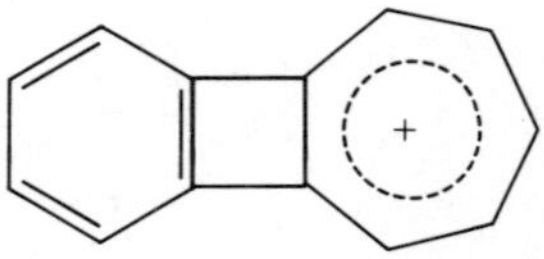 (**94**)

in trifluoroacetic acid.[134] There is little cross-conjugation between the two π-electron systems in the benzo[3′,4′]cyclobuta[1,2]cycloheptenyl cation (**94**): nucleophilic capture of this ion leads to 9-substituted fluorenes.[135] 7-Chloro- and 7-bromo-5*H*-benzocycloheptenylium salts have been prepared: kinetically controlled addition to the latter occurs at C(5), though 7-substituted products are favoured if the addition is reversible.[136]

A review has appeared of extended homoaromaticity.[137] Attempts to prepare the *anti*-1,3-bishomotropylium cation (**95**), a potentially double Möbius framework, result only in the formation of its *syn*-isomer (**96**). Protonation of the corresponding ketones leads, in the case of the *syn*-isomer, to a cation with some bishomotropylium character; however, the ion formed from the *anti*-ketone behaves as a hydroxyallyl cation (**97**).[138] Attempts to prepare the tricyclo[5.3.1.$0^{4,11}$]undecatrienyl cation (**98**), which is a potential bridged trishomotropylium ion, generated instead a species assigned the structure (**99**).[139] 1-Chloro-, 1,4-dichloro-, and 1,4,7-trichloro-triquinacene react with SbF_5 to yield bridgehead cations (**100**).[140] NMR-data indicate that the 10,11-homophenalenium cation (**101**) has little homoaromatic character.[141]

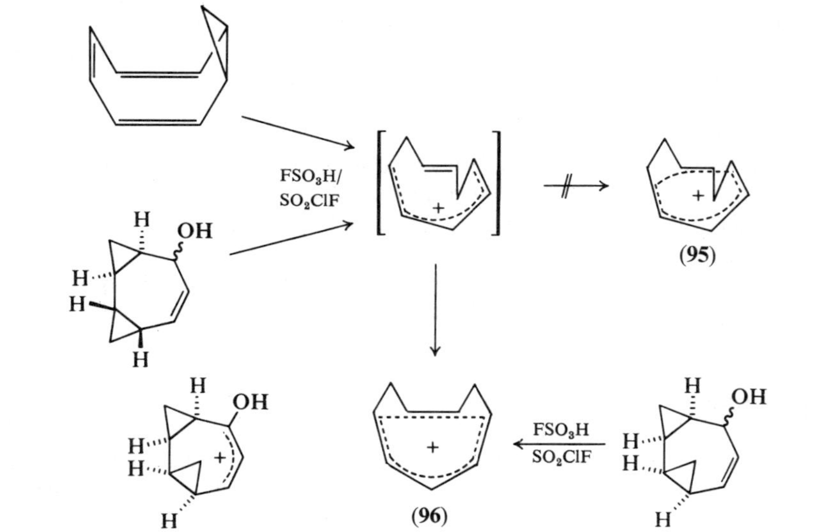

FSO$_3$H/ SO$_2$ClF

(95)

FSO$_3$H SO$_2$ClF

(96)

FSO$_3$D

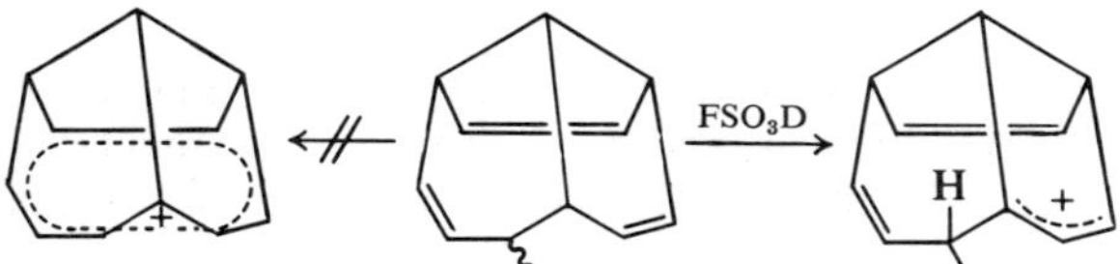

(98)

(99)

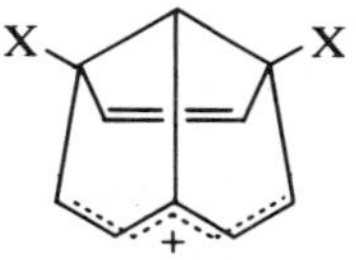

(100) X = H or Cl

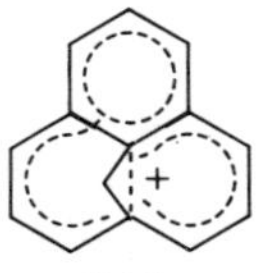

(101)

The relation between structure and ^{13}C-NMR shielding effects produced by substituents in a series of diarylmethyl cations has been investigated, it being concluded that both π-electron density and π-bond order terms are required to describe ^{13}C chemical shifts.[142] The selectivity of diarylmethyl cations in ethanol–water mixtures has been examined.[143] UV-spectroscopic and polarographic studies of 1,1-diarylethyl cations[144] have also been reported. The deactivation of electron excitation in acidic media in triarylmethyl cations,[145] and their reactions with triphenylphosphine[146] have been studied. 4-Hydroxytriarylmethanes, formed when 4-fluorotriarylmethanols are treated with boiling 90% HCOOH, arise from trapping of the intermediate triarylmethyl cation by water and subsequent elimination of hydrogen fluoride.[147] The rates of hydride-ion transfer from cyclic ethers to triphenylmethane have been determined by polarography.[148]

Intermediate oxocarbonium ions have been detected spectroscopically in the hydrolysis of certain acetals and ketals. The rates of hydrolysis of these intermediates have also been studied.[149] The rate constants for capture of the 1,3-dioxolan-2-ylium cation correlate with the basicity of the nuleophiles.[150] The regioselectivity of deprotonation of tetrahydropyranyl cations[151] and the electrochemical reduction of various stable carbonium ions in aqueous sulphuric acid[152] have also been reported.

Other Reactions

The reaction between crossed molecular beams of various alkyl halides and SbF_5 has been used to generate cations in the gas phase.[153] Electro-oxidation of alkyl bromides in acetonitrile is thought to lead to highly energetic carbonium ions which then react with the solvent directly, or after rearrangement through a series of 1,2-hydride shifts.[154,155] Enthalpy and entropy changes that occur upon protonation of alkenes have been measured by ICR (*i.e.* ion cyclotron resonance) spectrometry: the absolute entropy of, for example, the *tert*-pentyl cation has been estimated by this method.[156]

There is evidence for statistical deprotonation of the 1-methylcyclopentyl and *tert*-pentyl cation by triethylamine in the gas phase.[157] The reaction of *tert*-butyl cations, generated in the gas phase by radiolysis of neopentane, with water, alcohols, and ammonia has also been studied.[158] ICR-spectroscopic studies of the reactions of simple alkyl cations,[159] and mass-spectral measurements of the rate constants for reactions between alkyl cations and C_3–C_{10} cyclic and acyclic hydrocarbons[160] have also been reported.

Variation of the composition of ethanol–water mixtures has little effect on the selectivity of adamantyl derivatives towards the two solvent components, though addition of acetone increases this selectivity, probably through change in nucleophilicity of the ethanol and water.[161]

There is evidence that *ca.* 2.5% of the butan-2-ol obtained from nitrous acid deamination of butylamine-1-^{14}C in aqueous perchloric acid is formed through equilibrating corner-protonated methylcyclopropane intermediates.[162]

The substrate and positional selectivity of the tritritiomethyl cation, CT_3^+ (generated by decomposition of methane-$^3H_4$), in the gas or liquid phase for alkylation of benzene and toluene parallel those found for poorly solvated cations rather than conventional Friedel–Crafts reagents.[163] The high electrophilicity of CT_3^+ is also apparent from its reactions with alcohols.[164] A free acetylium ion has been prepared in the gas phase by methylation of carbon monoxide by CT_3^+: this species is more reactive to *n*- than to π-nucleophiles (it will not, for example, acylate benzene or toluene).[165]

A review has appeared concerning formation of cations through decomposition of diazonium salts.[166] Sodium metamolybdate decreases the rate of decomposition of phenyldiazonium tetrafluoroborate.[167]

A predictive model for the major unimolecular reactions of unsaturated carbonium ions has been described.[168]

There is evidence for an intermediate bromonium ion in the bromination of substituted styrenes bearing strongly electron-withdrawing aromatic substitutents.[169] A small equilibrium concentration of the 1,1-diphenylpropyl cation is formed when 1,1-diphenylpropene is treated with HCl–$SnCl_4$.[170] A quadratic linear free-energy relation has been formulated to describe the bromination of diphenylethylenes.[171]

The mechanistic and preparative aspects of vinyl cation chemistry[172] and the reactions of vinyl triflates[163] have been reviewed. *cis*- and *trans*-2-Phenyl-1,2-di-*p*-tolylvinyl-2-^{13}C bromide solvolyse through a free vinyl cation which then undergoes degenerate rearrangement by a 1,2-phenyl shift in competition with capture by solvent.[174] There is evidence for an intermediate cyclic vinyl cation in the trifluoroethanolysis of cyclo-octatetraenyl trifluoromethanesulphonate,[175] and for an α-(9-anthryl)vinyl cation in the solvolysis of 9-(α-chlorovinyl)anthracene.[176]

It has been suggested that an intermediate allyl cation occurs in the acid-catalysed epimerization of 15(*R*)- and 15(*S*)-methylprostaglandin-E_2.[177] The migratory aptitude of an allyl group is found to be only slightly greater than that of a methyl group in the acetolysis of substituted 3,3-dimethylhex-5-en-2-yl *p*-bromobenzenesulphonates.[178] Products have been obtained resulting from an effective [4+3]-cycloaddition of allenyl cations with cyclopentadiene.[179]

A carbonium-ion pathway is implicated in the aromatization of cyclohexa-1,3- and -1,4-dienes in liquid sulphur dioxide.[180] The ICR-dissociation spectra of protonated hexamethylbenzene and hexamethyl(Dewar benzene) ions are identical, indicating that the latter rearranges very quickly in the gas phase.[181] Whilst 4-trichloromethyl-4-methylcyclohexadienol undergoes a dienol–benzene rearrangement in concentrated mineral acids, 4,4-dimethylcyclohexadienol reacts similarly *ca.* 10^8 times faster in dilute buffers, and also shows an allylic rearrangement not observed for the former.[182]

Various studies of $C_7H_7^+$ ions in the gas phase have been described.[183–186] The bicyclic homotropylium cation (**102**) readily undergoes a circumambulation through the closed structure (**103**).[187]

I

AgOAc / acetone

(**102**)

(**103**)

$-H^+$

A detailed study of the potential-energy surface of $C_4H_9O^+$ ions in the gas phase,[188] the heats of formation of the ions CH_3O^+, $C_2H_5O^+$, etc.,[189] fragmentation of $C_2H_5S^+$ and $C_3H_7S^+$.[190] and ICR-spectroscopic studies of the gas-phase reactions of $MeOCH_2^+$ ions with alcohols, thiols, and amines,[191] have been reported. Oxonium ions (**104**) fragment in a non-concerted manner to give either alkenes or alkyl cations and formaldehyde.[192]

$$HCHO + Me(CH_2)_n^+ \xleftarrow{n>1} CH_2{=}\overset{+}{O}(CH_2)_nMe \xrightarrow{n=1} CH_2{=}\overset{+}{O}H + CH_2{=}CH_2$$

(**104**)

A method for estimating the heats of formation of some gaseous cations, *e.g.* $HO(CH_2)_2CH_2^+$, has been described.[193] Acetonides are formed in high yield when epoxides are treated with copper sulphate in acetone.[194] The rates of acid-catalysed hydrolysis of the Bay region epoxides of phenanthrene and chrysene have been measured.[195] There is evidence for a carbonium-ion intermediate in the acetolytic cleavage of 2β,19-epoxy-5α-cholestane.[196]

The presence of a free solvent-separated oxocarbonium-ion intermediate in the acid-catalysed hydrolysis of acetophenone dimethyl acetal has been demonstrated.[197]

An alkoxycarbonium-ion intermediate is formed in the hydrolysis of methyl 1-cyclopropylvinyl ether in the presence of a variety of proton sources.[198] Kinetic data for the acid-catalysed hydrolysis of 2,4-disubstituted-1,3-dioxolanes have been reported.[199] Anodic oxidation of β-oxocarboxylate ethylene acetals leads to 2-methoxy-1,4-dioxans by rearrangement of carbonium-ion intermediates.[200]

Dialkyl esters of (3-alkylalka-1,2-dienyl)phosphonic acids cyclize in the presence of acid *via* a carbonium-ion intermediate which results from protonation of the central allenic carbon atom.[201] The reaction of 2-methylallyl alcohol with aldehydes to give tetrahydropyranyl alcohols has also been studied.[202]

There is evidence for participation of chlorine *via* a chloronium ion in the silver ion-catalysed acetolysis of *E,Z*- and *Z,Z*-1-bromo-4-chloro-1,4-diphenylbutadiene;[203] and the binding energies and stabilities of chloronium ions in the gas phase have been determined.[204]

A series of 1,2,5-triazapentadienium salts has been prepared.[205] Carbonium-ion intermediates have been implicated in the cyclization of 1-(2-pyridyl)benzo-1,2,3-triazole in polyphosphoric acid,[206] the acid-catalysed racemization of (+)-halostachine and (−)-phenylephrine,[207] and the reaction of 9-(hydroxymethyl)carbazole with alcohols.[208] The reactions of hydroxypyrimidine cations[209] and di-indolylmethyl cations[210] have also been investigated. 1,3-Phenyl migration has been observed when *N*-substituted 2-hydroxy-2-phenyl-5-methylpiperidin-6-ones are heated in oxalic acid at 200 °C.[211] The kinetics of the Schmidt reactions of 4-methoxy- and 2,4,6-trimethyl-benzaldehyde in aqueous sulphuric acid have been studied.[212]

A cyclopropyl substituent has little effect on the rates of reaction of arylcarbonium ions, generated by pulse radiolysis, with trialkylamines and ammonia in chlorinated solvents.[213] The reactions of Malachite Green and tri-*p*-methoxyphenymethyl cation with 1-benzyldihydronicotinamide are catalysed by anionic micelles of sodium lauryl sulphate, but only weakly by cationic micelles of cetyltrimethylammonium bromide.[214] The equilibria and rates of reaction of *o*-methyl-(Malachite Green) with sulphite ions have also been investigated.[215]

$Me_3SiC{\equiv}C(CH_2)_nCOCl$ (**105**) $\xrightarrow[CH_2Cl_2]{AlCl_3}$ cyclic ynone $(CH_2)_n$; $n = 12$; Me_2CuLi → cyclic enone $(CH_2)_{12}$ (Me) + cyclic enone CH_2, Me $(CH_2)_{11}$ $\xrightarrow{\text{Reduction}}$ (**106**)

Large-ring ynones are readily synthesized by intramolecular acylation of ω-(trimethylsilylethynyl)alkanoyl chlorides, *e.g.* (**105**); a synthesis of (*d,l*)muscone (**106**) has been described which uses this reaction.[216]

The proton affinity of ketene has been measured by pulsed ICR-spectrometry.[217] The loss of carbon monoxide from the butanoylium cation in the gas phase requires a relatively low activation energy, probably because isomerization to the 1-methylethyl cation occurs before the fragmentation.[218] Gas-phase pinacol-type rearrangements are observed in the unimolecular decomposition of protonated butanone and protonated 2-methylpropanal but not for protonated butanal.[219]

Evidence has been obtained for an intermediate 4-membered ring oxonium ion in the $AgSbF_6$-catalysed dehalogenation of α-bromo-ketones in non-nucleophilic media.[220]

The asymmetric reduction of carbonium ions by chiral organosilicon hydrides has been investigated.[221] An intermediate silicenium ion is formed in the pyrolysis of 1-methyl-1-vinyl-1-silacyclobutane in phenol–benzene mixtures at 600 °C.[222] The reaction of the silyl cation (SiH_3^+) with ethene has been investigated by mass spectrometry.[223]

Theoretical Calculations[224, 225]

There is a rough correlation with proton affinities of alkyl chlorides predicted by the MINDO/3 method and their rates of solvolysis.[226] MINDO/3 calculations favour bridged structures for the ethyl and vinyl cations.[227] Calculations of potential-energy surfaces for open and cyclic forms of vinyl cations, $XCH{=}\overset{+}{C}H$, favour a bridged structure when X is SH or NH_2; however, this geometry constitutes an energy maximum when X is F or Cl.[228] Hyperconjugative stabilization of the 1-methylethyl and vinyl cations has also been examined.[229] The results of other *ab initio* calculations on the cations $C_2H_3^+$ and $C_2H_4X^+$ have also been reported.[230]

Ab initio calculations suggest that nucleophilic attack on propargylic↔allenylic cations should greatly favour formation of propargylic products as found experimentally but not as expected from NMR-spectroscopic studies of such species.[231]

MINDO/3 calculations favour a puckered geometry for the cyclobutyl cation, in which the cationic centre is not trigonal as a result of a non-classical bridging interaction with C(3).[232] *Ab initio* calculations suggest that the bisected conformation of the bicyclobut-1-ylmethyl cation is *ca.* 32 kcal mol^{-1} more stable than the eclipsed conformer.[233] MINDO/3 calculations on the bicyclo[1.1.1]pent-1-yl cation favour a structure that may be considered as a complex between CH^+ and trimethylenemethane.[234]

An open structure is favoured for both the methanediazonium and the benzenediazonium cations: in the former case, rearrangement occurs through a symmetrical bridged structure, whereas for the latter an asymmetric transition state is favoured.[235]

Ab initio SCF calculations suggest that, whilst the parent cyclobutadiene and cyclo-octatetraene dications might be kinetically stable, both the cyclopentadienyl and the cyclononatetraenyl trications should be too unstable to be observed.[236] Whilst the tetrakis(trifluoromethyl)cyclobutadiene dication should be planar, the parent ion and its tetramethyl analogue are predicted to prefer rather highly puckered geometries.[237] The heats of formation of monosubstituted tropylium ions, their molecular geometries, and charge distributions have been predicted by the MINDO/3 method.[238] *Ab initio* calculations suggest that the heptalenium dication is significantly less stable than might be expected by analogy with naphthalene.[239]

It has been suggested that the high energy barriers previously noted for the conversion of CH_3O^+ into protonated formaldehyde arise because the parent ion is a complex ($H_2.HCO^+$).[240] The proton affinities of benzaldehyde, tropone, and 4-methylenecyclohexa-2,5-dien-1-one calculated by the MINDO/3 method are in close agreement with those measured by ICR-spectrometry.[241]

Electronic spectra of and charge distributions in the 2,4,6-trimethylpyrylium cation,[242] and the oxazolium cation and its mono-, di-, and tri-phenyl derivatives[243] have been calculated. There is evidence for an unusual carbocation-stabilizing effect by σ-bivalent sulphur.[244] Studies of the stability of tetra- and penta-methylene-chloronium ions,[245] and α-substituted silicienium ions[246] have also been described.

References

[1] Kelly, D. P., and Spear, R. J., *Aust. J. Chem.*, **31**, 1209 (1978).
[2] See *Org. Reaction Mech.*, **1977**, 337.
[3] Farnum, D. G., Botto, R. E., Chambers, W. T., and Lam, B., *J. Am. Chem. Soc.*, **100**, 3847 (1978).
[4] Lambert, J. B., and Mark, H. W., *J. Am. Chem. Soc.*, **100**, 2501 (1978).
[5] Brown, H. C., Ravindranathan, M., Chloupek, F. J., and Rothberg, I., *J. Am. Chem. Soc.*, **100**, 3143 (1978).
[6] Harris, J. M., Mount, D. L., and Raber, D. J., *J. Am. Chem. Soc.*, **100**, 3139 (1978).
[7] Liu, K.-T., *Tetrahedron Lett.*, **1978**, 1129.
[8] Goetz, D. W., Schlegel, B., and Allen, L. C., *J. Am. Chem. Soc.*, **99**, 8118 (1977).
[9] Clark, D. T., Cromarty, B. J., and Colling, L., *J. Am. Chem. Soc.*, **99**, 8120 (1977).
[10] Olah, G. A., Prakash, G. K. S., Donovan, D. J., and Yavari, I., *J. Am. Chem. Soc.*, **100**, 7085 (1978).
[11] Arnett, E. M., and Petro, C., *J. Am. Chem. Soc.*, **100**, 2563 (1978).
[12] Coxon, J. M., Steel, P. J., Coddington, J. M., Rae, I. D., and Jones, A. J., *Aust. J. Chem.*, **31**, 1223 (1978).
[13] Coxon, J. M., Pojer, P. M., Steel, P. J., Rae, I. D., and Jones, A. J., *Aust. J. Chem.*, **31**, 1747 (1978).
[14] Coxon, J. M., Pojer, P. M., Robinson, W. T., and Steel, P. J., *J. Chem. Soc., Chem. Commun.*, **1978**, 111.
[15] Brown, H. C., and Ravindranathan, M., *J. Am. Chem. Soc.*, **100**, 1865 (1978).
[16] Bégué, J. P., Charpentier-Morize, M., Pardo, C., and Sansoulet, J., *Tetrahedron*, **34**, 293 (1978).
[17] Christol, H., Coste, J., Pietrasanta, F., Plénat, F., and Renard, G., *J. Chem. Res.*, (*S*) **1978**, 62.
[18] Sonney, J.-M., Vogel, P., and Burger, U., *Tetrahedron Lett.*, **1978**, 825.
[19] Goering, H. L., Chang, C.-S., and Masilamani, D., *J. Am. Chem. Soc.*, **100**, 2506 (1978).
[20] Burger, U., Sonney, J.-M., and Vogel, P., *Tetrahedron Lett.*, **1978**, 829.
[21] Fleming, I., and Michael, J. P., *J. Chem. Soc., Chem. Commun.*, **1978**, 245.
[22] Furusaki, A., and Matsumoto, T., *Bull. Chem. Soc. Jpn.*, **51**, 16 (1978).
[23] Gregorcic, A., and Zupan, M., *Tetrahedron*, **33**, 3243 (1977).
[24] Akiyama, T., Fujii, T., Ishiwari, H., Imagawa, T., and Kawanisi, M., *Tetrahedron Lett.*, **1978**, 2165.
[25] Chamberlain, P. H., *Diss. Abstr. Int. B.*, **38**, 2177 (1977); *Chem. Abs.*, **88**, 36731 (1978).
[26] Chang, L. W.-K., *Diss. Abstr. Int. B*, **38**, 1714 (1977); *Chem. Abs.*, **88**, 36842 (1978).
[27] Vilkas, E., Vilkas, M., Joniaux, D., and Pascard-Billy, C., *J. Chem. Soc., Chem. Commun.*, **1978**, 125.
[28] Banks, R. M., and Maskill, H., *J. Chem. Soc., Perkin Trans.* 2, **1977**, 1991.
[29] Schipper, P., and Buck, H. M., *J. Am. Chem. Soc.*, **100**, 5507 (1978).
[30] Nomura, Y., Takeuchi, Y., and Tomoda, S., *Tetrahedron Lett.*, **1978**, 911.
[31] Volz, H., Shin, J.-H., Prinzbach, H., Babsch, H., and Christl, M., *Tetrahedron Lett.*, **1978**, 1247.
[32] Paquette, L. A., Klein, G., and Doecke, C. W., *J. Am. Chem. Soc.*, **100**, 1527 (1978).
[33] Inamoto, Y., Aigami, K., Takaishi, N., Fujikura, Y., Tsuchihashi, K., and Ikeda, H., *J. Org. Chem.*, **42**, 3833 (1977).
[34] Fujise, Y., Nakatsu, T., Nakamura, A., and Ito, S., *Tetrahedron Lett.*, **1978**, 4293.
[35] Mehta, G., Chaudhuri, B., and Duddeck, H., *Tetrahedron Lett.*, **1978**, 1603.
[36] Mehta, G., Singh, V., Duddeck, H., *Tetrahedron Lett.*, **1978**, 1223.
[37] Klein, G., and Paquette, L. A., *J. Org. Chem.*, **43**, 1293 (1978).
[38] Schadt, F. L., III, Lancelot, C. J., and Schleyer, P. von R., *J. Am. Chem. Soc.*, **100**, 228 (1978).

[39] Tsuno, Y., Fujio, M., Seki, Y., Mishima, M., and Kim, S.-G., *Nippon Kagaku Kaishi*, **1977**, 1673; *Chem. Abs.*, **88**, 36878 (1978).
[40] Kirmse, W., and Günther, B.-R., *J. Am. Chem. Soc.*, **100**, 3619 (1978).
[41] Fain, D., and Dubois, J. E., *Tetrahedron Lett.*, **1978**, 791.
[42] Brown, H. C., Ravindranathan, M., and Rao, C. G., *J. Am. Chem. Soc.*, **100**, 1218 (1978).
[43] Goering, H. L., and Anderson, R. P., *J. Am. Chem. Soc.*, **100**, 6469 (1978).
[44] Tanaka, M., Kimoto, M., and Tokuyama, K., *Chem. Pharm. Bull.*, **26**, 38 (1978); *Chem. Abs.*, **88**, 120270 (1978).
[45] Rotem, M., and Kashman, Y., *Tetrahedron Lett.*, **1978**, 63.
[46] Corey, E. J., and Boger, D. L., *Tetrahedron Lett.*, **1978**, 2461.
[47] Fetizon, M., and Ragoussis, N., *Tetrahedron*, **34**, 287 (1978).
[48] McCormick, J. P., and Barton, D. L., *Tetrahedron*, **34**, 325 (1978).
[49] Harding, K. E., Cooper, J. L., and Puckett, P. M., *J. Am. Chem. Soc.*, **100**, 1015 (1978).
[50] Mizyuk, V. L., and Semenovsky, A. V., *Tetrahedron Lett.*, **1978**, 3603.
[51] Johnson, W. S., Yarnell, T. M., Myers, R. F., and Morton, D. R., *Tetrahedron Lett.*, **1978**, 2549.
[52] Cane, D. E., and Nachbar, R. B., *J. Am. Chem. Soc.*, **100**, 3208 (1978).
[53] Friedrich, E. C., and Jassawalla, J. D. C., *Tetrahedron Lett.*, **1978**, 953.
[54] Paquette, L. A., Degenhardt, C. R., and Berk, H. C., *J. Am. Chem. Soc.*, **100**, 1599 (1978).
[55] Goldstein, M. J., and Warren, D. P., *J. Am. Chem. Soc.*, **100**, 6539, 6541 (1978).
[56] Trost, B. M., and Scudder, P. H., *J. Am. Chem. Soc.*, **99**, 7601 (1977).
[57] Takakis, I. M., and Rhodes, Y. E., *Tetrahedron Lett.*, **1978**, 2475.
[58] Rhodes, Y. E., Takakis, I. M., Schueler, P. E., and Weiss, R. A., *Tetrahedron Lett.*, **1978**, 2479.
[59] Kwantes, P. M., and Klumpp, G. W., *Tetrahedron Lett.*, **1978**, 4097.
[60] Warner, P., Lu, S.-L., and Chang, S.-C., *Tetrahedron Lett.*, **1978**, 1947.
[61] Detty, M. R., and Paquette, L. A., *J. Org. Chem.*, **43**, 1118 (1978).
[62] Applequist, D. E., and Sasaki, T., *J. Org. Chem.*, **43**, 2399 (1978).
[63] Diaz, A. F., and Miller, R. D., *J. Am. Chem. Soc.*, **100**, 5905 (1978).
[64] Paquette, L. A., and Carmody, M. J., *J. Org. Chem.*, **43**, 1299 (1978).
[65] Petty, R. L., Ikeda, M., Samuelson, G. E., Boriack, C. J., Onan, K. D., McPhail, A. T., and Meinwald, J., *J. Am. Chem. Soc.*, **100**, 2464 (1978).
[66] Yano, K., Isobe, M., and Yoshida, K., *J. Am. Chem. Soc.*, **100**, 6166 (1978).
[67] Yoshida, K., and Yano, K., *Tetrahedron Lett.*, **1978**, 3039.
[68] Hayano, K., Ohfune, Y., Shirahama, H., and Matsumoto, T., *Tetrahedron Lett.*, **1978**, 1991.
[69] Davies, S. G., Green, M. L. H., and Mingos, D. P. M., *Tetrahedron*, **34**, 3047 (1978).
[70] Hayakawa, Y., Yokoyama, K., and Noyori, R., *J. Am. Chem. Soc.*, **100**, 1791 (1978).
[71] Noyori, R., Shimizu, F., and Hayakawa, Y., *Tetrahedron Lett.*, **1978**, 2091.
[72] Birney, D. M., Crane, A. M., and Sweigart, D. A., *J. Organomet. Chem.*, **152**, 187 (1978).
[73] Gresham, D. G., Kowalski, D. J., and Lillya, C. P., *J. Organomet. Chem.*, **144**, 71 (1978).
[74] Olah, G. A., Liang, G., and Yu, S., *J. Org. Chem.*, **42**, 4262 (1977).
[75] Pearson, A. J., *J. Chem. Soc.*, *Perkin Trans.* 1, **1978**, 495.
[76] Birch, A. J., and Pearson, A. J., *J. Chem. Soc.*, *Perkin Trans.* 1, **1978**, 638.
[77] Bayond, R. S., Biehl, E. R., and Reeves, P. C., *J. Organomet. Chem.*, **150**, 75 (1978).
[78] Abram, T. S., Crawford, W., Knipe, A. C., and Watts, W. E., *Proc. Roy. Irish Acad.*, **77B**, 317 (1977); *Chem. Abs.*, **89**, 146040 (1978).
[79] Cais, M., Dani, S., Herbstein, F., and Kapon, M., *J. Am. Chem. Soc.*, **100**, 5554 (1978).
[80] Powell, P., Russell, L. J., Styles, E., Brown, A. J., Howarth, O. W., and Moore, P., *J. Organomet. Chem.*, **149**, C1 (1978).
[81] Top, S., Caro, B., and Jaouen, G., *Tetrahedron Lett.*, **1978**, 787.
[82] Seyferth, D., Merola, J. S., and Eschbach, C. S., *J. Am. Chem. Soc.*, **100**, 4124 (1978).
[83] Clark, D. W., and Kane-Maguire, L. A. P., *J. Organomet. Chem.*, **145**, 201 (1978).
[84] Schilling, B. E. R., and Hoffmann, R., *J. Am. Chem. Soc.*, **100**, 6274 (1978).
[85] Sokolov, V. I., *Tezisy Dokl.-Vses. Konf. Khim. Atsetilena*, *5th*, **1975**, 347; *Chem. Abs.*, **89**, 146286 (1978).
[86] Driessen, P. B. J., and Hogeveen, H., *J. Am. Chem. Soc.*, **100**, 1193 (1978).
[87] Sivaram, S., *J. Organomet. Chem.*, **156**, 55 (1978).
[88] Hehre, W. J., *Mod. Theor. Chem.*, **4**, 277 (1977); *Chem. Abs.*, **88**, 36707 (1978).
[89] Sorensen, T. S., and Rauk, A., *Org. Chem.* (*N.Y.*), **35** (Pt. 2), 1 (1977); *Chem. Abs.*, **88**, 5642 (1978).

[90] Nishimura, J., and Olah, G. A., *Mem. Fac. Ind. Arts, Kyoto Tech. Univ. Sci. Technol.*, **25**, 92 (1976); *Chem. Abs.*, **88**, 189397 (1978).
[91] Olah, G. A., and Mo, Y. K., *Carbonium lons*, **5**, 2135 (1976).
[92] Arnett, E. M., and Petro, C., *J. Am. Chem. Soc.*, **100**, 5402 (1978).
[93] Arnett, E. M., and Petro, C., *J. Am. Chem. Soc.*, **100**, 5408 (1978).
[94] Olah, G. A., and Donovan, D. J., *J. Am. Chem. Soc.*, **100**, 5163 (1978).
[95] Saunders, M., and Kates, M. R., *J. Am. Chem. Soc.*, **100**, 7082 (1978).
[96] Siskin, M., *Tetrahedron Lett.*, **1978**, 527.
[97] Siskin, M., *J. Am. Chem. Soc.*, **100**, 1838 (1978).
[98] Coustard, J.-M., Douteau, M.-H., Jacquesy, R., Longevialle, P., and Zimmerman, D., *J. Chem. Res.*, (*S*) **1978**, 16, (*M*) 337.
[99] Coustard, J.-M., Douteau, M.-H., and Jacquesy, R., *J. Chem. Res.*, (*S*) **1978**, 18, (*M*) 357.
[100] Saunders, M., and Lloyd, J. R., *J. Am. Chem. Soc.*, **99**, 7090 (1977).
[101] Takahashi, Y., Fujimori, H., Yoneda, N., and Suzuki, A., *Nippon Kagaku Kaishi*, **1977**, 1861; *Chem. Abs.*, **88**, 88735 (1978).
[102] Kirchen, R. P., Sorensen, T. S., and Wagstaff, K. E., *J. Am. Chem. Soc.*, **100**, 6761 (1978).
[103] Kirchen, R. P., and Sorensen, T. S., *J. Chem. Soc., Chem. Commun.*, **1978**, 769.
[104] Hart, H., and Willer, R., *Tetrahedron Lett.*, **1978**, 4189.
[105] Kirchen, R. P., Sorensen, T. S., and Wagstaff, K. E., *J. Am. Chem. Soc.*, **100**, 5134 (1978).
[106] Kirchen, R. P., and Sorensen, T. S., *J. Am. Chem. Soc.*, **100**, 1487 (1978).
[107] Farcaşiu, D., *J. Am. Chem. Soc.*, **100**, 993 (1978).
[108] Olah, G. A., Liang, G., Babiak, K. A., Ford, T. M., Goff, D. L., Morgan, T. K., Jr., and Murray, R. K., Jr., *J. Am. Chem. Soc.*, **100**, 1494 (1978).
[109] Okazawa, N., and Sorensen, T. S., *Can. J. Chem.*, **56**, 2355 (1978).
[110] Cox, A., Kemp, T. J., Payne, D. R., Symons, M. C. R., and de Moira, P. P., *J. Am. Chem. Soc.*, **100**, 4779 (1978).
[111] Forsyth, D. A., Vogel, D. E., and Stanke, S. J., *J. Am. Chem. Soc.*, **100**, 5215 (1978).
[112] Siskin, M., and Schlosberg, R. H., *J. Am. Chem. Soc.*, **100**, 1842 (1978).
[113] Alieva, S. L., Aliev, A. D., Koshevnik, A. Y., and Krentsel, B. A., *Tezisy Dokl.-Vses. Konf.* "*Stereokhim. Konform. Anal. Org. Neftekhim. Sint.*", *3rd*, **1976**, 17; *Chem. Abs.*, **88**, 151918 (1978).
[114] Olah, G. A., Asensio, G., and Mayr, H., *J. Org. Chem.*, **43**, 1518 (1978).
[115] Karpov, V. M., Platonov, V. E., and Yakobson, G. G., *Tetrahedron*, **34**, 3215 (1978).
[116] Olah, G. A., Staral, J. S., Asensio, G., Liang, G., Forsyth, D. A., and Mateescu, G. D., *J. Am. Chem. Soc.*, **100**, 6299 (1978).
[117] Shtark, A. A., Pozdnyakovich, Y. V., and Shteingarts, V. D., *Zh. Org. Khim.*, **13**, 1671 (1977).
[118] Shtark, A. A., and Shteingarts, V. D., *Zh. Org. Khim.*, **13**, 1662 (1977).
[119] Sazonova, L. I., Shakirov, M. M., and Shubin, V. G., *Zh. Org. Khim.*, **13**, 2456 (1977).
[120] Bodoev, N. V., Mamatyuk, V. I., Krysin, A. P., and Koptyug, V. A., *Izv. Akad. Nauk SSSR, Ser. Khim.*, **1978**, 1199; *Chem. Abs.*, **89**, 107534 (1978).
[121] Borodkin, G. I., Shakirov, M. M., Shubin, V. G., and Koptyug, V. A., *Zh. Org. Khim.*, **14**, 989 (1978).
[122] Borodkin, G. I., Shakirov, M. M., and Shubin, V. G., *Zh. Org. Khim.*, **14**, 374 (1978).
[123] Borodkin, G. I., Shakirov, M. M., Shubin, V. G., and Koptyug, V. A., *Zh. Org. Khim.*, **14**, 321 (1978).
[124] Komatsu, K., Tomioka, I., and Okamoto, K., *Tetrahedron Lett.*, **1978**, 803.
[125] Bruck, D., Dagan, A., and Rabinovitz, M., *Tetrahedron Lett.*, **1978**, 1791.
[126] Willner, I., and Rabinovitz, M., *J. Am. Chem. Soc.*, **100**, 337 (1978).
[127] Pozdnyakovich, Y. V., and Shteingarts, V. D., *Zh. Org. Khim.*, **14**, 603 (1978).
[128] Feigel, M., Kessler, H., and Walter, A., *Chem. Ber.*, **111**, 2947 (1978).
[129] Föhlisch, B., and Welt, G., *Tetrahedron Latt.*, **1978**, 4019.
[130] Cavazza, M., Morganti, G., and Pietra, F., *J. Chem. Soc., Chem. Commun.*, **1978**, 710.
[131] Föhlisch, B., and Haug, E., *J. Chem. Res.*, (*S*) **1978**, 385, (*M*) 4801.
[132] Föhlisch, B., and Braun, R., *Tetrahedron Lett.*, **1978**, 2735.
[133] Butler, D. N., and Gupta, I., *Can. J. Chem.*, **56**, 80 (1978).
[134] Nakazawa, T., Niimoto, Y., and Murata, I., *Tetrahedron Lett.*, **1978**, 569.
[135] Lombardo, L., and Wege, D., *Aust. J. Chem.*, **31**, 1569 (1978).
[136] Föhlisch, B., Fischer, C., Widmann, E., and Wolf, E., *Tetrahedron*, **34**, 533 (1978).
[137] Paquette, L. A., *Angew. Chem. Int. Edn.*, **17**, 106 (1978).

[138] Paquette, L. A., and Detty, M. R., *J. Am. Chem. Soc.*, **100**, 5856 (1978).
[139] Du Vernet, R. B., Glanzman, M., and Schröder, G., *Tetrahedron Lett.*, **1978**, 3071.
[140] Bosse, D., and de Meijere, A., *Chem. Ber.*, **111**, 2243 (1978).
[141] Nakasuji, K., Katada, M., and Murata, I., *Tetrahedron Lett.*, **1978**, 2515.
[142] Ancian, B., Membrey, F., and Doucet, J. P., *J. Org. Chem.*, **43**, 1509 (1978).
[143] Karton, Y., and Pross, A., *J. Chem. Soc., Perkin Trans.* 2, **1977**, 1860.
[144] Din, K., and Plesch, P. H., *J. Chem. Soc., Perkin Trans.* 2, **1978**, 892.
[145] Ivanov, V. L., Al-Ainen, A. S., and Kuz'min, M. G., *Khim. Vys. Energ.*, **12**, 48 (1978); *Chem. Abs.*, **88**, 120451 (1978).
[146] Bidan, G., and Genies, M., *Tetrahedron Lett.*, **1978**, 2499.
[147] Andrews, A. F., Vandenbulcke, B. M. J., and Walton, J. C., *J. Chem. Soc., Chem. Commun.*, **1978**, 389.
[148] Din, K., and Plesch, P. H., *J. Chem. Soc., Perkin Trans.* 2, **1978**, 937.
[149] McClelland, R. A., and Ahmad, M., *J. Am. Chem. Soc.*, **100**, 7027, 7031 (1978).
[150] Kubisa, P., *Bull. Acad. Pol. Sci., Ser. Sci. Chim.*, **25**, 627 (1977); *Chem. Abs.*, **88**, 49923 (1978).
[151] Arakelyan, A. A., Kazaryan, P. I., and Gevorkyan, A. A., *Tezisy Dokl.—Molodezhnaya Konf. Org. Sint. Bioorg. Khim.*, **1976**, 34; *Chem. Abs.*, **88**, 169276 (1978).
[152] Feldman, M. R., and Flythe, W. C., *J. Org. Chem.*, **43**, 2596 (1978).
[153] Auerbach, A., Cross, R. J., Jr., and Saunders, M., *J. Am. Chem. Soc.*, **100**, 4908 (1978).
[154] Becker, J. Y., *J. Org. Chem.*, **42**, 3997 (1977).
[155] Becker, J. Y., *Tetrahedron Lett.*, **1978**, 1331.
[156] Ausloos, P., and Lias, S. G., *J. Am. Chem. Soc.*, **100**, 1953 (1978).
[157] Marinelli, W. J., and Morton, T. H., *J. Am. Chem. Soc.*, **100**, 3536 (1978).
[158] Attina, M., Cacace, F., Ciranni, G., and Giacomella, P., *J. Chem. Soc., Chem. Commun.*, **1978**, 938.
[159] Fiaux, A., Smith, D. L., and Futrell, J. H., *Int. J. Mass Spectrom. Ion Phys.*, **25**, 281 (1977).
[160] Moet-Ner, M., and Field, F. H., *J. Am. Chem. Soc.*, **100**, 1356 (1978).
[161] Karton, Y., and Pross, A., *J. Chem. Soc., Perkin Trans.* 2, **1978**, 595.
[162] Lee, C. C., Reichle, R., and Weber, U., *Can. J. Chem.*, **56**, 658 (1978).
[163] Cacace, F., and Giacomello, P., *J. Chem. Soc., Perkin Trans.* 2, **1978**, 652.
[164] Zharov, V. T., Nefedov, V. D., Sinotova, E. N., Korsakov, M. V., and Antonova, V. V., *Zh. Org. Khim.*, **14**, 35 (1978).
[165] Giacomello, P., and Speranza, M., *J. Am. Chem. Soc.*, **99**, 7918 (1977).
[166] Zollinger, H., *Angew. Chem. Int. Edn.*, **17**, 141 (1978).
[167] Kozlov, V. V., Smirnova, V. G., and Sagalovich, V. P., *Zh. Obshch. Khim.*, **47**, 2244 (1977); *Chem. Abs.*, **88**, 61814 (1978).
[168] Stapleton, J., Bowen, R. D., and Williams, D. H., *Tetrahedron*, **34**, 259 (1978).
[169] Dubois, J.-E., Ruasse, M.-F., and Argile, A., *Tetrahedron Lett.*, **1978**, 177.
[170] Bywater, S., and Worsfold, D. J., *Can. J. Chem.*, **56**, 2093 (1978).
[171] O'Brien, M., and More O'Ferrall, R. A., *J. Chem. Soc., Perkin Trans.* 2, **1978**, 1045.
[172] Hanack, M., *Angew. Chem. Int. Edn.*, **17**, 333 (1978).
[173] Stang, P. J., *Acc. Chem. Res.*, **11**, 107 (1978).
[174] Lee, C. C., Paine, A. J., and Ko, E. C. F., *J. Am. Chem. Soc.*, **99**, 7267 (1977).
[175] Hanack, M., and Sproesser, L., *J. Am. Chem. Soc.*, **100**, 7066 (1978).
[176] Rappoport, Z., Shulman, P., and Thuval, M., *J. Am. Chem. Soc.*, **100**, 7041 (1978).
[177] Merritt, M. V., and Bronson, G. E., *J. Am. Chem. Soc.*, **100**, 1891 (1978).
[178] Dillenberger, Z., Schmid, H., and Hansen, H.-J., *Helv. Chim. Acta*, **61**, 1856 (1978).
[179] Mayr, H., and Grubmüller, B., *Angew. Chem. Int. Edn.*, **17**, 130 (1978).
[180] Masilamani, D., and Rogic, M. M., *Tetrahedron Lett.*, **1978**, 3785.
[181] Dunbar, R. C., Fu, E. W., and Olah, G. A., *J. Am. Chem. Soc.*, **99**, 7502 (1977).
[182] Vitullo, V. P., Cashen, M. J., Marx, J. N., Caudle, L. J., and Fritz, J. R., *J. Am. Chem. Soc.*, **100**, 1205 (1978).
[183] Jackson, J.-A. A., *Diss. Abstr. Int. B*, **38**, 2691 (1977); *Chem. Abs.*, **88**, 61874 (1978).
[184] McCrery, D. A., and Freiser, B. S., *J. Am. Chem. Soc.*, **100**, 2902 (1978).
[185] Jackson, J.-A. A., Lias, S. G., and Ausloos, P., *J. Am. Chem. Soc.*, **99**, 7515 (1977).
[186] Houriet, R., Elwood, T. A., and Futrell, J. H., *J. Am. Chem. Soc.*, **100**, 2320 (1978).
[187] Scott, L. T., and Brunswold, W. R., *J. Am. Chem. Soc.*, **100**, 6535 (1978).
[188] Bowen, R. D., and Williams, D. H., *J. Am. Chem. Soc.*, **99**, 6822 (1977).
[189] Lossing, F. P., *J. Am. Chem. Soc.*, **99**, 7526 (1977).

[190] van de Graaf, B., and McLafferty, F. W., *J. Am. Chem. Soc.*, **99**, 6806, 6810 (1977).
[191] Pace, J. K., Kim, J. K., and Caserio, M. C., *J. Am. Chem. Soc.*, **100**, 3838 (1978).
[192] Bowen, R. D., Stapleton, B. J., and Williams, D. H., *J. Chem. Soc., Chem. Commun.*, **1978**, 24.
[193] Bowen, R. D., and Williams, D. H., *Org. Mass. Spectrom.*, **12**, 475 (1977).
[194] Hanzlik, R. P., and Leinwetter, M., *J. Org. Chem.*, **43**, 438 (1978).
[195] Whalen, D. L., Ross, A. M., Yagi, H., Karle, J. M., and Jerina, D. M., *J. Am. Chem. Soc.*, **100**, 5218 (1978).
[196] Gall, R. E., and Taylor, J., *Aust. J. Chem.*, **30**, 2249 (1977).
[197] Young, P. R., and Jencks, W. P., *J. Am. Chem. Soc.*, **99**, 8238 (1977).
[198] Kresge, A. J., and Chwang, W. K., *J. Am. Chem. Soc.*, **100**, 1249 (1978).
[199] Chernousova, N. N., *Deposited Doc.*, **1975**, VINITI 67-76; *Chem. Abs.*, **88**, 61761 (1978).
[200] Lelandovis, D., Bacquet, C., and Einhorn, J., *J. Chem. Soc., Chem. Commun.*, **1978**, 194.
[201] Mikhailova, T. S., Ignat'ev, V. M., Ionin, B. I., and Petrov, A. A., *Zh. Obshch. Khim.*, **48**, 701 (1978).
[202] Kazaryan, P. I., Arakelyan, A. S., and Gevorkyan, A. A., *Tzesiy Dokl.—Molodezhnaya Konf. Org. Sint. Bioorg. Khim.*, **1976**, 30; *Chem. Abs.*, **88**, 169491 (1978).
[203] Reich, I. L., Haile, C. L., and Reich, H. J., *J. Org. Chem.*, **43**, 2402 (1978).
[204] Sen Sharma, D. K., and Kebarle, P., *J. Am. Chem. Soc.*, **100**, 5826 (1978).
[205] McNab, H., *J. Chem. Soc., Perkin Trans.* 1, **1978**, 1023.
[206] Kalinowski, J., Nantka-Namirski, P., and Szczerek, I., *Acta Pol. Pharm.*, **34**, 9 (1977); *Chem. Abs.*, **88**, 61733 (1978).
[207] Kisbye, J., and Madsen, N. B., *Arch. Pharm. Chemi., Sci. Ed.*, **5**, 97, (1977); *Chem. Abs.*, **88**, 6012 (1978).
[208] Zherebtsov, I. P., Golmacheva, V. T., Lopatinskii, V. P., and Sedel'nikova, T. I., *Deposited Doc.*, **1976**, VINITI 3783-76; *Chem. Abs.*, **89**, 128747 (1978).
[209] Al-Yamoor, K. Y., Garner, A., Ali, K. M. I., and Scholes, G., *Proc. Tihany Symp. Radiat. Chem.*, **4**, 845 (1976).
[210] Berti, C., Greci, L., and Marchetti, L., *J. Heterocycl. Chem.*, **15**, 433 (1978).
[211] Wrobel, J. T., Cybulski, J., Dabrowski, Z., and MacLean, D. B., *Bull. Acad. Pol. Sci., Ser. Sci. Chim.*, **25**, 775 (1977); *Chem. Abs.*, **88**, 88842 (1978).
[212] Poplavskii, V. S., Ostrovskii, V. A., Koldobskii, G. I., Gidaspov, B. V., and Moskvin, A. V., *Zh. Org. Khim.*, **14**, 121 (1978).
[213] De Palma, V. M., Wang, Y., and Dorfman, L. M., *J. Am. Chem. Soc.*, **100**, 5416 (1978).
[214] Bunton, C. A., Carrasco, N., Huang, S. K., Paik, C. H., and Romsted, L. S., *J. Am. Chem. Soc.*, **100**, 5420 (1978).
[215] Hine, J., and Thiagarajan, V., *J. Org. Chem.*, **42**, 3978 (1977).
[216] Utimoto, K., Tanaka, M., Kitai, M., and Nozaki, H., *Tetrahedron Lett.*, **1978**, 2301.
[217] Ausloos, P., and Lias, S. G., *Chem. Phys. Lett.*, **51**, 53 (1977).
[218] Williams, D. H., Stapleton, B. J., and Bowen, R. D., *Tetrahedron Lett.*, **1978**, 2919.
[219] Bowen, R. D., and Williams, D. H., *J. Chem. Soc., Perkin Trans.* 2, **1978**, 68.
[220] Bégué, J.-P., and Malissard, M., *Tetrahedron*, **34**, 2095 (1978).
[221] Adlington, M. G., *Diss. Abstr. Int. B*, **38**, 2669 (1977); *Chem. Abs.*, **88**, 61798 (1978).
[222] Bertrand, G., Manuel, G., and Mazerolles, P., *Tetrahedron Lett.*, **1978**, 2149.
[223] Allen, W. N., and Lampe, F. W., *J. Am. Chem. Soc.*, **99**, 6816 (1977).
[224] Clark, D. T., and Cromarty, B. J., *Prog. Theor. Org. Chem.*, **2**, 447 (1977).
[225] Aleksankin, M. M., Lobanov, V. V., and Kruglyak, Y. A., *Fiz. Mol. (Kiev)*, **4**, 26 (1977); *Chem. Abs.*, **88**, 151594 (1978).
[226] Jorgensen, W. L., *J. Am. Chem. Soc.*, **100**, 1049 (1978).
[227] Dewar, M. J. S., and Rzepa, H. S., *J. Am. Chem. Soc.*, **99**, 7432 (1977).
[228] Lucchini, V., and Modena, G., *Prog. Theor. Org. Chem.*, **2**, 268 (1977).
[229] Streitwieser, A., and Alexandratos, S., *J. Am. Chem. Soc.*, **100**, 1979 (1978).
[230] Lischka, H., and Köhler, H.-J., *J. Am. Chem. Soc.*, **100**, 5297 (1978).
[231] Mirejovsky, D., Drenth, W., and van Duijneveldt, F. B., *J. Org. Chem.*, **43**, 763 (1978).
[232] Bauld, N. L., Cessac, J., and Holloway, R. L., *J. Am. Chem. Soc.*, **99**, 8140 (1977).
[233] Greenberg, A., *Tetrahedron Lett.*, **1978**, 3509.
[234] Chandrasekhar, J., Schleyer, P. von R., and Schlegel, H. B., *Tetrahedron Lett.*, **1978**, 3393.
[235] Vincent, M. A., and Radom, L., *J. Am. Chem. Soc.*, **100**, 3306 (1978).
[236] Radom, L., and Schaefer, H. F., III, *J. Am. Chem. Soc.*, **99**, 7522 (1977).

[237] Krogh-Jespersen, K., Schleyer, P. von R., Pople, J. A., and Cremer, D., *J. Am. Chem. Soc.*, **100**, 4301 (1978).
[238] Dewar, M. J. S., and Landman, D., *J. Am. Chem. Soc.*, **99**, 7439 (1977).
[239] Craig, D. P., Radom, L., and Schaefer, H. F., III, *Aust. J. Chem.*, **31**, 261 (1978).
[240] Schleyer, P. von R., Jemmis, E. D., and Pople, J. A., *J. Chem. Soc., Chem. Commun.*, **1978**, 190.
[241] Dits, H., Nibbering, N. M. M., and Verhoeven, J. W., *Chem. Phys. Lett.*, **51**, 95 (1977).
[242] Vlahovici, N., and Frangopool, P. T., *Rev. Roum. Chim.*, **22**, 1379 (1977).
[243] Vysotskii, Y. B., Kovach, N. A., and Shvaika, O. P., *Khim. Geterotsikl. Soedin.*, **1977**, 1186; *Chem. Abs.*, **88**, 6093 (1978).
[244] Mayer, R., Fabian, J., and Schoenfeld, P., *Phosphorus Sulphur*, **2**, 147 (1976); *Chem. Abs.*, **88**, 189773 (1978).
[245] Beatty, S. D., Worley, S. D., and McManus, S. P., *J. Am. Chem. Soc.*, **100**, 4254 (1978).
[246] Apeloig, Y., and Schleyer, P. von R., *Tetrahedron Lett.*, **1977**, 4647.

CHAPTER 9

Nucleophilic Aliphatic Substitution

J. SHORTER

Department of Chemistry, The University, Hull HU6 7RX

Vinylic Systems

A long series of papers on vinyl cations has been continued in a study of the solvolyses of certain vinyl fluorides, (**1**)–(**4**), in various aqueous-organic solvents.[1] Product studies and rate measurements indicated a vinyl cation (S_N1) mechanism of solvolysis.

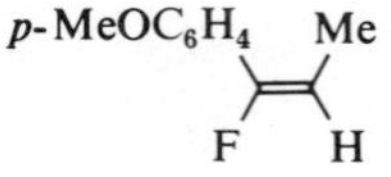

(1) 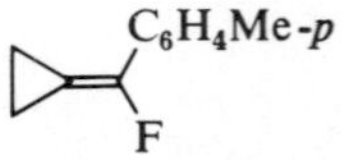 (2)

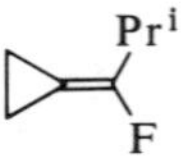

(3) 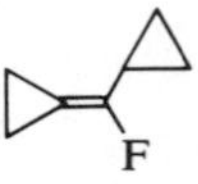 (4)

Nucleophilic attack at activated vinylic carbon usually proceeds with retention of configuration, *via* a tetrahedral transition state. Two families of vinylic substitutions proceed with inversion: the formation of charged rings, (**5**)→(**7**), and their subsequent ring-opening, (**8**)→(**10**) (*e.g.* Y = Ar, RS, I or Br); the stereochemistry of the overall process is thus retention *via* a double inversion. The transition state for each step involves planar tetraco-ordinate carbon. Rappoport has discussed the factors that may make a planar transition state more stable than the more common tetrahedral one.[2]

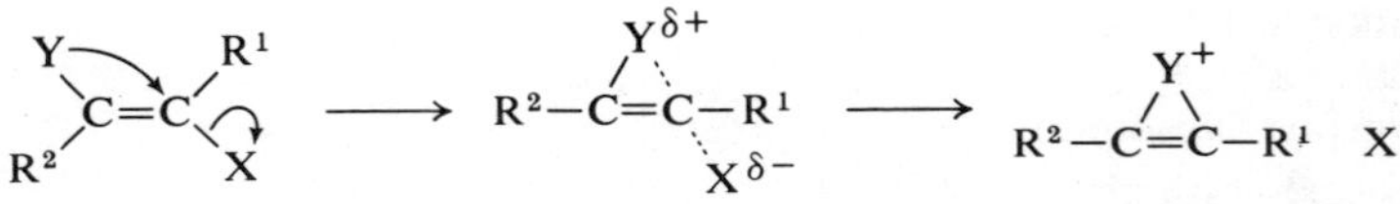

(**5**) (**6**) (**7**)

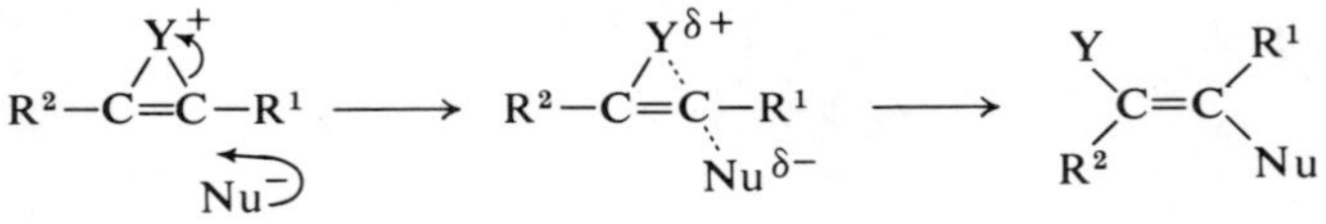

(**8**) (**9**) (**10**)

Rearrangement studies with ^{14}C have continued in work on the solvolysis of triphenylvinyl-2-^{14}C bromide in aqueous acetic acid and in 2,2,2-trifluoroethanol.[3] In the former solvent, scrambling of the label from C(2) to C(1) was observed in the reaction product but not in the unconsumed reactant. In TFE there was scrambling both in product and in reactant, which was unaffected by Et_4NBr, indicating occurrence of a 1,2-phenyl shift in the ion pair and intervention of ion-pair return.

$LiNEt_2$ or LiOEt reacts with $CF_3CF{=}CFR$ (R = Ph, $C_6H_4CF_3$, $CONEt_2$ or COOPr) preferentially at the α-position when R = Ph but exclusively at the β-position when R = COOPr.[4] For R = $CONEt_2$ the regioselectivity is not clear-cut.

A kinetic study of the reaction of tetracyanoethylene with *p*-$NH_2C_6H_4XMe$ (X = S or SO_2; the reaction is *N*-tricyanovinylation) indicated the formation of various intermediate complexes, with rate-determining formation of the final product.[5]

Kinetic studies of the reaction of (*Z*)-2-(2-bromovinyl)-5-nitrofuran with piperidine in various solvents showed that an addition–elimination mechanism was predominant.[6]

Nucleophilic substitution at the activated vinyl carbon atom has been reviewed in Russian.[7]

Allylic and Various Unsaturated Systems

There is continued interest in the behaviour of allylic systems, and particularly in the S_N2' mechanism (bimolecular nucleophilic substitution with allylic rearrangement).

The solvolyses of $RCH{=}CHCH_2Cl$ and $RCHClCH{=}CH_2$ (R = Me, Et, Pr^n, Bu^n, Bu^i or Bu^t) in alcohols (MeOH, EtOH or Pr^iOH) have been re-investigated.[8] Both (*Z*)- and (*E*)-$MeCH{=}CHCH_2Cl$, and the chiral compounds (+)-MeCHCl-$CH{=}CH_2$ and (+)-$Bu^iCHClCH{=}CH_2$ were included. Product and kinetic studies were made. The results were not consistent with any significant contribution from an S_N1 pathway and suggest a "process in which direct displacement by the solvent on a polarized molecule gives the secondary ether by a classical S_N2 reaction and indirect attack on the γ-carbon leads to the rearranged ether by the hitherto controversial and yet classical S_N2' reaction". However, in water–dioxan mixtures and in acetic acid "more ionized entities, intimate and solvent-separated ion-pairs, come into play".

The solvolysis of $HC{\equiv}CCHClCH{=}CH_2$ in ethanol–water or TFE–water mixtures does not involve nucleophilic participation of solvent in the rate-determining step, but is accompanied by some rearrangement to $HC{\equiv}CCH{=}CHCH_2Cl$ and there are various products.[9] The results were explained in terms of an ion pair, of tightness varying with the solvent. On the other hand, the solvolysis of $HC{\equiv}CCH{=}CHCH_2Cl$ involves nucleophilic participation of solvent in the rate-determining step, and the role of the S_N2' mechanism is enhanced by the electron-withdrawing properties of the ethynyl group.[9]

Products of solvolysis in various solvents and rates of ethanolysis have been determined for 2-(phenylethynyl)allyl, 2-cyclopropylallyl, and 2-phenylallyl toluene-*p*-sulphonates, and the results have been compared with those for their respective cyclopropyl toluene-*p*-sulphonate counterparts.[10] The theoretically expected ring-closure of 2-substituted allyl cations, with efficient electron-releasing groups, into stabilized cyclopropyl cations was not found, but a limitation of the anchimeric assistance of the double bond seems to result from the presence of such substituents.

An investigation[11] of the stereochemistry of the S_N2' reactions of acyclic allylic esters with a chiral primary amine found *syn*-displacement favoured over *anti*-displacement by a factor of 1.4–1.8. ". . . contrary to the long-held view that S_N2' reactions proceed with *syn*-stereochemistry, the whole spectrum spanned by the *syn*- and *anti*-extremes is to be expected depending, in any particular case, on the nature of the displacing and the displaced groups, counter-ions, and solvent." On the other hand, the S_N2' reaction of an acyclic allylic chloride with diethylamine shows *ca.* 97% *syn*-stereochemistry.[12]

2,3-Dichlorobuta-1,3-diene reacts with two equivalents of lithium diphenylphosphide to give $Ph_2PCH_2C{\equiv}CCH_2PPh_2$.[13] This can be rationalized as two S_N2'-type reactions involving phosphide anion. A similar reaction of chloroprene gives $Ph_2PCH_2CH{=}C{=}CH_2$.

Allylic alcohols react with N_3H/BF_3–OEt_2 in benzene *via* an ion-pair mechanism to give allylic azides.[14] Retention of configuration is enhanced when the concentration of N_3H is increased. Thus the reaction has characteristics different from those of the Ritter reaction of allylic alcohols with BF_3–OEt_2 in acetonitrile, in which an allylic cation is formed and is then attacked by MeCN as a nucleophile.[14]

The reactions of *cis*-3,4-dichlorocyclobutene (**11**) with OMe^- proceed exclusively with *syn*-stereochemistry and allylic rearrangement to the products (**12**) and (**13**).[15]

Solvolyses of cholest-4-en-3β- and -3α-yl trifluoroacetates give different mixtures of 3-substituted-Δ^4- and 5-substituted-Δ^3-products, probably *via* two distinct

Cl, Cl (**11**) —NaOMe/MeOH→ Cl, OMe (**12**) ⟶ OMe, OMe (**13**)

allylic carbonium ions.[16] When the same steroidal compounds are treated with NaN_3–HMPA, bimolecular substitution and configurational inversion occur.

Some examples of the little-studied S_Ni' reaction (the intramolecular version of S_N2') have been examined;[17] these involved the basic hydrolysis of four isomeric dichlorocyclohexenecarboxylic esters. The S_Ni' process was shown to be stereoselective. The occurrence of S_N2' and S_Ni' mechanisms in the reactions of cyclohexene systems has also been considered in terms of conformational analysis and the method of "*la notation des angles de torsion*".[18]

There is continued interest in the use of metal compounds to influence regio- or stereo-selectivity in the reactions of allylic systems. Grignard reagents are of particular interest in this connection, *e.g.* their reaction with allyl 2-pyridyl ethers in the presence of a considerable excess of $MgBr_2$.[19] With primary allyl ethers C(1)-alkylated products predominate (by S_N2 reaction), but with secondary or tertiary allyl ethers C(3)-alkylated products predominate (by S_N2' reaction). The N atom of the pyridine ring co-ordinates with $MgBr_2$ to form an active intermediate. In the case of primary allyl ethers the Grignard reagent can further co-ordinate to the $MgBr_2$ and attack C(1). However, for secondary and tertiary allyl ethers this process is subject to steric hindrance, and intermolecular attack by the Grignard reagent at C(3) is easier.

Several papers have dealt with the use of copper compounds to influence selectivity. A novel class of alkylating agent $RCuBF_3$ is said to effect substitution of allyl halides with complete allylic rearrangement.[20] Butylation of but-2-enyl chloride and its isomer 3-chlorobut-1-ene produces 3-methylhept-1-ene and oct-2-ene, respectively; it has therefore been suggested that the reactions do not involve a common intermediate such as an allylic radical; the reagent probably functions as $Cu^+RBF_3^-$. Allylic alcohols may also be attacked directly with this reagent, γ-alkylation predominating;[21] allylic acetates also react smoothly but allylic ethers do so only sluggishly.

Selective γ-alkylation of allylic alcohols may also be accomplished by using alkyl-lithium compounds in association with tributyl-(*N*-methyl-*N*-phenylamino)-phosphonium iodide and cuprous iodide.[22] The regioselectivity for γ-products is commonly 90–100%, and the stereochemistry of this S_N2' reaction is predominantly *anti*; the authors interpret these observations by means of the sequence (**14**)–(**16**). Two *cis–trans*-isomeric allylic ethers are methylated by MeMgI–CuI with allylic rearrangement and *anti*-stereochemistry.[23]

anti-1,3-Substitution also occurs predominantly when organocuprates react with the methylsulphinate of (*R*)-3-phenylprop-1-yn-3-ol to give substituted allenes.[24] A copper enolate species obtained by treating the lithium enolate of ethyl acetate with cuprous iodide reacts with propargyl bromide in a S_N2' manner to

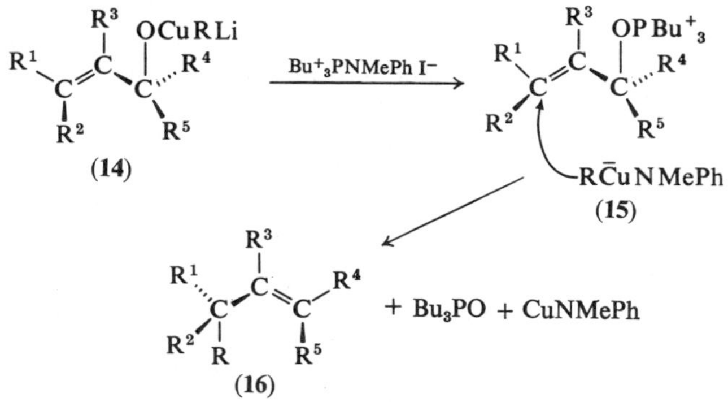

$$R^1R^2C{=}C(R^3){-}C(OCuRLi)R^4R^5 \ \textbf{(14)} \xrightarrow{Bu^t{}_3PNMePh\ I^-} R^1R^2C{=}C(R^3){-}C(OPBu^t{}_3{}^+)R^4R^5 \ \cdot \ R\bar{C}uNMePh \ \textbf{(15)}$$

$$\longrightarrow R^1R^2C(R){-}C(R^3){=}C{=}CR^4R^5 \ \textbf{(16)} + Bu_3PO + CuNMePh$$

give ethyl penta-3,4-dienoate;[25] under the same conditions the lithium enolate gives only ethyl pent-4-ynoate by S_N2 displacement. The rearrangement of propargyl compounds to allenes has also been discussed in two other papers.[26,27]

Nucleophilic substitution of allylic systems has also been discussed in relation to certain natural products and biological reactions.[28-30]

Norbornyl and Closely Related Systems

Interest in these systems appears unabated. The contributions of H. C. Brown's group to our knowledge and understanding of the 2-norbornyl system continue.[31-34]

The solvolysis of dimethyl(*endo*-2-norbornyl)carbinyl *p*-nitrobenzoate (**17**) in 80% aqueous acetone at 25 °C proceeds 18 times faster than that of the *exo*-isomer (**18**).[31] A similar factor is observed for the solvolysis of dimethylneopentylcarbinyl *p*-nitrobenzoate (**19**) compared with that of *tert*-butyl *p*-nitrobenzoate (**20**). These

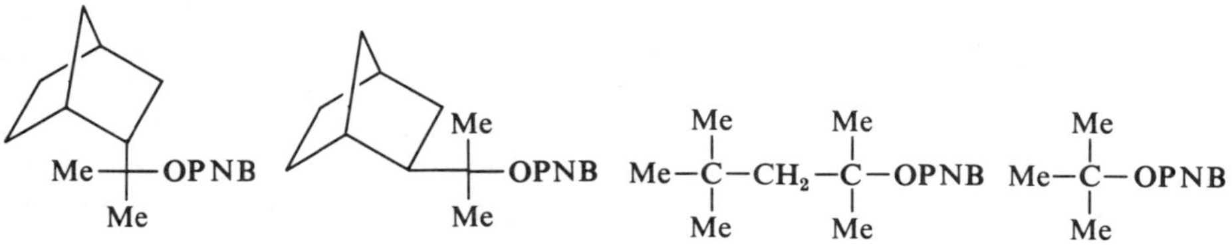

(17)   **(18)** $Me{-}C(Me)_2{-}CH_2{-}C(Me)_2{-}OPNB$ **(19)** $Me{-}C(Me)_2{-}OPNB$ **(20)**

OPNB = *p*-Nitrobenzoyloxy

results can be understood if relief of steric strain facilitates the solvolysis of both the *endo*-isomer and the dimethylneopentylcarbinyl derivative. On the other hand, the rate coefficient for *exo*-2-(dimethylamino)norbornane reacting with methyl iodide in nitrobenzene at 25 °C is 51 times that of the *endo*-isomer. For *n*-butyldimethylamine compared with dimethylneopentylamine in the same reaction the factor is 150. Thus reactions proceeding with an increase in steric strain are resisted by the systems which are initially the more strained. It is concluded that steric factors play an important role in governing the differences in behaviour between *exo*- and *endo*-2-norbornyl systems, whereas differences in electronic effects between the *exo*- and *endo*-moieties have never been convincingly demonstrated.

The last-mentioned matter is taken up again in a later paper.[32] The solvolyses of several aryl(*exo*- or *endo*-2-norbornyl)methylcarbinyl *p*-nitrobenzoates were studied in 80% aqueous acetone. The *endo*-derivatives react four or five times faster than the *exo*-derivatives, presumably because enhanced steric strain facilitates ionization of the former. The ρ^+ values (25 °C) for the effects of substituents in the aryl group are -4.44 for *exo*- and -4.47 for *endo*-derivatives. Thus H. C. Brown's "tool of increasing electron demand" fails to reveal any significant electronic factor in the *exo*-isomers that facilitates their ionization, an electronic factor not present in the corresponding *endo*-derivatives.

Rate coefficients (25 °C) have been determined for the solvolysis of *exo*- or *endo*-2-norbornyl toluene-*p*-sulphonate in a very wide variety of solvents, so that a reactivity range of about 10^5 is spanned.[33] Over the entire range of solvents, $\log k_{exo}$ is linearly related to $\log k_{endo}$. This is not consistent with the view that the high *exo* : *endo* rate ratio is due to the solvolysis of the *exo*-isomer involving a k_Δ process, insensitive to the nucleophilicity of the solvent, while the solvolysis of the *endo*-isomer involves a k_s process, sensitive to the nucleophilicity of the solvent. Further, $\log k_{exo}$ or $\log k_{endo}$ is linearly related to $\log k$ for the solvolysis of 2-adamantyl toluene-*p*-sulphonate (**21**) in the same range of solvents. The

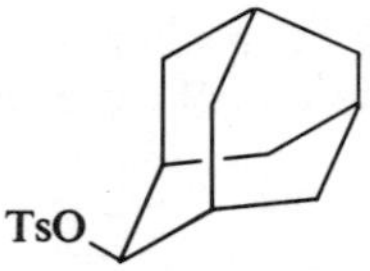

TsO

(**21**)

last-mentioned reaction is regarded as a k_c process (anchimerically and nucleophilically unassisted), and the authors suggest that the solvolyses of the *exo*- and *endo*-2-norbornyl sulphonates should both be regarded as essentially k_c processes.

A fourth paper is devoted to the solvolysis of *exo*- and *endo*-1,2-diphenyl-2-norbornyl and -1,2-dimethyl-2-norbornyl *p*-nitrobenzoates and chlorides.[34] Detailed comparisons are made with the corresponding tertiary 2-phenyl and 2-methyl derivatives and with the secondary parent 2-norbornyl compounds. In the space available we can but quote from the conclusions: the results ". . . fail to support the presence of a major non-classical resonance contribution in the *exo* secondary, absent in the *endo* secondary and in the *exo* and *endo* tertiary derivatives, as postulated in some current proposals". Steric hindrance to ionization of the *endo*-derivatives, secondary and tertiary, is held to be the factor responsible for the high *exo* : *endo* rate ratios.

The role of nucleophilic solvent assistance in the solvolysis of *endo*-2-norbornyl derivatives has also been examined by other workers.[35] Nine probes of nucleophilic solvent assistance have been applied to *endo*-norbornyl solvolysis and it has been determined that the amount of solvent assistance is small,[33] and the mechanism is essentially a k_c process.

An electron-withdrawing group at the 3-position of 2-norbornyl toluene-*p*-sulphonate (**22**) does not enhance the *exo* : *endo* rate ratio in solvolysis, in contrast to what is found in the 2-norborn-5-enyl system (**23**);[36] in the latter, the

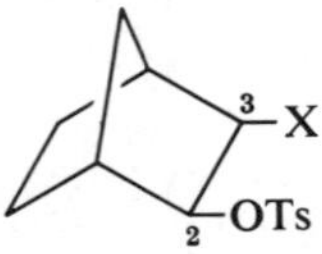

3 X
2 OTs
(**22**)

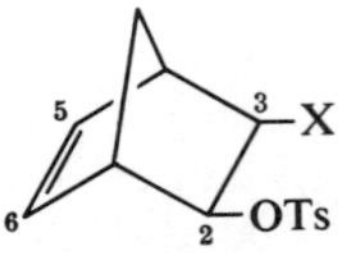

5 3 X
6 2 OTs
(**23**)

enhancement is due to increased participation by the 5,6-double bond and the lack of enhancement in the former system is consistent with the absence of σ-participation. The authors comment: "Our conclusions refer only to participation in the solvolytic transition state. These experiments do not speak to the issue of delocalization in the intermediate 2-norbornyl cation, which has been studied in fluorsulfonic acid and other media".

Comparison of product analysis of the acid-catalysed ring-opening of the epoxynorbornanes (**24**) and (**25**) with that of the corresponding reaction of nortricyclanol (**26**) indicates that the intermediates in the former are classical secondary carbenium ions.[37]

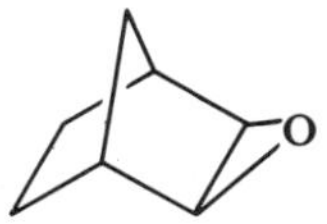

(**24**)  (**25**) 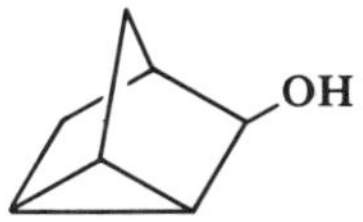 (**26**)

Work on solvolysis in carboxamides has continued with kinetic, product, and deuterium-tracer studies of the elimination of *exo*-2-norbornyl arenesulphonates.[38] It is suggested that the intimate ion-pair from which norbornene is formed is unsymmetrical, with a "classical" norbornyl-cation structure.

Other studies of the 2-norbornyl system have included the examination of stereo-electronic effects on the protonation and properties of 2-norbornyl chlorides through MINDO/3 calculations[39] and acetolyses of *exo*- and *endo*-5,6-bis-(methylene)-2-norbornyl *p*-bromobenzenesulphonates,[40] which closely resemble those of the corresponding 2-norborn-5-enyl compounds.[41] Kinetic and product studies have been made for the acid-catalysed hydrolyses of 7-*syn*- and 5-*endo*-diazoacetyl-2-norbornene, and the results have been compared with those of the corresponding saturated analogues.[42] The acetolysis of dibenzonorbornadienyl-1-carbinyl trifluoromethanesulphonate has been compared with that of the norbornyl-1-carbinyl system and its unsaturated analogues.[43] Studies of secondary deuterium isotope effects have confirmed the occurrence of π-participation in the solvolysis of the unsaturated fused-ring derivative of the *anti*-7-norbornyl system (**27**).[44]

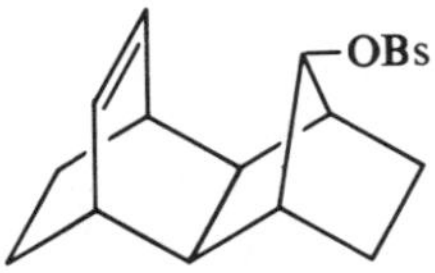

(**27**)

Miscellaneous Polycyclic Systems

Rate coefficients have been measured for the solvolysis of 1-substituted 3-bromo-adamantanes (**28**) in 80% ethanol.[45] When R is an alkyl or electron-attracting substituent, $\log k$ is well correlated with σ_I^q (inductive constant based on quinuclidine). Rates are higher than predicted by this correlation when R is a $+M$ substituent or electrofugal group. This has been attributed to CC hyper-conjugative relay of positive charge from the cationic centre at C(3) to the substituent at C(1). When R is a strong electron-donor (*e.g.* O^- or S^-) heterolytic fragmentation occurs, rates and products being controlled by the "frangomeric" effect.'

Inductive and hyperconjugative effects have also been studied for the solvolysis of 4-substituted bicyclo[2.2.2]oct-1-yl *p*-nitrobenzenesulphonates.[46] Here alkyl groups follow the inductive order, but all are less electron-releasing than H, *i.e.* $H > Bu^t > Pr^i > Et > Me$.

For the solvolysis of 1-alkyl-2-adamantyl substrates, empirical force-field calculations show that the acceleration in such series as $Me < Et < Pr^i < Bu^t$ may be fully accounted for by relief of steric strain.[47]

For the competitive ethanolysis and cyclization of stereoisomeric 4-benzamido-5-(methanesulphonyloxy)twistanes[48] and for the Hofmann reaction of the quaternary salts of 4-dimethylamino-5-twistanols,[49] the dependence of the course of the reaction on the torsion angle has been investigated.

Steric and conformational effects have been studied in the acetolysis of the ring-fused tertiary cyclopropyl derivatives (**29**) and (**30**), giving only allylic acetates.[50]

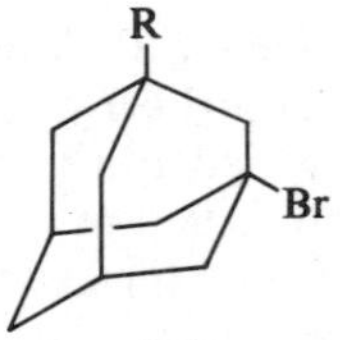

(**28**)

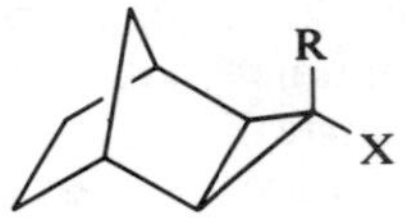

(**29**) R = Me; X = OTs
(**30**) R = Ph; X = OTFA

Kirmse's series of papers on deamination reactions has been extended with studies of the decomposition of bicyclo[4.1.0]heptane-7-,[51] bicyclo[4.1.0]hept-2-ene-7-,[52] bicyclo[5.1.0]oct-2-ene-8-,[53] and bicyclo[5.1.0]octa-2,4-diene-8-diazonium ions.[54]

Rate coefficients and secondary deuterium isotope effects have been measured for the solvolyses of *endo*- and *exo*-bicyclo[3.2.1]octan-3-yl toluene-*p*-sulphonates.[55] The reactions are close to the limiting (S_N1) extreme in almost all the solvents used, although for the *endo*-compound in 98% ethanol some direct S_N2 reaction of solvent with covalent sulphonate has been invoked; mechanisms of solvolysis generally have been discussed, including the roles of intimate and of solvent-separated ion-pairs, and solvent-induced S_N2.

A series of papers on 9-decalyl and related cations has been extended with kinetic and product studies of the generation of *cis*- and *trans*-2-*tert*-butyl-9-decalyl cations through σ-routes[56] and through π-routes[57] by acetolysis; anchimeric assistance, stereospecificity, and the role of ion pairs are discussed.

Molecular mechanics (the Schleyer method) has been used to predict the solvolysis rates of polycyclic secondary derivatives.[58] The predictions are good for acetolysis rates of rigid systems reacting by the k_c process. Calculated rates have been compared with experimental rates for substrates which may potentially react with σ assistance, and such assistance has been shown to be important for several reactions. The formation of σ-bridged non-classical intermediates appears to be involved in several cases. Similar methods have also been used to interpret the solvolytic behaviour (Cope rearrangement) of the highly hindered *exo*-tricyclo[4.4.1.1^{2,5}]dodeca-3,7,9-trien-11-yl toluene-*p*-sulphonates and derivatives.[59]

Stereochemical studies on 3,4-benzobicyclo[4.1.0]hept-3-en-2-ol systems and solvolytic studies on the corresponding *p*-nitrobenzoates have been carried out;[60] a homonaphthalenium ion intermediate (**31**) has been postulated.

The alcohol-to-chloride transformations of several isomeric benzobicyclo-octadienyl alcohols and dibenzobicyclononatrienyl alcohols upon reaction with

(**31**)

PPh_3–CCl_4 reagent[61] involve competition between S_N2, S_N1 and S_Ni' or S_N2' processes. The photochemical and thermal methanolyses of the epimeric 4-bromo-5-bromomethyldibenzobicyclo[3.2.1]octadienes have also been studied.[62]

Solvolytic studies of hypostrophene derivatives[63] and of *syn*- and *anti*-tricyclo-[4.2.0.0^{2,5}]octa-3,5-diene derivatives[64] have been made to elucidate the impact of high-lying σ orbitals and extensive through-bond interaction on chemical reactivity. Other systems studied with respect to solvolytic reactivity and mechanism include chloro-derivatives of dibenzobicyclo[2.2.2]octatriene (in H_2SO_4–HOAc),[65] *anti*-pentacyclo[5.4.0.0.^{2,6}0.^{3,9}0^{5,8}]undec-10-en-4-yl (a highly-strained cage compound) and related toluene-*p*-sulphonates (interest in conformational requirements),[66] and chloro-derivatives of tricyclo[5.2.1.0^{4,10}]deca-2,5,8-triene ("Triquinacene").[67] The last-mentioned is of interest in connection with the reactivity of the bridgehead chloro-compounds; these also react with SbF_5 in SO_2ClF to give allyl-stabilized "free" bridgehead carbenium ions, *e.g.* (**32**).

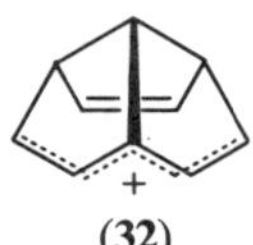

(**32**)

Epoxide Reactions

The kinetics of the hydrolytic ring-opening of tetramethyloxirane in aqueous buffers to form pinacol have been investigated by NMR-spectroscopy using a repetitive timed integration technique.[68] Two concurrent processes contribute to the catalytic term k_{HA}: one involves attack by the nucleophile A^- on protonated epoxide, the other involves general acid catalysis. Isotope effects and activation parameters indicated nucleophilic attack by water upon protonated epoxide. The mechanism was discussed in the light of MO theory and "a new view of the orientation of epoxide ring-opening, which may apply to other cyclic onium compounds" was put forward.

Six related cycloalkene epoxides were subjected to ring-opening by acidic solutions containing charged or uncharged nucleophiles.[69] By means of product analysis, nucleophilic attacks on different oxiranes were compared directly; the results were considered to provide the most extensive demonstration of the effects of charge orientation yet available.

The ring-opening of oxiranes (and also of aziridines and thi-iranes) may be accomplished by phenyl acetate in the presence of base (*e.g.* pyridine).[70]

The addition of salt increases markedly the *syn*-stereochemistry of the acid-catalysed ring-opening of 1-phenyl- and 1-(*m*-chlorophenyl)-cyclohexene oxide.[71]

Thus the acid-catalysed ethanolysis of the former normally gives predominantly *trans*-2-ethoxy-2-phenylcyclohexanol,[72] but in the presence of 3M-$LiClO_4$ gives predominantly the *cis*-compound.

The influence of 3-substituents on the rates of reaction of cyclohexene oxide with hydrogen chloride in aprotic solvents of low polarity has been studied.[73] Satisfactory second-order kinetics were obtained in toluene or carbon tetrachloride in the presence of small amounts of water or THF.

The epoxidation of 1-Bu^t-, 1-Me_3Si-, and 1-Me_3Ge-derivatives of 4,4-dimethylcyclohexenes has been described and the stereochemistry of ring-opening of the resulting epoxides by $LiAlH_4$, $MeOH/H^+$ or H_2O/H^+ has been examined.[74] The regiochemistry of the reactions of the Bu^t-epoxide is completely reversed in the case of the Me_3Si and Me_3Ge compounds by the electronic effects of these groups.

Product analysis for the action of borane on epoxymethylenecyclohexane has been carried out and the mechanism discussed.[75]

Solvent effects have been studied for simultaneous substitution and elimination in the reaction of allyl-lithium and of lithium acetylide with unsaturated cyclic epoxides.[76]

The hydration of 2-pyridyloxirane to the corresponding diol is strongly catalysed by copper(II) in dilute aqueous solution.[77] Kinetic studies are presented and the results (including previously reported regiospecificity) support a mechanism in which a copper(II) chelate of the oxirane undergoes attack by water or other nucleophile present. A bearing on the toxic biological effects of epoxides was seen.

The acid-catalysed ethanolysis of *trans*-2,3-diphenyloxirane in "pure" ethanol at 50 °C proceeds with 26% retention and 74% inversion.[78] While addition of hexane or benzene produces almost no change in the stereochemistry, acetonitrile, nitromethane or sulpholane increases the degree of retention, and DMSO, DMF or HMPA increases the degree of inversion. The results have been discussed in terms of the solvation-shell concept.

Rate coefficients were measured for the ring cleavage of the oxides (**33**) of substituted benzylideneacetophenones with morpholine.[79] For variation of R,

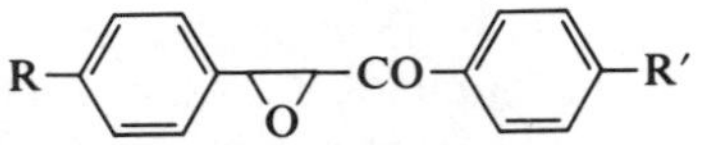

R–C6H4–(epoxide, O)–CO–C6H4–R′

(**33**)

$\log k$ *versus* σ^+ gave $\rho^+ = -0.64$ (50 °C), and for variation of R′ there was correlation with σ^0 and $\rho^0 = 0.455$ (70 °C).

p-Nitrostyrene oxides bearing α-acetoxy-, α-trifluoroethoxy- or α-fluoro-substituents (potential leaving groups) have been synthesized *via* peracid epoxidation of the corresponding olefins.[80] In pH 8.0 buffer each is cleanly converted into α-hydroxy-*p*-nitroacetophenone, the half-lives being 35, 0.4, and 0.4 min, respectively.

The stereochemistry of the opening of substituted 3,4-epoxybicyclo[4.1.0]-heptanes in the presence of nucleophilic reagents has been studied.[81]

The HBr-induced opening of the epoxide ring in the stereoisomeric 1-hydroxy-(or -acetoxy)-2,3-epoxy- and 3-hydroxy(or -acetoxy)-1,2-epoxy-5-α-cholestanes has been examined.[82] Product studies are evaluated with regard to the tendency to "diaxial opening" of the epoxide ring and the tendency of the nucleophile to attack at the C-atom β with respect to the neighbouring group. The same authors have also studied the cleavage of the epoxide ring in steroidal *trans*-$\alpha\beta$- and $\beta\gamma$-epoxy-alcohols by sodium borohydride in refluxing methanol.[83] The *cis*-compound is unchanged under these conditions.

Oxetane formation from 3,4-epoxy-alcohols under aqueous conditions proceeds regioselectively "beyond the expectation" and may be rationalized mainly in terms of the "collinearity requirement" in the transition state, although it depends to some extent on a steric effect.[84]

A convenient route to optically active hydrobenzoins lies in the *syn*-opening of a racemic stilbene epoxide by an optically active carboxylic acid.[85]

Kinetic studies have shown that the reaction of $Me(CH_2)_7SCH_2CH_2OH$ with methyloxirane to give $Me(CH_2)_7SCH_2CHMeOH$ and oxirane proceeds *via* a sulphonium intermediate.[86]

Rate coefficients and activation parameters have been measured for the reaction of hexafluoropropene oxide with various alkali-metal carboxylates in a variety of aprotic solvents.[87]

Two papers have dealt with epoxycarbinyl solvolyses: rate coefficients and activation parameters have been measured for the solvolyses of a pair of diastereomeric epoxycarbinyl *p*-bromobenzenesulphonate esters[88] and rates have also been measured for solvolytic reactions of *syn*- and *anti*-9-oxabicyclo[6.1.0]-non-2-yl *p*-bromobenzenesulphonates.[89] The latter reactions are slower by *ca.* 10^7 than those of the corresponding cyclopropyl compounds (*i.e.* CH_2 instead of O).

There have been various papers of biological interest in this field. Bruice's group has studied "structural features which determine the carcinogenesis, mutagenesis, and rates of acid- and water-mediated solvolysis of and nucleophilic attack upon diol epoxides, bay-region and non-bay-region tetrahydro-epoxides, and *K*-region and non-*K*-region arene oxides".[90] They have also studied the inactivation of glyceraldehyde 3-phosphate dehydrogenase and yeast alcohol dehydrogenase by arene oxides.[91] Other studies of biological interest include an investigation of stereo-electronic factors in the solvolysis of bay-region diol epoxides of polycyclic aromatic hydrocarbons[92] and a study of biomimetic reactions of epoxygermacrene D.[93]

As part of a study of the catalysis of organic reactions at alumina surfaces, the nucleophilic opening of structurally diverse epoxides has been performed under very mild conditions;[94] the reagents include alcohols, thiols, amines, and acetic acid. The ring-opening occurs stereospecifically (*trans*) and regioselectively, with nucleophilic incorporation preferentially at the less-substituted epoxide C-atom; the mechanism appears to be concerted. The *trans*-opening of *K*-region and non-*K*-region arene oxides, and of vinyl and aryl epoxides has also been studied.[95]

Substitution at Elements other than Carbon

Substitution at Silicon

The relationship between the stereochemistry of nucleophilic substitution at silicon and the electronic character of the attacking nucleophile has been studied;[96] hard nucleophiles favour retention (a charge-controlled mechanism), while soft nucleophiles favour inversion (a frontier orbital process).

α-Naphthylferrocenyl-fluorosilane and -chlorosilane have been synthesized as the first bifunctional enantiomeric silicon compounds.[97] A high selectivity between the two functional groups is observed in substitution reactions with organolithium compounds and Grignard reagents, the more polarizable group being specifically replaced. The stereochemistry of the reactions has been examined.

Rate measurements have been made for the hydrolysis of some chlorosilanes in weakly polar solvents, at a very low concentration of water and in the presence of nucleophiles.[98] The kinetics are of the third order and hydrolysis appears to be promoted by nucleophilic assistance localized on the Si atom. The slow step is the addition of H_2O to a pentaco-ordinate intermediate, to form a hexaco-ordinate intermediate. In a later paper a stereochemical proof of the existence of nucleophilic activation at a Si atom is given.[99] The reactions proceed with retention of configuration at Si, instead of the inversion always previously observed in the reactions of acyclic chlorosilanes.

New stereochemical and kinetic results have been reported for reactions between organolithium reagents and some fluoro- and methoxy-silanes.[100] They reveal the dominance of ion-pair dissociation, and thus of the electronic character of the nucleophile, and eliminate a mechanism involving complexation control, the S_Ni–Si mechanism.

The influence of the nature of the leaving group on the stereochemistry has been examined.[101] Results for optically active bromo- and thiophenyl-silanes are compared with those reported for Cl, F, OR, and H leaving groups. The predominant stereochemistry changes in the direction inversion→retention in the order of decreasing lability of the leaving groups: $Br \sim Cl \gg SR \sim F \gg OMe > H$.

Heterogeneous catalysis, by salts, of the alcoholysis of the Si—H bond has also been studied;[102] the reaction is highly selective and by choice of conditions it is possible to prepare mono-, di- or tri-alkoxysilanes. The salts include potassium, phthalate, thiocyanate, formate, and fluoride and caesium fluoride and acetate; the anion of the salt is thought to form a pentaco-ordinate Si complex in which the Si–H electron pair is displaced towards H and nucleophilic attack of alcohol on Si is thereby promoted.

The extraordinary reactivity of the ring Si—C bonds in silacyclopropanes (siliranes) has been investigated for 1,1-dimethyl-*trans*-2,3-bis(2′,2′-dimethylcyclopropylidene)-1-silacyclopropane (**34**).[103] Reaction with oxygen, water,

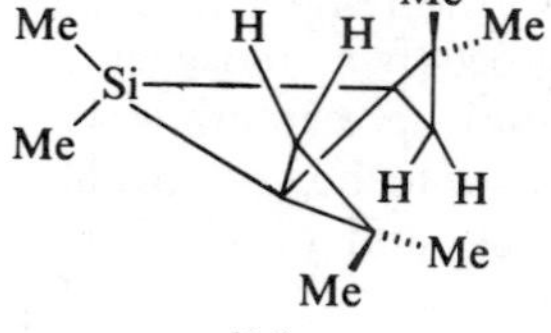

Me Me Si H H Me Me H H Me Me

(**34**)

alcohols, primary and some secondary amines, hydrogen sulphide, carboxylic acids or hydrazoic acid proceeds exothermically at room temperature with opening of the Si—C bond of the SiC_2 ring. Hydrogen chloride, methyl- or phenyl-lithium, or lithium dialkylamide reacts rapidly at −78 °C. Ring-cleavage can also be effected with acetyl chloride, boron trichloride or lithium aluminium hydride. The hyper-reactivity of these Si—C bonds is due to high ring strain, as is shown by a comparison with the reactivity of 1,1-dimethyl-1-silacyclobutane. The mechanism and stereochemistry of these ring-opening reactions are not yet clearly established.

Secondary phosphines, R_2PH, with R = Me, Ph or Me_3Si, cleave the ring of hexamethylsilirane exothermally at room temperature to form silylphosphines, $R_2PSiMe_2(CMe_2CHMe_2)$.[104] $MePH_2$ reacts similarly, but reaction of this silirane with $RPCl_2$ (R = Me, Et or Ph) gives Me_2SiCl_2, $Me_2C{=}CMe_2$, and cyclo-$[RP]_5$; with Ph_2PCl, the products include $Ph_2P{-}PPh_2$. Again, mechanistic details are not clear. The reactions of hexamethylsilirane and of 1,1-dimethyl-2,3-bis(trimethylsilyl)-1-silirene (**35**) with DMSO have also been examined.[105] In both cases nucleophilic attack of DMSO on ring-Si appears to generate dimethylsilanone, Me_2SiO, which then inserts into the strained and very reactive three-membered rings to give the cyclic siloxanes (**36**) and (**37**) respectively. Various mechanistic schemes have been considered.

Me_3Si, $SiMe_3$, C=C, Si, Me, Me
(**35**)

$Me_2C{-}CMe_2$, Me_2Si, $SiMe_2$, O
(**36**)

Me_3Si, $SiMe_3$, C=C, Me_2Si, $SiMe_2$, O
(**37**)

The kinetics of cleavage by butanol of the carboranes 1-$RMe_2SiCB_{10}H_{10}CH$-7 (R = Me, $ClCH_2$, MeO, EtO, Bu or $CF_3CH_2CH_2$) have been examined and the results have been correlated with the Taft substituent constant;[106] an S_N2 mechanism is involved. A long series of papers on silicon-containing heterocyclic compounds has continued with a study of the reactivity of 1,3-dichloro-1,3-diphenyl-2-oxa-1,3-disilaindan in hydrolytic polymerization;[107] inclusion of the SiOSi group in the indan nucleus increases the reactivity of the Si—Cl bond.

The group $(Me_3Si)_3C$ causes very large steric hindrance to nucleophilic attack at a Si atom to which the group is attached.[108] Thus $(Me_3Si)_3CSiMe_2Cl$ is even less reactive than $Bu^t{}_3SiCl$ towards base. The former chloro-compound, and also the corresponding bromo- and iodo-compounds, are cleaved by MeOH–MeONa to give $(Me_3Si)_2CHSiMe_2OMe$, possibly *via* the silaolefin $(Me_3Si)_2C{=}SiMe_2$. This reaction and other examples in the paper illustrate the reluctance of silicon to form siliconium ions in solution, *i.e.* if ordinary nucleophilic substitution is prevented, unusual reactions are preferred to reaction *via* a siliconium ion.

The alkaline hydrolysis of 3-(2-methoxyethoxy)dimethylpropylsilane (**38**) has been studied in a range of dioxan–water mixtures.[109] Statistical analysis of the kinetic data indicates that about $3H_2O$ are involved in the transition state of the rate-determining step, see (**39**). Rate constants have been determined for all

$CH_3O(CH_2)_2O(CH_2)_3SiMe_2H$
(**38**)

[HO—Si···H···H···O—H, HOH, HOH]
(**39**)

steps of the 3-step consecutive-parallel methanolysis of $RSiH_3$ [R = Me_2CHCH_2, $Ph(CH_2)_2$, $PhCH_2$, $NC(CH_2)_3$, Ph or $NC(CH_2)_2$].[110] Linear correlations with Taft substituent constants were obtained. In the methanolysis of the intermediates

$RSiH_2OMe$, the inductive effect of the OMe group was about 40% compensated by an opposing p_π–d_π conjugative effect. The kinetics of acidic hydrolysis of organosilicon hydrides have been used to estimate the steric constants, E_s, of certain substituents.[111]

For the reactions of alkoxysilanes, alkoxygermanes, and thioalkoxygermanes with various unsaturated reagents a four-centre transition state (**40**) was previously postulated; this has been confirmed by demonstrating retention of configuration at the atom M.[112] The kinetics of the alkaline hydrolysis of tetraethoxysilane have been studied.[113] In the esterification of tetrachlorosilane by ethanol, a computer method was used to determine the rate constants for the individual mono-, di-, tri-, and tetra-ethoxylation steps, and the reverse of the last step.[114]

A new and efficient cleavage of carboxylic esters with Me_3SiI or a mixture of $PhSiMe_3+I_2$ has been described.[115] Ethers also can be dealkylated by the latter reagent. Transition states have been postulated, *e.g.* (**41**), and it has been suggested that cleavage by $PhSiMe_3+I_2$ does not necessarily involve intermediate Me_3SiI.

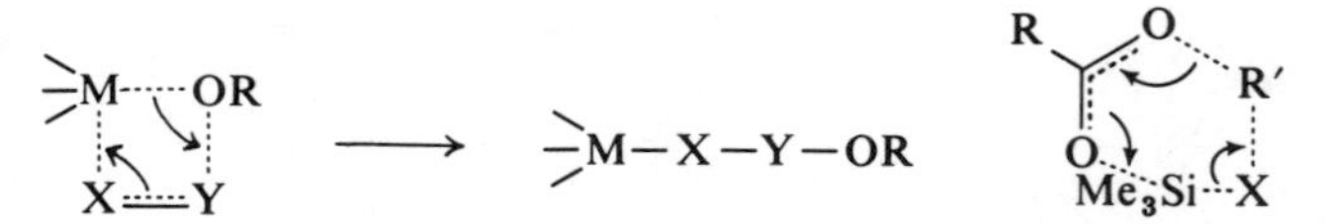

>M····OR
X═Y ⟶ >M—X—Y—OR

(**40**)

R, O, R′, O, Me_3Si···X

(**41**)

A kinetic study of the reactivity of acids, amines, and phenols in the cleavage of Si—S bonds in $(Me_2RSi)_2S$ has been carried out.[116]

Substitution at Phosphorus

The reactions of $C_6H_{11}P(OR)_2$ with phenol and of $C_6H_{11}P(NRR')_2$ with decanol at 363–463 K are of the second order.[117] Spectroscopic studies indicated that halide exchange occurs between (**42**) and (**43**) *via* a four-membered transition state with inversion at P; the solvent effect was examined.[118] CNDO/2 MO calculations have been carried out for three possible reaction routes for the attack of F^- on PHClF or SOClF, as exemplifying nucleophilic substitution at phosphorus or sulphur in phosphinyl or sulphinyl compounds.[119]

Kinetics, medium, and deuterium isotope effects have been studied for the alkaline decomposition of tetraphenylphosphonium chloride in DMSO–water mixtures.[120] The reaction is first-order in phosphonium cation and second-order in OH^- or OD^-. The rate of reaction increases dramatically as the DMSO content is increased. This has been attributed both to initial-state and to transition-state effects. The kinetic solvent isotope effect in the medium containing 60% DMSO is strongly dependent on temperature, since the activation energies differ by about 10 kcal mol^{-1}. However, the alkoxide-promoted decomposition of several phosphonium compounds[121] has a mechanism different from that of the hydroxide-promoted reaction and proceeds *via* a hexacovalent intermediate.

The *cis*- and the less stable *trans*-5-chloromethyl-5-methyl-2-*p*-nitrophenyl-2-oxo-1,3,2-dioxaphosphorinans (**44** and **45**, respectively) undergo substitution with an oxyanion to give both inversion and retention products.[122] The product isomer ratio depends strongly upon the degree of anion–cation association, a high degree of association favouring retention. The results have been interpreted in

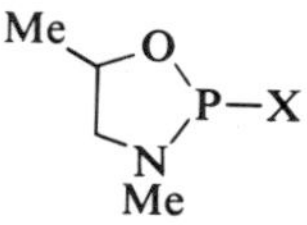

(42) X = Br
(43) X = Cl

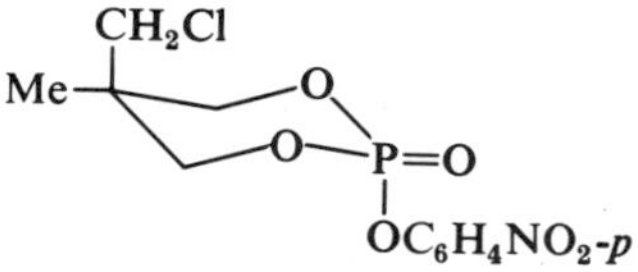

(44)

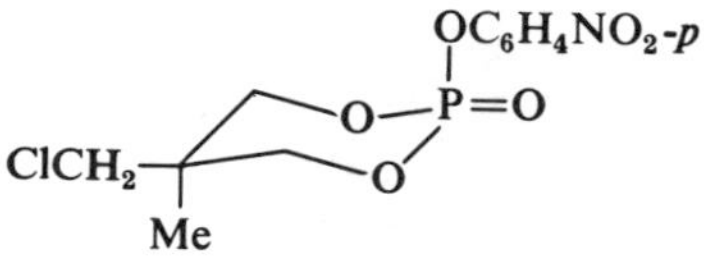

(45)

terms of a square-pyramidal intermediate, with the anion–cation complex spanning the phosphoryl—oxygen bond.

Nucleophilic substitution at tetraco-ordinated phosphinoyl phosphorus is believed to involve a pentaco-ordinate, trigonal-bipyramidal intermediate, (**46**).[123] It has now been claimed that such a structure, (**47**), has been detected in solution and isolated as its pure sodium salt.

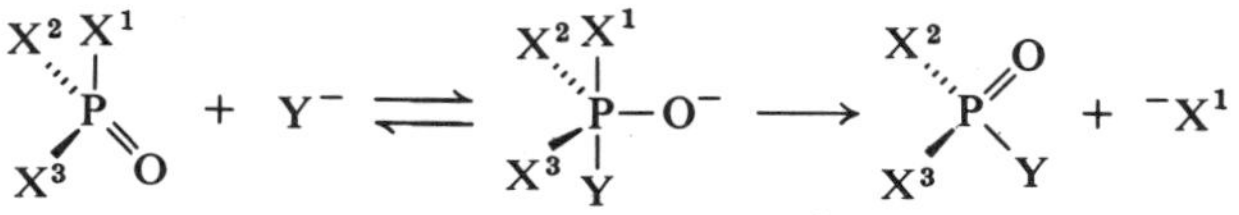

(**46**)

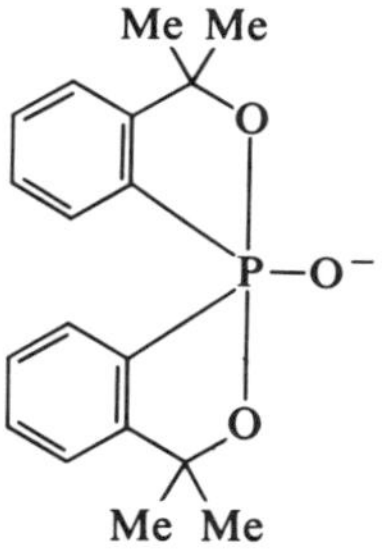

(**47**)

Finally, in their extensive correlation analysis of steric effects, M. and B. I. Charton have dealt with substituents at phosphorus.[124] Rate constants for 22 series of compounds XZ(PO)Cl reacting with water are best correlated by the steric constants v', and this agrees with a transition state having trigonal-bipyramidal geometry, with the nucleophile and leaving group at opposite positions. (v' values are derived from the reaction of alkylcarbinyl bromides with bromide ion.)

Substitution at Sulphur

There are several papers concerning the attack of disulphides by nucleophiles. Thus, secondary and primary amines react with 2,2′-dinitro-5,5′-dithiobenzoic acid (Ellman's reagent).[125] The reactions are not base-catalysed and the primary amines show a β_{nuc} value of +0.45, consistent with only partial development of the bond between N and S in the transition state. Proton transfer to yield the stable sulphenamide occurs in a fast step after the N—S bond has been formed. Tertiary amines catalyse the disulphide cleavage, showing the high β value of 0.8, and Me_3N has a solvent deuterium isotope effect of 1.72, consistent with general base-catalysed attack of water.

Hydrophobically alkylated polyethyleneimine derivatives also cleave Ellman's reagent.[126] A polyethyleneimine benzylated to the extent of 10% of its nitrogen atoms exhibited a 10^6-fold catalytic enhancement over the corresponding efficiency of a simple amine nucleophile of similar pK_a value. The process is considered to resemble enzymic catalysis, insofar as it involves a binding step, followed by a bond-forming step between S and N.

The dianion of Ellman's reagent undergoes two-stage attack by thiolate anions RS^- to form a new S—S link and release the dianion of 5-mercapto-2-nitrobenzoic acid, the final product being the disulphide RSSR.[127] Rate constants k_1 and k_2 have been measured for the two stages of attack by the anions of various alkyl and aryl thiols. Plots of $\log k_1$ *versus* pK_a generate separate but parallel straight lines for alkyl and aryl compounds, the aryl being the more reactive for a given pK_a value. The mechanism is believed to be S_N2, with no kinetically distinguishable metastable intermediate. The separate lines are ascribed either to a hydrophobic interaction between aryl groups on nucleophile and on substrate, or to "hard–soft" acid–base behaviour; thiophenoxides are softer nucleophiles than thio-alkoxides.

Another S_N2 reaction involving disulphides is attack by $Me_2CNO_2^-$;[128] RSSR sulphenylates the C-centre of the carbanion to give $Me_2C(NO_2)SR$ which undergoes a further reaction with the carbanion, apparently by a chain process.

Substitution reactions of 3-thiacycloalkyl derivatives, such as hydrolysis and aminolysis, have been shown to involve rate-determining formation of an episulphonium intermediate.[129] Disulphides and cyclic sulphides are also involved in further work on the *S*-alkylation of sulphides by an activated carbohydrate epimine under acidic catalysis.[130]

Methyl toluene-*p*-sulphenate undergoes rapid acid-catalysed hydrolysis in moist solvents to give various products.[131] A mechanism has been suggested in which thiosulphinate, usually suggested as intermediate in the hydrolysis of sulphenyl derivatives, combines with unchanged sulphenate ester to yield an intermediate sulphonium salt. The latter then reacts rapidly with either methanol or water to yield the products.

The sulphenate *p*-$MeOC_6H_4CH_2OSCF_3$ does not decompose in three days at 100 °C in hexane.[132] It undergoes ethanolysis with S—O bond cleavage, whereas the corresponding CCl_3 compound shows C—O cleavage.

The chemical and the optical instability of thiosulphinates RS(O)SR is due to ease of nucleophilic attack on the sulphenyl-sulphur atom, and can be reduced by introducing steric hindrance around one or both S atoms, *e.g.* $Bu^tS(O)SBu^t$ is chemically and optically very stable.[133]

The mechanisms of the reactions of *N*-bromo-*S*,*S*-diphenylsulphilimine Ph_2SNBr with sulphides, phosphines, and tertiary amines have been investigated;[134] a variety of S_N2 processes is indicated.

Nucleophilic substitution at trico-ordinate sulphur is discussed in a thesis.[135]

Miscellaneous Substitutions

The reaction of LiD with ammonium trifluoroacetate in a melt to give NH_3, NH_2D, HD, H_2, and D_2 has again been discussed with respect to the possible occurrence of S_N2 attack of D^- on NH_4^+.[136] The conclusion is that the results provide no evidence for such a process.

The conditions for stereospecificity in S_N2 reactions at a tetrahedrally substituted 3rd(4th, 5th, ...)-row atom have been discussed in terms of graphical or matrix description of the trigonal-bipyramidal transition state.[137]

Intramolecular Substitution

Work on the kinetics of formation of 3- to 23-membered lactones from ω-bromoalkanoate anions in 99% DMSO has been completed with the measurement of precise rate coefficients and activation parameters for the formation of four- to six-membered rings.[138] The enthalpies of activation closely follow the strain energies of the rings being formed. The entropies of activation are governed by several factors, but a trend towards more negative $\Delta S^{\ddagger}$ values with increasing ring size is apparent.

The kinetics of the base-induced cyclization of a series of substrates $HOZO(CH_2)_{12}Br$, in which Z denotes various phenylene and naphthylene groups, have been reported.[139] When allowance is made for the slight influence of Z on the nucleophilicity of the anionic end, the ease of ring closure appears as practically unaffected by the greatly varying geometry and size of the rigid components, constituting a very remarkable result.

The kinetics of dehydrochlorination of four chlorohydrins (to epoxides) in the presence of triethylamine have been studied.[140]

Methyl 4-bromobut-2-enoate may react with lithium *sec*- or *tert*-butyl sulphide to form much *trans*-disubstituted cyclopropane (**48**), rather than the expected S_N2 reaction product $RSCH_2CH{=}CHCOOMe$.[141] The yield of cyclopropane product is 70–80% in dichloromethane, diethyl ether, THF, benzene, or pentane, but practically zero in DMF or HMPA. The sodium or potassium compounds yield open-chain rather than cyclic product. The difference has been attributed to co-ordination of Li holding the attacking S in close proximity to the β-carbon (except in very basic solvents) [see (**49**)].

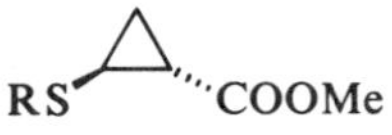

RS COOMe

(**48**)

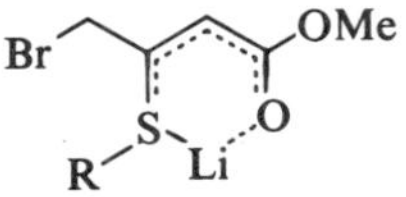

Br OMe S O R Li

(**49**)

Ba^{2+} or Sr^{2+} ions (but not K^+, Na^+, Li^+ or Et_4N^+) show a strong accelerating effect on the formation of benzo-18-crown-6 ether from *o*-$^-OC_6H_4(OCH_2CH_2)_5Br$ in aqueous solution.[142] It has been suggested that this is a "template effect": the co-ordination of the O^- and the five oxygen atoms to the M^{2+} brings the O^- and CH_2Br into a favourable configuration for nucleophilic substitution, (**50**).

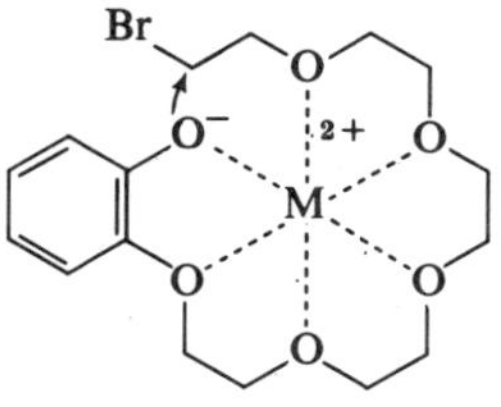

Br O O⁻ 2+ O M O O O

(**50**)

The kinetics of the S_Ni ring closure of 3-bromopropyldimethylamine (to a four-membered ring) and of 6-bromohexyldimethylane (to a seven-membered ring) have been studied.[143] Polymerization competes with cyclization but the latter is dominant in solutions more dilute than 0.1M. According to considerations of ring strain, the formation of the four-membered ring should be slower than that of the seven-membered ring by a factor of about 10^{12}, but in fact the rates are about equal.

Perfluoroalkyl groups are inert to direct nucleophilic attack (although S_N2' processes occur readily), but examples of cyclization have now been found in which F^- is displaced from a saturated position in a perfluoroalkyl group.[144] The structural features governing the intramolecular nucleophilic displacement of fluorine have been surveyed.[145]

Kinetic studies have been carried out for the formation and decomposition of the intermediate in the reaction of triphenylphosphine, carbon tetrachloride, and an alcohol.[146] The intermediate is Ph_3PClOR and it is decomposed to Ph_3PO and RCl through intramolecular nucleophilic attack of Cl on R. The decomposition occurs more easily with simple primary R groups than with secondary R groups, but the neopentyl intermediate is rather unreactive. Reaction between $MeSCH_2{}^{13}CH_2OH$ and CCl_4–R_3P (R = Ph, Pr^i or *n*-octyl) gives a 1 : 1 mixture of $MeSCH_2{}^{13}CH_2Cl$ and $MeS^{13}CH_2CH_2Cl$;[147] in this case the intramolecular attack in the above intermediate evidently involves S as nucleophile and the formation of 1-methylthi-iranium ion and chloride ion as further intermediates.

Kinetic and stereochemical aspects of the thermal decomposition of aralkyl carbonates, thiocarbonates and carbamates have been examined.[148] The reactions probably involve heterolysis of the aralkyl—oxygen bond, followed by breakdown of the resultant ion pair by a cyclic mechanism. The dependence of decomposition rate and loss of optical activity on the nature of the heteroatom of the anionic fragment has been discussed.

Anchimeric Assistance

The reaction of the sulphonate $PhSCH_2CH_2SO_3{}^- Na^+$ with PCl_5, $POCl_3$ or $SOCl_2$ gives $PhSCH_2CH_2Cl$ rather than the expected $PhSCH_2CH_2SO_2Cl$.[149] The latter type of product is formed when the β-substituent is Br, PhO or $PhSO_2$. It has been suggested that the anomalous reaction of the phenylthio-compound involves neighbouring-group participation to form an intermediate 1-phenylthi-iranium ion.

The solvolysis of 2-(phenylseleno)ethyl chloride is anchimerically assisted in methanol and particularly in aqueous ethanol, the rate in the latter solvent being over 200 times that for the corresponding sulphur compound.[150]

In a series of papers on neighbouring-group participation in solvolysis, the trifluoroacetolysis of ω-phenylalkyl-6-methylnaphthalene-2-sulphonates $Ph(CH_2)_n$-OMns ($n = 2$–6) has been studied.[151] Reactivity varies remarkably with n: $2 \gg 3 \ll 4 > 5 > 6$. This order was attributed to rate enhancement through anchimeric assistance by a remote phenyl group, competing with rate depression caused by the electron-withdrawing inductive effect of phenyl. Product studies for the compound where $n = 4$ showed that the only reaction pathway is the one with δ-phenyl participation.

To elucidate the role of cyclopropane as a neighbouring group, a detailed kinetic analysis has been carried out for the trifluoroethanolysis of 2-cyclopropylethyl toluene-p-sulphonate[152] and of the acetolysis and trifluoroethanolysis of 2-(1-methylcyclopropyl)ethyl arenesulphonates.[153]

In the hydrolysis of oxirylmethyl toluene-p-sulphonates, participation by the oxirane C—C bond gives rise to 2-oxocyclobutyl cations, which are further hydrolysed to β-ketols, see **(51–54)**;[154] stereospecificity, non-bonded interactions in the transition state, and a Taft–Streitwieser treatment have been discussed.

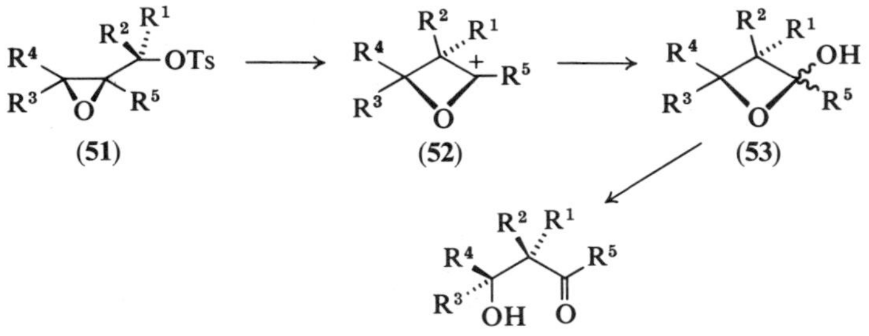

(51) **(52)** **(53)** **(54)**

A powerful anchimeric effect of the *N*-nitroso-group has been detected in the acetolysis of derivatives of β-hydroxydialkylnitrosamines; the compound **(55)** was conveniently reactive for kinetic studies but the rate for **(56)** was too great to measure.[155]

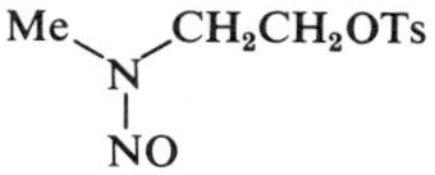

(55) 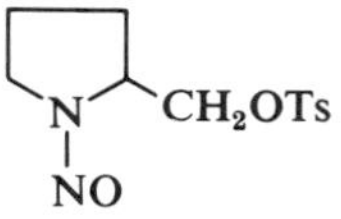 **(56)**

Evidence has been obtained for a new mode of neighbouring-group participation by the azo-group.[156] $Bu^tN{=}NCMe_2CH_2OBs$ readily undergoes ethanolysis for which a cationic mechanism, involving stabilization of the carbocationic centre by donation of an electron pair by the nearer nitrogen, has been suggested.

According to one group of authors their work "provides the most decisive evidence available for the covalent participation of hydroxyl as a neighbor group in the solvolysis of secondary systems (in acid solution) and contributes also to the study of the reverse reaction, the acid-catalysed scission of epoxides".[157]

In the reaction of *ortho*-substituted aryl methyl sulphides and diaryl sulphides with chloramine-T, substituent groups containing CO show an anchimeric effect in the fast nucleophilic product-controlling steps, which diminishes the yield of sulphilimine in solvents containing water.[158]

In a search for possible neighbouring group (direct *d*-orbital) participation by metal, formolysis rates of various methanesulphonates, "free" and complexed with chromium carbonyl, have been measured.[159] The results seem to indicate differing ion–dipole interactions in the transition state rather than direct *d*-orbital participation by Cr.

Ambident Nucleophiles

The alkylation of phosphorus monothioacids, *e.g.* $(EtO)_2P(S)OH$ or $Ph_2P(S)OH$, by aliphatic diazo-compounds, *e.g.* diazomethane or diazodiphenylmethane, in benzene, ether, THF, acetone, acetonitrile or nitromethane has been studied;[160] there is an inverse and linear relation between the logarithm of the product ratio of *O*-ester to *S*-ester and the dielectric constant of the solvent. *O*-Alkylation involves decomposition of an alkanediazonium cation and subsequent collapse along the O bond of the P(S)O triad, while *S*-alkylation involves an S_N2 mechanism.

A kinetic study has indicated that the reaction of tetrakis(trifluoromethyl)-cyclotetraphosphine with 1-iodoheptafluoropropane proceeds *via* an intermediate complex, whose decomposition is the rate-determining step.[161]

Methylation of a range of protomeric ambident nucleophiles, *e.g.* $PhCONH_2$, PhCSNHMe, or 2-aminopyridine, with methyl fluorosulphonate, has been found to occur regiospecifically at the heteroatom remote from the mobile proton.[162] In most cases the fluorosulphonate salts thus obtained can be isolated and converted into the corresponding neutral methylated derivatives by aqueous base; some variant behaviour is found, *e.g.* 2-piperidone forms the expected salt but the final product is 1-methyl-2-piperidone.

Ethylation of thiazolidine-2,4-diones (**57**) (R = H, Me, Me_2N or NO_2), *via* their sodium or triethylamine salts, with ethyl iodide in solvents of low polarity

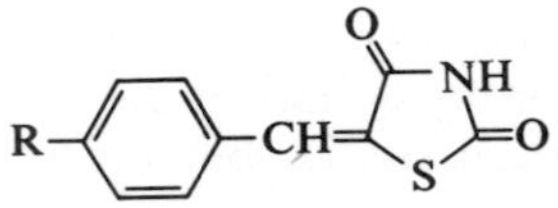

(**57**)

takes place at the N(3) centre of the ambident O—C^2—N^3—C^4—O system to give 3-ethyl derivatives.[163] Increasing solvent polarity or activity of the alkylating agent increases the yield of 2-ethoxy-derivative.

Advances in the chemistry of ambident enolate and phenolate ions have been reviewed in Russian.[164]

The α-Effect

α-Chloropolyfluoro-ketones are subject to displacement of Cl^- by relatively weak bases, notably fluoroalkoxide ions, under mild conditions.[165] Such S_N2 attack on sp^3-carbon in highly fluorinated systems is unusual and appears to be a case of the well-recognized α-effect, in which a CO group enhances the rate of displacement of an α-substituent. It has been suggested that the nucleophile adds (reversibly) first to the carbonyl-carbon and then attacks the α-carbon atom *via* a "triangular" transition state.

A comparison of the second-order rate constants for 2-bromo-1-phenylethanone (phenacyl bromide) and for methyl iodide reacting with various nucleophiles in acetonitrile at 25 °C reveals that the rate enhancement due to a carbonyl group in the position adjacent to the reacting carbon in RX is not a general effect but is very dependent upon the nucleophile.[166] There is substantial rate enhancement for nucleophiles giving tight transition states with essentially sp^2-C, *e.g.* Cl^-, but when the transition state is "early", with essentially sp^3-C, there is little effect of CO, *e.g.* the reactions with amines as nucleophiles.

In a quantum-mechanical study of the α-effect, *ab initio* SCF calculations with extended basis sets were made to obtain wave functions for a variety of nucleophiles: Cl^-, OH^-, NH_3, $MeNH_2$, NH_2OH, NH_2NH_2, ClO^-, and OOH^-.[167] In a very detailed discussion it emerges that most of the α-nucleophiles are characterized by an asymmetric antibonding HOMO, in which the charge density is more diffuse on the nucleophilic centre and which has a nodal plane perpendicular to the bond. The electrons of this HOMO are readily polarized by the electrophile, and their donation stabilizes the composite system along the potential-energy surface.

The α-effect in the chemistry of organic compounds has been reviewed in Russian.[168]

Isotope Effects

By ^{14}C-labelling it has been shown that the prochiral methyl groups in the (*RS*)-dimethyl(1-methylpropyl)sulphonium ion are transferred to an acceptor molecule, the 4-methylbenzenethiolate ion, with greatly different rates.[169]

$$EtCH(Me)S^+Me_2 + p\text{-}^-SC_6H_4Me \longrightarrow EtCH(Me)SMe + p\text{-}MeSC_6H_4Me$$

The observed rate difference is only partly due to the diastereotopic character of the two methyl groups, and is mainly due to a surprisingly large composite isotope effect of 1.16 ± 0.02, which is slightly greater than the highest previously reported $^{12}C/^{14}C$ primary isotope effect.

In a series of papers on dissected isotope effects (DIE), α-deuterium effects in the solvolysis of 1-methylheptyl *p*-bromobenzenesulphonate in 65% aqueous ethanol have been analysed.[170] The yields of seven products and the corresponding α-DIE are tabulated; the latter range from 1.103 ± 0.009 (ethyl 1-methylheptyl ether) to 1.193 ± 0.017 (oct-1-ene), at 54 °C. The results are interpreted in terms of the authors' assumption that the rate- and the product-determining steps of the reaction are the same. The same authors have also obtained β-DIE for the solvolysis of the same substrate at two temperatures.[171] While the overall isotope effect is temperature-independent, the β-DIE usually decrease slightly with increase in temperature.

The precision of recent results for the solvent isotope effect on the solvolysis of methyl bromide in H_2O/D_2O is sufficient to permit a separation into two contributions:[172] one is associated with the attacking nucleophile, the other with the developing bromide ion. Thus, the degree of nucleophilic participation in the transition state can be quantified and can be measured for other nucleophilic reactions from the gradient of the Swain–Scott plot.

Product isotope effects RH : RD have been determined for the cleavage of substituted benzyltrimethylsilanes $RSiMe_3$ by NaOMe in 1 : 1 MeOH–MeOD.[173] The values are believed to represent kinetic isotope effects for interactions of the carbanions R^- with MeOH and MeOD. As expected, except when there is unusual steric hindrance, they show an overall increase with increasing reactivity of $RSiMe_3$, and thus with the pK_a of RH.

In the exchange of OMe between methyl borate and methanol, there is a pH-independent phase of exchange which shows a large solvent isotope effect, $k(CH_3OH)/k(CH_3OD) = 11.0$ at 25 °C;[174] there is also catalysis by lyonium and lyate ions, the former showing an isotope effect of 1.4.

Gas-phase Reactions

In the current fashion for studying gas-phase reactivity, little attention seems to have been accorded recently to nucleophilic substitution.

Transaminations of silanamines have been studied by ICR.[175] The dominant pathway involves bimolecular nucleophilic substitution at Si. Under suitable conditions intermediate reaction complexes may be observed. The possible relationship of these to intermediates of the S_N2 reactions of organosilicon compounds has been discussed.

The nature and the isomer distribution of the neutral products obtained from gas-phase attack of Brönsted acids (formed radiolytically) on certain vicinal halo-alcohols provide evidence for gas-phase neighbouring-group participation.[176] This involves the intermediate formation of an *O*-protonated alkene oxide [see (**58**) and (**59**)].

Rates of reaction have been obtained for halogen exchange of MeF with $SiCl_3^+$ and $TiCl_3^+$, and several other halosilicon and halotitanium ions;[177] $GeCl_3^+$, $SnCl_3^+$, and CCl_3^+ are unreactive towards MeF. The exchange reaction involves ion-induced polarization of the C—F bond, as in (**60**). Halogen-exchange reactions of MeF with solid $AlCl_3$ and $ScCl_3$ have also been described.

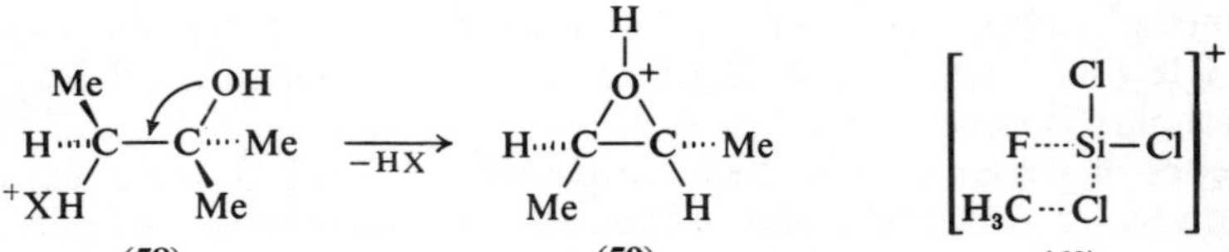

(**58**) (**59**) (**60**)

One-electron Processes

The reactions of alkali-metal diorganophosphides with a variety of organic halides have been studied by product analysis and ^{31}P-CIDNP.[178] Radical and non-radical paths are in competition. Alkyl, allyl or benzyl iodides or bromides all react by the radical path to some extent; alkyl chlorides follow an S_N2 mechanism. There is no evidence for radical processes when dialkyl phosphides react with aryl halides, or diaryl phosphides with alkyl halides. When the radical mechanism occurs, the CIDNP results are consistent with an electron-transfer step, followed by coupling of dialkylphosphinyl and organic radicals.

The alkylation of lithium anthracene with optically active 1-methylheptyl halides or methanesulphonate occurs with partial inversion of configuration, dependent on the leaving group;[179] the results may be rationalized most simply in terms of competition between electron-transfer and S_N2 processes.

A search has been made for nucleophilic substitution involving electron-transfer for compounds which possess low-lying easily accessible LUMO and for which the HOMO of the corresponding radical would be bonding.[180] Bicyclo[3.2.2]-nonatrienyl derivatives (**61**) were expected to be suitable for distinguishing between S_N1, S_N2, and electron-transfer processes. Evidence for the radical (**61**, X = electron) was obtained in the reactions between the ion (**61**, X = OSO_3^-)

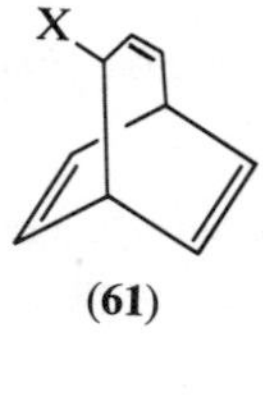

(**61**)

and polarizable, easily oxidized, nucleophiles (*e.g.* EtS^-). With nucleophiles of higher ionization potential (*e.g.* N_3^-) a cationic pathway becomes increasingly competitive with electron-transfer; with KOAc/18-crown-6 ether, or with KOH, the reaction occurs exclusively by the cationic pathway.

Reactions of arylchloromethanes with organolithium compounds, to give radical pairs, have been studied by ^{1}H, ^{19}F, and ^{13}C-CIDNP;[181] the effect of a magnetic field on the nature of the recombination products has also been studied.[182]

The role of single-electron transfer in substitution reactions has been reviewed in Russian.[183]

Solvent Effects

The correlation of solvent effects on rates of solvolysis and of S_N2 reactions has been discussed.[184] The solvent effects depend mainly on the anion-solvating properties of the solvents, although cation-solvation may play a significant role in S_N1 but not in S_N2 reactions. Various relationships have been devised and applied, in which transfer free energies of activation are related to free energies of transfer of K^+, Cl^-, and RX, and to solvent acceptor number and donor number.

The concept that water contains an equilibrium concentration of "free" OH and "free" lone-pair groups has been used to explain qualitatively the large rate changes observed for certain hydrolyses when co-solvents are added.[185]

As a solvent for S_N2 reactions, dimethyl sulphoximine has been shown by kinetic studies to have features characteristic of both protic and polar aprotic solvents.[186]

The effect of DMSO on the rate of, and the products of, hydrolysis of some longer-chain alkyl chlorides has been studied.[187] The accelerating effect of DMSO decreases with increase in the carbon number of the alkyl group, and high DMSO concentrations promote the formation of olefinic or ethereal product at the expense of alcoholic product.

Solvent effects on the structure of S_N2 transition states have been discussed in an attempt to resolve the apparent disagreement between various authors as to the solvent-dependence or -independence of the leaving-group kinetic isotope effect.[188] A model has been devised for the interaction between solvent molecules and S_N2 transition state, and a Solvation Rule has been developed to explain the observed kinetic isotope effects: "A change in solvent will not lead to a change in the structure of an S_N2 transition state if the charges on the two nucleophiles in the transition state are the same (Type I reaction) but will lead to a change in the structure of the transition state when a negatively charged nucleophile and a neutral nucleophile are present in the transition state (Type II reaction)".

The kinetics of the reaction of 5-bromopentan-2-one with diethylamine,[189] and of 5-chloropentan-2-one with sodium methoxide,[190] have been studied in various solvents. The former resembles the Menschutkin reaction, while the latter is a γ-elimination which gives cyclopropyl methyl ketone, probably *via* an *E*1*cb* process.

The plot of $\log k$ values for sulphonic esters reacting with halide ion in a given solvent against the corresponding data for the reaction in water gives a straight line for which the slope is characteristic of the solvent.[191] For a variety of dipolar aprotic solvents and methanol the slopes plotted against the E_T values of the solvents give a smooth curve. The reactions of crown-ether complex ion-pairs with the esters in benzene do not show the above LFER. The reactivity order for the halide ion-pairs in benzene was $I^- > Br^- > Cl^-$, the same as in the protic solvents water and methanol.

Kinetics have been studied for two tertiary phosphines reacting with three aliphatic halides in ten solvents (non-polar, dipolar aprotic, or aprotic) and the results have been compared with those from studies of tertiary amines.[192] The solvent effects are broadly similar for the reactions of the two classes of nucleophile, but protic solvents show specific effects connected with hydrogen-bonding to the amines. The applicability of various correlation equations has been examined.

The interaction of epichlorohydrin with Cl^-, Br^-, and I^- has been studied in various aprotic solvents and mixtures.[193]

Rate constants have been determined for the hydrolysis of secondary alkyl methanesulphonates ROMs in water[194] and a corresponding value for the 2-adamantyl sulphonate has been estimated by extrapolation of results in acetone–water mixtures. A measure of nucleophilic solvent assistance, calculated for each substrate ROMs, showed a dependence on R but the values were comparable with those obtained for toluene-*p*-sulphonate solvolysis in formic acid. This and other evidence suggests that methanesulphonate hydrolysis involves S_N2-like transition states, with high carbocation character.

The solvolysis rates of 1-adamantyl bromide and toluene-*p*-sulphonate, 2-adamantyl toluene-*p*-sulphonate, and *tert*-butyl chloride have been measured in several carboxamides;[195] their role as solvolytic solvents has been discussed on the basis of LFER.

In one of a series of papers on reactivity–selectivity relations, the influence of solvent composition on the selectivity (k_W/k_E) of adamantyl derivatives towards ethanol and water is discussed.[196] Variation in the composition of the binary mixture has little effect but addition of acetone to 60% aqueous ethanol increases substrate selectivity. The results have been interpreted as evidence for a change in the relative nucleophilicity of EtOH and H_2O in different solvent compositions. In a further paper in the same series, the selectivity of substituted benzyl chlorides towards ethanol and water has been considered.[197] The results provide a detailed description of the solvolytic intermediates, the role of intimate ion-pairs and solvent-separated ion-pairs being dependent on substituent and on solvent. A leaving-group effect on selectivity has been proposed as a new diagnostic tool for the identification of solvent-separated ion-pairs.

The solvolysis of 4,4′-dichlorobenzhydryl chloride in the nearly isodielectric mixtures of ethanol and 2,2,2-trifluoroethanol has a Grunwald–Winstein *m* value of 1.30, and is subject to common-ion rate depression.[198] A k_{TFE}/k_{EtOH} value of about 0.025 has been calculated from the product distribution by assuming reaction of the solvent components with a free cationic intermediate.

In 60% dioxan–water or in 60% acetone–water the solvolyses of *tert*-alkyl *p*-nitrobenzoates or chlorides containing chains of at least 8 carbon atoms in a row are slowed down by self-micellization.[199]

A paper on the influence of the solvent on rates of S_N1-type solvolyses of metal complexes is also of interest in connection with solvolyses of organic halides.[200]

Phase-transfer Catalysis and Other Intermolecular Effects

Ammonioalkanesulphonate esters ("betylates") $[R'Me_2N^+(CH_2)_nSO_2OR]\ X^-$ ($R' = H$ or Me; $n = 2$ or 3) are valuable as intermediates in the conversion of alcohols ROH into RNu by nucleophilic substitution with Nu^-.[201] Betylates are not remarkably reactive towards nucleophiles in a homogeneous system (*e.g.* acetone–water), but are extremely so in an aqueous suspension, or in a heterogeneous solvent system such as dichloromethane–water. [3]Betylates (*i.e.* $n = 3$) are about ten times less reactive than [2]betylates (*i.e.* $n = 2$), but the latter tend to eliminate the trialkylammonio-group in an undesirable side-reaction. The use of betylates illustrates two approaches to inducing reaction between a lipophilic substrate (*e.g.* ROH, particularly if R is a long-chain group) and a hydrophilic reagent (Nu^-): (i) "substrate phase transfer", in which the originally lipophilic substrate is made hydrophilic by attachment of a suitable group, and (ii) "substrate–reagent ion-pair reaction", in which the substrate is made into an ion, for which the reagent can act as counter-ion. In both approaches, substrate and reagent are thereby brought into intimate contact. Through the use of betylates, substitution may be accomplished under mild conditions even with poor nucleophiles such as ClO_4^- or $CF_3SO_3^-$.

When the compound $[n\text{-}C_{16}H_{33}N^+Me_2(CH_2)_2OSO_2R]\,X^-$ reacts with a nucleophile Y^-, the leaving group is RSO_3^- and the product $[n\text{-}C_{16}H_{33}N^+Me_2(CH_2)_2Y]\,X^-$ is also a surfactant (*cf.* behaviour of betylates above); initial studies of the reactivity of such compounds towards various nucleophiles have been reported.[202]

The reaction of substituted benzyl halides with aqueous NaCN or KCN gives only $ArCH_2OH$, but in the presence of a quaternary salt as a phase-transfer catalyst only $ArCH_2CN$ is produced.[203] The reaction between benzyl chloride and potassium acetate in acetonitrile is catalysed by tetramethylethylenediamine as a phase-transfer catalyst.[204] "Monoquat", $[PhCH_2N^+Me_2(CH_2)_2NMe_2]Cl^-$ (but not "diquat", $[(PhCH_2N^+Me_2CH_2)_2]2Cl^-$), acts comparably and benzyltriethylammonium chloride is an even better catalyst. Methyltrioctylammonium chloride (Aliquat 336) has been shown to catalyse the reactions of benzyl chloride and of butyl bromide or methanesulphonate with OAc^-, F^-, Br^-, NO_2^- or CN^- in acetonitrile or dichloromethane.[205]

The kinetics of reaction between *n*-octyl methanesulphonate and inorganic anions, catalysed by quaternary salts in a chlorobenzene–water two-phase system, have been studied;[206] reaction occurs in the organic phase and the methanesulphonate ion is irreversibly transferred to the aqueous phase. A relatively narrow reactivity range in the order $N_3^- > CN^- > Br^- \approx I^- > Cl^- > SCN^-$ is largely determined by specific solvation of the anion by a limited number of water molecules; rates are up to a power of 10 faster in anhydrous chlorobenzene, and the reactivity order becomes that normally found in dipolar aprotic solvents: $CN^- > N_3^- > Cl^- > Br^- > I^- > SCN^-$. In the two-phase systems, when the different concentrations of the various catalysts in the organic phase are allowed for, their catalytic activity varies only by a factor of about 2.5.

The stability of biphase- and triphase-transfer catalysts towards anions such as phenoxide and thiophenoxide has been examined.[207] The biphase catalysts included benzyltriethylammonium chloride and tetrabutylammonium bromide, and the triphase catalysts included Dowex 11 anion-exchange resin. In many cases the catalyst tends to decompose by alkylating the anion (particularly when this is a soft nucleophile) and this accounts for the origin of side-products and low yields (and in the case of supported catalysts, non-constancy of performance) when such materials are used to catalyse the reaction of an anion with an alkylating agent.

Reaction between (±)-2-bromoalkanoates and potassium phthalimide, under solid–liquid phase-transfer conditions, using either (−)-1-benzylcinchonidinium chloride or (+)-1-benzylcinchoninium chloride as chiral phase-transfer catalysts gives optically active 2-phthalimido-esters.[208] Optical yields are greatly affected by solvent polarity: dioxan or THF is best. There is evidence that the reaction takes place with partial inversion of configuration and that it seems to be a kinetic racemate resolution induced by the catalyst.

Solvent effects have been studied for the kinetics of the reaction of phenol with phenyl glycidyl ether (catalysed by $Et_3N^+CH_2CH(OH)CH_2OPh\ OPh^-$) and with *N,N*-diethylglycidylamine.[209]

Phase-transfer catalysis by quaternary ammonium salts has been reviewed in Japanese.[210]

Polystyrene-supported oligoethylene oxides surprisingly act as efficient catalysts in the displacement reactions of solid alkali-metal phenoxides on 1-bromobutane in toluene.[211]

The conversion of alcohols into alkyl chlorides and of acids into acyl chlorides by arylphosphines and CCl_4 proceeds more rapidly when the phosphine is polymer-supported;[212] the main pathway appears to involve co-operative reaction of two phosphorus-containing residues on the polymer. *C*- and *O*-Alkylations, sulphenylations, and Michael additions are facilitated by polymer-immobilized fluoride ion.[213]

$E2/S_N2$ ratios for the reactions of non-1-yl derivatives (leaving-group OTs^-, Cl^-, Br^- or I^-) with $KOBu^t$ in benzene, Bu^tOH, DMF or DMSO, in the absence of and in the presence of dicyclohexyl-18-crown-6 ether, have been found to lie in the range 0.02–24.0.[214] In benzene or Bu^tOH the crown ether raises $E2/S_N2$ in all cases but there is no effect in DMSO or DMF. In contrast, 18-crown-6 ether exerts a catalytic effect on reaction of alkyl halides with alkali-metal cyanides in the very polar aprotic solvent HMPA.[215]

Catalytic amounts of 18-crown-6 ether enhance the nucleophilicity of Bu^tO^- in THF, Bu^tOH, and benzene solutions (in the case of benzene the crown ether acts as a phase-transfer catalyst, as well as cation-solvator and anion-activator);[216] nucleophilicity seems to be enhanced more than basicity.

Rate constants have been measured for S_N1 and S_N2 components of the reaction of benzyl bromide (in CH_2Cl_2) with potassium *p*-nitrobenzoate (in water), with dicyclohexyl-18-crown-6 ether as a phase-transfer agent.[217] Even at very low crown ether concentrations, the S_N1 component is not greater than about 20% of the total reaction, while at the higher crown ether concentrations S_N1 is less than 3% of the total; the S_N2 reaction is of benzyl bromide with a KNB–crown-ether complex.

In a series of studies on the chemical behaviour of charge-transfer complexes, rates of solvolysis of several ethyl and isopropyl arenesulphonates having π-donor leaving-groups have been measured in the presence, and absence, of π-acceptors (1,3,5-trinitrobenzene or 2,4,7-trinitrofluorenone);[218] added acceptor produced small but measurable rate enhancement in four of the nine combinations investigated.

Activation parameters suggest that 1-bromoadamantane hydrolyses by an S_N1 mechanism at an unstirred toluene–water interface.[219]

A book on phase-transfer catalysis in organic synthesis[220] and a review of micellar reactions[221] have been published.

Structural Effects

Linear Free-energy Relations (LFER)

For the kinetics of solvolysis of substituted ω-bromo-1-acetonaphthones in aqueous ethanol, the Grunwald–Winstein $m = 0.112$ (55–80 °C) and the Hammett $\rho = 0.41$ (80% EtOH, 80 °C).[222] The mechanism is probably simple displacement of bromide by a solvent molecule. For reaction of the same substrates with benzoate ion in 80% aqueous acetone, $\rho = 1.33$ (30 °C) and activation parameters reveal an isokinetic temperature, $\beta = 380$ K.[223]

Extensive studies have been made of the Menschutkin reaction of substituted pyridines with methyl iodide.[224] Multiparameter equations have been used to correlate reactivity and various structural features, including the presence of *ortho*-substituents: the parameters include pK_a values, van der Waals radii, indicator variables, and dummy parameters; the correlations, at a high significance level, involve as many as 81 data points.

In their series of papers on structural effects in solvolytic reactions, Brown's group has presented a study of 1-aryl-1-cyclopropyl 3,5-dinitrobenzoates, containing activating substitutents in the aryl group;[225] the corresponding $\rho^+ = -5.19$ (80% acetone, 25 °C) is more negative than the "unusual" value observed for solvolysis of related cyclobutyl derivatives ($\rho^+ = -4.91$), and approaching that for 7-norbornyl ($\rho^+ = -5.27$) derivatives New σ^+ constants have been determined for *p*-MeS (-0.542) and 5-coumaranyl (-0.984). A further study from the same group is of the solvolysis of cycloalkyl(or alkenyl)arylmethylcarbinyl *p*-nitrobenzoates:[226] values of ρ^+ for cyclopropyl, -2.78; cyclobutyl, -3.94; cyclopentyl, -4.48; cyclohexyl, -4.71; cyclohex-1-enyl, -2.35; and cyclohex-2-enyl, -4.83 have been compared with $\rho^+ = -4.76$ for the corresponding isopropyl system. Thus, cyclopropyl and cyclohex-1-enyl groups show strong conjugative stabilization of the carbocationic centre.

The results of a study of the reactions of 5,5-dimethylcyclohexane-1,3-dione with substituted benzyl bromides are in accordance with the Kornblum rule (1953), *i.e.* electron-accepting and electron-releasing substituents favour the S_N2 and S_N1 mechanisms, respectively;[227] log (ratio of *C*- to *O*-benzylation) is a linear function of σ^+, with a positive reaction constant.

1-Aryl-3-(trimethylstannyl)propyl 3,5-dinitrobenzoates, $Me_3Sn(CH_2)_2CHAr$-(ODNB), are solvolysed in TFE to give arylcyclopropanes;[228] $\log k$ is correlated with σ^+, and $\rho^+ = -3.63$ at 100 °C. The corresponding 1-*tert*-butyl derivatives give $\rho^+ = -4.90$. The Grunwald–Winstein *m* values in aqueous acetic acid at 100 °C are 0.41 and 0.46, respectively, for the two series. This and other evidence indicates direct participation of the C—Sn σ electrons in the transition state, and the reaction is a concerted 1,3-elimination.

The kinetics of neutral and of basic solvolysis of substituted allyl benzenesulphonates in various methanol–water mixtures have been studied;[229] solvent and temperature effects on Hammett ρ values were examined.

Rate constants and activation parameters have been measured for the reaction of thiourea with some *N*-(2-substituted 3-bromopropyl)phthalimides;[230] $\log k$ is correlated satisfactorily with substituent inductive constants and the retarding effect of vicinal electron-acceptor substituents has been discussed.

For pyridine and some of its α-alkyl derivatives, constants have been proposed which characterize the inductive effect ($\Sigma\sigma^*$) of the hydrocarbon part and the steric effect of the whole molecule (E_N) on the *N*-nucleophilic reactivity of these amines.[231]

Rate constants of solvolysis of 3′- or 4′-substituted 1-(4-biphenylyl)ethyl chlorides in 80% v/v aqueous acetone have been measured.[232] The results were excellently correlated by means of the LArSR (Yukawa–Tsuno) equation:

$$\log(k/k_0) = \rho(\sigma^0 + r^+\Delta\bar{\sigma}_R{}^+)$$

with $\rho = -1.56$ and $r^+ = 0.84$; under identical conditions the ρ and r^+ values for the corresponding phenyl system are -4.95 and 1.15, respectively. The difference in the r^+ values has been interpreted in terms of the twisting of the two phenyl groups around the pivot bond in biphenyl. In a related study, rate constants have been determined for the solvolysis of 1-(7-substituted fluoren-2-yl)ethyl chlorides in aqueous acetone (both 90 and 80%) and for alkaline hydrolysis of ethyl 7-substituted fluorene-2-carboxylates in 85% aqueous ethanol;[233] correlation by the LArSR equation gives r^+ values (0.82 and 0.47, respectively) comparable with those found for the corresponding biphenylyl systems.

In a paper on the status of the Yukawa–Tsuno equation, it has been argued that the "extended selectivity plot" for log ($k_{OMe}k_H$) *versus* ρ for S_N1 hydrolysis of tertiary carbinyl systems is a good straight line of slope -0.78, *i.e.* σ^+ for *p*-OMe.[234] Accordingly, it has been concluded that variation of electron demand from that of the defining reaction for σ^+, as supposed in the usual interpretation of the Yukawa–Tsuno equation, does not occur. Genuine deviations from the above plot are due to twisting effects or mechanistic complexities.

In what is described as "a reply to recent criticism", the relationship between the Yukawa–Tsuno equation and the Grunwald–Winstein equation for solvent effects:

$$\log(k_{S_2}/k_{S_1}) = m(Y_{S_2} - Y_{S_1})$$

has been explored.[235] For the solvolysis of the aryldi-*tert*-butylcarbinyl *p*-nitrobenzoates (**62**) in acetic acid/water mixtures it is shown that $m_1/m_2 = \rho_1/\rho_2$; there is no mechanistic change involved, and the results are consistent with the idea that m_i and ρ_i are expressions of the ionic character of the transition state, although ρ_i for a given series is an averaged measure of a continuously varying quantity.

The reaction of pentamethyleneiodonium hexafluoroantimonate (**63**) with aqueous solutions of nucleophiles, or pairs of nucleophiles, has been used to generate a nucleophilicity scale encompassing both anionic and uncharged nucleophiles;[236] it shows a linear relationship to that based on reactions with methyl iodide. The charge type of both nucleophiles and substrates thus appears to be unimportant (in this respect) for S_N2 reactions in water.

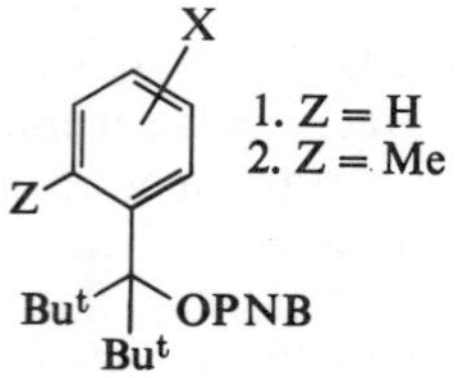

X, Z, Bu^t, OPNB, Bu^t

1. Z = H
2. Z = Me

(**62**)

SbF_6^-

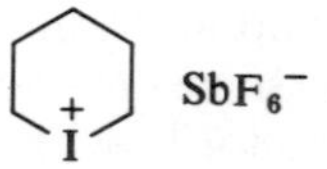

(**63**)

The rates of quaternization of p-$RC_6H_4NMe_2$ (R = H, I, Me, OMe, F, Cl or Br) and of m-$RC_6H_4NMe_2$ (same R) at 35, 45 or 55 °C indicate the possibility of participation of p-Cl, p-Br and p-I by d-orbital resonance.[237]

A study of the solvolysis of substituted benzylidene chlorides in dioxan–water mixtures has revealed evidence for the existence of a spiro intermediate in the solvolysis of p-$KOOCCH_2OC_6H_4CHCl_2$.[238]

Reactivity–Selectivity Principle (RSP)

For reactions of various alkylating agents (RX) with 3- or 4-substituted pyridines, plots of $\log k_{RX}$ against $\log k_{MeSO_3F}$ (as standard) generate a set of parallel straight lines, even though intrinsic reactivity varies a million-fold, *i.e.* the reactivity–selectivity principle does not apply;[239] the authors warn against assuming the RSP as a mechanistic tool.

In a series of studies on RS relationships, the solvolytic behaviour of octyl, 1-methylheptyl, and benzyl derivatives towards the competing nucleophiles EtOH and H_2O has been examined.[240] Low selectivity values k_E/k_W are found and the effects of changes in substrate, solvent polarity, leaving group, and temperature are consistent with the formation of ion-pair intermediates in these S_N2 reactions. In a further paper from the same research group, a breakdown is found in the RSP for S_N2 reactions of 1-octyl derivatives towards the competing nucleophiles m-chloroaniline and ethanol.[241] The results are to be understood in terms of HOMO and LUMO, and orbital interactions between nucleophile and electrophile.

Rate measurements have been made for reaction of various methylating agents (MeX) with p-nitrophenoxide and several substituted thiophenoxides in sulpholane.[242] The reactivity order (of X) with thiophenoxide is: $O_3SMe < p\text{-}O_3SC_6H_4Me < O_3SOCH_3 < I < O_3SF < O_3SCF_3 < \overset{+}{O}Me_2$; the rate constants range from 10^2 to $>10^8\ \text{M}^{-1}\,\text{s}^{-1}$. Substituent effects for reaction of substituted thiophenoxides with methyl toluene-p-sulphonate and methyl iodide are identical, and slightly greater than those for reaction with methyl trifluoroacetate. The trimethyloxonium ion is even less selective, but the rate approaches the diffusion limit. Considering the enormous range of intrinsic reactivity, the variation in selectivity is not really very large.

Alkyl aryl sulphates have been investigated as inhibitors of papain and ficin activity;[243] correlation analysis of the results indicates that the RSP is applicable to this system.

Theoretical Treatments

Semi-empirical MO calculations on the reaction of substituted benzyl chlorides with Cl^-, as a model S_N2 exchange reaction, have been used to investigate the electronic origins of substituent effects.[244] CNDO/2-calculated substituent effects on activation energy have been correlated with σ values and discussed in terms of mutual energy and charge perturbation of substituent, molecular framework, and reaction site. Similar calculations have also been done on 4-X-bicyclo[2.2.2]-octylmethyl and cubylmethyl chlorides.[245] The energetic trends produced by substitution in the two sets of alicyclic compounds correlate with each other and with related experimental data; substituent effects in the two series are comparable.

An *ab initio* FSGO study of the C_{3v} mechanism of the S_N2 reaction of Cl^- with MeCl indicates that the departure of the leaving group begins at an early stage;[246]

the transition state involves a tight ion triplet, $Cl^-[Me^+]Cl^-$, rather than a C with five distorted covalent bonds.

Fragmentation of $CH_3X_2^-$ into X^- and CH_3X, and of CH_3XY^- into X^- and CH_3Y, or Y^- and CH_3X has been examined by a recently devised method for the PMO analysis of an *ab initio* SCF–MO wave function.[247] The HOMO of the S_N2 transition states consist of a doubly occupied *p*-orbital of the fragmented X or Y and doubly occupied σ- and unoccupied σ^*-orbitals of CH_3X or CH_3Y. The four-electron destabilizing interaction between *p*- and σ-orbitals, and the two-electron stabilizing interaction between *p*- and σ^*-orbitals vary as X and Y are varied. The trends are mainly in agreement with experimental results for S_N2 reactions in the gas phase.

An *ab initio* MO study of organic reactions has included the exchange reactions of CH_4 with H^- and CF_4 with F^-.[248] Energy component analyses for the transition states indicate that the latter system is favoured relative to the former by greater electrostatic stabilization and intermolecular interaction, and by reduced exchange repulsion and intramolecular deformation.

Steric effects for S_N2 reactions of the lower alkyl compounds have been treated by molecular mechanics, with full relaxation and a force field based on reasonable estimates for the several special constants required.[249] The results differ appreciably from those of earlier calculations, which were based on static models and questionably weak non-bonded functions. In the β-substituted series (Et, Pr^n, Bu^i, neopentyl) the relative rates may be wholly explained in terms of steric effects, but in the α-series (Me, Et, Pr^i, Bu^t) agreement between calculation and experiment is poor.

S_N1 Reactions (Miscellaneous)

As part of a long series of studies in the chemistry of small ring compounds, it has been found that a cyclopropyl cation, stabilized by a *p*-anisyl group, is an intermediate in the alcoholysis of 1-(*p*-anisyl)cyclopropyl halides.[250] In various reactions of these substrates with nucleophiles, substitution products are in the main formed without ring rupture. Influence of solvent, nucleophile, and added salt on rate and on product distribution support an S_N1 mechanism. A *p*-tolyl group also has a stabilizing influence.[251] With poor nucleophiles such as NO_3^- in aqueous acetone, only polymeric ring-opened products are formed.

Dimethyl, diethyl, and di-isopropyl 2-(*N*,*N*-dimethylamino)ethyl phosphates are transformed spontaneously in polar solvents into their tetramethylpiperazinium salts.[252] The rate decreases in the order Me > Et > Pr^i and is about 25 times greater in water than in ethanol. The rate-determining step is the ionization of the substrate to give $(RO)_2PO_2^-$ and the 1,1-dimethylaziridinium cation (**64**).

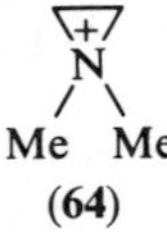

(**64**)

In neutral or weakly acidic 80% ethanol, 3-chloro-3-methylbutan-1-ol, 3-chloro-2,2,3-trimethylbutan-1-ol, and 3-chloroadamantan-1-ol form the typical products of unimolecular substitution (S_N1) and elimination (*E*1) mechanisms.[253] In the presence of NaOH the compounds are partially converted into their conjugate bases; this is reflected in greatly enhanced rates and the appearance of fragmentation products. Detailed product and kinetic studies of the above and related compounds have been presented.

First-order rate constants for hydrolysis of inden-1-yl and fluoren-9-yl, 3,5-dinitrobenzoates in 80% aqueous acetone at 80 °C have been indirectly determined;[254] the reactions of these substrates are estimated to be retarded by factors of about 10^{11} and 10^{8}, respectively, by the presence of destabilizing antiaromatic effects in their activated complexes for ionization.

Solvolytic rate constants of 1-arylethyl acetates and *p*-nitrobenzoates, in which aryl is 2-heteroaryl or phenyl, have been determined;[255] the reactivity order is 2-furyl > 2-pyrrolyl > 2-thienyl > phenyl > 2-selenienyl > 2-tellurienyl.

Product studies of the nitrous acid deamination of *n*-butylamine-1-^{14}C and the decomposition of 3-(*n*-butyl-1-^{14}C)-1-phenyltriazine have been carried out.[256] There is extensive scrambling of the ^{14}C-label but the pattern is different in the two reactions. This does not agree with the view that both reactions involve butyl carbocations in exactly the same way and it is suggested that there may be a more ready initial formation of an ion pair in the triazene decomposition. The deamination results suggest some role of protonated cyclopropane intermediates.

S_N2 Reactions (Miscellaneous)

Rate, product distribution, and stereochemical studies have been carried out for the reactions of sodium phenoxide with 1-phenylethyl chloride or butyl bromide, respectively, in acetone and in TFE as a function of [NaOPh].[257] The kinetics are second-order but the rate coefficients increase with dilution, indicating that OPh^- is more reactive than an ion pair or higher aggregate. Product analysis shows that OPh^- gives only *O*-alkyl product, but that the ion pair gives both *C*- and *O*-derivatives. Alkylations with optically active 1-phenylethyl chloride and [1-^{2}H]butyl bromide show that *O*- and *para*-alkylation proceed with complete inversion, but *ortho*-derivatives are formed with considerable racemization, despite the S_N2 conditions. The racemization is explained by assuming the *ortho*-reaction to proceed partly by front-side attack of the ion pair, probably *via* a cyclic six-membered transition state. The bearing of these matters on Sneen's mechanism has been discussed.

When 3-chlorobut-1-ene reacts with LiOEt in ethanol, the secondary ether formed is partially racemized;[258] inversion alone occurs when the Li^+ is complexed in the presence of crown ether or cage compounds and a four-centre transition state (**65**) has been suggested for the S_N2 process with retention.

The stereochemistry of nucleophilic displacement reactions of a vicinal primary and secondary sulphonic diester system has been examined.[259] Inversion occurs at both carbon atoms; this supports a "dual displacement" mechanism rather than an "acyl migration" mechanism.

Product stereochemistry has been established for the reactions of Me_3SnLi in THF, and for Me_3GeLi in HMPA, with 4-alkylcyclohexyl bromides and toluene-*p*-sulphonates.[260] For each nucleophilic unit there is evidence both for processes involving inversion in an S_N2 mechanism and for a process involving retention. The latter may involve a four-centre transition state, but the same result would be achieved by formation of a cyclohexyl lithium, followed by *electrophilic* attack of Me_3MBr on the C—Li bond.

The stereoselectivity of the Menschutkin reactions of nicotine and related compounds with methyl iodide has been investigated.[261]

The rapid nucleophilic step of the electrophilic bromination of alkenes has been interpreted quantitatively for the first time:[262] this involved the opening of seven methyl-substituted ethylenebromonium ions (**66**) by the competitive nucleophilic

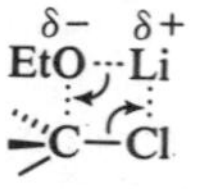

(65) 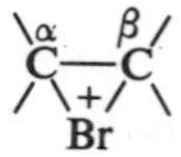 **(66)**

attack of MeOH and Br^- in a MeOH–O.2M-NaBr medium. Product analysis yielded information about chemoselectivity and regioselectivity and the results were interpreted in terms of the charge distribution on C_α and C_β of the bromonium ions (determined from ^{13}C-NMR measurements on the ions as stabilized in super-acid).

Improved methods of preparing diarylchloronium and diarylbromonium ions have enabled a study to be made of their nucleophilic nitrolysis with $NaNO_2$, giving nitroarenes.[263] The results are best accommodated by an S_N2-like mechanism controlling the collapse of ionic diarylhalonium nitrites (formed by displacement of the original counter-ion) to give nitro- and halo-benzenes.

Leaving-group abilities in the reactions of powerful methylating agents have been studied under four sets of reaction conditions;[264] the order of reactivity is always $Me_3O^+ > MeOSO_2CF_3 > MeOSO_2F > MeOClO_3$ but the range is more compressed in aprotic than in protic media.

Substrate–nucleophile association occurs between substituted chloromethanes ($ArXCH_2Cl$; X = none, S, SO, SO_2 or CO) and halide ions (Cl^-, Br^- or I^-) in dipolar aprotic solvents.[265] The thermodynamic characteristics of the equilibria have been studied by NMR-spectroscopy; spin–lattice relaxation times indicate that the halide ions place themselves close to the methylene protons, at the site of a potential S_N2 Finkelstein reaction.

Br exchange between LiBr and BuBr in acetone–methanol or acetone–ethanol mixtures has been examined;[266] kinetic studies suggest an S_N2 mechanism but Li^+Br^- ion pairs are kinetically inactive.

For the reaction $PhCOCH_2X + Y^-$ (X and Y may be Cl, Br, NCS or NCSe at 25 °C) a basicity order $NCSe^- \geqslant NCS^- > Cl^- \gg Br^-$ towards the CH_2–carbon is indicated by the equilibrium constants, while the second-order rate constants reveal a nucleophilicity order $Cl^- \sim Br^- \sim NCSe^- \gg NCS^-$; the leaving-group ability is $Br^- \gg Cl^- > NCSe^- > NCS^-$.[267]

Second-order kinetics are shown by the reaction $R_3SnSAr + R'Hal \rightarrow R_3SnHal + R'SAr$ for which substituent effects in arylthio, solvent effects, the relative reactivities of different halogens, and steric effects with various haloalkanes have been studied;[268] inversion occurs at α-C of R′Hal and a mechanism involving bimolecular nucleophilic attack of S on the haloalkane is indicated.

Kinetics of the reaction of 3-butyltetrahydro-1,3-oxazin-2-one with aromatic amines have been explained by theories of coulombic and frontier orbital control;[269] attack on CO is favoured by strong nucleophiles while attack on CH_2 is favoured by weak nucleophiles.

The substitution of Cl in 1,1-dichloro-2-(2- or 4-nitrophenyl)cyclopropanes by a variety of nucleophiles,[270] the reactions of several 2-aryl-1-bromo-1-fluoro-cyclopropanes with MeOH in the presence of $AgNO_3$ or Me_3COK,[271] and the reaction of arenediazotates with MeI have also been studied.[272]

Reactions of aromatic trifluoromethyl compounds with nucleophilic reagents have been reviewed;[273] the mechanisms include S_N1 and S_N2.

A book on the HSAB principle in organic chemistry has appeared.[274]

Kinetic Studies (Miscellaneous)

The kinetics of the quaternization of 4-methyl- and 4-ethyl-pyridine with *n*-propyl and *n*-butyl bromide in sulpholane have been studied for comparison with the quaternization of poly(4-vinyl)pyridine; the polymer reactions are slower at 319 K by a factor of about 0.8.[275]

A detailed study has been made of the volume of activation (and its pressure-dependence) for hydrolysis of benzyl chloride at 30 °C.[276] Pressures from 1 to 6895 bar were employed and a large body of data was accumulated, in order to test the various equations which have been proposed for the dependence of rate upon pressure for reactions in solution, and to extract the values of $\Delta V^{\ddagger}$ and $d(\Delta V^{\ddagger})/dp$.

A large compilation of activation volumes and reaction volumes in solution includes a considerable section on bimolecular nucleophilic substitution and on solvolysis.[277]

Other kinetic studies have included the following: the Acree relation for the reaction of alkoxide with alkyl halide;[278, 279] nitrous acid deamination of *sec*-butylamine;[280] the alkaline hydrolysis and the imidazole-catalysed hydrolysis of a series of *o*- or *p*-acetoxybenzyl bromides and *p*-nitrophenyl ethers;[281] the reactions of α-oxides with aliphatic amines;[282] the base-promoted dechlorination of methyl 9(10)-chloro-octadecanoate and sodium 9(10)-chloro-10(9)-oxo-octa-decanoate;[283] the S_N2 reactions of esters of As(III) acid;[284] and the reactivity of 2- or 3-indolylcarbinyl compounds as models for mitosane action.[285]

References

[1] Eckes, L., and Hanack, M., *Chem. Ber.*, **111**, 1253 (1978).
[2] Rappoport, Z., *Tetrahedron Lett.*, **1978**, 1073.
[3] Lee, C. C., and Ko, E. C. F., *Can. J. Chem.*, **56**, 2459 (1978).
[4] Nguyen, T., Rubinstein, M., and Wakselman, C., *J. Fluorine Chem.*, **11**, 573 (1978); *Chem. Abs.*, **89**, 107281 (1978).
[5] Gossen, L. P., Krulikovskaya, E. A., and Kumok, V. N., *Zh. Fiz. Khim.*, **52**, 1817 (1978); *Chem. Abs.*, **89**, 146084 (1978).
[6] Litvinenko, L. M., Popov, A. F., Kostenko, L. I., Kravchenko, V. V., and Vegh, D., *Dokl. Akad. Nauk SSSR*, **238**, 622 (1978) [Phys. Chem.]; *Chem. Abs.*, **88**, 120267 (1978).
[7] Kostenko, L. I., Popov, A. F., and Perel'man, L. A., *Deposited Doc.*, 1976, VINITI 1248–76, 45 pp. (Russ.). Avail. VINITI. *Chem. Abs.*, **88**, 73742 (1978).
[8] Georgoulis, C., and Ville, G., *J. Chem. Res.*, **1978**, (S) 248, (M) 3344.
[9] Rahimizadeh, M., and Waight, E. S., *J. Chem. Res.*, **1978**, (S) 186, (M) 2340.
[10] Salaün, J., *J. Org. Chem.*, **43**, 2809 (1978).
[11] Oritani, T., and Overton, K. H., *J. Chem. Soc., Chem. Commun.*, **1978**, 454.
[12] Magid, R. M., and Fruchey, O. S., *J. Am. Chem. Soc.*, **99**, 8368 (1977).
[13] Arthurs, M., Nelson, S. M., and Walker, B. J., *Tetrahedron Lett.*, **1978**, 1153.
[14] Kaboré, I. Z., Khuong-Huu, Q., and Pancrazi, A., *Tetrahedron*, **34**, 2807, 2815 (1978).

[15] Kirmse, W., Scheidt, F., and Vater, H.-J., *J. Am. Chem. Soc.*, **100**, 3945 (1978).
[16] Ortar, G., Paradisi, M. P., Morera, E., and Romeo, A., *J. Chem. Soc., Perkin Trans.* 1, **1978**, 4.
[17] Chiche, L., Coste, J., Christol, H., and Plenat, F., *Tetrahedron Lett.*, **1978**, 3251.
[18] Toromanoff, E., *Tetrahedron*, **34**, 1665 (1978).
[19] Mukaiyama, T., Yamaguchi, M., and Narasaka, K., *Chem. Lett.*, **1978**, 689.
[20] Maruyama, K., and Yamamoto, Y., *J. Am. Chem. Soc.*, **99**, 8068 (1977).
[21] Yamamoto, Y., and Maruyama, K., *J. Organomet. Chem.*, **156**, C9 (1978).
[22] Tanigawa, Y., Ohta, H., Sonoda, A., and Murahashi, S.-I., *J. Am. Chem. Soc.*, **100**, 4610 (1978).
[23] Claesson, A., and Olsson, L.-I., *J. Chem. Soc., Chem. Commun.*, **1978**, 621.
[24] Tadema, G., Everhardus, R. H., Westmijze, H., and Vermeer, P., *Tetrahedron Lett.*, **1978**, 3935.
[25] Amos, R. A., and Katzenellenbogen, J. A., *J. Org. Chem.*, **43**, 555 (1978).
[26] Mukaiyama, T., and Kawata, K., *Chem. Lett.*, **1978**, 785.
[27] Le Noble, W. J., Chiou, D.-M., and Okaya, Y., *Tetrahedron Lett.*, **1978**, 1961.
[28] Cane, D. E., and Murthy, P. P. N., *J. Am. Chem. Soc.*, **99**, 8327 (1977).
[29] Ikota, N., and Ganem, B., *J. Am. Chem. Soc.*, **100**, 351 (1978).
[30] Cherry, P. C., Gregory, G. I., Newall, C. E., Ward, P., and Watson, N. S., *J. Chem. Soc., Chem. Commun.*, **1978**, 467.
[31] Brown, H. C., and Ravindranathan, M., *J. Am. Chem. Soc.*, **100**, 1865 (1978).
[32] Brown, H. C., and Ravindranathan, M., *J. Org. Chem.*, **43**, 1709 (1978).
[33] Brown, H. C., Ravindranathan, M., Chloupek, F. J., and Rothberg, I., *J. Am. Chem. Soc.*, **100**, 3143 (1978).
[34] Brown, H. C., Ravindranathan, M., Rao, C. G., Chloupek, F. J., and Rei, M.-H., *J. Org. Chem.*, **43**, 3667 (1978).
[35] Harris, J. M., Mount, D. L., and Raber, D. J., *J. Am. Chem. Soc.*, **100**, 3139 (1978).
[36] Lambert, J. B., and Mark, H. W., *J. Am. Chem. Soc.*, **100**, 2501 (1978).
[37] Christol, H., Coste, J., Pietrasanta, F., Plénat, F., and Renard, G., *J. Chem. Res.*, **1978**, (S) 62, (M) 762.
[38] Saito, S., Moriwake, T., Takeuchi, K., and Okamoto, K., *Bull. Chem. Soc. Jpn.*, **51**, 2634 (1978).
[39] Jorgensen, W. L., and Munroe, J. E., *J. Am. Chem. Soc.*, **100**, 1511 (1978).
[40] Burger, U., Sonney, J.-M., and Vogel, P., *Tetrahedron Lett.*, **1978**, 829.
[41] Sonney, J.-M., and Vogel, P., *Tetrahedron Lett.*, **1978**, 825.
[42] Malherbe, R., and Dahn, H., *Helv. Chim. Acta*, **60**, 2539 (1977).
[43] Irie, T., and Tanida, H., *J. Org. Chem.*, **43**, 3274 (1978).
[44] Malojčić, R., Borčić, S., and Sunko, D. E., *Croat. Chem. Acta*, **49**, 743 (1977); *Chem. Abs.*, **89**, 128734 (1978).
[45] Fischer, W., and Grob, C. A., *Helv. Chim. Acta*, **61**, 1588 (1978).
[46] Grob, C. A., and Rich, R., *Tetrahedron Lett.*, **1978**, 663.
[47] Fărcasiu, D., *J. Org. Chem.*, **43**, 3878 (1978).
[48] Tichý, M., and Kniežo, L., *Collect. Czech. Chem. Commun.*, **43**, 2154 (1978).
[49] Tichý, M., Kniežo, L., and Svoboda, M., *Collect. Czech. Chem. Commun.*, **43**, 2165 (1978).
[50] Creary, X., Keller, M., and Dinnocenzo, J. P., *J. Org. Chem.*, **43**, 3874 (1978).
[51] Kirmse, W., and Jendralla, H., *Chem. Ber.*, **111**, 1857 (1978).
[52] Kirmse, W., and Jendralla, H., *Chem. Ber.*, **111**, 1873 (1978).
[53] Kirmse, W., and Richarz, U., *Chem. Ber.*, **111**, 1883 (1978).
[54] Kirmse, W., and Richarz, U., *Chem. Ber.*, **111**, 1895 (1978).
[55] Banks, R. M., and Maskill, H., *J. Chem. Soc., Perkin Trans.* 2, **1977**, 1991.
[56] Gream, G. E., Laffer, M. H., and Serelis, A. K., *Aust. J. Chem.*, **31**, 835 (1978).
[57] Gream, G. E., and Serelis, A. K., *Aust. J. Chem.*, **31**, 863 (1978).
[58] Smith, M. R., and Harris, J. M., *J. Org. Chem.*, **43**, 3588 (1978).
[59] Itô, S., Itoh, I., Fujise, Y., Nakatsu, T., Senkler, C. A., and Schleyer, P. von R., *Bull. Chem. Soc. Jpn.*, **51**, 2379 (1978).
[60] Ogawa, Y., Matsusaki, H., Hanaoka, K., Ohkata, K., and Hanafusa, T., *J. Org. Chem.*, **43**, 849 (1978).
[61] Cristol, S. J., Strom, R. M., and Stull, D. P., *J. Org. Chem.*, **43**, 1150 (1978).
[62] Cristol, S. J., Stull, D. P., and McEntee, T. E., *J. Org. Chem.*, **43**, 1756 (1978).
[63] Klein, G., and Paquette, L. A., *J. Org. Chem.*, **43**, 1293 (1978).

[64] Paquette, L. A., and Carmody, M. J., *J. Org. Chem.*, **43**, 1299 (1978).
[65] Miettinen, T., *Acta Chem. Scand., Ser. B*, **31**, 818 (1977).
[66] Yoshida, K., and Yano, K., *Tetrahedron Lett.*, **1978**, 3039.
[67] Bosse, D., and de Meijere, A., *Chem. Ber.*, **111**, 2243 (1978).
[68] Pocker, Y., and Ronald, B. P., *J. Am. Chem. Soc.*, **100**, 3122 (1978).
[69] Bovenkamp, J. W., Moir, R. Y., and Bannard, R. A. B., *Can. J. Chem.*, **55**, 4144 (1977).
[70] Funahashi, K., *Chem. Lett.*, **1978**, 1043.
[71] Battistini, C., Crotti, P., Ferretti, M., and Macchia, F., *J. Org. Chem.*, **42**, 4067 (1977).
[72] Battistini, C., Crotti, P., and Macchia, F., *Gazz. Chim. Ital.*, **107**, 153 (1977); *Chem. Abs.*, **87**, 167091 (1977).
[73] Bellucci, G., Berti, G., Ingrosso, G., Vatteroni, A., Conti, G., and Ambrosetti, R., *J. Chem. Soc., Perkin Trans. 2*, **1978**, 627.
[74] Richer, J.-C., Poirier, M.-A., Maroni, Y., and Manuel, G., *Can. J. Chem.*, **56**, 2049 (1978).
[75] Uzarewicz, A., and Segiet-Kujawa, E., *Rocz. Chem.*, **51**, 2343 (1977); *Chem. Abs.*, **88**, 135913 (1978).
[76] Apparu, M., and Barrelle, M., *Bull. Soc. Chim. Fr. II*, **1977**, 947.
[77] Hanzlik, R. P., and Hamburg, A., *J. Am. Chem. Soc.*, **100**, 1745 (1978).
[78] Inoue, M., Taguchi, Y., Sugita, T., and Ichihawa, K., *Bull. Chem. Soc. Jpn.*, **51**, 2098 (1978).
[79] Semenova, S. M., El-Sadani, S. K., Karavan, V. S., and Temnikova, T. I., *Zh. Org. Khim.*, **14**, 1268 (1978); *Chem. Abs.*, **89**, 107686 (1978).
[80] Hanzlik, R. P., and Hilbert, J. M., *J. Org. Chem.*, **43**, 610 (1978).
[81] Karaseva, A. N., Arbuzov, B. A., and Isaeva, Z. G., *Tezisy Dokl.—Vses. Konf. "Stereokhim. Konform. Anal. Org. Neftekhim. Sint."* 3*rd*, 78 (1976); *Chem. Abs.*, **89**, 42209 (1978).
[82] Glotter, E., and Krinsky, P., *J. Chem. Soc., Perkin Trans. 1*, **1978**, 413.
[83] Weissenberg, M., Krinsky, P., and Glotter, E., *J. Chem. Soc., Perkin Trans. 1*, **1978**, 565.
[84] Masamune T., Ono, M., Sato, S., and Murai, A., *Tetrahedron Lett.*, **1978**, 371.
[85] Brienne, M.-J., and Collet, A., *J. Chem. Res.*, **1978**, (S) 60, (M) 772.
[86] Malievskii, A. D., and Vints, V. V., *Izv. Akad. Nauk SSSR, Ser. Khim.*, **1977**, 2702; *Chem. Abs.*, **88**, 88746 (1978).
[87] Gubanov, V. A., Tyul'ga, G. M., Yatsenko, L. A., Karavan, V. S., Brettske, E. B., and Dolgopol'skii, I. M., *Zh. Org. Khim.*, **13**, 1809 (1977); *Chem. Abs.*, **88**, 21611 (1978).
[88] Whalen, D. L., Brown, S., Ross, A. M., and Russell, H. M., *J. Org. Chem.*, **43**, 428 (1978).
[89] Whalen, D. L., and Cooper, J. D., *J. Org. Chem.*, **43**, 432 (1978).
[90] Becker, A. R., Janusz, J. M., Rogers, D. Z., and Bruice, T. C., *J. Am. Chem. Soc.*, **100**, 3244 (1978).
[91] Bruice, P. Y., Wilson, S. C., and Bruice, T. C., *Biochemistry*, **17**, 1662 (1978).
[92] Whalen, D. L., Ross, A. M., Yagi, H., Karle, J. M., and Jerina, D. M., *J. Am. Chem. Soc.*, **100**, 5218 (1978).
[93] Niwa, M., Iguchi, M., and Yamamura, S., *Tetrahedron Lett.*, **1978**, 4043.
[94] Posner, G. H., and Rogers, D. Z., *J. Am. Chem. Soc.*, **99**, 8208 (1977).
[95] Posner, G. H., and Rogers, D. Z., *J. Am. Chem. Soc.*, **99**, 8214 (1977).
[96] Corriu, R. J. P., and Guerin, C., *J. Organomet. Chem.*, **144**, 165 (1978).
[97] Breliere, C., Corriu, R. J. P., and Royo, G., *J. Organomet. Chem.*, **148**, 107 (1978).
[98] Corriu, R. J. P., Dabosi, G., and Martineau, M., *J. Organomet. Chem.*, **150**, 27 (1978).
[99] Corriu, R. J. P., Dabosi, G., and Martineau, M., *J. Organomet. Chem.*, **154**, 33 (1978).
[100] Corriu, R. J. P., Fernandez, J. M., and Guerin, C., *J. Organomet. Chem.*, **152**, 21 (1978).
[101] Corriu, R. J. P., Fernandez, J. M., and Guerin, C., *J. Organomet. Chem.*, **152**, 25 (1978).
[102] Boyer, J., Corriu, R. J. P., Perz, R., and Reye, C., *J. Organomet. Chem.*, **157**, 153 (1978).
[103] Seyferth, D., Haas, C. K., Lambert, R. L., and Annarelli, D. C., *J. Organomet. Chem.*, **152**, 131 (1978).
[104] Hölderich, W., and Seyferth, D., *J. Organomet. Chem.*, **153**, 299 (1978).
[105] Seyferth, D., Lim, T. F. O., and Duncan, D. P., *J. Am. Chem. Soc.*, **100**, 1626 (1978).
[106] Korol'ko, V. V., Vecherskaya, V. I., Mileshkevich, V. P., and Il'ina, N. D., *Zh. Obshch. Khim.*, **48**, 596 (1978); *Chem. Abs.*, **89**, 41834 (1978).
[107] Chernyshev, E. A., Belkina, T. V., Nikitin, V. S., Komalenkova, N. G., Shapatin, A. S., and Zhinkin, D. Ya., *Zh. Obshch. Khim.*, **48**, 630 (1978); *Chem. Abs.*, **89**, 23364 (1978).
[108] Eaborn, C., Happer, D. A. R., Safa, K. D., and Walton, D. R. M., *J. Organomet. Chem.*, **157**, C50 (1978).
[109] Steward, O. W., Lutkus, A. G., and Greenshields, J. B., *J. Organomet. Chem.*, **144**, 147 (1978).

[110] Mileshkevich, V. P., and Novikova, N. F., *Zh. Obshch. Khim.*, **48**, 1121 (1978); *Chem. Abs.*, **89**, 42185 (1978).
[111] Mileshkevich, V. P., and Novikova, N. F., *Zh. Obshch. Khim.*, **48**, 1125 (1978); *Chem. Abs.*, **89**, 42186 (1978).
[112] Dubac, J., Dousse, G., Barrau, J., Cavezzan, J., Satgé, J., and Mazerolles, P., *Tetrahedron Lett.*, **1978**, 4499.
[113] Antipin, L. M., Krylov, V. V., Borisenko, A. I., Klygina, R. V., and Shaulov, Yu. Kh., *Izv. Akad. Nauk SSSR, Neorg. Mater.*, **14**, 935 (1978); *Chem. Abs.*, **89**, 41909 (1978).
[114] Ukhtomoskii, V. G., Utkin, O. V., Frolov, A. F., Musabekov, Yu. Yu., and Shapiro, Yu. E., *Zh. Prikl. Khim. (Leningrad)*, **51**, 1114 (1978); *Chem. Abs.*, **89**, 41913 (1978).
[115] Ho, T.-L., and Olah, G. A., *Proc. Natl. Acad. Sci. U.S.A.*, **75**, 4 (1978).
[116] Baburina, V. A., and Lebedev, E. P., *Zh. Obshch. Khim.*, **48**, 125 (1978); *Chem. Abs.*, **88**, 151676 (1978).
[117] D'yakonov, A. N., Zavlin, P. M., and Al'bitskaya, V. M., *Zh. Obshch. Khim.*, **47**, 2178 (1977); *Chem. Abs.*, **88**, 49916 (1978).
[118] Pudovik, M. A., Kibardina, L. K., Gol'dfarb, E. I., Zinin, V. N., and Pudovik, A. N., *Dokl. Akad. Nauk SSSR*, **238**, 1116 (1978) [Chem.]; *Chem. Abs.*, **89**, 41815 (1978).
[119] Minyaev, R. M., Minkin, V. I., and Kletskii, M. E., *Zh. Org. Khim.*, **14**, 449 (1978); *Chem. Abs.*, **89**, 42121 (1978).
[120] Khalil, F. Y., and Aksnes, G., *Phosphorus Sulfur*, **3**, 27 (1977); *Chem. Abs.*, **88**, 169376 (1978).
[121] Aksnes, G., Khalil, F. Y., and Majewski, P. J., *Phosphorus Sulfur*, **3**, 157 (1977); *Chem. Abs.*, **88**, 151694 (1978).
[122] Bauman, M., and Wadsworth, W. S., *J. Am. Chem. Soc.*, **100**, 6388 (1978).
[123] Granoth, I., and Martin, J. C., *J. Am. Chem. Soc.*, **100**, 5229 (1978).
[124] Charton, M., and Charton, B. I., *J. Org. Chem.*, **43**, 2383 (1978).
[125] Al-Rawi, H., Stacey, K. A., Weatherhead, R. H., and Williams, A., *J. Chem. Soc., Perkin Trans. 2*, **1978**, 663.
[126] Weatherhead, R. H., Stacey, K. A., and Williams, A., *J. Chem. Soc., Perkin Trans. 2*, **1978**, 800.
[127] Wilson, J. M., Bayer, R. J., and Hupe, D. J., *J. Am. Chem. Soc.*, **99**, 7922 (1977).
[128] Bowman, W. R., and Richardson, G. D., *Tetrahedron Lett.*, **1977**, 4519.
[129] Morita, H., and Oae, S., *Heterocycles*, **6**, 1593 (1977); *Chem. Abs.*, **88**, 5816 (1978).
[130] Bannister, B., *J. Chem. Soc., Perkin Trans. 1*, **1978**, 274.
[131] Ciuffarin, E., Gambarotta, S., Isola, M., and Senatore, L., *J. Chem. Soc., Perkin Trans. 2*, **1978**, 554.
[132] Braverman, S., and Manor, H., *Phosphorus Sulfur*, **2**, 215 (1976); *Chem. Abs.*, **89**, 42092 (1978).
[133] Mikołajczyk, M., and Drabowicz, J., *J. Am. Chem. Soc.*, **100**, 2510 (1978).
[134] Akasaka, T., Yoshimura, T., Furukawa, N., and Oae, S., *Phosphorus Sulfur*, **4**, 211 (1978); *Chem. Abs.*, **89**, 23374 (1978).
[135] Jacobine, A. F., Thesis, Univ. of N.H., Durham, N.H., 1977, 122 pp.; *Chem. Abs.*, **88**, 36885 (1978).
[136] Johnson, R. W., and Holm, E. R., *J. Am. Chem. Soc.*, **99**, 8077 (1977).
[137] Gielen, M., and Willem, R., *Phosphorus Sulfur*, **3**, 339 (1977); *Chem. Abs.*, **88**, 151930 (1978).
[138] Mandolini, L., *J. Am. Chem. Soc.*, **100**, 550 (1978).
[139] Mandolini, L., Masci, B., and Roelens, S., *J. Org. Chem.*, **42**, 3733 (1977).
[140] Markevich, M. A., Rogovina, S. Z., Nepomnyashchii, A. I., and Enikolopyan, N. S., *Dokl. Akad. Nauk SSSR*, **241**, 865 (1978) [Phys. Chem.]; *Chem. Abs.*, **89**, 162714 (1978).
[141] Little, R. D., and Dawson, J. R., *J. Am. Chem. Soc.*, **100**, 4607 (1978).
[142] Mandolini, L., and Masci, B., *J. Am. Chem. Soc.*, **99**, 7709 (1977).
[143] DeTar, D. F., and Brooks, W., *J. Org. Chem.*, **43**, 2245 (1978).
[144] Chambers, R. D., Lindley, A. A., Philpot, P. D., Fielding, H. C., Hutchinson, J., and Whittaker, G., *J. Chem. Soc., Chem. Commun.*, **1978**, 431.
[145] Hudlicky, M., *Isr. J. Chem.*, **17**, 80 (1978).
[146] Jones, L. A., Sumner, C. E., Franzus, B., Huang, T. T.-S., and Snyder, E. I., *J. Org. Chem.*, **43**, 2821 (1978).
[147] Billington, D. C., and Golding, B. T., *J. Chem. Soc., Chem. Commun.*, **1978**, 208.
[148] Verrinder, D. J., Hourigan, M. J., and Prokipcak, J. M., *Can. J. Chem.*, **56**, 2582 (1978).
[149] McManus, S. P., Smith, M. R., Herrmann, F. T., and Abramovitch, R. A., *J. Org. Chem.*, **43**, 647 (1978).

[150] McManus, S. P., and Lam, D. H., *J. Org. Chem.*, **43**, 650 (1978).
[151] Ando, T., Yamawaki, J., and Saito, Y., *Bull. Chem. Soc. Jpn.*, **51**, 219 (1978).
[152] Takakis, I. M., and Rhodes, Y. E., *Tetrahedron Lett.*, **1978**, 2475.
[153] Rhodes, Y. E., Takakis, I. M., Schueler, P. E., and Weiss, R. A., *Tetrahedron Lett.*, **1978**, 2479.
[154] Santelli, M., and Viala, J., *Tetrahedron*, **34**, 2327 (1978).
[155] Michejda, C. J., and Koepke, S. R., *J. Am. Chem. Soc.*, **100**, 1959 (1978).
[156] Allred, E. L., Oberlander, J. E., and Ranken, P. F., *J. Am. Chem. Soc.*, **100**, 4910 (1978).
[157] Cathcart, R. C., Bovenkamp, J. W., Moir, R. Y., Bannard, R. A. B., and Casselman, A. A., *Can. J. Chem.*, **55**, 3774 (1977).
[158] Ruff, F., Kapovits, I., Rábai, J., and Kucsman, Á., *Tetrahedron*, **34**, 2767 (1978).
[159] Bly, R. S., and Maier, T. L., *J. Org. Chem.*, **43**, 614 (1978).
[160] Mastryukova, T. A., Butorina, L. S., Uryupin, A. B., Svoren, V. A., and Kabachnik, M. I., *Zh. Obshch. Khim.*, **48**, 257 (1978); *Chem. Abs.*, **89**, 23549 (1978).
[161] Lavrent'ev, A. N., Maslennikov, I. G., and Sochilin, E. G., *Zh. Obshch. Khim.*, **48**, 942 (1978); *Chem. Abs.*, **89**, 23390 (1978).
[162] Beak, P., Lee, J., and McKinnie, B. G., *J. Org. Chem.*, **43**, 1367 (1978).
[163] V'yunkov, K. A., Ginak, A. I., and Sochilin, E. G., *Zh. Org. Khim.*, **14**, 1075 (1978); *Chem. Abs.*, **89**, 107257 (1978).
[164] Reutov, O. A., and Kurts, A. L., *Usp. Khim.*, **46**, 1964 (1977); *Chem. Abs.*, **88**, 73739 (1978).
[165] Krespan, C. G., *J. Org. Chem.*, **43**, 637 (1978).
[166] Halvorsen, A., and Songstad, J., *J. Chem. Soc., Chem. Commun.*, **1978**, 327.
[167] Heaton, M. M., *J. Am. Chem. Soc.*, **100**, 2004 (1978).
[168] Grekov, A. P., and Veselov, V. Ya., *Usp. Khim.*, **47**, 1200 (1978); *Chem. Abs.*, **89**, 106961 (1978).
[169] Grue-Sørensen, G., Kjaer, A., Norrestam, R., and Wieczorkowska, E., *Acta Chem. Scand., Ser. B*, **31**, 859 (1977).
[170] Gregoriou, G. A., and Varveri, F. S., *Tetrahedron Lett.*, **1978**, 287.
[171] Gregoriou, G. A., and Varveri, F. S., *Tetrahedron Lett.*, **1978**, 291.
[172] Albery, W. J., and Curran, J. S., *Finn. Chem. Lett.*, **1978**, 3.
[173] Macciantelli, D., Seconi, G., and Eaborn, C., *J. Chem. Soc., Perkin Trans.* 2, **1978**, 834.
[174] Hutton, W. C., and Crowell, T. I., *J. Am. Chem. Soc.*, **100**, 6904 (1978).
[175] Shea, K. J., Gobeille, R., and Wolf, J. F., *J. Organomet. Chem.*, **156**, 323 (1978).
[176] Angelini, G., and Speranza, M., *J. Chem. Soc., Chem. Commun.*, **1978**, 213.
[177] Kinser, R., Allison, J., Dietz, T. G., deAngelis, M., and Ridge, D. P., *J. Am. Chem. Soc.*, **100**, 2706 (1978).
[178] Bangerter, B. W., Beatty, R. P., Kouba, J. K., and Wreford, S. S., *J. Org. Chem.*, **42**, 3247 (1977).
[179] Malissard, M., Mazaleyrat, J. P., and Welvart, Z., *J. Am. Chem. Soc.*, **99**, 6933 (1977).
[180] Washburn, W. N., *J. Am. Chem. Soc.*, **100**, 6235 (1978).
[181] Leshina, T. V., Podoplelov, A. V., and Sagdeev, R. Z., *Tezisy Dokl. Vses. Konf. "Polyariz. Yader Elektronov Eff. Magn. Polya Khim. Reakts."*, 21 (1975); *Chem. Abs.*, **89**, 23437 (1978).
[182] Sagdeev, R. Z., Leshina, T. V., Podoplelov, A. V., Molin, Yu. N., Sarvarov, F. S., Salikhov, K. M., and Grishin, Yu. A., *Tezisy Dokl. Vses. Konf. "Polyariz. Yader Elektronov Eff. Magn. Polya Khim. Reakts."*, 34 (1975); *Chem. Abs.*, **89**, 23439 (1978).
[183] Todres, Z. V., *Usp. Khim.*, **47**, 260 (1978); *Chem. Abs.*, **88**, 151599 (1978).
[184] Parker, A. J., Mayer, U., Schmid, R., and Gutmann, V., *J. Org. Chem.*, **43**, 1843 (1978).
[185] Symons, M. C. R., *J. Chem. Soc., Chem. Commun.*, **1978**, 418.
[186] Furukawa, N., Takahashi, F., Yoshimura, T., and Oae, S., *Chem. Lett.*, **1977**, 1359.
[187] Furukawa, Y., and Komatsu, Y., *Sekiyu Gakkai Shi*, **17**, 773 (1974); *Chem. Abs.*, **87**, 167093 (1977).
[188] Westaway, K. C., *Can. J. Chem.*, **56**, 2691 (1978).
[189] Oncescu, T., and Dimofte, L., *Rev. Chim. (Bucharest)*, **28**, 1046 (1977); *Chem. Abs.*, **88**, 120271 (1978).
[190] Rădulescu, N., and Sârbu, C., *Rev. Roum. Chim.*, **23**, 197 (1978).
[191] Lemmetyinen, H., Edelmann, K., and Koskikallio, J., *Finn. Chem. Lett.*, **1978**, 169.
[192] Quéméneur, F., Bariou, B., and Kerfanto, M., *C. R. Hebd. Séances Acad. Sci., Sér. C*, **285**, 155 (1977).
[193] Vorobyov, B., Skop, V., and Shapiro, A., *Org. React. (Tartu)*, **14**, 444, 453 (1977).
[194] Bentley, T. W., and Bowen, C. T., *J. Chem. Soc., Perkin Trans.* 2, **1978**, 557.

195 Saito, S., Doihara, K., Moriwake, T., and Okamoto, K., *Bull. Chem. Soc. Jpn.*, **51**, 1565 (1978).
196 Karton, Y., and Pross, A., *J. Chem. Soc., Perkin Trans.* 2, **1978**, 595.
197 Aronovitch, H., and Pross, A., *J. Chem. Soc., Perkin Trans.* 2, **1978**, 540.
198 Rappoport, Z., Ben-Yacov, H., and Kaspi, J., *J. Org. Chem.*, **43**, 3678 (1978).
199 Rüchardt, C., Hertel, R., and Duismann, W., *Chem. Ber.*, **111**, 2179 (1978).
200 Wells, C. F., *J. Chem. Soc., Faraday Trans.* 1, **73**, 1851 (1977).
201 King, J. F., Loosmore, S. M., Lock, J. D., and Aslam, M., *J. Am. Chem. Soc.*, **100**, 1637 (1978).
202 Moss, R. A., and Sanders, W. J., *J. Am. Chem. Soc.*, **100**, 5247 (1978).
203 Fukunaga, K., Ide, S., Mori, M., and Kimura, M., *Nippon Kagaku Kaishi*, **1977**, 1379; *Chem. Abs.*, **88**, 5818 (1978).
204 Gokel, G. W., and Garcia, B. J., *Tetrahedron Lett.*, **1978**, 1743.
205 Jończyk, A., Ludwikow, M., and Mąkosza, M., *Angew. Chem. Int. Edn.*, **17**, 62 (1978).
206 Landini, D., Maia, A., and Montanari, F., *J. Am. Chem. Soc.*, **100**, 2796 (1978).
207 Dou, H. J.-M., Gallo, R., Hassanaly, P., and Metzger, J., *J. Org. Chem.*, **42**, 4275 (1977).
208 Juliá, S., Ginebreda, A., and Guixer, J., *J. Chem. Soc., Chem. Commun.*, **1978**, 742.
209 Sorokin, M. F., Shode, L. G., and Stokozenko, V. N., *Deposited Doc.*, 1975, VINITI 3734–75, 9 pp. Avail. VINITI. *Chem. Abs.*, **88**, 61759 (1978).
210 Seno, M., and Nanba, T., *Sekiyu Gakkai Shi*, **19**, 730 (1976); *Chem. Abs.*, **88**, 88564 (1978).
211 McKenzie, W. M., and Sherrington, D. C., *J. Chem. Soc., Chem. Commun.*, **1978**, 541.
212 Harrison, C. R., and Hodge, P., *J. Chem. Soc., Chem. Commun.*, **1978**, 813.
213 Miller, J. M., So, K.-H., and Clark, J. H., *J. Chem. Soc., Chem. Commun.*, **1978**, 466.
214 Závada, J., and Pánková, M., *Collect. Czech. Chem. Commun.*, **43**, 1080 (1978).
215 Shaw, J. E., Hsia, D. Y., Parries, G. S., and Sawyer, T. K., *J. Org. Chem.*, **43**, 1017 (1978).
216 DiBiase, S. A., and Gokel, G. W., *J. Org. Chem.*, **43**, 447 (1978).
217 Wong, K.-H., *J. Chem. Soc., Chem. Commun.*, **1978**, 282.
218 Colter, A. K., and Turkos, R. E. C., *Can. J. Chem.*, **56**, 585 (1978).
219 Regen, S. L., and Besse, J. J., *J. Am. Chem. Soc.*, **100**, 7117, (1978).
220 Weber, W. P., and Gokel, G. W., *Reactivity and Structure, Vol. 4: Phase Transfer Catalysis in Organic Synthesis*, Springer-Verlag, Berlin, 1977, 280 pp.
221 Bunton, C. A., *Pure. Appl. Chem.*, **49**, 969 (1977); *Chem. Abs.*, **88**, 88594 (1978).
222 Nadar, P. A., and Gnanasekaran, C., *J. Chem. Soc., Perkin Trans.* 2, **1978**, 277.
223 Nadar, P. A., and Gnanasekaran, C., *Indian J. Chem., Sect. B*, **16**, 139 (1978); *Chem. Abs.*, **89**, 41871 (1978).
224 Schaper, K.-J., *Arch. Pharm. (Weinheim, Ger.)*, **311**, 641, 650 (1978).
225 Brown, H. C., Rao, C. G., and Ravindranathan, M., *J. Am. Chem. Soc.*, **99**, 7663 (1977).
226 Brown, H. C., Ravindranathan, M., and Rao, C. G., *J. Am. Chem. Soc.*, **100**, 1218 (1978).
227 Hrnciar, P., and Fabianova, K., *Chem. Zvesti*, **30**, 520 (1976); *Chem. Abs.*, **87**, 200388 (1977).
228 McWilliam, D. C., Balasubramanian, T. R., and Kuivila, H. G., *J. Am. Chem. Soc.*, **100**, 6407 (1978).
229 Sendega, R. V., Makitra, R. G., and Protsailo, T. A., *Dopov. Akad. Nauk Ukr. RSR, Ser. B: Geol., Khim. Biol. Nauki*, **1977**, 711; *Chem. Abs.*, **87**, 200402 (1977).
230 Fedoseyev, V., Lis, J., Rjazantsev, G., and Sukhov, L., *Org. React. (Tartu)*, **14**, 140 (1977).
231 Popov, A., and Gelbina, Zh., *Org. React. (Tartu)*, **14**, 145 (1977).
232 Tsuno, Y., Chong, W.-Y., Tairaka, Y., Sawada, M., and Yukawa, Y., *Bull. Chem. Soc. Jpn.*, **51**, 596 (1978).
223 Nadar, P. A., and Gnanasekaran, C., *Indian J. Chem., Sect. B*, **16**, 139 (1978); *Chem. Abs.*, **89**, 41871 (1978).
234 Johnson, C. D., *J. Org. Chem.*, **43**, 1814 (1978).
235 Lomas, J. S., *Tetrahedron Lett.*, **1978**, 1783.
236 Peterson, P. E., Vidrine, D. W., Waller, F. J., Henrichs, P. M., Magaha, S., and Stevens, B., *J. Am. Chem. Soc.*, **99**, 7968 (1977).
237 Baliah, V., and Kanagasabapathy, V. M., *Indian J. Chem., Sect. B*, **16**, 64 (1978); *Chem. Abs.*, **88**, 120445 (1978).
238 Vitullo, V. P., and Capp, R. J., *J. Chem. Soc., Chem. Commun.*, **1978**, 98.
239 Arnett, E. M., and Reich, R., *J. Am. Chem. Soc.*, **100**, 2930 (1978).
240 Pross, A., Aronovitch, H., and Koren, R., *J. Chem. Soc., Perkin Trans.* 2, **1978**, 197.
241 Karton, Y., and Pross, A., *Tetrahedron Lett.*, **1978**, 3827.
242 Lewis, E. S., and Vanderpool, S. H., *J. Am. Chem. Soc.*, **100**, 6421 (1978).
243 Buncel, E., Chuaqui, C., Lynn, K. R., and Raoult, A., *Bioorg. Chem.*, **7**, 1 (1978).

[244] Davidson, R. B., and Williams, C. R., *J. Am. Chem. Soc.*, **100**, 73 (1978).
[245] Davidson, R. B., and Williams, C. R., *J. Am. Chem. Soc.*, **100**, 2017 (1978).
[246] Simons, G., and Talaty, E. R., *Chem. Phys. Lett.*, **56**, 554 (1978); *Chem. Abs.*, **89**, 128758 (1978).
[247] Wolfe, S., and Kost, D., *Nouveau J. de Chimie*, **2**, 441 (1978).
[248] Nagase, S., and Morokuma, K., *J. Am. Chem. Soc.*, **100**, 1666 (1978).
[249] DeTar, D. F., McMullen, D. F., and Luthra, N. P., *J. Am. Chem. Soc.*, **100**, 2484 (1978).
[250] van der Vecht, J. R., Dirks, R. J., Steinberg, H., and de Boer, Th. J., *Recl. Trav. Chim. Pays-Bas*, **96**, 309 (1977).
[251] van der Vecht, J. R., Steinberg, H., and de Boer, Th. J., *Recl. Trav. Chim. Pays-Bas*, **96**, 313 (1977).
[252] Manninen, P. A., *Acta Chem. Scand., Ser. B*, **32**, 269 (1978).
[253] Fischer, W., and Grob, C. A., *Helv. Chim. Acta*, **61**, 2336 (1978).
[254] Friedrich, E. C., and Taggart, D. B., *J. Org. Chem.*, **43**, 805 (1978).
[255] Clementi, S., Fringuelli, F., Linda, P., Marino, G., Savelli, G., Taticchi, A., and Piette, J. L., *Gazz. Chim. Ital.*, **107**, 339 (1977); *Chem. Abs.*, **88**, 36849 (1978).
[256] Lee, C. C., Reichle, R., and Weber, U., *Can. J. Chem.*, **56**, 658 (1978).
[257] Okamoto, K., Kinoshita, T., Oishi, T., and Moriyama, T., *J. Chem. Soc., Perkin Trans.* 2, **1978**, 453.
[258] Cayzergues, P., Georgoulis, C., and Ville, G., *J. Chem. Res.*, **1978**, (S) 325, (M) 4045.
[259] Kakinuma, K., *Tetrahedron Lett.*, **1978**, 765.
[260] Kitching, W., Olszowy, H., Waugh, J., and Doddrell, D., *J. Org. Chem.*, **43**, 898 (1978).
[261] Seeman, J. I., Secor, H. V., Whidby, J. F., and Bassfield, R. L., *Tetrahedron Lett.*, **1978**, 1901.
[262] Dubois, J.-E., and Chrétien, J. R., *J. Am. Chem. Soc.*, **100**, 3506 (1978).
[263] Olah, G. A., Sakakibara, T., and Asensio, G., *J. Org. Chem.*, **43**, 463 (1978).
[264] Kevill, D. N., and Lin, G. M. L., *Tetrahedron Lett.*, **1978**, 949.
[265] Hayami, J., Koyanagi, T., Hihara, N., and Kaji, A., *Bull. Chem. Soc. Jpn.*, **51**, 891 (1978).
[266] Holmgren, A., and Beronius, P., *Acta Chem. Scand., Ser. A*, **31**, 849 (1977); *Chem. Abs.*, **88**, 88750 (1978).
[267] Thorstenson, T., and Songstad, J., *Acta Chem. Scand., Ser. A*, **32**, 133 (1978); *Chem. Abs.*, **89**, 41903 (1978).
[268] Kozuka, S., and Ohya, S., *Bull. Chem. Soc. Jpn.*, **51**, 2651 (1978).
[269] Le Thi Nhut Hoa, Chu Pham Ngoc Son, Le Khac Huy, and Nguyen Huu Tinh, *Tap Chi Hoa Hoc*, **14**, 25 (1976); *Chem. Abs.*, **88**, 88867 (1978).
[270] Novokreshchennykh, V. D., Mochalov, S. S., and Shabarov, Yu. S., *Zh. Org. Khim.*, **14**, 546 (1978); *Chem. Abs.*, **89**, 41845 (1978).
[271] Aksenov, V. S., and Terent'eva, G. A., *Izv. Akad. Nauk SSSR, Ser. Khim.*, **1978**, 1344; *Chem. Abs.*, **89**, 107324 (1978).
[272] Voropaeva, A. P., Gladysheva, L. M., Ketlinskii, V. A., Bagal, I. L., Bryuske, Ya. E., and El'tsov, A. V., *Zh. Org. Khim.*, **13**, 2455 (1977); *Chem. Abs.*, **88**, 73803 (1978).
[273] Kobayashi, Y., and Kumadaki, I., *Acc. Chem. Res.*, **11**, 197 (1978).
[274] Ho, T.-L., *Hard and Soft Acids and Bases Principle in Organic Chemistry*, Academic, New York, 1977, 210 pp.
[275] Boucher, E. A., Khosravi-Babadi, E., and Mollett, C. C., *J. Chem. Soc., Faraday Trans.* 1, **74**, 427 (1978).
[276] Lohmüller, R., Macdonald, D. D., Mackinnon, M., and Hyne, J. B., *Can. J. Chem.*, **56**, 1739 (1978).
[277] Asano, T., and le Noble, W. J., *Chem. Rev.*, **78**, 407 (1978).
[278] Cayzergues, P., Georgoulis, C., and Papanastasiou, G., *C. R. Hebd. Séances Acad. Sci., Sér. C*, **285**, 163 (1977).
[279] Cayzergues, P., Georgoulis, C., and Papanastasiou, G., *J. Chim. Phys. Phys.-Chim. Biol.*, **74**, 1103 (1977); *Chem. Abs.*, **88**, 120454 (1978).
[280] Cachaza, J. M., Castro, A., and Sanchez Zas, J. L., *Acta Cient. Compostelana*, **14**, 515 (1977); *Chem. Abs.*, **89**, 75107 (1978).
[281] Remuzon, P., and Wakselman, M., *Tetrahedron*, **33**, 3097 (1977).
[282] Mutin, I. I., Arutyunyan, Kh. A., Davtyan, S. P., and Rozenberg, B. A., *Izv. Akad. Nauk SSSR, Ser. Khim.*, **1977**, 2828; *Chem. Abs.*, **88**, 73827 (1978).
[283] Ketola, M., Pihlaja, K., and Grönlund, K., *Finn. Chem. Lett.*, **1978**, 29.
[284] Valiullina, V. A., and Chernokal'skii, B. D., *Zh. Obshch. Khim.*, **48**, 1083 (1978); *Chem. Abs.*, **89**, 42184 (1978).
[285] Lown, J. W., and Weir, G. L., *Can. J. Chem.*, **56**, 249 (1978).

CHAPTER 10

Carbanions and Electrophilic Aliphatic Substitution

I. GOSNEY

Department of Chemistry, University of Edinburgh

Carbanion Structure and Stability

The methyl anion has been definitively observed in the gas phase and assigned a pyramidal geometry by photoelectron spectroscopy.[1] Application of the theory of self-consistent electron pairs to the methyl anion using Duke's large and flexible basis shows that the effects of electron correlation on geometrical predictions are essentially negligible.[2] In the first of two theoretical studies dealing with the stability of alkyl anions published this year, calculations including electron correlation provided evidence that alkyl anions do not exist as stable compounds (*i.e.* that alkyl radicals have at most very small electron affinities);[3] the second provides a model that enables more general prediction of the effect of a methyl substituent on the stability of anions.[4] The experimentally observed inversion at the RCH_2Li carbon within alkyl-lithium aggregates is indicated, by *ab initio* molecular-orbital calculations on models compound, to involve novel RCH_2Li_2 fragments in which two lithium atoms stabilize a planar RCH_2 group.[5] *Ab initio* SCF MO calculations with Pople's 4.31 basis set predict a planar structure as the most stable form of the nitromethyl anion and a monotonic increase in energy as the hydrogen atoms are bent out of the molecular plane; in contrast to previous *ab initio* calculations, the derived structure has high C=N double-bond character and N—O has single-bond character.[6] Complete optimization of classical and non-classical allyl-lithium structures at the RHF/STO-3G level with subsequent RHF/4-31G and RHF/6-31G* calculations confirms the C_S bridged species as having the lowest energy.[7] An assignment, on the basis of *ab initio* molecular orbital calculations, of the allylic structure to the carbanion produced by the reaction of OH^- with propene is supported, albeit indirectly, by flowing afterglow measurements with deuteriated propene.[8] According to *ab initio* SCF calculations, allyl anion has a larger ∠CCC skeletal angle than allyl cation, apparently so as to minimize the antiaromatic character of interactions between electrons in terminal p_π orbitals.[9] Users of the hex-5-enyl probe as a detector of radical intermediates should be aware that hex-5-enyl anions can isomerize prototropically to give 1-propylallyl-derived products which must be included in the reckoning of the yield of hex-5-enyl anions.[10] With BuLi–tetramethylethylene diamine, 3,7-dimethylocta-1,6-diene (**1**) is cyclized to the allylic carbanion of 3-isopropenyl-1,2-dimethylcyclopentane (**3**) either *via*

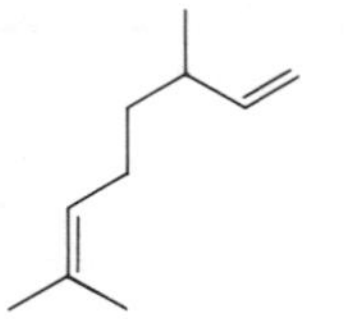

(**1**) 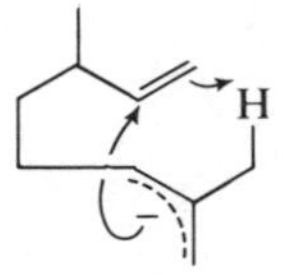 (**2**) 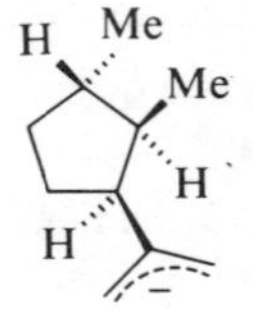 (**3**)

hydrogen transfer as in (**2**) or by transfer of Li rather than H followed by inter- or intra-molecular hydrogen exchange.[11] *Ab initio* MO calculations on the prop-2-ynl/allenyl anion suggest that it prefers an "allene-like" geometry and it seems likely that this is associated with a concentration of charge at the less substituted (CH) end of the molecule.[12] As a model for nucleophilic attack on acetylenes, the formation of the vinyl anion $C_2H_3^-$ and its rearrangements have been studied with *ab initio* SCF and correlated wave functions; unlike the vinyl cation, its rearrangement barriers are high in energy, the lowest being 40 kcal mol^{-1}.[13] Deprotonation of the alkylvinyl ether (**5**) with *tert*-butyl-lithium at −65 °C in the presence of HMPA yields the vinyl anion (**6**; equivalent to **7**) in complete preference to the

BuOCH—C̄H—CHOBu ↚ BuOCH=CHCH$_2$OBu ⟶ BuOC̄=CHCH$_2$OBu

(**4**) (5a) (**6**)

BuȮ=CH—C̄HCH$_2$OBu

(5b) (**7**)

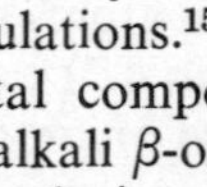

allyl anion (**4**).[14] This surprising selectivity has been attributed to the polarization of the vinyl double bond (see **5b**) which perturbs the relative acidities of the methylene and α-vinyl hydrogens, as well as destabilizing the allylic anion (**4**); support for this explanation has come from *ab initio* MO calculations.[15]

A review of skeletal rearrangements of organoalkali-metal compounds has appeared,[16] and the effect of ion-pairing on the reactivity of alkali β-oxoenolates has been assessed.[17] Caution has been advised in drawing conclusions regarding interpretation of small gegenion effects on ^{13}C-chemical shifts in enolates, especially in conformationally mobile systems (such as acetyl acetonates) where conformational factors can overwhelm any possible π-density shifts.[18] ^{1}H- and ^{13}C-Chemical shift considerations of benzyl alkaline-earth metals in THF indicate that the carbon–metal bond increases in ionic nature as the counterion varies, in the order: magnesium, calcium, strontium, lithium, barium and potassium.[19] The nature of the anion produced from 1,1-di-*p*-tolylethane is found to depend upon the metallating agent; thus, use of *tert*-butyl-lithium affords the benzyl type of anion (kinetic product), whereas benzyl-lithium affords the diphenylmethyl anion (thermodynamic product).[20] NMR observations on the delocalized anion (**8**) have shown a clear-cut differential substituent effect on the rates of rotation for the two rings.[21] Examination of the absorption spectra for diphenylmethyl-lithium and di-*p*-tolylmethyl-lithium in THF at different temperatures revealed a variation that could be interpreted on the basis of an equilibrium between contact and solvent-separated ion-pair species, with the latter predominating at low temperature.[22]

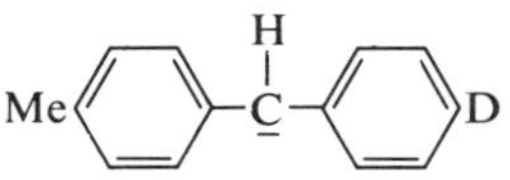

(**8**) 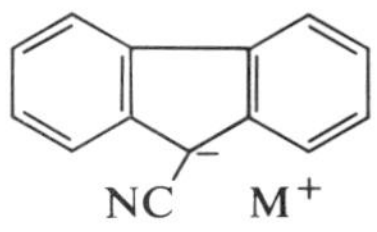 (**9**)

For the alkylation of (**9**; M = Cs, Li or Na) with ethyl iodide, the reactivity decreases with increasing solvation; notably, in Et_2O only (**9**·Li) contact ion pairs react, whilst in MeOH the reaction appears to be slower because MeO^- exists in equilibrium with (**9**) and undergoes competing Williamson reaction.[23] In the reaction of $Ph_3C^-M^+$ (M = Cs, Li or Na) with methyl chloride in THF at 25 °C, the free ion is more reactive than the contact ion pair.[24] Triphenylmethyl and diphenylmethyl alkali-metal salts in ethereal solvents exhibit visible absorption spectra which are dependent on the cation, solvent and temperature, indicative of equilibrium formation of contact and solvent-separated ion-pair species.[25] The normal *trans,trans*-conformation of the ion pairs of the 1,3-diphenylallyl and 1,3-diphenylbutenyl carbanions is converted into *cis,trans* by photolysis with white light; kinetic results obtained for the reverse dark reaction show that the rate of this process is much greater for the tight than for the loose ion pair.[26] Analysis of the ^{13}C-NMR spectra for 5,5-dimethylhexen-2-ylmetals and 2,5,5-trimethylhexen-2-ylmetals, where the metal is Li, Na, K, Rb or Cs, suggests that all these compounds possess essentially delocalized ionic structures with the charge in the caesium compound favouring slightly the α-position over the γ-position; however, as the counterion becomes smaller the charge resides more on the α-position.[27] A study of the IR spectral data for metallated acetonitrile (counterions Li^+, Na^+, and K^+ in THF and HMPA) and its D_3- and ^{15}N-derivatives, together with CNDO/2 and normal co-ordinate calcultions, showed that the mesomeric ion

$$H_2C{\cdots}C{\equiv}N$$

has a favoured planar structure and that the carbon–metal bond has a pronounced ionic character.[28]

During unsuccessful attempts to detect carbanion inversion by Me-topomerization, 1-lithio-2-methyl-1-phenylpropene (**11**) was found to "rearrange" to the isomeric allyl-lithium derivative (**12**); this unprecedented vinyl-to-allyl conversion was traced back to an intermolecular transmetallation mechanism (Scheme 1) involving small (< 10%) amounts of β,β-dimethylstyrene (**10**), the hydrolysis product of (**11**).[29] The deprotonation reactions of (**10**) with vinyl-lithium derivatives exhibit kinetic features that are very similar to those for *E*/*Z*-isomerization (topomerization) of the parent hydrocarbons.[30] The difference between topomerization and isomerization of the *cis,cis,cis,trans*-[9]annulene anion (**13**) (to **14**) is mainly

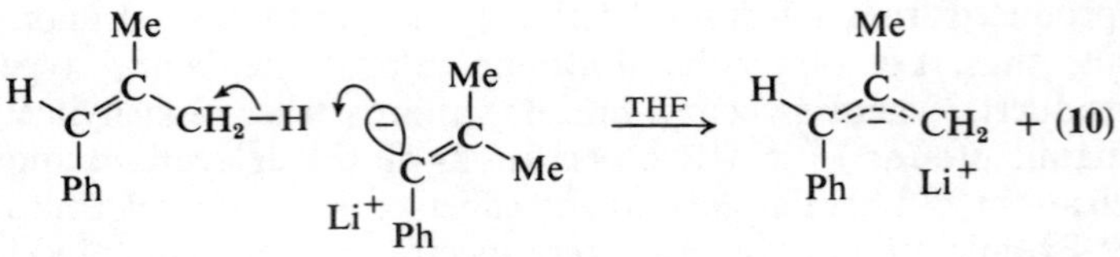

(10) (11) (12)

SCHEME 1

caused by a different degree of localization of charge in the transition states of both reactions: in contrast to the topomerization,[31] the isomerization requires complete localization of charge.[32] Determination of the regioselectivities in the

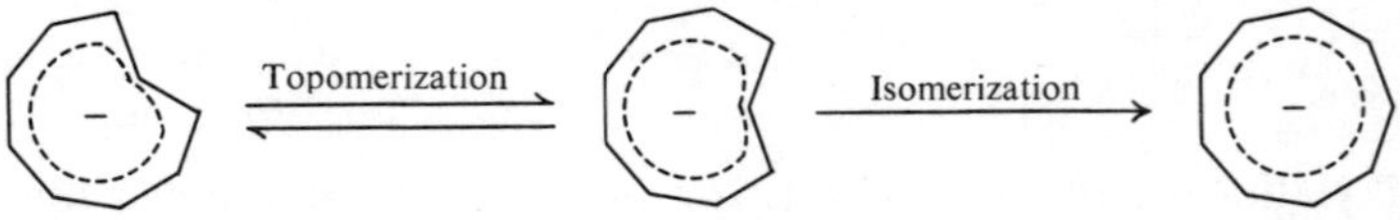

Topomerization Isomerization

(13) (13) (14)

base-promoted ring opening of unsymmetrically substituted methyl derivatives of *cis*-bicyclo[6.1.0]nona-2,4,6-triene (**15**) and its *trans*-isomer has led to the conclusion that the rate-limiting transition state for the conversion of (**15**) to (**17**)

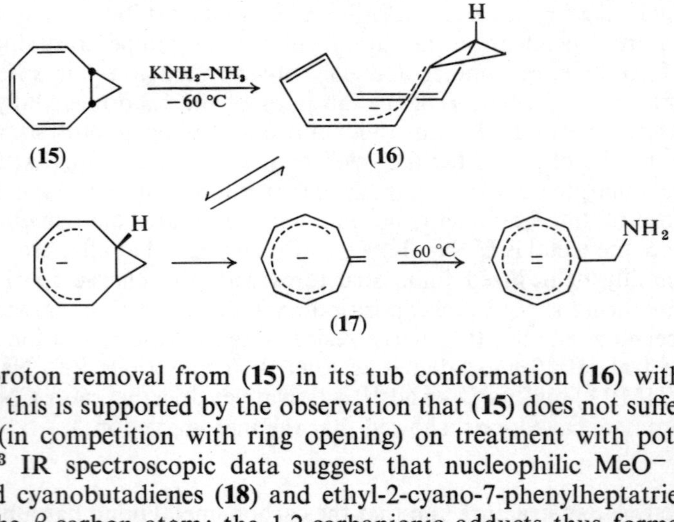

KNH_2–NH_3, −60 °C

(15) (16)

−60 °C

(17)

involves proton removal from (**15**) in its tub conformation (**16**) with little ring-flattening; this is supported by the observation that (**15**) does not suffer deuterium exchange (in competition with ring opening) on treatment with potassium *tert*-butoxide.[33] IR spectroscopic data suggest that nucleophilic MeO^- addition to substituted cyanobutadienes (**18**) and ethyl-2-cyano-7-phenylheptatrienoate takes place at the β-carbon atom; the 1,2-carbanionic adducts thus formed isomerize spontaneously into the thermodynamically more stable 1,4-adducts of the butadiene derivatives and the 2,7-adduct of the heptatrienoate.[34] Compared with the benzenoid

Ph—CH=CH—C=C(CN)X

X = CN, COOEt or p-$O_2NC_6H_4$

(18) (19) 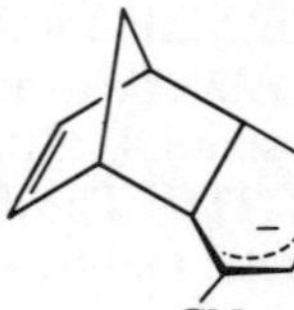 (20)

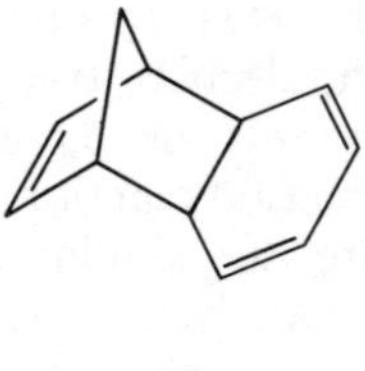

analogue (**19**), the anion (**20**) cycloreverts (to the cyclopentadienide ion and cyclopentadiene) at least 10^3 times faster; this can be rationalized in terms of FO-theory, since the HOMO of the cyclopentadienide ion (0.618) is closer in energy to the LUMO of cyclopentadiene (−0.618) than is that of benzene (1.000).[35] When treated with a large excess of potassium hydride (4–5 equivalents) in THF, the azabicyclo[2.2.1]heptene (**21**) undergoes a novel molecular reorganization to ethyl oxindole-4-carboxylate (**22**) by a mechanism envisioned to involve carbanion formation followed by β-elimination of the heteroatom bridge.[36]

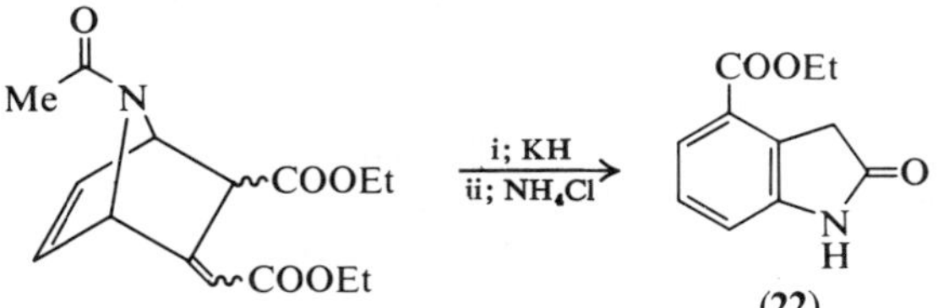

Me, O, N, COOEt, COOEt; i; KH / ii; NH_4Cl; COOEt, O, N H

(**21**) (**22**)

Pulsed radiolysis of THF solutions of $RSnMe_3$ (R = 9-methylfluorenyl or Ph_3C) generated solvated electrons (from the THF) which reacted with the organotin compounds to form R^-; the lifetime of R^- was 15–30 μs.[37] Addition of alkyl-lithiums to alk-2-enyl-1,3-dioxanes, *e.g.* (**23**), occurs readily to give, after hydrolysis,

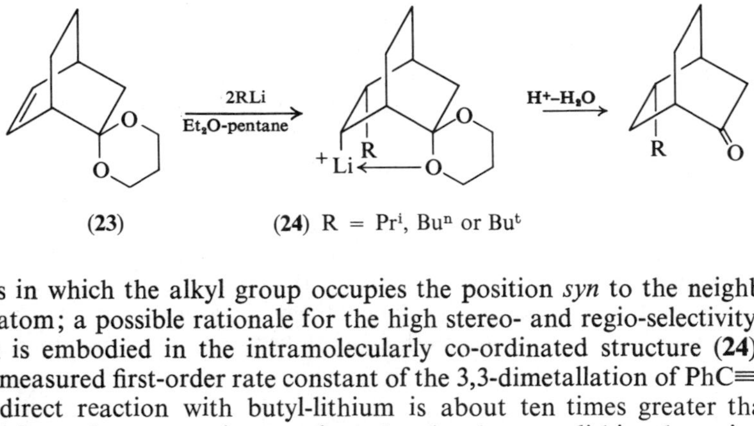

2RLi / Et_2O-pentane; H^+–H_2O; $^+$Li, R, O

(**23**) (**24**) R = Pr^i, Bu^n or Bu^t

products in which the alkyl group occupies the position *syn* to the neighbouring oxygen atom; a possible rationale for the high stereo- and regio-selectivity of this reaction is embodied in the intramolecularly co-ordinated structure (**24**).[38] The directly measured first-order rate constant of the 3,3-dimetallation of $PhC{\equiv}CCH_2$-Me by direct reaction with butyl-lithium is about ten times greater than that obtained from the consecutive reaction whereby the monolithiated species is first prepared and then metallated.[39] Substituent effects on the high rate of metallation of acetylene compounds by magnesiumbiphenyls suggest that the reactions proceed through a four-centred reagent-like transition state.[40]

Aromaticity

A critical survey of the major new developments in homoaromaticity has appeared, with particular emphasis on higher order (bis, tris, etc.) phenomena.[41] Unlike the related cation (**26**$^+$) which is capable of retaining its structural integrity under diverse conditions, the tetracyclic anion (**26**$^-$) exhibits a proclivity for skeletal rearrangement to (**25**$^-$) despite the energetic rewards of homoaromatic character (trishomocyclopentadienide character).[42] First-order analysis of the ^{1}H-NMR spectrum of the anionic species derived from 10,11-homophenalene by treatment with butyl-lithium is in full accord with the structure (**27**) rather than the *a priori* expected delocalized homophenalenide ion (**28**).[43]

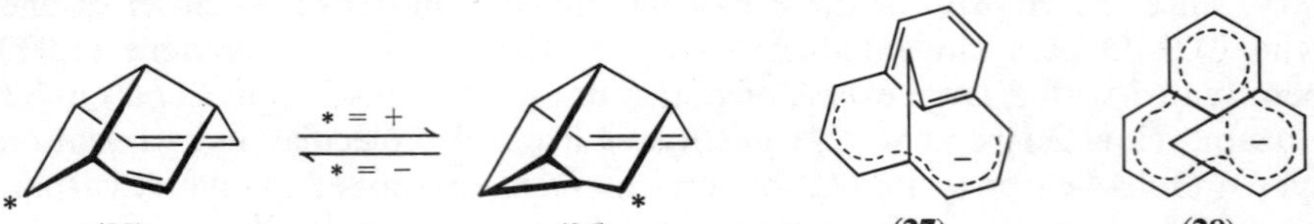

(**25**) (**26**) (**27**) (**28**)

NMR Spectroscopy has indicated the absence of homoaromaticity in the cyclohexadienyl anion, a conclusion that agrees with MINDO/3 calculations.[44] The products obtained from the lithium–ammonia reduction of bridged dicyclopropyl ketones, *e.g.* (**29**), support the notion that pericyclic processes lead to double ring-opening by two-electron reduction whenever two cyclopropane rings and a carbonyl group are conjugated in a pericyclic mode.[45] In the case of (**29a**) and (**29b**)

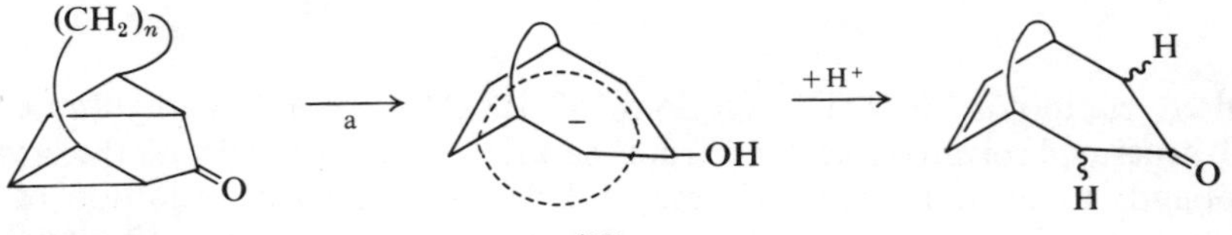

(**29**) a: $n = 0$
b: $n = 1$

(**30**)

Reagents: a: i, $+e^-$; ii, $+H^+$; iii, $+e^-$

it is most likely that the ring-opened product originates from the bishomocyclopentadienyl anions (**30**), whose formation is triggered by electron-transfer to intermediate protonated ketyls.[46] The extensive experimental evidence for absolute antiaromaticity (particularly for cyclobutadiene and cyclopropenide) has been critically examined and judged inconclusive in view of the existence of alternate, plausible rationalizations.[47] Reduction of pyrene and its isomers by metal under carefully controlled conditions yields the corresponding dianions which on the basis of their ^{1}H-NMR spectra can be regarded essentially as $(4n)\pi$-perimeters.[48] Information from ^{1}H- and ^{13}C-NMR spectroscopy shows that the methano-bridged fluorenyl anions (**31a** and **31b**) exist in "cycloheptatriene" rather than "norcaradiene" forms.[49] Deprotonation of *N,N*-dimethyl 7-cycloheptatrienylcarboxamide (**32**) with

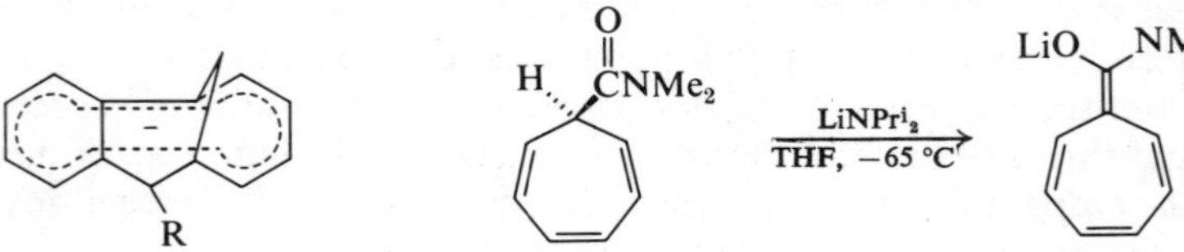

(**31**) a: R = H
b: R = Me

(**32**) (**33**)

lithium di-isopropylamide in THF at -65 °C generated an anionic species which, according to ^{1}H- and ^{13}C-NMR spectroscopy, is best viewed as an electron-rich heptafulvene system, *viz.* (**33**).[50] The aromaticity of the anion derived from 9-acetyl-*all-cis*-cyclonona-1,3,5,7-tetraene is found to depend dramatically on ion-pair character; in THF, the potassium compound features the (solvent-separated) acyl-substituted aromatic cyclononatetraenyl anion, whereas its lithium counterpart contains the corresponding non-aromatic enolate anion within a contact ion pair.[51] The reaction of 9-*anti*-methoxy- or 9-*anti*-chloro-*cis*-bicyclo[6.1.0]nona-2,4,6-triene with lithium, sodium or potassium in THF at -30 °C to -20 °C leads stereoselectively ($>97\%$ in the former case) to the aromatic *cis,cis,cis,trans*-[9]annulene anion, which can be isolated in the form of crystalline salts.[52] The dark red solution which is formed when 2, 3 : 6, 7 : 8, 9 : 12, 13-tetrabenzo[13]-annulene is treated with butyl-lithium in THF-d_8 contains the planar delocalized aromatic [13]annulenyl anion which is considerably less diatropic than the parent non-benzannelated component.[53] From the ^{1}H-NMR shifts of the benzene rings fused to the pentalene unit, the charged species derived from 1,9-dihydrodibenzo-[*b,f*]pentalene by treatment with butyl-lithium in THF-d_8 may be considered as a diatropic ("aromatic") perturbed [16]annulene dianion (**34**).[54] PMR spectroscopy

(34) **(35)**

has shown that the [17]annulenyl anion (**35**)[55] and its aza-analogue[56] are both strongly diatropic, the difference in chemical shifts of the inner and outer protons being *ca.* 17 and 14.2 ppm, respectively; interestingly, methylation of the aza-analogue is preceded by an isomerization which brings nitrogen from an internal (as shown) into an external position. Inspection of the ^{13}C-chemical shifts of the dianion derived from 1,2,3,4-dibenzocyclo-octatetraene on reduction by metal indicates that the excess charge is essentially localized in the eight-membered ring, thereby creating a bonding situation which approaches the border-line structure (**36**).[57] Reduction of the polycyclic compound (**37**) by metal yields a diamagnetic dianionic species in which the molecular framework of the starting material is surprisingly preserved.[58] Protonation of the tryptycene-cyclopentadienide anion analogue (**38**) occurs specifically at C-14, indicating incyclic double-bond character in the cyclopentadieno two-carbon arm of the dibenzobicyclo[2.2.2]octadiene skeleton to be thermodynamically unfavourable.[59] Failure to obtain the

(36) **(37)** **(38)**

(39) **(40)** **(41)**

bicyclo[4.3.2] undecatetraenyl anion (**39**) under conditions closely similar to those that had earlier provided (**40**) and (**41**) corresponds to an explicitly predicted limit of the longicyclic stabilization rule when its homologation reaches $C_{11}H_{11}$.[60]

Reactions of Carbanions

A review illustrating the synthetic value of dipole-stabilized carbanions has appeared.[61] Evidence that the carbonyl group plays a crucial role in providing thermodynamic (dipole) stabilization of a lithiomethyl thioester group relative to a lithiomethyl ether function has come from the observation that methyl 2,4,6-tri-isopropylthiobenzoate is completely metallated by (methylthio)methyl-lithium-TMEDA at −78 °C in THF-d_8.[62]

Reaction of 3,3-dimethylallenyl-lithium with aldehydes and ketones gives rise to both acetylenic and allenic alcohols in varying amounts, thereby exemplifying the ambident nature of this nucleophile.[63] 3,3-Diphenylacrylonitrile reacts with butyl-lithium in THF at −78 °C to give the derived vinyl carbanion $Ph_2C{=}\bar{C}CN$,[64] which acts as a nucleophile in addition and in S_N-type reactions with the following electrophiles: 3,3-diphenylacrylonitrile, benzophenone, *trans*-α-cyanostilbene, CO_2, and alkyl iodides yielding the expected products, along with some by-products.[65] 1,1-Dibromo-2-ethoxycyclopropane has been used as the synthon $CH_2{=}\bar{C}CHO$ in formation of 2-substituted propenal diethyl acetals,[66] which are particularly useful intermediates for the synthesis of α,β-unsaturated aldehydes of homoterpenoid structure.[67] 2,4,6-Tri-isopropylbenzenesulphonylhydrazones, *e.g.* (**42**), appear to offer distinct advantages over tosylhydrazones for vinyl anion generation; for example, when (**43**) is generated from the tosylhydrazone by butyl-lithium in hexane/TMEDA, only alkylation products derived from the rearranged 1-phenylallyl anion (**45**) could be obtained, whereas decomposition of

$PhCH_2C(Me){=}N{-}NHSO_2C_6H_2Pr^i_3$ (**42**) ⟶ $PhCH_2\bar{C}{=}CH_2$ (**43**) + $PhCH{=}\bar{C}Me$ (**44**)

(**43**) ⟶ $(PhCH{\cdots}CH{\cdots}CH_2)^-$ (**45**)

the dianion of trisylhydrazone (**42**) is much faster and the unrearranged anions (**43**) and (**44**) (ratio *ca* 9 : 1) can be trapped by rapid quenching with reactive electrophiles.[68] In this fashion, α-methylene-lactones have been prepared from acetone 2,4,6-tri-isopropylbenzenesulphonylhydrazone, an aliphatic ketone or aldehyde, and carbon dioxide in a "one-pot" good-yield reaction.[69] Br/Li-exchange in vinyl bromides with two equivalents of *tert*-butyl-lithium leads directly to haloalkane-free solutions of vinyl-lithium compounds which are useful in the synthesis of carbonyl compounds; for example, sulphenylation of (**46**) followed by hydrolysis affords cyclo-octanone in 63% yield.[70] Reaction of α-bromo-α,β-unsaturated ketals with *n*-butyl-lithium in THF at −76 °C, followed by trapping of the resultant α-ketovinyl anions with various electrophiles, provides an exceptionally efficient method for the construction of α-substituted α,β-unsaturated ketones directly from the parent enone, without the intervention of the thermodynamic dienolate.[71] A useful chelating effect, which should have wide applicability, has been used to stabilize the anion (**48**) from the Birch reduction product (**47**) and permit formation of 2-alkenyl-3-methylcyclohex-2-en-1-ones (**49**).[72]

e.g. R = $CH_2{=}CHCH_2$ (88%)
$CH_2{=}CH(CH_2)_2$ (86%)
p-$MeOC_6H_4(CH_2)_2CH{=}CH(CH_2)_2$ (89%)

A brief survey of the formation and electrophilic reactions of vinyl anions stabilized by heterotams has appeared.[73] Deprotonation of *trans*-3-(pyrrolidin-1-yl)-acrylonitrile (**50**) at −105 °C with lithium di-isopropylamide occurs exclusively at the 3-position to give the functional vinyl-lithium derivation (**51**) (a β-acylvinyl anion equivalent) which rearranges at higher temperatures, fast and almost quantitatively, into the thermodynamically more stable vinyl-lithium derivative (**52**), an equivalent of the monoanion of formylacetonitrile.[74] In the case of ethyl 3-(pyrrolidin-1-yl)acrylate (**53**), kinetic deprotonation with *tert*-butyl-lithium below −100 °C generates the corresponding β-(alkoxycarbonyl)vinyl-lithium derivative (**54**) which reacts with α,β-unsaturated carbonyl compounds to give functionalized cyclopent-2-en-1-ones (**55**).[75] Under appropriate conditions, ethyl 3-nitropropionate also serves as a β-acylvinyl anion equivalent in synthesis and provides, for example, useful routes to 4-hydroxyacrylates from aldehydes, and 5-ketoenones from α,β-unsaturated ketones.[76]

The use of complex bases and complex reducing agents in organic synthesis has been surveyed.[77] The aromatization of dihydroarenes, which in some cases (e.g. synthesis of 9,10-dialkylanthracenes) was formerly only possible by devious means, can now be accomplished by deprotonation and hydride elimination promoted by the combined action of potassium fencholate as base and fenchone as hydride acceptor; only catalytic amounts of the base are required since it is regenerated.[78,79]

The reactivity of carbanions, with particular emphasis on kinetic and mechanistic studies, has been reviewed.[80] The multitude of experimental data[81] relating to the sodium hydroxide-catalysed condensation of furfural (F) with acetone (Ac) has now been integrated into a general reaction mechanism which explains the formation of the main reaction products FAc and F_2Ac (difurfurylideneacetone),[82] as well as the formation of secondary products and polymers.[83] The rate of condensation of formaldehyde with ethyl cyanoacetate in absolute ethanol in the presence of basic catalysts may be satisfactorily described by the two-parameter equation (1) which takes into account both the basicity and polarity of the catalysts:[84]

$$\log k = -5.943 + 0.858 \frac{\varepsilon - 1}{2\varepsilon + 1} + 0.00183\text{B} \tag{1}$$

Quantitative considerations of structure, medium and temperature effects on the kinetics of solvolysis of β-polynitroalkylarylamines have been reported.[85] The base-catalysed condensation of substituted benzaldehydes with 3,5-dimethyl-4-nitroisoxazole in methanol follows a bimolecular rate law at high catalyst concentrations.[86,87]

Further studies into the factors affecting the stereochemistry of alkylation reactions of 9-alkyl-10-lithio-9,10-dihydroanthracene have appeared; kinetic measurements now show that for reactions with primary halides, the preponderant attack by anion is in the preferred conformer of axial–axial orientation (**56**),

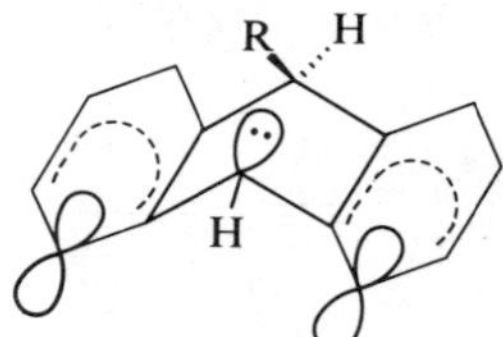

(**56**) R = H, Et, Pr^i or Bu^t

whereas in the case of secondary halides the data are more consistent with attack by the anion in its quasiequatorial conformation.[88] With (**56**; R = Bu^t), deuteriation by MeOD occurs stereospecifically *cis* in dioxan and Et_2O, but *trans* in HMPA.[89] Investigation of the formation of the carbanion of 9-trimethylsilyl-9,10-dihydroanthracene (Me_3Si DHA) has revealed an interesting competition in the abstraction by base of H-9 and H-10; although the latter hydrogen is less crowded, the 9-carbanion is stabilized by the vicinity of the silicon atom and alkylation (alkyl = Me, Et or Pr^i) leads exclusively to 9,9-disubstituted derivatives: 9-Me_3Si-9-alkyl-DHA.[90] The alkylation of lithium anthracene with optically active oct-2-yl halides and mesylate occurs with partial inversion of configuration the extent of which depends on the leaving group; the simplest rationalization of these results is a competition between electron transfer and S_N2 processes.[91] Trimethylsilylpotassium, readily prepared in a non-polar solvent by the reaction of hexamethyldisilane with potassium methoxide in the presence of 18-crown-6, is a useful reagent for efficient homogeneous electron-transfer.[92] The contention that the reaction of alkyl halides with Me_3Sn-Li, -Na, and -K proceeds by two (or more) competing reaction pathways, one of which involves intermediate free alkyl radicals, is supported by the observation that the reaction of lithium, sodium, and potassium trimethyl tin with cyclopropylmethyl bromide and iodide, but not chloride or tosylate, yields two alkylation products: trimethyl(cyclopropylmethyl)-tin and trimethyl(allylmethyl)tin.[93] Alkylation reactions of arylacetic esters can be greatly improved by the use of the chromium tricarbonyl [$Cr(CO)_3$] unit as a temporary complexing group of the aromatic ring; thus the electron-withdrawing effect of the $Cr(CO)_3$ group enhances the acidity of the esters, allowing a ready alkylation with different alkylating agents (either by phase-transfer catalysis or by sodium hydride in DMF), whilst its steric bulk may induce stereospecific alkylations, *e.g.* (**57**→*exo*-**58**).[94]

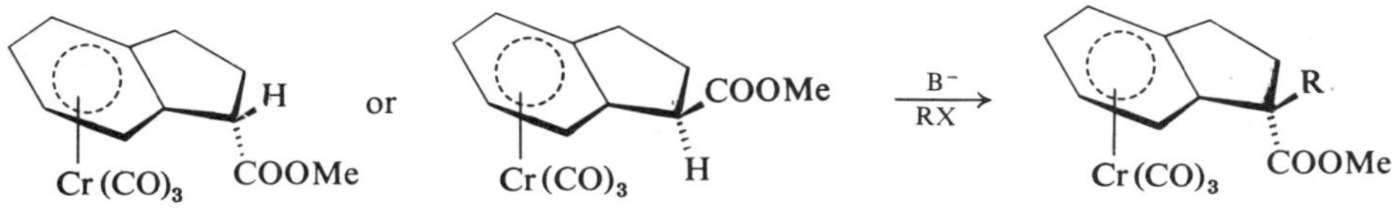

H or COOMe $\xrightarrow[RX]{B^-}$ R

Cr(CO)$_3$ COOMe Cr(CO)$_3$ H Cr(CO)$_3$ COOMe

(**57**a) (**57**b) (**58**)

R = Me, $PhCH_2$, $CH_2CH{=}CH_2$ or $CH_2C{\equiv}CH$

Reactions of π-allylpalladium compounds with sodium acetylacetonate, sodium malonitrile, and ethyl sodioacetoacetate in DMSO solution give diallyl substitution products.[95] Given a carbanionic nucleophile and an activated haloalkyne it has been shown that conditions which favour substitution and minimize addition and halogen abstraction are a relatively low pK of the parent carbon acid and an aprotic medium, *e.g.* glyme.[96] Steric factors are more important than electronic factors in the substitution of halogens and alkoxy-groups in vinylacetylene systems by carbanions.[97] When treated with LiCl/HMPA in THF, methyl chlorodifluoroacetate readily undergoes demethoxycarbonylation to generate the chlorodifluoromethide ion $ClCF_2^-$ (or the complexed carbenoid $ClCF_2Li$/HMPA) which may be trapped by a variety of electrophilic reagents.[98] Difluorochloromethane acts as a remarkably efficient alkylating agent towards the anion (**59**) giving rise to the

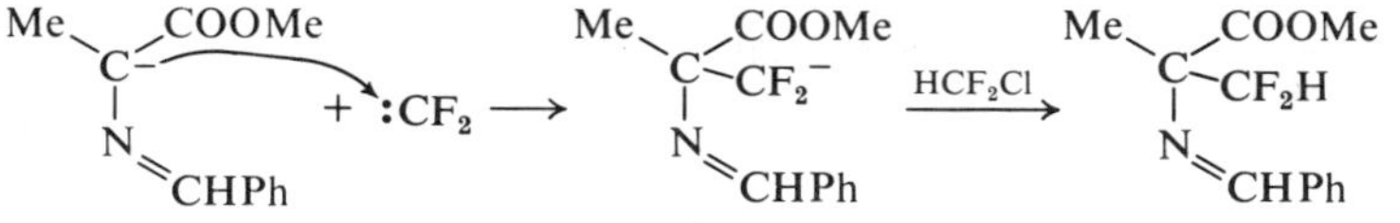

Me, COOMe, C^-, N=CHPh + :CF_2 → Me, COOMe, C–CF_2^-, N=CHPh $\xrightarrow{HCF_2Cl}$ Me, COOMe, C–CF_2H, N=CHPh

(**59**) (**60**)

+ :CF_2 + Cl^-

SCHEME 2

difluoromethyl adduct (**60**) by the mechanism of Scheme 2; subsequent removal of the benzylidene group followed by hydrolysis then completes a convenient synthesis of the corresponding fluorinated amino-acid without resorting to the use of toxic fluorinating agents.[99] Selected alkylation experiments with the cyclohexenyl acetamide derivatives (**61a**) and (**61b**) demonstrate that the derived mono- and di-lithiated species provide deconjugative α-substituted products (**62a**) and (**62b**), respectively, using primary, secondary, allylic and benzylic halides as well as benzaldehyde, methyl benzoate, and phenylselenyl bromide.[100]

NR^1R^2 (61) —i: $LiNPr^i_2$ or Bu^nLi; ii: E^+→ (62)

a; $R^1 = R^2 = Me$
b; $R^1 = H, R^2 = o\text{-}MeC_6H_4$

The addition of organometallics to enones is a widely employed reaction in organic synthesis which can lead to problems through lack of regioselectivity. One method of circumventing this difficulty in the case of acetylides is by the use of a catalyst prepared from $Ni(acac)_2$ and di-isobutylaluminium hydride (DiBAH) (1 : 1) which directs the process towards conjugate addition; for example, conversion of *tert*-butylacetylene into the aluminium derivative (**63**) followed by reaction with cumyloxycyclopentenone in the presence of the nickel catalyst

$Bu^tC{\equiv}CH$ —i; BuLi; ii; Me_2AlCl→ $Bu^tC{\equiv}C{-}AlMe_2$ (63) —a, b→ (64) $PhCMe_2O$ $C{\equiv}CBu^t$

Reagents: a: $Ni(AcAc)_2$, DiBAH

b: $PhCMe_2O$

preferently gives compound (**64**) in 85% isolated yield.[101] By contrast, methoxyallylcadmium or zinc reagents add regiospecifically 1,2 to enones, affording dienols which, as their potassium salts, serve as useful precursors to monoprotected 1,6-dicarbonyl derivatives (*via* an oxy-Cope rearrangement).[102] Sometimes control of the regioselectivity can be achieved by proper choice of experimental conditions. Thus $R^1SeCHLiCOOR^2$ ($R^1 = Me$ or Ph; $R^2 = Me$ or Bu^t) adds kinetically to cyclohexenone in THF at −78 °C to give the 1,2-adduct, but at higher temperature (or on addition of HMPA to the medium) thermodynamic control takes over, leading to the formation of the 1,4-adduct.[103] Similar behaviour is observed with the tin reagent Me_3SnLi, but even under thermodynamic control the silicon analogue reacts *via* the 1,4-mode.[104]

Although dimethyl sodiomalonate either is unreactive or gives many products, a variety of organolithium and organomagnesium compounds, possessing harder carbanion character, react with α,β-unsaturated thioamides to give selectively 1,4-addition products.[105] β-Sulphinyl-enones such as (**65**) readily undergo 1,4-addition of stabilized anions and enolates at room temperature with elimination of the sulphoxide substituent (see **66**) to form β-substituted enones (**67**) in excellent yield.[106] 2,3-Dimethylthi-iren 1,1-dioxide (**68**) behaves as an ambident electrophile in its reactions with α-metallated nitriles (**69a**) and (**69b**), affording two kinds of dihydrothiophen derivatives (**71**) and (**72**), respectively; thus the vinyl carbons are the electrophilic sites in the reaction with (**69a**), but the sulphur atom is attacked by the anion (**69b**) followed by ring-opening and recyclization to the intermediate (**70**).[107] For the Michael reaction of α-chloro-esters with α,β-unsaturated esters, leading to cyclopropane 1,2-diesters by cyclization of the intermediate carbanions,

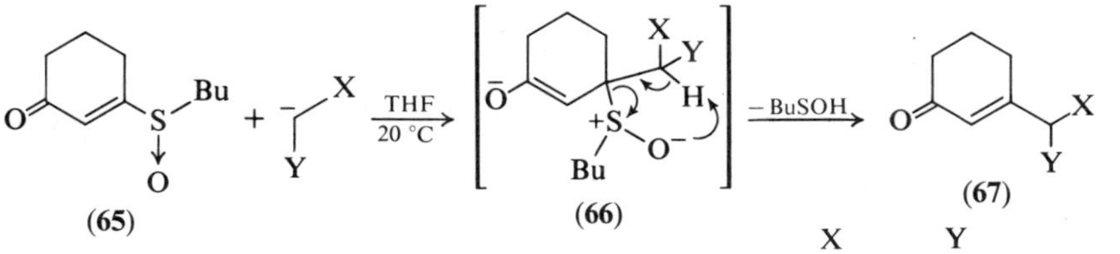
THF
20 °C
−BuSOH
(65)
(66)
(67)

	X	Y	
e.g.	COOEt	COOEt	(78%)
	CN	COOEt	(89%)
	MeCO	COOEt	(54%)

addition of macrocyclic polyethers to the reaction mixture leads to a remarkable increase in the proportion of *trans*-diester; this has been attributed to alternative conformational orientations of the γ-halogenocarbanion in its ion-aggregate and free form.[108]

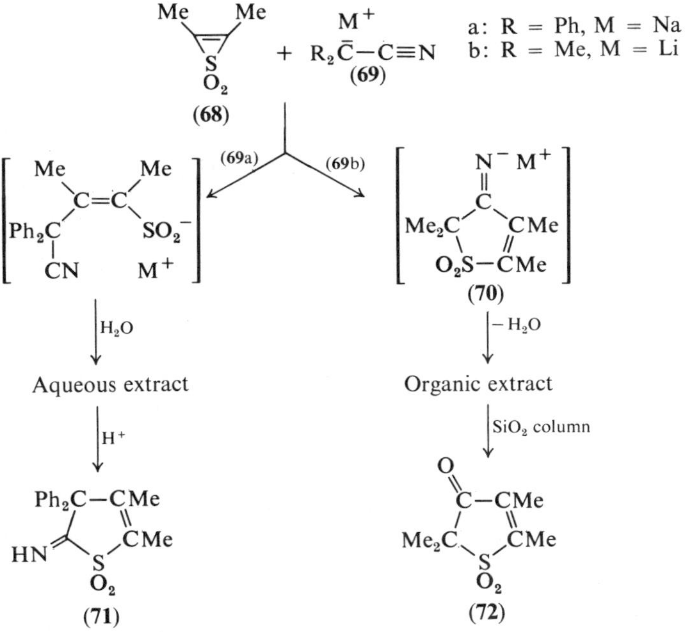
a: R = Ph, M = Na
b: R = Me, M = Li
(68)
(69)
(69a)
(69b)
(70)
H₂O
− H₂O
Aqueous extract
Organic extract
H⁺
SiO₂ column
(71)
(72)

A new carbon–carbon bond synthesis has been described which involves the addition of carbon nucleophiles to olefins activated as dicarbonyl η^5-cyclopentadienyl(olefin)iron cations.[109]

The reactions of carbanions and halocarbenes in two-phase systems have been reviewed.[110] Conversion of phenylacetonitrile (PAN) to dialkylated products occurs almost quantitatively when an excess of alkylating agent (molar ratio; alkyl bromide : PAN = 5 : 1) is used under standard "phase-transfer" conditions in the presence of tertiary amines.[111] Alkylation of isobutyraldehyde (and methyl acetoacetate) on use of a solid/liquid phase-transfer system (powdered NaOH–$Bu^t_4N^+I^-$) takes place readily even with less reactive halides, thereby providing a convenient route for the preparation of certain substituted aldehydes (and ketones) which are otherwise difficult to prepare.[112] Cyanohydrins of aliphatic aldehydes are etherified, by phase-transfer catalysis with allylic bromides, to β,γ-olefinic α′-cyano-ethers; these compounds on treatment with lithium di-isopropylamide give β,γ-olefinic ketones *via* a [2,3]-sigmatropic rearrangement of the resulting carbanions and elimination of lithium cyanide.[113] Under the same conditions, cyanoacetals rearrange to enolic monoethers of γ-keto-aldehydes or of γ-diketones; this forms the basis of a useful synthesis of dihydrojasmone.[114] Use of anion-exchange resins (triphase catalysis) is possible in the *C*-alkylation of phenyl acetonitrile, but the yields obtained are not as good as with ammonium or phosphonium salts under classical conditions of phase-transfer catalysis.[115] The condensation of α-chloro-α-phenylacetonitrile with benzaldehyde, carried out in the presence of 50% aqueous sodium hydroxide and a quaternary ammonium catalyst, gives mostly *trans*-2,3-diphenylglycidonitrile **(74a)**, but the *cis*-isomer **(74b)** predominates if the catalyst is omitted; the reason for this seems to be a preference for transition states **(73a)** in organic solvent and **(73b)** at the phase boundary.[116]

PhC̄ClCN + PhCHO

(73a)

(73b)

(74a)

(74b)

The reaction of fluorenylsodium with *N*-acetylaziridines in THF under nitrogen leads to ring-opening of the aziridines and formation of mono- or bis-9-amido-ethylated fluorene derivatives, the relative quantities of both products depending on experimental details.[117] Attempts to cyclize benzyl *o*-chlorophenyl ether, sulphide, sulphoxide, and sulphone by way of intermediates of the type **(75)** failed,

KNH_2 / $NH_3(l)$

X = O, S, SO or SO_2

(75)

but a similar reaction of 4- and 2-(*o*-chlorophenethyl)pyridine afforded products comprised of benzisoquinolines and 1-pyridylbenzocyclobutenes.[118] Other reactions of synthetic interest that have been studied include the unusual formation of an octalin derivative from the reaction of β-nitrostyrene with 1-morpholinoethene under conditions of kinetic control,[119] the palladium(II)-assisted alkylation of olefins with stabilized carbanions,[120] the stereospecific formation of functionalized 1,3-cyclohexadienes by alkylation of anions derived from 3,6-dihydrobenzoic acid esters,[121] and a novel *trans*-carboxylation reaction involving carbanionic attack on CO_2 generated from sodium fluorene-9-carboxylate in DMSO.[122]

In contrast to previous findings with the organometallic reagents (**76**), there is no cation or solvent effect on the stereoselectivity (>40% axial) of the addition of the charged-localized anions (**77**), (**78**), and (**79**) to 4-*tert*-butylcyclohexanone.[123] In spite of certain difficulties, it has been possible to use *cis*-2,2,6,6-tetramethylhept-4-en-3-one (**80**) as a probe to gain insight into the nature of Grignard reactions with ketones; for example, the reaction of *tert*-butylmagnesium chloride with an excess of enone shows almost complete isomerization of both starting material and products, indicative of a single-electron transfer pathway, whereas the corresponding reaction with allylmagnesium bromide gives indications of only a polar mechanism (*i.e.* it produces almost all 1,2-*cis*-addition product and does not significantly isomerize the remaining *cis*-enone starting material).[124] Methylmagnesium bromide (four equivalents, 3M in THF–benzene) adds to 1-trimethylsilyloct-1-yne (**81**) in the presence of 10 mol-% of 1 : 1 $Ni(acac)_2$-trimethylaluminium

MCH_2COO-alkyl

(**76**) M = MgX or ZnBr

$[ClCRCOO\text{-alkyl}]^-M^+$

(**77**)

$[ClC(R)CN]^-M^+$

(**78**) R = H or Me
Me = Li, Na or K

$[CH_2CN]^-M^+$

(**79**)

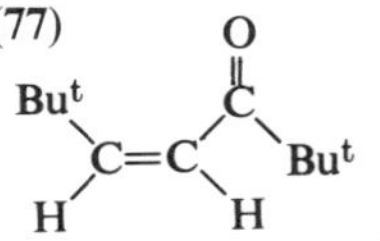

$Bu^tCH{=}CHCOBu^t$ (*cis*)

(**80**)

$C_6H_{13}C{\equiv}CSiMe_3 \xrightarrow{a,\ b,\ c}$ $(C_6H_{13})(Me)C{=}C(SiMe_3)(MgBr(Ni))$ + $(C_6H_{13})(Me)C{=}C(MgBr(Ni))(SiMe_3)$

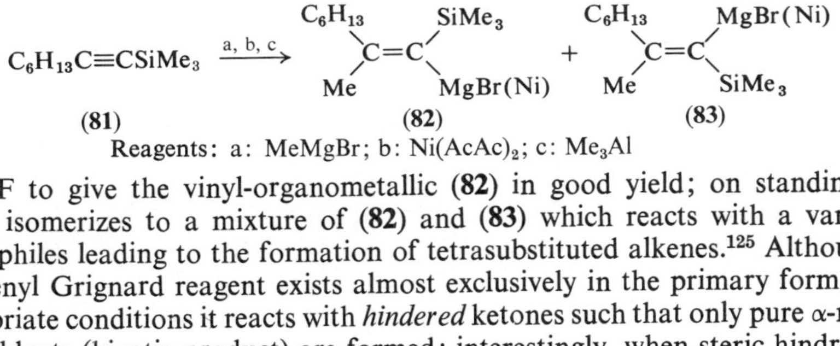

(**81**) (**82**) (**83**)

Reagents: a: MeMgBr; b: $Ni(AcAc)_2$; c: Me_3Al

in THF to give the vinyl-organometallic (**82**) in good yield; on standing, (**82**) slowly isomerizes to a mixture of (**82**) and (**83**) which reacts with a variety of electrophiles leading to the formation of tetrasubstituted alkenes.[125] Although the but-2-enyl Grignard reagent exists almost exclusively in the primary form, under appropriate conditions it reacts with *hindered* ketones such that only pure α-methylallyl adducts (kinetic product) are formed; interestingly, when steric hindrance in the carbonyl substrate is extremely severe (as with di-*tert*-butyl ketone), the thermodynamic butenyl adducts are obtained *both* from a reversal process involving the α-methylallyl adduct (Scheme 3) and directly from the ketone and the butenyl Grignard reagent *via* a four-centre transition state.[126]

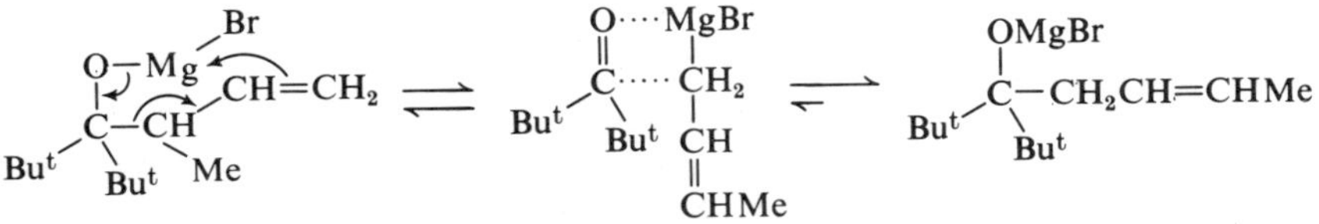

$Bu^t_2C(OMgBr)CH(Me)CH{=}CH_2 \rightleftharpoons [Bu^t_2C{=}O\cdots MgBr\cdots CH_2CH{=}CHMe] \rightleftharpoons Bu^t_2C(OMgBr)CH_2CH{=}CHMe$

SCHEME 3

The seemingly contradictory results previously reported for the reactions of terminal propargyl halides with alkyl Grignard reagents have been clarified with the discovery that the formation of allenes can occur not only on thermal isomerization of the initially formed alkynes, but also by a catalysed process involving transition-metal impurities in the magnesium used to prepare the Grignard reagents.[127,128] Dialkylcuprates also react with propargyl chlorides to form allenes in generally excellent yield.[129] The readily available dithioesters (**84**), known to undergo *thiophilic* addition with Grignard reagents, furnish magnesio-dithioacetals (**85**) which react

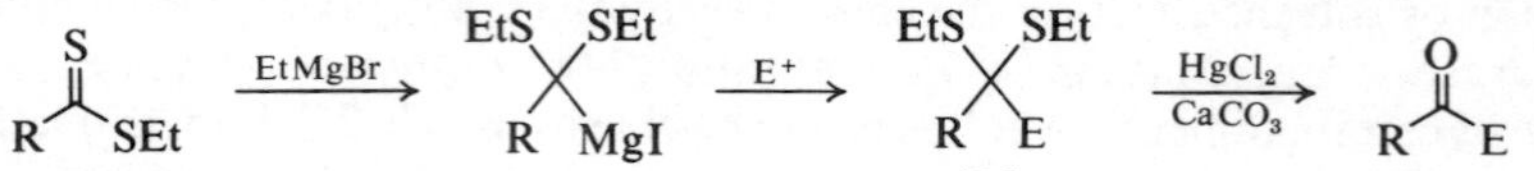

(**84**) (**85**) (**86**)

with a variety of electrophilic reagents (E) to give dithioketals (**86**) (masked carbonyl derivatives); in the case of (**84**; R = Me), prior metallation and treatment with carbonyl compounds leads to an overall reaction sequence that is tantamount to the double homologation of the novel E_1,E_2-synthon $\bar{C}H_2—\bar{C}=O$.[130]

Another paper has described the synthetic potential of *carbophilic* addition of allylic Grignard reagents to dithioesters leading to β,γ-unsaturated ketones.[131] The palladium-catalysed cross-coupling reactions of alkenyl iodides with a variety of Grignard reagents have been examined and found to proceed with retention of configuration (>97%) (*cf.* cuprates).[132] Reaction of 2-alkyl-2,3-epoxynaphthoquinone (R = Me, Et, Pr or Bu^t) with allyltributyltin occurs specifically at the sterically less crowded carbonyl group to give the corresponding tetrahedral adduct in which the allyl group is situated *cis* to the oxirane ring.[133]

According to kinetic and spectral data, the rearrangement of 1-chloroindole to 3-chloroindole in alkaline alcoholic media is a pseudo-first-order process preceded by an induction period which is seen as involving an equilibrium between 1-chloroindole, 3-chloro-3*H*-indole and the conjugate base of 3-chloroindole.[134]

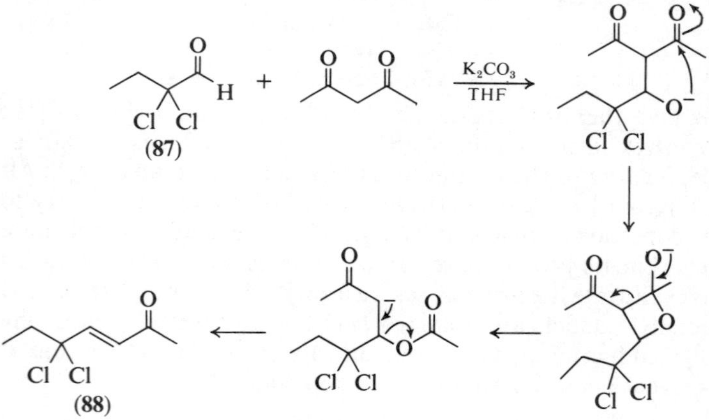

(**87**) (**88**)

SCHEME 4

Reaction of 2,2-dichlorobutanal (**87**) with pentane-2,4-dione in THF in the presence of K_2CO_3 leads cleanly to *trans*-5,5-dichlorohept-3-en-2-one (**88**), presumably by the mechanism of Scheme 4.[135] Kinetic studies of the mechanism of the anomalous formation of 4,4′-dinitrostilbene from the reaction of *tert*-butoxide with 4-nitrobenzyl halides are consistent with a chain reaction in which radical anions act as the chain carrier and exclude a bimolecular displacement reaction.[136] Kinetic and other evidence supports a free-radical mechanism for the oxidation of α-carbanions of alcohols, such as 9-hydroxyfluorene, by flavins.[137] Other carbanionic processes to have received mechanistic attention include the addition of benzene to α-substituted chalcone derivatives in the presence of catalytic amounts of palladium(II) acetate,[138] the unexpected formation of 1-methyl-3,4-benzo-7-thia-2-azabicyclo[3.3.1]nonane 7-oxide during the methylation of quinaldine with dimsyl anion,[139] the reaction of silylynamines with active triple bonds,[140] and a novel coupling–condensation reaction of *O*-methylarenesulphonates and *O*-methylarenesulphonanilides.[141]

1,1-Diaryl-2,2-dihaloethylenes are reduced under polarographic and cyclic voltametric conditions in non-aqueous DMF to 1,1-diarylethylenes in two two-electron waves *via* intermediary 1,1-diaryl-2-haloethylenes.[142] In the case of 2,2-dichloro-1,1-bis-(*p*-tolyl)ethylene, controlled potential electrolysis in non-aqueous DMF containing D_2O (0.2M) and lithium bromide produced the corresponding 2-chloro,2-deuterio-1,1-diarylethene as the only reduction product into which deuterium had been incorporated (68%), apparently by way of the electrochemically produced intermediate carbanion $Ar_2C{=}\bar{C}Cl$.[143] Trichloromethyl anion, cathodically generated from carbon tetrachloride, adds to carbonyl compounds to form trichloromethyl carbinols in yields that are comparable to those achieved by phase-transfer catalysis.[144] (2*S*,2′*S*)-2-Hydroxymethyl-1-[(1-methylpyrrolidin-2-yl)methyl]pyrrolidine is a new and efficient chiral ligand for effecting asymmetric induction in the addition reactions of alkyl-lithiums to carbonyl compounds;[145] for example, in the case of the reaction of butyl-lithium with benzaldehyde in a 1 : 1 mixture of dimethoxymethane and dimethyl ether at −123 °C, the corresponding alcohol (77% yield) is obtained in 95% optical yield.[146] Tri-*tert*-butylcarbinol is converted completely and instantly into di-*tert*-butyl ketone and isobutane when treated with the potassium salt of DMSO in that solvent at 25 °C; whether this reaction occurs through the expulsion of a *tert*-butyl carbanion or by a radical-cleavage pathway is not entirely clear, but the lack of coupling products in the product mixture and the initiation of the reaction by strong base makes the carbanion mechanism seem much more likely.[147] The absence of a detectable pH-independent pathway for breakdown of benzaldehyde cyanohydrins in aqueous solution suggests that complete proton removal, to form the oxy-anion, is a prerequisite for the CN^- departure.[148]

Metallation of cyclo-octa-1,5-diene with phenylsodium in the presence of TMEDA, followed by oxidation with oxygen, provides an exceptionally convenient synthesis of cyclo-octatetraene on a preparative scale.[149] Other topics that have been studied include the cleavage of tetrahydrofuran by butyl-lithium,[150] the scope of the DBU–$BrCCl_3$-induced bromination of active methylene compounds,[151] and their dimerization by treatment with dimethylbromosulphonium bromide in the presence of a strong base.[152]

Addition of *tert*-butyl α-chloro-α-trimethylsilyl-acetate to a THF solution of lithium di-isopropylamide at −78 °C produced the lithium reagent (**89**) which reacts with aldehydes and ketones to give, after quenching with two equivalents of thionyl chloride, reasonable yields of α-chloro-α,β-unsaturated esters (**90**).[153] In a sequence similar to that reported last year, reaction of 1-trimethylsilylvinyl-lithium (**91**) with ketones afforded the 2-hydroxyvinylsilanes (**92**) which could be converted by the novel methods indicated into a mixture of tetra-substituted alkenes (**93**) and (**94**), the major isomer being the one with the bulky trimethylsilyl group *trans* to the larger of the two alkyl groups R^1 and R^2.[154] Treatment of (**92**)

$Cl-C(SiMe_3)(Li)-COOBu^t$ **(89)** $\xrightarrow{\text{i; } R^1R^2CO \text{; ii; } 2SOCl_2}$ $R^1R^2C{=}C(Cl)COOBu^t$ **(90)**

e.g.

R^1	R^2	Yield %
Ph	H	44
Pr^i	H	25
$CH_2{=}CHCH_2CH$	H	49
Pr^i	Me	17
$PhCH_2$	Me	40

$Me_3Si(Li)C{=}CH_2$ **(91)** + R^1R^2CO ⟶ $R^1-C(R^2)(OH)-C(SiMe_3){=}CH_2$ **(92)** $\xrightarrow{a}$ $R^2(R^1)C{=}C(Et)SiMe_3$ **(93)** + $R^1(R^2)C{=}C(Et)SiMe_3$ **(94)**

Reagents: a: i; AcCl–AgCN; ii; Me_2CuMgI

R^1	R^2	Yield %	$E:Z$
Me	Et	80	65 : 35
Me	Bu	70	80 : 20
Me	Pr^i	75	92 : 8
Et	$PhCH_2$	75	76 : 24

derived from aldehydes (*i.e.*, $R^2 = H$) or aryl ketones ($R^1 = R^2 = Ar$) with thionyl chloride followed by β-elimination of halosilane with fluoride ion gives terminal allenes in good yields.[155] In the reactions of the (1-trimethylsilylallyl) carbanion with aldehydes and ketones, prior addition of $MgBr_2$ causes a remarkable shift in regioselectivity from the γ- to the α-site, thus allowing access to 1,3-dienes (by the subsequent elimination of Me_3SiO^-).[156] *gem*-Chloro(trimethylsilyl)allyl-lithium, $Li(Me_3SiCClCH{=}CH_2)$, prepared by the transmetallation between Ph_3Pb-$CH_2CH{=}C(Cl)SiMe_3$ and *n*-butyl-lithium in THF at −90 °C, is a novel ambident nucleophile which in coupling reactions with Me_3SiCl and Me_3SnCl forms the new bond to carbon exclusively at the CH_2 terminus; on the other hand, addition of $Li(Me_3SiCClCH{=}CH_2)$ to carbonyl compounds is less selective: its reaction with cyclohexanone, for example, gives a 38 : 62 mixture of products derived from reaction at the $CClSiMe_3$ and CH_2 termini, respectively.[157] *n*-Butyl-lithium reacts with 3,3-dichloroallyltrimethylsilane to metallate the vinyl proton, but under the reaction conditions the $Me_3SiCH_2C(Li){=}CCl_2$ formed undergoes β-elimination of LiCl to give $ClC{\equiv}CCH_2SiMe_3$.[158] Evidence from both trapping and competition experiments indicates that silaethylene intermediates can be produced by the elimination of lithium chloride from α-lithiochlorosilanes under appropriate reaction conditions.[159] The mechanism of the alkaline cleavage of the Si—H bond in penta- or tetra-methyldisiloxane involves either a fast equilibrium between hydroxide ion and the silane, followed by approach of a hydroxylated molecule in the rate-determining step, or a concerted attack on the Si atom.[160] Reaction of trimethylsilylmethylenedimethylsulphurane $Me_3SiCHSMe_2$ (prepared from Me_3Si-$CH_2S^+Me_2I^-$ with one equivalent of Bu^tOK) with carbonyl compounds leads in a Peterson-like way to a non-isolable vinylsulphoniumdimethylsilanoate intermediate which undergoes, amongst other processes, elimination of trimethylsilanol followed by a 2,3-sigmatropic rearrangement to give 2,3,3-trialkyl-5-thiahex-1-enes.[161]

The mechanism of the Wittig reaction has been discussed from the point of view of hard and soft acids and bases and compared with the analogous reactions of trivalent phosphorus compounds with carbonyl compounds.[162] Two new, but related, approaches to the preparation of stable sulphines, namely the alkylidenation of SO_2 by use of reactive phosphorus ylides[163] and α-silyl carbanions,[164] have been reported. When treated with singlet oxygen, methoxycarbonylphosphoranes are cleanly converted into α-keto-carboxylic esters.[165] Both benzyltri-(3-furyl)- and benzyltri-(3-thienyl)-phosphonium bromides (**95**; X = O or S, R = $PhCH_2$) undergo alkaline hydrolysis with preferential loss of the benzyl group, indicating that the latter is better able to support the forming carbanionic centre than are the 3-heteroaryl groups; this is in contrast to the situation with the corresponding 2-heteroaryl isomers.[166] Reaction of the five-membered cyclic phosphonate (**96b**) (mixture of *meso* and racemic) with 5,5-dimethyl-1-pyrroline *N*-oxide (**97**) in DME with sodium hydride as base gave, as with (**96a**), the aziridine (**98**) (30%) as the only detectable product, whereas after reaction of the six-membered phosphonate (**96c**) with (**97**) under the same conditions only (**99**) could be detected.[167]

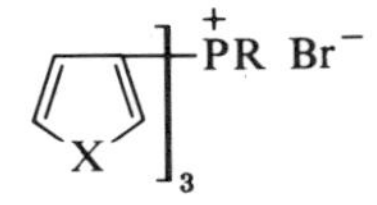

(**95**)

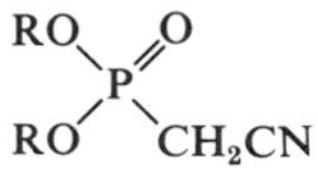

(**96**) a: R = Et
b: R, R = —CHMeCHMe—
c: R, R = —$CH_2CMe_2CH_2$—

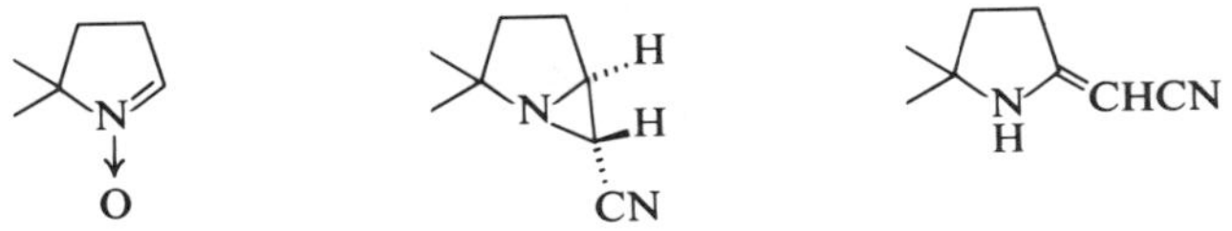

(**97**) (**98**) (**99**)

The course of the reaction between *N*-methylisatoic anhydride (**100**) and the phosphonate anion (**101**) is drastically solvent-dependent, that in benzene giving mainly the benz[*d*][1,3]oxazin-4-one (**102**), whilst that in DMF affords only the

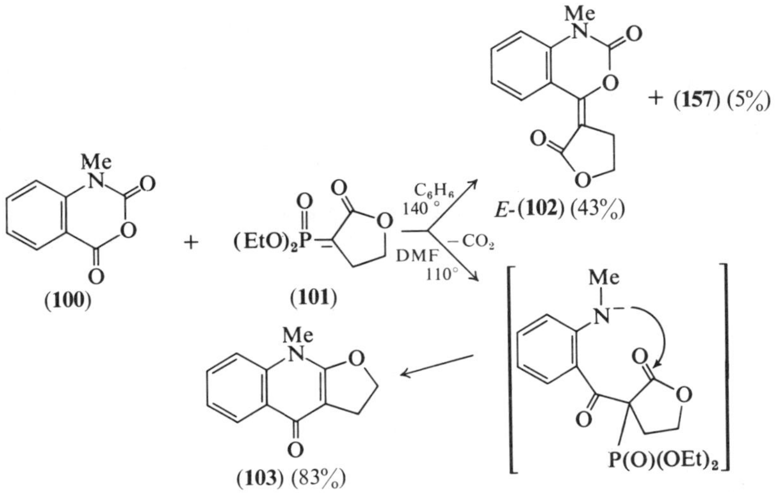

(**100**) + (**101**) → (C_6H_6, 140 °) *E*-(**102**) (43%) + (**157**) (5%)

(DMF, 110°, $-CO_2$) → (**103**) (83%)

furoquinoline (**103**) in high yield.[168] Treatment of 1,2-epoxyalkylphosphonates, *e.g.* (**104**) with two equivalents of butyl-lithium in THF at −78 °C, resulted in the unexpected formation of β-substituted allylic alcohols, presumably by a rearrangement of the type shown in Scheme 5.[169] The allyl anion (**106**), derived from the enol phosphate (**105**) by treatment with *tert*-butyl-lithium, reacts with MeI at C(1) to

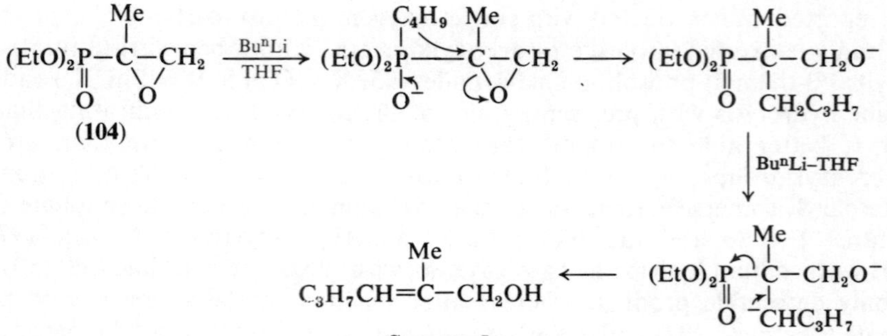

SCHEME 5

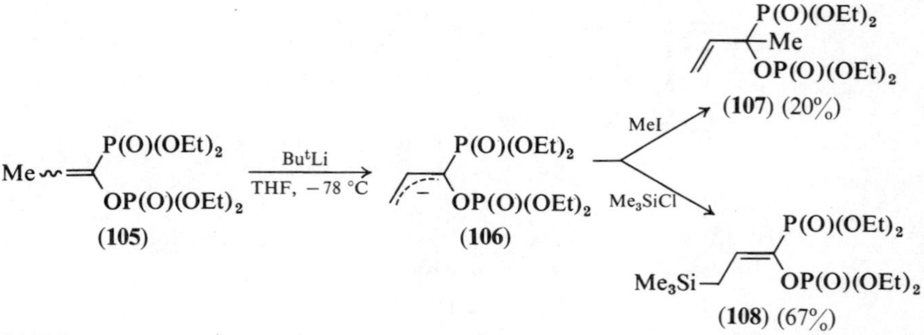

give (**107**), but with Me_3SiCl at C(3) to give (**108**); deprotonation of the latter followed by reaction with carbonyl compounds leads exclusively to the products of the Wittig–Horner olefination.[170] Metallation of *N*-allyl-*N*-methylphosphoramides with butyl-lithium generates an ambident anion which reacts regiospecifically with alkylating agents at the γ-position; hydrolysis of the resulting enephosphoramides then affords aldehydes in good yield.[171] Kinetic data for the reaction of trimethyl phosphite with α-bromo- and α-iodo-acetophenones leading to both oxophosphates $(MeO)_2P(O)CH_2COAr$ (Arbusov reaction) and vinyl phosphonates $(MeO)_2P(O)OCArCH_2$ (Perkov reaction) support a mechanism involving slow formation of an enolate halophosphonium ion pair produced by attack of the phosphite on the halogen.[172]

Addition of one equivalent of $LiClO_4$ to a solution of the aziridine (**109**) in acetonitrile leads to the practically quantitative formation of the complex (**110**)

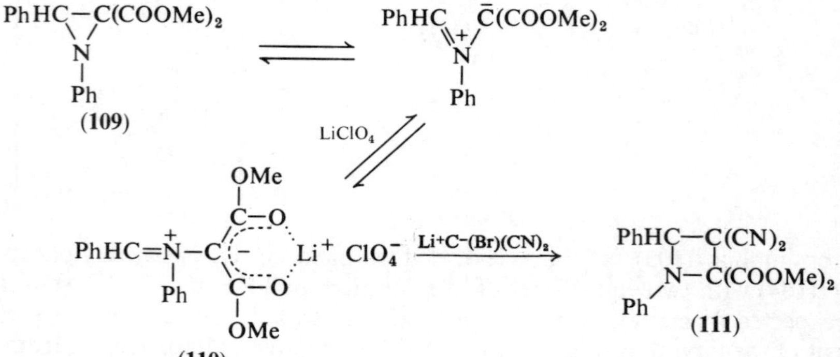

which reacts with bromomalonitrile carbanion to give the azetidine (**111**) in 71% yield.[173] Reaction of perfluorocyclobutene (**112**) with pyridine gives a mixture of dimers (**114a**) and (**114b**), and a trimer (**115**) whose formation can be rationalized on the basis of a reaction of an ylide (**113**) with the dimer (**114b**).[174] Metallation

Pyridine

(**112**) (**113**)

(114b)

(**114a**) (**114b**) (**115**)

of nitrosoamines with potassium *tert*-butoxide/butyl-lithium/di-isopropylamine (KDA) takes place more rapidly than with lithium di-isopropylamide (LDA) and leads to more reactive potassium derivatives for C—C bond formation with alkyl halides.[175] Silyl nitronates (**116**), easily prepared from both primary and secondary nitroalkanes, react readily with aliphatic and aromatic aldehydes under fluoride ion catalysis to give the derived 2-nitroalkyl *O*-silyl ethers (**117**) which can be smoothly reduced to 2-amino-alcohols (**118**) by addition to $LiAlH_4$ in refluxing ether.[176] The 2,4,6-tri-*tert*-butylphenoxyl group is a new sterically effective carbonyl-protecting group which provides access to the $Me_2NCH_2^-$ synthon from the *N,N*-dimethylcarbamate (**119**) by deprotonation with *sec*-butyl-lithium; thus reaction of the derived lithium compound (**120**) with electrophiles (E = alkyl halides and carbonyl compounds) leads to "higher" carbamates, which can be reductively cleaved with $LiAlH_4$ to tertiary amines of the type (**121**).[177]

RCHO, $Bu^n_4\overset{+}{N}F^-$; i; $LiAlH_4$, Et_2O ii; Aq. Na_2SO_4

(**116**) (**117**) R = alkyl or aryl (**118**)

sec-BuLi THF–TMEDA

(**119**) (**120**)

i; E^+ ii; $LiAlH_4$

Me_2NCH_2E
(**121**)

Enolates

When treated with an excess of lithium di-isopropylamide in THF, substituted cyclohexenones afford cross-conjugated dienolate anions which react quickly with methyl acrylate by a sequential Michael mechanism to give bicyclo[2.2.2]-octanones in excellent yields.[178] Contrary to earlier calculations concluding that α-attack on dienolates would always be favoured, quenching of the salt (**122**) with bromine, benzenesulphonyl chloride, or benzeneselenyl bromide leads *exclusively* to products of γ-attack.[179] Equally surprising is the propensity of the lithium dienolates (**123**), derived from 3(2*H*)-furanones, to undergo selective γ-alkylation.[180]

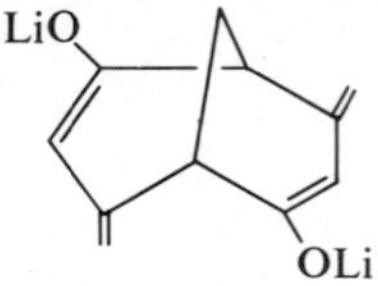

(**122**) 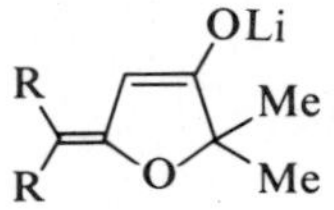 (**123**) R = H or Me 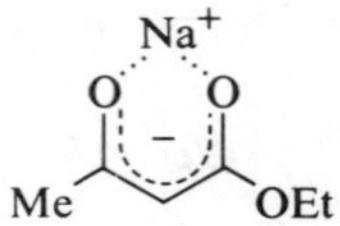 (**124**)

Perturbation theory has been used to interpret the breakdown in the reactivity–selectivity principle for the ethylation of the ambident enolate anion (**124**) on changing the leaving group (EtX, X = I, Br, OTS, OSO_2Et, OSO_2F or OSO_2CF_3).[181] The kinetics of iodination of RC_6H_4COMe (R = H, *p*-Cl, *m*- or *p*-NO_2) have been studied in various carboxylate buffers in 50 vol.-% aqueous methanol; the reaction was first-order in substrate and catalyst and zero-order in iodine.[182] In the alkylation of tetrabutylammonium acetylacetonate with butyl iodide, the *C*/*O*-alkylation ratio changes neither with the solvents (C_6H_6, $CHCl_3$, CH_2Cl_2 or DMF) nor with the enolate concentrations.[183] Silylation of acyclic ketones with ethyl trimethylsilylacetate in the presence of 1 mol-% of tetrabutylammonium fluoride proceeds highly stereoselectively, giving (*Z*)-enol silyl ethers which can be transformed into the corresponding lithium enolates with retention of stereochemistry.[184] Platinum(0) complexes which are effective in catalytic allylation of β-dicarbonyl enolates are also catalysts for the conversion of allyl ethers derived from phenols and ketoenolic compounds into the corresponding *C*-allylated derivatives.[185] Investigation of the reaction between sodium salts of α-formylcycloalkanones and 1-ethoxycarbonylcyclopropyltriphenylphosphonium tetrafluoroborate has led to the development of a new route to a variety of 2-substituted spiro[4,5]decanes [including the sesquiterpenes, (±)-β-vetivone, (±)-hinesol, (±)-β-vetispirene and (±)-α-vetispirene] *via* a pivotal intermediate.[186] A useful synthetic route to cyclohexanone derivatives has also been developed based upon the intramolecular cyclization of terminal enolates formed by the kinetic deprotonation of methyl ω-bromoalkyl ketones with the hindered base, lithium di-isopropylamide.[187, 188] Similar intramolecular cyclization by the bromo-ketone (**125**) produced the perhydroazulene derivative (**126**) (77–84%) accompanied by only 2% of the alternative cyclization product (**127**).[189]

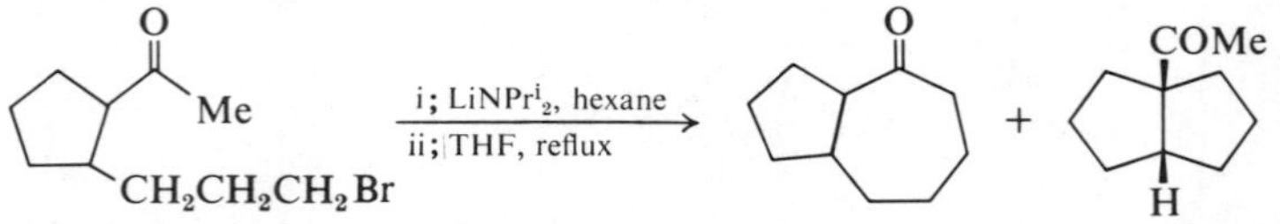

i; $LiNPr^i_2$, hexane
ii; THF, reflux

(**125**) (**126**) (**127**)

In the reaction of methyl dithioacetate with trimethyl phosphite, the phosphite behaves as a base (not as a nucleophile) and promotes formation of a mixture of the geometric isomers of Me(MeS)C=$CHCS_2Me$ by an aldol-type mechanism.[190] A stable intermediate (**129**) has been detected by ^1H-NMR spectroscopy during reaction of organic isothiocyanates and isoselenocyanates with glutacondialdehyde anion (**128**), leading to 1-substituted 3-formyl-2(1*H*)-pyridine-thiones (**130a**) and -selones (**130b**), respectively.[191]

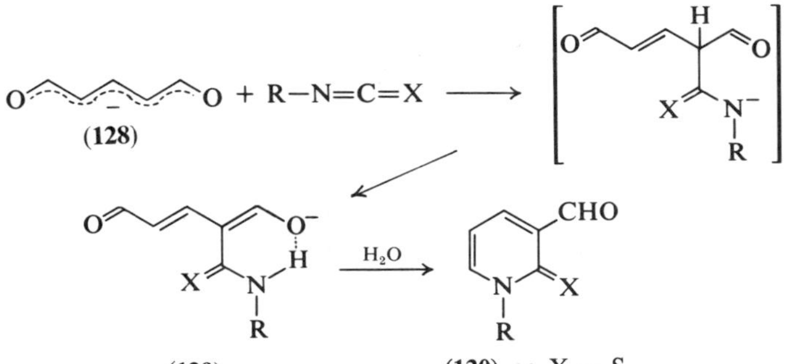

(**128**)

(**129**)

(**130**) a: X = S
b: X = Se

Either transmetallation of *cis*-1-tri-*n*-butylstannyl-2-ethoxyethylene[192] or halogen–metal exchange with *cis*-2-ethoxyvinyl bromide[193] leads directly to the hitherto elusive *cis*-2-ethoxyvinyl-lithium; this compound serves as a synthetically useful equivalent of the acetaldehyde anion, providing access to α,β-unsaturated aldehydes from carbonyl compounds, and a simple method for the two-carbon homologation of halides to aldehydes. Carbonyl substrates can also be efficiently converted into conjugated dienals by an extension of this methodology, using the vinyl analogue of *cis*-2-ethoxyvinyl-lithium.[194] Doubly deprotonated 4-nitrobut-1-ene (**131**) is a crotonaldehyde enolate equivalent (cf. **132**) which preferentially affords γ-substituted products (**133**) upon treatment with electrophiles, rather than the α-substituted products (**134**) formed from the normal dienolates; (**133**) can be subsequently

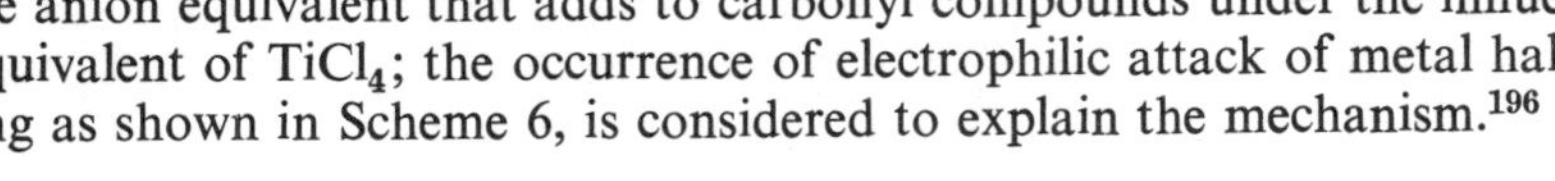

(**131**) (**132**) (**133**) (**134**)

transformed into α,β-unsaturated aldehydes by Nef reaction with $TiCl_3$.[195] 1-Ethoxy-1-trimethylsiloxycyclopropane is the first example of an ester homoenolate anion equivalent that adds to carbonyl compounds under the influence of one equivalent of $TiCl_4$; the occurrence of electrophilic attack of metal halide on the ring as shown in Scheme 6, is considered to explain the mechanism.[196]

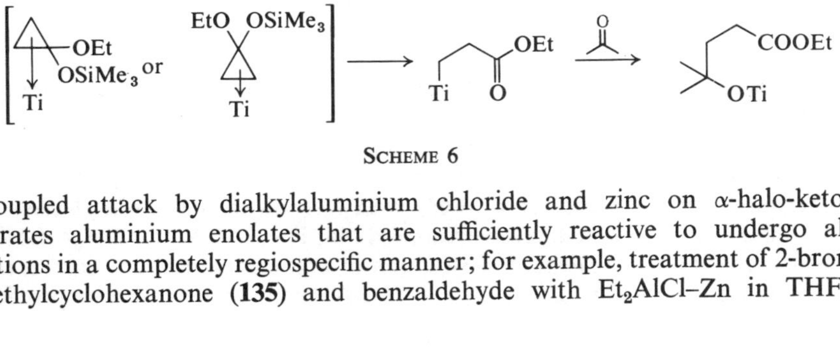

Scheme 6

Coupled attack by dialkylaluminium chloride and zinc on α-halo-ketones generates aluminium enolates that are sufficiently reactive to undergo aldol reactions in a completely regiospecific manner; for example, treatment of 2-bromo-2-methylcyclohexanone (**135**) and benzaldehyde with Et_2AlCl–Zn in THF at

(135) + PhCHO $\xrightarrow[\text{THF, }-20\,^\circ\text{C}]{Et_2AlCl-Zn}$ (136)

(137) $\xrightarrow[\text{ii; RCHO}]{\text{i; LiNPr}^i_2}$ (138) $\xrightarrow{HIO_4}$ (139)

R = Ph (77%), Pr^i (50%) or $PhCH_2$ (76%)

−20 °C produces the β-hydroxy-ketone (**136**) in quantitative yield, uncontaminated by the regioisomer.[197] In a sequence similar to that reported last year, condensation of the enolate of 2-methyl-2-trimethylsilyloxypentan-3-one (**137**) with aldehydes affords the diastereomerically pure β-hydroxy-ketones (**138**) which can be converted into β-hydroxy-acids (**139**) upon treatment with periodic acid.[198] Chiral β-ketols have been prepared in good chemical yields and enantiomeric excesses of 31–62% by the intermolecular aldol reaction of ketones R^2R^3CO with chiral hydrazones of ketones R^1COMe.[199]

Isomeric enamines derived from unsymmetrical methyl ketones and morpholine are readily converted into the least substituted isomer by regioselective deprotonation of the corresponding immonium salts with one equivalent of *tert*-butylamine.[200] Deprotonation of the tertiary enamine (**140**) with butyl-lithium in THF–TMEDA at 0 °C leads to the highly nucleophilic 1-aminoallylic anion (**141**)

(140) ⟶ (141) $\xrightarrow[\text{ii: }H_2O]{\text{i: }E^+}$

(141) ≡ (142)

which serves as a synthetic equivalent of the homoenolate anion (**142**) in reactions with electrophiles (alkyl halides, carbonyl compounds, epoxides, imines, etc.).[201] Asymmetric synthesis of α-alkylated cyclic ketones *via* chiral chelated lithioenamines continues to attract attention; in the present method, alkylation of cyclohexanone through the imine (**143**) with L-*tert*-leucine *tert*-butyl ester as a chiral source leads to 2-alkylcyclohexanones (**144**; R = Me, Pr^n or $CH_2{=}CHCH_2$) in 84–98% enantiomeric purity.[202] Reduction of imine derivatives of enones to the corresponding metalloenamines followed by alkylation and hydrolysis provides a

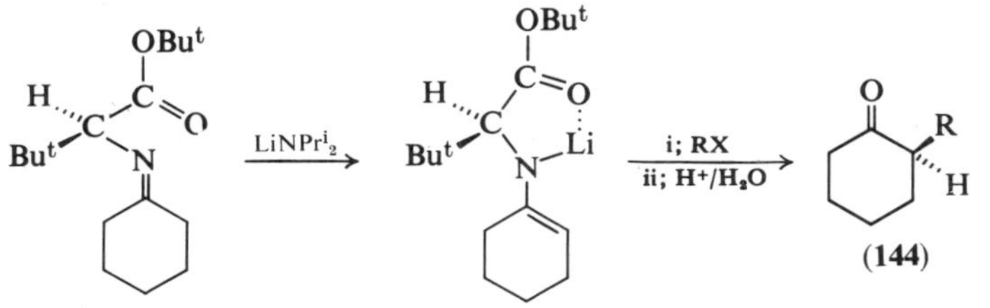

OBu^t H C C=O Bu^t N LiNPr$^i_2$ OBu^t C=O H C Li Bu^t N i; RX ii; H^+/H_2O O R H

(143) (144)

highly effective route to regiospecifically α-alkylated ketones.[203] Methylation of the dianion of (**145**) takes place exclusively at the *syn*-carbon and in a conformationally specific manner to give the axially substituted oxime (**146**) in 88% yield.[204] Whereas

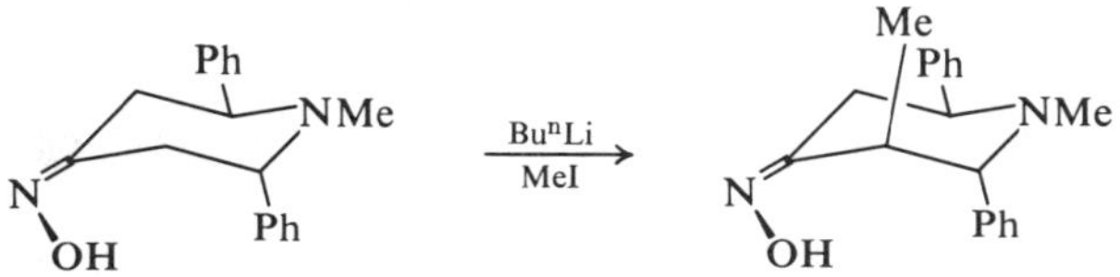

Ph NMe N OH Ph Bu^nLi MeI Me Ph NMe N OH Ph

(145) (146)

N,N-dimethylhydrazones undergo kinetic *anti*-deprotonation and rapid isomerization (about the C—N bond) to give the more stable *syn*-anion, formation of the *syn*-*O*-tetrahydropyranyl oxime anion is both *kinetically* and *thermodynamically* favoured.[205] Full details have appeared of the use of metallated dimethylhydrazones as equivalents of enolate anions in synthesis.[206, 207] The mixed Gilman reagent (**147**), obtained from α-lithioacetone *N,N*-dimethylhydrazone and cuprous thiophenoxide, adds Michael fashion to enones, *e.g.* (**148**), giving products such as (**149**) which can be used for the preparation of a variety of annulated fused-ring cyclohexenones.[208]

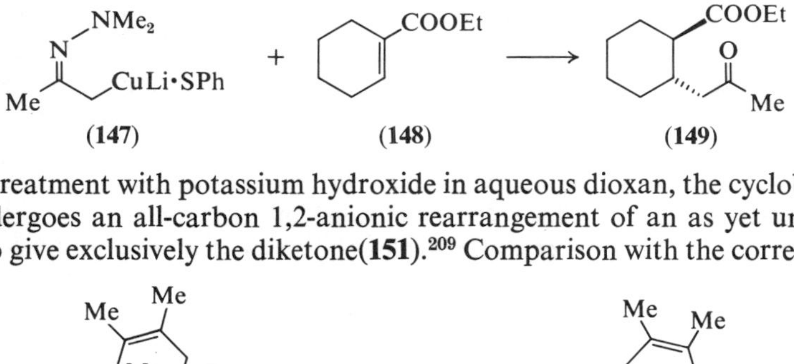

NMe_2 N Me CuLi·SPh + COOEt ⟶ COOEt O Me

(147) (148) (149)

Upon treatment with potassium hydroxide in aqueous dioxan, the cyclobutanone (**150**) undergoes an all-carbon 1,2-anionic rearrangement of an as yet unspecified nature, to give exclusively the diketone(**151**).[209] Comparison with the corresponding

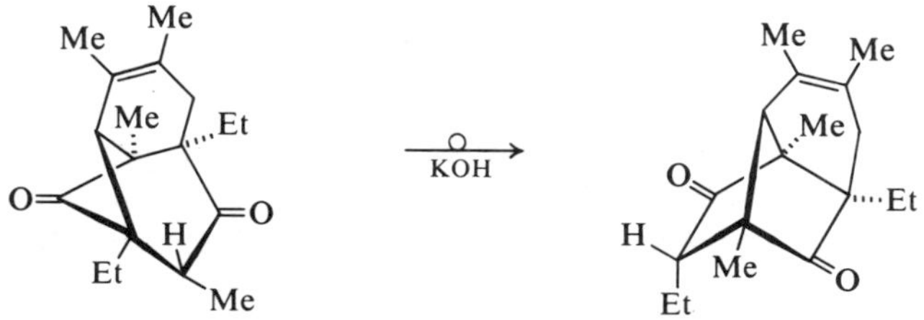

Me Me Me Et O O H Et Me KOH Me Me Me O Et H Me O Et

(150) (151)

saturated ketone shows that the presence of the double bond in α,α-dimethyl bicyclo[2.2.2]-octenone enhances its reactivity toward β-enolate formation and subsequent rearrangement to bicyclo[3.2.1]oct-2-en-6-one and its Δ^3-isomer.[210]

Through application of a new mechanistic criterion, which involves a log–log plot of the *exo* and *endo* H→D exchange rate constants,[211] it is apparent that, for bicyclic ketones such as (**152**), charge-transfer to oxygen is greater in the transition state for *endo*- than for *exo*-deprotonation.[212] The rate ratio (290 : 1) in the base-

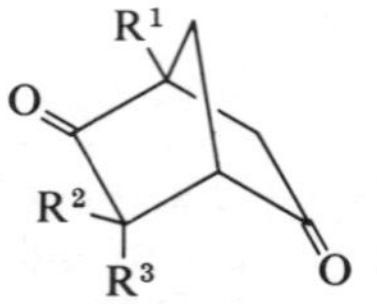

(**152**) $R^1 = R^2 = R^3$ = H or Me
R^1 = H, $R^2 = R^3$ = Me

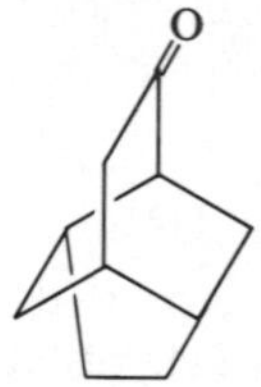

(**153**)

catalysed H–D exchange of the diastereotopic protons α to the carbonyl group in twistan-4-one (**153**) is in marked contrast to the lack of selectivity observed with cyclohexanones, and clearly shows that the principle of stereoelectronic control can be of major importance in the reaction of an α-keto-carbanion provided the ketone produces the requisite geometry.[213] Comparison of the base-catalysed deuterium exchange rates of the protons at C(3) in a series of 4-substituted bornanones (**154**; X = H, Me, Ac, Cl, Br or NO_2) shows that there is steric inhibition to

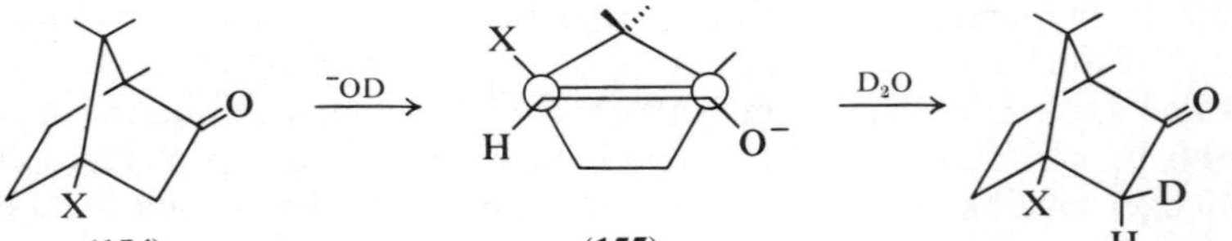

(**154**) $\xrightarrow{^-OD}$ (**155**) $\xrightarrow{D_2O}$

reaction of the enolate (**155**) with D_2O at its *exo*-face on account of a non-bonded interaction between D_2O and the substituent X at C(4) which increases with the size of X.[214] Base-catalysed H–D exchange (α-thioenolization) of the *exo*- and *endo*-protons of thiocamphor occurs much more readily than the corresponding α-enolization of camphor.[215] A review article concerned with bifunctional catalysis of α-hydrogen exchange of aldehydes and ketones has appeared.[216] Comparison of the rates of deprotonation of some cyclopropyl methyl ketones with those of suitable model compounds shows that a cyclopropyl group exerts little stabilization on an adjacent carbanionic centre, whereas the effect of a vinyl group is considerable.[217]

Dianions

Low-temperature matrix photoionization of the condensed cyclobutadiene dianion (**156**) leads *reversibly* to the corresponding radical anion which undergoes irreversible, and as yet unexplained, structural changes, on photoejection of a second electron.[218] Two separate and reversible one-electron transfers are observed in the electrochemical reduction of $[2_4]$-paracyclophanetetraene (**157**) to its dianion, which can be quenched with acetic acid to give $[2_4]$-paracyclophanetriene.[219] In contrast to the behaviour of 1,3-diphenylallyl-lithium, reduction of triphenylallyl-lithium with lithium, sodium, potassium or caesium leads to the exclusive formation of the allyl trianions (**158**) with no apparent dimerization of the intermediate

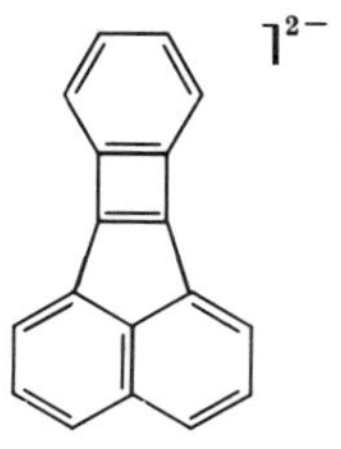

(156)

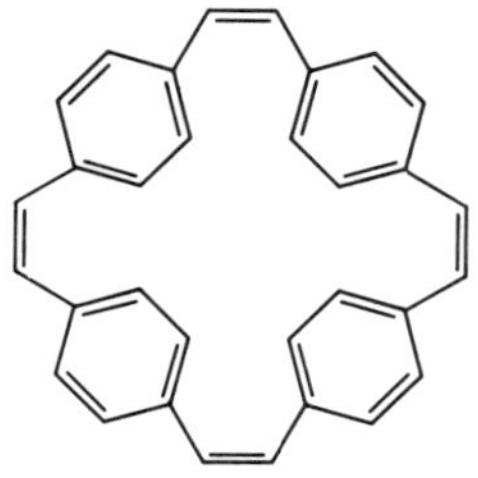

(157)

dianion radicals.[220] Irradiation of dilithioacetylene in liquid ammonia at −45 °C affords a monomeric substance of empirical formula C_4Li_4 whose NMR spectrum and field-desorption mass spectrum are compatible with the face-centred tetralithiotetrahedrane structure **(159)**.[221] Infrared spectra of the dianion **(160)** (as

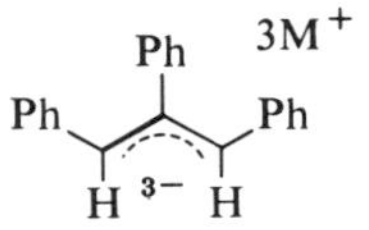

(**158**) M = Li, Na, K or Cs  (**159**) 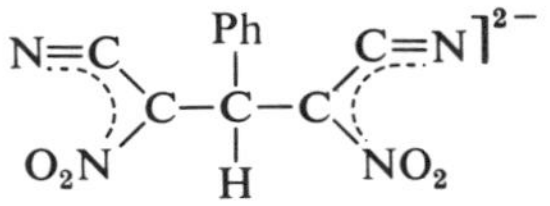 (**160**)

alkylammonium, deuteriated alkylammonium, and potassium salts) indicate the anionic charge to be localized mainly over the nitro-groups.[222] Organic substrates with second pK_a's well below 35, *e.g.* $Ph_2CHCOOH$, $PhSCH_2COOH$, $PhSO_2CH_2COOH$, $EtOOCCH_2COCH_2COOEt$, and indan-2-one, can be converted by dimsyl anion/DMSO into dianions, whose structures may be proved unambiguously by splitting patterns and chemical shifts of their ^{13}C-NMR spectra.[223] Attempts to observe spectroscopically the intermediate monolithium derivative during dimetallation of $CH_2{=}C(CH_2CH{=}CH_2)_2$ failed since the second step of metallation, leading to a ten-electron cross-conjugated dianion, is surprisingly much faster than the first one, despite the additional charge introduced in the π-system.[224]

Counter to predictions based solely on pK_a considerations, alkylation of the phenylacetone dianion is non-regioselective, carbon–carbon bond formation taking place at either the α- or α'-position.[225] Unlike the reactions of dianions of β-keto-esters with electrophilic reagents, those involving dianions of β-keto-amides occur exclusively at the γ-position without the problems associated with self-condensation.[226] When treated with ethyl bromoacetate, the dianion **(161)** of

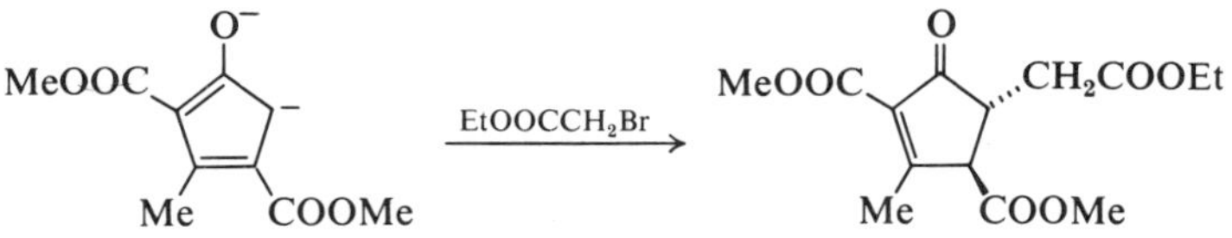

(**161**) (**162**) (61%)

2,4-bis(methoxycarbonyl)-3-methylcyclopent-2-enone undergoes alkylation exclusively on carbon to give the *trans*-substituted product **(162)** which can be used in the synthesis of Corey-type prostaglandin precursors.[227] Double deprotonation of readily available α-mercapto-γ-butyrolactone with lithium di-isopropylamide in THF/TMEDA at −78 °C yields the hitherto unknown dianion **(163)** which reacts with a variety of carbonyl compounds to give, after treatment with ethyl

(163) i: R^1R^2CO; ii: ClCOOEt → (164)

e.g.

R^1	R^2	$E:Z$
Ph	H	100 : 0 (54%)
Bu^n	H	95 : 4 (69%)
Ph	Me	100 : 0 (40%)

chloroformate, α-alkylidene γ-butyrolactones (**164**) with high *E*-stereoselectivity.[228] The stilbenediol dianion (**165**), generated by stirring a benzene solution of benzoin with 50% aqueous sodium hydroxide in the presence of a phase-transfer catalyst, *e.g.* $Bu^n_4N^+Br^-$, is easily dialkylated by various alkylating agents, in a manner typical of an ambident nucleophile.[229] The dianion prepared from the urethane (**166**) by using an excess of lithium di-isopropylamide in HMPA undergoes regioselective alkylation at carbon with a variety of electrophiles, thereby providing general synthetic access to α-acetylenic α-amino-acids.[230] Double deprotonation of methallyl alcohol with potassium *tert*-butoxide/*n*-butyl-lithium complex gives the

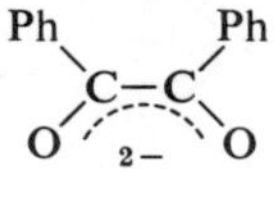

(165) 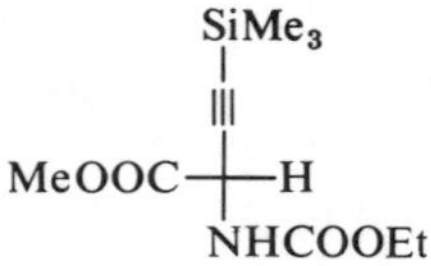 (166) 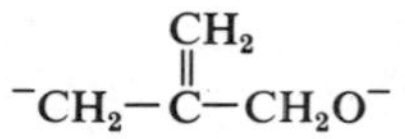 (167) 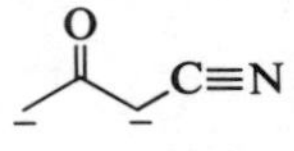 (168)

dianion (**167**) which reacts with aldehydes and ketones to afford diols; subsequent oxidation by activated manganese dioxide then completes an easy synthesis of α-methylene-γ-lactones.[231] Treatment of 5-methylisoxazole with two equivalents of lithium di-isopropylamide in THF at −10 °C provides a convenient method for the preparation of the synthetically useful acetoacetonitrile dianion (**168**).[232]

Anions α to Sulphur

According to recent CNDO/2-calculations, the higher carbanion stabilization effect of sulphur than of oxygen is related to the greater capability of the σ-bivalent sulphur to take up excess charge into the *sp* valence shell rather than to *d*-orbital conjugation.[233] ^{13}C-NMR studies on the structure of sulphur-stabilized carbanions show that the metallated carbon is nearly pyramidal in $PhSCH_2^-M^+$ and nearly planar in $PhSOCH_2^-M^+$, whatever the cation or solvent, whereas $PhSO_2CH_2^-M^+$ and $PhSO(NMe)CH_2^-M^+$ are in an intermediate hybridization state which is cation- and solvent-dependent.[234] The anion (**169**) of 2*H*-thiopyran reacts with

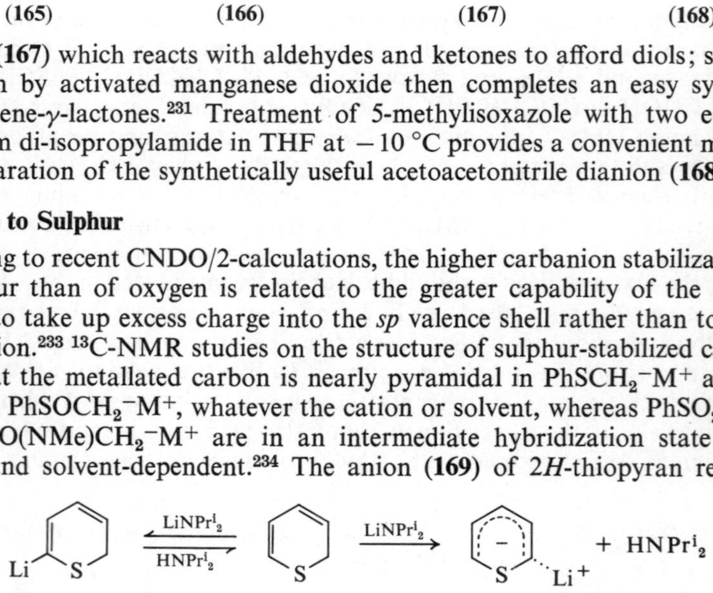

(169) ⇌ ($LiNPr^i_2$ / $HNPr^i_2$) → ($LiNPr^i_2$) (170) + $HNPr^i_2$

tert-butyl bromide and cyclohexyl bromide in liquid ammonia to give 2-*tert*-butyl-2*H*-thiopyran and a mixture of 2-cyclohexyl-2*H*-thiopyran and 4-cyclohexyl-4*H*-thiopyran in 53 and 82% yield, respectively.[235] Conversion of 6-lithio-2*H*-thiopyran (**169**) into the charge-delocalized species (**170**) takes place in the presence of di-isopropylamine, probably by a process of proton donation and abstraction.[236]

Exchange experiments on 2,2-dimethyl-1,3-dithiane and *ab initio* SCF on MeSEt anions show that while hyperconjugation may control the stereochemistry of

C—H acidification, C—S bond polarization accounts for the regiochemistry.[237] Previous conclusions as to the origins of the high degree of stereoselectivity in the reactions of 2-lithio-1,3-dithanes with electrophiles have been revised; it is now suggested (in accord with theoretical arguments) that an intrinsically preferred equatorial orientation of the lone pair in the carbanions may be responsible for such stereoselectivity.[238] A virtually completely asymmetric synthesis of (*S*)-(+)-atrolactic acid methyl ether has been described: the extremely high (~100%) optical yield results from a stereoselective electrophilic attack of benzaldehyde on the lithium derivative of an anancomeric 1,3-oxathiane, leading to equatorial substitution; this is followed by oxidation of the product alcohol to give a ketone which undergoes an asymmetric Grignard reaction according to Cram's rule where the rigid (cyclic) model applies.[239] The protonation (deuteriation and tritiation) of the 2-anion of 2-nonyl-1,3-dithian shows an inverse isotope effect, the magnitude of which has been explored for the anions of several 2-alkyl- and 2-aryl-1,3-dithianes.[240] The differential kinetic acidity of diastereotopic protons α to the sulphonium function has been studied in a series of conformationally biased or rigid thianium cations and found to increase as the ring becomes more rigid; that this change of stereoselectivity obtains from a monotonic decrease of the reactivity of the axial proton, whilst that of the equatorial proton remains constant, indicates that the conformation of the transition state for exchange of the axial proton, but not of the equatorial one, tends to differ considerably from the chair ground state.[241]

α-Chlorination of alkyl aryl sulphoxides whilst using optically active methyl phenyl *N*-chlorosulphoximide occurs with low asymmetric induction at the sulphinyl-sulphur; kinetic results suggest that the proton-removal step from the α-carbon is a cyclic concerted process (Scheme 7), which also explains the asymmetric induction.[242]

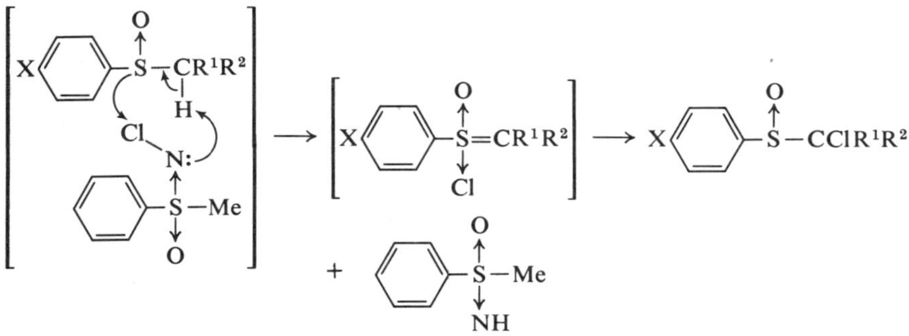

$$\left[X\text{-}C_6H_4\text{-}S(\to O)\text{-}CR^1R^2H \cdots Cl\text{-}N\text{:}\text{-}S(\to O)(Ph)\text{-}Me\right] \longrightarrow \left[X\text{-}C_6H_4\text{-}S(\to O)(\to Cl)\text{=}CR^1R^2\right] \longrightarrow X\text{-}C_6H_4\text{-}S(\to O)\text{-}CClR^1R^2$$

$$+\ Ph\text{-}S(\to O)(\to NH)\text{-}Me$$

SCHEME 7

$$R\text{-}C(=O)\text{-}O\text{-}Men \xrightarrow[\text{ii; }H^+]{\text{i; }2\ p\text{-}MeC_6H_4S(\to O)CH_2Li} p\text{-}MeC_6H_4\text{-}\overset{*}{S}(\to O)\text{-}CH_2\text{-}C(=O)\text{-}R + p\text{-}MeC_6H_4\text{-}\overset{*}{S}(\to O)\text{-}Me + Men\text{-}OH$$

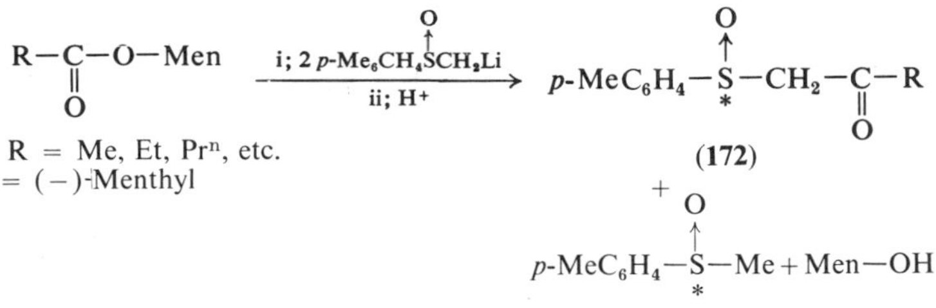

(171) R = Me, Et, Pr^n, etc.
Men = (−)-Menthyl

(172)

When two equivalents of (±)-*p*-tolylsulphinyl carbanion are allowed to react with (−)-menthyl carboxylates **(171)** in THF, the corresponding optically active β-keto-sulphoxides (**172**) are produced together with optically active methyl *p*-tolyl sulphoxide which has the opposite configuration to (**172**).[243]

A total synthesis of biotin based on the stereoselective alkylation of sulphoxides has been achieved.[244] Convincing evidence that the alkylation stereochemistry of α-lithio-sulphoxides is governed by their chelated structures has come from independent observations[245, 246] that the stereoselectivity of methylation (*e.g.* of MeSOCHLiPh)[245] is reversed on going from methyl iodide (inversion) to trimethyl phosphate (retention), a reagent needing electrophilic assistance by the Li^+ cation to be reactive.

2-Lithiobenzothiazole readily adds to carbonyl compounds to give, after dehydration, vinylbenzothiazoles that form the basis for a broad range of synthetic applications, including their conversion into α,β-unsaturated aldehydes and ketones,[247] and the stereoselective formation of fused and spiro-five- and spiro-six-membered rings.[248, 249]

A review dealing with nucleophilic eliminative ring-fission has appeared.[250] Studies of activated alkene-forming 1,2-eliminations have been extended to 2-arylsulphonylethyl- and 2-cyanoethyl-ammonium and -sulphonium salts in ethanolic triethylamine buffers which minimize ion-pairing; the reactions show buffer saturation kinetics: at low buffer-base concentrations ionization to form the intermediate carbanion is rate-determining, but at high buffer-base concentrations the intermediate carbanion is formed in a rapidly established pre-equilibrium step and the observed rate constant does not change with increasing base concentration at constant buffer ratio.[251] Orientation in 1,2-elimination from a carbanion in substrates of the type $GCHPhCH_2G$ [G = PhS or $(EtO)_2PO$] is directed by a phenyl group to give the alkene derived by the more rapid deprotonation at the phenyl-bearing site, whereas in the disulphone $PhSO_2CHPhCH_2SO_2Ph$ the opposite orientation of elimination is observed owing to steric repression of deprotonation.[252]

Deprotonation of 9-cyano-10-methylthioxanthenium perchlorate with sodium hydride in THF generated an orange-yellow crystalline product that was identified as 9-cyano-10-methyl-10-thia-anthracene, the first example of a stable 1,4-ylide in the thia-anthracene system;[253] when heated, the ylide undergoes a thermal Steven-type 1,4-rearrangement *via* radical intermediates to give 9-cyano-9-methylthioxanthene in excellent yield.[254]

[*n*,4]-Spiroannelation of α,β-epoxyketones utilizing diphenylsulphonium cyclopropylides followed by fragmentation constitutes a novel chain-extension procedure, *e.g.* (**173**)→(**174**), which can produce either olefin geometry in the chain-extended product.[255] Dimethylsulphoxonium allylides, derived from β-chlorocycloalkenones

i: $\triangleright\!=\!\overset{+}{S}Ph_2$; ii: $LiBH_4$, C_6H_6; NaOMe

(**173**) (**174**a) (**174**b)

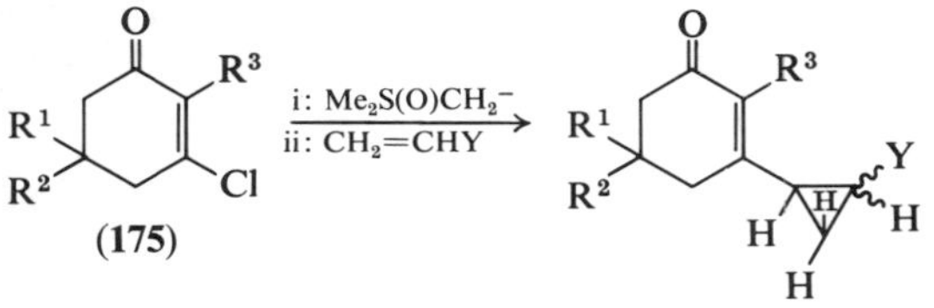

(175) —[i: $Me_2S(O)CH_2^-$; ii: $CH_2{=}CHY$]→ (176)

R^1	R^2	R^3	Y	Yield %
Me	Me	Me	COOMe	60
H	H	H	CN	35
Me	Me	H	CHO	40

(175), react readily with electrophilic olefins to give functionalized *cis*- and *trans*-cycloalkenyl cyclopropanes **(176)** in fair-to-good yields.[256] Compounds $R_2\overset{+}{S}{-}\bar{C}(NO_2)_2$ (R not specified) have been prepared and their reactions with electrophiles examined.[257]

Nitriles (RCN; R = ethyl, phenyl or β-naphthyl) react with dimsyl anion to give, after quenching with water, β-imino-sulphoxides which can be hydrolysed to the corresponding methyl ketone RCOMe in nearly quantitative yield.[258] On treatment with metallic sodium in the presence of trimethylsilyl chloride, 5-substituted 2-furaldehyde diphenyl dithioacetals undergo facile ring-opening to yield, after aqueous work-up, polysilylated ketones.[259] In the reaction of (phenylthio)acetic acid dianion, $PhS\bar{C}HCO_2^-$, and its ester monoanion, $PhS\bar{C}HCO_2Me$, with conjugated enones, the former gives predominantly the 1,2-adducts (β-hydroxy-acids) while the latter affords the 1,4-adducts (δ-oxo-esters).[260]

Lithium naphthalenide is a very useful reagent for the reductive cleavage of C—S bonds, allowing formation of sulphur-stabilized cyclopropyl anions from cyclopropanone dithioketals,[261] and primary alkyl-lithiums from alkyl phenyl sulphides.[262] Deprotonation of α-(phenylthio)alkaneboronic esters **(177)** with lithium di-isopropylamide followed by acylation of the resultant carbanions **(178)** with methyl esters provides an efficient, regiospecific, general synthesis of α-(phenylthio)-ketones **(179)**, the synthetic utility of which is well established.[263]

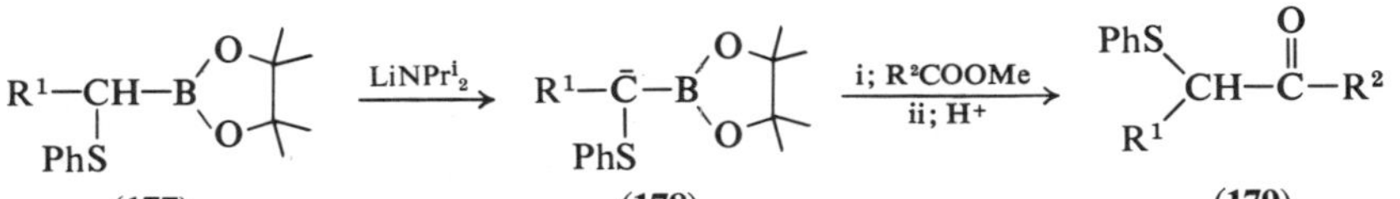

(177) —[$LiNPr^i_2$]→ (178) —[i; R^2COOMe; ii; H^+]→ (179)

α-Selenocarbanions $MeSeCLi(R^2)CH_2R^1$, easily prepared from carbonyl compounds $R^1CH_2COR^2$, react with monoalkyl (R^3) epoxides exclusively at the least substituted carbon to produce γ-hydroxy-selenides **(180)** which can be specifically

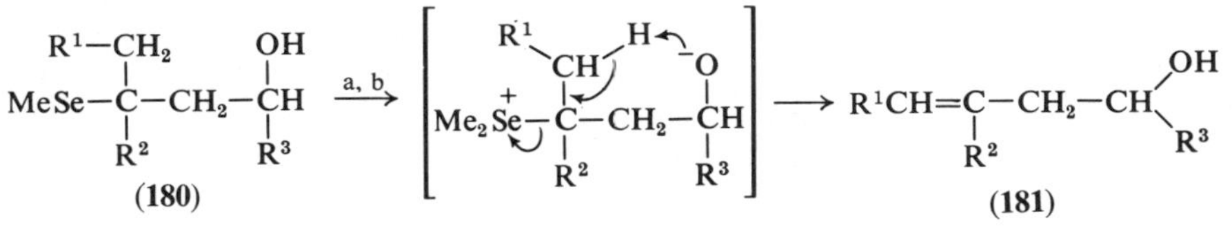

(180) —[a, b]→ [intermediate] → (181)

SCHEME 8

Reagents: a: MeI
b: $KOBu^t$–DMSO

converted by the novel methods indicated into homoallyl alcohols **(181)** in high yield; the high regiospecificity of this transformation can be explained by the cyclic elimination depicted in Scheme 8.[264] Diastereoisomeric β-hydroxyselenides, obtained by reaction of the α-selenocarbanions $RSeCLi(R^1)COOR^2$ with aldehydes and ketones R^3COR^4, have been separated by chromatography and converted into stereochemically pure α,β-unsaturated esters by *trans*-deoxy-selenation promoted by phosphorus oxychloride.[265]

Proton-transfer and Hydrogen Isotope Effects

The nitroalkanes are a group of compounds that continue to attract widespread attention as a result of their anomalous behaviour in various aspects of proton-transfer reactions. It is now apparent that for systems such as $G(CH_2)_nNO_2$, where G is a heteroatom substituent and $n = 1, 2$ or 3, the Taft relationship can be applied only in a limited sense because (*a*) σ^*_G constants ($n = 1$) cannot be mixed with $\sigma^*_{CH_2G}$ constants ($n = 2$); (*b*) "methylene transmission coefficients" which relate $\sigma^*_{CH_2G}$ to $\sigma^*_{CH_2CH_2G}$ vary with the geometry; and (*c*) σ^* constants in general give only a rough measure of polar effects because they depend on geometry.[266] It also seems likely that such deprotonations, with the exception of that of nitrocyclopropane,[267] involve an essentially pyramidal, singly H-bonded intermediate or virtual intermediate (**182** in Scheme 9) rather than a single transition state.[268]

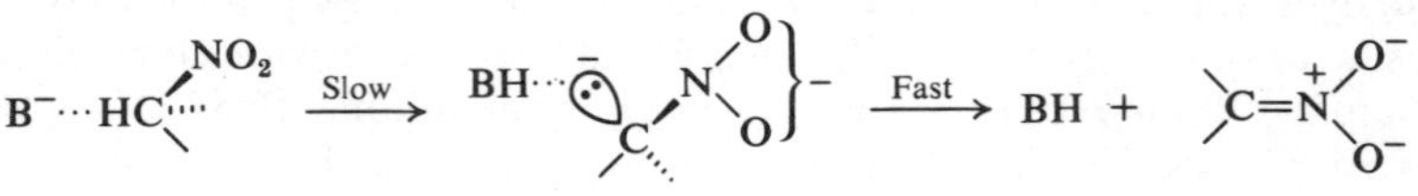

(**182**)

SCHEME 9

Proton-transfer from $MeCH(NO_2)_2$ to trioctylamine showed none of the characteristics that would be expected if tunnelling were important.[269] The kinetic isotope effect for the reaction of 4-nitrophenylnitromethane (4-NPNM) with pentamethylguanidine in toluene is 13.7 ± 0.4, a value significantly lower than that reported for tetramethylguanidine (TMG), but about the same as those observed with other bases in toluene and for TMG in the more polar solvents THF, CH_2Cl_2, and MeCN; this value, together with the corresponding activation enthalpy difference $(\Delta H_f^{\ddagger D} - \Delta H_f^{\ddagger})$ of 2.2 ± 0.4 kcal mol^{-1}, suggests that there is a significant tunnelling correction.[270] Tunnelling would also account for the considerable isotope effects and very low Arrhenius activation energies observed in the proton-transfer reactions of 4-NPNM with some alkylamidines $HN{=}C(NEt_2)R$,[271] and the cyclic amidine DBU (1,5-diazabicyclo[5,4,0]undec-5-ene) in toluene.[272] The suggestion has been made that the effective potential energy barrier used in calculating tunnelling corrections should include a mass-dependent term consisting of the zero-point energies of vibrational modes orthogonal to the reaction co-ordinate.[273] Just as with ketones, close agreement is observed between the Brønsted β-coefficient determined for the deprotonation of phenylnitromethane by hydroxide ion in aqueous DMSO solutions (on the basis of the enhanced basicity of hydroxide ion in these mixtures) and that obtained in the standard manner (for reaction with bases in aqueous solution).[274] Proton transfer between 2,4,6-trinitrotoluene and TMG is faster in PhCN than in MeCN owing to the more ordered (solvated) transition state in the latter solvent.[275] NMR data show absence of hydrogen bonding in the $CF_3CH_2NO_2$–Bu_3N system, but that reversible proton-transfer and ion-pair formation does occur.[276] Details have appeared of a simple method, applicable to base-catalysed ionization of nitroalkanes, for the empirical characterization of transition states in terms of structure–reactivity parameters on a three-dimensional energy-contour diagram.[277]

Kinetic data for the deprotonation of 4-nitrophenylacetonitrile by hydroxide ion in DMSO–water mixtures are consistent with a reaction sequence involving two steps:[278]

$$HA + OH^- = A^- + H_2O$$

$$A^- + H_2O = A^- \cdot H_2O$$

in which two coloured forms of the anion are produced, one unspecifically solvated (purple; λ_{max} 535 nm) and the other hydrogen-bonded to the nitro-group of the anion (orange; λ_{max} 495 nm).[279] Comparison of the rates of deprotonation of cyclopentadienes with related data for simple nitro- and carbonyl compounds suggests that structural reorganization on going from the acid to its conjugate base, and not electronic delocalization in the carbanion, is the main cause of the activation barrier against proton-transfer involving carbon acids.[280] Further studies on the isomerization and protophilic H–D exchange reactions of phenylated cyclopropanes have appeared.[281] The energies required for deprotonation of the substituted benzenes XC_6H_5 have been calculated by an *ab* initio method; the results show that substituent effects in base cleavage of $XC_6H_4SiMe_3$ compounds (and possibly those in base-catalysed hydrogen exchange in substituted benzenes) are consistent with the formation of the aryl carbanion $XC_6H_4^-$ in the rate-determining step.[282] Phase-transfer catalysis seems to be an efficient tool for deuterium exchange and isomerization of acidic hydrocarbons.[283]

Linear free-energy relationships are observed between the (log) rate constants for H–D exchange of vinyl hydrogen atoms in substituted ethylenes and $\sigma_p{}^0$ constants of the α and β substituents: $\rho_\alpha{}^0 = 27.9$ and $\rho_\beta{}^0 = 13.9$.[284] Linear correlations also exist between the H–D exchange rate constants and calculated deprotonation energies for methyl derivatives of five-membered aromatic heterocycles[285] and of benzene derivatives.[286] In strongly alkaline media, 1-alkylimidazoles undergo deuterium exchange *only* at C(5) by a carbanion pathway in which a proton is abstracted from the neutral molecule in the rate-determining step; the resistance to carbanion formation at C(4) is ascribed to the *adjacent lone pair* (ALP) effect, *i.e.* a significant electrostatic repulsion between lone pairs in the coplanar sp^2-orbitals at N(3) and C(4).[287] The same effect is also seen in the lack of exchange at C(4) in 1-methyl-2- and -5-X-imidazoles (X = Me, F or NO_2).[288] In the deuterium exchange reactions of 3-hydroxypyridine, only the anionic form underwent exchange at H(4).[289] No relation is found between the rates of base-catalysed H–D exchange in a variety of perimidines and naphthimidazoles and their reactivity toward butyl-lithium; for example, (**183**) undergoes exchange most readily, but is not metallated.[290] Treatment of the protoporphyrin IX (**184**) dimethyl ester with MeONa–MeOD in DMF for five days leads to deuterium exchange not only in the methylene groups adjacent to the carbonyl, but also in the methyl groups at positions 1 and 3 in the ring; the latter clearly exhibit a greater reactivity.[291] The methyl group of 5-methyl-2-phenyltetrazole undergoes H–D exchange some 20 times faster than that of the corresponding 1-phenyl derivative.[292]

In the presence of D_2O in the gas phase, highly basic anions, such as allyl, hexenyl, and benzyl, undergo sequential H–D exchange, a process that promises to be of value as a diagnostic tool for the determination of the structure of an ion as well as for distinguishing among isomeric anions.[293] In the case of less basic anions the extent of exchange with different deuteron sources can be used not only to give structural information, but also to probe the details of the mechanisms

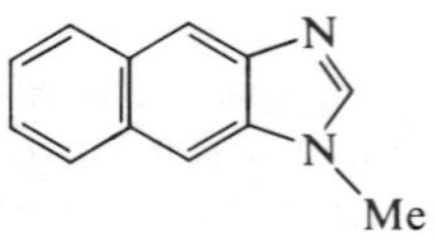

(183)

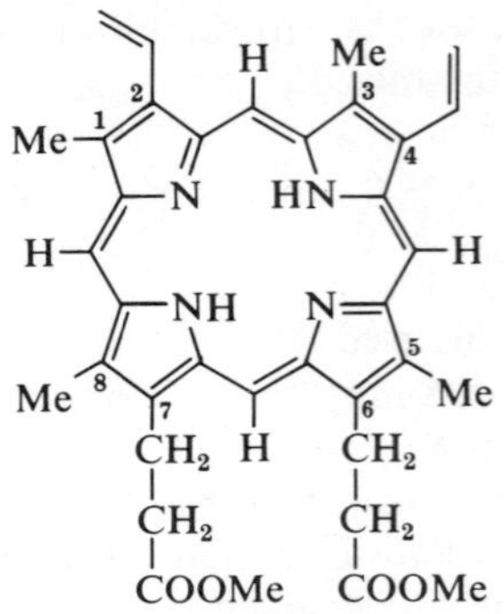

(184)

of gas phase ion–molecule reactions; for example, the penta-1,3-dienyl anion exchanges a maximum of four hydrogens with MeOD, thereby showing that reprotonation of the central carbon atom, which could lead to the intermediacy of a nonconjugated diene (and a maximum of five exchanges), does not occur.[294]

A review in Japanese entitled "Kinetic Isotope Effects and Organic Reactions" has appeared.[295] Kinetic isotope effects have been determined by a polarimetric differential method in which the dextrorotatory form, (+)-A, of a substance A is mixed with the laevorotatory form, (−)-A*, of the isotopically substituted substance A* such that the net optical rotation is zero; thus in the ensuing reaction which transforms A and A* into non-chiral products the kinetic isotope effect induces optical activity in the reaction mixture and from the maximum optical rotation its magnitude can easily be calculated.[296] For proton transfer from O_2NCH_2COOEt to bases (HO^-, PhO^-, *o*-$ClC_6H_4O^-$, AcO^-, $ClCH_2COO^-$, H_2O), the calculated plot of kinetic isotope effect *vs.* free energy of reaction agrees with experimental data.[297] A reinvestigation of the cleavage of Ph_3CSiMe_3 by methanolic sodium methoxide has shown that the product isotope effect, given by the product ratio RH : RD obtained in 1 : 1 MeOH–MeOD, is in fact in the region of 1.3, a value consistent with those for substituted benzyl compounds of comparable reactivity, but incompatible with a reported value (*ca.* 8.1) of the kinetic isotope effect for proton abstraction from Ph_3CH by sodium methoxide in methanol.[298]

Evidence for bifunctional catalysis of the ionization of dihydroxyacetone phosphate (DHAP) by several diamines has been obtained, but evidence is marginal for trifunctional catalysis by monoprotonated *N*-[*N*′-(2-pyridylalkyl)aminoalkyl]-guanidines in which the guanidium component interacts electrostatically with the phosphate of DHAP, the secondary amino-group forms an imininium bond with the carbonyl group of DHAP, and the 2-pyridyl component abstracts the α-proton of DHAP to generate the carbanion.[299] The Brønsted plot for the rate-determining enolization of 4-(*p*-nitrophenoxy)butan-2-one with a series of oxyanion bases exhibits substantial curvature, the cause of which has been ascribed to a solvation effect rather than a "Hammond postulate" type of change in structure of the transition-state.[300] It seems likely that solvation perturbation also dominates the structure–reactivity correlations for nucleophilic reactions of the same oxyanions with pivalate esters.[301] Contrary to expectations, the rates of racemization and H_α exchange of a series of pyridoxal–amino-acid Schiff's bases do not accord with predictions based solely on electronic or steric effects, but instead parallel the proportions of the reactive conformers (*i.e.* conformers with the C_α—H_α bond orthogonal to the π-system) estimated by CPK models.[302] A new model for water, based on the simple equilibrium:

$$(H_2O)_{bound} \rightleftharpoons (OH)_{free} + (LP)_{free} \quad (LP = \text{lone pair})$$

has been proposed to explain electrolyte and cosolvent effects on the rates of water-catalysed deprotonation reactions such as the hydrolysis of arylsulphonylmethyl perchlorates.[303]

Absolute and Relative Acidity

A novel method for determining thermodynamic pK_a's of weakly acidic hydrocarbons has been described; the basis of the method which involves the establishment of the equilibrium:

$$RH + MeO^- \rightleftharpoons R^- + MeOH$$

is the measurement of the concentration of RT after the addition of a given volume of the equilibrated solution to a quench solution with a relative high total concentration of protons and tritons of known isotopic ratio.[304] More data have appeared concerning the acidity of acetylenic compounds in DMSO; the most striking feature is the following order of decreasing acidities: phenylacetylene (22.6) > acetylene (25.5) > *tert*-butylacetylene (27.6).[305] From the observation that 2-benzylbiphenyl (2-BBP) has essentially the same acidity as diphenylmethane (*ca.* 1.7 p*K* units less acidic than **4-BBP**), it appears that any steric interactions that prevent conjugation in the anion of 2-BBP are just balanced by the inductive effect of the *o*-phenyl group.[306] A previous measurement of the equilibrium acidity of 4-methylbiphenyl (toward caesium cyclohexylamide, CsCHA) has been found to be in error because of an unrecognized solvent-levelling effect; the corrected pK_{CsCHA} is 38.95, 0.22 units higher than the old value.[307] Comparison of the acidity of the 9-phenylfluorenyl system (**185**) with fluorene itself shows that a fully conjugating phenyl group has about a 7 p*K* unit acidifying effect, of which about one-third is attributed to polar effects and two-thirds to conjugation.[308] Brønsted correlations between p*K*'s and relative kinetic acidities have been obtained for some alkyl-aromatic hydrocarbons; the kinetic values were measured by means of competitive reactions between two alkylaromatics and the tertiary aliphatic polyamine ionized lithium salt of a third hydrocarbon and the p*K* values derived from the equilibrium constants of the same reactions.[309]

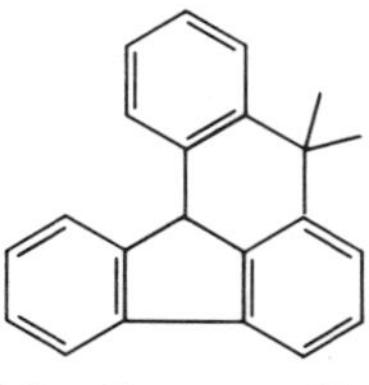

(**185**) $pK_{CsCHA} = 15.45$

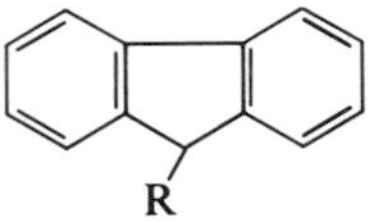

R

(**186**)

MO calculations for the electron configuration of the 9-fluorenyl derivatives (**186**; R = H, allyl or allyloxy) show that their C—H acidity decreases in the order given.[310] Alkyl substituent effects on the acidity of 9-substituted fluorenes can be satisfactorily described by an equation of the type: $pK = f(\rho_\alpha, E_{s\alpha}, E_{s\beta})$ where the parameters ρ_α and $E_{s\alpha}$ characterize α-crowding and $E_{s\beta}$ characterizes

β-crowding.[311] Substitution of various groups for a hydrogen atom at the α, β, and γ positions of 3-nitropropene causes a complex variation of kinetic and equilibrium acidities that can be interpreted by assuming that the kinetic acidities are governed primarily by polar effects and steric inhibition of solvation in the transition state, and the equilibrium acidities by steric strains in the product nitronate ions.[312]

Gas-phase acidities, which are valuable in that they allow intrinsic and solvent-induced substituent effects to be separated, have been determined for some simple carbonyl compounds from the photodetachment thresholds for the corresponding enolate anions.[313] The effects of alkyl substituents on equilibrium acidities of various weak carbon acids in protic and dipolar aprotic media and in the gas phase have been assessed; for the methyl group these may be divided into four types: (a) acid-weakening hyperconjugative and polar effects; (b) acid-strengthening hyperconjugative effects (on ketones, nitroalkanes, and 9-methylfluorene); (c) acid-weakening polar methyl effects (on sulphones and nitriles); and (d) acid-weakening steric effects.[314] Comparison of the gas-phase acidities of a number of sulphones with similarly substituted carbonyl and nitro-compounds shows that there is less stabilization by conjugation in the sulphone anions.[315]

The equilibrium acidities of numerous fluorinated diarylacetonitriles, arylmalonate esters, and similar compounds have been determined by transmetallation reactions and ^{19}F-NMR spectroscopy.[316, 317] On the basis of equation (2), the pK_a values of different acidic sites in carbon compounds can be estimated provided that the corresponding rate coefficients of detritiation are known.[318, 319] The effect of increasing methylation on the acidities of the enol form of β-keto-aldehydes

$$\mathrm{p}K_a = (-2.23 \pm 0.25)\log k + (15.62 \pm 0.62) \qquad (2)$$

has been assessed.[320] pK_a values of 44 enolizable compounds have been determined and collated with the half-wave oxidation potentials of the corresponding anions.[321] Taft plots for the ionization, in ethanolic sodium ethoxide, have been constructed for three series of carbon acids activated by a phenylsulphonyl, cyano, or benzoyl group; comparison with earlier work on nitro-compounds, fluorenes, and esters shows that the sensitivity of the ionization rates of carbon acids (G—CH⟨) is in the order G=$PhSO_2$ > CN > PhCO > fluorenyl > NO_2 ~ COOEt, a trend that is strikingly consistent with the degree of delocalization of negative charge in the derived carbanion.[322] According to CNDO/2 calculations, solvation effects are approximately half as important as intrinsic effects in accounting for the non-additive behaviour of the acidities of nitromethanes.[323] Interpretation of the acidity of carbon acids using ϕ-constants has been extended to *gem*-dinitroalkanes with polar substituents.[324] That the ∠CCC skeletal angle is not the only factor that governs the kinetic acidity of cycloalkenes is shown by the unusual reversal in the acidity order, namely $C_5 < C_6 < C_7 > C_8$, as the ring size increases.[325] Comparison of the kinetic acidities of 4-alkyl-1-methylpyridinium ions with those of non-aromatic ammonium ions has revealed the full effect, on reactivity, of a reduction in resonance energy in a transition state and the extent to which this is offset by charge neutralization.[326]

Other Reactions

A review on the mechanisms of electrophilic substitution at saturated carbon[327] and a brief account of electrophilic replacement of olefinic hydrogen atoms have appeared.[328] Whereas 9-allyltryptycene gives the expected dibromide when treated with an excess of bromine in chloroform at room temperature, the 1,2,3,4-tetrachloro-derivative (**187**; R = H) affords a mixture of geometrical isomers in the substitution product (**187**; R = Br); examination of the reactions of model compounds suggests that this result is probably due to the through-space electronic effect of the chloro-substituent in the *peri*-position which stabilizes the positively charged intermediate resulting from electrophilic attack by Br^+.[329] The kinetics of chlorination of metal acetylacetonates $[M^{III}(acacH)_3]$ by *N*-chlorosuccinimide have been investigated by the relatively new technique of stopped-flow pulse Fourier-transform ^{1}H-NMR spectroscopy; the solvent-dependence of the second-order rate constants, the normal absence of CIDNP, the absence of a hydrogen isotope effect, and the simple reaction kinetics all support an S_E2 mechanism involving a transition state of the type (**188**) which probably forms a transient ion pair with the succinimide anion.[330]

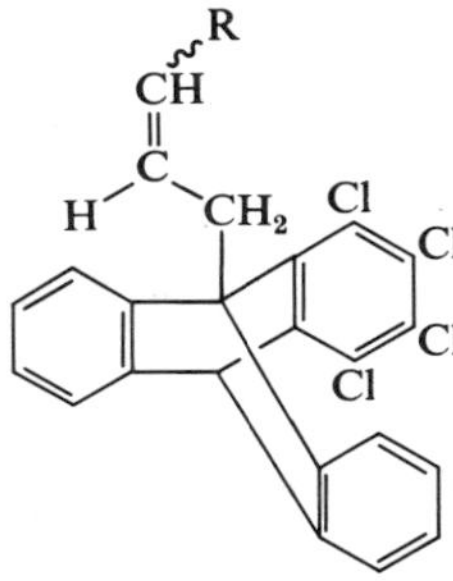

(**187**)

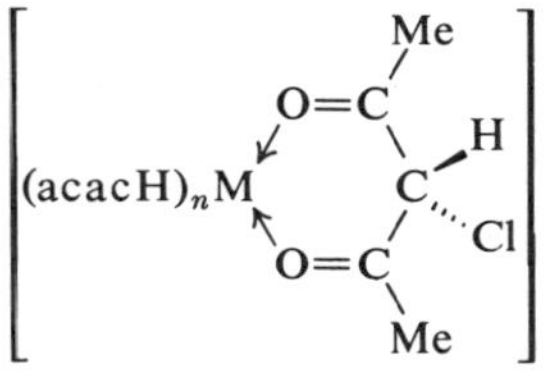

$M = Co^{III}, Rh^{III}, Ir^{III}, Al^{III}$ or Ga^{III}

(**188**)

The reaction of $RSnMe_3$ (R = Me, allyl, PhC≡C, 1-indenyl or CN) with Ph_3CX (X = BF_4 or $SbCl_6$) gives Ph_3CR by an electrophilic mechanism.[331] I^- accelerates the reaction of Ph_2Hg with $(PhC{\equiv}C)_2Hg$ but retards that with $Hg(CN)_2$, $HgBr_2$ or HgI_2, and has no effect on that with $(CCl_3)_2Hg$; this has been explained in terms of the opposing effect of I^- which decreases the electrophilicity of Hg in the electrophilic reagent and increases the nucleophilicity of the C in Ph_2Hg.[332] For the acidolysis of $C_5H_5HgBr_4^{3-}$ the product of isotopic fractionation factors for untransferred protons is 3.8, which is larger than the upper value 2.1 thought to be allowed for direct proton transfer from H_3O^+; this result may be due to an accumulation of errors, but, if it is real, it suggests that the rate- and product-determining steps may not be the same.[333] In the reaction of mercury(II) bromide with trialkyltin (SnR_3) derivatives of Mn, Fe, and Mo, the general pattern of reactivity is (R =) Me < Et < Bu^n < C_6H_{11}, indicating that mercuration proceeds *via* an S_E2 (open) transition state similar to structure (**189**) or possibly *via* a cyclic structure (**190**) with little interaction between the bromine and the tin atoms.[334]

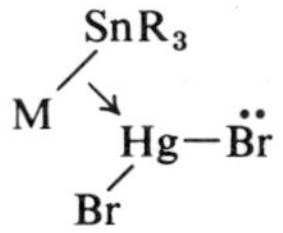

S_E2 (open)

(**189**)

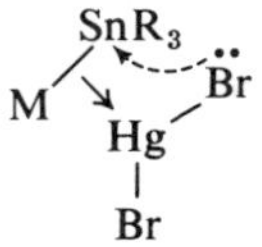

S_E2 (cyclic)

(**190**)

$M = Mn(CO)_5$, $Fe(cp)(CO)_2$ or $Mo(cp)(CO)_3$

Cleavage reactions of the manganese–carbon bond of *cis*-[(*threo*-PhCHDCHD)-$Mn(CO)_4(PEt_3)$] by halogens and mercury(II) halides proceed with both retention and inversion of configuration at the α-carbon atom depending on the reagent used and the solvent conditions; the occurrence of competitive processes involving initial electrophilic attack at both the α-carbon and the metal atoms is inferred.[335] Reactions of methyl (4-tolyl)metal derivatives with various electrophilic reagents give preferential cleavage (79%) of the methyl–metal bond in *cis*-[PtMe(4-MeC_6H_4)-$(PMe_2Ph)_2$] but the 4-tolyl–metal bond in [PtMe(4-MeC_6H_4)(cyclo-octa-1,5-diene)] and in *cis*-[$AuMe_2$(4-MeC_6H_4)(PPh_3)].[336]

References

1 Ellison, G. B., Engelking, P. C., and Lineberger, W. C., *J. Am. Chem. Soc.*, **100**, 2556 (1978).
2 Dykstra, C. E., Hereld, M., Lucchese, R. R., Schaefer, H. F., and Meyer, W., *J. Chem. Phys.*, **67**, 4071 (1977).
3 Kollmar, H., *J. Am. Chem. Soc.*, **100**, 2665 (1978).
4 Pross, A., and Radom, L., *J. Am. Chem. Soc.*, **100**, 6572 (1978).
5 Clark, T., and Schleyer, P. von R., *J. Chem. Soc., Chem. Commun.*, **1978**, 137.
6 Murdoch, J. R., Streitwieser, A., and Gabriel, S., *J. Am. Chem. Soc.*, **100**, 6338 (1978).
7 Clark, T., Jemmis, E. D., Schleyer, P. von R., Binkley, J. S., and Pople, J. A., *J. Organomet. Chem.*, **150**, 1 (1978).
8 Mackay, G. I., Lien, M. H., Hopkinson, A. C., and Bohme, D. K., *Can. J. Chem.*, **56**, 131 (1978).
9 Streitwieser, A., and Boerth, D. W., *J. Am. Chem. Soc.*, **100**, 750 (1978).
10 Garst, J. F., Pacifici, J. A., Felix, C. C., and Nigam, A., *J. Am. Chem. Soc.*, **100**, 5974 (1978).
11 Edwards, J. H., and McQuillin, F. J., *J. Chem. Soc., Chem. Commun.*, **1977**, 838.
12 Bushby, R. J., Patterson, A. S., Ferber, G. J., Duke, A. J., and Whitham, G. H., *J. Chem. Soc., Perkin Trans. 2*, **1978**, 807.
13 Dykstra, C. E., Arduengo, A. J., and Fukunaga, T., *J. Am. Chem. Soc.*, **100**, 6007 (1978).
14 Gould, S. J., and Remillard, B. D., *Tetrahedron Lett.*, **1978**, 4353.
15 Rossi, A. R., Remillard, B. D., and Gould, S. J., *Tetrahedron Lett.*, **1978**, 4357.
16 Grovenstein, E., Jr., *Angew. Chem. Int. Edn.*, **17**, 313 (1978).
17 DePalma, V. M., and Arnett, E. M., *J. Am. Chem. Soc.*, **100**, 3514 (1978).
18 Raban, M., and Haritos, D., *J. Chem. Soc., Chem. Commun.*, **1978**, 965.
19 Takahashi, K., Kondo, Y., and Asami, R., *J. Chem. Soc., Perkin Trans. 2*, **1978**, 577.
20 Bank, S., and Sturges, J. S., *J. Organomet. Chem.*, **156**, 5 (1978).
21 Bushweller, C. H., Sturges, J. S., Cipullo, M., Hoogasian, S., Gabriel, M. W., and Bank, S., *Tetrahedron Lett.*, **1978**, 1359.
22 Menon, B., and Buncel, E., *J. Organomet. Chem.*, **159**, 357 (1978).
23 Solov'yanov, A. A., Dem'yanov, P. I., Beletskaya, I. P., and Reutov, O. A., *Zh. Org. Khim.*, **13**, 2246 (1977); *Chem. Abs.*, **88**, 61749 (1978).
24 Solov'yanov, A. A., Karpyuk, A. D., Beletskaya, I. P. and Reutov, O. A., *Dokl. Akad. Nauk. SSR*, **237**, 360 (1977); *Chem. Abs.*, **88**, 49935 (1978).
25 Buncel, E., and Menon, B., *J. Chem. Soc., Chem. Commun.*, **1978**, 758.
26 Parkes, H. M., and Young, R. N., *J. Chem. Soc., Perkin Trans. 2*, **1978**, 249.
27 Bywater, S., and Worsfold, D. J., *J. Organomet. Chem.*, **159**, 229 (1978).
28 Juchnovski, I. N., Dimitrova, J. S., Binev, I. G., and Kaneti, J., *Tetrahedron*, **34**, 779 (1978).
29 Knorr, R., and Lattke, E., *Tetrahedron Lett.*, **1977**, 4655.
30 Knorr, R., and Lattke, E., *Tetrahedron Lett.*, **1977**, 4659.
31 Boche, G., Weber, H., and Bieberbach, A., *Chem. Ber.*, **111**, 2833 (1978).
32 Boche, G., and Bieberbach, A., *Chem. Ber.*, **111**, 2850 (1978).
33 Staley, S. W., Linkowski, G. E., and Fox, M. A., *J. Am. Chem. Soc.*, **100**, 4818 (1978).
34 Juchnovski, I. N., and Binev, I. G., *Bull. Soc. Chim. Belg.*, **86**, 793 (1977).
35 Neukam, W., and Grimme, W., *Tetrahedron Lett.*, **1978**, 2201.
36 Kozikowski, A. P., and Kuniak, M. P., *J. Org. Chem.*, **43**, 2083 (1978).
37 Pikaev, A. K., Artamkina, G. A., and Beletskaya, I. P., *Izv. Akad. Nauk SSSR, Ser. Khim.*, **1977**, 2171; *Chem. Abs.*, **88**, 21696 (1978).

[38] Kool, M., and Klumpp, G. W., *Tetrahedron Lett.*, **1978**, 1873.
[39] Becker, J. Y., and Klein, J., *J. Organomet. Chem.*, **157**, 1 (1978).
[40] Kurrikoff, S., and Tuulmets, A., *Org. React.* (*Tartu*), **15**, 124 (1978).
[41] Paquette, L. A., *Angew. Chem. Int. Edn.*, **17**, 106 (1978).
[42] Paquette, L. A., Degenhardt, C. R., and Berk, H. C., *J. Am. Chem. Soc.*, **100**, 1599 (1978).
[43] Nakasuji, K., Katada, M., and Murata, I., *Tetrahedron Lett.*, **1978**, 2515.
[44] Olah, G. A., Asensio, G., Mayr, H., and Schleyer, P. von R., *J. Am. Chem. Soc.*, **100**, 4347 (1978).
[45] Bos, H., and Klumpp, G. W., *Tetrahedron Lett.*, **1978**, 1863.
[46] Bos, H., and Klumpp, G. W., *Tetrahedron Lett.*, **1978**, 1865.
[47] Bauld, N. L., Welsher, T. L., Cessac, J., and Holloway, R. L., *J. Am. Chem. Soc.*, **100**, 6920 (1978).
[48] Müllen, K., *Helv. Chim. Acta*, **61**, 2307 (1978).
[49] Hunadi, R. J., and Helmkamp, G. K., *J. Org. Chem.*, **43**, 1586 (1978).
[50] Rapp, K. M., Burgemeister, T., and Daub, J., *Tetrahedron Lett.*, **1978**, 2685.
[51] Boche, G., and Heidenhain, F., *Angew. Chem. Int. Edn.*, **17**, 283 (1978).
[52] Boche, G., Weber, H., Martens, D., and Bieberbach, A., *Chem. Ber.*, **111**, 2480 (1978).
[53] Willner, I., Gamliel, A., and Rabinovitz, M., *Chem. Lett.*, **1977**, 1273.
[54] Willner, I., and Rabinovitz, M., *J. Am. Chem. Soc.*, **100**, 337 (1978).
[55] Hildenbrand, P., Plinke, G., Oth, J. F. M., and Schröder, G., *Chem. Ber.*, **111**, 107 (1978).
[56] Rottele, H., Heil, G., and Schröder, G., *Chem. Ber.*, **111**, 84 (1978).
[57] Müllen, K., *Helv. Chim. Acta*, **61**, 1296 (1978).
[58] Müllen, K., *Helv. Chim. Acta*, **61**, 1305 (1978).
[59] Butler, D. N., and Gupta, I., *Can. J. Chem.*, **56**, 80 (1978).
[60] Goldstein, M. J., *J. Am. Chem. Soc.*, **100**, 4899 (1978).
[61] Beak, P., and Reitz, D. B., *Chem. Rev.*, **78**, 275 (1978).
[62] Reitz, D. B., Beak, P., Farney, R. F., and Helmick, L. S., *J. Am. Chem. Soc.*, **100**, 5428 (1978).
[63] Creary, X., *J. Am. Chem. Soc.*, **99**, 7632 (1977).
[64] Feit, B. A., and Melamed, U., *J. Chem. Soc.*, *Perkin Trans*, 1, **1978**, 1228.
[65] Melamed U., and Feit, B. A., *J. Chem. Soc.*, *Perkin Trans.* 1, **1978**, 1232.
[66] Hiyama, T., Kanakura, A., Yamamoto, H., and Nozaki, H., *Tetrahedron Lett.*, **1978**, 3046.
[67] Hiyama, T., Kanakura, A., Yamamoto, H., and Nozaki, H., *Tetrahedron Lett.*, **1978**, 3051.
[68] Chamberlin, A. R., Stemke, J. E., and Bond, F. T., *J. Org. Chem.*, **43**, 147 (1978).
[69] Adlington, R. M., and Barrett, A. G. M., *J. Chem. Soc. Chem. Commun.*, **1978**, 1071.
[70] Neumann, H., and Seebach, D., *Chem. Ber.*, **111**, 2785 (1978).
[71] Guaciaro, M. A., Wovkulich, P. M., and Smith, A. B., III, *Tetrahedron Lett.*, **1978**, 4661
[72] Amupitan, J., and Sutherland, J. K., *J. Chem. Soc. Chem. Commun.*, **1978**, 852.
[73] Takei, H., *Yuki Gosei Kagaku Kyokaishi*, **35**, 755 (1977); *Chem. Abs.*, **88**, 21447 (1978).
[74] Schmidt, R. R., and Talbiersky, J., *Angew. Chem. Int. Edn.*, **16**, 853 (1977).
[75] Schmidt, R. R., and Talbiersky, J., *Angew. Chem. Int. Edn.*, **17**, 204 (1978).
[76] Bakuzis, P., Bakuzis, M. L. F., and Weingartner, T. F., *Tetrahedron Lett.*, **1978**, 2371.
[77] Caubère, P., *Top. Curr. Chem.*, **73**, 49, 105 (1978).
[78] Reetz, M. T., and Eibach, F., *Angew. Chem. Int. Edn.*, **17**, 278 (1978).
[79] Reetz, M. T., and Eibach, F., *Justus Liebigs Ann. Chem.*, **1978**, 1598.
[80] Solov'yanov, A. A., and Beletskaya, I. P., *Usp. Khim.*, **47**, 819 (1978); *Chem. Abs.*, **89**, 89772 (1978).
[81] Isăcescu, D. A., and Avramescu, F., *Rev. Roum. Chim.*, **23**, 661 (1978).
[82] Isăcescu, D. A., and Avramescu, F., *Rev. Roum. Chim.*, **23**, 865 (1978).
[83] Isăcescu, D. A., and Avramescu, F., *Rev. Roum. Chim.*, **23**, 873 (1978).
[84] Protsaylo, T., Makitra, R., and Poljanski, I., *Org. React.* (*Tartu*), **15**, 78 (1978).
[85] Pivovarov, S., Perepletchikova, G., Selivanov, V., and Gidaspov, B., *Org. React.* (*Tartu*), **14**, 211, 226 (1977).
[86] Jagannadham, V., Kandlikar, S., Sethuram, B., and Rao, T. N., *Natl. Acad. Sci. Lett.* (*India*), **1**, 207 (1978); *Chem. Abs.*, **89**, 162906 (1978).
[87] Jagannadham, V., Sethuram, B., and Rao, T. N. *Curr. Sci.*, **46**, 704 (1977); *Chem. Abs.*, **88**, 88880 (1978).
[88] Bank, S., Bank, J., Daney, M., Labrande, B., and Bouas-Laurent, H., *J. Org. Chem.*, **42**, 4058 (1977).

[89] Daney, M., Lapouyade, R., and Bouas-Laurent, H., *Tetrahedron Lett.*, **1978**, 783.
[90] Daney, M., Labrande, B., Lapouyade, R., Bouas-Laurent, H., *J. Organomet. Chem.*, **159**, 385 (1978).
[91] Malissard, M., Mazaleyrat, J. P., and Welvart, Z., *J. Am. Chem. Soc.*, **99**, 6933 (1977).
[92] Sakurai H., Kira, M., and Umino, H., *Chem. Lett.*, **1977**, 1265.
[93] San Filippo, J., Jr., Silbermann, J., and Fagan, P. J., *J. Am. Chem. Soc.*, **100**, 4834 (1978).
[94] des Abbayes, H., and Boudeville, M.-A., *J. Org. Chem.*, **42**, 4104 (1977).
[95] Jackson, W. R., and Strauss, J. U., *Aust. J. Chem.*, **31**, 1073 (1978).
[96] Izumi, T., and Miller, S. I., *J. Org. Chem.*, **43**, 871 (1978).
[97] Sargsyan, M. S., and Badanyan, I. O., *Tezisy Dokl.-Vses. Konf. Khim. Atsetilena* 5th, **1975**, 384; *Chem. Abs.*, **89**, 59403 (1978).
[98] Wheaton, G. A., and Burton, D. J., *J. Org. Chem.*, **43**, 2643 (1978).
[99] Bey, P., and Vevert, J. P., *Tetrahedron Lett.*, **1978**, 1215.
[100] Oakleaf, J. A., Thomas, M. T., Wu, A., and Snieckus, V., *Tetrahedron Lett.*, **1978**, 1645.
[101] Hansen, R. T., Carr, D. B., and Schwartz, J., *J. Am. Chem. Soc.*, **100**, 2244 (1978).
[102] Evans, D., Baillargeon, D. J., and Nelson, J. V., *J. Am. Chem. Soc.*, **100**, 2242 (1978).
[103] Luchetti, J., and Krief, A., *Tetrahedron Lett.*, **1978**, 2697.
[104] Still, C. W., and Mitra, A., *Tetrahedron Lett.*, **1978**, 2659.
[105] Tamaru, Y., Harada, T., Iwamoto, H., and Yoshida, Z., *J. Am. Chem. Soc.*, **100**, 5221 (1978).
[106] Bryson, T. A., Dardis, R. E., and Gammill, R. B., *Tetrahedron Lett.*, **1978**, 743.
[107] Agawa, T., Yoshida, Y., Komatsu, M., and Ohshiro, Y., *J. Chem. Soc., Chem. Commun.*, **1977**, 931.
[108] Akabori, S., and Yoshii, T., *Tetrahedron Lett.*, **1978**, 4523.
[109] Lennon, P., Rosan, A. M., and Rosenblum, M., *J. Am. Chem. Soc.*, **99**, 8426 (1977).
[110] Makosza, M., *Usp. Khim.*, **46**, 2174 (1977); *Chem. Abs.*, **88**, 61659 (1978).
[111] Chiellini, E., and Solaro, R., *J. Org. Chem.*, **43**, 2550 (1978).
[112] Purohit, V. G., and Subramanian, R., *Chem. Ind. (London)*, **1978**, 731.
[113] Cazes, B., and Julia, S., *Bull. Soc. Chim. Fr. II*, **1977**, 925.
[114] Cazes, B., and Julia, S., *Bull. Soc. Chim. Fr. II*, **1977**, 931.
[115] Komeili-Zadeh, H., Dou, H. J.-M., and Metzger, J., *J. Org. Chem.*, **43**, 156 (1978).
[116] Jończyk, M., Kwast, A., and Makosza, M., *J. Chem. Soc. Chem. Commun.*, **1977**, 902.
[117] Stamm, H., and Wiesert, W., *Chem. Ber.*, **111**, 2665 (1978).
[118] Kessar, S. V., Nadir, U. K., Singh, P., and Gupta, Y. P., *Tetrahedron*, **34**, 449 (1978).
[119] Pitaco, G., Risaliti, A., Trevisan, M. L., and Valentin, E., *Tetrahedron*, **33**, 3145 (1977).
[120] Hayashi, T., and Hegedus, L. D., *J. Am. Chem. Soc.*, **99**, 7093 (1977).
[121] Boeckman, R. K., Ramaiah, M., and Medwid, J. B., *Tetrahedron Lett.*, **1977**, 4485.
[122] Felicioli, M. G., Bottaccio, G., and Chiusoli, G. P., *Gazz. Chim. Ital.*, **106**, 1127 (1976).
[123] Idriss, N., Perry M., Maroni-Barnaud, Y., Roux-Schmitt, M.-C., and Seyden-Penne, J., *J. Chem. Res.*, **1978** (S) 128; (M) 1601.
[124] Ashby, E. C., and Wiesemann, T. L., *J. Am. Chem. Soc.*, **100**, 3101 (1978).
[125] Snider, B. B., Karras, M., and Conn, R. S. E., *J. Am. Chem. Soc.*, **100**, 4624 (1978).
[126] Benkeser, R. A., Siklosi, M. P., and Mozdzen, E. C., *J. Am. Chem. Soc.*, **100**, 2134 (1978).
[127] Pasto, D. J., Schults, R. H., McGath, J. A., and Waterhouse, A., *J. Org. Chem.*, **43**, 1832 (1978).
[128] Pasto, D. J., Chou, S.-K., Waterhouse, A., Schults, R. H., and Hennion, G. F., *J. Org. Chem.*, **43** 1385 (1978).
[129] Pasto, D. J., Chou, S.-K., Fritzen, E., Schults, R. H., Waterhouse, A., and Hennion, G. F., *J. Org. Chem.*, **43**, 1389 (1978).
[130] Meyers, A. I., Tait, T. A., and Comins, D. L., *Tetrahedron Lett.*, **1978**, 4657.
[131] Gosselin, P., Masson, S., and Thuiller, A., *Tetrahedron Lett.*, **1978**, 2717.
[132] Dang, H. P., and Linstrumelle, G., *Tetrahedron Lett.*, **1978**, 191.
[133] Maruyama, K., and Naruta, Y., *Chem. Lett.*, **1978**, 431.
[134] De Rosa, M., and Triana Alonos, J. L., *J. Org. Chem.*, **43**, 2639 (1978).
[135] Verhé, R., De Kimpe, N., De Buyck, L., Swyngedouw, C., and Schamp, N., *Bull. Soc. Chim. Belg.*, **86**, 893 (1977).
[136] Bethell, D., and Bird, R., *J. Chem. Soc., Perkin Trans. 2*, **1977**, 1856.
[137] Novak, M., and Bruice, T. C., *J. Am. Chem. Soc.*, **99**, 8079 (1977).
[138] Yamamura, K., *J. Org. Chem.*, **43**, 724 (1978).
[139] Kato, H., Takeuchi, I., Hamada, Y., Ono, M., and Hirota, M., *Tetrahedron Lett.*, **1978**, 135.

[140] Sato, Y., Kobayashi, Y., Sugiura, M., and Shirai, H., *J. Org. Chem.*, **43**, 199 (1978).
[141] Truce, W. E., and VanGemert, B., *J. Am. Chem. Soc.*, **100**, 5525 (1978).
[142] Merz, A., and Thumm, G., *Justus Liebigs Ann. Chem.*, **1978**, 1526.
[143] Merz, A., and Thumm, G., *Tetrahedron Lett.*, **1978**, 679.
[144] Karrenbrock, F., and Schäfer, H. J., *Tetrahedron Lett.*, **1978**, 1521.
[145] Mukaiyama, T., Soai, K., and Kobayashi, S., *Chem. Lett.*, **1978**, 219.
[146] Soai, K., and Mukaiyama, T., *Chem. Lett.*, **1978**, 491.
[147] Arnett, E. M., Small, L. E., McIver, R. T., Jr., and Miller, J. S., *J. Org. Chem.*, **43**, 815 (1978).
[148] Ching, W.-M., and Kallen, R. G., *J. Am. Chem. Soc.*, **100**, 6119 (1978).
[149] Gausing, W., and Wilke, G., *Angew. Chem. Int. Edn.*, **17**, 371 (1978).
[150] Baryshnikov, Yu. N., Kaloshina, N. N., and Vesnovskaya, G. I., *Zh. Obshch. Khim.*, **47**, 2790 (1977); *Chem. Abs.*, **88**, 104321 (1978).
[151] Furukawa, N., Inoue, T., Aida, T., Akasaka, T., and Oae, S., *Phosphorus Sulphur*, **4**, 15 (1978); *Chem. Abs.*, **88**, 189302 (1978).
[152] Hori, Y., Nagano, Y., Uchiyama, H., Yamada, Y., and Taniguchi, H., *Chem. Lett.*, **1978**, 73.
[153] Chan, T. H., and Moreland, M., *Tetrahedron Lett.*, **1978**, 515.
[154] Amouroux, R., and Chan, T. H., *Tetrahedron Lett.*, **1978**, 4453.
[155] Chan, T. H., Mychajlowskij, W., Ong, B. S., and Harpp, D. N., *J. Org. Chem.*, **43**, 1526 (1978).
[156] Lau, P. W. K., and Chan, T. H., *Tetrahedron Lett.*, **1978**, 2383.
[157] Seyferth, D., and Mammarella, R. E., *J. Organomet. Chem.*, **156**, 279 (1978).
[158] Seyferth, D., and Mammarella, R. E., *J. Organomet. Chem.*, **156**, 299 (1978).
[159] Jones, P. R., and Lim, T. F. O., *J. Am. Chem. Soc.*, **99**, 8447 (1977).
[160] Tondeur, J. J., Vandendunghen, G., and Xhigne, M., *React. Kinet. Catal. Lett.*, **7**, 327 (1977); *Chem. Abs.*, **88**, 61737 (1978).
[161] Fleischmann, C., and Zbiral, E., *Tetrahedron*, **34**, 317 (1978).
[162] Nesterov, L. V., Krepysheva, N. E., Sabirova, R. A., and Lipkina, G. N., *Zh. Obshch. Khim.*, **48**, 790 (1978); *Chem. Abs.*, **89**, 41888 (1978).
[163] Zwanenburg, B., Venier, C. G., Porskamp, P. A. T. W., and Van der Leij, M., *Tetrahedron Lett.*, **1978**, 807.
[164] Van der Leij, M., Porskamp, P. A. T. W., Lammerink, B. H. M., and Zwanenburg, B., *Tetrahedron Lett.*, **1978**, 811.
[165] Jefford, C. W., and Barchietto, G., *Tetrahedron Lett.*, **1977**, 4531.
[166] Allen, D. W., and Hutley, B. G., *J. Chem. Soc. Perkin Trans.* 1, **1978**, 675.
[167] Zbaida, S., and Breuer, E., *J. Chem. Soc. Chem. Commun.*, **1978**, 6.
[168] Minami, T., Matsumoto, M., Suganuma, H., and Agawa, T., *J. Org. Chem.*, **43**, 2149 (1978).
[169] Sturtz, G., and Pondaven-Raphalen, A., *Tetrahedron Lett.*, **1978**, 629.
[170] Ahlbrecht, H., König, B., and Simon, H., *Tetrahedron Lett.*, **1978**, 1191.
[171] Coutrot, P., and Savignac, P., *J. Chem. Res.*, **1977** (S) 308; (M) 3401.
[172] Toke, L., Petnehazy, I., and Szakal, G., *J. Chem. Res.*, **1978**, (S) 155; (M) 1975.
[173] Vaultier, M., and Carrie, R., *Tetrahedron Lett.*, **1978**, 1195.
[174] Chambers, R. D., Taylor, G., and Powell, R. L., *J. Chem. Soc. Chem. Commun.*, **1978**, 433.
[175] Renger, B., and Hügel, H., *Chem. Ber.*, **111**, 2630 (1978).
[176] Colvin, E. W., and Seebach, D., *J. Chem. Soc. Chem. Commun.*, **1978**, 689.
[177] Seebach, D., and Hassel, T., *Angew. Chem. Int. Edn.*, **17**, 274 (1978).
[178] White, K. B., and Reusch, W., *Tetrahedron*, 1978, 2439.
[179] Lam, C. N., Mellor, J. M., Picard, P., Rawlins, M. F., and Stibbard, J. H. A., *Tetrahedron Lett.*, **1978**, 4103.
[180] Smith, A. B., and Scarborough, R. M., *Tetrahedron Lett.*, **1978**, 4193.
[181] Lefour, J. M., Sarthou, P., Bram, G., Guibé, F., Loupy, A., and Seyden-Penne, J., *Tetrahedron Lett.*, **1978**, 3831.
[182] Satyanarayana, N., and Sundaram, E. V., *J. Indian Chem. Soc.*, **54**, 886 (1977); *Chem. Abs.*, **89**, 162711 (1978).
[183] D'Incan, E., and Viout, P., *Tetrahedron*, **34**, 2469 (1978).
[184] Nakamura, E., Hashimoto, K., and Kuwajima, I., *Tetrahedron Lett.*, **1978**, 2079.
[185] Balavoine, G., Bram, G., and Guibé, F., *Nouveau J. de Chimie* **2**, 207 (1978).
[186] Dauben, W. G., and Hart, D. J., *J. Am. Chem. Soc.*, **99**, 7307 (1977).
[187] House, H. O., Phillips, W. V., Sayer, T. S. B., and Yau, C.-C., *J. Org. Chem.*, **43**, 700 (1978).

[188] House, H. O., and Phillips, W. V., *J. Org. Chem.*, **43**, 3851 (1978).
[189] House, H. O., Sayer, T. S. B., and Yau, C.-C., *J. Org. Chem.*, **43**, 2153 (1978).
[190] Yoshida, Z., Yoneda, S., Kawase, T., and Inaba, M., *Tetrahedron Lett.*, **1978**, 1285.
[191] Becher, J., Frandsen, E. G., Dreier, C., and Henriksen, L., *Acta Chem. Scand., Ser. B*, **31**, 843 (1977).
[192] Wollenberg, R. H., Albizati, K. F., and Peries, R., *J. Am. Chem. Soc.*, **99**, 7365 (1977).
[193] Kreisler, S. Y. L., and Schlosser, M., *J. Org. Chem.*, **43**, 1595 (1978).
[194] Wollenberg, R. H., *Tetrahedron Lett.*, **1978**, 717.
[195] Seebach, D., Henning, R., and Lehr, F., *Angew. Chem. Int. Edn.*, **17**, 458 (1978).
[196] Nakamura, E., and Kuwajima, I., *J. Am. Chem. Soc.*, **99**, 7360 (1977).
[197] Maruoka, K., Hashimoto, S., Kitagawa, Y., Yamamoto, H., and Nozaki, H., *J. Am. Chem. Soc.*, **99**, 7705 (1977).
[198] Buse, C. T., and Heathcock, C. H., *J. Am. Chem. Soc.*, **99**, 8109 (1977).
[199] Eichenauer, H., Friedrich, E., Lutz, W., and Enders, D., *Angew. Chem. Int. Edn.*, **17**, 206 (1978).
[200] Carlson, R., Nilsson, L., Rappe, C., Babadjamian, A., and Metzger, J., *Acta Chem. Scand., Ser. B*, **32**, 85 (1978).
[201] Ahlbrecht, H., *Chimia*, **31**, 391 (1977).
[202] Hashimoto, S., and Koga, K., *Tetrahedron Lett.*, **1978**, 573.
[203] Wender, P. A., and Eissenstat, M. A., *J. Am. Chem. Soc.*, **100**, 292 (1978).
[204] Lyle, R. E., Fribush, H. M., and Lyle, G. G., *J. Org. Chem.*, **43**, 1275 (1978).
[205] Ensley, H. E., and Lohr, R., *Tetrahedron Lett.*, **1978**, 1415.
[206] Corey, E. J., and Enders, D., *Chem. Ber.*, **111**, 1337 (1978).
[207] Corey, E. J., and Enders, D., *Chem. Ber.*, **111**, 1362 (1978).
[208] Corey, E. J., and Boger, D. L., *Tetrahedron Lett.*, **1978**, 4597.
[209] Scheffer, J. R., Gayler, R. E., Zakouras, T., and Dzakpasu, A. A., *J. Am. Chem. Soc.*, **99**, 7726 (1977).
[210] Cheng, A. K., and Stothers, J. B., *Can. J. Chem.*, **56**, 1342 (1978).
[211] Werstiuk, N. H., Taillefer, R., and Banerjee, S., *Can. J. Chem.*, **56**, 1140 (1978).
[212] Werstiuk, N. H., Taillefer, R., and Banerjee, S., *Can. J. Chem.*, **56**, 1148 (1978).
[213] Fraser, R. R., and Champagne, P. J., *J. Am. Chem. Soc.*, **100**, 657 (1978).
[214] Brown, F. C., Casadevall, E., Metzger, P., and Morris, D. G., *J. Chem. Res.*, **1977**, (S) 335; (M) 3588.
[215] Werstiuk, N. H., and Andrews, P., *Can. J. Chem.*, **56**, 2605 (1978).
[216] Hine, J., *Acc. Chem. Res.*, **11**, 1 (1978).
[217] Perkins, M. J., Peynircioglu, N. B., and Smith, B. V., *J. Chem. Soc., Perkin Trans. 2*, **1978**, 1025.
[218] Dvorak, V., Manzara, A. P., and Michl, J., *Tetrahedron*, **34**, 2433 (1978).
[219] Ankner, K., Lamm, B., Thulin, B., and Wennerström, O., *Acta Chem. Scand., Ser. B*, **32**, 155 (1978).
[220] Boche, G., and Buckl, K., *Angew. Chem. Int. Edn.*, **17**, 284 (1978).
[221] Rauscher, G., Clark, T., Poppinger, D., and Schleyer, P. von R., *Angew. Chem. Int. Edn.*, **17**, 276 (1978).
[222] Andreev, G., Demireva, Z., Yuchnovski, I., Kuzmanova, R., and Binev, I., *Izv. Khim.*, **10**, 379 (1977); *Chem. Abs.*, **89**, 41717 (1978).
[223] Lambert, J. B., and Wharry, S. M., *J. Chem. Soc., Chem. Commun.*, **1978**, 172.
[224] Klein, J., and Medlik-Balan, A., *Tetrahedron Lett.*, **1978**, 279.
[225] Bays, J. P., *J. Org. Chem.*, **43**, 38 (1978).
[226] Hubbard, J. S., and Harris, T. M., *Tetrahedron Lett.*, **1978**, 4601.
[227] Ide, J., Inoue, K., and Sakai, K., *Chem. Lett.*, **1978**, 747.
[228] Tanaka, K., Uneme, H., and Yamagishi, N., *Chem. Lett.*, **1978**, 653.
[229] Merz, A., and Tomahogh, R., *J. Chem. Res.*, **1977**, (S) 273; (M) 3070.
[230] Casara, P., and Metcalf, B. W., *Tetrahedron Lett.*, **1978**, 1581.
[231] Carlson, R. M., *Tetrahedron Lett.*, **1978**, 111.
[232] Vinick, F. J., Pan, Y., and Gschwend, H. W., *Tetrahedron Lett.*, **1978**, 4221.
[233] Fabian, J., Schoenfeld, P., and Mayer, R., *Phosphorus Sulphur*, **2**, 151 (1976).
[234] Chassaing, G., and Marquet, A., *Tetrahedron*, **34**, 1399 (1978).
[235] Gräfing, R., Verkuijsse, H. D., and Brandsma, L., *J. Chem. Soc., Chem. Commun.*, **1978**, 596.
[236] Gräfing, R., and Brandsma, L., *Recl. Trav. Chim. Pays-Bas*, **97**, 208 (1978).

[237] Borden, W. T., Davidson, E. R., Andersen, N. H., Denniston, A. D., and Epiotis, N. D., *J. Am. Chem. Soc.*, **100**, 1604 (1978).
[238] Abatjoglou, A. G., Eliel, E. L., and Kuyper, L. F., *J. Am. Chem. Soc.*, **99**, 8262 (1977).
[239] Eliel, E. L., Koskimies, J. K., and Lohri, B., *J. Am. Chem. Soc.*, **100**, 1614 (1978).
[240] Golding, B. T., Ioannou, P. V., and Eckhard, I. F., *J. Chem. Soc., Perkin Trans.* 2, **1978**, 774.
[241] Barbarella, G., Dembech, P., Garbesi, A., Bernardi, F., Bottoni, A., and Fava, A., *J. Am. Chem. Soc.*, **100**, 200 (1978).
[242] Morita, H., Itoh, H., Furukawa, N., and Oae, S., *Chem. Lett.*, **1978**, 817.
[243] Kunieda, N., Motoki, H., and Kinoshita, M., *Chem. Lett.*, **1978**, 713.
[244] Lavielle, S., Bory, S., Moreau, B., Luche, M. J., and Marquet, A., *J. Am. Chem. Soc.*, **100**, 1558 (1978).
[245] Biellmann, J. F., and Vicens, J. J., *Tetrahedron Lett.*, **1978**, 467.
[246] Chassaing, G., Lett, R., and Marquet, A., *Tetrahedron Lett.*, **1978**, 471.
[247] Corey, E. J., and Boger, D. L., *Tetrahedron Lett.*, **1978**, 5.
[248] Corey, E. J., and Boger, D. L., *Tetrahedron Lett.*, **1978**, 9.
[249] Corey, E. J., and Boger, D. L., *Tetrahedron Lett.*, **1978**, 13.
[250] Stirling, C. J. M., *Chem. Rev.*, **78**, 517 (1978).
[251] Barlow, K. N., Marshall, D. R., and Stirling, C. J. M., *J. Chem. Soc., Perkin Trans.* 2, **1977**, 1920.
[252] Thomas, P. J., and Stirling, C. J. M., *J. Chem. Soc., Chem. Commun.*, **1978**, 975.
[253] Hori, M., Kataoka, T., Shimizu, H., Ohno, S., and Narita, K., *Tetrahedron Lett.*, **1978**, 251.
[254] Hori, M., Kataoka, T., Shimizu, H., and Ohno, S., *Tetrahedron Lett.*, **1978**, 255.
[255] Trost, B. M., Bogdanowicz, M. J., Frazee, W. J., and Salzmann, T. N., *J. Am. Chem. Soc.*, **100**, 5512 (1978).
[256] Chalchat, J.-C., Garry, R., Michet, A., and Vessière, R., *C. R. Hebd. Séances Acad. Sci., Sér. C*, **286**, 329 (1978).
[257] Shevelev, S. A., and Semenov, V. V., *Tezisy Dokl. Nauchn. Sess., Khim. Tekhnol. Org. Soedin. Sery Sernistykh Neftei*, 14th, **1975**, 164; *Chem. Abs.*, **88**, 189831 (1978).
[258] Yokoyama, M., and Takeshima, T., *Tetrahedron Lett.*, **1978**, 147.
[259] Atsumi, K., and Kuwajima, I., *Chem. Lett.*, **1978**, 387.
[260] Yamagiwa, S., Hoshi, N., Sato, H., Kosugi, H., and Uda, H., *J. Chem. Soc., Perkin Trans*, 1, **1978**, 214.
[261] Cohen, T., Daniewski, W. M., and Weisenfeld, R. B., *Tetrahedron Lett.*, **1978**, 4665.
[262] Screttas, C. G., and Micha-Screttas, M., *J. Org. Chem.*, **43**, 1064 (1978).
[263] Matteson, D. S., and Arne, K., *J. Am. Chem. Soc.*, **100**, 1325 (1978).
[264] Sevrin, M., and Krief, A., *Tetrahedron Lett.*, **1978**, 187.
[265] Lucchetti, J., and Krief, A., *Tetrahedron Lett.*, **1978**, 2693.
[266] Bordwell, F. G., and Bartmess, J. E., *J. Org. Chem.*, **43**, 3101 (1978).
[267] Bordwell, F. G., Bartness, J. E., and Hautala, J. A., *J. Org. Chem.*, **43**, 3113 (1978).
[268] Bordwell, F. G., Barmess, J. E., and Hautala, J. A. *J. Org. Chem.*, **43**, 3107 (1978).
[269] Golubev, N. S., Safarov, N. A., and Tanasiichuk, A. S., *Kinet. Katal.*, **19**, 302 (1978); *Chem. Abs.*, **89**, 5790 (1978).
[270] Heggen, I., Lindstrøm, J., and Rogne, O., *J. Chem. Soc., Faraday Trans.* 1, **74**, 1263 (1978).
[271] Caldin, E. F., Rogne, O., and Wilson, C. J., *J. Chem. Soc., Faraday Trans.* 1, **74**, 1796 (1978).
[272] Caldin, E. F., and Rogne, O., *J. Chem. Soc., Faraday Trans.* 1, **74**, 2065 (1978).
[273] Bell, R. P., *Finn. Chem. Lett.*, **1978**, 7.
[274] Slater, C. D., and Chan, D., *J. Org. Chem.*, **43**, 2423 (1978).
[275] Pruszynski, P., and Jarczewski, A., *Rocz. Chem.*, **51**, 2171 (1977); *Chem. Abs.*, **88**, 151666 (1978).
[276] Golubev, N. S., and Safarov, N. A., *React. Kinet. Catal. Lett.*, **7**, 451 (1977); *Chem. Abs.*, **88**, 104326 (1978).
[277] Jencks, D. A., and Jencks, W. P., *J. Am. Chem. Soc.*, **99**, 7948 (1977).
[278] Walters, E. A., *J. Phys. Chem.*, **82**, 1219 (1978).
[279] Walters, E. A., *J. Phys. Chem.*, **81**, 1995 (1977).
[280] Okuyama, T., Ikenouchi, Y., and Fueno, T., *J. Am. Chem. Soc.*, **100**, 6162 (1978).
[281] Leonova, T. V., Shapiro, I. O., Ranneva, Yu. I., Vasyanina, L. K., Shatenshtein, A. I., and Shabarov, Yu. S., *Zh. Org. Khim.*, **13**, 2290 (1977); *Chem. Abs.*, **88**, 73801 (1978).
[282] Eaborn, C., Stamper, J. G., and Seconi, G., *J. Organomet. Chem.*, **150**, C23 (1978).
[283] Willner, I., Halpern, M., and Rabinovitz, M., *J. Chem. Soc., Chem. Commun.*, **1978**, 155.

[284] Zatsepina, N. N., Tupitsyn, I. F., Belyashova, A. I., and Sudakova, G. N., *Zh. Org. Khim.*, **13**, 1802 (1977); *Chem. Abs.*, **87**, 200622 (1977).
[285] Zatsepina, N. N., Tupitsyn, I. F., Kane, A. A., and Sudakova, G. N., *Khim. Geterotsikl. Soedin.*, **1977**, 1192; *Chem. Abs.*, **88**, 6094 (1978).
[286] Zatsepina, N. N., Kane, A. A., and Tupitsyn, I. F., *Zh. Org. Khim.*, **13**, 1793 (1977); *Chem. Abs.*, **88**, 6060 (1978).
[287] Takeuchi, Y., Yeh, H. J. C., Kirk, K. L., and Cohen, L. A., *J. Org. Chem.*, **43**, 3565 (1978).
[288] Takeuchi, Y., Kirk, K. L., and Cohen, L. A., *J. Org. Chem.*, **43**, 3570 (1978).
[289] Lezina, V. P., Stepanyants, A. U., Smirnov, L. D., and Vinnik, M. I., *Izv. Akad. Nauk SSSR, Ser. Khim.*, **1978**, 317; *Chem. Abs.*, **88**, 151699 (1978).
[290] Belyashova, A. I., Zatsepina, N. N., Malysheva, E. N., Pozharskii, A. F., Smirnova, L. P., and Tupitsyn, I. F., *Khim. Geterotsikl. Soedin.*, **1977**, 1544; *Chem. Abs.*, **88**, 73917 (1978).
[291] Evans, B., Smith, K. M., La Mar, G. N., and Viscio, D. B., *J. Am. Chem. Soc.*, **99**, 7070 (1977).
[292] Zatsepina, N. N., Zyryanov, V. A., Kirova, A. V., Rusinov, V. L., Postovskii, I. Ya., and Tupitsyn, I. F., *Khim. Geterotsikl. Soedin.*, **1978**, 127; *Chem. Abs.*, **88**, 169262 (1978).
[293] Stewart, J. H., Shapiro, R. H., DePuy, C. H., and Bierbaum, V. M., *J. Am. Chem. Soc.*, **99**, 7650 (1977).
[294] DePuy, C. H., Bierbaum, V. M., King, G. H., and Shapiro, R. H., *J. Am. Chem. Soc.*, **100**, 2921 (1978).
[295] Ando, T., *Kagaku No Ryoiki*, **31**, 339 (1977); *Chem. Abs.*, **88**, 49768 (1978).
[296] Bergson, G., Matsson, O., and Sjöberg, S., *Chem. Scr.*, **11**, 25 (1977).
[297] German, E. D., *Izv. Akad. Nauk. SSSR, Ser. Khim.*, **1978**, 959; *Chem. Abs.*, **89**, 42158 (1978).
[298] Macciantelli, D., Seconi, G., and Eaborn, C., *J. Chem. Soc., Perkin Trans.* 2, **1978**, 834.
[299] Gettys, G. A., and Gutsche, C. D., *Bioorg. Chem.*, **7**, 141 (1978).
[300] Hupe, D. J., and Wu, D., *J. Am. Chem. Soc.*, **99**, 7653 (1977).
[301] Hupe, D. J., Wu, D., and Shepperd, P., *J. Am. Chem. Soc.*, **99**, 7659 (1977).
[302] Tsai, M.-D., Weintraub, H. J. R., Byrn, S. R., Chang, C., and Floss, H. G., *Biochemistry*, **17**, 3183 (1978).
[303] Symons, M. C. R., *J. Chem. Res.*, **1978**, (S) 140.
[304] Stoffer, J. O., Straib, D. R., Filger, D. L., Lloyd, E. T., and Crain, C., *J. Org. Chem.*, **43**, 1812 (1978).
[305] Chrisement, J., and Delpuech, J.-J., *J. Chem. Res.*, **1978**, (S) 340; (M) 3701.
[306] Juaristi, E., and Streitwieser, A., *J. Org. Chem.*, **43**, 2704 (1978).
[307] Streitwieser, A., and Guibé, F., *J. Am. Chem. Soc.*, **100**, 4532 (1978).
[308] Streitwieser, A., and Nebenzahl, L. L., *J. Org. Chem.*, **43**, 598 (1978).
[309] Veracini, S., and Gau, G., *Nouveau J. de Chimie*, **2**, 523 (1978).
[310] Akperov, O. G., Gyul'maliev, A. M., Dzhafarova, E. A., and Ibragimova, D. S., *Teor. Eksp. Khim.*, **13**, 806 (1977); *Chem. Abs.*, **88**, 88918 (1978).
[311] Talvik, A., *Org. React.* (*Tartu*), **15**, 91 (1978).
[312] Bordwell, F. G., and Hautala, J. A., *J. Org. Chem.*, **43**, 3116 (1978).
[313] Zimmerman, A. H., Reed, K. J., and Brauman, J. I., *J. Am. Chem. Soc.*, **99**, 7203 (1977).
[314] Bordwell, F. G., Bartmess, J. E., and Hautala, J. A., *J. Org. Chem.*, **43**, 3095 (1978).
[315] Cumming, J. B., and Kebarle, P., *J. Am. Chem. Soc.*, **100**, 1835 (1978).
[316] Vlasov, V. M., Zakharova, O. V., and Yakobson, G. G., *Izv. Sib. Otd. Akad. Nauk SSSR, Ser. Khim. Nauk*, **1977**, 127; *Chem. Abs.*, **88**, 6108 (1978).
[317] Vlasov, V. M., Zakharova, O. V., and Yakobson, G. G., *Zh. Org. Khim.*, **13**, 2372 (1977); *Chem. Abs.*, **88**, 104464 (1978).
[318] Kankaanpera, A., Salomaa, P., Oinonen, L., and Mattsen, M., *Finn. Chem. Lett.*, **1978**, 25.
[319] Kankaanpera, A., Oinonen, L., and Salomaa, P., *Acta Chem. Scand., Ser. A*, **31**, 551 (1977).
[320] Terpinski, J., Zajaczkowska-Terpinska, E., and Kozerski, L., *Bull. Acad. Pol. Sci., Ser. Sci. Chim.*, **26**, 197 (1978); *Chem. Abs.*, **89**, 128939 (1978).
[321] Kern, J. M., and Federlin, P., *Tetrahedron*, **34**, 661 (1978).
[322] Thomas, P. J., and Stirling, C. J. M., *J. Chem. Soc., Perkin Trans.* 2, **1977**, 1909.
[323] Niemeyer, H. M., *Tetrahedron*, **34**, 1369 (1978).
[324] Talvik, A., *Org. React.* (*Tartu*), **14**, 187 (1977).
[325] Streitwieser, A., and Boerth, D. W., *J. Am. Chem. Soc.*, **100**, 755 (1978).
[326] Zoltewicz, J. A., and Jacobson, H. L., *J. Org. Chem.*, **43**, 19 (1978).
[327] Gielen, M., *Actual Chim.*, **1976**, 6; *Chem. Abs.*, **88**, 21460 (1978).

[328] Hojo, M., *Yuki Gosei Kagaku Kyokaishi*, **36**, 473 (1978); *Chem. Abs.*, **89**, 162584 (1978).
[329] Hatakeyama, S., Mitsuhashi, T., and Oki, M., *Chem. Lett.*, **1978**, 599.
[330] Brown, A. J., Howarth, O. W., and Moore, P., *J. Am. Chem. Soc.*, **100**, 713 (1978).
[331] Kashin, A. N., Bumagin, N. A., Beletskaya, I. P., and Reutov, O. A., *Zh. Org. Khim.*, **14**, 1141 (1978); *Chem. Abs.*, **89**, 107303 (1978).
[332] Beletskaya, I. P., Shishkin, V. N., Butin, K. P., and Reutov, O. A., *Zh. Org. Khim.*, **13**, 2241 (1977); *Chem. Abs.*, **88**, 61748 (1978).
[333] Ibrahim, S. E., and Kreevoy, M. M., *J. Phys. Chem.*, **81**, 2143 (1977).
[334] Chipperfield, J. R., Hayter, A. C., and Webster, D. E., *J. Chem. Soc., Dalton Trans.*, **1977**, 921.
[335] Dong, D., Hunter, B. K., and Baird, M. C., *J. Chem. Soc., Chem. Commun.*, **1978**, 11.
[336] Jawad, J. K., and Puddephatt, R. J., *J. Chem. Soc., Chem. Commun.*, **1977**, 892.

CHAPTER 11

Elimination Reactions

A. F. HEGARTY

Chemistry Department, University College, Cork, Ireland

Stereochemistry, Orientation, and Isotope Effects in *E*2 Reactions

Kinetic isotope effects[1] and primary, secondary, and solvent isotope effects[2] in *E*2 eliminations have been reviewed.

Dehydrohalogenation of 1-aryl-1-bromo-, and -1-chloro-ethanes promoted by 2,6-di-*tert*-butylphenoxide ion in DMSO–DMF (1 : 9 v/v) is characteristically *E*2 with $\rho = +2.44$ and $k_H/k_D = 9.0$;[3] the k_{Br}/k_{Cl} ratio of 146 is similar to that obtained when sodium phenoxide is used as base. It has been concluded that steric effects on rates of *E*2 elimination are not very large, in spite of the large effect often observed on positional and geometrical orientation, and the effect of crown ether complexation of the Bu^{t}OK has been commented upon. The results are consistent with the "Thornton model", there being less H^+ transfer in the transition state for reaction with the stronger base.

Preferential *anti*-elimination of HBr from the *meso*- and *dl*-isomers (**1**) and (**4**) occurs when the reaction is carried out at 100° in the presence of solid Bu^{t}OK.[4] The reactivity of (**5**) exceeds that of (**2**) which is not completely inert; however, base-catalysed isomerization of (**2**) to (**5**) could not be ruled out. From a reinvestigation[5] of the elimination reaction of *trans*-2-methylcyclo-octyltoluene-*p*-sulphonate (first studied by Brown and Kliminch in 1966), it appears that formation of 1-methylcyclo-octene occurs by competing *E*1 reaction and does not constitute an exception to the antiperiplanar rule for *E*2 reactions.

An extensive study[6] of base-catalysed eliminations from quaternary hydroxides formed from piperidine, morpholine, and decahydroquinolines generally confirms classical work; conformational effects are important and elimination occurs when,

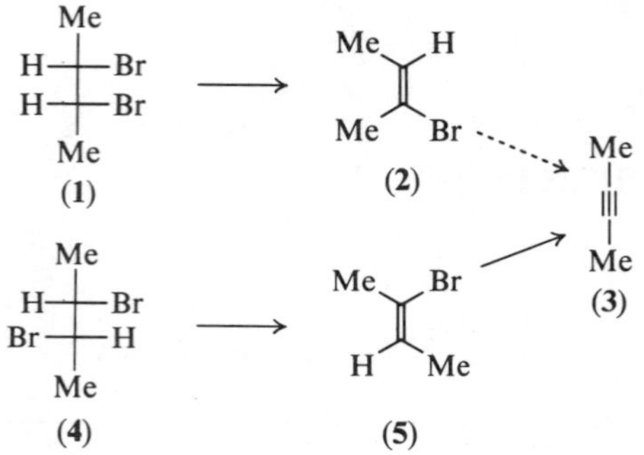

(1) (2) (3)

(4) (5)

in the six-membered rings, anti-coplanarity of H_β, C_β, C_α, and N^+ can be achieved. A reverse Hofmann elimination involving a transannular antiperiplanar addition of an amine to an alkene has been described; steric strain is relieved on cyclization.[7] On irradiation of (+)-glaucine methiodide ($\lambda > 300$ nm) in methanol a photo-Hofmann elimination occurs to give 1-[2′-(*N*,*N*-dimethylaminoethyl)]-3,4,6,7-tetramethoxyphenanthrene in good yield.[8] Hofmann elimination from *N*,*N*,*N*-trimethyl-(*p*-nitrophenethyl)ammonium iodide in 0.1 M-NaOH is inhibited by anionic micelles; some cationic and zwitterionic micelles have little effect but *N*,*N*-dimethyl-*N*-hexadecyl-*N*-(2-hydroxyethyl)ammonium bromide is an effective catalyst at high pH when it is converted into (**6**), which participates as a base in elimination.[9]

$R\overset{+}{N}Me_2CH_2CH_2O^-$ (**6**) $RCH_2CH(OTs)C_5H_{11}$ (**7**) $RCH(OTs)CH_2C_5H_{11}$ (**8**)

Zavada and his co-workers have investigated the rates[10] of formation and the isomer distribution of the alkene products[11] of *anti*-*E*2 eliminations of the toluenesulphonate series (**7**) and (**8**). *tert*-Butoxide was used as base and the reactions were studied with solvent Bu^tOH (where the base is associated) and with DMF (dissociated base). Changes in the structure of R produce a relatively small effect (*e.g.* there is < 10-fold variation in the rate of the *E*2 reaction when R is changed from Me to Bu^i under any conditions) and this has been attributed to a combination of steric and polar effects, the latter being of secondary importance; the rate variation is even smaller when the dissociated base is used. The preferential formation of *cis*-alkenes from (**7**) and (**8**) has been attributed to the importance of interactions between the cation of the associated base and the departing toluene sulphonate group, and it has been suggested that large groups R hinder this by preventing optimum tilting of the base.

The effect of ion association on competing *E*2 and S_N2 reactions has been examined, for 1-nonyl toluenesulphonates, chlorides, bromides, and iodides, by carrying out reactions of Bu^tO^- in benzene, Bu^tOH, DMF, and DMSO in the presence and absence of crown ethers.[11,12] The $E2/S_N2$ ratios are always lower in the absence of the crown, but the effect is negligible in the more dipolar solvents; the ratios follow the usual pattern and increase in the order $OTs < Cl \leqslant Br \leqslant I$.

anti-Saytzeff orientation has been reported for the elimination (in dilute HCl or on alumina) of ferrocenyl alcohols and attributed to a special effect of the preferred conformation (**9**);[13] however, it has already been pointed out[14] that terminal alkenes are always formed preferentially from alcohols of type (**10**), when R is a bulky group.

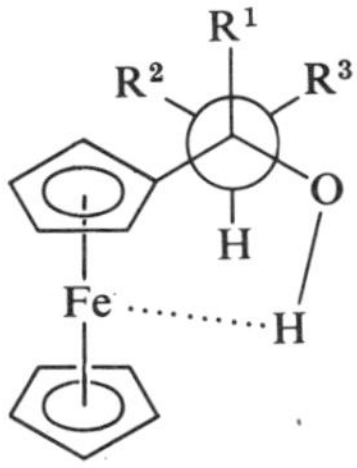

(9)

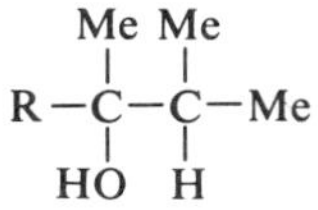

$$R-C(Me)(OH)-C(Me)_2H \quad \text{i.e. } R-\underset{HO}{\overset{Me}{C}}-\underset{H}{\overset{Me}{C}}-Me$$

(10)

Elimination from the *erythro*-sulphonyl system $MeCHXCHMeSO_2Ph$ with amines gives the *E*-alkene whereas varying amounts of *E*- and *Z*-alkenes are obtained from the *threo*-isomer.[15] When the leaving group X is changed from Cl to Br the *threo*-substrate gives different ratios of *E*- and *Z*-alkenes; other differences are noted upon temperature variation and it has been concluded that an *E*1*cB* mechanism may be operative when X = Cl.

Ahlberg's elegant work on the effect of leaving-groups on competing elimination and 1,3-proton transfer reactions of indenes has now been reported in full.[16]

The heavy-atom kinetic isotope effects on alkoxide-induced eliminations from $Ar_2CHCHCl_2$ (k_{35}/k_{37} and k_{12}/k_{14} for α- and β-carbons) have been calculated and compared with experimentally determined values,[17] with the ultimate aim of deriving a transition-state map. The results imply a half-broken C_α–Cl bond in the transition state and are consistent with the known variation of *E*2 transition-state structure. The primary and secondary kinetic isotope effects for the elimination of HCl (or DCl) from substituted 1,2-diaryl-1-chloroethanes agree with theoretical predictions but give significantly low A_H/A_D values, implying that either proton tunnelling or a competing internal-return mechanism is occurring.[18] Primary isotope effects on the *E*2 elimination reaction of HO^- with *N,N,N*-trimethylphenethylammonium ion in DMSO–H_2O have also been reported.[19]

The *E*1*cB* Mechanism

Base-catalysed elimination from 4-(*p*-substituted phenoxy)-2-oxobutanoic acids (**11**) is not a simple *E*1*cB* reaction.[20]. General base catalysis of proton transfer by amine bases is observed, but the dependence of rate on base concentration is non-linear, indicating a change in rate-determining step. Exchange of the α-hydrogen is faster than elimination at high amine concentration; therefore, under these conditions elimination occurs by spontaneous decomposition of the enolate and general base-catalysed reaction of the enol.

$$\underset{\textbf{(11)}}{ArOCH_2CH_2COCO_2^-} \underset{k_3BH}{\overset{k_1B}{\rightleftharpoons}} [ArOCH_2CHCOCO_2]^- \xrightarrow[k_3'BH]{k_3B} CH_2{=}CHCOCO_2^- + ArO^-$$

In an extensive series of papers Stirling and his co-workers have tackled the difficult problem of pinpointing the effect of varying the leaving group in eliminations, uncomplicated by other reactions (such as proton transfer pre-equilibria).[21–26] The general system used is shown in equation (1); Z is the leaving group and G is an electron-withdrawing group (such as $PhSO_2$) which stabilizes the carbanion (**13**). For those substrates which react by the $(E1cB)_R$ mechanism (as shown by deuterium–hydrogen exchange at C_β), the equilibrium constants k_1/k_{-1} were estimated. There is no simple correlation between k_2 (the leaving-group ability

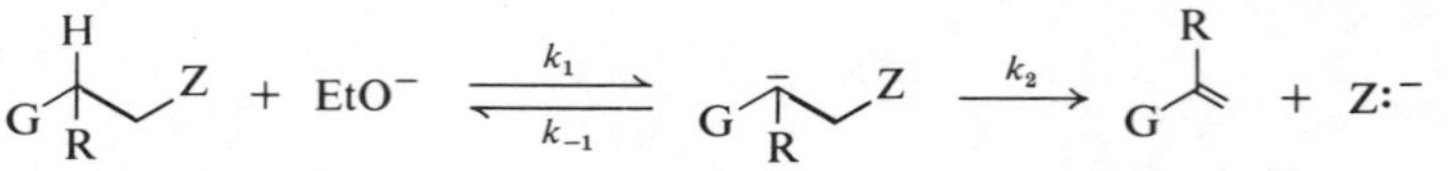

(12) + EtO^- $\underset{k_{-1}}{\overset{k_1}{\rightleftharpoons}}$ (13) $\xrightarrow{k_2}$ (14) + $Z:^-$ (1)

from the carbanion) and the pK_a of ZH.[21] All positively charged groups are good leaving groups; for neutral groups the order is PhSe > PhO > PhS > $PhSO_2$ > PhSO > NMeTs > NMeAc > CN, and covers a reactivity range of 10^{16}. When the group G is changed the rate of ionization (k_1) can become rate-determining with poor leaving groups; the ability of G to stabilize the carbanion (**13**) varies in the order G = $PhSO_2$ > CN > PhCO > fluorenyl ~ NO_2 ~ CO_2Et, and does not follow the Taft σ^* order in all cases.[22] The mechanism changes from $(E1cB)_R$, when G is $PhSO_2$ or −CN, to $(E1cB)_I$, when G is PhCO; a further change to E2 probably occurs with leaving groups such as Br.[23] Buffer catalysis by triethylamine in ethanol is observed with (**12**; G = $ArSO_2$ or CN, and Z = $^+NR_3$ or $^+SR_2$). Saturation kinetics are observed; at low amine concentrations ionization is rate-determining but at high buffer concentrations decomposition of the carbanion becomes rate-limiting. Although formation of (**13**) is very sensitive to polar and steric effects the ratio k_2/k_{-1} (elimination–reprotonation ratio) is more or less insensitive to the structure of the leaving group.[24]

Elimination of PhO^- from *erythro*- and *threo*-1,2-diphenyl-2-phenoxyethyl phenyl sulphone occurs by an $(E1cB)_R$ mechanism in EtO^-/EtOH with a stereospecificity that depends on the configuration of the carbanion; explusion of the leaving group occurs more readily from the ion derived from the *erythro*-isomer[25] since the antiperiplanar arrangement of C^- and the leaving group can be more readily achieved. In reactions of (**12**; G = Z and R = Ph) the 1,2-alkene (**14**; R = Ph) is formed when G = Z = SPh or $(EtO)_2PO$, being derived from the more rapidly deprotonated site; however, when G = Z = SO_2Ph the 2,1-alkene (Ph—CH=CH—SO_2Ph) is formed since the coplanar arrangement normally required for stabilization of the carbanion is obviated.[26]

The β-elimination of (**15**; R = Me) in MeOD or MeOH exhibits kinetic induction due to competing hydrogen exchange between the substrate and solvent;[27] the kinetic analysis is consistent with an *E*1*cB* mechanism, rather than E2, as might be expected on the basis of previous work on (**15**; R = H). An *E*1*cB* mechanism has

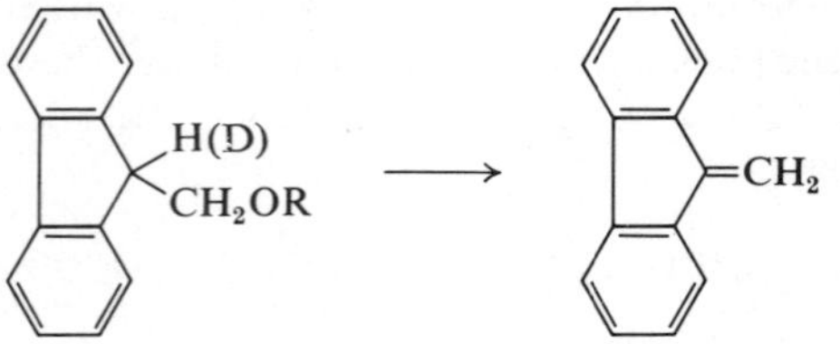

(15)

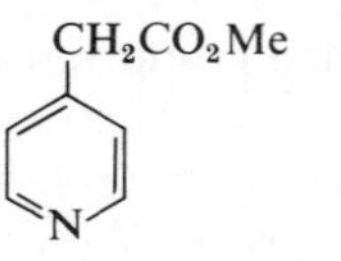

(16) 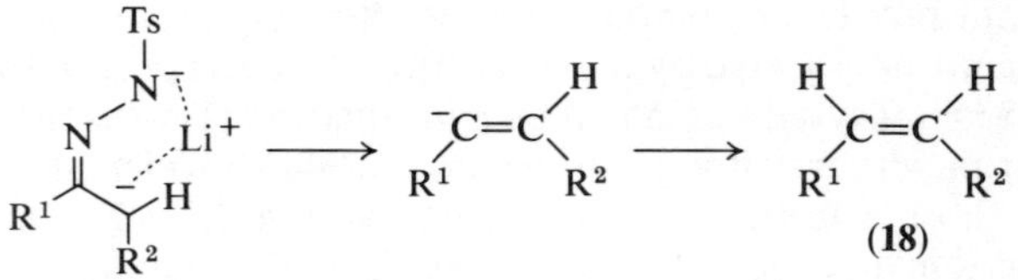 (17) (18)

also been proposed for the hydrolysis of esters of type (**16**) in DMSO–H_2O mixtures;[28] the rather weak evidence is based mainly on a rate decrease observed as the fraction of DMSO is increased.

Regiospecific alkylation and elimination from tosylhydrazones continue to attract interest. The dianion (**17**), whose configuration is maintained by coordinato Li^+, is the reactive species (BuLi is used as base); high *cis/trans* ratios are observed and a k_H/k_D ratio of 7.5 for the α-position suggests an $(E1cB)_I$ mechanism *via* (**18**).[29] The reaction can be extended to the production of trisubstituted alkenes by using lithium di-isopropylamide as base.[30] The effect of pressure on elimination of HCl (thought to occur by an *E*1*cB* mechanism) from chloromaleic acid has been described.[31]

Pyrolytic Elimination Reactions

Acetates and Carbamates

The rates of gas-phase eliminations of acetic acid from 1-arylethyl acetate (equation 2) and benzoates have been measured and used to determine accurate σ^+ values for a variety of substitutents including *m*-OAc, *p*- and *m*-NHAc,[32,33] and benzo[*b*]-thiophens.[34] The transition state for pyrolysis of the benzoates ($\rho^+ = -0.66$) is

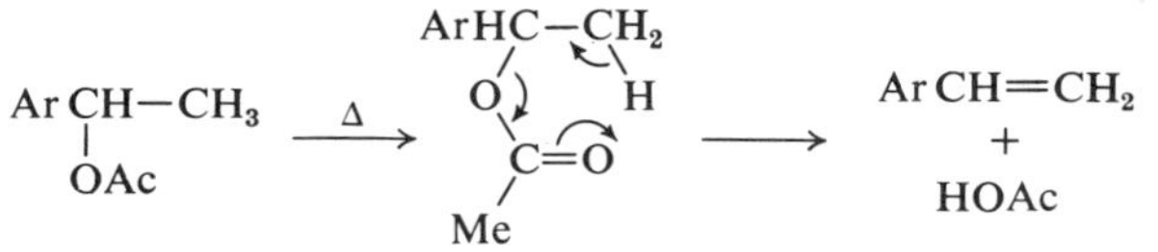

$$\mathrm{ArCH(OAc)CH_3} \xrightarrow{\Delta} [\text{ArHC–CH}_2\text{, O–C(Me)=O, H six-membered cyclic transition state}] \longrightarrow \mathrm{ArCH{=}CH_2} + \mathrm{HOAc} \quad (2)$$

more polar[35] than that of the acetates (and thus more suitable for the determination of substituent effects) but less so than that of the corresponding carbamates.

The rate of gas-phase pyrolysis of ispropyl α-haloacetates[36] at 330° is sensitive to the nature of the halo-substituents (I > Br > Cl > F) and a good correlation (also involving alkyl and α-cyano[6] substituents) is obtained[37,38] with σ^* where $\rho^* = -1.85$.

The minor product, 4-protoadamantene, from gas-phase pyrolysis of 2-adamantyl methanesulphonate arises from a concerted homoretroene reaction (rather than from the major product, 2,4-dehydroadamantane) which is formed predominantly by 1,3-elimination (92%) rather than *via* a carbene intermediate (8%). The failure to observe cross-over products upon pyrolysis of equatorial and axial 2-adamantyl methanesulphonates favours the concerted rather than an ion-pair mechanism; in general, product stability tends to govern the nature of the product formed.[39]

Lactone pyrolysis occurs in the range 520–590° and the rate of reaction varies with ring size (5, 7, 10, and 12 atom ring sizes were studied)[40]. The reaction is apparently concerted since only when the 6-membered transition state can be achieved without ring strain (*e.g.* **19**) is (**20**) formed rapidly; small lactones cannot achieve this geometry and thus undergo pyrolysis considerably more slowly. Thermolysis[41] of β-lactones (**21**) (formed by cycloaddition of dichloroketene and

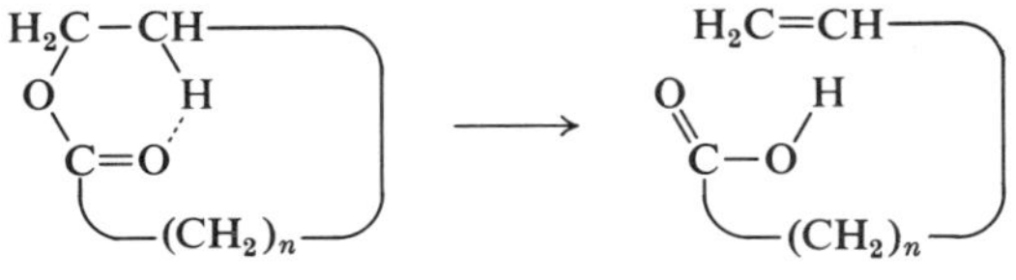

$$\text{(19)}\ \mathrm{H_2C{-}CH{-}(CH_2)_n{-}C(=O){-}O}\ \text{(cyclic, H transferred to C=O)} \longrightarrow \text{(20)}\ \mathrm{H_2C{=}CH{-}(CH_2)_n{-}C(=O)OH}$$

(**19**) (**20**)

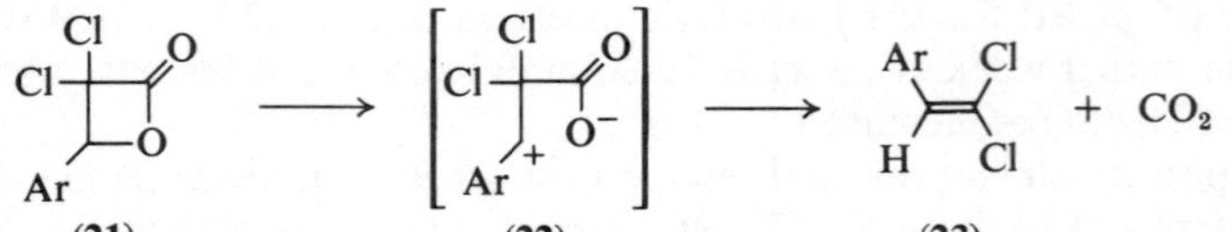

(21) → [(22)] → (23) + CO_2

the aldehyde) leads to loss of CO_2 and formation of the alkene (**23**). The reaction is of the first order and probably occurs *via* a highly polarized transition state that resembles (**22**) since $\rho = -3.07$ (correlation with σ^+ for substitutents on Ar); however, attempts to trap (**22**) were unsuccessful. β-Peroxylactones (**24**) undergo thermal decomposition *via* the diradical (**25**).[42] Solvent and secondary isotope effects are consistent with this mechanism although the lifetime of (**25**) must be $<10^{-7}$ s.

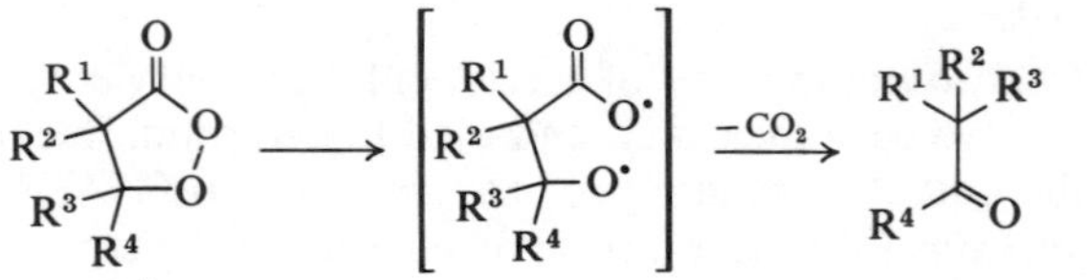

(24) → [(25)] $\xrightarrow{-CO_2}$ (26)

Methyleneketene (propa-1,2-dien-1-one) ($CH_2{=}C{=}C{=}O$) is generated by flash-vacuum pyrolysis of acrylic anhydride at 510–560°;[43] gas-phase pyrolysis of *tert*-pentyl nitrite[44] and elimination of acetic acid and ethylene from ethyl acetate induced by a pulsed CO_2 laser[45] have also been described.

The transition state for gas-phase thermal decomposition of 1-arylethyl phenyl carbonates[46] is more polar ($\rho = -0.84$) and *E*1-like than that for acetate pyrolysis (where $\rho = -0.66$); this is supported by the kinetic isotope effect ($k_H/k_D = 2.11$) which is also substantially smaller (2.32) than that observed for the acetates. Tri-carbonates (**27**) give mixtures of di- (**28**) and mono-carbonates (**29**) (R = adamantyl) when heated at 150°;[47] the conversion of (**28**) into (**29**) at 170° in benzonitrile was investigated in some detail and is believed to involve initial scission to give $ROCO^+$ and $ROCO_2^-$. The products formed upon pyrolysis of carbonate tosylhydrazone

ROC(O)OC(O)OC(O)OR (27) ROC(O)OC(O)OR (28) ROC(O)OR (29)

salts have been rationalized[48] in terms of the formation of partially equilibrating radicals; this stepwise process is favoured rather than an allowed $\{{}_\sigma 2_s + {}_\sigma 2_s + {}_\omega 2_s\}$ linear cheleotropic process.

The ^{2}H-kinetic isotope effect on the thermal decomposition of 1,1-diphenylethyl and 1-phenylethyl *N*-(*p*-tolyl) carbamates is approximately additive for each deuterium at the β-carbon; thus hyperconjugative stabilization of the incipient carbocation is apparently the most important contributor (a secondary effect) to the isotope effect (consistent with rate-determining C—O bond cleavage).[49–50]

Extrusion of CO_2, CO, and SO_2

The decarboxylation of 6-nitrobenzisoxazole to 6-nitrosalicylonitrile continues to attract attention as a sensitive probe of solvent structure; decarboxylation in water containing poly-vinylbenzo-18-crown-6) occurs 2300-fold faster than in water itself[51] since the isoxazole is drawn into the inner core of the tightly coiled polymer before undergoing decomposition.

Theoretical calculations (by the MINDO/3 method) of primary kinetic isotope effects and activation parameters for the decarboxylation of but-3-enoic acid indicate that this retroene reaction is synchronous.[52] An interesting point is that the same method predicts a non-symmetric transition state for the Diels–Alder reaction.[53]

Decarbonylation of 2,2-dimethylbut-3-enal is intramolecular with transfer of hydrogen atom concerted with extrusion of CO (c.f. **30**); a kinetic isotope effect ($k_H/k_D = 2.8$) is observed and radicals that can be trapped by a scavenger are

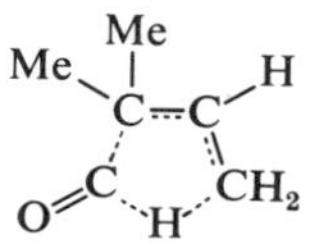

(30)

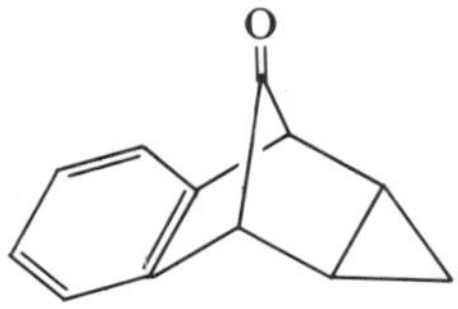

(31)

absent.[54] The stereoelectronic effect of a cyclopropane ring has been investigated in the thermal decarboxylation of *endo*-6,7-benzotricyclo[3.2.1.0^{2,4}]octen-8-one (**31**) and several analogues.[55] Pyrolytic decarbonylation of cyclopent-4-ene-1,2,3-trione has also been reported.[56]

The 3,4-disubstituted 1,2,3,5-oxathiadiazole 2-oxides (**32**) loose SO_2 under mild thermal conditions, to yield carbodi-imides (**33**); substituent effects in R^1 and R^2 were examined and these, together with the failure to trap intermediates, suggest that the fragmentation occurs by a concerted mechanism[57] with simultaneous rearrangement.

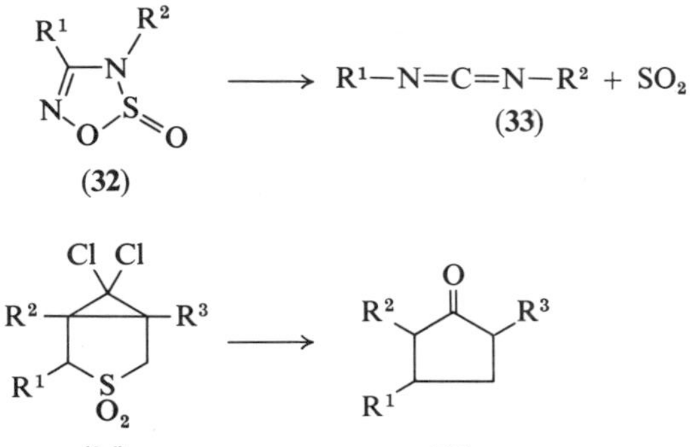

$$\text{(32)} \longrightarrow R^1{-}N{=}C{=}N{-}R^2 + SO_2 \quad \text{(33)}$$

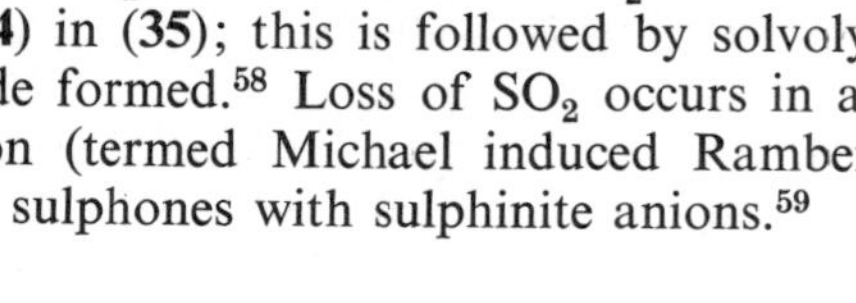

(34) ⟶ (35)

A thermal cheleotropic elimination of SO_2 also occurs in the slow step of the conversion of (**34**) in (**35**); this is followed by solvolysis and cyclization of the dienylic dichloride formed.[58] Loss of SO_2 occurs in a key step in a new chain-extension reaction (termed Michael induced Ramberg–Backland synthesis) of halomethyl vinyl sulphones with sulphinite anions.[59]

Extrusion of Nitrogen

The thermal decomposition of *meso*- and *dl*-forms of the diazenes (**36**) is *cis*-stereospecific yielding *cis*- and *trans*-3,4-dimethylhex-3-enes, respectively. This favours a concerted two-bond cleavage for these cyclic azoalkanes; if stepwise reaction occurs then C—C bond rotation is slow relative to N_2 loss (which is possible but speculative).[60]

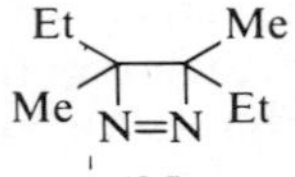

(36) 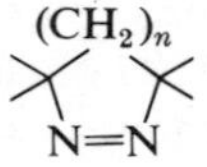 (37)

By comparing the rates of (thermal and photochemically induced) extrusion of nitrogen from azoalkanes (**37**) of varying ring size it has been shown[61,62] that (**37**; $n = 0$) and its bicyclo-analogues react surprisingly slowly despite their high ring strain. It has been concluded that the transition state retains most of the strain present in the starting azoalkane; the small effect of methyl groups α- to the azo-linkage supports this view. Loss of N_2O from bicyclic azoxy-compounds occurs considerably more slowly than nitrogen loss from the corresponding azo-compounds (*e.g.* activation energies of 37 and 14 kcal mol^{-1} have been reported for similarly substituted azoxy- and azo-analogues).[63]

Decomposition of cyclic *cis*-tetrazenes (**38**) occurs with negative entropy change, suggesting a concerted reaction; also the high yield (>90%) of the diaza-product (**39**) is not concordant with a biradical intermediate with a significant lifetime.[64] Unsymmetrically substituted diazenes, on the other hand, generally undergo

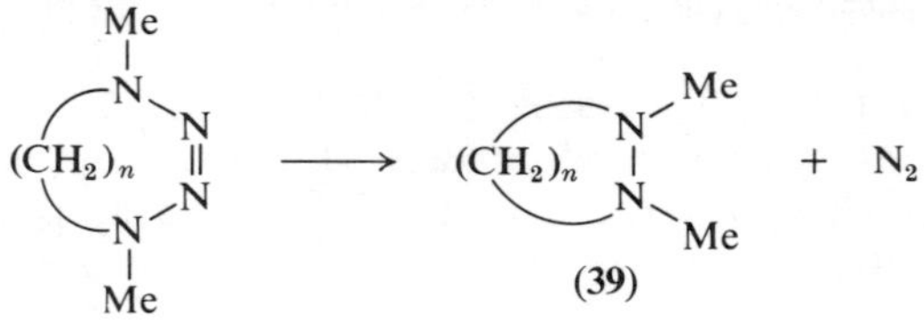

(38) → (39) + N_2

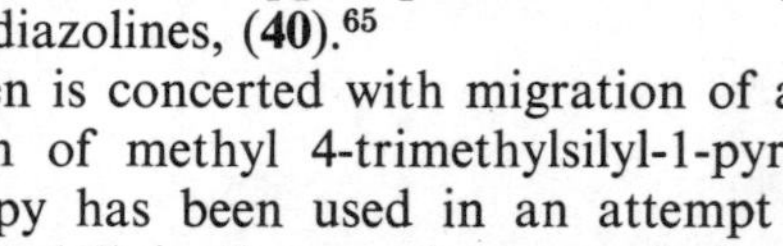

(40) → (41)

stepwise decomposition; for example, an intermediate 1,5-diradical (**41**) has been proposed (on the basis of the trapped products formed) for the thermolysis of 2-acetoxy-Δ^3-1,3,4-oxadiazolines, (**40**).[65]

Explusion of nitrogen is concerted with migration of a trimethylsilyl group in thermal decomposition of methyl 4-trimethylsilyl-1-pyrazoline-3-carboxylates.[66] Microwave spectroscopy has been used in an attempt to detect a potentially antiaromatic 1*H*-azirine (**43**) in the gas-phase pyrolysis of 1,2,3-triazole (**42**); no evidence for the intermediacy of (**43**) was found.[67]

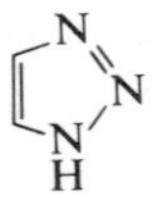

(42) 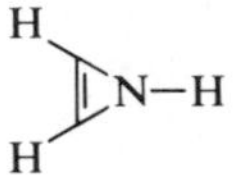 (43) 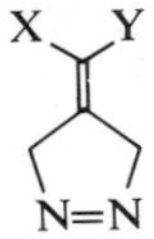 (44) 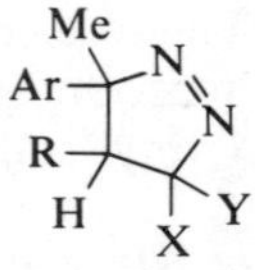 (45)

The effects of substitutents X and Y on the rate of nitrogen elimination from 4-alkylidene-Δ^1-pyrazolines (**44**) show that these do not correlate with the expected radical-stabilizing power of X and Y. This suggests either a concerted mechanism or the formation of radicals that are orthogonal rather than allyl in type.[68] Intermediate ion pairs have been proposed to account for retention of stereochemistry in cyclopropanes formed by nitrogen extrusion from pyrazolines (**45**) bearing substituents Y = COOAr.[69] Pyrazolines (**45**; X, Y = cyano, alkoxycarbonyl, or halogen etc.) yield both alkenes and cyclopropanes, the relative amounts of which can be rationalized in terms of abilities of the substituent X,Y to stabilize dipolar intermediates.[70]

The fragmentation of the pyrazoline epoxide (**46**) to the unsaturated ketone (**49**) suggests the intermediacy of an unstable oxabicyclobutane (**47**);[71] (**48**), which is formed along with (**49**) (in the ratio 2.8 : 1), may also be formed by way of (**47**) or directly *via* a diradical intermediate.

(**46**) (**47**) (**48**) (**49**)

Photochemically induced loss of nitrogen from unstable tricyclic azo-compounds has also been investigated.[72]

Other Pyrolyses

Deuterium kinetic isotope effects are in the range 2–3 for reaction of the *N*-oxides of *N*,*N*-dimethyl-2-phenylethylamine and -2-phenylpropylamine in DMSO–H_2O, –Bu^tOH, and –THF solvent mixtures.[73] The effective basicity of the solvent was changed by varying the solvent composition, and the fact that the isotope effect was small and did not vary greatly was taken as evidence for a non-linear proton transfer in the transition state (in confirmation of theoretical predictions). The skew transition state can also account for the temperature-insensitive value; k_H/k_D ~2 is constant for thermolysis of *N*-oxides over a 100° range; this implies that $A_H/A_D > 1.2$ (the theoretical limit for proton transfer *via* a linear transition state). The isotope effect for thermolysis of sulphoxides on the other hand shows the temperature-dependence expected of a linear proton transfer from carbon to oxygen.[74]

The ^{37}Cl isotope effects on the pyrolysis of alkyl halides[75] is such as to suggest a cyclic transition state involving migration of H from C to Cl.

A number of studies have also been reported in which proton transfer is to a neighbouring basic heterocycle; in general, these reactions appear to be concerted and analogous to ester pyrolyses. Thus, 2-ethoxypyridine (**50**) undergoes thermal elimination of ethylene and formation of 2-pyridone when heated at >400°.[76]

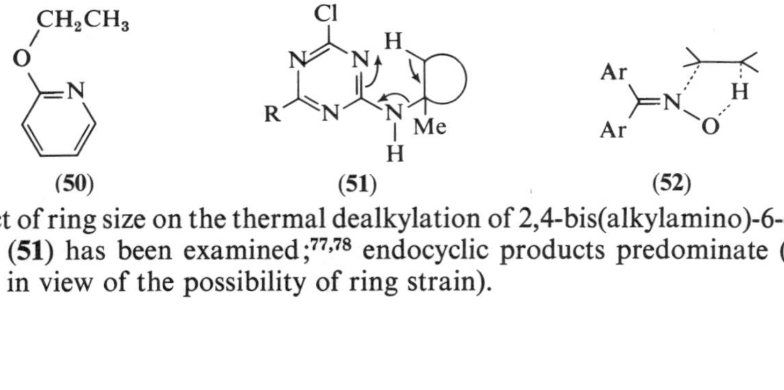

(**50**) (**51**) (**52**)

The effect of ring size on the thermal dealkylation of 2,4-bis(alkylamino)-6-chloro-*s*-triazines (**51**) has been examined;[77,78] endocyclic products predominate (which is unusual, in view of the possibility of ring strain).

Substituent effects[79] ($\rho = +0.22$, correlation with σ^+) on the thermal elimination of isobutene from *N*-(diarylmethylene)-*tert*-butylamine *N*-oxides indicate a concerted process (**52**), although the stepwise reaction is sometimes favoured when Ar are highly electron-donating groups.

A direct comparison[80] of the pyrolysis of ketones with that of the corresponding esters shows that the ketones, in general, require temperatures up to 150° higher for pyrolysis. For example, heptane-2,4-dione gives typically 30–40% conversion into acetone and but-3-enal on pyrolysis at 600–700°; because of the higher temperatures required, ketones tend to give also side-products of competing free-radical reactions.

The reteroene conversion of ethyl vinyl ether into acetone and ethylene, which occurs on CO_2TEA-laser photolysis, competes with simple bond fission.[81] 3-Vinylcyclobutanone is pyrolysed smoothly to butadiene and ketene; a twisted activated complex has been suggested.[82] Unimolecular homolysis occurs when isopropyl peroxide is heated, in competition with an electrocyclic reaction [for which the rate constants correlate with E_t (or Z) parameters of solvent polarity].[83]

In the acetylenic analogue of the retroene reaction the propargyl ethers (**53**) are converted into allenes (**54**), which retain stereochemical integrity;[84] this result is unexpected in view of the linear rigidity of (**53**). In general, vinyl ethers are pyro-

(53) ⟶ $R^1R^2C{=}O$ + (54)

lysed at 40–50° lower temperatures than are the corresponding esters; methyl vinyl ether is considerably more stable than its homologues and decomposes by an entirely different mechanism, in which an ionic intermediate is probably involved.[85]

The pyrolysis of butyric acid is catalysed by HBr and proceeds rapidly at >633° (to give propene, carbon monoxide, and water); the reaction is of the second order and the entropy of activation, *ca.* 0, does not suggest a cyclic mechanism.[86] Amine elimination from (N^2-(alk-1-enyl)hydrazines to give pyrrolinones (the Brunner oxindole synthesis) is shown, by specific ^{2}H-labelling, to involve a prior {3,3}-sigmatropic rearrangement.[87]

The negligible electronic effect of substituents, on Ar, on the thermolysis of *N*-alkylidenearenesulfinamides (**55**) suggests that the transition state is cyclic; the arenesulfinic acids (**56**) were trapped by reaction with methyl propiolate.[88] Regiospecific exchange of deuterium for hydrogen in alkenes can be achieved by carrying

$$\text{Ar–S(=O)–N=CH–R} \longrightarrow \text{ArSOH} + \text{RC}{\equiv}\text{N}$$

(55) (56)

out the reversible reaction with SO_2 in the presence of D_2O.[89] A mechanism has been proposed which involves the intermediacy of an allylic sulfinic acid; the competing rearrangement of the alkene is suppressed by the presence of D_2O. Substituent effects on the competing [2,3]-sigmatropic rearrangement and *syn*-elimination of allylic aryl sulphoxides have also been reported.[90] The oxidation of alkyl iodides that have strongly electron-withdrawing substituents (*e.g.* methoxycarbonyl or sulphonyl) at the α-carbon initially yields an iodoso-compound which undergoes *syn*-elimination; an example is given in equation (3).[91]

$$\mathrm{PhSO{-}C(I)Me_2} \xrightarrow[\mathrm{CH_2Cl_2}]{m\text{-CPBA}} \left[\mathrm{PhSO_2{-}C(IO)Me_2}\right] \longrightarrow \underset{(87\%)}{\mathrm{PhSO_2{-}C(Me){=}CH_2}} \quad (3)$$

$$\mathrm{Ph{-}C(OPO_2^-OMe)Br{-}CHBrMe}$$

(57)

The thermal cleavage of β-bromo-phosphonates (*e.g.* **57**) to yield alkenes and methyl metaphosphate (which is trapped by reaction with diethylaniline) is strictly trans-stereospecific.[92]

On pyrolysis of 1-methylcyclohex-1-ene at 1000–1180° in a pulse shock tube a reverse Diels–Alder reaction occurs; a biradical mechanism is consistent with the observed Arrhenius parameters;[93] under the same conditions propene yields methane, ethylene, allene, and propyne.[94] Also reported are: low-pressure pyrolysis of 3-chloropropionitrile;[95] thermal and IR-induced unimolecular decomposition of pairs of alkenes, esters, and alkyl halides;[96] single-pulse shock-tube decomposition of simple alkyl chlorides;[97] thermal decomposition of biurea;[98] and participation by neighbouring groups in gas-phase eliminations.[99,100]

Rates of gas-phase elimination from α-substituted alkyl chlorides, $ZCHClCH_3$, have been correlated with σ^* values of σ_p^+ (when the group Z has available lone pair-electrons).[101] *cis*- and *trans*-4-Chloropent-2-enes give HCl and penta-1,3-diene at 520–640°, the *trans*-isomer reacting by the usual four-centre process.[102] Most (>90%) of the product formed from pyrolyses of cyanides involves C(2)–C(3) bond fission.[103] Increased size and branching of the alkoxy-group increased the thermal stability of 2-alkoxytetrahydropyrans;[104] homolytic and cyclic mechanisms compete. Pyrolysis of tetralin[105] (giving mainly naphthalene but also considerable amounts of decomposition products such as indene, methane *etc.*) is first-order in the range 500–670°. A convenient method has been described for preparation of *trans*-alkenes by reaction of 1,2-diols with $Me_2NCH(OMe)_2$ followed by methylation and pyrolysis in refluxing toluene.[106]

Activation energies of 26 and 32 kcal mol^{-1} have been determined for the thermal evolution of NO_2 from 1-chloro- and 1-amino-2,4,6-trinitrobenzene, respectively.[107] Decomposition of 2-nitrophenols, anilines, and nitrobenzenes occurs (as shown by k_H/k_D values) *via* a concerted mechanism.[108] Fulven-6-one, which has been postulated as an intermediate in the thermolysis of *o*-substituted aromatic compounds, has been generated and characterized by low-temperature IR spectroscopy and by trapping experiments.[109]

Other Topics

The preparation of unsaturated ketones by elimination reactions has been reviewed.[110] Reviews have also dealt with β-eliminations,[111,112] and alkene-forming hydride eliminations.[113]

The structural and solvent effects on the rates of *syn*-eliminations from alkyl selenoxides show[114] that polar solvents reduce the rate of elimination; Cl or Ph substitutents (α- or β-) or α-alkyl substituents increase the rate of elimination but β-alkyl or methoxy-substituents retard elimination. A side-reaction is the addition of the benzeneselenic acid produced to the alkene, but this can be suppressed by the addition of a secondary amine. The procedure, involving *in situ* oxidation to the selenoxide, is a particularly mild method for the formation of alkenes.[115]

α-Haloselenides (**58**) undergo rapid elimination (10 min) in DMSO to give the aldehyde (**59**; $R^2 = H$); in DMF, the selenoalkene (**61**) is obtained (*via* **60**).[116] The synthesis of 1,3-diketones by elimination of selenium from selenocarboxylates has also been described.[117]

$R^1_2CH-C(R^2)(SeR^3)-Br$ (**58**) $\xrightarrow{DMSO}$ $R^1_2CH-C(R^2)(SeR^3)-O\overset{+}{S}Me_2$ (Br^-) $\longrightarrow$ $R^1_2CH-C(R^2)=O$ (**59**)

(**58**) $\xrightarrow{DMF}$ $R^1_2CH-C(R^2)(SeR^3)-O-CH=\overset{+}{N}Me_2$ (**60**) $\longrightarrow$ $R^1_2C=C(R^2)SeR^3$ (**61**)

α,β-Unsaturated ketones are formed on treatment of the silanes (**63**) with bromine or with copper(II) bromide (followed by desilylbromination); the latter reaction is reversible since the silanes from (**62**) are prepared in the presence of

(**62**) $\underset{Cu(II)}{\overset{Cu(I),\ R_3SiLi}{\rightleftharpoons}}$ (**63**) (SiR_3)

copper(I) iodide.[118,119] The synthesis of alkenes, α,β-unsaturated esters, aldehydes, and nitriles from β-functionalized organosilicon compounds has been reviewed.[120]

Allyl alcohols can be converted into 1,3-dienes *via* selenoxide or sulphoxide intermediates.[121] The reagent of choice is 4-methyl-2-nitrobenzeneselenyl chloride and the inital reaction at the alcohol function is followed by a [2,3]-sigmatropic rearrangement and *syn*-elimination of the selenoxide. The effects of the size of alkyl groups on the reactivity of β-alkyl-9-borabicyclo[3.3.1]nonanes with aldehydes and ketones show[122] that high reactivity occurs only when the adduct can achieve *syn*-coplanar alignment of B–C–C–H. The reaction is second-order (first-order in the organoborane and in the carbonyl compound) and is accelerated by increasing electrophilicity in the aldehyde. An alkene is produced in the rate-determining step; the overall reaction can be used to reduce aldehydes under mild conditions.

A new reductive elimination from α,β-dinitroalkanes to give alkenes (**64**; X = CN or COOEt) has been described; a radical chain (rather than an *E*2 mechanism) involving one-electron transfers has been proposed.[123] Other reductive

$R^2R^1C(NO_2)-CR^3(NO_2)X \longrightarrow R^1R^2C=CR^3X$ (**64**)

eliminations include: formation of biaryls from diarylbis(phosphine)Pt(II) complexes in solution (a concerted unimolecular reductive elimination with conformational restriction in the transition state has been suggested to account for the substituent effects and large negative entropy of reaction);[124] regiospecific addition followed by *syn*-elimination of hydridopalladium salt;[125] β-hydroxy-sulphides with 2-fluoropyridinium ion and lithium iodide as reagents;[126] addition of allylic acetates to zerovalent palladium followed by elimination of acetic acid, which liberates a diene and regenerates the palladium which recycles.[127]

The enolization of oxaloacetic acid (**65**; R = COO^-)[128] and its diethyl ester[129] (**65**; R = COOEt) is catalysed by tertiary amines with $pK_a > 8$; plots of observed rate against amine concentration show a break, and k_{obs} increases less rapidly at high amine concentration. The kinetic scheme (equation 4) proposed involves a novel amine-promoted *E*2 elimination from the adduct (**66**). The rates of solvolysis of

$$\underset{(\mathbf{65})}{R-\overset{O}{\overset{\|}{C}}-CH_2-R} + N{<} \rightleftharpoons R-\overset{O^-}{\overset{|}{C}}(-\overset{+}{N}{<})-CH_2R \rightleftharpoons \underset{(\mathbf{66})}{R-\overset{OH}{\overset{|}{C}}(-\overset{+}{N}{<})-CH_2-R} \xrightarrow{N<} R-\overset{OH}{\overset{|}{C}}=CH-R \quad (4)$$

R = COO^- or COOEt

4-aryl-2-bromo-4-oxobutanoic acids in DMSO are correlated with ordinary σ values ($\rho = +0.65$); no H–D exchange occurs and a primary deuterium isotope effect is observed; the corresponding chlorides give the alkene *ca.* three-fold more slowly. The proposed transition state resembles a solvent-separated ion pair which begins to lose the proton and generate a small amount of negative charge at C_β.[130]

3,3,4,4-Tetrafluoropyrazoline is a good catalyst for α-deprotonation of ketones by intermediate formation of a detectable enimmonium ion.[131] The mechanism of dehydrohalogenation of hydroxamoyl chlorides(**67**; Ar = 2,4,6-$Me_3C_6H_2$) and bromides to the corresponding nitrile oxides (**68**) in acetonitrile is catalysed by

$$\underset{(\mathbf{67})}{\overset{Ar}{\underset{Cl}{>}}C=NOH} \longrightarrow \underset{(\mathbf{68})}{Ar-C\equiv\overset{+}{N}-\bar{O}}$$

tertiary amines and by $AgNO_3$;[132] leaving-group effects (k_{Br}/k_{Cl}) of 10^3 and 1.6 have been reported for catalysis by $AgNO_3$ and by 4-methylmorpholine, respectively. A primary isotope effect (k_H/k_D) of 2 was estimated (which might suggest an unusual concerted mechanism of elimination) but, since the isotopic purity of the substrate was only 18%, the value is obviously subject to error. The thermal decomposition of furoxazans (nitrile oxide dimers) has also been reported.[133]

The mechanism of amine elimination from *gem*-diamines (**69**) (formed in situ by $NaBH_4$ reduction of amidines) has been studied by the clever technique of reductive trapping of the iminium cation so formed, at pH 2–11;[134] the reaction proceeds by elimination of the less basic amine regardless of whether it gives rise to an aromatic or an aliphatic iminium cation (**70**). The rates of solvolysis of the Mannich bases $XC_6H_4NHCH_2C(NO_2)_2C_6H_4Y$ have been investigated[135] and correlated with a four-parameter equation.

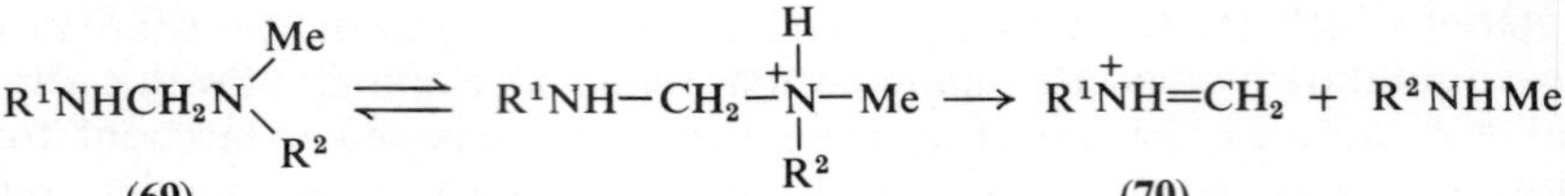

$$R^1NHCH_2N(Me)R^2 \rightleftharpoons R^1NH{-}CH_2{-}\overset{+}{N}H(Me)R^2 \longrightarrow R^1\overset{+}{N}H{=}CH_2 + R^2NHMe$$

(69) (70)

The secondary deuterium isotope effects on the competing elimination and substitution of 1,1-dihaloethylbenzenes increase (1.187–1.375) as the dioxan fraction of the solvent is increased from 35 to 75%;[136] the percentage of elimination also increases at high dioxan concentration and is thought to occur from a tight ion pair (whereas substitution occurs from a solvent-separated ion pair). Dehydration of 4-hydroxyphenyl-3-methoxypropanol in acid and in base has been studied,[137] as have the solvolyses of 7β-methylbicyclo[3.3.1]non-3β-yl-toluene-*p*-sulphonate[138] and *exo*-2-norbornylarenesulphonates[139] in a series of carboxamides including DMF. The products are formed from a carbocation–carboxamide adduct which undergoes competing *syn*-*E*1 (the major pathway) and S_N1 processes; the *E*1 reaction occurs *via* an intimate ion pair and can be suppressed by hydrogen-bonding of the amide with the toluenesulphonate anion.

The rate constant for acid-catalysed dehydration of the *anti*-periplanar (*ap*)-isomer of the alcohol (**71**) is 10^4-fold greater than that of the *syn*-periplanar (*sp*)-isomer;[140] this has been attributed to ground-state steric strain in the (*ap*)-isomer together with smaller contributions from steric effects on resonance stabilization of the forming carbonium ion together with entropy effects. Solvolysis of 2,2-

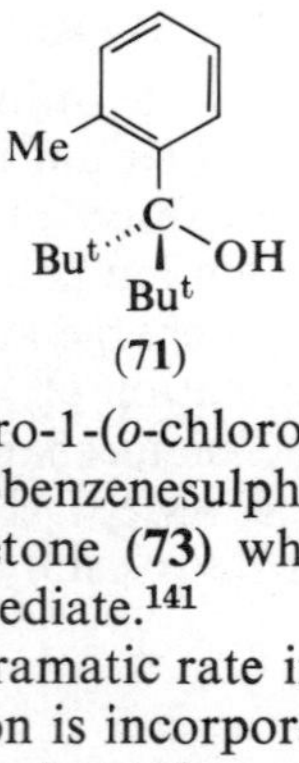

(71): o-Me-C6H4–C(OH)(But)2

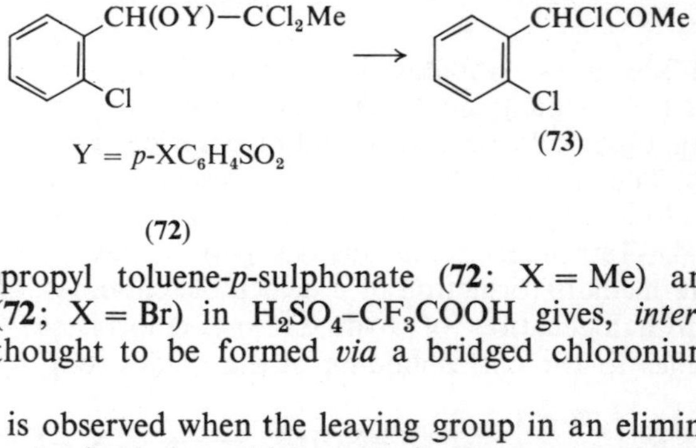

(72): o-Cl-C6H4–CH(OY)–CCl2Me → (73): o-Cl-C6H4–CHClCOMe

Y = p-$XC_6H_4SO_2$

(71) (72) (73)

dichloro-1-(*o*-chlorophenyl)propyl toluene-*p*-sulphonate (**72**; X = Me) and *p*-bromobenzenesulphonate (**72**; X = Br) in H_2SO_4–CF_3COOH gives, *inter alia*, the ketone (**73**) which is thought to be formed *via* a bridged chloronium ion intermediate.[141]

A dramatic rate increase is observed when the leaving group in an elimination reaction is incorporated in a strained ring which opens in the course of reaction. This is shown by comparing the reactivities of (**74–76**); in fact, ring strain accelerates eliminative ring fission to such an extent that β-deprotonation becomes rate-limiting.[142]

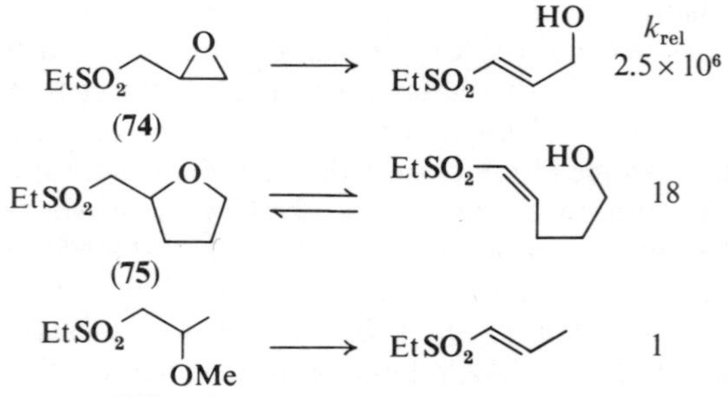

Substrate		Product	k_{rel}
$EtSO_2CH_2$-oxiranyl (74)	⟶	$EtSO_2CH{=}CHCH_2OH$	2.5×10^6
$EtSO_2CH_2$-tetrahydrofuranyl (75)	⇌	$EtSO_2CH{=}CH(CH_2)_3OH$	18
$EtSO_2CH_2CH(OMe)CH_3$ (76)	⟶	$EtSO_2CH{=}CHCH_3$	1

Nucleophilic substitution of Cl^- in 1,1-dichloro-2-(nitrophenyl)cyclopropanes (**77**) by, *e.g.*, MeO^- or PhS^- involves the initial formation of cyclopropenes (**78**);[143,144] ring-opening of (**79**) occurs in $AgNO_3$/MeOH or in Bu^{t}OK;[145] electron-

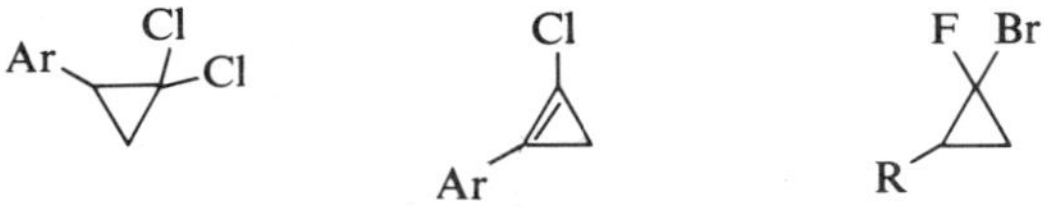

(**77**) (**78**) (**79**)

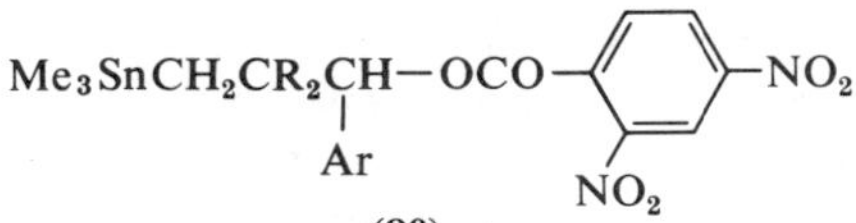

$Me_3SnCH_2CR_2CH(Ar)–OCO–C_6H_3(NO_2)_2$

(**80**)

donating substituents increase the rate of reaction with $AgNO_3$/MeOH but retard reaction with Bu^{t}OK. The kinetics of reaction of Bu^{t}O$^-$ with epoxides have also been examined,[146] as has the reaction of epoxides with amines.[147] The solvolysis of (**80**) in CF_3COOH leads to cyclopropanes by 1,3-deoxystannylation;[148] a kinetic isotope effect ($k_H/k_D = 0.94$; R = H or ^{2}H) and substituent effect ($\rho = -3.63$ correlation with σ^+) indicate direct participation of C–Sn σ-electrons in the transition state.

The four diastereomeric 2-bromo-3-(phenylsulphinyl)butanes (e.g. **81**) react with tributyltin radicals, ultimately to give but-2-enes stereoselectively.[149] The stereoselectivity is thought to arise from rapid loss of PhSO from the intermediate radical (**82**) before appreciable C(2)–C(3) rotation can occur. The comparable

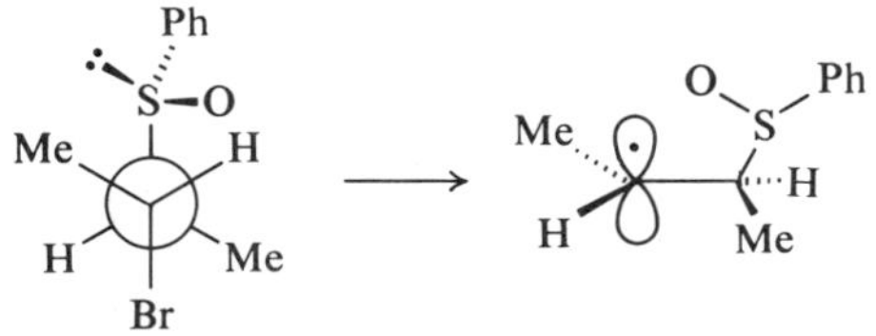

(**81**) (**82**)

rates of reaction for the various isomers indicate, at best, a modest stabilization of the intermediate radical by the neighbouring sulphur.

Alkaline dehydrochlorination of 1-arylnaphthalene tetrachlorides shows the following order of reactivity for *E*2 elimination: 1,2 or 3,4 > 2,3 (all *anti*) > *syn* 3,4 > *syn* 2,3.[150] Elimination, catalysed by pyridine, HOAc, or Zn in HOAc, of androstanone gives *trans*-β-elimination of an 11α-acetoxy-group[151] (rather than a thermal *cis*-elimination); acetoxy-group elimination from steroid derivatives, promoted by PCl_5 or methylenesulphonyl chloride, has also been reported.[152]

A *syn*-elimination of H_2O occurs in the dehydration of (3*R*)-hydroxyalkyl *S*-thioesters by yeast fatty acid synthetase.[153] The D-amino-acid transaminase from *Bacillus sphaericus* catalyses β-elimination with β-bromo-D-alanine or β-bromopyruvate;[154] there is direct evidence for an α-amino-acrylate–Schiff's base intermediate. Ethanolamine–ammonia lyase-catalysed deamination of ethanolamine and 2-aminopropanol has been described.[155] A stereospecific *syn*-1,4-elimination from 4-deuteriated cyclohexanols has been used as a model for the chorismate synthetase reaction;[156] model studies for β-eliminations of amino-acids catalysed by pyridoxal have also been described,[157] as has base-catalysed elimination of cysteine from adducts with 5-bromo-2′-deoxyuridine.[158]

Dehydrohalogenations of the following compounds have been reported: 5-chloropentan-2-one (solvent effects, with NaOMe as base)[159] $PrCCl_2Me$ (and interconversion of three chloropentenes formed);[160] *N*-1-(2,2,2-trichloroethylidene)-*tert*-butylamine,[161] $MeCO(CH_2)_3Br$ (ratio of substitution to elimination as a function of amine bases),[162] *meso*- and *dl*-PhCHBrCHBrPh (loss of Br_2 by a radical chain mechanism),[163] dibromocyclohexanediones (to give ultimately cyclopentenediones *via* cyclopropane intermediates),[164] 2-bromo-2-nitrobornane (which gives 4-nitrocamphene selectively on treatment with $AgSbF_6$);[165] chlorinated hydrocarbons,[166] and α,β-dibromo-ketones.[167]

1,2-Diols can be converted into alkenes by a mild route that involves preliminary conversion into the bis(dithiocarbamates) followed by reaction with $Bu^n{}_3SnH$; the elimination step is probably radical in nature with $Bu^n{}_3Sn$ as chain carrier.[168] Dehydration of acetylenic alcohols has also been reported,[169] as has elimination of EtOH from the diastereomers of p-$O_2NC_6H_4CH(OEt)CH(CO_2Et)$-COMe.[170]

Several competing elimination reactions occur when dialkyl ethers are treated with alkyl-lithium compounds,[171] as shown by specific ^2H-labelling; thus, MeOMe

$$R{-}CH_2CH_2OR' \xrightarrow{\beta} R{-}CH{=}CH_2 + LiOR'$$

$$\xrightarrow{\alpha} RCH_2CH(Li)OR' \longrightarrow RCH_2\ddot{C}H \longrightarrow RCH{=}CH_2$$

$$\xrightarrow{\alpha,\beta'} R{-}CH_2{-}CH(Li){-}O{-}CH_2{-}CH(H)R'' \longrightarrow R''CH{=}CH_2 + LiOCH_2CH_2R$$

reacts only by α-elimination while ethyl ethers undergo mainly β-elimination with, however, significant amounts of α,β′-elimination. When these eliminations are hindered, Wittig rearrangement (up to 21% of total reaction) can compete.

Reduction of methylsulphonylmethyl ethers (**83A**; Z = $MeSO_2CH_2O$–) or even the simple methyl ethers (**83A**; Z = MeO–) with sodium naphthalenide in HMPA results in a reductive decyanation with formation of the corresponding *cis*-alkene.[172]

$$(CH_2)_{10}\,C(Me)(Z){-}C(Me)(CN) \longrightarrow (CH_2)_{10}\,C(Me){=}C(Me)$$

(83A) **(83B)**

The conversion of a cyclic ketal into a lactone by *m*-chloroperbenzoic acid in the presence of BF_3 is thought to involve *syn*-elimination of a peracylal in a key step.[173] Piperidine reacts with (*Z*)-2-(2-bromovinyl)-5-nitrofuran by an addition–elimination rather than an elimination–addition mechanism.[174] A novel mode of base-induced reaction of *N*-alkoxypyridinium salts (**84**) (with a β-hydrogen atom), to give alkenes (**85**) competes with the classical aldehyde formation;[175] the formation of alkenes is particularly favoured when the pyridine nucleus is unsubstituted in the 2- and 6-positions.

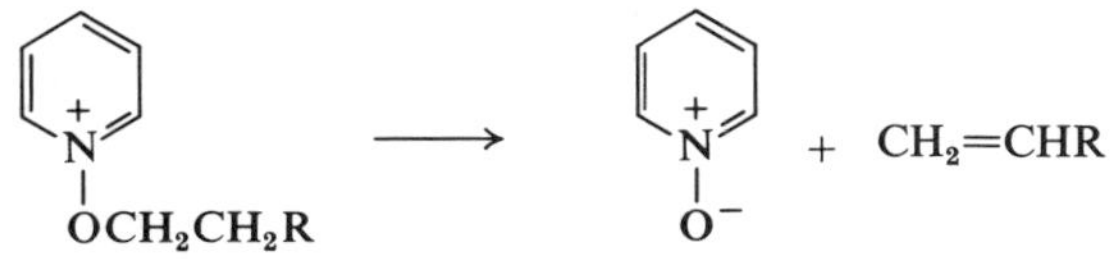

(84) → (85) + $CH_2{=}CHR$

MO calculations on elimination of MeOH from $MeOCH_2CH(CHO)COOH$ have been carried out[176] and hydrosulphurization of thiophene has been examined by Hückel theory.[177] Other topics reported include: kinetics of retroisomerization of retinol and retinyl acetate,[178] deamination of 4-amino-3,4-dihydro-2*H*-thiopyrans,[179] trapping of thiaziridinimines with *N*-sulfinyl-amines,[180] methanolysis of acetals,[181] formation of alkenes from dithiocarbamates *via* *S*-alkylation,[182] and fragmentation of α,β-epoxycyclopentanone.[183]

Dioxetanes and Dioxetanones

The chemiluminesence of the dioxetanone (**86**) is initiated by electron-transfer to an aromatic electron-acceptor followed by rapid loss of CO_2 from the reduced

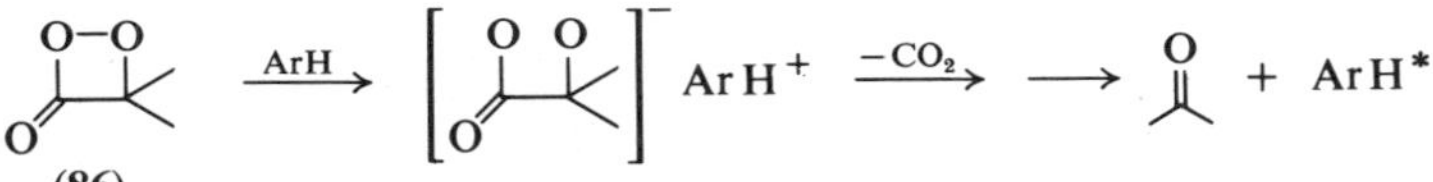

(86) —ArH→ [ ]$^-$ ArH$^+$ —$-CO_2$→ → + ArH*

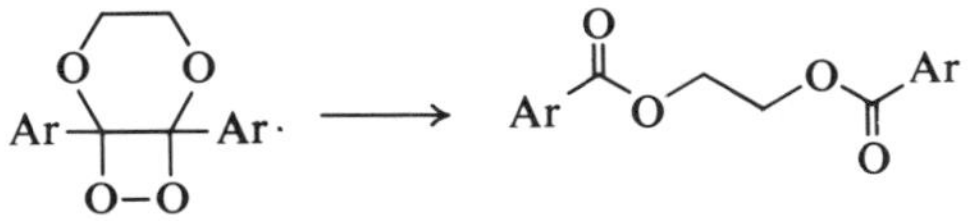

(87) → (88)

species.[184] Chemiluminescence also occurs during an uncatalysed thermal fragmentation.[185] A non-linear Hammett plot is observed for substituent effects on the decomposition of the 1,2-dioxetanes (**87**) to (**88**); the rate constants were obtained from chemiluminesence measurements.[186] The decarboxylative thermolysis of the *o*-xylylene peroxide (**89**) is chemiluminescent,[187] although thermolysis of *trans*-4,4-dimethyl-2,3,5-trioxabicyclo[4,4,0]decane (to acetone and adipaldehyde) could give chemiluminesence on energetic grounds, yet none was detected.[188] The chemistry of 1,2-dioxetanes has been reviewed,[189] and activation energies for the reaction $O_2^*(\Delta g) + R^1R^2C{=}CR^3R^4 \rightarrow R^1R^2CO + R^3R^4CO$ (which involves a dioxetane intermediate) have been measured.[190]

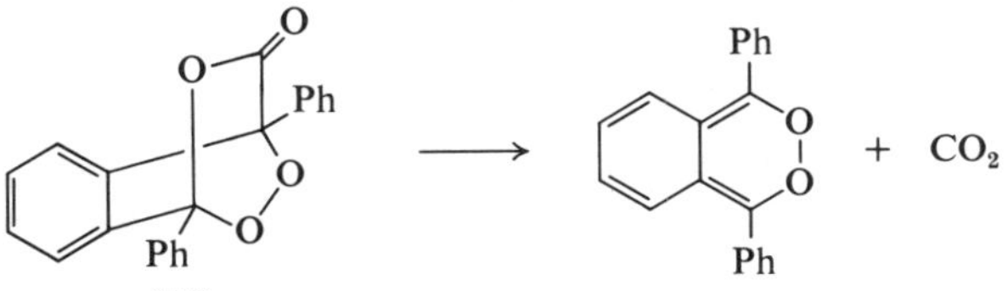

(89) → + CO_2

Heterogeneous Catalysis

The mechanism of eliminations catalysed by heterogeneous catalysts has been reviewed.[191]

Dehydration and isomerization of but-2-enols occur over UO_2 catalysts at 250–500°. The mechanism of the eliminations was studied by using specific ^{2}H-labelling; the *syn-E2*/*anti-E2* ratio is 67 : 33; on CeO_2 below 400° and on ThO_2 an

*E*1*cB* mechanism is operative, while on CeO_2 above 400° and on UO_2, the *E*1 and *E*2 mechanisms are operative, respectively.[192] It has been concluded that both but-2-ene and butan-2-one are formed from butan-2-ol over La_2O_3 and ThO_2 at ~440°, *via E*1*cB* reactions of a common carbanionic intermediate.[193]

Hofmann alkenes (but-1-enes) are formed upon dehydration of 3,3- and 2,3-dimethylbutan-2-ol over boron phosphate catalysts of high surface acidity; lower acidity of the catalyst tends to favour but-2-ene formation.[194] The asymmetric dehydration of racemic butane-1,3-diol gives >98% of MeCOEt in PhOH with chiral catalysts $RhCl_3$–(+)-(neomenthyldiphenylphosphine) but there is enrichment in the R-(−)-isomer of the unreacted diol, which increases with increasing conversion.[195] Other systems studied include: dehydration of alcohols on PbO_4, $Ca_3(PO_4)_2$ and SmO_3;[196] β-alkyl-substituted alcohols on basic oxides;[197] competing dehydration and dehydrogenation of alcohols.[198]

The stereochemistry of 1,3-dehalogenation of (R)-*meso*- and (*S*)-*meso*-2,4-dibromo-3-methylpentane (**90** and **92**) over Zn, Na, and chromous sulphate

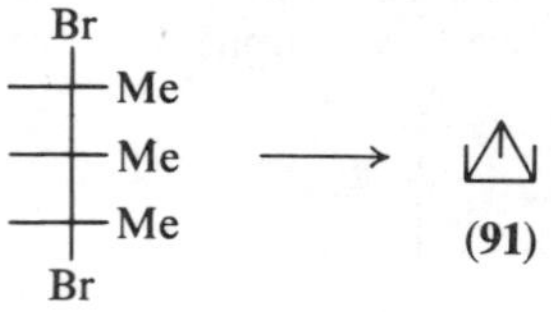

Br
Me
Me
Me
Br
(90) ⟶ **(91)**

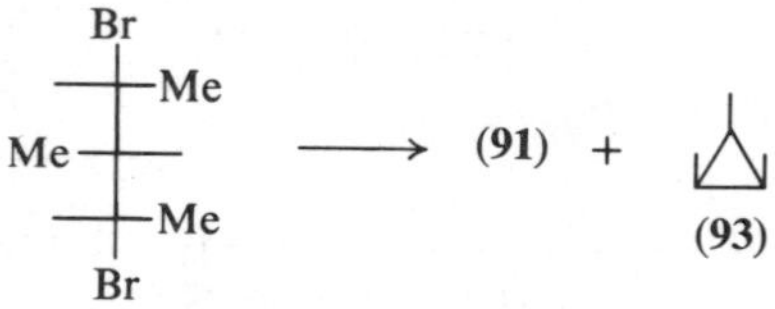

Br
Me
Me
Me
Br
(92) ⟶ **(91)** + **(93)**

catalysts has been studied;[199] the principal products are the *trans*- and *cis*-1,2,3-trimethylcyclopropanes (**91**) and (**93**). The fact that no *cis*-product (**93**) was obtained from (**90**) with any of the reducing agents indicated that at least one centre undergoes inversion on ring closure, while the formation of (**93**) and (**91**) from (**92**) means that the remaining centre undergoes either inversion (giving the *cis*-) or retention (giving the *trans*-product). The intermediate that does not preserve the stereochemistry of the second carbon may be of radical, carbanionic or organometallic character.

Electron Impact

The effects of substituents on the elimination of HCN from the aldehydic oximes *m*-$XC_6H_4CH{=}NOH$ under electron impact are negligible; two elimination routes (involving a 4- or 5-membered ring intermediate) have been suggested.[200] 1,4-Elimination of H_2O from *o*-substituted benzyl alcohols (in the mass spectrometer) occurs with either large ("late" transition state) or small ("early" transition state) kinetic energy release in the metastable species.[201] The three *C*-bromo-1,2,4-triazole tautomers, which can be generated by elimination of ethylene from the corresponding *N*-ethyl isomers, retain their structure prior to further fragmentation.[202] Deuterium-labelling has shown that under electron impact 1,2-elimination of HCl from 2-$(Cl(CH_2)_2)C_6H_4COOMe$ involves neighbouring-group participation by the ester function.[203] Ketene loss from o-$(NO_2)C_6H_4NHOAc$ gives a deuterium isotope effect of 2–2.5 when measured by negative ion mass spectrometry.[204] Ketene elimination from a quinone acetate molecular anion is thought to occur by

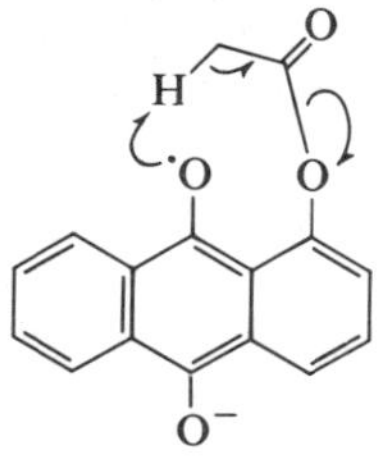

(94)

hydrogen atom abstraction (**94**); deuterium isotope effects (1.4–2.4) have been reported.[205] A reversible base-catalysed antiperiplanar elimination of methanol from 3-methoxycycloalkanones has been studied.[206]

References

1 Chiao, W. B., *Diss. Abstr. Int. B*, **38**, 683 (1977); *Chem. Abs.*, **88**, 88751 (1978).
2 Schroeder, G., Jarczewski, A., and Dorozalska, A., *Wiad. Chem.*, **31**, 293 (1977); *Chem. Abs.*, **87**, 166929 (1977).
3 Alunni, S., Baciocchi, E., Perucci, P., and Renzziconi, R., *J. Org. Chem.*, **43**, 2414 (1978).
4 Tremelling, M. J., Hopper, S. P., and Mendelowitz, P. C., *J. Org. Chem.*, **43**, 3076 (1978).
5 Whitesell, J. K., and Matthews, R. S., *J. Org. Chem.*, **42**, 3443 (1977).
6 Booth, H., Bostock, A. H., Franklin, N. C., Griffiths, D. V., and Little, J. H., *J. Chem. Soc., Perkin Trans.* 2, **1978**, 899.
7 Kirby, A. J., and Logan, C. J., *J. Chem. Soc., Perkin Trans.* 2, **1978**, 642.
8 Bremmer, J. B., and Winzenberg, K. N., *Aust. J. Chem.*, **31**, 313 (1978).
9 Minch, M. J., Chen, S.-S., and Peters, R., *J. Org. Chem.*, **43**, 31 (1978).
10 Pánková, M., and Závada, J., *Collect. Czech. Chem. Commun.*, **42**, 3439 (1977).
11 Závada, J., and Pánková, M., *Collect. Czech. Chem. Commun.*, **42**, 3421 (1977).
12 Závada, J., and Pánková, M., *Collect. Czech. Chem. Commun.*, **43**, 1080 (1978).
13 Chen, S., and Elofson, R. M., *Chem. Ind.* (London), **1978**, 64.
14 Kieboom, A., *Chem. Ind. (London)*, **1978**, 434.
15 Philips, J. C., and Hernandez, L. C., *Tetrahedron Lett.*, **1977**, 4461.
16 Ahlberg, P., and Thibblin, A., *J. Am. Chem. Soc.*, **99**, 7926 (1977).
17 Burton, W. B., Sims, L. B., and McLennan, D. J., *J. Chem. Soc., Perkin Trans.* 2, **1977**, 1847.
18 Fouad, F. M., and Farrell, P. G., *Tetrahedron Lett.*, **1978**, 4735.
19 Kaldor, S. B., and Saunders, W. H., *J. Chem. Phys.*, **68**, 2509 (1978); *Chem. Abs.*, **89**, 41860 (1978).
20 Hilbert, J. M., and Fedor, L., *J. Org. Chem.*, **43**, 452 (1978).
21 Marshall, D. R., Thomas, P. J., and Stirling, C. J. M., *J. Chem. Soc., Perkin Trans.* 2, **1977**, 1898.
22 Thomas, P. J., and Stirling, C. J. M., *J. Chem. Soc., Perkin Trans.* 2, **1977**, 1909.
23 Marshall, D. R., Thomas, P. J., and Stirling, C. J. M., *J. Chem. Soc., Perkin Trans.* 2, **1977**, 1914.
24 Barlow, K. N., Marshall, D. R., and Stirling, C. J. M., *J. Chem. Soc., Perkin Trans.* 2, **1977**, 1920.
25 Redman, R., Thomas, P., and Stirling, C. J. M., *J. Chem. Soc., Chem. Commun.*, **1978**, 43.
26 Stirling, C. J., and Thomas, P. J., *J. Chem. Soc., Chem. Commun.*, **1978**, 975.
27 More O'Ferrall, R. A., and Warren, P. J., *Proc. R. Irish Acad., Sect. B*, **77B**, 513 (1977); *Chem. Abs.*, **89**, 146042 (1978).
28 Rao, G. V., Balakrishnan, M., Venkatasubramanian, N., Subramanian, P. V., and Subramanian, V., *J. Chem. Soc., Perkin Trans.* 2, **1978**, 8.
29 Lipton, M. F., and Shapiro, R. H., *J. Org. Chem.*, **43**, 1409 (1978).
30 Kolonko, K. J., and Shapiro, R. H., *J. Org. Chem.*, **43**, 1404 (1978).

[31] Sen, A. K., Palit, S. K., and Mezumder, S., *Indian J. Chem.*, *Sect. B*, **15B** (7), 649 (1977); *Chem. Abs.*, **88**, 21627 (1978).
[32] Taylor, R., *J. Chem. Soc.*, *Perkin Trans.* 2, **1978**, 755.
[33] Amin, H. B., and Taylor, R., *J. Chem. Soc.*, *Perkin Trans.* 2, **1978**, 1095.
[34] Amin, H. B., and Taylor, R., *J. Chem. Soc.*, *Perkin Trans.* 2, **1978**, 1053.
[35] Amin, H. B., and Taylor, R., *Tetrahedron Lett.*, **1978**, 267.
[36] Chuchani, G., Hernandez, J. A., Yepez, M., and Alonso, M. E., *Int. J. Chem. Kinet.*, **9**, 811 (1977).
[37] Chuchani, G., and Fraile, G., *React. Kinet. Catal. Lett.*, **8**, 13 (1978), *Chem. Abs.*, **88**, 189743 (1978).
[38] Chuchani, G., Martin, I., Fraile, G., Lingstuyl, O., and Diaz, M. J., *Int. J. Chem. Kinet.*, **10**, 893 (1978).
[39] Fărcasiu, D., Slutsky, J., Schleyer, P. V. R., Overton, K. H., Luk, K., and Stothers, J. B., *Tetrahedron*, **33**, 3269 (1977).
[40] Bailey, W. J., and Bird, C. N., *J. Org. Chem.*, **42**, 3895 (1977).
[41] Krabbenhoft, H. O., *J. Org. Chem.*, **43**, 1305 (1978).
[42] Adam, W., Cueto, O., Guedes, L. N., and Rodriguez, L. O., *J. Org. Chem.*, **43**, 1466 (1978).
[43] Blackman, G. L., Brown, R. D., Brown, R. F. C., Eastwood, F. W., McMullen, G. L., and Robertson, M. L., *Aust. J. Chem.*, **31**, 209 (1978).
[44] Batt, L., Islam, T. S. A., and Rattray, G. N., *Int. J. Chem. Kinet.*, **10**, 931 (1978).
[45] Danen, W. C., Munslow, W. D., and Setser, D. W., *J. Am. Chem. Soc.*, **99**, 6961 (1977).
[46] Amin, H. B., and Taylor, R., *J. Chem. Soc.*, *Perkin Trans.* 2, **1978**, 1090.
[47] Pope, B. M., Sheu, S.-J., Stanley, R. L., Tarbell, D. S., and Yamamoto, Y., *J. Org. Chem.*, **43**, 2410 (1978).
[48] Borden, W. T., and Hoo, L. H., *J. Am. Chem. Soc.*, **100**, 6274 (1978).
[49] Thorne, M. P., *J. Chem. Res.*, (*S*) **1978**, 222.
[50] Thorne, M. P., *J. Chem. Soc.*, *Perkin Trans.* 2, **1978**, 716.
[51] Shah, S. C., and Smid, J., *J. Am. Chem. Soc.*, **100**, 1426 (1978).
[52] Dewar, M. J. S., and Ford, G. P., *J. Am. Chem. Soc.*, **99**, 8343 (1977).
[53] Dewar, M. J. S., Olivella, S., and Rzepa, H. S., *J. Am. Chem. Soc.*, **100**, 5650 (1978).
[54] Crawford, R. J., Lutemer, S., and Tokunaga, H., *Can. J. Chem.*, **55**, 3951 (1977).
[55] Battiste, M. A., and Visnick, M., *Tetrahedron Lett.*, **1978**, 4771.
[56] Kasai, M., Oda, M., and Kitahara, Y., *Chem. Lett.*, **1978**, 217.
[57] Dondoni, A., Barbaro, G., and Battaglia, A., *J. Org. Chem.*, **42**, 3372 (1977).
[58] Ganoi, Y., *Tetrahedron Lett.*, **1978**, 3277.
[59] Chen, T. B. R. A., Burger, J. J., and de Waard, E. R., *Tetrahedron Lett.*, **1977**, 4527.
[60] White, D. K., and Greene, F. D., *J. Am. Chem. Soc.*, **100**, 6760 (1978).
[61] Engel, P. S., Hayes, R. A., Keifer, L., Szilagyi, S., and Timberlake, J. W., *J. Am. Chem. Soc.*, **100**, 1876 (1978).
[62] Snyder, J. P., and Olsen, H., *J. Am. Chem. Soc.*, **100**, 2566 (1978).
[63] Oth, J. F., Olsen, H., and Snyder, J. P., *J. Am. Chem. Soc.*, **99**, 8505 (1977).
[64] Michejda, C. J., Koepke, S. R., and Campbell, D. H., *J. Am. Chem. Soc.*, **100**, 5979 (1978).
[65] Yeung, D. W. K., MacAlpine, G. A., and Warkentin, J., *J. Am. Chem. Soc.*, **100**, 1962 (1978).
[66] Cunico, R. F., and Lee, H. M., *J. Am. Chem. Soc.*, **99**, 7613 (1977).
[67] Winnewisser, M., Vogt, J., Ahlbrecht, H., *J. Chem. Res.* (*S*) **1978**, 298.
[68] Bushby, R. J., and Pollard, M. D., *Tetrahedron Lett.*, **1978**, 3859.
[69] Begley, M., Dean, F. M., Houghton, L. E., Johnson, R. S., and Park, B. K., *J. Chem. Soc.*, *Chem. Commun.*, **1978**, 461.
[70] Nagai, W., and Hirata, Y., *J. Org. Chem.*, **43**, 626 (1978).
[71] Friedrich, L. E., de Vera, N. L., and Lam, Y.-K. S., *J. Org. Chem.*, **43**, 34 (1978).
[72] Turro, N. J., Liu, K. C., Cherry, W., Liu, J. M., and Jacobson, B., *Tetrahedron Lett.*, **1978**, 555.
[73] Chiao, W.-B., and Saunders, W. H., *J. Am. Chem. Soc.*, **100**, 2802 (1978).
[74] Kwart, H., George, T. J., Louw, R., and Ultee, W., *J. Am. Chem. Soc.*, **100**, 3927 (1978).
[75] Christie, J. R., *Mol. Rate Processes*, *Symp.*, **1975**, B5; *Chem. Abs.*, **88**, 88935 (1978).
[76] Taylor, R., *J. Chem. Soc.*, *Chem. Commun.*, **1978**, 732.
[77] Muškatirović, M. D., Jovanović, B. Ž., Bončić-Caričič, G. A., and Tadic, Ž. D., *J. Chem. Soc.*, *Perkin Trans.* 2, **1978**, 948.

[78] Jurcu, S., Cirstea, M., Chivulescu, E., and Dragusa, E., *Rev. Chim. (Bucharest)* **29**, 109 (1978), *Chem. Abs.*, **89**, 107554 (1978).
[79] Boyd, D. R., Neill, D. C., and Stubbs, M. E., *J. Chem. Soc., Perkin Trans.* 2, **1978**, 30.
[80] Bailey, W. J., and Cesare, F., *J. Org. Chem.*, **43**, 1421 (1978).
[81] Rosenfeld, R. N., Brauman, J. I., Barker, J. R., and Golden, D. M., *J. Am. Chem. Soc.*, **99**, 8063 (1977).
[82] Frey, H. M., and Smith, R. A., *J. Chem. Soc., Perkin Trans.* 2, **1977**, 2082.
[83] Hiatt, R., and Rahimi, P. M., *Int. J. Chem. Kinet.*, **10**, 185 (1978).
[84] Viola, A., Dudding, G. F., and Proverb, R. J., *J. Am. Chem. Soc.*, **99**, 7390 (1977).
[85] Bailey, W. J., and Di Pietro, J., *J. Org. Chem.*, **42**, 3899 (1977).
[86] Ahonkhai, S. I., and Emovon, E. U., *J. Chem. Res. (S)* **1978**, 114.
[87] Von Rohrscheidt, C., and Fritz, H., *Justus Liebigs Ann. Chem.*, **1978**, 680.
[88] Davis, F. A., Friedman, A. J., and Nadir, U. K., *J. Am. Chem. Soc.*, **100**, 2844 (1978).
[89] Masilamani, D., and Rogic, M. M., *J. Am. Chem. Soc.*, **100**, 4634 (1978).
[90] Isobe, M., Iio, H., Kitamura, M., and Goto, T., *Chem. Lett.*, **1978**, 541.
[91] Reich, H. J., Peake, S. L., *J. Am. Chem. Soc.*, **100**, 4888 (1978).
[92] Satterthwait, A. C., and Westheimer, F. H., *J. Am. Chem. Soc.*, **100**, 3197 (1978).
[93] Simmie, J. M., *Int. J. Chem. Kinet.*, **10**, 227 (1978).
[94] Yano, T., *Int. J. Chem. Kinet.*, **9**, 725 (1977).
[95] King, K. D., *J. Chem. Soc., Faraday Trans.* 1, **74**, 912 (1978).
[96] Gutman, D., Breuen, W., and Tsang, W., *J. Chem. Phys.*, **67**, 4291 (1977); *Chem. Abs.*, **88**, 50041 (1978).
[97] Evans, P. J., Ichimura, T., and Tschuikowroux, E., *Int. J. Chem. Kinet.*, **10**, 855 (1978).
[98] Russell, P. R., and Strachan, A. N., *J. Chem. Soc., Perkin Trans.* 2, **1978**, 323.
[99] Hernandez, J. A., and Chuchani, G., *Int. J. Chem. Kinet.*, **10**, 923 (1978).
[100] Chuchani, G., Martin, I., and Bigley, D. B., *Int. J. Chem. Kinet.*, **10**, 649 (1978).
[101] Chuchani, G., Martin, I., and Alonso, M. E., *Int. J. Chem. Kinet.*, **9**, 819 (1977).
[102] Robinson, P. J., Skelhorne, G. G., and Waller, M. J., *J. Chem. Soc., Perkin Trans.* 2, **1978**, 349.
[103] King, K. D., and Goodard, R. D., *Mol. Rate Processes, Peptide Symp.*, **1975**, E1; *Chem. Abs.*, **88**, 104439 (1978).
[104] Klyaulin, M. S., and Rakhmankulov, D. L., *Neftekhim. Sint. Tekh. Prog.*, **28** (1976); *Chem. Abs.*, **88**, 189740 (1978).
[105] Tominega, H., Yahagi, N., and Ishiwateri, M., *J. Fac. Eng., Univ. Tokyo, Ser. A*, **1977**, 68; *Chem. Abs.*, **89**, 128862 (1978).
[106] Hanessien, S., Bergiotti, A., and Lakua, M., *Tetrahedron Lett.*, **1978**, 737.
[107] Hara, Y., Kawano, F., and Osado, H., *Kogyo Kayaku*, **38**, 266 (1977); *Chem. Abs.*, **89**, 59506 (1978).
[108] Matueev, V. G., Dubikhin, V. V., and Nazin, G. M., *Izv. Akad. Nauk SSSR, Ser. Khim.*, **2**, 474 (1978); *Chem. Abs.*, **88**, 151803 (1978).
[109] Block, B., *Tetrahedron Lett.*, **1978**, 1071.
[110] Bayer, O., *Methoden Org. Chem.*, **7**, 2113 (1977); *Chem. Abs.*, **88**, 135741 (1978).
[111] Aleskerov, M. A., Yufit, S. S., and Kucherov, V. F., *Ush. Khim.*, **47**, 235 (1978); *Chem. Abs.*, **88**, 151598 (1978).
[112] Fanega, S. M., *Agham*, **1975**, 3, 203; *Chem. Abs.*, **89**, 23238 (1978).
[113] Reetz, M. T., *Nachr. Chem. Tech. Lab.*, **25**, 594 (1977); *Chem. Abs.*, **88**, 36711 (1978).
[114] Reich, H. J., Wollowitz, S., Trend, J. E., Chow, F., and Wendelborn, D. F., *J. Org. Chem.*, **43**, 1697 (1978).
[115] Laber, D., Hevesi, L., Dumont, W., and Krief, A., *Tetrahedron Lett.*, **1978**, 1141.
[116] Dumont, W., Sevrin, M., and Krief, A., *Tetrahedron Lett.*, **1978**, 183.
[117] Ishihara, H., and Hirabayashi, Y., *Chem. Lett.*, **1978**, 1007.
[118] Ager, D. J., and Fleming, I., *J. Chem. Soc., Chem. Commun.*, **1978**, 177.
[119] Fleming, I., and Goldhill, J., *J. Chem. Soc., Chem. Commun.*, **1978**, 176.
[120] Chan, T.-H., *Acc. Chem. Res.*, **10**, 442 (1977).
[121] Reich, H. J., Reich, I. L., and Wollowitz, S., *J. Am. Chem. Soc.*, **100**, 5981 (1978).
[122] Midland, M. M., Tramontano, A., and Zderic, S. A., *J. Organomet. Chem.*, **156**, 203 (1978).
[123] Ono, N., Tamura, R., Hyami, J., and Kaji, A., *Tetrahedron Lett.*, **1978**, 763.
[124] Braterman, P. S., Cross, R. J., and Young, G. B., *J. Chem. Soc., Dalton Trans.*, **1977**, 1892.
[125] Arai, I., and Doyle, D. G., *J. Am. Chem. Soc.*, **100**, 287 (1978).

[126] Mukaiyama, T., *Chem. Lett.*, **1978**, 413.
[127] Tsuji, J., Yamakawa, T., Kaito, M., and Mandai, T., *Tetrahedron Lett.*, **1978**, 2075.
[128] Bruice, P. Y., and Bruice, T. C., *J. Am. Chem. Soc.*, **100**, 4793 (1978).
[129] Bruice, P. Y., and Bruice, T. C., *J. Am. Chem. Soc.*, **100**, 4802 (1978).
[130] Bolognese, A., and Scherillo, G., *J. Org. Chem.*, **42**, 3867 (1977).
[131] Roberts, R. D., and Spencer, T. A., *Tetrahedron Lett.*, **1978**, 2557.
[132] Beltrame, P., Dondoni, A., Burbaro, G., Gelli, G., Loi, A., and Steffé, S., *J. Chem. Soc., Perkin Trans.* 2, **1978**, 607.
[133] Zverev, V. U., Saifullin, I., and Shernin, G. P., *Izv. Akad Nauk SSSR, Ser. Khim.*, **2**, 313, (1978); *Chem. Abs.*, **88**, 169377 (1978).
[134] Muad, G., and Benkovic, S. J., *J. Am. Chem. Soc.*, **100**, 5495 (1978).
[135] Riikoja, J., Timotheus, H., and Palm, V., *Org. React.* (*Tartu*), **14**, 507 (1977).
[136] Sridharan, S., and Vitullo, V. P., *J. Am. Chem. Soc.*, **99**, 8093 (1977).
[137] Reznikov, V. M., and Chirich, L. V., Deposited Doc., 1976 VINITI 2132–76, *Chem. Abs.*, **89**, 107336 (1978).
[138] Saito, S., Yaluki, T., Moriwake, T., and Okamoto, K., *Bull. Chem. Soc. Jpn.*, **51**, 529 (1978).
[139] Saito, S., Moriwake, T., Takeuchi, K., and Okamoto, K., *Bull. Chem. Soc. Jpn.*, **51**, 2634 (1978).
[140] Lomas, J. S., and Dubois, J. E., *Tetrahedron*, **34**, 1597 (1978).
[141] Jensen, B. L., and Peterson, P. E., *J. Org. Chem.*, **42**, 4052 (1977).
[142] Palmer, R. J., and Stirling, C. J. M., *J. Chem. Soc., Chem. Commun.*, **1978**, 338.
[143] Novokreschennykh, V. D., Mochalov, S. S., and Shabarov, Yu. S., *Zh. Org. Khim.*, **14**, 546 (1978); *Chem. Abs.*, **89**, 41845 (1978).
[144] Arct, J., Migaj, B., and Zych, J., *Bull. Acad. Pol. Sci., Ser. Sci. Chim.*, **25**, 697 (1977).
[145] Aksenov, V. S., and Terent'eva, G. A., *Izv. Akad. Nauk SSSR, Ser. Khim.*, **6**, 1344 (1978); *Chem. Abs.*, **89**, 107324 (1978).
[146] Hassan, M., Nour, A. R. O. A., and Satti, A. M., *Rev. Roum. Chim.*, **23**, 747 (1978).
[147] Mukhamedova, L. A., Kursheva, L. I., and Anoshina, W. P., *Khim. Geterotsikl. Soedin.*, **1**, 31 (1978); *Chem. Abs.*, **88**, 169261 (1978).
[148] McWilliam, D. C., Balasubramanian, T. R., and Kuivila, H. G., *J. Am. Chem. Soc.*, **100**, 6407 (1978).
[149] Boothe, T. E., Greene, J. L., Shevlin, P. B., Willcott, III, M. R., Inners, R. R., and Cornelis, A., *J. Am. Chem. Soc.*, **100**, 3874 (1978).
[150] Bedford, K. R., and de la Mare, P. B. D., *J. Chem. Soc., Perkin Trans.* 2, **1978**, 932.
[151] Murphy, W. S., and Cocker, D., *J. Chem. Soc. Perkin Trans.* 1, **1977**, 2565.
[152] Hanson, J. R., Johnson, A. W., and Kaplen, M. A., *J. Chem. Soc., Perkin Trans.* 1, **1978**, 263.
[153] Sedgwick, B., Morris, C., and French, S. J., *J. Chem. Soc., Chem. Commun.*, **1978**, 193.
[154] Soper, T. S., and Manning, J. M., *Biochemistry*, **17**, 3377 (1978).
[155] Krouwer, J. S., Schultz, R. M., and Babior, B. M., *J. Biol. Chem.*, **253**, 1041 (1978).
[156] Hill, R. K., and Bock, M. G., *J. Am. Chem. Soc.*, **100**, 637 (1978).
[157] Karube, Y., and Matsushima, Y., *J. Am. Chem. Soc.*, **99**, 7356 (1977).
[158] Pal, B. C., *J. Am. Chem. Soc.*, **100**, 5170 (1978).
[159] Radulescu, N., and Sarbu, C., *Rev. Roum. Chim.*, **23**, 197 (1978).
[160] Shishinova, V. I., Levanova, V. A., Shevtsova, L. A., Rozhnov, A. M., and Tokareuskii, V. A., *Zr. Fiz. Khim.*, **52**, 497 (1978); *Chem. Abs.*, **88**, 169289 (1978).
[161] R. Verhé, De Kimpe, N., De Buyck, L., Tilley, M., and Schamp, N., *Bull. Soc. Chim. Belg.*, **86**, 879 (1977).
[162] Oncescu, T., and Dimofte, L., *Rev. Chim* (*Bucharest*), **28**, 1139 (1977); *Chem. Abs.*, **88**, 104341 (1978).
[163] Itoh, K., *Nippon Kezaku Kaishi*, **9**, 1359 (1977); *Chem. Abs.*, **88**, 104291 (1978).
[164] Theobald, D. W., *Tetrahedron*, **34**, 1567 (1978).
[165] Bégué, J.-P., Pardo, C., and Sansoulet, J., *J. Chem. Res.*, (*S*) **1978**, 52.
[166] Kovařik, P., Valka, L., and Gatial, A., *Collect. Czech. Chem. Commun.*, **42**, 3265 (1977).
[167] Rosnati, V., and Salimbani, A., *Gazz. Chim. Ital.*, **107**, 271 (1977); *Chem. Abs.*, **88**, 104306 (1978).
[168] Barrett, A., Barton, D., Bielski, R., and McCombie, S., *J. Chem. Soc., Chem. Commun.*, **1977**, 866.
[169] Glazunova, E. M., Grigina, I. W., Marupov, R. M., Samiev, M., and Fedulov, V. P., *Tezisy Dokl.-Vses. Konf. Khim. Atsetilana 5th*; *Chem. Abs.*, **88**, 169471 (1978).

[170] Yufit, S. S., and Esikova, I. A., *React. Kinet. Catal. Lett.*, **4**, 121 (1976); *Chem. Abs.*, **88** 21592 (1978).
[171] Maercker, A., and Demuth, W., *Justus Liebigs Ann. Chem.*, **1977**, 1909.
[172] Marshall, J. A., and Karas, L. J., *J. Am. Chem. Soc.*, **100**, 3615 (1978).
[173] Grieco, P. A., Oguri, J., and Yokoyama, Y., *Tetrahedron Lett.*, **1978**, 419.
[174] Litvinenko, L. M., Popov, A. F., Kostenko, L. I., Krauchenko, U. V., and Vegh, D., *Dokl. Akad. Nauk SSSR*, **238**, 622 (1978); *Chem. Abs.*, **88**, 120267 (1978).
[175] Sbiva, H., and Tarter, A., *Tetrahedron*, **33**, 3111 (1977).
[176] Faustov, V. I., Fundyler, I. N., Esikova, I. A., and Yufit, S. S., *Izv. Akad. Nauk SSSR, Ser. Khim.*, **12**, 2694 (1977); *Chem. Abs.*, **88**, 104491 (1978).
[177] Duben, A. J., *J. Phys. Chem.*, **82**, 348 (1978).
[178] Pekkarinen, L., *Finn. Chem. Lett.*, **1977**, 261.
[179] Pradere, J.-P., and Hadjukovic, G., *C. R. Hebd. Séances Acad. Sci. Sér. C*, **286**, 553 (1978).
[180] L'abbé, G., Van Asch, A., Declercq, J. P., Germain, G., and Van Meerssche, M., *Bull. Soc. Chim. Belg.*, **87**, 285 (1978).
[181] Zakoshanskii, V. M., Blazhin, Y. M., Idlis, G. S., and Ogorodnikov, S. K., *Zh. Org. Khim.*, **14**, 1359 (1978); *Chem. Abs.*, **89**, 128871 (1978).
[182] Hayashi, T., Sakurai, A., and Oishi, T., *Chem. Lett.*, **1977**, 1483.
[183] Tsuzuki, K., Hashimoto, H., Shirahama, H., and Matsumoto, T., *Chem. Lett.*, **1977**, 1469.
[184] Schmidt, S. P., and Schuster, G. B., *J. Am. Chem. Soc.*, **100**, 1966 (1978).
[185] Schmidt, S. P., and Schuster, G. B., *J. Am. Chem. Soc.*, **100**, 5559 (1978).
[186] Zaklika, K. A., Thayer, A. L., and Schaap, A. P., *J. Am. Chem. Soc.*, **100**, 4916 (1978).
[187] Smith, J. P., and Schuster, G. B., *J. Am. Chem. Soc.*, **100**, 2564 (1978).
[188] Schuster, G. B., and Bryant, L. A., *Gov. Rep. Announce.* Index (U.S.), **77**, 101 (1977); *Chem. Abs.*, **87**, 183766 (1977).
[189] Horn, K. A., Schmidt, S. P., Koo, J., and Schuster, G. B., U.S.NTIS, AD Rep. 1978, AD–A053290 from *Gov. Rep. Announce.* Index (U.S.) 1978, **78** (15), 108; *Chem. Abs.*, **89**, 162587 (1978).
[190] Alben, K. T., Auerbach, A., Ollison, W. M., Weiner, J., and Cross, Jr., R. J., *J. Am. Chem. Soc.* **100**, 3274 (1978).
[191] Kraus, M., *Chem. Listy*, **72**, 449 (1978); *Chem. Abs.*, **89**, 75045 (1978).
[192] Thomke, K., *Proc. Int. Congr. Catal. 6th 1976* (1977), 303; *Chem. Abs.*, **88**, 5821 (1978).
[193] Thomke, K., *Z. Phys. Chem.* (*Frankfurt am Main*), **106**, 225 (1977).
[194] Jewur, S. S., and Moffat, J. B., *J. Chem. Soc., Chem. Commun.*, **1978**, 801.
[195] Ohkulo, K., Hori, K., and Yoshinaga, K., *Inorg. Nucl. Chem. Lett.*, **13**, 637 (1977); *Chem. Abs.*, **89**, 23602 (1978).
[196] Thomke, K., *Z. Phys. Chem.* (*Wiesbaden*), **106**, 295 (1977).
[197] Thomke, K., *Z. Phys. Chem.* (*Wiesbaden*), **107**, 99 (1977).
[198] Tomke, K., *Z. Phys. Chem.* (*Wiesbaden*), **106**, 225 (1977).
[199] Applequist, D. E., and Pfohl, W. F., *J. Org. Chem.*, **43**, 867 (1978).
[200] Vijfhuizen P. C., and Dijkstra, G., *Org. Mass Spectrom.*, **12**, 241 (1977); *Chem. Abs.*, **87**, 200295 (1977).
[201] Gross, M. L., De Roos, F. L., and Hoffman, M. K., *Org. Mass Spectrom.*, **12**, 258 (1977); *Chem. Abs.*, **87**, 200298 (1977).
[202] Mequestiau, A., Van Haverbeke, Y., Flammeng, R., and Mispreuve, H., *Org. Mass Spectrom.*, **12**, 205 (1977; *Chem. Abs.*, **87**, 200289 (1977).
[203] Borchers, F., Heimback, H., Leusen, K., Morhann, H., and Schwarz, H., *Org. Mass Spectrom.*, **12**, 573 (1977); *Chem. Abs.*, **88**, 104218 (1978).
[204] Benbow, J. A., Wilson, J. C., and Bowie, J. H. *Int. J. Mass Spectrom. Ion Phys.*, **26**, 173 (1978); *Chem. Abs.*, **88**, 135830 (1978).
[205] Wilson, J. C., Benbow, J. A., Bowie, J. H., and Prager, R. H., *J. Chem. Soc., Perkin Trans.* 2, **1978**, 498.
[206] Hert, H. and Dunkelblum, E., *J. Am. Chem. Soc.*, **100**, 5141 (1978).

CHAPTER 12

Addition Reactions: Polar Addition

C. BROWN

Chemical Laboratory, University of Kent, Canterbury

Electrophilic Additions

A review of addition reactions has appeared.[1] Attempts have been made to identify suitable model reactions for electrophilic additions: for example, sulphenylation and hydration of olefins have been proposed[2] as models for distinguishing between electrophilic addition reactions proceeding *via* bridged and open carbonium-ion-like transition states, respectively. In particular, the large difference in range of reactivities of a series of cyclopropyl-substituted alkenes towards acid-catalysed hydration ($k_{max}/k_{min} \approx 10^{16}$) and towards addition of 4-chlorobenzenesulphenyl chloride ($k_{max}/k_{min} \approx 200$) has been pointed out.[3] The method has been applied to the bromination of styrene,[4] and it has been concluded that this reaction proceeds *via* a bridged rate-limiting transition state. The regio- and stereo-specificities of the addition of a variety of electrophilic reagents to 3-methyl- and 3-*tert*-butylcyclohexene have been reported, together with results of ring-opening reactions of the corresponding epoxides, taken as models for the second stage of the reaction.[5] *Ab initio* calculations have also been carried out[6] on model electrophilic addition reactions. Participation by 19-hydroxyl[7] and by 19-methoxyl[8] groups in the electrophilic reactions of some unsaturated steroids has been investigated.

Halogen and Related Addition

In continued work with xenon difluoride, Stavber and Zupan have shown[9] that BF_3-catalysed reaction of this reagent with 1-substituted pentafluorobenzenes gives >80% yields of the 1-substituted heptafluorocyclohexa-1,4-dienes. The same reagent gives vicinal difluorides on reaction with phenyl-substituted 5-, 6-, and 7-membered cycloalkenes; both *syn*- and *anti*-addition are observed,[10] the ratio depending on ring size. The regio- and stereo-specificity of cohalogenation of olefins in liquid HF has been noted[11] and the reaction of benzofuran and *N*-acylindoles with trifluoro(fluoro-oxy)methane (CF_3OF) has been shown to involve electrophilic fluorination of C(2) followed by nucleophilic capture of the C(3) cation so formed.[12]

Solvent effects on the rates of addition of chlorine to a variety of substituted alkenes have been reported;[13] π-complex intermediates were invoked, together with competing polar and molecular pathways, to explain the results. The chlorination of 3-substituted prop-1-enes has been studied,[14] and a Taft ρ^* value of -1.20 was found[15] for the chlorination of vinyl derivatives in the dark in the presence of oxygen. The factors affecting the ratio of electrophilic attack at the α- and β-face of cholest-5-enes by chlorine have been investigated.[16] Addition of chlorine to some acrylic acids and esters in acetic acid solution has been shown to be a second-order process with a highly polar transition state,[17] and anti-Markownikoff addition is observed[18] in the reaction of Cl_2, Br_2, ClBr, ClI, and BrI to the vinylphosphonate $CH_2{=}CHP(O)(OCHMe_2)_2$. The mechanism of the chlorination of 2-phenylnaphthalene has been studied and interpreted in terms of intermediate chlorodialins,[19] and the isolation of tetrachloro-isomers (**1**) from the reaction of the benzotriazole (**2**) with gaseous chlorine has been cited[20] as evidence for the quinonoidal nature of such compounds.

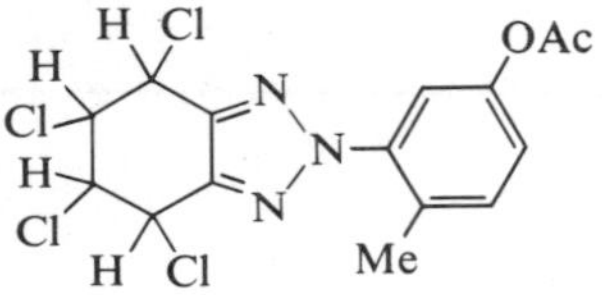

H Cl H Cl H Cl H Cl N N N OAc Me

(**1**)

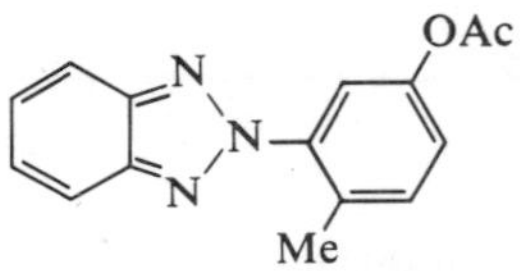

N N N OAc Me

(**2**)

Product studies indicate[21] that the addition of methyl hypochlorite to cyclopentadiene under ionic conditions involves the same intermediate carbonium ion as that involved in addition of chlorine, and the addition of *tert*-butyl hypochlorite to 2-alkoxy- and 2-phenoxy-3,4-dihydro-2*H*-pyrans in methanol has been shown to give mainly the 1,2- addition product.[22] *cis*-Chlorination of alkenes can be achieved in high yield[23] by treatment with tetrabutylammonium octamolybdate and acetyl chloride. A mechanism was proposed involving *cis*-insertion of the C=C double bond into a Mo—Cl bond of an (unspecified) polychlorinated Mo(VI) species, followed by reductive elimination.

The utility of quadratic linear free-energy relationships in analysing non-additive substituent effects in several reactions, including the bromination of diphenylethenes, has been critically examined.[24] A very careful study of the bromination of substituted styrenes at very low reactant concentrations in methanol–sodium bromide mixtures has been carried out; the dangers of over-simplified interpretation of kinetic parameters in attempts to delineate transition-state structure were pointed out.[25] *Ab initio* MO calculations give no clear-cut answer[26] to the problem of the shape of the energy profile for bromination of ethylene. However, despite recently expressed doubts about the applicability of bromonium-ion-like transition states in alkene bromination reactions, it seems likely (for alkyl-substituted alkenes at least) that this model best represents the situation.[27] Likewise, on the basis of substituent effects, it can be concluded that the bromination of styrenes also proceeds to a large extent *via* bromonium ions, especially when the aromatic ring carries electron-withdrawing substituents.[28]

A detailed study of the addition of bromine to a series of alkenes in acetic acid and in TCE has been reported.[29] In the less polar solvent, third-order kinetics and a large structure-dependence were observed, consistent with a bromine-assisted Ad_E2 mechanism involving a cyclic bromonium ion. Charge distribution in such bromonium-ion intermediates plays an important part in determining both chemoselectivity and regioselectivity in the bromination of substituted alkenes in methanol solution.[30] Bromine gives products of addition at the C(5)=C(6) and C(11)=C(12) double bonds of hexahelicene.[31] The addition of BrCl to hex-1-ene in CCl_4 appears to involve a symmetrically bridged bromonium ion.[32] As expected, the main products from reaction of 3-substituted cholest-5-enes with sources of electrophilic bromine are the 5α-bromo-6β-substituted adducts, as the result of electrophilic attack on the α-face,[33] while cholest-5-en-3-one gives products of attack by bromine chloride at both the α- and the β-face of the molecule.[34] Ionic addition of Cl_2, Br_2, CH_3OCl, CH_3OBr, or NBS to ethyl sorbate in methanol solution gave predominantly products of addition to the γ—δ-bond.[35] The kinetics of bromination of crotonic acid by NBS have been measured:[36] the reaction is first-order in crotonic acid and in hydrogen-ion concentration, and zero-order in NBS. Substituents in the *peri*-position of 9-allyltriptycenes suppress addition of bromine to the non-conjugated C=C bond at the expense of electrophilic substitution at the olefinic carbon atoms.[37] *trans*-Stereospecific methoxybromination of olefins has been carried out by using bromine in methanol in the presence of $AgNO_3$, $Pb(NO_3)_2$, or yellow lead oxide.[38] In an intriguing paper, it has been shown[39] that bromine preadsorbed on a 5A zeolite selectively adds to styrene in the presence of a two-fold excess of cyclohexene: the absence of any bromoacetate product when the reaction is carried out in the presence of acetic acid, the slowness of the reaction relative to homogeneous bromination, and the lack of selectivity of bromine preadsorbed on 13X sieve, all point to the reaction taking place within a sieve capacity, whose window dimension is critical. Electrophilic addition of Cl_2, Br_2, HBr, or HI to dispiro[2.0.2.2]oct-7-ene (**3**) gives a variety of products,[40] including in addition to adducts such as (**4**) (from Br_2) and (**5**) (from HBr), ring-expansion and ring-opened derivatives such as (**6**) (from HI) and (**7**) (from HBr), respectively.

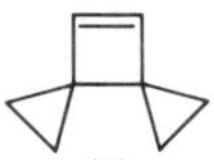

Br Br Br

(**3**) 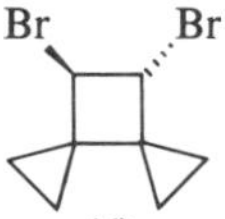 (**4**) 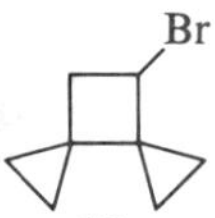 (**5**)

I CH_2CH_2Br

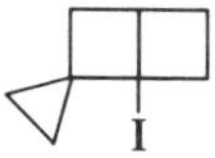

(**6**) 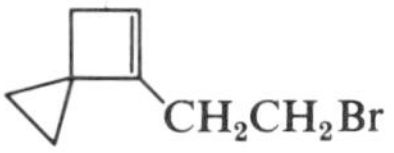 (**7**)

Both the stereo- and regio-chemistry of addition of the elements of Br_2 and BrCl to conjugated dienes can be controlled by suitable choice of halogenating agent.[41] Salt effects have been studied for the bromination of alk-1-en-3-ynes.[42]

X-ray crystallographic measurements have been used[43] to determine the structures of the products (**9**) and (**10**) from addition of bromine and of iodine to the bicyclic olefin (**8**). These products emphasize the importance of π-participation by the second double bond in such reactions.

Neighbouring-group participation by both acetoxy- and hydroxy-groups has been observed in the addition of HOBr to some unsaturated steroids,[44] and

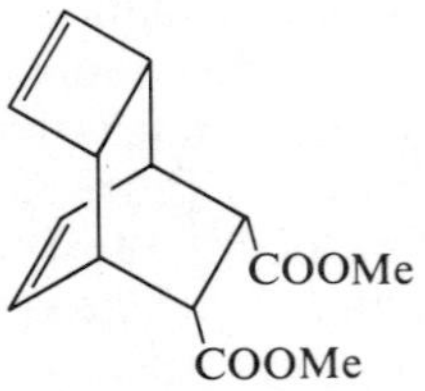

(8) 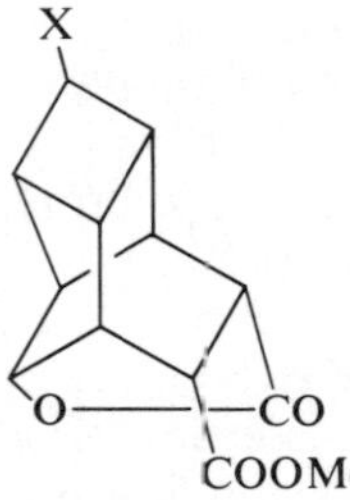 (9) 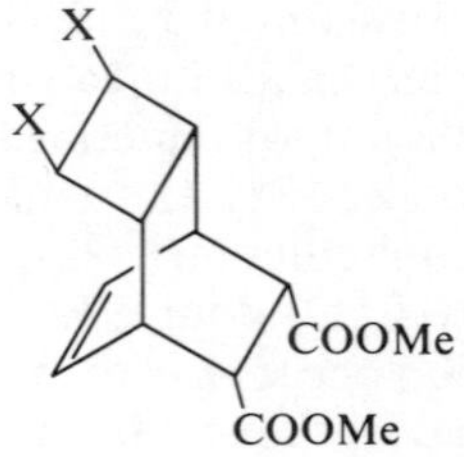 (10)

X = Br or I

participation by an allylic-OH group has been suggested to account for the unusual stereochemistry of the dibromides obtained from addition of bromine to 1α- and 1β-hydroxy-2,3-dehydro-5α-santanolide.[45] Evidence has been obtained for both double-bond and hydroxyl-group transannular interactions in the reaction of (*Z*,*Z*)-cyclonona-1,5-diene with NBS and water or methanol.[46,47] (Scheme 1).

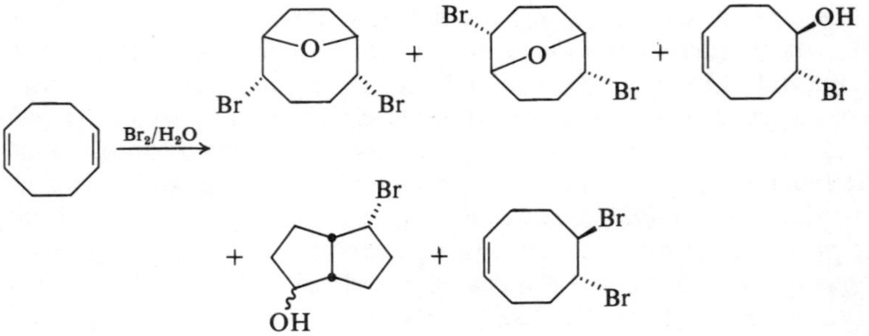

Br_2/H_2O

SCHEME 1

Ethyl phenylpropiolate undergoes reaction with bromine both by a bimolecular process and by a bromide-ion-catalysed termolecular one, as evidenced by the observed rate law:

$$-\mathrm{d}[Br_2]/\mathrm{d}t = k_2[\text{acetylene}][Br_2]+k_3[\text{acetylene}][Br_2][Br-]$$

However, phenylpropiolate and its anion react by the bimolecular pathway only ($k_3 \sim 0$). These results, taken together with the product compositions observed, were interpreted in terms of an intermediate vinyl carbonium ion, which in the ester is stabilized by bromide-ion participation, but which in the acid or anion may be stabilized by intramolecular carboxyl participation.[48] An intermediate vinyl-carbonium ion was also invoked to account for the effect of added LiBr on the nature and rate of formation of the products derived from bromination of phenyl-acetylene in methanol.[49] The rates of bromination of a wide range of alkynes in a variety of solvents have been reported: substituent effects in phenyl- and diphenyl-acetylenes indicate a highly asymmetrically charged transition state;[50] a weakly bridged bromonium ion, rather than an open vinyl cation, best explains the nature of the products of addition of BrCl to hex-1-yne.[32] Evidence for phenyl participation to give the ion (**11**) (R = 4-OMe) in the addition of bromine to 3-arylpropynes has been presented.[51] With less electron-releasing substituents, (H, 4-Me, 3-CF_3), an unsaturated bromonium ion (**12**) is preferred.

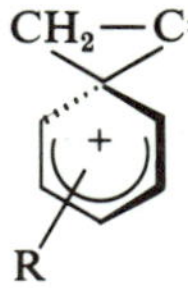

$CH_2-C=CHBr$ (ring with + ; R)

(11)

$RC_6H_4CH_2-\overset{\overset{+}{Br}}{C}=CH$

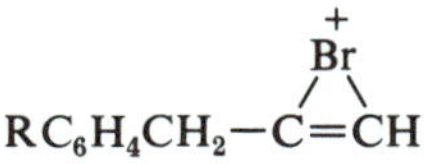

(12)

Cyclic iodonium ions were postulated as intermediates in the iodination of pyruvic acid enol phosphate.[52] The addition of iodine to the C=C double bond in enol acetates in the presence of thallium(I) acetate gives the 2-iodo-ketone, presumably *via* deacylation of the intermediate iodonium ion by acetate ion. In the case of the enol acetate of cyclohexanone, the iodo-*gem*-diacetate was also formed.[53] The iodoacetoxylation of alkenes with iodine and peracetic acid in acetic acid follows the rate law:

$$\text{rate} = k_1[I_2][MeCO_3H] + k_2[\text{alkene}][I_2][MeCO_3H]$$

Only electron-rich alkenes react, and a mechanism involving two separate pathways was proposed.[54] *trans*-Opening of asymmetric iodonium ions by azide ion, with kinetic rather than thermodynamic control, best explains the product distribution in the addition of iodine(I) azide to alkenes.[55]

The use of formic acid as a solvent in sulphenyl halide addition reactions has been advocated.[56] The rates of addition of MeSCl and of PhSCl to cinnamic acid, its methyl ester, and *N*-phenylamide have been measured. The reaction is both regiospecific [giving PhCHClCH(SR)COX] and stereospecifically *trans*.[57] Compelling evidence has been presented[58] to support the claim that the addition of arenesulphenyl chlorides to polyisoprene occurs by addition to successive double bonds as the result of activation by neighbouring sulphur participation. Similar participation, especially by sulphur attached to electron-rich benzene rings is invoked to account for the ready loss of HCl from the modified polymers (Scheme 2).

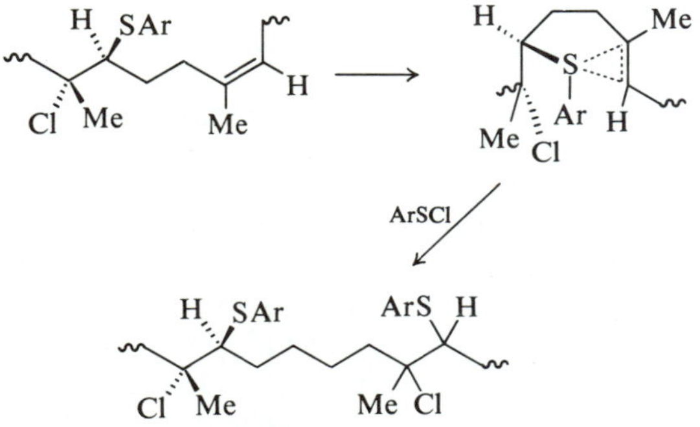

H SAr, Cl Me, Me, H → H, Me, S, Ar, H, Me Cl

ArSCl

H SAr, ArS H, Cl Me, Me Cl

SCHEME 2

Addition of benzenesulphenyl chloride to α,β-unsaturated ethers and sulphides is completely regiospecific (Markownikoff addition) but not stereospecific: rate-limiting formation of a bridged ion, which is in equilibrium with an open ion, was proposed to account for these observations.[59] Evidence (based on product studies) for the reaction of preformed episulphonium ions with a variety of nucleophiles has been cited[60] as favouring a σ-sulphurane rather than an episulphonium ion as the more likely intermediate in the reaction of alkenes with sulphenyl derivatives under non-polar conditions; SCF-MO calculations on the hypothetical reaction between HSCl and ethylene favour a cyclic sulphurane as intermediate, although it is conceded that in polar solvents episulphonium ions may also be involved.[61] A bridged thiiranium ion, however, seems best to rationalize the rates and products of addition of arenesulphenyl chlorides to allene and its methyl-substituted derivatives.[62] Other experimental evidence seems to support this general conclusion.[63] Addition of benzenesulphenyl chloride to the olefinic alcohol, $CH_2{=}CMe(OH)Ar$, gives not only the two isomeric 1 : 1 adducts but also the 1,2- rearrangement products, $PhSCH_2CMe(Ar)COMe$, the proportion of the latter increasing on addition of Et_4NCl or $LiClO_4$: these observations, together with kinetic data, indicate a two-step process.[64] Addition of 2,4-dinitrobenzenesulphenyl chloride to the olefin (**8**) was shown[65] by X-ray crystallography to give the *trans*-adduct (**13**). The reaction of sulphur mono- and di-chloride with phenylpropiolic acid and its methyl ester has been examined in detail.[66] The trifluoroacetoxysulphenylation of olefins by use of diaryl disulphides and lead tetra-acetate in the presence of an excess of trifluoroacetic acid is an electrophilic process, probably involving episulphonium intermediates, as deduced from regio- and stereochemical studies.[67] The addition of thiocyanogen to cyclohexene in acetic acid gives a mixture of products (**14**). A relatively polar cyclic transition state was proposed for this reaction,[68] and an ionic mechanism was similarly proposed for the addition of electrochemically generated thiocyanogen and selenocyanogen to alkenes.[69]

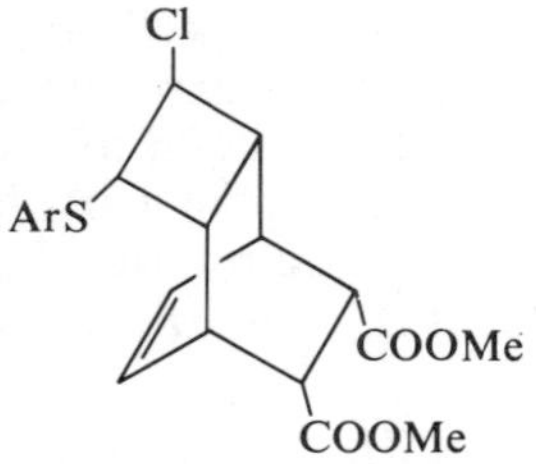

Cl
ArS
COOMe
COOMe

(**13**) Ar = 2,4-$(NO_2)_2C_6H_3$

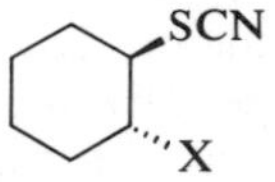

SCN
X

(**14**) **a**; X = SCN
b; X = NCS
c; X = OAc

The rate of addition of benzeneselenenyl chloride to some olefins has been measured; evidence of significant differences between this reaction and that of addition of benzenesulphenyl chlorides in both the rate-limiting and product-forming steps was detected.[70] Also, while the addition of benzeneselenenyl chloride to 3,4-dihydro-2*H*-pyran appears to proceed through an initially-formed seleniranium ion, it is postulated that this opens to an oxacarbonium ion before the product-forming step.[71] (*Z*,*Z*)-Cyclonona-1,5-diene (**15**) reacts in a stereoselective, transannular fashion with PhSeCl to give (**16**), affording another example[72] of cyclofunctionalization: the reaction of benzeneselenenyl chloride with Δ^4-unsaturated alcohols leads to tetrahydrofurans by 5-*exo-trig* closure,[73] via hydroxyl participation. Attempts to isolate a stable episelenurane or seleniranium salt by addition of benzeneselenenyl chloride to the highly crowded alkene, adamantylideneadamantane (**17**), failed; only the rearranged dichloride (**18**) was obtained.[74]

The addition of benzeneselenenic acid (generated *in situ*) to olefins gives essentially the Markownikoff β-hydroxyselenide adduct.[75]

Halofluorination of norbornadiene with a variety of reagents involves both *exo*- and *endo*-attack by the electrophile.[76]

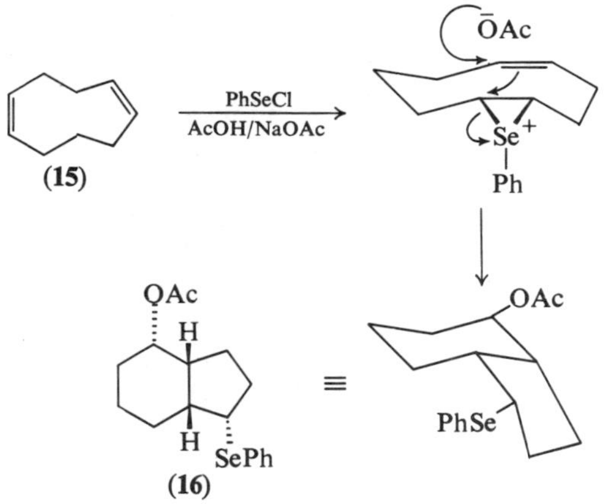

PhSeCl
AcOH/NaOAc
(15)
OAc
Se+
Ph
OAc
H
H
SePh
(16)
≡
OAc
PhSe
Cl 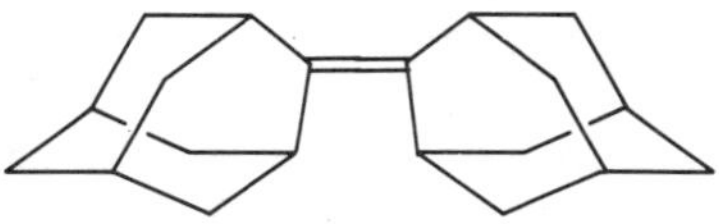 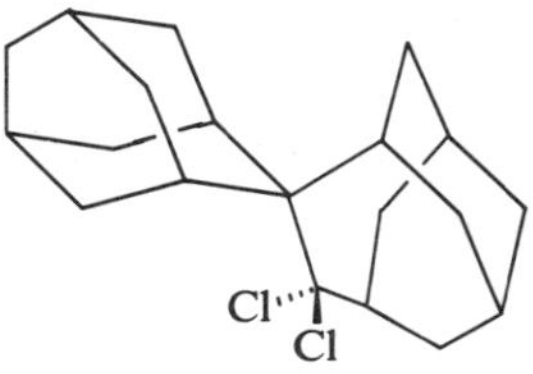
Cl
(17)
(18)

Hydrogen Halide addition

Conformational and substituent effects on the role of lactonization during treatment of cyclohexene-4-carboxylic acids with HCl have been investigated,[77] and the kinetics of HCl addition to dimethylbutadiene have been examined.[78] In addition to the expected products, *viz.* 2-chloropropene and 2,2-dichloropropane, a further ten cyclic trimerization products and one linear dimerization product have now been isolated after reaction of propyne with HCl.[79] The dimethylcyclobutenyl cation (**19**), formed from reaction of the initially formed methylvinyl cation with a second mole of propyne, is given a key rôle in the reaction scheme proposed (Scheme 3).

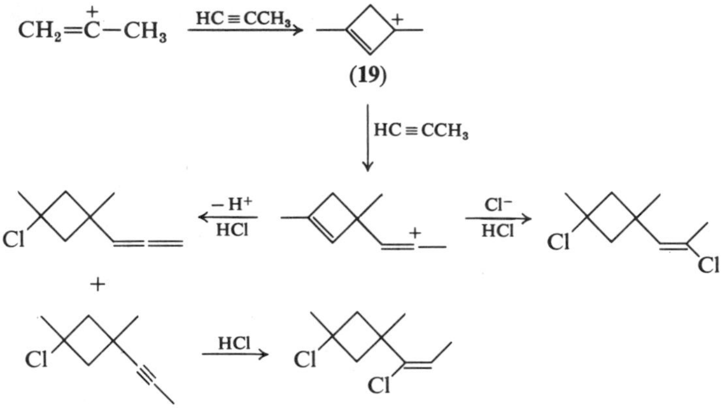

$CH_2{=}\overset{+}{C}{-}CH_3$ $\xrightarrow{HC\equiv CCH_3}$ (19)

$HC\equiv CCH_3$

$-H^+$
HCl
Cl^-
HCl
Cl
Cl
Cl
+
Cl
HCl
Cl
Cl

SCHEME 3

Addition of HCl to the triple bond of vinylacetylene is catalysed by CuCl;[80] complexation of the triple bond with Cu ion, followed by electron transfer, was suggested[81] to account for the result. Extensive rearrangement of the carbon skeleton was observed,[82] as well as normal addition, in the reaction of 1,4-dihydro-1,1,4,4-tetramethylnaphthalene with HBr; deuterium-tracer studies indicated (*a*) competing concerted *anti*- and (*b*) carbenium ion-mediated *syn*-AdE3 mechanisms for this reaction.[83] An electrophilic mechanism involving rate-limiting formation of $^{+}CPh{=}CH_2$ followed by attack of Br^- ion, HBr, or solvent has been suggested to account for the formation of $PhCBr{=}CH_2/PhCOCH_3$ mixtures from addition of HBr to phenylacetylene in 90% aqueous acetic acid.[84] A novel method for the anti-Markownikoff hydrohalogenation of alkenes involves catalysed addition of trichlorosilane to the alkene to give the alkyltrichlorosilane;[85] reaction with potassium fluoride gives the corresponding pentaflurosilicate which reacts with Cl_2, Br_2, or I_2 to give the alkyl halide.

Hydration and Related Reactions

Rate studies indicate that a wide variety of olefins, from ethylene itself to 1-methylcyclohexene, undergo acid-catalysed hydration by way of open carbonium ions. A caveat concerning the mechanistic significance of solvent isotope effects was issued.[86] The acid-catalysed addition of acetic acid to (*E*)- and (*Z*)-but-2-ene proceeds by a predominantly (*ca.* 84%) *anti*-AdE3 mechanism when catalysed by DCl, DBr, or $MeSO_3D$, but by the AdE2 pathway when catalysed by the stronger acid, CF_3SO_3D.[87] The relative rates of hydration of *Z*- and *E*-isomers of but-2-ene, hex-3-ene, 2,5-dimethylhex-3-ene and 2,2,5,5-tetramethylhex-3-ene can be interpreted[88] in terms of rate-determining protonation involving a late transition state with considerable residual double-bond character, and minimal twisting about the central C—C bond. Deuterium-solvent-isotope and buffer-catalysis effects indicate that the rate-limiting step in the hydration of (**20**) is a slow proton transfer: the effects of cationic and anionic micelles on the rate of this process have been examined.[89] The mechanism of hydration of the bridged, α,β-unsaturated ketone, norborn-2-enone, differs from that reported for the isomerization of unbridged β,γ-unsaturated cyclic ketones in that slow protonation of the C=C double bond constitutes the rate-limiting step.[90] The hydrolysis of vinyl sulphides has been examined.[91] Although the disubstituted allene (**21a**) unexpectedly undergoes protonation at the central carbon atom and subsequent cyclization to the

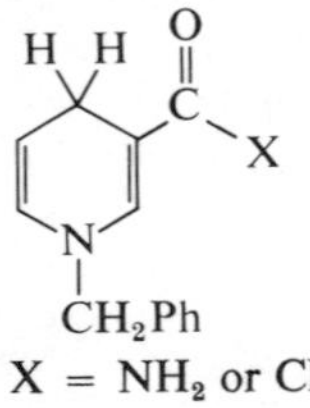

(**20**) X = NH_2 or CH_3

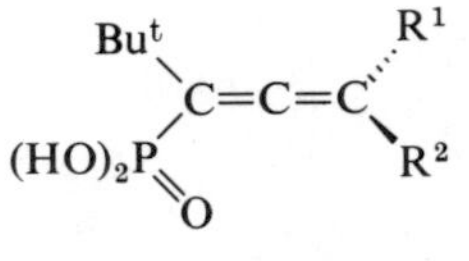

(**21**)

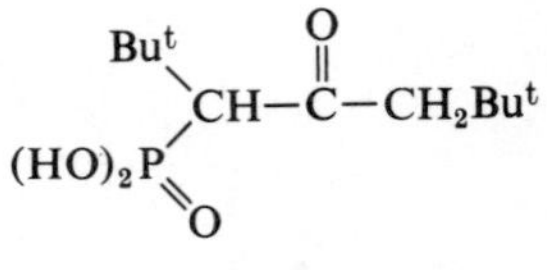

(**23**)

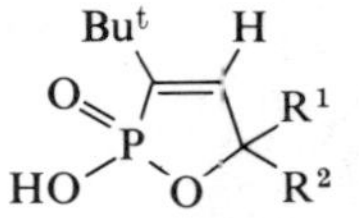

(**22**) **a**; $R^1 = R^2 = Me$
b; $R^1 = Bu^t$, $R^2 = H$

oxaphospholene (**22a**) in TFA, monosubstituted allene (**21b**) gave the normal terminal protonation product (**23**).[92]

Rate-limiting formation of a vinyl cation has been demonstrated in the acid-catalysed hydration of ethynylferrocenes,[93] but Hammett ($\rho = 0.33$) and Taft ($\rho^* = 1.78$) correlations for the hydration of aromatic and aliphatic nitriles, respectively, indicate a weakly polar transition state for the reaction in the presence of copper-containing catalysts.[94]

Miscellaneous Electrophilic Additions

The Prins reaction has been reviewed.[95] The results of MINDO/3 calculations are in agreement with the Stork–Eschenmoser hypothesis governing D/C ring-closure in steroids.[96]

An issue of the *Journal of Organometalic Chemistry* [**156** (1978)], dedicated to Professor H. C. Brown, contains papers on hydroboration and related reactions.[97] *Ab initio* calculations indicate that the gas-phase reaction of BH_3 with ethylene proceeds *via* a π-complex but displays no activation barrier. It is concluded, therefore, that the selectivity observed in hydroboration in solution arises from the introduction of an activation barrier on co-ordination of solvent.[98] MNDO calculations for the reaction of alkenes and alkynes with borane indicate, however, that the orientation of addition is controlled in the main by steric effects,[99] and a CNDO/2 study indicates that a 3-centre rather than a 4-centre transition state is to be preferred for the reaction of BH_3 with ethylene.[100] Extremely high regio-selectivity is observed[101] in the hydroboration of but-3-enyl derivatives using 9-BBN. Diphenylborane adds to alkenes to give alkyldiphenylboranes, which are useful reagents for alkylation of α,β-unsaturated ketones in the β-position.[102] Careful NMR spectral analysis has provided direct evidence for the predominantly *cis*-nature of the hydroboration of alkenes.[103] The rôle of isomerization of dialkyl-borane dimers in determining the effectiveness of asymmetric hydroboration using these compounds has been emphasized.[104] Exclusive monohydroboration of certain conjugated dienes can be achieved with the highly selective reagent, 9-BBN.[105] Alkyldibromoboranes can now be synthesized as their Me_2S addition compounds, $RBBr_2 \cdot SMe_2$, conveniently prepared in high yield by the surprisingly ready addition of $HBBr_2 \cdot SMe_2$ to olefins, a process which does not even require Lewis-acid catalysis.[106] Addition of dialkylboranes to 1-bromoalk-1-ynes, followed by treatment of the bromovinylboranes so produced with lead tetra-acetate or iodosobenzene diacetate, affords good yields of vinyl bromides.[107] Hydroboration and H_2O_2 oxidation of β,γ-unsaturated sulphones with 9-BBN or borane gives mixtures of the isomeric hydroxysulphones, the ratio of which varies with olefin and hydroboration reagent.[108] The system, Cp_2TiCl_2–$LiAlH_4$, hydroaluminates alkenes at the terminal carbon atom; dienes with one terminal double bond are selectively attacked at this position.[109]

Examination of the proton-thallium coupling constants in the products of oxythallation of acyclic alkenes has provided direct evidence that the process gives the *trans*-adduct.[110] Oxythallation of α,β-unsaturated ketones, ArCH=CHCOPh, is accompanied by 1,2-aryl migration.[111]

The hydroformylation reaction continues to attract attention.[112] Hydroformylation of ethylene in homogeneous phase with rhodium-complex catalysts has been studied,[113] and a mechanism for ethylene hydroformylation over a rhodium(I) complex adsorbed on γ-alumina has been proposed.[114] Deuterium-tracer studies of $Rh_4(CO)_{12}$-catalysed hydroformylation of olefins have been used to investigate the mechanism of this process,[115] and hydroformylation of oct-1-ene at 100 atm., using transition-metal complexes, has been examined.[116] Reaction temperature has been shown to have a marked effect on product distribution in hydroformylation of styrene, α-methylstyrene, allylbenzene, and *trans*-propenylbenzene with octacarbonyldicobalt as catalyst and pyridine as activator.[117] Asymmetric induction is observed when olefins are hydroformylated by means of chiral platinum-complex catalysts.[118]

Hydrosilylation of olefins, catalysed by $(Ph_3P)_4Pt$, has been reported,[119] and the rôle of diplatinum complexes in the hydrosilylation of alkenes[120] and of alkynes[121] has been examined. The H_2PtCl_6-catalysed hydrosilylation of isoprene with Cl_3SiH, Cl_2MeSiH, and Me_3SiH has been examined. Contrary to literature reports, the addition occurred from the less hindered side of the diene.[122] Hydrosilylation of a variety of activated olefins with Me_2SiH_2 yields products of addition of only one Si—H bond when $Ni(acac)_2/AlR_3/Ph_3P$ mixtures are used as catalysts.[123] Asymmetric hydrosilylation, using homogeneous catalysts with chiral ligands, has been reviewed,[124] and the 1,4-addition of disilanes to a variety of substrates has been reported.[125] The regioselectivity of hydrosilylation of 1,3-dienes has been shown to depend quite markedly on the catalyst used.[126] Aminosilylation of acetylenedicarboxylate has been studied.[127]

Epoxidation of unsaturated compounds by peracids[128] and of alkenes by organic hydroperoxides[129] has been reviewed. Both semiempirical and *ab initio* MO calculations indicate that epoxidation of alkenes by peroxy-acids involves nucleophilic attack of the alkene π-MO on the O—O antibonding orbital,[130] and *ab initio* calculations using STO-2G and STO-4G basis sets on the transition state for the reaction of ethylene with performic acid indicate that unsymmetric transition states are energetically more favourable than symmetric ones. This is in keeping with the notion that peracids behave as electrophiles towards alkenes.[131] Epoxidation of propene by peracetic acid has been shown to be a second-order process occurring twice as fast in ethyl acetate as in acetone,[132] and it has been found that basic solvents retard the rate of peracetic acid epoxidation of olefins, allegedly by complexations of the peracid.[133] The importance of a sound understanding of the structural features of reactants in the interpretation of product distributions in epoxidation reactions has been emphasized.[134] Epoxidation of dicyclopentadiene by monoperoxyphthalic acid has also been studied.[135]

Epoxidations with ethylbenzene hydroperoxide have been studied by a number of workers: alkenes examined include propene and isobutene,[136] and cyclohexene.[137,138] The kinetics of liquid-phase epoxidation of isobutene with cumene hydroperoxide catalysed by molybdenum acetylacetonate have been measured, and a mechanistic scheme has been proposed.[139] Rates of epoxidation of fused-ring cyclopropenes, namely the 8,8-disubstituted bicyclo[5.1.0]oct-1(7)-enes (**24**), have been reported.[140] It was concluded, from a consideration of relative rates, that the ethyl group has a significantly larger steric requirement, in comparison with a methyl group, than is suggested by tabulated E_s values, as a result of electronic as well as steric-effect contributions to E_s values as obtained from ester hydrolysis data. Epoxidation of α-substituted styrenes (**25**) leads to α-substituted epoxides (**26**) which may be isolated in the pure state when $R^1 = NO_2$. These compounds are useful substrates for photometric studies with epoxide hydrase.[141]

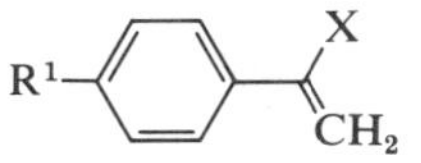

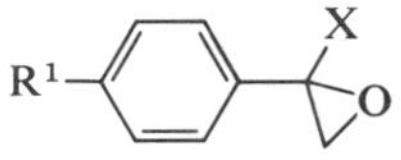

(24) **a**; $R^1 = R^2 = Me$
b; $R^1 = Me, R^2 = Et$
c; $R^1 = R^2 = Et$
d; $R^1 = Me, R^2 = Pr^i$

(25)
R^1 = H or NO_2
X = F, Cl, Br, $OSiR_3$, OAlk, or OAc

(26)

The mechanism of the reaction of *cis*- and *trans*-1,2-difluoroethylene with ozone has been investigated,[142] as has the mechanism of the gas-phase ozonolysis of vinyl chloride, using IR-matrix and microwave spectroscopy.[143]

Steric effects appear to dominate the stereochemistry of the products of oxymercuration of methylenecyclohexanes and methylenecyclopentanes.[144] The rôle of ligand electronegativity in determining the regiochemistry of intramolecular phenoxymercuration of 2-allylphenols has been studied. Benzofuran derivatives are favoured over chroman derivatives as products with less electronegative ligands and *vice versa*.[145] A number of other oxymercurations has been studied, including the oxymercuration of a range of alkenylsilanes,[146] and of 2- and 4-nitrostyrenes: the latter reaction gives products resistant to reductions by $NaBH_4$.[147] Oxymercuration of alkynes is largely a *trans*-process, with the regiospecificity depending on alkyne structure: an unsymmetrical mercury-bridged intermediate was proposed to account for the results.[148] The allylic hydroperoxide (**27**) appears to give the 1,2-dioxetane (**28**) on treatment with mercuric trifluoroacetate in rigorously anhydrous $CDCl_3$ solution.[149]

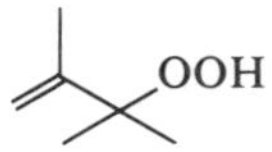

(27) 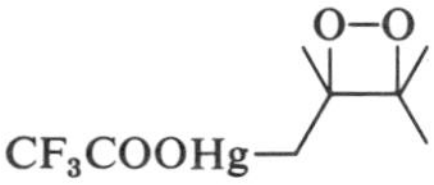 **(28)**

The reaction of alkenylboranes with palladium acetate in the presence of Et_3N gives generally good yields of olefins with greater than 96% *E*-stereochemistry, probably *via* the acetoxypalladation adduct;[150] alkynes can be carbonylated regiospecifically by using palladium complexes under very mild conditions.[151] Vicinal diamination of olefins on use of $(PhCN)_2PdCl_2$ and amines involves prior *trans*-aminopalladation, followed by an oxidant-promoted displacement of the palladium by a second mole of amine.[152] Evidence has been obtained for intramolecular addition of alkoxycarbonyl palladium species to carbon—carbon triple bonds in the synthesis of α-methylene-lactones from carbon monoxide and acetylenic alcohols.[153] Reaction of the organopalladium species derived from 1,3-dimethylpyrimidine-2,4-dion-5-ylmercuric acetate and palladium acetate with cyclic enol ethers appears to involve (the expected) *syn*-addition of the aryl-palladium.[154] Hydroxypalladation of *cis*-1,2-dideuterioethylene in the presence of carbon monoxide is a *trans*-addition process[155] as shown in Scheme 4.

Carbometallation of acetylenes has been shown to give the *cis*-adduct;[156] carbometallation of alkynylsilanes by Ziegler–Natta alkylating agents has also been studied.[157] Cobalt carbonyl-catalysed carbonylation of acetylenes with phase-transfer catalysts affords a good route to but-2-enolides: dienes give acylated

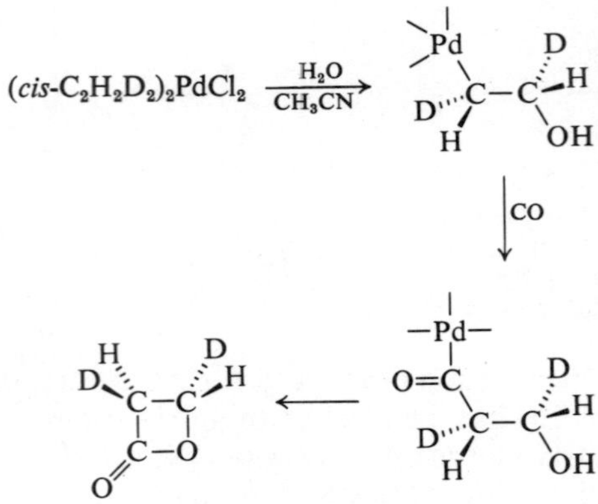

$(cis\text{-}C_2H_2D_2)_2PdCl_2 \xrightarrow[CH_3CN]{H_2O}$

Scheme 4

derivatives under similar conditions.[158] Ene addition of chloral or bromal to 1,2-dialkylethylenes has been investigated.[159] The isomerization of pent-1-ene to *trans*- and *cis*-pent-2-ene on using ruthenium cluster carbonyl hydrides in toluene has been shown to involve a metal hydride-elimination process.[160]

Addition of carbon tetrachloride to 1,3-dienes to give 1,4-adducts proceeds readily under mild conditions when dichlorotris(triphenylphosphine)ruthenium(II) is used.[161]

The ionic addition of halomethanes to fluoro-olefins has been reviewed.[162] Solvent effects on the rate of reaction of triethylaluminium with olefins have been attributed to changes in the equilibrium concentrations of different forms of the organometallic reagent giving rise to different reactivities.[163] A hydride shift in the intermediate cation is proposed[164] to account for the formation of (**29**) in addition to the expected product (**30**) in the $AlCl_3$-catalysed cyclization of (**31**).

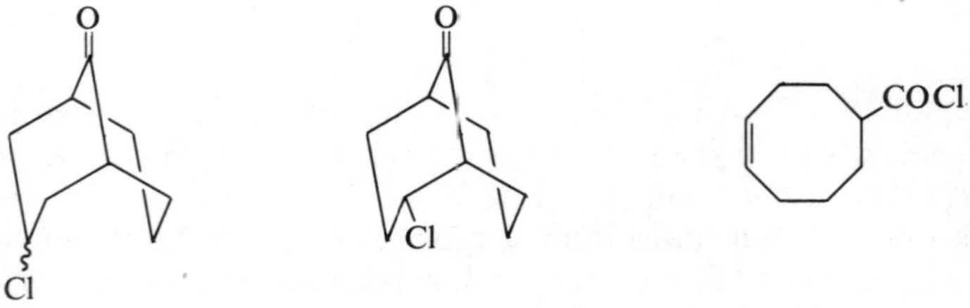

(**29**) (**30**) (**31**)

The reaction of enol ethers with *N*-haloamides, catalysed by $CrCl_2$, to give α-acylamino-acetals and -ketals has been extended to some sugar derivatives.[165] Electron-transfer processes appear to be implicated in the CuCl-catalysed addition of sulphonyl chlorides to a variety of olefins and acetylenes.[166]

Nucleophilic Additions

Reduction of the diene ester (**32**) with $LiAlH_4$, followed by treatment with D_2O or $LiAlD_4$, followed further by treatment with H_2O, gives (**33**) and (**34**), respectively, evidently by the highly unusual *endo*-conjugate addition of hydride followed by protonation on the *exo*-face.[167]

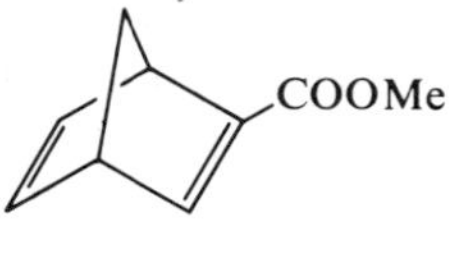

COOMe D CH₂OD H H H CD₂OH H D

(32) 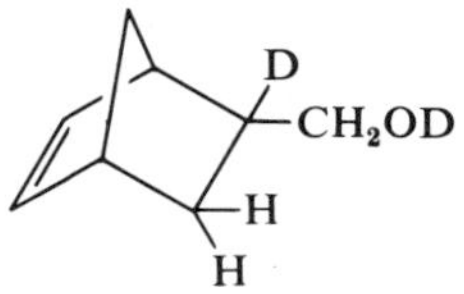 **(33)** 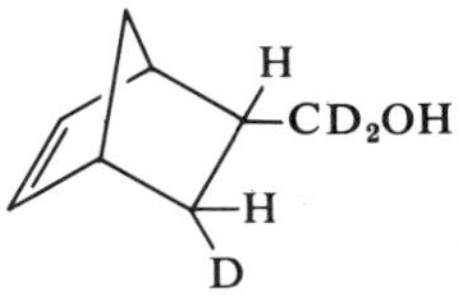 **(34)**

The regio- and stereo-specificity of the reactions of the series of copper hydrides Li_nCuH_{n+1} with enones and cyclic ketones indicate that the species with $n = 1-5$ behave as discrete entities.[168]

Strong support for the Ritchie N_+ correlation has been presented, from the rates of addition of a series of nucleophiles to **(35)**, a reaction covering eleven orders of magnitude in rate.[169]

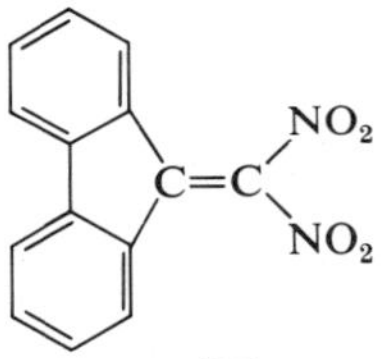

NO_2 C=C NO_2 O O Ph N Me O Me

(35) 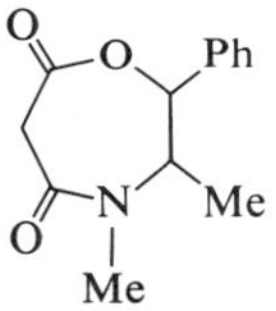 **(36)**

Fluoride ion, immobilized on strongly basic anion-exchange resins, acts as an effective catalyst for Michael additions;[170] the use of preformed enolates in such reactions has been advocated.[171]

It has been claimed that consideration of torsion angle allows prediction as well as interpretation of the stereochemistry of Michael addition to cyclopent-2-enones.[172] The conjugate addition of lithium dialkylcuprates to carbohydrate α-enones has been demonstrated: stereoselectivity depends on the substrate structure.[173] The preparation of optically active compounds by the Michael reaction has been approached in two ways: thus, the chiral oxazepine **(36)**, prepared from (−)-ephedrine, undergoes Michael addition to achiral α,β-unsaturated ketones with considerable asymmetric induction,[174] and improvements (up to 23% enantiomeric excess) in the degree of asymmetric induction in the Michael addition of nitromethane to chalcones have been achieved by using chiral ammonium fluorides in aprotic solvents.[175] The enantiomeric excess observed in such reactions was found to be inversely proportional to the dielectric constant of the solvent employed.[176]

The rôle of cations in controlling the regioselectivity of nucleophilic additions to α,β-unsaturated ketones has been investigated, and a distinction made between complexation-controlled and association-controlled reactions:[177] results from addition of acetonitrile anion and phenylacetonitrile anion to a series of cyclic and acyclic α-enones support the general conclusions reached.[178] A series of Group IV anions (namely, Me_3SnLi, Me_3SiLi, and Me_3CLi) has been added to cyclohex-2-enones and the kinetically- and the thermodynamically-controlled products for each reaction have been established. The results were contrary to those expected from HSAB theory, in that the Me_3SnLi gave mostly the 1,2-adduct under kinetic control. However, a range of product distributions from 100% of 1,2-addition to 90% of 1,4-addition was observed for reaction of Me_3CLi with cyclohexenone on changing the solvent from Et_2O to THF + 20% HMPA.[179]

Selenium-stabilized enolates likewise added to the carbonyl-carbon of cyclohexenone,[180] but a continuum of 1,2 : 1,4 product ratios was observed for addition of dilithium carboxylates to α-enones when substituents in the dianion or in the α-enone were varied.[181] Reaction temperature was also shown to be important[182] in determining the ratio of 1,2- to 1,4-addition of $CH_3SCH_2S(O)CH_3$ anion to α-enones. β-Sulphinyl enones (**37**) undergo Michael addition exclusively at the carbon bearing the sulphinyl group to give (**38**) from which RSOH is immediately lost to give (**39**).[183] 1,2- *versus* 1,4- Addition of lithiomethyl-nitrogenous heterocycles to α,β-unsaturated carbonyl systems has also been studied.[184] Benzylmagnesium chloride undergoes conjugate 1,4-addition to furan-2-aldehyde,[185] and benzylmagnesium halides add to the dienone (**40**) to give a mixture of 1,2-, 1,4-, and 1,6-addition products, in the presence of 10 mol % of copper(I) bromide; however, the product was almost entirely the 1,6-adduct.[186]

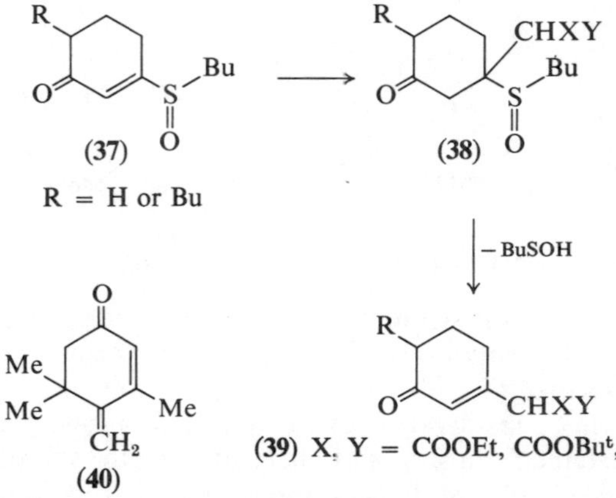

(**37**) R = H or Bu

(**38**)

− BuSOH

(**40**)

(**39**) X, Y = COOEt, $COOBu^t$, or CN

Addition of alk-2-enylmagnesiums to styrene[187] gives a mixture of primary and secondary alkylmagnesium chlorides; with 1,4-dienes the addition gives the 2-metalated product, with inversion of the alk-2-enyl group.[188]

STO-3G *ab initio* calculations on the transition state for nucleophilic addition to the C≡C triple bond indicate optimum attack along an axis making an angle of *ca.* 120° with the bond.[189] The addition of Bu^t_2Zn to terminal acetylenes is regio- but not stereo-selective, giving exclusively the 1,2-disubstituted ethenes (*cis* and *trans*) in fair yield.[190] Addition of zirconium alkenyls to α,β-unsaturated ketones proceeds efficiently in the presence of catalytic quantities of $Ni(acac)_2$; nickel(II) rather than nickel(0) appears to be involved in this process.[191] An improved procedure for the addition of trialkylsilyl-lithiums to α,β-unsaturated, ketones has been reported.[192]

Kinetic data indicate an addition–elimination mechanism for the reaction of piperidine with (*Z*)-2-(2-bromovinyl)-5-nitrofuran to give the 2-(2-piperidinovinyl) analogue.[193] Rates of addition of piperidine to a series of β-substituted acrylonitriles have been measured and shown to be rather insensitive to the nature of β-alkyl substituents.[194] A kinetic study of the addition of amines to tetracyanoethene has also been made, and a mechanistic scheme has been proposed.[195] Oxaphospholenes have been postulated[196] as possible intermediates in the Michael addition of certain silicon phosphite esters to α,β-unsaturated carbonyl compounds to explain the preferential formation of the *Z*-adducts (Scheme 5).

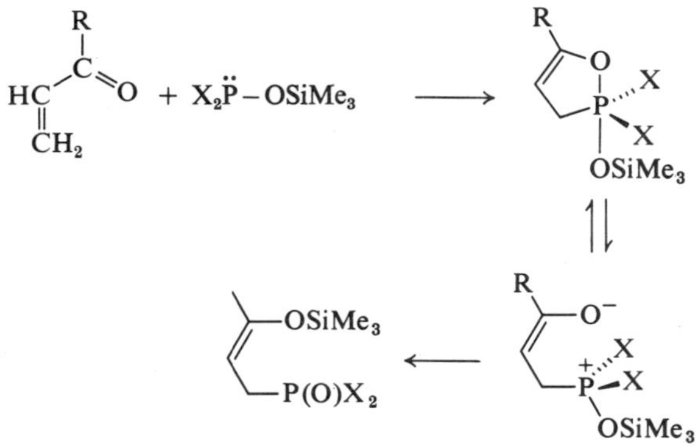

SCHEME 5

Angle- and torsional-strain effects are postulated to account for the unusual stability of the hydroperoxy-ketone (**41**) derived from Michael addition of alkaline hydrogen peroxide to the indenone (**42**) and related compounds; such strain effects prevent either reversal of the Michael addition or base-promoted cyclization to the epoxide.[197]

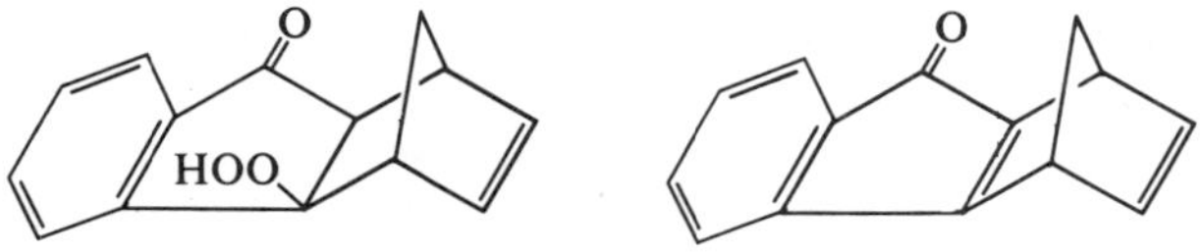

(**41**) (**42**)

The rate of addition of alkoxide ions to cyclohexenone correlates well with the basicity of the nucleophile.[198] Competition between simple nucleophilic substitution and Michael addition by mercaptide ions on methyl 4-bromocrotonate, $BrCH_2CH{=}CHCOOMe$, depends heavily on solvent and gegenion.[199] The intermediate bromoenolate cyclizes to a sulphur-substituted cyclopropane. The addition of butanethiol to the *N*-arylmaleimides (**43**) increases with the electron-attracting effect of R; 2-chloro-substituents retard the reaction because the two rings are thereby prevented from being coplanar.[200]

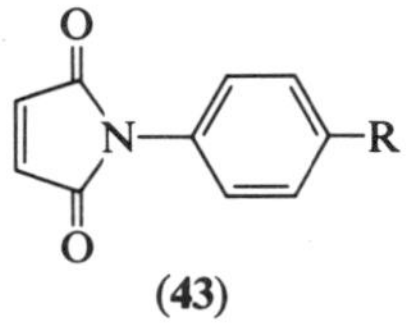

(**43**)

Michael addition of aqueous sodium sulphide to the dienone (**44**) gives the 1 : 2 adduct (**47**). A pathway involving the intermediates (**45**) and (**46**) was suggested to account for this.[201]

The reaction of phosphorus dithio-acids with β-alkoxyvinyl phosphonates has also been studied.[202]

A full account has appeared of experiments designed to test Baldwin's rules as applied to ring closure by conjugate addition of oxygen nucleophiles to double and triple bonds.[203] The extent to which 5-*endo-trig* cyclizations are disfavoured appears (perhaps not unexpectedly) to depend on electronic factors modifying the reactivity

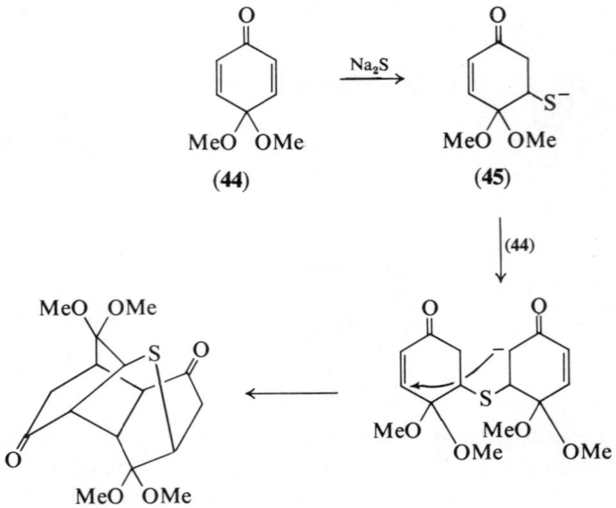

(44) $\xrightarrow{Na_2S}$ (45) $\xrightarrow{(44)}$ (46) → (47)

of the nucleophilic centre. Thus, the cyclization of the hydroxylamine derivative, PhCH=CHCOONHMe, to (**48**) takes place under quite mild conditions.[204] Intramolecular Michael addition is favoured over vinylogous aldol cyclization in (**49**), giving the angularity substituted *cis*-perhydroindene-2,5-diones (**50**).[205] With an excess of piperidine in toluene or propanol solution, the quinone (**51**) gives (by a pseudo-first-order process) the enamine (**52**) which cyclizes to a mixture of 3-piperidino-9,10-phenanthraquinone and 2-piperidino-9,10-anthraquinone.[206] The acetylenic amine, $Me_2NCH_2CMe(OH)C{\equiv}CPh$, cyclizes in methanol to give (**53**).[207]

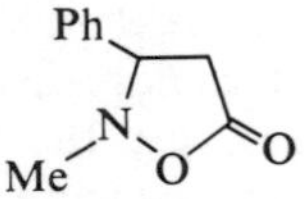

(48) 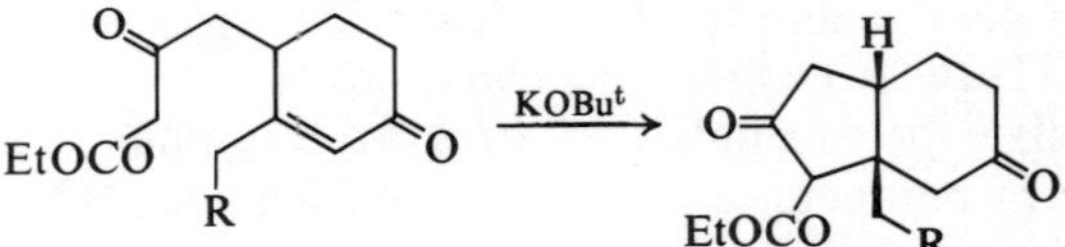 (49) $\xrightarrow{KOBu^t}$ (50)

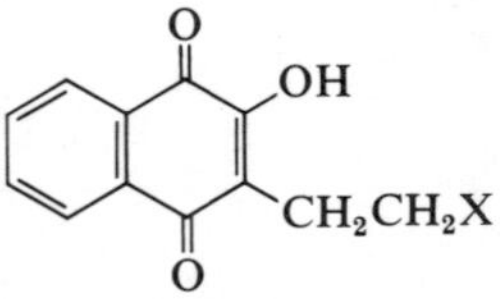

(**51**) X = $COCH_3$
(**52**) X = 1-piperidinovinyl

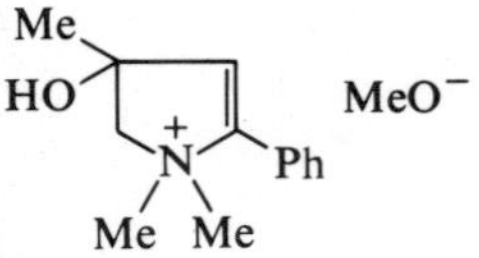

(53)

The allene (**54**) can be isolated from the reaction of (**55**) with sodium isopropoxide, lending support to the suggestion[208] that it is the intermediate involved in the eventual formation of (**56**). In what is claimed to be the first example of such a process,[209] cyclization of the allylphenol (**57**) to optically active (**58**) has been achieved in 12% optical yield by using palladium(II) acetate in the presence of catalytic amounts of (−)-β-pinene. The catalytic species is suggested to be one in which π-bonded β-pinene exists, since *bis*[acetoxy-(7,1,2-η-pinene)palladium(II)] also gives optically active (**58**) from (**57**) in similar optical yield.

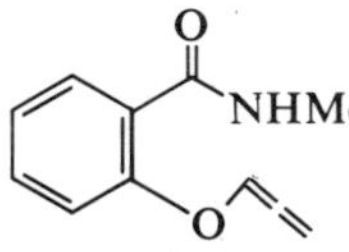

(54) 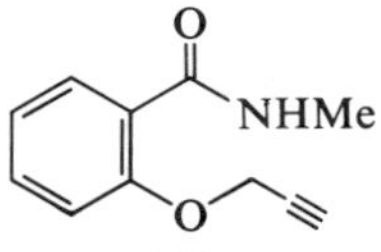 (55) 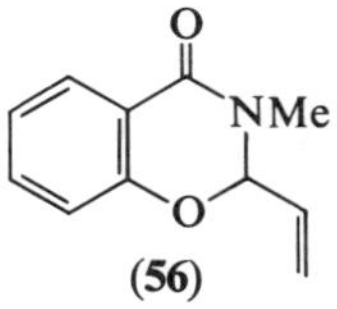 (56)

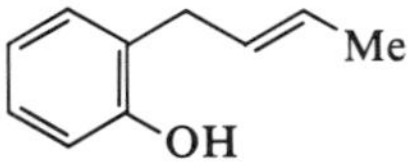

(57) 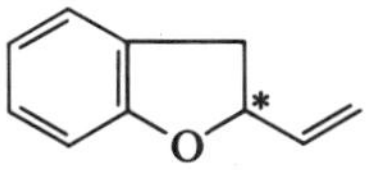 (58)

Divinyl ketone has been shown to be a potentially useful C_5 synthon for terpene and steroid syntheses, by way of stepwise Michael addition reactions.[210] The stereochemistry of the adducts (**59**) formed from thermal [to give (**59a**)] or photochemical [to give (**59b**)] addition of methanol to the benzoxazepine (**60**) can be explained in terms of *trans*- and *cis*-Michael addition to (**60**) respectively, followed by retro-ene reaction.[211] The *trans*-isomer (**62**) has been postulated[212] as the species involved in the photoaddition of methanol to (**61**) to give (**63**).

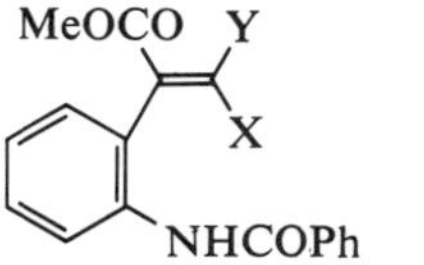

(59) **a**; Y = H, X = OMe
b; Y = OMe, X = H

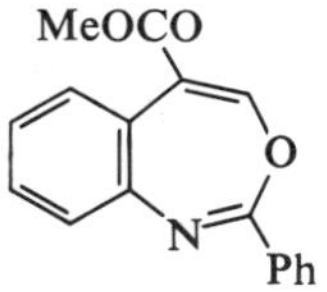

(60)

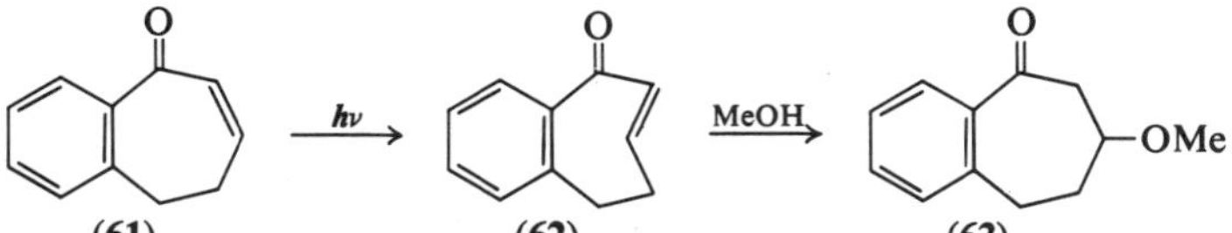

(61) $\xrightarrow{h\nu}$ (62) $\xrightarrow{MeOH}$ (63)

A mechanism involving concerted nucleophilic attack by nitrogen on one end of the central C=C double bond in (**64**) and proton transfer from general acids to the other is proposed for the cyclization of (**64**) to (**65**): significant release of ground-state strain on cyclization appears to be necessary for the occurrence of such nucleophilic addition to unactivated double bonds.[213] A similar intramolecular addition of an amine to an unactivated *triple* bond has also been reported.[214]

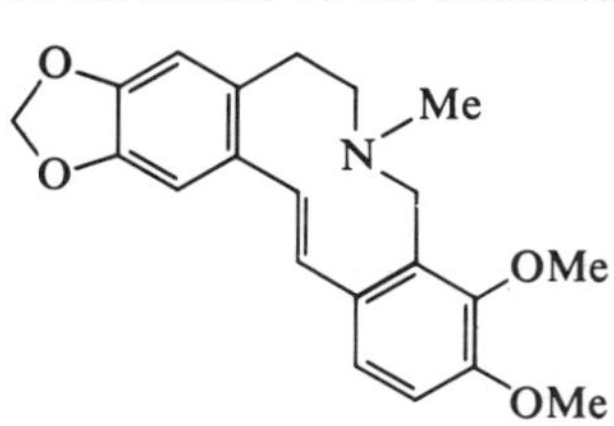

(64) 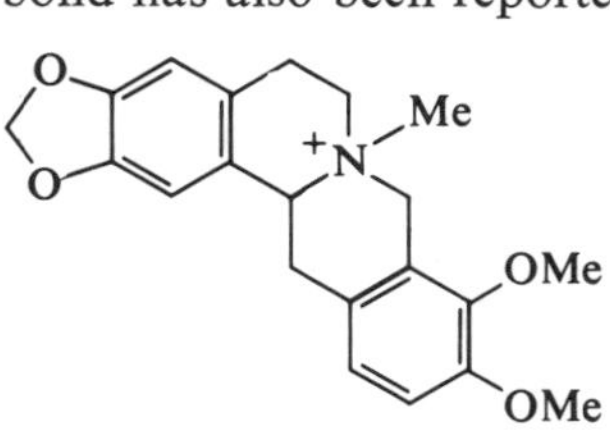 (65)

Extra-thermodynamic parameters for the addition of diphenylsilane and diphenylgermane to acetylenes have been determined calorimetrically.[215] The rates of addition of a range of primary and secondary amines to PhC≡CCOPh have been shown to be controlled by entropy factors.[216] Substituted derivatives, $RC_6H_4C{\equiv}CCOC_6H_4R'$, have also been examined.[217] Organocopper(I) reagents add to vinylacetylene in a regiospecific *syn*-fashion, but addition of the reagents to $CH_2{=}CHC{\equiv}CCH_2OEt$ gives a mixture of the regioisomers together with large quantities of the allene, $CH_2{=}CHCBu{=}C{=}CH_2$.[218] A pair of Michael-type reactions is proposed[219] to account for the formation of complexes from *cis*-$M(CO)_4(PPh_2H)_2$ and acetylenes as shown in Scheme 6.

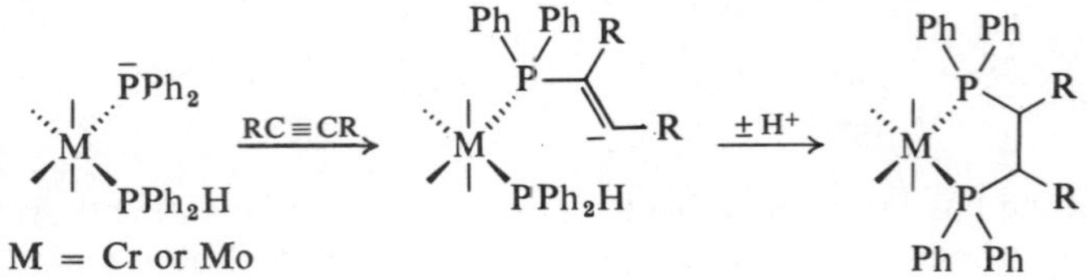

M = Cr or Mo

SCHEME 6

The X-ray structure determination of the adduct from $CF_3C{\equiv}CCF_3$ and $Cr(CO)_4(PPh_2H)_2$, however, shows it to be a 2 : 1 adduct of 1,4-difluorobutadiene.[220] Lithium dimethylcuprate adds to $CH_2{=}C{=}CHCOCH_2Ph$ at the activated carbon—carbon double bond to give the α-cuprio-ketone, which slowly equilibrates with the copper enolate.[221] Similar results have been obtained for other allenic carbonyl compounds, and for phosphine oxides,[222] sulphoxides, and sulphones.[223]

The reaction of trialkyl phosphites with nitroalkenes has been studied: the products can be rationalized in terms of a common dipolar intermediate formed by Michael addition of the phosphite to the β-carbon atom.[224] The Michael addition of lithium alkynylborates, $Li^+R_3\bar{B}C{\equiv}CR'$, to nitroalkenes[225] occurs with migration of an alkyl group from boron to carbon to give adducts (**66**) which may be hydrolysed to alkenes or oxidized to ketones (Scheme 7).

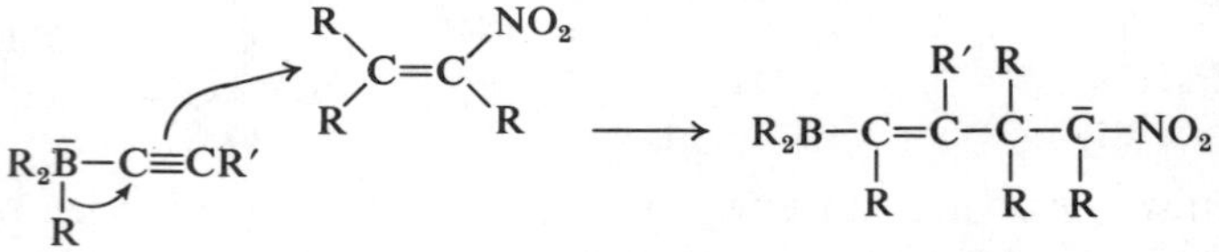

(**66**)

SCHEME 7

Nucleophilic addition of cysteine to 5-chlorouracil has been studied.[226] α,β-Unsaturated thioamides serve as Michael acceptors towards a variety of organolithium and organomagnesium compounds; the intermediate thioenolate anions may be trapped with electrophiles such as diphenyl disulphide or allyl bromide, or quenched in methanol.[227]

Rates of addition of triethylsilane to aromatic azomethines, $ArN{=}CHPh$, have been measured and a Hammett correlation has been given;[228] also the mechanism of the addition of SO_2 to Schiff's bases has been clarified.[229]

Addition reactions of thiocumulenes of the general structure $X{=}S{=}Y$ have been reviewed.[230] Kinetic data, supported by CNDO/2 calculations, indicate that attack of alkoxide ion on sulphur constitutes the rate-limiting step in the Cu^{2+}-catalysed ethanolysis of *N*-sulphinylamines, ArNSO.[231] Chimishkyan and his co-workers have continued their study of addition of amines to isocyanates,[232,233] and the kinetics of addition of ethanol to 1-naphthylisocyanate have been measured in various solvents.[234]

Addition of pyridine to ethene π-bonded to platinum(II) gives a σ-bonded adduct.[235] However, addition of malonate to the π-allylpalladium derivatives of steroidal 4-en-3-ones gives the 6β-derivatives stereospecifically.[236] This type of addition of stabilized carbanions to π-allylpalladium complexes is greatly facilitated by the presence of phosphine or phosphite ligands.[237] The method has been elegantly exploited in prenylation procedures[238] and in the elaboration of steroid side-chains.[239] Addition of sodium alkanesulphinates to 1,3-dienes in the presence of $PdCl_2$, followed by treatment with dimethylglyoxime (DMG), affords the corresponding anti-Markownikoff 1,2-hydrosulphonylation products, γ,δ- unsaturated sulphones (**68**), *via* the π-allylpalladium complexes (**67**).[240]

R^1, R^2, R^3; R^4SO_2Na, $PdCl_2$ → R^1, R^2, R^3, SO_2R^4, Pd, Cl, 2 (**67**); DMG → R^1, R^2, R^3, SO_2R^4 (**68**)

Nickel acetylacetonate–trimethylaluminium mixture catalyses the *cis*-addition of Grignard reagents to silylacetylenes,[241] and rhodium carbonyl cluster complexes catalyse the addition of benzene and monosubstituted derivatives to diphenylketene and to aryl isocyanates to give aryl diphenylmethyl ketones and *N*-arylbenzamides, respectively.[242]

References

1 Boyd, G. V., *Aromat. Heteroaromat. Chem.*, **5**, 359 (1977); *Chem. Abs.*, **88**, 5620 (1978).
2 Schmid, G. H., and Tidwell, T. T., *J. Org. Chem.*, **43**, 460 (1978).
3 Cerksus, T. R., Csizmadia, V. M., Schmid, G. H., and Tidwell, T. T., *Can. J. Chem.*, **56**, 205 (1978).
4 Schmid, G. H., *J. Org. Chem.*, **43**, 777 (1978).
5 Bellucci, G., Berti, G., Ferretti, M., Ingrosso, G., and Mastrorilli, E., *J. Org. Chem.*, **43**, 422 (1978).
6 Nagase, S., and Morokuma, K., *J. Am. Chem. Soc.*, **100**, 1666 (1978).
7 Kocovsky, P., and Cerny, V., *Collect. Czech. Chem. Commun.*, **43**, 327 (1978).
8 Kocovsky, P., and Cerny, V., *Collect. Czech. Chem. Commun.*, **43**, 1924 (1978).
9 Stavber, S., and Zupan, M., *J. Chem. Soc., Chem. Commun.*, **1978**, 969.
10 Zupan, M., and Sket, B., *J. Org. Chem.*, **43**, 696 (1978).
11 Boguslavskaya, L. S., Morozova, T. V., and Voronin, A. P., *Zh. Org. Khim.*, **14**, 1442 (1978); *Chem. Abs.*, **89**, 146307 (1978).
12 Barton, D. H. R., Hesse, R. H., Jackman, G. P., and Pechet, M. M., *J. Chem. Soc., Perkin Trans.* 1, **1977**, 2604.
13 Serguchev, Y. A., Konyushenko, V. P., Sergeev, G. B., and Bitko, S. A., *Ukr. Khim. Zh.*, **44**, 736 (1978); *Chem. Abs.*, **89**, 107744 (1978).

[14] V'yunov, K. A., Garabadzhiu, A. V., and Ginak, A. I., *Zh. Org. Khim.*, **14**, 1598 (1978); *Chem. Abs.*, **89**, 162716 (1978).
[15] Shinoda, K., *Bull. Chem. Soc. Jpn.*, **51**, 913 (1978).
[16] de la Mare, P. B. D., and Wilson, R. D., *J. Chem. Soc., Perkin Trans.* 2, **1977**, 2062.
[17] Subramanian, R., and Ganesan, R., *Indian J. Chem.*, **15B**, 645 (1977); *Chem. Abs.*, **88**, 21626 (1978).
[18] Haegele, G., and Dolhaine, H., *Phosphorus Sulfur*, **3**, 47 (1977); *Chem. Abs.*, **88**, 169280 (1978).
[19] Aversa, M. C., Cum, G., Giannetto, P., and Romeo, G., *J. Chem. Res. (S)*, **1978**, 292.
[20] Norris, T., Payne, A., and Southby, D. T., *J. Chem. Soc., Chem. Commun.*, **1978**, 932.
[21] Heasley, G. E., Emery, W. E., Hinton, R., Shellhamer, D. F., Heasley, V. L., and Rodgers, S. L., *J. Org. Chem.*, **43**, 361 (1978).
[22] Hall, S. S., Weber, G. F., and Duggan, A. J., *J. Org. Chem.*, **43**, 667 (1978).
[23] Nugent, W. A., *Tetrahedron Lett.*, **1978**, 3427.
[24] O'Brien, M., and More O'Ferral, R. A., *J. Chem. Soc., Perkin Trans.* 2, **1978**, 1045.
[25] Dubois, J.-E., and de Ficquelmont-Loizos, M. M., *Tetrahedron*, **34**, 2247 (1978).
[26] Yates, K., *Prog. Theor. Org. Chem.*, **1977**, 261; *Chem. Abs.*, **88**, 169277 (1978).
[27] Bienvenue-Goetz, E., and Dubois, J.-E., *Tetrahedron* **34**, 2021 (1978).
[28] Dubois, J.-E., Ruasse, M.-F., and Argile, A., *Tetrahedron Lett.*, **1978**, 177.
[29] Modro, A., Schmid, G. H., and Yates, K., *J. Org. Chem.*, **42**, 3673 (1977).
[30] Dubois, J.-E., and Chretien, J. R., *J. Am. Chem. Soc.*, **100**, 3506 (1978).
[31] op den Brouw, P. M., and Laarhoven, W. H., *Recl. Trav. Chim. Pays-Bas*, **97**, 265 (1978).
[32] Heasley, V. L., Shellhamer, D. F., Iskikian, J. A., Street, D. L., and Heasley, G. E., *J. Org. Chem.*, **43**, 3139 (1978).
[33] de la Mare, P. B. D., and Wilson, R. D., *J. Chem. Soc., Perkin Trans.* 2, **1977**, 2048.
[34] de la Mare, P. B. D., and Wilson, R. D., *J. Chem. Soc., Perkin Trans* 2, **1977**, 2055.
[35] Shellhamer, D. F., Heasley, V. L., Foster, J. E., Luttrull, J. K., and Heasley, G. E., *J. Org. Chem.*, **43**, 2652 (1978).
[36] Sukla, U. K. S., and Sharma, J. P., *Monatsh. Chem.*, **109**, 451 (1978).
[37] Hatakeyama, S., Mitsuhashi, T., and Oki, M., *Chem. Lett.*, **1978**, 559.
[38] Vishwakarma, L. C., and Walia, J. S., *J. Indian Chem. Soc.*, **53**, 156 (1978); *Chem. Abs.*, **88**, 5622 (1978).
[39] Risbood, P. A., and Ruthven, D. M., *J. Am. Chem. Soc.*, **100**, 4919 (1978).
[40] Bottini, A. T., and Cabral, L. J., *Tetrahedron*, **34**, 3195 (1978).
[41] Heasley, G. E., Bundy, J. M., Heasley, V. L., Arnold, S., Gipe, A., McKee, D., Orr, R., Rodgers, S. L., and Shellhamer, D. F., *J. Org. Chem.*, **43**, 2793 (1978).
[42] Mel'nikov, G. D., Mel'nikova, S. P., and Porfir'eva, Y. I., *Zh. Org. Khim.*, **13**, 2482 (1977); *Chem. Abs.*, **88**, 73822 (1978).
[43] Kondo, A., Yamane, T., Ashida, T., Sasaki, T., and Kanematsu, K., *J. Org. Chem.*, **43**, 1180 (1978).
[44] Glotter, E., and Krinsky, P., *J. Chem. Soc., Perkin Trans.* 1, **1978**, 408.
[45] Yamakawa, K., Azusawa, K., and Nishitani, K., *Chem. Lett.*, **1977**, 1159.
[46] Haufe, G., Muehlstaedt, M., and Graefe, J., *Monatsh. Chem.*, **108**, 1431 (1977).
[47] Haufe, G., Kleinpeter, E., Mühlstadt, M., and Graefe, J., *Monatsh. Chem.*, **109**, 575 (1978).
[48] Ehrlich, S. J., and Berliner, E., *J. Am. Chem. Soc.*, **100**, 1525 (1978).
[49] Mel'nikov, G. D., and Mel'nikova, S. P., *Tezisy Dokl.-Vses. Konf. Khim. Atsetilena 5th*, **1975**, 355; *Chem. Abs.*, **89**, 59397 (1978).
[50] Modena, G., Rivetti, F., and Tonellato, U., *J. Org. Chem.*, **43**, 1521 (1978).
[51] Pincock, J. A., and Somawardhana, C., *Can. J. Chem.*, **56**, 1164 (1978).
[52] Volkova, N. V., and Kanivets, N. P., *Ukr. Khim. Zh.*, **43**, 1221 (1977); *Chem. Abs.*, **88**, 61745 (1978).
[53] Cambie, R. C., Hayward, R. C., Jurlina, J. L., Rutledge, P. S., and Woodgate, P. D., *J. Chem. Soc., Perkin Trans.* 1, **1978**, 126.
[54] Urasaki, I., *Numazu Kogyo Koto Semmon Gakko Kenkyu Hokoku*, **11**, 87 (1976); *Chem. Abs.*, **88**, 21642 (1978).
[55] Cambie, R. C., Rutledge, P. S., Smith-Palmer, T., and Woodgate, P. D., *J. Chem. Soc., Perkin Trans.* 1, **1978**, 997.
[56] Zefirov, N. S., Sadovaya, N. K., Novgorodtseva, L. A., and Bodrikov, I. V., *Tetrahedron*, **34**, 1373 (1978).

[57] Rasteikiene, L., Vidugiriene, V., Taliene, V., and Talaikyte, Z., *Zh. Org. Khim.*, **13**, 2099 (1977); *Chem. Abs.*, **88**, 36862 (1978).
[58] Buchan, G. M., and Cameron, G. G., *J. Chem. Soc., Perkin Trans.* 1, **1978**, 783.
[59] Toyoshima, K., Okuyama, T., and Fueno, T., *J. Org. Chem.*, **43**, 2789 (1978).
[60] Smit, W. A., Gybin, A. S., Bogdanov, V. S., Krimer, M. Z., and Vorobieva, E. A., *Tetrahedron Lett.*, **1978**, 1085.
[61] Csizmadia, V. M., *Prog. Theor. Org. Chem.*, **1977**, 280; *Chem. Abs.*, **88**, 169438 (1978).
[62] Schmid, G. H., Garratt, D. G., and Yeronshalmi, S., *J. Org. Chem.*, **43**, 3764 (1978).
[63] Zefirov, N. S., Smit, V. A., Bodrikov, I. V., and Krimer, M. Z., *Dokl. Akad. Nauk SSSR*, **240**, 858 (1978); *Chem. Abs.*, **89**, 89866 (1978).
[64] Kartashov, V. R., Bodrikov, I. V., Skorobogatova, E. V., and Zefirov, N. S., *Phosphorus Sulfur*, **3**, 213 (1977); *Chem. Abs.*, **88**, 169363 (1978).
[65] Zefirov, N. S., Kirin, V. N., Potekhin, K. A., Koz'min, A. S., Sadovaya, N. K., Kurkutova, E. N., and Bodrikov, I. V., *Zh. Org. Khim.*, **14**, 1224 (1978); *Chem. Abs.*, **89**, 107841 (1978).
[66] Lok, W. N., and Ward, A. D., *Aust. J. Chem.*, **31**, 605 (1978).
[67] Trost, B. M., Ochiai, M., and McDougal, P. G., *J. Am. Chem. Soc.*, **100**, 7103 (1978).
[68] Kartashov, V. R., Akimkina, N. F., Skorobogatova, E. V., and Sanina, N. L., *Kinet. Katal.*, **19**, 785 (1978); *Chem. Abs.*, **89**, 162686 (1978).
[69] Cauquis, G., and Pierre, G., *Tetrahedron*, **34**, 1475 (1978).
[70] Schmid, G. H., and Garratt, D. G., *Tetrahedron*, **34**, 2869 (1978).
[71] Garratt, D. G., *Can. J. Chem.*, **56**, 2184 (1978).
[72] Clive, D. L. J., Chittattu, G., and Wong, C. K., *J. Chem. Soc., Chem. Commun.*, **1978**, 441.
[73] Clive, D. L. J., Chittattu, G., and Wong, C. K., *Can. J. Chem.*, **55**, 3894 (1977).
[74] Garratt, D. G., *Tetrahedron Lett.*, **1978**, 1915.
[75] Hori, T., and Sharpless, K. B., *J. Org. Chem.*, **43**, 1689 (1978).
[76] Gregorcic, A., and Zupan, M., *Tetrahedron*, **33**, 3243 (1977).
[77] Ismailov, A. G., Rustamov, M. A., and Akhmedov, A. A., *Tezisy Dokl.-Vses. Konf. "Stereokhim. Konform. Anal. Org. Neftekhim. Sint."*, 3rd, **1976**, 106; *Chem. Abs.*, **88**, 169493 (1978).
[78] Volens, T., *Tezisy Dokl.–Resp. Konf. Molodykh. Uch.-Khim.*, 2nd, **1977**, 59; *Chem. Abs.*, **89**, 89874 (1978).
[79] Griesbaum, K., Singh, A., and El Abed, M., *Tetrahedron Lett.*, **1978**, 1159.
[80] Shestakov, G. K., Bel'skii, F. I., Airyan, S. M., and Temkin, O. N., *Kinet. Katal.*, **19**, 334 (1978); *Chem. Abs.*, **89**, 5641 (1978).
[81] Shestakov, G. K., Airyan, S. M., and Temkin, O. N., *Tezisy Dokl.-Vses. Konf. Khim. Atsetilena*, 5th, **1975**, 365; *Chem. Abs.*, **89**, 59400 (1978).
[82] Staab, H. A., Wittig, C. M., and Naab, P., *Chem. Ber.*, **111**, 2965 (1978).
[83] Naab, P., and Staab, H. A., *Chem. Ber.*, **111**, 2982 (1978).
[84] Mel'nikov, G. D., Mel'nikova, S. P., and Kovaleva, L. A., *Zh. Org. Khim.*, **14**, 701 (1978); *Chem. Abs.*, **89**, 41887 (1978).
[85] Tamao, K., Yoshida, J., Takahashi, M., Yamamoto, H., Kakui, T., Matsumoto, H., Kurita, A., and Kumada, M., *J. Am. Chem. Soc.*, **100**, 290 (1978).
[86] Chwang, W. K., Nowlan, V. J., and Tidwell, T. T., *J. Am. Chem. Soc.*, **99**, 7233 (1977).
[87] Pasto, D. J., and Gadberry, J. F., *J. Am. Chem. Soc.*, **100**, 1469 (1978).
[88] Chwang, W. K., and Tidwell, T. T., *J. Org. Chem.*, **43**, 1904 (1978).
[89] Bunton, C. A., Rivera, F., and Sepulveda, L., *J. Org. Chem.*, **43**, 1166 (1978).
[90] Lajunen, M., and Sura, T., *Tetrahedron*, **34**, 189 (1978).
[91] Volkov, A. N., Khudyakova, A. N., Kudyakova, R. N., Keiko, V. V., and Trofimov, B. A., *Izv. Akad. Nauk SSSR, Ser. Khim.*, **1978**, 153; *Chem. Abs.*, **88**, 151686 (1978).
[92] Macomber, R. S., *J. Org. Chem.*, **42**, 3297 (1977).
[93] Abram, T. S., Crawford, W., Knipe, A. C., and Watts, W. E., *Proc. Roy. Ir. Acad.*, **77B**, 317 (1977); *Chem. Abs.*, **89**, 146040 (1978).
[94] Zil'berman, E. N., Trachenko, V. I., Danov, S. M., and Shipunova, N. R., *Izv. Vyssh. Uchebn. Zaved., Khim. Khim. Tekhnol.*, **20**, 1141 (1977); *Chem. Abs.*, **87**, 183830 (1977).
[95] Adams, D. R., and Bhatnagar, S. P., *Synthesis*, 1977, 661; *Chem. Abs.*, **88**, 5618 (1978).
[96] Corvers, A., Scheers, P. C. H., Castenmiller, W. A. M., and Buck, H. M., *Tetrahedron*, **34**, 457 (1978).
[97] Various authors, *J. Organomet. Chem.*, **156** (1978).
[98] Clark, T., and Schleyer, P. von R., *J. Organomet. Chem.*, **156**, 191 (1978).

[99] Dewar, M. J. S., and McKee, M. L., *Inorg. Chem.*, **17**, 1075 (1978); *Chem. Abs.*, **88**, 135956 (1978).
[100] Dasgaupta, S., Datta, M. K., and Datta, R., *Tetrahedron Lett.*, **1978**, 1309.
[101] Chi-San Chen, J., *J. Organomet. Chem.*, **156**, 213 (1978).
[102] Jacob, P., *J. Organomet. Chem.*, **156**, 101 (1978).
[103] Kabalka, G. W., Newton, R. J., and Jacobus, J., *J. Org. Chem.*, **43**, 1567 (1978).
[104] Mandal, A. K., and Yoon, N. M., *J. Organomet. Chem.*, **156**, 183 (1978).
[105] Brown, H. C., Liotta, R., and Kramer, G. W., *J. Org. Chem.*, **43**, 1058 (1978).
[106] Brown, H. C., and Ravindran, N., *J. Am. Chem. Soc.*, **99**, 7097 (1977).
[107] Masuda, Y., Arase, A., and Suzuki, A., *Chem. Lett.*, **1978**, 665.
[108] Moreau, R., Adam, Y., and Loiseau, P., *C. R. Hebd. Séances Acad. Sci.*, **287C**, 39 (1978).
[109] Isagawa, K., Tatsumi, K., and Otsuji, Y., *Chem. Lett.*, **1977**, 1117.
[110] Kurosawa, H., Kitano, R., and Sasaki, T., *J. Chem. Soc., Dalton Trans.*, **1978**, 234.
[111] Leon, M. A., and Cabaleiro, M. C., *An. Asoc. Quim. Argent.*, **64**, 331 (1976); *Chem. Abs.*, **89**, 107486 (1978).
[112] Kang, H., Maudlin, C. H., Cole, T., Slegeir, W., Cann, K., and Pettit, R., *J. Am. Chem. Soc.*, **99**, 8323 (1977).
[113] Rosas, N., Gomez-Lara, J., Cabrera, A., and Alvarez, C., *Rev. Latinoam. Quim.*, **8**, 212 (1977); *Chem. Abs.*, **88**, 6062 (1978).
[114] Tjan, P. W. H. L., and Scholten, J. J. F., *Proc. Int. Congr. Catal.*, 6th, **1**, 488 (1977); *Chem. Abs.*, **88**, 5823 (1978).
[115] Consiglio, G., von Bezard, D. A., Morandini, F., and Pino, P., *Helv. Chim. Acta*, **61**, 1703 (1978).
[116] Cesarotti, E., Fusi, A., Ugo, R., and Zanderighi, G. M., *J. Mol. Catal.*, **4**, 205 (1978); *Chem. Abs.*, **89**, 89913 (1978).
[117] Botteghi, C., Branca, M., Marchetti, M., and Saba, A., *J. Organomet. Chem.*, **161**, 197 (1978).
[118] Kawabata, Y., Susuki, T. M., and Ogata, I., *Chem. Lett.*, **1978**, 361.
[119] Reikhsfel'd, V. O., Khvatova, T. P., and Astrakhanov, M. I., *Zh. Obshch. Khim.*, **47**, 2625 (1977); *Chem. Abs.*, **88**, 49933 (1978).
[120] Green, M., Spencer, J. L., Stone, F. G. A., and Tsipis, C. A., *J. Chem. Soc., Dalton Trans.*, **1977**, 1519.
[121] Green, M., Spencer, J. L., Stone, F. G. A., and Tsipis, C. A., *J. Chem. Soc., Dalton Trans.*, **1977**, 1525.
[122] Benkeser, R. A., Merritt, F. M., and Roche, R. T., *J. Organomet. Chem.*, **156**, 235 (1978).
[123] Salimgareeva, I. M., Kaverin, V. V., and Jurjev, V. P., *J. Organomet. Chem.*, **148**, 23 (1978).
[124] Ojima, I., Yamamoto, K., Kumada, M., *Aspects Homogeneous Catal.*, **3**, 185 (1977); *Chem. Abs.*, **88**, 189277 (1978).
[125] Tamao, K., Okazaki, S., and Kumada, M., *J. Organomet. Chem.*, **146**, 87 (1978).
[126] Ojima, I., and Kumagai, M., *J. Organomet. Chem.*, **157**, 359 (1978).
[127] Srivastava, G., *J. Organomet. Chem.*, **152**, 39 (1978).
[128] Gershanova, E. L., Stratonova, E. I., Sorokin, M. F., and Mikhitarova, Z. A., *Deposited Doc. 1976, VINITI 792*; *Chem. Abs.*, **88**, 61683 (1978).
[129] Grinenko, S. B., and Belousov, V. M., *Metallokompleksnyi Katal.*, **1977**, 40; *Chem. Abs.*, **89**, 23239 (1978).
[130] Bach, R. D., Willis, C. L., and Domagala, J. M., *Prog. Theor. Org. Chem.*, **1977**, 221; *Chem. Abs.*, **88**, 169436 (1978).
[131] Plesnicar, B., Tasevski, M., and Azman, A., *J. Am. Chem. Soc.*, **100**, 743 (1978).
[132] Sarancha, V. N., Abadzhev, S. S., and Romanyuk, I. M., *Kinet. Katal.*, **18**, 1612 (1977); *Chem. Abs.*, **88**, 120336 (1978).
[133] Makitra, R. G., and Pirig, Y. N., *Zh. Fiz. Khim.*, **52**, 785 (1978); *Chem. Abs.*, **88**, 189839 (1978).
[134] Pizzala, L., Aycard, J.-P., and Bodot, H., *J. Org. Chem.*, **43**, 1013 (1978).
[135] Sorokin, M. F., Gershanova, E. L., Stratonova, E. I., and Bulavintseva, T. G., *Izv. Vyssh. Uchebn. Zaved., Khim. Khim. Tekhnol.*, **20**, 1781 (1977); *Chem. Abs.*, **88**, 120340 (1978).
[136] Gedra, A., *Magy. Kem. Lapja*, **33**, 259 (1978); *Chem. Abs.*, **89**, 162778 (1978).
[137] Sapunov, V. N., Litvintsev, I. Y., Magomedov, G. I., and Margitfal'vi, I., *Deposited Doc., 1976, VINITI 1451*; *Chem. Abs.*, **88**, 61797 (1978).

[138] Sapunov, V. N., Litvintsev, I. Y., Margitfal'vi, I., and Lebedev, N. N., *Deposited Doc., 1976, VINITI 1446*; *Chem. Abs.*, **88**, 50009 (1978).
[139] Costa Novella, E., Martinez de la Cuesta, P. J., Rus Martinez, E., and Galleja Pardo, G., *An. Quim.*, **73**, 1198 (1977); *Chem. Abs.*, **89**, 128828 (1978).
[140] Friedrich, L. E., Leckonby, R. A., Stout, D. M., and Lam, Y.-S. P., *J. Org. Chem.*, **43**, 604 (1978).
[141] Hanzlik, R. P., and Hilbert, J. M., *J. Org. Chem.*, **43**, 610 (1978).
[142] Gillies, C. W., *J. Am. Chem. Soc.*, **99**, 7239 (1977).
[143] Vaccani, S., Kuehne, H., Bauder, A., and Guenthard, H. H., *Chem. Phys. Lett.*, **50**, 187 (1977); *Chem. Abs.*, **87**, 200513 (1977).
[144] Senda, Y., Kamiyama, S., and Imaizumi, S., *J. Chem. Soc., Perkin Trans.* 1, **1978**, 530.
[145] Hosokawa, T., Miyagi, S., Murahashi, S.-I., and Sonada, A., *J. Org. Chem.*, **43**, 719 (1978).
[146] Soderquist, J. A., and Thompson, K. L., *J. Organomet. Chem.*, **159**, 237 (1978).
[147] Shabarov, Y. S., Mochalov, S. S., Oretskaya, T. S., and Karpova, V. V., *J. Organomet. Chem.*, **150**, 7 (1978).
[148] Spear, R. J., and Jensen, W. A., *Tetrahedron Lett.*, **1977**, 4535.
[149] Adam, W., and Sakanishi, K., *J. Am. Chem. Soc.*, **100**, 3935 (1978).
[150] Yatagai, H., Yamamoto, Y., Muruyama, K., Sonada, A., and Murahashi, S.-I., *J. Chem. Soc., Chem. Commun.*, **1977**, 852.
[151] Knifton, J. F., *J. Mol. Catal.*, **2**, 293 (1977); *Chem. Abs.*, **87**, 183851 (1977).
[152] Backvall, J.-E., *Tetrahedron Lett.*, **1978**, 163.
[153] Murray, T. F., Varma, V., and Norton, J. R., *J. Am. Chem. Soc.*, **99**, 8085 (1977).
[154] Arai, I., and Daves, G. D., Jr., *J. Am. Chem. Soc.*, **100**, 287 (1978).
[155] Stille, J. K., and Divakaruni, R., *J. Am. Chem. Soc.*, **100**, 1303 (1978).
[156] Van Horn, D. E., Valente, L. F., Idacavage, M. J., and Negishi, E.-I., *J. Organomet. Chem.*, **156**, C20 (1978).
[157] Eisch, J. J., Manfre, R. J., and Kamar, D. A., *J. Organomet. Chem.*, **159**, C13 (1978).
[158] Alper, H., Currie, J. K., and des Abbayes, H., *J. Chem. Soc., Chem. Commun.*, **1978**, 311.
[159] Gill, G. B., Parrott, S. J., and Wallace, B., *J. Chem. Soc., Chem. Commun.*, **1978**, 655.
[160] Vaglio, G. A., Osella, D., and Valle, M., *Transition Met. Chem. (Weinheim)*, **2**, 94 (1977); *Chem. Abs.*, **87**, 167250 (1977).
[161] Matsumoto, H., Nakano, T., Nikaido, T., and Nagai, Y., *Chem. Lett.*, **1978**, 115.
[162] Paleta, O., *Fluorine Chem. Rev.*, **8**, 39 (1977); *Chem. Abs.*, **88**, 5668 (1978).
[163] Allen, P. E. M., and Lough, R. M., *Mol. Rate Processes, Pap. Symp.*, **1975**, C3; *Chem. Abs.*, **88**, 135973 (1978).
[164] Heumann, A., and Kraus, W., *Tetrahedron*, **34**, 405 (1978).
[165] Driguez, H., Vermes, J.-P., and Lessard, J., *Can. J. Chem.*, **56**, 119 (1978).
[166] Tanaskov, M. M., and Stadnichuk, M. D., *Zh. Obshch. Khim.*, **48**, 1140 (1978); *Chem. Abs.*, **89**, 41911 (1978).
[167] Jackson, W. R., Norman, J. W., and Rae, I. D., *Tetrahedron Lett.*, **1978**, 2061.
[168] Ashby, E. C., Lin, J.-J., and Goel, A. B., *J. Org. Chem.*, **43**, 183 (1978).
[169] Hoz, S., and Spiezman, D., *Tetrahedron Lett.*, **1978**, 1775.
[170] Miller, J. M., So, K.-H., and Clark, J. H., *J. Chem. Soc., Chem. Commun.*, **1978**, 466.
[171] Dionne, G., and Eugel, Ch. R., *Can. J. Chem.*, **56**, 419 (1978).
[172] Toromanoff, E., *Tetrahedron*, **34**, 2105 (1978).
[173] Yunker, M. B., Plaumann, D. E., and Fraser-Reid, B., *Can. J. Chem.*, **55**, 4002 (1977).
[174] Mukaiyama, T., Hirako, Y., and Takeda, T., *Chem. Lett.*, **1978**, 461.
[175] Colonna, S., Hiemstra, H., and Wynberg, H., *J. Chem. Soc., Chem. Commun.*, **1978**, 238.
[176] Wynberg, H., and Greijdanus, B., *J. Chem. Soc., Chem. Commun.*, **1978**, 427.
[177] Lefour, J.-M., and Loupy, A., *Tetrahedron*, **34**, 2597 (1978).
[178] Sauvetre, R., Roux-Schmitt, M.-C., and Seyden-Penne, J., *Tetrahedron*, **34**, 2135 (1978).
[179] Still, W. C., and Mitra, A., *Tetrahedron Lett.*, **1978**, 2659.
[180] Luchetti, J., and Krief, A., *Tetrahedron Lett.*, **1978**, 2697.
[181] Mulzer, J., Harz, G., Kuhl, U., and Brüntrup, G., *Tetrahedron Lett.*, **1978**, 2949.
[182] Ogura, K., Yashita, M., and Tsuchihashi, G., *Tetrahedron Lett.*, **1978**, 1303.
[183] Bryson, T. A., Dardis, R. E., and Gammill, R. B., *Tetrahedron Lett.*, **1978**, 743.
[184] Kaiser, E. M., Knutson, P. L., and McLure, J. R., *Tetrahedron Lett.*, **1978**, 1747.
[185] Sjöhlm, R., *Acta Chem. Scand.*, **32B**, 105 (1978).
[186] Davis, B. R., and Johnson, S. J., *J. Chem. Soc., Chem. Commun.*, **1978**, 614.

[187] Lehmkuhl, H., and Bergstein, W., *Justus Liebigs Ann. Chem.*, **1978**, 1436.
[188] Lehmkuhl, H., Reinehr, D., Mehler, K., Schomburg, G., Kötter, H., Henneberg, D., and Schroth, G., *Justus Liebigs Ann. Chem.*, **1978**, 1449.
[189] Eisenstein, O., Procter, G., and Dunitz, J. D., *Helv. Chim. Acta*, **61**, 2538 (1978).
[190] Courtois, G., and Miginiac, L., *C. R. Hebd. Séances Acad. Sci.*, **285C**, 207 (1977).
[191] Loots, M. J., and Schwartz, J., *J. Am. Chem. Soc.*, **99**, 8045 (1977).
[192] Ager, D. J., and Fleming, I., *J. Chem. Soc., Chem. Commun.*, **1978**, 177.
[193] Litvinenko, L. M., Popov, A. F., Kostenko, L. I., Kravchenko, V. V., and Vegh, D., *Dokl. Akad. Nauk SSSR*, **238**, 622 (1978); *Chem. Abs.*, **88**, 120267 (1978).
[194] Adomeniene, O., *Tezisy Dokl.-Resp. Konf. Molodykh Uch.-Khim.*, 2nd, **1977**, 37; *Chem. Abs.*, **89**, 107313 (1978).
[195] Gossen, L. P., Krulikovskaya, E. A., and Kumok, V. N., *Zh. Fiz. Khim.*, **52**, 1817 (1978); *Chem. Abs.*, **89**, 146084 (1978).
[196] Evans, D. A., Hurst, K. M., and Takacs, J. M., *J. Am. Chem. Soc.*, **100**, 3467 (1978).
[197] Paquette, L. A., Carr, R. V. C., and Bellamy, F., *J. Am. Chem. Soc.*, **100**, 6764 (1978).
[198] Fellons, R., Luft, R., and Vellutini, M.-J., *Bull. Soc. Chim. Fr. II*, **1977**, 993.
[199] Little, R. D., and Dawson, J. R., *J. Am. Chem. Soc.*, **100**, 4607 (1978).
[200] Augustin, M., Werudl, R., Koehler, M., and Ruettinger, H. H., *Pharmazie*, **33**, 191 (1978); *Chem. Abs.*, **89**, 107676 (1978).
[201] Foster, C. H., and Payne, D. A., *J. Am. Chem. Soc.*, **100**, 2834 (1978).
[202] Kutyrev, G. A., Kashina, N. V., Ishmaeva, E. A., Cherkasov, R. A., and Pudovik, A. N., *Zh. Obshch. Khim.*, **47**, 2460 (1977); *Chem. Abs.*, **88**, 88727 (1978).
[203] Baldwin, J. E., Thomas, R. C., Kruse, L. I., and Silberman, L., *J. Org. Chem.*, **42**, 3846 (1977).
[204] Fountain, K. R., and Gerhardt, G., *Tetrahedron Lett.*, **1978**, 3985.
[205] Stork, G., Taber, D. F., and Marx, M., *Tetrahedron Lett.*, **1978**, 2445.
[206] Santos, J., Schalscha, D., and Valderrama, J., *An. Quim.*, **73**, 855 (1977); *Chem. Abs.*, **87**, 200408 (1977).
[207] Ivachnyuk, M. S., and Perveev, F. Y., *Zh. Org. Khim.*, **14**, 214 (1978); *Chem. Abs.*, **88**, 169272 (1978).
[208] Scherrer, V., Jackson-Mully, M., Zsindely, J., and Schmid, H., *Helv. Chim. Acta*, **61**, 716 (1978).
[209] Hosokawa, T., Miyagi, S., Murahashi, S.-I., and Sonada, A., *J. Chem. Soc., Chem. Commun.*, **1978**, 687.
[210] Spitzner, D., *Angew. Chem. Int. Edn.*, **17**, 197 (1978).
[211] Somei, M., Kitamura, R., Fujii, H., Hashiba, K., Kawai, S., and Kaneko, C., *J. Chem. Soc., Chem. Commun.*, **1977**, 899.
[212] Hart, H., and Dunkelblum, E., *J. Am. Chem. Soc.*, **100**, 5141 (1978).
[213] Kirby, A. J., and Logan, C. J., *J. Chem. Soc., Perkin Trans. 2*, **1978**, 642.
[214] Szmuszkovicz, J., Musser, J. H., and Laurian, L. G., *Tetrahedron Lett.*, **1978**, 701.
[215] Lebedev, B. V., Milov, V. I., Rabinovich, I. B., Luneva, L. K., and Sladkov, A. M., *Tezisy Dokl.-Vses. Konf. Khim. Atsetilina*, 5th, **1975**, 453; *Chem. Abs.*, **88**, 169472 (1978).
[216] Korzhova, N. V., Pisareva, V. S., Slyusareva, O. M., and Korshunov, S. P., *Zh. Org. Khim.*, **13**, 2555 (1977); *Chem. Abs.*, **88**, 73823 (1978).
[217] Slyusareva, O. M., Pisareva, V. S., Korshunov, S. P., and Kazantseva, V. M., *Zh. Org. Khim.*, **13**, 2285 (1977); *Chem. Abs.*, **88**, 49931 (1978).
[218] Scott, F., Cahiez, G., Normant, J. F., and Villieras, J., *J. Organomet. Chem.*, **144**, 13 (1978).
[219] Treichel, P. M., and Wong, W. K., *J. Organomet. Chem.*, **157**, C5 (1978).
[220] Treichel, P. M., Wong, W. K., and Calabrese, J. C., *J. Organomet. Chem.*, **159**, C20 (1978).
[221] Berlan, J., Battioni, J.-P., and Koosha, K., *J. Organomet. Chem.*, **152**, 359 (1978).
[222] Berlan, J., and Koosha, K., *J. Organomet. Chem.*, **153**, 99 (1978).
[223] Berlan, J., and Koosha, K., *J. Organomet. Chem.*, **153**, 107 (1978).
[224] Krueger, W. E., McLean, M. B., Rizwaniuk, A., Maloney, J. R., Behelfer, G. L., and Boland, B. E., *J. Org. Chem.*, **43**, 2877 (1978).
[225] Pelter, A., and Hughes, L., *J. Chem. Soc., Chem. Commun.*, **1977**, 913.
[226] Pal, B. C., *J. Am. Chem. Soc.*, **100**, 5170 (1978).
[227] Tamaru, Y., Harada, T., Iwamoto, H., and Yoshida, Z., *J. Am. Chem. Soc.*, **100**, 5221 (1978).

[228] Andrianov, K. A., Sidorov, V. I., and Filimonova, M. I., *Izv. Nauk SSSR, Ser. Khim.*, **1978**, 460; *Chem. Abs.*, **88**, 151700 (1978).
[229] Clark, D. R., and Miles, P., *J. Chem. Res.* (*S*), **1978**, 124.
[230] Inagaki, Y., and Okazaki, R., *Yuki Gosei Kagaku Kyokaishi*, **36**, 1 (1978); *Chem. Abs.*, **89**, 41580 (1978).
[231] Glass, W. K., King, I. J., and Shiels, A., *Inorg. Chim. Acta*, **25**, 157 (1977); *Chem. Abs.*, **88**, 104528 (1978).
[232] Chimishkyan, A. L., Dragalov, V. V., Kelekhsaeva, E. A., and Dvornikova, E. E., *Deposited Doc.*, *1976*, *VINITI 3300*; *Chem. Abs.*, **89**, 128751 (1978).
[233] Chimishkyan, A. L., Dragalov, V. V., and Sirovskii, F. S., *Deposited Doc. 1976*, *VINITI 3299*; *Chem. Abs.*, **89**, 107726 (1978).
[234] Griger'eva, V. A., Baturin, S. M., and Entelis, S. G., *Kinet. Katal.*, **18**, 1404 (1977); *Chem. Abs.*, **88**, 135853 (1978).
[235] Al-Najjar, I. M., and Green, M., *J. Chem. Soc., Chem. Commun.*, **1977**, 926.
[236] Collins, D. J., Jackson, W. R., and Timms, R. N., *Aust. J. Chem.*, **30**, 2167 (1977).
[237] Trost, B. M., Weber, L., Strege, P. E., Fullerton, T. J., and Dietsche, T. J., *J. Am. Chem. Soc.*, **100**, 3416 (1978).
[238] Trost, B. M., Weber, L., Strege, P. E., Fullerton, T. J., and Dietsche, T. J., *J. Am. Chem. Soc.*, **100**, 3426 (1978).
[239] Trost, B. M., and Verhoeven, T. R., *J. Am. Chem. Soc.*, **100**, 3435 (1978).
[240] Tamaru, Y., Kagotani, M., and Yoshida, Z., *J. Chem. Soc., Chem. Commun.*, **1978**, 367.
[241] Snider, B. B., Karras, M., and Conn, R. S. E., *J. Am. Chem. Soc.*, **100**, 4624 (1978).
[242] Hong, P., Yamazaki, H., Sonogashira, K., and Hagihara, N., *Chem. Lett.*, **1978**, 535.

CHAPTER 13

Addition Reactions: Cycloaddition

C. Brown

Chemical Laboratory, University of Kent, Canterbury

Introduction

Cycloaddition reactions in general have been reviewed,[1] together with gas-phase polar cycloaddition reactions,[2] the synthesis of carbocyclic spiro-compounds by cycloaddition reactions,[3] and the use of ketens and ketenimines in the synthesis of heterocyclic compounds by way of cycloaddition processes.[4] A comprehensive review of reaction and activation volumes has also appeared:[5] this includes data on 2+2-, 2+4- and 2+3-cycloaddition reactions. The consistency of the conclusions reached about reaction feasibility using the Woodward–Hoffmann rule or the Fukui frontier orbital approach has been rationalized in terms of HOMO–HOMO interactions,[6] and predictions of "allowedness" of pericyclic reactions using calculated resonance energy differences (RED) between ground and transition states for such reactions have been shown to give reasonable correlations with experimental findings.[7] A perturbation density matrix method for the study of intramolecular cycloaddition processes has been described: the application of the method to systems bearing heteroatoms was particularly stressed.[8] A simple model for predicting qualitatively the effect of substituents on a variety of thermal pericyclic reactions, including 2+2- and 2+4-cycloadditions, has been proposed,[9] and 1,1-difluoroallene has been suggested as a highly reactive probe for concertedness in cycloaddition reactions.[10]

2+2-Cycloaddition

The linear combination of fragment configurations (LCFC) method has been applied to $\pi^2+\pi^2$-cycloadditions: contrary to current thinking, the $\pi^2s+\pi^2s$ mechanism is suggested for non-ionic 2+2-cycloadditions, and also for ionic 2+2-cycloadditions. Furthermore, the possible intermediacy of exciplexes in 2+2-photocycloadditions is predicted by this method.[11] Ionization potentials have been shown experimentally to play a key role in determining the reactivity of olefins in 2+2-cycloaddition reactions.[12]

In an elegantly conceived and executed set of experiments, Doering and Guyton have convincingly eliminated both the $\pi^2s+\pi^2a$-cycloaddition and stereorandom mechanisms for dimerization of *cis*-1,2-dideuterioacrylonitrile:[13] the results are more consistent with a diradical process, although the exact structures of those diradical species still await clarification. Diradical species were also invoked to rationalize the formation of (**2**) and (**3**) from (**1**) at 190 °C in dilute solution.[14] The methylenecyclopropanes (**4**) undergo self- and cross-dimerization reactions at temperatures near 200 °C to give mixtures of the *cis*- and *trans*-7,8-disubstituted dispiro[2.0.2.2]octanes (**5**). With acrylonitrile the 2+2-cycloadduct (**6**) is formed.[15]

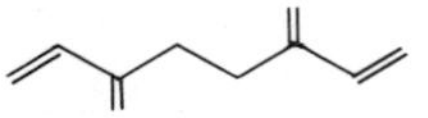

(**1**) 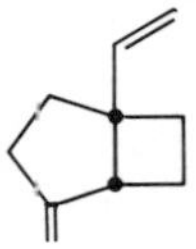 (**2**) 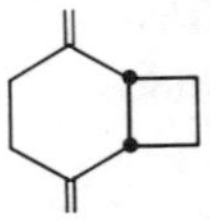 (**3**)

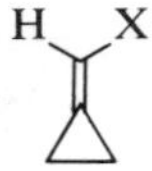

H X

(**4**)

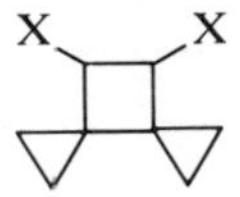

X X

(**5**)

X = Cl, Br, or OEt

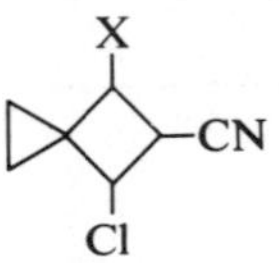

X CN Cl

(**6**)

Cyclodimerization of the *O*-silylated enol of biacetyl has been reported.[16] The 2+2-cycloadduct (**8**) is among the various products formed from the reaction of a thiophenamine (**7**) with dimethyl acetylenedicarboxylate.[17] Addition of TCNE to the *trans*-fixed 1,3-diene (**9**) gives the adduct (**10**) *via* a dipolar intermediate.[18]

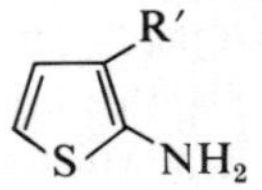

R′ S NH_2

(**7**)

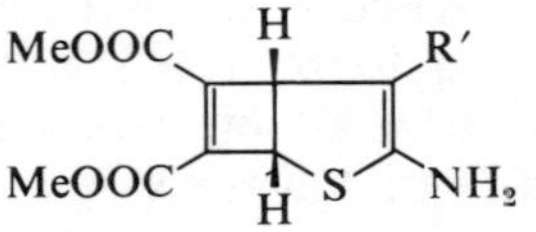

MeOOC MeOOC H H R′ S NH_2

(**8**)

R′ = COOMe or COOEt

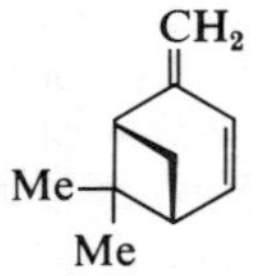

CH_2 Me Me

(**9**)

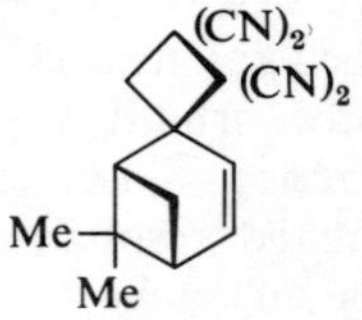

$(CN)_2$ $(CN)_2$ Me Me

(**10**)

It has been proposed that the polarity of the exocyclic double bond in 7-methylenenorbornadiene accounts for its selective participation in 2+2-cycloadditions of this hydrocarbon with diphenylketen and tetracyanoethylene.[19] A variety of analogous *exo*-methylene compounds give similar products with diphenylketen, there being apparently no correlation between the rate of addition and the ionization potential of the double bond,[20] despite the conclusions referred to above.[12]

There have been several studies of 2+2-cycloadditions to vinyl ethers and related compounds: for example, activation parameters for addition of vinyl ethers to TCNE in a variety of solvents were measured, and the results interpreted in terms of solvation-energy differences in the reactions.[21] The same group concluded that donor–acceptor interactions are largely responsible for the reactivity order $PhSCH{=}CH_2 > PhSeCH{=}CH_2 > PhOCH{=}CH_2$ for addition to TCNE;[22] the effects of pressure on such cycloadditions to TCNE have been measured.[23]

The nature of the substituents X in the vinylcyclopropyl derivatives (**11**) markedly affects the reactivity of the double bond in 2+2-cycloaddition reactions.[24] Thus (**11a**) and (**11b**) failed to react with TCNE even at 60 °C, whereas (**11c**) reacted smoothly with TCNE to give (**12**).

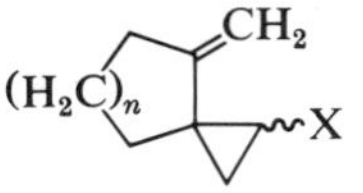

(**11a**) X = CN
(**11b**) X = COOMe
(**11c**) X = CH_2OH

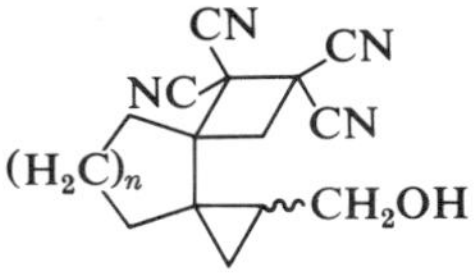

(**12**)

Radical anions derived from olefins such as (**13**) and (**14**) have been shown to undergo $\pi^2+\pi^2$-cycloaddition[25] (Scheme 1).

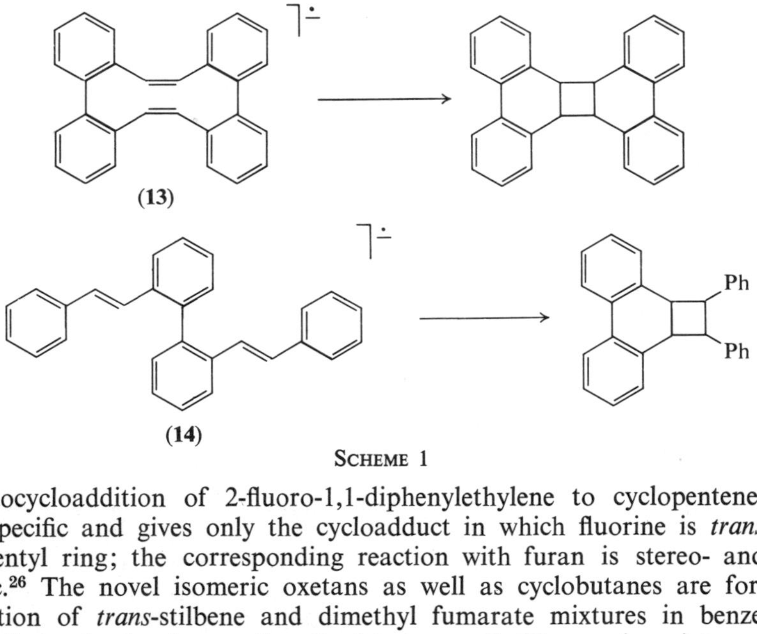

(**13**)

(**14**)

SCHEME 1

Photocycloaddition of 2-fluoro-1,1-diphenylethylene to cyclopentene is *cis*-stereospecific and gives only the cycloadduct in which fluorine is *trans* to the cyclopentyl ring; the corresponding reaction with furan is stereo- and regio-specific.[26] The novel isomeric oxetans as well as cyclobutanes are formed on irradiation of *trans*-stilbene and dimethyl fumarate mixtures in benzene; the oxetan-formation has been rationalized in terms of stilbene-eximer interception by ground-state fumarate.[27]

Evidence for a 1,4-diradical intermediate in the sensitized 2+2-cycloaddition of phenanthrene to dimethyl fumarate has been presented.[28] The photocycloaddition of cyclopenten-2-one to electron-rich alkenes has been studied: with sufficiently powerful π-donor substituents on the alkene, very high degrees of stereo- and regio-specificity can be realized.[29] Dramatic effects on the quantum yield for acenaphthene photodimerization have been observed when liquid-crystal solvents are employed.[30]

2+2-Cycloaddition accompanies the ene reaction of 2-methylpenta-2,3-diene with hexafluorobut-2-yne or dimethyl acetylenedicarboxylate,[31] and the nature of the products obtained from the reaction of enamines with the latter reagent depends on the solvent used: this has been rationalized in terms of the nature of the initial 1 : 1 adduct formed.[32] In polar solvents a zwitterionic species is favoured, but in less polar solvents a "forbidden" $\pi^2s+\pi^2s$ concerted cycloaddition may take place.

The regioselectivity observed in enone–alkyne photoannelation is probably largely controlled by dipole–dipole interactions between the alkyne ground state and the enone triplet (π, π^*) state.[33] The rates of photochemical cycloaddition of cyclohex-2-enone and some 3-substituted derivatives to hex-2-yne at 25 °C ($\rho = 1.554$) are lower than that for addition of cyclopent-2-enone: a non-concerted mechanism involving electrophilic attack of triplet state enone on the triple bond was proposed.[34] The $\pi^2+\sigma^2$-cycloadduct (**15**) has been proposed as an intermediate in the photoaddition of 1,3-dienes to *N*-alkylphthalimides.[35] The first case of a $\pi^2s+\pi^2s$-photocycloaddition between non-activated C=C and N=N bonds has been reported [(**16**)→(**17**)]: photoelectron spectra of the thermally stable (**16**) (up to 250 °C) indicate strong π–π interaction between the C=C and N=N bonds.[36] Dehydrogenation of 2,2′-dihydroxybiphenyls does *not* give benzoxetes, but gives oxepinobenzofurans.[37]

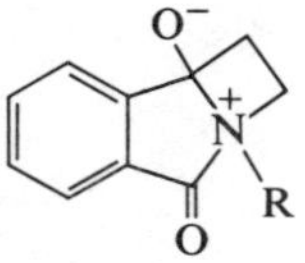

(**15**) 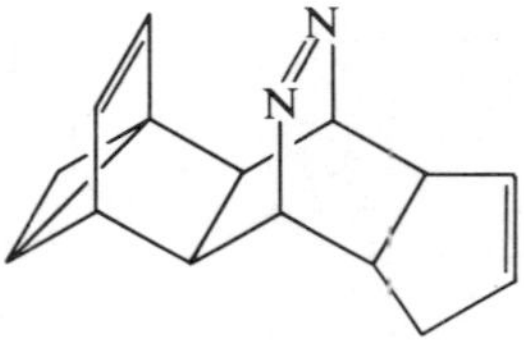 (**16**) 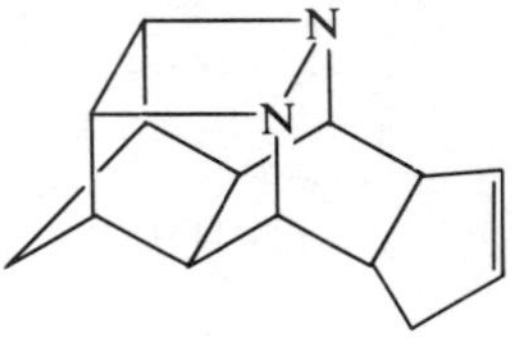 (**17**)

Kinetic studies indicate that the reaction of the *S*,*S*-disubstituted *N*-imidoyl-sulphimides with CS_2 proceeds by way of a 2+2-cycloaddition.[38] *Ab initio* calculations indicate a planar structure for $CH_2{=}SiH_2$ with a π-bonded ground state: despite the relatively high π-bond strength, however, the polarity of the bond and the low energy of the π^* MO predict a low barrier (*ca.* 14 kcal mol^{-1}) for 2+2-cyclodimerization, in keeping with experimental results.[39] Similar dimerization of $Me_2Si{=}CHNp$ and 2+2-cycloaddition to one of the double bonds in 1,3-butadiene has been noted.[40] Despite earlier reports to the contrary, sulphinylamines do not undergo cycloaddition to aldehydes in aprotic solvents.[41] It is still not clear whether the formation of the dioxetan (**18**) from singlet oxygen and 2,3-dihydro-4-methylpyran involves a concerted 2+2-cycloaddition or a stepwise process *via* a perepoxide.[42] On reaction with *C*,*N*-diphenylnitrone the perfluoroalkene, $CF_2{=}CF{-}CF_3$, surprisingly gives the β-lactam (**19**), rather than a 5-membered heterocycle.[43]

Keten fails to dimerize in the gas phase below 1 atm; however, a thorough study of the reverse reaction, the thermolysis of diketen, has been made.[44] Both the C=C

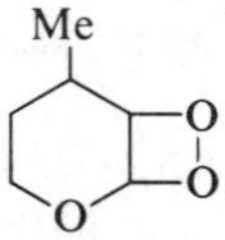

(**18**) 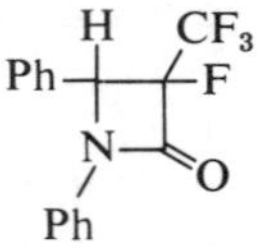 (**19**) 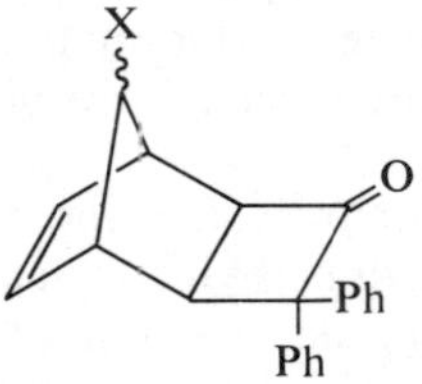 (**20**) (**21**)

X = OBu^t or OAc

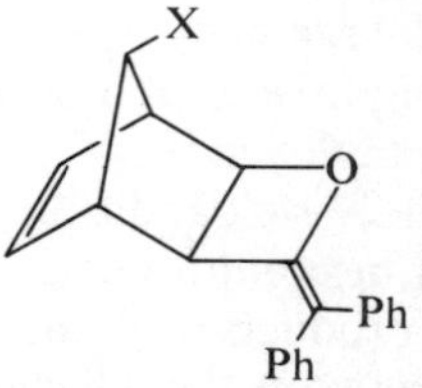

and C=O bonds of diphenylketen add to quadricyclene and its derivatives to give the adducts (**20**) and (**21**).[45]

Carbon-14 and deuterium isotope effects[46] on the rate of addition of styrene to diphenylketen support the hyperconjugation interaction previously proposed for this reaction.[47] The stereochemistry of the cyclobutanones produced on reaction of *tert*-butylcyanoketen with a variety of bicyclo[2.2.1]heptene derivatives was rationalized in terms of a $\pi^2s+\pi^2a$-cycloaddition pathway, with the *tert*-butyl group taking up the *exo*-position in order to minimize steric interactions in the transition state.[48] On reaction with *tert*-butylcyanoketen 2-methylbut-2-ene gives only the 2+2-adduct 2-*tert*-butyl-2-cyano-3,4-dimethylcyclobutanone, in which the alkyl groups are mutually of *cis*-configuration; the regioselectivity is controlled by the coefficients of the AOs making up the olefin HOMO.[49]

Although solvent effects and measurements of $\Delta S^{\ddagger}$ have proved to be mechanistically inconclusive, other product studies on the reaction of *tert*-butylcyanoketen with activated olifins suggest a zwitterionic intermediate for this reaction.[50]

The spiro β-lactone (**23**) is suggested as the key intermediate in the conversion of (**22a**) into (**24**) on reaction with diphenylketen;[51] with the cycloheptatriene complex (**22b**), the 2+2-cycloadduct (**25**) is obtained.[52]

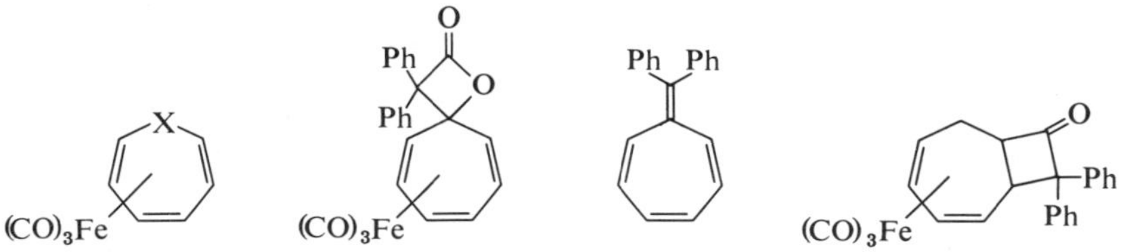

X
(CO)$_3$Fe
(**22a**) X = CO
(**22b**) X = CH_2

O
Ph
O
Ph
(CO)$_3$Fe
(**23**)

Ph Ph
(**24**)

O
Ph
Ph
(CO)$_3$Fe
(**25**)

^{13}C-NMR methods, including NOE determinations, were used to establish the stereochemistry of the 2+2-cycloadducts formed by addition of cyanoketens to formimidate esters.[53] Convincing evidence for the intermediacy of zwitterionic intermediates in the cycloaddition of cyanoketens and imidate esters to give β-lactams has been adduced,[54] and similar independent generation and trapping of the zwitterionic intermediate in keten + keten cycloadditions have been successfully carried out.[55] 2+2-Addition of methyl(phenylthio)keten to alkenes and to imines has been reported; oxidation of the product to the sulphoxide followed by pyrolytic elimination gives products that are equivalent to 2+2-cycloadducts of methyleneketen, CH_2=C=C=O [56] Ketones add to phosphazenes to give crystalline 2+2-cycloadducts,[57] presumably stabilized by the incorporation of the five-membered ring into the phosphorane; this small-ring effect on phosphorane-phosphorus stability has also been exploited in the isolation of the Standinger-type intermediate 2+2-cycloadducts from the reaction of azaphospholes with isocyanates. Similar results were obtained by using azaphospholines.[58]

2+2-Cycloadducts as well as 4+2-cycloadducts were obtained from reaction of *N*-(2-pyridyl)-substituted azomethines with diphenylketen, phenyl isocyanate or diphenylcarbodi-imide.[59] In contrast to the usual finding that aryl isocyanates react with 2π-addends at the C=N π-bond, it has been shown[60] that in CCl_4 solution phenyl isocyanate reacts with MeC≡CNEt$_2$ at the C=O π-bond to give a ketenimine intermediate; it was not possible to distinguish clearly the ıôle of the cyclic and zwitterionic precursors. Isopropylideneketen, Me_2C=C=C=O, reacts with ketens by 2+2-cycloaddition to give both oxetanones and cyclobutane-1,3-diones with ketens, together with its dimer.[61]

2+3-Cycloaddition

A comprehensive and critical review of the various criteria applied to the elucidation of the mechanisms of 1,3-dipolar cycloadditions has been presented: overall, the evidence favours the diradical rather than the concerted type of mechanism for such reactions.[62] However, Harcourt has attempted to reconcile this view with that of Huisgen, and presents a "concerted-diradical" description of these reactions in terms of valence-bond theory.[63]

The success of the frontier molecular orbital approach in rationalizing the regioselectivity of a variety of cycloadditions is undisputed: attention has now been focussed on anomalous results, and more sophisticated methods have been devised to explain these. Thus, a rationale for the apparently "schizophrenic" nature of substituents in cycloaddition reactions has been presented: a key feature of this is an allowance for the possibility of LUMO polarization inversion on formation of reaction transition state complexes.[64] The importance of overlap-repulsion and of overlap-induced polarization in determining the symmetry of transition states for reaction involving the interaction of the three electron pairs in six-electron thermal pericyclic reactions has also been emphasized by Godfrey.[65] Unique long-range interactions between the primary-carbon HOMO of 1,3-dipoles and the secondary adjacent atom LUMO of electron-deficient alkyne dipolarophiles in the 1,3-cycloaddition transition state has been proposed[66] to account for the "anomalous" regioselectivities often encountered in these reactions, but not in the corresponding reactions with olefinic dipolarophiles. However, in the case of intramolecular 1,3-dipolar cycloadditions, steric effects rather than HOMO–LUMO interactions appear to dominate the regioselectivity.[67]

The applicability of the "*exo*-rule" has been explored for a variety of bicyclic olefinic dipolarophiles and 1,3-dipoles.[68–70] Additions of $PhCO\overset{+}{C}{=}N{-}\bar{O}$, $Ph\overset{+}{C}{=}N{-}\bar{N}Ph$, or of *C*-4-nitrophenyl-*N*-phenylnitrone to the norbornadiene (**26**) do not follow the "*exo*-rule"; NMR spectral studies indicate that the *endo*-side of (**26**) is homoconjugated, and this is postulated to be responsible for the formation of *endo*- as well as *exo*-1,3-dipolar addition products.[68] The additions of the nitrile oxides $PhCO\overset{+}{C}{=}N{-}\bar{O}$, $Ph\overset{+}{C}{=}N{-}\bar{O}$ and phenyl azide to the alkenes (**27**) to (**29**) have also been studied: only *endo*-adducts are obtained with (**27**), but with (**28**) only *exo*-adducts are formed. $Ph\overset{+}{C}{=}N{-}\bar{N}Ph$ adds to both bonds of (**27**) to give *endo*-adducts and to those of (**28**) to give *exo*-adducts; on the other hand, (**29**) gives both *exo*- and *endo*-adducts with $PhCO\overset{+}{C}{=}N{-}\bar{O}$. These results have been

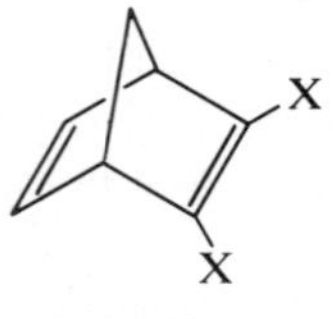

X = H, COOMe, or CN
(**26**)

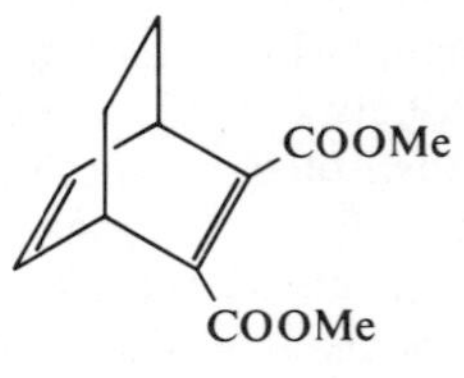

(**27**)

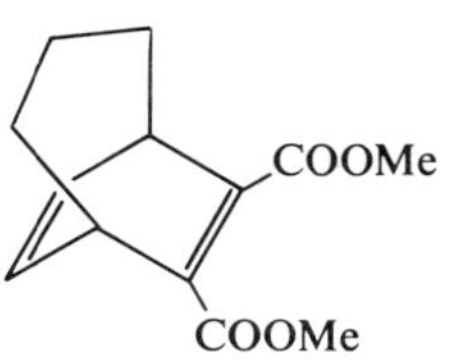

COOMe
COOMe
(28)

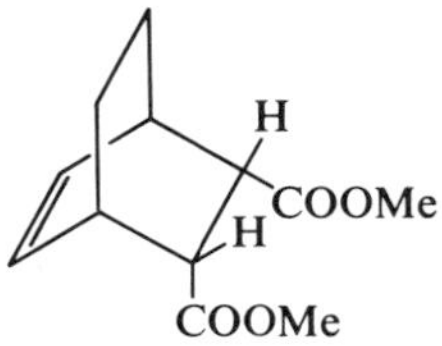

H
COOMe
H
COOMe
(29)

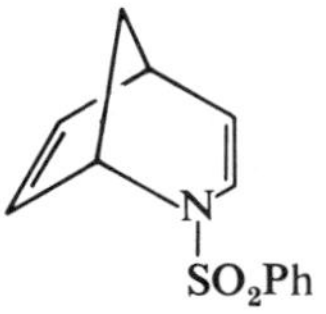

SO_2Ph
(30)

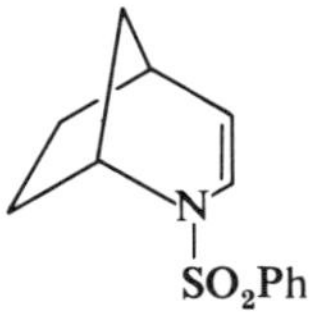

SO_2Ph
(31)

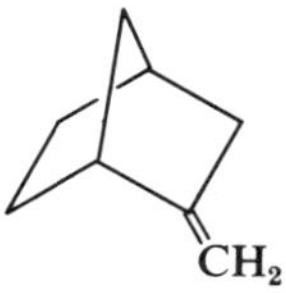

CH_2
(32)

rationalized in terms of the orbital mixing rule.[69] In the case of (**30**) to (**32**) the rule does apply, again as the result of the operation of the orbital-mixing rule.[70]

With customary elegance, Berson and his co-workers[71] have put forward a frontier molecular-orbital explanation of the regioselectivity observed in singlet cycloadditions of 2-methylenecyclopentane-1,3-diyls (**33**), generated from the diazenes (**34**). Cyclopropenones (**35**) and (**36**) react with ketimines to give 2-pyrrolin-4-ones (**37**), *via* the 1,3-dipole (**38**).[72]

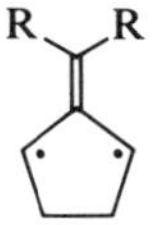

(**33**) 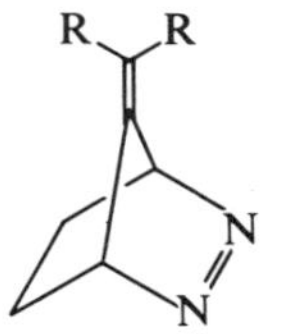 (**34**) R = H, Me, or OMe

(**35**) R = Me
(**36**) R = Ph

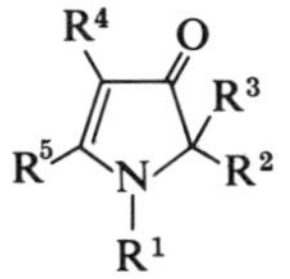

(**37**)
R^3 = Me or Aryl
R^2 = Me or Et
R^1 = Me or Et

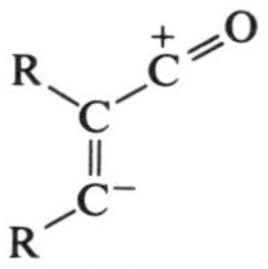

(**38**)

Oxallyl-Fe(II) cations, generated from iron carbonyl and α,α′-dibromo-ketones, add to olefinic dipolarophiles to give cyclopentanones or tetrahydrofuran derivatives[73] *via* catonic intermediates. The addition of diphenylcyclopropenone to azines affords 2,3-diphenyl-2-pyrrolin-4-one derivatives.[74]

The ketiminium salt (**39**) has been synthesized. Deprotonation gives the azomethine ylide (**40**), which dimerizes to a piperazine or adds to dipolarophiles such as norbornene to give (**41**).[75] Thermal opening of a range of 2-alkoxycarbonylaziridines has been found to give the corresponding azomethine ylides, which add to isocyanates and isothiocyanates regiospecifically to produce epimeric imidazolidones and thioimidazolidones, respectively.[76] Reaction of carbon dioxide with propylimine in the presence of Lewis acids in several solvents has been studied: the best system was I_2 in acetone or CH_2Cl_2; a mechanism has been proposed to account for the kinetics, which are of first order in CO_2, second order in imine for oxazolidone formation, and first order in I_2 and second order in imine for polymerization.[77]

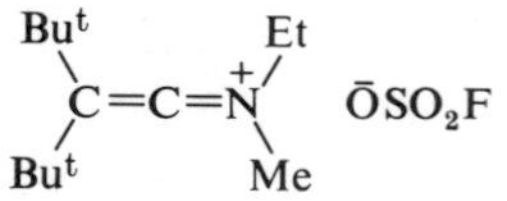

$(Bu^t)_2C{=}C{=}\overset{+}{N}(Et)Me\ \ \bar{O}SO_2F$ **(39)**

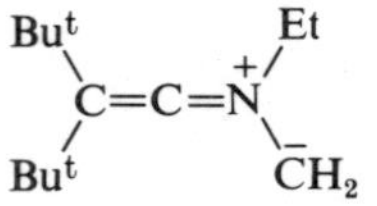

$(Bu^t)_2C{=}C{=}\overset{+}{N}(Et)\bar{C}H_2$ **(40)**

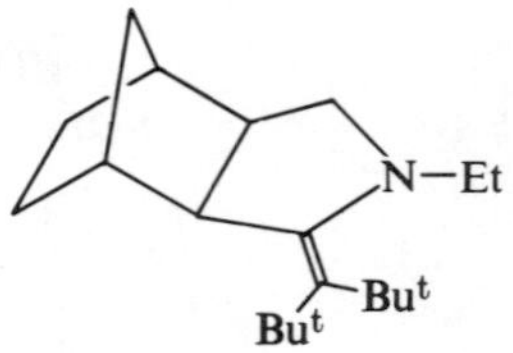

N–Et, Bu^t, Bu^t **(41)**

The rate of reaction of aziridines (**42**) with DMAD depends only on aziridine concentration, and the ρ-value for change in the *N*-aryl substituents is -0.80. These results, taken together with the stereochemistry of the adducts of aziridine (**42a**) to some acetylenes and olefins, indicate a mechanism involving slow conrotatory opening of the aziridine to the azomethine ylide, followed by rapid *cis–cis* dipolar addition. With the fused ring aziridines (**43**) a slow disrotatory opening appears to operate.[78] The ylide derived from thermal ring-opening of *N*-lithiated *cis*-diphenylaziridine adds to a variety of double and triple carbon–carbon bonds to give five-membered nitrogen-containing heterocycles.[79]

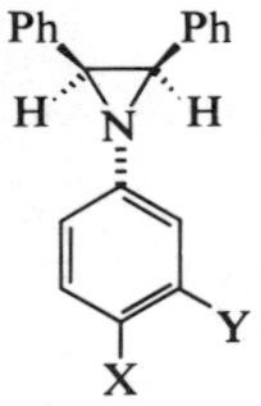

Ph, Ph, H, N, H, Y, X

(**42**) a, X = Y = H
b, X = Me, Y = H
c, X = Cl, Y = H
d, X = H, Y = Cl

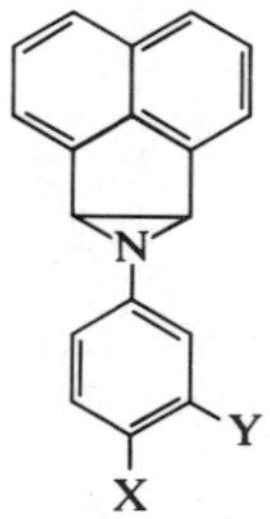

N, Y, X

(**43**) a, X = Y = H
b, X = Me, Y = H
c, X = OMe, Y = H
d, X = Br, Y = H
e, X = NO_2, Y = H
f, X = Y = Cl

The thermal reactions of 2-allyl-2*H*-azirines (**44**), in contrast to the photochemical reactions which are believed to involve discrete nitrile ylide intermediates, probably involve nitrene intermediates, the final pyridine product (**45**) arising from thermal sigmatropic rearrangement of the initially formed bicycloaziridine (**46**).[80] On photolysis the 2-allylphenyl-2*H*-azirines (**47**) yield nitrile ylides; however, the geometric restraints imposed by the system preclude attainment of the "two-plane orientation approach" generally recognized as the one normally involved in 1,3-dipolarcycloadditions of such species.[81] A non-concerted pathway of the type advocated by Firestone[62] probably operates in such cases. The effect of substituents on both the mode[82] and rate[83] of these nitrile ylide cycloadditions has been examined in detail; substituents may affect the preference for bent over linear geometry in the ylide, and thus change the mechanism dramatically.

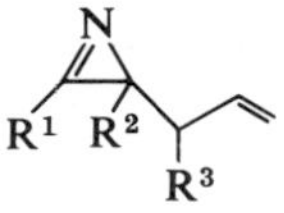

(44) 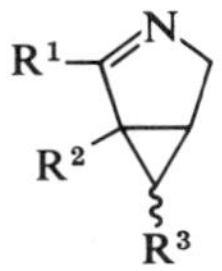 (46) 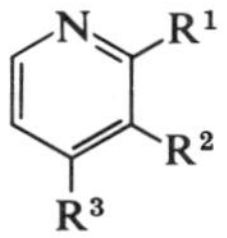 (45)

R^1, R^2 = Ph or Me, R^3 = H or Me

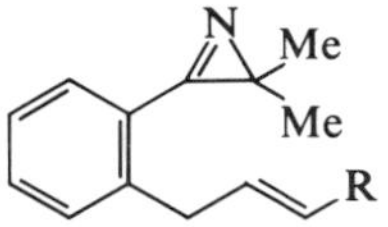

(47) R = H or Me

Imines of α-amino-esters undergo a variety of 1,3-cycloadditions: the 1,3-dipolar tautomers (**48**) have been suggested as likely reactive intermediates in such reactions.[84] The amide (**49**) on treatment with $SOCl_2$ and then with Et_3N gives (**50**), apparently by intramolecular 1,3-cycloaddition of the intermediate ylide (**51**).[85] Phenyl isocyanate traps the nitrile ylide $(CF_3)_2\bar{C}{-}N{=}\overset{+}{C}Ar$ by attack at the C=O bond to give (**52**) regiospecifically and at the C=N bond to give both regio-isomers, (**53**) and (**54**).[86]

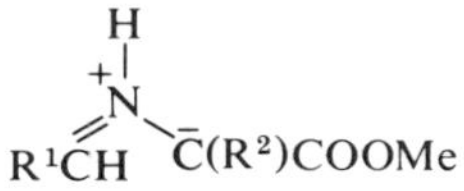

(48) R^1, R^2 = aryl or alkyl

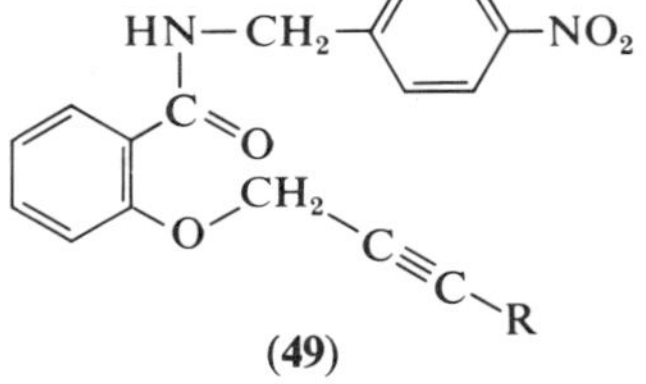

(49)

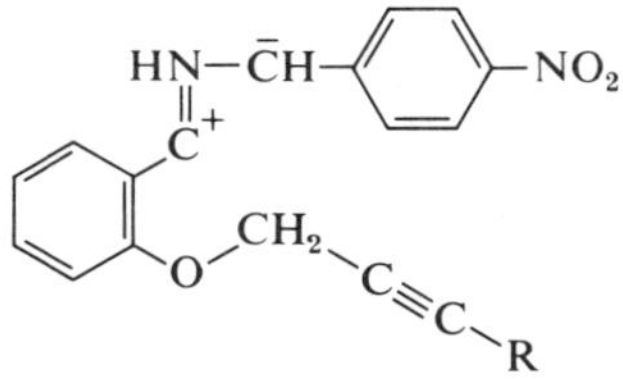

(51)

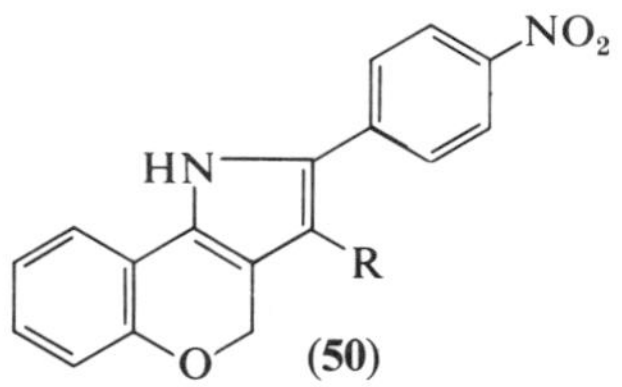

(50)

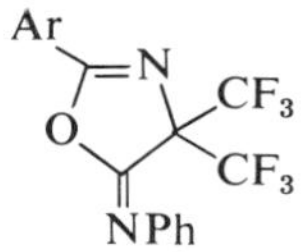

(52) 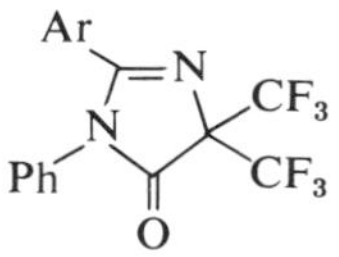 (53) 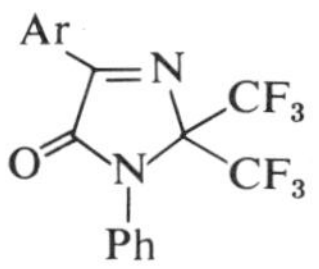 (54)

Low-temperature matrix flash-photolysis of α-cyano-*cis*- and -*trans*-stilbene oxides gives the carbonyl ylides. Careful kinetic analysis allowed construction of a free-energy profile for *cis–trans* isomerization and recyclization of the ylides.[87]

Nitrilimines bearing *N*-alkenyl substituents normally underwent intramolecular 1,3-dipolar cycloaddition; however, in the presence of active dipolarophiles, exclusively intermolecular cycloaddition was observed.[88] Second-order perturbation methods were successfully employed to rationalize the regioselectivity observed in

1,3-dipolar cycloaddition of diarylnitrilimines to *N*-methylindole.[89] The nitrilimine, $CH_3COC{\equiv}\overset{+}{N}{-}\overset{-}{N}Ph$ adds to pyrroles to give a variety of 2 : 1 bis-adducts,[90] including spiro-adducts. Diphenylnitrilimine adds regiospecifically to the C=S bond of sulphines, $R^1R^2C{=}S{=}O$, to give adducts (**55**). Isomeric sulphines give the same product, apparently as the result of stereomutation of the cycloadduct.[91] Some interesting intramolecular cyclizations of nitrilimines have been reported.[92]

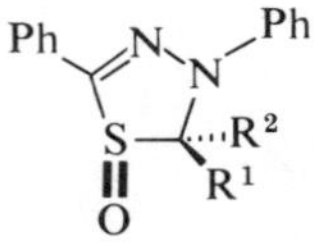

R^1, R^2 = alkyl, aryl, or SPh

(**55**)

Both 1 : 1 and 2 : 1 adducts are observed on reaction of diazomethane with cinnamylidenecyanoacetic esters and nitriles.[93] Addition of diaryl- and arylmethyl-diazomethanes to penta- and hexa-fluoroacetone at −20 °C gives 2,5-dihydro-1,3,4-oxadiazoles but, at higher temperatures, loss of nitrogen takes place, forming oxiranes; evidence for an intermediate carbonyl ylide has been presented.[94] It has been shown that the (unexpected) specificity of addition of diazoethane to 7-chloronorbornadiene to give exclusively *endo,anti*-isomer (**56**) is not general for diazo-compounds: diphenyldiazomethane gives three adducts, (**57**)–(**59**),[95] The 1,3-dipolar addition of 2-diazopropane to the diesters (**60a**) and (**60b**) has been re-examined, and the products have been reassigned the structures (**61a**) and (**61b**).[96]

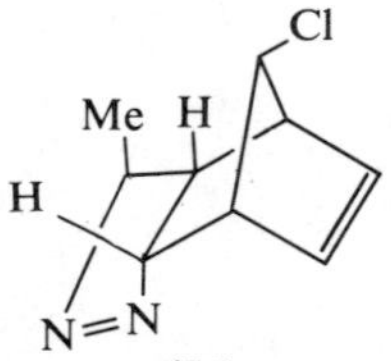

(**56**) 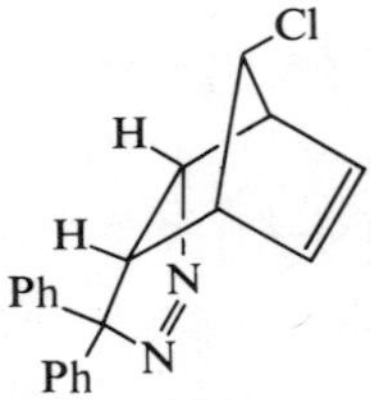 (**57**)

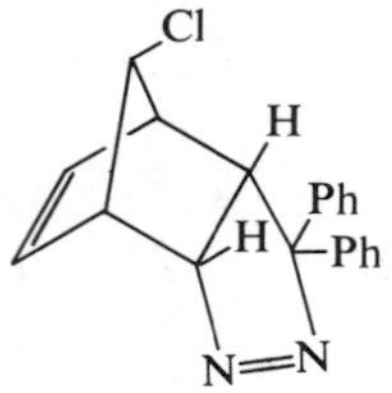

(**58**) 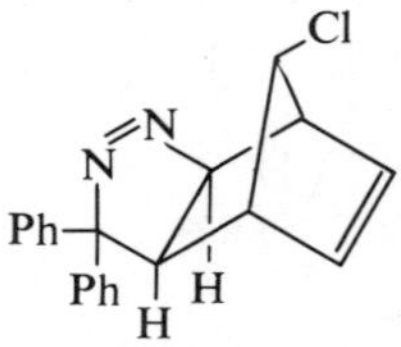 (**59**)

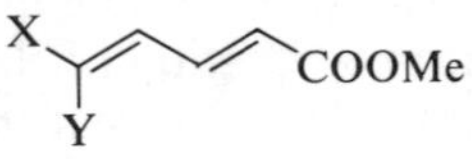

(**60**) 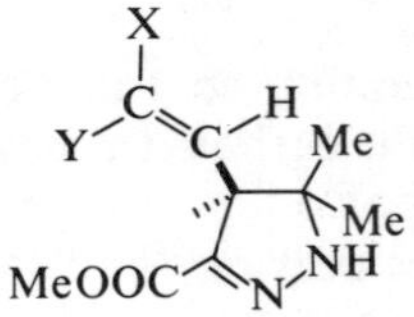 (**61**)

a, X = Me, Y = CO_2Me
b, X = CO_2Me, Y = Me

Addition of diazoalkanes to allenic derivatives has been studied,[97] as have the additions of nitrilimines, nitrile oxides and azides.[98] The stereochemistry of the pyrazolines obtained from cyclopentadiene on reaction with acetylenic esters and diazoalkanes depends on the order of addition of the reagents: the *endo*-isomer is the only product when addition of the acetylene precedes diazoalkane addition, whereas, when the order is reversed, the *exo*-isomer is the major product.[99] 1,3-Dipolar addition of diazomethanes to tetrachlorocyclopropene gives the bicyclic adducts (**62**) which decompose thermally to give pyridazines (**62a**, **62b**, **62c**), dihydropyridazines (**62d**) or dienes (**62e**) at temperatures which vary with the nature of the groups R^1 and R^2. Analogous adducts (**63**) are postulated to account for the formation of the products of reaction of tetrachlorocyclopropene with aryl azides.[100]

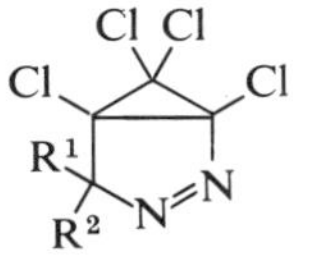

(**62**) **a** $R^1 = R^2 = H$
b $R^1 = H, R^2 = Me$
c $R^1 = H, R^2 = COOMe$
d $R^1 = R^2 = Me$
e $R^1 = R^2 = Ph$

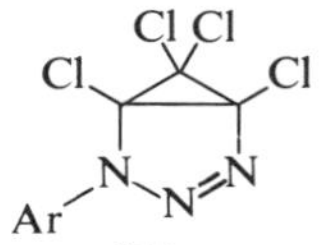

(**63**)

Electrochemical oxidation of trisubstituted hydrazines gives diazenium cations, which on treatment with Et_3N yield azomethinimines, $RR^1\overset{+}{C}—N(R^2)—\overset{-}{N}Ar$; these species give 1,3-dipolar cycloadducts with electron-rich olefins, but not with electron-depleted ones. Frontier molecular-orbital explanations for the regio- and stereo-chemistry of these reactions have been given.[101]

The Dewar-thiophen (**64**) reacts with phenyl or cyclohexyl azide to give the corresponding 2+3-cycloadducts (**65**); these adducts give Dewar pyrroles (**66**) on UV-irradiation followed by treatment with Ph_3P.[102]

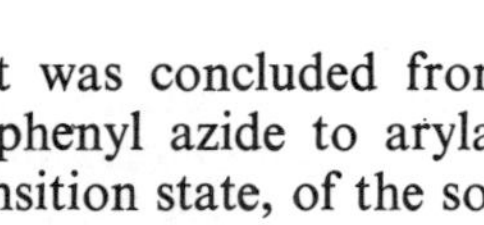

(**64**) 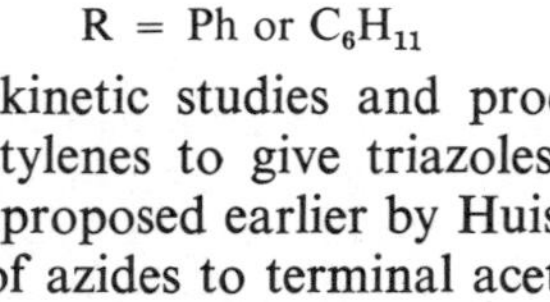 (**65**) 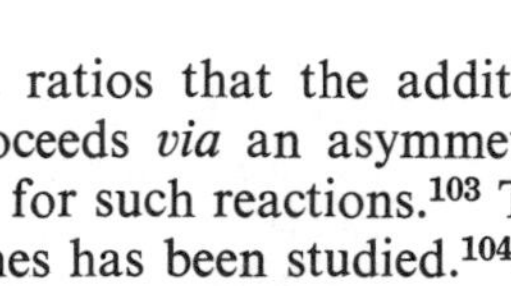 (**66**)

R = Ph or C_6H_{11}

It was concluded from kinetic studies and product ratios that the addition of phenyl azide to arylacetylenes to give triazoles proceeds *via* an asymmetric transition state, of the sort proposed earlier by Huisgen for such reactions.[103] The cycloaddition of a variety of azides to terminal acetylenes has been studied.[104, 105] Vinyl azides with neighbouring double (**67**) and triple (**68**) bonds undergo intramolecular 1,3-dipolar addition to give the triazo-derivatives (**69**) and (**70**), respectively.[106] The highly strained diazo-ketone (**71**) reacts with $CH_2{=}CHCN$ to give the *spiro*-2-pyrazoline (**72**) apparently *via* the tautomeric 1-pyrazoline (**73**).[107]

Lewis acids such as $AlCl_3$ are effective catalysts for the 2+3-cycloaddition of diazocarbonyls to nitriles, to give oxazoles.[108]

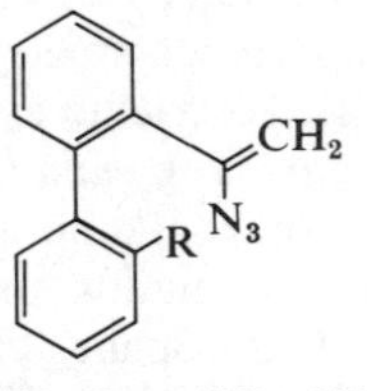

(67) R = $CH{=}CH_2$
(68) R = $C{\equiv}CH$

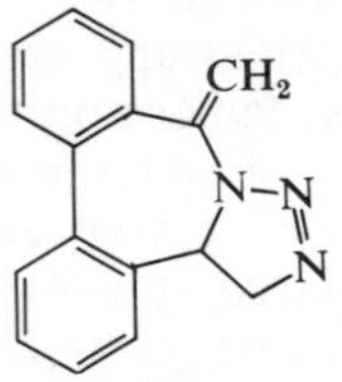

(69) 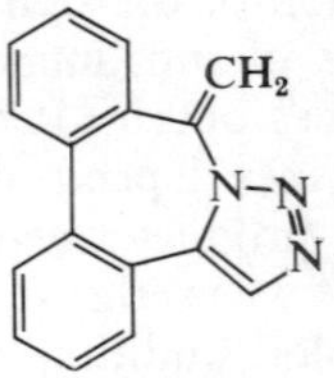 **(70)**

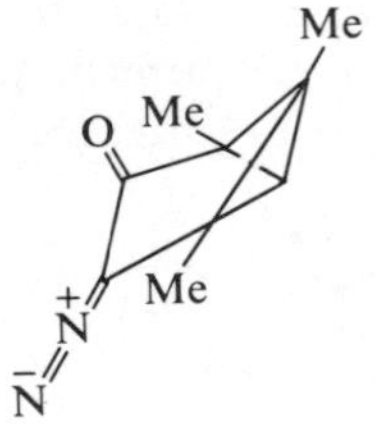

(71) 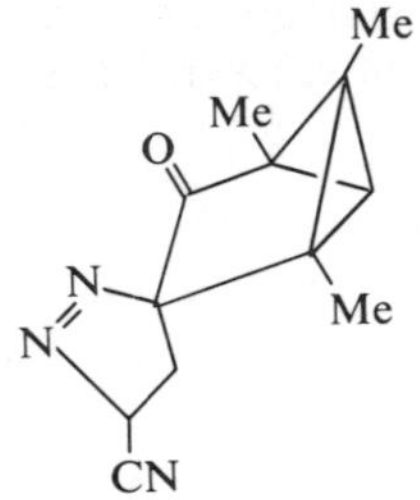 **(72)** 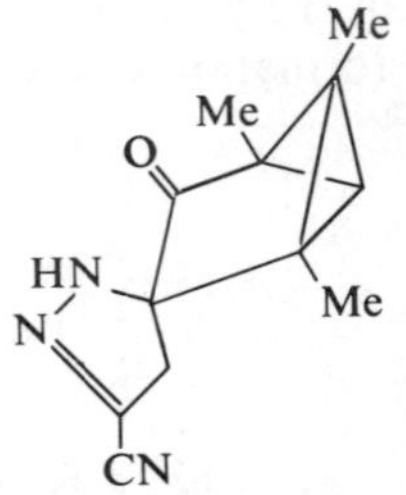 **(73)**

1,3-Dipolar addition of 2,3,4,5-tetrahydropyridine-1-oxide to a variety of monosubstituted alkenes, $CH_2{=}CHX$, gives stereoisomeric isoxazolidines *via* transition states with the substituents oriented *exo* and *endo* with respect to the heterocyclic ring; the *exo*-addition is favoured in all cases, and it has been suggested that unfavourable steric interactions between the substituent and ring hydrogen atoms in the *endo*-transition state are mainly responsible for this preference.[109]

1,3-Dipolar addition of *N*-(4-substituted benzylidene) 4-substituted benzylamine *N*-oxides, $ArCH{=}\overset{+}{N}(Ar)\overset{-}{O}$, to *trans*-1,2-diaroylethenes gives isoxazolidines in a process controlled by HOMO(*N*-oxide)–LUMO(diaroylethyene) interactions,[110] and MO-methods predicted correctly the effect of nitrone ionization potential on the regioselectivity of their 1,3-dipolar addition to substituted alkenes.[111] The LCAO–SCF–MO method has been applied to the mapping of the potential energy hypersurface for the nitrone–ethylene 1,3-dipolar cycloaddition.[112] The kinetics of the addition of amido-nitrones, $RR'NCOCH{=}N(O)R''$ to styrene to give the corresponding 5-phenyl oxazolidines have been measured, and the activation energies have been determined.[113] The ratio of *exo*- to *endo*-approach in reactions between nitrones and olefins bearing electron-attracting substituents, to give oxazolidines, has been interpreted in terms of HOMO(dipole)–LUMO(olefin) interactions;[114] the importance (in some cases) of secondary interactions, not taken into account in the calculations, has been pointed out.[115] Secondary orbital interactions were also shown to be important in determining the orientation of addition of nitronic esters, $MeOCOCH{=}N(O)OMe$, to deactivated olefins.[116] An MO explanation has also been given for the differences between the mode of addition of nitrones and nitrile oxides to alkenes and to alkynes: distortions in the transition-state geometry may have quite significant effects.[117] Unusual temperature effects on the *exo*/*endo* ratio for addition of nitrile oxides to 2,3-bismethoxycarbonylnorbornadiene have been explained in terms of solvent cage effects.[118]

A kinetic study of 1,3-cycloaddition of benzonitrile oxides to aliphatic nitriles, giving 1,2,4-oxadiazoles, shows that this regiospecific reaction is a concerted process, with little response of rate to polar effects.[119] A number of perfluoroalkylisoxazolines and isoxazoles have been prepared by cycloaddition of nitrile oxides to perfluoroalkyl ethenes or ethynes.[120] Electrophilic nitrile oxides (Ar = Ph) tend to add to give adducts of type **(74)** whereas more nucleophilic nitrile oxides (Ar = mesityl) give adducts of type **(75)** with benzofuran; this regiochemical inversion stems from the similarity of the two frontier interactions, which results in a high regio-dependence on minor changes in the structure of the 1,3-dipole.[121] Using photoelectron spectra, Houk and his co-workers have determined the ionization potentials of some cyclopentenes and have established a reasonable correlation between these and the regioselectivities of their reactions with benzonitrile oxide.[122] The effect of solvents on the *exo*/*endo* ratio for addition of $PhCO\overset{+}{C}{=}N{-}\overset{-}{O}$ to norbornadienes **(76)** can be explained in terms of an *endo*-transition state with a greater dipole moment than that of the *exo*-forms.[123]

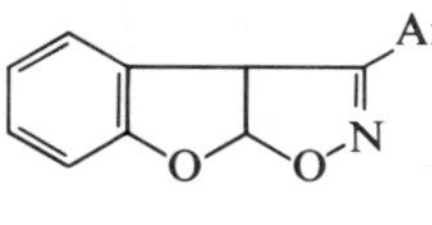

(74) 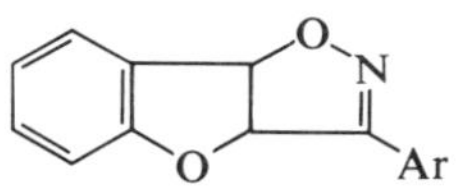 **(75)** **(76)**

Ar = Ph or mesityl

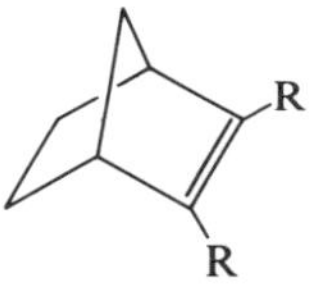

R = H, CN, or COOMe

Rate data for 1,3-dipolar additions of a variety of 3-oxidopyridiniums to olefinic dipolarophiles have been collected; the results of Hammett studies are consistent with a synchronous mechanism.[124] The 1,ω-bis-(3-oxidopyridinio)alkanes **(77)** fail to give any products arising from intramolecular dimerization, although they do give rise to a variety of bis-adducts with 2π-addends.[125] The reactivity of 1-methyl-3-oxidopyridinium is greatly enhanced by the introduction of a 5-methoxy-group.[126]

Mesoionic 1,3-dithiolylium-4-olates **(78)** react with unsymmetrical alkynes ($R^3C{\equiv}CR^4$) to give thiophenes **(79, 80)** and COS. The regiospecificities of these

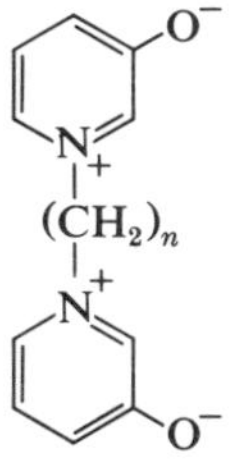

(77) 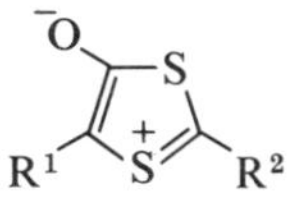 **(78)** 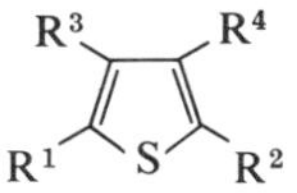 **(79)** 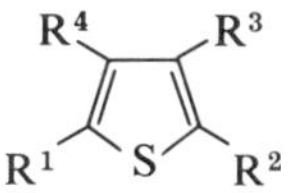 **(80)**

R^1, R^2 = aryl
R^3 = H, Me, or Ph
R^4 = H, Ac, COOMe, COOEt, or CH_2OH

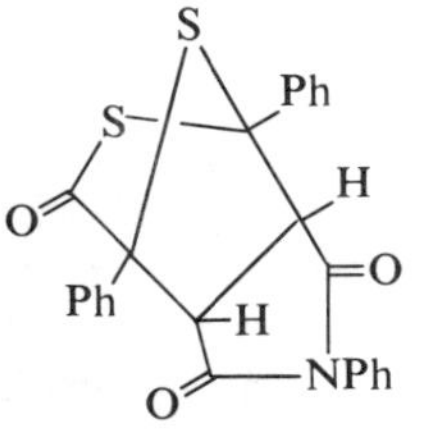

(81) 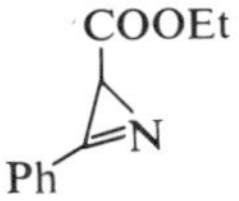 **(82)** 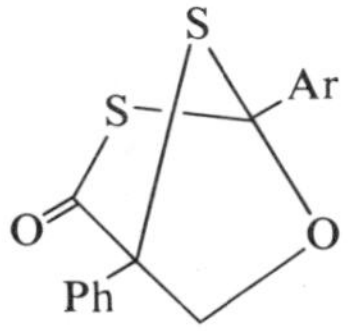 **(83)** 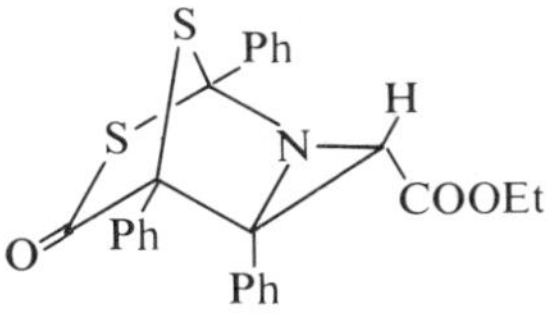 **(84)**

reactions can be accounted for in terms of perturbation theory;[127] with *N*-phenylmaleoylimide, bicyclic adducts such as (**81**) and its *exo*-isomer could be isolated,[128] and with triphenylcyclopropene the analogous fused cyclopropane was also obtained (100% *exo*).[129] The reaction was also extended to double bonds involving heteroatoms: for example formaldehyde and the azirine (**82**) gave (**83**) and (**84**), respectively.[130]

The first examples of cycloadducts of mesoionic compounds with carbonyl derivatives has been claimed: reaction of (**85**) with (**86**) gave adducts (**87**) or (**88**), depending on the nature of the carbonyl component.[131] 1,3-Dipolar addition of mesoionic oxazolium-5-olates to CS_2 is highly regiospecific, leading eventually to thiazoles,[132] and mesoionic triazapentalenes (**89**) give good yields of the 1,3-cycloadduct (**90**) with DMAD.[133]

Azomethinimine intermediates have been detected in "criss-cross" cycloaddition of perfluoroazines with HC≡CCOOEt.[134] Addition of CS_2 to Hector's base gives the adduct (**91**).[135] Dioxygen platinum complexes $Pt(PR_3)_2(O_2)$ react with hexafluorobut-2-yne and dimethyl acetylenedicarboxylate to give[136] the adducts (**92**).

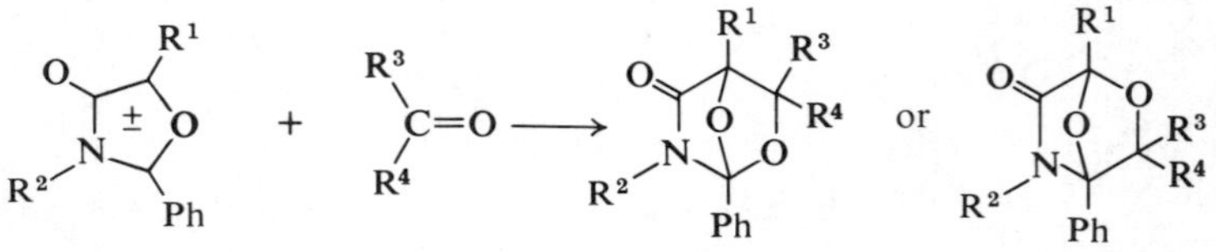

(**85**) (**86**) (**87**) (**88**)

R^1, R^2 = alkyl, aryl, or H
R^3 = aryl or COPh
R^4 = H, alkyl, or Ph

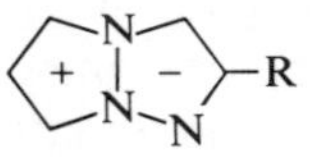

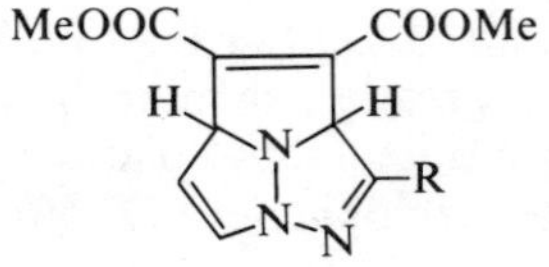

(**89**) R = Me or Ph (**90**)

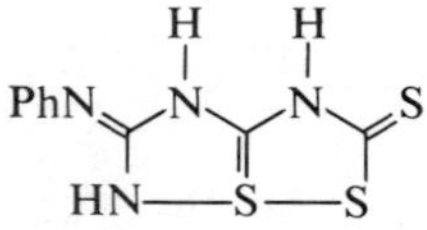

(**91**) 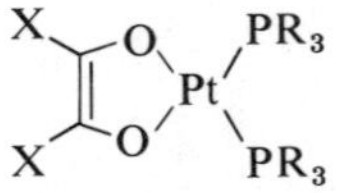 (**92**)

X = CF_3 or COOMe
R = Me, Bu^n, Bu^t, or Ph

2+4-Cycloaddition

A comprehensive review of the applications of retro-Diels–Alder reactions for the period 1970 to 1976 has been published.[137]

MINDO/3 studies of the retro-Diels–Alder reaction of cyclohexene, the reverse of the archetypal Diels–Alder reaction, indicate a highly unsymmetrical, biradicaloid transition state,[138] and the product distributions observed in 2+4-cyclodimerization of the electronegatively substituted dienes, CH_2=CHCH=CHCN,

$CH_2{=}CHC(CN){=}CH_2$ and $CH_2{=}CHCF{=}CH_2$, can readily be explained in terms of diradical intermediates.[139] A diradical pathway was also proposed for the cyclodimerizations of acrylonitrile and styrene.[140] However, dipolar character in the transition state for addition of the C=N bond of 1-arylimidazoline-2,5-dione to buta-1,3-diene derivatives has been invoked[141] to explain the regioselectivity observed in such reactions.

Other workers have used MINDO/2 methods to obtain information on the potential surface for the 4+2-cycloaddition of ethylenes to monosubstituted butadienes.[142]

Evidence has been presented[143] which supports the involvement of charge-transfer complexes in cationic polar 4+2-cycloaddition. Further studies of such reactions have uncovered, in the reactions of acridizinium ion with ethyl acrylate, *cis*-crotononitrile, and 2-stilbazole, the first examples of non-regiospecific polar 4+2-cycloaddition; the importance of charge-transfer processes was again emphasized.[144] Configuration interaction calculations also showed that charge-transfer effects were important in determining regiochemistry in Diels–Alder reactions.[145] The enthalpies (ΔH) and rate constants (k) for Diels–Alder reactions of anthracene and its 9-chloro-, 9-methyl-, 9-methoxy-, 9,10-dimethyl and 9,10-dimethoxy-derivatives with TCNE were determined. Linear relations were found between ΔH and anthracene localization energies, and between log k and the energy of the donor–acceptor interaction.[146]

Calculations indicate that the greatest rate enhancements in Diels–Alder reactions are to be expected when donor substituents are placed on the addend with the higher HOMO, and attractor substituents are placed on the addend with the lower HOMO.[147] The two pairs of formal Diels–Alder adducts of cyanocyclopentadiene with cyclopentadiene and with benzene undergo cycloreversion to cyanocyclopentadienide ion on treatment with the base LiTMP, some 10^3 times faster than does the analogous benzene/cyclopentadiene Diels–Alder adduct. Frontier orbital considerations, rather than thermochemical ones best explain this result.[148] *N*-(Acylamino)-1,3-alkadienes react readily with methyl acrylate in a higher regio- and stereo-selective manner. FMO theory using diene IP data from PE spectroscopy also readily explains these experimental results.[149] Although a refreshingly sceptical view of the usefulness of applying FMO theory to systems displaying relatively small regioselectivities has been expressed,[150] following an examination of the reactions of the dienes **(93)** with a variety of dienophiles, explanations continue to be found for a host of "anomalous" results. Thus, it was shown by a variety of methods (CNDO/2, INDO, CNDO/S and Hückel) that secondary orbital interactions have to be considered in order to predict the regioselectivity of 1,2-disubstituted butadienes in Diels–Alder reactions;[151] examination of the stereochemistry, as well as regiochemistry, of Diels–Alder reactions of 1-substituted dienes also led to the conclusion that both primary and secondary overlap are important in controlling the distribution of regioisomers.[152] Likewise, inclusion of subjacent and superjacent as well as frontier orbitals in the PMO treatment of the 2+4-cycloadditions of methyl 2-pyronecarboxylates to 1,3-dienes dramatically improves the agreement between theoretical and experimental results.[153] 2-Azabicyclo[2.2.2]oct-5-enes **(94)** and **(95)** were formed from Diels–Alder reactions of protonated imines, $RCH{=}\overset{+}{N}HR^1$, with cyclohexa-1,3-diene: in all cases a preference for **(94)** over **(95)** is observed: an explanation involving secondary orbital interactions between the diene and aldimine substituent was suggested.[154] Secondary overlap also appears to affect the regioselectivity of addition of acetylenes to 2-pyrones.[155] The reaction rate sequence **(96a)** > **(96b)** ~ **(96c)** ~ **(96d)** > **(96e)** towards **(97)** to give **(98)** and nitrogen provides convincing evidence of orbital interactions through space (OITS).[156]

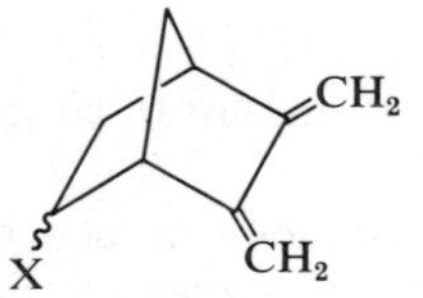

(93)
X = OH or OAc

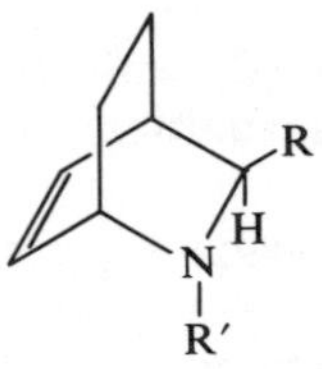

(94) 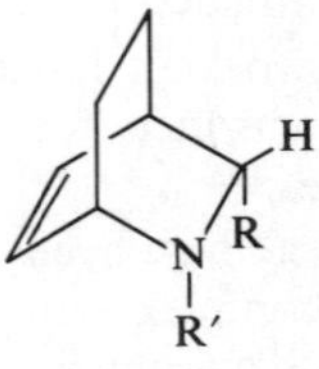 **(95)**
R = COOMe, COOEt, $COOCH_2Ph$, COMe, Ph, or $4\text{-}NO_2C_6H_4$

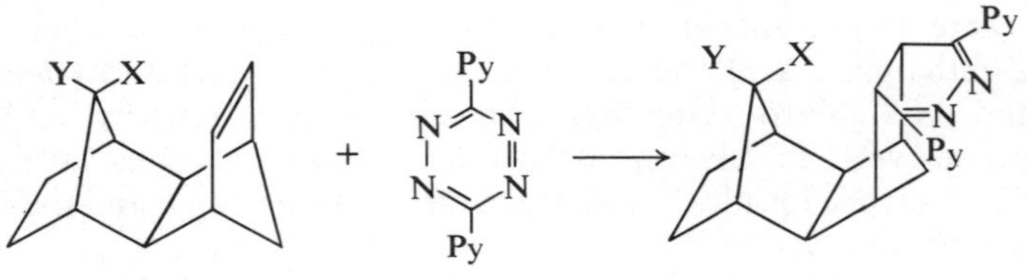

(96) + **(97)** ⟶ **(98)**

a X = OMe, Y = H
b X = Y = H
c X = H, Y = OH
d X = H, Y = OMe
e X, Y = O

Perturbation molecular orbital theory was used to generate predictions of *syn/anti* selectivity in the $\pi^4s+\pi^2s$ reactions of 7-substituted norbornadienes:[157] through space interactions between the *syn*-double bond and the substituent were shown to be an important factor. Again in apparent contradiction of FMO predictions, cyclopentadienones act both as dienes and as dienophiles in Diels–Alder reactions with 6,6-dimethylfulvene, to give the adducts **(99)** and **(100)**, which interconvert in a Cope rearrangement process. This is in contrast to the behaviour of cyclopentadienones with cyclopentadiene, but can be reasonably explained in terms of angle strain in the transition states involved.[158]

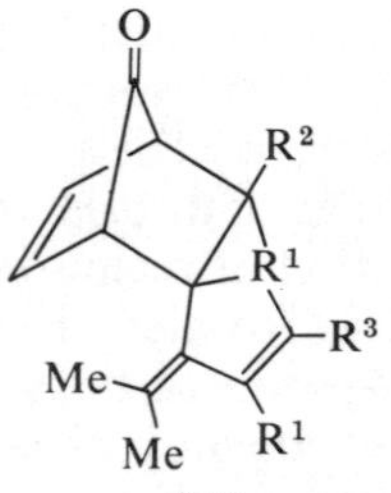

(99) 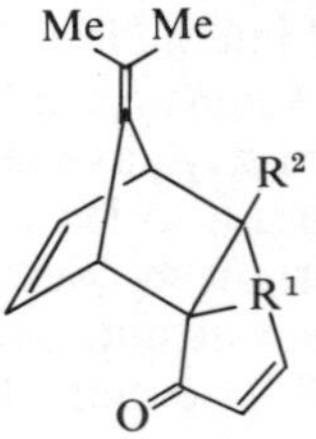 **(100)**

R^1 = H, Me, Ph, or CF_3
R^2 = H, Ph, or CF_3
R^3 = H, Bu^t, Ph, or CF_3

The rate and stereochemistry of Diels–Alder addition of α,β-unsaturated aldehydes to cyclopentadiene are both affected by reaction temperature and by the use of BF_3—OEt_2 as catalyst.[159] The synthetic utility of dienes such as CH_2:C(SAr)C(OR):CH_2 (R = Me or Ac) and related compounds has been further enhanced by the discovery that, depending on the substituents present, addition of Lewis acids may either enhance or oppose the orientational effect of sulphur in Diels–Alder reactions of these dienes.[160] Both the nature of the Lewis acid employed as catalyst in this way and the concentration employed have, however, been shown to be important in determining the regiochemistry of the product obtained.[161] Methods for "manipulating" the orientation of addition of dienes to oxygenated naphthaquinones have been put forward.[162] The importance of careful choice of diene component in influencing regiochemical control in Diels–Alder reactions has been emphasized.[163] Regioselectivity in Diels–Alder reactions of 1-(trimethylsilyl)-butadienes $R^1CH{=}C(R^2)CH{=}CHSiMe_3$, R^1 or R^2 = Me is controlled by the 3- or 4-substituent rather than by the trimethylsilyl group.[164] Methyl *trans*-β-nitro-acrylate provides a useful dienophile for the preparation of "meta"-cyclohexenes by Diels–Alder reaction with nucleophilic dienes.[165] Perturbation calculations show that the enhanced reactivity of *o*-quinonoidal heterocycles (**101**) relative to their counterparts (**102**) can be explained in terms of non-interaction of the two π-systems present in these heterocycles, the butadiene system of the carbocyclic ring, and the six-electron system of the fused heterocyclic ring.[166]

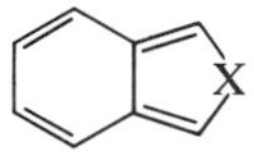

(**101**) 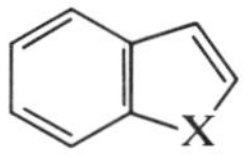 (**102**)

X = NH, O, S, or Se

The "*endo*-rule" largely prevails in the 4+2-cycloaddition of cyclobutenes to the 4π-system of cyclopentadiene,[167] and in the addition of allyl esters of chloroacetic acids to cyclopentadiene and its perchloro-derivative,[168] but apparently not completely for reaction of 1-methoxy-1,3-butadiene with aliphatic aldehydes.[169] An explanation for the apparently anomalous nature of the reaction between furan and and maleic anhydride has been discovered: although this reaction satisfies all the usual criteria for expecting *endo*-selectivity, mainly the *exo*-adduct (**103**) is obtained on work-up since formation of both *exo*- and *endo*-adducts (**104**) is reversible under these conditions; the kinetic product is the expected *endo*-isomer which is formed *ca.* 500-fold faster than the *exo*-isomer.[170]

A small amount (*ca.* 8%) of asymmetric induction was observed on reaction of cyclopentadiene with allyl (−)-menthyl ether to give norbornenylmethyl menthyl ethers.[171]

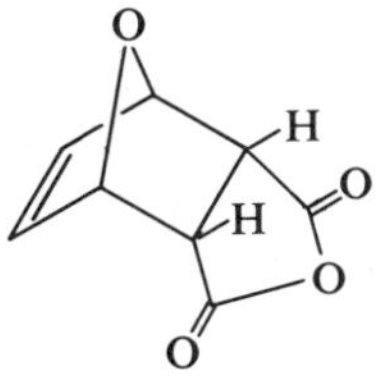

(**103**) 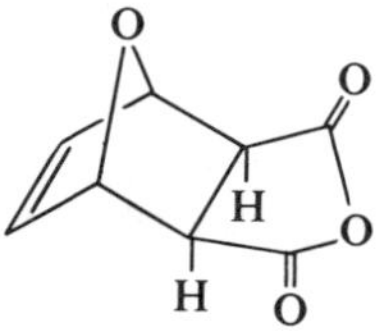 (**104**) 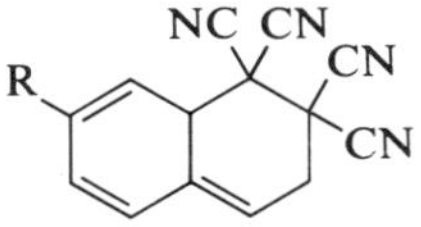 (**105**)

R = H, Cl, or OMe

High-pressure reaction of TCNE with styrenes affords adducts (**105**).[172]

Kinetic studies of Diels–Alder dimerization reactions at high pressures indicate a product-like transition state and exclude a two-step mechanism;[173] high-pressure studies of the Diels–Alder reaction of 1-alkoxy-1,3-butadienes with aldehydes indicate a transition state that resembles the plane-parallel pre-reaction complex more than the chair-form adduct.[174] Although some difficulties due to partial polymerization were encountered in certain cases, nevertheless, high pressure (3000 bar) was effective in bringing about 2+4-cycloadditions between acrolein, methyl vinyl ketone or crotonaldehyde with acrylic or methacrylic dienophiles to give substituted dihydropyrans.[175] The application of pressures at around 20–40 kbar not only accelerates the addition of the 4π-system of 3-hydroxy-2-pyrone to a wide range of dienophiles, but also suppresses extrusion of CO_2 from the bicyclic adducts, except in the case of the DMAD adduct, which loses CO_2 when the reaction vessel is opened.[176]

The configurations of the Diels–Alder adducts of maleic anhydride with cyclopentadiene have been determined by using NMR.[177] The reaction of (**106a**) with hexachlorocyclopentadiene is faster than that of (**106b**).[178] While 1-hydroxy- and 1-acetoxy-indenes fail to give Diels–Alder products, indenone readily gives 2+4-cycloadducts with a range of dienes.[179]

Addition of α-pyrone to *trans*-benzazonine (**107**) gives only (**108**) and not the other possible 2+4-regioisomer, apparently because of a greater influence of the benzene ring than of the urethane moiety on the 2π-fragment HOMO.[180]

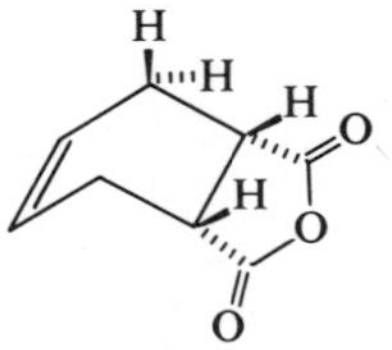

(**106a**) 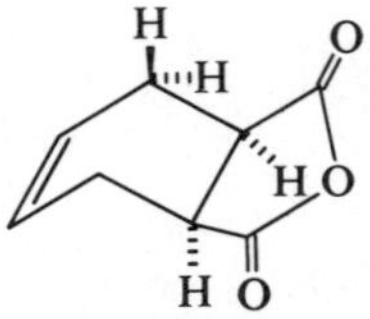 (**106b**)

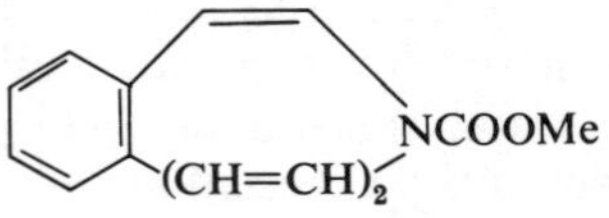

NCOOMe
(CH=CH)$_2$
(**107**)

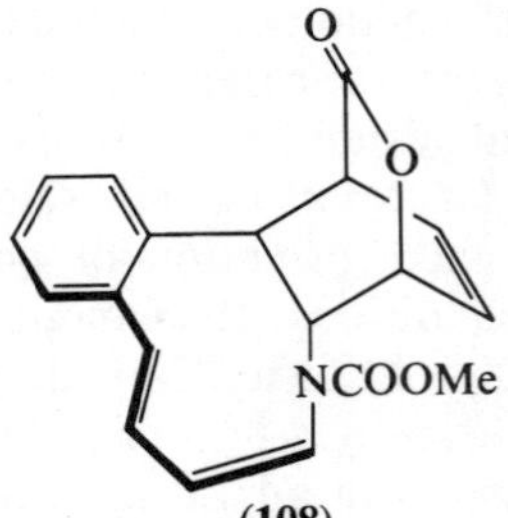

NCOOMe
(**108**)

Thermal intramolecular Diels–Alder reaction of [3,3]paracyclo(9,10)anthracenophane has been reported.[181] The Diels–Alder reactions of some propellane derivatives have also been studied:[182–184] it would appear that steric effects can at least partly offset electronic influences on the stereochemistry of such reactions.[185,186] The methylenecyclopropanes (**4**) act as 2π-addends in conventional Diels–Alder reactions[15] with furan, cyclopentadiene and 1,3-cyclohexadiene; with 2,3-dimethylbutadiene unusual products are formed, presumably from the expected 2+4-cycloadduct *via* cyclopropylcarbinyl rearrangement.

Diels–Alder reactions of benzvalene with a variety of electron-deficient dienes and heterodienes have been reported.[187] The product formed with *ortho*-benzoquinone (**109**) can be photodecarbonylated to give the new $C_{10}H_{10}$ isomer (**110**). The 2+4-cycloaddition reactions of the bicyclobutane-bridged diene (**111**) give

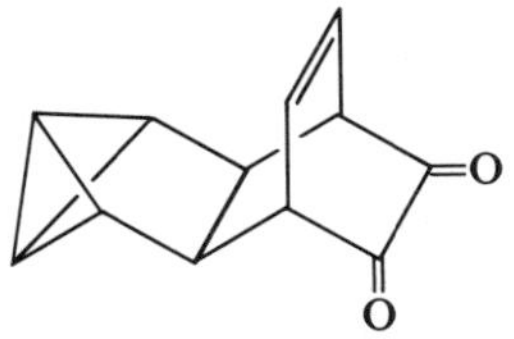

(**109**) 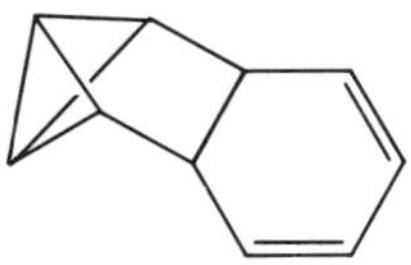 (**110**) 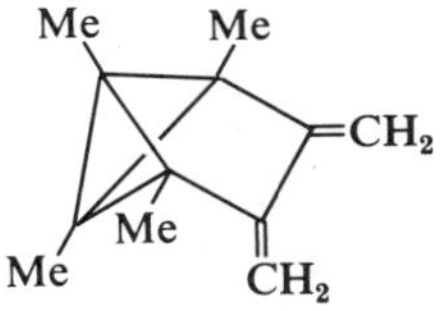 (**111**)

rise to different types of product, depending on the dienophile used:[188] thus, TCNE gives the benzvalene (**112**), maleic anhydride the homofulvene (**113**) and more reactive dienophiles (such as dibenzoylethylene) the tetralin (**114**); the last type of reaction probably proceeds *via* benzvalene adducts. This type of reaction occurs extremely readily, in contrast to the analogous reactions of the cyclobutane-bridged diene (**115**), despite the fact that *ab initio* calculations indicate a reasonable level of activity for (**115**): the low reactivity was attributed to repulsive interactions between occupied FMOs of the cyclobutane ring and the C=C bridge in Diels–Alder adducts of (**115**).[189] 3,4-Dimethylenecyclohepta-1,5-diene has been prepared and shown to take part in a variety of 4+2-cycloadditions, including dimerization.[190]

Cyclophanes with small numbers of bridging atoms, such as [7]- and [8]-paracyclophane and [7]metacyclophane, behave as dienes towards $CF_3C{\equiv}CCF_3$ in Diels–Alder reactions at around 160 °C; NCC≡CCN requires Lewis acid catalysis, and methyl acetylenedicarboxylate fails to react.[191]

Substituted arynes add to hexamethylcyclohexa-2,4-dienone to give isomeric benzobarrelenones;[192] the novel cycloalkyne, 4,5-dehydrotropone, acts as an acetylenic dienophile in Diels–Alder reactions with cyclopentadiene and cyclohexadiene.[193] The interesting hydrocarbon (**116**) has been trapped as the [4.1.1]propellane adduct (**117**) by reaction with anthracene.[194]

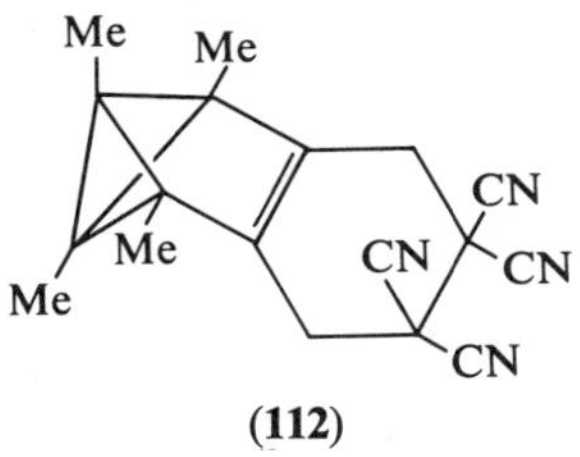

(**112**) 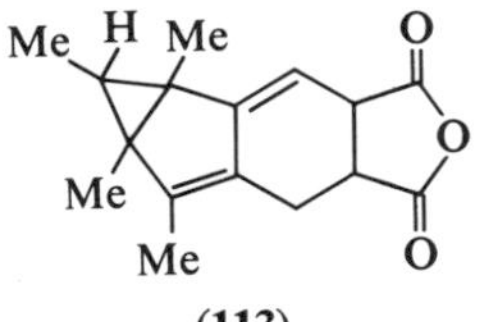 (**113**) 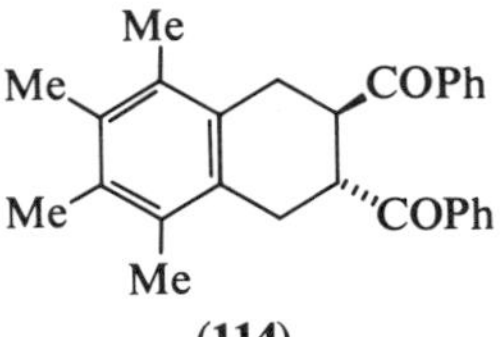 (**114**)

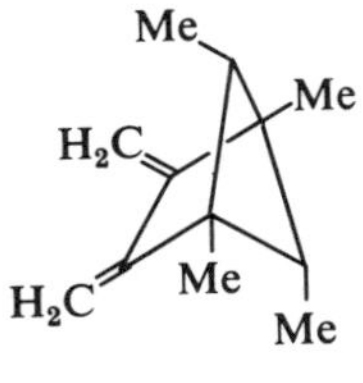

(**115**)  (**116**) 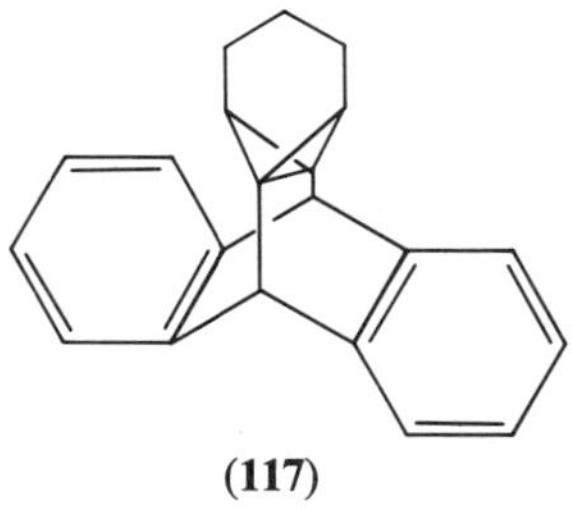 (**117**)

Intramolecular Diels–Alder reactions have been successfully employed in the synthesis of hydrocarbons with bridgehead double bonds.[195] Although the benzenesulphenic acid produced in the final stage of the proposed reaction scheme may pose some problems, nonetheless, the use of phenyl vinyl sulphoxide as an acetylene synthon in Diels–Alder reactions looks very promising.[196] (Methylthio)maleic anhydride undergoes a wide range of Diels–Alder reactions with the sulphur exercising a high degree of regiocontrol; these reactions are effectively Diels–Alder reactions of methoxycarbonylketen.[197]

It has been shown that the complex $[Ru_4H_4(CO)_{12}]$ catalyses the dimerization of COT to give the new dimer (**118**).[198] Although the Cu(I)-catalysed cyclization of buta-1,3-diene to cyclobutene failed, contrary to earlier reports, copper(I) trifluoromethanesulphonate did catalyse the 4+2-cycloaddition reaction of this diene with cyclohexene.[199] Diels–Alder-like additions of TCNE and $CF_3C{\equiv}CCF_3$ to the cyclopentadienyl portion in (cyclopentadienyl)nitrosylmolybdenum complexes have been reported.[200] A clever device for the generation of inaccessible but potentially useful systems such as (**119**), which should cyclize to the five-membered lactam precursors (**120**), has been discovered. Oxidation of $CH_3CONHOH$ in the presence of 9,10-dimethylanthracene gives (**121**) by 4+2-cycloaddition of the acylnitroso-intermediate: this can be elaborated by standard procedures to give (**122**) which on thermolysis in benzene gives 9,10-dimethylanthracene and the intramolecular 4+2-cycloadducts (**120a**) and (**120b**) derived from intermediate (**119**).[201]

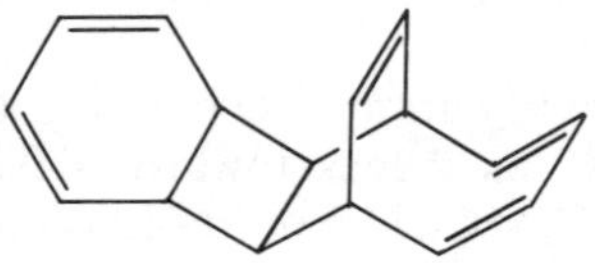

(**118**)

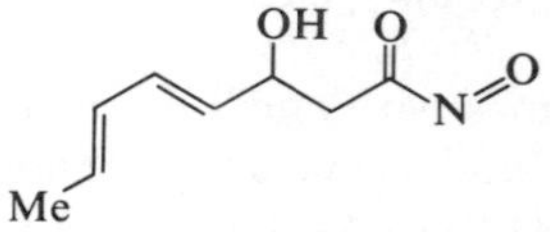

(**119**)

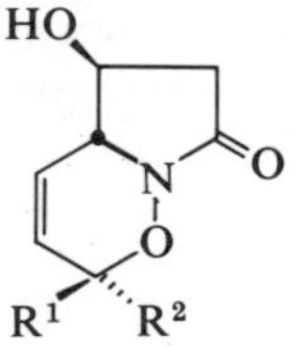

(**120**) (**a**) R^1 = H, R^2 = Me
(**b**) R^2 = H, R^1 = Me

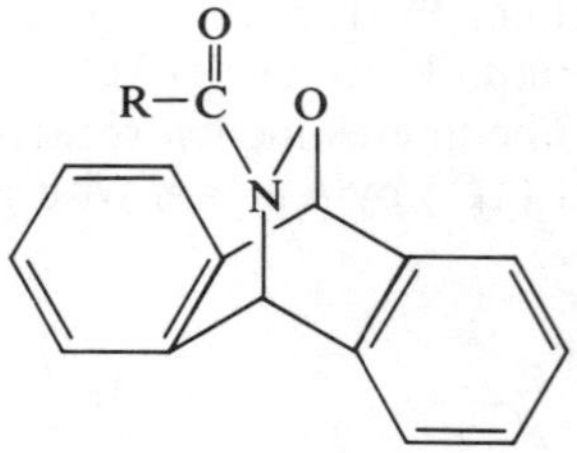

(**121**) R = Me

(**122**) R = $CH_3(CH{=}CH)_2\overset{\mathrm{OH}}{\overset{|}{C}}HCH_2$

Kinetic studies[202] of the 2+4-cycloaddition reaction between 1,2-diarylbuta-1,3-dienes and nitrosobenzenes to give dihydro-oxazines indicate that, for the *E*-isomers at least, charge-transfer complexes play a significant rôle in the direction and rate of the reaction; aryl substituent effects (ρ) are relatively large for those on the nitrosobenzene but rather small for those on the diene.[203] Aryldiazenium salts, $Ph(R)\overset{+}{N}{=}NH\ \overset{-}{Cl}O_4$ (R = Me, Et or Ph), react with *cis*-PhCH=CHD to give (**123**) and with *trans*-PhCH=CHD to give (**124**); the unlabelled 1,2,3,4-tetrahydrocinnolines oxidize on standing to give (**125**).[204] 2+4-Cycloaddition of two moles of (cyclobutadiene)tricarbonyliron to 1,2,4,5-tetrazines gives the azo-bridged pterodactylanes (**126**).[205]

Diels–Alder reactions of triphenylvinylphosphonium salts lead to cyclohex-3-enyltriphenylphosphonium salts, from which the corresponding methylene compounds can be synthesized by the Wittig reaction. The central C=O bond in diethyl oxomalonate acts as a dienophile to produce compounds that can be degraded to valerolactones by using either the Curtius reaction or $Pb(OAc)_4$. These two processes are therefore equivalent to allene and CO_2 Diels–Alder reactions, respectively.[206]

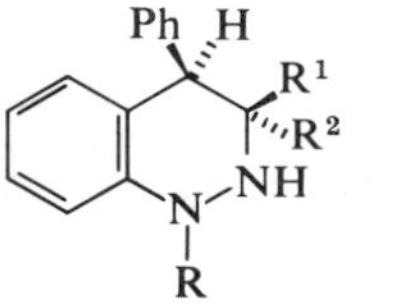

(**123**) R^1 = D, R^2 = H
(**124**) R^1 = H, R^2 = D

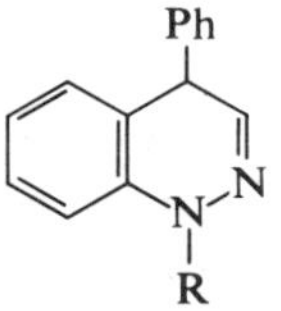

(**125**)

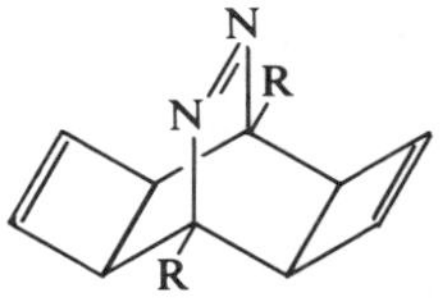

(**126**)
R = COOMe, Me, or Ph

The Diels–Alder reaction between butyl glyoxalate and the butadienyl ethers of two tetra-*O*-benzyl-α-D-glucopyranosides has been examined. Although in each case a strong preference for one face of the prochiral ether was observed, no particular preference for *endo*-addition could be discerned.[207] 3-Aryl-2-cyanothioacrylamides, ArCH=C(CN)C(S)NH_2, furnish both the 2π- and 4π-components of a 4+2-cycloaddition process in dimerizing to 3,4-dihydro-2*H*-thiopyrans; complete regiospecificity is observed, in keeping with predictions from FMO theory.[208]

Diels–Alder reactions of thiocumulenes of the X=S=Y type are included in a review of the reactions of this type of species.[209] The 4+2-cycloaddition of SO_2 to *o*-quinodimethane is accompanied by competitive η+ππ-cycloaddition (*ca.* 10%).[210]

X-Ray structural data have shown that the reaction product from phencyclone and *N*-ethoxycarbonylazepine is in fact the 4+2-cycloadduct, rather than the previously reported 6+2-adduct.[211] The mesoionic species (**127**) add to acetylenes to give the unstable adducts (**128**) which lose COS to yield furans.[212] The *cis*-stereospecificity and high regioselectivity in reaction of 4-tosylisonitrosomalonodinitrile to yield acyclic and cyclic dienes indicate a concerted 2+4-cycloaddition mechanism.[213] *N,N'*-Dibenzoyl- and *N,N'*-di(phenylsulphonyl)-*o*-benzoquinone di-imines act as the 4π-component in Diels–Alder reactions with inverse electron demand with olefins[214] and with fulvenes.[215,216] The hetero-Diels–Alder reaction with inverse electron demand between azodibenzoyl and enamines has been further studied.[217] Thioacylketen thioacetals and *N*-thioacyldithioimidocarbonates act as heterodienes in Diels–Alder reactions with active acetylenes and olefins to give six-membered heterocycles.[218] Similar reactions were observed for *o*-thioquinonemethides with enamines,[219] and α,β-unsaturated ketones and amides.[220] The reaction of $PhP(OMe)_2$ with $RCH{=}CHNO_2$ (R = Et, Pr or Pr^i) at −50 °C gave (**129**), but in ether at room temperature gave $PhP(O)(OMe)_2$ and $Ph(MeO)P(O)CR{=}CH_2$, also formed by thermolysis of (**129**); (**129**) with H_2O gave diastereomeric $MeO(Ph)P(O)CH(CHMe_2)CH_2NO_2$. $Me_3SiOPPhOMe$ and $RCH{=}CHNO_2$ gave diastereomeric $MeO(Ph)P(O)CHRCH{=}N(O)OSiMe_3$ which, with water, gave diastereomeric $MeO(Ph)P(O)CHRCH_2NO_2$.[221]

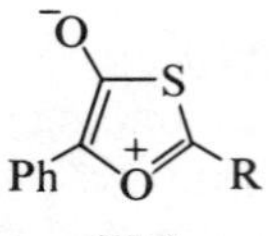

(127)

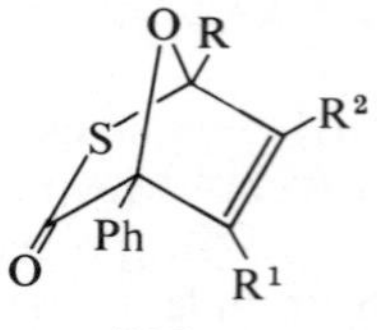

(128)

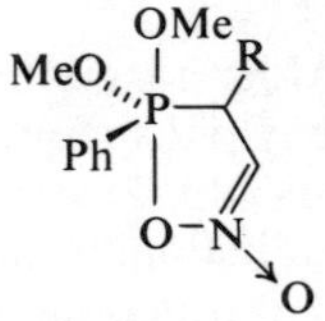

R = *s*-alkyl, piperidino, or morpholino
$R^1 = R^2$ = H, Ph, COMe, COPh, or COOMe

(129)

Miscellaneous Cycloadditions

Complexation between the hydroxyl group and the reagent in transition states involving an eclipsed carbon–carbon double bond has been invoked to account for the stereospecificity of the Simmons–Smith reactions of *cis*- and *trans*-allyl alcohols.[222]

The conversions of (**130**) into (**131**) by use of DDQ and into (**132**) by using TCNE provide examples of a $(\sigma^2+\pi^2+\pi^2)+\pi^2$-cycloaddition.[223] The 2+2+2-cycloaddition of unactivated acetylenes $R^1C{\equiv}CR^2$ (R^1, R^2 = H, alkyl or aryl) to norbornadiene to give adducts (**133**) can be promoted by using cobalt complexes as catalysts.[224] An analogous reaction with norbornene as the 2π-component yields (**134**).[225]

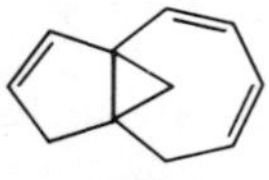

(130)

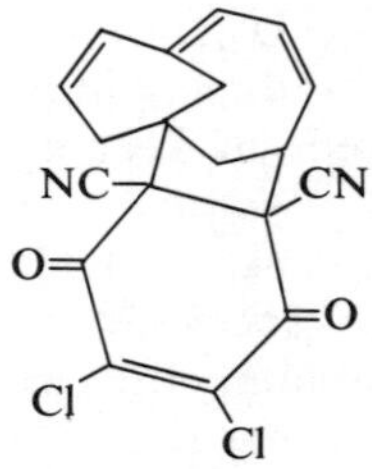

(131)

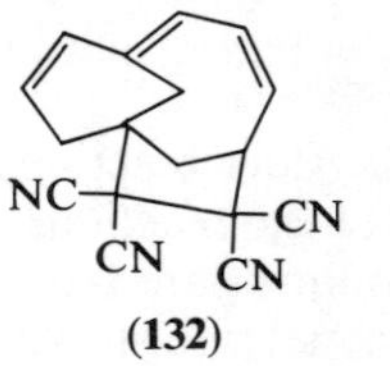

(132)

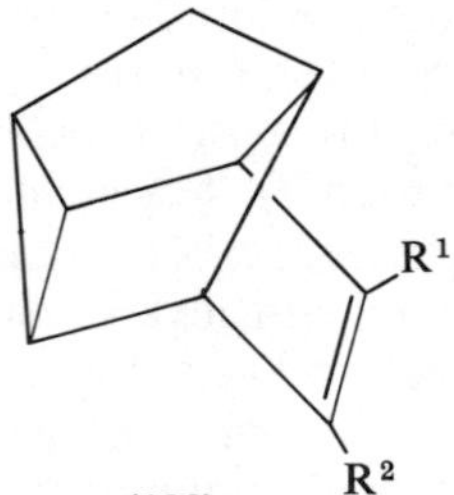

(133)

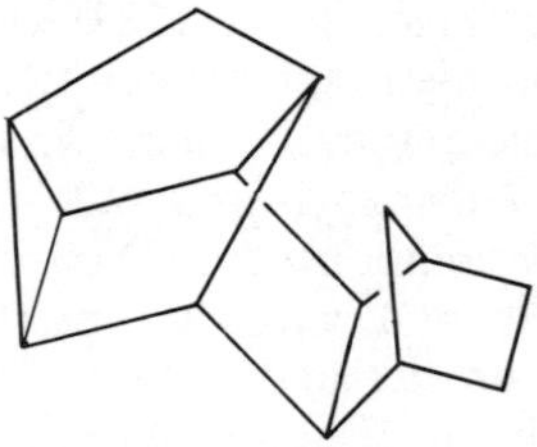

(134)

Nitrilimines, $ArCO\overset{+}{C}{=}N{-}\overset{-}{N}Ar$, are generated by reaction of benzyl- and phenacyl-arsonium salts with arenediazonium salts, and dimerize to give 1,4-dihydro-1,2,4,5-tetrazines.[226] The choice of solvent plays an important part in determining the regiochemical outcome of the 4+3-cycloaddition between *trans*-pentadiene and 1,1-dimethyloxallyl cation.[227] The related (iron carbonyl)-promoted reactions of polybromo-ketones with 1,3-dienes and with enamines have been given considerable attention by Noyori and his collaborators.[228–231] Allenyl cations add to cyclopentadiene to give bicyclic products *via* 4+3-cycloaddition reaction.[232] The

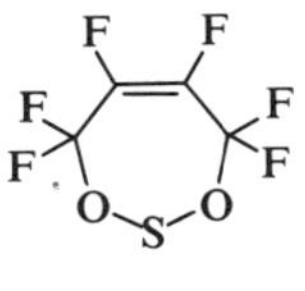

(135) 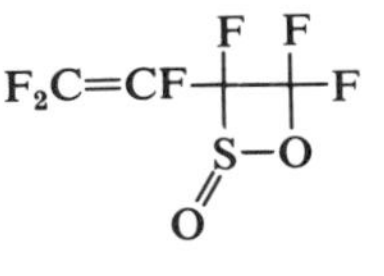 **(136)** 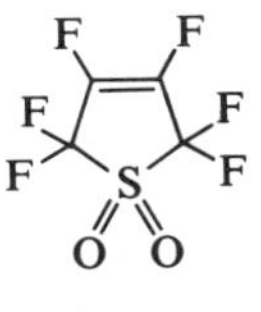 **(137)**

reaction of SO_2 with hexafluorobutadiene gives mostly **(135)** together with some **(136)** and **(137)**.[233]

Photoexcited *N*-methyl-2-pyridone has been shown to undergo exclusive 4+4-cycloaddition, giving the *trans-anti-*, *cis-anti-*, *cis-syn-* and *trans-syn-*adducts in yields of 51, 11.2, 6.8 and 0.6% respectively.[234] Treatment of semicarbazones of general formula EtOCOCH=NNHCONR^{1}R^2 where R^1 = H, Me or Ph, and R^2 = Me or Ph, on treatment with Br_2 and NaOAc gave the bromo-derivatives, which when heated gave high yields of oxadiazoles by intramolecular cyclization of the 1,5-dipole.[235]

X-Ray structural determination has established that the product formed on cycloaddition of *N*-(ethoxycarbonyl)azepine to nitrosobenzene is a 6+2-cyclo-adduct.[236] *N,N'*-Diarylsulphonyl-*o*-benzoquinone di-imines give both unstable 6+4-cycloadducts and benzoquinoxaline 4+4-cyclo-adducts with fulvenes.[237] A careful examination[238] of the reaction between tropone and monocyclic fulvenes indicates strongly that the initial product of such reactions is of the 4^F+6^T-cyclo-adduct type. The regioselectivity of 6+4-cycloadditions of dienamines to fulvenes to yield dihydroazulenes has also been investigated.[239] The first example of a 6+4-cycloaddition of a nitrile ylide has been reported,[240] viz., reaction of 3-phenyl-2,2-dimethyl-2*H*-azirine with fulvenes. A low yield of a 6+4-cycloadduct has also been detected in the reaction of diphenylnitrilimine with tropone.[241]

The reaction of sterically hindered 1,2-dicyclopropylethenes with tetracyano-quinodimethane to give [10]paracyclopha-4,6-dienes is formally a $\pi^6+\sigma^2+\pi^2+\sigma^2$-cycloaddition: radical cations have been proposed as intermediates in this reaction.[242] Analogous reactions have been observed with 1-substituted 1-cyclo-propylbuta-1,3-diene.[243] The related dispiro[2.2.2.2]deca-4,9-diene undergoes a variety of interesting cycloaddition reactions with conjugated dienes, with dimethyl maleate and with TCNQ; the first two processes appear to involve diradical intermediates, but the last process seems to proceed *via* a dipolar species which may be trapped.[244]

1,8-Dipolar addition intermediates have been postulated in the reaction of azafulvenes with dimethyl acetylenedicarboxylate,[245,246] phenylsulphene, and benzoylsulphene.[247] The cycloaddition reactions of fulvenoid dipoles seem to be limited to those with electron-deficient dipolarophiles. One of the products with phenyl vinyl sulphone, a forbidden, formally 10+2-cycloadduct, was shown to be formed *via* an unstable intermediate.[248]

A 14+2-cycloaddition reaction between [15]annulenone dioxide and dimethyl acetylenedicarboxylate gave the adduct **(138)** which on oxidation with DDQ yielded **(139)**.[249]

Anthraquinones have been obtained from the 2 : 1 additions of keten acetals to naphthaquinones.[250] Similar reactions with benzoquinones gave 1,4-naphtha-quinones.[251] *peri*-Annelation reactions of heptalene derivatives afford good entries into penta- and hexa-cyclic conjugated hydrocarbons.[252] A variety of heterocyclic

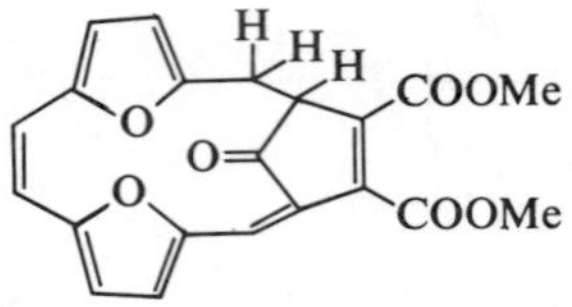

(**138**) 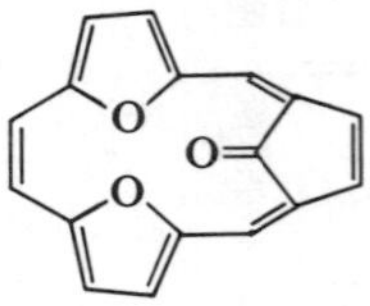 (**139**)

products is obtained on reaction of substituted-1,2,4-triazines with DMAD.[253] Reversible formation of an SO_2 insertion product with some cycloheptatriene–iron complexes has been observed.[254]

References

[1] Boyd, G. V., *Aromat. Heteroaromat. Chem.*, **5,** 123 (1977); *Chem. Abs.*, **88,** 21418 (1978).
[2] Russel, D. H., and Gross, M. L., *Lect. Notes Chem.*, **7.** 209 (1978); *Chem. Abs.*, **89,** 89781 (1978).
[3] Krapcho, A.P., *Synthesis*, **1978,** 77; *Chem. Abs.*, **88,** 151562 (1978).
[4] Ghosez, L., *Med. Chem., Proc. Int. Symp., 5th, 1976*, 363; *Chem. Abs.*, **87,** 183555 (1977).
[5] Asano, T., and Le Noble, W. J., *Chem. Rev.*, **78,** 407 (1978).
[6] Chu, S.-Y., *Tetrahedron*, **34,** 645 (1978).
[7] Aihara, J., *Bull. Chem. Soc. Jpn.*, **51,** 1788 (1978).
[8] El-Basil, S., and Hilal, R., *Bull. Chem. Soc. Jpn.*, **51,** 2749 (1978).
[9] Carpenter, B. K., *Tetrahedron*, **34,** 1877 (1978).
[10] Dolbier, W. R., Piedrahita, C., Houk, K. N., Strozier, R. W., and Gandour, R. W., *Tetrahedron Lett.*, **1978** 2231.
[11] Epiotis, N. D., and Shaik, S., *J. Am. Chem. Soc.*, **100,** 9 (1978).
[12] Ishigami, T., Uehara, M., Murata, T., and Endo, T., *J. Chem. Soc., Chem. Commun.*, **1978,** 786.
[13] von E. Doering, W., and Guyton, C. A., *J. Chem. Soc.*, **100,** 3229 (1978).
[14] Shea, K. J., and Wise, S., *Tetrahedron Lett.*, **1978,** 2283.
[15] Bottini, A. T., and Cabral, L. J., *Tetrahedron*, **34,** 3187 (1978).
[16] Murai, S., Ryu, I., Kadono, Y., Katayama, H., Kondo, K., and Sonoda, N., *Chem, Lett.*, **1977,** 1219.
[17] Biere, H., Herrmann, C., and Hoyer, G.-A., *Chem, Ber.*, **111,** 770 (1978).
[18] Huisgen, R., and Ortega, J. P., *Tetrahedron Lett.*, **1978,** 3975.
[19] Hoffmann, R. W., and Becherer, J., *Tetrahedron*, **34,** 1187 (1978).
[20] Becherer, J., and Hoffmann, R. W., *Tetrahedron*, **34,** 1193 (1978).
[21] Solomonov, B. N., Antipin, I. S., and Konovalov, A. I., *Zh. Org. Khim.*, **13,** 2491 (1977); *Chem. Abs.*, **88,** 73929 (1978).
[22] Solomonov, B. N., Arkhireeva, I. A., Mannafov, T. G., and Konovalov, A. I., *Zh. Obshch. Khim.*, **47,** 2624 (1977); *Chem. Abs.*, **88,** 49932 (1978).
[23] Von Jouanne, J., and Kelm, H., *High Temp.–High Pressures*, **9,** 515 (1977); *Chem. Abs.*, **89,** 128771 (1978).
[24] Langbeheim, M., and Sarel, S., *Tetrahedron Lett.*, **1978,** 1219.
[25] Müllen, K., and Huber, W., *Helv. Chim. Acta*, **61,** 1310 (1978).
[26] Šket, B., and Zupan, M., *Tetrahedron Lett.*, **1978,** 2607.
[27] Lewis, F. D., and Johnson, D. E., *J. Am. Chem. Soc.*, **100,** 983 (1978).
[28] Caldwell, R. A., and Creed, D., *J. Am. Chem. Soc.*, **99,** 8360 (1977).
[29] Termont, D., De Keukeleire and Vandewalle, M., *J. Chem. Soc., Perkins Trans.*, 1, **1977,** 2349.
[30] Nerbonne, J. M., and Weiss, R. G., *J. Am. Chem. Soc.*, **100,** 2571 (1978).
[31] Lee, C. B., Newman, R. J. J., and Taylor, D. R., *J. Chem. Soc., Perkins Trans.* 1, 1978, 1161.
[32] Reinhoudt, D. N., Geevers, J., and Trompenaars, W. P., *Tetrahedron Lett.*, **1978,** 1351.
[33] Burstein, K. Y., and Serebryakov, E. P., *Tetrahedron*, **34,** 3233 (1978).
[34] Serebryakov, E. P., Margaryan, A. K., and Kucherov, V. F., *Izv. Akad. Nauk SSSR, Ser. Khim.*, 1978, 416; *Chem. Abs.*, **88,** 169284 (1978).
[35] Mazzocchi, P. H., Bowen, M. J., and Narain, N. K., *J. Am. Chem. Soc.*, **99,** 7063 (1977).

[36] Berning, W., and Hünig, S., *Angew. Chem. Int. Edn.*, **16,** 777 (1977).
[37] Meier, H., Schneider, H.-P., Rieker, A., and Hitchcock, P. B., *Angew. Chem. Int. Edn.*, **17,** 121 (1978).
[38] Yoshida, H., Ogata, T., and Inokawa, S., *Bull. Chem. Soc. Jpn.*, **50,** 3302 (1977).
[39] Ahlrichs, R., and Heinzmann, R., *J. Am. Chem. Soc.*, **99,** 7452 (1977).
[40] Jones, P. R., and Lim, T. F. O., *J. Chem. Soc.*, **99,** 8447 (1977).
[41] Mirek, J., and Rachwal, S., *Phosphorus Sulfur*, **3,** 333 (1977); *Chem. Abs.*, **88,** 169282 (1978).
[42] Frimer, A. A., Bartlett, P. D., Boschung, A. F., and Jewett, J. G., *J. Am. Chem. Soc.*, **99,** 7977 (1977).
[43] Tada, K., and Toda, F., *Tetrahedron Lett.*, **1978,** 563.
[44] Chickos, J. S., Sherwood, D. E., and Jug, K., *J. Org. Chem.*, **43,** 1146 (1978).
[45] Becherer, J., Hauel, N., and Hoffmann, R. W., *Justus Liebigs Ann. Chem.*, **1978,** 312.
[46] Collins, C. J., Benjamin, B. M., and Kabalka, G. W., *J. Am. Chem. Soc.*, **100,** 2570 (1978).
[47] Baldwin, J. E., and Kapecki, J. A., *J. Am. Chem. Soc.*, **92,** 4874 (1970).
[48] Gheorghiu, M. D., Drăghici, C., and Pârvulescu, L., *Tetrahedron*, **33,** 3295 (1977).
[49] Brook, P. R., Eldeeb, A. M., Hunt, K., and McDonald, W. S., *J. Chem. Soc., Chem. Commun.*, 1978, 10.
[50] Becker, D., and Brodsky, N. C., *J. Chem. Soc., Chem. Commun.*, 1978, 237.
[51] Goldschmidt, Z., and Antebi, S., *Tetrahedron Lett.*, **1978,** 1225.
[52] Goldschmidt, Z., and Antebi, S., *Tetrahedron Lett.*, **1978,** 271.
[53] Chambers, R., Kunert, D., Hernandez, L., Mercer, F., and Moore, H. W., *Tetrahedron Lett.*, 1978, 933.
[54] Moore, H. W., Hernandez, L., and Chambers, R., *J. Am. Chem. Soc.*, **100,** 2245 (1978).
[55] Moore, H. W., and Wilbur, D. S., *J. Am. Chem. Soc.*, **100,** 6523 (1978).
[56] Minami, T., Ishida, M., and Agawa, T., *J. Chem. Soc., Chem. Commun.*, **1978,** 12.
[57] Schmidpeter, A., and von Criegern, T., *Angew. Chem. Int. Edn.*, **17,** 55 (1978).
[58] Schmidpeter, A., and von Criegern, T., *J. Chem. Soc., Chem. Commun.*, **1978,** 470.
[59] Bödeker, J., and Courault, K., *Tetrahedron*, **34,** 101 (1978).
[60] Piper, J. U., Allard, M., Faye, M., Hamel, L., and Chow, V., *J. Org. Chem.*, **42,** 4261 (1977).
[61] Baxter, G. J., Brown, R. F. C., Eastwood, F. W., Gatehouse, B. M., and Nesbit, M. C., *Aust. J. Chem.*, **31,** 1757 (1978).
[62] Firestone, R. A., *Tetrahedron*, **33,** 3009 (1977).
[63] Harcourt, R. D., *Tetrahedron*, **34,** 3125 (1978).
[64] Houk, K. N., Domelsmith, L. N., Strozier, R. W., and Patterson, R. T., *J. Am. Chem. Soc.*, **100,** 6531 (1978).
[65] Godfrey, M., *J. Chem. Res.* (S) **1978,** 258.
[66] Gordon, M. D., Alston, P. V., and Rossi, A. R., *J. Am. Chem. Soc.*, **100,** 5701 (1978).
[67] Padwa, A., Ku, H., and Mazzu, A., *J. Org. Chem.*, **43,** 381 (1978).
[68] Taniguchi, H., Ikeda, T., and Imoto, E., *Bull. Chem. Soc. Jpn.*, **50,** 2694 (1977).
[69] Taniguchi, H., Ikeda, T., Yoshida, Y., and Imoto, E., *Bull. Chem. Soc. Jpn.*, **51,** 1495 (1978).
[70] Taniguchi, H., Ikeda, T., and Imoto, E., *Bull. Chem. Soc. Jpn.*, **51,** 1859 (1978).
[71] Siemionko, R., Shaw, A.. O'Connell, G., Little, R. D., Carpenter, B. K., Shen, L., and Berson, J. A., *Tetrahedron Lett.*, **1978,** 3529.
[72] Eicher, T., Weber, J. L., and Chatila, G., *Justus Liebigs Ann. Chem.*, **1978,** 1203.
[73] Hayakawa, Y., Yokoyama, K., and Noyori, R., *J. Am. Chem. Soc.*, **100,** 1791 (1978).
[74] Takahashi, M., Yamaguchi, S., and Igari, N., *Chem. Lett.*, **1977,** 1497.
[75] Deyrup, J. A., and Kuta, G. S., *J. Org. Chem.*, **43,** 501 (1978).
[76] Benhaona, H., Texier, F., Guenot, P., Martelli, J., and Carrié, R., *Tetrahedron*, **34,** 1153 (1978).
[77] Soga, K., Hosoda, S., and Ikeda, S., *Nippon Kagaku Kaishi*, **1978,** 246; *Chem. Abs.*, **88,** 135877 (1978).
[78] Oehlschlager, A. C., Yim, A. S., and Akhtar, M. H., *Can. J. Chem.*, **56,** 273 (1978).
[79] Vo-Quang, L., and Vo-Quang, Y., *Tetrahedron Lett.*, **1978,** 4679.
[80] Padwa, A., and Carlsen, P. H. J., *J. Org. Chem.*, **43,** 2029 (1978).
[81] Padwa, A., and Ku, A., *J. Am. Chem. Soc.*, **100,** 2181 (1978).
[82] Padwa, A., Carlsen, P. H. J., and Ku, A., *J. Am. Chem. Soc.*, **100,** 3494 (1978).
[83] Padwa, A., and Carlsen, P. H. J., *J. Org. Chem.*, **43,** 3757 (1978).
[84] Grigg, R., Kemp, J., Sheldrick, G., and Trotter, J., *J. Chem. Soc., Chem. Commun.*, **1978,** 109.

[85] Garanti, L., Padova, G., and Zecchi, G., *J. Heterocycl. Chem.*, **14**, 947 (1977).
[86] Burger, K., Goth, H., and Hoehenberger, W., *Chem–Ztg.*, **101**, 453 (1977); *Chem. Abs.*, 88, 49909 (1978).
[87] Huisgen, R., Markowski, V., and Hermann, H., *Heterocycles*, **7**, 61 (1977).
[88] Padwa, A., Nahm, S., and Sato, E., *J. Org. Chem.*, **43**, 1664 (1978).
[89] Laude, B., Soufiaoui, M., and Arriau, J., *J. Heterocycl. Chem.*, **14**, 1183 (1977).
[90] Ruccia, M., Vivona, N., and Cusmano, G., *J. Heterocycl. Chem.*, **15**, 293 (1978).
[91] Bonini, B. F., Maccagnani, G., Mazzanti, G., Thijs, L., Veenstra, G. E., and Zwanenburg, B., *J. Chem. Soc., Perkins Trans.*, 1, **1978**, 1218.
[92] Könnecke, A., Dörre, R., and Lippmann, E., *Tetrahedron Lett.*, **1978**, 2071.
[93] Martelli, J., and Carrié, R., *Tetrahedron*, **34**, 1163 (1978).
[94] Shimizu, N., and Bartlett, P. D., *J. Am. Chem. Soc.*, **100**, 4260 (1978).
[95] Wilt, J. W., and Roberts, W. N., *J. Org. Chem.*, **43**, 170 (1978).
[96] Scharf, H.-D., and Mattay, J., *Chem. Ber.*, **111**, 2206 (1978).
[97] Battioni, P., Vo-Quang, L., and Vo-Quang, Y., *Bull. Soc. Chim. Fr. II*, **1978**, 401.
[98] Battioni, P., Vo-Quang, L., and Vo-Quang, Y., *Bull. Soc. Chim. Fr. II*, **1978**, 415.
[99] Dietrich-Buchecker, C., Martina, D., and Franck-Neumann, M., *J. Chem. Res.*, (S) **1978**, 78.
[100] Dehmlow, E. V., and Naser-ud-Din, *J. Chem. Res.*, (S) **1978**, 40.
[101] Cauquis, G., and Chabaud, B., *Tetrahedron*, **34**, 903 (1978).
[102] Kobayashi, Y., Kumadaki, I., Ohsawa, A., and Ando, A., *J. Am. Chem. Soc.*, **99**, 7350 (1977).
[103] Stephan, E., *Bull. Soc. Chim. Fr. II*, **1978**, 364.
[104] Tyspin, G. I., Timofeeva, T. N., Mel'nikov, V. V., and Gidaspov, B. V., *Zh. Org. Khim.*, **13**, 2275 (1977); *Chem. Abs.*, **88**, 73911 (1978).
[105] Tyspin, G. I., Mel'nikov, V. V., Timofeeva, T. N., and Gidaspov, B. V., *Zh. Org. Khim.*, **13**, 2281 (1977); *Chem. Abs.*, **88**, 49930 (1978).
[106] Padwa, A., Ku, A., Ku, H., and Mazzu, A., *J. Org. Chem.*, **43**, 66 (1978).
[107] Elzinga, J., Heldeweg, R. F., Hogeveen, H., and Schudde, E. P., *Tetrahedron Lett.*, **1978**, 2107.
[108] Doyle, M. P., Oppenhuizen, M., Elliott, R. C., and Boekins, M. R., *Tetrahedron Lett.*, **1978**, 2247.
[109] Tufariello, J. J., and Ali, S. A., *Tetrahedron Lett.*, **1978**, 4647.
[110] Tada, K., Yamada, T., and Toda, F., *Bull. Chem. Soc. Jpn.*, **51**, 1839 (1978).
[111] Houk, K. N., Bimanand, A., Mukherjee, D., Sims, J., Chang, Y.-M., Kaufman, D. C., and Domelsmith, L. N., *Heterocycles*, **7**, 293 (1977); *Chem. Abs.*, **88**, 21984 (1978).
[112] Leroy, G., Nguyen, M. T., and Sana, M., *Tetrahedron*, **34**, 2459 (1978).
[113] Akmanova, N. A., Shaul'skii, Y. M., and Svetkin, Y. V., *Org. Khim.*, **1976**, 88; *Chem. Abs.*, **88**, 169274 (1978).
[114] Joucla, M., Tonnard, F., Greé, D., and Hamelin, J., *J. Chem. Res.*, (S) **1978**, 240.
[115] Joucla, M., Hamelin, J., and Greé, D., *J. Chem. Res.*, (S) **1978**, 276.
[116] Greé, R., and Carrié, R., *J. Heterocycl. Chem.*, **14**, 965 (1977).
[117] Houk, K. N., Chang, Y.-M., Strozier, R. W., and Caramella, P., *Heterocycles*, **7**, 793 (1977).
[118] Taniguchi, H., and Imoto, E., *Bull. Chem. Soc. Jpn.*, **51**, 2405 (1978).
[119] Beltrame, P., Gelli, G., and Loi, A., *J. Chem. Res.*, (S) **1978**, 420.
[120] Gallucci, J., LeBlanc, M., and Riess, J. G., *J. Chem. Res.*, (S) **1978**, 192.
[121] Caramella, P., Cellerino, G., Houk, K. N., Albini, F. M., and Santiago, C., *J. Org. Chem.*, **43**, 3006 (1978).
[122] McAlduff, E. J., Caramella, P., and Houk, K. N., *J. Am. Chem. Soc.*, **100**, 105 (1978).
[123] Taniguchi, H., Yoshida, Y., and Imoto, E., *Bull. Chem. Soc. Jpn.*, **50**, 3335 (1977).
[124] Katritzky, A. R., El-Osta, B., Musumarra, G., and Öğretir, C., *J. Chem. Res.*, (S) **1978**, 322.
[125] Dennis, N., Dowlatshahi, H. A., and Katritzky, A. R., *J. Chem. Res.*, (S), **1978**, 102.
[126] Tamura, Y., Akita, M., Kiyokawa, H., Chen, L.-C., and Ishibashi, H., *Tetrahedron Lett.*, **1978**, 1751.
[127] Gotthardt, H., and Weisshuhn, C. M., *Chem. Ber.*, **111**, 2028 (1978).
[128] Gotthardt, H., and Christl, B., *Chem. Ber.*, **111**, 3029 (1978).
[129] Gotthardt, H., Weisshuhn, C. M., and Christl, B., *Chem. Ber.*, **111**, 3037 (1978).
[130] Gotthardt, H., and Weisshuhn, C. M., *Chem. Ber.*, **111**, 3171 (1978).
[131] Hamaguchi, M., *J. Chem. Soc., Chem. Commun.*, **1978**, 247.
[132] Huisgen, R., and Schmidt, T., *Justus Liebigs Ann. Chem.*, **1978**, 29.
[133] Koga, H., Hirobe, M., and Okamoto, T., *Tetrahedron Lett.*, **1978**, 1291.

[134] Burger, K., and Hein, F., *Chem.-Ztg.*, **102**, 152 (1978); *Chem. Abs.*, **89**, 107289 (1978).
[135] Butler, A. R., *J. Chem. Res.*, (S) **1978**, 50.
[136] Clark, H. C., Goel, A. B., and Wong, C. S., *J. Am. Chem. Soc.*, **100**, 6241 (1978).
[137] Ripoll, J. L., Rouessac, A., and Rouessac, F., *Tetrahedron*, **34**, 19 (1978).
[138] Dewar, M. J. S., Olivella, S., and Rzepa, H. S., *J. Am. Chem. Soc.*, **100**, 5650 (1978).
[139] Weigert, F. J., *J. Org. Chem.*, **42**, 3859 (1977).
[140] Shuraeva, V. N., and Katsobashvili, V. Y., *React. Kinet. Catal. Lett.*, **7**, 267 (1977); *Chem. Abs.*, **88**, 36852 (1978).
[141] Kim, D., and Weinreb, S. M., *J. Org. Chem.*, **43**, 121 (1978).
[142] Oliva, A., Fernandez-Alonso, J. I., and Bertran, J., *Tetrahedron*, **34**, 2029 (1978).
[143] Bradsher, C. K., Carlson, G. L. B., Porter, N. A., Westerman, I. J., and Wallis, T. G., *J. Org. Chem.*, **43**, 822 (1978).
[144] Westerman, I. J., and Bradsher, C. K., *J. Org. Chem.*, **43**, 3002 (1978).
[145] Sordo, J. A., and Bertran, J., *An. Quim.*, **74**, 333 (1978); *Chem. Abs.*, **89**, 146262 (1978).
[146] Kiselev, V. D., Konovalov, A. I., Veisman, E. A., and Ustyugov, A. N., *Zhur. Org. Khim.*, **14**, 128 (1978); *Chem. Abs.*, **88**, 169270 (1978).
[147] Anh, N. T., Canadell, E., and Eisenstein, O., *Tetrahedron*, **34**, 2283 (1978).
[148] Neukam, W., and Grimme, W., *Tetrahedron Lett.*, 1978, 2201.
[149] Overman, L. E., Taylor, G. F., Houk, K. N., and Domelsmith, L. N., *J. Am. Chem. Soc.*, **100**, 3182 (1978).
[150] Carrupt, P.-A., Avenati, M., Quarroz, D., and Vogel, P., *Tetrahedron Lett.*, 1978, 4413.
[151] Alston, P. V., Ottenbrite, R. M., and Cohen, T., *J. Org. Chem.*, **43**, 1864 (1978).
[152] Fleming, I., Michael, J. P., Overman, L. E., and Taylor, G. F., *Tetrahedron Lett.*, **1978**, 1313.
[153] Imagawa, T., Haneda, A., Nakagawa, T., and Kawanishi, M., *Tetrahedron*, **34**, 1893 (1978).
[154] Krow, G. R., Johnson, C., and Boyle, M., *Tetrahedron Lett.*, **1978**, 1971.
[155] Kahl, R. T., Katto, T., Braham, J. N., and Stille, J. K., *Macromolecules*, **11**, 340 (1978); *Chem. Abs.*, **89**, 41872 (1978).
[156] Paddon-Row, M, N., Patney, H. K., and Warrener, R. N., *J. Chem. Soc., Chem. Commun.*, 1978, 296.
[157] Alston, P. V., and Ottenbrite, R. M., *J. Heterocycl. Chem.*, **14**, 1443 (1977).
[158] Paddon-Row, M. N., Patney, H. K., and Warrener, R. N., *Aust. J. Chem.*, **30**, 2307 (1977).
[159] Lindsay Smith, J. R., Norman, R. O. C., and Stillings, M. R., *Tetrahedron*, **34**, 1381 (1978).
[160] Trost, B. M., Ippen, J., and Vladuchick, W. C., *J. Am. Chem. Soc.*, **99**, 8116 (1977).
[161] Kelly, T. R., and Montury, M., *Tetrahedron Lett.*, **1978**, 4311.
[162] Kelly, T. R., *Tetrahedron Lett.*, **1978**, 1387.
[163] Boeckman, R. K., Dolak, T. M., and Culos, K. O., *J. Am. Chem. Soc.*, **100**, 7098 (1978).
[164] Fleming, I., and Percival, A., *J. Chem. Soc., Chem. Commun.*, **1978**, 178.
[165] Danishefsky, S., Prisbylla, M. P., and Hiner, S., *J. Am. Chem. Soc.*, **100**, 2918 (1978).
[166] Chacko, E., Bornstein, J., and Sardella, D. J., *J. Am. Chem. Soc.*, **99**, 8248 (1977).
[167] Martin, H.-D., Iden, R., and Schiwek, H.-J., *Tetrahedron Lett.*, **1879**, 3337.
[168] Kyazimova, T. G., Babaev, R. S., and Bairamov, A. A., *Dokl. Akad. Nauk Az. SSR.* **33**, 34 (1977); *Chem. Abs.*, **89**, 23612 (1978).
[169] Raifel'd, Y. E., El'yanov, B. S., and Makin, S. M., *Tezisy Dokl-Vses. Konf. "Stereokhim. Konform. Anal. Org. Neftekhim. Sint"., 3rd*, 1976, 35; *Chem. Abs.*, **89**, 23619 (1978).
[170] Lee, M. W., and Herndon, W. C., *J. Org. Chem.*, **43**, 518 (1978).
[171] Akhmedov, I. M., Mamedov, E. G., Guseinov, M. M., and Mamedov, A. A., *Zhur. Org. Khim.*, **14**, 1197 (1978); *Chem. Abs.*, **89**, 128970 (1978).
[172] Nakahara, M., Uosaki, Y., Sasaki, M., and Osugi, J., *Rev. Phys. Chem. Jpn.*, **47**, 119 (1977); *Chem. Abs.*, **88**, 151684 (1978).
[173] Jenner, G., and Rimmelin, J., *High Temp.–High Pressures*, **9**, 517 (1977); *Chem. Abs.*, **89**, 128772 (1978).
[174] El'yanov, B. S., Makin, S. M., and Raifel'd, Y. E., *Ref. Dokl. Soobshch.–Mendeleevsk. S'ezd Obshch. Prikl. Khim., 11th*, **2**, 59 (1978); *Chem. Abs.*, **88**, 135880 (1978).
[175] Jenner, G., Abdi-oskoui, H., and Rimmelin, J., *Bull. Soc. Chim. Fr. II*, **1977**, 983.
[176] Gladysz, J. A., Lee, S. J., Tomasello, J. A. V., and Yu., Y. S., *J. Org. Chem.*, **42**, 4170 (1977).
[177] Verma, S. M., and Singh, A. K., *Indian J. Chem., Sect. B*, **15**, 700 (1977); *Chem. Abs.*, **88**, 104241 (1978).
[178] Musaeva, N. F., Salakhov, M. S., Guseinov, M. M., Suleimanov, S. N., and Poladov, P. M., *Tezisy Dokl.–Vses. Konf. "Stereokhim. Konform. Anal. Org. Neftekhim. Sint.," 3rd, 1976*, 107; *Chem. Abs.*, **88**, 151693 (1978).

[179] Szmant, H. H., and Nanjundiah, R., *J. Org. Chem.*, **43**, 1835 (1978).
[180] Anastassiou, A. G., and Badri, R., *Tetrahedron Lett.*, **1977**, 4465.
[181] Shinmyozu, T., Inazu, T., and Yoshino, T., *Chem. Lett.*, **1978**, 405.
[182] Kalo, J., Ginsburg, D., and Bloomfield, J. J., *Tetrahedron*, **34**, 2153 (1978).
[183] Ashkenazi, P., Vogel, E., and Ginsburg, D., *Tetrahedron*, **34**, 2167 (1978).
[184] Ashkenazi, P., Oliker, J., and Ginsburg, D., *Tetrahedron*, **34**, 2171 (1978).
[185] Kalo, J., and Ginsburg, D., *Tetrahedron*, **34**, 2155 (1978).
[186] Ashkenazi, P., Kalo, J., Rüttimann, A., and Ginsburg, D., *Tetrahedron*, **34**, 2161 (1978).
[187] Christl, M., Lüddeke, H.-J., Nagyrevi-Neppel, A,. and Freitag, G., *Chem. Ber.*, **110**, 3745 (1977).
[188] Hogeveen, H., and Huurdeman, W. F. J., *J. Chem. Soc.*, **100**, 860 (1978).
[189] Hogeveen, H., Huurdeman, W. F. J., and Kok, D. M., *J. Am. Chem. Soc.*, **100**, 871 (1978).
[190] Oda, M., M., Kato, K., Kitahara, Y., Kuroda, S., Morita, N., and Asao, T., *Chem. Lett.*, **1978**, 961.
[191] Noble, K.-L., Hopf, H., Jones, M., and Kammula, S. L., *Angew. Chem. Int. Edn.*, **17**, 602 (1978).
[192] Oku, A., and Matsui, A., *Bull. Chem. Soc. Jpn.*, **50**, 3338 (1977).
[193] Nakazawa, T., Niimoto, Y., and Murata, I., *Tetrahedron Lett.*, **1978**, 569.
[194] Szeimies-Seebach, U., and Szeimies, G., *J. Am. Chem. Soc.*, **100**, 3966 (1978).
[195] Shea, K. J., and Wise. S., *J. Am. Chem. Soc.*, **100**, 6519 (1978).
[196] Paquette, L. A., Moerck, R. E., Harirchian, B., and Magnus, P. D., *J. Am. Chem. Soc.*, **100**, 1597 (1978).
[197] Trost, B. M., and Lunn, G., *J. Am. Chem. Soc.*, **99**, 7079 (1977).
[198] Humphries, A. P., and Knox, S. A. R., *J. Chem. Soc., Dalton Trans.*, **1978**, 1514.
[199] Evers, J. T. M., and Mackor, A., *Tetrahedron Lett.*, **1978**, 2317.
[200] Hunt, M. M., and McCleverty, J. A., *J. Chem. Soc., Dalton Trans.*, **1978**, 480.
[201] Keck, G. E., *Tetrahedron Lett.*, **1978**, 4767.
[202] Häussinger, P., and Kresze, G., *Tetrahedron*, **34**, 689 (1978).
[203] Kresze, G., and Mavromatis, A., *Tetrahedron*, **34**, 697 (1978).
[204] Zelenin, K. N., Verbov, V. N., and Matveeva, Z. M., *Khim. Geterotsikl. Soedin.*, **1977**, 1658; *Chem. Abs.*, **88**, 104567 (1978).
[205] Martin, H.-D., and Hekman, M., *Tetrahedron Lett.*, **1978**, 1183.
[206] Bonjouklian, R., and Ruden, R. A., *J. Org. Chem.*, **42**, 4095 (1977).
[207] David, S., Lubineau, A., and Thieffry, A., *Tetrahedron*, **34**, 299 (1978).
[208] Brunskill, J. S. A., De, A., and Ewing, D. F., *J. Chem. Soc., Perkin Trans.*, 1, **1978**, 629.
[209] Inagaki, Y., and Okazaki, R., *Yuki Gosei Kagaku Kyokaishi*, **36**, 1 (1978); *Chem. Abs.*, **89**, 41580 (1978).
[210] Durst, T., and Tétreault-Ryan, L., *Tetrahedron Lett.*, **1978**, 2353.
[211] Harano, K., Ban, T., Yasuda, M., and Kanematsu, K., *Tetrahedron Lett.*, **1978**, 4037.
[212] Gotthard, H., Weisshuhn, C. M., and Dörhöfer, K., *Chem. Ber.*, **111**, 3336 (1978).
[213] Fleury, J.-P., Desbois, M., and See, J., *Bull, Soc. Chim. Fr. II*, **1978**, 147.
[214] Friedrichsen, W., and Schmidt, R., *Justus Liebigs Ann. Chem.*, **1978**, 1129.
[215] Friedrichsen, W., and Oeser, H.-G., *Justus Liebigs Ann. Chem.*, **1978**, 1139.
[216] Friedrichsen, W., and Oeser, H.-G., *Justus Liebigs Ann. Chem.*, **1978**, 1146.
[217] Marchetti, L., *J. Chem. Soc., Perkins Trans.*, 2, **1978**, 382.
[218] O-oka, M., Kitamura, A., Okazaki, R., and Inamoto, N., *Bull. Chem. Soc. Jpn.*, **51**, 301 (1978).
[219] Okazaki, R., Ishii, F., and Inamoto, N., *Bull. Chem. Soc. Jpn.*, **51**, 309 (1978).
[220] Okazaki, R., Kang, K.-T., Sunagawa, K., and Inamoto, N., *Chem. Lett.*, **1978**, 55.
[221] Gareev, R. D., Loginova, G. M., Samitov, Y. Y., and Pudovik, A. N., *Zh. Obshch. Khim.*, **47**, 2663 (1977); *Chem. Abs.*, **88**, 135852 (1978).
[222] Ratier, M., Castaing, M., Godat, J.-Y., and Pereyre, M., *J. Chem. Res.*, (S) **1978**, 179.
[223] Scott, L. T., and Brunsvold, W. R., *J. Chem. Soc., Chem. Commun.*, **1978**, 633.
[224] Lyons, J. E., Myers, H. K., and Schneider, A., *J. Chem. Soc., Chem. Commun.*, **1978**, 636.
[225] Lyons, J. E., Myers, H. K., and Schneider, *J. Chem. Soc., Chem. Commun.*, **1978**, 638.
[226] Bansal, R. K., and Sharma, S. K., *J. Organomet. Chem.*, **155**, 293 (1978).
[227] Hoffman, H. M. R., and Chidgey, R., *Tetrahedron Lett.*, **1978**, 85.
[228] Takaya, H., Makino, S., Hayakawa, Y., and Noyori, R., *J. Am. Chem. Soc.*, **100**, 1765 (1978).

[229] Takaya, H., Hayakawa, Y., Makino, S., and Noyori, R., *J. Am. Chem. Soc.*, **100,** 1778 (1978).
[230] Hayakawa, Y., Baba, Y., Makino, S., and Noyori, R., *J. Am. Chem. Soc.*, **100,** 1786 (1978).
[231] Hayakawa, Y., Yokoyama, K., and Noyori, R., *J. Am. Chem. Soc.*, **100,** 1799 (1978).
[232] Mayr, H., and Grubmüller, B., *Angew. Chem. Int. Edn.*, **17,** 130 (1978).
[233] Kaz'mina, N. B., Knunyants, I. L., Mysov, E. I., and Kuz'yants, G. M., *Izv. Akad. Nauk SSSR, Ser. Khim.*, **1978,** 163; *Chem. Abs.*, **88,** 151687 (1978).
[234] Nakamura, Y., Kato, T., and Morita, Y., *J. Chem. Soc., Chem. Commun.*, **1978,** 620.
[235] Werber, G., Buccheri, F., Noto, R., and Gentile, M., *J. Heterocycl. Chem.*, **14,** 1385 (1977).
[236] Murphy, W. S., Raman, K. P., and Hathaway, B. J., *J. Chem. Soc. Perkins Trans.*, 1, **1977,** 2521.
[237] Friedrichsen, W., and Oeser, H.-G., *Justus Liebigs Ann. Chem.*, **1978,** 1161.
[238] Tegmo-Larsson, I.-M., and Houk, K. N., *Tetrahedron Lett.*, **1978,** 941.
[239] Dunn, L. C., and Houk, K. N., *Tetrahedron Lett.*, **1978,** 3411.
[240] Padwa, A., and Nobs, F., *Tetrahedron Lett.*, **1978,** 93.
[241] Mukherjee, D., Watts, C. R., and Houk, K. N., *J. Org. Chem.*, **43,** 817 (1978).
[242] Kataoka, F., and Nishida, S., *J. Chem. Soc., Chem. Commun.*, **1978,** 864.
[243] Katoaka, F., and Nishida, S., *Chem. Lett.*, **1978,** 1053.
[244] Tsuji, T., Shibata, T., Hienuki, Y., and Nishida, S., *J. Am. Chem. Soc.*, **100,** 1806 (1978).
[245] Abe, N., Tanaka, Y., and Nishiwaki, T., *J. Chem. Soc., Perkins Trans.*, 1, **1978,** 429.
[246] Hafner, K., Römer, M., and aus der Fünten, W., *Justus Liebigs Ann. Chem.*, **1978,** 376.
[247] Iwasaki, T., Kajigaeshi, S., and Kanemasa, S., *Bull. Chem. Soc. Jpn.*, **51,** 229 (1978).
[248] Mukherjee, D., Domelsmith, L. N., and Houk, K. N., *J. Am. Chem. Soc.*, **100,** 1954 (1978).
[249] Ogawa, H., and Chisaka, A., *Tetrahedron Lett.*, 1978, 4811.
[250] Cameron, D. W., Crossley, M. J., Feutrill, G. I., and Griffiths, P. G., *Aust. J. Chem.*, **31,** 1335 (1978).
[251] Cameron, D. W., and Crossley, M. J., *Aust. J. Chem.*, **31,** 1353 (1978).
[252] Hafner, K., Diesel, H. D., and Richarz, W., *Angew. Chem. Int. Edn.*, **17,** 763 (1978).
[253] Ewald, H., Lehman, B., and Neunhoeffer, H., *Justus Liebigs Ann. Chem.*, **1977,** 1718.
[254] Cunningham, D., McArdle, P., Sherlock, H., Johnson, B. F. G., and Lewis, J., *J. Chem. Soc., Dalton Trans.*, **1977,** 2340.

CHAPTER 14

Molecular Rearrangements

A. W. MURRAY

Department of Chemistry, University of Dundee

Aromatic Rearrangements

Benzene Derivatives

A general review on thermal rearrangements[1] and a review that draws attention to the excellent correlation that exists between activation and reaction volumes in solution, and the main mechanistic features of various rearrangement reactions,[2] have appeared. The use of the semidione spin probe to study molecular rearrangements[3] has also been reviewed.

^{13}C-Fourier transform NMR spectroscopy has been used to follow, *in situ*, the course of a Fischer indole reaction and, interestingly, imine intermediates such as (**1**) (corresponding to the stage of the reaction after the benzidine-type rearrangement, but before closure of the five-membered heterocycle) have been observed and characterized spectroscopically for the first time.[4] The ring-protonation mechanism of the benzidine rearrangement has been discussed on the basis of a detailed product analysis of the acid-catalysed rearrangements of 4-chloro- and 4-methoxyhydrazobenzenes[5] while a reinvestigation of the solvolysis of hydrazobenzene in methyl iodide has established[6] that the product obtained is aniline hydriodide and not hydrazobenzene dihydriodide, as previously reported.

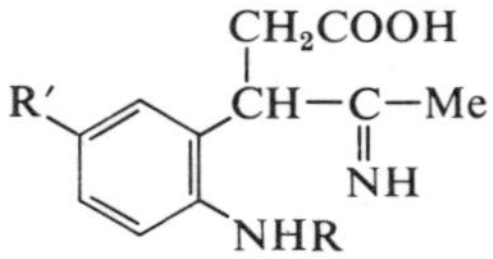

CH_2COOH / R′ / CH—C(=NH)—Me / NHR

(**1**)

R—S—N(Me)—Ph

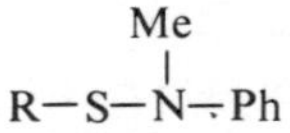

(**2**)

A paper has appeared on the abnormal Wallach rearrangement of 4-alkylazoxybenzenes,[7] and the rearrangement of *N*-(chloroacyl)arylamides in carboxylic acids or toluene-*p*-sulphonic acids has been interpreted[8] as an intermolecular chlorination in which the protonated *N*-chloro-compound acts as an electrophilic chlorinating agent. A study of the rearrangement of sulphenanilides (**2**) demonstrates[9] that the process, which is partly intermolecular, may provide a convenient route to *ortho*- and *para*-(alkylthio)anilines.

The thermal interconversion of phenylhydrazine and *o*-phenylenediamine in a molten mixture of $AlCl_3$–NaCl–KCl and, furthermore, the conversion of various aromatic compounds into the corresponding aromatic amines when heated in the same mixture in the presence of phenylhydrazine suggests a novel method for the direct preparation of aromatic amines.[10]

A study of solvent effects on the dediazoniation of benzene- and 2,4,6-trimethylbenzene-diazonium ions has been interpreted[11] in terms of (at least) two intermediates involved in the course of dediazoniation; a tight nitrogen–(aryl cation) molecule-ion pair through which the rearrangement occurs, and a solvated aryl cation free of, or solvent-separated from, nitrogen. The results further indicate that the exchange reaction with external nitrogen proceeds through the second intermediate, and a statistical treatment of the kinetic data[12] provides further evidence for the validity of the proposed two-intermediate mechanism.

Homolysis of the amide C—N bond in α-phenylacetanilide, into anilino and phenylacetyl free radicals, is considered to be the initiation step in its pyrolysis which results in migration of the benzyl group to the *ortho*- and *para*-positions of the aniline nucleus,[13] while free-radical mechanisms have also been proposed to account for the formation of *N′*,*N′*-diarylhydrazides on the thermolysis of aryl alkanehydrazonates,[14] and for the photorearrangement of bis-(2-nitrophenyl) disulphide to 2-aminobenzenesulphonanilides.[15] Radical intermediates have also been postulated for the formation of 10-aryl-9-arsaanthracenes (**4**) on the pyrolysis of (**3**).[16]

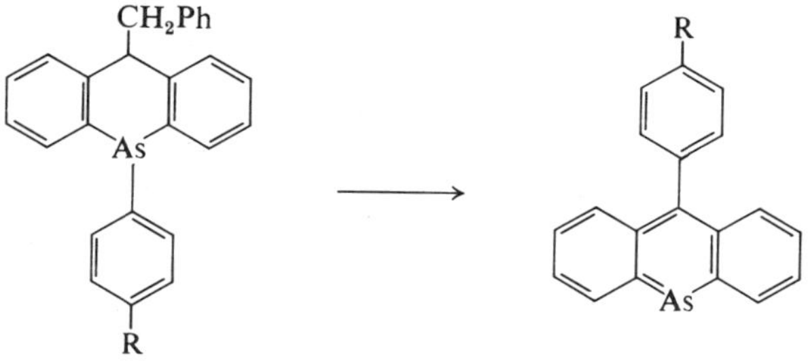

Spectroscopic confirmation of the existence of a spiro anionic σ-complex has been obtained[17] in the base-catalysed Smiles rearrangement of 2-(acetylamino)ethyl 2,4-dinitrophenyl ether, while base-catalysed cyclization of 1-(3-hydroxy-2,2-dimethylpropyloxy)-2,4,6-trinitrobenzene (**5**) was found to yield the Meisenheimer complex (**6**), which itself rearranged further in refluxing dimethyl sulphoxide containing triethylamine to yield the benzodioxepine (**7**).[18] Likewise, a study of the photo-Smiles rearrangement of a homologous series of *N*-[ω-(*p*-nitrophenoxy)]-alkylanilines (**8**) to the corresponding *N*-(ω-hydroxyalkyl)-*p*-nitrodiphenylamines (**10**) has indicated[19] the involvement of an ionic intermediate (**9**) in the rearrangement.

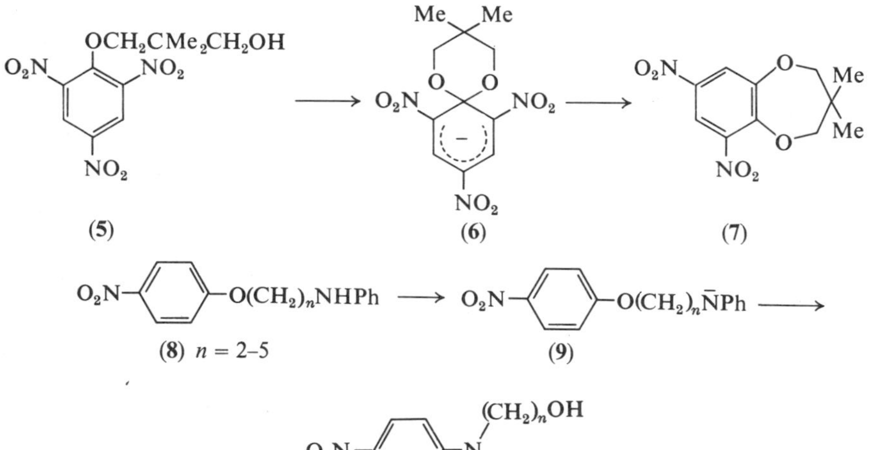

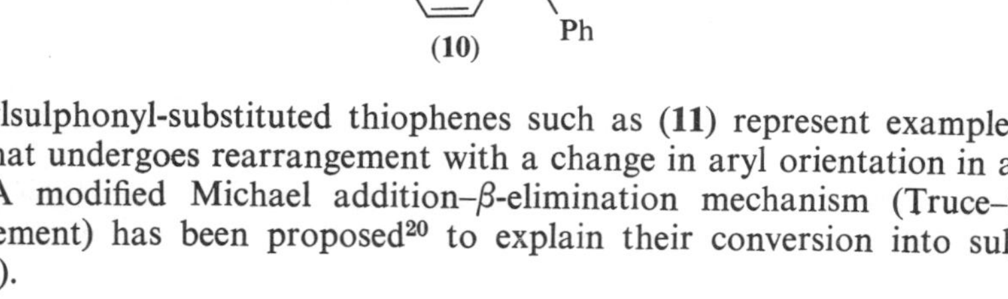

Mesitylsulphonyl-substituted thiophenes such as (**11**) represent examples of a system that undergoes rearrangement with a change in aryl orientation in aprotic media. A modified Michael addition–β-elimination mechanism (Truce–Smiles rearrangement) has been proposed[20] to explain their conversion into sulphinic acids (**12**).

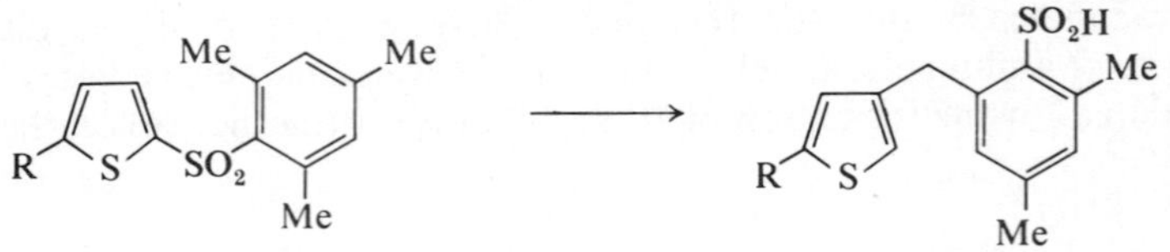

(11) (12)

It has been shown that spirocyclization is often more favourable than the alternative *ortho*-cyclization, even when the latter leads to a six-membered ring;[21] thus, methoxybenzylamino-acetonitriles with benzyl or phenethyl substituents attached to the nitrile α-carbon, *e.g.* (**13**), have been shown to undergo spirocyclization in concentrated H_2SO_4 followed by rearrangement with cyclization;[22] see Scheme 1.

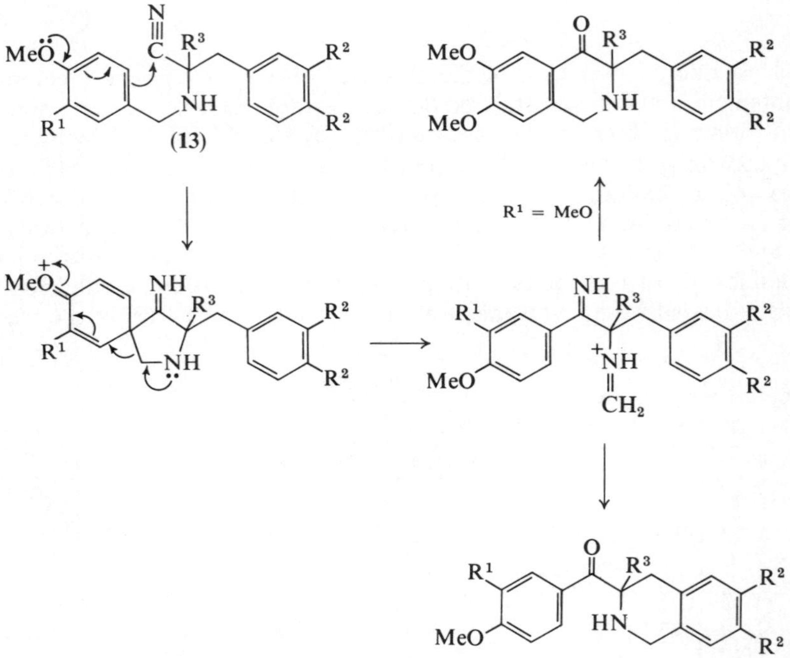

(13)

R^1 = MeO

SCHEME 1

Rearranged products arising from 1,4- or 1,5-aryl shifts from Si to C have been reported to occur on the treatment of γ- and δ-halosilanes, *e.g.* (**14**), with di-*tert*-butyl peroxide.[23] It is postulated that these shifts proceed *via* spiro-radical transition states or intermediates, *e.g.* (**15**), thus inferring that α-silyl radical stability and/or transition-state steric strain may be the principal anti-rearrangement factors blocking the 1,2-shift. Evidence has been obtained to substantiate the formation and rearrangement of an *ipso*-intermediate (**16**) in the photoinitiated chlorination of *p*-bromonitrobenzene to (**17**).[24]

An interesting example of a *para*- to *meta*-bromine shift has been observed during Friedel–Crafts acylations and alkylations of bromobenzene with cinnamoyl chloride.[25] The phenomenon of reversibility of Friedel–Crafts acylation has been

$ArSiMe_2CH_2CH_2CH_2Cl \rightleftharpoons$ (15) $\rightleftharpoons Ar(CH_2)_3SiHMe_2$

(14) (15)

(16) (17)

exemplified by the *para-*$\rightleftharpoons$*ortho*-acyl rearrangements of fluorofluorenones in polyphosphoric acid,[26] while the kinetics of $AlCl_3$-catalysed acetyl exchange between acetylmesitylene and acetyl chloride have provided proof of a further example of reversibility in a classical Friedel–Crafts acylation.[27] Interestingly, 2-acetyl-3-ethylbenzofuran was obtained along with 6-acetyl-2,3-dimethylbenzofuran on acetylation of 2,3-dimethylbenzofuran.[28] This "non-conventional" Friedel–Crafts reaction probably involves attack of the electrophile at the 2-position of the benzofuran and subsequent rearrangement of the 2-methyl group. On treatment with SbF_5–HF, ethers of *p*-alkylated or 2,6-dialkylated phenols have been found to afford the corresponding *m*-alkylated phenols by a mechanism that implies alkylation of *O*-protonated ethers or phenols.[29]

A variety of papers concerned with the Fries rearrangement has appeared. One has established[30] that the product arising from the Fries rearrangement of trimethylhydroquinone diacetate (**18**) is 1-(2,6-dihydroxy-3,4,5-trimethylphenyl)-ethanone (**23**) and not the expected, and previously reported, 1-(2,5-dihydroxy-3,4,6-trimethylphenyl)ethanone. A mechanistic rationale, depicted in Scheme 2, envisages that complex (**20**) derived from (**18**) *via* the first-formed (**19**) undergoes an intramolecular conjugate addition promoted by chelation with aluminium to yield the zwitterionic enolate species (**21**); regeneration of the ketone grouping could then occur with methyl migration, providing (**22**) in which the original *o*-methyl and *m*-acetoxyl substituents have been transposed; finally deacetylation and hydrolysis would then yield (**23**). The Fries rearrangements of *O*-acetyl and *O*-benzyl derivatives of 4-hydroxy-3-nitrobenzylated polystyrene and 5-polystyrylmethylquinolin-8-ol,[31] and of 4-hydroxyfluoranthene esters[32] and *N*-benzenesulphonylaniline[33] have been studied and it has been shown that the formation of *m*-acylphenols in the Fries rearrangement is practically limited to *o*-alkoxyphenol esters.[34]

MO calculations of the energy-transfer activities of organic π-structures in the photo-Fries rearrangement have been made.[35] The photo-Fries rearrangement of polyacetoxynaphthalenes has been applied[36] to the synthesis of acetylnaphthazarin derivatives, and seleno-[37] and telluro-[38] photo-Fries products have been reported in new aromatic photosubstitution reactions. In the course of attempted electrophilic cyclization, *o*-(phenyltelluro)benzoyl chloride (**24**) did not cyclize to telluroxanthone, but isomerized to 2-chlorotellurobenzophenone (**25**) through intramolecular carbodetelluration.[39]

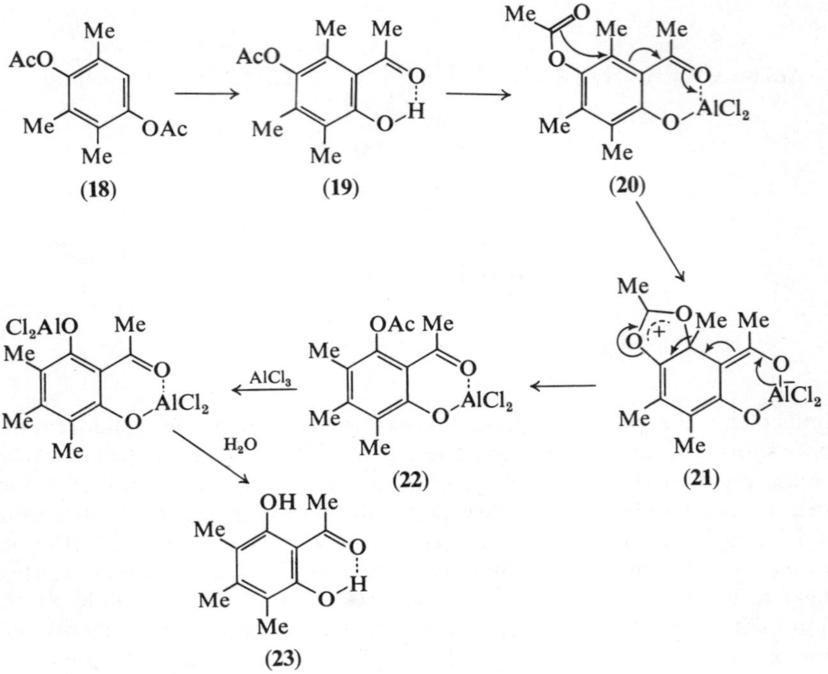

(18) (19) (20) (21) (22) (23)

SCHEME 2

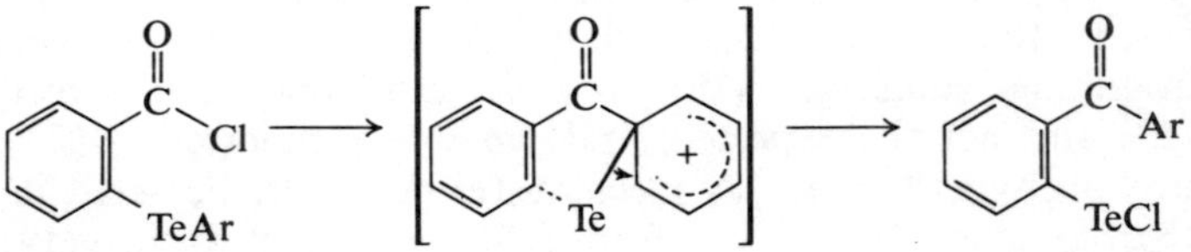

(24) (25)

Intramolecular migration of an amino group *via* a transannular process during the alkaline hydrolysis of *O*-salicyloylglycolamides has been observed;[40] evidence has been obtained to substantiate the 1→4 migration of an anilino-group in the alkaline reaction of *O*-aroyl-*N*-arylglycolamides,[41] and the novel rearrangement of 2-thiophenoxybenzhydroxamic acid to 2-carboxamidodiphenyl sulphoxide (**27**) has been rationalized[42] on the basis of an initial rearrangement in the hydroxamic acid side-chain leading to (**26**) which immediately undergoes transformation into the sulphoxide (**27**).

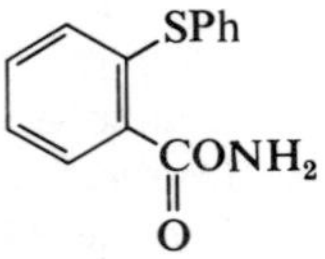

(26) 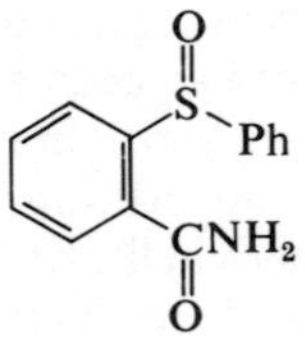 (27)

A mechanism involving a "cage radical pair" has been proposed[43] for the photochemical rearrangement of phenyl benzenesulphonates to *o*- and *p*-hydroxyphenyl phenyl sulphones and phenol, while the conversion of the *N*-aryl-*N*-quinolylamine sulphonamide (**28**) into the isomeric diaryl sulphone (**29**) has been achieved by proton catalysis.[44] Extended Hückel calculations of the potential surface for the sold-state rearrangement of *p*-$Me_2NC_6H_4SO_3Me$ to *p*-$Me_2\overset{+}{N}C_6H_4SO_3^-$ together with crystal potential-energy calculations have indicated[45] a two-step mechanism involving a molecular ion-pair intermediate for the process.

The reactivity order for the isomerization of biphenyldisulphonic acids in aqueous H_2SO_4 at elevated temperatures has been established[46] as $2,3' > 2,2' > 2,4' \gg 4,4' > 3,3'$; the same conditions have been shown to cause disproportionation and isomerization of iodoanisoles.[47] Other aromatic rearrangements include the gas-phase rearrangements of anisole and methylanisoles over a perfluorinated resin-sulphonic acid catalyst,[48] the photorearrangement of 4-nitrophenylnitromethane into 4-nitrobenzaldehyde[49] (a reaction shown to involve oxygen-transfer from the nitro- to the aldehyde group), and the high-yield conversion of 2,4,6-tris(chloromethyl)mesitylene (**30**) into hexaradialene (**31**).[50]

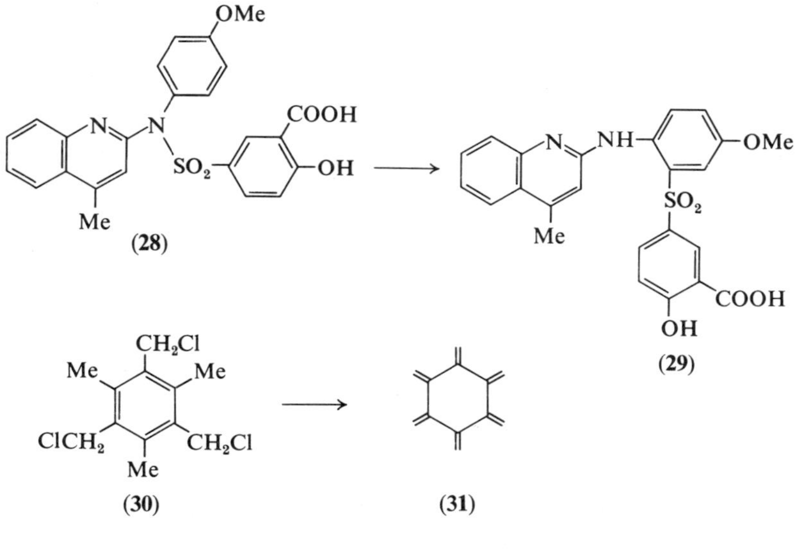

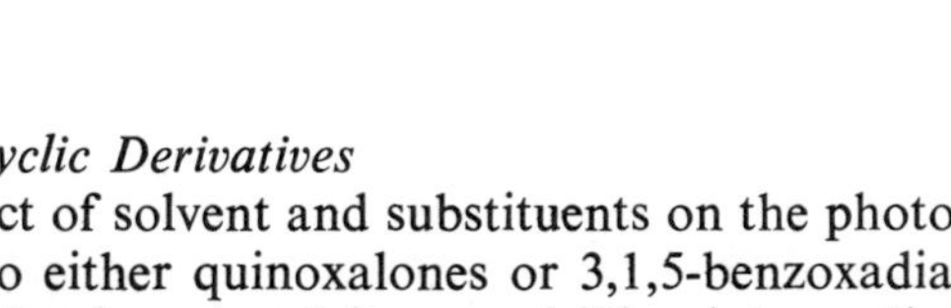

(**28**) → (**29**)

(**30**) → (**31**)

Heterocyclic Derivatives

The effect of solvent and substituents on the photoisomerization of quinoxaline-1-oxides to either quinoxalones or 3,1,5-benzoxadiazepines has been investigated[51] and mechanisms involving oxaziridine intermediates have been proposed for the photochemical rearrangement of 2-unsubstituted imidazole 3-oxides to imidazolones,[52] and the photolysis of 1-hydroxybenzimidazole 3-oxides to *N*-acyl-*o*-nitroso(or nitro)-anilines.[53] A new type of base-catalysed ring transformation of pyrazolo[4,3-*d*]pyrimidine 1-oxides (**32**) into pyrimido[5,4-*d*]pyrimidines (**35**) has been described.[54] A reasonable mechanism for this transformation is shown in Scheme 3. Proton abstraction from the active *N*(2)-methylene followed by cleavage of the $N(1)-N(2)$ bond would give an open-chain azahexatriene (**33**) which could then undergo intramolecular cycloaddition to furnish the dihydro-intermediate (**34**), aromatization of which would lead to (**35**). Δ^3-1,2,3-Thiadiazoline 1-oxides have been obtained[55] by acid-catalysed ring-contraction of 6-dihydro-2*H*-1,2,3,6-thiatriazine 1-oxides.

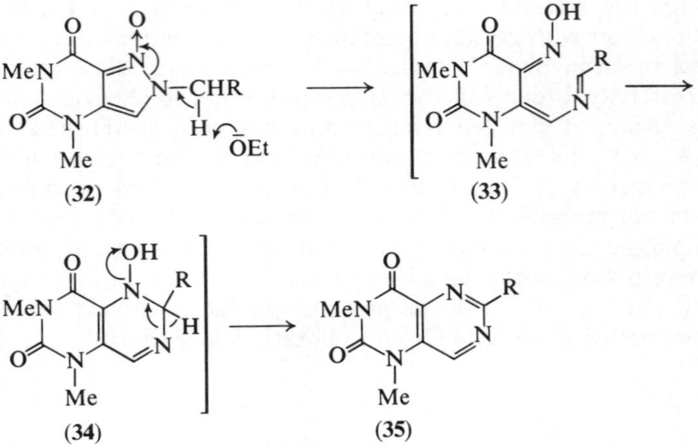

(**32**) (**33**) (**34**) (**35**)

SCHEME 3

It has been reported[56] that inclusion of SO_2 in the reaction medium circumvents the previously unattainable conversion of strained furoxans of the norbornane series into the corresponding currently inaccessible di-isocyanates, namely (**36**) to (**38**). A bis-dioxazathiole (**37**) has been suggested as a likely intermediate in this route.

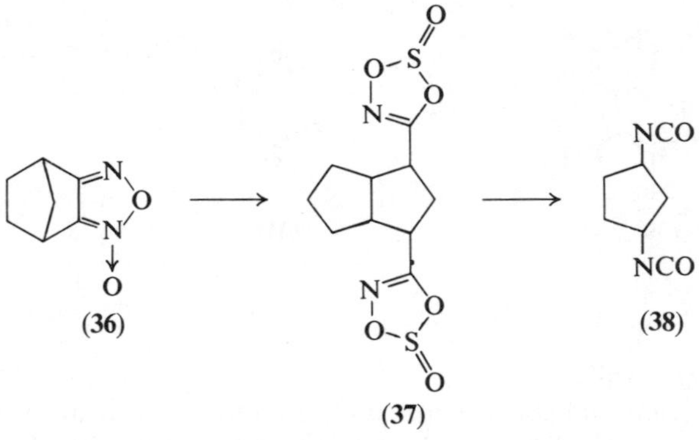

(**36**) (**37**) (**38**)

In the presence of alkylamines, 3-cyanopyridinium salts undergo a double rearrangement,[57] while it has been shown[58] that alkyl-group migration, as well as hydrogen–deuterium exchange, take place during the Ladenburg rearrangement of *N*-(trideuteriomethyl)quinolinium iodide. Photolysis of oxidopyrido[1,2-*b*]pyridazinium compounds (**39**) results in rearrangement to pyrido[1,2-*a*]pyrimidinones (**40**), which themselves undergo photochemical ring-contraction in alcohol to give alkoxydiazaspiro[4,5]decenones (**41**).[59]

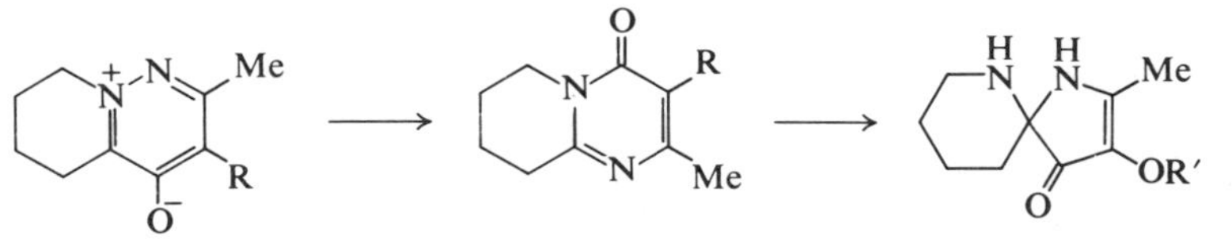

(39) (40) (41)

Various other novel rearrangements involving pyrimidines have been published; thus, it has been reported that heating 2-alkylpyrimidinium salts **(42)** in a sealed tube with an excess of alcoholic methylamine furnishes the corresponding 2-methylaminopyridines **(43)**[60] (see Scheme 4); heating 5-nitropyrimidine in dilute

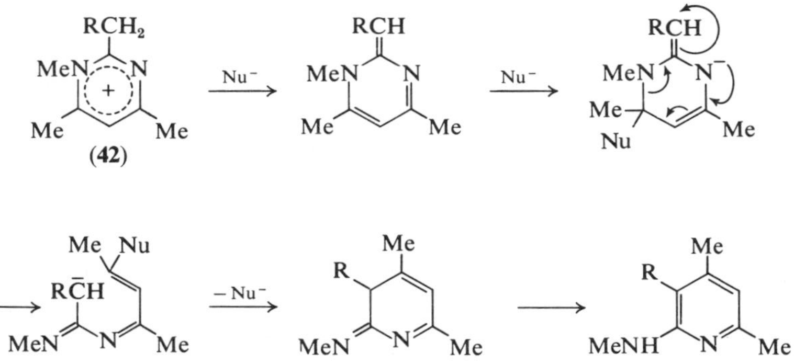

(42) (43)

SCHEME 4

acetic acid affords 3,5-dinitropyridine,[61] apparently by fragmentation of the pyrimidine ring and recondensation, while the first example of the conversion of a pyrimidine into a 1,3,5-triazepine, as illustrated by the conversion of 6-azido-1,3-dimethylthymine **(44)** into **(45)**, has been achieved[62] by irradiation in various amines. A similar ring-expansion, *viz.* the photolytic conversion of 3-azido- and

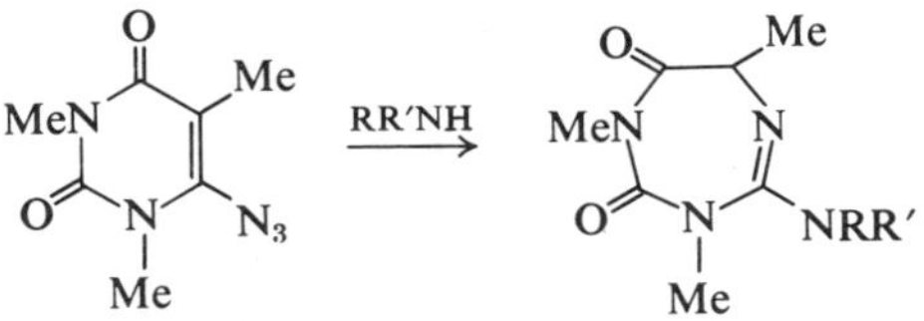

(44) (45)

4-azido-quinoline into benzo-1,4-diazepines in the presence of amines or methoxide ions, has also been reported.[63] An unusual rearrangement has been found[64] in the 8-substituted pyrido[2,3-*d*]pyrimidine series; in this case, rearrangement of the uracil-methyl component from N(8) to N(3) of the pyridopyrimidine **(46)** was observed and considered to proceed by the route outlined in Scheme 5.

An isomerization akin to the Dimroth rearrangement has been reported[65] in the conversion of simple *s*-triazolo[4,3-*a*]pyrimidines into their [1,5-*a*]-isomers. A number of unstable intermediates, namely the anionic σ-complex **(48)**, the anionic,

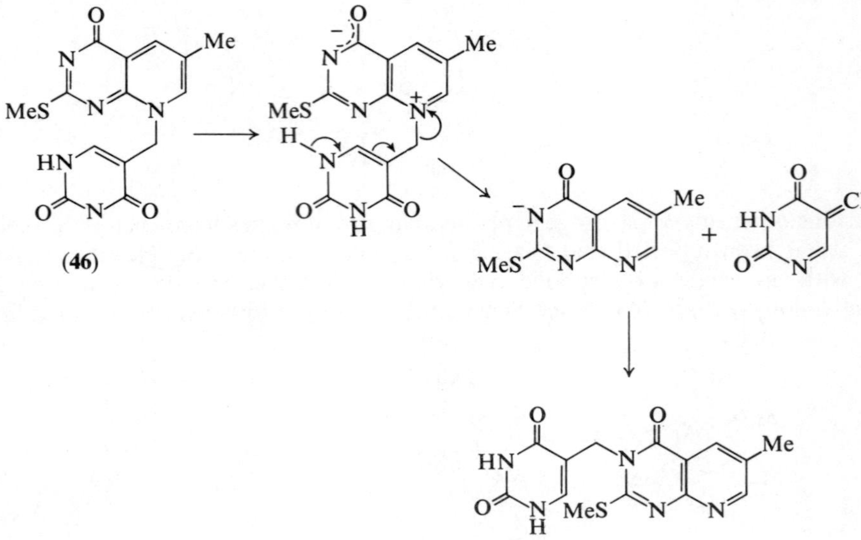

(46)

SCHEME 5

open-chain aminoethynyldiazabutadiene (**49**), and the cyanoaminoazabutadiene (**50**) have been identified[66] (by ^{13}C-NMR spectroscopy) as participating in the potassamide/liquid ammonia-catalysed ring-transformation of 4-chloro-2-dimethylamino-5-phenylpyrimidine (**47**) into 4-benzyl-2-dimethylamino-*s*-trazine (**51**).

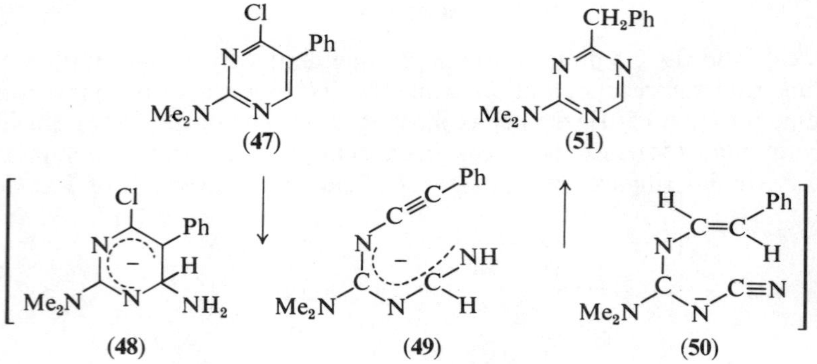

(47) (51)

(48) (49) (50)

The reaction pathway proposed[67] for the lead tetra-acetate-oxidative rearrangement of tetrahydroisoqunolines such as (**52**) is outlined in Scheme 6. It is thought to involve a retro-Mannich reaction of a *p*-quinol acetate (**53**) to generate an iminium ion (**54**) which recyclizes *ortho* to the phenolic function, to supply the tetrahydroisoquinoline (**55**). A study has been made[68] of the thermal rearrangements of 1,2-dihydropyridines and, interestingly, contrary to previous speculation regarding the relative stability of these ring systems, results suggest that the more extensively conjugated 1,2-dihydropyridines are less stable than the 2,3-dihydropyridines. Additional rearrangements of six-membered heterocycles that have been reported include the intramolecular rearrangement of bicyclic 2,2-disubstituted 4-oxo-1,2,3,4-tetrahydropyrimidines in the presence of polyphosphoric acid,[69] the

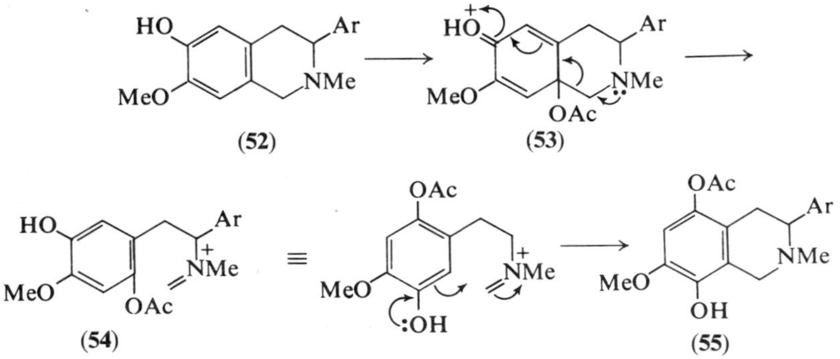

(52) (53)

(54) (55)

SCHEME 6

thermal rearrangement of 2,3,4,4a-tetrahydro-3,3,4a-trimethyl-1-thioxo-1*H*,6*H*-pyrimido[1,6-*a*][3.1]benzoxazin-6-one,[70] the ring transformations (to 2-pyridones) that accompany the hydrolysis of 6-(2-dimethylaminovinyl)uracils,[71] the decarboxylative rearrangement of 4-phenyl-4*H*-cyclopenta[*c*]cinnolinecarboxylic acids,[72] and the thermal decomposition of 2-methylthio-4-(2-chloroethoxy)-6-(dialkylamino)-*s*-triazines.[73]

An intermolecular parallel for the key alkylation reaction proposed in the biosynthetic conversion of secoyohimbine to *Strychnos* and picraline alkaloids has been observed[74] in the rearrangement of 3-chloroindolenine derivatives such as **(56)** to the 3-spiro-2-alkylideneindoline **(57)** on treatment with dimethyl thalliomalonate; the importance of reaction conditions in determining the nature of the products obtained from the rearrangement of chloroindolenine derivatives of 4-oxoindolo[2,3-*a*]quinolizidines has been emphasized.[75] A series of *as*-triazino-[4,5-*a*]indolones **(59)** has been prepared by the base-catalysed rearrangement of 2-(oxadiazolyl)indoles **(58)**.[76]

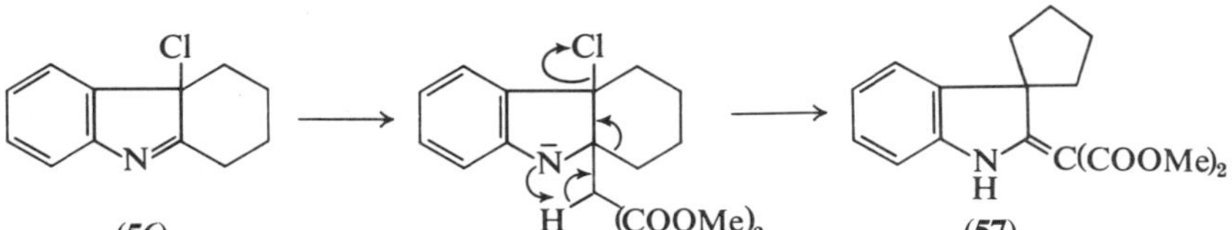

(56) (57)

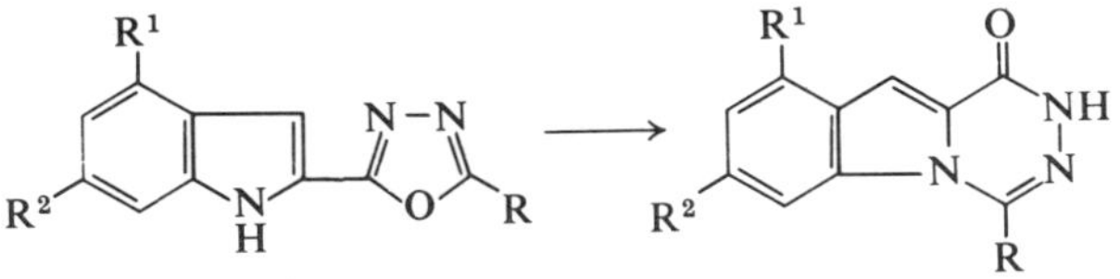

(58) (59)

An examination of the Fischer-indole rearrangement of sterically hindered systems has been made[77] and unusual ring-expanded products have been obtained from the reaction of strained indoles with acetylenedicarboxylate.[78] Interestingly, the gas-phase, acid-catalysed elimination of NH_3 from protonated phenylhydrazones has been shown[79] to be analogous in detail to the solution reaction, *i.e.* to the Fischer indole reaction.

A study of the asymmetric synthesis of a number of *Araliopsis* alkaloids has established[80] the occurrence of a dual mechanism for the dihydrofuroquinolone–dihydropyranoquinolone acetate rearrangement involved in the conversion of, for example, balfourodine (**60**) into isobalfourodine acetate (**61**). The reorganization of functional groups that occur during the transformation of furans (**62**; R = CN, COOEt, COMe) into pyrroles (**64**) is assumed[81] to proceed through the open dicyano-diketone (**63**). An analogy to this rearrangement has been found in the base-induced rearrangement of 2-aminothiophenes to 3-cyanothiophenes.[82]

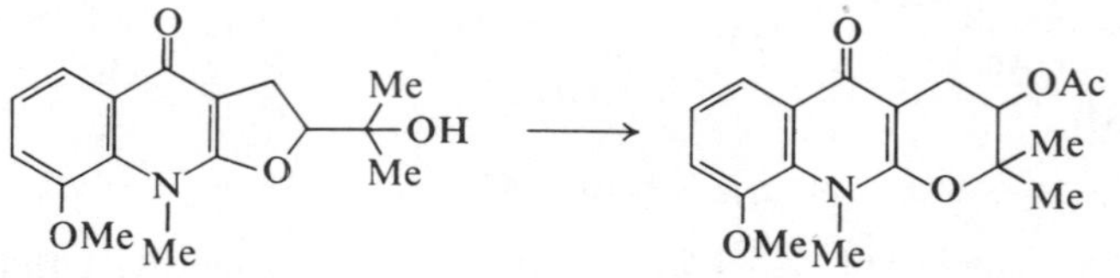

(**60**) (**61**)

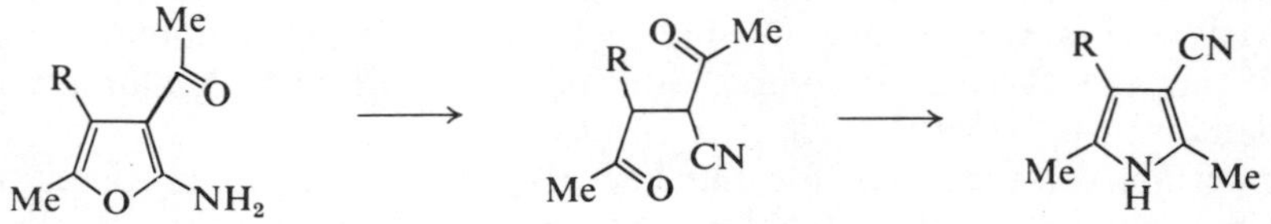

(**62**) (**63**) (**64**)

A mechanism analogous to a Lewis acid-catalysed Fries rearrangement has been proposed[83] for the $BF_3 \cdot OEt_2$-catalysed rearrangement of acyloxyfuran (**65**) to the previously unreported class of keto-lactones represented by (**66**); the same catalyst has proved effective[84] in converting 2,9-dioxa-1-azabicyclo[4.3.0]nonane derivatives (**67**) into *cis*- and *trans*-dioxa-azaspirononanes (**68**).

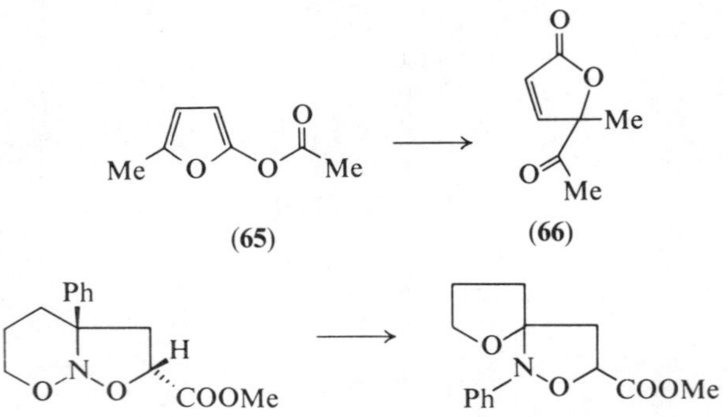

(**65**) (**66**)

(**67**) (**68**)

Whereas anils (**69**) derived from 3-nitrothiophene-2-carbaldehyde undergo reductive cyclization with triethyl phosphite to give the expected 2-arylthieno-[3.2-*c*]pyrazoles (**70**), the isomeric anils (**71**; Ar = Ph or $C_6H_4NMe_2$-*p*) give the 1-arylpyrrole-3-carbonitriles (**72**).[85] A mechanism which accounts for the latter products is shown in Scheme 7. It has been noted[86] that 2,5-bis(alkylthio)thiophenes undergo intramolecular disproportionation–rearrangement to furnish the 2,4-isomers, on protonation with HCl–$AlCl_3$, while attempted cyclization of ethyl

2-(2-amino-4-fluoroanilino)-5-ethylthiophene-3-carboxylatè with sodium hydride at elevated temperatures has been shown[87] to result in rearrangement to an *N*-thienylbenzimidazolone.

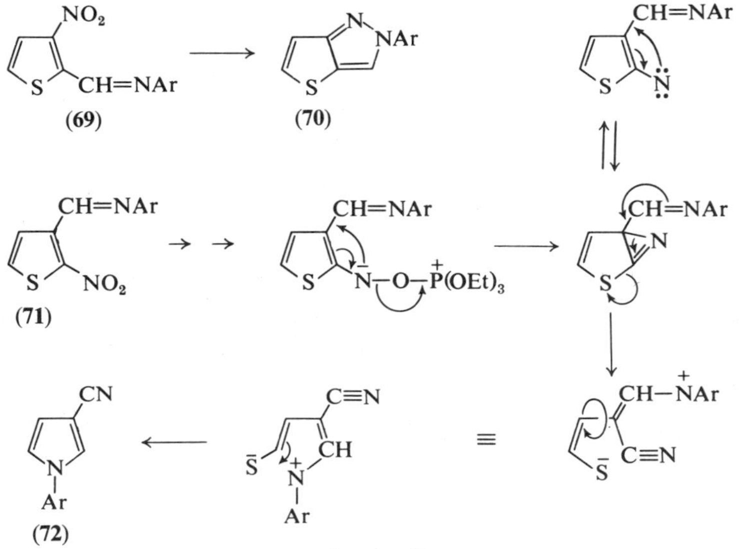

(69) (70) (71) (72)

SCHEME 7

The pyrazolo[3,4-*c*]pyrazole, which was reported previously as the product of thermal rearrangement of 4-acetyl-3-methyl-1-phenylpyrazol-5-one phenylhydrazone **(73)**, has since been shown[88] to be **(74)**, presumably arising from a Boulton–Katritzky-type rearrangement of **(73)**. Moreover, the kinetics of the above type of rearrangement of *meta*- and *para*-substituted phenylhydrazones of 3-benzoyl-5-phenyl-1,2,4-oxadiazole into 2-aryl-4-benzoylamino-5-phenyl-1,2,3-triazoles show[89] that the reaction rates are moderately affected by the nature of the substituent present in the aryl fragment of the phenylhydrazone portion, and are consistent with an internal nucleophilic displacement by nitrogen on nitrogen.

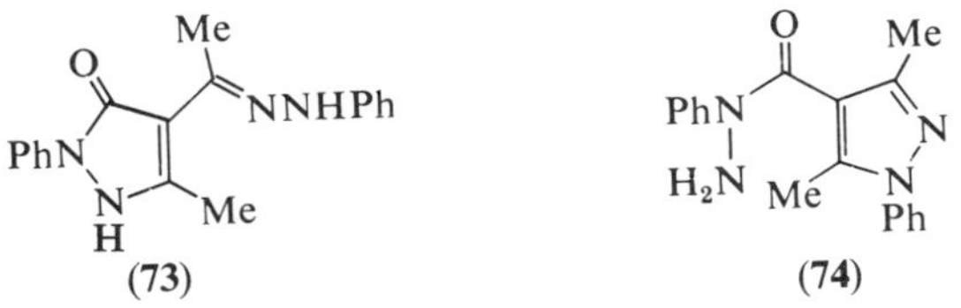

(73) (74)

The reported rearrangement of acyloxazolinones **(75)** to oxazoles **(76)**[90] represents an aza analogy of the α-acyl-lactone rearrangement, while the reported photo-chemical conversion of 3-methyl-4-phenylisoxazoles bearing differently substituted hydrazinic groups at the 5-position, *e.g.* **(77)**, into 3-methylamino-pyrazolones **(79)**, exemplifies an unusual methyl shift from the 3-position to the nitrogen of the isoxazole ring.[91] The intermediate of the reaction is believed to be the ketenimine **(78)** which is considered to undergo nucleophilic intramolecular attack by the β-hydrazinic nitrogen leading to the final product.

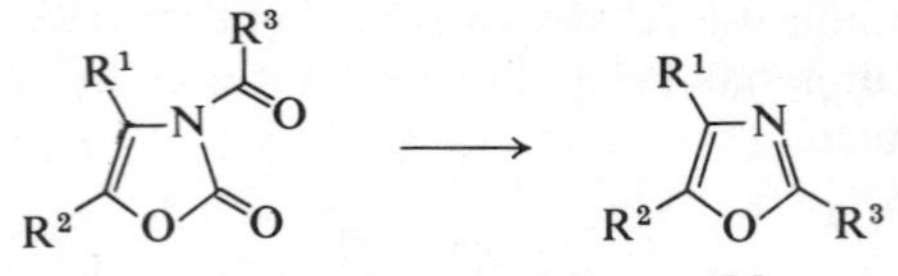

(75) → (76)

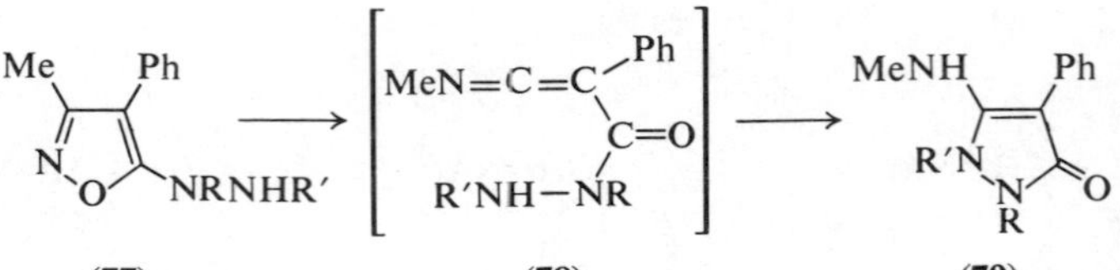

(77) → [(78)] → (79)

2,1-Benzisoxazoles bearing a vinylic side-chain at the 3-position, *e.g.* (**80**), have been found[92] to undergo a series of intriguing transformations which are considered to involve ring-opening of the system to the nitreno-ketone (**81**) followed by rearrangement of the charged intermediate (**82**) (see Scheme 8) in a process that closely resembles the production of acridones from 3-arylanthranils. Further ring-expansions are represented by the base-catalysed conversion of 2-substituted 2*H*-1,2-benzisoxazolin-3-ones into 2-substituted 2*H*-1,3-benzoxazin-4-ones[93] and the conversion of dimorpholinoimidazolines (**83**) into 5-arylpyrimidine-3,4-diamines (**85**), presumably by the intermediacy of (**84**).[94] The generality of C—N transposition in the pyrolysis of benzisoxazoles and 2,3-benzoxazin-1-ones, and the involvement of azirine intermediates in these processes, have been examined.[95]

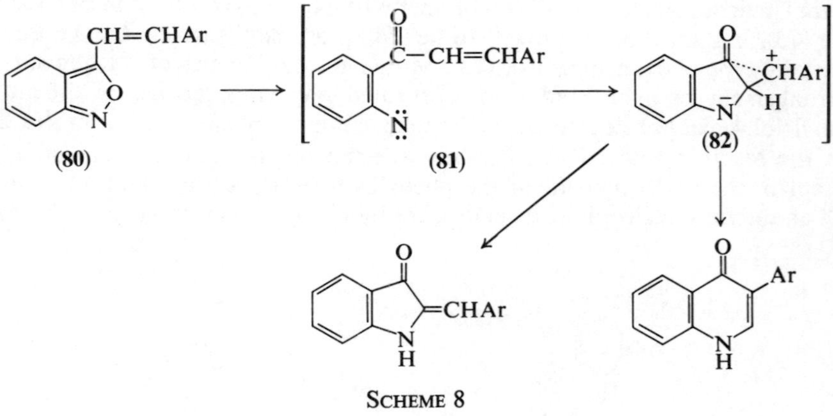

(80) → [(81) → (82)]

Scheme 8

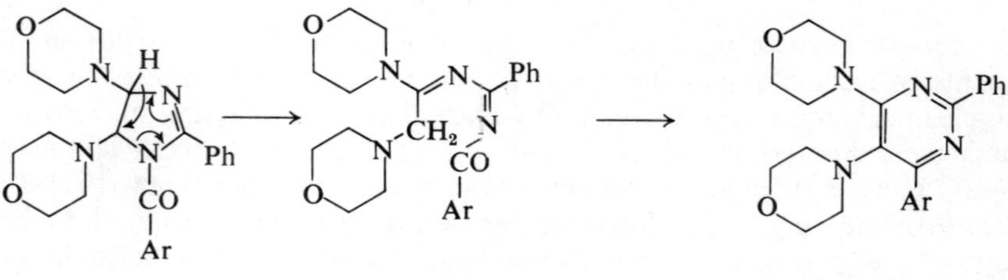

(83) → (84) → (85)

A novel type of hexa-arylbi-imidazole has been obtained by the thermal rearrangement of 2,4,4-triaryl-5-methylthio-4*H*-imidazoles;[96] the conversion of 3,3-disubstituted pyrazolidines into pyrazoles,[97] and the peroxide-initiated conversion of 2-pyrazolines into 1-pyrazolines[98] have been reported.

2-Acetyl-2-ethyl-2-phenylbenzothiazoline (**86**) reacts with sulphuryl chloride to give the 1,4-benzothiazine dye (**88**). A tentative suggestion[99] for the mechanism of this rearrangement invokes the initial ring-expansion of the chlorosulphonium salt of (**86**) with S—C(2) cleavage leading to the benzothiazine (**87**) and thence (**88**). Treatment of 1,2-thiazine *S*-oxides (**89**) with concentrated H_2SO_4 gives oxazines (**90**).[100]

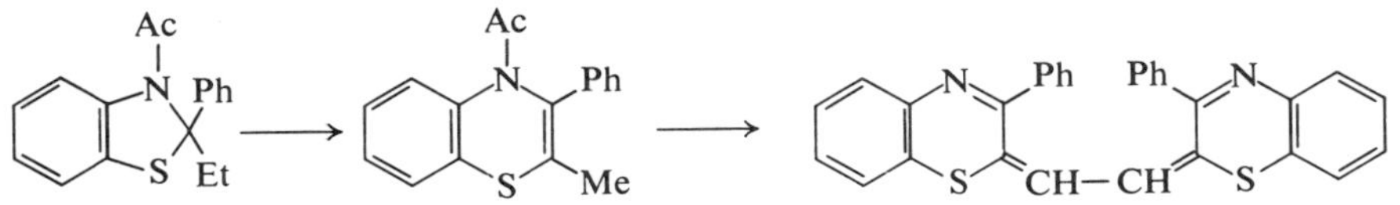

(**86**) (**87**) (**88**)

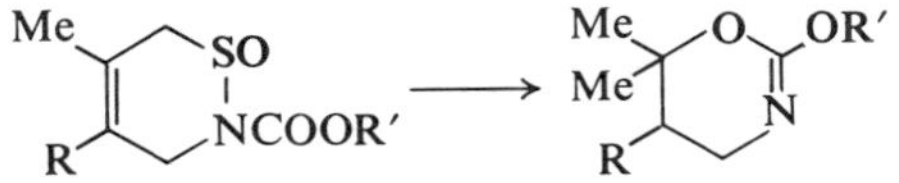

(**89**) (**90**)

An interesting approach to benzoxazaboroles, benzothiazaboroles, and benzoselenazaboroles has been achieved[101] by the reaction of diborane with benzoxazole and its sulphur and selenium analogues. It has been suggested that the pathway proceeds *via* BH_3-adduct formation, intramolecular hydroboration, and rearrangement *via* a hydride transfer from B to C (see Scheme 9). Boron derivatives of the type (**92**) have been obtained[102] by the thermal rearrangement of oxaborapyrimidine derivatives of type (**91**).

The reported rearrangement of 1,4-dihydro-5*H*-1,3,4-benzotriazepin-5-ones (**93**) to 1-methyl-2-(methylamino)-4(1*H*)-quinazolinones (**94**) exemplifies a unique

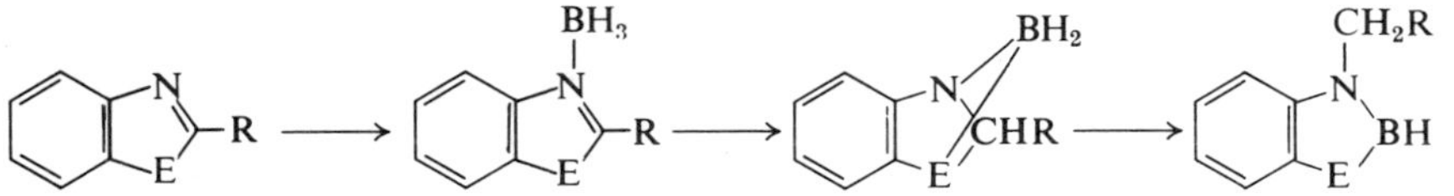

E = O, S or Se.

SCHEME 9

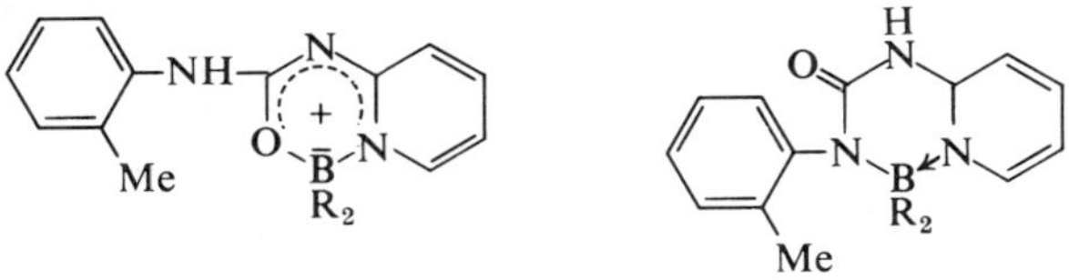

(**91**) (**92**)

ring-contraction which proceeds *via* an intermediate Dimroth-like species (Scheme 10).[103] Again, the occurrence of the observed products from the reaction of the 5-hydrazinotriazepine with an excess of ethyl orthoformate has been explained[104] on the basis of a thermal Dimroth-type rearrangement. The thermal isomerization of a 1,2-diazepine to a 1,3-diazepine has been reported[105] and the acid-catalysed rearrangement of the novel bicyclo[2.2.2]octanyl[1.4]benzodiazepinone ring

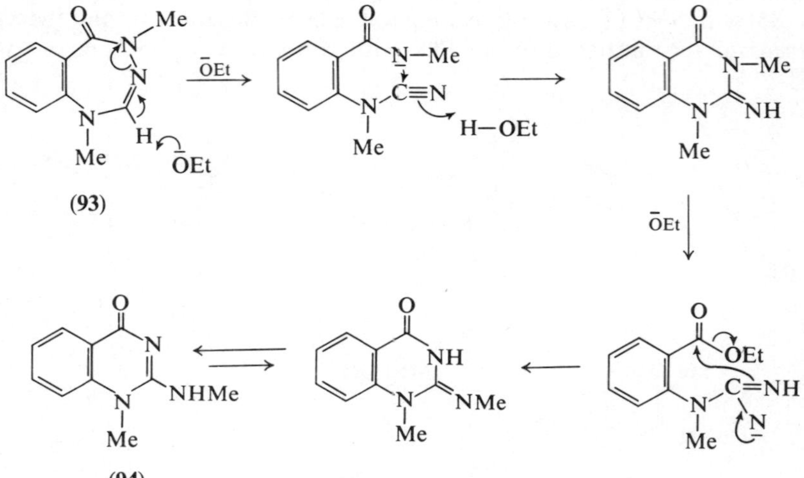

(93)

(94)

SCHEME 10

system to the corresponding bicyclo[2.2.2]oct-2-enylbenzimidazole system (**96**) has been described.[106] The mechanism of the latter rearrangement probably involves protonation followed by proton loss from the bicyclic structure (**95**), thus effecting ring-opening which is followed by acid-catalysed condensation. The thermal ring-contractions of 2,3,4,5-tetrahydro-1,3-oxazepin-5-ones and 4,5-dihydro-2*H*-1,3-dioxepin-5-ones have been reported.[107]

Finally, reference must be made to a number of other heterocyclic rearrangements that have appeared; these include the base-catalysed rearrangement of

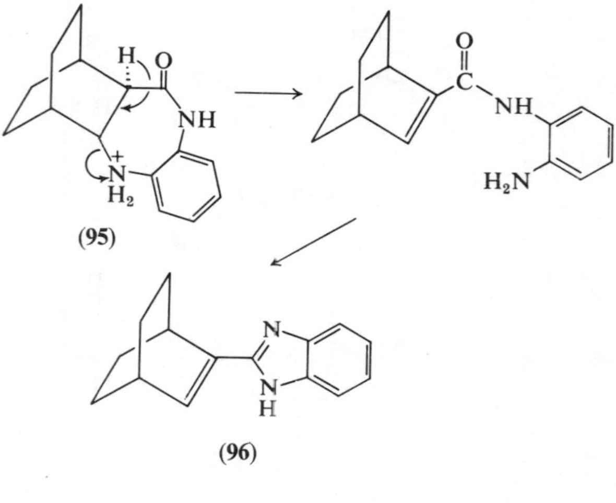

(95)

(96)

aminoindanediones into isoquinolines,[108] rearrangements of chloroethynylpyrazoles[109] and of methylthiopyrazoles,[110] thermal rearrangements of 1,3,4-oxadiazoles[111] and 1,2,4-oxadiazol-5(4*H*)-one,[112] novel rearrangements of certain isoxazole derivatives,[113] photochemical conversions of 3-phenyl-2-isoxazolines fused with cycloalkane rings,[114] the photo-Fries-type rearrangement of *N*-acetylbenzimidazolones,[115] the *v*-triazole–thiadiazole rearrangement of 4,6-dimethyl-*v*-triazolo[4,5-*d*]pyrimidine-5,7[4*H*,6*H*]-dione,[116] thermal rearrangement of 1-alkoxy-5-phenyltetrazoles to 5-phenyl-3-alkyltetrazole 1-oxides,[117] and thermolysis of 5-aryl-2-(2,4-dinitrophenyl)tetrazoles.[118]

Cyclohexadiene Derivatives

The mechanisms of photochemical transformations of cross-conjugated cyclohexadienones have been reviewed.[119] The abnormal tricyclodecenones (**99**), which resulted from irradiation of 2-methoxy 6/6-fused cross-conjugated dienones of type (**97**), have been considered[120] to arise either by a symmetry-allowed 1,4-sigmatropic rearrangement in a cyclopropyl intermediate of type (**98**) produced by electronic excitation and β,β-bonding in the dienone system, or by a photoinduced $[{}_\sigma 2_a + {}_\pi 2_a]$ cycloaddition process involving the 5,10-σ and the 1,2-π bond of the dienone. A photo-induced rearrangement of 2-epoxycyclohexane-1,4-diones to γ-alkylidene-γ-butyrolactones has been reported,[121] and a novel method based on the above rearrangement has been devised for the construction of the γ-alkylidene-$\Delta^{\alpha,\beta}$-butenolide skeleton.[122]

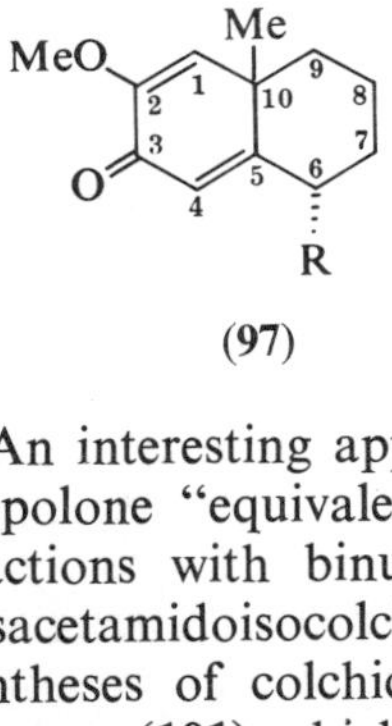

(**97**) 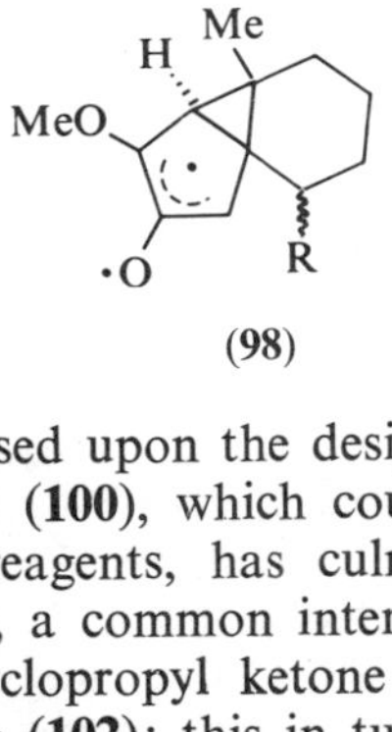 (**98**) 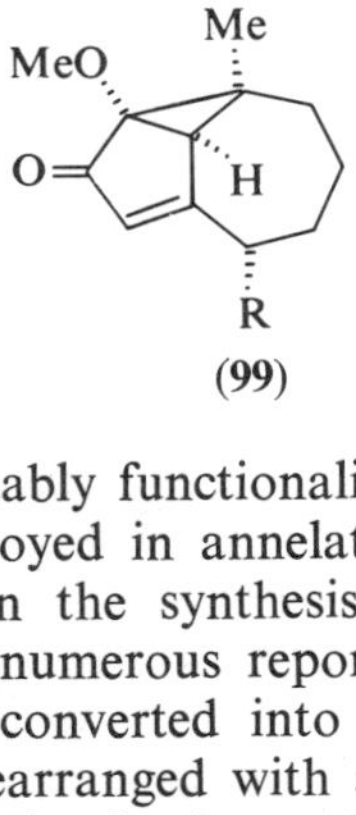 (**99**)

An interesting approach focused upon the design of a suitably functionalized tropolone "equivalent", namely (**100**), which could be employed in annelation reactions with binucleophilic reagents, has culminated[123] in the synthesis of desacetamidoisocolchicine (**104**), a common intermediate in numerous reported syntheses of colchicine: the cyclopropyl ketone (**100**) was converted into the dienone (**101**) which cyclized to (**102**); this in turn slowly rearranged with aryl migration, to afford the dihydrotropolone ether (**103**). A synthesis of (±)-grandisol (**106**) from 4-methoxy-3,6,6-trimethylcyclohexa-2,4-dienone (**105**) has also been described;[124] the route (see Scheme 11) includes regiospecific cyclopropanation of the dienone, acid-catalysed hydrolytic rearrangement of the product, and five-membered ring-opening of the resultant bicyclo[3.2.0]heptanone system. A reported rearrangement of hydrobenzofuranoid neolignans has provided[125] a semisynthetic entry into the bicyclo[3.2.1]octanoid series; depsidones have been obtained by the thermal rearrangement of grisadienediones,[126] while photochemical transformation of the latter apparently lead to dibenzofurans;[127] both reactions are considered to proceed by radical ring-opening.

Studies that confirm aromatization of 4-methyl-4-nitrocyclohexadienones by a formal 1,3-shift of the nitro-group, and indicate a radical dissociation–recombination mechanism for the process, have been reported.[128] Further studies of the

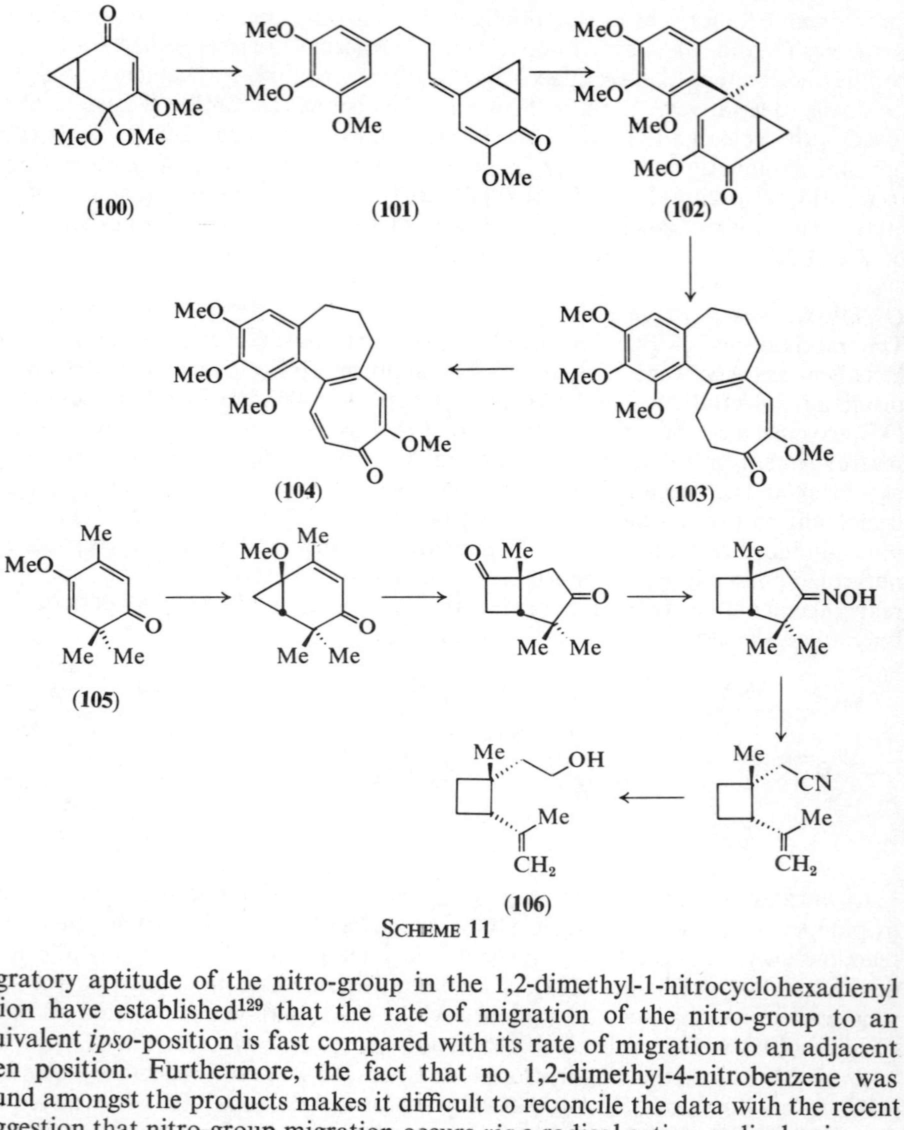

SCHEME 11

migratory aptitude of the nitro-group in the 1,2-dimethyl-1-nitrocyclohexadienyl cation have established[129] that the rate of migration of the nitro-group to an equivalent *ipso*-position is fast compared with its rate of migration to an adjacent open position. Furthermore, the fact that no 1,2-dimethyl-4-nitrobenzene was found amongst the products makes it difficult to reconcile the data with the recent suggestion that nitro-group migration occurs *via* a radical cation–radical pair.

A detailed kinetic examination of the acid-catalysed dienol–benzene rearrangements of 4-methyl-4-(trichloromethyl)cyclohexadienol and 4,4-dimethylcyclohexadienol has appeared,[130] and an attempt has been made[131] to clarify the mechanism of the *ortho*-selective hydroperoxylation of 2,4,6-tri-*tert*-butylphenol (**107**) on base-catalysed oxygenation. The exclusive migration of (**108**) to (**110**) has been rationalized by assuming chelate formation in a non-radical, π-complex transition state (**109**), as depicted in Scheme 12. The unexpected cyclization of 6-acetonyl-6-hydroxy-2,4-di-*tert*-butylcyclohexa2,4-dienone oxime (**111**) with HCl has been interpreted[132] as involving initial rearrangement to an isomeric oxime (**112**) followed by cyclization to the oxazine (**113**).

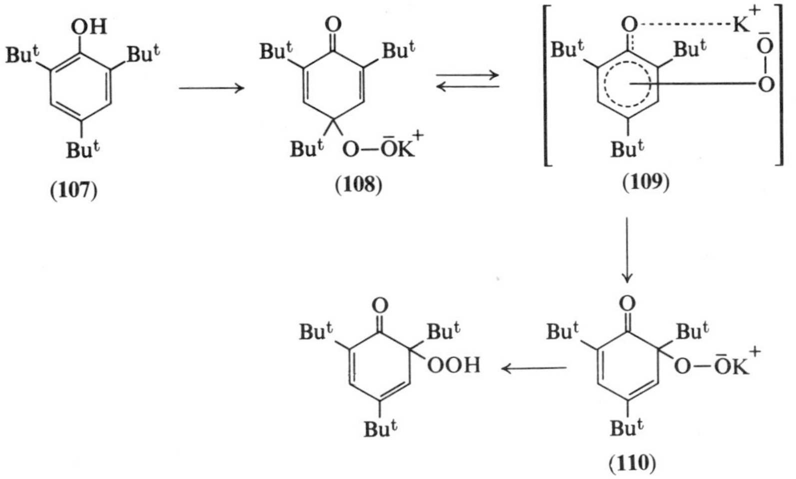

(107) (108) (109) (110)

SCHEME 12

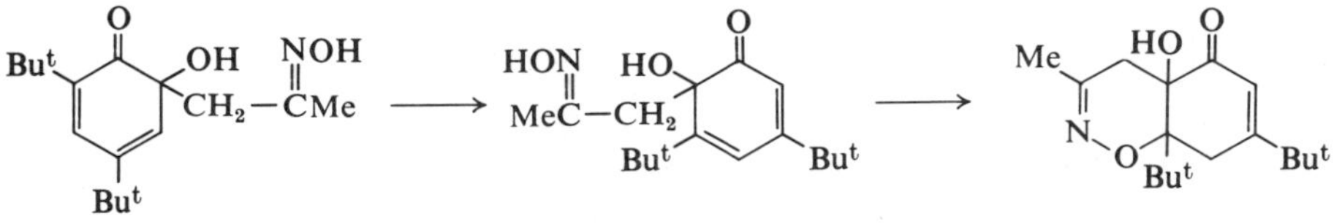

(111) (112) (113)

The preparation of 2-substituted aminoarsabenzenes from the oximes of 1-substituted 1-arsacyclohexa-2,5-dienones has been achieved[133] by a rearrangement analogous to the arsacyclohexadienone–arsaphenol, and Beckmann rearrangements; a study has been made of arsaphenol–arsacyclohexadienone tautomerism.[134] In a similar manner, entry to *C*-monosubstituted λ^5-phosphorins (**115**) has been provided by rearrangement of the phosphacyclohexadienes (**114**).[135] The arsenic compounds corresponding to (**114**) form λ^3-arsenins under analogous conditions.

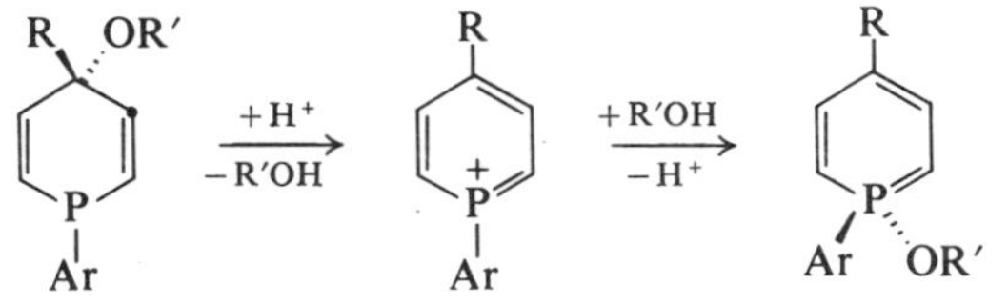

(114) (115)

Products obtained from the photolysis of perfluorinated derivatives of aza- and diaza-cyclohexadienes show that, in this reaction, a variety of skeletal processes occurs, including ring-opening, fluorine migration, and retro-Diels–Alder reactions.[136]

Sigmatropic Rearrangements

An operationally useful model has been proposed[137] for predicting the qualitative effect of substituents on the rate of thermal pericyclic reactions; this has provided a unified interpretation of several previously unrelated experimental observations and enables substituent effects to be predicted in pericyclic reactions for which experimental data are not yet available. A review of the skeletal rearrangements of organoalkali-metal compounds has appeared,[138] where this pointed out that, while organometallic compounds of the alkali metals undergo [1,*n*]-sigmatropic migrations of aryl, vinyl, carbonyl, alkoxycarbonyl, silyl, germyl, and related groups, sigmatropic migrations of alkyl, benzyl, and hydrogen occur only under special circumstances and then often by an alternative fragmentation–recombination pathway. Allyl groups, on the other hand, may migrate by the latter process or by a concerted [2,3]-sigmatropic reaction. The review includes a discussion of these molecular rearrangements from the viewpoint of simple MO theory and it is shown how, by suitable choice of the alkali metal, temperature, ligand of the alkali cation, and/or the solvent, considerable control may be exercised over the rate of rearrangement. The Hückel–Hubbard theory has been applied to thermal and photochemical [1,3]-, [1,4]-, [1,5]-, and [1,6]-sigmatropic shifts;[139] inclusion of electron correlation is explicit in this treatment which is shown to give considerably more detailed information than the usual molecular-orbital theories.

Phase-transfer catalysis has proved[140] to be an easy and efficient method for deuterium exchange and isomerization of acidic hydrocarbons such as indene and fluorene, while a polarimetric differential method has been used[141] to determine kinetic isotope effects in various sigmatropic rearrangements.

[3,3]-*Migrations*

Claisen and related rearrangements. A study of the effect of substituents on the Claisen rearrangements of allyl [*o*(or *p*)-substituted phenyl] ethers has been carried out;[142] the same authors have shown[143] that the magnitude of the catalytic effect of $ZnCl_2$ on the rearrangement is governed by the strength of co-ordination of the catalyst at the ether oxygen, and this in turn is controlled by the inductive effect, steric hindrance, and chelating effect of the *ortho*-substituent.

It has been shown that the [3,3]-sigmatropic rearrangement of benzyl vinyl ethers is markedly facilitated by an ethoxycarbonyl group at the benzylic position;[144] thus, the reaction of ethyl mandelate derivatives (**116**) in the Claisen orthoester rearrangement with (**117**) provides an extremely convenient method for the regiospecific synthesis of substituted arenes (**118**).

COOEt OH X + RR′CHC(OR″)$_3$ ⟶ COOEt COOR″ R R′

(**116**) (**117**) (**118**)

It has been established[145] that the presence of a β-substituent on the allyl unit of *p*-tolyl allyl ethers does not effect the pathway by which the catalytic Claisen rearrangement proceeds (intramolecular with inversion of the allyl group, intramolecular without inversion of the allyl group). Moreover, a careful study of the Lewis acid-catalysed Claisen rearrangement of *p*-tolyl [2,3,3-D_3]-allyl ether to the *p*-cresol diallyl derivative (**119**) has indicated[146] that the inverted allyl group in (**119**) arises by an intramolecular process, and the uninverted one by an intermolecular process. Allylquinols lacking substituents at C(1) of the allyl group have been shown[147] to undergo [3,3]- or [1,3]-rearrangements readily under neutral conditions while, under acid conditions, [1,2]-rearrangements become important. Moreover, when C(1) is disubstituted, [3,3]-rearrangements become completely suppressed and [1,2]-rearrangements, accompanied by fragmentation, result under both neutral and acid conditions. The unexpected elimination of a 2-allyloxy-group from the Claisen rearrangement of 1,2-di(allyloxy)anthraquinone has been reported.[148]

Conclusive evidence has been obtained[149] for the occurrence of [3,3]-sigmatropy in the solvolysis of bridged tricyclic tosylates of the type (**120**); the intramolecular deuterium kinetic isotope effect observed in the Claisen-type rearrangement of 2-deuteriophenyl propargyl ether has been interpreted[150] in terms of a non-synchronous mechanism, and the Claisen rearrangements of the 7-cinnamyloxy-derivatives of 3-methoxyflavone and 4-methylcoumarin have been reported.[151]

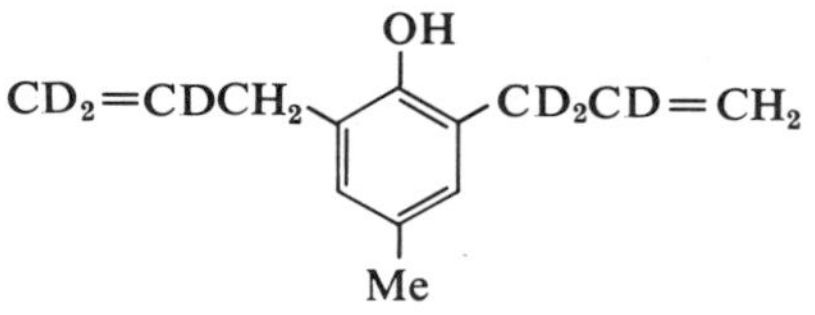

OH
$CD_2{=}CDCH_2$ $CD_2CD{=}CH_2$
Me

(**119**)

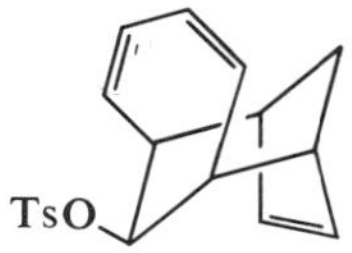

TsO

(**120**)

It has been established[152] that *endo*-arylbicyclo[5.1.0]octa-2,4-dienes (**121**) isomerize thermally to the benzo-condensed tricycles (**124**). The reaction has been interpreted as a carbo-Claisen rearrangement of (**121**) to (**123**); the latter compound is subsequently stabilized in a thermally allowed [1,5]-homodienyl hydrogen shift with regeneration of the aromatic system to give (**124**). However, experimental data do not permit a decision as to whether the formation of (**123**) is concerted or proceeds *via* the diradical (**122**), although the authors consider the two-step mechanism to be the more likely.

The report of the selective rearrangement of diallyl ethers (**125**) catalysed by tris(triphenylphosphine)ruthenium(II) dichloride has opened up a valuable new route

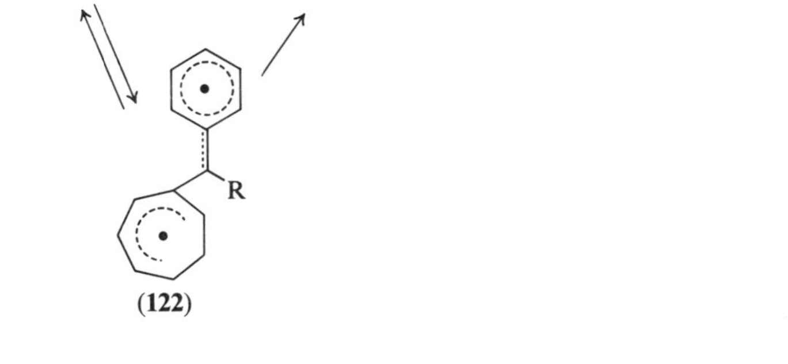

R

(**121**)

R

(**123**)

R

(**124**)

R

(**122**)

for the synthesis of γ,δ-unsaturated aldehydes (**126**) (and ketones).[153] The Claisen–Eschenmoser [3,3]-sigmatropic rearrangement of appropriately functionalized 3-arylcyclohex-2-enols has provided a ready synthesis of mesembrane alkaloids, and the total synthesis of *rac-O*-methyljoubertiamine has been described[154] as part of the development of a general synthetic route to these alkaloids involving the aforementioned rearrangement as the key synthetic step. The synthetic utility of the Claisen rearrangement has also been used[155] in the synthesis of (*S*)-(*Z*)-2-(5-hydroxy-2-methylenecyclohexylidene)ethanol, of interest as a ring A intermediate for a synthesis of vitamin D, while the [3,3]-sigmatropic rearrangement of glycals has been used to prepare branched-chain sugars as is illustrated by the conversion of (**127**) into (**128**).[156] A new synthesis of (2*R*,4′*R*,8′*R*)-α-tocophenyl acetate has been achieved[157] by application of a stereoselective Claisen rearrangement, further demonstrating the wide potential applicability of the [3,3]-sigmatropic process in the synthesis of optically active substances.

O → CHO

(**125**) (**126**)

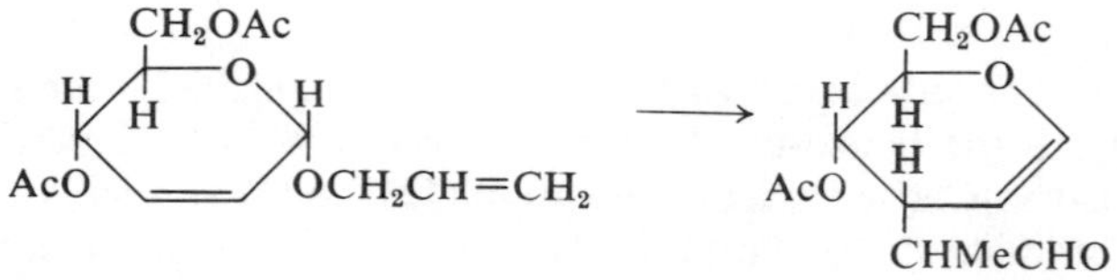

CH_2OAc O H H H AcO $OCH_2CH{=}CH_2$ → CH_2OAc O H H H AcO CHMeCHO

(**127**) (**128**)

A mechanism involving the mercuric ion-catalysed Claisen rearrangement of bis-(3-aryloxyprop-1-ynyl)mercury has been postulated[158] to account for the formation of the novel 4,4′-bis-(2*H*-chromenyl)mercury derivatives on treatment of a series of aryl prop-2-ynyl ethers with mercuric trifluoroacetate followed by quenching with alkaline $NaBH_4$. In a similar manner, 1,4-bis-(aryloxy)but-2-ynes (**129**) have been selectively converted into 4-(aryloxymethyl)-2*H*-chromenes (**130**), 6*H*-benzofuro[3,2-*c*]-6a,11a-dihydro-11a-methylbenzopyrans (**131**), or 5a,10b-dihydro-5a,10b-dimethylbenzofuro[2,3-*b*]benzofurans (**132**) by treating a dichloromethane solution of (**129**) with mercuric trifluoroacetate, silver tetrafluoroborate, or anhydrous aluminium chloride, respectively.[159] A mechanism involving a charge-induced Claisen rearrangement, triggered by π-complex formation between the heavy-metal ion and the C—C multiple bond, is postulated for formation of (**130**) and (**131**). Sequential charge-induced Claisen rearrangement of (**129**) into (**131**) by co-ordination of $AlCl_3$ with the oxygen atoms of (**129**) and (**130**), followed by ionic rearrangement of (**131**) into (**132**), is also postulated. In this paper the author also discusses the differing efficacy of metal ions in promoting these isomerizations. Mercuric trifluoroacetate has also been used as an effective catalyst for equilibrating allylic carbamates, namely (**133**) and (**135**).[160] The experimental observations for this transposition of allylic oxygen functionality are most consistent with a mechanism in which the mercuric catalyst interacts with the carbon–carbon π-bond to promote an intramolecular cyclization to yield an intermediate 1,3-dioxanium ion (**134**). The reported titanium tetrachloride catalysis of the

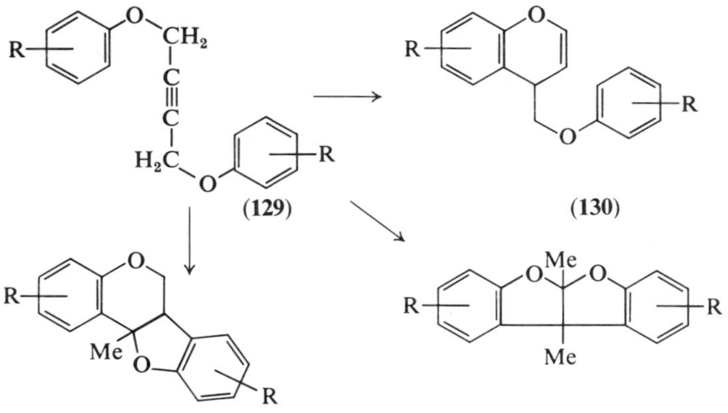

(129) (130) (131) (132)

aza-Claisen rearrangement represents a further example of such a "charge-induced" pericyclic reaction.[161] This new-found procedure makes it possible (see Scheme 13) to prepare and rearrange *N*-allylenamines of aldehydes (**136**) to alkylated aldehydes (**137**) in a single step. Moreover, results show that titanium tetrachloride catalysis does not change the characteristic chair-like transition state, with substituents predominantly equatorial, typical of Claisen and other [3,3]-sigmatropic rearrangements.

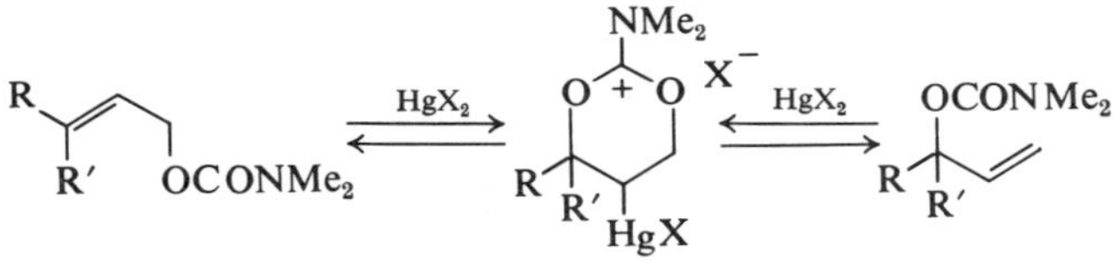

(133) (134) (135)

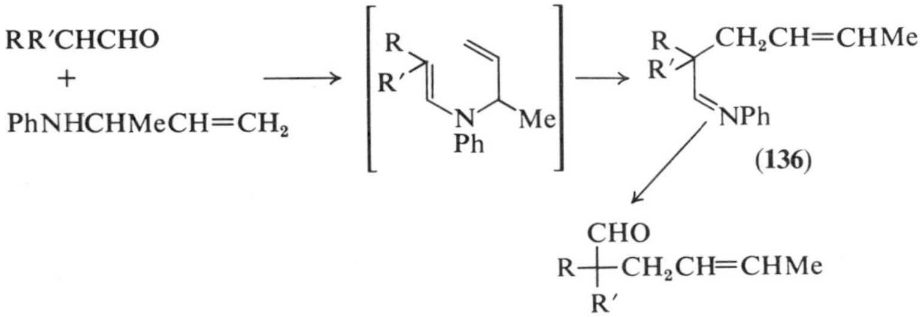

(136) (137)

SCHEME 13

N-Mesyl- and *N*-tosyl-*o*-allylanilines have been synthesized[162] by the amido-Claisen rearrangement, and kinetic data have been correlated with bond-localization energy for the rearrangement of the *N*-allyl-*N*-mesyl derivatives of aniline, 1- and 2-naphthylamines, and 2-, 3-, and 9-aminophenanthrenes.[163] The pyrolysis of *N*-alkylanilinium salts such as (**138**) has led to the development of a new type of amino-Claisen rearrangement, a reaction which allows the introduction of functional groups into the *ortho*-position of aniline derivatives and enables the preparation of potential precursors for indolines and indoles.[164] A versatile procedure for the elaboration of allylic thiolcarbamates (**139**) to α,β-unsaturated carboxylic esters (**140**) has been reported.[165] The thiolcarbamates are readily derived from allylic alcohols *via* a [3,3]-sigmatropic rearrangement.

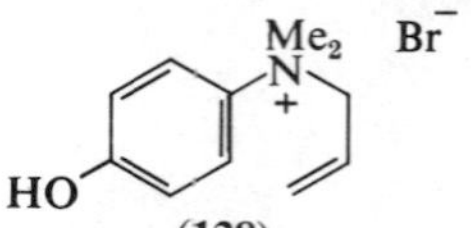

(**138**)

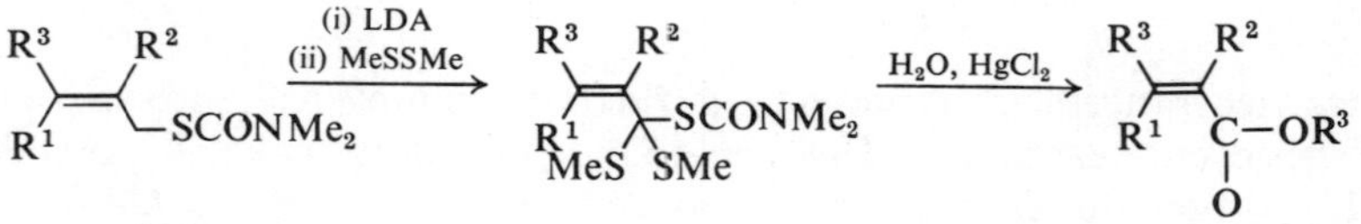

(**139**) → (i) LDA, (ii) MeSSMe → H_2O, $HgCl_2$ → (**140**)

A new synthesis of conjugated dienones has also been described:[166] it involves addition of a primary or secondary allyl alcohol to an allenyl sulphoxide to form an enol ether that undergoes Claisen rearrangement and subsequent elimination of benzenesulphenic acid, as portrayed in Scheme 14.

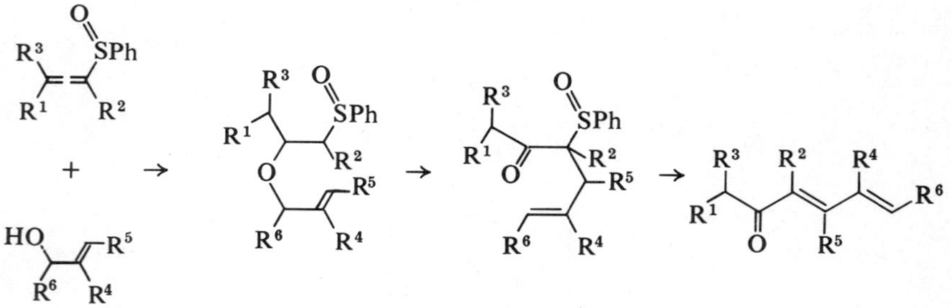

SCHEME 14

The Claisen rearrangement of allylthienyl sulphides to the corresponding allylthiophenethiols,[167] and the thio-Claisen rearrangements of a variety of 2-allylthiopyrroles[168] have been reported. A novel method has been found[169] for the introduction of the isoprene skeleton into chloromethyl-arenes and -heteroarenes in a three-step sequence involving a solvent-assisted Claisen–Cope rearrangement. Two consecutive Claisen–Cope rearrangements have been used[170] as key steps in an easy synthesis of 3,7-unsaturated esters, and the first example has been reported[171] of the double thio-Claisen and Cope rearrangements as a means of furnishing α,β-ε,ξ-unsaturated thioamides possessing a fundamental terpene skeleton.

Cope and related rearrangements. An attempt to clear the issue as to whether the mechanism of the [3,3]-sigmatropic shift of hexa-1,5-dienes is a concerted process *via* a single chair-like transition state with partial bonding between C(1) and C(6) and between C(3) and C(4), or whether cyclohexane-1,4-diyl (a biradicaloid) is an intermediate, has demonstrated[172] that the rearrangement proceeds by the former process. The same authors have further attempted to resolve the magnitude of these partial bonds between the allylic moieties; although they have failed to answer the question in an absolute sense, they have shown by the measurement of secondary deuterium kinetic isotope effects that dramatic changes in the magnitudes of these partial bonds result as a function of substituents.[173] The rates of Cope rearrangement of various 2-arylhexa-1,5-dienes have been measured[174] and the relationship between the results and the mechanism of the rearrangement have been discussed. The rate data do not give a linear Hammett plot and the evidence concerning the mechanism of rearrangement of 2-arylhexa-1,5-dienes seems to be inconclusive. On balance, the available rate data can be quite reasonably interpreted by the concerted, single-transition-state route although the evidence presented does reduce the latitude of structural possibilities for the transition state to those having more resemblance to a diradical than to two separated allyl radicals. The >90-fold rate enhancement found in the Cope rearrangement of methylenecyclobutanes (**141**) to (**142**) when a vinyl group ($R = CH_2{=}CH$) is substituted for a methyl group ($R = Me$) is entirely consistent with, and has been interpreted in terms of,[175] a biradicaloid transition state that leads to a 1,4-diradical intermediate.

R R

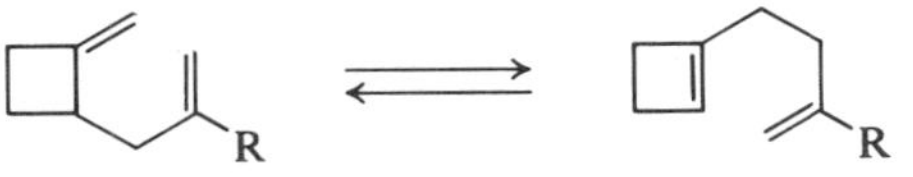

(**141**) (**142**)

Kinetic and product data for the Cope rearrangement of allylcyclopropenes (**143**) are consistent with the interpretation that the process is not a true pericyclic reaction but involves derivatives of 1,4-cyclohexylene as intermediates.[176] Thus, the formation of (**146**) and its analogues occurs *via* the biradical intermediate (**144**) which can either fragment to give the Cope product or undergo ring inversion to (**145**) followed by radical coupling to yield the tricyclo[2.2.0.0^{2,6}]hexane ring system.

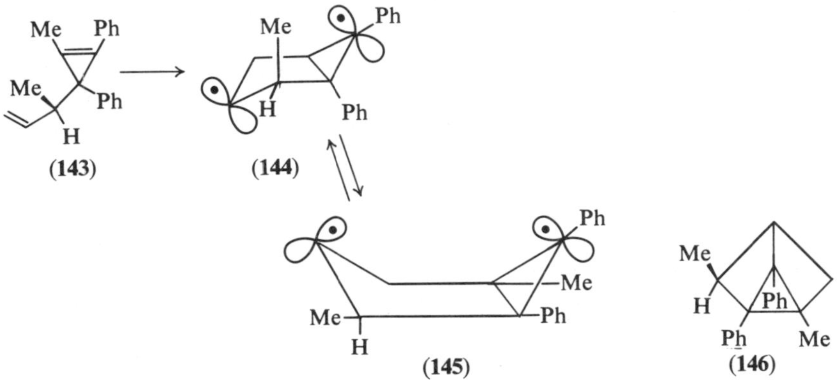

Me Ph Me Ph Me Ph H H Ph

(**143**) (**144**)

Ph Me Me Ph H Me H Ph Ph Me

(**145**) (**146**)

Although previously there existed no unequivocal evidence for participation of an aromatic double bond in the Cope rearrangement, it has now been shown[177] that a 4-phenylbut-1-ene system, albeit activated by an appropriately substituted cyclopropane ring as in (**147**), rearranges to (**148**) in a process which is considered to be a Cope rearrangement involving an aromatic ring.

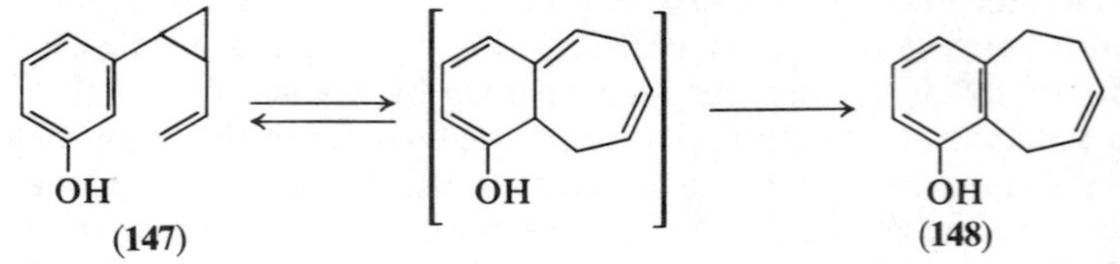

(**147**) (**148**)

The Cope rearrangement of 2-methyl-3-(2,2-dimethyl-*cis*-3-vinylcyclopropyl)-cyclohex-2-en-1-one has been observed to be a remarkably sluggish reaction, particularly when compared with the rapid rearrangement of structurally very similar compounds. This striking difference in reactivity has been rationalized[178] by proposing that a severe steric interaction between the vinyl methyl group and the *cis*-methyl group on the cyclopropane ring develops as the reaction proceeds *via* the boat-like transition state characteristic of the Cope rearrangement of *cis*-divinylcyclopropanes. Interestingly, an examination of the thermal isomerization of cyclopropyl imidates, *e.g.* (**149**)→(**150**), has indicated[179] that imidates do not

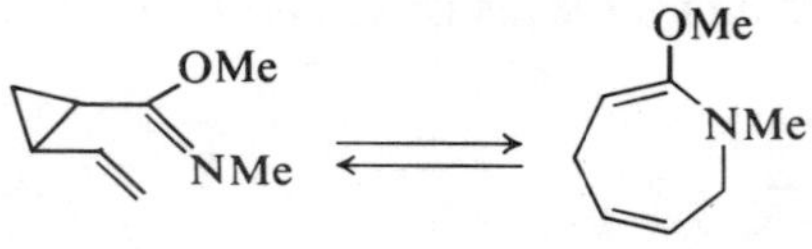

(**149**) (**150**)

enter into electrocyclic rearrangements with the same facility as their isoconjugate olefinic counterparts. This information has been used further to explain the mechanism of thermal isomerization of homoazocines.[180] The influence of complexation to transition metals on the Cope rearrangement of *cis*-divinylcyclopropanes has also been studied, using 4-azabicyclo[5.1.0]octa-2,5-diene-4-carboxylate as a substrate;[181] a study has been made of the steric and electronic effects in the Cope rearrangement of homotropilidenes.[182]

Preliminary results from an investigation of the thermal rearrangement of bis(methylenecycloalkenes) have permitted[183] independent observation of high and low conformations of the Cope rearrangement. A detailed product and kinetic study of the pyrolysis of 6-methylenebicyclo[3.2.1]oct-2-ene has established that the major portion of the reaction is a Cope rearrangement, but that two products of formal [1,3]-sigmatropic rearrangement are also formed at rates, the sum of which corresponded to *ca.* 20% of the process.[184]

A labelling study of the interconversion of 1-methylene-3-vinylcyclopentane and 5-methylenecycloheptene established the reaction as a concerted Cope process.[185] The unexpected formation of hydrocarbon (**153**) on thermal decarbonylation of ketone (**151**) is considered to occur by a ready [3,3]-sigmatropic Cope rearrangement of the intermediate diene (**152**),[186] while a study of the cycloaddition of tropone to 8-phenylisobenzofulvene implies[187] that the production of (**155**) is consistent with initial formation of (**154**) and subsequent Cope rearrangement.

The remarkable difference in thermal behaviour observed in the pyrolysis of the pentacyclic bishomotropilidenes (**156**) and (**158**) has been ascribed[188] to the difference in the ring strain imposed on the two systems. An initial retro-Diels–Alder reaction followed by a [1,3]-sigmatropic shift is considered to produce (**157**)

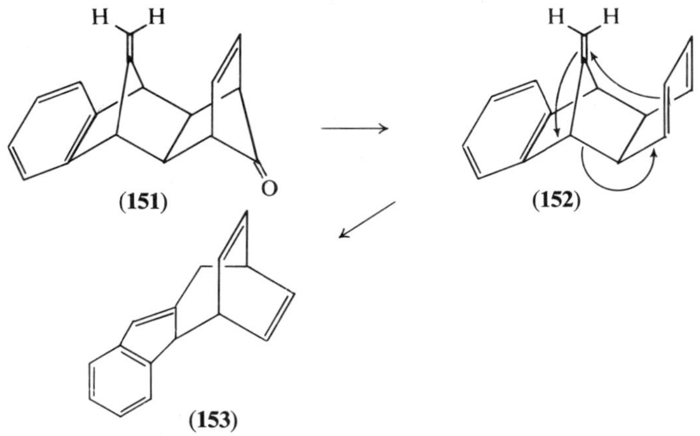

H H O (151) (152) (153)

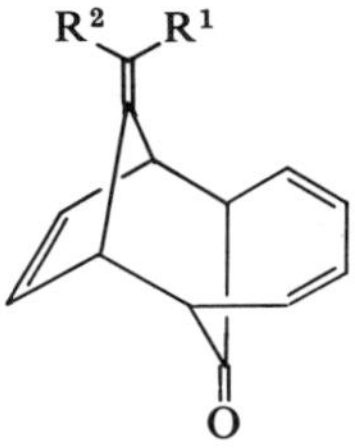

R2 R1 O (154) O R2 R1 (155)

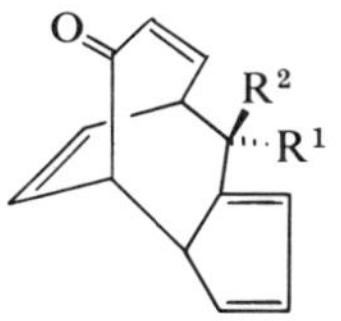

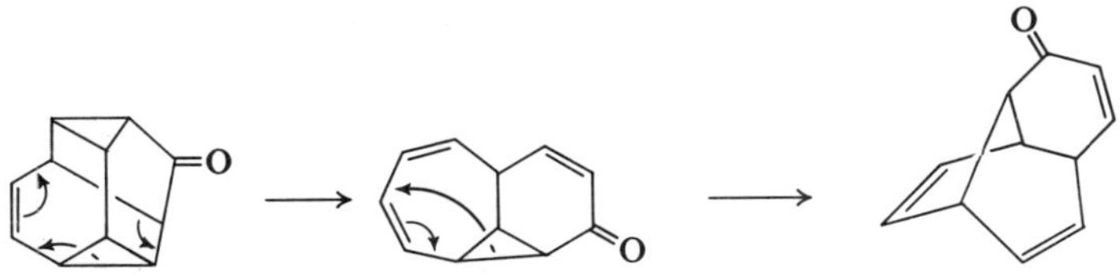

O (156) O O (157)

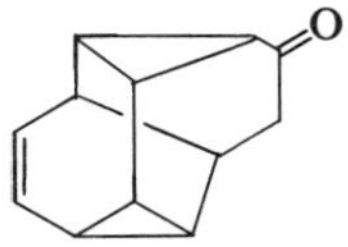

O (158) HO (159) HO (160)

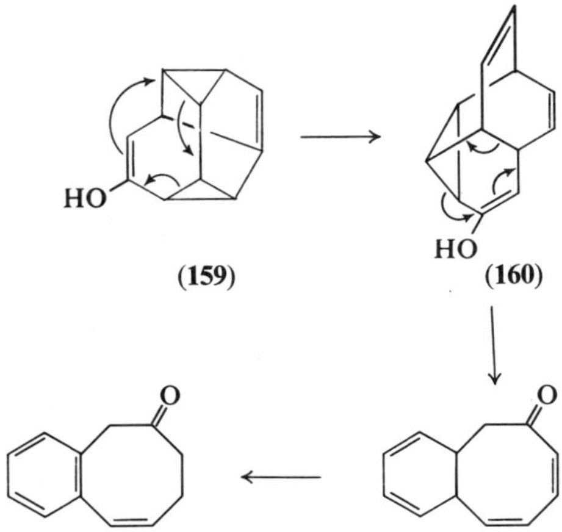

O (162) O (161)

from (**156**), while (**162**) is thought to arise from (**158**) by a homo-Cope rearrangement in the enol form (**159**), subsequent retro-Diels–Alder reaction of (**160**) to (**161**), and successive [1,5]-hydrogen shifts. A series of sequential Cope rearrangements, or their equivalent, as depicted in Scheme 15, is thought to provide the mechanistic connection between cyclododecatriyne (**163**) and hexaradialene (**164**).[178]

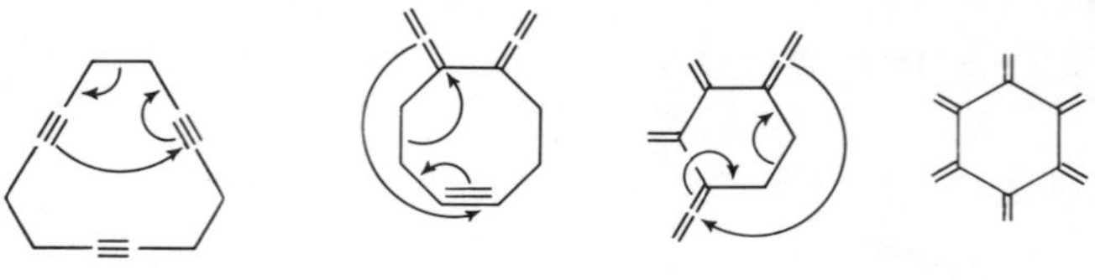

(**163**) (**164**)

SCHEME 15

Novel thermal transformations of (+)-pin-2-en-9-al (**165**) and its analogues to allyl vinyl ethers (**166**) have been described;[190] the ether (**166**; R = Me) underwent further rearrangement in an acid-catalysed step to the (−)-2,6-dimethylbicyclo-[3.3.1]non-6-en-2-ones (**167**). The Cope rearrangement of (**168**) has been utilized in an interesting approach to 3-acetylcyclohexanone,[191] while a new application of the base-catalysed oxy-Cope rearrangement of hexa-1,5-dien-3-ols (**169**; X = OR′) has opened up a general approach to the synthesis of 1,6-dicarbonyl substrates (**170**).[192]

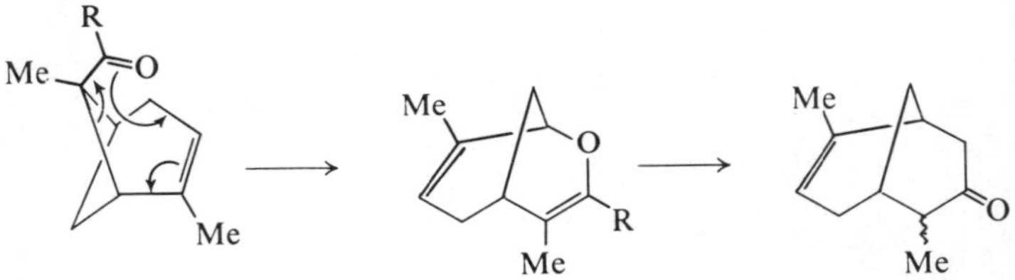

(**165**) (**166**) (**167**)

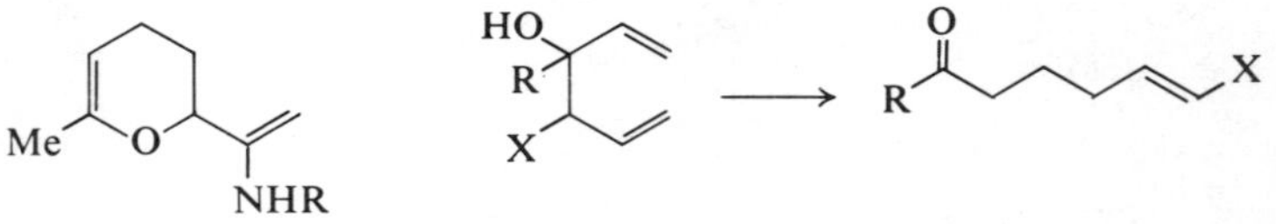

(**168**) (**169**) (**170**)

A paper on the oxy-Cope rearrangement of 1,5,6-trien-3-ols has appeared;[193] the first reported instance of the anionic oxy-Cope rearrangement in bicyclo[2.2.1]-heptene systems has been used to afford a rapid synthesis of 18,19-bisnorsteroids;[194] the acetylenic oxy-Cope rearrangement of cyclohex-2-enylprop-2-ynols, a new type of synthetic equivalent of ionone and its analogues, has been investigated[195] in neat and solvent-assisted systems. The effect of solvent on the oxy-Cope rearrangement of the 4-isopropenyl-3,7-dimethylocta-1,6-dien-3-ol diastereomeric system has also been studied.[196] Interestingly, experimental evidence[197] has suggested that the epimerically unfavourable anionic oxy-Cope rearrangement of (**171**) to (**172**) requires a thermally allowed [1,5]-sigmatropic epimerization and then conventional Cope rearrangement. Papers have appeared describing the generation of 4,5-

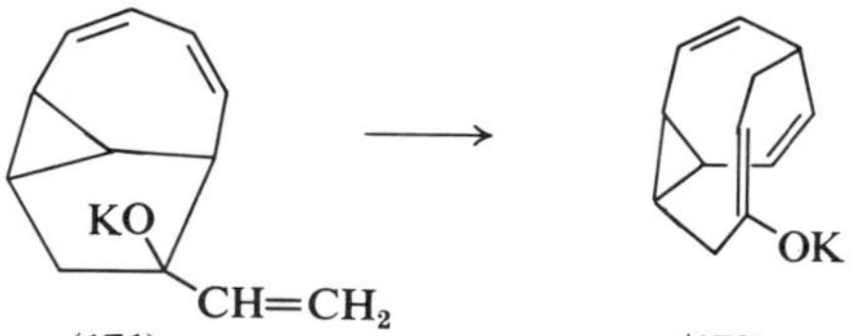

(171) (172)

dimethyl-*sym*-oxepin oxide and its Cope rearrangement to 2,7-dimethyl-*sym*-oxepin oxide,[198] the epimerically controlled thermolysis of synthetically useful azatricyclic compounds,[199] and stereospecific carbon–carbon coupling in tetra-arylmonoaza-Cope systems.[200] New synthetic routes to pyrroles, tetrahydropyridazines, and pyrazoles by way of a novel Cope-type rearrangement of ketazine dianions, have been uncovered.[201] The main factor governing the course of this reaction appears to be the electron density on the carbon termini; thus, ketazine dianions bearing electron-releasing groups on their carbon termini rearrange to pyrroles, while dianions without substituents or with electron-withdrawing substituents rearrange to tetrahydropyridazines and pyrazoles, respectively. The first example of stereo-controlled α-coupling of thioamides to give *d,l*-1,4-dithioamides, taking advantage of the stereoselectivity of the Cope-type rearrangement of divinyl sulphides, has been described by the same authors.[202]

Variable-temperature ^1H-NMR spectroscopic studies of a number of 2,8-pentamethylenesemibullvalenes having a heteroaromatic unit (X = O, NCH_2Ph, S, SO_2, *syn*-SO, and *anti*-SO) at the central linkage have indicated[203] that the ground-state equilibrium is shifted substantially in the direction of the tautomer (**173**) where the cyclopropane ring is in the periphery of the structure. Progression through the order listed above was shown to lead to an enhanced partiality for tautomer (**174**).

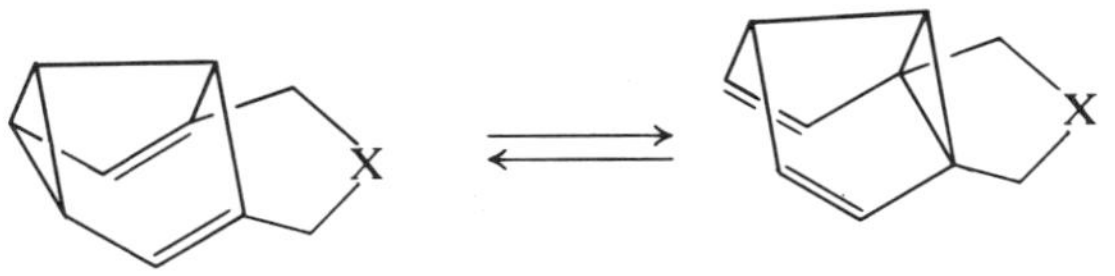

(173) (174)

The authors discuss the possible origin of the factors which contribute to the enhanced concentration of one tricyclic tautomer relative to that of the other.

[2,3]-Migrations
The [2,3]-sigmatropic rearrangements of allylic selenoxides and of selenium ylides have been discussed in a general review on modern organoselenium chemistry,[204] and a new and efficient procedure for isomerizing (*E*)- and (*Z*)-allylic alcohols of the type (**175**) to those of type (**178**) has been reported.[205] It is proposed that the formation of alcohol (**178**) is the result of two consecutive and spontaneous processes, *viz.* a [2,3]-sigmatropic rearrangement of the selenoxide (**176**) and hydrolysis of the selenenic ester (**177**). Since the corresponding sigmatropic rearrangement of allylic sulphoxides is not spontaneous, it is clear that this use of selenium chemistry permits the contrathermodynamic transformation, (**175**) to (**178**), to be carried out under extremely mild conditions. The rearrangement of *N*-arylbenzohydroxamic acids to *p*-hydroxybenzanilides in the presence of catalytic amounts of seleninic acids or selenium(II) species such as phenylselenenyl chloride has been interpreted[206]

in terms of the initial formation of an *O*-seleninylhydroxamic acid, reduction to the corresponding selenenylhydroxamic acid by some as yet unknown reductant, followed by [2,3]-sigmatropic shifts.

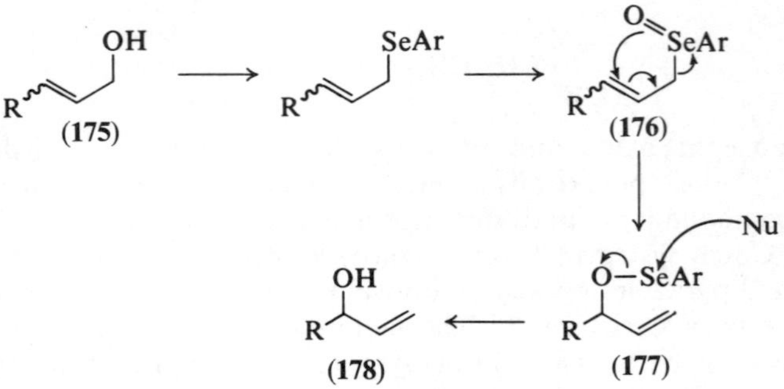

(175) (176) (177) (178)

The unexpected formation of 4-methylthio-2-phenylbut-1-ene derivatives **(181)** on treatment of β-hydroxy-β-phenylsulphonium iodide **(179)** with base has been explained[207] by proposing dehydration of **(179)** followed by concomitant [2,3]-sigmatropic rearrangement of the *S*-ylide intermediates **(180)** as shown in Scheme 16. Even more surprising is the fact that the reaction of methanesulphenyl salts with 3-methylthiopropene to give 1-chloro-2,3-di(methylthio)propane and its regioisomer is not a straightforward electrophilic addition to the double bond, but involves an unsuspected sigmatropic rearrangement by way of attack of the sulphenating agent at sulphur.[208] Such a result emphasizes the remarkable generality of symmetry-allowed [2,3]-sigmatropic rearrangements whereby allylic groups migrate to and from various atom types and charge types (in this case from $\overset{+}{S}$ to $\ddot{S}$).

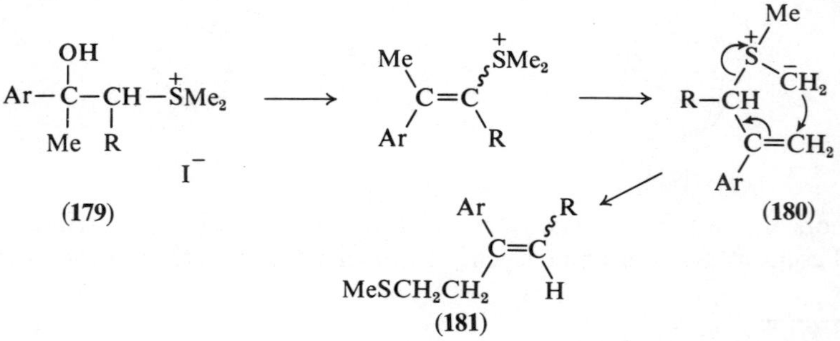

(179) (180) (181)

SCHEME 16

Evidence has been obtained[209] for the participation of a nitrile group in a [2,3]-sigmatropic rearrangement of sulphonium ylides. Thus, the formation of *N*-(methylthiomethyl)ketenimines **(184)** on reaction of (α-cyanoalkyl)dimethylsulphonium salts **(182)** with *n*-butyl-lithium or triethylamine is considered to proceed by way of a [2,3]-sigmatropic rearrangement of intermediate ylides **(183)**.

$$RR'C(CN)\overset{+}{S}Me_2\ X^- \longrightarrow RR'C(CN)\overset{+}{S}Me(\overset{-}{C}H_2) \longrightarrow RR'C{=}C{=}NCH_2SMe$$

(182) $X = BF_4$ or PF_6 **(183)** **(184)**

The first example of a [2,3]-sigmatropic rearrangement having a transition state which prefers pseudo-axial substitution has opened the way to a highly stereo-selective synthesis of *Z*-trisubstituted olefins[210] (see Scheme 17), while β,γ-olefinic ketones have been synthesized[211, 212] by [2,3]-sigmatropic rearrangement of the lithiated carbanions of β,γ-unsaturated α'-cyanoethers, followed by elimination of lithium cyanide; see **(185)**→**(186)**→**(187)**.

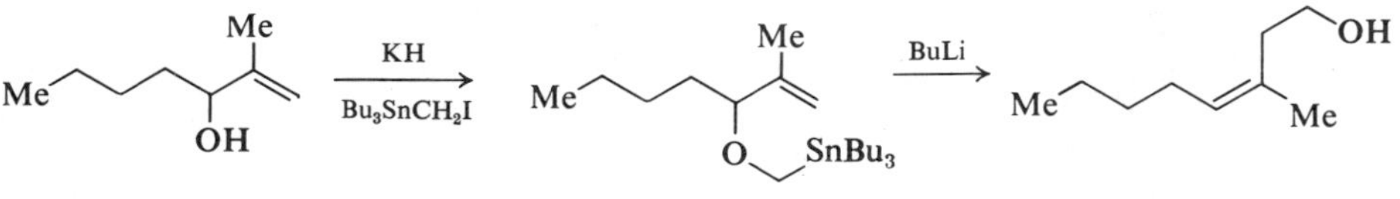

SCHEME 17

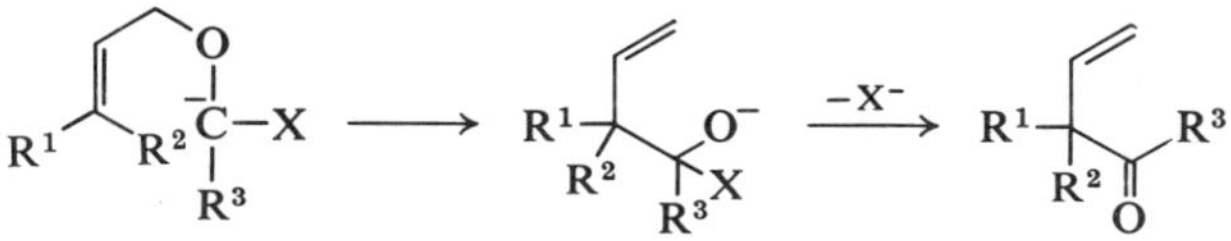

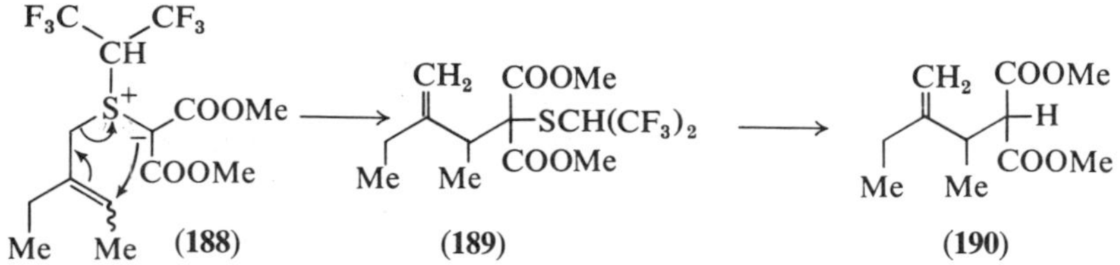

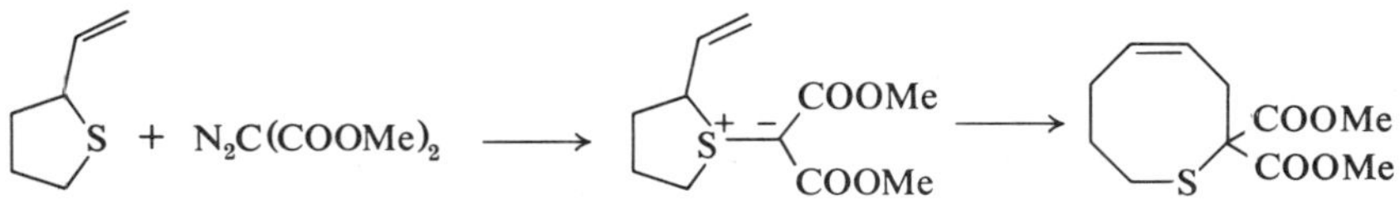

A most useful preparation of allylic malonates from alkenes has been developed,[213] utilizing as a key step the [2,3]-sigmatropic rearrangement of ylides **(188)**. Desulphurization of the resulting allyl(hexafluoropropylthio)malonates **(189)** leads to allylic malonates **(190)**. 2-Thienylmalonic esters have been prepared similarly[214] by thermolysis of thiophenium bismethoxycarbonylmethylides. A closely related [2,3]-sigmatropic shift has been very neatly used[215] for the conversion of five-membered heterocycles into eight-membered heterocycles, for example **(191)**→**(192)**, and in much the same way successive [2,3]-sigmatropic shifts of α-vinyl-*S*-allylsulphonium salts have been employed[216] for the systematic ring enlargements of α-vinyl sulphides. The synthesis of *trans*-2-methylthiacyclo-oct-4-ene has been achieved[217] by an analogous ring expansion procedure involving rearrangement of the allylsulphonium ylide derived from *cis*- and *trans*-1-ethyl-2-vinylthiolanium hexafluorophosphate; the authors present evidence indicating that this sulphur-containing eight-membered cyclic olefin is an intrinsically asymmetric molecule whose enantiomeric conformers most likely are capable of separate existence. [2,3]-Shifts in six-membered cyclic systems have been shown to take place[218] with remarkable specificity by either the *transoid* or the *cisoid* geometry, depending on ring conformation and stereochemistry.

[2,3]-Sigmatropic rearrangements have been found to be commonplace in the reactions of trimethylsilylethylene-dimethylsulphurane with carbonyl compounds.[219] Similar shifts are also involved in the preparation of carbonyl compounds by cupric sulphate-catalysed hydrolysis of α(dialkylamino)nitriles,[220] in a new synthesis of β-lactams from α-diazothioesters,[221] and in the benzylsulphonyl chloride-initiated rearrangement of cephalosporins to penicillins.[222] A complete transfer of chirality has been effected in a [2,3]-sigmatropic rearrangement of allylic alcohols to β,γ-unsaturated amides[223] and the procedure has been used to prepare optically active 9- and 14-carbon saturated isoprenoid synthons. Moreover, a procedure utilizing the [2,3]-sigmatropic rearrangements of ylides derived from azasulphonium salts in the key functionalization step has been developed for the specific *ortho*-alkylation of polycyclic aromatic amines.[224]

A study has been made[225] of the effect of substituents on the ratio of [2,3]-sigmatropic rearrangement *versus syn*-elimination in a series of allylic aryl sulphoxides and a fairly detailed investigation has been made[226] of the stereochemistry of an allyl sulphoxide-allyl sulphenate rearrangement, namely **(193)**, *via* **(194)**, to 2-methylenecyclohexanol **(195)**. An interesting [2,3]-sigmatropic rearrangement of allyl sulphoxides **(196)** to allyl sulphenates **(197)**, followed by

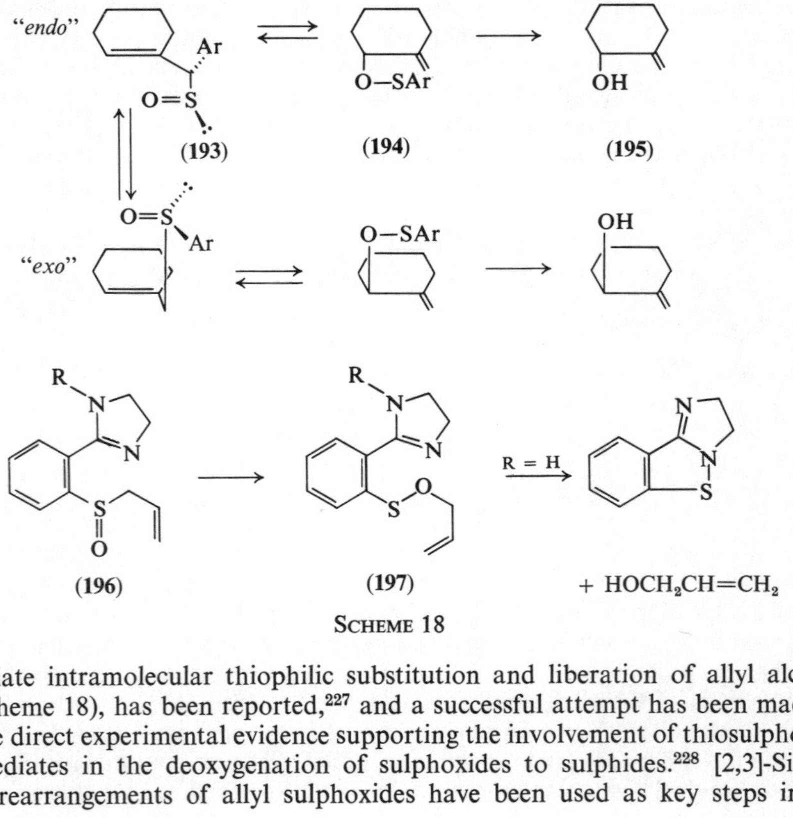

"*endo*" Ar O=S **(193)** O—SAr **(194)** OH **(195)**

"*exo*" O=S Ar O—SAr OH

R N N S O **(196)** R N N S O **(197)** R = H N N S + $HOCH_2CH{=}CH_2$

SCHEME 18

immediate intramolecular thiophilic substitution and liberation of allyl alcohol (see Scheme 18), has been reported,[227] and a successful attempt has been made to provide direct experimental evidence supporting the involvement of thiosulphoxide intermediates in the deoxygenation of sulphoxides to sulphides.[228] [2,3]-Sigmatropic rearrangements of allyl sulphoxides have been used as key steps in the synthesis of verrucarol[229] and in a stereoselective synthesis of 11-deoxy-10-oxa-prostaglandin E.[230]

[1,3]-Migrations

The spirotriene (**198**) has been thermally equilibrated[231] *via* a formal [1,3]-sigmatropic rearrangement with the highly strained 6,6-dimethylenefulvene (**199**).

(**198**) (**199**)

Various medium-ring 1-vinyl alcohols, having either a double bond or a benzo-group at the 3-position, have been shown[232] to undergo unusually fast [1,3]-sigmatropic shifts to the ring-expanded ketones at room temperature, under the influence of potassium hydride, or in the presence of hexamethylphosphoric triamide or dimethoxyethane/18-crown-6 media. The authors postulated that the enormous rate enhancements observed for these anionic shifts are partly due to the naked anion becoming delocalized in the transition state, thus leading to the resonance-stabilized enolate, and partly due to the better donor properties of the anion unit relative to the neutral compounds. Moreover, the reported room-temperature, potassium hydride-catalysed isomerization of *exo*-bicyclo[3.2.0]hept-2-en-7-ol to *exo*-norbornenol[233] is further confirmation that many thermal rearrangements may be accelerated by alkoxide substituents on the C—C bond undergoing cleavage.

Insufficient evidence has as yet been collected[234] to establish whether the thermal degenerate rearrangement of bicyclo[3.2.1]oct-2-en-7-one, (**200**)→(**201**), can be

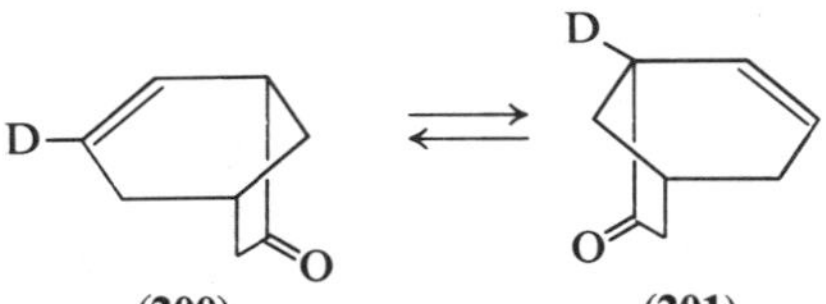

(**200**) (**201**)

interpreted as a truly concerted [1,3]-sigmatropic acyl shift or as a stepwise reaction passing over an allylic–acyl biradical intermediate derived by cleavage of the C(1)—C(7) bond, but it has been established[235] that [1,3]-alkyl migration in electron-rich olefins, as exemplified by 3,3′-dibenzyl-$\Delta^{2,2'}$-bibenzothiazoline, proceeds by a dissociative mechanism. The key experiment performed with a mixture of protio- and deuterio-starting materials showed intermolecular crossover to the extent of 28 ± 2%, a finding typical of radical dissociation–recombination reactions in which a "cage-effect" operates. A new synthetic route to various asymmetrical *N,N,N′,N′*-tetrasubstituted 1,1-diaminoethylenes by reaction of silylynamines with secondary amines has been reported.[236] At elevated temperatures, these compounds further rearrange to amidines by a thermal [1,3]-alkyl rearrangement which is also considered to proceed by a radical dissociation–recombination path.

A detailed analysis of the thermal rearrangement of homoazocines has been made[237] and the findings, although not allowing knowledge of the precise extent, if any, to which [1,3]-sigmatropy contributes to the thermal bond relocation of *cis*-bicyclo[6.1.0]nona-2,4,6-trienes to *cis*-8,9-dihydroindenes, do indicate that such a process must now be given explicit consideration in any explanation of the thermolysis of the bicyclononatrienes.

A series of 2-allyl-substituted 2*H*-azirines (**202**) has been found to undergo smooth thermal rearrangement to azabicyclo[3.1.0]hex-3-enes (**205**) in high yield. The reaction was rationalized[238] (see Scheme 19) in terms of an equilibration of the 2*H*-azirine ring with a transient vinyl-nitrene (**203**), which subsequently adds to the adjacent π-bond, the initially formed bicycloaziridine (**204**) rearranging to the 3-azabicyclohexene system (**205**) by means of a [1,3]-sigmatropic shift. In the case of 2-(but-2-ynyl)-2-methyl-3-phenyl-2*H*-azirine (**206**), thermolysis resulted in a novel rearrangement and production of 2,5-dimethyl-6-phenylpyridine (**209**) as the major product; the formation of pyridine (**209**) from azirine (**206**) is also considered to be compatible with a vinyl-nitrene intermediate (**207**), in this case, attack by the neighbouring acetylenic functionality on the nitrene would lead to carbene (**208**), which would be expected to undergo easily a hydrogen migration to produce (**209**).

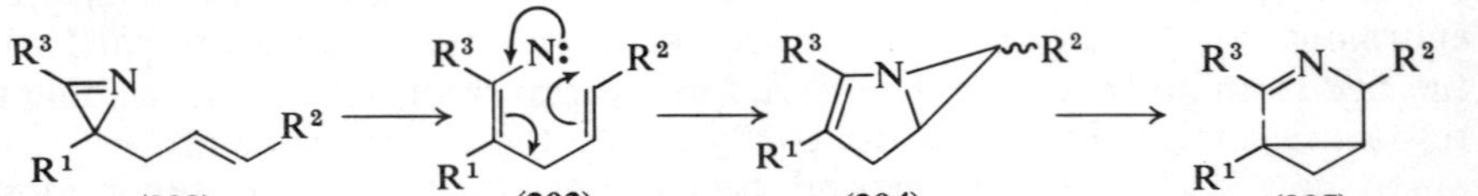

(**202**) (**203**) (**204**) (**205**)

SCHEME 19

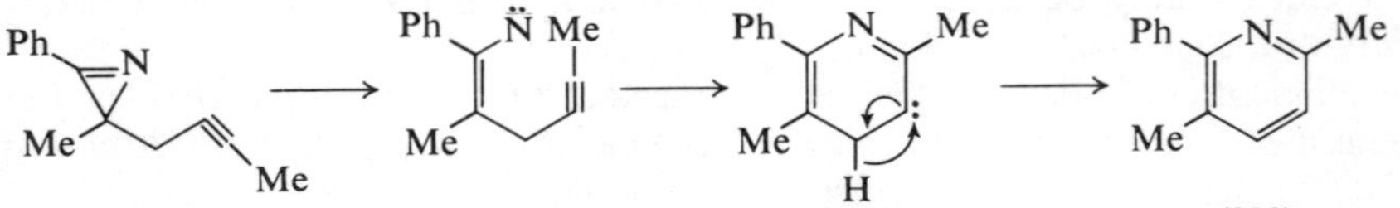

(**206**) (**207**) (**208**) (**209**)

Consecutive [1,2]- and [1,3]-migrations of sulphur functionality as outlined in Scheme 20 have been used[239] to convert the hydroxysulphoxide terpene building block (**210**) into the charge-affinity-inverted synthon (**211**). [1,3]-Phenylthio-shifts have been observed[240] in the photolysis of the trimethylsilyl enol ethers of α,α-dialkyl α-phenyl thioketones, in the acetolysis of 1-chloro-4,4-dimethyl-3-phenylthiopentan-2-one[241] and of α'-arylthio-α-diazoketones[242] – reactions which may well go through cyclic intermediates formed by *S* neighbouring-group participation; the [1,3]-shift of a $BEt_2 \cdot C_5H_5N$ fragment, and incidentally the [1,5]-shift of an $SiMe_3$ fragment, have been observed[243] on using NMR spectroscopy of organometallic derivatives of alkanenitronic acids $RR'C{=}N(O)(OX)$ ($X = BEt_2 \cdot C_5H_5N$ and $SiMe_3$, respectively).

[1,5]-Migrations

Recent work[244] has shown that isoindenes can be generated as transient intermediates by aryl migration during the photolysis of suitable indenes. Pyrolysis studies of spiro[4.4]nona-1,3,6-triene have established[245] that the resulting [1,5]-sigmatropic shift proceeds with negligible primary deuterium isotope effect, thus eliminating a rate-determining hydrogen shift and establishing that the vinyl substituent is the migrating group. Moreover, by extending the study to the

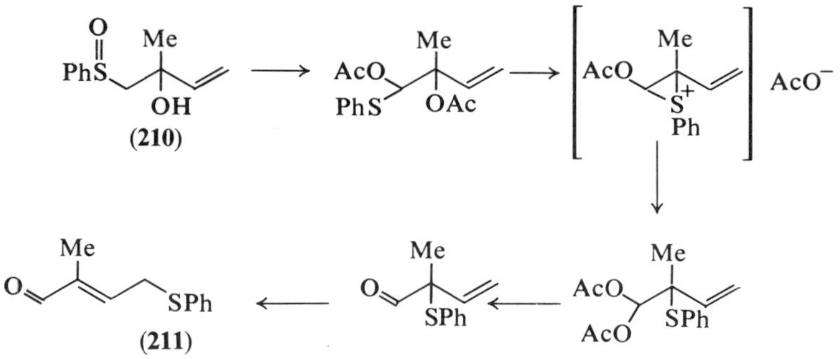

(210)

(211)

SCHEME 20

1,2,3,4-tetradeuteriospiro[4.4]nonatriene isomer (**212**), which rearranges to (**213**), it was concluded that [1,5]-vinyl shifts are favoured over [1,5]-alkyl migrations in such spirocycles. On the other hand, the unprecedented low-temperature ring expansion of the unstable spirotriene (**214**) to the bicyclic triene (**215**), which process is thought to be assisted by strain relief of the methylenecyclopropane, is considered to proceed by the [1,5]-migration of the allylic rather than the vinylic σ-bond.[246]

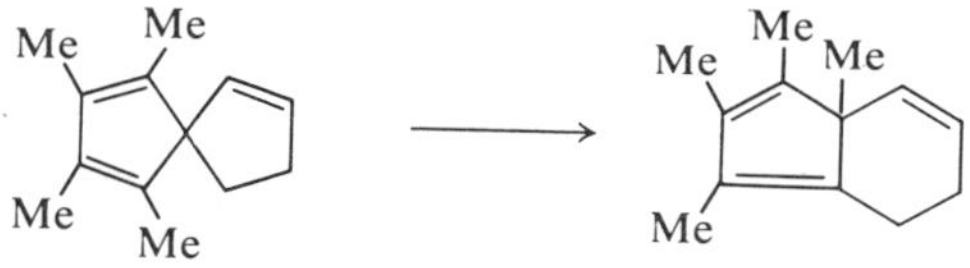

(212) (213)

(214) (215)

By studying the pyrolysis of 7,7-dideuterio-*anti*-1,5-bishomocycloheptatriene, evidence has been obtained[247] indicating that thermal rearrangement of this ring system to *all cis*-cyclonona-1,3,6-triene proceeds by two-fold 1,5-homodienyl rearrangement, the second stage of which involves prior conformational ring inversion and geminal hydrogen interchange. An interesting kinetic study of the stereomutation of (+)-2-deuterio-3,7-dimethyl-7-methoxymethylcyclohepta-1,3,5-triene has demonstrated[248] a sigmatropic [1,5]-carbon shift with allowed stereochemistry in a type of molecule previously thought to isomerize according to a "forbidden" least-motion stereochemical route. The system also provides an instance in which one-centred epimerization in a cyclopropane does not involve a simple diradical intermediate. An attempt has been made[249] to distinguish between competing pericyclic and diradical processes in the thermal rearrangements of bicyclo[3.2.1]octa-2,6-dienes (**216**). The interpretation of the results is outlined in Scheme 21; the formation of (**217**) is thought to start with a homo-[1,5]-hydrogen shift while the formation of (**219**) is depicted as proceeding from *endo*-(**216**) to diradical (**218**) carrying R in the *trans* position.

Tricyclic compounds such as (**220**) have been shown[250] to undergo ready and often completely site-selective homo-[1,5]-sigmatropic hydrogen migrations to

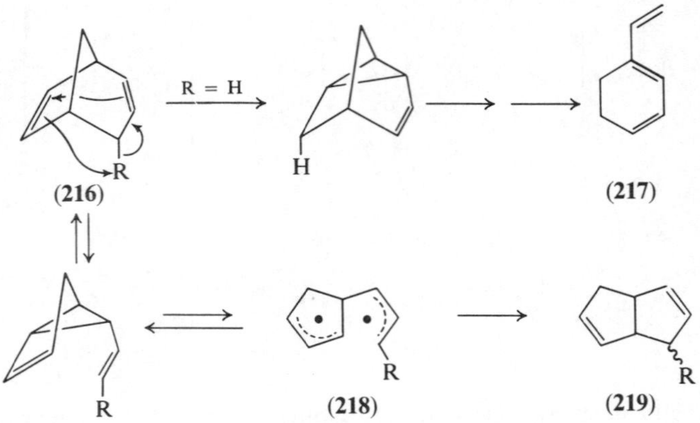

R = H

H

(216)

(217)

R

R

R

(218)

(219)

SCHEME 21

afford products such as (**221**); [1,5]-sigmatropic rearrangement of a vinylallene system and subsequent [1,7]-shifts have been used[251] in a novel route to the 1-hydroxyvitamin D skeleton; and allo-ocimene, obtained *via* ocimene by heating α-pinene, has been further pyrolysed[252] to a complex mixture of products derived primarily from antarafacial [1,7]- and suprafacial [1,5]-hydrogen migrations, and σ-electron disrotatory electrocyclizations.

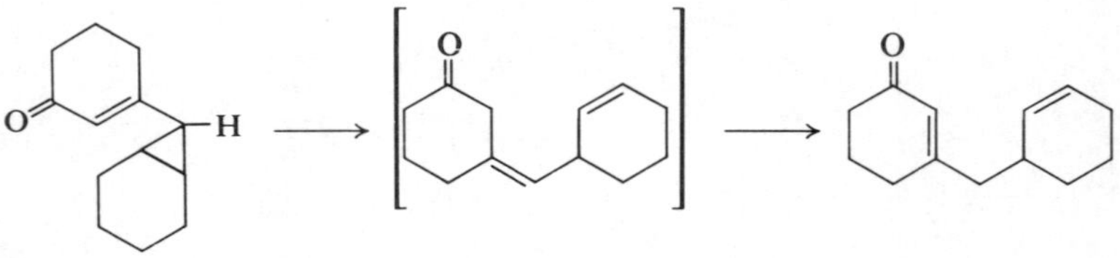

(**220**) (**221**)

Treatment of cycloheptatrienone with either ethane-1,2-dithiol or propane-1,3-dithiol under typical conditions for dithioketalization of ketones has been shown[253] to result instead in a fast 1,2-annelation of dithiols. One likely route to the 1,2-annelated products (see Scheme 22) is considered to involve the dithioketals (**222**) as labile intermediates which are rapidly rearranged to (**223**) by either two consecutive [1,5]- or a single [1,7]-sigmatropic sulphur migration. Such a pathway has broad implications, suggesting the exploitation, for synthetic purposes, of the presumed extremely high aptitude of bico-ordinated sulphur towards sigmatropic migrations.

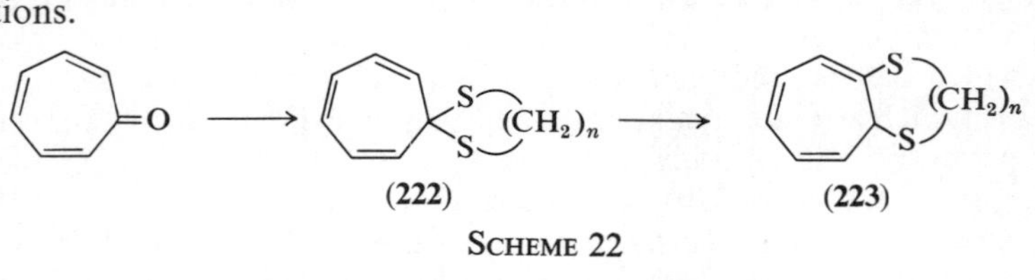

(**222**) (**223**)

SCHEME 22

Although photolysis of 3*H*-thieno-1,2-diazepines has been shown to afford 3-vinylthienopyrazoles, deuterium-labelling studies have shown that their thermolysis gives thienylpyrazoles *via* a [1,5]-hydrogen shift in the diazepine ring.[254]

Various 6,6-diphenyl-3-phosphonyl-3*H*-pyrazoles have been isomerized by a [1,5]-sigmatropic shift of the aryl group to give *N*-phenyl-substituted phosphonylpyrazoles,[255] while two examples of the [1,5]-shift of a methyl group have been observed during studies of Fischer indole syntheses.[256] Moreover, one of the few authenticated cases of a [1,5]-alkyl rearrangement occurring in a mass spectrometer has been observed[257] on an examination of the electron-impact spectra of some trimethylsilylated di- and poly-amines; it appears that the primary impetus for the rearrangement is the high reactivity of the silicon centre, although the proximities of the various amine centres must also affect the fragmentation process.

The easy thermal and photochemical rearrangement of the 8-azabicyclo[5.1.0]-octadienone derivative (**224**; X = NR) by C(4)—C(5) bond cleavage to (**225**) has been contrasted both with the failure of (**224**; X = O) to rearrange by a related path, and with the rearrangement by C—N bond cleavage of simple 2-vinylaziridines.[258] Although a formal [1,3]-shift converts (**224**) into (**225**) (see arrows), it is suggested that the ease of this particular reaction is possibly due to it proceeding *via* a biradical or a zwitterionic intermediate.

X = NR

(**224**) (**225**)

The mechanism of photo-deconjugation of α,β-unsaturated nitro-compounds has been investigated[259] by studying the photochemistry of 6-nitrocholest-5-ene and its 3β-acetoxy-derivative (**226**). A pathway analogous to that accepted for the photo-deconjugation of ketones has been suggested – namely, [1,5]-hydrogen shift from C(4) to oxygen to yield the *aci*-nitro-compounds (**227**) which then tautomerize to the mixture of nitro-olefins (**228**) and (**229**).

An unusually facile thermal [1,5]-hydrogen shift found for a nitroarene with *ortho*-benzylic hydrogens, see (**230**), has resulted in the reported formation of benzisoxazoles from *o*-alkylnitrosobenzenes;[260] 2-ethoxypyridine has been shown[261] to undergo thermal elimination of ethylene to give pyrid-2-one in a formally similar manner, *i.e. via* a six-centred cyclic process which is a nitrogen analogue of ester pyrolysis – indeed, this represents the first reported reaction of this type in which aromatic π-electrons participate.

[1,5]-Suprafacial sigmatropic migrations have been observed during the thermolysis of anthracene photo-oxide;[262] a fascinating rearrangement of phenanthraquinone monoimine (**231**) has been uncovered[263] which involves a ready [1,5]-shift in the intermediate (**232**), to give (**233**), while an NMR study of cyclopentadienyldimethylborane and (pentamethylcyclopentadienyl)dimethylborane has indicated,[264] in each of these compounds, fluxional behaviour formally described as a [1,5]-sigmatropic migration of the Me_2B group.

Miscellaneous

Dyotropic rearrangements and related σ–σ exchange processes have been reviewed.[265]

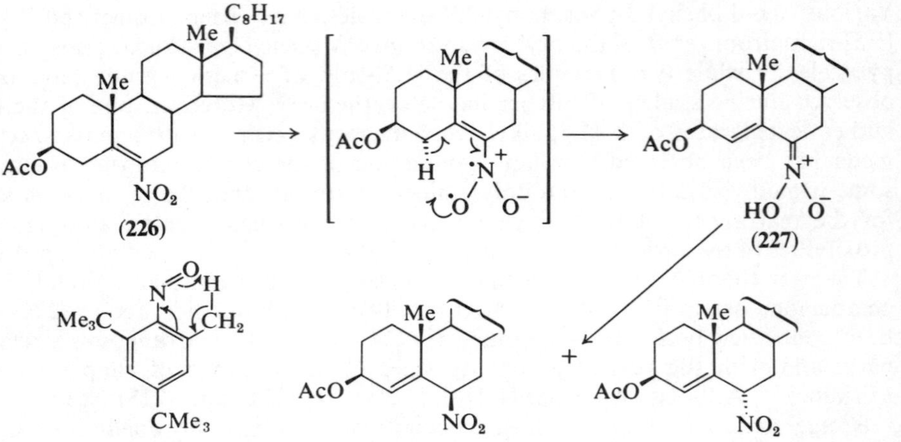

(226) (227) (228) (229) (230)

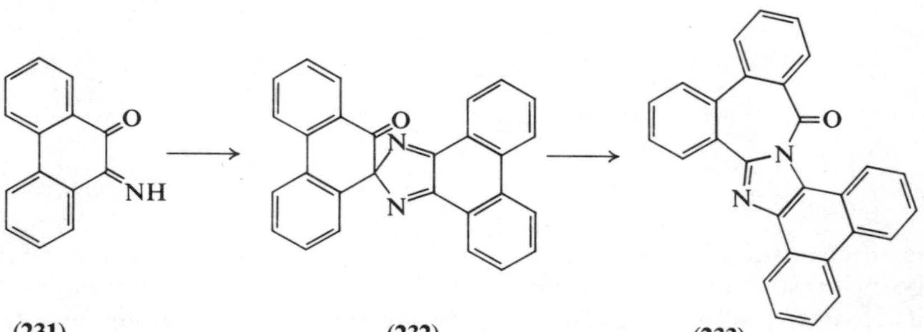

(231) (232) (233)

Since the thermal rearrangement of alkyl(silylmethyl) ethers appears to proceed with anchimeric assistance from the silyl group,[266] it is not surprising that it is only those compounds which potentially liberate resonance-stabilized ketyl and alkyl radicals that have been shown to rearrange.[267]

Although an enamine tautomer (**235**) has not yet been detected in the reaction, it is considered to be the reactive form through which the dyotropic-like thermal rearrangement of *C*-silylimines (**234**) to *N*-silylenamines (**236**) proceeds.[268] Certainly the observed entropy of activation for the reaction indicates considerable loss of freedom of motion in the transition state and is consistent with a concerted process rather than the alternative route *via* the silylaminocarbene (**237**). It has been noted[269] that the stereospecific transformations of methyl 4-trimethylsilyl-1-pyrazoline-3-carboxylates to 4-trimethylsilylbut-2-enoates, in which the silyl group participates in the expulsion of nitrogen [see (**238**) to (**239**)], bears a similarity to a dyotropic process. Consequently the term "triotropic" has been used to describe this general class of transformations which involve simultaneous rearrangement of three σ-bonds.

Organic and organometallic haptotropic shifts in cyclopentadienyl compounds (CpX) have been analysed[270] by studying the motion of an X^+ across the face of a $(C_5H_5)^-$ ring. Kinetics of the metallotropic rearrangement of η^1-idenyltrimethyllead and the relative migratory ability of the trimethyl-lead group have been

$R^3_3Si-C(=NR^2)-CHR^1_2$ → $R^3_3Si-C(NHR^2)=CR^1_2$ → $R^2-N(H)-C(SiR^3_3)=CR^1_2$ → $R^2(R^3_3Si)N-CH=CR^1_2$

(234) (235) (236)

$R^2(R^3_3Si)N-\ddot{C}-CHR^1_2$

(237)

MeOOC, N_2, Me_3Si, $SiMe_3$

(238) ⟶ (239)

examined,[271] and [1,3]-boranyl shifts in carbon–carbon–oxygen skeletons have been reported.[272]

It has been shown that the thermal rearrangements of the ammonium ylides, $PhCO\bar{C}R\overset{+}{N}Me_2R'$, give products formally derived from competing [1,2]-, [1,3]- and [1,4]-rearrangements.[273] Furthermore, the intramolecularities of these processes are compatible with a radical-pair mechanism involving initial homolysis of the ylide followed by three possible modes of radical-pair recombination. A further study using a chiral ylide has demonstrated[274] the stereoselectivity of the competing [1,2]- and [1,3]-processes, while an examination of the thermolysis of pentadienylammonium ylides[275] has indicated that the rearrangement proceeds by two distinct reaction pathways, the first a forbidden sigmatropic rearrangement that involves an interacting radical pair having some translational freedom and giving predominantly the products of [1,2]- and [5,2]-rearrangements, and the second a formally allowed [3,2]-sigmatropic rearrangement that involves an interacting radical pair indistinguishable from the transition state of a concerted pericyclic process.

The first example of a [1,2]-hydrogen shift accompanying acid-catalysed hydrogen exchange has been observed[276] during detritiation of methyl-substituted 9-tritiated phenanthrenes, and it has been shown that magnesium reacts in diethyl ether with 5-chloro-4,4-diphenylpent-1-ene, 2-chloro-1,1,1-triphenylethane, and 1-chloro-2,2,3-triphenylpropane, to give Grignard reagents in which the phenyl groups have undergone an apppreciable amount of [1,2]-sigmatropic rearrangement.[277]

It appears[278] that 3-methylenebicyclo[3.2.2]nona-6,8-diene (**240**) isomerizes under relatively mild conditions *via* a one-bond cleavage, first to the highly stabilized diradical (**241**) which, after a formal [1,4]-hydrogen shift (or consecutive [1,3]- and [1,2]-shifts), closes to produce (**242**) which then isomerizes and gives the observed mixture of (**242**) and (**243**).

The calculated values of the activation volumes for the concerted [1,4]-shift of the 2-alkoxypyridine *N*-oxide (**244**; R = CH_2Ph) to the 1-alkoxypyrid-2-one (**245**)

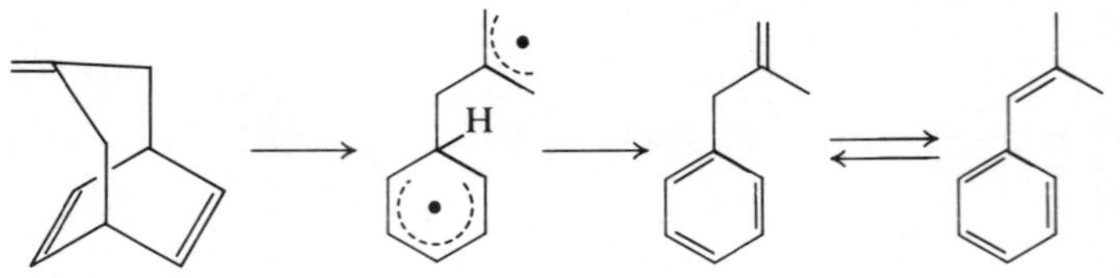

(240) (241) (242) (243)

and for the stepwise, diradicaloid reaction of the diphenylmethyl analogue (**244**; R = $CHPh_2$), namely -30 ± 5 cm^3 $mole^{-1}$ and $+10 \pm 2$ cm^3 $mole^{-1}$, respectively, clearly confirm[279] that this activation parameter provides a criterion for concertedness in sigmatropic shifts. A benzo[*f*]quinoline alcohol has been converted[280] into azasteroids by application of the recently discovered [1,4]-aryl radical rearrangement of α-halomethyl-*N*-arylsulphonylpiperidines; see (**246**)→(**247**). In the case considered, R contains additional functionality and further cyclization was envisaged and realized. Consequently the method can be considered as an attractive pathway for the construction of bridgehead nitrogen heterocycles.

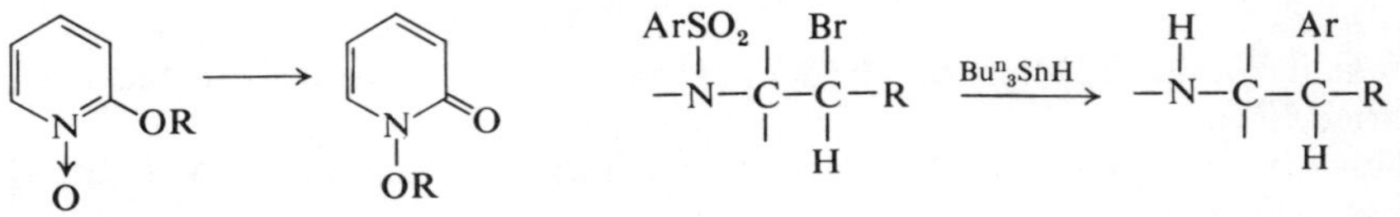

(244) (245) (246) (247)

A concerted [1,4]-sigmatropic rearrangement *via* a thia-anthracene anion has been proposed[281] as a possible mechanism for the base-promoted rearrangement of thioxanthene *N*-(*p*-toluenesulphonyl)sulphilimine although an intermolecular, dissociation–recombination process involving a thioxanthylium ion is also considered as a possibility. On the other hand, 1,4-ylidic thia-anthracenes undergo [1,4]-rearrangement *via* 9-arylthioxanthyl radicals.[282]

A [3,4]-allyl shift has been observed[283] in the thermolysis of 1-allyl-1-methoxynaphthalen-2-one. It appears that such a shift is markedly dependent on the presence of a methoxyl group at C(1), since the corresponding naphthalenones with methyl or allyl groups in place of the methoxyl at C(1) only undergo the expected equilibration *via* reverse Claisen migration. The intermediacy of a dipolar species and its effective stabilization by the methoxyl group are considered to account for the favourable influence of such a group in facilitating the [3,4]-allyl migration.

The formation of biphenyls (**249**) instead of Fischer indolization products on treatment of arylhydrazones (**248**) with polyphosphoric acid has been explained[284] on the basis of a [5,5]-sigmatropic rearrangement of the protonated substrate;

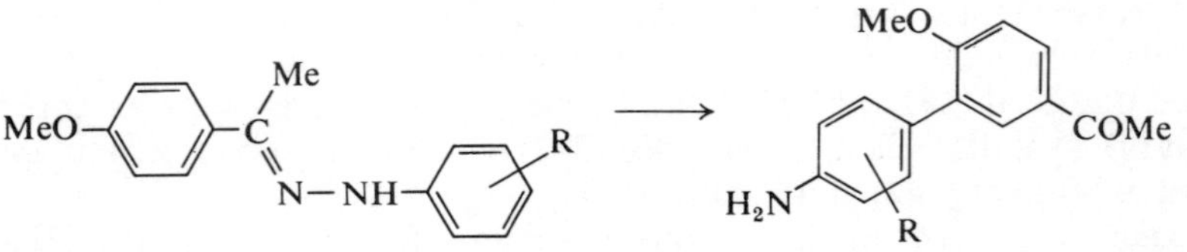

(248) (249)

(3*S*,10*R*)-(*Z*,*Z*)-9,10-secocholesta-5,7,14-trien-3-ol and its 10*S*-epimer have been prepared[285] from *cis*-isotachysterol *via* facile [1,7]-sigmatropic rearrangements, while a rapid [1,7]-interchange of unhindered alkylthio-groups has been observed[286] in the 1-alkyl-7-[alkyl(aryl)thio]cycloheptatrienes obtained by the addition of alkene- or arene-thiolates to 1-alkylthiotropenylium ions.

The thermal and photolytic decomposition of phenyl(diphenylthiophosphor-amyl)methyl azide to benzylideneaminodiphenylthiophosphorane and aminodi-phenylthiophosphorane appears to be the first reported migration of a thio-phosphorus group;[287] the authors favour a concerted mechanism for the process rather than one involving a nitrene intermediate. A thiophosphonate–thiophosphate rearrangement has been observed[288] in the reactions of incompletely esterified phosphorus acids with thiobenzophenone. Finally, the accumulated results from a mechanistic study of the aluminium chloride-catalysed rearrangement of 3-chloro-4′-fluoro-2-methylpropiophenone have indicated[289] that its conversion into 5-fluoro-3-methylindan-1-one and 2-(4′-fluorophenyl)-1-oxoniacyclopent-1-enyl cation occurs by means of alkyl and hydride shifts and not by acyl migration; although preceding those or any other carbon bond reorganization, there appears to be a degenerate 1,3-chloride–hydride transposition within the first-formed $AlCl_3$-complex.

Electrocyclic Reactions

It has been suggested[290] that the degree of aromaticity in a conjugated transition state is an important factor in determining the allowedness of pericyclic reactions. The MINDO/3-UHF method has been used to investigate the potential surface behaviour of some doublet systems;[291] the calculations support the expected strong similarity between the transition states of forbidden electrocyclic radical rearrange-ments and the ground state of Jahn–Teller distorted radicals.

It has been shown that the n-orbital basis employed in the conventional analysis of n-carbon atom electrocyclic isomerizations is not sufficient for a complete symmetry analysis of the problem. Selection rules for the variously charged species have been obtained by using a five-orbital basis in OCAMS and Hückel–Hubbard calculations. The additional pair of σ-orbitals behaves as a closed shell throughout the reactions and thus a sound theoretical foundation is provided for the con-ventional treatment of σ-orbitals in electrocyclic reactions.[292] MNDO calculations have been made for the electrocyclic ring-opening of 1- and 2-azabicyclo[2.2.0]-hexa-2,5-dienes,[293] while kinetic data and MO calculations have indicated that the intramolecular non-mutual exchange of 5-halobenzofuroxans, *i.e.* **(250)** ⇌ **(252)**, proceeds *via* species **(251)** which are generated by electrocyclic ring-opening.[294] Electrocyclic ring-opening (to isocyanates) is observed too, on the flash-vacuum pyrolysis of 1,3-oxazin-2-ones.[295]

A report has appeared[296] that simple aliphatic polyenes such as 3,6-dimethylidene-octa-1,7-diene undergo thermal rearrangement to products that result from a

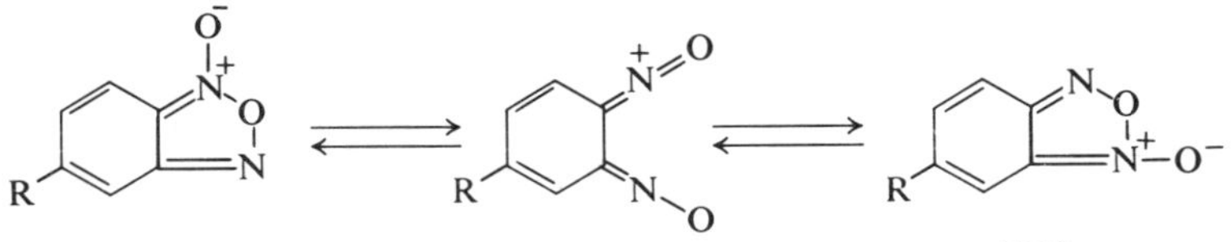

(250) ⇌ **(251)** ⇌ **(252)**

"formal" $_{\pi}2+_{\pi}2$ intramolecular cycloaddition, while the 14-electron-electrocyclization of the fulvene (**253**) has been shown to yield (**254**) by a symmetry-forbidden conrotatory pathway.[297] A kinetic study of the cycloaromatization of

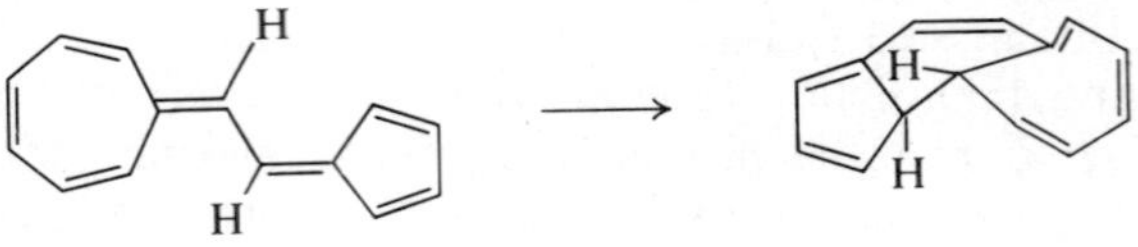

(**253**) (**254**)

bridged diallenes,[298] *e.g.* (**255**), in which the bridge contributes an aromatic character and consequently an extra driving force for the cyclization, is compatible with a two-step process involving an electrocyclic six-electron reaction in the first step, (**255**)→(**256**), followed by a [1,5]-hydrogen shift to give (**257**). The observable products from electrocyclization of conjugated tetraenes with both central double bonds of *cis*-configuration have been shown[299] to be markedly dependent on the nature of the terminal substituents, while electron-withdrawing groups, *e.g.* R = COOMe, have been shown[300] considerably to accelerate the intramolecular ene reaction, (**258**)→(**259**), of 1,6-enynes.

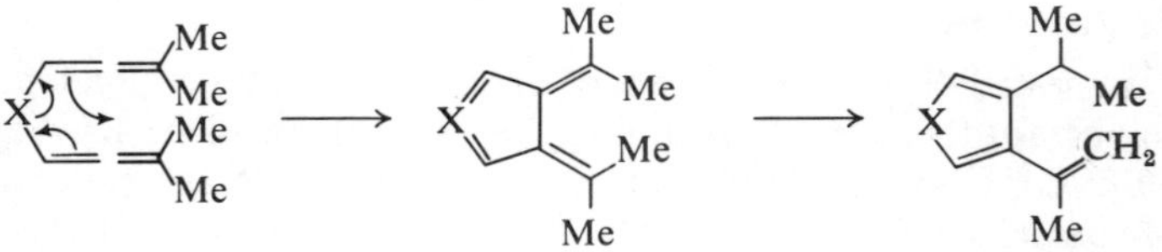

(**255**) (**256**) (**257**)

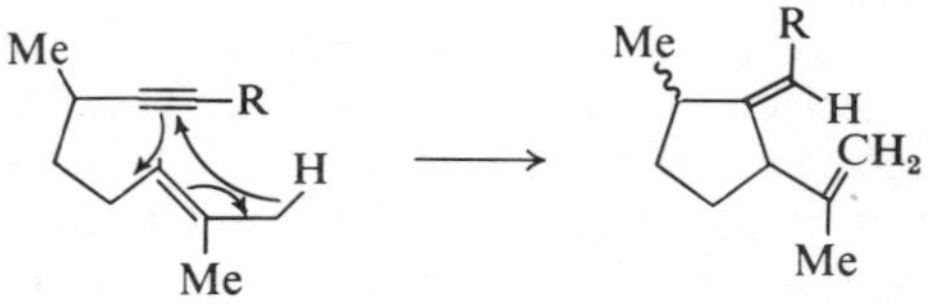

(**258**) (**259**)

An example of a remarkably easy symmetry-forbidden thermal conversion of a polycyclic butadiene unit into a cyclobutene is observed[301] in the room-temperature conversion of the "biradicaloid" hydrocarbon, 7,12-dimethylpleiadene (**260**), to 6b,10b-dimethyl-6b,10b-dihydrobenzo[*j*]cyclobut[*a*]acenaphthylene (**261**). The intriguing possibility has been suggested[302] that *exo*-tetracyclo[4.3.0.0^{2,4}0^{5,7}]non-8-ene (**262**), on thermolysis, undergoes a retro-Diels–Alder cleavage to vibrationally excited (**263**) in its extended conformation, which thereupon suffers immediate symmetry-allowed electrocyclic opening to *cis, cis, cis, trans*-cyclononatetraene, and thence to *trans*-dihydroindene.

Evidence has been presented[303] for a new mechanism for the formation of cyclopropane derivatives in the thermolysis of pyrazolines. In this mechanism ionization of the pyrazoline is considered to permit a ring-expansion and then extrusion of nitrogen in an electrocyclic process, consistent with retention of stereochemistry by the resulting cyclopropane. Electrocyclic elimination of CO_2 from the cycloadduct (**264**) of α-pyrone and mono-*trans*-3-benzazonine yields the

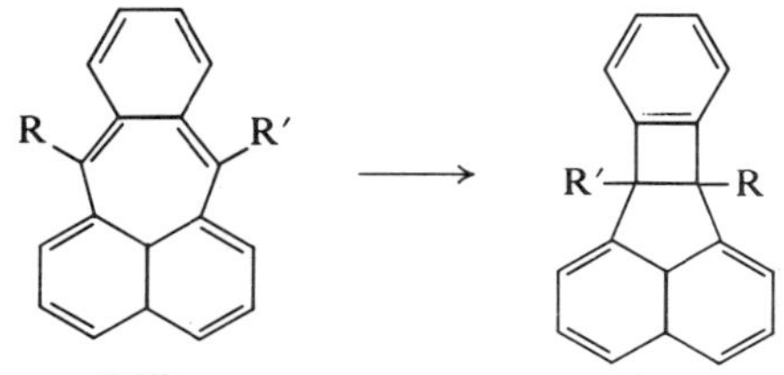

(260) **(261)**

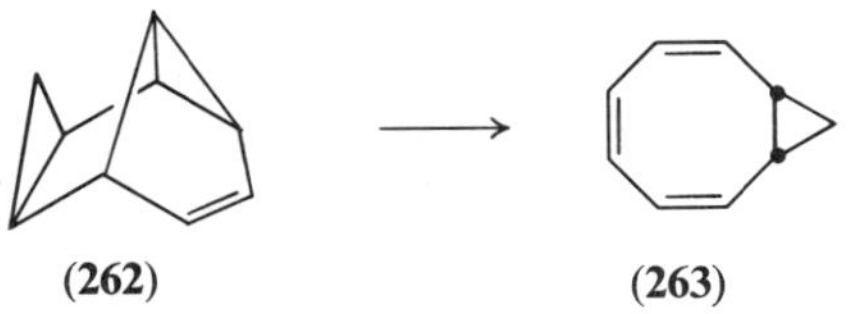

(262) **(263)**

ester (**265**).[304] Interestingly, the primary cycloadducts from 2-phenylbenzazete and diarylnitrilimines have been found[305] to rearrange spontaneously to 1,2,4-triarylbenzo[*f*]-1,3,5-triazepines (**268**). The postulated mechanism involves initial ring-opening of the adduct (**266**) to the benzo[*e*]-1,2,4-triazepine (**267**) which rearranges to (**268**) by electrocyclic ring closure to a diazaridine, a [1,5]-NAr shift, and ring-opening.

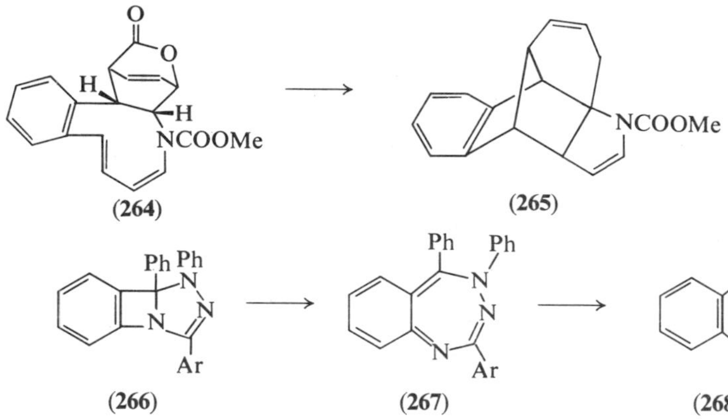

(264) **(265)**

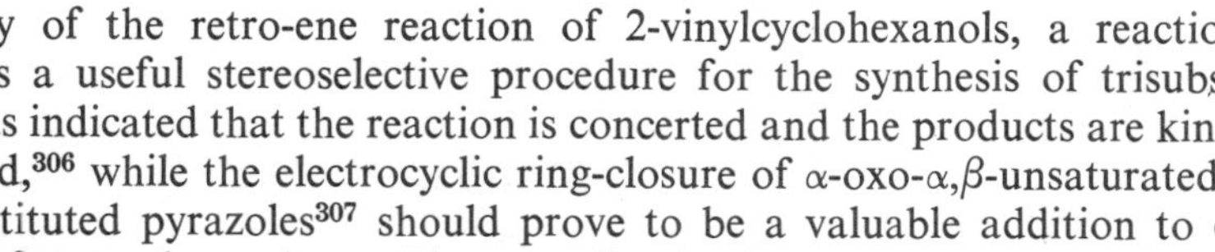

(266) **(267)** **(268)**

A study of the retro-ene reaction of 2-vinylcyclohexanols, a reaction that constitutes a useful stereoselective procedure for the synthesis of trisubstituted olefins, has indicated that the reaction is concerted and the products are kinetically determined,[306] while the electrocyclic ring-closure of α-oxo-α,β-unsaturated azines to *N*-substituted pyrazoles[307] should prove to be a valuable addition to current methods of pyrazole syntheses. Electrocyclization is also considered to be involved in the thermally induced cyclization of 1,16-dimethylhexahelicenes into 5*H*-benzo-[*c,d*]pyrene-5-spiro-1′-indane derivatives,[308] in the isomerization of 1,6-benzodithiocins,[309] in the synthesis of 9,12-dithia-*trans*-bicyclo[6,4,0]dideca-2.4.6,10-tetraenes by reaction of 1,2-dithiols with cyclo-octatetraene *trans*-dihalides,[310] in the syntheses of 2,3,4,5,6,7-hexachloro-8,8-dicyanoheptafulvene[311] and *N*-(ethoxycarbonyl)aza[17]annulenes,[312] respectively, and in the thermal transformations of naphthacenic *endo*-peroxides.[313]

Anionic Rearrangements

The Ramberg–Bäcklund rearrangement has been reviewed[314] and a review has appeared that describes in detail the 1,2-anionic rearrangements of organosilicon and organogermanium compounds.[315] The modes of rearrangement in the ionic compounds RPF_5^- have been elaborated upon[316] and the rearrangements of η^5- and η^6-anionic complexes of the chromium-group metals have been studied.[317] 1,3-Migration of the trimethylsilyl group has been observed to occur in the anions of bromophenoxysilanes,[318] and dialkyl α-siloxybenzylphosphonates have been prepared[319] by reaction of silyl phenyl ketones with trialkyl phosphites by a pathway that involves the migration of a silyl group from carbon to anionic oxygen.

Wittig-type rearrangements have been observed[320] in the course of an investigation of reactions of alkyl diphenylphosphinates and related thioesters with organolithiums and Grignard reagents and it has been shown[321] that, in the cleavage of dialkyl ethers with alkyl-lithium compounds, Wittig rearrangements become important only when all possible competing elimination reactions are completely hindered.

Several studies of the Stevens rearrangement have been made. Thus, the effect of different bases on the competition of Stevens rearrangement and Hofmann degradation with ring-opening of *N*-methallyl-*N*-methyl-piperidinium and -morpholinium iodides has been investigated,[322] the primary products of the Stevens rearrangement of trialkyl-(2-methylalk-2-enyl)ammonium iodides have been shown to arise as a result of [3,2]-rearrangement,[323] while transesterification has been observed during the Stevens rearrangement of ammonium salts containing alkoxycarbonylmethyl groups.[324] A Stevens rearrangement accompanied by an intramolecular aldol-type condensation has been employed[325] in the preparation of various cyclopentenone derivatives and the Stevens rearrangement of benzylsuccinimidosulphonium salts provides a simple synthesis of *N*-(1-chloroalkyl)-succinimides.[326] Hydantoin-2-sulphonium salts (**269**), in the presence of tertiary amines, undergo Stevens rearrangement to yield unstable imidates (**270**) which are changed into their isomers (**271**) by heat,[327] while the same rearrangement of the N^1,N^3-disulphonium salts of hydantoin takes place with the formation of bis-(alkylthio)- or -(arylthio)-methylated hydantoins.[328] The observed product ratios obtained from an examination of the thermal rearrangement of ylides of the type (**272**) and (**273**) are consistent with limited rotational equilibration of diradical intermediates.[329]

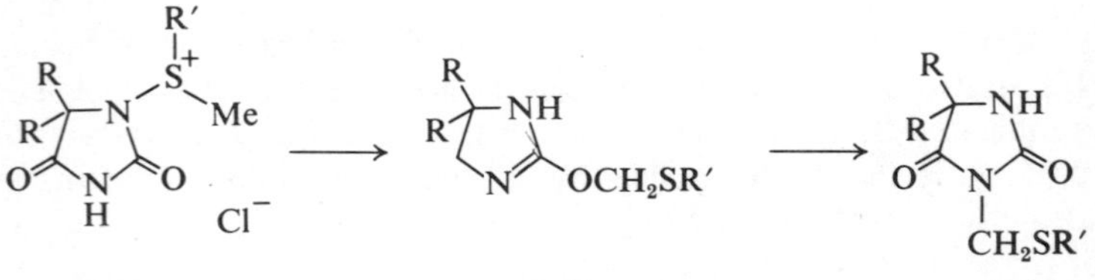

(**269**) (**270**) (**271**)

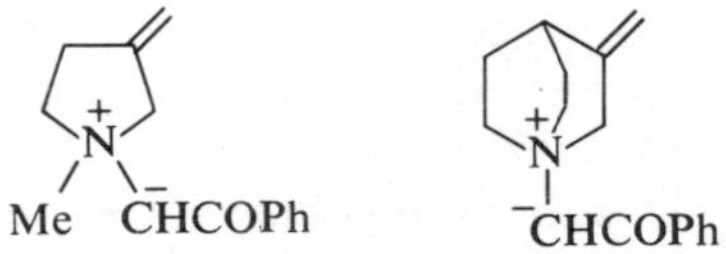

(**272**) (**273**)

Pummerer rearrangements of the 3-(methylsulphinyl) derivatives of quinolinones, cinnolinones, chromanones, and chromones have been reported;[330] 4-methoxycoumarin has been converted into 2-formylchromone in a series of steps, the key reaction of which is a Pummerer rearrangement of 2-[(methanesulphinyl)methyl]-chromone;[331] and the details of several synthetically useful Pummerer-type reactions of α-phosphoryl sulphoxides have been presented.[332] The last-mentioned report also describes a new example of the transfer of chirality from sulphur to carbon. The Pummerer-type mechanism of the acetic anhydride-induced rearrangement of benzyl phenylmethanethiosulphinate to a 1 : 1 mixture of *erythro*- and *threo*-1-(thioacetoxy)benzyl benzyl sulphoxides has been revised[333] in the light of a recent product analysis study for ^{13}C- and ^{18}O-labelled substrates. Recently obtained evidence[334] has led to the conclusion that a 2,5-dienone intermediate pathway, followed by a [1,3]-trifluoroacyl migration, probably accounts for the observed Pummerer rearrangement of 4-methylsulphinyl-3,5-xylenol with trifluoroacetic anhydride. The first example of a reaction sequence involving the functionalization of an allylic methyl group proceeding *via* an additive Pummerer-rearranged intermediate has been reported.[335] This novel reaction (see Scheme 23)

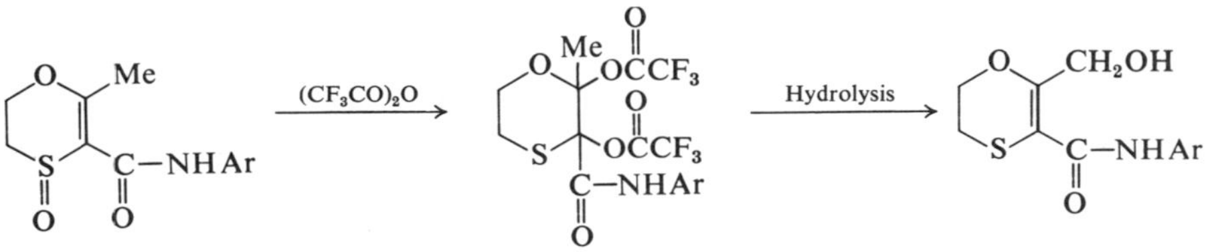

$(CF_3CO)_2O$ → Hydrolysis →

SCHEME 23

provides ready access to a number of previously unavailable 2-oxymethyl analogues of the highly active and widely used systemic fungicide, carboxin. The first example of a carbothioate acyl anion equivalent which is capable of homologation of alkyl halides, permitting their conversion into thiol ester derivatives (**275**), has also been reported.[336] The key step in this transformation is the silicon-Pummerer rearrangement of a silylated α-chloro-sulphoxide (**274**) as illustrated in Scheme 24.

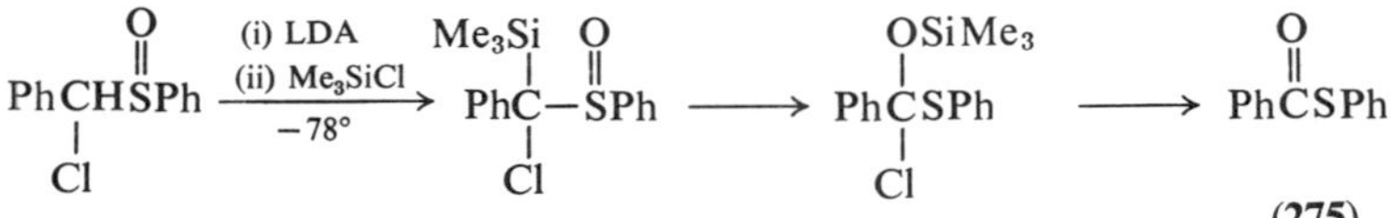

PhCH(Cl)S(O)Ph (**274**) —(i) LDA, (ii) Me_3SiCl, −78°→ PhC(Me_3Si)(Cl)S(O)Ph → PhC($OSiMe_3$)(Cl)SPh → PhC(O)SPh (**275**)

SCHEME 24

A full report has now appeared[337] concerning the reaction of cyanoacetamides with α-oxo- and α-cyano-*SS*-acetals in the presence of base to afford 4-(alkylthio)-3-cyano-2(1*H*)-pyridones and the corresponding 6-aminopyridones, respectively. A novel potassium hydride-induced reorganization of an azabicyclo[2.2.1]heptene has produced[338] ethyl oxindole-4-carboxylate, a product which may serve as a valuable precursor for the synthesis of various members of the ergot family. The dimsylsodium-induced rearrangement of alkylimidates to *N*-alkyl-lactams has been reported,[339] and the base-catalysed chlorohydrin rearrangement of 1,1-dichloro-3-piperidinylpropan-2-ol (**276**) and subsequent reaction of the intermediate α-chloro-epoxide (**277**) have been used[340] to prepare quinuclidine aldehydes (**278**). The structure of the product obtained from the base-catalysed rearrangement of *exo*-2,3-epoxynorbornane-*endo*-5,6-carbonate, which was previously assigned as *endo*-epoxide (**279**), has now been elucidated[341] as the oxetane (**280**). A study of

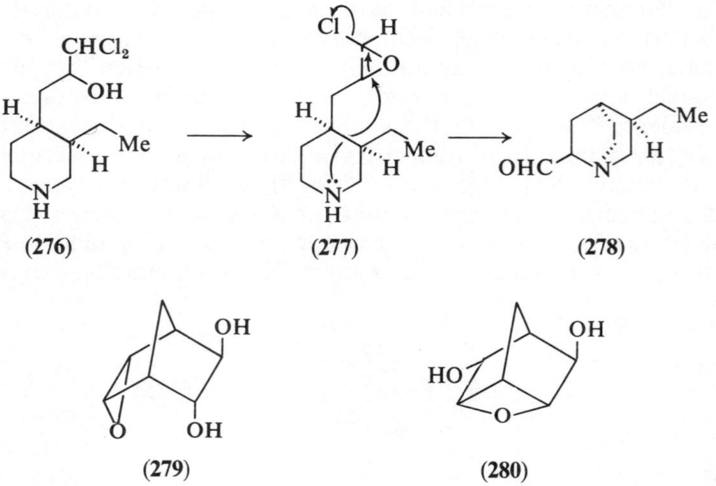

(**276**) (**277**) (**278**)

(**279**) (**280**)

the thermal rearrangement of the 3-β-D-glucoside of the cytokinin, 6-benzylamino-purine, has shown[342] that the major process is an intramolecular [1,3]-shift of the glucosyl grouping to the 9-position, with retention of configuration at the migrating centre, together with a smaller contribution from an intermolecular [1,3]-rearrangement which gives rise to both the 9α- and the 9β-anomers.

The formation and rearrangements of the vinyl anion have been studied[343] with *ab initio* molecular wave functions, as a model for nucleophilic attack on acetylenes. Unlike the case for a vinyl cation, the rearrangement barriers are high, the lowest being 40 kcal mol^{-1}. The kinetics of an apparent vinyl-to-allyl anion rearrangement in 1-phenyl-2,2-dimethylvinyl-lithium have been studied[344] and 1,1-diaryl-2-halo-2-carbanions have been observed as intermediates in the electrochemically induced rearrangement of 1,1-diaryl-2,2-dihaloethylenes.[345] Unexpected products obtained[346] on reaction of hex-5-enyl chloride with alkali metals point to isomerization of the hex-5-enyl anion by an intramolecular [1,4]-proton transfer; an attempted metalation of anion (**281**) resulted in a novel rearrangement and the formation of (**282**),[347] while a study of the base treatment of 1-(1-X-1-methylethyl)indenes, using different leaving groups (X = Cl, OAc, OMe, Me), shows[348] that 1,2-elimination and [1,3]-

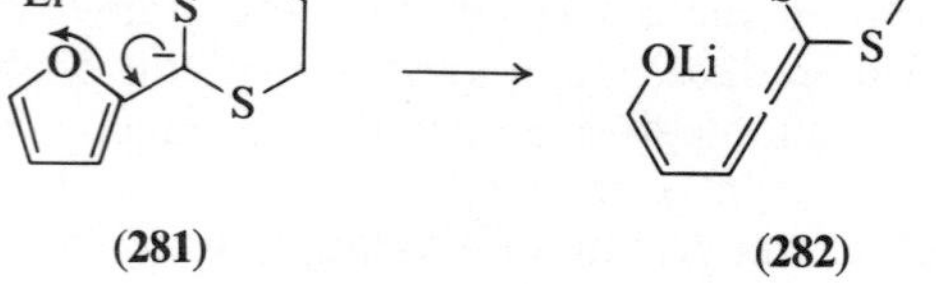

(**281**) (**282**)

proton transfer are coupled *via* common intermediates which presumably are of the ion-pair and carbanion type, respectively.

The prediction that the energetic disadvantage inherent in the 1,2-migration of saturated carbon to an anionic centre could be overcome, at least in part, by designing a system in which relief of ring strain provides the driving force for the rearrangement and in which the starting and final carbanions maintain an equal degree of resonance stabilization, has been realized[349] in a report of the base-catalysed rearrangement of (**283**) to (**284**). A report that concerns the thermodynamic stability of the nortriquinacenyl anion relative to other $(CH)_9^-$ isomers

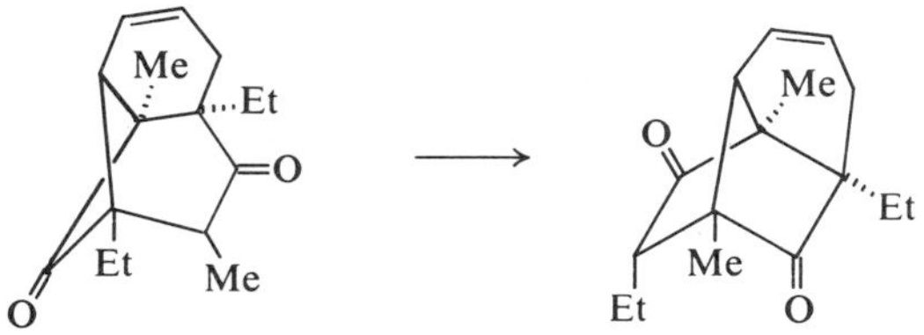

(**283**) (**284**)

and, further, describes the full complement of reversible charge control criteria for this system has appeared,[350] while failure to obtain the bicyclo[4.3.2]undecatetraenyl anion because of spontaneous rearrangement, even under conditions closely similar to those that had earlier provided the bicyclo[3.2.2]nonatrienyl anion and the bicyclo[3.3.2] decatrienyl dianion, seems to confirm the explicitly predicted limit of longicyclic stabilization.[351]

The general base-catalysed rearrangement of methyl- and phenyl-glyoxal glutathionyl hemithiol acetals to *S*-lactoyl- and *S*-mandeloyl-glutathiones, respectively, which was examined as a model for the glyoxalase enzyme system, has been shown[352] to require enediol proton transfer, while a mechanistic examination[353] of the ready thioester migration (**285**)→(**286**), a model reaction for the methylmalonyl coenzyme

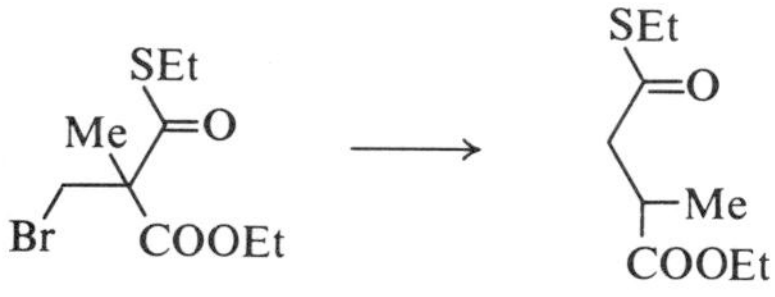

(**285**) (**286**)

A mutase conversion, has shown that the overall mechanism for the reaction is anionic; the skeletal rearrangement probably takes place at the radical (or radical anion) stage, followed by a rapid second-electron reduction to give a carbanionic intermediate which is then protonated by the reaction medium.

Enolate-induced Rearrangements (Favorskii Types)

The attempted Favorskii reaction of 4-biphenylyl 1-bromocyclobutyl ketone gives 4-(*p*-phenylbenzoyl)butanoic acid instead of 1-(4-biphenylyl)cyclobutane-1-carboxylic acid.[354] The first application of Favorskii rearrangement conditions to a chloro-enol-lactone (**287**) and the resulting synthesis of a 4-epigiberellin system (**288**) has been reported,[355] while a Favorskii-type ring contraction of 5*β*,6*β*-epoxyandrostane-4,17-dione has been shown[356] to afford a rapid route to A-norsteroids. A plausible mechanism is outlined in Scheme 25.

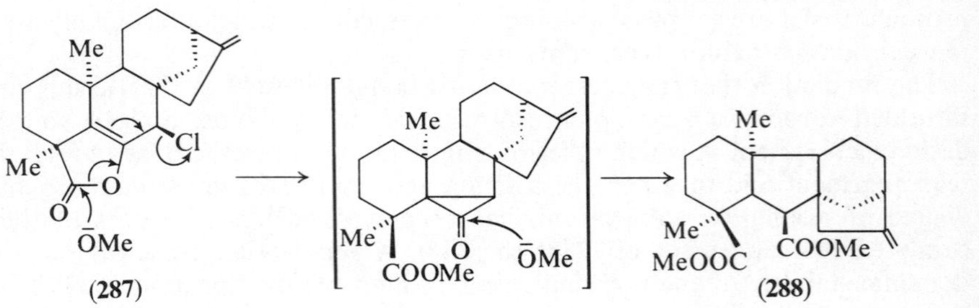

(287) (288)

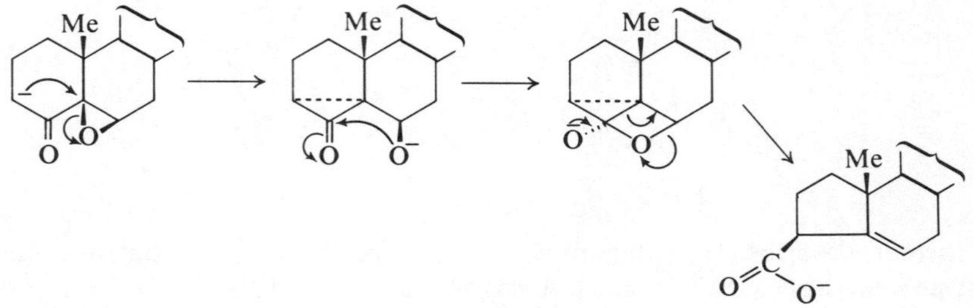

Scheme 25

1-Hydroxy-6-isopropylbicyclo[3.2.0]heptan-7-one (**290**), which appears to be in equilibrium with 1-hydroxy-7-isopropylbicyclo[4.1.0]heptan-2-one (**291**), has been formed from the corresponding 6-*exo*-chlorobicycloheptan-7-one (**289**) by ciné-substitution which presumably involves near-concerted loss of the diaxially disposed H(1) and Cl(6) from (**289**) to give the zwitterionic Favorskii intermediate which is then attacked either by solvent or by hydroxide ion at C(1) to yield the most stable enolate ion.[357] Base-catalysed rearrangement of 2β-hydroxy-1β-methyl-2α-vinyl-tricyclo[4.3.0]non-6-en-8-one was observed to yield a transannular reaction product, 8α-hydroxy-2-methyltricyclo[6.3.0.0^{1,8}]undec-2-en-4-one,[358] and the formation of the epimeric rearrangement product (**293**) by treatment of 6-*endo*-hydroxy-9-thiabicyclo[3.3.1]nona-3,7-dien-2-one (**292**) has been explained[359] in terms of a common enolate intermediate formed by retro-aldol cleavage of (**292**) and (**293**).

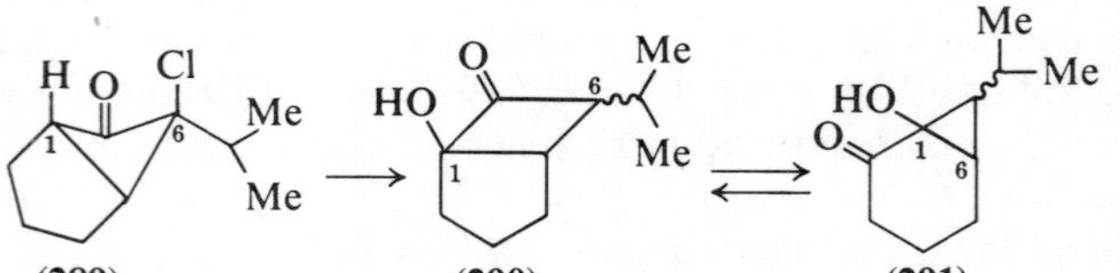

(289) (290) (291)

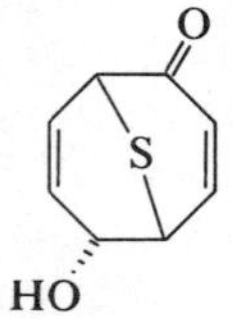

(292) 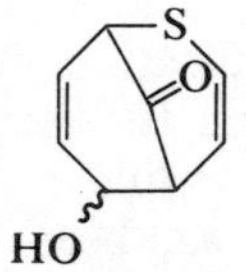 (293)

Cationic Rearrangements

A review-type discussion concerning degenerate equilibration *versus* degenerate rearrangement of carbocations has appeared,[360] as have several quantitative descriptions of the rearrangements of various carbonium ions. These include calculations of the activation energies for the degenerate rearrangements of 1,2,3,4,5,6-hexamethyl-1-phenylbenzenonium,[361] and 1,1,2-trimethyl- and 1,2-dimethyl-1*H*-acenaphthylenonium ions,[362] kinetic studies of the degenerate rearrangements of ($2H^+$)-biphenylenonium[363] and 1-benzyl-1,2,3,4,5,6-hexamethylbenzenonium ions,[364] and the establishment of relationships between the structures of rearranging carbonium ions and the rates of [1,2]-phenyl[365] and [1,2]-hydrogen shifts.[366] Rates for degenerate [1,2]-hydride and [1,2]-methide shifts have been measured from the high-field ^{13}C-NMR spectra of a variety of tertiary alkyl cations.[367]

Ab initio calculations with minimal and split-valence basis sets have been carried out for the open and bridged structures of the methane- and benzene-diazonium ions.[368] Interestingly, NMR data for the 1-methylcyclobutyl cation are not in accordance with a classical 1-methylcyclobutyl cation containing an sp^3-hybridized carbenium centre. Rather they indicate a σ-delocalized 1-methylcyclopropylcarbinyl cation containing two different methylene groups, one of which is highly shielded, suggesting a five-co-ordinated carbon.[369] A description of the very unusual features of both the ^1H- and ^{13}C-NMR spectra of the cyclodecyl cation has led to the conclusion that this cation possesses a 1,6-μ-hydrido-structure,[370] while a study of the temperature-dependent ^1H- and ^{13}C-NMR spectra of the α,1-dimethylcyclopropylcarbinyl cation shows that at temperatures < -100 °C it is a static carbocation, at -70 °C it undergoes equilibration through the 1,2-dimethylcyclobutyl cation, whereas at -39 °C it rearranges irreversibly to the 1,1,3-trimethylallyl cation.[371] A similar study of methyl-substituted 2-arylbicyclo[2.2.1]hept-2-yl cations shows[372] them to rearrange at elevated temperatures to 2-aryl-5,5-*endo*,6-trimethylbicyclo[2.2.1]hept-2-yl cations, thereby relieving the unfavourable C(1)methyl–C(2)aryl steric interaction present in the parent cations. At higher temperatures, the bicyclic skeletons collapse to monocyclic allylic cations. The technique of photodissociation spectroscopy indicates that the gas-phase protonated hexamethyl(Dewar benzene) cation rearranges within less than a few seconds to the protonated hexamethylbenzene structure.[373] Several 8,8-dimethyl-substituted homotropylium cations have been prepared and shown[374] to undergo a variety of molecular rearrangements including circumambulation, while two unexpected skeletal reorganizations of 1,7-bridged homotropylium cations, which too can reasonably be explained by circumambulatory rearrangements, have been discovered.[375]

A detailed study of multiple ring-expansion and ring-contraction rearrangements in observable cycloalkyl cations has been made[376] and, in an attempt to learn more of the structural and electronic requirements for cyclopropylcarbinyl–cyclopropylcarbinyl cation rearrangements, the effects of individual 1,2,6, or *exo*-7-methyl substitution on the rearrangement of the bicyclo[4.1.0]hept-2-yl system, namely **(294)**$\rightarrow$**(295)**, have been studied;[377] results appear to indicate that significant amounts of charge delocalization at C(1) are critical for the rearrangement to take place. Any substituent effects that increase charge elsewhere and thereby decrease charge delocalization at C(1) decrease the likelihood of rearrangement. Such findings suggest that the most favourable pathway for the rearrangement is *via* a puckered cyclobutyl (or related delocalized) species (**296**) rather than through another cyclopropylcarbinyl species such as (**297**). The solvolytic behaviour of the strained 1-, 4-, and 5-methylbenzo[*b*]bicyclo[3.1.0]hex-2-en-4-yl cations has also been examined.[378]

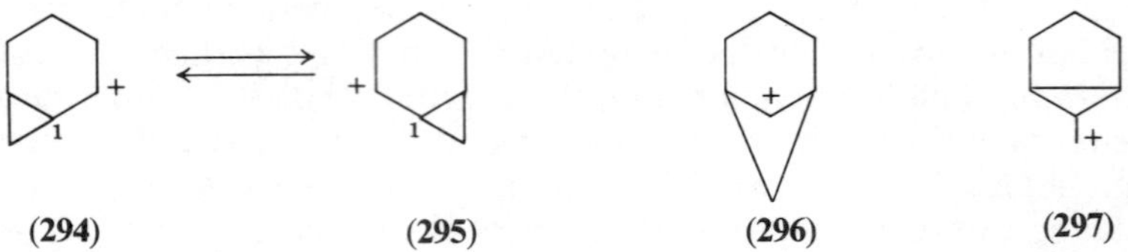

(**294**) (**295**) (**296**) (**297**)

The permethyl-substituted $AlCl_3$ σ-complex of cyclobutadiene (a cyclobutenyl cation) is involved in a dynamic process which has been investigated[379] by NMR spectroscopic techniques and found to consist predominantly, if not exclusively, of consecutive [1,2]-shifts of the $AlCl_3$ group, and the mechanism of interconversion of the dynamic η^1- and η^3-benzylbis(triethylphosphine)palladium(II) cations has been investigated[380] by NMR spectroscopy.

The exclusive formation of only eight-membered ring products on solvolysis of cyclo-octatetraenyl trifluoromethanesulphonates is presented[381] as evidence for the generation of the intermediate cyclic vinyl cation (**298**), and its capture by solvent.

The rearrangement of 1-phenylallyl cations to indanyl cations in superacidic media has been described.[382] Results from the acetolysis of *cis*- and *trans*-2-phenyl-1,2-di-*p*-tolylvinyl-2-^{13}C bromides point to[383] the formation, without phenyl participation, of a free 2-phenyl-1,2-di-*p*-tolyvinyl cation which can then undergo competitively degenerate rearrangement by a [1,2]-phenyl shift and solvent capture to give product – the less nucleophilic the solvent, the greater the extent of isotopic scrambling. Furthermore, a recent study of the solvolysis of triphenylvinyl-2-^{14}C bromide in aqueous acetic acid and in 2,2,2-trifluoroethanol has provided much additional information on the degenerate rearrangement arising from [1,2]-phenyl shifts in the triphenylvinyl cationic system.[384] An important rôle is attributed to ion pairs, ion-pair return being shown to occur in the reaction of substrate with 2,2,2-trifluoroethanol, but not with aqueous acetic acid. A detailed study[385] of the electrophilic addition of benzenesulphenyl chloride to olefinic systems produced data that are consistent only with a [1,2]-aryl migration in a fast non-limiting step, and a two-step mechanism involving an intermediate with a low degree of positive charge development on the carbon atoms in the rate-determining step. It has been demonstrated[386] that phenonium ions can undergo sequential rearrangements in competition with attack by external nucleophiles, while the [1,3]-rearrangement of hydroxypiperidinones, *e.g.* (**299**) to (**302**) and (**303**), has been shown[387] to involve protonation to afford (**300**) and then phenyl migration *via* (**301**).

The tool of increasing electron-demand, combined with ^{13}C-NMR as a probe, has been used[388] to examine several 2-arylnorbornyl cations. Interestingly, the data cannot be interpreted in terms of equilibrating classical cations, but are readily interpreted in terms of σ-bonding in the more electron-demanding derivatives. Comparison of these results with related solvolysis studies has allowed a comparison of the electron demands of the transition states with those of the carbocations to which they are presumed to lead. Results obtained from an acetolysis of *exo*- and *endo*-5,6-bis(methylene)-2-norbornyl *p*-bromobenzenesulphonates are

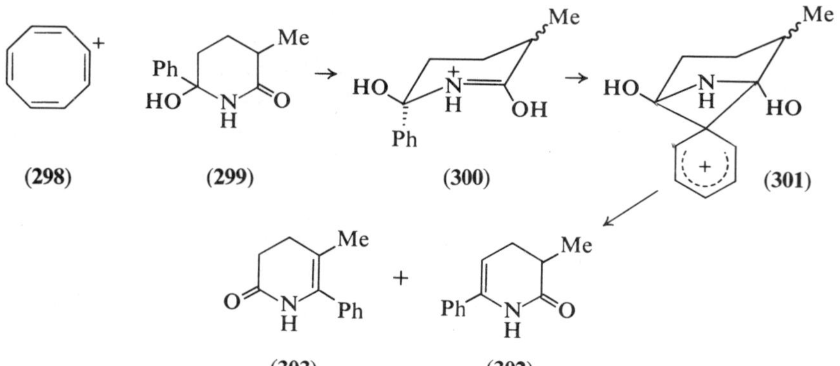

(298) (299) (300) (301)

(303) (302)

most simply explained by invoking the intermediacy of the symmetrical cation **(304)** (or corresponding symmetrical ion-pairs) which may be common to the solvolysis of both the *exo*- and *endo*-derivatives.[389]

A report has appeared[390] showing how the presence of a 7-silyl group encourages carbonium-ion rearrangements in the norbornyl system so that 7-functionalized norbornenes are the products. Thus, methyl 7-(trimethylsilyl)bicyclo[2.2.1]hept-5-ene-2-carboxylate **(305)** and its epoxide **(306)** rearrange on treatment with Br_2 or BF_3 to give 7-bromonorbornene (**308**; X = Br) and norbornen-7-ol (**308**; X = OH), respectively. In both cases the silyl group encourages rearrangement because the cations (**307**; X = Br) and (**307** X = OBF_3^-) are stabilized by the neighbouring

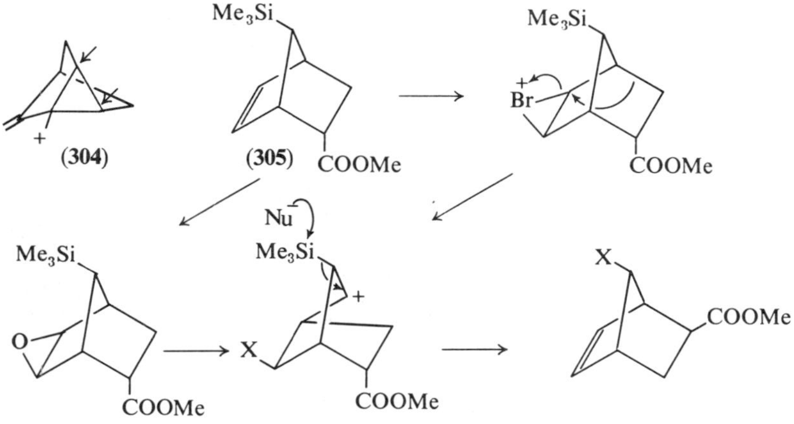

(304) (305) (306) (307) (308)

C—Si bond and the outcome is controlled by the ready loss of the silyl group. Evidence has been presented[391] to support the intermediacy of a carbonium ion in the rearrangement of enol silyl ether epoxides to α-hydroxy-ketones.

Various 1,4,5,6-tetramethylbicyclo[2.2.0]hex-5-ene-2-*endo*-carboxylic esters have been reported to undergo $AlCl_3$-induced rearrangement to a number of bicyclo-[3.1.0]hexenecarboxylic esters.[392] The isomerization seems to proceed by stereo-specific *endo*-protonation and subsequent rearrangement to 2-methoxycarbonyl-1,4,5,6-*exo*-tetramethylbicyclo[3.1.0]hex-4-yl carbocations.

A study of the reactions of several isomeric benzobicyclo-octadienyl alcohols and dibenzobicyclononatrienyl alcohols with triphenylphosphine and carbon tetrachloride has demonstrated[393] that the alcohol-to-chloride transformations are much more complex than originally proposed; direct displacements, allylic rearrangement displacements, and Wagner–Meerwein rearrangement displacements all occur. Halofluorination studies of norbornadiene have shown[394] that only the *exo*-β-halocarbocation (and not its *exo*-isomer) undergoes Wagner–Meerwein rearrangement. An attempt to establish whether enantioselective syntheses could be induced in a chiral medium by employing high pressures proved successful by the isolation of optically active 2-(4-methoxyphenyl)-2-phenylpropiophenone from the Wagner–Meerwein rearrangement of racemic 1-(4-methoxyphenyl)-1-methyl-2,2-diphenyloxirane in chiral solvents at high pressure.[395]

Derivatives of 2-azabicyclo[2.2.2]oct-5-enes substituted on nitrogen by a hydroxy, tosyloxy, or acyloxy group have been rearranged to 1-azabicyclo[3.2.1]-oct-2-enes by a concerted Wagner–Meerwein rearrangement;[396] addition of norbornene to 2,4-dinitrophenylsulphenyl chloride to give 2-*exo*-acetoxy-7-*anti*-(dinitrophenylthio)norbornane takes place through a Wagner–Meerwein rearrangement accompanied by a 1,6-hydride shift;[397] treatment of 5-methylenebicyclo[2.2.2]oct-2-ene with organic acids has been shown to yield Wagner–Meerwein rearrangement products.[398] Photo-Wagner–Meerwein rearrangements of halogenated dibenzobicyclononatriene systems have been reported[399] and a Wagner–Meerwein rearrangement appears to be involved in the ring-expansion of 4*H*-cyclopenta[*def*]phenanthrene to pyrene.[400] Moreover, it has been shown[401] that the highly strained σ-bond in pterodactyladienes (**309**; R = OH, R′ = OTs) is prone to Wagner–Meerwein rearrangement with virtually concomitant 1,2-shift of the leaving group, thereby offering an easy access to tetracyclo[$4.4.1.0^{2,5}0^{7,10}$]-undeca-3,8-dienes (**310**). An interesting communication has appeared[402] reporting a unique stereoselective method for synthesizing 1,2,3-trisubstituted cyclopentanes (**313**) based on the Wagner–Meerwein rearrangement of the 7-oxabicyclo[2.2.1]-hept-2-yl cation (**311**) into the 2-oxabicyclo[2.2.1]hept-3-yl cation (**312**), by virtue of the participation of the oxygen atom (see Scheme 26).

The results of a solvolytic study of *exo*- and *endo*-bicyclic[3.2.1]oct-3-en-2-yl *p*-nitrobenzoates have been explained[403] by assuming that the first-formed ion-pair intermediate is unsymmetrical.

Evidence has been obtained[404] for the formation of a γ-rearranged allylic phosphonium salt in the reaction between an allyl bromide derivative and triphenylphosphine. Other examples of allylic rearrangements have been reported; these include rearrangements in the 6,6-dimethylbicyclo[3.1.1]heptane series,[405] rearrangements of *p*-mentha-1(7),8-dien-2-ols involving a synthesis of perilla alcohol,[406] while complete allylic rearrangement is achieved on reacting allyl halides with a new alkylating agent of the type $RCu\cdot BF_3$.[407] Bromination of α-benzylidene-γ-butyrolactone also proceeds with allylic rearrangement to give 3-(α-bromobenzyl)-2(5*H*)-furanone.[408]

In direct contrast to the deamination of 2-hydroxy-2-phenyl-2-(2-methoxyphenyl)ethylamine, no evidence was found[409] for *o*-MeO-5 participation in the deamination of 2-phenyl-2-(2-methoxyphenyl)ethylamine. Acid-catalysed rearrangements of the spiro-epoxy-alcohol (**314**),[410] leading to ring-expanded keto-alcohols (**315**) and (**316**), are reminiscent of pinacol and Demjanov rearrangements, while gas-phase pinacol-type rearrangements have been reported[411] for the

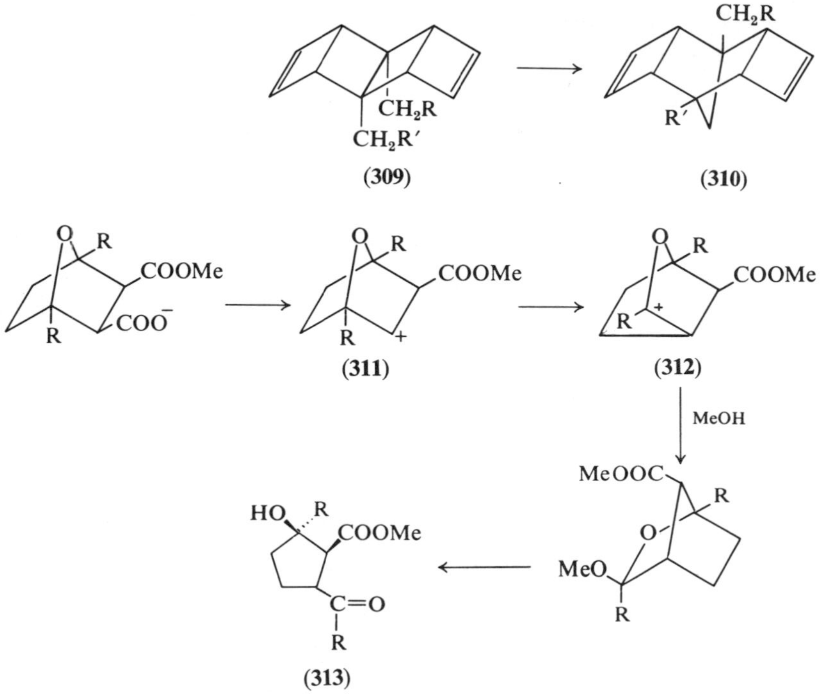

(309) (310)

(311) (312) (313)

SCHEME 26

unimolecular decomposition of some isomers of $C_4H_9O^+$. For two such isomers, *viz.* protonated butanone and protonated isobutyraldehyde, equilibration *via* a pinacol-type rearrangement precedes dissociation. ^{1}H-NMR spectroscopy has revealed the nature of the intermediates participating in the superacid-catalysed isomerization of allyl alcohols into aldehydes.[412]

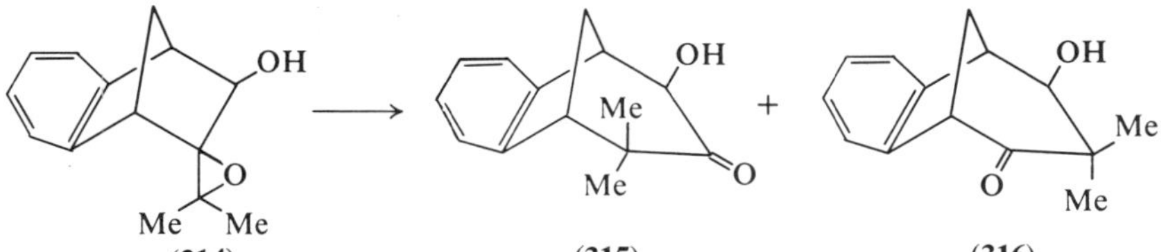

(314) (315) (316)

Various pinacolic and carbonium-ion rearrangements have been observed[413] on treating (+)-carane-3β,4α-diol with sulphuric acid, a convenient new procedure for achieving pinacolone rearrangements by using perfluorinated resin-sulphonic acids as catalyst has been reported,[414] while ferric chloride adsorbed on silica gel has also been utilized[415] as an effective reagent for pinacol- and acyloin-type rearrangements. A mechanism involving an acyloin rearrangement, followed by an intramolecular nucleophilic displacement, has been proposed[416] to explain the base-induced rearrangement of sphaerococcenol A and bridgehead hydroxyl-substituted benzobicyclo[3.2.1]-octenes and -octadienes have been prepared by the easy acyloin rearrangement of a benzobicyclo[2.2.2]octene.[417] A variety of bicyclo-[3.2.1]oct-3(and 6)-enes has also been prepared[418] by the acid-catalysed rearrangement of bicyclo[2.2.2]octene derivatives.

The rates of hydrolysis of a series of isomeric vinyl ethers have been measured.[419] The results show that the *cis*- and *trans*-isomers do not interconvert under the hydrolysis conditions, implying that formation of the intermediate alkoxy-carbonium ion in these reactions is not reversible. A mechanism involving a rate-limiting hydride migration to an α-alkoxy-cation has been suggested[420] for the acid-catalysed conversion of α-hydroxy-ketals into alkoxy-ketones, while a one-pot reaction between 1,2- or 1,3-diols, sulphonic acids, and carboxylic acid derivatives has been used[421] to prepare mixed carboxylate–sulphonate esters by a process believed to involve intermediate 1,3-dioxolan-2-yl or 1,3-dioxan-2-yl cations.

An ion-cyclotron-resonance study of the low-pressure gas-phase reactions of methoxymethyl cation ($CH_3OCH_2^+$) with alcohols, thiols, and amines has been described.[422] The results underscore the difference in behaviour of gaseous and solvated ions; indeed it has been suggested that their structures are not strictly comparable and that the reactions of the gaseous ions can be attributed to the effects of internal solvation and product stability. Contrary to a recent report, it has been shown[423] that the "perpendicular" hydroxymethyl cation (**317**) is not responsible for the high isomerization barrier between "CH_3O^+" and "$CH_2{=}\overset{+}{O}H$". A self-consistent explanation of the main reactions undergone in metastable transitions by some isomers of the $C_4H_9O^+$ ion has been given in terms of the potential-energy surfaces over which the reactions are considered to occur.[424] The rearrangements and fragmentations of $[C_2H_5S]^+$ ions have been examined,[425] as has the mechanism of hydrogen migrations in the molecular ions of 1,3-diphenylpropane,[426] and the mechanism of the isomerization of the *n*-heptyl ion in the mass spectrum of *n*-heptyl iodide has been simulated by a computer model which takes carbon atom rearrangements into account. It has been shown[427] that isomerization mechanisms in the gas or solution phase are similar and involve protonated cyclopropane intermediates.

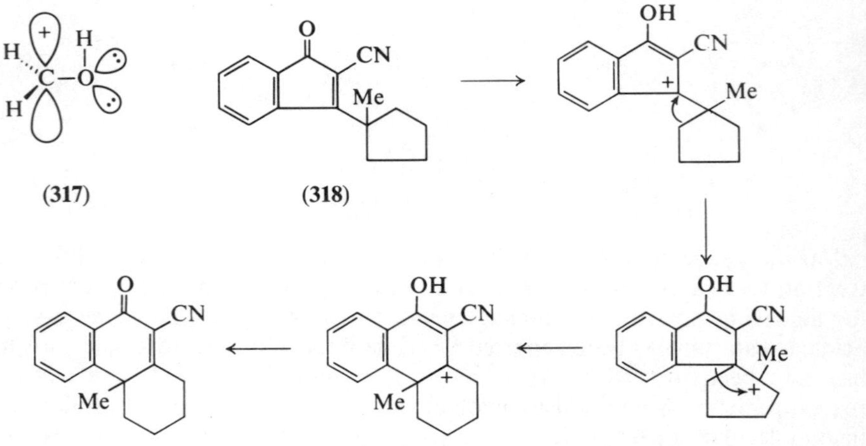

(**317**) (**318**) (**319**)

SCHEME 27

The possibility of preparing partially saturated phenanthrenes with 2-aminoethyl side-chains at the 4a-position (key precursors in the synthesis of the morphinan ring system) by a complex carbonium-ion rearrangement of 2-cyano-3-(1-methylcyclopentyl)indenone (**318**) has been realized.[428] Thus, (**318**) with sulphuric acid gives 1,2,3,4,4a,9-hexahydro-4a-methyl-9-oxophenanthrene-10-carbonitrile (**319**) by a rearrangement involving migration of a cyclopentyl carbon followed by a phenyl migration (see Scheme 27). An account has appeared[429] of a new general and efficient method for cyclopentenone synthesis principally based on the thermal conrotatory ring-closure of a chloropentadienyl cation followed by hydrolysis. Although heterolytic cleavage of selenoxides to carbonium ions has not been heretofore recognized, a recent report demonstrates that under special circumstances such a process can be very favourable indeed.[430] Furthermore, this new rearrangement has been used to expand the scope of the geminal alkylation procedure and the γ-butyrolactone annelation.

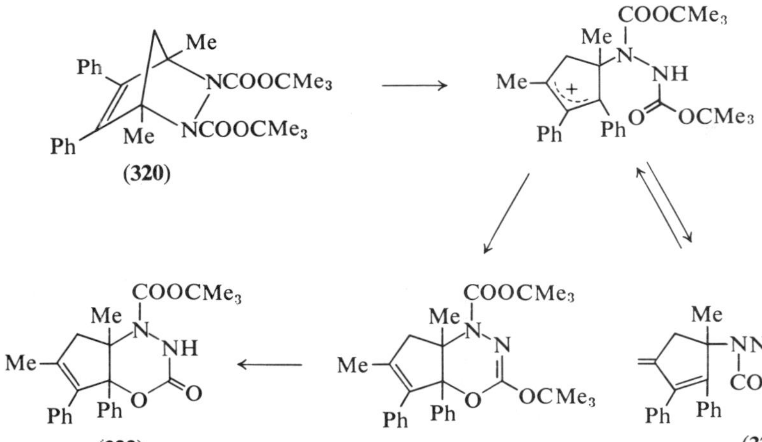

Cationic rearrangement have also been implicated in the novel acid-catalysed isomerization of the adducts of 1,4-dimethyl-2,3-diphenylcyclopentadiene and dialkyl azodicarboxylates, namely (**320**), to a mixture of the exocyclic alkene (**321**) and the oxadiazinone (**322**),[431] in the acid-catalysed rearrangement and cyclization of the cycloadducts of 2-acetonyl-5-benzylfuran with alkyl acetylenedicarboxylates to furo[2,3-*f*]naphtho[2,3,4-*i,j*]phthalazine derivatives,[432] in the salicyclic acid-catalysed rearrangements of α- and β-phellandrenes and α-terpinene,[433] and in the nitrosation of *cis*- and *trans*-1,3,3-trimethyl-2-oxabicyclo[2.2.2]octan-6-amine to 1-[3-(1-hydroxy-1-methylethyl)cyclopentyl]ethanone and *cis*-4,7,7-trimethyl-6-oxabicyclo[3.2.1]octan-4-ol, respectively.[434] A mechanism involving a stabilized carbocation has been proposed[435] to account for the ene epoxide-to-dihydrofuran rearrangement of vinyl epoxides derived from bicyclo[4.2.0]octa-2,4,7-trienes; and skeletal rearrangements of carbonium ions are implied[436] in the photosensitized oxygenation of olefins such as α-vinylcyclopropane.

Rearrangements in Polycyclic Systems

A review which details the pathways of the acid-catalysed, multiple-step isomerizations of tetramethylenenorboranes has appeared.[437] 1,3-Bishomoadamantane has been synthesized *via* the ring expansion of homoadamant-4-en-2-one,[438] and 9-substituted derivatives of triamantane have been prepared by the $AlBr_3$-catalysed rearrangement of $C_{19}H_{26}$ polycycles.[439] It has been suggested[440] that the acid-catalysed isomerization of 2-protoadamantenone (**323**) to 8,9-dehydroadamantan-2-one (**326**) involves the initial protonation of the carbonyl group in (**323**) to give the homoallyl cation (**324**) which then undergoes a homoallyl–cyclopropylcarbinyl rearrangement to cation (**325**) and ketone (**326**). Mechanisms have been proposed to account for the degenerate rearrangements of 8,9-dehydroadamant-2-yl and 2,5-dehydro-4-protoadamantyl cations;[441] the rearrangements of tricyclo[4.1.0.0^{2,7}]-heptenyl cations have been studied;[442] and a convenient and simple technique has been devised for low-temperature synthesis of reactive carbocations in superacid media.[443]

(**323**) (**324**)

(**326**) (**325**)

The ease of rearrangement of (**327**) to (D_3)-trishomocubane (**329**) has been rationalized[444] by considering the various cations derivable from (**327**); see (**328**). Cations at C(3) and C(4) are analogous to the 1- and 7-norbornyl cations, respectively, and should be quite unfavourable. Cations at C(1) and C(2) would appear to be even more unstable. A cation at C(9) may not be prohibitively strained, but all possible 1,2-alkyl shifts lead to products having four-membered rings. Thus, the most likely positively charged species to be involved is at C(8). Of the four possible 1,2-alkyl shifts from this cation, only paths *a* and *b* do not form new four-membered rings. However, it has been calculated that the product from path *b* is considerably less stable than (**397**), so this path should not compete with path *a* which leads to (D_3)-trishomocubane with considerable strain relief.

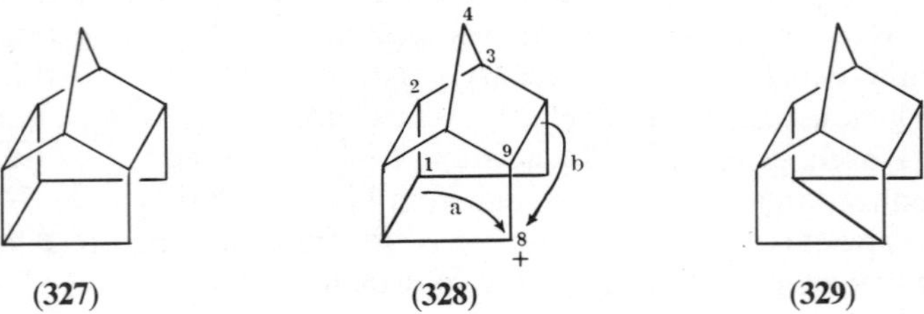

(**327**) (**328**) (**329**)

The formation of the novel lactone (**331**) on ceric ion oxidation of diketone (**330**; R = R′ = H) is considered[445] to involve a cyclobutyl-to-cyclopropylcarbinyl type of carbonium-ion rearrangement as depicted in Scheme 28, while a deep-seated rearrangement of the hexacyclic dione (**330**; RR′ = —CH=CH—CH=CH—) to benzo-annulated tricyclo[5.2.1.0^{3,8}]decane and tricyclo[4.2.1.0^{2,5}]nonane ring systems, under the same conditions, has been described.[446] Trishomocubane derivatives have been prepared[447] by the interaction of diketone (**330**; R = R′ = H) and chlorosulphonic acid; and the characterization[448] and quantitative analysis[449] of two $C_{11}H_{11}$ homoallyl cationic rearrangements have been described.

(**330**) (**331**)

SCHEME 28

Acetolyses of bicyclo[3.2.0]hept-6-en-2-yl and bicyclo[3.2.0]hept-2-yl derivatives have been examined[450] in an attempt to obtain a direct insight into anchimeric assistances of cyclobutane and cyclobutene rings. Apparently homoallylic participation (**332**) is more effective for stabilization of the transition state than σ-participation (**333**), a result which is contrary to that observed in the bicyclic system.

A convenient synthesis of hypostrophene, *i.e.* tetracyclo[5.3.0.0^{2,6}0^{3,10}]deca-4,8-diene (**334**), has been developed and a possible correlation between the high-lying σ-orbitals present in (**334**), its marked preference for through-bond instead of

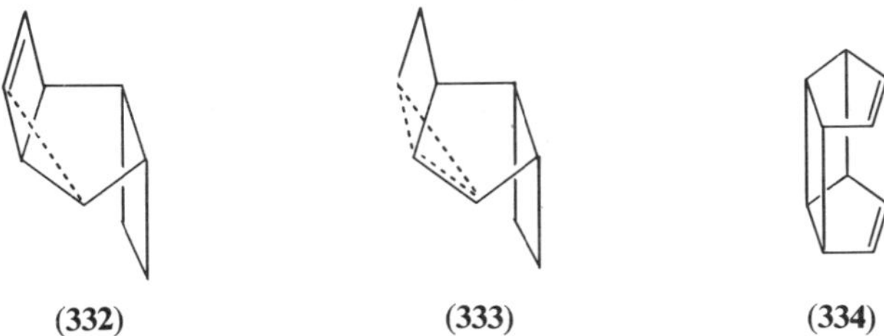

(**332**) (**333**) (**334**)

through-space interaction, and strain relief have been presented[451] in rationalization of its susceptibility to rearrangement under electrophilic conditions. Moreover, the findings obtained[452] from a solvolytic study of hypostrophene derivatives demonstrate that the frequently observed ability of a proximate double bond to engage in transannular participation can be overridden by lateral σ-bond participation while a comparison of the results of a similar study[453] of *syn*- and *anti*-tricyclo[4.2.0.0^{2,5}]octa-3.5-diene *p*-toluenesulphonates, with that available from other cyclobutane-containing sulphonate esters, has allowed conclusions to be made concerning the geometric requirements for effective intervention of bicyclobutonium ion. The extensive skeletal reorganization observed[454] during the solvolysis of *exo*- and *endo*-3-tosyloxytricyclo[4.2.1.0^{2,5}]nonane derivatives illustrates further the importance of orbital alignment in the ground state for participation in the course of a solvolysis reaction. Rearrangement reactions of derivatives of *syn*-tricyclo[4.2.1.1^{2,5}]decane have been described[455] and it has been proposed that the driving force behind the entire range of rearrangements examined is steric strain relief, since they involve passage of a boat cyclohexane conformation to a chair form.

The unusual acid-catalysed isomerization of tricyclo[6.3.0.0^{1,6}]undecan-2-one (**335**) to tricyclo[3.3.3.0]undecan-2-one (**336**)[456] is thought to be related to the biosynthesis of the sesquiterpene modhephene, while isobarbatene, the stable tricyclic end-point of the bazzanene–barbatene sesquiterpenes, has been obtained from barbatene by a bicyclo-[3.2.1]→-[2.2.2]→-[3.2.1] transformation followed by an *exo*-methyl shift and deprotonation.[457]

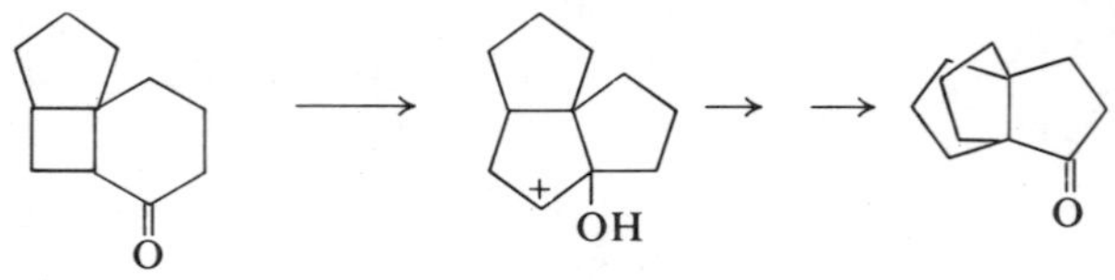

OH

(**335**) (**336**)

The relative kinetic and thermodynamic factors controlling the Wagner–Meerwein inter-relationship of perhydromethenocyclopenta[a]pentalenyl brosylates have been determined,[458] pathways have been outlined for the acid-catalysed rearrangement of *cis*-2,3-trimethylenebicyclo[2.2.2]octane to 4-homoisotwistane,[459] and concomitant ring enlargement and skeletal isomerization in 4-homobrendylcarbinols have been interpreted[460] in terms of the relative stabilities of the intermediate bridged cations. The factors controlling formation of the products variously obtained from the acetolysis of 8-bromoneoisolongifolene have been discussed and the possible rôle of energetics of the incipient carbonium ion at the migration origin has been highlighted.[461]

Tricyclo[4.1.0.0^{2,7}]hept-3-enes have been prepared by thermal rearrangement of 7-*endo*-chloro- and 7-*endo*-bromo-tetracyclo[4.1.0.0^{2,4}0^{3,5}]heptanes,[462] solvolysis of dibenzobullvalene bromide,[463] and chloro-derivatives of dibenzobicyclo[2.2.2]-octatriene[464] have been investigated, and a mechanism (Scheme 29) has been tentatively proposed[465] for the thermolysis of the bicyclo[4.2.1]nona-2,4,7-trien-9-yl system (**337**). In this mechanism, the interaction between the monoene and diene begins synchronously with C—O bond ionization as partial positive charge develops at C(9). Subsequently, sufficient interaction is achieved between C(2) [C(5)] and C(8) [C(7)] to generate an intermediate like (**338**) (not a discrete cation), which then rapidly collapses into *endo*-(**339**), the observed stereospecific product.

It has been shown[466] that the stability of benzvalene derivatives to acid and to thermal treatment is enhanced by the presence of strongly electron-withdrawing groups at the carbon–carbon double bond of the benzvalene skeleton; two types of rearrangement have been observed[467] on treating multilayered [2,2]paracyclophanes with Friedel–Crafts acids, depending on the catalyst used, and an unusual acid-catalysed rearrangement of a β,γ-unsaturated ketone has been used[468] to convert a hexahydrophenanthrene into a dihydroanthracene.

Rearrangements in Natural-product Systems

These studies are classified thus because they involve rearrangements of natural-product systems that probably proceed *via* carbonium ions.

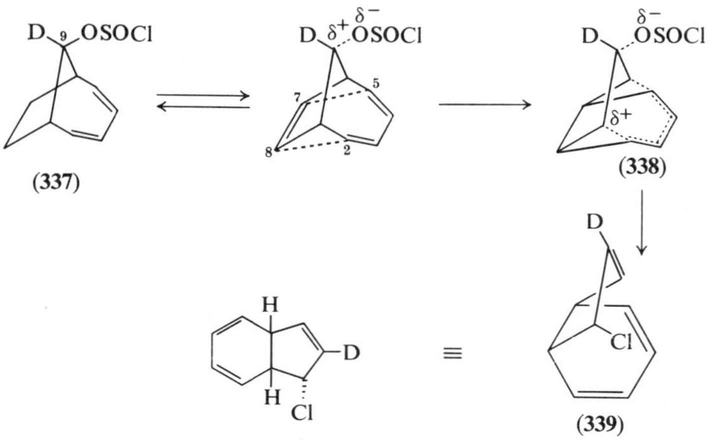

(337)

(338)

(339)

SCHEME 29

A study of the biosynthesis of illudin sesquiterpenoids from [1,2-C^{13}(2)]acetate has established[469] that the illudin-M-to-isoilludin-M rearrangement is a simple 1,2-methyl shift rather than a pinacolic shift; further, transformation of $\Delta^{7(13)}$-protoilludene (**340**) to hirsutene (**341**), through a process that involves a hitherto unknown biogenetic type of skeletal rearrangement as the key step (see Scheme 30),

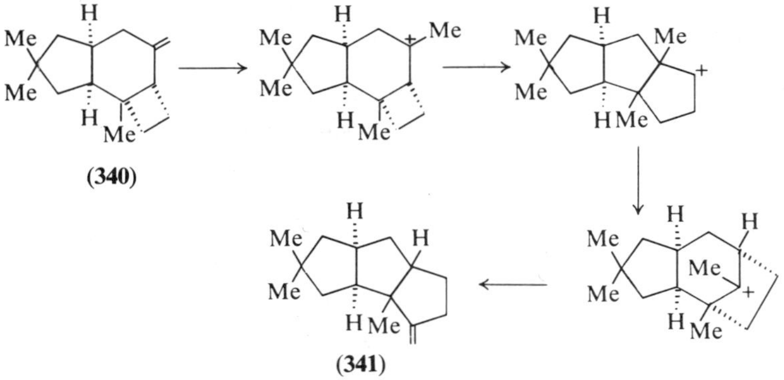

(340)

(341)

SCHEME 30

has been described.[470] A Wagner–Meerwein-type of rearrangement is involved as a key step in a recently reported synthesis of (−)-hibaene.[471] The mechanism of the geranyl–geranyl pyrophosphate cyclization in fusicoccin biosynthesis has been studied[472] with ^{13}C-NMR spectroscopy by use of deuterium labelling, and the first non-enzymic generation of cations (**342**) and (**343**), possible intermediates in the intricate series of cyclizations and rearrangement steps leading to a variety of polycyclic sesquiterpenes, has been achieved and their reactivity has been examined.[473] Phosphoric acid (85%) has been established as a medium particularly well suited for the examination of carbocations in processes initiated by ionization, and the fate of the α-terpinyl cation in this solvent has been examined.[474] The fate of various monoterpene alcohols and their acetates in aqueous citric acid has also been investigated.[475]

(**342**) (**343**)

(X = + charge or its functional equivalent)

The stereochemistry of hydrogen migration from C(24) to C(25) during the biosynthesis of isofucosterol from lanosterol has been established,[476] and the ^{13}C-labelling patterns of β-sitosterol isolated from *Isodon japonicus Hara* tissue cultures fed with [4-^{13}C] mevalonic acid and [1,2-^{13}C]acetate, provide confirmatory evidence for the postulated backbone rearrangement during biosynthesis of β-sitosterol.[477] Proton-catalysed rearrangement of 4,4-dimethyl-5α, 13α-androst-7-ene has been shown to proceed with methyl migration to give 4,4,14α-trimethyl-18-nor-5α-androst-13(17)-ene.[478] The mechanism of this reaction has been compared with the rearrangement of various other triterpenoids. Likewise, protic and Lewis acids cause partial backbone rearrangement involving the C—D ring junction in the normal steroid series with ring contraction and formation of 12,14α-cyclo-12,13-seco-5α-cholest-13(17)-ene compounds.[479] Further backbone rearrangements of the steroid skeleton have been reported; oleanolic acid has been transformed by methyl-group shift and by aromatization into 3-*O*-acetyl-28-norursa-9(11),12,15,-17,19,21-hexaen-3β-ol,[480] the acid-catalysed rearrangements of cholesterol, cholest-5-ene,[481] and friedo-oleanenes[482] have been described, and solvent effects on the backbone rearrangement of 3β,4β-epoxyfriedelane have been studied.[483] Labelling studies have shown[484] that the acid-catalysed rearrangement of 2α-hydroxycholest-4-en-3-one to the aromatic 2-methoxy-4-methyl-19-norcholesta-1,3,5(10)-triene occurs by a dienone–phenol-type of rearrangement, thus supporting previous mechanistic proposals describing the reaction, while oxidation of 11-oxolanostan-3β-yl acetate with selenium dioxide in acetic acid is accompanied by a C-nor-D-homo-rearrangement of the steroid skeleton with aromatization of ring D and formation of 9β-hydroxy-4,4,15-trimethyl-11-oxo-14(13$\rightarrow$12)*abeo*-5α-cholesta-12,14,16-trien-3β-yl acetate.[485]

A new structure, namely (**344**), has been proposed[486] for the transition state in the BF_3-catalysed D-homoannulation of 17α-hydroxypregnan-20-ones; a proton as the species responsible for holding the two oxygen atoms of the ketol in a *syn*-orientation, and catalysis by co-ordination of each oxygen atom with a molecule of Lewis acid, are the novel features in the proposal. Selection of C(16) as the

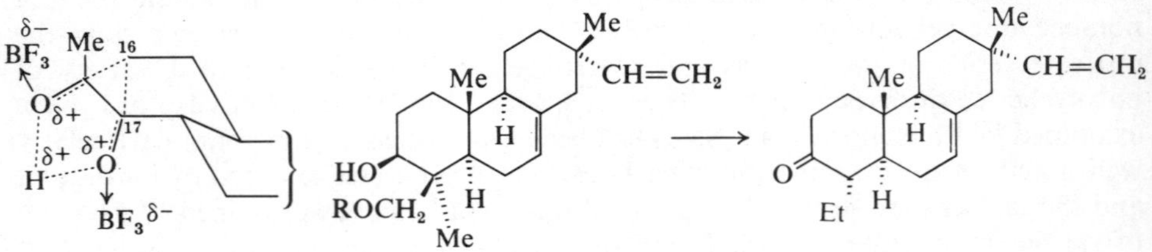

(**344**) (**345**) (**346**)

migrating centre is attributed to steric effects associated with the conformational rigidity of the hydrogen-bridged transition state.

It has been shown[487] that the α-epoxide derived from $BF_3 \cdot OEt_2$ treatment of *l*-abietic acid yields a methyl ketone by stereoselective methyl migration to the 13β-position, whereas the corresponding β-epoxide undergoes a methylene migration (ring expansion) to give cycloheptenone derivatives. The formation of ketone (**346**) as one of the products from the solvolytic rearrangement of virescenol B 19-tosylate (**345**; R = *p*-$MeC_6H_4SO_2$)[488] has been considered with much interest from a biosynthetic point of view since it represents a structural type found among natural pimaradiene-derived diterpenes.

A reinvestigation[489] of the acid-catalysed rearrangement of the dihydroxyacetone side-chain of hydrocortisone during ketal exchange has shown that the by-products obtained are not bisketals but instead result from Mattox rearrangement of an intermediate in the ketal exchange; and an examination of the acid-catalysed ring contraction of methyl 3,4-*O*-benzylidene-β-D-ribopyranoside to the diastereomeric forms of methyl 2,3-*O*-benzylidene-β-D-ribofuranoside has been described[490] as part of a programme concerned with the formation and migration of cyclic acetals of carbohydrates.

Careful labelling studies[491] have proved beyond doubt that the biosynthesis of natural porphyrins (and relatives) occurs by head-to-tail assembly of four porphobilinogen units to form an unrearranged bilane, or its enzyme-bound equivalent, which is then converted into uro'gen-III by an intramolecular process. Observed results further limit to two the processes that are possible for the rearrangement; that involving a protonated spiro-intermediate is preferred.

Finally, the *O*-cyclohexyl group has been designed[492] as a useful new protecting group that minimizes rearrangement of tyrosine ethers to 3-alkyl derivatives, a process that normally prevails during acidolysis in peptide synthesis.

Miscellaneous

Carbene–carbene rearrangements in solution have been reviewed,[493] [1,2]-shifts in alkoxycarbenes have been reported[494] to resemble in character the Wittig rather than the Wagner–Meerwein rearrangement, and the outcome of flash-vacuum pyrolysis of cycloalkylidene derivatives (**347**) has been rationalized[495] in terms of formation of cycloalkylidene carbenes (**348**) and their rearrangements to cycloalkynes, to bicyclic cyclopropenes, or to further products. It has been proposed[496] that a carbene–carbene rearrangement of phenyltrimethylsilylcarbene (**349**) to (**350**) followed by a conventional C—H insertion reaction explains the formation of 1,1-dimethyl-1-silabenzocyclopentene (**351**) on the gas-phase pyrolysis of phenyltrimethylsilyldiazomethane. Two competitive mechanisms, one concerted and one proceeding *via* a delocalized carbocation, have been proposed[497] in order to explain the structure of cyclic and linear products formed in the thermolysis of 2-sila- and 3-sila-6,6-dichlorobicyclo[3.1.0]hexanes. Contrary to expectations, β-oxoalkyltrimethylsilanes (**354**) are obtained when α-haloacylsilanes (**352**) are treated with Grignard reagents. The reaction is considered to proceed through initial removal of chloride ion from alkoxide (**353**) followed by migration of the trimethylsilyl group to the cationic site.[498]

In an attempt to develop synthetic operations which reverse the normal polar reactivity pattern of the carbonyl function, a general study of the 1,2- and 1,4-addition reactions of trialkylsilyl tervalent phosphorus esters to saturated and

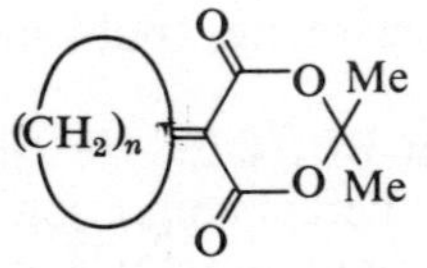

(347) 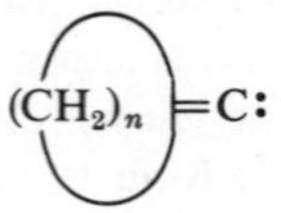 (348)

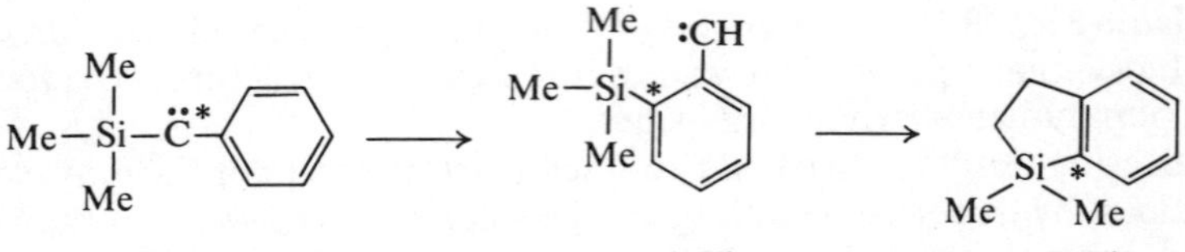

(349) (350) (351)

unsaturated aldehydes and ketones has been carried out and some of the mechanistic details of these addition reactions have been elucidated;[499] a general accounting of the observations associated with 1,2-addition is illustrated in Scheme 31.

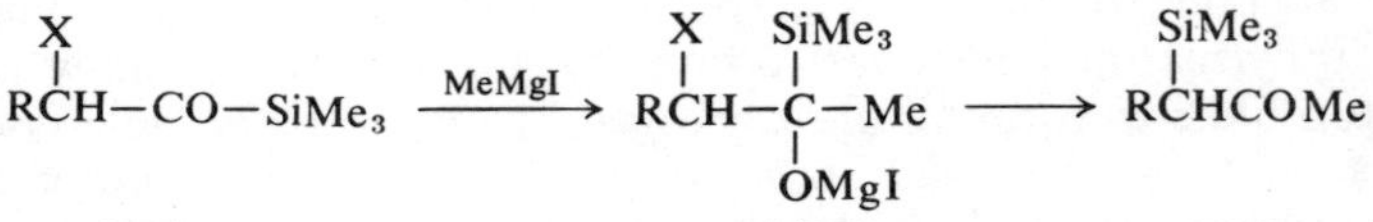

(352) (353) (354)

X=Cl or Br

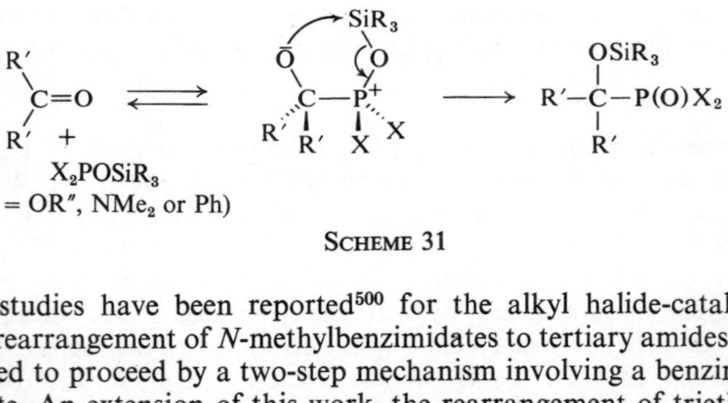

X_2POSiR_3
(X = OR″, NMe_2 or Ph)

SCHEME 31

Kinetic studies have been reported[500] for the alkyl halide-catalysed pseudomolecular rearrangement of *N*-methylbenzimidates to tertiary amides. The reaction is considered to proceed by a two-step mechanism involving a benzimidonium ion intermediate. An extension of this work, the rearrangement of triethyl *N*-phenylphosphorimidate to diethyl *N*-ethyl-*N*-phenylphosphoramidate, has likewise been subjected to a kinetic study;[501] the results are compared with the former rearrangement and discussed in relation to the ambident nucleophilic properties of phosphylamidates; thus it is suggested that alkylation occurs most readily at the oxygen atom of neutral phosphylamidates to give a phosphylimidate (kinetic product) which then rearranges in the presence of electrophilic catalysts to an *N*-substituted phosphylamidate (thermodynamic product). Kinetic data and an observed special salt effect have led to the conclusion[502] that the thermal isomerization of *O*-but-2-enyl *O*,*O*,-dimethyl thiophosphate to the corresponding *S*-but-2-enyl derivatives proceeds by a dissociative mechanism in which ion pairs are proposed as intermediates. The mechanistic aspects of the photorearrangements of phenylthiazoles have been discussed[503] and an interesting mechanism invoking the intermediacy of a tricyclic sulphonium cation has been proposed for the rearrangement.

A vicinal phenyl-group shift has been reported[504] in lead tetra-acetate and related oxidations of 5,5,5-triphenylpentan-1-ol, and a new mechanism of γ-cleavage has been characterized in the mass spectral fragmentation of aliphatic carbonyl compounds.[505] Moreover, an ion-dipole intermediate undergoing nucleophilic attack by water as the rate-determining step has been proposed as participant in the Meyer–Schuster rearrangement of tertiary arylpropargyl alcohols to α, β-unsaturated carbonyl compounds.[506]

An attempt to define the mechanism of the acid-catalysed conversion of 1-(*o*-chlorophenyl)-2,2-dichloropropan-1-ol and its derivatives into 1-(*o*-chlorophenyl)-2,2-dichloropropan-2-one has provided evidence to suggest the intermediacy of a chloronium ion in the process.[507] A more recent attempt to delineate this mechanism using improved solvolytic conditions has further implicated the existence of an intermediate halonium ion during rearrangement.[508] A report has appeared[509] pointing out that α-haloketones can rearrange by either a reductive dehalogenation–rehalogenation sequence involving loss of cationic halogen, or by an α'-enol allylic rearrangement involving loss of anionic halogen. Moreover, contrary to general opinion, it appears that the latter pathway is the common path for migration of bromine.

A further example of reversible intramolecular transfer of hydride from hydroxylated methine to carbonyl-carbon (in this case a [3,9]-shift) has been reported[510] in the base-induced rearrangement of *endo*-3-hydroxy-7,7-dimethylbicyclo[3.3.1]nonan-9-one to *anti*-9-hydroxy-7,7-dimethylbicyclo[3.3.1]nonan-3-one, and a possible mechanism for the formation of 5,5-diphenylhydantoins from the reaction of ureas and benzils has been deduced[511] from a kinetic study.

A novel Michael reaction that proceeds with migration of an alkyl group from boron to carbon,[512] and the first direct synthesis of carboxylic acids from organoboranes, by a process that similarly involves migration of an alkyl group from boron to carbon[513] (see Scheme 32), have been reported. The discovery of the oxidative rearrangement of boron-methylated alk-1-ene-1,1-diboronic esters to methyl ketones illustrates[514] the point that structures of organoborane intermediates

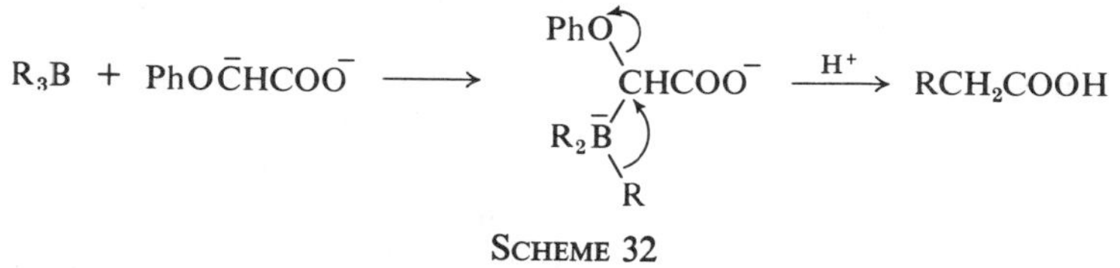

$$R_3B + PhO\bar{C}HCOO^- \longrightarrow \begin{matrix} PhO \\ \bar{R_2B}(R)-CHCOO^- \end{matrix} \xrightarrow{H^+} RCH_2COOH$$

SCHEME 32

$$\overset{+}{Li}[R_3B-C(OMe)=CH_2]^- \xrightarrow{I_2} \left[R_2B(R)-C(MeO)\overset{+}{\underset{I}{=}}CH_2 \right] \longrightarrow R_2B(R)(MeO)C-CH_2I \xrightarrow{-R_2BI} R-C(OMe)=CH_2$$

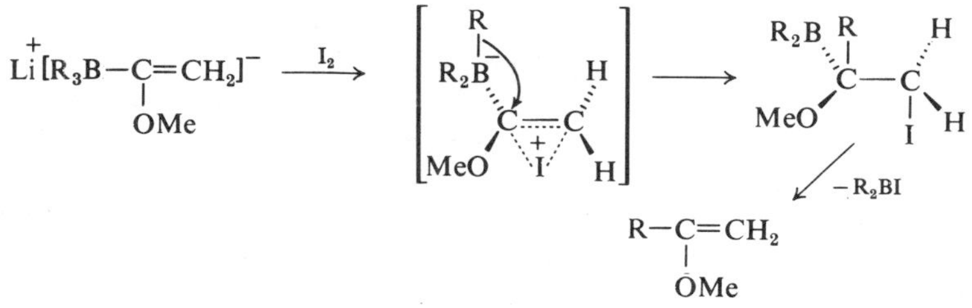

SCHEME 33

that contain reactive neighbouring groups cannot be assigned solely on the basis of their peroxide oxidation products. The products obtained from the reaction of 1-methoxyvinyl-lithium with trialkylboranes, followed by treatment with an electrophile, have been rationalized[515] by invoking electrophilic attack at the β-carbon of an alkenylborate salt and subsequent "α-transfer" (*i.e.* transfer of an alkyl group from tetraco-ordinate boron to an adjacent atom bearing a leaving group); see Scheme 33. A mechanistic investigation has been made[516] of the Lewis acid-catalysed alkyl migration reactions of $[RFe(CO)_4]^-$.

Rearrangements Involving Electron-deficient Heteroatoms

The Beckmann rearrangements of aldonitrones to carboxamides have been reviewed.[517] The effect of various substituted benzyl groups in the 1-position of the oximes of 2-tetralones and 2-indanones on the course of their Beckmann rearrangement, has been investigated.[518] In the former series, a trend to a preferred configuration of the oxime is found as the benzyl group becomes increasingly large, whereas 2-indanones form only one oxime regardless of the size of the substituent.

A variety of catalysts has been used to effect Beckmann rearrangements; thus strongly acidic styrene–divinylbenzene cation-exchange resin has proved an efficient catalyst in the liquid-phase rearrangement of cyclohexanone oxime to ε-caprolactam;[519] copper(II) acetate has been used to catalyse the rearrangement of aldoximes to amides,[520] and B_2O_3-based binary metal oxides have been used to catalyse the rearrangement of acetoxime;[521] further, 3-aza-A-homo-steroids have been prepared by an improved regiospecific Beckmann rearrangement of oxime *p*-toluenesulphonates of 3-oxo-4-ene steroids.[522] A study of the photo-Beckmann rearrangement of a variety of cholestanone oximes has established[523] that the chirality of the migrating group is retained in the rearrangement and that the predominant lactam in the product mixture is the one resulting from migration of the more substituted carbon centre. The path of this photo-rearrangement best fits a scheme involving transformation of excited lactams in a fully concerted manner.

N-Cyanato(diphenylmethanimine) isomerizes rapidly in a Beckmann rearrangement to *N*-phenylbenzimidoyl isocyanate,[524] and the unexpected formation of *N*-benzylethane-1,2-diamine and *N*-benzyl-*N'*-methylethane-1,2-diamine on reduction of *N*-benzyl-2-cyano-2-(hydroximo)acetamide has been accounted for[525] by invoking the migration of a cyano-group in a Beckmann rearrangement.

The 10π-electron heterocyclic methano-bridged system, 4-methoxy-1,6-methanoisoquinoline, has been prepared by utilizing a Beckmann rearrangement of tricyclo[4.3.1.0^{1,6}]deca-2,4-dien-8-one oxime;[526] Beckmann rearrangements have been used to prepare lactams of 1,5,6,7-tetrahydro-4*H*-indol-4-ones,[527] homopiperazines,[528] and 5,11-dihydro-6,11-dioxo-11-phenyldibenzo[*b*,*c*]aza-5,11-phosphepine, a new carbon–phosphorus heterocycle (**355**),[529] while the Beckmann rearrangement of bishomocubanone oximes has been examined.[530]

It has been shown[531] that, under acid conditions, *N*-(arylsulphonyloxy)-phthalimides undergo isomerization followed by Beckmann rearrangement to yield a mixture of 4-aryl-1*H*-2,3-benzoxazin-1-ones and diaryl sulphones whereas, under base-catalysed conditions (amines), they undergo Lossen rearrangement to produce *N*,*N'*-diarylureas and amine salts. The reported rearrangement of 2-phenyl-1,2-benzoisoxazol-3-ones (**356**) to 3-phenyl-benzoxazol-2-ones (**358**) represents an *N*-phenyl version of the Lossen rearrangement.[532] Acylnitrenium ion (**357**) formed from (**356**) by heterolysis of the protonated oxygen–nitrogen bond is probably the pertinent intermediate; aryl migration and cyclization of the resulting carbonium ion would then lead to the oxazolones.

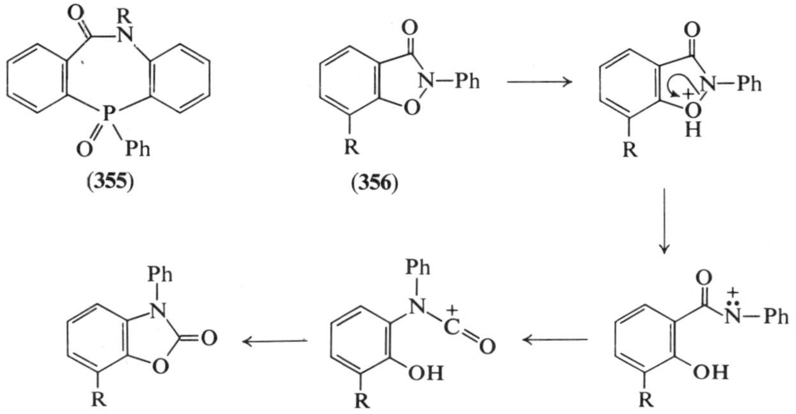

(**355**) (**356**)

(**358**) (**357**)

Kinetic data and the results of product analyses have been presented[533] for the reactions of phosphylated amide oximes in alkaline solution and probable mechanisms for the processes have been discussed in terms of Lossen and Tiemann rearrangements and in the light of current theories of substitution at phosphorus. Evidence which implicates the formation and polar decomposition of hydroxylamine *O*-arylsulphonates with concomitant aryl rearrangement to electron-deficient nitrogen has been reported[534] for the reaction of amines with arenesulphonyl peroxides, and formation of *N*-acetyl-1,3,8-trichloro-4-azahomoadamantane on carbon tetrachloride–aluminium chloride treatment of *N*-acetyl-*N*-chloroadamantane-1-amine has been rationalized[535] mechanistically on the basis of two known reactions, *viz.* the rearrangement of *N*,*N*-dichloroadamantane-1-amine to the azahomoadamantyl system, apparently *via* electron-deficient nitrogen, and the chlorination of C—H bonds by the alkyl halide–aluminium chloride combination.

A kinetic study has been carried out on the Curtius rearrangement of 5-substituted 2-furoyl azides,[536] a Curtius rearrangement has been used[537] as a key step in the development of a new route for the synthesis of pyrrolo[3,2-*d*]- and pyrrolo-[2,3-*d*]pyrimidine systems starting from 3-ethoxycarbonyl-4-methylpyrrole-2-carboxylic acid, while a Hofmann rearrangement of 5-carbamylpyridazine-4-carboxylic acid, itself obtained from pyridazine-4,5-dicarboxylic acid through the corresponding anhydride, provides an easy route to 5-aminopyridazine-4-carboxylic acid.[538] Disubstituted chloroformiminium chlorides (**359**) have been shown[539] to react with tetrabutylammonium azide (and other azide sources) to give the corresponding disubstituted cyanamides (**361**), presumably *via* the "Curtius-like" rearrangement of an intermediate azidoformiminium salt (**360**).

A detailed kinetic study of the Baeyer–Villiger reaction of acetophenones with permonophosphoric acid has been undertaken[540] and it has been shown that, in the Baeyer–Villiger reaction of strained cage compounds such as the 1,3-bishomocubanones, unusual mechanisms different from the established one, and involving cyclobutyl and cyclopropylcarbinyl cations, act predominantly, to give rearranged lactones.[541]

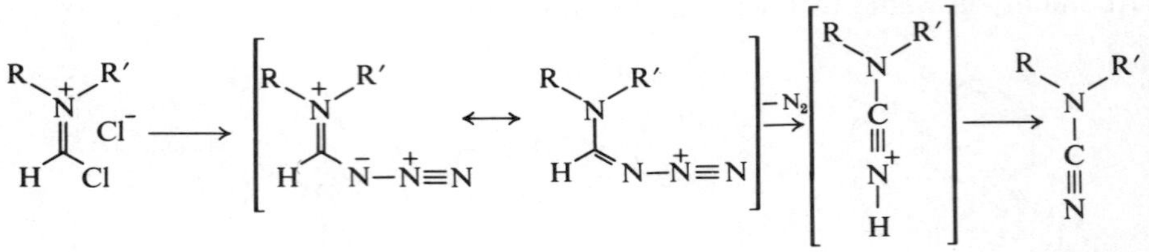

(359) (360) (361)

Metal-catalysed Rearrangements

Acyclic Systems

Several topics related to metal-catalysed rearrangements have been reviewed. These include the rearrangements of organoaluminium compounds and their Group III analogues,[542] organomagnesium rearrangements,[543] aryl migration in organometallic compounds of the alkali metals,[544] fluxional and non-rigid behaviour of organo-transition metal π-complexes,[545] σ-π-rearrangements of organo-transition metal complexes,[546] and molecular rearrangements in polynuclear transition metal complexes,[547] while a coverage of skeletal rearrangements also appears in a review that discusses organic reactions at alumina surfaces.[548] A new methodology involving the use of low-melting silver salt eutectics as both reaction media and catalysts has been developed[549] to enhance the efficiency of silver(I)-promoted rearrangements.

A kinetic and labelling study has been carried out on the isomerization of *n*-butenes over zinc oxide and alumina;[550] the isomerization of butenes on titanium dioxide has been shown to proceed through three different mechanisms involving, respectively, protonic sites, Lewis acid sites, and basic sites, depending on the pretreatment temperature of the catalyst,[551] while the abnormal dependence of the rate constant for but-1-ene isomerization in the presence of μ-dichlorotetraethylenedirhodium has been explained[552] by the inhibiting effect of ethylene liberated from the catalyst. Transition-metal salts have proved effective catalysts in olefin photorearrangements.[553]

Results have been presented[554] indicating that olefin isomerization by $(Ph_3P)_nNiX$ occurs by a metal-hydride pathway, it being suggested that the hydride is found as one product of an unusual reaction involving oxidative addition of an α-C—H bond of the olefin to two nickel(I) centres, as in Scheme 34. Furthermore, azo-compounds produced by the nickel-catalysed reaction of butadiene with hydrazones have been further isomerized in the presence of the same nickel catalyst.[555]

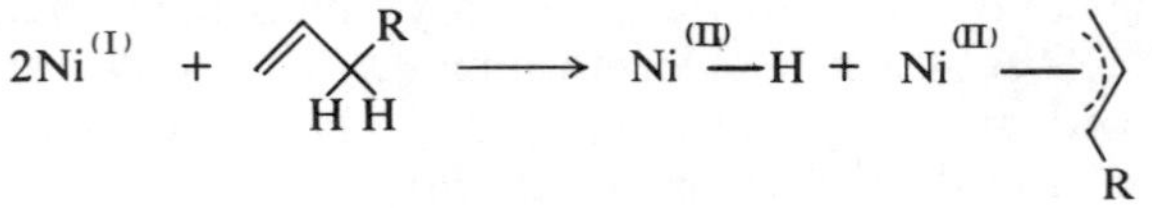

$$2Ni^{(I)} + \text{[olefin]} \longrightarrow Ni^{(II)}\text{—}H + Ni^{(II)}\text{—[allyl]}$$

SCHEME 34

The first reported stereoselective conversion of allyl into *trans*-propenyl ethers has been achieved[556] by hydrogen-activated $[Ir(\text{cyclo-octa-1,5-diene})(PMePh_2)_2]PF_6$, a π-allyl hydrido-species being proposed as the rearrangement intermediate; and a report has appeared[557] of an enantioselective isomerization of prochiral allylamines to chiral enamines, reputedly the first stereospecific hydrogen migration observed in olefin catalysis.

A novel palladium-catalysed exchange reaction of tertiary amines has been reported.[558] The inital step in this reaction seems to be an insertion of palladium into a carbon–hydrogen bond adjacent to nitrogen leading to a highly active iminium ion complex.

The kinetics of the allene–methylacetylene isomerization over silica-supported cobalt and iron catalysts have been studied,[559] and a mechanism has been formulated[560] to account for the stereochemistry of the conversion of 1-phenylprop-2-yn-1-ol to 1-chloro-3-phenylpropa-1,2-diene *via* the cuprous chloride-catalysed rearrangement of 3-chloro-3-phenylprop-1-yne. Rhodium(I)-catalysed rearrangements of 3,4-bis(acyloxy)hexa-1,5-diynes (**362**) have resulted in cyclization to (*E*)-4-[(acyloxy)methylidene]cyclopent-2-en-1-ones (**363**) regardless of the configuration of the starting material, and a mechanism which accounts for the stereospecificity of the process has been proposed[561] (see Scheme 35).

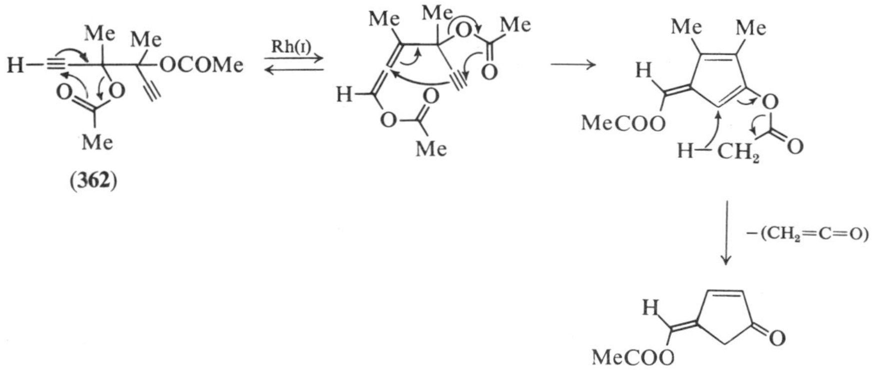

(**362**) Rh(I) $-(CH_2{=}C{=}O)$ (**363**)

SCHEME 35

The results of a study of prototropic rearrangements of alkenes and dienes promoted by iron carbonyls have led to the suggestion[562] that diene isomerization can emanate directly from iron tricarbonyl through a pathway in which co-ordination and hydride abstraction occur either simultaneously or in rapid succession to yield the intermediate π-allyliron tricarbonyl hydride necessary for rearrangement. The mechanism of the reaction of α,α'-dibromo-ketones and iron carbonyls has been examined[563] with various substrates and reaction conditions. The findings indicate that the reaction first produces enolate intermediates and subsequently oxyallyliron(II) species which, depending on their structures and reaction conditions experience a variety of cationic rearrangements, nucleophilic trapping, and prototropy to form α,β-unsaturated ketones.

Monocyclic Systems

The similarity between cyclopropanes and olefins as manifested by the ability of small rings to enter into conjugation with adjacent π-electrons and to insert metal residues into the carbon–carbon σ-bond has been discussed[564] in a review which also outlines the novel features characterizing the interaction between zerovalent transition metals and multi-σ,π-electron systems.

The first observation of enhanced chemiluminescence associated with the catalytic decomposition of a 1,2-dioxetane has been made using silica gel as the catalyst.[565] The reported formation of 1,2,3-trimethylbenzene in the palladium choride–acetic acid treatment of apopinene (**364**) represents the first ever opening of a cyclobutane ring by fission of the C(1)—C(7) bond.[566] A concerted reaction of the palladium chloride complex, as outlined in Scheme 36, is a probable course for the reaction.

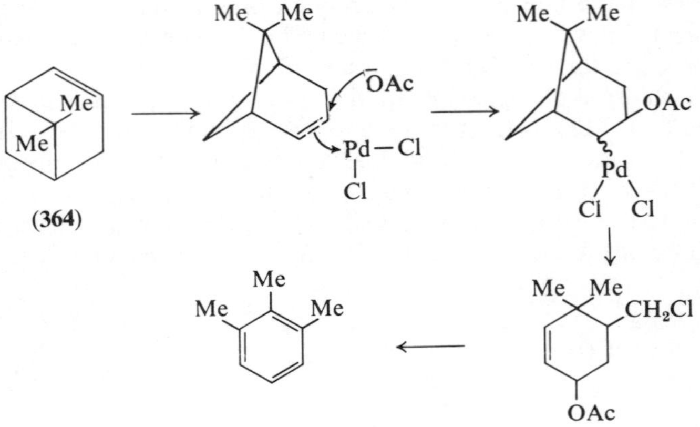

(**364**)

SCHEME 36

An attempt has been made[567] to determine the stability of rhodium complexes derived from simple divinylcyclopropanes. *cis*-Divinylcyclopropane reacts with bis(ethylene)hexafluoroacetylacetonatorhodium(I) to give a 1 : 1 complex with displacement of ethylene. This material, which still contains an intact cyclopropane ring, slowly rearranges to the hexafluoroacetylacetonatorhodium(I) complex of cyclohepta-1,4-diene. In contrast, *trans*-divinylcyclopropane reacts with the same catalyst to give a 1 : 1 complex in which the cyclopropane ring has opened to form a bis-π-allylrhodium complex. The iron carbonyl-induced vinylcyclopropane–cyclopentene rearrangement in the 4,5-homotropilidene series has been shown to involve successive cleavage of two bonds of the cyclopropane ring in the ligand sphere of a $Fe_2(CO)_6$ unit,[568] while in the presence of $Fe_2(CO)_9$, (*Z*)-2-phenyl-3-(2,3-diphenyl-2*H*-azirin-2-yl)acrylophenone yields 2,3,5,6-tetraphenylpyridine in a novel transformation which interestingly implies loss of an oxygen atom.[569]

Evidence has been presented[570] showing that 7,7-bis(dialkylaluminio)hept-1-enes (**365**) readily rearrange by intramolecular olefin exchange to the 1,7-bis(dialkylaluminio)hept-1-enes (**366**), which cyclize in a very fast subsequent reaction to [bis(dialkylaluminio)methyl]cyclohexanes (**367**). Metal–η^2-alkenesulphinato-products containing a 1,3-rearranged $MSO_2CR'R''CR{=}CH_2$ and/or an unrearranged

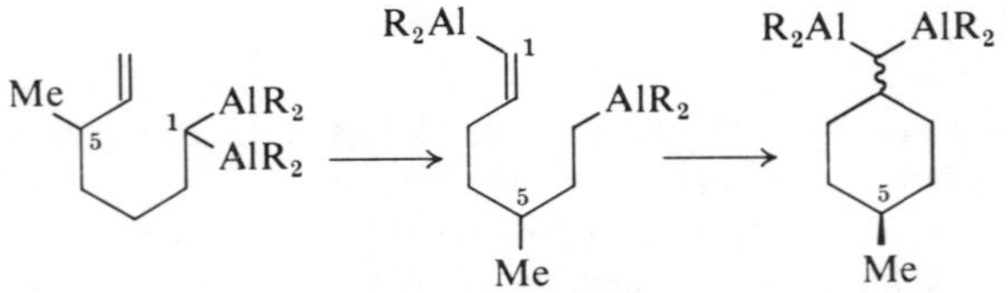

(**365**) (**366**) (**367**)

$MSO_2CH_2CR{=}CR'CR''$ allylic fragment (M = Fe, Mo, W, *etc.*) have been obtained from the reaction of cyclopentadienylmetal(2-alkenyl)carbonyl complexes with sulphur dioxide.[571] These alkene-*S*-sulphinates exhibit *cis–trans*-isomerism the incidence of which has been rationalized[572] in terms of a mechanism that entails the formation and subsequent dissociation of a metal-η^2-alkene-sulphinate zwitterion, a 1,3-rearrangement *via* internal return of the dissociated alkenesulphinate anion, and a recombination of the rearranged and/or unrearranged anion with the cationic cyclopentadienylmetalcarbonyl species. The factors favouring formation of the rearranged metal *S*-sulphinates over the unrearranged species have been discussed.[573]

In a study of the action of various transition-metal complexes on the isomerization of dihydropyridines,[574] only $RhCl(PPh_3)_3$ was found effective in converting 1,2-dihydropyridines into their 1,4-isomers; moreover, it has been shown[575] that rhodium chloride trihydrate induces aromatization *via* remote double-bond migration in unsaturated cyclohexenones; thus, the remote double bond of 4-(pent-4-enyl)cyclohex-2-enone (**368**) migrates to afford the phenol (**369**). Similarly, unsaturated imines are aromatized to substituted aniline derivatives; see (**370**)→(**371**).

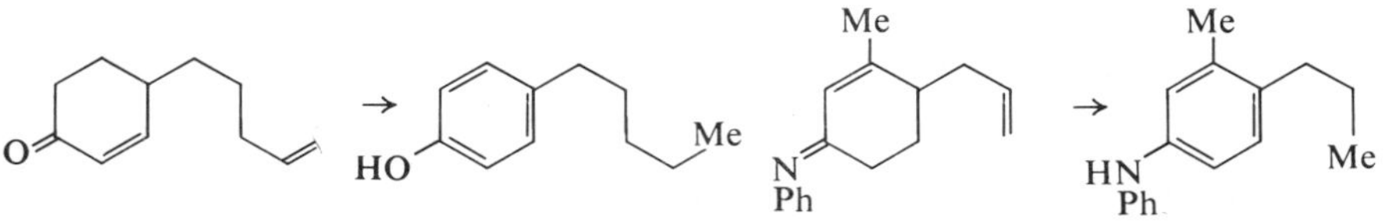

(**368**) (**369**) (**370**) (**371**)

High-performance liquid chromatography has been used[576] to study the thermal rearrangement of tricarbonyl(phenylcycloheptatriene)iron isomers, and a tricarbonylnorcaradieneiron complex has been postulated[577] as intermediate in the 1,3-iron shifts in tricarbonylcycloheptatrieneiron complexes. The activation energies for these shifts are found to be higher than those associated with 1,2-metal shifts in conjugated cyclic polyene and polyenyl complexes. The fluxional behaviour of (1-5-η-cycloheptadienyl)(1-5-η-cycloheptatrienyl)iron has been explained[578] by invoking a 1,2-shift of the 1-5-η-cycloheptatrienyl moiety with respect to the central iron atom while an examination of the fluxional behaviour of bis(cyclopentadienyl)-dithiocarbamatomolybdenum complexes has shown[579] conclusively that the rate of 1,2-shifts of the σ-ring is significantly faster than that of 1,3-shifts.

By selection of suitable ligands (L), 1,3-dithiol-2-ylideneiron(0) complexes (**372**) have been converted with ease into the isomeric heterometallocycles (**373**),[580] while ring-contraction of the five-membered chelate ring in cation (**374**) to afford three-membered ring species surprisingly occurs on its reaction with methyl-lithium.[581] When 2,6-bis(2-benzothiazolinyl)pyridine (**375**) was treated with divalent zinc salts, {2-(2-benzothiazolyl)-6-[2-(2-thiolophenyl)-2-azaethyl]pyridine}acetatozinc (**376**) was obtained.[582] In order to account for this unusual ligand rearrangement, the ring-opening of one of the thiazolinyl rings to an imino-thiolate chelate must occur, while a net "internal hydrogen transfer" reaction is required to explain dehydrogenation of the other benzothiazolinyl ring and the hydrogenation of the imino-linkage. Thus, the formation of (**376**) provides a rare example of a dehydrogenation/hydrogenation reaction which is promoted by a bivalent metal ion which

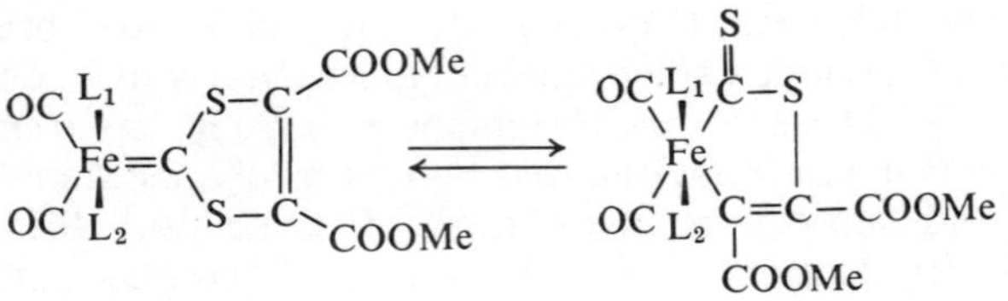

(372) (373) 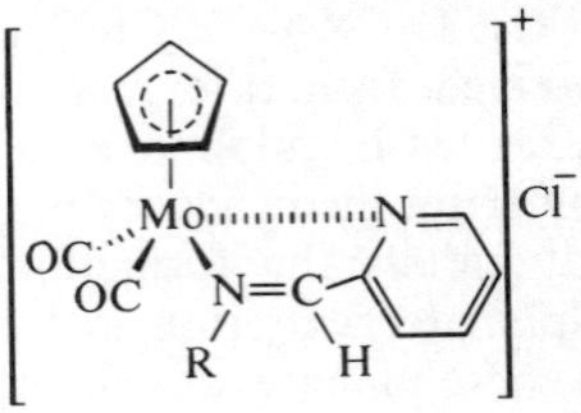 (374)

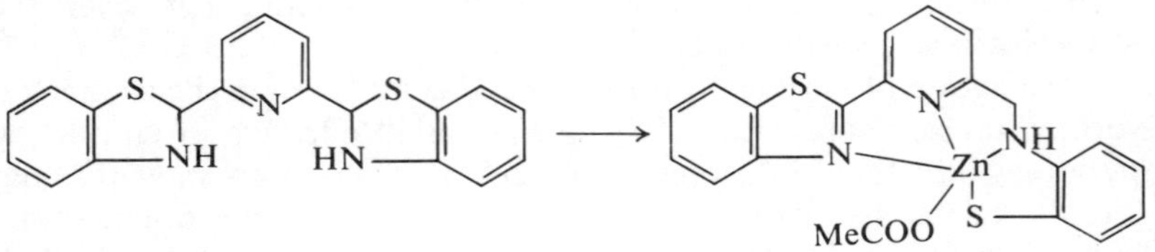

(375) (376)

does not easily show changes in its oxidation state; in this respect it provides an analogy to the zinc-containing enzyme, liver dehydrogenase.

An interesting O → As(Sb) methyl shift observed[583] in triethylphosphite-arsenido and -antimonido complexes of transition metals transforms a phosphite into a phosphonate ligand, with concomitant formation of a trimethylarsane (-stibane) ligand; see **(377)** → **(378)**. Moreover, compared with the few known examples of direct P=O bond formation resulting from methyl migration, this reaction proceeds under astonishingly mild conditions.

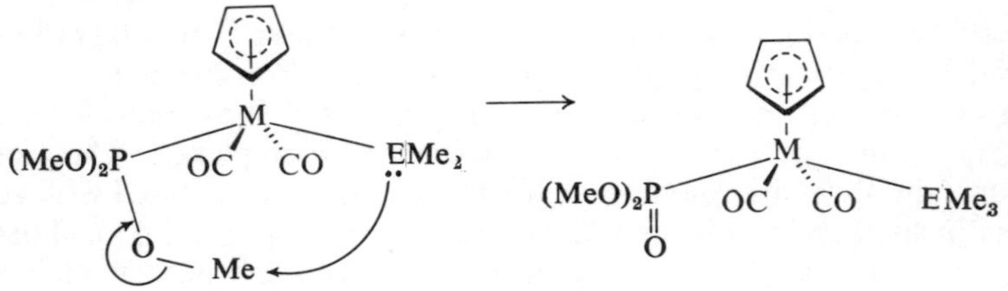

(377) (378)

(M = Mo or W; E = As or Sb)

Interestingly, co-ordination of a nitrile-nitrogen with silver(I) appears to render a nitrile-carbon more susceptible to attack under conditions near neutral pH, a fact that is considered instrumental in explaining the ready rearrangement of cyanodihydroazepines to furo[2,3-*b*]pyridines;[584] finally, monometalated and bimetalated ketenes have been prepared[585] by the copper-catalysed Wolff rearrangement of α-metalated diazocarbonyl compounds.

Polycyclic Systems

The dienone–iron complex **(379)**, prepared by cycloaddition of diphenylketene to tricarbonylcycloheptatrieneiron, has been found to rearrange in refluxing benzene to the σ,π-allyliron complex **(380)** in a process which seems to be without precedent;[586] formally, it represents a unique [2,2]-dyotropic rearrangement in which the iron attached to C(2) of (**379**) migrates to the neighbouring carbon forming the Fe—C(9) σ-bond in (**380**), while the diphenylcarbinyl-carbon migrates from C(1) to C(2) [in (**379**)] to form the C(1)—C(8) σ-bond of (**380**). The rearrangement of 9,9-dichlorobicyclo[6.1.0]non-4-ene to its bicyclonon-3-ene isomer and ultimately to the bicyclonon-2-ene, in the presence of $Fe_2(CO)_9$, has been examined,[587] and an interesting example of a rhodium(I)-promoted six-electron bond reorganization, namely *syn*-1,3-bishomocycloheptatriene to bicyclo[6.1.0]nona-2,5-diene (**383**), has been reported[588] and represents a previously unobserved and novel facet of organo-rhodium chemistry. A hypothetical mechanism proposed for this reaction involves rate-determining oxidative addition of rhodium into the fused bond of the central cyclopropane ring, as in (**381**), followed in a subsequent more rapid step by a [3,3]homodienyl rhodium shift with relief of structural strain to give (**382**) and ultimate reductive extrusion of the metal to construct a new three-membered ring.

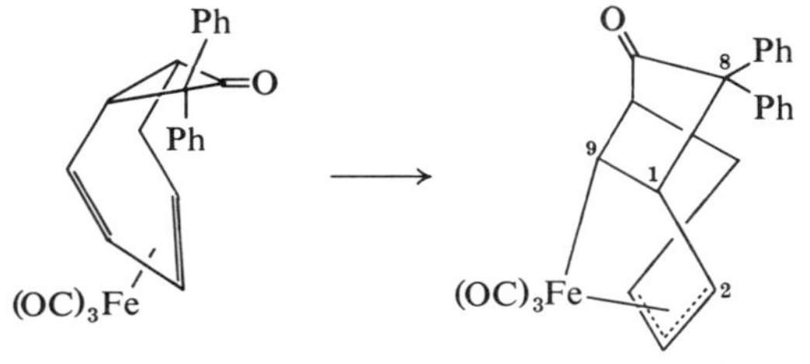

(**379**) (**380**)

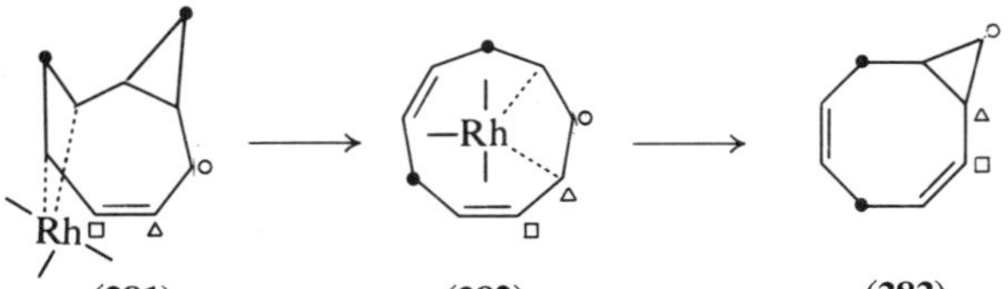

(**381**) (**382**) (**383**)

The same catalyst effects the rearrangement of cycloprop[*a*]acenaphthylene to phenalene by oxidative addition of rhodium(I) to the central bond of the substrate followed by migration of the C(7)-*exo*-hydrogen to give a η^3-allylrhodium(III) complex.[589] Rhodium(I) complexes of *syn*-tricyclo[4.2.1.1^{2,5}]deca-3,7-dienes have been prepared readily by rhodium(I)-induced cleavage of strained secopenta-prismanes,[590] while a study has been made of the rhodium(I)-catalysed cubane to *syn*-tricyclo-octadiene-type of rearrangement; the critical use of this kind of metal-induced rearrangement for the tactical synthesis of octahydro- and perhydro-[0.0]paracyclophanes has been described.[591] Chiral rhodium chloride complexes have been used to provide the first example of a kinetic resolution of a strained polycyclic compound,[592] the effect of substituents on the rhodium-catalysed conversion of bicyclo[3.1.0]hex-2-enes to cyclohexadienes has been investigated,[593] the reactivity of 1,5-bismethoxycarbonyl-3-oxaquadricyclanes towards different transition-metal catalysts has been found to vary strongly with the type of catalyst used,[594] and an examination has been made of the rhodium(I)-catalysed transformations of the cycloaddition products of 3,4-dimethoxyfuran with acetylenedicarboxylates.[595]

Tricyclo[4.2.0.0^{2,5}]octa-1,3,5,7-tetraene(η^5-cyclopentadienyl)cobalt has been suggested as a possible intermediate in the remarkable gas-phase rearrangement of

1,2-bis(trimethylsilyl)-3,4-diethynylcyclobutadiene(η^5-cyclopentadienyl)cobalt to 1,2-bis(trimethylsilylethynyl)cyclobutadiene(cyclopentadienyl)cobalt,[596] and a metalocyclic intermediate has been isolated in the $[Ir(CO)_3Cl]_2$-catalysed rearrangement of 1,3-bishomocubane to dicyclopentadiene.[597]

The ability of variously unsaturated [4.4.2]propellanes to co-ordinate with transition-metal carbonyl fragments has been assessed experimentally,[598] and it has been demonstrated that hexacarbonyl molybdenum is capable of transforming suitably strained propellanes into isomeric hydrocarbons under relatively mild conditions. In the specific case of [4.4.2]propella-2,4,11-trienes(**384**), these complexes (**385**) undergo further [1,5]-sigmatropic shifting of the four-membered ring to deliver 1,2-annulated cyclo-octatetraenes (**386**; R = Me, R′ = H and R = H, R′ = Me), although the catalyst is apparently also capable of promoting more extended circumambulation when C(11) and C(12) are both substituted. Results of kinetic and deuterium-labelling studies unequivocally establish[599] that C(11) of the cyclobutene ring, and not C(7) of the cyclohexane moiety, is directly involved in the formal [1,5]-sigmatropic migration.

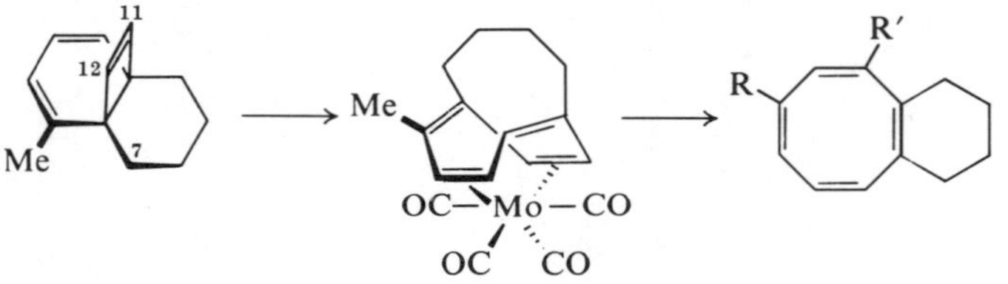

(**384**) (**385**) (**386**)

A study of the intramolecular rearrangements of square-planar β-diketonate complexes of palladium(II) and platinum(II) indicates a dissociative rather than an associative solvolysis,[600] and an unusual gas-phase rearrangement–cyclization reaction on a platinum–silica catalyst must be considered in the synthesis of *anti*-tetramantane.[601] C-5 allyl, propyl, and propenyl uracil and cytosine nucleosides have been prepared[602] from unprotected pyrimidine nucleosides *via* organopalladium intermediates by a route that has several advantages over any method appearing in the literature to date, while an examination of the effectiveness of various transition-metal complexes of poly(styryl)bipyridine, as catalyst for the quadricyclene to norbornadiene isomerization, has shown that only the palladium(0) species is a suitable catalyst.[603]

Rearrangements have been observed in the mercuration of bornylene,[604] a number of methylenecyclopropanes have been shown to undergo a variety of novel carbon skeletal reorganizations upon photolysis in the presence of copper(I) trifluoromethanesulphonate,[605] and the silver bromide-catalysed retro-rearrangement of a dibenzotricyclo[4.3.1.0^{2,5}]decadiene has been reported to yield the [4.2.2.0^{2,5}] system.[606] Silver ion also catalyses the cyclobutyl–cyclopropylcarbinyl rearrangement of substituted 4,4-dichlorotetracyclo[3.3.0.0^{2,8}0^{3,6}]octanes (**387**) to mixtures of epimeric tetracyclo[3.3.0.0^{2,8}0^{4,6}]octan-3-ols (**388**),[607] and the effect of Group IV substituents (*e.g.* trimethylsilyl, trimethylgermyl, *etc.*) on the course of silver(I)-catalysed rearrangement reactions of the tricyclo[4.1.0.0^{2,7}]heptane ring system has been investigated.[608]

Treatment of chalcones (ArCH=CHCOAr′) with thallium(III) nitrate in acidic methanol was found to yield the propanones [ArCH(COAr′)CH(OMe)$_2$] resulting

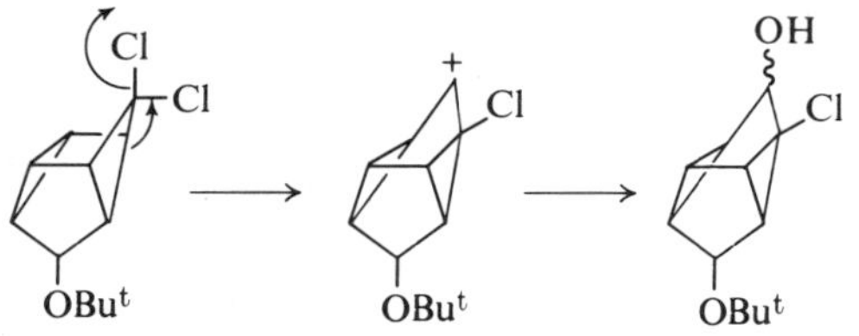

(387) (388)

from migration of the Ar group. However, prior conversion of the chalcones into their dimethyl ketals, followed by reaction with thallium(III) nitrate, yields methyl 2,3-diaryl-3-methoxypropanoates [ArCH(OMe)CH(COOMe)Ar′] by rearrangement of the Ar′ group.[609] Natural flavone and isoflavone glycosides[610] and prenylated isoflavonoids[611] have been obtained by way of the oxidative rearrangement of chalcones by thallium(III) nitrate; thallium nitrate-induced oxidative rearrangements of chalcones have been utilized in the synthesis of odoratin and 7-hydroxy-2′,4′,5′,6-tetramethoxyisoflavone,[612] and the same reagent has initiated the stereoselective transformation of the 16-exocyclic methylene group into a carboxy-group in *ent*-kaurene and its 19-oic acid.[613] Moreover, 17-nor-13β-kauran-16-one (**389**) on treatment with thallium(III) nitrate yields an oxidatively rearranged product, 9,10-*friedo*-17-norkaur-5(10)-en-12-one (**390**), which further rearranges in the presence of acid to give the α,β-unsaturated ketone, 17-norkaur-9(11)-en-12-one (**391**).[614] A reasonable mechanism for this rearrangement is shown

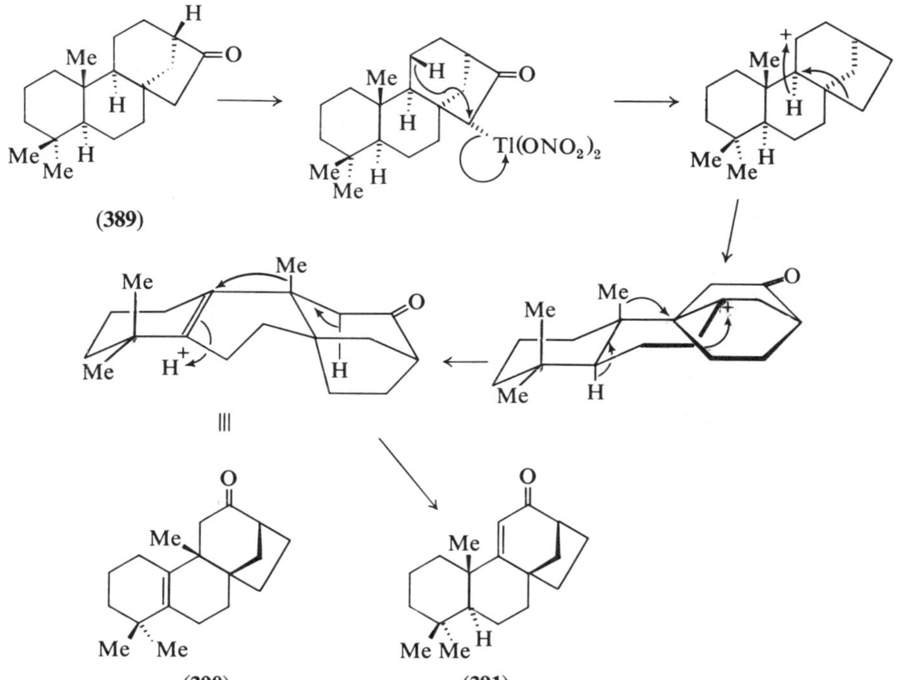

(389)

(390) (391)

Scheme 37

in Scheme 37, while a mechanism involving the formation of a bridged organothallium intermediate is proposed for the formation of 3α,10α-epoxy-5-methyl-19-nor-5β-cholestan-6β-yl acetate on treatment of epicholesterol with thallium triacetate.[615]

Rearrangements Involving Ring-opening and Ring-closure

Three-membered Rings

Structural isomerization of cyclopropane behind reflected shock waves has been studied[616] and a complete kinetic analysis of the thermal stereomutation of (+)-(1*S*,2*S*,3*S*)-*r*-1-cyano-*t*-2-phenyl-1,*c*-3-dideuteriocyclopropane has been made.[617] It has been demonstrated that α-methylstyrene is not the major product of thermal structural isomerization of phenylcyclopropane, the main products being *trans*- and *cis*-β-methylstyrene and allylbenzene. Moreover, it has been shown that the reaction proceeds by the synchronous double rotation of the phenyl-bearing carbon and one of the methylene groups.[618]

A novel regiospecific 2-chlorotropone synthesis has been achieved involving acid-catalysed rearrangement of 2,2-dichlorocyclopropan-1-ols,[619] while a skeletal rearrangement involving ring cleavage of an intermediate 1-benzylidene-2-chlorocyclopropane has been observed during the dihydrochlorination of 2-benzyl-1,1-dichlorocyclopropane.[620] The relative amounts of various fluoropropenes resulting from the thermal isomerization of fluorocyclopropane have been shown to be pressure-dependent,[621] while preliminary experimental kinetic and thermodynamic results on the thermal isomerization of 2,2-difluoro-1-methylenecyclopropane to (difluoromethylene)cyclopropane have at last provided experimental support for the prediction that the bonds adjacent to the substituted carbon, as well as the bond opposite the fluorine-substituted carbon, are weakened.[622] The study also provides the first quantitative evaluation of the relative thermodynamic stability of cyclopropane-bound fluorine *versus* vinylic fluorine.

The surface for the degenerate thermal rearrangement of methylenecyclopropane has been investigated by a molecular-orbital method (PRDDO). The observed low barriers to internal rotation in the singlet states of the bisallyl intermediates are considered to be important in explaining the racemization observed in the thermal rearrangements of optically active, substituted methylenecyclopropanes and 1,2-dimethylenecyclobutanes.[623] The effect of substituents on the thermal rearrangement of the former compounds has been interpreted[624] in terms of an early transition state in the formation of the intermediate perpendicular trimethylenemethane, or alternatively by a concerted process having only partial radical character in the transition state, and an NMR spectroscopic study has shown that base treatment of α-methylenecyclopropane esters, ketones, and alcohols specifically leads to vinylcyclopropane compounds in which the substituent groups always show a *trans*-relationship.[625] Thermolysis of 1-phenyl-2-vinylcyclopropane leads to the rapid formation of an equilibrium mixture of the *cis*- and *trans*-isomers and the slow but irreversible production of 4-phenylcyclopent-1-ene.[626] Evidence has been presented to show that vinylcyclopropylidene to cyclopentenylidene rearrangements such as (**392**) to (**393**) do not involve carbenes, but rather species in which bonding to lithium is necessary[627] (see Scheme 38), and a quantum-mechanical investigation has been carried out on the parent vinylcyclopropylidene to cyclopentenylidene rearrangement.[628]

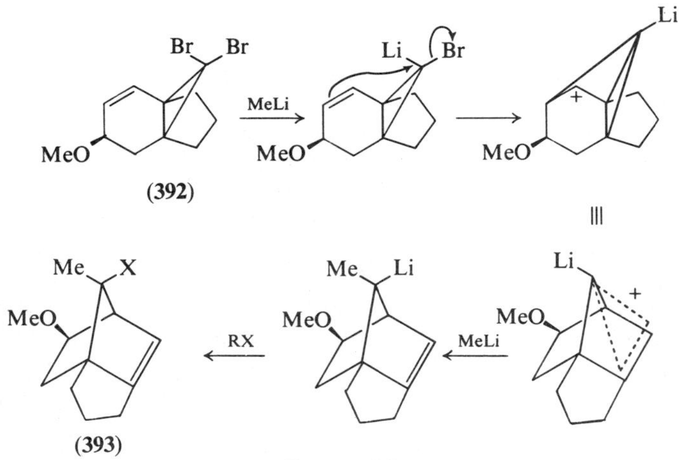

(392)

(393)

SCHEME 38

Methylacetylene appears to be the major product of gas-phase pyrolysis of cyclopropene, the minor product, allene, being formed by two pathways, one molecular and the other radical by nature.[629] The thermal and thermocatalytic isomerizations of methyl 2,3,3-triphenylcyclopropene-1-carboxylate to methyl 2,3-diphenylindene-1-carboxylate have been studied,[630] photolytic rearrangements of triphenylcyclopropenes to the corresponding 1,2-diphenylindenes have been described,[631] the photochemical rearrangement of 3-vinylcyclopropenes to cyclopentadienes has been rationalized in terms of both carbene and diradical mechanisms,[632] and a careful study of the products and mechanism of the photolysis of several unsymmetrical 3-aryl- and 3-vinyl-substituted cyclopropenes has been undertaken; and a consistent and satisfying mechanistic rationale for the ring-opening reaction has been presented.[633]

A useful synthesis of monoterpenes has been developed from the ring-opening of a number of C_{10}-alkenylidenecyclopropanes under acid- and base-catalysed conditions;[634] thus, treatment of the acetylenic alkenylidenecyclopropane **(394)** with $BF_3 \cdot OEt_2$ in methanol leads directly to the head-to-tail terpenoid allene–acetylene structure **(395)**, while dispiro[2.2.2.2]deca-4,9-diene, where the two cyclopropane rings can be cleaved and the central cyclohexadiene ring can be

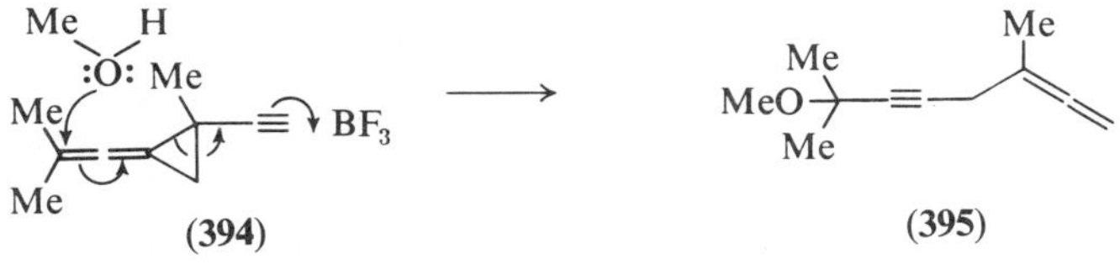

(394) (395)

aromatized, has proved to be an excellent precursor for paracyclophane derivatives.[635] Rate constants have been evaluated for the thermally induced geometric isomerization of 1,2,4-trimethylspiropentane,[636] while a product analysis of the deamination of 2-deuteriospiropentylamine seems to rule out incursion of CH^+-trimethylenemethane, or the bicyclo[1.1.1]pent-1-yl cation, as intermediates in the spiropentyl-to-3-methylenecyclobutyl cation rearrangement, and instead verifies an earlier proposal of unsymmetrical involvement of C(4) and C(5) in the deamination.[637]

The thermal labilities of a series of spiro[*n*,2]alkan-2-ones have been examined; in most cases, significant oxovinylcyclopropane to dihydrofuran rearrangements are observed.[638] The thermal rearrangement of an aminomethyl cyclopropyl ketone has been used in a synthesis of pentazocine[639, 640] while a recently reported synthesis of 2-(6-carboxyhexyl)cyclopent-2-en-1-one, a prostanoid synthon, involves the acid-catalysed cyclopropane rearrangement of 9-cyclopropyl-9-oxononanoic acid as the key step.[641]

The formation of, and opening of, the cyclopropane ring in tricyclo[3.2.0.0^{1,4}]-heptan-2-one has been proposed to account for the novel base-catalysed rearrangement of bicyclo[3.2.0]heptan-6-one to the corresponding bicyclo[3.2.0]heptan-2-one;[642] the formation and ring-opening of a cyclopropane moiety have been invoked to explain the formation of 1,6-dimethylcyclohepta-1,3,5-triene from 2,7-dimethyloxepine on treatment of the latter with dimethyl azodicarboxylate,[643] while the 1,4-dihydrocinnoline derivatives (**397**) [obtained on heating 1a,7b-dihydro-1*H*-cyclopropa[*c*]cinnolines (**396**)] are thought to arise through concerted cleavage of the cyclopropane ring and [1,5]-hydrogen shift from carbon to heteroatom.[644]

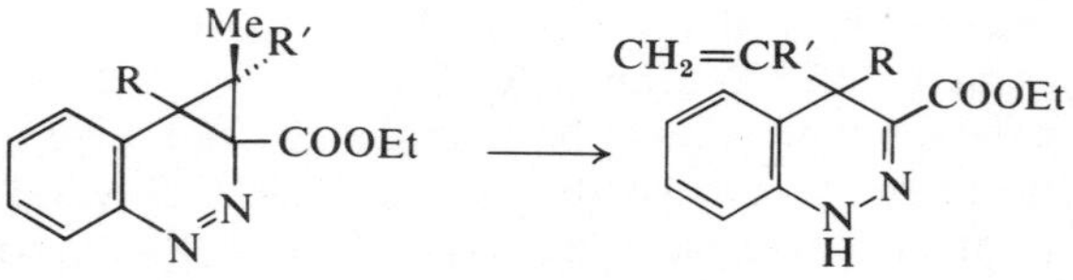

(**396**) (**397**)

A comparison has been made of the Berson–Willcott rearrangement of 11,11-dimethyltricyclo[4.4.1.0^{1,6}]undeca-2,4,7,9-tetraene and of its radical anion,[645] and a kinetic and product study of the photo-induced Berson–Willcott rearrangement of methyl 3-isopropyl-1a,7b-dihydro-1*H*-cyclopropa[*a*]naphthalene-1-*exo*- and -1-*endo*-acetate supports a slither motion of C(1), with inversion of configuration, for the process.[646] The thermolysis of *endo,exo*-tetracyclo[6.1.0.0^{2,4}0^{5,7}]nonane to *trans*-bicyclo[4.3.0]nona-3,7-diene is considered to proceed by cleavage of the cyclopropane rings to yield *cis,trans,trans*-cyclonona-1,4,7-triene which then undergoes a [$_{\pi}2_s + _{\pi}2_s$]-cycloaddition between the two strained *trans* double bonds followed by rearrangement with retention of the *trans*-linkage.[647] Mechanistic schemes that concisely account for the transformations of tricyclo[4.1.0.0^{2,7}]hept-3-enes have been proposed;[648] such schemes emphasize the interdependence of structural features and chemical reactivity in bicyclobutane systems; thus, for example, a marked preference for electrophilic attack by D^+ at the edge bicyclobutane bonds in these molecules is apparent, in contradistinction to benzvalene which is π-olefinic reactive.

Rearrangements of cyclopropylmethyl- and but-3-enyl-cobaloximes have been studied,[649] and it has been observed that 1-methylbut-3-enylcobaloxime and 2-methylbut-3-enylcobaloxime rearrange to an equilibrium mixture in a manner which clearly implicates cyclopropylcarbinyl intermediates.[650] This ability of cobaloxime(II) to induce both ring-closure and ring-opening suggests a simple but novel explanation for related coenzyme B_{12}-catalysed rearrangements.

In addition to generating a ring B-enlarged ketone (**400**), which product almost certainly arises by incorporation of the C(10) angular methyl group and ring-opening of the intermediate cyclopropane (**399**), the $BF_3 \cdot OEt_2$ treatment of 3β-acetoxy-9β,11β-epoxylanostan-7-one (**398**) has also been shown to afford cucurbitane derivatives, *e.g.* (**401**), by rearrangement of the β-epoxide ring.[651]

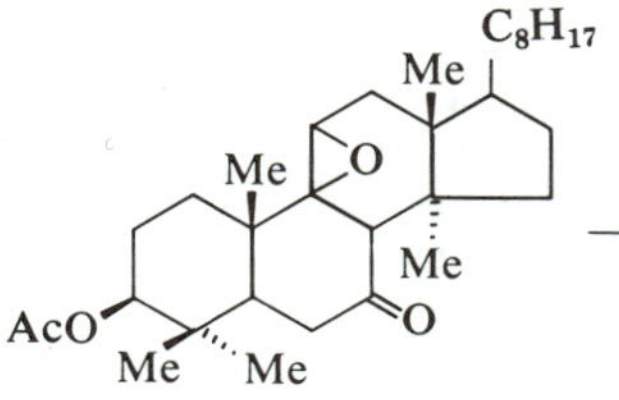

(**398**) 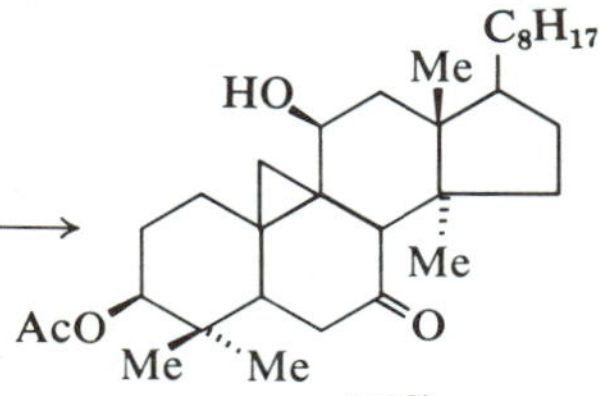 (**399**)

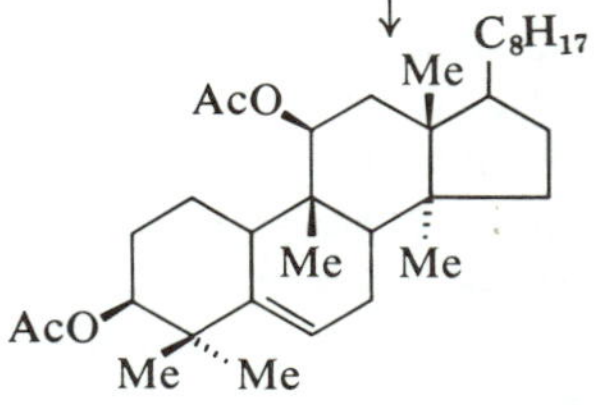

(**401**) 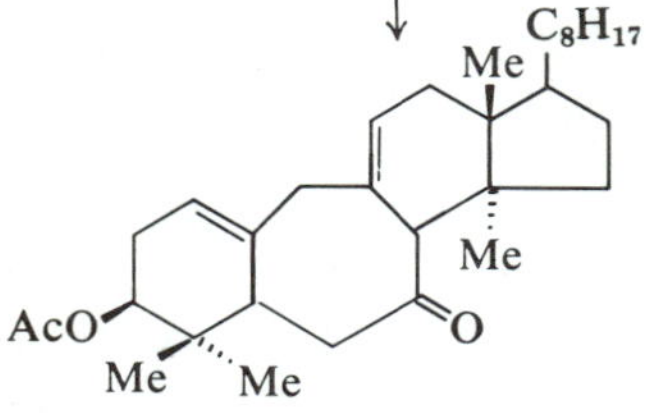 (**400**)

An investigation of the aromatization of 2-carboxy- and 2-methoxycarbonyl-oxepine/benzene oxide supports the suggestion that 1,2-oxides of benzoic acids are intermediates in biological oxidative decarboxylation reactions and further suggests that 2-carboxyoxepin/benzene oxide is itself an intermediate in the *ortho*-hydroxylation of benzoic acid. The observed aromatization of the 2-methoxycarbonyl derivative provides the first case of alkoxycarbonyl migration in the aromatization of arene oxides.[652] The novel *m*-chloroperbenzoic acid-induced contraction of the seven-membered ring in mutagenic azuleno[1,2,3-*c*,*d*]phenalene is considered to occur by formation and subsequent opening of an arene ring which in turn activates the seven-membered ring in the substrate to nucleophilic attack by *m*-chloroperbenzoic acid. The authors suggest[653] that this reaction might provide a model for the mutagenic and carcinogenic activity of the phenalene.

A new methodology for accomplishing aromatic ring annelation with high regiochemical control has been described;[654] the annelation reagents are epoxides derived from cycloalk-2-en-1-ones containing n ring atoms, from which $(n-1)$ atoms are incorporated into the new ring. Moreover, having realized that the use of a vinyl ether as the nucleophilic component in cationic olefin cyclizations would offer the advantage of retaining useful functionality in the cyclization products, a study has been initiated of the Lewis acid-induced cyclization of vinyl ether epoxides of the type (**402**) in an attempt to demonstrate the feasibility and utility of such cationic cyclization reactions.[655] Such a study has indeed shown that exposure of (**402**; $R' = R^2 = H$, $R^1 = Me$, $n = 2$) to basic Al_2O_3 affords the hydroxydihydropyran (**403**) which itself can be transformed into the functionalized carbocyclic ketone (**404**). Five- and six-membered rings have also been formed from olefinic α,β-epoxy-ketones and hydrazine;[656] thus, epoxy-ketone (**405**) cyclizes to the hydrindenol (**406**).

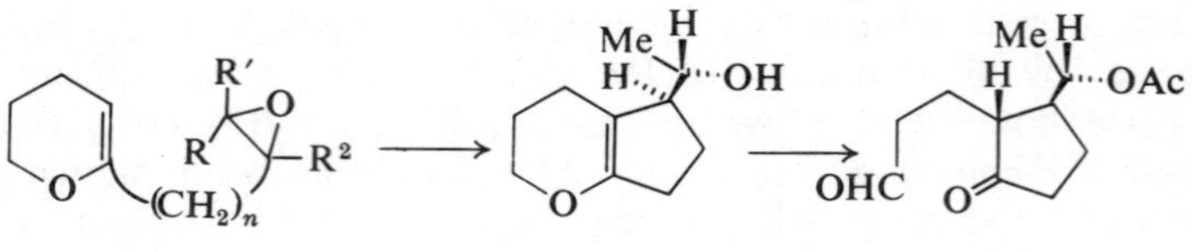

(402) (403) (404)

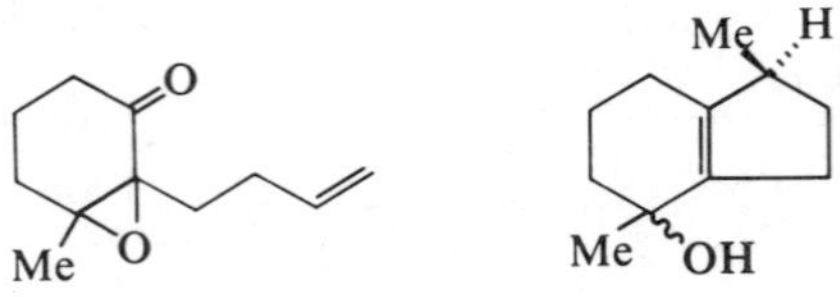

(405) (406)

It has been shown[657] that the course of the reaction of allene oxides with nucleophiles depends critically on the nature of the substituents on C(1). The relevance of these observations to the mechanism of the allene oxide–cyclopropane isomerization process has been discussed and a consistent picture has emerged in which a rate-determining step leading to the formation of oxyallyl is proposed for the isomerization. The mechanism that has been proposed for the conversion of various vinyloxiranes into the corresponding dihydrofurans involves a two- or multi-step reaction sequence with 2-oxapentadienyl dipoles as intermediates.[658] Furthermore, it appears that in most cases the first (carbon–carbon bond cleavage of the oxirane ring) and the last step (cyclization of the 1,5-dipole) take place in a stereospecific manner. A recent investigation of the thermal isomerization of *cis*-2-ethynyl-3-vinyloxiranes (**407**), substituted at C(2) by an alkyl group, lends support to the proposed formation of a highly strained seven-membered allenic intermediate (**408**) in the process, and brings a new insight into the reactivity of these molecules.[659]

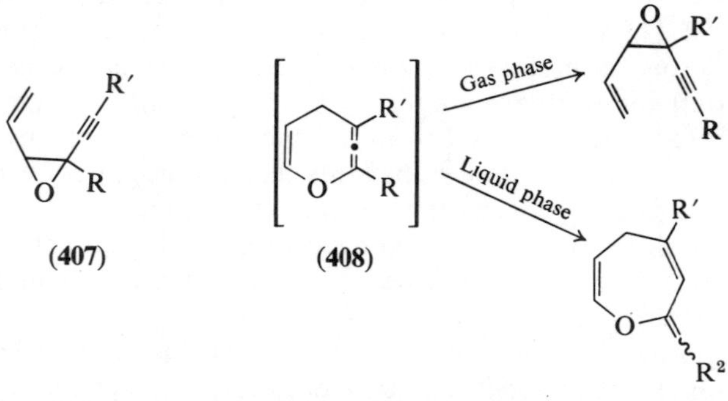

Gas phase

Liquid phase

(407) (408)

Ring-opening of the epoxide ring in 6*H*-dibenz[*b,f*]oxireno[*d*]azepines appears to lead to normal cleavage of the ring or to a rearrangement involving acridine ring formation, depending on the nature of the substituents on N(6).[660]

Direct evidence for the covalent participation of hydroxyl as a neighbouring group has been presented in a paper[661] which also contributes to a study of the reverse reaction, the acid-catalysed scission of epoxides; a study has been made of the isomerization of car-2- and -3-ene oxides over several solid acid and base catalysts,[662] and the kinetics of the thermal gas-phase decomposition of 1,2-epoxypropane has been studied.[663] Treatment of various 1,2-epoxy-3-phenylpropanes with a base appears to proceed by proton abstraction to a carbanion which, in a concerted reaction, affords the corresponding cinnamyl alcohol;[664] the rearrangement of chlorinated β-diketone monomethyl enol ethers with sodium methoxide has been shown to proceed by a mechanism which involves the formation and opening of an oxirane ring;[665] and an improved procedure for the transformation of silyloxiranes to α-trimethylsilyl-ketones, *via* iodohydrin intermediates, has been described.[666] An interesting but unsuccessful attempt[667] to produce an oxirene by thermal rearrangement of 1,4,7,8-tetramethyldibenzobarrelene epoxide produced only 1-acetyl-1,4,5-trimethyl-2,3 : 6,7-dibenzocyclohepta-2,4,6-triene.

Irradiation of a series of tetra-aryl-substituted imidazole *N*-oxides (**409**) in both polar and non-polar media has been shown to afford unsymmetrical benzil di-imine derivatives (**411**), suggesting that fused oxaziridines (**410**) are intermediates.[668] Oxaziridine intermediates have also been invoked for the photochemical isomerization of quinoline *N*-oxides to oxazepines.[669]

(**409**) (**410**) (**411**)

The kinetics of thermal isomerization of 3,3-dihalo-1,2-diphenylaziridines have been determined[670] and the reaction shown to take place by a co-ordinated mechanism involving a polar cyclic transition state with σ-migration of a halogen atom and conrotatory ring scission to give PhCHXCY=NPh (X, Y = various halogens), while a pyrolysis study of some *gem*-dichloroaziridines has delineated the factors controlling the ring-opening reaction and demonstrated the synthetic utility of this reaction.[671] Efforts to prepare phosphorus-substituted heterocycles, by rearrangement of activated aziridine synthons such as [2-aziridin-1-ylalkenyl)-triphenylphosphonium bromides, have been described,[672] 1,2,4-oxadiazoline (**413**) has been obtained from the reaction of sodium nitrite with aziridine (**412**) in the

(**412**) (**413**)

presence of benzoic acid, by a route that is considered to involve formation and decarboxylation of a β-lactam,[673] and strong evidence has been presented for the intermediacy of 5-azabicyclo[2.1.0]pent-2-enes; this in turn indicates a transposition mechanism involving 2,5-bonding, followed by "walk" of the aziridine nitrogen, in the photochemical rearrangement of 2-cyano- to 3-cyano-pyrroles.[674] Under appropriate gas-chromatographic conditions, 1,2,2-trisubstituted aziridines have been shown to undergo rearrangement to the corresponding secondary *N*-1-(alkylidene)amines.[675] The rearrangement products of 2-vinylaziridines have been noted to vary considerably with the nature of substituents on the carbon and nitrogen atoms,[676] a high degree of stereospecificity has been observed[677] in the methyleneaziridine–cyclopropanimine rearrangement, (**414**) to (**415**), and an aziridinium ion has been proposed as intermediate in the oxidative piperidine-ring contraction of the aspidospermane alkaloid, 14-hydroxycathovaline.[678] Aziridines

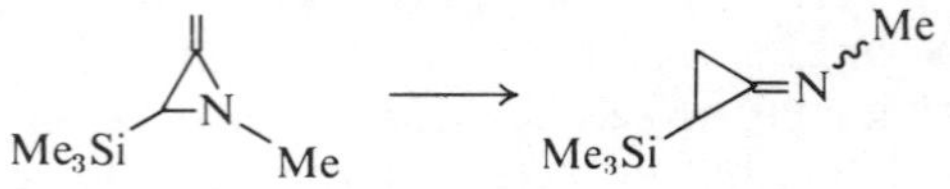

(**414**) (**415**)

have also been invoked to explain the rearrangements of 5,9-methano-6,7,8,9-tetrahydro-5*H*-benzocycloheptenes (*via* their oximes) to the 1,4-ethano-1,2,3,4-tetrahydronaphthalene system.[679]

A mechanism corresponding to the Neber rearrangement, and involving an azirine intermediate, has been discussed for the phenylhydrazine-amination of 3-hydroxyphenalenone to 2-amino-3-hydroxy-1-phenalenone;[680] azirine intermediates have been invoked to account for the pyrolytic transpositions of benzoisoxazoles into benzoxazoles and 2,3-benzoxazin-1-ones into 2,4-benzoxazin-1-ones;[681] and a novel pathway in azirine chemistry, namely a concerted mechanism [see arrows in (**416**)], has been proposed[682] to account for the molybdenum hexacarbonyl-induced rearrangement of (*Z*)-oxovinylazirine (**416**) to the 1,3-oxazepine (**417**). The mechanistic details of the thermal transformation of allyl-substituted 2*H*-azirines to pyridines have been probed[683] and the unusual formation of oxazoles by base- or acid-catalysed ring-opening of 2-acyl-2*H*-azirines has been reported.[684]

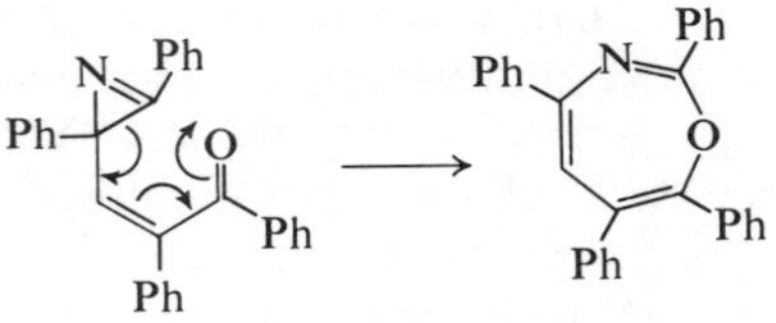

(**416**) (**417**)

The kinetics of the base-catalysed rearrangement of 2-methyl-*N*-nitrosoamino)-acetonitrile (**418**) to α-isonitroso-*N*-methylaminoacetonitrile (**419**) have been studied by differential pulse polarography.[685] Results of the study lend support to the mechanism shown in Scheme 39 for the process.

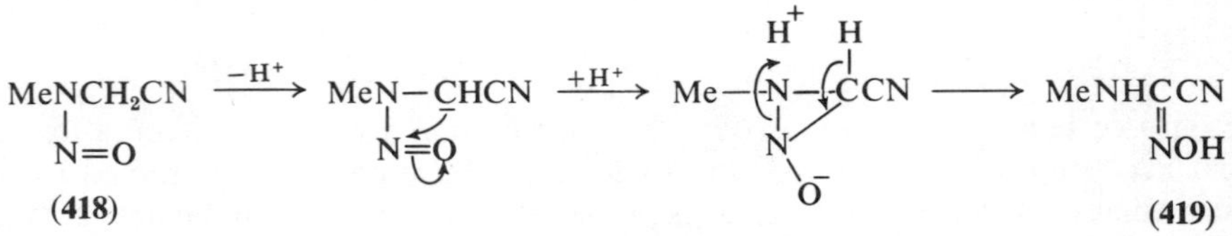

(**418**) (**419**)

Scheme 39

Disilacyclopropane (**421**), which can undergo α-elimination to produce the β-silylsilylene (**422**) and hence the 1,3-disilacyclobutane (**423**), has been propounded[686] as an intermediate in the gas-phase rearrangement of tetramethyldisilene (**420**), and evidence has been presented for the transient formation of 1,1-dimethyl-2-phenylgermirane in the gas-phase reactions of phenyl(trimethylgermyl)carbene.[687]

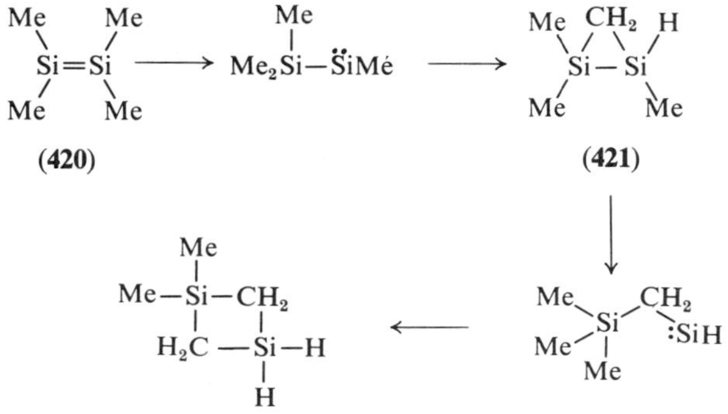

(420) (421)

(423) (422)

Four-membered Rings
The rearrangements of 3-thia- and 3-oxa-nonam analogues of penam and cepham have been reviewed[688] and a critical report on the development of β-lactam chemistry, including their rearrangement reactions, has appeared.[689] A new rearrangement resulting from cleavage of the S—C(5)-bond of penicillin sulphoxide, and possibly involving an episulphonium ion as an intermediate, has been reported,[690] and the rearrangements of a penicillin-derived 4-mercaptoazetidin-2-one have been outlined.[691] A variety of heterocycles have been prepared by ring-transformations of 1-arylazetidin-2-ones,[692] while an attempt has been made[693] to account for the diversity of rearrangement products arising from 1,2-diazabicycloheptenones, such as (**424**). A recent investigation of the cyclization of some 3-(2-pyrrolyl)propionic acids clearly disproves the normally implicit assumption that arylpropionic acids of the type (**425**) cyclize without rearrangement; thus (**425**), as well as giving the expected 4*H*-cyclopenta[*b*]pyrrol-4-ones (**427**), also afford the corresponding 6-ones (**428**) by a single rearrangement, and the cyclopenta[*c*]pyrrol-4-ones (**429**) by a double rearrangement, depending upon the substituents. The spirocyclic intermediate (**426**) is considered to be involved in these transformations.[694]

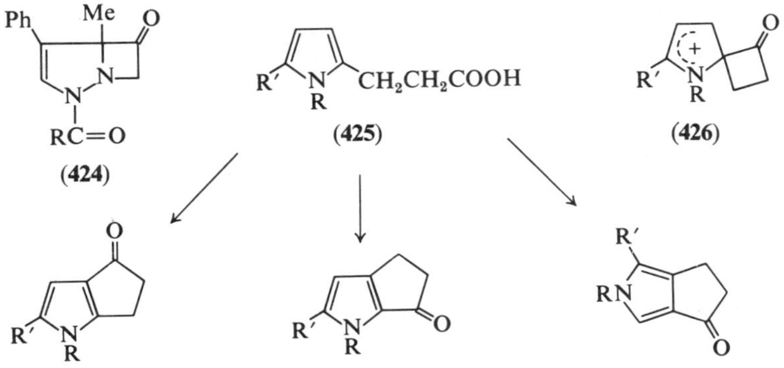

(424) (425) (426)

(427) (428) (429)

A new annelation of cyclobutanones, allowing the development of a net synthon in which a carbanion is β to a carbonyl group, has been described,[695] and its application to the synthesis of steroids has been reported. Moreover, 2-phenylcyclobutanones (**430**) bearing an activating group in the 3-position have been shown readily to undergo acid-catalysed rearrangement to tetral-2-ones (**431**),[696] and this process, too, has been extended[697] to synthesis of polycyclic systems such as steroids. Interestingly, rearrangement of the β,γ-unsaturated ketone (**432**) in an acidic medium affords[698] in two steps the tetracyclic system common to various diterpenes such as phyllocladene (**433**) and its derivatives.

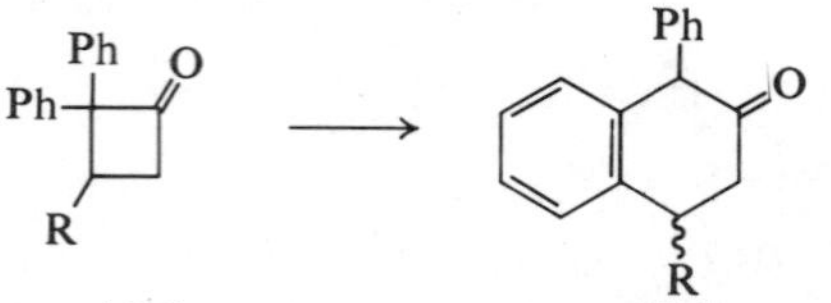

(**430**) (**431**)

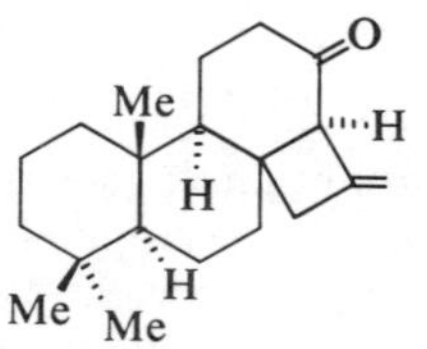

(**432**) 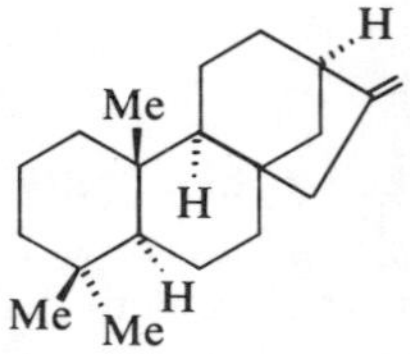 (**433**)

The nature of the rearrangement products obtained by the action of nucleophiles on *trans*-8-tosyloxybicyclo[4.2.0]oct-3-en-7-one has been established;[699] it has been demonstrated[700] that the reaction of *cis*-3,4-dichlorocyclobutene with methoxide proceeds exclusively with *syn*-stereochemistry and with allylic rearrangement to produce *cis*-3-chloro-4-methoxycyclobutene and *cis*-3,4-dimethoxycyclobutene; and activation volumes have been determined for the ring-opening isomerizations of dimethyl 1-*cis*-3-methyl-1,2-diphenylcyclobutene-3,4-dicarboxylate and dimethyl 1-*cis*-3,4-dimethyl-1,2-diphenylcyclobutene-3,4-dicarboxylate.[701]

A rare example of a ready vinylcyclobutane–cyclohexene rearrangement has been provided by the conversion of 1-(2,2-dicyclopropylvinyl)-2,2,3,3-tetracyanocyclobutane (**434**; $R^2 = R^1 =$ cyclopropyl) into the cyclohexene (**436**) in polar solvents. The authors conclude[702] that this rearrangement is a heterolytic process involving the zwitterionic intermediate (**435**), while both a study of the products of the rearrangement of cyclobutenylidene (**437**) and molecular-orbital calculations of

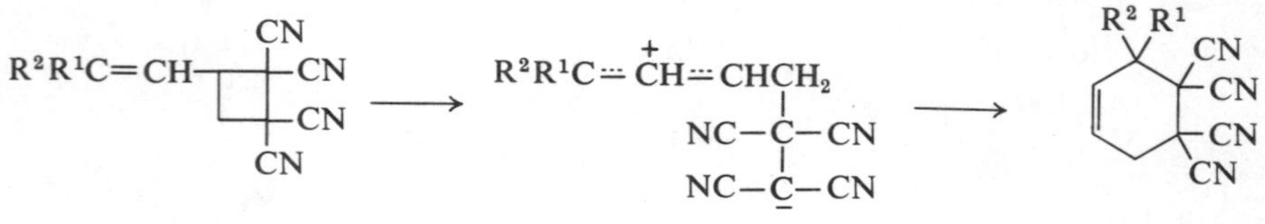

(**434**) (**435**) (**436**)

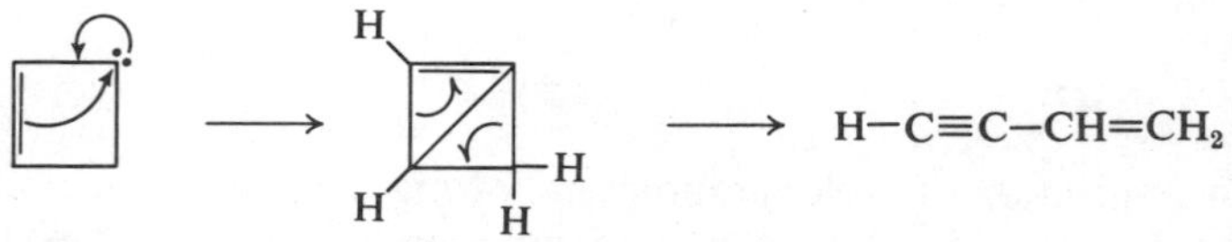

$H-C{\equiv}C-CH{=}CH_2$

(**437**) (**438**)

SCHEME 40

possible reaction co-ordinates indicate[703] an interesting bicyclobutene-like transition state (**438**) and a process as depicted in Scheme 40 for the reaction.

It has been reported[704] that rate-enhancing effects of alkoxy-groups in cyclopropane rearrangements are much smaller in the analogous thermal reorganizations of alkoxycyclobutanes, while a recent kinetic study[705] has emphasized the low reactivity of acenaphthylene-fused cyclobutanes and further demonstrated that acetolysis products obtained from the methanesulphonate of *endo*-7-hydroxy-6b,7,8,8a-tetrahydrocyclobut[*a*]acenaphthylene result from cleavage of an "interior" cyclobutane bond, while in the *exo*-isomer it is an "external" cyclobutane bond that cleaves.

Five-membered Rings

It has been suggested[706] that thermolysis of 2-(1,2,5-triphenyl-1λ^5-phosphol-1-ylidene)acenaphthen-1-one (**439**) gives 7,10-diphenylfluoranthene (**440**) by intramolecular rearrangement and elimination. A key step in the proposed mechanism, which is outlined in Scheme 41, involves ring-enlargement of the strained phosphole ring (a precedented reaction), followed by extrusion of the PhPO component.

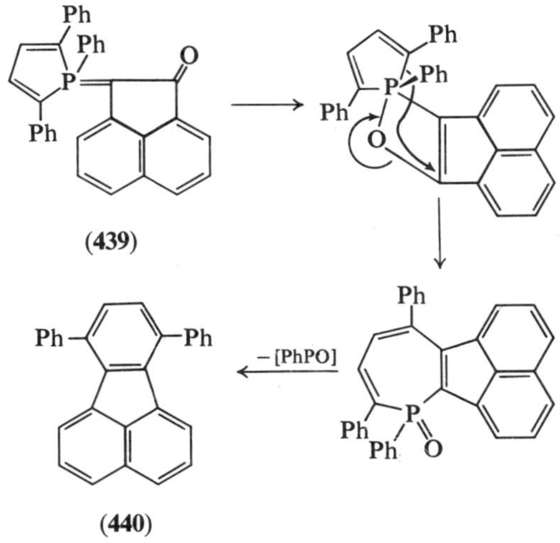

(**439**)

(**440**)

– [PhPO]

SCHEME 41

Another example of ring expansion, this time by migration of sulphur, has been reported for 1,3-benzodithioles (**441**);[707] thus, when (**441**) are subjected to thionyl chloride–triethylamine treatment, they yield the ring-expanded 2-alkylidene-1,4-benzodithians (**442**) by a [1,2]-sulphur migration, while treatment of (**441**; $R^3 = H$) with acid affords 1,4-benzodithiins (**443**). The ring-opening rates of some substituted 3-thienyl-lithium derivatives have been measured,[708] activation parameters have been calculated, and an attempt has been made to explain the unexpected large negative entropy of activation obtained for the ring-opening. Interestingly, the photolysis of dihydrothiophen-3(2*H*)-ones has been shown to follow a different course from that of their oxygen analogues, the resulting 5,6-dithiadecane-2,9-diones being formed by cleavage of the S—C(2) bond (β-cleavage) rather than by α-cleavage.[709]

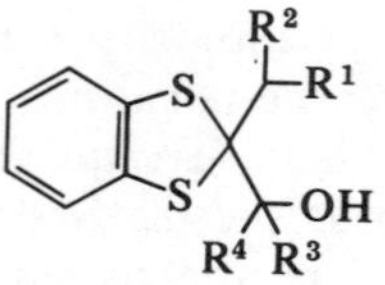

(441) 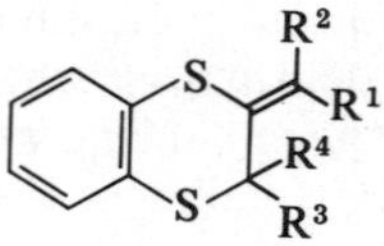 (442) 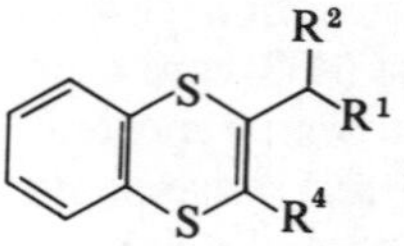 (443)

Several *N*-benzoyl-*N'*-oxamoylureas have been observed to undergo facile skeletal rearrangement easily when heated in methanol to yield 1-aryl-3-(ω-benzamidoalkyl)imidazolidine-2,4,5-triones,[710] while the synthesis of 3-acyltetramic acids *via* an intramolecular rearrangement of α-[*N*-(acylacetyl)amino] cyclic imides has provided[711] a new source of these biologically active compounds. The intramolecular photoaddition of bridged tricyclic alkenyl nitrosamines such as (**444**) and the subsequent rearrangements of the cyclization products have been described,[712] a typical reaction pattern being represented in Scheme 42.

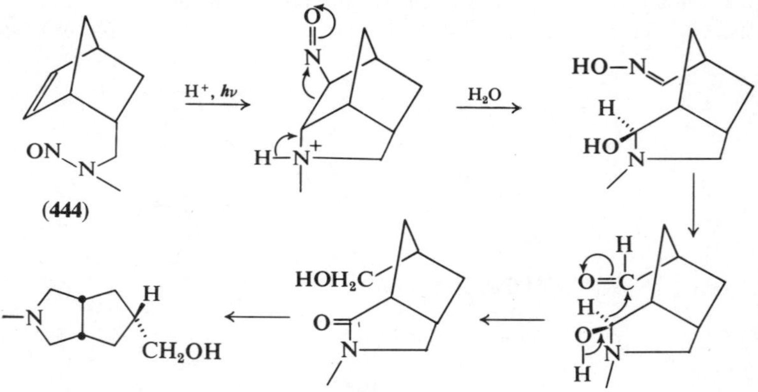

(**444**)

SCHEME 42

Treatment of 2-acetoxy-5-nitro-2,5-dihydro-2-furylmethylene diacetate with methanolic HCl has been shown[713] to yield methyl 5,5-dimethoxy-4-oxopentanoate rather than methyl 5,5-dimethoxy-2-oxopent-3-enoate, as previously suggested. Benzofuran derivatives have been obtained from $\Delta^{\beta,\gamma}$-butenolides[714] and base-treatment of 4-ylidenebutenolides (**445**) has resulted in rearrangement to the corresponding cyclopentene-1,3-diones and the method has been applied to a synthesis of calythrone (**446**) and related compounds.[715] The results of lead tetra-acetate-induced oxidative rearrangement of unsymmetrical 1,3,4,6-tetraketones, *e.g.* (**447**), into dehydroacetic acid analogues (**449**; R = Ph, R' = Me and R = Me, R' = Ph) have been rationalized[716] in terms of a mechanism involving ring-opening of cyclic intermediates such as (**448**). Base-catalysed rearrangement of 2-alkyl-5-hydroxycyclopent-2-enones,[717] general synthesis of 4-hydroxy-2-alkylcyclopentenones from furyl acetate,[718] one-pot conversion of readily available furfuryl alcohols into maltol and related γ-pyrones,[719] syntheses of chrysanthemum-monocarboxylic acid and related compounds from appropriate dihydrofurans,[720] synthesis of 1,1,2,2-tetracyclopropane derivatives by photochemical rearrangement of 2,2,4-triacyl-2,3-dihydrofurans,[721] synthesis of hydrindane derivatives by

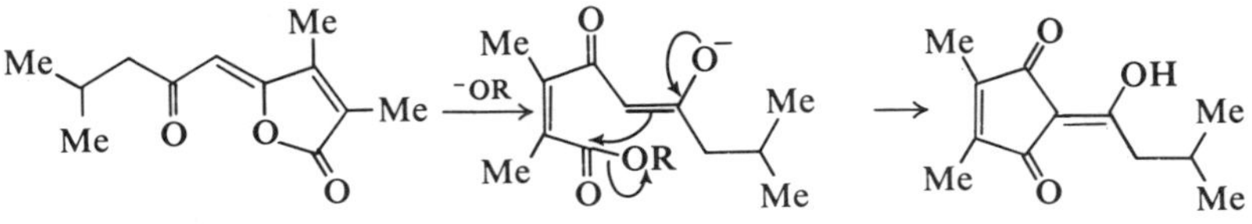

(445) (446)

$PhCOCH_2COCOCH_2COMe$ (447) (448) (449)

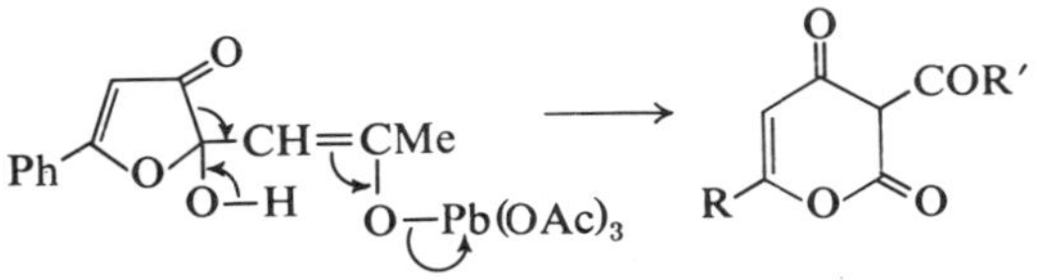

rearrangement of nardofuran,[722] synthesis of 2-aryl-3-hydroxy-1,4-naphthoquinones from 2,2-disubstituted indane-1,3-diones,[723] acid-catalysed isomerization of 1,3-dioxolanes to 3-acetyltetrahydrofurans,[724] thermal rearrangement of 2,3-dioxabicyclo[2.2.1]heptane,[725] rearrangement of 2-(α-bromoalkyl)-1,3-dioxacyclanes,[726] rearrangement of 3-arylidene-2-thiophthalides to bis-(2-arylinden-1-on-3-yl) sulphides,[727] rearrangement of a modified penicillin-derived thiazoline to yield a 6-amino-2-chloromethylpenam ester,[728] and preparation of bis(silyl)-enamines on thermal rearrangement of triazolines[729] also involve reactions that proceed by the opening of a five-membered ring.

Six-membered and Larger Rings

The mechanism and synthetic utility of the base-catalysed opening and reclosure of the adenine ring have been reviewed.[730]

A scheme involving a retro-Friedel–Crafts alkylation followed by aromatization has been proposed[731] in order to explain the acid-catalysed conversion of 1,5,8-trimethoxybenzobarrelene **(450)** into 2,4′,5-trimethoxybiphenyl **(451)**.

The mechanism of the base-catalysed oxygenation of 4-aryl-2,6-di-*tert*-butylphenols **(452)** to 3-aryl-2,5-di-*tert*-butylcyclopenta-2,4-dienones **(454)** has been clarified[732] and a new ring-contraction process for the conversion of the intermediate

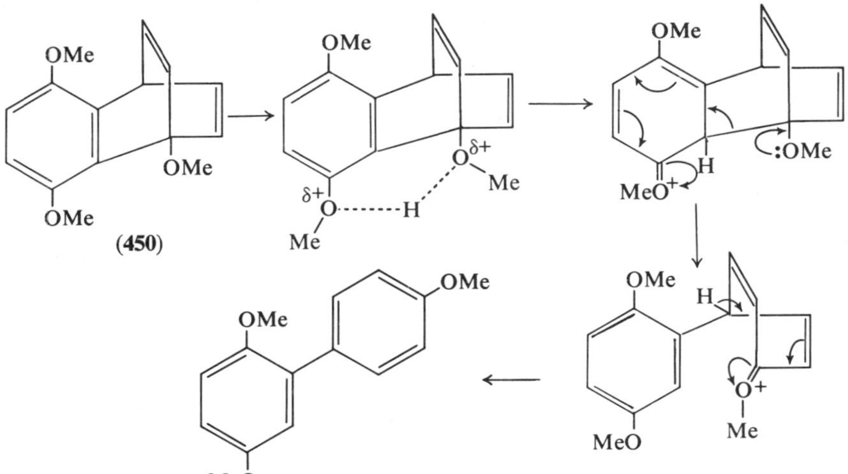

(450) (451)

epoxy-*o*-quinols (**453**) into cyclopentadienones (**454**) has been revealed; see Scheme 43. An "abnormal" base-catalysed isomerization of a humulone to a cyclopentenone,[733] and the Lewis acid-induced ring-contraction of chromanones to

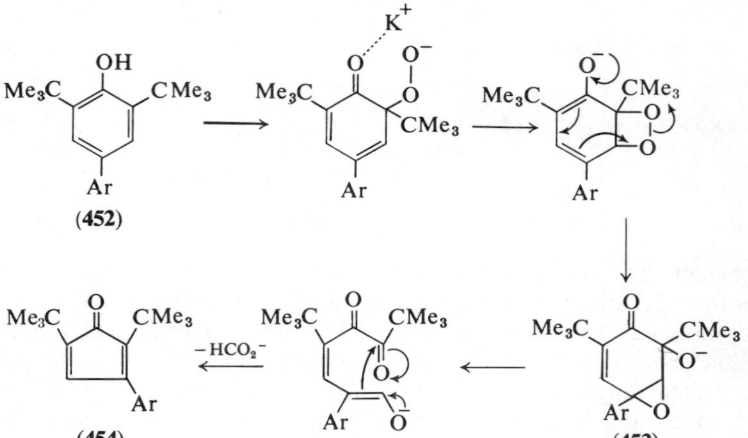

SCHEME 43

4,7-dimethoxyindan-1-ones[734] have been reported, while the observed rearrangement of vinyl-*β*-hydroxy-ketones (**456**) to (**457**) emphasizes the importance of readily available hydrogenated indenes of type (**455**) as intermediates in the synthesis of the tricyclo[6.3.0.0^{2,8}]undecane ring system.[735]

The conversion of semicarbazones, *N*-benzoylhydrazones, and *p*-nitrophenylhydrazones of 4-oxo-4*H*-[1]benzopyran-3-carboxaldehydes into (hydroxybenzoyl)-pyrazoles has been reported[736] and a novel ring-opening/ring-closure sequence has been described which allows the synthesis of pyridaz[4,5-*b*]carbazoles from 5-acylpyrido[4,3-*a*]carbazoles.[737]

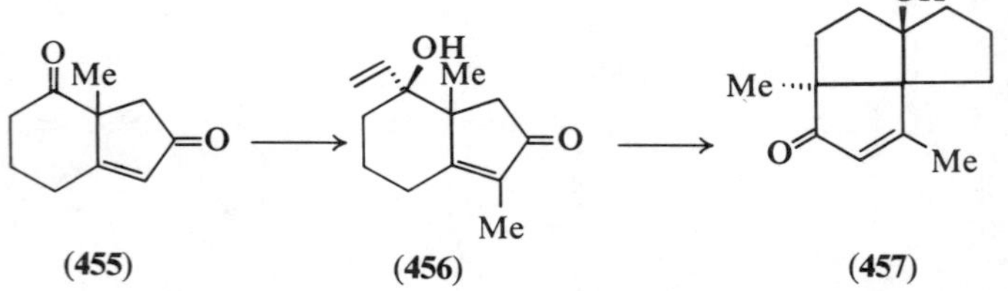

S-Alkyl and *S*-aryl *Z*-perchloropenta-2,4-dienethioates (**458**; R = SR′) have been shown to rearrange thermally *via* the pyran intermediates (**459**) and (**460**) into the 5-thiosubstituted acyl chlorides (**461**) and (**462**).[738] Likewise, the pentadienone (**458**; R = Ph) rearranges *via* pyran intermediates with the predominant formation of the *Z,Z*-acyl choride (**461**; R = Ph). Increasing temperature appears to accelerate the establishment of an equilibrium with the *Z,E*-acyl chloride (**462**; R = Ph), a fact which is interpreted on the basis of different ring-opening directions of the common pyran intermediate (**460**).[739]

Ring-opening and -closure sequences are also implicated in the intramolecular rearrangements of 6-acetyl-2,9-dioxa-1-azabicyclo[4.3.0]nonanes[740] and in the

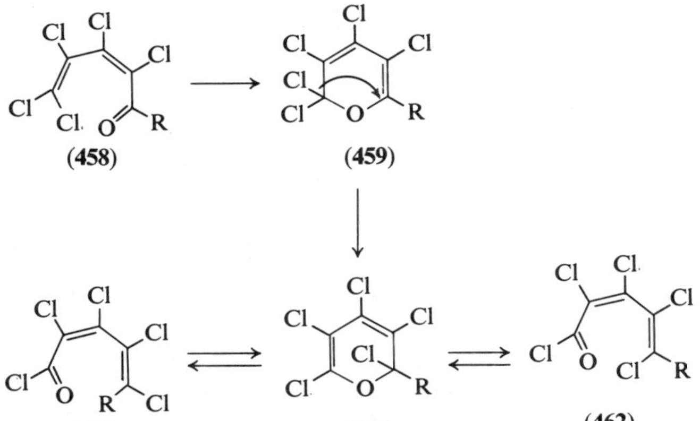

(458) (459) (461) (460) (462)

synthesis of alkyl- and aryl-substituted dihydropyrans from 4-methyl-4-phenyl- and 2,4,4,6-tetramethyl-1,3-dioxans.[741] The effect of acid, hence the reversibility of ring-opening and colour formation, on a series of thermochromic spiropyrans, has been examined.[742]

The ring-contraction/ring-expansion transformation of 6β-bromo-4,4-dimethyl-1α,2α-epoxy-3α,5α-oxidocholestane into 6β-bromo-4,4-dimethyl-3α,5α-oxido-5(10$\rightarrow$1βH)abeo-10(19)-cholesten-2α-ol has provided[743] a model for the possible biosynthetic route of the grayanane carbon skeleton from (−)-kaurene. Oxa[17]-annulenes have been obtained by photolysis of the monoepoxides of the tricyclic dimer of cyclo-octatetraene[744] and it has been shown that, in a similar manner, UV light cleaves all the endocyclic σ-bonds of the tetracyclic compounds formed on reaction of cyclo-octatetraene with ethoxycarbonylcarbene, to produce cyclo-heptadecaoctaenes.[745]

The reported[746] internal translactonization of the type **(463)**$\rightarrow$**(464)** represents a useful new approach to the generation of macrocyclic lactones. A similar trans-amidation, which converts structures of the type **(465)** to **(466)**, has been utilized[747] in the synthesis of 17- and 21-membered polyamino-lactams.

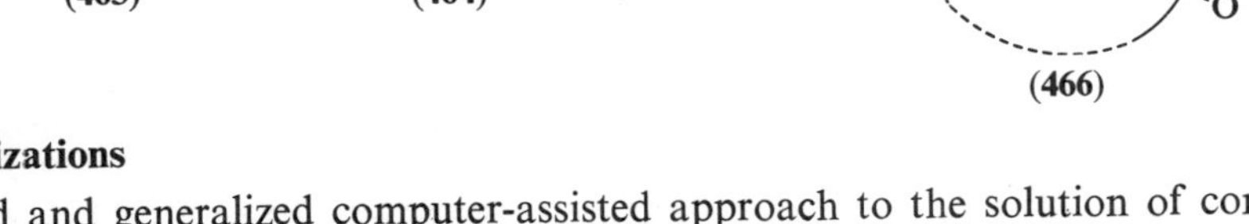

(463) (464) (465) (466)

Isomerizations

A rapid and generalized computer-assisted approach to the solution of complex problems in dynamic stereochemistry has been described,[748] graphs have been derived for intramolecular rearrangements of tetragonal-pyrimidal complexes,[749] and a theoretical study concerning mechanisms of the molecular rearrangement of carbonyl compounds to oxacarbenes in the lowest triplet state has been carried out.[750] By use of the self-consistent electron-pair method, the unimolecular rearrangement of vinylidene into acetylene has been studied[751] as a model for a potentially important pathway for the disappearance of carbenes and nitrenes. Single-pulse shock-tube measurements have been reported[752] in an attempt to resolve conflicting values for the rate of isomerization of allene to methylacetylene, while the fact that $HC{\equiv}^{13}CD$ eliminated from ethenoanthracene at 600 °C undergoes rearrangement at higher temperatures to $DC{\equiv}^{13}CH$ without appreciable intermolecular hydrogen or deuterium exchange suggests that unlabelled acetylene might also undergo degenerate rearrangement at similar temperatures. The authors suggest[753] that the rearrangement may go through methylenecarbene, $H_2C{=}C{:}$ as an intermediate although they hint that a species of higher energy may be involved in the intervening pathway.

A new type of halotropy has been observed in a buta-triene–enyne rearrangement.[754] Prototropic rearrangements have been observed in the reactions of 4-alkoxy-3-chlorobutan-1-ols with bases,[755] prototropic isomerizations of unsaturated sulphides, sulphoxides, and sulphones have been examined and the thermodynamic stability of the isomers calculated and interpreted in terms of the electronic effect of alkyl groups and of the steric demands of sulphide, sulphoxide, and sulphone groups,[756] while a similar study has been made of the prototropic isomerization of variously substituted isopentene derivatives.[757]

Evidence has now been provided[758] to confirm the rôle of allylic sulphinic acids as intermediates in the rapid, regiospecific isomerization of olefinic double bonds in liquid sulphur dioxide, while a study of the degree of double-bond shift during ketalization of cycloalkenones has shown[759] that the shift is dependent on ring-size and location of substituents; for example, it has been shown that the double bond in bicyclic systems containing enones in a six-membered ring has a strong tendency to shift to the β,γ-position on ketalization.

Double-bond migration in alkenes has been effected by nitrogen dioxide adsorbed on various adsorbents,[760] by ZnX zeolites in the presence of adsorbed sulphur dioxide,[761] and the mechanisms of double-bond migrations, initiated by superbasic or one-electron centres, have been discussed.[762] In all these cases, *cis*$\rightarrow$*trans*-isomerization appears as a competing reaction although, interestingly, a recently reported study[763] of the treatment of cinnamic acids with polyphosphoric acid has provided no evidence to support the previously claimed contrathermodynamic isomerization (*trans*$\rightarrow$*cis*-isomerization) of these acids. The presence of *cis–trans*-isomerism at ambient temperatures in polymethylbis(trihalogenoacetyl)benzenes has been demonstrated[764] for the first time by the observation of separate ^{1}H-NMR signals, emphasis has been drawn to the rôle of steric effects in the fast thermal *cis*$\rightarrow$*trans*-isomerization of overcrowded ethylenes as represented by *N,N'*-dimethylbiacridans,[765] while a study of geometric isomerization in model annulated azonines has firmly established the crucial influence that lone-pair mobility has in determining a system's preference for a specific geometric arrangement.[766] The interesting protolytic/photochemical *cis*$\rightarrow$*trans*-isomerization of a stilbazolium betaine (represented by M) has been shown to constitute a complete molecular reaction cycle,[767] namely; $M_{trans} \rightleftarrows MH^+_{trans} \rightleftarrows MH^+_{cis} \rightleftarrows M_{cis} \rightarrow M_{trans}$, which may well serve as a chemical model for the storage of information and subsequent regeneration of the information carrier in, for instance, biological systems.

Other examples of *cis*$\rightarrow$*trans*-isomerizations have been reported; these include the photochemically-induced isomerizations of α-(4*H*-3,1-benzothiazin-4-ylidene) ketones and esters,[768] the base-catalysed isomerization of bis-(4-benzylideneaminocyclohexyl)methane,[769] and the ready isomerization of 1,2-dialkoxy-1-haloethenes;[770] the gas-phase reactions of nuclear recoil generated ^{38}Cl atoms with 2,3-dichlorohexafluorobut-2-ene have been shown to result predominantly in chlorine-for-chlorine substitution in a reaction that proceeds with geometric isomerization.[771] The isomerization rate of the iodine-catalysed *cis–trans*-stereomutation of optically active 1,4-dialkylbutadienes has been shown to depend on the substituents of the diene component.[772]

A study of the rotational propensity of substituents in the thermal *cis–trans* isomerization of cyclopropanes has provided[773] a further example of the principle of additivity of substituent effects as a quantitative method for estimating the activation energies of not-obviously-concerted thermal rearrangements.

Polarimetry and ^{1}H-NMR spectroscopy have shown that the thermal epimerization of (−)-(2*S*)- and (+)-(2*R*)-2-[(*R*)-α-methylbenzyl]-3,3-diphenyloxaziridines involves neither bond cleavage nor racemization, but only a nitrogen inversion mechanism.[774] The *ab initio* molecular-fragment method has been used to calculate two possible intermediate states in the *syn–anti*-isomerization of formaldoxime and the authors suggest[775] that the intermediate is stabilized by a hybridization change that results in the formation of a linear C–N–O π-system. Energy barriers for the *cis–trans* isomerization of a series of *N*-sulphenylimines have been interpreted in terms of hyperconjugation involving the nitrogen lone-pair in an inversion mechanism,[776] while it has been shown that azides (**467**) are stabilized in the open-chain form by a slow *cis–trans* isomerization about the C=N bond and that electron-withdrawing substituents in Ar′ can either lower (by the operation of a mesomeric effect) or raise (by the inductive effect) this inversion barrier.[777]

Cyclization of *O*-acylazidoximes to tetrazoles appears to be initiated by protonation of the C-bonded azide-nitrogen atom which bears the highest negative charge,[778] while *O*-sulphinylated oximes (**468**), obtained from the reaction of oximes with sulphinyl chlorides, readily isomerize to the thermally stable *N*-sulphonylimines (**469**) in a process that has been shown to involve homolysis of the N—O bond to iminyl and sulphonyl radicals followed by radical recombination with N—S bond formation.[779]

Ar(N$_3$)C=N̈–N̈=CHAr′ (**467**)

RR^{1}C=N–O–S(=O)R^2 (**468**) ⟶ RR^{1}C=N· + ·O–S(=O)R^2 ⟶ RR^{1}C=N–S(=O)$_2$R^2 (**469**)

A mechanistic appraisal of the intra- or inter-molecular thermal isomerization of allylic alcohols to saturated ketones has appeared,[780] while evidence has been presented[781] to suggest that the mechanism of thermal isomerization of 1,2-diolates is compatible only with a redox pathway and not with the reverse pinacol-coupling pathway originally proposed.

Labelling studies have shown[782] that pentadienyl hydroperoxides undergo thermal rearrangement by a process in which the oxygen atoms of the hydroperoxy-group exchange with atmospheric oxygen, while the results of a study[783] of the effect of high pressure on the iosmerization of organosilicon peroxides, $R_3SiOOSiR_3$, to $ROSiR_2OSiR_3$ are consistent with earlier mechanistic proposals.

Tautomerism

A review of tautomeric rearrangements of Group IV and V element-substituted carbonyl compounds has appeared,[784] and *ab initio* molecular-orbital theory has been used to examine the effect of simple π-electron-accepting substituents (Li, BeH, BH_2) on the keto–enol equilibrium in the acetaldehyde–vinyl alcohol system.[785] A theoretical study of the influence of conjugative effects on the enol–enethiol tautomerism of β-thioxoketones has been carried out,[786] and an NMR spectroscopic investigation in various solvents has indicated that the enol tautomer of thioacetylacetone is favoured in less polar solvents while the enethiol tautomer predominates in more polar solvents.[787] Tautomerism of the enedione system of 15-oxoprostaglandin D_2 has been examined[788] and ring–chain tautomerism has been observed on acid treatment of certain iridoid aglycones.[789]

It is proposed[790] that preparation of bicyclo[2.2.2]octan-2-ones and bicyclo-[2.2.1]heptan-2-ones by thermal cyclization of 3-alkenylcyclohexanones **(470)** proceeds *via* the enol tautomer (**471**) by a process in which the enol-hydrogen is transferred to the terminus of the double bond resulting in the formation of a new carbon–carbon bond; see (**471**)→(**472**).

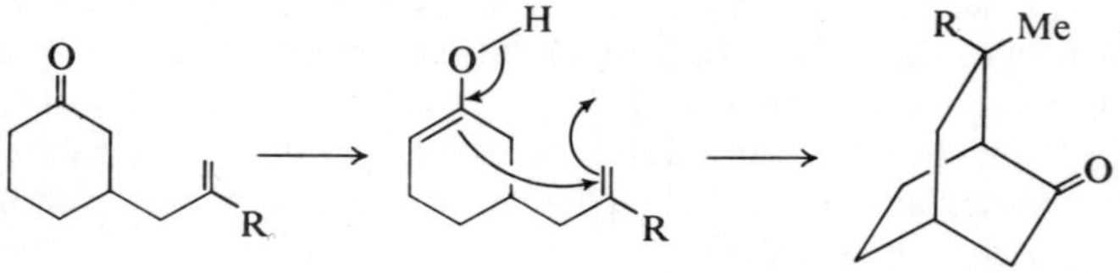

(470) **(471)** **(472)**

Ring–chain tautomerism of various aldehyde semicarbazones, namely **(473)** ⇄ **(474)** has been studied by NMR spectroscopy, and the results have indicated that protonation of the C=N nitrogen atom and steric hindrance of the 2-methyl group (R^1 = Me) with an aromatic group at the C=N carbon atom (R^3 = Ar) in the

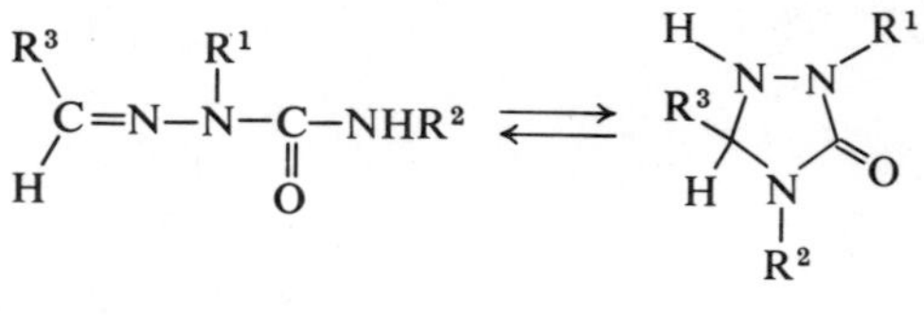

(473) **(474)**

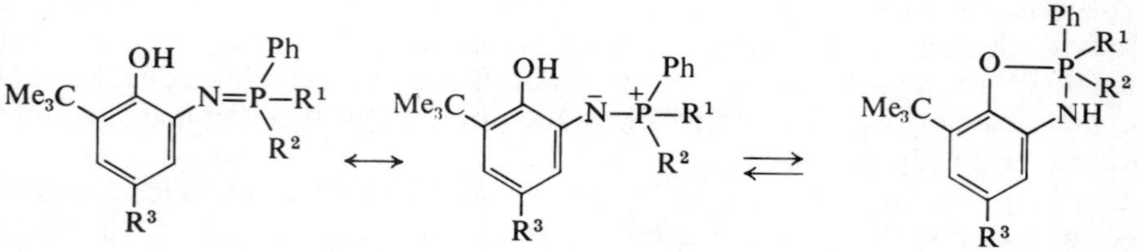

(475) **(476)**

chain form promotes cycloisomerization to ring tautomers;[791] it has also been shown that the position of equilibrium of the two valence tautomeric forms (**475**) and (**476**) of (2-hydroxyphenylimino)phosphoranes depends strongly on the substituents as well as on the solvents used; consequently, it is possible to prepare either iminophosphoranes or benzoxazaphospholines.[792]

Kinetic data obtained for both the base- and acid-catalysed rearrangements of a series of methyl pseudo-8-benzoyl-1-naphthoates and pseudo-2-benzoylbenzoates in methanol to the corresponding normal esters are consistent with a stepwise mechanism in which the formation or decomposition of tetrahedral intermediates is rate-determining.[793]

The strength of intramolecular hydrogen-bonding in various 2-(arylaminomethylene)cycloalkanones has been found to be dependent on the size, and substitution pattern, of the cycloalkanone unit;[794] the effects of substituents and solvents on the tautomerism of 4-arylazonaphth-1-ols have been discussed;[795] in aqueous solution, hydroxy-1,5-naphthyridines have been shown to exist predominantly as the pyridine tautomers.[796] It has been reported[797] that under certain conditions cyclopenta[*c*]quinolines show a pronounced preference to exist as the NH-tautomer, while their 5-methyl derivatives, in sharp contrast to the cyclopenta[*b*]quinoline system, show no resemblance whatsoever in visible spectra to that of the azulene analogues. A resonance explanation is offered to account for the spectral differences between the two systems and for the varying chromoisomerism of cyclopentaquinolines themselves. The first example of pyrimidine–pyrimidylidene tautomeric equilibrium in the pyrimidinyl-2-methane system has been observed.[798]

Use of the temperature-jump relaxation technique has shown that the tautomeric interconversion of 6-methoxypyrid-2-one involves a concerted proton-transfer.[799] Apparently, this transfer requires an association between the substrate and an acid–base bifunctional catalyst that can be another substrate molecule or a hydroxylated molecule. Thus, in anhydrous media, the interconversion-limiting step is a diffusion-controlled dimerization of the substrate: when water is present, competition between this mechanism and a water bifunctional catalysis is observed. Moreover, it has been shown that variations of the tautomeric equilibrium constant of heteroaromatic systems in water–aprotic solvent media are directly related to the activity of water.[800] The same group of workers have further shown that cation-binding strongly favours the lactam tautomer of 2-hydroxypyridines,[801] a result which, on extrapolation to the nucleic acid bases, may cast a new light on the theory of spontaneous mutagenesis. Temperature-jump relaxation spectrophotometry has shown that in contrast with the findings for α-pyridones, the tautomeric interconversion of 5-methyl-3-phenylpyrazole implies two successive intermolecular proton transfers.[802]

Spectroscopic examinations have been made of amino–imino tautomerism in several 1-alkyladenines[803] and in neutral and cationic *N*-substituted 4-aminopyrazolo[3,4-*d*]pyrimidines.[804] In the latter instance the biochemical implications of the location of the basic sites in these compounds, and of the existence of the imino-tautomers, have been discussed. Interestingly, it has been demonstrated[805] that the temperature-dependent tautomeric equilibrium between N(1)—H and N(2)—H forms in 7-amino-3β-D-ribofuranosyl-1*H*-pyrazolo[4,3-*d*]pyrimidine (formycin), an important analogue of adenosine, is faster than the turnover rates for many enzymes. This means that, even though the 1-H tautomer is predominant, the quick tautomerization to the 2-H form occurs sufficiently rapidly to obviate its limiting the rate of an enzymic process that is specific for the 2-H species.

A theoretical study of the vinyl azide-*v*-triazole isomerization has permitted the reaction to be classed as a 1,5-dipolar electrocyclic reaction.[806] The effect of substituents on the azido–tetrazolo equilibrium[807] in azido-thiazoles, -benzothiazoles, -thiadiazoles, and -isoxazoles, and in 5-aryl-9-methyl-*s*-triazolo[4,3-*c*]tetrazolo-[1,5-*a*]pyrimidines[808] has been examined, while failure to induce thermal or photochemical valence tautomerization of substituted 2-phenyl-4-phenylazo-benzotriazoles[809] provides an unexpected limitation to the Boulton–Katritzky rearrangement.

An NMR-spectroscopic examination of a series of recently prepared alkyl- and aryl-substituted thieno[3,2-*b*]thiophene-2,5-diols showed that all existed as dithio-lactones, no evidence being found for the presence of the hydroxy-forms of these compounds.[810] NMR spectroscopy has also been used to investigate 2-aryl-iminothiazolidine–2-arylaminothiazoline tautomerization,[811] the tautomerism of 3-acetyl-5-isopropylpyrrolidine-2,4-dione,[812] and of 2,5-disubstituted pyrroles;[813] and the rate of intramolecular exchange of 5- and 6-fluorobenzofuroxans has been measured by analysing the line shapes of their NMR signals.[814]

A reversible photovalence tautomerism has been reported in diaryl-substituted cycloheptatrienes;[815] see (**477**) ⇄ (**478**). It has been postulated that the observed spontaneous racemization of 1,2- and 3,4-epoxydihydrophenanthrene proceeds by an epoxide–oxepin valence tautomerism,[816] while an interesting report has

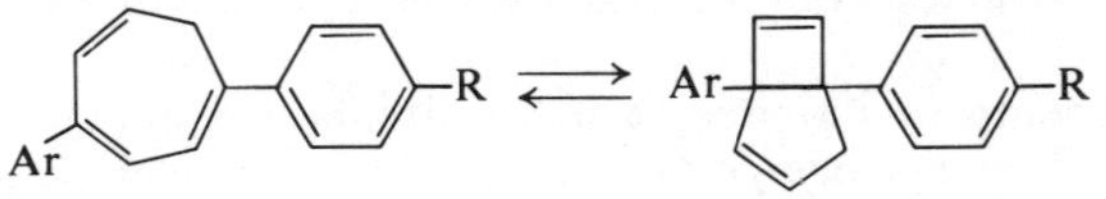

Ar R Ar R

(**477**) (**478**)

Ph Me₃C Me N O O CMe₃ NO₂ Me N Ph X X

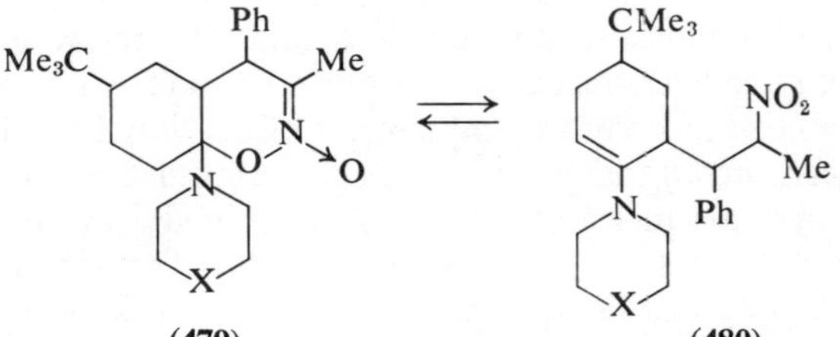

(**479**) (X = O or CH_2) (**480**)

appeared of the unusual ring–chain tautomerism of cyclic nitronic esters of the type (**479**) to trisubstituted enamines (**480**).[817] This represents the first example of reversible formation of an enamine from a heterocyclic compound.

Acyl Derivatives

A review that places special emphasis on mechanistic studies deals with the 1,2-intramolecular shift of alkoxycarbonyl groups to carbonium sites.[818]

It has been proposed that the formation of aldehydes (**482**) and (**483**) on photolysis of citral (**481**; X = CHO) at elevated temperatures requires the unusual 1,2-shift of a formyl group in a free-radical reaction.[819] Moreover, further exploration of this process has shown that the photochemical isomerization of geranonitrile (**481**; X = CN) to (**484**) occurs in a process that may be mechanistically

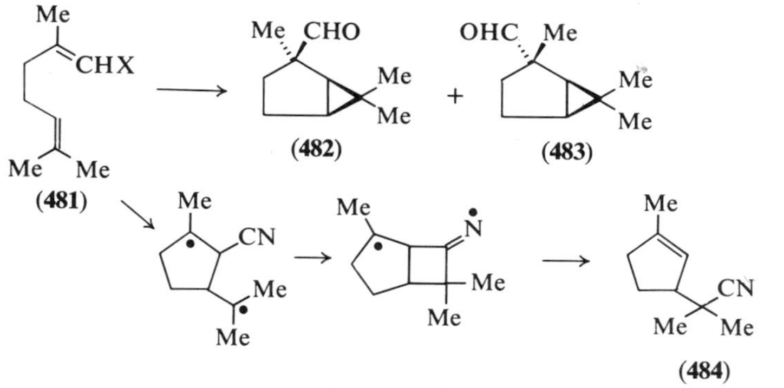

(481) (482) (483) (484)

SCHEME 44

related to the isomerization of citral, and thus provides an example of a new type of 1,6-rearrangement of a cyano-group in a biradical intermediate;[820] see Scheme 44.

The recent identification of appropriately labelled gentisic acid as a product formed by strains of *Bacillus stearothermophilus* utilizing labelled 4-hydroxybenzoate has been shown to be in accord with a reaction mechanism involving an intramolecular *ortho*-migration of a carboxy-group as a consequence of C(1)-hydroxylation.[821] The migration of an ester group has been observed during the irradiation of a pyrido[2,1-*b*]thiazoletetracarboxylate,[822] and the migratory aptitudes of a variety of acyl groups have been measured by examining 1,5-acyl shifts (see Scheme 45) in optically active indenes such as (**485**).[823] The fast migration of formyl,

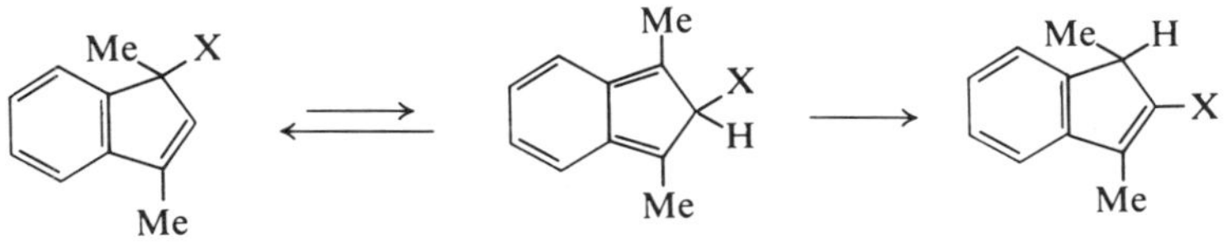

(**485**) SCHEME 45

acetyl, and benzoyl groups has been explained, at least in part, by envisaging a secondary interaction (between the π-system of the migrating group and the diene unit) acting in support of the primary interaction between the migrating σ-bond and the diene system. It has been reported[824] that Lewis acid-catalysed 1,2-benzoyl migration in both (*Z*)- and (*E*)-1,3-diphenylbut-2-en-1-one oxide proceeds in non-polar solvents with 100% inversion of configuration at the migration terminus, thus excluding a long-lived, freely rotating carbenium-ion intermediate on the reaction pathway, and supporting a concerted migration in which the carbonyl moiety migrates concurrently with oxirane ring-opening. The acylation of 3-alkylindoles has been envisaged as involving initial attack at the 3-position of the indole nucleus followed by migration of the acyl group to afford 2-acyl-3-alkylindoles,[825] and formation of 5-acetoxy-4-hydroxybenzo[*g*]indole derivatives from the reaction of 2-acetoxy-1,4-naphthoquinone with enamines has been shown to be the result of acyl migration, the course of which appears to be partially intermolecular with participation of the carboxylic acid used as solvent.[826] Carbamoyl group migration has been demonstrated to occur in synthesis of methyl 3-*O*-carbamoyl-α-D-mannopyranoside,[827] while a recent study of the reaction of a series of tertiary glycidamides with boron trifluoride etherate has established[828] that amide migration is not as general a phenomenon as ketone, ester, or thiol ester migration. The apparent reluctance of the amide group to migrate has been explained as being due to its high relative basicity. Interestingly, the thermal conversion of the stearamido-alcohol, $C_{17}H_{35}CONHCH_2CH_2OH$, to the ester $C_{17}H_{35}CONHCH_2$-$CH_2OCOC_{17}H_{35}$ does not involve a direct cleavage of the amide but rather appears to involve an N→O acyl transfer between two molecules of the amido-alcohol,[829] while attempted *O*-benzylation of (**486**) has been shown to yield (**487**) in what appears to be the first example of an ethoxycarbonyl group migration from an amide-nitrogen to oxygen.[830] The reactions of *N*,*N*-dimethylcyclohex-1-enylamine *N*-oxide with various acylating agents have been found to afford products derived from β-substitution of an intermediate acyloxyenammonium salt,[831] 10-acylphenothiazine 5-imide 5-oxides (**488**; R = COR^2, R′ = H) and 10-acyl-10,11-dihydrodibenzo[*b*,*f*][1,4]thiazepine 5-imide 5-oxides (**488**; R = COR^2, R′ = H) have been shown to undergo base-induced transannular migration of acyl groups to provide the 5-acylimide compounds (**489**; R = H, R′ = COR^2) and (**489**; R = H, R′ = COR^2), respectively,[832] and the effect of substituents on the kinetics of non-degenerate acyl rearrangements of *N*-picryl and *N*-4-nitrobenzoyl derivatives of

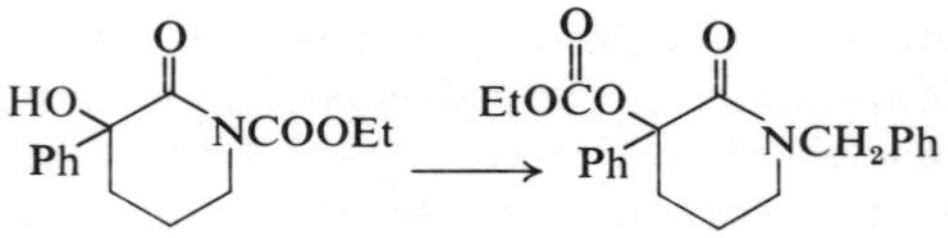

HO, Ph, O, NCOOEt → EtOCO, O, O, Ph, NCH_2Ph

(**486**) (**487**)

R, N, S, O, NR′ R, N, S, O, NR′

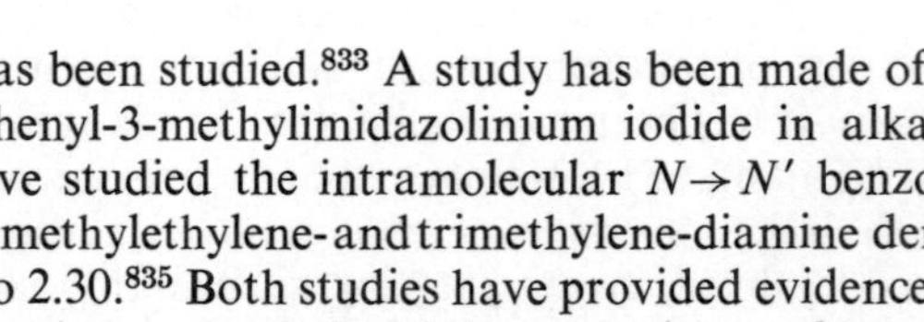

(**488**) (**489**)

diarylbenzamidines has been studied.[833] A study has been made of *N*→*N*′ benzoyl migration of 1,2-diphenyl-3-methylimidazolinium iodide in alkaline solution;[834] the same authors have studied the intramolecular *N*→*N*′ benzoyl migration of *N*′-aryl-*N*-benzoyl-*N*-methylethylene- and trimethylene-diamine derivatives over the H_0-pH range −2.28 to 2.30.[835] Both studies have provided evidence for mechanisms that involve the formation and catalysed decomposition of tetrahedral addition intermediates. The effect of detergents on the *S*→*N*-acyl transfer of *S*-acyl-β-mercaptoethylamines has been examined.[836]

Quantitative data for the rapid 1,2-shift of acyl groups in the acid-catalysed decomposition of α-hydroperoxy ketones, $RCOC(OOH)R^1R^2$, have been published.[837] The results suggest that the order of acyl migration is determined by the electron-releasing power of R, while the ease of migration is related to the resonance stabilization with lone-pair electrons of carbonyl-oxygen. Finally, the application of photochemical 1,3-acyl migrations to the synthesis of a series of tris-σ-homobenzene ketones has been reported.[838]

Polycyclic Rearrangements

An investigation of the valence isomerism between cage compounds (**490**) and dienes (**492**) has been carried out.[839] An attractive pathway proposed for the process involves initial homolysis of the C(3)—C(4) bond to give the intermediate radical (**491**). Interestingly, the rates of diene formation from (**490**) increase as the length of the bridge "X" increases. On the basis of a detailed kinetic study, an intermediate 1,4-dihydronaphthalene-1,4-diyl diradical has been postulated[840] for the thermal isomerization of benzobicyclo[2.2.0]hexa-2,5-diene to naphthalene, and correlation diagrams have been produced for the cyclization of bicyclobutane-2,4-diyl to tetrahedrane and cyclobutadiene.[841]

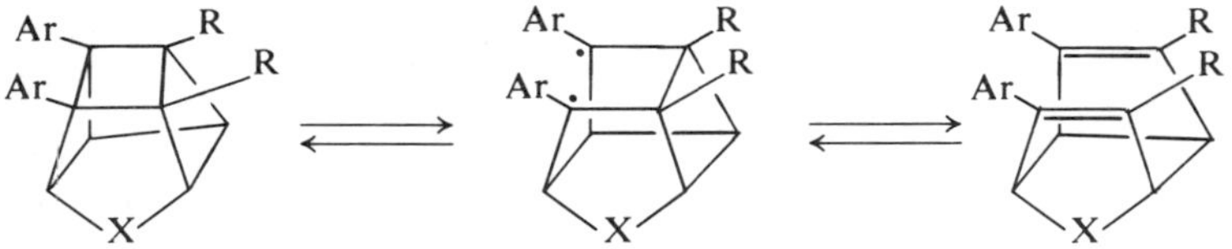

(**490**) (**491**) (**492**)

Application of ESR spectroscopy has shown that bicyclo[4.2.0]octa-3,7-diene-2,5-semidione is a more stable valence isomer than the monocyclic cycloocta-2,5,7-triene-1,4-semidione,[842] line-shape analysis of ^{1}H- and ^{13}C-NMR spectra has been used to determine the activation parameters of the valence isomerization of substituted spironorcaradienes/spirocycloheptatrienes, (**493**) ⇄ (**494**),[843] while NMR spectral data have shown that the mutual interconversion of *cis*-bicyclo-[6.1.0]nona-2,4,6-triene, [(**495**) ⇄ (**496**) ⇄ (**497**) ⇄ (**498**)], proceeds stereospecifically with inversion at the migrating carbon atom C(9).[844]

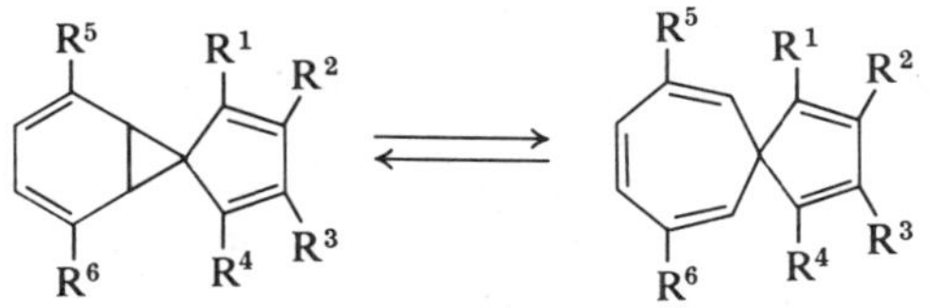

(**493**) (**494**)

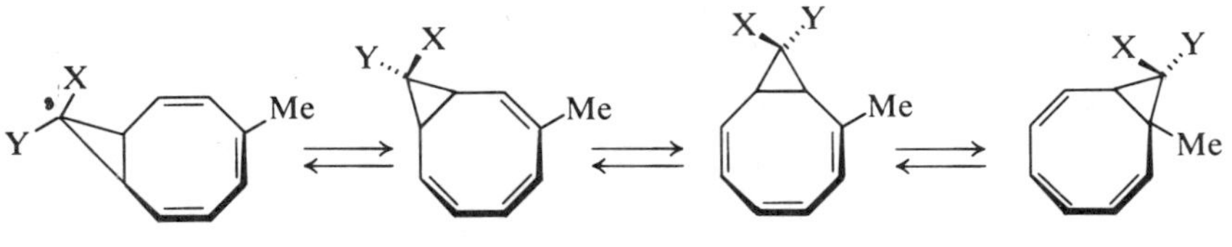

(**495**) (**496**) (**497**) (**498**)

(X, Y = CN or Me)

The generality of the slither, or bicycle, rearrangement has been examined in some new systems.[845] In these, the 3-ring carbon of a 2-methylenebicyclo[3.1.0]-hex-3-ene has been shown to "walk" around the 5-ring and also out onto the exocyclic π-bond. The easy thermal isomerization of *syn*-tricyclo[4.2.1.1^{2,5}] deca-3,7-diene (**499**) to (**500**) to the exclusion of thermodynamically more favoured products can be regarded as an exploitable benefit of orbital symmetry.[846] Thermal isomerization of tetraene (**501**) has provided a novel route to bridgehead bicyclo-[3.4.1]alkenes such as (**504**). The reaction has been formally viewed[847] as an intramolecular Diels–Alder reaction to give (**502**) followed by cleavage of the bond between C(7) and C(8) to produce diradical (**503**) which then undergoes transannular ring closure to (**504**).

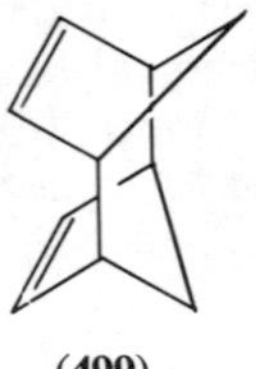

(**499**)  (**500**)

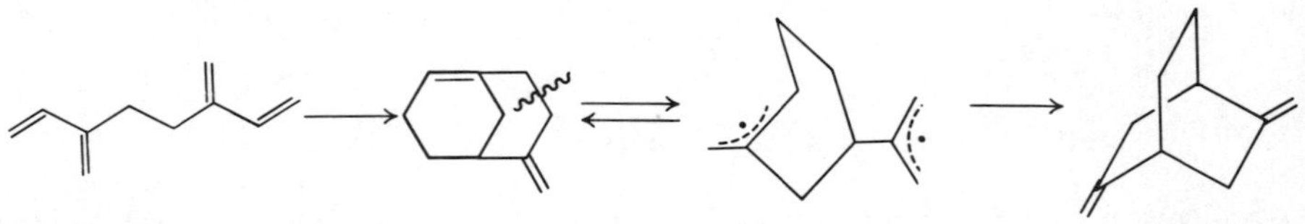

(**501**) (**502**) (**503**) (**504**)

Recent crystallographic findings for (−)-camphene-8-carboxylic acid have been related[848] to earlier explanations of the *exo/endo* Nametkin rearrangement. These new results support the *exo*-migration pathway for the rearrangement. Other polycyclic isomerizations studied include the base-catalysed rearrangement of 7-methylenebicyclo[4.1.0]hept-2-ene[849] and the reversible thermal reaction of the photoisomer of 6,7,14,15-tetrahydro-15,16[1′,2′]8,13[1″,2″]dibenzenodibenzo[*a*,*g*]-cyclododecene;[850] surprisingly, adamantene obtained from the gas-phase isomerization of 3-noradamantylcarbene[851] possesses properties different from those previously attributed to the material.

MINDO/3 calculations have substantiated the view that the thermal isomerization of *cis-trans-σ*-homobenzenes to cyclonona-1,4,7-trienes can be regarded as a $[_{\sigma}2_s + _{\sigma}2_s + _{\sigma}2_s]$ cycloreversion.[852] Isomerization of 1,2,3,5,6,7-hexachloro-4-methyl-enebicyclo[3.2.0]hepta-2,6-dienes (**505**) to heptafulvenes (**506**) is assumed to account for the various products obtained from the pyrolysis of (**505**).[853] The

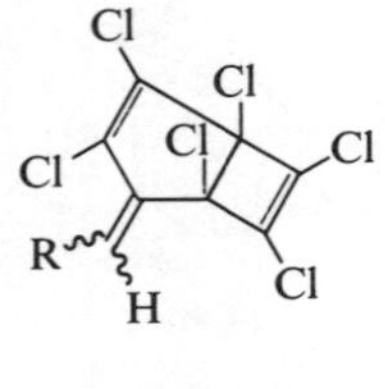

(**505**) 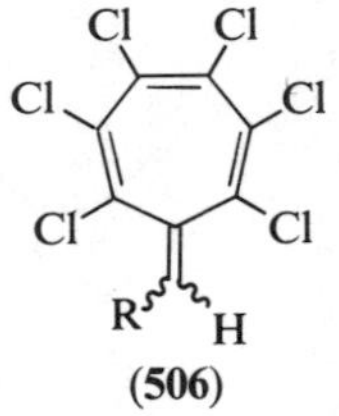 (**506**)

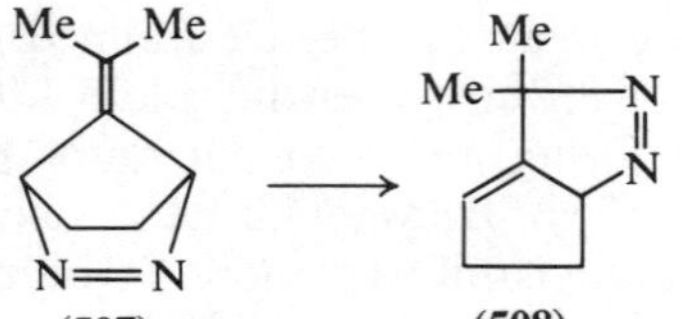

(**507**) (**508**) 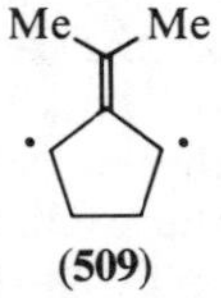 (**509**) 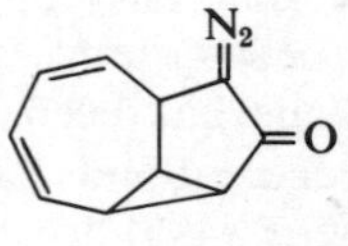 (**510**)

irreversible rearrangement of (**507**) to (**508**) has been shown to proceed directly, rather than *via* the intermediate trimethylenemethane biradical (**509**),[854] while the photochemically induced valence isomerization of α-diazocarbonyl compounds to give diazirines, a process previously thought to have been restricted to α-diazo-amides, has now been observed in α-diazo-ketones such as (**510**).[855]

Variable-temperature ^{13}C-NMR spectroscopy has indicated the occurrence of permutational isomerization in a variety of spirosulphuranes derived from catechols,[856] and studies have been made of the isomerization of anthracene photo-oxides[857] and naphthacene photo-oxide.[858] Thermolysis of halocineole derivatives (**511**; X = Br or Cl) offers the first synthetic entry to pinol derivatives (**513**) bearing a halogen at the C(4)-position.[859] The isomerization appears to involve ionization by the favourably located neighbouring oxygen atom to form an ion-pair intermediate (**512**), followed by internal return with attack of halide occurring at the more hindered tertiary position, yielding the thermodynamically more stable pinol (oxabicyclo[3.2.1]octane) ring system.

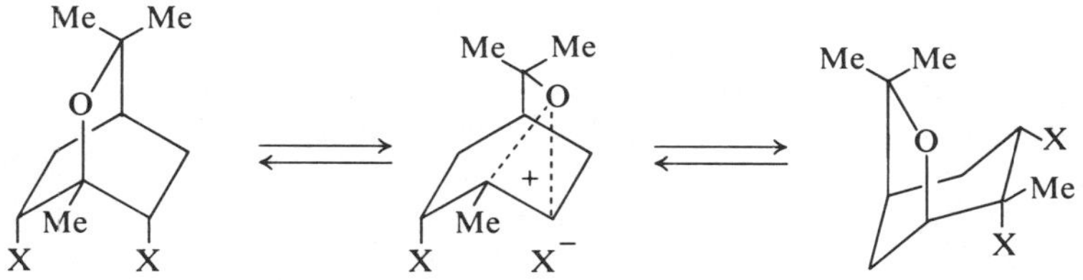

(**511**) ⇌ (**512**) ⇌ (**513**)

Addenda

Photochemical Studies

The following references are appended for general interest.

Photochemical rearrangements of alkenes and polyenes have been reviewed.[860]

The effect of substituents on rates and regioselectivities in the di-π-methane rearrangement has been examined.[861] Quantitative correlations between orbital shapes and energies in benzonorbornadienes and the regioselectivity of triplet di-π-methane rearrangements have been found,[862] reports of unusual regioselectivity in the di-π-methane rearrangement have appeared,[863,864] and a case of multiplicity-dependent regiospecificity in the di-π-methane rearrangement has been unearthed in a photochemical synthesis of chrysanthemic acid and cyclopropylacrylic acid derivatives.[865] Substituent effects on the di-π-methane rearrangement of 1,1-dicyano-2-methyl-3-phenylpropene have been examined,[866] the di-π-methane rearrangement of systems with simple vinyl components has been studied,[867] a di-π-methane rearrangement involving an oxime moiety has been reported,[868] and the effect of configuration in some bicyclic di-π-methanes with simple chromophores has been examined.[869] A report on the arylvinylmethane version of the di-π-methane rearrangement has concluded that its stereochemistry parallels that of the divinylmethane version (*i.e.* complete inversion of configuration at the methane-carbon) rather than that of the oxa-di-π-methane variety, and that singlet *vs.* triplet factors seem dominant in the process;[870] a detailed investigation of the rearrangement has revealed a pattern of substituent effects that can be understood readily on a theoretical basis.[871] In the triplet state, the tricyclo-[5.2.2.0^{2,6}]undeca-3,10-dien-9-one system has been shown to undergo an oxa-di-π-methane rearrangement,[872] the photochemical rearrangement of β-apolignans to tetrahydrocyclopropindenes has been characterized as a vinyl-aryl di-π-methane process[873] and a di-π-methane rearrangement has been observed in the photolysis of 17β-acetoxy-4-oxa-androsta-1,5-dien-3-one.[874]

A study has been undertaken on the mechanism of the photochemical "bicycle" reaction,[875] while a study of the photochemistry of disubstitued cyclic olefins such as 1,2-dimethylcyclohexene has provided evidence to show that, in addition to the intermediacy of the "expected" carbene which is formed by an initial migration of a C—C bond, a second carbene which results from the migration of a C—H bond is also operative.[876] The intriguing question that remains unanswered in the study is the factor (or factors) that determines the direction of the excited state between the two pathways.

Novel 1,4-substituent shifts have been observed in the photorearrangement of 2-hydroxymethyl-2*H*-azirine derivatives into *N*-vinylimines[877] and a dimethylsilyl imine has been obtained from the photolysis of trimethylazidosilane in inert matrices.[878]

Several rearrangements have been induced by photo-oxygenation; these include the photo-oxygenation of 8-methoxyberberinephenolbetaine to a spirobenzylisoquinoline[879] and the photosensitized oxidative rearrangements of indole derivatives.[880] Photochemical rearrangements of the following systems have been reported: α- and γ-methylallyl chorides,[881] cycloalkenes,[882] cyclic azoalkanes,[883] chiral cyclohex-2-enones,[884] tetra-*tert*-butylcyclopentadienone,[885] 2,3-epoxy-3-phenylpropiophenones,[886] spiro-epoxy-ketones,[887] diterpenoid quinones,[888] 7,12-dimethylbenz[*a*]-anthracene 7,12-epidioxide,[889] 2,2,4-triacyl-2,3-dihydrofurans,[890] 4-pyrones and 4-pyridones,[891,892] pyridinium-3-olates,[893] 1-(2-pyridyl)- and 1-(3-pyridyl)-2-nitropropene,[894] and 2-pyridyl acetate methides and their pyrimidine analogues.[895]

References

[1] Berson, J. A., *Ann. Rev. Phys. Chem.*, **28,** 111 (1977).
[2] Asano, T., and le Noble, W. J., *Chem. Rev.*, **78,** 407 (1978).
[3] Russell, G. A., Schmidt, K., Tanger, C., Goettert, E., Yamashita, M., Kosugi, Y., Siddens, J., and Senatore, G., *A.C.S. Symp. Ser.*, **1978,** 69.
[4] Douglas, A. W., *J. Am. Chem. Soc.*, **100,** 6463 (1978).
[5] Allan, Z. J., *Justus Liebigs Ann. Chem.*, **1978,** 705.
[6] Lin., C. T., and Byrn, S. R., *Tetrahedron Lett.*, **1978,** 1975.
[7] Shimao, I., and Matsumura, S., *Bull. Chem. Soc. Jpn.*, **51,** 926 (1978).
[8] Badr, M. Z. A., Aly, M. M., and Salem, S. S., *Tetrahedron*, **34,** 123 (1978).
[9] Ainpour, P., and Heimer, N. E., *J. Org. Chem.*, **43,** 2061 (1978).
[10] Imaizumi, H., Hashida, Y., and Matsui, K., *Bull. Chem. Soc. Jpn.*, **51,** 1507 (1978).
[11] Szele, I., and Zollinger, H., *J. Am. Chem. Soc.*, **100,** 2811 (1978).
[12] Hashida, Y., Landells, R. G. M., Lewis, G. E., Szele, I., and Zollinger, H., *J. Am. Chem. Soc.*, **100,** 2816 (1978).
[13] Badr, M. Z. A., Aly, M. M., and Salem, S. S., *Tetrahedron*, **33,** 3155 (1977).
[14] Shawali, A. S., Hassaneen, H. M., and Almousawi, S., *Bull. Chem. Soc. Jpn.*, **51,** 512 (1978).
[15] Pillai, V. N. R., *Chem. Ind.* (*London*), **1977,** 665.
[16] Weustink, R. J. M., Jongsma, C., and Bickelhaupt, F., *Recl. Trav. Chim. Pays-Bas*, **96,** 265 (1977).
[17] Okada, K., Matsui, K., and Sekiguchi, S., *Bull. Chem. Soc. Jpn.*, **51,** 2601 (1978).
[18] Knyazev, V. N., Drozd, V. N., and Minov, V. M., *Zh. Org. Khim.*, **14,** 105 (1978); *Chem. Abs.*, **89,** 43360 (1978).
[19] Mutai, K., Kanno, S.-i., and Kobayashi, K., *Tetrahedron Lett.*, **1978,** 1273.
[20] Truce, W. E., Van Gemert, B., and Brand, W. W., *J. Org. Chem.*, **43,** 101 (1978).
[21] Harcourt, D. N., Taylor, N., and Waigh, R. D., *J. Chem. Soc.*, *Perkin Trans.* 1, **1978,** 722.
[22] Harcourt, D. N., Taylor, N., and Waigh, R. D., *J. Chem. Soc.*, *Perkin Trans.* 1, **1978,** 1330.

[23] Wilt, J. W., Chwang, W. K., Dockus, C. F., and Tomink, N. M., *J. Am. Chem. Soc.*, **100,** 5534 (1978).
[24] Everly, C. R., and Traynham, J. G., *J. Am. Chem. Soc.*, **100,** 4316 (1978).
[25] Shotter, R. G., Johnston, K. M., and Jones, J. F., *Tetrahedron*, **34,** 741 (1978).
[26] Agranat, I., Bentor, Y., and Shih, Y.-S., *J. Am. Chem. Soc.*, **99,** 7068 (1977).
[27] Andreau, A. D., Gore, P. H., and Morris, D. F. C., *J. Chem. Soc., Chem. Commun.*, **1978,** 271.
[28] Baciocchi, E., Cipiciani, A., Clementi, S., and Sebastiani, G. V., *J. Chem. Soc., Chem. Commun.*, **1978,** 597.
[29] Gesson, J.-P., Di Giusto, L., and Jacquesy, J.-C., *Tetrahedron*, **34,** 1715 (1978).
[30] Cohen, N., Lopresti, R. J., and Williams, T. H., *J. Org. Chem.*, **43,** 3723 (1978).
[31] Warshawsky, A., Kalir, R., and Patchornik, A., *J. Am. Chem. Soc.*, **100,** 4544 (1978).
[32] Shenbor, M. I., Azarov, A. S., and Denisenko, S. N., *Zh. Org. Khim.*, **13,** 2411 (1977); *Chem. Abs.*, **88,** 89434 (1978).
[33] Srinivasan, C., *Acta Cienc. Indica*, **3,** 18 (1977); *Chem. Abs.*, **88,** 169736 (1978).
[34] Martin, R., *Bull. Soc. Chim. Fr.*, **1977,** 901.
[35] Mehlhorn, A., Schwenzer, B., Brückner, H.-J., and Schwetlick, K., *Tetrahedron*, **34,** 481 (1978).
[36] Escobar, C., Farina, F., Martinez-Utrilla, R., and Paredes, M. C., *J. Chem. Res.*, **1977,** (S) 266; (M) 3154.
[37] Martens, J., Praefcke, K., and Simon, H., *Chem.-Ztg.*, **102,** 108 (1978); *Chem. Abs.*, **89,** 41945 (1978).
[38] Hoehne, G., Lohner, W., Praefcke, K., Schulze, U., and Simon, H., *Tetrahedron Lett.*, **1978,** 613.
[39] Piette, J.-L., Thibaut, P., and Renson, M., *Tetrahedron*, **34,** 655 (1978).
[40] Boudreau, J. A., Farrar, C. R., and Williams, A., *J. Chem. Soc., Perkin Trans.* 2, **1977,** 1804.
[41] Boudreau, J. A., and Williams, A., *J. Chem. Soc., Perkin Trans.* 2, **1977,** 2028.
[42] Dhareshwar, G. P., and Hosangadi, B. D., *Indian J. Chem.*, **16B,** 143 (1978).
[43] Ogata, Y., Takagi, K., and Yamada, S., *J. Chem. Soc., Perkin Trans.* 2, **1977,** 1629.
[44] Srimali, S., S., Sharma, N. D., Kumar, S., and Joshi, B. C., *Rev. Roum. Chim.*, **23,** 613 (1978).
[45] Gavezzotti, A., and Simonetta, M., *Nouveau J. Chim.*, **2,** 69 (1978).
[46] Kortekaas, T. A., and Cerfontain, H., *J. Chem. Soc., Perkin Trans.* 2, **1978,** 742.
[47] Muramoto, Y., Asakura, H., and Sizuki, H., *Nippon Kagaku Kaishi*, **1978,** 259; *Chem. Abs.*, **89,** 59387 (1978).
[48] Kaspi, J., and Olah, G. A., *J. Org. Chem.*, **43,** 3142 (1978).
[49] Yamada, K., Shigehiro, K., Kiyozuka, T., and Iida, H., *Bull. Chem. Soc. Jpn.*, **51,** 2447 (1978).
[50] Harruff, L. G., Brown, M., and Boekelheide, V., *J. Am. Chem. Soc.*, **100,** 2893 (1978).
[51] Albini, A., Colombi, R., and Minoli, G., *J. Chem. Soc., Perkin Trans.* 1, **1978,** 924.
[52] Bartnik, R., and Mloston, G., *Rocz. Chem.*, **51,** 1747 (1977); *Chem. Abs.*, **88,** 62332 (1978).
[53] Haddadin, M. J., Hawi, A. A., and Nazer, M. Z., *Tetrahedron Lett.*, **1978,** 4581.
[54] Senda, S., Hirota, K., and Asao, T., *Tetrahedron Lett.*, **1978,** 2295.
[55] Sommer, S., and Schubert, U., *Chem. Ber.*, **111,** 1989 (1978).
[56] Barnes, J. F., Paton, R. M., Ashcroft, P. L., Bradbury, R., Crosby, J., Joyce, C. J., Holmes, D. R., and Milner, J. A., *J. Chem. Soc., Chem. Commun.*, **1978,** 113.
[57] Gromov, S. P., Kost, A. N., and Sagitullin, R. S., *Zh. Org. Khim.*, **14,** 1316 (1978); *Chem. Abs.*, **89,** 108989 (1978).
[58] Cervinka, O., Fabryova, A., and Snuparek, V., *Z. Chem.*, **18,** 137 (1978); *Chem. Abs.*, **89,** 23501 (1978).
[59] Yamazaki, T., Nagata, M., Hirokami, S.-I., and Miyakoshi, S., *Heterocycles*, **8,** 377 (1977).
[60] Sagitullin, R. S., Kost, A. N., and Danagulyan, G. G., *Tetrahedron Lett.*, **1978,** 4135.
[61] van der Plas, H. C., Jongejan, H., and Koudijs, A., *J. Heterocycl. Chem.*, **15,** 485 (1978).
[62] Senda, S., Hirota, K., and Asao, T., *Tetrahedron Lett.*, **1978,** 1531.
[63] Hollywood, F., Scriven, E. F. V., Suschitzky, H., Thomas, D. R., and Hull, R., *J. Chem. Soc., Chem. Commun.*, **1978,** 806.
[64] Srinivasan, A., Fagerness, P. E., and Broom, A. D., *J. Org. Chem.*, **43,** 828 (1978).
[65] Brown, D. J., and Nagamatsu, T., *Aust. J. Chem.*, **30,** 2515 (1977).
[66] Geerts, J. P., and van der Plas, H. C., *J. Org. Chem.*, **43,** 2682 (1978).
[67] Hara, H., Hosaka, M., Hoshino, O., and Umezawa, B., *Tetrahedron Lett.*, **1978,** 3809.
[68] Hasan, I., and Fowler, F. W., *J. Am. Chem. Soc.*, **100,** 6696 (1978).
[69] Upadysheva, A. V., Grigor'eva, D. N., and Znamenskaya, A. P., *Khim. Geterotsikl. Soedin.*, **1977,** 1549; *Chem. Abs.*, **88,** 105251 (1978).

[70] Zigeuner, G., Schweiger, K., Baier, M., and Fuchsgruber, A., *Monatsh. Chem.*, **109,** 113 (1978).
[71] Senda, S., Hirota, K., Asao, T., and Abe, Y., *Heterocycles,* **9,** 739 (1978).
[72] Barinov, I. V., and Chertkov, V. A., *Zh. Org. Khim.*, **13,** 2458 (1977); *Chem. Abs.*, **88,** 74361 (1978).
[73] Dovlatyan, V. V., Dovlatyan, A. V., Eliazyan, K. A., and Mirzoyan, R. G., *Khim. Geterotsikl. Soedin.*, **1977**, 1420; *Chem. Abs.*, **88,** 50810 (1978).
[74] Kuehne, M. E., and Hafter, R., *J. Org. Chem.*, **43,** 3702 (1978).
[75] Laronze, J.-Y., Laronze, J., Royer, D., Levy, J., and Le Men, J., *Bull. Soc. Chim. Fr. II*, **1977,** 1215.
[76] Robba, M., Maume, D., and Lancelot, J. C., *J. Heterocycl. Chem.*, **14,** 1365 (1977).
[77] Kollenz, G., *Monatsh. Chem.*, **109,** 249 (1978).
[78] Acheson, R. M., Letcher, R. M., and Procter, G., *J. Chem. Soc., Chem. Commun.*, **1978,** 332.
[79] Glish, G. L., and Cooks, R. G., *J. Am. Chem. Soc.*, **100,** 6720 (1978).
[80] Grundon, M. F., and Surgenor, S. A., *J. Chem. Soc., Chem. Commun.*, **1978,** 624.
[81] Blount, J. F., Coffen, D. L., and Katonak, D. A., *J. Org. Chem.*, **43,** 3821 (1978).
[82] Meth-Cohn, O., and Narine, B., *J. Chem. Res.*, **1977,** (S), 294; (M) 3262.
[83] Kraus, G. A., and Roth, B., *J. Org. Chem.*, **43,** 2072 (1978).
[84] Chlenov, I. E., Salamonov, Yu. B., Khasapov, B. N., Karpenko, N. F., Chizhov, O. S., and Tartakovskii, V. A., *Izv. Akad. Nauk SSSR, Ser. Khim.*, **1977,** 2830; *Chem. Abs.*, **88,** 121078 (1978).
[85] Colburn, V. M., Iddon, B., Suschitzky, H., and Gallagher, P. T., *J. Chem. Soc., Chem. Commun.*, **1978,** 453.
[86] Yakubov, A. P., Grigor'eva, N. V., and Belen'kii, L. I., *Zh. Org. Khim.*, **14,** 641 (1978); *Chem. Abs.*, **89,** 107512 (1978).
[87] Chakrabarti, J. K., Hicks, T. A., Hotten, T. M., and Tupper, D. E., *J. Chem. Soc., Perkin Trans.* 1, **1978,** 937.
[88] Elguero, J., Gonzalez, E., and Sarlin, R., *An. Quim.*, **74,** 527 (1978); *Chem. Abs.*, **89,** 128854 (1978).
[89] Spinelli, D., Frenna, V., Corrao, A., and Vivona, N., *J. Chem. Soc., Perkin Trans.* 2, **1978,** 19.
[90] Lautenschlaeger, H., *Justus Liebigs Ann. Chem.*, **1978,** 566.
[91] Adembri, G., Camparini, A., Donati, D., and Ponticelli, F., *Tetrahedron Lett.*, **1978,** 4439.
[92] Smalley, R. K., Smith, R. H., and Suschitzky, H., *Tetrahedron Lett.*, **1978,** 2309.
[93] Uno, H., and Kurokawa, M., *Chem. Pharm. Bull.*, **26,** 549 (1978).
[94] Citerio, L., Garufi, M., and Stradi, R., *Tetrahedron Lett.*, **1978,** 2175.
[95] Davies, K. L., Storr, R. C., and Whittle, P. J., *J. Chem. Soc., Chem. Commun.*, **1978,** 9.
[96] Domany, G., Nyitrai, J., Lempert, K., Voelter, W., and Horn, H., *Chem. Ber.*, **111,** 1464 (1978).
[97] Le Fevre, G., and Hamelin, J., *Tetrahedron Lett.*, **1978,** 4503.
[98] Abushanab, E., Sytwu, I.-I., Zabbo, A., and Goodman, L., *J. Org. Chem.*, **43,** 2017 (1978).
[99] Chioccara, F., Mangiacapra, V., Novellino, E., and Prota, G., *J. Chem. Soc., Chem. Commun.*, **1977,** 863.
[100] Slyusarenko, E. I., Pel'kis, N. P., and Lechenko, E. S., *Zh. Org. Khim.*, **14,** 1092 (1978); *Chem. Abs.*, **89,** 163520 (1978).
[101] Knapp, K. K., Keller, P. C., and Rund, J. V., *J. Chem. Soc., Chem. Commun.*, **1978,** 971.
[102] Dorokhov, V. A., Lavrinovich, L. I., Bochkareva, M. N., Zolotarev., B. M., and Mikhailov, B. M., *Izv. Akad. Nauk SSSR, Ser. Khim.*, **1978,** 253; *Chem. Abs.*, **88,** 170216 (1978).
[103] Leiby, R. W., Corley, E. G., and Heindel, N. D., *J. Org. Chem.*, **43,** 3427 (1978).
[104] Hasnaoui, A., Lavergne, J.-P., and Viallefont, P., *J. Chem. Res.*, **1978,** (S) 190; (M) 2501.
[105] Moore, J. A., Freeman, W. J., Gearhart, R. C., and Yokelson, H. B., *J. Org. Chem.*, **43,** 787 (1978).
[106] Fairhurst, J., Horwell, D. C., and Timms, G. H., *J. Heterocycl. Chem.*, **14,** 1199 (1977).
[107] Burger, K., and Meffert, A., *Justus Liebigs Ann. Chem.*, **1978,** 1052.
[108] Ozols, J., Liepins, E., Mazeika, I., Vigante, B., and Duburs, G., *Khim. Geterotsikl. Soedin.*, **1978,** 789; *Chem. Abs.*, **89,** 109014 (1978).
[109] Sinyakov, A. N., Vasilevskii, S. F., and Shvartsberg, M. S., *Izv. Akad. Nauk SSSR, Ser. Khim.*, **1977,** 2306; *Chem. Abs.*, **88,** 62331 (1978).
[110] Maquestiau, A., Van Haverbeke, Y., and Vanovervelt, J. C., *Bull. Soc. Chim. Belg.*, **86,** 961 (1977); *Chem. Abs.*, **88,** 136512 (1978).

[111] Mitra, R. B., Kulkarni, G. H., and Shirwaiker, G. S., *Indian J. Chem.*, **16B**, 146 (1978).
[112] Marquez, V. E., DiParsia, M. T., and Kelley, J. A., *J. Heterocycl. Chem.*, **14**, 1427 (1977).
[113] Jerzmanowska, Z., and Basinski, W., *Rocz. Chem.*, **51**, 2283 (1977); *Chem. Abs.*, **89**, 107505 (1978).
[114] Seshimoto, O., Kumagai, T., Shimizu, K., and Mukai, T., *Chem. Lett.*, **1977**, 1195.
[115] Bouchet, P., Joncheray, G., Jacquier, R., and Elguero, J., *J. Heterocycl. Chem.*, **15**, 625 (1978).
[116] Nishigaki, S., Shimizu, K., and Senga, K., *Chem. Pharm. Bull.*, **25**, 2790 (1977).
[117] Plenkiewicz, J., *Tetrahedron Lett.*, **1978**, 399.
[118] Koennecke, A., Lepom, P., and Lippmann, E., *Z. Chem.*, **18**, 214 (1978); *Chem. Abs.*, **89**, 146843 (1978).
[119] Schuster, D. I., *Acc. Chem. Res.*, **11**, 65 (1978).
[120] Caine, D., Deutsch, H., Chao, S. T., Van Derveer, D. G., and Bertrand, J. A., *J. Org. Chem.*, **43**, 1114 (1978).
[121] Kitamura, T., Imagawa, T., and Kawanisi, M., *Tetrahedron Lett.*, **1978**, 3443.
[122] Kitamura, T., Kawakami, Y., Imagawa, T., and Kawanisi, M., *Tetrahedron Lett.*, **1978**, 4297.
[123] Evans, D. A., Hart, D. J., and Koelsch, P. M., *J. Am. Chem. Soc.*, **100**, 4593 (1978).
[124] Wenkert, E., Berges, D. A., and Golob, N. F., *J. Am. Chem. Soc.*, **100**, 1263 (1978).
[125] De Alvarenga, M. A., Brocksom, U., Gottlieb, O. R., and Yoshida, M., *J. Chem. Soc., Chem. Commun.*, **1978**, 831.
[126] Sala, T., and Sargent, M. V., *J. Chem. Soc., Chem. Commun.*, **1978**, 1043.
[127] Sala, T., and Sargent, M. V., *J. Chem. Soc., Chem. Commun.*, **1978**, 1044.
[128] Barnes, C. E., and Myhre, P. C., *J. Am. Chem. Soc.*, **100**, 973 (1978).
[129] Barnes, C. E., and Myhre, P. C., *J. Am. Chem. Soc.*, **100**, 975 (1978).
[130] Vitullo, V. P., Cashen, M. J., Marx, J. N., Candle, L. J., and Fritz, J. R., *J. Am. Chem. Soc.*, **100**, 1205 (1978).
[131] Nishinaga, A., Shimizu, T., and Matsuura, T., *Tetrahedron Lett.*, **1978**, 3747.
[132] Forrester, A. R., Thomson, R. H., and Woo, S. O., *J. Chem. Soc., Perkin Trans.* 1, **1978**, 755.
[133] Maerkl, G., and Rampal, J. B., *Tetrahedron Lett.*, **1978**, 1175.
[134] Maerkl, G., and Rampal, J. B., *Tetrahedron Lett.*, **1978**, 1471.
[135] Maerkl, G., Liebl, R., and Hüttner, A., *Angew. Chem. Int. Edn.*, **17**, 528 (1978).
[136] Chambers, R. D., Hercliffe, R. D., and Middleton, R., *J. Chem. Soc., Chem. Commun.*, **1978**, 305.
[137] Carpenter, B. K., *Tetrahedron*, **34**, 1877 (1978).
[138] Grovenstein, E., *Angew. Chem. Int. Edn.*, **17**, 313 (1978).
[139] Seitz, W. A., Welsher, T. L., Yurke, B., and Matsen, F. A., *J. Am. Chem. Soc.*, **100**, 4679 (1978).
[140] Willner, I., Halpern, M., and Rabinovitz, M., *J. Chem. Soc., Chem. Commun.*, **1978**, 155.
[141] Bergson, G., Matsson, O., and Sjoberg, S., *Chem. Scr.*, **11**, 25 (1977).
[142] Hayashi, T., and Goto, M., *Nippon Kagaku Kaishi*, **1977**, 1512; *Chem. Abs.*, **88**, 21796 (1978).
[143] Hayashi, T., and Goto, M., *Nippon Kagaku Kaishi*, **1978**, 1007; *Chem. Abs.*, **89**, 128852 (1978).
[144] Raucher, S., and Lui, A. S.-T., *J. Am. Chem. Soc.*, **100**, 4902 (1978).
[145] Bunina-Krivorukova, L. I., Rossinski, A. P., and Bal'yan, Kh. V., *Zh. Org. Khim*, **14**, 586 (1978); *Chem. Abs.*, **89**, 42037 (1978).
[146] Feoktistov, V. M., Bunina-Krivorukova, L. I., and Bal'yan, Kh. V., *Zh. Org. Khim.*, **14**, 807 (1978); *Chem. Abs.*, **89**, 42037 (1978).
[147] Hegedus, L. S., and Evans, B. R., *J. Am. Chem. Soc.*, **100**, 3461 (1978).
[148] Bell, K. A., Flatman, I. J., Golborn, P., Pachl, A., and Scheinmann, F., *J. Chem. Soc., Chem. Commun.*, **1978**, 900.
[149] Fujise, Y., Nakatsu, T., Nakamura, A., and Itô, S., *Tetrahedron Lett.*, **1978**, 4293.
[150] Al-Sader, B. H., and Al-Fekri, D. M., *J. Org. Chem.*, **43**, 3626 (1978).
[151] Jain, A. C., Tuli, D. K., and Kumar, A., *Curr. Sci.*, **46**, 839 (1977).
[152] Maas, G., and Regitz, M., *Chem. Ber.*, **111**, 1733 (1978).
[153] Reuter, J. M., and Salomon, R. G., *J. Org. Chem.*, **42**, 3360 (1977).
[154] Strauss, H. F., and Wiechers, A., *Tetrahedron*, **34**, 127 (1978).
[155] Lythgoe, B., Manwaring, R., Milner, J. R., Moran, T. A., Nambudiry, M. E. N., and Tideswell, J., *J. Chem. Soc., Perkin Trans.* 1, **1978**, 387.
[156] Heyns, K., and Hohlweg, R., *Chem. Ber.*, **111**, 1632 (1978).
[157] Chan, K.-K., Specian, A. C., and Saucy, G., *J. Org. Chem.*, **43**, 3435 (1978).

[158] Bates, D. K., and Jones, M. C., *J. Org. Chem.*, **43,** 3775 (1978).
[159] Bates, D. K., and Jones, M. C., *J. Org. Chem.*, **43,** 3856 (1978).
[160] Overman, L. E., Campbell, C. B., and Knoll, F. M., *J. Am. Chem. Soc.*, **100** 4822 (1978).
[161] Hill, R. K., and Khatri, H. N., *Tetrahedron Lett.*, **1978,** 4337.
[162] Inada, S., Hirabayashi, S., Taguchi, K., and Okazaki, M., *Nippon Kagaku Kaishi*, **1978,** 86; *Chem. Abs.*, **88,** 151779 (1978).
[163] Inada, S., Sodeyama, Y., and Okazaki, M., *Nippon Kagaku Kaishi*, **1978,** 571; *Chem. Abs.*, **89,** 107515 (1978).
[164] Katayama, H., *Chem. Pharm. Bull.*, **26,** 2027 (1978).
[165] Nakai, T., Mimura, T., and Kurokawa, T., *Tetrahedron Lett.*, **1978,** 2895.
[166] Cookson, R. C., and Gopalan, R., *J. Chem. Soc., Chem. Commun.*, **1978,** 608.
[167] Anisimov, A. V., *I*onova, V. F., and Viktorova, E. A., *Khim. Geterotsikl. Soedin.*, **1978,** 186; *Chem. Abs.*, **89** 5715 (1978).
[168] Teo, K.-E., Barnett, G. H., Anderson, H. J., and Loader, C. E., *Can. J. Chem.*, **56,** 221 (1978).
[169] Fujita, Y., Onishi, T., and Nishida, T., *Synthesis*, **1978,** 612.
[170] Fujita, Y., Onishi, T., and Nishida, T., *Synthesis*, **1978,** 532.
[171] Tamaru, Y., Harada, T., and Yoshida, Z.-i., *Tetrahedron Lett.*, **1978,** 2167.
[172] Gajewski, J. J., and Conrad, N. D., *J. Am. Chem. Soc.*, **100,** 6268 (1978).
[173] Gajewski, J. J., and Conrad, N. D., *J. Am. Chem. Soc.*, **100,** 6269 (1978).
[174] Marvell, E. N., and Li, T. H.-C., *J. Am. Chem. Soc.*, **100,** 883 (1978).
[175] Shea, K. J., and Wise, S., *J. Org. Chem.*, **43,** 2710 (1978).
[176] Padwa, A., and Blacklock, T. J., *J. Am. Chem. Soc.*, **100,** 1321 (1978).
[177] Marvell, E. N., and Lin, C., *J. Am. Chem. Soc.*, **100,** 877 (1978).
[178] Piers, E., Nagakura, I., and Morton, H. E., *J. Org. Chem.*, **43,** 3630 (1978).
[179] Paquette, L. A., and Ewing, G. D., *J. Am. Chem. Soc.*, **100,** 2908 (1978).
[180] Ewing, G. D., Ley, S. V., and Paquette, L. A., *J. Am. Chem. Soc.*, **100,** 2909 (1978).
[181] Aumann, R., and Knecht, J., *Chem. Ber.*, **111,** 3429 (1978).
[182] Dyllick-Brenzinger, R., Oth, J. F. M., Fühlhuber, H. D., Gousetis, C., Troll, T., and Sauer, J., *Tetrahedron Lett.*, **1978,** 3907.
[183] Shea, K. J., and Phillips, R. B., *J. Am. Chem. Soc.*, **100,** 654 (1978).
[184] Janusz, J. M., and Berson, J. A., *J. Am. Chem. Soc.*, **100,** 2237 (1978).
[185] Rhoads, S. J., Watson, J. M., and Kambouris, J. G., *J. Am. Chem. Soc.*, **100,** 5151 (1978).
[186] Warrener, R. N., Russell, R. A., and Collin, G. J., *Tetrahedron Lett.*, **1978,** 4447.
[187] Tegmo-Larsson, I.-M., and Houk, K. N., *Tetrahedron Lett.*, **1978,** 941.
[188] Miyashi, T., Kawamoto, H., and Mukai, T., *Tetrahedron Lett.*, **1977,** 4623.
[189] Barkovich, A. J., Strauss, E. S., and Vollhardt, K. P. C., *J. Am. Chem. Soc.*, **99,** 8321 (1977).
[190] Bessiere, Y., Grison, C., and Boussac, G., *Tetrahedron*, **34,** 1957 (1978).
[191] Mundy, B. P., and Bornmann, W. G., *Tetrahedron Lett.*, **1978,** 957.
[192] Evans, D. A., Baillargeon, D. J., and Nelson, J. V., *J. Am. Chem. Soc.*, **100,** 2242 (1978).
[193] Doutheau, A., Balme, G., Malacria, M., and Goré, J., *Tetrahedron Lett.*, **1978,** 1803.
[194] Jung, M. E., and Hudspeth, J. P., *J. Am. Chem. Soc.*, **100,** 4309 (1978).
[195] Onishi, T., Fujita, Y., and Nishida, T., *J. Chem. Soc., Chem. Commun.*, **1978,** 651.
[196] Fujita, T., Onishi, T., and Nishida, T., *J. Chem. Soc., Chem. Commun.*, **1978,** 972.
[197] Miyashi, T., Hazato, A., and Mukai, T., *J. Am. Chem. Soc.*, **100,** 1008 (1978).
[198] Rastetter, W. H., and Richard, T. J., *Tetrahedron Lett.*, **1978,** 2995.
[199] Anastassiou, A. G., and Mahaffey, R. L., *J. Chem. Soc., Chem., Commun.*, **1978,** 915.
[200] Brombach, D., and Voegtle, F., *Synthesis*, **1977,** 800.
[201] Tamaru, Y., Harada, T., and Yoshida, Z.-i., *J. Org. Chem.*, **43,** 3370 (1978).
[202] Tamaru, Y., Harada, T., and Yoshida, Z.-i., *J. Am. Chem. Soc.*, **100,** 1923 (1978).
[203] Paquette, L. A., and Burson, R. L., *Tetrahedron*, **34,** 1307 (1978).
[204] Clive, D. L. J., *Tetrahedron*, **34,** 1049 (1978).
[205] Clive, D. L. J., Chittattu, G., Curtis, N. J., and Menchen, S. M., *J. Chem. Soc., Chem. Commun.*, **1978,** 770.
[206] Frejd, T., and Sharpless, K. B., *Tetrahedron Lett.*, **1978,** 2239.
[207] Kano, S., Yokomatsu, T., and Shibuya, S., *Tetrahedron Lett.*, **1978,** 4125.
[208] Kim, J. K., Kline, M. L., and Caserio, M. C., *J. Am. Chem. Soc.*, **100,** 6243 (1978).
[209] Morel, G., Khamsitthideth, S., and Foucaud, A., *J. Chem. Soc., Chem. Commun.*, **1978,** 274.
[210] Still, W. C., and Mitra, A., *J. Am. Chem. Soc.*, **100,** 1927 (1978).
[211] Cazes, B., and Julia, S., *Bull. Soc. Chim. Fr. II*, **1977,** 925.

[212] Cazes, B., and Julia, S., *Bull. Soc. Chim. Fr. II*, **1977**, 931.
[213] Snider, B. B., and Füzesi, L., *Tetrahedron Lett.*, **1978**, 877.
[214] Gillespie, R. J., Porter, A. E. A., and Willmott, W. E., *J. Chem. Soc., Chem. Commun.*, **1978**, 85.
[215] Vedejs, E., Hagen, J. P., Roach, B. L., and Spear, K. L., *J. Org. Chem.*, **43**, 1185 (1978).
[216] Vedejs, E., Mullins, M. J., Renga, J. M., and Singer, S. P., *Tetrahedron Lett.*, **1978**, 519.
[217] Ceré, V., Pollicino, S., Sandri, E., and Fava, A., *J. Am. Chem. Soc.*, **100**, 1516 (1978).
[218] Vedejs, E., Arco, M. J., and Renga, J. M., *Tetrahedron Lett.*, **1978**, 523.
[219] Fleischmann, C., and Zbiral, E., *Tetrahedron*, **34**, 317 (1978).
[220] Büchi, G., Liang, P. H., and Wüest, H., *Tetrahedron Lett.*, **1978**, 2763.
[221] Georgian, V., Boyer, S. K., and Edwards, B., *Heterocycles*, **7**, 1003 (1977).
[222] Kobayashi, T., and Hiraoka, T., *Chem. Lett.*, **1977**, 1399.
[223] Chan, K.-K., and Saucy, G., *J. Org. Chem.*, **42**, 3828 (1977).
[224] Gassman, P. G., and Schenk, W. N., *J. Org. Chem.*, **42**, 3240 (1977).
[225] Isobe, M., Iio, H., Kitamura, M., and Goto, T., *Chem. Lett.*, **1978**, 541.
[226] Hoffmann, R. W., Gerlach, R., and Goldmann, S., *Tetrahedron Lett.*, **1978**, 2599.
[227] Hoffmann, R. W., and Goldmann, S., *Chem. Ber.*, **111**, 2716 (1978).
[228] Baechler, R. D., Daley, S. K., Daly, B., and McGlynn, K., *Tetrahedron Lett.*, **1978**, 105.
[229] Trost, B. M., and Rigby, J. H., *J. Org. Chem.*, **43**, 2938 (1978).
[230] Tunemoto, D., Takahatake, Y., and Kondo, K., *Chem. Lett.*, **1978**, 189.
[231] Miller, R. D., and Kaufmann, D., *J. Chem. Soc., Chem. Commun.*, **1978**, 496.
[232] Thies, R. W., and Seitz, E. P., *J. Org. Chem.*, **43**, 1050 (1978).
[233] Wilson, S. R., and Mao, D. T., *J. Chem. Soc., Chem. Commun.*, **1978**, 479.
[234] Janusz, J. M., Gardiner, L. J., and Berson, J. A., *J. Am. Chem. Soc.*, **99**, 8509 (1977).
[235] Baldwin, J. E., Branz, S. E., and Walker, J. A., *J. Org. Chem.*, **42**, 4142 (1977).
[236] Sato, Y., Kato, M., Aoyama, T., Sugiura, M., and Shirai, H., *J. Org. Chem.*, **43**, 2466 (1978).
[237] Ewing, G. D., Ley, S. V., and Paquette, L. A., *J. Am. Chem. Soc.*, **100**, 2909 (1978).
[238] Padwa, A., and Carlsen, P. H. J., *J. Org. Chem.*, **43**, 2029 (1978).
[239] Nederlof, P. J. R., Moolenaar, M. J., De Waard, E. R., and Huisman, H. O., *Tetrahedron* **34**, 2205 (1978).
[240] Gerber, U., Widmar, U., Schmid, R., and Schmid, H., *Helv. Chim. Acta*, **61**, 83 (1978).
[241] Gladiali, S., Porcu, M. P., Rosnati, V., Saba, A., Soccolini, F., and Selva, A., *Gazz. Chim. Ital.*, **107**, 293 (1977).
[242] Gladiali, S., Pusino, A., Rosnati, V., Saba, A., Soccolini, F., and Selva, A., *Gazz. Chim. Ital.*, **107**, 535 (1977).
[243] Ioffe, S. L., Kalinin, A. V., Khasapov, B. N., Stepanyants, A. U., and Tartakovskii, V. A., *Izv. Akad. Nauk SSSR, Ser. Khim.*, **1978**, 1162; *Chem. Abs.*, **89**, 107532 (1978).
[244] de Fonseka, K. K., Manning, C., McCullough, J. J., and Yarwood, A. J., *J. Am. Chem. Soc.*, **99**, 8257 (1977).
[245] Semmelhack, M. F., Weller, H. N., and Clardy, J., *J. Org. Chem.*, **43**, 3791 (1978).
[246] Miller, R. D., Kaufmann, D., and Mayerle, J. J., *J. Am. Chem., Soc.*, **99**, 8511 (1977)
[247] Detty, M. R., and Paquette, L. A., *J. Chem. Soc., Chem. Commun.*, **1978**, 365.
[248] Baldwin, J. E., and Broline, B. M., *J. Am. Chem. Soc.*, **100**, 4599 (1978).
[249] Japenga, J., Klumpp, G. W., and Schakel, M., *Tetrahedron Lett.*, **1978**, 1869.
[250] Piers, E., Nagakura, I., and Shaw, J. E., *J. Org. Chem.*, **43**, 3431 (1978).
[251] Hammond, M. L., Mouriño, A., and Okamura, W. H., *J. Am. Chem. Soc.*, **100**, 4907 (1978).
[252] Crowley, K. J., and Traynor, S. G., *Tetrahedron*, **34**, 2783 (1978).
[253] Cavazza, M., Morganti, G., and Pietra, F., *Tetrahedron Lett.*, **1978**, 2137.
[254] Tsuchiya, T., Enkaku, M., and Sawanishi, H., *J. Chem. Soc., Chem. Commun.*, **1978**, 568.
[255] Heydt, H., and Regitz, M., *Justus Liebigs Ann. Chem.*, **1977**, 1766.
[256] Fusco, F., and Sannicolo, F., *Tetrahedron Lett.*, **1978**, 4827.
[257] Smith, R. G., and Daves, G. D., *J. Org. Chem.*, **43**, 2178 (1978).
[258] Jones, D. W., *J. Chem. Soc., Chem. Commun.*, **1978**, 404.
[259] Pinhey, J. T., Rizzardo, E., and Smith, G. C., *Aust. J. Chem.*, **31**, 97 (1978).
[260] Okazaki, R., Watanabe, M., Inagaki, Y., and Inamoto, N., *Tetrahedron Lett.*, **1978**, 3439.
[261] Taylor, R., *J. Chem. Soc., Chem. Commun.*, **1978**, 432.
[262] Defoin, A., Baranne-Lafont, J., Rigandy, J., and Guilhem, J., *Tetrahedron*, **34**, 83 (1978).
[263] Ranganathan, S., and Panda, C. S., *Heterocycles*, **7**, 529 (1977).
[264] Johnson, H. D., Hartford, T. W., and Spangler, C. W., *J. Chem. Soc., Chem. Commun.*, **1978**, 242.

[265] Reetz, M. T., *Adv. Organomet. Chem.*, **16**, 33 (1977).
[266] Reetz, M. T., and Kliment, M., *Chem. Ber.*, **111**, 1083 (1978).
[267] Reetz, M. T., Greif, N., and Kliment, M., *Chem. Ber.*, **111**, 1095 (1978).
[268] Brook, A. C., Golino, C., and Matern, E., *Can. J. Chem.*, **59**, 2286 (1978).
[269] Cunico, R. F., and Lee, H. M., *J. Am. Chem. Soc.*, **99**, 7613 (1977).
[270] Anh, N. T., Elian, M., and Hoffmann, R., *J. Am. Chem. Soc.*, **100**, 110 (1978).
[271] Grishin, Yu. K., Bakhbukh, M., Ustynyuk, Yu. A., Zemlyanskii, N. N., Kolosova, N. D., and Kocheshkov, K. A., *Dokl. Akad. Nauk SSSR*, **237**, 594 (1977); *Chem. Abs.*, **88**, 61811 (1978).
[272] Paetzold, P., and Biermann, H. P., *Chem. Ber.*, **110**, 3678 (1977).
[273] Chantrapromma, K., Ollis, W. D., and Sutherland, I. O., *J. Chem. Soc., Chem. Commun.*, **1978**, 670.
[274] Chantrapromma, K., Ollis, W. D., and Sutherland I. O., *J. Chem. Soc., Chem. Commun.*, **1978**, 672.
[275] Chantrapromma, K., Ollis, W. D., and Sutherland, I. O., *J. Chem. Soc., Chem. Commun.*, **1978**, 673.
[276] Ansell, H. V., Sheppard, P. J., Simpson, C. F., Stroud, M. A., and Taylor, R., *J. Chem. Soc., Chem. Commun.*, **1978**, 586.
[277] Grovenstein, E., Cottingham, A. B., and Gelbaum, L. T., *J. Org. Chem.*, **43**, 3332 (1978).
[278] Schneider, M. P., and Csacsko, B., *J. Chem. Soc., Chem. Commun.*, **1978**, 964.
[279] Le Noble, W. J., and Daka, M. R., *J. Am. Chem. Soc.*, **100**, 5961 (1978).
[280] Loven, R. P., and Speckamp, W. N., *Tetrahedron*, **34**, 1023 (1978).
[281] Tamura, Y., Nishikawa, Y., Sumoto, K., and Ikeda, M., *J. Org. Chem.*, **42**, 3226 (1977).
[282] Hori, M., Kataoka, T., Shimizu, N., and Onho, S., *Tetrahedron Lett.*, **1978**, 255.
[283] Miller, B., and Lin, W.-O., *Tetrahedron Lett.*, **1978**, 751.
[284] Fusco, R., and Sannicoló, F., *Tetrahedron Lett.*, **1978**, 1233.
[285] Onisko, B. L., Schnoes, H. K., and Deluca, H. F., *J. Org. Chem.*, **43**, 3441 (1978).
[286] Cavazza, M., Morganti, G., and Pietra, F., *J. Chem. Soc., Chem. Commun.*, **1978**, 945.
[287] Cauquis, G., and Divisia, B., *Bull. Soc. Chim. Fr. II.*, **1978**, 375.
[288] Pudovik, A. N., Konovalova, I. V., Zimin, M. G., and Dvoinishnikova, T. A., *Zh. Obshch. Khim.*, **48**, 490 (1978); *Chem. Abs.*, **89**, 43607 (1978).
[289] Pines, S. H., and Douglas, A. W., *J. Org. Chem.*, **43**, 3126 (1978).
[290] Aihara, J., *Bull. Chem. Soc. Jpn.*, **51**, 1788 (1978).
[291] Bischof, P., *J. Am. Chem. Soc.*, **99**, 8145 (1977).
[292] Halevi, E. A., Katriel, J., Pauncz, R., Matsen, F. A., and Welsher, T. L., *J. Am. Chem. Soc.*, **100**, 359 (1978).
[293] Dewar, M. J. S., Ford, G. P., Ritchie, J. P., and Rzepa, H. S., *J. Chem. Res.*, **1978**, (S) 26; (M), 484.
[294] Uematsu, S., and Akahori, Y., *Chem. Pharm. Bull.*, **26**, 25 (1978).
[295] Manley, P. W., Storr, R. C., Baydar, A. E., and Boyd, G. V., *J. Chem. Soc., Chem. Commun.*, **1978**, 902.
[296] Shea, K. J., and Wise, S., *Tetrahedron Lett.*, **1978**, 2283.
[297] Prinzbach, H., Babsch, H., and Hunkler, D., *Tetrahedron Lett.*, **1978**, 649.
[298] Braverman, S., and Duar, Y., *Tetrahedron Lett.*, **1978**, 1493.
[299] Marvell, E. N., Seubert, J., Vogt, G., Zimmer, G., Moy, G., and Siegmann, J. R., *Tetrahedron*, **34**, 1323 (1978).
[300] Snider, B. B., and Killinger, T. A., *J. Org. Chem.*, **43**, 2161 (1978).
[301] Steiner, R. P., and Michl, J., *J. Am. Chem. Soc.*, **100**, 6413 (1978).
[302] Battiste, M. A., and Fiato, R. A., *J. Org. Chem.*, **43**, 1282 (1978).
[303] Begley, M. J., Dean, F. M., Houghton, L. E., Johnson, R. S., and Park, B. K., *J. Chem. Soc., Chem. Commun.*, **1978**, 461.
[304] Anastassiou, A. G., and Badri, R., *Tetrahedron Lett.*, **1977**, 4465.
[305] Manley, P. W., Somanathan, R., Reeves, D. L. R., and Storr, R. C., *J. Chem. Soc., Chem. Commun.*, **1978**, 396.
[306] Marvell, E. N. and Rusay, R., *J. Org. Chem.*, **42**, 3336 (1977).
[307] Albright, T. A., Evans, S., Kim, C. S., Labaw, C. S., and Russiello, A. B., *J. Org. Chem.*, **42**, 3691 (1977).
[308] Borkent, J. H., Rouwette, P. H. F. M., and Laarhoven, W. H., *Tetrahedron*, **34**, 2569 (1978).
[309] Schroth, W., and Moegel, L., *Z. Chem.*, **17**, 441 (1977).
[310] Moegel, L., Schroth, W., and Werner, B., *J. Chem. Soc., Chem. Commun.*, **1978**, 57.

[311] Roedig, A., and Foersch, M., *Chem. Ber.*, **111,** 2738 (1978).
[312] Rottele, H., Heil, G., and Schroeder, G., *Chem. Ber.*, **111,** 84 (1978).
[313] Rigaudy, J., and Sparfel, D., *Tetrahedron*, **34,** 2263 (1978).
[314] Paquette, L. A., *Organic Reactions*, **25,** 1 (1977).
[315] West, R., *Adv. Organomet. Chem.*, **16,** 1 (1977).
[316] Musher, J. I., *Phosphorus Sulphur*, **3,** 247 (1977).
[317] Nesmeyanov, A. N., Ustynyuk, N. A., Makarova, L. G., Andre, S., Ustynyuk, Yu. A., Novikova, L. N., and Luzikov, Yu. N., *J. Organomet. Chem.*, **154,** 45 (1978).
[318] Muslin, D. V., Lyapina, N. Sh., Kut'in, A. P., and Vasileiskaya, N. S., *Izv. Akad. Nauk SSSR, Ser. Khim.*, **1978,** 890; *Chem. Abs.*, **89,** 43585 (1978).
[319] Sekiguchi, A., Ikeno, M., and Ando, W., *Bull. Chem. Soc. Jpn.*, **51,** 337 (1978).
[320] Goda, K., Okazaki, R., Akiba, K., and Inamoto, N., *Bull. Chem. Soc., Jpn.*, **51,** 260 (1978).
[321] Maercker, A., and Demuth, W., *Justus Liebigs, Ann. Chem.*, **1977,** 1909.
[322] Dietrich, W., Schulze, K., and Muehlstaedt, M., *J. Prakt. Chem.*, **319,** 799 (1977).
[323] Dietrich, W., Schulze, K., and Muehlstaedt, M., *J. Prakt, Chem.*, **320,** 143 (1978).
[324] Kocharyan, S. T., Ogandzhanyan, S. M., Razina, T. L., and Babayan, A. T., *Arm. Khim. Zh.*, **30,** 977 (1977); *Chem. Abs.*, **89,** 163026 (1978).
[325] Babayan, A. T., Kocharyan, S. T., Voskanyan, V. S., and Babayan, M. A., *Arm. Khim. Zh.*, **30,** 971 (1977); *Chem. Abs.*, **89,** 23812 (1978).
[326] Vilsmaier, E., Schuetz, J., and Moessel, W., *Chem. Ber.*, **111,** 400 (1978).
[327] Vilsmaier, E., Bayer, R., Laengenfelder, I., and Welz, U., *Chem. Ber.*, **111,** 1136 (1978).
[328] Vilsmaier, E., Bayer, R., Welz, U., and Dittrich, K.-H., *Chem. Ber.*, **111,** 1147 (1978).
[329] Ollis, W. D., Rey, M., and Sutherland, I. O., *J. Chem. Soc., Chem. Commun.*, **1978,** 675.
[330] Connor, D. T., Young, P. A., and von Strandtmann, M., *J. Heterocycl. Chem.*, **15,** 115 (1978).
[331] Connor, D. T., Young, P. A., and von Strandtmann, M., *Synthesis*, **1978,** 208.
[332] Mikoajczyk, M., Zatorski, A., Grzejszczak, S., Costisella, B., and Midur, W., *J. Org. Chem.*, **43,** 2518 (1978).
[333] Furukawa, N., Morishita, T., Akasaka, T., Oae, S., and Uneyama, K., *Tetrahedron Lett.*, **1978,** 1567.
[334] King, R. R., *J. Org. Chem.*, **43,** 3784 (1978).
[335] King, R. R., Greenhalgh, R., and Marshall, W. D., *J. Org. Chem.*, **43,** 1262 (1978).
[336] More, K. M., and Wemple, J., *J. Org. Chem.*, **43,** 2713 (1978).
[337] Rastogi, R. R., Kumar, A., Ila, H., and Junjappa, H., *J. Chem. Soc., Perkin Trans.* 1, **1978,** 549.
[338] Kozikowski, A. P., and Kuniak, M. P., *J. Org. Chem.*, **43,** 2083 (1978).
[339] Kano, S., Yokomatsu, T., Hibino, S., Imamura, K., and Shibuya, S., *Heterocycles*, **6,** 1319 (1977).
[340] Grethe, G., Lee, H. L., Mitt, T., and Uskokovic, M. R., *J. Am. Chem. Soc.*, **100,** 581 (1978).
[341] Saksena, A. K., Mangiaracina, P., Brambilla, R., McPhail, A. T., and Onan, K. D., *Tetrahedron Lett.*, **1978,** 1729.
[342] Eichholzer, J. V., MacLeod, J. K., and Summons, R. E., *Aust. J. Chem.*, **31,** 893 (1978).
[343] Dykstra, C. E., Arduengo, A. J., and Fukunaga, T., *J. Am. Chem. Soc.*, **100,** 6007 (1978).
[344] Knorr, R., and Lattke, E., *Tetrahedron Lett.*, **1977,** 4655.
[345] Merz, A., and Thumm, G., *Justus Liebigs Ann. Chem.*, **1978,** 1526.
[346] Garst, J. F., Pacifici, J. A., Felix, C. C., and Nigam, A., *J. Am. Chem. Soc.*, **100,** 5974 (1978).
[347] Taschner, M. J., and Kraus, G. A., *J. Org. Chem.*, **43,** 4235 (1978).
[348] Thibblin, A., and Ahlberg, P., *J. Am. Chem. Soc.*, **99,** 7926 (1977).
[349] Scheffer, J. R., Gayler, R. E., Zakouras, T., and Dzakpasu, A. A., *J. Am. Chem. Soc.*, **99,** 7726 (1977).
[350] Paquette, L. A., Degenhardt, C. R., and Berk, H. C., *J. Am. Chem. Soc.*, **100,** 1599 (1978).
[351] Goldstein, M. J., Nomura, Y., Takeuchi, Y., and Tomoda, S., *J. Am. Chem. Soc.*, **100,** 4899 (1978).
[352] Hall, S. S., Doweyko, A. M., and Jordan, F., *J. Am. Chem. Soc.*, **100,** 5934 (1978).
[353] Scott, A. I., Kang, J., Dalton, D., and Chung, S. K., *J. Am. Chem. Soc.*, **100,** 3603 (1978).
[354] Morand, P., and Samad, S. A., *Bangladesh J. Sci. Ind. Res.*, **12,** 236 (1977); *Chem. Abs.*, **89,** 23900 (1978).
[355] González, A. G., Fraga, B. M., Hernández, M. G., and Luis, J. G., *Tetrahedron Lett.*, **1978,** 3499.
[356] Hanson, J. R., Raines, D., and Wadsworth, H., *J. Chem. Soc., Perkin Trans.* 1, **1978,** 743.

[357] Brook, P. R., and Kitson, D. E., *J. Chem. Soc., Chem. Commun.*, **1978,** 87.
[358] Geetha, K. Y., Rajagopalan, K., and Swaminathan, S., *Tetrahedron*, **34,** 2201 (1978).
[359] Miyashi, T., Suto, N., and Mukai T., *Chem. Lett.*, **1978,** 157.
[360] Le Noble, W. J., *Chimia*, **32,** 31 (1978).
[361] Borodkin, G. I., Shakirov, M. M., and Shubin, V. G., *Zh. Org. Khim.*, **13,** 2152 (1977).
[362] Bushmelev, V. A., Shakirov, M. M., and Koptyug, V. A., *Zh. Org. Khim.*, **13,** 2161 (1977).
[363] Bodoev, N. V., Mamatyuk, V. I., Krysin, A. P., and Koptyug, V. A., *Izv. Akad. Nauk SSSR, Ser. Khim.*, **1978,** 1199; *Chem. Abs.*, **89,** 107534 (1978).
[364] Sazonova, L. I., Shakirov, M. M., and Shubin, V. G., *Zh. Org. Khim.*, **13,** 2456 (1977).
[365] Borodkin, G. I., Shakirov, M. M., Shubin, V. G., and Koptyug, V. A., *Zh. Org. Khim.*, **14,** 321 (1978).
[366] Borodkin, G. I., Shakirov, M. M., Shubin, V. G., and Koptyug, V. A., *Zh. Org. Khim.*, **14,** 989 (1978).
[367] Saunders, M. and Kates, M. R., *J. Am. Chem. Soc.*, **100,** 7082 (1978).
[368] Vincent, M. A., and Radom, L., *J. Am. Chem. Soc.*, **100,** 3306 (1978).
[369] Olah, G. A., Prakash, G. K. S., Donovan, D. J., and Yavari, I., *J. Am. Chem. Soc.*, **100,** 7085 (1978).
[370] Kirchen, R. P., Sorensen, T. S., and Wagstaff, K., *J. Am. Chem. Soc.*, **100,** 6761 (1978).
[371] Olah, G. A., Donovan, D. J., and Prakash, G. K. S., *Tetrahedron Lett.*, 1978, **4779.**
[372] Coxon, J. M., Steel, P. J., Coddington, J. M., Rae, I. D., and Jones, A. J., *Aust. J. Chem.*, **31,** 1223 (1978).
[373] Dunbar, R. C., Fu, E. W., and Olah, G. A., *J. Am. Chem. Soc.*, **99,** 7502 (1977).
[374] Childs, R. F., and Rogerson, C. V., *J. Am. Chem. Soc.*, **100,** 649 (1978).
[375] Scott, L. T., and Brunsvold, W. R., *J. Am. Chem. Soc.*, **100,** 6535 (1978).
[376] Kirchen, R. P., Sorensen, T. S., and Wagstaff, K. E., *J. Am. Chem. Soc.*, **100,** 5134 (1978).
[377] Friedrich, E. C., and Jassawalla, J. D. C., *Tetrahedron Lett*, **1978,** 953.
[378] Wertheimer, V., Glatz, A. M., and Razus, A. C., *Rev. Roum. Chim.*, **22,** 1505 (1977); *Chem. Abs.*, **88,** 104354 (1978).
[379] Driessen, P. B. J., and Hogeveen, H., *J. Am. Chem. Soc.*, **100,** 1193 (1978).
[380] Becker, Y., and Stille, J. K., *J. Am. Chem. Soc.*, **100,** 845 (1978).
[381] Hanack, M., and Sproesser, L., *J. Am. Chem. Soc.*, **100,** 7066 (1978).
[382] Olah, G. A., Asensio, G., and Mayr, H., *J. Org. Chem.*, **43,** 1518 (1978).
[383] Lee, C. C., Paine, A. J., and Ko, E. C. F., *J. Am. Chem. Soc.*, **99,** 7267 (1977).
[384] Lee, C. C., and Ko, E. C. F. *Can. J. Chem.*, **56,** 2459 (1978).
[385] Kartashev, V. R., Bodrikov, I. V., Skorobogatova, E. V., and Zefirov, N. S., *Phosphorus Sulphur*, **3,** 213 (1977).
[386] Kirmise, W., and Günther, B.-R., *J. Am. Chem. Soc.*, **100,** 3619 (1978).
[387] Wrobel, J. T., Cybulski, J., Dabrowski, Z., and MacLean D. B., *Bull. Acad. Pol. Sci, Ser. Sci. Chim.*, **25,** 775 (1977); *Chem. Abs.*, **88,** 88842 (1978).
[388] Farnum, D. G., Botts, R. E., Chambers, W. T., and Lam, B., *J. Am. Chem. Soc.*, **100,** 3847 (1978).
[389] Burger, U., Sonney, J.-M., and Vogel, P., *Tetrahedron Lett.*, **1978,** 829.
[390] Fleming, I., and Michael, J. P., *J. Chem. Soc., Chem. Commun.*, **1978,** 245.
[391] Rubottom, G. M., Gruber, J. M., Boeckman, R. K., Ramaiah, M., and Medwid, J. B., *Tetrahedron Lett*, **1978,** 4603.
[392] Van Rantwijk, F., Van Der Stoel, R. E., and Van Bekkum, H., *Tetrahedron*, **34,** 569 (1978).
[393] Cristol, S. J., Strom, R. H., and Stull, D. P., *J. Org. Chem.*, **43,** 1150 (1978).
[394] Gregorcic, A., and Zupan, M., *Tetrahedron*, **33,** 3243 (1977).
[395] Kraemer, H. P., and Plieninger, H., *Tetrahedron*, **34,** 891 (1978).
[396] Fleury, J.-P., and Desbois, M., *J. Heterocycl. Chem.*, **15,** 1005 (1978).
[397] Potekhin, K. A., Kurkutova, E. N., Antipin, M. Yu., Struchkov, Yu. T., Magerranov, A. M., Sadovaya, N. K., and Zefirov, N. S., *Zh. Org. Khim.*, **13,** 2093 (1977); *Chem. Abs.*, **88,** 62032 (1978).
[398] Belikova, N. A., Bobyleva, A. A., Dzhigirkhanova, A. V., Pehk, T., Lippmaa, E., and Plate, A. F., *Zh. Org. Khim.*, **13,** 1898 (1977); *Chem. Abs.*, **88,** 37310 (1978).
[399] Cristol, S. J., Stull, D. P., and Daussin, R. D., *J. Am. Chem. Soc.*, **100,** 6674 (1978).
[400] Kimura, T., Minabe, M., and Suzuki, K., *J. Org. Chem.*, **43,** 1247 (1978).
[401] Martin, H.-D., and Hekman, M., *Chem. Ber.*, **111,** 3004 (1978).
[402] Akiyama, T., Fujii, T., Ishiwari, H., Imagawa, T., and Kawanisi, M., *Tetrahedron Lett.*, **1978,** 2165.

[403] Goering, H. L., and Anderson, R. P., *J. Am. Chem. Soc.*, **100**, 6469 (1978).
[404] Font, J., and March, P., *Tetrahedron Lett.*, **1978**, 3601.
[405] Bessiere, Y., Reca, E., Chatzopoulos-Ouar, F., and Boussac, G., *J. Chem. Res.*, **1977**, (S) 302; (M) 3501.
[406] Bessiere, Y., and Derguini-Boumechal, F., *J. Chem. Res.*, **1977**, (S), 304; (M) 3522.
[407] Maruyama, K., and Yamamoto, Y., *J. Am. Chem. Soc.*, **99**, 8068 (1977).
[408] Zimmer, H., Hillstrom, W. W., Rothe, J., and Schmidt, J. C., *Justus Liebigs Ann. Chem.*, **1978**, 124.
[409] Hill, G., and Harris, F. L., *J. Org. Chem.*, **42**, 3306 (1977).
[410] Lee, T. S., Russell, R. A., and Warrener, R. N., *Aust. J. Chem.*, **31**, 1129 (1978).
[411] Bowen, R. D., and Williams, D. H., *J. Chem. Soc., Perkin Trans.* 2, **1978**, 68.
[412] Olah, G. A., Meidar, D., and Liang, G., *J. Org. Chem.*, **43**, 3890 (1978).
[413] Cocker, W., and Grayson, D. H., *J. Chem. Soc., Perkin Trans.* 1, **1978**, 155.
[414] Olah, G. A., and Meidar, D., *Synthesis*, **1978**, 358.
[415] Keinan, E., and Mazur, Y., *J. Org. Chem.*, **43**, 1020 (1978).
[416] Cafieri, F., De Napoli, L., Fattorusso, E., and Piattelli, M., *Tetrahedron*, **34**, 1225 (1978).
[417] Grunewald, G. L., Walters, D. E., and Kroboth, T. R., *J. Org. Chem.*, **43**, 3478 (1978).
[418] Monti, S. A., Chen, S.-C., Yang, Y.-L., Yuan, S.-S., and Bourgeois, O. P., *J. Org. Chem.*, **43**, 4062 (1978).
[419] Chiang, Y., Kresge, A. J., and Young, C. I., *Can. J. Chem.*, **56**, 461 (1978).
[420] Creary, X., and Rollin, A. J., *J. Org. Chem.*, **42**, 4231 (1977).
[421] Auteri, S. C., Cameron, D. W., and Drake, C. B., *Aust. J. Chem.*, **30**, 2479 (1977).
[422] Pau, J. K., Kim, J. K., and Caserio, M. C., *J. Am. Chem. Soc.*, **100**, 3838 (1978).
[423] Schleyer, P. von R., Jemmis, E. D., and Pople, J. A., *J. Chem. Soc., Chem. Commun.*, **1978**, 190.
[424] Bowen, R. D., and Williams, D. H., *J. Am. Chem. Soc.*, **99**, 6822 (1977).
[425] Broer, W. J., and Weringa, W. D., *Org. Mass Spectrom.*, **13**, 232 (1978).
[426] Gruetzmacher, H. F., *Org. Mass Spectrom.*, **13**, 90 (1978).
[427] Stahl, D., and Gaeumann, T., *Adv. Mass Spectrom.*, **7B**, 1190 (1978).
[428] Campaigne, E., and Forsch, R. A., *J. Org. Chem.*, **43**, 1044 (1978).
[429] Hiyama, T., Shinoda, M., and Nozaki, H., *Tetrahedron Lett.*, **1978**, 771.
[430] Trost, B. M., and Scudder, P. H., *J. Am. Chem. Soc.*, **99**, 7601 (1977).
[431] Dao, L. H., and Mackay, D., *Can. J. Chem.*, **56**, 1724 (1978).
[432] Gomès, L. M., and Cabarès, J., *C. R. Hebd. Séances Acad. Sci., Ser. C*, **73**, 287 (1978).
[433] Popova, L. A., *Tezisy Dokl. – Resp. Konf. Molodykh Uch.-Khim., 2nd*, **1**, 58 (1977); *Chem. Abs.*, **89**, 90226 (1978).
[434] Bondavalli, F., Ranise, A., Schenone, P., and Lanteri, S., *J. Chem. Soc., Perkin Trans.* 1, **1978**, 804.
[435] Warrener, R. N., Russell, R. A., and Tan, R. Y. S., *Tetrahedron Lett.*, **1978**, 1585.
[436] Takeshita, H., Hatsui, T., Kouno, I., and Mori, A., *Asahi Garasu Kogyo Gijutsu Shoreikai Kenkyu Hokoku*, **31**, 243 (1977); *Chem. Abs.*, **89**, 107547 (1978).
[437] Inamoto, Y., *Yuki Gosei Kagaku Kyokaishi*, **35**, 550 (1977); *Chem. Abs.*, **88**, 22240 (1978).
[438] Sasaki, T., Eguchi, S., and Hattori, S., *Tetrahedron*, **34**, 67 (1978).
[439] Hollowood, F., Karim, A., McKervey, M. A., and McSweeney, P., *J. Chem. Soc., Chem. Commun.*, **1978**, 306.
[440] Karlović, G., and Majerski, Z., *J. Org. Chem.*, **43**, 746 (1978).
[441] Olah, G. A., Liang, G., Babiak, K. A., Ford, T. M., Goff, D. L., Morgan, T. K., and Murray, R. K., *J. Am. Chem. Soc.*, **100**, 1494 (1978).
[442] Volz, H., Shin, J.-H., Prinzbach, H., Babsch, H., and Christl, M., *Tetrahedron Lett.*, **1978**, 1247.
[443] Ahlberg, R., and Engdahl, C., *Chem. Scr.*, **11**, 95 (1977).
[444] Kent, G. J., Godleski, S. A., Ōsawa, E., and Schleyer, P. von R., *J. Org. Chem.*, **42**, 3852 (1977).
[445] Mehta, G., Singh, V., and Duddeck, H., *Tetrahedron Lett.*, **1978**, 1223.
[446] Mehta, G., and Singh, V., *Tetrahedron Lett.*, **1978**, 4591.
[447] Tolstikov, G. A., Lerman, B. M., Galin, F. Z., and Struchkov, Yu. T., *Tetrahedron Lett.*, **1978**, 4145.
[448] Goldstein, M. J., and Warren, D. P., *J. Am. Chem. Soc.*, **100**, 6539 (1978).
[449] Golstein, M. J., and Warren, D. P., *J. Am. Chem. Soc.*, **100**, 6541, (1978).
[450] Yano, K., Isobe, M., and Yoshida, K., *J. Am. Chem. Soc.*, **100**, 6166 (1978).
[451] Paquette, L. A., James, D. R., and Klein, G., *J. Org. Chem.*, **43**, 1287 (1978).

[452] Klein, G., and Paquette, L. A., *J. Org. Chem.*, **43,** 1293 (1978).
[453] Paquette, L. A., and Carmody, M. J., *J. Org. Chem.*, **43,** 1299 (1978).
[454] Diaz, A. F., and Miller, R. D., *J. Am. Chem. Soc.*, **100,** 5905 (1978).
[455] Paquette, L. A., Klein, G., and Doecke, C. W., *J. Am. Chem. Soc.*, **100,** 1595 (1978).
[456] Cargill, R. L., Dalton, J. R., O'Connor, S., and Michels, D. G., *Tetrahedron Lett.*, **1978,** 4465.
[457] Andersen, N. H., Tseng, C.-L. W., Moore, A., and Ohta, Y., *Tetrahedron*, **34,** 47 (1978).
[458] Battiste, M. A., Timberlake, J. F., Paquette, L. A., Degenhardt, C. R., Martin, J. T., Hedaya, E., Su, T. M., and Theodorpulos, S., *J. Chem. Soc., Chem. Commun.*, **1977,** 941.
[459] Inamoto, Y., Aigami, K., Takaishi, N., Fujikura, Y., Tsuchihashi, K., and Ikeda, H., *J. Org. Chem.*, **42,** 3833 (1977).
[460] Fujikura, Y., Ohsugi, M., Inamoto, Y., Takaishi, N., and Aigami, K., *J. Org. Chem.*, **43,** 2608 (1978).
[461] Yadav. J. S., Chawla, H. P. S., and Dev, S., *Tetrahedron*, **34,** 475 (1978).
[462] Christil, M., Freitag, G., and Brüntrup, G., *Chem. Ber.*, **111,** 2320 (1978).
[463] Bancui, M., Vlad, R., Draghici, C., and Cioranescu, E., *Rev. Roum. Chim.*, **22,** 1017 (1977); *Chem. Abs.*, **88,** 49906 (1978).
[464] Miettinen, T., *Acta Chem., Scand.*, **31 B,** 818 (1977).
[465] Nomura, Y., Takeuchi, Y., and Tomoda, S., *Tetrahedron Lett.*, **1978,** 911.
[466] Hogeveen, H., and Huudeman, W. F. J., *J. Am. Chem. Soc.*, **100,** 860 (1978).
[467] Horita, H., Koizumi, Y., Otsubo, T., Sakata, Y., and Misumi, S., *Bull. Chem. Soc., Jpn.*, **51,** 2668 (1978).
[468] ApSimon, J. W., Badripersaud, S., Post, M. L., and Gabe, E. J., *Can. J. Chem.*, **56,** 2150 (1978).
[469] Bradshaw, A. P. W., Hanson, J. R., and Siverns, M., *J. Chem. Soc., Chem. Commun.*, **1978,** 303.
[470] Hayano, K., Ohfune, Y., Shirahama, H., and Matsumoto, T., *Tetrahedron Lett.*, **1978,** 1991.
[471] Duc, D. K. M., Fetizon, M., and Lazare, S., *Tetrahedron*, **34,** 1207 (1978).
[472] Banerji, A., Hunter, R., Mellows, G., Sim, K.-Y., and Barton, D. H. R., *J. Chem. Soc., Chem. Commun.*, **1978,** 843.
[473] Itoh, A., Nozaki, H., and Yamamoto, H., *Tetrahedron Lett.*, **1978,** 2903.
[474] McCormick, J. P., and Barton, D. L., *Tetrahedron*, **34,** 325 (1978).
[475] Baxter, R. L., Laurie, W. A., and McHale, D., *Tetrahedron*, **34,** 2195 (1978).
[476] Nicotra, F., Ronchetti, F., Russo, G., Lugaro, G., and Casellato, M., *J. Chem. Soc., Chem. Commun.*, **1977,** 889.
[477] Seo, S., Tomita, Y., and Tori, K., *J. Chem. Soc., Chem. Commun.*, **1978,** 319.
[478] Sheppard, R. C., and Turner, S., *J. Chem. Soc., Perkin Trans.* 1, **1977,** 2551.
[479] Anastasia, M., Soave, A. M., and Scala, A., *J. Chem. Soc., Perkin Trans.* 1, **1978,** 1131.
[480] Brieskorn, C. H., and Unger, G., *Chem. Ber.*, **111,** 1160 (1978).
[481] Alamaula, R. U., Trivedi, G. K., and Bhattacharyya, S. C., *Indian J. Chem.* **16B,** 257 (1978).
[482] Sengupta, P., Sen, M., and Sarkar, U., *J. Chem. Soc., Perkin Trans.* 1, **1978,** 384.
[483] Tori, M., Tsuyuki, T., and Takahashi, T., *Bull. Chem. Soc., Jpn.*, **50,** 3381 (1977).
[484] Davis, B. R., Rewcastle, G. W., and Woodgate, P. D., *J. Chem. Soc., Perkin Trans.* 1, **1978,** 735.
[485] Lawrie, W., Hamilton, W., McLean, J., and Meney, J., *J. Chem. Soc., Perkin Trans.* 1, **1978,** 471.
[486] Kirk, D. N., and McHugh, C. R., *J. Chem. Soc., Perkin Trans.* 1, **1978,** 173.
[487] Shimagaki, M., Anazawa, A., Oishi, T., and Tahara, A., *Heterocycles*, **8,** 237 (1977).
[488] Ceccherelli, P., Curini, M., Pellicciari, R., Baddeley, C. V., Raju, M. S., and Wenkert, E., *J. Org. Chem.*, **43,** 4244 (1978).
[489] Sloan, K. B., Little, R. J., and Bodor, N., *J. Org. Chem.*, **43,** 3405 (1978).
[490] Clode, D. M., *Can. J. Chem.*, **55,** 4071 (1977).
[491] Battersby, A. R., Fookes, C. J. R., McDonald, E., and Meegan, M. J., *J. Chem. Soc., Chem. Commun.*, **1978,** 185.
[492] Engelhard, M., and Merrifield, R. B., *J. Am. Chem. Soc.*, **100,** 3559 (1978).
[493] Jones, W. J., *Acc. Chem. Res.*, **10,** 353 (1977).
[494] Iwamura, H., and Tukada, H., *Tetrahedron Lett.*, **1978,** 3451
[495] Baxter, G. J., and Brown, R. F. C., *Aust. J. Chem.*, **31,** 327 (1978).
[496] Ando, W., Sekiguchi, A., Rothschild, A. J., Gallucci, R. R., Jones, M., Barton, T. J., and Kilgour, J. A., *J. Am. Chem. Soc.*, **99,** 6995 (1977).

[497] Bertrand, G., Manuel, G., and Mazerolles, P., *J. Organomet. Chem.*, **144,** 303 (1978).
[498] Sato, T., Abe, T., and Kuwajima, I., *Tetrahedron Lett.*, **1978,** 259.
[499] Evans, D. A., Hurst, K. M., and Takacs, J. M., *J. Am. Chem. Soc.*, **100,** 3467 (1978).
[500] Challis, B. C., and Frenkel, A. D., *J. Chem. Soc., Perkin Trans.* 2, **1978,** 192.
[501] Challis, B. C., Challis, J. A., and Iley, J. N., *J. Chem. Soc., Perkin Trans* 2., **1978,** 813.
[502] Markowska, A., and Nowicki, T., *Chem. Ber.*, **111,** 56 (1978).
[503] Maeda, M., and Kojima, M., *J. Chem. Soc., Perkin Trans.* 1, **1978,** 685.
[504] Mihailović, M. L., Milošević, G., and Milovanović, A., *Tetrahedron*, **34,** 2587 (1978).
[505] McAdoo, D. J., Witiak, D. N., and McLafferty, F. W., *J. Am. Chem. Soc.*, **99,** 7265 (1977).
[506] Edens, M., Boerner, D., Chase, C. R., Nass, D., and Schiavelli, M. D., *J. Org. Chem.*, **42,** 3403 (1977).
[507] Jensen, B. L., and Peterson, P. E., *J. Org. Chem.*, **42,** 4052 (1977).
[508] Jensen, B. L., Burke, S. E., and Thomas, S. E. *Tetrahedron*, **34,** 1627 (1978).
[509] Warnhoff, E., Rampersad, M., Raman, P. S., and Yerhoff, F. W., *Tetrahedron Lett.*, **1978,** 1659.
[510] Watt, I., *Tetrahedron Lett.*, **1978,** 4175.
[511] Butter, A. R., and Leitch, E., *J. Chem. Soc., Perkin Trans.* 2, **1977,** 1972.
[512] Pelter, A., and Hughes, L., *J. Chem. Soc., Chem. Commun.*, **1977,** 913.
[513] Hara, S., Kishimura, K., and Suzuki, A., *Tetrahedron Lett.*, **1978,** 2891.
[514] Moody, R. J., and Matteson, D. S., *J. Organomet. Chem.*, **152,** 265 (1978).
[515] Levy, A. B., Schwartz, S. J., Wilson, N., and Christie, B., *J. Organomet. Chem.*, **156,** 123 (1978).
[516] Collman, J. P., Finke, R. G., Cawse, J. N., and Brauman, J. I., *J. Am. Chem. Soc.*, **100,** 4766 (1978).
[517] Zinner, G., *Chem. Ztg.*, **102,** 58 (1978).
[518] Jensen, B. L., and Michaud, D. P., *J. Heterocycl. Chem.*, **15,** 321 (1978).
[519] Mostowicz, R., and Berak, J. M., *Przem. Chem.*, **57,** 119 (1978); *Chem. Abs.*, **88,** 190044 (1978).
[520] Zhdanov, Yu. A., Alekseev, Y. E., and Sudareva, T. P., *Dokl. Akad. Nauk SSSR*, **238,** 580 (1978); *Chem. Abs.*, **88,** 152880 (1978).
[521] Takamiya, N., Tsunoda, H., Suzuki, S., and Murai, S., *Nippon Kagaku Kaishi*, **1978,** 799; *Chem. Abs.*, **89,** 75231 (1978).
[522] Oka, K., and Hara, S., *J. Org. Chem.*, **43,** 3790 (1978).
[523] Suginome, H., and Yagihashi, F., *J. Chem. Soc., Perkin Trans* 1, **1977,** 2488.
[524] Holm, A., Christophersen, C., Ottersen, T., Hope, H., and Christensen, A., *Acta Chem. Scand.*, **31B,** 687 (1977).
[525] Darling, C. M., and Chen. C. P., *J. Pharm. Sci.*, **67,** 860 (1978).
[526] Lipa, W. J., Crawford, H. T., Radlick, P. C., and Helmkamp, G. K., *J. Org. Chem.*, **43,** 3813 (1978).
[527] Bardakos, V., and Sucrow, W., *Chem. Ber.*, **111,** 1780 (1978).
[528] Baliah, V., Lakshmanan, M., and Pandiarajan, K., *Indian J. Chem.*, **16B,** 72 (1978).
[529] Petrov, K. A., Chauzov, V. A., Lebedeva, N. Y., and Kostrova, S. M., *Zh. Obshch. Khim.*, **48,** 1187 (1978); *Chem. Abs.*, **89,** 109759 (1978).
[530] Hirao, K., Miura, H., and Yonemitsu, O., *Heterocycles*, **7,** 857 (1977).
[531] Fahmy, A. F. M., Aly, N. F., Nada, A., and Aly, N. Y., *Bull. Chem. Soc., Jpn.*, **50,** 2678 (1977).
[532] Sheradsky, T., and Avramovici-Grisaru, S., *Tetrahedron Lett.*, **1978,** 2325.
[533] Hudson, R. F., and Woodcock, R. C., *Justus Liebigs Ann. Chem.*, **1978,** 176.
[534] Hoffman, R. V., Cadena, R., and Poelker, D. J., *Tetrahedron Lett.*, **1978,** 203.
[535] Starewicz, P. M., Sackett, A., and Kovacic, P., *J. Org. Chem.*, **43,** 739 (1978).
[536] Saikachi, H., and Kitagawa, T., *Chem. Pharm. Bull.*, **26,** 1054 (1978).
[537] Garcia-Lòpez, M. T., de las Heras, F. G., and Stud, M., *J. Chem. Soc., Perkin Trans.* 1, **1978,** 483.
[538] Chimichi, S., and Nesi, R., *J. Heterocycl. Chem.*, **14,** 1099 (1977).
[539] Imhof, R., Ladner, D. W., and Muchowski, J. M., *J. Org. Chem.*, **42,** 3709 (1977).
[540] Ogata, Y., Tomizawa, K., and Ikeda, T., *J. Org. Chem.*, **43,** 2417 (1978).
[541] Miura, H., Hirao, K.-I., and Yonemitsu, O., *Tetrahedron*, **34,** 1805 (1978).
[542] Oliver, J. P., *Adv. Organomet. Chem.*, **16,** 111 (1977).
[543] Hill, E. A., *Adv. Organomet. Chem.*, **16,** 131 (1977).
[544] Grovenstein, E., *Adv. Organomet. Chem.*, **16,** 167 (1977).

[545] Faller, J. W., *Adv. Organomet. Chem.*, **16,** 211 (1977).
[546] Tsutsui, M., and Courtney, A., *Adv. Organomet. Chem.*, **16,** 241 (1977).
[547] Evans, J., *Adv. Organomet. Chem.*, **16,** 319 (1977).
[548] Posner, G. H., *Angew. Chem. Int. Edn.*, **17,** 487 (1978).
[549] Chamot, E., Sharma, A. K., and Paquette, L. A., *Tetrahedron Lett.*, **1978,** 1963.
[550] Lombardo, E. A., Conner, W. C., Madon, R. J., Hall, W. K., Kharlamov, V. V., and Minachev, Kh. M., *J. Catal.* **53,** 135 (1978).
[551] Lemberton, J.-L., Perot, G., and Guisnet, M., *J. Chem. Soc., Chem. Commun.*, **1977,** 883.
[552] Okura, I., Takahashi, N., and Keii, T., *J. Mol. Catal.*, **4,** 237 (1978).
[553] Salomon, R. G., *Adv. Chem. Ser.*, **168,** 174 (1978).
[554] D'Aniello, M. J., and Barefield, E. K., *J. Am. Chem. Soc.*, **100,** 1474 (1978).
[555] Blaser, H.-U., and Reinehr, D., *Helv. Chim. Acta*, **61,** 1118 (1978).
[556] Baudry, D., Ephritikhine, M., and Felkin, H., *J. Chem. Soc., Chem. Commun.*, **1978,** 694.
[557] Kumobayashi, H., Akutagawa, S., and Otsuka, S., *J. Am. Chem. Soc.*, **100,** 3949 (1978).
[558] Murahashi, S.-I., Hirano, T., and Yano, T., *J. Am. Chem. Soc.*, **100,** 348 (1978).
[559] Khulbe, C. P., and Mann, R. S., *Can. J. Chem.*, **56,** 2791 (1978).
[560] Muscio, O. J., Jun, Y. M., and Philip, J. B., *Tetrahedron Lett.*, **1978,** 2379.
[561] Purro, S., Pryde, A., Zsindely, J., and Schmid, H., *Helv. Chim. Acta*, **61,** 266 (1978).
[562] Gibson, D. H., Ong, T.-S., and Khoury, F. G., *J. Organomet. Chem.*, **157,** 81 (1978).
[563] Noyori, R., Hayakawa, Y., Takaya, H., Murai, S., Kobayashi, R., and Sonoda, N., *J. Am. Chem. Soc.*, **100,** 1759 (1978).
[564] Sarel, S., *Acc. Chem. Res.*, **11,** 204 (1978).
[565] Zaklika, K. A., Burns, P. A., and Schaap, A. P., *J. Am. Chem. Soc.*, **100,** 318 (1978).
[566] Giddings, R. M., and Whittaker, D., *Tetrahedron Lett.*, **1978,** 4077.
[567] Brown, J. M., Golding, B. T., and Stofko, J. J., *J. Chem. Soc., Perkin Trans.* 2, **1978,** 436.
[568] Aumann, R., Averbeck, H., and Krüger, C., *J. Organomet. Chem.*, **160,** 241 (1978)
[569] Bellamy, F., *J. Chem. Soc., Chem. Commun.*, **1978,** 998.
[570] Reinaecker, R., and Schwengers, D., *Justus Liebigs Ann. Chem.*, **1977,** 1633.
[571] Chen, L. S., Su, S. R., and Wojcicki, A., *Inorg. Chim. Acta*, **27,** 79 (1978).
[572] Downs, R. L., and Wojcicki, A., *Inorg. Chim. Acta*, **27,** 91 (1978).
[573] Ross, D. A., and Wojcicki, A., *Inorg. Chim. Acta*, **28,** 59 (1978).
[574] Eisner, U., and Sadeghi, M. M., *Tetrahedron Lett.*, **1978,** 299.
[575] Grieco, P. A., and Marinovic, N., *Tetrahedron Lett.*, **1978,** 2545.
[576] Pryde, A., *J. Chromatogr.*, **152,** 123 (1978).
[577] Karel, K. J., and Brookhart, M., *J. Am. Chem. Soc.*, **100,** 1619 (1978).
[578] Blackborow, J. R., Grubbs, R. H., Hildenbrand, K., von Gustorf, E. A. K., Miyashita, A., and Scrivanti, A., *J. Chem. Soc., Dalton Trans.*, **1977,** 2205.
[579] Hunt, M. M., Kita, W. G., Mann, B. E., and McCleverty, J. A., *J. Chem. Soc., Dalton Trans.*, **1978,** 467.
[580] Le Bozec, H., Gorgues, A., and Dixneuf, P., *J. Chem. Soc., Chem. Commun.*, **1978,** 573.
[581] Brunner, H., Schwâgerl, H., Wachter, J., Reisner, G. M., and Bernal, I., *Angew. Chem. Int. Edn.*, **17,** 453 (1978).
[582] Liles, D. C., McPartlin, M., and Tasker, P. A., *J. Am. Chem. Soc.*, **99,** 7704 (1977).
[583] Malisch, W., and Janta, R., *Angew. Chem. Int. Edn.*, **17,** 211 (1978).
[584] Gregory, B., Bullock, E., and Chen, T.-S., *Can. J. Chem.*, **55,** 4061 (1977).
[585] Kruglaya, O. A., Fedot'eva, I. B., Fedot'ev, B. V., Kalikhman, I. D., Brodskaya, E. I., and Vyazankin, N. S., *J. Organomet. Chem.*, **142,** 155 (1977).
[586] Goldschmidt, Z., and Antebi, S., *Tetrahedron Lett.*, **1978,** 271.
[587] Barborak, J. C., Dasher, L. W., McPhail, A. T., Nichols, J. B., and Onan, K. D., *Inorg. Chem.*, **17,** 2936 (1978).
[588] Paquette, L. A., and Detty, M. R., *Tetrahedron Lett.*, **1978,** 713.
[589] Paquette, L. A., and Gree, R., *J. Organomet. Chem.*, **146,** 319 (1978).
[590] Eaton, P. E., and Patterson, D. R., *J. Am. Chem. Soc.*, **100,** 2573 (1978).
[591] Eaton, P. E., and Chakraborty, U. R., *J. Am. Chem. Soc.*, **100,** 3634 (1978).
[592] Gassman, P. G., Sugawara, T., and Tillotson, L. G., *J. Org. Chem.*, **42,** 3785 (1978).
[593] Beach, D. L., and Barnett, K. W., *J. Organomet. Chem.*, **142,** 225 (1977).
[594] Hogeveen, H., and Nusse, B. J., *J. Am. Chem. Soc.*, **100,** 3110 (1978).
[595] Iten, P. X., and Eugster, C. H., *Helv. Chim. Acta*, **61,** 1134 (1978).
[596] Fritch, J. R., and Vollhardt, K. P. C., *J. Am. Chem. Soc.*, **100,** 3643 (1978).
[597] Blum, J., Zlotogorski, C., Schwarz, H., and Höhne, G., *Tetrahedron Lett.*, **1978,** 3501.

[598] Paquette, L. A., Photis, J. M., and Micheli, R. P., *J. Am. Chem. Soc.*, **99,** 7899 (1977).
[599] Paquette, L. A., Micheli, R. P., and Photis, J. M., *J. Am. Chem. Soc.*, **99,** 7911 (1977).
[600] May, C. J., and Powell, J., *Inorg. Chim. Acta*, **26,** L21 (1978).
[601] Burns, W., McKervey, M. A., Mitchell, T. R. B., and Rooney, J. J., *J. Am. Chem. Soc.*, **100,** 906 (1978).
[602] Ruth, J. L., and Bergstrom, D. E., *J. Org. Chem.*, **43,** 2870 (1978).
[603] Card, R. J., and Neckers, D. C., *J. Org. Chem.*, **43,** 2958 (1978).
[604] Skorobogatova, E. V., Povelikina, L. N., and Kartashov, V. R., *Zh. Org. Khim.*, **14,** 663 (1978).
[605] Salomon, R. G., Sinha, A., and Salomon, M. F., *J. Am. Chem. Soc.*, **100,** 520 (1978).
[606] Dinulescu, I. G., Pop, M. S., and Avram, M., *Rev. Roum. Chim.*, **22,** 1037 (1977); *Chem. Abs.*, **88,** 105002 (1978).
[607] Kwantes, P. M., and Klumpp, G. W., *Tetrahedron Lett.*, **1978,** 4097.
[608] Taylor, R. T., and Paquette, L. A., *J. Org. Chem.*, **43,** 242 (1978).
[609] Taylor, E. C., Conley, R. A., Johnson, D. K., and McKillop, A., *J. Org. Chem.*, **42,** 4167 (1977).
[610] Antus, S., Boross, F., Farkas, L., and Nogradi, M., *Flavonoids Bioflavonoids, Proc. Hung. Bioflavonoid Symp.*, 5th, **1977,** 171.
[611] Gottsegen, A., Antus, S., Farkas, L., Kardos-Balogh, Z., and Nogradi, M., *Flavonoids Bioflavonoids, Proc. Hung. Bioflavonoid Symp.*, 5th, **1977,** 181.
[612] Kardos-Balogh, Z., Farkas, L., and Wolfner, A., *Acta Chim. Acad. Sci., Hung.*, **94,** 75 (1977); *Chem. Abs.*, **88,** 105068 (1978).
[613] Fujita, E., and Ochiai, M., *Chem. Pharm. Bull.*, **25,** 3013 (1977).
[614] Fujita, E., and Ochiai, M., *Can. J. Chem.*, **56,** 246 (1978).
[615] Schwartz, A., and Glotter, E., *J. Chem., Soc., Perkin Trans.* 1, **1977,** 2470.
[616] Lewis, D. K., Giesler, S. E., and Brown, M. S., *Int. J. Chem. Kinet.*, **10,** 277 (1978).
[617] Baldwin, J. E., and Carter, C. G., *J. Am. Chem. Soc.*, **100,** 3942 (1978).
[618] Wood, J. T., Arney, J. S., Cortès, D., and Berson, J. A., *J. Am. Chem. Soc.*, **100,** 3855 (1978).
[619] Macdonald, T. L., *J. Org., Chem.*, **43,** 4241 (1978).
[620] Surmina, L. S., Formanovskii, A. A., and Bolesov, I. G., *Zh. Org., Khim.*, **14,** 883 (1978); *Chem. Abs.*, **89,** 42600 (1978).
[621] Flowers, M. C., *Can. J. Chem.*, **56,** 29 (1978).
[622] Dolbier, W. R., and Fielder, T. H., *J. Am. Chem., Soc.*, **100,** 5577 (1978).
[623] Dixon, D. A., Foster, R., Halgren, T. A., and Lipscomb, W. N., *J. Am. Chem. Soc.*, **100,** 1359 (1978).
[624] Creary, X., *J. Org. Chem.*, **43,** 1777 (1978).
[625] Vicens, M., Dumont, C., and Vidal, M., *C. R. Hebd. Séances Acad. Sci.*, **286C,** 717 (1978).
[626] Andrist, A. H., and Kalynchuk, D. G., *Spectrosc. Lett.*, **11,** 133 (1978).
[627] Warner, P., and Chang, S.-C., *Tetrahedron Lett.*, **1978,** 3981.
[628] Schoeller, W. W., and Brinker, U. H., *J. Am. Chem. Soc.*, **100,** 6012 (1978).
[629] Bailey, I. M., and Walsh, R., *J. Chem. Soc., Faraday Trans.* 1, **74,** 1146 (1978).
[630] Komendantov, M. I., Bekmukhametov, R. R., and Domnin, I. N., *Tetrahedron*, **34,** 2743 (1978).
[631] Stoffer, J. O., and Bohanon, J. T., *J. Chem. Soc., Perkin Trans.* 2, **1978,** 692.
[632] Zimmerman, H. E., and Aasen, S. M., *J. Org. Chem.*, **43,** 1493 (1978).
[633] Padwa, A., Blacklock, T. J., Getman, D., Hatanaka, N., and Loza, R., *J. Org. Chem.*, **43,** 1481 (1978).
[634] Crombie, L., Maddocks, P. J., and Pattenden, G., *Tetrahedron Lett.*, **1978,** 3479.
[635] Tsuji, T., Shibata, T., Hienuki, Y., and Nishida, S., *J. Am. Chem. Soc.*, **100,** 1806 (1978).
[636] Gajewski, J. J., and Weber, R. J., *J. Am. Chem. Soc.*, **99,** 8054 (1977).
[637] Gajewski, J. J., and Chang, M. J., *J. Org. Chem.*, **43,** 765 (1978).
[638] Khazanie, P. G., and Lee-Ruff, E., *Can. J. Chem.*, **56,** 808 (1978).
[639] Nemoto, H., Seto, H., and Kametani, T., *Heterocycles*, **6,** 1569 (1977).
[640] Kametani, T., Seto, H., Nemoto, H., and Fukumoto, K., *J. Org. Chem.*, **42,** 3605 (1977).
[641] Wada, E., Nakai, T., and Okawara, M., *Chem. Lett.*, **1977,** 1121.
[642] Grudzinski, Z., and Roberts, S. M., *Tetrahedron Lett.*, **1978,** 389.
[643] Rastetter, W. H., and Richard, T. J., *Tetrahedron Lett.*, **1978,** 2999.
[644] Garanti, L., and Zecchi, G., *J. Heterocycl. Chem.*, **15,** 509 (1978).
[645] Gerson, F., Huber, W., and Müllen, K., *Angew. Chem. Int. Edn.*, **17,** 208 (1978).
[646] Kato, M., Tsuji, M., and Miwa, T., *Bull. Chem. Soc. Jpn.*, **51,** 1450 (1978).

[647] Spielmann, W., Kaufmann, D., and de Meijere, A., *Angew. Chem. Int. Ed.*, **17,** 440 (1978).
[648] Bischof, P., Gleiter, R., Taylor, R. T., Browne, A. R., and Paquette, L. A., *J. Org. Chem.*, **43,** 2391 (1978).
[649] Atkins, M. P., Golding, B. T., and Sellars, P. J., *J. Chem. Soc., Chem. Commun.*, **1978,** 954.
[650] Bury, A., Ashcroft, M. R., and Johnson, M. D., *J. Am. Chem. Soc.*, **100,** 3217 (1978).
[651] Baddeley, G. V., Samaan, H. J., Simes, J. J. H., and Ai, T. H., *J. Chem. Soc., Chem. Commun.*, **1978,** 411.
[652] Boyd, D. R., and Berchtold, G. A., *J. Am. Chem., Soc.*, **100,** 3958 (1978).
[653] Nakasuji, K., Nakamura, T., and Mnrata, I., *Tetrahedron Lett.*, **1978,** 1539.
[654] Schultz, A. G., Erhardt, J., and Hagmann, W. K., *J. Org. Chem.*, **42,** 3458 (1977).
[655] Boeckman, R. K., Bruza, K. J., and Heinrich, G. R., *J. Am. Chem. Soc.*, **100,** 7101 (1978).
[656] Stork, G., and Williard, P. G., *J. Am. Chem. Soc.*, **99,** 7067 (1977).
[657] Chan, T. H., and Ong, B. S., *J. Org. Chem.*, **43,** 2994 (1978).
[658] Eberbach, W., and Burchardt, B., *Chem. Ber.*, **111,** 3665 (1978).
[659] Bourelle-Wargnier, F., Vincent, M., and Chuche, J., *Tetrahedron Lett.*, **1978,** 283.
[660] Kawashima, K., and Ishiguro, T., *Chem. Pharm. Bull.*, **26,** 951 (1978).
[661] Cathcart, R. C., Bovenkamp, J. W., Moir, R. Y., Bannard, R. A. B., and Casselman, A. A., *Can. J. Chem.*, **55,** 3774 (1977).
[662] Arata, K., Bledsoe, J. O., and Tanabe, K., *J. Org. Chem.*, **43,** 1660 (1978).
[663] Flowers, M. C., *J. Chem. Soc., Faraday Trans.* 1, **73,** 1927 (1977).
[664] Hassan, M., Nour, A. R. O., and Satti, A. M., *Rev. Roum, Chim.*, **23,** 747 (1978).
[665] Verhe, R., De Buyck, L., De Kimpe, N., De Rooze, A., and Schamp, N., *Bull. Soc., Chim. Belg.*, **87,** 143 (1978).
[666] Obayashi, M., Utimoto, K., and Nozaki, H., *Tetrahedron Lett.*, **1978,** 1383.
[667] Hart, H., Jiang, J. B.-C., and Sasaoka, M., *J. Org. Chem.*, **42,** 3840 (1977).
[668] Gainsford, G. J., and Woolhouse, A. D., *J. Chem. Soc., Chem. Commun.*, **1978,** 857.
[669] Hata, N., and Oguro, T., *Chem. Lett.*, **1978,** 597.
[670] Kostikov, R. R., Khlebnikov, A. F., and Ogloblin, K. A., *Khim. Geterotsikl. Soedin.*, **1978,** 48; *Chem. Abs.*, **88,** 169358 (1978).
[671] Meilahn, M. K., Olsen, D. K., Brittain, W. J., and Anders, R. T., *J. Org. Chem.*, **43,** 1346 (1978).
[672] Calcagno, M. A., and Schweizer, E. E., *J. Org. Chem.*, **43,** 4207 (1978).
[673] Vaultier, M., and Carrié, R., *J. Chem. Soc., Chem. Commun.*, **1978,** 356.
[674] Barltrop, J. A., Day, A. C., and Ward, R. W., *J. Chem. Soc., Chem. Commun.*, **1978,** 131.
[675] De Kimpe, N., Verhe, R., De Buyck, L., and Schamp, N., *Recl. Trav. Chim. Pays-Bas*, **96,** 242 (1977).
[676] Sauleau, J., Sauleau, A., and Huet, J., *Bull. Soc. Chim. Fr. II*, **1978,** 97.
[677] Quast, H., and Vélez, C. A. W., *Angew. Chem. Int. Edn.*, **17,** 213 (1978).
[678] Langlois, N., and Potier, P., *Bull. Soc. Chim. Fr. II*, **1978,** 144.
[679] Stoll, A. P., Loosli, H. R., Niklaus, P., and Zardin-Tartaglia, T., *Helv. Chim. Acta*, **61,** 648 (1978).
[680] Kollenz, G., *Justus Liebigs Ann. Chem.*, **1978,** 1670.
[681] Davies, K. L., Storr, R. C., and Whittle, P. J., *J. Chem. Soc., Chem. Commun.*, **1978,** 9.
[682] Bellamy, F. D., *Tetrahedron Lett.*, **1978,** 4577.
[683] Padwa, A., and Carlsen, P. H. J., *Tetrahedron Lett.*, **1978,** 433.
[684] Isomura, K., Hirose, Y., Shuyama, H., Abe, S., Ayabe, G., and Taniguchi, H., *Heterocycles*, **9,** 1207 (1978).
[685] Vohra, S. K., Harrington, G. W., and Swern, D., *J. Org., Chem.*, **43,** 3617 (1978).
[686] Wulff, W. D., Goure, W. F., and Barton, T. J., *J. Am. Chem., Soc.*, **100,** 6236 (1978).
[687] Norsoph, E. B., Coleman, B., and Jones, M., *J. Am. Chem., Soc.*, **100,** 994 (1978).
[688] Bose, A. K., Hoffmann, W. A., and Manhas, M. S., *Spec. Publ. Chem. Soc.* (Recent Adv. Chem. β-Lactam Antibiot)., **28,** 269 (1977).
[689] Mukerjee, A. K., and Singh, A. K., *Tetrahedron*, **34,** 1731 (1978).
[690] Chou, T. S., Spitzer, W. A., Dorman, D. E., Kukolja, S., Wright, I. G., Jones, N. D., and Chaney, M. O., *J. Org. Chem.*, **43,** 3835 (1978).
[691] Baldwin, J. E., and Christie, M. A., *J. Chem., Soc., Chem. Commun.*, **1978,** 239.
[692] Kano, S., Ebata, T., Denta, Y., Hibino, S., and Shibuya, S., *Heterocycles*, **8,** 411 (1977).
[693] Moore, J. A., and Staskun, B., *J. Org. Chem.*, **43,** 4021 (1978).
[694] Palmer, M. H., Leitch, D. S., and Greenhalgh, C. W., *Tetrahedron*, **34,** 1015 (1978).
[695] Trost, B. M., Bogdanowicz, M. J., Frazee, W. J., and Salzmann, T. N., *J. Am. Chem. Soc.*, **100,** 5512 (1978).

[696] Abegg, V. P., Hopkinson, A. C., and Lee-Ruff, E., *Can. J. Chem.*, **56,** 99 (1978).
[697] Dao, Lê H., Hopkinson, A. C., and Lee-Ruff, E., *Tetrahedron Lett.*, **1978,** 1413.
[698] Do, K. M., Fetizon, M., and Lazare, S., *J. Chem. Res.*, **1978,** (S) 22; (M) 167.
[699] Casadevall, E., and Pouet, Y., *Tetrahedron*, **34,** 1921 (1978).
[700] Kirmse, W., Scheidt, F., and Vater, H. J., *J. Am. Chem. Soc.*, **100,** 3945 (1978).
[701] Mündnich, R., and Plieninger, H., *Tetrahedron*, **34,** 887 (1978).
[702] Shimizu, N., and Nishida, S., *J. Chem. Soc., Chem. Commun.*, **1978,** 931.
[703] Dyer, S. F., Kammula, S., and Shevlin, P. B., *J. Am. Chem. Soc.*, **99,** 8104 (1977).
[704] Kirmse, W., and Murawski, H. R., *J. Chem. Soc., Chem. Commun.*, **1978,** 392.
[705] Petty, R. L., Ikeda, M., Samuelson, G. E., Boriack, C. J., Onan, K. D., McPhail, A. T., and Meinwald, J., *J. Am. Chem. Soc.*, **100,** 2464 (1978).
[706] Cadogan, J. I. G., Rowley, A. G., and Wilson, N. H., *Justus Liebigs Ann. Chem.*, **1978,** 74.
[707] Blatcher, P., Warren, S., Ncube, S., Pelter, A., and Smith, K., *Tetrahedron Lett.*, **1978,** 2349.
[708] Frejd, T., *J. Heterocycl. Chem.*, **14,** 1085 (1977).
[709] Yates, P., and Toong, Y. C., *J. Chem. Soc., Chem. Commun.*, **1978,** 205.
[710] Richter, R., Tucker, B., and Ulrich, H., *J. Org. Chem.*, **43,** 4150 (1978).
[711] Cartwright, D., Lee, V. J., and Rinehart, K. L., *J. Am. Chem. Soc.*, **100,** 4237 (1978).
[712] Lockhart, R. W., Kitadani, M., Einstein, F. W. B., and Chow, Y. L., *Can. J. Chem.*, **56,** 2897 (1978).
[713] Greene, B. B., and Lewis, K. G., *Aust. J. Chem.*, **31,** 627 (1978).
[714] Hashem, A. I., *J. Prakt. Chem.*, **319,** 689 (1977).
[715] Gedge, D. R., and Pattenden, G., *J. Chem. Soc., Chem. Commun.*, **1978,** 880.
[716] Poje, M., Gašpert, B., and Balenović, K., *Recl. Trav. Chim. Pays-Bas*, **97,** 242 (1978).
[717] Caton, M. P. L., Darnbrough, G., and Parker, T., *Synth. Commun.*, **8,** 155 (1977).
[718] Shono, T., Hamaguchi, H., and Aoki, K., *Chem. Lett.*, **1977,** 1053.
[719] Brennan, T. M., Weeks, P. D., Brannegan, D. P., Kuhla, D. E., Elliott, M. L., Watson, H. A., and Wlodecki, B., *Tetrahedron Lett.*, **1978,** 331.
[720] Ohkata, K., Isako, T., and Hanafusa, T., *Chem. Ind.* (*London*), **1978,** 274.
[721] Ohkata, K., Sakai, T., Kulo, Y., and Hanafusa, T., *J. Org. Chem.*, **43,** 3070 (1978).
[722] Ruecker, G., and Hembeck, H. W., *Arch. Pharm.* (*Weinheim*), **311,** 511 (1978).
[723] Shcherban, A. I., Aleksyuk, M. P., and Kharitonov, G. V., *Izv. Vyssh. Uchebn. Zaved., Khim. Khim. Tekhnol.*, **21,** 342 (1978); *Chem. Abs.*, **89,** 108783 (1978).
[724] Malardeau, C., and Mousset, G., *Bull. Soc. Chim. Fr. II*, **1977,** 988.
[725] Salomon, R. G., Salomon, M. F., and Coughlin, D. J., *J. Am. Chem. Soc.*, **100,** 660 (1978).
[726] Bogatskii, A. V., Kotlyar, S. A., Kamalov, G. L., and Savranskaya, R. L., *Zh. Org. Khim.*, **14,** 1112 (1978); *Chem. Abs.*, **89,** 108033 (1978).
[727] Muceniece, D., Vigante, B., and Rotbergs, J., *Zh. Org. Khim.*, **14,** 152 (1978); *Chem. Abs.*, **88,** 169806 (1978).
[728] Eagle, S. J., and Kitchin, J., *Tetrahedron Lett.*, **1978,** 4703.
[729] Bassindale, A. R., Brook, A. G., Jones, P. F., and Stewart, J. A. G., *J. Organomet. Chem.*, **152,** C25 (1978).
[730] Fuju, T., Itaya, T., and Saito, T., *Symp. Heterocycl.*, (Pap.), **1977,** 129; *Chem. Abs.*, **89,** 163445 (1978).
[731] Heaney, H., Hollinshead, J. H., and Sharma, R. P., *Tetrahedron Lett.*, **1978,** 67.
[732] Nishinaga, A., Itahara, T., Matsuura, T., Rieker, A., Koch, D., Albert, K., and Hitchcock, P. B., *J. Am. Chem. Soc.*, **100,** 1826 (1978).
[733] John, G. D., and Shannon, P. V. R., *Chem. Ind.* (*London*), **1977,** 876.
[734] Kasturi, T. R., and Parvathi, S., *Indian J., Chem. Soc.*, **15B,** 857 (1977).
[735] Dauben, W. G., and Hart, D. J., *J. Org. Chem.*, **42,** 3787 (1977).
[736] Ghosh, C. K., and Mukhopadhyay, K. K., *J. Indian Chem. Soc.*, **55,** 386 (1978).
[737] Baradarani, M. M., and Joule, J. A., *J. Chem. Soc., Chem. Commun.*, **1978,** 309.
[738] Roedig, A., Fleischmann, K., Frank, F., and Rettenberger, R., *Justus Liebigs Ann. Chem.*, **1977,** 2091.
[739] Roedig, A., Goepfert, H., and Renk, H., *Chem. Ber.*, **111,** 860 (1978).
[740] Chlenov, I. E., Salamonov, Y. B., Khasapov, B. N., Shitkin, V. M., Karpenko, N. F., Chizhov, O. S., and Tartakovskii, V. A., *Izv. Akad. Nauk SSSR, Ser. Khim.*, **1978,** 1149; *Chem. Abs.*, **89,** 109305 (1978).
[741] Rakhmankulov, D. L., Kantor, E. A., Musavirov, R. S., Romanov, N. A., and Paushkin, Ya M., *Dokl. Akad. Nauk SSSR*, **236,** 1383 (1977); *Chem. Abs.*, **88,** 89470 (1978).
[742] Menger, F. M., and Perinis, M., *Tetrahedron Lett.*, **1978,** 4653.
[743] Iriye, R., and Sasakura, M., *Agric. Biol. Chem.*, **41,** 2109 (1977).

[744] Schröder, G., Plinke, G., and Oth, J. F. M., *Chem. Ber.*, **111**, 99 (1978).
[745] Hildenbrand, P., Plinke, G., Oth, J. F. M., and Schröder, G., *Chem. Ber.*, **111**, 107 (1978).
[746] Corey, E., J., Brunelle, D. J., and Nicholou, K. C., *J. Am. Chem. Soc.*, **99**, 7359 (1977).
[747] Kramer, U., Guggisberg, A., Hesse, M., and Schmid, H., *Angew. Chem. Int. Edn.*, **16**, 861 (1977).
[748] Hutchings, M. G., Johnson, J. B., Klemperer, W. G., and Knight, R. R., *J. Am. Chem. Soc.*, **99**, 7126 (1977).
[749] Balaban, A. T., *Rev., Roum. Chim.*, **23**, 733 (1978).
[750] Altmann, J. A., Csizmadia, I. G., Robb., M. A., Yates, K., and Yates, P., *J. Am. Chem. Soc.*, **100**, 1653 (1978).
[751] Dykstra, C. E., and Schaefer, H. F., *J. Am. Chem. Soc.*, **100**, 1378 (1978).
[752] Simmie, J. M., and Melvin, D., *J. Chem. Soc., Faraday Trans.* 1, **74**, 1337 (1978).
[753] Brown, R. F. C., Eastwood, F. W., and Jackman, G. P., *Aust. J. Chem.*, **31**, 579 (1978).
[754] Sargsyan, M. S., and Badanyan, Sh. O., *Arm. Khim. Zh.*, **31**, 416 (1978); *Chem. Abs.*, **89**, 179443 (1978).
[755] Poiesel, K., Krejča, F., and Jonas, J., *Collect. Czech. Chem. Commun.*, **43**, 876 (1978).
[756] Saváta, V., Prochazka, M., and Bakos, V., *Collect. Czech. Chem. Commun.*, **43**, 2619 (1978).
[757] Siroký, M., and Prochažka, M., *Collect. Czech. Chem. Commun.*, **43**, 2635 (1978).
[758] Masilamani, D., and Rogic, M. M., *J. Am. Chem. Soc.*, **100**, 4634 (1978).
[759] Becker, D., Brodsky, N. C., and Kalo, J., *J. Org. Chem.*, **43**, 2557 (1978).
[760] Hattori, M., Wada, Y., Morikawa, A., and Otsuka, K., *J. Chem. Soc., Chem. Commun.*, **1977**, 830.
[761] Otsuka, K., Oouchi, R., and Morikawa, A., *J. Catal.*, **50**, 379 (1977).
[762] Kijenski, J., and Malinowski, S., *Bull. Acad. Pol. Sci,, Ser. Sci. Chim.*, **25**, 669 (1977); *Chem. Abs.*, **88**, 50030 (1978).
[763] Quick, J., and Meltz, C., *J. Chem. Soc., Chem. Commun.*, **1978**, 355.
[764] Pinkus, A. G., and Kalyanam, N., *J. Chem. Soc., Chem. Commun.*, **1978**, 201.
[765] Agranat, I., and Tapuhi, Y., *J. Am. Chem. Soc.*, **100**, 5604 (1978).
[766] Anastassiou, A. G., Sabahi, M., and Badri, R., *Tetrahedron Lett.*, **1978**, 4755.
[767] Steiner, U., Abdel-Kader, M. H., Fischer, P., and Kramer, H. E. A., *J. Am. Chem. Soc.*, **100**, 3190 (1978).
[768] Lohse, C., Pedersen, C. T., Ebel, M., and Callendret, R., *J. Chem. Soc., Perkin Trans.* 1, **1978**, 1432.
[769] Richter, R., and Temme, G. H., *J. Org. Chem.*, **43**, 1825 (1978).
[770] Pericás, M. A., and Serratosa, F., *Tetrahedron Lett.*, **1978**, 4969.
[771] Stevens, D. J., and Spicer, L. D., *J. Am. Chem. Soc.*, **100**, 3295 (1978).
[772] Giacomelli, G., Lardicci, L., and Saba, A., *J. Chem. Soc., Perkin Trans.* 1, **1978**, 314.
[773] Doering, W. von E., and Barsa, E. A., *Tetrahedron Lett.*, **1978**, 2495.
[774] Forni, A., Garuti, G., Moretti, I., Andreetti, G. D., Bocelli, G., and Sgarabotto, P., *J. Chem. Soc., Perkin Trans.* 2, **1978**, 401.
[775] Hwang, W. F., and Kuska, H. A., *J. Phys. Chem.*, **82**, 2126 (1978).
[776] Brown, C., Hudson, R. F., and Grayson, B. T., *J. Chem. Soc., Chem. Commun.*, **1978**, 156.
[777] Hegarty, A. F., Brady, K., and Mullane, M., *J. Chem. Soc., Chem. Commun.*, **1978**, 871.
[778] Plenkiewicz, J., *Tetrahedron*, **34**, 2961 (1978).
[779] Brown, C., Hudson, R. F., and Record, K. A. F., *J. Chem. Soc., Perkin Trans.* 2, **1978**, 822.
[780] Andrist, A. H., Slivon, L. E., and Graas, J. E., *J. Org. Chem.*, **43**, 634 (1978).
[781] McMurry, J. E., and Choy, W., *J. Org. Chem.*, **43**, 1800 (1978).
[782] Chan, H. W.-S., Levett, G., and Matthew, J. A., *J. Chem. Soc., Chem. Commun.*, **1978**, 756.
[783] Yablokov, V. A., Ganyushkin, A. V., Botnikov, M. Y., and Zhulin, V. M., *Izv. Akad. Nauk SSSR, Ser. Khim.*, **1978**, 484; *Chem. Abs.*, **88**, 151789 (1978).
[784] Lutsenko, I. F., Baukov, Y. I., Foss, V. L., and Novikova, Z. S., *Vestn. Mosk, Univ., Ser 2: Khim.*, **18**, 504 (1977); *Chem. Abs.*, **88**, 49765 (1978).
[785] Bouma, W. J., and Radom, L., *Aust. J. Chem.*, **31**, 1167 (1978).
[786] Carlsen, L., and Duus, F., *J. Am. Chem. Soc.*, **100**, 281 (1978).
[787] Yusupov, I. A., Gindin, V. A., Ershov, B. A., and Kol'tsov, A. I., *Zh. Org. Khim.*, **14**, 210 (1978); *Chem. Abs.*, **88**, 135835 (1978).
[788] Jones, R. L., and Wilson, N. H., *J. Chem. Soc., Perkin Trans.* 1, **1978**, 209.
[789] Bianco, A., Bonini, C. C., Guiso, M., Iavarone, C., and Trogolo, C., *Tetrahed. on Lett.*, **1978**, 4829.
[790] Conia, J. M., and Lange, G. L., *J. Org. Chem.*, **43**, 564 (1978).

[791] Uda, M., and Kubota, S., *J. Heterocycl. Chem*, **15,** 807 (1978).
[792] Stegmann, H. B., Haller, R., and Scheffler, K., *Chem. Ber.*, **110,** 3817 (1977).
[793] Bowden, K., and El-Kaissi, F. A., *J. Chem. Soc., Perkin Trans.*, 2, **1977,** 1927.
[794] Gogte, V. N., Jose, C. I., Namjoshi, A. G., Vankar, Y. D., and Tilak, B. D., *Indian J. Chem.*, **15B,** 778 (1977).
[795] Kishimoto, S., Kitahara, S., Manabe, O., and Hiyama, H., *J. Org. Chem.*, **43,** 3882 (1978).
[796] Brown, S. B., and Dewar, M. J. S., *J. Org. Chem.*, **43,** 1331 (1978).
[797] Eisch, J. J., Gopal, H., and Kuo, C. T., *J. Org. Chem.*, **43,** 2190 (1978).
[798] Lapachov, V. V., Zagulayeva, O. A., Bichkov, S. F., and Mamaev, V. P., *Tetrahedron Lett.*, **1978,** 3055.
[799] Bensaude, O., and Dubois, J.-E., *C. R. Hebd. Séances Acad. Sci.*, **285C,** 503 (1977).
[800] Bensaude, O., Chevrier, M., and Dubois, J.-E., *Tetrahedron Lett.*, **1978,** 2221.
[801] Bensaude, O., Chevrier, M., and Dubois, J.-E., *J. Am. Chem. Soc.*, **100,** 7055 (1978).
[802] Bensaude, O., Chevrier, M., and Dubois, J.-E., *Tetrahedron*, **34,** 2259 (1979).
[803] Dreyfus, M., Dodin, G., Bensaude, O., and Dubois, J.-E., *J. Am. Chem. Soc.*, **99,** 7027 (1977).
[804] Dodin, G., Dreyfus, M., Bensaude, O., and Dubois, J.-E., *J. Am. Chem. Soc.*, **99,** 7257 (1977).
[805] Cole, F. X., and Schimmel, P. R., *J. Am. Chem. Soc.*, **100,** 3957 (1978).
[806] Burke, L. A., Leroy, G., Nguyen, M. T., and Sana, M., *J. Am. Chem. Soc.*, **100,** 3668 (1978).
[807] Faure, R., Galy, J.-P., Vincent, E.-J., and Elguero, J., *J. Heterocycl. Chem.*, **14,** 1299 (1977).
[808] Könnecke, A., Dörre, R., Kleinpeter, E., and Lippmann, E., *Tetrahedron Lett.*, **1978,** 1311.
[809] Balli, H., and Gunzenhauser, S., *Helv. Chim. Acta*, **61,** 2628 (1978).
[810] Testaferri, L., Tiecco, M., Zanirato, P., and Martelli, G., *J. Org. Chem.*, **43,** 2197 (1978).
[811] Toth, G., and Hemela, J., *Acta Chim. Acad. Sci. Hung.*, **95,** 71 (1977); *Chem. Abs.*, **89,** 42035 (1978).
[812] Steyn, P. S., and Wessels, P. L., *Tetrahedron Lett.*, **1978,** 4707.
[813] Flitsch, W., and Kappenberg, F., *Chem. Ber.*, **111,** 2401 (1978).
[814] Uematsu, S., and Akahori, Y., *Chem. Pharm. Bull.*, **25,** 3261 (1977).
[815] Abraham, W., Buck, K., and Kreysig, D., *Tetrahedron Lett.*, **1978,** 4109.
[816] Boyd, D. R., Neill, J. D., and Stubbs, M. E., *J. Chem. Soc., Chem. Commun.*, **1977,** 873.
[817] Pitacco, G., and Valentin, E., *Tetrahedron Lett.*, **1978,** 2339.
[818] Rozwadowska, M., *Wiad. Chem.* **31,** 575 (1977); *Chem. Abs.*, **88,** 73732 (1978).
[819] Barany, F., Wolff, S., and Agosta, W. C., *J. Am. Chem. Soc.*, **100,** 1946 (1978).
[820] Wolff, S., and Agosta, W. C., *J. Org. Chem.*, **43,** 3627 (1978).
[821] Keenan, S. L., and Chapman, P. J., *J. Chem. Soc., Chem. Commun.*, **1978,** 731.
[822] Ito, S., Maeda, M., and Kojima, M., *Heterocycles*, **8,** 455 (1977).
[823] Field, D. J., Jones, D. W., and Kneen, G., *J. Chem. Soc., Perkin Trans.* 1, **1978,** 1050.
[824] Domagala, J. M., and Bach, R. D., *J. Am. Chem. Soc.*, **100,** 1605 (1978).
[825] Jackson, A. H., Naidoo, B., Smith, A. E., Bailey, A. S., and Vandrevala, M. H., *J. Chem. Soc., Chem. Commun.*, **1978,** 779.
[826] Kuckländer, U., *Justus Liebigs, Ann. Chem.*, **1978,** 129.
[827] Rahman, M. M. A. A., *Indian J. Chem.*, **15B,** 1032 (1977).
[828] Butke, G. P., Jimeniz, F., Michalik, J., Gorski, R. A., Rossi, N. F., and Wemple, J., *J. Org. Chem.*, **43,** 954 (1978).
[829] Butler, R. N., O'Regan, C. B., and Moyniham, P., *J. Chem. Soc., Perkin Trans.* 1, **1978,** 373.
[830] Grunewald, G. L., and Brouillette, W. J., *J. Org. Chem.*, **43,** 1839 (1978).
[831] Krouwer, J. S., and Richmond, J. P., *J. Org. Chem.*, **43,** 2464 (1978).
[832] Stoss, P., and Satzinger, G., *Chem. Ber.*, **11,** 1453 (1978).
[833] Olekhnovich, L. P., Mikhailov, I. E., Ivanchenko, N. M., Metlushenko, V. P., Zhdanov, Y. A., and Minkin, V. I., *Zh. Org. Khim.*, **14,** 340 (1978); *Chem. Abs.*, **88,** 169365 (1978).
[834] Fernández, B., Perillo, I., and Lamdan, S., *J. Chem. Soc., Perkin Trans.* 2, **1978,** 545.
[835] Fernández, B., Perillo, I., and Lamden, S., *J. Chem. Soc., Perkin Trans.* 2, **1978,** 550.
[836] Cuccovia, I. M., Schröter, E. H., de Baptista, R. C., and Chaimovich, H., *J. Org. Chem.*, **42,** 3400 (1977).
[837] Sawaki, Y., and Ogata, Y., *J. Am. Chem. Soc.*, **100,** 856 (1978).
[838] Prinzbach, H., Schal, H.-P., and Hunkler, D., *Tetrahedron Lett.*, **1978,** 2195.
[839] Mukai, T., and Yamashita, Y., *Tetrahedron Lett.*, **1978,** 357.
[840] Grimmie, W., and Heinze, U., *Chem. Ber.*, **111,** 2563 (1978).
[841] Boehm, M. C., and Gleiter, R., *Tetrahedron Lett.*, **1978,** 1179.
[842] Russell, G. A., Tanger, C. M., and Kosugi, Y., *J. Org. Chem.*, **43,** 3278 (1978).
[843] Kausch, M., and Dürr, H., *Chem. Ber.*, **111,** 1 (1978).

[844] Klaerner, F. G., and Wette, M., *Chem. Ber.*, **111**, 282 (1978).
[845] Zimmerman, H. E., and Cutler, T. P., *J. Org. Chem.*, **43**, 3283 (1978).
[846] Doecke, C. W., Klein, G., and Paquette, L. A., *J. Am. Chem. Soc.*, **100**, 1596 (1978).
[847] Shea, K. J., and Wise, S., *J. Am. Chem. Soc.*, **100**, 6519 (1978).
[848] Moews, P. C., Knox, J. R., and Vaughan, W. R., *J. Am. Chem. Soc.*, **100**, 260 (1978).
[849] Kovner, O. Y., Ranneva, Y. I., Mil'vitskaya, E. M., Shapiro, I. O., Luzikov, Y. N., Shatenshtein, A. I., and Plate, A. F., *Zh. Org. Khim.*, **14**, 1562 (1978); *Chem. Abs.*, **89**, 129097 (1978).
[850] Mau, A. W. H., *J. Chem. Soc., Faraday Trans.* 1, **74**, 603 (1978).
[851] Martella, D. J., Jones, M., and Schleyer, P. von R., *J. Am. Chem. Soc.*, **100**, 2896 (1978).
[852] Spanget-Larsen, J., and Gleiter, R., *Angew. Chem. Int. Edn.*, **17**, 441 (1978).
[853] Roedig, A., and Försch, M., *Justus Liebigs Ann. Chem.*, **1978**, 1406.
[854] Cichra, D., Platz, M. S., and Berson, J. A., *J. Am. Chem. Soc.*, **99**, 8507 (1977).
[855] Muyashi, T., Nakajo, T., and Mukai, T., *J. Chem. Soc., Chem. Commun.*, **1978**, 442.
[856] Belkind, B. A., Denney, D. B., Denney, D. Z., Hsu, Y. F., and Wilson, G. E., *J. Am. Chem. Soc.*, **100**, 6327 (1978).
[857] Rigaudy, J., Baranne-Lafont, J., Defoin, A., and Cuong, N. K., *Tetrahedron*, **34**, 73 (1978).
[858] Rigaudy, J., and Sparfel, D., *Tetrahedron*, **34**, 113 (1978).
[859] Wolinsky, J., Thorstenson, J. H., and Vogel, M. K., *J. Org. Chem.*, **43**, 740 (1978).
[860] Kaupp, G., *Angew. Chem. Int. Edn.*, **17**, 150 (1978).
[861] Zimmerman, H. E., Gruenbaum, W. T., Klun, R. T., Steinmetz, M. G., and Welter, T. R., *J. Chem. Soc., Chem. Commun.*, **1978**, 228.
[862] Santiago, C., McAlduff, E. J., Houk, K. N., Snow, R. A., and Paquette, L. A., *J. Am. Chem. Soc.*, **100**, 6149 (1978).
[863] Zimmerman, H. E., and Gruenbaum, W. T., *J. Org. Chem.*, **43**, 1997 (1978).
[864] Alexander, D. W., Pratt, A. C., Rowley, D. H., and Tipping, A. E., *J. Chem. Soc., Chem. Commun.*, **1978**, 101.
[865] Baeckström, P., *Tetrahedron*, **34**, 3331 (1978).
[866] Ferreira, A. B. B., and Salisbury, K., *J. Chem. Soc., Perkin Trans.* 2, **1978**, 995.
[867] Zimmerman, H. E., and Klun, R. T., *Tetrahedron*, **34**, 1775 (1978).
[868] Nitta, M., Inoue, O., and Tada, M., *Chem. Lett.*, **1977**, 1065.
[869] Weiss, R. G., and Hammond, G. S., *J. Am. Chem. Soc.*, **100**, 1172 (1978).
[870] Zimmerman, H. E., Gannett, T. P., and Keck, G. E., *J. Am. Chem. Soc.*, **100**, 323 (1978).
[871] Zimmerman, H. E., Steinmetz, M. G., and Kreil, C. L., *J. Am. Chem. Soc.*, **100**, 4146 (1978).
[872] Hirao, K., Unno, S., Miura, H., and Yonemitsu, O., *Chem. Pharm. Bull.*, **25**, 3354 (1977).
[873] Momose, T., Kanai, K., and Nakamura, T., *Chem. Pharm. Bull.*, **26**, 1592 (1978).
[874] Vallet, J. A., Boix, J., Bonet, J. J., Brianso, M. C., Miravitlles, C., and Brianso, J. L., *Helv. Chim. Acta*, **61**, 1158 (1978).
[875] Zimmerman, H. E., and Cutler, T. P., *J. Chem. Soc., Chem. Commun.*, **1978**, 232.
[876] Srinivasan, R., and Brown, K. H., *J. Am. Chem. Soc.*, **100**, 4602 (1978).
[877] Padwa, A., Carlsen, P. H. J., and Tremper, A., *J. Am. Chem. Soc.*, **100**, 4481 (1978).
[878] Perutz, R. N., *J. Photochem.*, **9**, 195 (1978).
[879] Hanaoka, M., and Mukai, C., *Heterocycles*, **6**, 1981 (1977).
[880] Nakagawa, M., and Hino, T., *Heterocycles*, **6**, 1675 (1977).
[881] Cristol, S. J., and Micheli, R. P., *J. Am. Chem., Soc.*, **100**, 850 (1978).
[882] Inoue, Y., Takamuku, S., and Sakurai, H., *J. Chem. Soc., Perkin Trans.* 2, **1977**, 1635.
[883] Turro, N. J., Cherry, W. R., Mirbach, M. F., and Mirbach, M. J., *J. Am. Chem. Soc.*, **99**, 7388 (1977).
[884] Schuster, D. I., Brown, R. H., and Resnick, B. M., *J. Am. Chem. Soc.*, **100**, 4504 (1978).
[885] Maier, G., Pfriem, S., Schaefer, U., and Matusch, R., *Angew. Chem. Int. Edn.*, **17**, 520 (1978).
[886] Dewar, D. J., and Sutherland, R. G., *J. Chem. Soc., Perkin Trans.* 2, **1977**, 1522.
[887] Muzart, J., and Pete, J.-P., *Tetrahedron*, **34**, 1179 (1978).
[888] Edwards, O. E., and Ho, P.-T., *Can. J. Chem.*, **56**, 733 (1978).
[889] Logani, M. K., Austin, W. A., and Davies, R. E., *Tetrahedron Lett.*, **1978**, 511.
[890] Ohkata, K., Sakai, T., Kubo, Y., and Hanafusa, T., *J. Org. Chem.*, **43**, 3070 (1978).
[891] Ishibe, N., and Yutaka, S., *J. Org. Chem.*, **43**, 2138 (1978).
[892] Ishibe, N., Yutaka, S., Matsui, J., and Ihda, N., *J. Org. Chem.*, **43**, 2144 (1978).
[893] Laerum, T., and Undheim, K., *Acta Chem. Scand.*, **32B**, 68 (1978).
[894] Hunt, R. G., and Reid, S. T., *J. Chem. Soc., Perkin Trans.* 1, **1977**, 2462.
[895] Takagi, K., and Ogata, Y., *J. Chem. Soc., Perkin Trans.* 2, **1977**, 1410.

Author Index

In this index bold figures relate to chapter numbers, roman figures are reference numbers

Subject Index

Erratum for Organic Reaction Mechanisms 1977

P. 60: The three-phase test referred to should be attributed to Rebek and Gaviña who first described the test in *J. Am. Chem. Soc.* **96**, 7112 (1974).